Studies in Logic
Volume 61

Philosophical Applications of Modal Logic

Volume 50
Intuitionistic Set Theory
John L. Bell

Volume 51
Metalogical Contributions to the Nonmonotonic Theory of Abstract Argumentation
Ringo Baumann

Volume 52
Inconsistency Robustness
Carl Hewitt and John Woods, eds.

Volume 53
Aristotle's Earlier Logic
John Woods

Volume 54
Proof Theory of N4-related Paraconsistent Logics
Norihiro Kamide and Heinrich Wansing

Volume 55
All about Proofs, Proofs for All
Bruno Woltzenlogel Paleo and David Delahaye, eds

Volume 56
Dualities for Structures of Applied Logics
Ewa Orłowska, Anna Maria Radzikowska and Ingrid Rewitzky

Volume 57
Proof-theoretic Semantics
Nissim Francez

Volume 58
Handbook of Mathematical Fuzzy Logic, Volume 3
Petr Cintula, Petr Hajek and Carles Noguera, eds.

Volume 59
The Psychology of Argument. Cognitive Approaches to Argumentation and Persuasion
Fabio Paglieri, Laura Bonelli and Silvia Felletti, eds

Volume 60
Absract Algebraic Logic. An Introductory Textbook
Josep Maria Font

Volume 61
Philosophical Applications of Modal Logic
Lloyd Humberstone

Studies in Logic Series Editor
Dov Gabbay   dov.gabbay@kcl.ac.uk

Philosophical Applications of Modal Logic

Lloyd Humberstone

© Individual author and College Publications 2015
All rights reserved.

ISBN 978-1-84890-196-4

College Publications
Scientific Director: Dov Gabbay
Managing Director: Jane Spurr
Department of Informatics
King's College London, Strand, London WC2R 2LS, UK

http://www.collegepublications.co.uk

Original cover design by Orchid Creative   www.orchidcreative.co.uk
Printed by Lightning Source, Milton Keynes, UK

All rights reserved. No part of this publication may be reproduced, stored in a retrieval system or transmitted in any form, or by any means, electronic, mechanical, photocopying, recording or otherwise without prior permission, in writing, from the publisher.

Contents

	Preface	ii
1	**Modal Logic**	**1**
1.1	Introductory Overview	1
1.2	Non-Modal Background	5
1.3	Modal Logics: A Hierarchy of Classes	16
1.4	Refinements and Generalizations	30
2	**Normal Modal Logics**	**33**
2.1	Some Candidate Axioms	33
2.2	Models and Truth: Simplified Semantics	53
2.3	Models and Truth: Kripke Semantics	58
2.4	Canonical Models and Generated Submodels	61
2.5	From Models to Frames	71
2.6	The Rule of Disjunction	112
2.7	Revision Exercises	129
2.8	Supplement: Matsumoto's Embedding	142
2.9	Supplement: Matsumoto's Embedding (Concluded)	149
2.10	Semantical Postscript	152
	A Quick History	152
	Frames with Functions	157
	An Application of Neighbourhood Semantics	171
3	**Applications: Tense Logic**	**177**
3.1	Axiomatizing the Basic Logic	177
3.2	Extensions of $\mathbf{K}_t$	183
3.3	Temporally Motivated Concerns: Density and Discreteness	194
4	**Applications: Alethic, Nomic, Deontic**	**203**
4.1	Introduction	203
4.2	Nomic Necessity I: Pargetter	217
4.3	Nomic Necessity II: Bacon	226
4.4	Deontic Logic: Main Themes	234
4.5	Deontic Logic: More Translations, More Issues	278
4.6	Logics Which Are Fully Modalized	290
4.7	"Nothing in Between": A Remark by A. N. Prior	304
4.8	The Fatal Disjunction: Danielsson on Ross's Paradox	324

5	**Applications: Doxastic and Epistemic Logic**	**331**
	5.1 The Logical Omniscience Issue	331
	5.2 Introspection Issues	370
	5.3 Negative Introspection: Van der Hoek *et al.*	380
	5.4 Negative Introspection: Halpern	385
	5.5 Logics Between **S4** and **S5**	402
	5.6 Concept-Possession Problems	420
	5.7 Another Use for B^{-1}	441

6	**Coming to Stand in a Relation**	**449**
	6.1 Introduction	449
	6.2 The Logic of Coming About	452
	6.3 Relational Change and Coming About	456
	6.4 Multiple Inchoativity	459
	6.5 Cross-Predicability	460
	6.6 Cross-Predicability and Multiple Inchoativity	467
	6.7 Postscript: Completeness for the Logic of §6.2.	468

7	**Appendix: Natural Deduction for S4 and S5**	**471**
	7.1 Non-Modal Rules	471
	7.2 Natural Deduction Rules for $\Box$	475
	7.3 Adding Rules for $\Diamond$	478
	7.4 The Semantics of **S5** and Some Rules for **S4**	481
	7.5 Natural Deduction for Weaker Normal Modal Logics	486

References	**494**
Index	**554**

Philosophical Applications of Modal Logic

Preface

Modal logic, especially the range of what are called normal modal logics, taken collectively, has been applied to many areas of philosophical interest: to the representation of moral claims and principles ('deontic logic'), to questions of knowledge and belief ('epistemic logic'), and so on. It is to such philosophically motivated uses of modal logic – and specifically modal propositional logic (modal predicate being mentioned only in passing from time to time) – that the present work is devoted, though naturally the purely formal elaboration of the subject will require some attention first, before candidate applications can be reviewed. No fixed line is taken on the plausibility of the applications in question, but those wanting to make up their own minds will have many of the pertinent issues aired here in order to help them do so, together with plenty of references to the relevant literature. This should give the general idea of what is covered, though the following paragraph goes into more detail. In particular, while taking a great interest in the application of modal logic to various areas of philosophy, this book pays little attention to the philosophy of modal logic (or philosophy of logic more generally) and none to the philosophy (the metaphysics and epistemology, that is) of modality. The latter area has been treated extensively in many works, some of them (such as Forbes [319] and Lewis [726]) already in our list of references because of the logical issues they touch on, as well as some mentioned only here to redirect readers wanting such material to them: Jubien [633], and – for a more logic-oriented approach – Rini and Cresswell [959], as well as Stalnaker [1084], or Hale and Hoffmann [425] for a useful anthology.

Chapters 1 and 2 give the necessary technical background on the taxonomy of modal logics, their semantic treatment, especially for the normal modal logics, completeness and (what is called 'global') modal definability, the rule of disjunction and variations, etc., again, mostly as they arise for normal modal logics. Indeed somewhat more is provided than the bare technical prerequisites for subsequent chapters, when matters of logical interest arise naturally in the course of supplying such a background. And further, some technical matters are found in the later chapters when the application under discussion naturally calls for them or for this or that reason are most conveniently discussed there. For example, local (as opposed to global) modal definability is addressed in the 'Applications' chapter on tense logic – more specifically, in Section 3.2 – while a brief introduction to two-dimensional modal semantics can be found in the course of the discussion of deontic logic (after Example 4.4.45), in a chapter which also considers (Section 4.6) the issue of what have been called fully modalized logics, we well as in Section 5.7. Similarly, quite a bit of the discussion of non-normal modal logics is deferred to a point (p. 360$f\!f$.) at which it becomes particularly relevant – in the discussion of deductive omniscience in epistemic logic (Section 5.1). Moral: in some cases the index will be of greater assistance than the contents pages for locating a topic of interest. It goes without saying, perhaps, that the topics chosen for discussion here reflect the author's own interests, which are in broadly philosophical and conceptual matters, and those wanting material on, for example, questions of computational complexity, should look elsewhere. The same goes for those wanting an overview of the last thirty years' advances in technical modal logic. Further prefatory material on notation, pre-requisites, etc., can be found at the end of Section 1.1; this includes suggestion for reading in the areas just described as not lying on the present agenda. In addition, I should mention that philosophically motivated departures from the standard language with one or more primitive one-place modal operators, such as the two-place conditional obligation operator often urged in deontic logic, are remarked on in passing but not given a detailed treatment – though plenty of pointers to the pertinent literature are supplied on such occasions.

This material started life as a lecture course in modal logic delivered to philosophy undergraduates at Princeton University in 2007. I am grateful to the Princeton Philosophy Department for inviting me to visit on that occasion and to those students who participated in the course for their feedback and their questions. I have since also had the benefit of assistance on this project from many people, including Rohan French, Allen Hazen, Wolfgang Lenzen, David Makinson and Evgeny Zolin, all of whom provided information or advice on specific topics. For their proof-reading assistance, I am greatly indebted to Rohan French and Sam Butchart.

Chapter 1

Modal Logic

1.1 Introductory Overview

Modal logic might, somewhat narrowly construed, be defined as the logic of necessity and possibility. Already, this encompasses several projects since we can distinguish various types of necessity and possibility: physical, logical, metaphysical, etc. The reason for the word "narrowly" is that once we develop the apparatus for a study of these notions, we find it can be brought to bear on the logical treatment of some others also: the logic of obligation ('deontic logic'), of knowledge ('epistemic logic'), and of certain temporal constructions ('tense logic') – to mention only three. What do such applications have in common? In each case, we take an interest in one or more operators which apply to a sentence to make a new sentence – 1-ary connectives, that is, syntactically like "¬" – which are not, as "¬" is, amenable to a truth-functional semantic treatment. Often the term *alethic* is used to modify the phrase "modal logic" in order to indicate that it is necessity and possibility that are under consideration rather than these other notions; this will be our focus for the moment. (The etymological source of "alethic" is a Greek word roughly translatable as *truth*; note that in some cases necessity is spoken of in such a way that it does not imply truth, as when something is said to be morally necessary or morally required. This places us in the deontic rather than the alethic realm. The same is even more frequently the case with the corresponding modal auxiliary verb *must*, which has in addition, an epistemic interpretation. See the index entry "modal auxiliaries" for further treatment of such issues.)

Although the general notion of truth-functionality will receive closer attention in Section 1.2, we can say something here as to how it fails to apply to these modal notions. To see that this description ('non-truth-functional') covers the case of necessity and possibility (of whatever type) notice that there is no plausible way to complete the partial truth-tables of Figure 1.1, in which "□" is read "it is necessary that" and "◇" is read "it is possible that" (a common alternative symbolism uses "L" for "□", and "M" for "◇"), respectively:

$\Box A$	A
?	T
F	F

$\Diamond A$	A
T	T
?	F

Figure 1.1: Truth-tables – floundering

If, as in *deontic* logic, mentioned above, we read the "□" as "it ought to be the case that" (or as "it is obligatory that") and the "◇" as "it is morally permissible that" – for which readings it is customary to write "O" and "P" – then we cannot even fill in *this* much of the would-be truth-tables, since what ought to be the case, unlike what is necessarily the case, need not after all be the case. The same applies if we read $\Box$ as "*a* believes that __", for some cognitive agent a – the *doxastic* application of modal logic – since our agent's beliefs needn't be true, though it is again unproblematic for a reading as "*a* knows that __", for *epistemic* logic, as in the alethic case. For the moment, we will not worry about what should replace the method of truth-tables as a semantic account of the language we are developing, and judge the acceptability or otherwise of formulas by simply reading the "$\Box$" as "(it is) necessarily (the case that)", and similarly with "$\Diamond$" and possibility. A semantic approach with greater promise – variously known as 'possible worlds' semantics, Kripke semantics or the relational semantics for modal logic – will be presented in Sections 2.2 and 2.3. It is the unavailability of a complete truth-table description of the semantic behaviour of these operators makes them *non-truth-functional* connectives. (See p. 10 below for a more precise definition of truth-functionality w.r.t. a class of valuations; if we are thinking of the connectives as individuated by their inferential behaviour the claim of non-truth-functionality amounts to the claim that the logic codifying this behaviour is not sound and complete w.r.t. any class of valuations w.r.t. which the connectives concerned are truth-functional.[1])

The alethically (and epistemically) uncontroversial T and F entries in our partial truth-tables above indicate an informal judgment of validity for arguments of the forms "$\Box A \therefore A$" and "$A \therefore \Diamond A$". In treating such issues formally, we may decide to work entirely with formulas of a formal language, treating "A" here as a schematic letter for such formulas (a metalinguistic variable ranging over the class of formulas), and represent the argument forms as implicational principles: $\Box A \to A$ and $A \to \Diamond A$, or we may treat them more directly as sequents (formal analogues of arguments with a premiss and conclusion structure) $\Box A \succ A$ and $A \succ \Diamond A$. We mostly take the former route, for conformity with the bulk of the literature, but include several asides and an appendix (Chapter 7) in the latter style, explaining the idea of using sequents in Section 1.2.

Since doxastic and deontic logic are no less cases of applying modal logic (to *belief* and *obligation*) the epistemic and alethic cases (for *knowledge* and *necessity*), the deep analogies between the various cases – first articulated by G. H. von Wright in the seminal publications [1204], [1205], [1206] – cannot have anything to do with whether A follows from $\Box A$, however we choose to record this. (For more on this veridical/non-veridical contrast, see p. 204. The present cursory remarks do not attend to distinctions among various things that go under the names of necessity. It might be thought for example that logical and metaphysical necessity need to be distinguished, and certainly *nomic* (or nomological necessity) – the kind of non-accidentality often held to characterize the laws of nature – raises its own questions; we will encounter some of them next in Section 4.2.) They pertain instead to how $\Box$ and $\Diamond$ relate to each other, the logical relations among sequences of these operators, and to how they interact with conjunction and disjunction, considerations which will lead us by the end of this section to isolate the class of what are called normal modal logics, studied further in Chapter 2, before we return to the various applications mentioned above (as well as some others) in later chapters. These are what we have in mind as 'philosophical applications' of modal logic, though the basic machinery of modal logic has numerous further applications to areas not traditionally regarded as being of special philosophical interest – modal provability logic, the modal logic of programs ('dynamic' logic), and tense logic, to mention only three; here only the third of these gets any discussion to itself (Chapter 3) and even there the motivating concerns from a historical perspective – specifically temporal matters – are given a low prominence. It should also be understood that, as was indicated in the Preface, this talk of applications of modal logic to all such areas is meant to suggest the attempted or intended application of modal logic on the part of various theorists. In many cases there will be others who claim the putative application is not appropriate for one reason or another. Pointers to the relevant polemical literature will be given. But it

---

[1]Soundness here means that everything provable in the logic is true/truth-preserving on each valuation in the class in question, and completeness means that everything with the latter property is provable in the logic. We are having to be a little vague as to what kind of things a logic provides proofs of to give a description which is neutral between thinking of these things as *formulas* (making the talk of truth appropriate) or instead – see the following paragraph – *sequents* (making the talk of truth-preservation appropriate).

1.1. INTRODUCTORY OVERVIEW

would be unwise to venture into such debates without a prior familiarity with the formal developments in question, which is what will be supplied here. Further: we concentrate almost exclusively on modal propositional logic, with only occasional asides on modal predicate logic (quantified modal logic), which raises numerous technical questions of its own and would certainly warrant more extensive treatment had our interest been in the philosophy of modality (to use a phrase from the Preface). See 'Reading Around' below for some pertinent references. The formal techniques for studying modal logic that we shall use were nearly all available forty years ago and can be found in Lemmon and Scott [705], which was circulated in the mid 1960s, though some themes from the 1980s will arise (cf. van Benthem [73]). Nothing too mathematically demanding will come up. (See 'Conceptual Prerequisites', below.)

Notation for schematic letters. The notation for schematic letters for formulas used above, "A", "B",... will be the default choice, but when modal operators are themselves being written as capital (Roman) letters, it is easier on the eye to switch to a different style, and the Greek letters "φ", "ψ", "χ" when this happens, in the expectation that the reader will develop a certain bilingualism over this. For example, since the discussion of tense logic (Chapter 3) uses A. N. Prior's notation G, F, H, P for the tense operators, we switch to the Greek style of schematic letters there. Likewise in Section 4.3, where an operator written as N (for nomic-but-not-logical necessity is in play), and in Chapter 5, where Hintikka's notation K, P, B, C for the epistemic and doxastic operators is used, as well as in Chapter 5 where an operator written as D is used (for "it comes about that" – or initially, for "the agent brings it about that"). Otherwise, when the bulk of the attention is on modal operators written as (perhaps decorated, e.g. by subscripts and the like) "$\Box$" and "$\Diamond$", we use the schematic letters $A, B, C \ldots$. One exception to this rule: in deontic logic (Sections 4.4 and 4.5, where the letters O and P appear as the local incarnations of $\Box$ and $\Diamond$ (signifying roughly obligation and permissibility) Roman $A, B, C, \ldots$ have been retained, since there is little danger of confusion and not too much aesthetic distraction. In fact the danger of confusion is at its greatest in the specifically doxastic application of modal logic, in which the $\Box$ and $\Diamond$ operators are written as B and C (signifying roughly belief and belief-compatibility), so in that case there is a particularly pressing need to avoid using early Roman capitals as schematic letters. This spreads over to the epistemic case because of the need to consider mixed principles involving both epistemic and doxastic notions, such as the principle that whatever is known by a given subject is believed by that subject, where the principle in question appears as the schema $K\varphi \to B\varphi$. (Though this is widely endorsed, Section 5.4 will look at a dissenting opinion from J. Y. Halpern..) Very occasional resort to other parts of the Greek alphabet: $\alpha, \beta, \gamma \ldots$ will be made as necessary. However schematic letters are notated, it is important that they be clearly distinguished from propositional variables (for which we be using p, q, r, abbreviating the first three elements of an infinite sequence, $p_1, \ldots, p_n, \ldots$ of such variables – also called sentence letters). The latter are symbols of the object language – the language under discussion, that is – rather than the meta-language (the language in which that discussion is cast, i.e., English, with a few technical terms and special devices added). Most authors reserve the letters just parenthetically mentioned to serve as propositional variables, though one exception is Jaakko Hintikka, the founder of epistemic logic, whose book [490] uses precisely these symbols as schematic letters for formulas, instead, and is followed in this practice by some others discussing his work (e.g., Lenzen [707], Chapter 6 of Girle [379]); the same usage is also to be found in Forbes [322], though it tends to be seen more in older work – for example Fitch [306] uses $p, q, r \ldots$ schematically, while in the later Fitch [308], they have been replaced in this rule by $\varphi, \psi, \ldots$. (It should be noted also that some writers use $p, q, r \ldots$ as schematic letters not for arbitrary formulas but specifically for sentence letters, which in practice amounts to the same as the present policy – of taking them to *be* sentence letters – in that the notation alerts us to the fact that we dealing with atomic formulas rather than formulas in general.) Finally, regardless of what notation is in play for individual formulas, we most often use capital Greek letters as metalinguistic variables over *sets* of formulas: Γ, Δ, Θ, except when such sets of formulas are logics in their own right, in which case we tend to use "S", perhaps decorated, as the variable in question: think S̲ystem in this case. (The letter "L" will be sometimes used for languages and so is not available to stand for logics; the present "S" is not to be confused with a special use of boldface "**S**" as part of the names of certain normal modal logics, explained early in Section 2.1.)

Notation for Binary Relations. Especially when we get to the semantics of modal logic (in Section 2.3), we will be using the letter "R", sometimes subscripted or otherwise decorated, to stand for a binary relation. There are two options here: infix notation, writing (e.g.) "xRy" to mean that x bears the relation in question to y, and prefix notation, writing instead "Rxy" for the same thing. Each has its advantages. The second fits in the with systematic practice in predicate logic of writing the predicate letter first, so that if we have S as a three-place relation symbol, for instance, then we write "$Sxyz$" and so on; but the infix notation has the advantage (by analogy with symbols like "=" and "$\leq$" which are always used in the infixed position) that we can allow ourselves the convenience of writing such things as "$wRxRyRz$" to mean "$wRx \,\&\, xRy \,\&\, yRz$". For the most part, we will use the infix notation despite its slightly clunky appearance, but sometimes, e.g., when discussing passages in which other writers use the prefix notation, or when the connection with first-order logic is worth emphasizing, we use the prefix notation instead. (In the sentence before last, "&" is just an abbreviation in the metalanguage for "and"; if we were emphasizing the connection with first-order logic we might use the object language conjunction connective "$\wedge$" instead.[2]) So, as in the case of schematic letters, a certain degree of bilingualism is being expected of the reader.

Abbreviations. In addition to abbreviations found in general writing, such as the occasional replacement of the word "Section", when followed by a section number, by "§", the following abbreviations, found in logical and mathematical writing, will also be used: "iff" for "if and only if"; "w.r.t." for "with respect to"; "l.h.s." ("r.h.s.") for "left-hand side" ("right-hand side"). Further, "Prop.", "Exc.", "Thm." and "Coro." abbreviate "Proposition", 'Exercise", "Theorem", and "Corollary".[3] "*Nat*" denotes the set of natural numbers $\{0, 1, \ldots\}$, sometimes called $\mathbb{N}$. In the latter vein, the sets of integers and of rational numbers are denoted by $\mathbb{Z}$ and $\mathbb{Q}$, respectively.

Roadsigns. An occurrence of "■" marks the end of proofs, with "✠" marking the end of exercises, and "◀" the end of examples and (numbered) remarks. (There is no need to mark the end of lemmas, theorems, etc., since these stand out from the surrounding text by being set in italic.) More precisely, "■" marks the end of a metalogical proof we are providing; sometimes a proof in this or that proof system is itself given as an example of how the system in question functions, in which case the example ends, like any other, with a "◀". (The main proof systems to be found here are axiom systems, except for the natural deduction systems in the Appendix, Chapter 7.)

Conceptual Prerequisites, Set-theoretic Notation. Readers are assumed to be familiar with the general idea of proof by (mathematical) induction, but any who are not will be able to pick up the idea from the first few examples of this proof technique in action (as in the proof of Proposition 1.2.5). I am also assuming some familiarity with basic set-theoretic ideas and notation, such as that of the union, intersection, and cartesian product of sets X and Y, denoted respectively $X \cup Y$ and $X \cap Y$, and $X \times Y$, as well as the operation of set difference ($X \smallsetminus Y$: the set of elements of X which are not elements of Y); unions and intersections of indexed collections of sets are denoted using "$\bigcup$", "$\bigcap$", respectively, with appropriate sub- and possibly superscripting to show the range of indexing. We also have the relations of membership ($\in$) and inclusion ($\subseteq$), the notation "$\{x \in X \mid \Phi(x)\}$" for the set of elements of X which satisfy the condition $\Phi(\cdot)$, with the idea of n-ary relations on a set X as sets of ordered n-tuples of elements of X, the concept of a function, as well as other similar ideas. For the ordered n-tuple with elements $x_1, \ldots, x_n$ in that order, we use the notation: $\langle x_1, \ldots, x_n \rangle$. The notation "$\{x \in X \mid \Phi(x)\}$" may have the "$\in X$" part suppressed or replaced by something else, and to avoid an ugly proliferation of vertical lines caused by a particular way of spelling out the $\Phi(\cdot)$ in question, have its "such that" vertical

---

[2]Similarly, we use "$\Rightarrow$" as an informal metalinguistic analogue of the formal implication connective "$\rightarrow$".

[3]"Exercise" does not begin with the letters "Exc", of course; but then nor does "Theorem" begin with "Thm". (Compare also: "Sgt." for "Sergeant".)

line "|" replaced by ":". An unrelated but standard use of vertical lines – one that can result in the proliferation just alluded to – is that according to which, for a set X, $|X|$ is the cardinality of X (number of elements of X). The power set of a set X – the set of all subsets of X, that is, including X itself and also $\varnothing$ (the empty set) – is denoted by $\wp(X)$. No set theory beyond obvious properties of these concepts is involved; the terms "class" and "set" (and "collection") are used interchangeably.

Index. To find such special symbols as were just in play ("$\subseteq$" etc.), and non-alphanumeric symbols in general, look them up as subentries under "notation" in the index. Surnames consisting of a prepositional prefix and a name will be found in the index under the first portion of the name with a capital letter (when occurring other than at the start of a sentence). Thus "von Wright" is with the "W"s rather then the "V"s, for example.

Italics. Italics are used for foreign words or phrases and for emphasis and also – a convention of mathematical English – to indicate that the term or phrase is being defined in the surrounding text, as well as for letters of the English alphabet used as symbols. In addition, results (lemmas, theorems, etc.) are set in italics, as was mentioned under "roadsigns" above.

Reading Around. There are good articles on various areas to be touched on here in the collection Goble [385]. In particular, a quick overview of modal logic is provided by Cresswell [217], and there are surveys of epistemic, deontic, and tense logic (Meyer [806], Hilpinen [487], Venema [1129], respectively). Articles on these subjects in the Stanford Encyclopedia of Philosophy are also reliable and regularly updated. There are many excellent textbooks on modal logic and its applications. Carnielli and Pizzi [152] is a good recent example, pitched at about the same level as this one, though going into less detail in respect of philosophical criticism or defence when it comes to applications – the proposed deontic, epistemic, etc. interpretations of modal logic. Garson [351] supplies some such discussion but with fewer references to the literature for those wanting to follow matters up, and is pitched at a somewhat lower level of technical sophistication. Several other texts, typically including more advanced material, will be referred to in the course of our exposition. Many such books make it their business to discuss not only modal logic but also – something not covered here – intuitionistic logic (and intermediate logics), because of various similarities between the semantic treatments of the two areas; these include Chagrov and Zakharyaschev [166] and van Benthem [79]. And even in respect of (classically based) modal logic itself, some of the texts mentioned here go into the details of quantified modal logic. As already mentioned, this is a topic touched on only in passing in what follows; further references to this area are given just before Exercise 2.2.8 on p. 58 below.

Skimming Advice. For a briefer pass through this material, Sections 1.4, 2.8, 2.9 and 4.7 could be omitted without much loss of pre-requisites for other (sub)sections. Passages explicitly marked with the word "Digression" can be skimmed on quick reading, but may need to be consulted when they are referred back to. Exercises should be at least skimmed by readers not intending to attempt them, as concepts that will be used later are sometimes introduced within them.

1.2 Non-Modal Background

If we want to do some modal logic, then we should start with some non-modal logic and add some modal vocabulary. In the narrowest sense of the term *modal* this means adding some vocabulary for talking about what is necessary, impossible, and possible. (This is sometimes called *alethic* modal logic, or, more accurately, the alethic interpretation or application of modal logic.) The simplest and most widely familiar non-modal basis to add the new vocabulary to would be the language of classical propositional logic, which is what we will choose here. But if you wanted to do modal predicate logic instead, you would want to begin perhaps with classical predicate logic (alias *first-order* or *elementary* logic). For occasional references to this topic, see 'modal predicate logic' in the index. We avoid the area for the most part for the two reasons – relative insignificance for the heart of most applications of modal logic, combined with considerable complexity – indicated by Melvin Fitting in the following quotation from the opening page of [311]:

> Propositional modal logic is a well-known tool, since possible worlds can represent computational states or moments of time or ways an agent thinks the world is. The addition of quantifiers, however, opens the door to a labyrinth full of twists and problems (...) and comparatively few have been willing to enter.

Fitting goes on to list some of the issues involved, but even since the semantics in terms of possible worlds has yet to be explained, these would not make a lot of sense at this stage. And the reaction described is perhaps slightly exaggerated, as our occasional allusions in what follows to quantified modal logic will indicate. (Some bibliographical references can be found in the paragraph before Exercise 2.2.8, p. 58 below.) At any rate, we cannot cover everything, and have chosen to omit detailed consideration of modal predicate logic for roughly the reasons gestured at in Fitting's remarks. Propositional modal logic it is, then.

Of course you may not be happy with the account specifically *classical* propositional logic provides of the connectives it takes to represent natural language expressions *not, and, or, if __ then ...* – written throughout in what follows as $\neg, \wedge, \vee, \rightarrow$; you may even think that no coherent concepts are embodied in these connectives as they behave according to classical logic, say, because you are an adherent of (the weaker) intuitionistic logic. Then by all means start with intuitionistic propositional logic – or indeed intuitionistic predicate logic – or with whatever else you are comfortable with (*relevant* logic, perhaps). There is a considerable literature on intuitionistic modal logic and a somewhat smaller literature on relevant modal logic. But we are going to make life simple by keeping the underlying non-modal logic *classical*. There is also some literature on using a less than functionally complete stock of non-modal connectives (this terminology being explained below, after Exercise 1.2.4), so that we are looking at fragments of the full classically based language with $\Box$ (and perhaps $\Diamond$ too) – such as Schumm and Edelstein [1010] (also Schumm [1008]), Humberstone [560], and Dunn [265]. But again, interesting though it is, this is not an area into which we shall be venturing here. Our non-modal apparatus will be *full* rather than *fragmentary* classical logic.

The intended reader of the present work may, by way of non-modal experience, be taken to have been introduced to classical (propositional) logic one way or another – semantically (via truth-tables, say) and or perhaps also syntactically via some proof-system (axioms, natural deduction rules, or sequent calculus rules, to cite the most common formats – and tableau/proof-tree methods should be mentioned too,[4] though these are inimical to an appreciation of the contrast between proof theory and semantics). Indeed, the reader has probably encountered both a semantic characterization and a matching syntactic characterization. The match here is described in more detail by saying that the chosen proof system provides proofs only for what is valid according to the semantics, which amounts to its being *sound* with respect to the semantics, and it provides a proof for everything that is valid, making it *complete* w.r.t. the semantics. For the usual bivalent truth-table semantics, the validity of a formula – also put by calling the formula in question a *tautology* – is a matter of its receiving the value True, or T for short, when evaluated by the usual tables in T and F(alse), regardless of how these values are assigned to the propositional variables (or 'sentence letters'). The latter we assume come in infinite (more precisely, countably infinite) supply, as $p_1, p_2, \ldots, p_n, \ldots$, though for convenience we take the liberty of referring to the first three of them as p, q and r. The formulas of the language of propositional logic are the propositional variables, together with the result of applying the binary connectives, $\rightarrow, \wedge, \vee$, to any pair of formulas, or the singulary (i.e., 1-ary or one-place) connective $\neg$ to any formula.[5] Note that

---

[4]Such methods are used – and moreover used for modal logic – in Fitting [310], Priest [909], Girle [378] and Garson [351], for example. For an overview of this approach, see Goré [401].

[5]You might like to say "unary" here (for singulary/1-ary). On etymological grounds, I can't quite bring myself to, though. For a vigorous reaction against such qualms (as expressed by Quine and Church) see Curry [226], correcting – a

1.2. NON-MODAL BACKGROUND

we sometimes say "truth-functional tautology", the adjective included purely for emphasis here; also we reserve the term "tautology" for a formula constructed using the familiar Boolean connectives (see below), and not for formulas with additional connectives present, even if the formula in question is a substitution instance (see Section 1.3) of a tautology in this narrow sense. (Either use is perfectly legitimate, but that chosen is more convenient here.)

In the last paragraph, soundness for a proof system with respect to a notion of validity was rather vaguely said to consist in providing for proofs "only for things which are valid", and the completeness of such a system was a matter of the availability of a proof "for everything that is valid". What are these *things*? Well, if you are used to the axiomatic approach to proofs, the most obvious things that have proofs are formulas, and it was with these in mind that the soundness and completeness of a proof system was cashed out as the coincidence of its provable formulas (or "theorems") and the tautologies. But if you are familiar with a natural deduction system or with a sequent calculus, then the objects that are proved are not formulas but things called sequents, which themselves come in various shapes and sizes, the simplest of which is as ordered pairs comprising a finite set, multiset or sequence of formulas, Γ, say, and a single formula A. Various notations are found for such pairs, usually more suggestive of the idea that we are thinking of the formulas in Γ as like the premisses of an argument and the formula A as the conclusion. Thus instead of the bare ordered-pair notation $\langle \Gamma, A \rangle$, for instance, it has become common to right "$\Gamma \Rightarrow A$".[6] Note the use of "$\Rightarrow$" as a sign for separating the premiss-side of the sequent from the conclusion side. Since we are using $\to$ as our implicational connective, it would be very confusing to use *that* symbol to play this separator role. (Some people, including Gerhard Gentzen, who pioneered these ideas in the 1930s, have used $\to$ as a separator, but then they typically use another symbol for the implicational connective so that, again, no confusion arises. Gentzen used $\supset$ for that purpose, for example. Even this last symbol has been used by others – Hösli and Jäger [529], for instance – as sequent separator.) The implicational connective constructs formulas from formulas, whereas the sequent separator constructs sequents from formulas. Other notations one sees for this separator are the colon notation used by M. Dummett ("$\Gamma : A$" for the above sequent) and the arrow tail notation of Stephen Blamey ("$\Gamma \succ A$"); Humberstone [594] uses the latter notation, and this is what is used in some of the practice exercises below, and throughout the Appendix on Natural Deduction.[7] Such complications need not detain us here. In fact, for most of what follows, we don't even *need* the present simple notion of a sequent. The present point is that just as there are proof systems which provide proofs of sequents, so there is a natural notion of truth-table validity to go with them to: we can say that $\Gamma \succ A$ is valid by truth-tables, or to have a single word for this, *tautologous* when every assignment of truth-values to the propositional variables occurring in A and (the formulas in) Γ which results in every formula in Γ getting the value T results in the formula A getting the value T. (We also describe a formula A as tautologous when A is a tautology, which is equivalent to the tautologousness of the sequent $\succ A$.) Note than when Γ is empty this coincides with A's being a tautology, so the 'formula' treatment to logic is a special case of the 'sequent' treatment. Indeed we can also go the other way, and instead of asking about whether the sequent $A_1, \ldots, A_n \succ B$ is tautologous, we can trade this in for a formula which is guaranteed to be a

---

typo – the word "first" to "second" in the second last line of the first column of p. 246.

[6]Sequent calculus – or 'Gentzen system' – rules governing a connective insert an occurrence of the connective formula on the left or the right of the conclusion sequent, whereas natural deduction rules either insert into or remove such an occurrence from the formula on the right, and are called respectively introduction or elimination rules for that connective. The right insertion sequent calculus rules coincide with the natural deduction introduction rules, and the left insertion rules do the work of the elimination rules, something most evident in the presence of certain rules – in particular the Cut Rule – not governing any particular connective. But such a rule has disadvantages of its own and much effort is expended to show that it can be dropped without reducing the set of provable sequents. We are not concerned with the sequent calculus approach here.

[7]Warning: sometimes you may see things like "$\Gamma \succ \Delta$", where not only Γ but also Δ represents a set or multiset of formulas. See p. 311 for some explanation, and the discussion in 1.21 in [594] of *logical frameworks* there to get a feel for what is going on; when sets (as opposed to multisets) of formulas are involved, these sequents with several (or no) formulas on the right stand in the same relation to the generalized consequence relations mentioned on p. 14 as ordinary sequents with a single formula on the right stand to consequence relations.

tautology under the same conditions, namely the formula $(A_1 \wedge \ldots \wedge A_n) \to B$. Instead, we could have used the formula

$$A_1 \to (A_2 \to \ldots \to (A_n \to B)\ldots)$$

or again

$$\neg A_1 \vee \neg A_2 \vee \ldots \vee \neg A_n \vee B.$$

In this last example I have left out the internal parentheses which should really be there to reflect the fact that $\vee$ is a binary connective, but can be inserted in any way without making a difference to the result (because $\vee$ is associative in classical propositional logic). The same liberties have been taken in writing "$(A_1 \wedge \ldots \wedge A_n)$" for the antecedent of the earlier conditional. Although we don't need to work with sequents for the bulk of what follows, they figure extensively in the Appendix on natural deduction (Chapter 7) so some practice exercises on classical propositional logic cast in terms of them are included here. Of the several options mentioned above, the simplest is to think of the formulas on the left of the separator $\succ$ as constituting a *set* of formulas, rather than a sequence of formulas (where the order matters) or a multiset of formulas (where the number of occurrences matters). (In the terminology of my [594], this means we are in the logical framework SET-FMLA, rather than SEQ-FMLA or MSET-FMLA.)

EXERCISE 1.2.1 Which of the following sequents are tautologous? For (iii) and (iv) think of $\leftrightarrow$ as introduced by a definition: "$A \leftrightarrow B$" abbreviates the formula $(A \to B) \wedge (B \to A)$.

(i) $\quad \neg(q \wedge p), \neg(r \wedge \neg p) \succ \neg(q \wedge r)$
(ii) $\quad p \to (q \vee r) \succ (p \to q) \vee r$
(iii) $\quad p \leftrightarrow (q \to r) \succ (p \to q) \leftrightarrow (p \to r)$
(iv) $\quad p \to (q \leftrightarrow r) \succ (p \leftrightarrow q) \to (p \leftrightarrow r)$
(v) $\quad (p \to q) \to (q \to r) \succ (p \to q) \to (p \to r)$
(vi) $\quad (p \to q) \to (p \to r) \succ (p \to q) \to (q \to r)$. ✠

A slightly more abstract practice exercise:

EXERCISE 1.2.2 In each case, either provide a formula A (in the language we have been working with) which has the property in question, or show that there can be no such A. In the interests of brevity, I have said 'valid' here, to mean 'tautologous'.

(i) Is there a formula A for which the sequent $A \to p \succ q$ is valid?
(ii) Is there a formula A for which the sequent $p \succ A \to q$ is valid?
(iii) Is there a formula A for which the sequent $p \to A \succ q$ is valid?
(iv) Is there a formula A for which the sequent $p \succ q \to A$ is valid?
(v) Is there a formula A for which the sequents $p \wedge q \succ A$ and $A \succ p$ are valid but neither of the 'converse' sequents $A \succ p \wedge q$ or $p \succ A$, is valid? (We return to a reformulation of this question in two paragraphs, and answer it in Section 4.7 below; it would be helpful for you to address it now.) ✠

There is another way of asking all these questions which does not introduce sequents as objects in their own right, and which does not have us talking directly about the corresponding formulas (implications with conjunctive antecedents, say, as earlier). This is to introduce the idea of a *consequence relation*, which is to say, a relation between sets of formulas and individual formulas meeting some simple conditions, given below. We use the symbol "$\vdash$" (and similar looking notations) for consequence relations; the defining conditions to be satisfied are that:

$$\{A\} \vdash A \text{ for all formulas } A \text{ of the language concerned,}$$

and that

$$\Gamma \vdash A \text{ implies } \Gamma \cup \Delta \vdash A, \text{ for all sets of formulas } \Gamma, \Delta,$$

and of course all formulas A; and finally that for all Γ, Δ, B,

$$\text{if } \Gamma \vdash A \text{ for each } A \in \Delta \text{ and } \Gamma \cup \Delta \vdash B, \text{ then } \Gamma \vdash B.$$

These conditions are known by several names; in terminology due to Dana Scott they are called, respectively, the conditions of reflexivity, motononicity and transitivity – though the first and third of these labels are not to be taken literally, for obvious enough reasons.[8] Actually this last 'transitivity' condition can be replaced by the simpler condition that (for all Γ, A, B), $\Gamma \vdash A$ and $\Gamma \cup \{A\} \vdash B$ should together imply $\Gamma \vdash B$, in the case in which $\vdash$ is *finitary*, which is to say that for any Γ, A, if $\Gamma \vdash A$, then there exists a finite subset Γ_0 of Γ, for which $\Gamma_0 \vdash A$. (It may have seemed, in view of the way Γ was used in writing sequents above, that we were understanding only finite sets of formulas to be involved, but actually that would be an inconvenient restriction to impose in the discussion of consequence relations.) In case the terminology of reflexivity and transitivity deployed here is unfamiliar, its use is explained as applied to binary relations on p. 14; the present use is adapted to consequence relations $\vdash$ via the special case of the binary relation holding between formulas A and B when $\{A\} \vdash B$.

The particular consequence relation associated with classical propositional logic we may call the relation of *tautological consequence*, or alternatively *truth-functional consequence*, and denote by $\vdash_{CL}$. ("CL" for "Classical Logic".) We can define this semantically (in terms of truth-values) or syntactically (in terms of some particular proof system, that is). As the terminology just mentioned suggests, we take the former route: $\Gamma \vdash_{CL} A$ if and only if every truth-value assignment to the variables in the formulas in $\Gamma \cup \{A\}$ which, when the usual truth-functions are used to evaluate compound formulas in terms of the values of their components, assigns the value T to all formulas in Γ, assigns the value T to the formula A. This consequence relation turns out to be finitary – something we are not showing here. (The result in question is called the Compactness Theorem for classical propositional logic, "compact" being another word for "finitary", used especially when a consequence relation has been characterized, as here, semantically.) Further, when Γ is itself a finite set of formulas, the claim that $\Gamma \vdash_{CL} A$ is (trivially) equivalent to the claim that the sequent $\Gamma \succ A$ is tautologous.

Both in connection with sequents and in talking about consequence relations certain notational shortcuts are taken. In particular, instead of writing carefully, as above "$\Gamma \cup \{A\} \vdash B$" and "$\Gamma \cup \Delta \vdash A$", one would normally write simply "$\Gamma, A \vdash B$" and "$\Gamma, \Delta \vdash A$", respectively. Some further abbreviations that are common: "$\Gamma \nvdash A$" for: it is not the case that $\Gamma \vdash A$; and "$A \dashv\vdash B$" for: both $A \vdash B$ and $B \vdash A$.[9] Also, when Γ is empty, instead of writing explicitly "$\varnothing \vdash A$", one usually just writes "$\vdash A$". Using some of this notation, Exercise 1.2.2(v), for example, could be reformulated thus:

Does there exist a formula A such that $p \wedge q \vdash_{CL} A$ and $A \vdash_{CL} p$ while $A \nvdash_{CL} p \wedge q$ and $p \nvdash_{CL} A$?

Incidentally, some authors use the symbol "$\vdash$" itself as a sequent separator, instead of as a metalinguistic symbol to stand for a consequence relation. Indeed you may be having some trouble seeing what this distinction amounts to, and if the above example of reformulating part (v) of Exercise 1.2.2 hasn't helped, then I would advise you to ignore the distinction – no great harm will come of that – and regard it as something to come back to at some stage in the future.

For many purposes a slightly different presentation of the semantic side of classical propositional logic is preferable. Instead of talking about truth-value assignments to the propositional variables in a given set

---

[8] In non-monotonic logic, a topic which will come up only tangentially in what follows, one works with consequence-like relations in which the monotonicity condition is dropped or restricted.

[9] If "$\vdash$" is subscripted – e.g. with "CL" – then we append the subscript only to the "$\vdash$" part of the composite "$\dashv\vdash$" notation.

of formulas and the extension of such assignments to arbitrary formulas constructed from those variables, let us work with the undifferentiated notion of a *valuation*, which is just a function assigning truth-values to formulas (variables or otherwise), i.e., a mapping from the set of formulas to the two-element set $\{T, F\}$. When $v(A) = T$, we describe the formula A as *true on* the valuation v. There are conventions associated with various primitive connectives as to how a valuation should respect the intended meaning of the connective; we call the connectives in question *Boolean* connectives, and the valuations that respect the intended meaning *Boolean valuations*. We are not supposing that in any given choice of language, all of these connectives are present, but we count as Boolean only such valuations as abide by the conditions for as many of these connectives as are present. For the binary connectives $\wedge$, $\vee$, $\rightarrow$ and $\leftrightarrow$ (in case we wish to treat this a primitive connective rather than as defined *à la* Exc. 1.2.1) we count a valuation v as a Boolean valuation when for all formulas A, B:

$$v(A \wedge B) = T \text{ iff } v(A) = T \text{ and } v(B) = T;$$
$$v(A \vee B) = T \text{ iff } v(A) = T \text{ or } v(B) = T \text{ (or both)};$$
$$v(A \rightarrow B) = T \text{ iff } v(A) \neq T \text{ or } v(B) = T \text{ (or both)};$$
$$v(A \leftrightarrow B) = T \text{ iff } v(A) = v(B).$$

When $\vee$ is being interpreted semantically via the above condition as opposed to a similar condition with "or both" replaced by "but not both", the phrase *inclusive disjunction* (as opposed to *exclusive disjunction*) is used. Similarly the connectives $\rightarrow$ and $\leftrightarrow$ when treated as here are sometimes called the *material* conditional and biconditional respectively – or as connectives of material implication and material equivalence. The terminology recalls an older form/matter contrast ('formal implication' etc.), with "matter" amounting to *truth-value* here. More saliently in the setting of modal logic the contrast would be with strict implication (understood as necessary material implication: see p. 46). For the 1-ary negation connective $\neg$, a valuation is Boolean when for all A:

$$v(\neg A) = T \text{ iff } v(A) = F,$$

while for the 0-ary connectives $\top$ and $\bot$, propositional constants for truth and falsity, we require of Boolean v, as this description suggests, that:

$$v(\top) = T \quad \text{and} \quad v(\bot) = F.$$

$\top$ and $\bot$ are sometimes called the *Verum* and the *Falsum*, respectively, or *top* and *bottom* (again respectively), or *tee* and *eet*.

REMARK 1.2.3 It is sometimes useful to apply the concept of being Boolean more selectively, and say that a valuation v is, specifically, $\wedge$-Boolean, $\vee$-Boolean, $\rightarrow$-Boolean or $\leftrightarrow$-booolean according as it satisfies for all A, B, the first or the second or third or fourth condition inset above, and similarly in the 1-ary and 0-ary cases just reviewed. ◀

An n-ary truth-function is just a function from $\{T, F\}^n$ to $\{T, F\}$ – i.e., taking any n-tuple of truth-values to a truth-value. An n-ary connective $\#$ is *truth-functional* with respect to a class $\mathcal{V}$ of valuations just in case there exists an n-ary truth-function f such that for all formulas $A_1, \ldots, A_n$, and all $v \in \mathcal{V}$:

$$v(\#(A_1, \ldots, A_n)) = f(v(A_1), \ldots, v(A_n)).$$

Here we have used the prefix rather than infix notation for $\#$ since it doesn't require a convention as to where inside the $\#$-compound to place the "$\#$". With binary connectives, we continue to write them in between the two components – using the infix notation, that is. Thus for binary $\#$, we can write the above condition as demanding the existence of a function $f_\#$ ("of two variables" as they say) – for which we still use prefix notation – such that for all $v \in \mathcal{V}$ and for all formulas A, B:

1.2. NON-MODAL BACKGROUND

$$v(A \# B) = f_{\#}(v(A), v(B)).$$

Note that all the Boolean connectives are truth-functional w.r.t. the class of all Boolean valuations. For example, in the case of $\vee$, we have, as the associated truth-function, $f_{\vee}$, let's call it: $f_{\vee}(T, T) = f_{\vee}(T, F) = f_{\vee}(F, T) = T$, while $f_{\vee}(F, F) = F$. In view of this, since the only valuations of interest to us in what follows are Boolean valuations, we will simply use the term 'truth-functional connective' for 'Boolean connective'.

For the following exercise, we recall some familiar terminology. A binary operation $\cdot$ on a set X (i.e. a function of two arguments under which X is closed) is said to be *idempotent* when for all $a \in X$, $a \cdot a = a$, to be *commutative* when for all $a, b \in X$, $a \cdot b = b \cdot a$, and *associative* when for all $a, b, c \in X$, $(a \cdot b) \cdot c = a \cdot (b \cdot c)$. A binary truth-function is accordingly a binary operation on the two-element set $\{T, F\}$.

EXERCISE 1.2.4 (*i*) Show that any binary truth-function which is idempotent is associative. (*Suggestion*: since we have to show that for all choices of a, b, c from our two-element set, $(a \cdot b) \cdot c = a \cdot (b \cdot c)$, on the hypothesis that $\cdot$ is idempotent, we can use a brute force method and consider all possible cases, since each of b, c, must be either a itself or else the only other element of the set, which we will call $\bar{a}$ (so $\bar{T} = F$ and $\bar{F} = T$, for our example), giving four cases to check: $(a \cdot a) \cdot a = a \cdot (a \cdot a)$, $(a \cdot a) \cdot \bar{a} = a \cdot (a \cdot \bar{a})$, $(a \cdot \bar{a}) \cdot a = a \cdot (\bar{a} \cdot a)$, and $(a \cdot \bar{a}) \cdot \bar{a} = a \cdot (\bar{a} \cdot \bar{a})$. By way of illustration, we work this last case, which divides into two subcases, in the first of which $a \cdot \bar{a} = a$ and in the second of which $a \cdot \bar{a} = \bar{a}$. In the first subcase, the l.h.s. of the equation we need to check, $(a \cdot \bar{a}) \cdot \bar{a} = a \cdot (\bar{a} \cdot \bar{a})$, reduces to a $a \cdot \bar{a}$ on replacing the bracketed term with a in accordance with the (sub)case hypothesis (i.e., that $a \cdot \bar{a} = a$) and the r.h.s. also reduces to $a \cdot \bar{a}$, on simplifying the bracketed term to $\bar{a}$ by the assumption of idempotence. So we see that the two sides are equal – even without further reducing them to a in accordance with the case hypothesis. For the second subcase, the bracketed term in the l.h.s. reduces to $\bar{a}$ by the case hypothesis, so the l.h.s. itself reduces to $\bar{a}$ by idempotence, while the bracketed term in the r.h.s. reduces to $\bar{a}$ by idempotence and then the r.h.s. also simplifies to $\bar{a}$ by the case hypothesis. The other three (main) cases need to be checked similarly.)

(*ii*) Is it the case that every binary truth-function which is idempotent is commutative? (Give a proof or a counterexample, according as you answer *Yes* or *No*.) ✠

A set of truth-functional connectives is *functionally complete* if for any n-ary truth-function (where $n \geqslant 1$) f, there is a formula $A = A(p_1, \ldots, p_n)$ constructed from the variables exhibited and using connectives in the given set, with the property that for any Boolean valuation v, $v(A) = f(v(p_1), \ldots, v(p_n))$. A stronger notion of functional completeness is obtained by requiring this for all $n \geqslant 0$ rather than $n \geqslant 1$ (understanding by a 0-ary truth-function, a truth-*value*). Whereas, for example, the sets $\{\vee, \neg\}$, $\{\wedge, \neg\}$, $\{\rightarrow, \neg\}$ are all functionally complete in the sense defined, though not 'strongly functionally complete', the sets $\{\rightarrow, \bot\}$ and $\{\rightarrow, \neg, \top\}$ are functionally complete in this stronger sense. You might think that this last set involves a redundancy, since we could define $\top$ as $p \rightarrow p$, but this elides the distinction between the truth-value T and the constant 1-ary truth-function which assumes this value for all arguments.[10] For the most part, functional completeness in the weaker sense introduced above is all that is called for. A truth-functional connective – or the associated truth-function – whose unit set is functionally complete in this weaker sense is often called a Sheffer function, after the special case ('the Sheffer stroke', also called *nand*) of negated conjunction.[11]

---

[10]This distinction (see Humberstone [565]) between a stronger and a weaker notion of functional completeness is not widely drawn in the literature.

[11]A similar notion of expressive completeness has also been considered in modal logic, though we do not go into details here, along with an analogous notion of Sheffer status. The latter is discussed with various modal logics in mind in Massey [779] (and Sobociński [1072] for some simplifications); also Hendry and Massey [472] and references there given, as well as Dubikajtis and de Moraes[257]. A related exercise – not restricting the primitive connectives to one, but insisting that those employed are in a certain sense *self-dual* – is pursued in Rose [965].

The next observation we need to make does not depend on the definability of *all* truth-functional connectives, concerning instead what happens when *only* such connectives are involved, namely that in this case the truth-value of any formula is fixed by the truth-values of the propositional variables used in its construction:

PROPOSITION 1.2.5 *For any formula A constructed using only Boolean connectives, and any Boolean valuations v, v', with $v(p_i) = v'(p_i)$ for each propositional variable p_i occurring in A, $v(A) = v'(A)$.*

Proof. By induction on the complexity of A.

Basis Case. The complexity of A is 0, so A is a propositional variable, so $v(A) = v'(A)$ if v, v', agree on any variables in A (A itself being the only such variable).

Inductive Step. Suppose A is $\#(B_1, \ldots, B_n)$ and all connectives involved are Boolean. For Boolean valuations v, v' we have

$$v(A) = f_\#(v(B_1), \ldots, v(B_n)) \qquad v'(A) = f_\#(v'(B_1), \ldots, v'(B_n)),$$

where $f_\#$ is the truth-function associated with $\#$. Suppose that v, v', agree on the variables in A. Then they agree on the variables in $B_1, \ldots, B_n$ and so by the inductive hypothesis they agree on the formulas $B_1, \ldots, B_n$ themselves, i.e., $v(B_i) = v'(B_i)$ for $i = 1, \ldots, n$. From this and the two inset equalities above, we get $v(A) = v'(A)$. ∎

The formulas which are tautologies (or sequents which are tautologous) in the sense of always receiving (resp., preserving, from left to right of the "≻") the value T on every truth-value assignment to the variables involved, coincide with the formulas true on (resp. sequents truth-preserving on) every Boolean valuation, despite the overwhelmingly greater number – there being uncountably many – of the latter valuations. If a formula (or sequent) is constructed from k distinct propositional variables, then there are only 2^k assignments to these variables, and the truth-table decision procedure for validity checks that the formula is true (the sequent is truth-preserving) in each of these 2^k cases. But there are uncountably many Boolean valuations – one for each subset of the countably infinite set $\{p_1, p_2, \ldots, p_n, \ldots\}$ (since putting $v(p_i) = T$ iff p_i belongs to a given subset uniquely determines v). But Prop. 1.2.5 reassures us that most (to put it mildly) of the differences between these valuations do not matter, since they pertain to formulas constructed out of variables not occurring in the formula (or sequent) we are considering at any given moment. Suppose this formula (or sequent) involves only variables in a (finite) set P, and say that valuations are P-equivalent if the only propositional variables on which they differ lie outside P. The resulting equivalence classes then correspond to the lines in a truth-table as used for testing the formula (or sequent) in question for tautologousness. We could arbitrarily select one valuation from each equivalence class – for instance the unique such valuation falsifying every propositional variable lying outside P: and there are only 2^k of these (where $k = |P|$).

The following observation is a version of the $\wedge/\vee$ distribution equivalences of classical propositional logic, i.e., the truth-functional equivalence of $A * (B \# C)$ with $(A \# B) * (A \# C)$, where $*$ and $\#$ are respectively $\wedge$ and $\vee$, or else respectively $\vee$ and $\wedge$. A more general form could be given, with more than two conjuncts (here A_i, B_i) per disjunct, so that the jth disjunct appeared as $A_1^j \wedge \ldots \wedge A_m^j$, say, but the present binary form ($m = 2$) will suffice for our later appeal to this equivalence (for Theorem 4.5.12):

PROPOSITION 1.2.6 *For a formula $D = (A_1 \wedge B_1) \vee \ldots \vee (A_n \wedge B_n)$, let Φ_D be the set of formulas $C_1 \vee \ldots \vee C_n$, where each C_i is either A_i or B_i ($1 \leq i \leq n$), and let D' be the conjunction of the formulas in Φ_D. Then D' is truth-functionally equivalent to D.*

Proof. Suppose D is true on some Boolean valuation. Then at least one of its disjuncts, $A_i \wedge B_i$ is true (on that valuation: words which will be omitted from now on). So for each $C_1 \vee \ldots \vee C_n \in \Phi_D$, C_i is true,

1.2. NON-MODAL BACKGROUND

in which case so is their conjunction D'. On the other hand, suppose that D is false (on some Boolean valuation). Then each of D's disjuncts, $A_i \wedge B_i$, is false, so for each i we can take C_i as a false conjunct, A_i or B_i, and then at least one conjunct of D' is false, namely $C_1 \vee \ldots \vee C_n$, is false, so D' is false. ∎

We turn to a point of contrast (Prop. 1.2.7) between classical propositional logic and its modal extensions, as we shall remark in Proposition 2.5.36 (p. 86). We need the notion of a (sentential) *context*, and specifically that of a 1-ary context, by which is meant a formula containing a distinguished propositional variable, p_i say. Putting another formula into this context is simply substituting the formula for all occurrences of p_i. We indicate this by writing the original formula as $C(p_i)$ and the result of making the substitution as $C(A)$, say, where A is the formula substituted for p_i. Sometimes, to emphasize the idea of the original formula as a formula with gaps in it (where p_i occurs) waiting to be filled by A, we call it $C(\cdot)$.

PROPOSITION 1.2.7 *(i) For any 1-ary Boolean connective # we have:*

$$\vdash_{CL} (\#p \wedge \#q) \to \#(p \wedge q),$$

and in fact (ii) for any 1-ary context $C(\cdot)$, we have:

$$\vdash_{CL} (C(p) \wedge C(q)) \to C(p \wedge q).$$

Proof. (i) is easily proved as there are only four cases to check, corresponding to the four 1-ary truth-functions. It is in any case just the special case of (ii) in which $n = 1$, which we now proceed to address. Let $r_1, \ldots, r_{n-1}$ be the propositional variables in the n-variable formulas $C(p)$ and $C(q)$ other than p and q, and let v be a Boolean valuation with $v(C(p) \wedge C(q)) = T$, with a view to showing that $v(C(p \wedge q)) = T$. Put z_i for $v(r_i)$ $(i = 1, \ldots, n-1)$, x for $v(p)$ and y for $v(q)$. Our formulas $C(p)$, $C(q)$, give an n-ary truth-function f with $v(C(p)) = f(x, z_1, \ldots, z_{n-1})$ and $v(C(q)) = f(y, z_1, \ldots, z_{n-1})$, the right-hand sides of which we will now abbreviate to $f(x, \vec{z_i})$ and $f(y, \vec{z_i})$. Since $v(C(p) \wedge C(q)) = T$, we have

$$f(x, \vec{z_i}) = T \quad \text{and} \quad f(y, \vec{z_i}) = T.$$

To be able to conclude, as desired, that $v(C(p \wedge q)) = T$, it suffices to show that $f(x \wedge y, \vec{z_i}) = T$, where here we have used $\wedge$ to denote the truth-function associated on any Boolean valuation, with the connective of the same name. Consider two cases. Case (1): $x = y$. Then $x \wedge y = x = y$ and so, in virtue of either of the hypotheses inset above, we are done. Case (2): $x \neq y$. Then since one of x, y, is T and the other is F, we have

$$\text{(2a)} \ f(T, \vec{z_i}) = T \quad \text{and} \quad \text{(2b)} \ f(F, \vec{z_i}) = T.$$

Since in case (2) $x \wedge y = F$, (2b) implies that $f(x \wedge y, \vec{z_i}) = T$. ∎

If we run a similar argument but with $\vee$ in place of $\wedge$ (on the right), we find that at a corresponding point in the proof, (2a) rather than (2b) is appealed to. Leaving the interested reader to reconstruct the proof, we just state the corresponding result:

PROPOSITION 1.2.8 *For any C as in Prop. 1.2.7 we have* $\vdash_{CL} (C(p) \wedge C(q)) \to C(p \vee q)$.

We can reduce some of the clutter in the formulation of Propositions 1.2.7(ii) and 1.2.8 by speaking in consequence relation terms:

$$C(p), C(q) \vdash_{CL} C(p \wedge q) \quad \text{and} \quad C(p), C(q) \vdash_{CL} C(p \vee q),$$

for any context $C(\cdot)$.

EXERCISE 1.2.9 Show that for any context C:
(i) $\vdash_{CL} C(p \wedge q) \to (C(p) \vee C(q))$;
(ii) $\vdash_{CL} C(p \vee q) \to (C(p) \vee C(q))$. ✠

To produce uncluttered versions of these last two examples, it helps to use not consequence relations but *generalized consequence relations*, also called 'multiple-conclusion' consequence relations, which relate pairs of sets of formulas. We do not give the general definition here (for which see, e.g., Shoesmith and Smiley [1050], p. 29, or Segerberg [1032], in which the defining conditions are (Refl), (Mono) and (Cut) from §2.2, which are suitable multiple-conclusion versions of the conditions given on p. 8 above). Instead we simply give a semantic definition of the generalized consequence relation of classical logic, notated $\Vdash_{CL}$: $\Gamma \Vdash_{CL} \Delta$ if and only if for all Boolean valuations v, if $v(C) = T$ for all $C \in \Gamma$ then $v(D) = T$ for some $D \in \Delta$. As in the case of Γ on the left with a consequence relation, here when spelling out a particular Γ, Δ, the formulas are usually listed without the accompanying set-collection brackets "{" and "}". Note that when $\Delta = \{A\}$ for some formula A, we have the same verdicts from $\vdash_{CL}$ and $\Vdash_{CL}$: $\Gamma \Vdash_{CL} A$ if and only if $\Gamma \vdash_{CL} A$. Now we can streamline (i) and (ii) of Exercise 1.2.9:

(i) $C(p \wedge q) \Vdash_{CL} C(p), C(q)$ (ii) $C(p \vee q) \Vdash_{CL} C(p), C(q)$.

The situation with modal logic illustrated for classical non-modal logic in Propositions 1.2.7 and 1.2.8, and Exercise 1.2.9 is very different. Any sublogic of the famous logic S5, which we shall meet in due course, suffices to provide counterexamples in the shape of contexts $C(\cdot)$ for which the results just cited do not hold. (Most obviously, for Prop. 1.2.7(ii), take $C(p)$ as $\Diamond p$.)

Since our final topic under the non-modal preliminaries rubric is less intimately connected with propositional logic, we include at this point an exercise to assess familiarity with the latter topic.

EXERCISE 1.2.10 Say for each of these "if and only if" claims whether it is true (in both directions), or true in the "if" direction and not the "only if" direction, or vice versa, or false in both directions. For all formulas A, B, C, and all sets of formulas Γ:
(i) $\Gamma \vdash_{CL} A \wedge B$ if and only if $\Gamma \vdash_{CL} A$ and $\Gamma \vdash_{CL} B$;
(ii) $\Gamma \vdash_{CL} A \vee B$ if and only if $\Gamma \vdash_{CL} A$ or $\Gamma \vdash_{CL} B$;
(iii) $\Gamma \vdash_{CL} \neg A$ if and only if $\Gamma \nvdash_{CL} A$;
(iv) $\Gamma, A \vee B \vdash_{CL} C$ if and only if $\Gamma, A \vdash_{CL} C$ and $\Gamma, B \vdash_{CL} C$;
(v) $\Gamma, A \wedge B \vdash_{CL} C$ if and only if $\Gamma, A \vdash_{CL} C$ or $\Gamma, B \vdash_{CL} C$. ✠

We turn to the subject of relations. Many properties of binary relations will be defined in the course of discussing the semantics of modal logic, but for the record the three best-known such properties are recalled here. A binary relation R on a set U is

- *reflexive* when for all $x \in U$, xRx
- *symmetric* when for all $x, y \in U$, if xRy then yRx
- *transitive* when for $x, y, z \in U$, if xRy and yRz, then xRz.

We will sometimes use a more formal notation and write the quantifiers in the style of first-order logic, in which case the conditions are understood as interpreted in a structure with universe U; thus, transitivity amounts to the following condition: $\forall x \forall y \forall z ((Rxy \wedge Ryz) \to Rxz)$. (A related condition appears as the first of two mentioned in Proposition 1.2.12 below: it diversifies the two occurrences of "y" in the condition of transitivity.) Conditions like those above associated with corresponding negative terminology – irreflexivity, asymmetry, and intransitivity – are obtained from those given by negating the rightmost atomic subformula. (Thus, in the case of the third of these, replace "xRz" – or "Rxz" –

1.2. NON-MODAL BACKGROUND

with its negation.) Asymmetry in this sense is to be distinguished from *antisymmetry*, which says instead that $\forall x \forall t (Rxy \land Ryx) \to x =, y$. Some combinations of these conditions have names of their own; in a particular a *pre-order* (or "quasi-order") is a transitive and reflexive binary relation. Antisymmetric pre-orders are called *partial orders*, while symmetric pre-orders are called *equivalence relations*.

For our final topic, we consider some ways of using a pair of (not necessarily distinct) subsets X, Y of some non-empty set U to induce binary relations on U. The first of these ways is familiar from standard discussions of set theory (and was mentioned under 'Conceptual Prerequisites' in Section 1.1); the second and third are less so, and the notation used here is far from standard.

- $X \times Y = \{\langle x, y \rangle \mid x \in X \text{ and } y \in Y\}$
- $X + Y = \{\langle x, y \rangle \mid x \in X \text{ or } y \in Y\}$
- $X \sim Y = \{\langle x, y \rangle \mid x \in X \text{ if and only if } y \in Y\}$

We define a binary relation R on U to be $\land$-*representable*, $\lor$-*representable*, or $\leftrightarrow$-*representable* just in case there exist $X, Y \subseteq U$ with $R = X \times Y$, $R = X + Y$, or $R = X \sim Y$, respectively.[12] We will be concerned especially with the first two of these, the Cartesian product and (what we might call) the Cartesian sum of X and Y, but before saying goodbye to the third, let us pause to observe that if any binary Boolean connective (of the *metalanguage*) $\#$ is employed in place of those appearing on the right in the bulleted definitions above, we get a corresponding notion of an $\#$-representable relation; for example, if we use (material) implication, we have $R \to$-representable just in case there exist $X, Y \subseteq U$ such that for all $x, y \in U$, xRy if and only if $x \in X \Rightarrow y \in Y$. Clearly, any such relation is also $\lor$-representable: just choose a new X as the complement (relative to U) of the old X (and keep Y the same). More generally, the three cases singled out above suffice for the subsumption of all such cases, in that – as you may check, if necessary, by working through all 16 cases (one for each binary truth-function), calling a binary relation *monadically representable* when it is similarly $\#$-representable for some binary Boolean $\#$ – a relation turns out to be monadically representatble just in case it is $\land$-representable, $\lor$-representable, or $\leftrightarrow$-representable. The following material relates only to $\lor$- and $\land$-representability, since these are the cases we shall encounter in later chapters (in Sections 5.5 and 6.3, respectively).

EXERCISE 1.2.11 Can one and the same binary relation be $\land$-representable and also $\lor$-representable? Can a binary relation be both $\lor$-representable and $\leftrightarrow$-representable? Give an example, in each case, or show why there can be no such examples (in the case of a negative answer). ✠

The notions of $\#$-representability for $\# = \land, \lor, \leftrightarrow$, are higher-order in form: they involve (existential) quantification over subsets of the given underlying set U. But they have simple first-order equivalents, as we indicate here for the first two. The proofs can be found in my [552], along with a first-order formulation of $\leftrightarrow$-representability.[13]

PROPOSITION 1.2.12 *A relation* $R \subseteq U \times U$ *is $\land$-representable just in case the following condition, written as an explicit first order condition, with quantifiers understood as ranging over U, is satisfied:*

$$\forall x \forall y \forall u \forall z ((Rxy \land Ruz) \to Rxz),$$

and R is $\lor$-representable just in case the following condition is satisfied:

$$\forall x \forall y \forall u \forall z (Rxz \to (Rxy \lor Ruz)).$$

---
[12]What we are calling $\land$-representable relations here are often referred to as *rectangular* relations in the literature, but the present terminology has been chosen to emphasize the parallel with the other cases, for which there is no corresponding established nomenclature.

[13]See Humberstone [562] for some typographical corrections to [552] as well as further related discussion, and for more of the latter, see also [566].

EXERCISE 1.2.13 Show that for $X, Y, X', Y' \subseteq U$:

(i) If $X \times Y = X' \times Y'$ and none of X, Y, X', Y' is $\varnothing$, then $X = X'$ and $Y = Y'$;

(ii) If $X + Y = X' + Y'$ and none of X, Y, X', Y' is U, then $X = X'$ and $Y = Y'$. ✠

PROPOSITION 1.2.14 *If for $R \subseteq U \times U$, $R = X + Y$, then R is transitive if and only if $X = U$ or $Y = U$ or $X \cap Y = \varnothing$.*

Proof. 'Only if': Suppose, arguing contrapositively, $X \neq U$ and $Y \neq U$ and $X \cap Y \neq \varnothing$. Since $X \neq U$, for some $a \in U$, $a \notin X$, and since $Y \neq U$, for some $b \in U$, $b \notin Y$. As $X \cap Y \neq \varnothing$, we may choose $c \in X \cap Y$. Since $R = X + Y$ and $c \in X$, cRb, and since $c \in Y$, aRc. Thus R is not transitive since these two conclusions would then imply aRb, contradicting the fact that $a \notin X$ and $b \notin Y$, as $R = X + Y$.

'If': Having $X = U$ or $Y = U$ would imply $R = U \times U$, the universal relation on U, which is certainly transitive. So it remains only to show that the transitivity of $R = X + Y$ follows also from the supposition that $X \cap Y = \varnothing$. So, making this supposition, assume further, for a contradiction, that R is not transitive. Thus there are $a, b, c \in U$ with aRb, bRc, but not aRc. Since not aRc, $a \notin X$ and $c \notin Y$. As $a \notin X$ while aRb, we must have $b \in Y$. As $c \notin Y$ while bRc, we must have $b \in X$. Thus $b \in X \cap Y$, contradicting our supposition that $X \cap Y = \varnothing$. ■

EXERCISE 1.2.15 Show that for $R \subseteq U \times U$ of the form $X + Y$, R is reflexive if and only if $X \cup Y = U$. ✠

1.3 Modal Logics: A Hierarchy of Classes

Now to obtain the language of modal logic we need to add to the language described in the previous section a new primitive 1-ary connective, written as $\Box$. The original motivation for this, as mentioned there, was that we want a way of expressing (alethic) modal concepts: saying that things are necessary, impossible or possible. Heuristically, for the moment, we may follow the *alethic* strand from Section 1.1 and think of $\Box A$ as saying "It is necessary that A", or just "Necessarily, A". (However, in general the safest procedure is to just read "$\Box$" aloud as "Box", in order to avoid the distractions of any particular intended interpretation; similarly, "$\Diamond$", introduced below, should be pronounced "Diamond".)

But wait – what about impossibility and possibility? Taking the latter first, we can indeed have a separate symbol for possibility, another modal operator alongside $\Box$, this companion operator being customarily written as $\Diamond$, but we don't need it to be taken as another primitive. (For the history of the symbols $\Box$ and $\Diamond$, see Hughes and Cresswell [535], Appendix 4. Those authors themselves use "L" and "M" for $\Box$ and $\Diamond$.) Instead we can define it in terms of $\Box$ and $\neg$:

$$\Diamond A =_{\mathsf{Df}} \neg \Box \neg A.$$

(A small aside: in intuitionistic modal logic, mentioned in Section 1.2, this would not be a sensible definition, for the same reason that one cannot define the existential quantifier ("$\exists$") in terms of the universal quantifier ("$\forall$") and negation in intuitionistic predicate logic.) We will usually drop the subscripted "Df" and generally record such definitions by saying that $\Diamond$ is to be understood in such a way that (for all formulas A), $\Diamond A = \neg \Box \neg A$.

Now that we have possibility as well as necessity, we can obviously express the idea that it is impossible that A, just by taking this as the negation of $\Diamond A$ – which amounts, on the above definition of $\Diamond$, to taking the claim that it is impossible that A as the claim that $\Box \neg A$: it is necessarily not the case that A. Note the order here: "Necessarily not", as opposed to "Not necessarily" ($= \neg \Box$), which means something quite

1.3. MODAL LOGICS: A HIERARCHY OF CLASSES

different. To get clear about the logical relations between necessity, possibility and impossibility, it helps to draw a square of opposition, and maybe even – an idea from Robert Blanché, adding the notions of contingency and noncontingency (see Exercises 1.3.1 and 1.3.8) – a *hexagon* of opposition; Humberstone [582] gives an introduction to modality concentrating on such figures, which raise interesting questions not entered into here, together with references to Blanché's work.

Digression. Note that we could equally well have taken $\Diamond$ as primitive, as many texts do, defining $\Box$ in terms of it: by putting $\Box A = \neg \Diamond \neg A$. While this would be slightly less convenient for present purposes, since several rules and conditions on modal logics are more succinctly formulated using $\Box$ – see for example what is called Scott's rule on p. 22 below – this alternative would arguably be more natural psychologically, to judge by such linguistic evidence as the following. According to de Haan [418], p. 55, "There are languages in which strong modality is expressed only by means of a weak modal and a double negation. That is, there is no separate strong modal in the language, and $\neg \Diamond \neg p$ is the only way to express $\Box p$." (Malagasy, Classical Tibetan and Japanese are cited as examples). De Haan continues: "There is also a logical equivalence between $\neg \Box \neg p$ and $\Diamond p$, but there seem to be no languages that have a weak modal that is made up of a strong modal and a double negation." Incidentally there is a slight danger in the use of the phrase 'double negation' here, because of its association with an immediately doubled negation "$\neg\neg$" which for classical (though not intuitionistic) logic could be dropped entirely. Matters are entirely different when a modal operator intervenes between the two occurrences of "$\neg$". There may be some confusion on this front in Cinque [180], on p. 126 of which two examples are given:

- He couldn't (possibly) not have accepted.
- He couldnae have no been no working. [One geolect of Scots English.]

Cinque glosses the second example with a paraphrase in standard English: "It is impossible that he has not been out of work", and goes on to say: "The fact that sentences containing more than three negations (canceling each other out) are in general unacceptable may be due to processing difficulties rather than to..." Quite how *three* negations could manage to cancel each other out – or what this would even *mean* – in the way two immediately adjacent negations (classically) do is anything but clear. From the gloss preferred, the relevant initial prefix is "$\neg\Diamond\neg\neg$" and it is the last two negations that might be described as cancelling each other out, rather than the first and either of the other two. (Of course we could rephrase "$\neg\Diamond\neg\neg$" to "$\Box\neg\neg\neg$" and reduce the last three "$\neg$"s to one – but again this doesn't seem to be describable as their all cancelling each other out.) We will in due course encounter a modal logic (called **KD!**) in which the two negations surrounding a $\Diamond$ (or for that matter, a $\Box$) cancel each other out in the sense that $\neg\Diamond\neg A$ is always equivalent to $\Diamond A$, but this fact itself shows that **KD!** does not do a very good job of reflecting the logical relations between any everyday notions of necessity and possibility – despite its theoretical importance in modal logic broadly conceived, as witnessed by the large number of times we shall have occasion to refer to it (which are listed in the index). Before leaving the subject of *could*, we should pause to note that this modal auxiliary, which superficially looks as though it simply stands in for either of *was able to* or *would be able to* (in standard English) in fact raises a few mysteries of its own. For example, in "I ran fast and was able to catch the bus" the *was able to* cannot be replaced by *could*, even though had the original been "I ran fast but was not able to catch the bus" the replacement of *was not able to* by *couldn't* is fine. And further, even with the "was able" formulation, there is a strong suggestion (to say the least) that the ability in question was not just possessed but *exercised*. Otherwise the preferred formulation would be "would have been able to". These examples come from Palmer [863], where some reflections at to what might explain them can be found. **End of Digression.**

EXERCISE 1.3.1 (*i*) Explain, in ordinary (but *clear*) English, the difference between saying "It is contingent that A" and saying "It is contingent whether A".
(*ii*) How would the two notions mentioned under (*i*) be represented in the language with $\Box$ and $\Diamond$ (alongside the various truth-functional connectives)?

(*iii*) Writing ∇A for "it is contingent whether A" (cf. the discussion in and after Exercise 1.3.8, p. 21 below), and supposing this is taken as primitive, can you supply a definition of $\Box$ in terms of ∇ and the Boolean connectives? (Assume here that for any A, A follows from $\Box A$. The more general situation is discussed in Cresswell [211] and Kuhn [680]; The present author's first moves in the area can be found in [569] and [576]. See further note 217 on p. 253 below.)

(*iv*) Let us make informal use of the consequence relation notation $\vdash$, as meaning that what is on the right should follow logically from what is on the left. Which of the following four would-be equivalences are plausible for necessity and possibility (and conjunction and disjunction)? If an equivalence fails, say which direction it fails in and why.

$$\Box(p \wedge q) \dashv\vdash \Box p \wedge \Box q \qquad \Box(p \vee q) \dashv\vdash \Box p \vee \Box q$$
$$\Diamond(p \wedge q) \dashv\vdash \Diamond p \wedge \Diamond q \qquad \Diamond(p \vee q) \dashv\vdash \Diamond p \vee \Diamond q.$$

✠

REMARK 1.3.2 Apropos of Parts (*i*) and (*ii*) of the above exercise, do not make the mistake of thinking that "it is contingent whether A" is just represented by $\Diamond A$. Certainly saying that something was possible would invite a hearer to infer that its negation was also possible, and hence that it was contingent whether it was the case, but this invitation is not based on an inference from the *content* of what is said, and thinking otherwise is like thinking that it is part of the *meaning* of "Some Fs are Gs" that some Fs are not Gs, as opposed to something suggested by making an assertion to the effect that some Fs are Gs and thereby inviting the hearer to infer that not all Fs are Gs, since otherwise that's what would have been said. This inference is what is called by those who attend to these matters a *conversational implicature*. See further Example 5.1.11, p. 349. ◀

In the class of arbitrary modal logics, the plausible principles from among those listed under Exercise 1.3.1(*iv*) will not be forthcoming (as provable implications, since we are treating modal logics as sets of formulas), but they will in some special classes, and in particular the class – of greatest interest to us – of *normal* modal logics. (In fact this is also true of a broader class: the *regular* modal logics.) But let's begin with the most general case.

A *modal logic* is a set S of formulas in a language with a functionally complete set of truth-functional connectives and the additional 1-ary connective $\Box$ (with $\Diamond$ taken as defined, as above) with the following properties:

- S contains all tautologies;

- S is closed under uniform substitution of arbitrary formulas for propositional variables;

- S is closed under Modus Ponens: i.e., if $A \to B \in S$ and $A \in S$ then $B \in S$.

The conspicuously missing feature of this definition is any reference to the specifically modal item of (primitive) vocabulary, $\Box$. That is because we are here looking at the broadest possible definition of what a modal logic might be, and not requiring any particular behaviour on the part of $\Box$. Well, by the first condition, we do need to have $p \vee \neg p$, for instance, in any modal logic, and therefore, by the second condition, substituting $\Box p$ uniformly in this formula for p (which means that since p occurs twice in the formula, we must make the substitution for both occurrences), we also have $\Box p \vee \neg \Box p$ in any modal logic. But this isn't really a special demand on $\Box$ – it's something that applies to whatever formula we substitute for p in the original tautology: for any modal logic S, and any formula A, we have $A \vee \neg A \in S$.

Instead of this "$\in$" notation, we will more often use a provability turnstile and write "$\vdash_S A$" for $A \in S$". But why are we talking about provability all of a sudden? Provability in what proof system? Well, the above characterization of what it takes to be a modal logic is, in effect, an axiomatization of the smallest modal logic. Take as axioms all tautologies (in the non-modal language), or perhaps in the

1.3. MODAL LOGICS: A HIERARCHY OF CLASSES

interests of economy, any finite axiomatic basis for this set (given the rule Modus Ponens, mentioned in the third condition), and, as rules, the following two: Uniform Substitution and Modus Ponens. (These are respectively the one-premiss rule that allows us to pass from a formula A and any substitution s to the formula $s(A)$ – called a *substitution instance* of A – which results from applying the substitution s uniformly to A, and the two-premiss rule which allows us to pass from A and $A \to B$ to B, for any formula B.) We will allow simultaneous uniform substitutions for several variables. Thus for example, if we have a formula A containing perhaps amongst others, the variables p, q and r, then for any formulas B, C, D, we count the result of replacing all occurrences of p, q and r in A by B, C, D, respectively, as a substitution instance of A, and hence as a possible conclusion for the application of the rule Uniform Substitution with A as premiss. (It is well known that in the presence of an infinite supply of propositional variables, one can get the effect of such simultaneous substitutions by successive single substitutions, but let us take the rule in this more generous form to start with.)

The provable formulas, or theorems, of this system are those that can be obtained from the axioms by applying the rules any number of times. There is another way of axiomatizing the same system which does not use the rule of Uniform Substitution, namely by allowing the axioms to be arbitrary substitution instances (in the language with $\Box$) of truth-functional tautologies – or again in the interests of finiteness, the instances of finitely many axiom-schemata which would suffice for the provability of all tautologies when Modus Ponens is used as the sole rule. Then the only rule we need is Modus Ponens. Note that although we haven't used Uniform Substitution as a rule in the axiomatization in this version, our set of provable formulas is still closed under the uniform substitution of arbitrary formulas for propositional variables – the rule of Uniform Substitution is still *admissible*, as it is put, so we are indeed dealing with a modal logic in the sense of the definition above.[14]

EXERCISE 1.3.3 Consider the smallest modal logic containing the formula (often called **T**): $\Box p \to p$. Along the lines of the discussion above, this is the same logic as the smallest modal logic containing all instances of the schema (also called **T**) $\Box A \to A$. Which of the following formulas are also provable? (For (vi) onward, recall that $\Diamond A$ is the formula $\neg \Box \neg A$; the formula at (viii) – or the corresponding schema – is often called **D**. We will see these and many other labels for modal principles, together with etymological explanations for most of them, in Section 2.1.)

(i) $\Box p \to \neg \neg p$ (ii) $\Box \neg \neg p \to p$ (iii) $\Box p \to \Box \neg \neg p$
(iv) $\Box p \to \neg \neg \Box p$ (v) $\Box \neg \neg \Box p \to p$ (vi) $\Box \neg \neg \Box p \to \Diamond \Box p$
(vii) $p \to \Diamond p$ (viii) $\Box p \to \Diamond p$ (ix) $\Diamond (p \vor \neg p)$.

☧

If S is a modal logic in the sense already defined, then we say S is

- *congruential* if for all formulas A, B, if $\vdash_S A \leftrightarrow B$ then $\vdash_S \Box A \leftrightarrow \Box B$,
- *monotone* if for all formulas A, B, if $\vdash_S A \to B$ then $\vdash_S \Box A \to \Box B$,
- *regular* if S is monotone and for all A, B, we have $\vdash_S (\Box A \land \Box B) \to \Box(A \land B)$, and
- *normal* if S is regular and in addition $\vdash_S \Box\top$.

The smallest normal modal logic is called **K**, after Kripke, a label which is also, as we shall see shortly, given to a particular axiom (or axiom schema), the normal modal logics also being definable as the modal logics containing this axiom/all instances of this schema and closed under the rule of *Necessitation*, taking us from A to $\Box A$. Because we are concentrating mostly on normal modal logics in what follows the above list of properties does not include one which will receive only occasional mention, namely the property of

---

[14] More generally, a logic is said to be *closed under* a rule or to have that rule *admissible* if whenever premisses for an application of the rule are provable in the logic, so is the conclusion.

being *antitone*, meaning that whenever the logic contains $A \to B$, it contains $\Box B \to \Box A$. If one wanted to give as an informal reading for $\Box$, 'it is impossible that', or more simply, just interpret $\Box$ as $\neg$, such a property would be expected – even if the use of the notation "$\Box$" might be less so. (Instead of *monotone* and *antitone*, the phrases *upward monotone* and *downward monotone*, respectively, are sometimes used.)

REMARK 1.3.4 Note that the second part of the official definition of normality above could be replaced by the condition of closure under Necessitation (the rule just described). Evidently for $A = \top$ we have the antecedent of this conditional, whose consequent is the second part of the above definition of normality. Conversely, if we have $\vdash_S \Box \top$ then for any modal logic S with $\vdash_S A$ we have $\vdash_S \top \to A$, so on the assumption that S is regular (indeed on the weaker assumption that A is monotone) $\vdash_S \Box \top \to \Box A$, and therefore since $\vdash_S \Box \top$, $\vdash_S \Box A$. ◀

There are many equivalent alternatives to each of these properties of modal logics – see Chellas [171] and Segerberg [1032] in particular. (The term "congruential" here was introduced by David Makinson; in Segerberg [1025], the term "classical" was used instead for this property. The latter term survives in Segerberg [1032] for a related but different property. (See again note 34, p. 39.) Congruentiality is evidently a matter of the interreplaceability of provably equivalent formulas in the scope of $\Box$; it is sometimes called the Replacement Property.)

Part (*iii*) of the following exercise should be looked at even by those skipping the exercises, because of the terminology introduced there.

EXERCISE 1.3.5 (*i*) Show that if S is a modal logic, then S is regular in the sense of the above definition just in case S is congruential and $\vdash_S \Box(A \wedge B) \leftrightarrow (\Box A \wedge \Box B)$, for all formulas A, B.
(*ii*) Show that if S is a modal logic, then S is regular just in case S is closed under the following rule, for all $n \geqslant 1$:

$$\frac{(A_1 \wedge \ldots \wedge A_n) \to B}{(\Box A_1 \wedge \ldots \wedge \Box A_n) \to \Box B}$$

(*iii*) Show that if S is a modal logic, then S is normal just in case S is closed under the above rule but now understood as applying for all $n \geqslant 0$. (In the $n = 0$ case, we understand the antecedent as not existing, so that the schematically represented condition just amounts to the formula B.) This rule is often described as the *rule of normality*, and with the $n \geqslant 1$ restriction, as the *rule of regularity*. With the restriction to $n = 1$ we have the *rule of monotony*, and with the restriction to $n = 0$, the *rule of Necessitation* – or, more frequently, just *Necessitation* or *Nec*. (Warning: some older – and even some not so old – sources use "rule of regularity" to mean *rule of monotony*.)
(*iv*) Show that any modal logic closed under the $n = 2$ case of the rule of regularity is closed under the full rule (i.e., also the $n = 1$ and $n \geqslant 3$ cases). (Thus a normal modal logic can be alternatively defined as a modal logic closed under the $n = 2$ and $n = 0$ cases of the rule of normality – i.e., as a Necessitation-closed regular modal logic.) ✠

REMARK 1.3.6 Monotone modal logics are also called *monotonic* modal logics, though the latter terminology is potentially confusing, especially if you are in the company of advocates of "non-monotonic logic," as explained in note 8 – which has nothing to do with a failure of a modal logic to be monotone in the present sense; for a textbook presentation of this topic, see Makinson [765]. (We will not be concerned with non-monotonic logic in that sense here, though it certainly comes up in discussions of belief revision and doxastic logic – see the Digression on p. 125 – as well as apropos of deontic logic: see Horty [525].) This also explains why the word "monotony" was used rather than "monotonicity", despite the awkward fact that this is the abstract noun not only for the technical adjective "monotone" but also the everyday adjective "monotonous". (The trouble is that there are three adjectives but only two nouns to go around.) ◀

1.3. MODAL LOGICS: A HIERARCHY OF CLASSES

Note that a modal logic S's being closed under the $n = 2$ case of the rule of regularity is not equivalent to (though it does imply) S's proving, for all A, B, the schema – often[15] labelled as here – mentioned under the original definition of regularity, namely:

Aggregation $\qquad \vdash_S (\Box A \land \Box B) \to \Box(A \land B)$,

because this leaves the monotone property out of account. To see this, interpret "$\Box$" as "$\neg$": now, even with only one of the conjuncts of the antecedent, and so certainly with both, the consequent is provably implied (in classical propositional logic), but of course we do not have the monotone property in this case. (Proposition 1.2.7 on p. 13 showed that all 1-ary contexts are similarly 'aggregative' – though this word was not used there – in classical (non-modal) propositional logic, something that we won't be surprised not to see in general for modal logics: consider putting $\Diamond$ for $\Box$ in the above aggregation schema. See further Humberstone [597], [600].)

EXAMPLE 1.3.7 Can the above point be illustrated by providing a non-monotone interpretation of $\Box$ for which the aggregation schema above holds, but in which neither $\Box A \to \Box(A \land B)$ nor the corresponding schema with "$\Box B$" in the antecedent holds? We certainly cannot do this with a truth-functional interpretation of $\Box$, since there are only four 1-ary truth-functions and one can check that none of them qualifies. Suppose, however, we pick a propositional variable p_i and interpret $\Box A$, for any formula A, as the formula $p_i \leftrightarrow A$. This is readily seen not to be monotone, but by Proposition 1.2.7(ii) it satisfies the aggregation schema, which now amounts to the following, for arbitrary A, B:

$$((p_i \leftrightarrow A) \land (p_i \leftrightarrow B)) \to (p_i \leftrightarrow (A \land B)).$$

Further, we do not have, for all formulas A, B, that the result of dropping the first or the second conjunct from the antecedent of the inset formula here still yields a tautologous implication. ◂

The above example interprets $\Box A$ as a certain context of A (namely "$p_i \leftrightarrow \_$") but not as a primitive or definable 1-ary connective – because of the intrusion of the formula p_i – by contrast with the earlier case of $\neg$ (mentioned before Example 1.3.7). An exercise illustrates that possibility, which, in view of the above point about the four 1-place truth-functions, will require a non-Boolean connective:

EXERCISE 1.3.8 Consider the 1-ary connective $\triangle$ with the intended reading of $\triangle A$ (for which this notation is fairly standard – cf. Humberstone [576] and references there cited) as "It is noncontingent whether A". (Recall Exc. 1.3.1.) We can define $\triangle$ in any modal logic with a suitable alethic interpretation for $\Box$ thus: $\triangle A = \Box A \lor \Box \neg A$. Show that $\triangle$ satisfies the aggregation principle (i.e., putting $\triangle$ for $\Box$ in the above formulation of the principle) – but *not* the version we get by omitting one of the conjuncts of the antecedents – while at the same time not being monotone. ✠

Apropos of this last exercise, it should be mentioned that in the extensive literature on using contingency or noncontingency as modal primitives rather than possibility or necessity – see the references in [576] and in the references they in turn cite – $\triangle$ (as in the preceding exercise) and $\triangledown$ (as in Exercise 1.3.1(iii)) are used respectively to symbolize noncontingency and contingency, meaning here its being (non)contingent *whether* such-and-such, rather than its being (non)contingent *that* such and such. Some further noncontingency-like notions will come up in the discussion in and on either side of Exercise 2.5.60 (p. 102) below.

---

[15] See Schotch and Jennings [1000] for an early occurrence of the word in this connection. Another term for the same principle is *agglomeration*. The latter term was used in a 1965 article by Bernard Williams for a deontic version of the principle; see Williams [1170]. See also Sorensen [1074] for a similar usage (apropos of *a priori* knowability). MacPherson [756] uses the term *adjunction* for aggregation. Warning: the term *aggregation* is also used in an unrelated to refer to a hypothetical process for combining ('aggregating') the beliefs or preferences of individuals in a group to arrive at a group belief or preference; see List and Pettit [736], for example.

Digression. The text Blackburn *et al.* [98] uses these symbols for something quite different, as does Zolin (occasionally in) [1226]. Bílková, Palmigiano, and Venema (in [94]) use "∇" – and often Δ and ∇ are written in place of $\triangledown$ and $\triangle$ – for something of a different syntactic category altogether: it takes a (finite) set of formulas and makes a formula, $\nabla(\{A, B, C\})$, to take a representative case by way of example, amounting to the conjunction of $\Box(A \vee B \vee C)$ with $\Diamond A \wedge \Diamond B \wedge \Diamond C$. In an otherwise useful survey article, de Rijke and Wansing [958] tell readers on their opening page that "it is contingent that A" can be represented as "$\neg \Box A$". In answer to Exercise 1.3.1 above, "it is contingent that A" should be represented by "$A \wedge \neg \Box A$" (or by "$A \wedge \Diamond \neg A$") and "it is contingent whether A" by "$\Diamond A \wedge \neg \Box A$" (or by "$\Diamond A \wedge \Diamond \neg A$"). De Rijke and Wansing's "$\neg \Box A$" expresses no kind of contingency, since it is entailed by A's being impossible ($\Box \neg A$, that is). Aristotle is often claimed to have confused possibility with contingency: see Ackrill [4], p. 152. While some caution is called for in view of the potential for difficulties over translation from Aristotle's Greek into contemporary English, it is still somewhat surprising to read in a contemporary discussion the claim – described as debatable but endorsed nonetheless – that "the possibility of φ entails the possibility of $\neg\varphi$ (and vice versa)", only to find this backed up in a footnote with a citation of Aristotle "as an ally of highest authority here". (Kahle, [634], p. 107.) Certainly von Wright [1205], p. 32, gets into a muddle along similar lines when discussing the epistemic analogues of these notions. ***End of Digression.***

A somewhat less obvious characterization of regularity can be given in terms of the schema **K** (alluded to in connection with the normal modal logic – the smallest such logic – of the same name, mentioned above):

K $\quad \Box(A \to B) \to (\Box A \to \Box B)$ (schematic form)

We give the same name to a representative instance (this phrase being explained in the terminological notes below) of this schema, namely:

K $\quad \Box(p \to q) \to (\Box p \to \Box q)$ (formula form)

Since we are assuming for all modal logics that they are closed under uniform substitution, it makes no difference whether we require a modal logic to contain all instances of the schema above, or simply to contain the single formula listed below it. A terminological point here: in the interests of clarity, we distinguish *instances* and *substitution instances*. A formula – or if we were working with them, a sequent – has formulas (or sequents) as *substitution instances*, while a schema (here, a formula schema, but we could consider sequent-schemata too) has formulas as *instances*. A *representative* instance of a schema (or scheme) is a formula resulting by replacing any distinct schematic letters by distinct propositional variables.

Digression on Sequent Formulations. The above distinction between instances of a schema and substitution instances of a formula applies at the level of sequents too. For example $(p \wedge \neg r) \to q, p \wedge \neg r \succ q$ is an *instance* of the sequent-schema $A \to B, A \succ B$ and a *substitution instance* of the sequent $p \to q, p \succ q$. The most natural way to use sequents in modal logic is one that preserves the non-modal equivalence between sequents and formulas got by conjoining their left-hand formulas as the antecedent of an implication whose consequent is the right-hand formula. Thus the rules of monotony, regularity and normality mentioned in Exc. 1.3.5(iii) can be more succinctly described in terms of the sequent-to-sequent rule:

$$\frac{A_1, \ldots, A_n \succ B}{\Box A_1, \ldots, \Box A_n \succ \Box B}$$

with appropriate restrictions on n (e.g., no restrictions at all for normality – this case of the rule sometimes being called 'Scott's Rule', after Dana Scott). In a similar vein it is reasonable – though unfortunately in some quarters a different policy has emerged – to call a consequence relation $\vdash$ (on the present language) monotone, regular or normal according as $\Gamma \vdash B$ implies $\Box \Gamma \vdash \Box B$ for any B and any Γ with $|\Gamma| = 1$,

1.3. MODAL LOGICS: A HIERARCHY OF CLASSES

or $|\Gamma| \geq 1$, or $|\Gamma| \geq 0$, respectively.[16] (Here $\Box\Gamma = \{\Box C \,|\, C \in \Gamma\}$ and $|\Gamma|$ is, as explained at p. 5, the cardinality of Γ.) One advantage of the sequent and consequence relation approaches is that they still provide a way of saying what follows from what, even in the absence of the connective $\to$, so that we can study modal logics in restricted (not functionally complete) fragments. For example, understood as consequence relations, the $\{\land, \top, \Box\}$-fragments of the logics **S4** and **S5**, introduced in Section 2.1, turn out to coincide, while this is no longer the case if one admits $\to$ in the language. **End of Digression.**

REMARKS 1.3.9 (*i*) Whether one thinks of normality in terms of Scott's Rule or in terms the corresponding condition on consequence relations, or – our main focus – on logics as sets of formulas, what it amounts to, somewhat loosely speaking, is that the conclusions of valid arguments with necessary premisses are themselves necessary. This idea has seldom been contest, at least for alethic interpretations of $\Box$. (An exception is mentioned under (*ii*) below.) There is occasional sympathy for a matching principle to the effect that the conclusions of valid arguments with contingent premisses are themselves contingent. Examples of the frequent sympathy expressed on this score as well as many counterexamples to the principle, taken at face value, can be found in Routley and Routley [974]. But perhaps the principle can be tweaked a little to capture what has won it so many adherents while avoiding the counterexamples a simple-minded interpretation of it faces. Efforts in this direction are made in Anderson and Belnap [20], §§5.2.1 and 22.2.1 (the latter contributed by Alberto Coffa), in Humberstone [548], and in (§5.2 of) Restall and Russell [956]. One important observation made by Anderson and Belnap is that discussion of this topic is greatly facilitated by a distinction between a statement which is necessarily true and a statement to the effect that something is necessarily true – a distinction between necessary and *necessitative* statements, as we may put it. (Anderson and Belnap actually say "necessitive" with the suggestion they are following the analogy of "negative", but since the verbs involved in the two cases are *negate* and *necessitate*, they have not actually followed through on the analogy.) This allows us to distinguish the claim, illustrated here in the one-premiss case, that no contingent statement entails a necessary statement from the claim that no non-necessitative statement entails a necessitative statement, though it has to be conceded that outside of the special setting in which Anderson, Belnap and Coffa discuss the matter (their favoured logic of entailment) the latter claim is no more immune to counterexamples than the former. The notion of necessativity is, however, quite interesting in its own right and we shall return to it briefly after we have seen enough modal logic to be able to say something about it (Example 4.1.13, p. 213).

(*ii*) Although the idea that necessary premisses themselves only necessarily imply necessary conclusions is generally accepted, Lazerowitz [686] was an early (1936) dissident voice, which because it concerns only a single premiss amounts to an attack on the axioms **K** – or at least its necessitated form. The argument turns on a further modal axiom, that called **T**: $\Box p \to p$ (introduced officially below in Section 2.1), as well as on the contrast between material and strict implication mentioned on p. 10 above. Substituting into **T** $p \to q$ for p gives us:

$$\Box(p \to q) \to (p \to q),$$

which so by an application of Necessitation gives us the following, in which the strict implication connective $\strictif$ of C. I. Lewis (see p. 46) is used to emphasize the point, with formulas $\Box(A \to B)$ written as $A \strictif B$:

---

[16]The unfortunate alternative policy, found in the influential Blok and Pigozzi [101], for instance, requires for the normality of $\vdash$ that we should have $A \vdash \Box A$ for all A, putting this 'horizontal' condition in place of the 'vertical' form of Necessitation which emerges from the Scott's Rule condition when $n = 0$ (or $|\Gamma| = 0$, in the formulation just given). For more of this complaint, see Humberstone [581] and [592]. Here we are concentrating on $\vdash$ as signalling the local as opposed to global consequence relation determined by a class of frames – see notes 115 and 117 (pp. 146, 148) – because it is the former that corresponds to argument validity, for which reason these relations are contrasted in subsection 2.23 of Humberstone [594] as inferential consequence and model consequence, respectively (alongside two other relations – point consequence, i.e., preservation of valdity at a point, another 'local' matter, and frame consequence, preservation of validity on a frame, another 'global' matter – this one introduced below on p. 80). These refinements will not concern us here.

$$(p \strictif q) \strictif (p \to q).$$

Then Lazerowitz argues that this represents a violation of the claim that necessary truth strictly imply only further necessary truths since it has a strict conditional strictly implying the corresponding material conditional and a merely material conditional is precisely the kind of thing that is not necessary but only contingent (thus, even when true, being such that it could have failed to be true). The rhetorical trick here – which captures the heart of Lazerowitz's reasoning in [686], although it does not purport to be a direction quotation and ignores various aspects of the discussion there (such as the use of propositional quantifiers) – is the use of the word "merely". This invites us to pass from the correct claim that the material conditional $p \to q$ does not say that it is necessary that if its antecedent is true then its consequent is true, to the incorrect claim that it says that it is not necessary that if its antecedent is true then its consequent is true. If one wanted to inject such a modal ingredient and introduce a special connective for *merely material* implication, say written as $\to_{\mathsf{mm}}$, one could of course do so, with $A \to_{\mathsf{mm}} B$ defined as $(A \to B) \land \neg \Box (A \to B)$ (or, for the second conjunct here, $\Diamond (A \land \neg B)$). But then the implication $(p \strictif q) \strictif (p \to_{\mathsf{mm}} q)$ would no longer be the necessitation of a substitution instance of the **T** axiom. ◀

Having rescued the **K** axiom from Lazerowitz's criticism in (*ii*) above, we return to investigation of its properties:

PROPOSITION 1.3.10 *A modal logic is regular if and only if it is monotone and contains the formula* **K**.

Proof. 'If': Suppose S is a monotone modal logic containing the formula **K**. Given the definition of regularity above, we have to show that the aggregation condition is satisfied: $\vdash_S (\Box A \land \Box B) \to \Box (A \land B)$ for any formulas A, B. S contains all substitution instances of tautologies and so in particular for any given A, B, we have

$\vdash_S A \to (B \to (A \land B))$;

thus by the monotone rule:

$\vdash_S \Box A \to \Box (B \to (A \land B))$.

Now because the formula concerned is a substitution instance of the formula **K**, we have:

$\vdash_S \Box (B \to (A \land B)) \to (\Box B \to \Box (A \land B))$.

A tautological consequence of this and the formula previously cited is $\Box A \to (\Box B \to \Box (A \land B))$, so reformulating this slightly (again just using truth-functional manipulations), we conclude that

$\vdash_S (\Box A \land \Box B) \to \Box (A \land B))$,

as required.

'Only if': Suppose S is a regular modal logic. We must show that S is monotone and that $\vdash_S$ **K**. Being monotone was built into the definition of regularity so it is only the second half of the claim that requires work. Since it is a truth-functional tautology, we have

$\vdash_S ((p \to q) \land p) \to q$.

So by the rule of regularity

$\vdash_S (\Box(p \to q) \land \Box p) \to \Box q$,

from which the formula **K** is only a step of truth-functional reformulation away. ■

The above proof assumes toward the end that regular modal logics are closed under the rule of regularity, so strictly speaking the proof is not self-contained, relying on Exc. 1.3.5(ii).

Another useful terminological convention, explicit in Segerberg [1025] though going back to McKinsey in the 1940s and 1950s, involves the prefix "quasi-". If "X" is an adjective for a class of modal logics which has a least element (a smallest modal logic to which the adjective applies, that is), and in particular, one

1.3. MODAL LOGICS: A HIERARCHY OF CLASSES

of the adjectives "congruential", "monotone", "regular", "normal", then we use "quasi-X" to apply to all modal logics extending the smallest X logic. For example, the smallest modal logic extending **K** (i.e., including amongst its theorems all the theorems of **K**) and the formula $\Box p \to p$ – a formula often known as **T** because of its role in axiomatizing a logic originally known as "t"[17] but now more commonly known as **T** (again) or **KT** (the label we follow Chellas [171] in using) – contains all the substitution instances $\Box A \to A$ of this formula, because closure under uniform substitution is part of the definition of what it takes to be a modal logic, but it does not include the necessitation $\Box(\Box p \to p)$ of this formula, since closure under Necessitation[18] was not part of the definition of a modal logic, only of a normal modal logic. Thus the modal logic just described is quasi-normal but not normal. (This is not the logic **KT**, the normal extension of **K** by **T**, i.e., the smallest normal modal logic containing the formula **T**. In the notation in play in Exercise 2.5.25 below, **KT** is the logic **K** ⊕ **T** whereas the logic we have just been considering is **K** + **T**. The "⊕" notation itself is introduced before Exc. 2.5.10, p. 76.)

EXERCISE 1.3.11 (*i*) Show that any quasi-normal modal logic is normal if and only if it is congruential. (*ii*) Is it similarly true that any quasi-regular modal logic is regular if and only if it is congruential. ✣

The "hierarchy" terminology of the title of this section is intended to draw attention to the linear arrangement evident in the fact that the normal modal logics comprise a proper subset of the regular modal logics, which compromise a proper subset of the monotone modal logics, which in turn comprise a proper subset of the congruential modal logics (themselves properly included within the class of all modal logics). But the class of quasi-normal modal logics and that of the congruential modal logics are incomparable with respect to inclusion: **K** + **T**, mentioned before the previous exercise, is quasi-normal without being congruential (equivalently, for quasi-normal logics: without being normal), for example, while the smallest congruential modal logic (**E** – see note 32) is not quasi-normal. What the various classes of modal logics we have been attending to have in common is not (speaking loosely) that they are linearly ordered with respect to inclusion, but rather that they are all 'families' of logics in the sense of the following definition:[19] a set Λ of modal logics is a *family* if and only if for any $\Lambda_0 \subseteq \Lambda$, we have

$$(1) \quad \bigcap \Lambda_0 \in \Lambda \qquad \text{and} \qquad (2) \quad \bigcap \{S \in \Lambda \mid S \supseteq \bigcup \Lambda_0\} \in \Lambda.$$

Condition (1) here on Λ implies that for any $S, S' \in \Lambda$, we have $S \cap S' \in \Lambda$ (take $\Lambda_0 = \{S, S'\}$), and also tells us that there is a weakest logic in Λ – one included in all of them, that is (take Λ_0 as Λ itself). Condition (2) demands that there is always a weakest logic in the family stronger than all those in a given set of logics in the family. (In lattice-theoretic terminology this is their least upper bound, or join, in the lattice; see further the Digression on p. 39.) Again this guarantees the 'binary' version, that for any pair of logics in the family there is a weakest logic stronger than both which is in the family.[20]

EXERCISE 1.3.12 (*i*) Explain why the class of normal modal logics constitutes a family of modal logics in the sense just defined; similarly for the class of quasi-normal modal logics.
(*ii*) (This question presumes some familiarity with the names for modal logics and modal principles introduced in Section 2.1.) Which of the following classes of modal logics are families in the current sense? (Give reasons.)

---

[17] Another early name for this logic was "M" – e.g. in von Wright [1205].
[18] In this sentence we use *necessitation* for the result of prefixing a $\Box$ to a formula and *Necessitation* (with a capital N) specifically for the rule which derives this result as provable from the provability of the original formula.
[19] This terminology is not standard. What we are calling a family of logics is the universe (i.e., underlying set) of a complete lattice of logics, with inclusion ($\subseteq$) as the associated partial order. (Here familiarity with basic concepts concerning lattices is assumed.) We return to this in the Digression on p. 39.
[20] The join of a pair of logics S_1, S_2, belonging to more than one family will depend on which lattice we consider; we use the notations $S_1 + S_2$ for their join in the lattice of all modal logics, and $S_1 \oplus S_2$ for their join in the lattice of all normal modal logics. In fact, as officially introduced below (see Exercise 2.5.25, p. 81) the notation "$S_1 \oplus S_2$" makes sense even for non-normal modal logics S_1, S_2 (and denotes the smallest normal modal logic containing $S_1 \cup S_2$); indeed, it makes sense for arbitrary sets of formulas – not even specifically modal logics – S_1 and S_2.

(a) The set of normal modal logics S such that $\mathbf{S4} \subseteq S \subseteq \mathbf{S5}$.
(b) The set of consistent modal logics.
(c) The set of normal modal logics in which $\mathbf{T}$ is not provable.

We continue to take a special interest in the family of normal modal logics. To give an idea of what a formal proof from the axioms and rules of a particular axiomatization of a normal modal logic might look like we consider the axiomatization of the smallest such logic $\mathbf{K}$ along the following lines:

Axioms:

- All substitution instances (in the present language) of truth-functional tautologies
- all instances of the schema $\mathbf{K}$: $\Box(A \to B) \to (\Box A \to \Box B)$

Primitive Rules

- Modus Ponens $(A, A \to B / B)$
- Necessitation $(A / \Box A)$.

Note that the slash notation, as with "$A/\Box A$" for Necessitation, is simply a more compact version of the 'rule display' notation with a horizontal line separating premiss(es) from conclusion – as in Exercise 1.3.5(ii), where this rule first appeared (see part (iii) of that exercise). If one wanted an axiomatization along similar lines of the normal modal logic $\mathbf{KT}$, one would simply add to the list of axioms that not only all instances of the schema $\mathbf{K}$, but also all instances of the schema $\mathbf{T}$ (i.e., $\Box A \to A$, mentioned above), and so on for other such logics.

As *derived* rules – to expedite proofs – we use a family of rules collectively dubbed "TF" (for "truth-functional reasoning"), comprising the following (for which we revert to the full 'rule display' notation), for each $n \geqslant 0$:

$$(\text{TF}) \quad \frac{A_1 \quad \ldots \quad A_n}{B}$$

whenever the formula $(A_1 \wedge \ldots \wedge A_n) \to B$ is a substitution instance of a tautology. By allowing $n = 0$, we subsume the first batch of axioms listed above (i.e., substitution instances of tautologies), as applications of this rule. Note that Modus Ponens is a special case of this derived rule, but for old time's sake we record its effects separately, with the aid of the annotation "MP" (rather than "TF"). Let us also make use of the derived rules $\Box$Mono and $\Diamond$Mono:

$$(\Box\text{Mono}) \quad \frac{A \to B}{\Box A \to \Box B} \qquad (\Diamond\text{Mono}) \quad \frac{A \to B}{\Diamond A \to \Diamond B}$$

The first rule is derivable by the same considerations as show that normality implies monotony; for the second rule, see Prop. 1.4.1 below.[21] We illustrate these derived rules in action to give short proofs in Examples 1.3.13 and 1.3.16.

EXAMPLE 1.3.13 A proof in $\mathbf{K}$, as axiomatized here, of the formula $(\Box p \vee \Box q) \to \Box(p \vee q)$:

(1) $p \to (p \vee q)$ TF
(2) $\Box p \to \Box(p \vee q)$ 1 $\Box$Mono
(3) $q \to (p \vee q)$ TF
(4) $\Box q \to \Box(p \vee q)$ 3 $\Box$Mono
(5) $(\Box p \vee \Box q) \to \Box(p \vee q)$ 2, 4 TF

Thus by Uniform Substitution, we have $\vdash_{\mathbf{K}} (\Box A \vee B) \to \Box(A \vee B)$, for all formulas A, B.

---

[21]These derived rules are called DR1 and DR2, respectively, in the classic text Hughes and Cresswell [535].

1.3. MODAL LOGICS: A HIERARCHY OF CLASSES

EXAMPLE 1.3.14 Let us show that the following rule, with a conclusion of the same form as that of $\Box$Mono but a different premise, is also derivable on the basis of our axiomatization of **K**:

$$\frac{\Diamond A \to \Box B}{\Box A \to \Box B}$$

Derivation:

(1)	$\Diamond A \to \Box B$	Given, by hypothesis.
(2)	$\neg\Box\neg A \to \Box B$	1 Def. $\Diamond$
(3)	$\Box\neg A \lor \Box B$	2 TF
(4)	$(\Box\neg A \lor \Box B) \to \Box(\neg A \lor B)$	See 1.3.13
(5)	$\Box(\neg A \lor B)$	3, 4 MP
(6)	$(\neg A \lor B) \to (A \to B)$	TF
(7)	$\Box(\neg A \lor B) \to \Box(A \to B)$	6 $\Box$Mono
(8)	$\Box(A \to B)$	5, 7 MP
(9)	$\Box(A \to B) \to (\Box A \to \Box B)$	**K**
(10)	$\Box A \to \Box B$	9, 10 MP

◀

EXERCISE 1.3.15 Which (if any) of the following three rules, variations on the theme of Example 1.3.14, is/are also derivable? (Justification not required; in fact a negative answer would be hard to justify definitively on the basis of the material so far presented.)

$$\frac{\Box A \to \Diamond B}{\Box A \to \Box B} \qquad \frac{\Box A \to \Diamond B}{\Diamond A \to \Diamond B} \qquad \frac{\Diamond A \to \Box B}{\Diamond A \to \Diamond B}$$

✠

EXAMPLE 1.3.16 A proof from this axiomatization of **K** for the formula $\Box(p \lor q) \to (\Diamond p \lor \Box q)$:

(1)	$\Box(\neg p \to q) \to (\Box\neg p \to \Box q)$	**K**
(2)	$\Box(\neg p \to q) \to (\neg\Box\neg p \lor \Box q)$	1, TF
(3)	$\Box(\neg p \to q) \to (\Diamond p \lor \Box q)$	2, Def. $\Diamond$
(4)	$(p \lor q) \to (\neg p \to q)$	TF
(5)	$\Box(p \lor q) \to \Box(\neg p \to q)$	4, $\Box$Mono
(6)	$\Box(p \lor q) \to (\Diamond p \lor \Box q)$	3, 5 TF

◀

EXERCISE 1.3.17 (*i*) Could line (6) in the above proof be obtained directly from line (3) by appeal to the derived rule (TF)? (Explain why or why not.)

(*ii*) Provide a proof using the current axiomatization of **K** (in the style of Example 1.3.16) for the three formulas

$$(\Box p \land \Diamond q) \to \Diamond(p \land q) \qquad \Diamond(p \lor q) \leftrightarrow (\Diamond p \lor \Diamond q) \qquad \Diamond(p \to q) \leftrightarrow (\Box p \to \Diamond q).$$

✠

We have seen numerous equivalent characterizations of normality for modal logics, which could equally well be taken as alternative axiomatizations of **K**, e.g., in the original definition at p. 19 and in Exercise 1.3.5(*iii*), to say nothing of that of our latest axiomatization of **K**: normal modal logics as those modal logics containing the **K** axiom and closed under Necessitation. But here we will have a look at a failed characterization.

EXAMPLE 1.3.18 Font and Jansana [318], p. 448, purport to characterize **K** as the smallest set of formulas containing all substitution instances of truth-functional tautologies, all instances of the schema

$$\Box(A \wedge B) \leftrightarrow (\Box A \wedge \Box B)$$

and closed under the rules Modus Ponens and Necessitation. This amounts to saying that normal modal logics are modal logics containing all instances of the inset schema – a biconditional version of the aggregation schema – and closed under Necessitation. (The same mistake appears at the base of p. 276 of Pelletier and Urquhart [879]; the 'correction' details in our bibliographical entry for this paper concern the correction of an unrelated error.)

However, this is not an alternative axiomatization of **K** at all, but of some non-normal modal logic which is not congruential (and *a fortiori* not monotone). To see this, we isolate a notion of validity which is possessed by the axioms and preserved by the rules, but is not possessed by the **K**-provable formula $\Box\neg\neg p \to \Box p$. First, we need an auxiliary notion applying to conjuncts of conjuncts of ... of conjuncts of (conjunctive) formulas. Define the relation 'is a conjunct* of' to be the smallest binary relation between formulas satisfying the following conditions, for all formulas C, D and E: C is a conjunct* of C, and if C is a conjunct* of D then C is a conjunct* of $D \wedge E$ and also a conjunct* of $E \wedge D$. (This makes the conjunct* relation the *ancestral* – see the discussion before Theorem 2.4.10 below (p. 70) – of the usual 'is a conjunct of' relation.) For example, q is a conjunct* of $p \wedge (q \wedge (r \vee p))$, but q is not a conjunct* of $p \wedge \neg\neg q$ or of $p \wedge (q \vee q)$: being equivalent, according to this or that logic, to a conjunct* of a formula is not the same as being such a conjunct*, which is a purely syntactical matter.

Let $\mathcal{V}$ be the class of Boolean valuations v satisfying the further condition that for all formulas A, $v(\Box A) = \mathrm{T}$ iff no propositional variable p_i is a conjunct* of A. For purposes of this Example only, understand a formula A to be *valid* just in case for all $v \in \mathcal{V}$ ($\mathcal{V}$ as just fixed), we have $v(A) = \mathrm{T}$. To check that all axioms of the present axiomatization are valid in this sense, note that these are either substitution instances of truth-functional tautologies, in which case they are valid because all valuations in $\mathcal{V}$ are Boolean (i.e. #-Boolean for each primitive Boolean connective # in the language, recalling Remark 1.2.3), or instances of the biconditional inset above: in which case we use the fact that the definition of *conjunct** above implies that $A \wedge B$ has a propositional variable (or indeed any other formula) as a conjunct* if and only if either A does or B does, so $v(\Box(A \wedge B)) = \mathrm{F}$ iff $v(\Box A) = \mathrm{F}$ or $v(\Box B) = \mathrm{F}$, if $v \in V$; thus for all such v, $v(\Box(A \wedge B)) = \mathrm{T}$ iff $v(\Box A) = \mathrm{T}$ and $v(\Box B) = \mathrm{T}$, which means that $v(\Box(A \wedge B)) = \mathrm{T}$ iff $v(\Box A \wedge \Box B) = \mathrm{T}$, since v is $\wedge$-Boolean.

It remains to check the rules. Modus Ponens preserves validity in the current sense because the valuations in $\mathcal{V}$ are $\to$-Boolean. For Necessitation, suppose that $v(\Box A) = \mathrm{F}$ for some $v \in \mathcal{V}$. Then A has some propositional variable p_i as a conjunct*. Let v' be any valuation in $\mathcal{V}$ with $v'(p_i) = \mathrm{F}$; since v' is a Boolean valuation, $v'(A) = \mathrm{F}$, and so A is not valid. Thus all theorems of the present system are valid, but $\Box\neg\neg p \to \Box p$ is not valid, since for any $v \in \mathcal{V}$, $v(\Box\neg\neg p) = \mathrm{T}$, as $\neg\neg p$ has no propositional variable as a conjunct*, while $v(\Box p) = \mathrm{F}$, as p has p as a conjunct*. Thus $v(\Box\neg\neg p \to \Box p) = \mathrm{F}$. This implicational formula is therefore not provable from the present axiomatization, which accordingly gives a modal logic strictly weaker than **K**.

While that disposes of the claim that the biconditional schema at the start of this example serves over truth-functional logic with the aid of Necessitation and Modus Ponens to axiomatize **K**, a conceptually interesting point arises over the fact alluded to in the previous paragraph, to the effect that all valuations in $\mathcal{V}$ treat $\Box$-formulas (and indeed all fully modalized formulas, this notion being defined at p. 56) in the same way: for any such formula A, $v(A) = v'(A)$ for any $v, v' \in V$. In particular, to pick up on the illustration just used, all valuations in $\mathcal{V}$ verify $\Box\neg\neg p$ and all falsify $\Box p$. Now this means that $\neg\Box p$ is valid (in the sense of the present discussion) while $\neg\Box\neg\neg p$ is not.[22] Thus the class of valid formulas is not a

---

[22] We could equally well choose $\neg\Box(p \to p)$ in place of the latter formula. This variant is an illustration of the failure of a system – (propositional) S13 – due to David Kaplan, and described in §6 of Cocchiarella [182], to be closed under Uniform Substitution.

modal logic, not being closed under Uniform Substitution (in the present instance, of $\neg\neg p$ for p in $\neg\Box p$). The interest of this point is that Font and Jansana [318] actually uses the axiom $\Box(p \wedge q) \leftrightarrow (\Box p \wedge \Box q)$ rather than (all instances of) the axiom schema given above, so that along with Modus Ponens and Necessitation, they use Uniform Substitution as an additional primitive rule. In this setting the analogue of the above inductive argument (by induction on the length of proofs) to the conclusion that all provable formulas are valid (in our current sense) would falter in the inductive step for the case in which a formula is derived from one having a shorter proof by an application of Uniform Substitution, since, as just noted, this rule does not preserve validity. This makes no difference to our conclusion that the logic axiomatized is strictly weaker than **K**, however, since (as noted in the discussion before Exc. 1.3.3) the axiomatization using the axiom-schema has the same theorems as that using the axiom with Uniform Substitution, the earlier argument shows that [318] errs in claiming to have produced an axiomatization of (the set of theorems of) **K**. However, if one wants to give an inductive argument along the above lines for the axiomatization using $\Box(p \wedge q) \leftrightarrow (\Box p \wedge \Box q)$ and Uniform Substitution, this can easily be done by changing the property one shows is possessed by the axioms and preserved by the rules (but lacked by $\Box\neg\neg p \to \Box p$) from validity as defined above to validity$^+$, where a formula is defined to be *valid*$^+$ just in case (it and) all its substitution instances are valid in the original sense (i.e., true on every $v \in \mathcal{V}$). The argument is as before except that we now need to check that Uniform Substitution preserves validity$^+$, which requires only the observation that the relation 'is a substitution instance of' is transitive. (Even further from being a satisfactory axiomatization of **K**, or a satisfactory definition of normality, than that offered by Font and Jansana in [318] is that given in Miyazaki [812], p. 304, or [813], p. 382, in which only the same rules as are cited in [318] but with only one half (the "$\to$" direction) of the biconditional axiom $\Box(p \wedge q) \leftrightarrow (\Box p \wedge \Box q)$.) The oversight here drawn attention to is easily corrected without jeopardising the results of Font and Jansana (or Pelletier and Urquhart,[879]; mentioned earlier, or Miyazaki) – which go well beyond the technical level of anything explained in the present survey; it simply shows how easy it is to make a mistake in this elementary part of the subject. ◀

EXERCISE 1.3.19 Correcting the axiomatization described with Necessitation and $\Box(A \wedge B) \leftrightarrow (\Box A \wedge \Box B)$ in 1.3.18 by explicitly adding the Congruentiality rule (from $A \leftrightarrow B$ to $\Box A \leftrightarrow \Box B$) would give us two modal (i.e. $\Box$-involving) proper rules. Show that we could get an equivalent axiomatization with only one such rule by dropping Necessitation in favour of the additional modal axiom $\Box\top$ (as in the definition of normality at p. 19). ✠

EXERCISE 1.3.20 Consider the following claim and the putative proof of it given below:

CLAIM. If S is a modal logic closed under Necessitation and containing all instances of the aggregation schema $(\Box A \wedge \Box B) \to \Box(A \wedge B)$, then S is normal.

Proof. It will suffice to show that for all $n \geq 0$, if $\vdash_S (A_1 \wedge \ldots A_n) \to B$ then $\vdash_S (\Box A_1 \wedge \ldots \Box A_n) \to \Box B$. We prove the implication here by induction on n.

Basis Case. $n = 0$: this case is given to us by the supposition S closed under Necessitation.

Inductive Step. Suppose that we have the result for a given n (the inductive hypothesis) and want to show that it holds for $n+1$, to which end suppose further that $\vdash_S (A_1 \wedge \ldots A_n \wedge A_{n+1}) \to B$. The inductive hypothesis, taking the conjunction $A_n \wedge A_{n+1}$ as the nth conjunct, that $\vdash_S (\Box A_1 \wedge \ldots \wedge \Box(A_n \wedge A_{n+1})) \to \Box B$. The supposition that $\Box$ is aggregative in S tells us that $\vdash_S (\Box A_n \wedge \Box A_{n+1}) \to \Box(A_n \wedge A_{n+1})$. But from by truth-functional reasoning from these last two S-theorems we get the desired conclusion that $\vdash_S (\Box A_1 \wedge \ldots \wedge \Box A_n \wedge \Box A_{n+1}) \to \Box B$. ∎

The discussion provided by Example 1.3.18 shows us that the Claim in the above exercise is in fact false, since we saw in 1.3.18 that even supposing a modal logic S was closed under Necessitation and contained all instances of the aggregation schema *as well as the converse schema* was not enough to secure normality. So: where does the proof go wrong? ✠

The would-be inductive proof in the above exercise is not unlike a standard example of fallacious reasoning by induction to the conclusion that everyone in a given room has the same age (by induction on the number – assumed finite – of people in the room in question), or that all billiard balls have the same colour, to cite the form this example takes at p. 47 of Bunch [129], *q.v.* for further details; for an interesting variation, see Ramsamujh [942] and again, Chapter 5 (esp. 5.3.2) of Gunderson [417].

1.4 Refinements and Generalizations

Instead of considering 1-ary $\Box$ and $\Diamond$ operators some authors have instead consider the more general case of n-ary such operators, with

$$\Diamond(A_1, \ldots, A_n) = \neg \Box(\neg A_1, \ldots, \neg A_n).$$

They are amenable to a semantic treatment similar to that reported for the 1-ary case in Section 2.3, using $(n+1)$-ary accessibility relations in place of the binary relations there in play.[23] A more refined terminology would be called for in the classification of such modal logics, since (for a given choice of $n > 1$) the normality condition might be satisfied with respect to one position, but only the weaker condition of congruentiality, with respect to another. For instance, suppose that $n = 2$; congruentiality in the second position and normality in the first, for a logic S, amounts to closure under the following pair of rules, addressing these conditions respectively; the second rule is meant to apply for any $m \geqslant 0$ (as in Exc. 1.3.5(iii), though we have changed 'n' to 'm' to avoid a collision with n as the currently supposed arity of $\Box$):

$$\frac{A \leftrightarrow B}{\Box(C,A) \leftrightarrow \Box(C,B)} \qquad \frac{(A_1 \wedge \ldots \wedge A_m) \to B}{(\Box(A_1,C) \wedge \ldots \wedge \Box(A_m,C)) \to \Box(B,C)}$$

We will not go further into such developments here or their (informal) semantical motivation, on which see the references cited in note 23.

Another generalization of greater relevance to subsequent chapters would be to have several primitive (1-aet) $\Box$ operators rather than just one. To keep things simple, let us consider this possibility for the case in which each such operator is, as on the original definition of a modal logic, 1-ary. But now we have n of them: $\Box_1, \ldots, \Box_n$, say. We have on our hands an n-modal logic (usually called monomodal, bimodal, trimodal, etc., if $n = 1, 2, 3$, etc.). If we impose in respect of each of them the same restrictions the earlier hierarchical terminology transfers unproblematically. The logic is congruential if the condition imposed on S's treatment of $\Box$ for congruentiality in the case of monomodal S is satisfied in the n-modal case by each of $\Box_1, \ldots, \Box_n$. Likewise with monotony, regularity, and normality. If we want to consider knowledge and belief in a logic rich enough to represent both of them (e.g., to record logical relations between the two notions), we will need a bimodal logic; in the case of the first candidate bimodal logic for such an epistemic–doxastic interpretation, from Hintikka [490], this was a normal bimodal logic – and we will see more of it below, especially in Chapter 5. Another famous "binormal" (for short) case is that of tense logic, in which one $\Box$ operator is used for talking about the future and the other for talking about the past; again we return to this below (Section 3.1). But the possibility just canvassed in the case of a single n-ary $\Box$ operator arises here too, since our various operators might not behave uniformly in respect to the properties we have been considering. We might have one of them being normal, and the other not even being congruential, for example. (This is what happens with a mixed alethic–epistemic logic treated in Williamson [1179].) This description is assumed to be readily intelligible. When we say that $\Box_i$ is normal in S, for instance, what is meant is that S is closed under the rule of normality – from Exc. 1.3.5, p. 20 – as formulated with $\Box_i$ in place of $\Box$.

---

[23]See Schotch and Jennings [1000], p. 271, where the idea is credited to earlier unpublished work of R. Goldblatt. We touch on the semantic side of this in note 210, p. 245.

1.4. REFINEMENTS AND GENERALIZATIONS

Having introduced this last refinement – of a given modal operator's being congruential, monotone, regular or normal, according to a modal logic – for the sake of multimodal logics, we can see it as being of use in the monomodal case, since although there we have only one primitive modal connective, we have also various other 1-ary connectives derivable by definition from the primitives. For example, in any monotone (mono)modal logic – which in this operator-specific way of speaking, means a logic in which $\Box$ is monotone, then $\Diamond$ is also monotone:

PROPOSITION 1.4.1 *If S is a monotone modal logic, then $\Diamond$ is monotone according to S.*

Proof. Suppose $\vdash_S A \to B$. Then by truth-functional reasoning $\vdash_S \neg B \to \neg A$, so since S is monotone, $\vdash_S \Box \neg B \to \Box \neg A$, whence by truth-functional reasoning again $\vdash_S \neg \Box \neg A \to \neg \Box \neg B$, i.e., $\vdash_S \Diamond A \to \Diamond B$. ■

Thus in any regular – or, further specializing, any normal – modal logic, $\Diamond$ is monotone. However, the analogue of Prop. 1.4.1 with both occurrences of "monotone" replaced by "regular" (or by "normal") would not be correct: in any modal logic intended for a plausible alethic interpretation we would not want to be able to prove the aggregation principle for $\Diamond$ – $(\Diamond A \land \Diamond B) \to \Diamond(A \land B)$, because we would want to distinguish the weaker claim that two things are each possible from the claim that the two things are compatible (or *compossible*, as it is sometimes put).[24]

We can also apply the monotone/regular/normal terminology usefully to strings of modal operators. An *affirmative modality* is a string of "$\Box$"s and "$\Diamond$"s, which can be regarded as a composite 1-ary connective (or 'operator') in its own right. (Since we are not taking $\Diamond$ as primitive, a string of $\Box$'s and $\Diamond$'s is really a string of occurrences of "$\Box$" and "$\neg \Box \neg$". A *modality* – without the "affirmative" – is usually taken to be any string of "$\Box$"s, "$\Diamond$"s, and "$\neg$"s, which, alluding to the primitive vocabulary, means any string of $\Box$'s and $\neg$'s. The term "modality" is understood more liberally in Zolin [1226], to mean any formula constructed from a single propositional variable and thought of as a context into which a formula may be fitted by substituting it uniformly for that variable – these are the 'modal functions' of, e.g., Hughes and Cresswell [535] – and modalities in the sense defined are called *linear* modalities.) In the proof below, we use numerically subscripted "O" to stand for a "$\Box$" or "$\Diamond$"; "O" here is mnemonic for (modal) Operator, and has no connection with the italic "O" – itself suggestive of Ought or Obligatory – for a deontically interpreted $\Box$ operator.

PROPOSITION 1.4.2 *In a monotone modal logic, any affirmative modality is monotone.*

Proof. We argue by induction on k that where $O_1 O_2 \ldots O_k$ is an affirmative modality of length k, then $\vdash_S A \to B$ implies $\vdash_S O_1 O_2 \ldots O_k A \to O_1 O_2 \ldots O_k B$.

Basis Case: $k = 0$. Then $O_1 O_2 \ldots O_k A \to O_1 O_2 \ldots O_k B$ is just the formula $A \to B$ and the result holds automatically.

Inductive Step: Assume (the "inductive hypothesis") that we have the result for affirmative modalities of length $< k$ and want to show that it holds for $O_1 O_2 \ldots O_k$. Suppose that $\vdash_S A \to B$, for our monotone modal logic S. Then by the inductive hypothesis $\vdash_S O_2 \ldots O_k A \to O_2 \ldots O_k B$. If O_1 is $\Box$ we have the result that $\vdash_S O_1 O_2 \ldots O_k A \to O_1 O_2 \ldots O_k B$ immediately since the definition of monotony for S means we can prefix a $\Box$ to the antecedent and consequent of any provable implication. If on the other hand O_1 is $\Diamond$ then we appeal to Prop. 1.4.1 to justify prefixing O_1 in the same way. ■

---

[24] Although one would not want to endorse aggregation-for-$\Diamond$ for many applications, we should still ponder its significance; a question about a representative instance of the aggregation schema is raised in Exc. 2.5.28(*ii*), p. 82 below, after the semantics for modal logic has been introduced. Early efforts in modal logic – by Łukasiewicz, Moisil and Curry, for example, tended – one would like to be able to say inadvertently – to have aggregation for $\Diamond$ as a by-product: see the discussion before (for Łukasiewicz) and within (for Curry and Moisil) Examples 2.1.20, p. 50 below.

Chapter 2

Normal Modal Logics

2.1 Some Candidate Axioms

What is special about the normal modal logics is that they all admit of a simple semantic characterization in terms of Kripke models – though we are not going into that until Section 2.3, and should quickly point out that Kripke also developed a model-theoretic semantics for some non-normal (though regular) modal logics which included the systems S2 and S3 (see p. 362) of C. I. Lewis, who developed a range of modal logics of which these two are at the weaker end, with the stronger systems **S4** and **S5** being normal. Here we are following a definite convention in always using boldface type for the names of normal modal logics, and in general these labels always begin with the letter "**K**", followed by labels for particular axioms. (We will explain how "**S4**", "**S5**" manage to fall under this description in a moment.) To have a whole host of them in one place, we list some (though not all) to which we shall have occasion to refer later, in Table 2.1. Most of the labels come from Lemmon and Scott [705] or Segerberg [1025].

Label	Schema	Dual (or other alt've) Form
T	$\Box A \to A$	$A \to \Diamond A$
4	$\Box A \to \Box\Box A$	$\Diamond\Diamond A \to \Diamond A$
5	$\Diamond A \to \Box\Diamond A$	$\Diamond\Box A \to \Box A$
D	$\Box A \to \Diamond A$	$\Diamond\top$
B	$A \to \Box\Diamond A$	$\Diamond\Box A \to A$
U	$\Box(\Box A \to A)$	—
.2 (or G)	$\Diamond\Box A \to \Box\Diamond A$	—
.3	$\Box((\Box A \land A) \to B) \lor \Box(\Box B \to A)$	—
.4	$\Diamond A \to (A \lor \Box\Diamond A)$	$(A \land \Diamond\Box A) \to \Box A$
Ver	$\Box\bot$	$\Box A$
M	$\Box\Diamond A \to \Diamond\Box A$	$\Diamond(\Box A \lor \Box\neg A)$
End	$\Box\bot \lor \Diamond\Box\bot$	$\Box A \lor \Diamond\Box B$
W	$\Box(\Box A \to A) \to \Box A$	$\Diamond A \to \Diamond(A \land \Box\neg A)$

Table 2.1: Some Candidate Axioms

So, for instance, if we want to consider the smallest normal modal logic in which all instances of schemas labelled **X** and **Y** are provable, we refer to this as **KXY**. To subsume Lewis's nomenclature for some special cases, we agree to write number labels after the letter labels **D** and **T** (thus "**KD4**" rather than "**K4D**", for instance), and – a neat suggestion of Segerberg's – when "**KT**" is followed by a numerical label not (which in practice will either be "**4**" or "**5**" – we do not include quasi-decimal labels ".**2**", ".**3**", and the like – we abbreviate "**KT**" to "**S**". (No relation to the use of "*S*" as a variable ranging over modal logics, normal or otherwise.) Similarly, amongst the logics between **S4** and **S5**, the latter being a proper extension of the former, are two particularly well known logics, **S4.2** and a stronger logic **S4.3**. **S4.4** is a still stronger sublogic of **S5**. The label "S4.1" (with or without the boldface) was also used historically for another extension of **S4**, namely **S4M**, but, as has often been recognised since, this is confusing since this logic is not a sublogic of **S4.2** – or even of **S5**. In the older literature, especially, labels like "S4.3.1" and "S4.3.2", or "**S4.3.1**" and "**S4.3.2**", are sometimes encountered – see Zeman [1221] or the index entries in Zeman [1224] or Hughes and Cresswell [535] – for logics felt to be slightly stronger than **S4.3** (and neither of these logics is included in the other,[25] making the nomenclature more confusing still) but we avoid them here (following Segerberg [1025]) as not conducive to clarity.[26] In particular, the second 'decimal point' entries have nothing to do with the axioms .**2** and .**3**. The labels **D** and **B** were coined to call to mind 'deontic' and 'Brouwersche' (i.e., Brouwerian), in the former case because **D** is naturally thought of as a deontically suitable weakening of **T** and in the latter because if intuitionistic negation is thought of as analogous to impossibility, the principle says that the double negation of A is implied by A: the intuitionistically acceptable half of the classical equivalence of A with its double negation.

Digression. The "M" is for McKinsey, who first considered this axiom – its necessitation appearing on line 6 of McKinsey [789], p. 93 – this paper being the source of the unfortunate label "S4.1" mentioned above – and then in essentially the form $\Diamond(\Box p \lor \Box \neg p)$ in McKinsey and Tarski [791], in the latter case not apropos of normal modal logics. Quite the opposite, in fact: he wanted to exhibit a non-normal extension of **S4**, and did so with the example of the smallest modal logic with all of **S4**'s theorems and also (what we, amongst many others, call) **M**. The logic McKinsey and Tarski produced was 'properly quasi-normal' – quasi-normal but not normal, that is. In the notation of Exercise 2.5.25, p. 81 below, it is the logic **S4** + **M**, whereas **S4M** is the logic **S4** ⊕ **M**. See Proposition 3.2.27 on p. 193 below, for this result, and the discussion leading up to it for other relevant material. Scroggs [1020] showed that there are no such logics amongst the extensions of **S5**, a result improved in Segerberg [1028] where it is shown that in fact all modal logics extending **S4.3** were normal. The present use of "M" has nothing to do with von Wright's use of "M" for **KT** (and "M′" and "M″" for **S4** and **S5**, respectively) in [1205], or, for that matter, with the use of "M" to recall "Monotone" in Chellas [171], a use appearing in the label "EM" in the discussion after Theorem p. 160 below. ***End of Digression.***

As explained toward the end of Section 1.3 in connection with **K**, we do not notationally distinguish between (labels for) the schemata listed here and (labels for) the corresponding formulas which are representative instances thereof. Thus, when convenient, we also refer (for example) to the formulas $\Box p \to \Box\Box p$ and $\Diamond p \to \Box \Diamond p$ as **4** and **5** respectively.

The last column in Table 2.1 gives dual forms or other alternative versions[27] of the modal principles listed. They are interdeducible, or more precisely, **K**-interdeducible, with the initially given forms.[28] Without going too much into the theoretical background to explain how this comes to be an appro-

---

[25]The inclusion relations between these and numerous other logics between **S4** and **S5** are conveniently depicted in a diagram at the end of Lenzen [708]. Lenzen's philosophical interest in such systems will emerge in Section 5.5.

[26]The logic referred to as **S4.3.2** is in fact **S4F**, for the axiom **F** mentioned on p. 66, as one learns from [1025], p. 164.

[27]While the phrase "other alternative" would normally involve a confusion – since the idea of otherness is already built into the word *alternative* – I do really mean to use both words together here: alternatives to the original axioms which are *not* the dual forms thereof.

[28] Formulas A and B are *S-interdeducible* (S typically a normal modal logic) when the smallest normal modal logic extending S and containing A coincides with the smallest normal modal logic extending S and containing B; in the notation of the previous Digression, this means: $S \oplus A = S \oplus B$. In more dynamic terms, this means that we can move from A to B or back making use of theorems of S and the rules Uniform Substitution, Modus Ponens, and Necessitation. The definition is biased for convenience in what is mainly a discussion (as ours is) of normal modal logics in including Necessitation here; without it, we have a smaller interderivability relation ($S + A = S + B$ in the notation alluded to in the Digression), in which no special interest will be taken here.

2.1. SOME CANDIDATE AXIOMS

priate terminology, let us just say that against the backdrop of classical logic, an n-ary connective $\#'$ is regarded as the *dual* of an n-ary connective $\#$ if in classical logic or whatever extension with new vocabulary of classical logic we are considering, any formula $\#'(A_1, \ldots, A_n)$ is equivalent to the formula $\neg\#(\neg A_1, \ldots, \neg A_n)$. The relation of duality is symmetric in view of properties of classical $\neg$, in classical non-modal logic and in any congruential modal logic. $\wedge$ and $\vee$ are duals, as are (in predicate logic, making the obvious adaptation of the definition to cover quantifiers) $\forall$ and $\exists$. For us the main point is that this relation obtains between $\Box$ and $\Diamond$, as is evident from the way $\Diamond$ was defined. What is meant by the dual (sometimes called "inverted") form of a schema with $\to$ as its main connective is the schema in which the antecedent and consequent are interchanged and any $\wedge$ or $\vee$ or $\Box$ or $\Diamond$ is replaced by its dual in the process.[29] Thus if we started with a schema, such as

$$\Diamond A \to (\Box\Diamond\Diamond A \vee \Box\Box A),$$

the new antecedent will be a conjunction – since the original consequent was a disjunction – and the first conjunct will attach the modality $\Diamond\Box\Box$ to A, since that is coordinatewise dual to the modality in the old first disjunct. The second conjunct similarly replaces "$\Box\Box$" with "$\Diamond\Diamond$". That finishes what is to be the new antecedent. The new consequent is simply $\Box A$, replacing the $\Diamond$ of the old antecedent by its dual operator. Thus we end up with:

$$(\Diamond\Box\Box A \wedge \Diamond\Diamond A) \to \Box A.$$

To see that it makes no difference if we replace a schema by its dual form in the axiomatization of a congruential modal logic (and thus *a fortiori* in the case of any normal modal logic), let us show that if all instances of the first schema above are provable in such a logic, so are all instances of the second schema. Of course this is only one example, but it is thoroughly representative and saves us all the machinery establishing the general point rigorously would require.

So we begin by assuming that all instances of the first schema above are provable in a congruential modal logic S. These include of course all instances of the following more specific form:

$\Diamond\neg A \to (\Box\Diamond\Diamond\neg A \vee \Box\Box\neg A).$

By truth-functional reasoning (specifically "by contraposition"), we also have in S anything of this form

$\neg(\Box\Diamond\Diamond\neg A \vee \Box\Box\neg A) \to \neg\Diamond\neg A.$

By further truth-functional processing ("De Morgan"), S proves

$(\neg\Box\Diamond\Diamond\neg A \wedge \neg\Box\Box\neg A) \to \neg\Diamond\neg A.$

To arrive finally at the dual form of the schema, all that is needed is the observation that in any congruential modal logic, for any affirmative modality $O_1 O_2 \ldots O_k$ and any formula A, $\neg O_1 O_2 \ldots O_k A$ is equivalent to $\widetilde{O}_1 \widetilde{O}_2 \ldots \widetilde{O}_k \neg A$, where $\widetilde{O}_i$ is the dual of O_i ($1 \leqslant i \leqslant k$), already implicit in the above remarks about dual modalities.

The dual of an implicational principle (and we use the term 'principle' so as to remain neutral between axiom-schemata, as above, and particular implicational axioms, to which an analogous account applies) is of course to be distinguished from its converse, and a handy notation from Chellas (e.g., [171]) enables us to economize on new labels by using $\mathbf{X}_c$ as the label for the converse of the principle labelled by $\mathbf{X}$, provided the latter is of implicational form. Beware: we are really considering the explicit syntactic form of the principle, and that two equivalent implicational principles can have non-equivalent converses. Thus the schema $\mathbf{T}_c$ is $A \to \Box A$. If we are thinking, instead, of the formula $\mathbf{T}$, then $\mathbf{T}_c$ is the formula

---

[29]For simplicity, we are ignoring occurrences of $\to$ and $\neg$, though it will be clear how to deal with these along similar lines, and so apply the procedure to the case of $\mathbf{W}$. The dual form of this schema, when that procedure is mechanically followed, would be "$\Diamond A \to \Diamond(\Box\neg A \wedge A)$". These last conjuncts appear in reverse order in Table 2.1, for better readability in idiomatic English – see the tense-logical gloss on $\mathbf{W}$ below (before Exc. 2.1.5).

$p \to \Box p$. Let's continue with the schematic orientation here: $\mathbf{4}_c$ is $\Box\Box A \to \Box A$ (or in the dual version: $\Diamond A \to \Diamond\Diamond A$), while $\mathbf{5}_c$ is $\Box\Diamond A \to \Diamond A$ (or $\Box A \to \Diamond\Box A$, in the dual form). $\mathbf{D}_c$ and $\mathbf{B}_c$ are respectively $\Diamond A \to \Box A$ (which, like $\mathbf{D}$ itself, is its own dual form) and $\Box\Diamond A \to A$. (One point at which we have not followed this convention is in using the label "$\mathbf{M}$" in its own right, rather than writing ".$\mathbf{2}_c$" – or "$\mathbf{G}_c$" – for this.)

EXERCISE 2.1.1 (i) What is the dual form of $\mathbf{B}_c$, as just given?
(ii) Two other schemata listed in Table 2.1 already stand in the $\mathbf{X}$-to-$\mathbf{X}_c$ relationship; which ones?
(iii) Do any of the schemata listed in Table 2.1 have themselves as their dual forms? If so, which ones?
(iv) True or false: "The dual form of the converse of an implicational principle always coincides with the converse of the dual form of that principle"? ✠

Here is a simple application duality-related ideas:

EXAMPLE 2.1.2 For a congruential modal logic S, let $Fm_S^\Box$ be the set of formulas provably equivalent in S to a $\Box$-formula (i.e. one of the form $\Box A$) and $Fm_S^\Diamond$ be the set of formulas provably equivalent in S to a $\Diamond$-formula. As a moment's thought will reveal the claim that $Fm_S^\Box \subseteq Fm_S^\Diamond$ amounts to the claim that every $\Box$-formula is (provably) equivalent to a $\Diamond$-formula in S'. Similarly, the converse inclusion amounts to the claim that $\Diamond$-formula is (provably) equivalent to a $\Box$-formula in S'. In fact the two claims are equivalent for any congruential S. Suppose for instance that $Fm_S^\Box \subseteq Fm_S^\Diamond$ we have $A \in Fm_S^\Diamond$, wanting to show that $A \in Fm_S^\Box$. Since $A \in Fm_S^\Diamond$, there is some formula B with $\vdash_S A \leftrightarrow \Diamond B$, i.e., $\vdash_S A \leftrightarrow \neg\Box\neg B$, in which case we have:

$\vdash_S \neg A \leftrightarrow \Box\neg B$, and so since $Fm_S^\Box \subseteq Fm_S^\Diamond$, for some C with $\vdash_S \Diamond C \leftrightarrow \Box\neg B$

$\vdash_S \neg A \leftrightarrow \Diamond C$, i.e.,

$\vdash_S \neg A \leftrightarrow \neg\Box\neg C$, so

$\vdash_S A \leftrightarrow \Box\neg C$,

which places A in $Fm_S^\Box$, completing the demonstration that

$$Fm_S^\Box \subseteq Fm_S^\Diamond \Rightarrow Fm_S^\Diamond \subseteq Fm_S^\Box.$$

The converse implication has a similar proof with one twist: we did not need to actually use the background assumption of congruentiality for the above demonstration, by contrast with the earlier demonstration that any congruential modal logic containing a formula (or all instances of some schema) contains the dual formula (or all instances of the dual schema), since we need not need to make replacements of provably equivalent formulas within the scope of $\Box$. But running a similar argument for the converse implication requires us to use the equivalence of the prefixes "$\neg\Diamond\neg$" and $\Box$ as above we used that of $\neg\Box\neg$ with $\Diamond$, an while our definitional conventions give us that for free, this new equivalence requires some congruentiality, since a direct unwinding into our chosen primitives gives us and equivalence between $\neg\Diamond\neg A$ and $\neg\neg\Box\neg\neg A$, forcing us to replace $\neg\neg A$ with A inside the scope of $\Box$. For more on $Fm_S^\Box$ and $Fm_S^\Diamond$, see Revision Exercise 2.7.11, p. 131. ◀

REMARK 2.1.3 Although, as we have used the word 'schema', any number of schematic letters ("A", "B",...) may be involved, as we see in the case of $\mathbf{D}$ in the final column of Table 2.1, with $\Diamond\top$ as an alternative form, and of $\mathbf{Ver}$ and $\mathbf{End}$ in the main column, that number can be 0. It is worth wondering, when faced with a candidate axiom schema, such as $\mathbf{T}$, $\mathbf{B}$, $\mathbf{M}$ or $\mathbf{W}$, or, to include some cases not explicitly included in that table, $\mathbf{T}_c$ or $\mathbf{D}_c$, whether or not an alternative form of the axiom exists without any schematic letters – alternative in the sense that the smallest normal modal logic containing the proposed formula coincides with the smallest normal modal logic containing all instances of the original schema. A negative answer in the case of $\mathbf{D}_c$ is an immediate consequence of the observation in Revision Exercise

2.1. SOME CANDIDATE AXIOMS

2.7.39 on p. 139; part (*ii*) of the exercise addresses the case of **T**. (The discussion there is couched not in terms of schemata without schematic letters but using the (coextensive) notion of 'pure formulas' introduced in Revision Exercise 2.7.21.) ◀

As already mentioned, duals are interdeducible as additions to any congruential modal logic, so duality is useful even for principles not considered as potential further axioms to add to **K**.

EXAMPLE 2.1.4 The dual of the Aggregation schema above (p. 21), $(\Box A \wedge \Box B) \to \Box(A \wedge B)$ is thus

$$\Diamond(A \vee B) \to (\Diamond A \vee \Diamond B),$$

and this should sound as plausible when "$\Diamond$" is given its alethic reading of "it is possible that" as Aggregation itself does when $\Box$ is read as "necessarily". As has often been observed, if we are dealing with the logic of ability (dynatic logic) and read "$\Diamond$" as "it is possible for agent a to make it the case that" then dual aggregativity is no longer plausible since the agent a's degree of control over outcomes will not be unlimited. For example by tossing a coin I can bring it about that it lands heads or tails, while I have neither the ability to bring it about that the coin lands heads and I also lack the ability to bring it about that the coin lands tails. Dynatic logic may accordingly be better accommodated among the broader range of monotone modal logics rather specifically those that are regular (let alone normal): see Brown [121]. This issue has made its way into the philosophical literature on compatibilism (the view that the existence of free action is compatible with the laws of nature being deterministic): see Humberstone [582], mid p. 559, for further references. The issue is further complicated by the option of rejecting even monotony for this dynatic reading of $\Diamond$ on the grounds that "*a* beings it about that __" is itself not monotone, an option that will be canvassed in Section 6.1, p. 449. ◀

Chellas [171] introduces a related and similarly convenient notation **X!** for the conjunction of an implicational schema **X** with its converse (i.e., with $\mathbf{X}_c$). This amounts to strengthening the main "$\to$" of **X** to a "$\leftrightarrow$"; thus **T!**, **D!** and **4!** are respectively $\Box A \leftrightarrow A$, $\Box A \leftrightarrow \Diamond A$ and $\Box A \leftrightarrow \Box\Box A$.[30] It might be wondered why anyone would want to consider such a schema as **T!**, or even the weaker $\mathbf{T}_c$. These may be implausible as principles governing necessity (or obligation, knowledge or belief, for that matter) but **KT!** is a normal modal logic nonetheless and in fact one playing an important structural role: in this logic $\Box$ is behaving as a truth-functional connective, with associated truth-function the identity function. So the modal logics all of whose theorems become purely truth-functional tautologies on erasing all occurrences of "$\Box$" and "$\Diamond$" inside them are precisely the sublogics of **KT!**. Hughes and Cresswell [535] followed a precedent set by B. Sobociński of calling such logics "regular", which would of course clash with the sense in which this term has been used here (and elsewhere). But we can perhaps without causing too much confusion call them *Sobociński-regular*. The Sobociński-irregular normal modal logics axiomatizable by a selection of schemata from Table 2.1 are the extensions of **KVer**, **KW** and **KEnd**. The first of these, sometimes called the 'Verum' logic, enjoys a status like that of **KT!**, and we return to it presently (in discussing Post completeness after the Exercise below). The second has an important place in the application of modal logic to reasoning about provability (see Boolos [112]), and both it and the third have been used as axioms in tense logic (see Section 3.1). One can see this in the case of the second (the case of **W**) most easily by giving the tense-logical interpretation "it will always be the case that __" and "it will at some time be the case that __", to $\Box$ and $\Diamond$, and considering the dual form listed in Table 2.1, which then says: If it will ever be the case that A then it will at some time be the case that A for the last time. So we have a kind of end-of-change principle. By contrast, what we have called **End** is more of an end-of-time principle when $\Box$ is given a similar future tense reading. (See Prior [925] for more information.)

---

[30]Another device like Chellas's $\mathbf{X}_c$ and **X!** we shall occasionally employ, though on a less systematic basis, is to write $\mathbf{X}^\Box$ for the result of prefixing an occurrence of $\Box$ to the formula or schema labelled by **X**. Thus $\mathbf{B}^\Box$ would be a label for the schema $\Box(A \to \Box\Diamond A)$. The label **U**, however, is retained rather than being replaced by $\mathbf{T}^\Box$.

EXERCISE 2.1.5 (*i*) The above glosses on **W** and **End** suggest that we should have **End** provable in **KW**; show that this is indeed the case.

(*ii*) Do we similarly have $\vdash_{\mathbf{KEnd}}$ **W**, understanding **End** here as the axiom rather than the schema of the same name (i.e., pick a representative instance)? Justify your answer. (This last part will be difficult if the correct answer is negative, since as yet no method had been suggested for showing that something is not provable; but anything that makes a negative answer plausible will do, in this case.) ✠

The special status of **KVer** and **KT!** among normal modal logics is that these are the only two such logics enjoying a certain maximality property called Post completeness. Logics in the current set-of-formulas sense are called *consistent* when they do not contain every formula.[31] Let Λ be a collection – typically a *family* in the sense explained on p. 25 – of logics in the same language which includes the inconsistent logic in that language (i.e., the set of all formulas of the language). Then $S \in \Lambda$ is *Post complete* in Λ if S is consistent but every proper extension of S in Λ is inconsistent. Since there is only one inconsistent logic in a given language (namely the set of all formulas of that language), we could equally well put this by saying that the unique proper extension of S lying in Λ is the inconsistent logic in the given language. One speaks of Post complete modal logics, Post complete regular modal logics, Post complete normal modal logics, etc., meaning logics which are Post complete in Λ for Λ as the family of all modal logics, all regular modal logics, all normal modal logics, etc. (Warning: this practice is far from universal, and some would confine talk of Post completeness to the case of the lattice of all modal logics, using such phrases as "maximal consistent normal" for the lattice of normal modal logics, and so on. See Humberstone [601].) So we can more accurately put the opening point of the present paragraph like this: what is special about **KVer** and **KT!** is that they are the only two Post complete normal modal logics, and we can add – without, however, giving the justification (by a well known argument due to Lindenbaum) for this here – that every consistent normal modal logic has one or other (or both) of these two as a consistent extension. In each of these logics, $\Box$ behaves like a truth-functional connective: in **KT!** interpretable as the identity truth-function and in **KVer** as the constant-true truth-function. See further Exercise 2.7.22 (p. 134), which, however, requires the semantic apparatus of later sections.

REMARK 2.1.6 The translations $(\cdot)^{\mathbf{T!}}$ and $(\cdot)^{\mathbf{Ver}}$ of the exercise just referred to, with $A^{\mathbf{T!}}$ being truth-functionally equivalent to erasing all occurrences of $\Box$ (and $\Diamond$) in A, and $A^{\mathbf{Ver}}$ being the result of replacing any $\Box B$ subformula of A with $\top$ have the following property: for any consistent modal logic, at least one of them maps all of its theorems to truth-functional tautologies. Translations which work on the internal structure of the formula to be translated eliminating occurrences of $\Box$ as they proceed, as these do, are called "$\Box$-definitional" in the discussion preceding Example 4.4.27 on p. 266 below (as well as Example 4.4.25 for background), where a more precise explanation is given. For any such translation τ, we might usefully generalize some of the preceding discussion by calling a modal logic S τ-*regular* if $\tau(S) \subseteq \mathsf{CL}$, where $\tau(S) = \{\tau(A) \,|\, A \in S\}$ and CL is the set of truth-functional tautologies of classical logic. (One might prefer to write "A^τ" in place of "$\tau(A)$".) Note that this amounts to saying that τ embeds S into CL, where the embedding is most definitely not required to be faithful – see Remark 2.8.1, p. 142. Thus, the Sobociński-regular logics are those which are τ-regular for τ as $(\cdot)^{\mathbf{T!}}$, and any consistent normal modal logic which is not Sobociński-regular (as well as many which are) is τ-regular for $\tau = (\cdot)^{\mathbf{Ver}}$. This raises the question of whether there are consistent modal logics that are not τ-regular for any ($\Box$-definitional translation) τ: such logics are called *contra-classical* modal logics in Humberstone [573], where an example of such a logic (the non-normal S6: see p. 362) is given. [573] asked whether every such examples must be (as S6 is) non-congruential. The answer turns out to be negative: see Example 2.10.33, p. 174 below. ◀

---

[31]Analogously, logics as sets of sequents are consistent when they do not have every sequent as a member; logics as consequence relations are consistent when for some Γ and B, we have $\Gamma \nvdash B$, where $\vdash$ is the consequence relation concerned. By the definition of "consequence relation" this is equivalent to the existence of a formula B for which $\nvdash B$, i.e., $\varnothing \nvdash B$.

2.1. SOME CANDIDATE AXIOMS

Digression. The various families of logics alluded to in the discussion before Remark 2.1.6 – all modal logics, all regular modal logics, etc. – are usefully considered as endowed with a lattice structure with the partial order being the relation of inclusion ($\subseteq$, that is), as was mentioned in the preamble to Exercise 1.3.12. The *join*, or least upper bound, of a set of logics is then the smallest logic in the lattice including all of them, and their *meet*, or greatest lower bound, is just their intersection. (Note that because of the 'Horn form' – see Hodges [504] – of their definitions, the classes of all modal logics, all congruential modal logics, all monotone modal logics, all regular modal logics, and all normal modal logics, are closed under intersections – though not under unions, which is why we can't just take the union as the join in the lattice in question.) The top element of these lattices is always the inconsistent logic – the set of all formulas – while the bottom element varies from lattice to lattice; for example, if Λ is the set of all normal modal logics, then in the lattice of logics in Λ, we have **K** as the bottom element, whereas if we take Λ as the set of congruential modal logics, we get a much weaker logic as bottom element, often called E or **E**.[32] Elements below the top element but with nothing between them and it – "co-atoms" (or *dual atoms*), in lattice-theoretic terminology – are the Post complete logics (of the lattice in question).[33] Makinson [760] drew attention to a somewhat surprising anomaly: the structure of the lattice of all modal logics is different depending on different choices as to what the primitive Boolean connectives are to be, even setting aside functionally incomplete candidate choices. As [760] observes, if we were to choose $\{\bot, \rightarrow\}$ then the least normal modal logic – the bottom element of the resulting lattice – is the meet (intersection) of two of its proper extensions, while this is not so if we chose instead $\{\neg, \wedge\}$ or $\{\neg, \rightarrow\}$ (or any other set containing neither $\top$ nor $\bot$). Thus, here the usually inconsequential distinction (see p. 11) between functionally complete and strongly functionally complete sets of truth-functions seems to be making a difference.

Makinson [760] also cites in his discussion Lemmon's paper [701], in which is it shown that a modal logic is the intersection of two of its proper extensions just in case it is *Halldén-incomplete*, where the latter means that some disjunction $A \vee B$ is provable in the logic for which A and B have no propositional variables in common and neither A nor B is itself provable; otherwise it is called *Halldén-complete*. In this terminology, the least modal logic in the language with Boolean primitives $\{\bot, \rightarrow\}$ is Halldén-incomplete – with, for instance $\Box\top \vee \neg\Box\top$ as a disjunction with the features required. (For this last case, think of $\top$ as defined by $\bot \rightarrow \bot$.) On the other hand, the least modal logic in the language with Boolean primitives $\{\neg, \wedge\}$ (or $\{\neg, \rightarrow\}$) is Halldén-complete. For more information on Halldén completeness, see Chapter 15 of Chagrov and Zakharyaschev [166] and also my [588] – where Halldén incompleteness is seen as a desirable feature for encapsulating certain philosophical positions – as well as references cited in both these sources. The result attributed to Lemmon above was also proved independently in Kripke [670]; note that the modal logics whose intersection yields a given *normal* Halldén-incomplete logic are not themselves guaranteed to be normal. For a contrast here: see Exercise 5.5.17(*iii*), p. 415. As Makinson notes, the phenomenon highlighted in [760] disappears once we pass from the lattice of all modal logics to the lattice of *congruential* modal logics;[34] however, many proposed axiomatizations even of *normal* modal

---

[32]Thus an axiomatic description of this logic might run as follows: for axioms, take all substitution instances of truth-functional tautologies; for rules, take Modus Ponens and the rule $A \leftrightarrow B / \Box A \leftrightarrow \Box B$. See Theorem 2.10.12, p. 160 below, and the discussing preceding it.

[33]If one element is strictly greater than another in a lattice and there is nothing properly between them, the former is said to *cover* – or be a cover of – the latter. Thus the dual atoms in a lattice with a top (greatest) element are elements covered by the top element. *Atoms* themselves are elements in a lattice with a bottom which cover that element. The present lattice-relative account of Post completeness is employed, for example, in Humberstone [573], p. 444, though some restrict the terminology to the case in which the lattice concerned comprises all modal logics and use a different terminology for other cases – e.g. for the lattice of all normal modal logics, in connection with which Wiliamson [1187] says "maximal consistent normal". Fritz [336] has a convenient variation: relativize talk of Post completeness to sets of rules, so that relative to a given such set, we are looking at the dual atoms in the lattice of logics closed under those rules.

[34]The same applies if we consider the lattice of quasi-congruential modal logics – modal logics containing all theorems of **E**, that is. In all such logics formulas which are truth-functionally equivalent are interreplaceable in modal contexts, which is to say that, in the terminology of Segerberg [1032] these are *classical* logics. In the terminology of Segerberg [1025], "classical" just means *congruential*, thus the logics which would there – as well as in Blok and Köhler [100], where they are

logics obtained by transcribing the axioms given for a different choice of primitives (typically $\Diamond$ taken as primitive in place of $\Box$) end up not axiomatizing the intended logic. See Humberstone [580] and note 17 of [588] for several examples and references; the *locus classicus* here is Hiż [503]. Makinson's warning about the (sensitivity of claims about the structure of the lattice of logics to the) choice of primitives receives further discussion in §4.6 of Segerberg [1032] and the final section of Humberstone [565]. **End of Digression.**

Proofs from the axiomatizations suggested by the various labels for the normal modal logics are just like the **K** proofs illustrated in the preceding chapter (for instance, in Example 1.3.16, p. 27), except that we cite the axioms in question to justify some lines in the proof. In the first example below, we have use the name for an axiom (or axiom schema) to stand in for any substitution instance (resp. instance) of the axiom (schema):

EXAMPLES 2.1.7 (*i*) A proof in $\mathbf{K4}_c$ for the formula $((\Box q \to p) \land \Box(p \to \Box q)) \to (\Box p \to p)$:

(1) $\Box(p \to \Box q) \to (\Box p \to \Box\Box q)$ **K**
(2) $\Box\Box q \to \Box q$ $\mathbf{4}_c$
(3) $\Box(p \to \Box q) \to (\Box p \to \Box q)$ 2, TF
(4) $((\Box q \to p) \land \Box(p \to \Box q)) \to (\Box p \to p)$ 3, TF

Of course the two TF steps could be combined into one, but taking them separately makes the proof more readily intelligible. In the second example in which we add **4** to $\mathbf{4}_c$, boosting the logic to **K4!**, we proceed more informally with a proof-sketch rather than a rigorous proof dotting all "i"s and crossing all "t"s.

(*ii*) Showing that $\vdash_{\mathbf{K4!}} \Box(p \leftrightarrow \Box q) \to \Box(\Box p \to p)$: By regularity and the $\mathbf{K4}_c$-provability of the formula featuring in (*i*), the following formula is **K4!**-provable:

$$(\Box(\Box q \to p) \land \Box\Box(p \to \Box q)) \to \Box(\Box p \to p).$$

Appealing to **4** we can reduce the doubled "$\Box$" on the second conjunct of the antecedent to a single "$\Box$":

$$(\Box(\Box q \to p) \land \Box(p \to \Box q)) \to \Box(\Box p \to p).$$

and then by the fact that $\Box$ commutes with conjunction:

$$(\Box((\Box q \to p) \land (p \to \Box q))) \to \Box(\Box p \to p).$$

Thus by TF and congruentiality, we have the desired formula

$$\Box(p \leftrightarrow \Box q) \to \Box(\Box p \to p)$$

as a theorem of **K4!**. Aside from illustrating the method of proof from additional axioms, both formal (as in (*i*)) and informal (as here), the **K4!**-provability of this particular formula will be of service to us later on (in Example 4.1.13, p. 213). ◀

In fact for the use we need to make of the second of the above examples, we will also need the result of the following:

EXERCISE 2.1.8 Give a formal proof that $\vdash_{\mathbf{K4}} \Box(p \leftrightarrow \Box q) \to \Box(p \to \Box p)$. ✠

For convenient back reference, let us parcel up a combined consequence of Example 2.1.7(*ii*) and Exercise 2.1.8:

PROPOSITION 2.1.9 $\vdash_{\mathbf{K4!}} \Box(p \leftrightarrow \Box q) \to \Box(p \leftrightarrow \Box p)$.

---

studied using algebraic techniques – be called quasi-classical are quasi-congruential in the current terminology.

2.1. SOME CANDIDATE AXIOMS

We turn to some more famous tourist destinations in the modal landscape. For alethic application the most popular modal logics are the normal logics **S4** and **S5**. The latter is amenable to an especially simple semantic treatment in terms of possible worlds (see Section 2.2), which is reflected syntactically by the fact that in it, every affirmative modality is equivalent to either $\Box$, to $\Diamond$, or to the null modality ($O_1 O_2 \ldots O_n$ with $n = 0$, that is, in the notation of the proof of, e.g., Proposition 1.4.2), and every modality is equivalent to one of these three or the negation of one of them. Further, there is a very simple way of working out which equivalence class a given $O_1 O_2 \ldots O_n$ belongs to: if $n = 0$ we have, as just remarked, the null modality, and if $n \geqslant 1$ then the whole thing is equivalent to O_n, its final operator. A proof of this fact can be found in Hughes and Cresswell [535] or [537], but we give our own below (Coro. 2.1.12), placing the result in a more general setting. These are also convenient sources for the corresponding point about **S4**: in this logic there are 7 non-equivalent affirmative modalities, and so 14 modalities in all (taking their negations), appearing together with the implicational relations between them in Fig. 2.1, formulas provably implying those to which they are linked going *upward* on the diagram by 0 or more edges. Essentially this diagram appears on its side at p. 48 of Hughes and Cresswell [535], where detailed historical information can also be found, and upside down at p. 56 of the same authors' [537].[35]

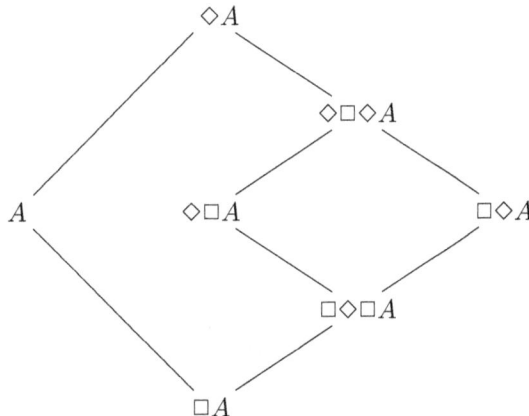

Figure 2.1: *Implications Between Affirmative Modalities in* **S4**

For good measure, Figure 2.2 is included – a 'multiplication table' summarizing the effect of composing one affirmative modality with another in **S4**. That is, for modalities, X, Y, Z, having Z as the entry in the table with the row labelled X and column labelled Y means that $\vdash_{\mathbf{S4}} XYp \leftrightarrow Zp$, or, as recorded in the table, $X \circ Y = Z$ (X composed with Y is – to within provable equivalence in **S4** – Z). There are only six rows (and columns) because the null modality is not included in the table. Its inclusion among the rows and columns would not make the table more informative, and in the case of **S4** composing non-null

---

[35] A horizontal version of the diagram appeared already on the second page of the 1959 paper Dummett and Lemmon [263] and indeed the diagram appears on p. 124 of Prior [919] published in 1957 and delivered orally two years before that; all these authors certainly mention (p. 149 of) Parry [866] as a place where all the relevant information is represented, in a somewhat clunky quasi-diagrammatic form. The orientation of the diagram chosen here reflects the fact that the usual partial ordering associated with the Lindenbaum algebra of a logic, as conventionally depicted in a Hasse diagram (see Figure 2.13a at p. 225 of Humberstone [594], for example), has the implying formula's equivalence class lower than ($\leqslant$) the implied formula's equivalence class. (Don't worry if that terminology is unfamiliar. It will not be used again, except briefly in Section 4.7.) An inverted version of the diagram is attempted at p. 81 of Mints [810], where it appears in a garbled form. Chellas [171], pp. 147–157 gives diagrams depicting the relations between affirmative modalities for numerous normal modal logics.

affirmative modalities X and Y never yields the null modality, so such an entry is never needed in the body of the table either. For a study of normal modal logics which contrast with **S4** – and for that matter with **K**, **KT**, **S5**, ... – in this respect, in which such an XY 'collapses' in the sense that XYp and p are provably equivalent, see my [591].[36] Such a (finite) table would not be possible for **K** or **KT**, each of which provides infinitely many non-equivalent affirmative modalities, and would be a very dull affair for **S5**, as we shall see in Coro. 2.1.12.

∘	□	◇	□◇	◇□	□◇□	◇□◇
□	□	□◇	□◇	□◇□	□◇□	□◇
◇	◇□	◇	◇□◇	◇□	◇□	◇□◇
□◇	□◇□	□◇	□◇	□◇□	□◇□	□◇
◇□	◇□	◇□◇	◇□◇	◇□	◇□	◇□◇
□◇□	□◇□	□◇	□◇	□◇□	□◇□	□◇
◇□◇	◇□	◇□◇	◇□◇	◇□	◇□	◇□◇

Figure 2.2: Composition of Affirmative Modalities in **S4**

Working our way toward the result for **S5** just alluded to, let us say that a normal modal logic S has the *Innermost Reduction Property* if for any $n \geq 1$ and any affirmative modality $O_1 O_2 \ldots O_n$,

$$\vdash_S O_1 O_2 \ldots O_n p \leftrightarrow O_n p,$$

and that S has the *Outermost Reduction Property* if for any $n \geq 1$ and any affirmative modality $O_1 O_2 \ldots O_n$,

$$\vdash_S O_1 O_2 \ldots O_n p \leftrightarrow O_1 p.$$

The above definitions would make sense for any modal logic and be of interest for any congruential modal logic, but our formulations are tailored to the case of normal modal logics. The Innermost Reduction Property (under whatever name) is considerably more familiar than the Outermost Reduction Property. For the former we need a preliminary observation which relates the principles **4!** and **5!**, which (by way of reminder) are, when formulated as schemata, respectively:

$$\Box A \leftrightarrow \Box\Box A \qquad \text{and} \qquad \Diamond A \leftrightarrow \Box \Diamond A,$$

with respective duals:[37]

$$\Diamond A \leftrightarrow \Diamond\Diamond A \qquad \text{and} \qquad \Box A \leftrightarrow \Diamond \Box A.$$

For the proof of this preliminary lemma, we use the following congruentiality rule for $\Diamond$, derivable from the axiomatization of **K** given above in much the same way as the rule ◇Mono:

$$(\Diamond\text{Cong}) \qquad \frac{A \leftrightarrow B}{\Diamond A \leftrightarrow \Diamond B}$$

LEMMA 2.1.10 $\vdash_{\mathbf{K5!}}$ **4!**, *and thus* **K4!5!** = **K5!**.

---

[36]In terms of the principles mentioned in Table 2.1 (p. 33) and variations thereon, in **KT!** all affirmative modalities collapse, while – less obviously – in **KB**$_c$ all affirmative modalities of length 2 collapse.

[37]Note that we are speaking here of the dual form of a ↔-schema, whereas this terminology was introduced strictly only for →-schemata. The intention is (of course) to apply the given definition to the two implications (→ and ←) in the ↔-schema and then combine the duals thus obtained into a new single ↔-schema.

2.1. SOME CANDIDATE AXIOMS

Proof. We give a formal proof of **4!** from **5!** and the rest of the basis of **K** assisted by the derived rule $\Diamond$Cong just mentioned; note that available to us here is not only **5!** $= \Diamond p \leftrightarrow \Box \Diamond p$ but also the dual form $\Box p \leftrightarrow \Diamond \Box p$, of which we shall use the substitution instance (*) $\Diamond \Box \Diamond p \leftrightarrow \Box \Diamond p$ obtained by substituting $\Diamond p$ (uniformly) for p:

(1) $\Diamond p \leftrightarrow \Box \Diamond p$ **5!**
(2) $\Diamond \Diamond p \leftrightarrow \Diamond \Box \Diamond p$ 1 $\Diamond$Cong
(3) $\Diamond \Box \Diamond p \leftrightarrow \Box \Diamond p$ (*) above
(4) $\Diamond \Diamond p \leftrightarrow \Box \Diamond p$ 2, 3 TF
(5) $\Diamond p \leftrightarrow \Diamond \Diamond p$ 1, 4 TF

Well, this derivation ends with the dual form of **4!** rather than with **4!** itself, but that will be enough. Note how little of the normality of **K** was used here: the proof shows that **4!** is a theorem of any congruential modal logic containing **5!**, as is pointed out in Chellas [171], Exercise 8.38. ∎

Digression. The derivation (1)–(5) given in the course of the proof of Lemma 2.1.10 is perhaps not the most perspicuous representation of what is going on, for which reason an equational version is given here instead.

Recall that a *semigroup* consists of a non-empty set together with an associative binary operation under which that set is closed. (Associativity was defined just before Exercise 1.2.4 on p. 11.) Now suppose that a, b, are elements of a semigroup standing in the relations indicated in (1) and (2),

$$(1)\ ab = b \qquad \text{and} \qquad (2)\ ba = a,$$

in which the operation is indicated by juxtaposition. Then for these elements we have $aa = a$ and $bb = b$.[38] The justification is given as follows, in which "(1)" or "(2)" is written under an equation involving replacements licensed by (1) or (2) respectively, with the undecorated "=" involving an appeal to associativity :

$$a \underset{(2)}{=} ba \underset{(1)}{=} (ab)a = a(ba) \underset{(2)}{=} aa$$

Similarly, interchanging "a" and "b" (and "(1)" and "(2)"), we see that $b = bb$. Think of a and b as $\Box$ and $\Diamond$ respectively, with "$\Box\Diamond = \Diamond$" meaning that $\Box\Diamond p$ and $\Diamond p$ are provably equivalent (in the modal logic concerned). Naturally, one could expose further structure by making $\neg$ explicit (satisfying suitable conditions), so that only one of equations (1) and (2) above is needed as a hypothesis, the other following from it. ***End of Digression.***

PROPOSITION 2.1.11 *A normal modal logic S has the Innermost Reduction Property if and only if $S \supseteq$* **K5!**.

Proof. "Only if": The equivalence **5!** is one special case of what the Innermost Reduction Property (with $n = 2$) requires, so if (normal) S has this property, we must have $S \supseteq$ **K5!**.

"If": By induction on n (in the definition of the Innermost Reduction Property).

<u>Basis Case</u>. $n = 1$. This case of the condition for S to have the Innermost Reduction Property is automatically satisfied by any modal logic, since the affirmative modality in question coincides with its final modal operator. For the inductive step, we need to exploit the hypothesis that $S \supseteq$ **K5!**, however.

<u>Inductive Step</u>. Suppose – the inductive hypothesis – that $\vdash_S O_1 \ldots O_n p \leftrightarrow O_n p$, for any choice of the O_i, with a view to showing that $\vdash_S O_1 \ldots O_n O_{n+1} p \leftrightarrow O_{n+1} p$, for any choice of O_{n+1}. From the supposition, uniformly substituting $O_{n+1} p$ for p, we get as S-provable

---

[38]These would usually be written as $a^2 = a$ and $b^2 = b$, and a and b would be described as idempotent elements of the semigroup, in an adaptation of the terminology introduced for binary operations just before Exc. 1.2.4.

$$\vdash_S O_1 \ldots O_n O_{n+1} p \leftrightarrow O_n O_{n+1} p,$$

so it remains only to show that the r.h.s. here is equivalent to $O_{n+1}p$. If O_n and O_{n+1} are both $\Box$ or both $\Diamond$, the equivalence is given by **4!** or its dual form – available courtesy of Lemma 2.1.10 – while if one of them is $\Box$ and the other is $\Diamond$, the equivalence is given by **5!** or its dual form. ∎

COROLLARY 2.1.12 **S5** *has the Innermost Reduction Property; thus there are only three pairwise non-equivalent affirmative modalities in* **S5**: $\Box$, $\Diamond$ *and the null modality.*

Proof. From Proposition 2.1.11, as **S5** extends **K5!**, 5_c being a special case of (the dual form of) **T**. ∎

EXERCISE 2.1.13 Formulate and prove a result analogous to Prop. 2.1.11 for the Outermost Reduction Property. (Don't worry if the logic playing the role of **K5!** doesn't have a handy name – just specify it as the smallest normal modal logic containing some explicitly listed formula(s). We will encounter this logic again in Section 4.5 below: see the discussion following Exercise 4.5.20 there.) ✠

As Hughes and Cresswell (esp. [535]) emphasize, the fact that each of **S4** and **S5** has finitely many modalities, should not make us expect these logics to fare similarly in respect of how many non-equivalent formulas in even a single (fixed) propositional variable they provide. **S5** is what is often called *locally finite* or *locally tabular*, which means that for any n, there are only finitely many pairwise non-equivalent formulas constructed out of the variables $\{p_1, \ldots, p_n\}$, whereas even for $n = 1$, in **S4** there are infinitely many non-equivalent formulas that can be constructed from just the variable p_1 (alias p).[39]

A third schema equivalent to the two given for **5** in Table 2.1 is given here on the right, for comparison with **4** repeated here, on the left, from the earlier table:

$$\Box A \to \Box\Box A \qquad\qquad \neg\Box A \to \Box\neg\Box A.$$

So while **4** tells us that what is necessary, is so necessarily, **5** tells us that what is not necessary, is so necessarily. Putting them together they say that everything has its modal status (i.e., its necessity or non-necessity) necessarily. An interesting point is that while over **K** the two principles are independent, in that neither **K4** ⊆ **K5** nor **K5** ⊆ **K4**, in the presence of **T**, **4** can be derived from **5**, so **KT5** ⊆ **KT4**. Note that this last claim is just another way of saying that $\vdash_{\mathbf{KT5}}$ **4**. A semantic gloss which helps to make some sense of the syntactic facts here – including the striking detour through formulas of modal degree 3 (i.e., have a $\Box$ in the scope of a $\Box$ in the scope of a $\Box$) in lines (4) and (6) of the syntactic derivation we present as Example 2.1.14, in deducing **4**, a formula of degree 2, from **T** and **5**, formulas of modal degree 1 and 2 respectively – may be found in the discussion preceding Exercise 2.5.42 below. (The notion of modal degree employed in the remarks just made is more precisely explained at p. 70.) The fact that **4** could be derived from the **KT5** axiomatization was nevertheless known before the semantics which made this obvious had been introduced – for example in von Wright [1205] (in which, as remarked in the Digression on p. 34, **KT**, **KT4** and **KT5** appear as M, M', and M'').

EXAMPLE 2.1.14 A **KT5** proof (using **T** and **5**) of **4**, going via **B** (at line (3)); "US" abbreviates "Uniform Substitution":

---

[39]In the terminology of Zolin [1226], already mentioned, this would be put by saying that although there are only finitely many linear modalities in **S4**, there are infinitely many modalities altogether. In **S5**, by contrast, there are 16 non-equivalent formulas constructible from a single propositional variable, as Rudolf Carnap pointed out in the 1940s ([151]); you can see a diagrammatic representation of these formulas and their logical relationships in Figure 20.A1 (p. 605) of my [582], consulting the text before it for an explanation of some abbreviations.

2.1. SOME CANDIDATE AXIOMS

(1) $p \to \Diamond p$ **T**, dual form
(2) $\Diamond p \to \Box \Diamond p$ **5**
(3) $p \to \Box \Diamond p$ 1, 2 TF
(4) $\Box p \to \Box \Diamond \Box p$ 3, US ($\Box p$ for p)
(5) $\Diamond \Box p \to \Box p$ **5**, dual form
(6) $\Box \Diamond \Box p \to \Box \Box p$ 5, $\Box$Mono
(7) $\Box p \to \Box \Box p$ 4, 6 TF

◂

EXERCISE 2.1.15 Show that the axiomatizations suggested in the following labels in fact all yield the same logic (namely **S5**, that label being officially an abbreviation of the first of the following four):

KT5   **KTB4** (or '**KT4B**')   **KDB4**   **KDB5**,

and that in each case, none of the axioms alluded to is redundant – as would be the case with the label "**KT45**", for instance. (Note: if in difficulties, consult Carnielli and Pizzi [152], pp. 37–40.) ✠

Digression. Does the above exercise show that our notation for normal modal logics gives rise to *ambiguities*? From the following quotation from Green [407], p. 60; it is not necessary to know what a coset is in order to detect a confusion:

> It is at first rather difficult to realise that the notation Hx for a right coset is ambiguous, because the same coset may be written Hy, for some y different from x.

This is of course nonsense – unless the author's reason for saying that it's difficult to realise is that it's impossible to realise, in which case why the *"at first* rather difficult"? The point is that ambiguity would result from having the same notation denote two separate things, not – as in the above case – having two notations which denote the same thing. The latter is what we have in Example 2.1.15. Of course one may find this feature already objectionable in a system for naming modal logics – as Garson does, for example (see [351], p. 42). Indeed, as the formulation of Exercise 2.1.15 shows, we regard the 'anatomical' labels as primarily being labels for axiomatizations and derivatively as labels for the logics thus axiomatized, sometimes playing on the former association. One way of making this precise would be to invoke Frege's distinction between sense and reference (or denotation), regarding an axiomatization of a logic as a 'mode of presentation' of that logic: thus the names "**KT5**" and "**KTB4**" differ in sense but have the same reference. The first picks out a logic as the smallest normal modal logic containing all instances of the schemata **T** and **5**, and the second as the smallest such logic containing all instances of the schemata **T**, **B** and **4**; the fact that the same logic is thus picked out twice over is then potentially informative, as in the case of Frege's examples in [331]. (On the other hand, the labels "**KT4B**" and "**KTB4**" have not just the same reference but the same sense.) **End of Digression.**

There is some genuine ambiguity in the nomenclature, the above Digression notwithstanding. In the first place, we use labels like "**T**" ambiguously as between names for axioms ($\Box p \to p$, in the present case) and axiom-schemata ($\Box A \to A$). Secondly in the case of "**K**" this refers either to a modal principle (whether the candidate axiom $\Box(p \to q) \to (\Box p \to \Box q)$ or the corresponding schema) or else to a modal logic, namely the smallest modal logic. We have tried to reduce the latter kind of ambiguity to a minimum, by using "**KT**" and "**KD**" as names of modal logics, rather than following the fairly widespread practice of calling the logics concerned **T** and **D**. In some cases the latter usage can become quite confusing. For instance the logic referred to as **B** in such works as [535] is not **KB** but **KTB**; a dramatically simplified axiomatization of this logic is presented at Coro. 3.1.7. (The more explicit policy followed here is that of Chellas [171].)

REMARK 2.1.16 A number of papers on modal logic have been stimulated by questions related to nomenclature, settling the potential confusions that arise with one modal logic having been explored under two names, two logics having been given the same name, and so on. Examples include, in chronological order, Zeman [1223], Lenzen [709] ("The aim of this note is to show that Georgacarakos' S4.03 and my S4.03 are one and the same system"), Bernert and Biela [86], and Williamson [1180]. As Segerberg ([1025], p. 49) rather charmingly says: "Finding good names for modal systems is one of the more recalcitrant problems in the field." ◀

Some words are in order on C. I. Lewis, last mentioned at the beginning of this section. Lewis's interest in modal logic was not an interest in the study of necessity and possibility for their own sake, so much as in their utility in explicating the informal idea of *entailment*. His proposal was to represent p's entailing q as $\Box(p \to q)$, or equivalently as $\neg\Diamond(p \land \neg q)$. Lewis introduced a special binary connective to indicate when this relationship holds between p and q, called *strict implication*, and in place of either of the two formulas just cited, one writes, as already foreshadowed in Remark 1.3.9(*ii*) on p. 23, "$p \strictif q$". Not everyone has been convinced that entailment can be formalized as strict implication, since, for example, it is provable in all the usual systems of modal logic that $\Box A$ implies that B strictly implies A – in other words, the formula $\Box A \to \Box(B \to A)$, or $\Box A \to (B \strictif A)$ – whereas these critics have held that a necessary truth (A here) is not automatically entailed by absolutely any old statement (B). In particular, the school of *relevant logicians* (mentioned in passing in our discussion of non-modal propositional logic) have held that there must be some connection of meaning between two things one of which is said to entail the other. Formulas such as that mentioned above which seem, on these grounds, to present anomalies for the explication of entailment as strict implication are sometimes called *paradoxes of strict implication*; another example would be: $\Box\neg B \to \Box(B \to A)$, or $\Box\neg B \to (B \strictif A)$. Why should the fact that B is necessarily false mean that it entails everything? (Does the proposition that $2 + 3 = 4$, for example, entail that some tadpoles grow into partridges?) Several items in our bibliography address matters in this general area, the most useful of which is probably Anderson and Belnap [20], whose own bibliography includes many publications (such as the articles listed by Jonathan Bennett) dissenting from their opinion that no version of strict implication will do as an explication of entailment. (One application of these logics is as *paraconsistent* logics, intended to support theories in which a contradiction is provable but not every formula is. From this perspective, the exclusion of principles like $(p \land \neg p) \to q$ is of greater significance that the exclusion of – equally relevance-violating – principles like $p \to (q \lor \neg q)$. See Priest [910].)

EXERCISE 2.1.17 (*i*) Suppose we started with $\strictif$ rather than $\Box$ as primitive, along with the Boolean connectives, and wanted to give a definition of $\Box$, with the effect that, using $\Box$ as so defined, $A \strictif B$ and $\Box(A \to B)$ were equivalent (provably equivalent in a version of **K** in the new language, for definiteness). How might such a definition run?
(*ii*) In fact it is possible to answer (*i*) with a definition in which no Boolean connectives appear, only $\strictif$. This is a harder question. Give a definition of $\Diamond$ which would work for **S5**, though not necessarily for weaker normal modal logics, in which the only connective to appear is $\strictif$. (For this, it suffices to find a formula **S5**-equivalent to p in which the only connective appearing is $\strictif$, taken as primitive. There are only five **S5**-equivalence classes of formulas thus definable, and that appropriate for $\Diamond p$ is one of them. If in distress, consult Lemmon, Meredith *et al.* [704].) ✠

Although we are mainly interested in normal modal logics, a few words on their non-normal cousins are included here, especially since we have just been discussing C. I. Lewis and he had a soft spot for one of these, namely S2, mentioned among the (second batch of) examples that follow. A semantic treatment aimed at bringing out the appeal of this particular non-normal logic for Lewis is provided by Mares [773], while for the semantics devised by Kripke for these logics, see p. 360*ff*. below.

EXAMPLES 2.1.18 (*i*) There is a systematic nomenclature available for these, thanks to Chellas [171], and to some extent Segerberg [1025], but as we shall not have much occasion to consider these logics

2.1. SOME CANDIDATE AXIOMS

or many of them to consider, we follow the higgledy-piggledy *ad hoc* nomenclature that has been used in the historical literature. This includes "C2" (called C in [1025]) for the smallest regular modal logic, "E2" for a logic explored by Lemmon (see [694]), which is the smallest regular modal logic containing **K** (the axiom **K**, that is) and **T**. The "regular" in this description can be replaced by "monotone" without affecting the result. Thus more explicitly, one has the following axiomatization of E2, with the non-modal part provided as for normal modal logics, e.g., by taking all substitution instances (in the language with $\Box$) of truth-functional tautologies as axioms, and Modus Ponens as a primitive rule, and for the modal part, all instances of the following schemata:

K $\quad \Box(A \to B) \to (\Box A \to \Box B)$
T $\quad \Box A \to A$

and the rule, from p. 26:

($\Box$Mono) $\quad A \to B \,/\, \Box A \to \Box B$

(If we omit **T** here, we have an axiomatization of C2, already mentioned.) The idea is to approximate **KT** as far as is possible without having any provable $\Box$-formulas, and the "E" was suggestive to Lemmon of "epistemic" in part on the grounds that epistemic logic should not be telling us unconditionally of any A it was known that A – though other aspects of Lemmon's motivation were later shown not to stand up to scrutiny: see Example 4.3.1 (p. 227 below). Lemmon also considered stronger logics which he called E3 and E4, to axiomatize the first of which, **K** above is replaced by:

$$\Box(A \to B) \to \Box(\Box A \to \Box B),$$

from which **K** can be recovered via **T**; for E4 one adds instead **4**: $\Box A \to \Box\Box A$.

(*ii*) The numerical part of the nomenclature under (*i*) was intended to recall that of C. I. Lewis's S2, S3 and S4. This last is simply the normal **S4** ($=$ **KT4**), and S2 and S3 are are quasi-regular (but not regular) modal logics, extending E2, for which the following axiomatizations are among those to be found in the literature. In the case of S2, the non-modal part is as for E2, but we also have as axioms and rule(s):

- $\Box A$ for any truth-functional tautology A;
- axioms **K** and **T** as above;
- the necessitations of **K** and **T**;
- the rule $\Box(A \to B) \,/\, \Box(\Box A \to \Box B)$.

(If we change the modal part of this axiomatization by dropping the ingredients under the third and fourth bullet points, we have an axiomatization of the weak logic S0.5 from Lemmon [695] and [694], sometimes called "S.O5" in the literature – e.g. in Cresswell [200]: see further Example 5.1.28(*i*), p. 368.) There is some redundancy here: certainly we do not need the axiom **K** is we have its necessitation and also **T**. (On the other hand, we cannot get rid of **T** by similar reasoning, relying on its necessitation, alias **U**. Why not?) For S3, on the other hand, we replace the modal part of this axiomatization with:

- $\Box A$ for any truth-functional tautology A;
- axioms **K** and **T**;
- $\Box(A \to B) \to \Box(\Box A \to \Box B)$;
- the necessitations of **K**, **T** and $\Box(A \to B) \to \Box(\Box A \to \Box B)$.

This time there are no specifically modal rules. (Recall from note 18, p. 25, that the reference in these various descriptions to the necessitations of formulas are not meant to suggest that we are applying the rule of Necessitation.) One obtains an axiomatization of S4/**S4** by adding **4** to the above basis.

There are numerous variations – here illustrated for the case of S2 – to be found in the literature, aside from the above Lemmon-derived axiomatizations and those of C. I. Lewis himself (the latter being less perspicuous): see Kripke [670] for an axiomatization of S2 as the smallest modal logic extending E2 by $\Box(p \to p)$ – N.B., not the smallest monotone such extension: so $\Box$Mono cannot be used in the unrestricted manner one might have expected. Another is to be found in Schumm [1006], which replaces the third and fourth bullet-pointed ingredients (from the above axiomatization of S2) in its modal part with the rule $A \mathbin{/} \Box(\Box B \to A)$. (I am grateful to Allen Hazen for showing me that this does indeed succeed in axiomatizing S2, as Schumm claims. Note that we are concerned throughout only with capturing the same set of theorems, not requiring reaxiomatizations to respect the consequence relation generated by the original axiomatization – a theme famously explored in the case of the normal modal logic **S5** in Porte [895], though see also the Digression after Remark 2.1.19 below; a survey of the area can be found in Kracht [666].) Considerable energy is required to enumerate all the non-equivalent modalities of such non-normal modal logics as S2 and S3 – compare Figure 2.1 on p. 41 above (doubling that number, since only the affirmative modalities are there depicted) – and it has indeed been expended: see Parry [866], Feys [286], Pledger [886].

(iii) Numerous further systems are considered by C. I. Lewis and also by Lemmon, such as their respective S1 and E1 (on the first of which, see the references cited in note 304 on p. 362 below), and Lemmon also offered an E5 to match **S5**, on which see note 301, p. 360. An **S5**-like strengthening of S3 was isolated by Åqvist under the name S3.5: an informative discussion is provided by Cresswell [202] and [203]; this is the smallest modal logic extending S3 and containing $\Diamond p \to \Box\Diamond p$. See also Example 5.1.28(ii) (p. 368 below). For more information on all these logics, see Lemmon [694] and Kripke [670], and for textbook presentations, Zeman [1224] and Hughes and Cresswell [535], esp. Chapters 12 and 13 for details and references. For other devised Lewis-inspired logics S6, S7, S8, (introduced under those names by M. J. Alban and S. Halldén in the 1940s and '50s) see Chapters 14 and 15 of [535], in the section of which starting on p. 267, the "non-regular" in the section heading means "not Sobociński-regular" rather than not regular in the sense defined on p. 19. (Sobociński-regular modal logics were defined on p. 37; we return to S6 (mentioned already in Remark 2.1.6, p. 38) on p. 362 below. These three logics all have $\Diamond\Diamond p$, and thus also $\Diamond\Diamond\bot$, among their theorems. Hughes and Cresswell also discuss an 'S9', on which see further Bernert and Biela [86], mentioned in Remark 2.1.16 on p. 45 above.) Chapters 3 and 4 of Segerberg [1025] are a mine of information on non-normal modal logics; semantic remarks on some of these logics can be found on p. 362 below. (Further non-normal modal logics will receive occasional mention below; for example in Section 4.3 and in Chapter 6, in which see especially Section 6.1 and also Example 6.2.4.) Instead of going into any detail on the logics just mentioned by name here, we describe, as our final non-normal modal logic here, Łukasiewicz's 'Ł-modal system'. Here we take as the modal part of an axiomatic basis just two axioms (or axiom-schemata, to be more accurate), namely **T** and $\Box B \to (A \to \Box A)$, with no special modal rules. As you can see from the second schema, this logic has no plausibility on an alethic (or epistemic or deontic) reading of $\Box$, something realised early on by everyone except Łukasiewicz, Chapter 1 of Prior [919] being a solid reference in this connection. These philosophical shortcomings notwithstanding, the Ł-modal system has been the focus of much logical interest and we will have occasion to refer to it more than once in what follows, aside from just before Examples 2.1.20 below: see pp. 156, 261 and 363 – in the last case in Remark 5.1.23(ii). (For more information, see also the Digression beginning on p. 470 of Humberstone [594], together with the end of section notes and references – p. 484 there.) For this reason, it is worth mentioning a somewhat simpler recent axiomatization of the logic (from p. 231 of Tkaczyk [1108]), using for its modal part just the single schema:

$$(\Box A \wedge B) \to (A \wedge \Box B).$$

There is, incidentally, a mistaken characterization of this logic on p. 210 of Segerberg [1025] as arising from "adding to C [= the smallest regular modal logic] the schema

$$\text{L} \qquad \Box\top \to (A \to \Box A).\text{''}$$

It is clear that we can derive the earlier ('implausible') axiom, since C (alias C2 under (*i*) above) contains $\Box B \to \Box\top$, but equally clearly there is no way of deriving **T** from this axiomatization (since consistently with this basis, $\Box$ could be interpreted as expressing the constant-true truth-function, an interpretation invalidating **T**). ◀

REMARK 2.1.19 Note that we write "S2", etc., and not "**S2**", because the latter would conflict with the convention that labels of the form "**S**" followed by a numeral are reserved for normal extensions of **KT**. This would not prevent us from writing "**E2**" for Lemmon's system, but boldface **E** is used here in a different capacity, for the smallest congruential modal logic, itself non-normal: see also note 32 and Theorem 2.10.12. The latter use of this letter abbreviates *extensional*, employed to mean *congruential* in Segerberg [1025];[40] As already noted under Example 2.1.18(*i*), Lemmon's usage was derived from *epistemic*, with his E2 and E3 being 'epistemic' versions of Lewis's S2, S3. Lemmon also has 'D-for-deontic' systems D2, D3, etc., axiomatized by replacing **T** with **D** – for a note on which, see Routley and Montgomery [972] – as well as the sequence C2, C3,... whose first element has already been mentioned, which drops **T** without replacing it. (Another nomenclature introduced for such systems by Robert Feys is $S2^0, S3^0,\ldots$: for bibliographical details, see Hughes and Cresswell [535], p. 302. Zeman [1224] uses "T^0" for **K**.) Note, incidentally, that there is no connection between this use of the labels C2, C3, ... and the use of the same labels in Lemmon and Meredith *et al.*, [704], mentioned under Exercise 2.1.17(*ii*) above, where they stand for the strict implication fragments of S2, S3,... ◀

Digression. Somewhat further from the comparative mainstream of non-normal modal logics than Examples 2.1.18 are several logics, some of them in fact normal, considered as part of a systematic investigation, over several decades, by Jean Porte. (We can make sense of normality for consequence relations over the present language using the definition given in the Digression on p. 22 above.) The details are conveniently available in Porte [893] and [894] – but see Kielkopf [654] for corrections to some of the proofs in [894], as well as Kondo [660] for further information – and only a rough summary is given here, as follows. Porte begins with three non-normal logics S_a, S_b, S_c (increasing in strength in respect of a replacement or congruentiality-like property), and two ways of strengthening a consequence relation on the language of these logics represented by the Greek letters *nu* and *rho* (ν and ρ, though Porte uses the variant form 'ϱ' for the latter). For a consequence relation $\vdash$, $\nu(\vdash)$ is the smallest normal consequence relation extending $\vdash$ and $\rho(\vdash)$ is the smallest consequence relation extending $\vdash$ and satisfying the Deduction Theorem (i.e., having $A \to B$ as a consequence of Γ whenever B is a consequence of $\Gamma \cup \{A\}$). In the case of S_a, or as it might more explicitly be denoted, $\vdash_{S_a}$, we have six logics distinct from S_a and from each other: $\nu(S_a), \rho(S_a), \nu(\rho(S_a))$, and the further results of applying ν and ρ to these last two, respectively. (Any further applications of ν and ρ yield logics already on this list.) The inclusion relations among these are depicted in Figure 1 of [893], p. 15. The same paper describes similar extensions for S_b and S_c, in whose cases iterated application of ν and ρ yields fewer distinct results, and also details some relations with more familiar 'mainstream' logics: for example ([893], p. 26), $\rho(S_a)$ coincides with Lemmon's S0.5 mentioned parenthetically under Examples 2.1.18(*ii*) above, and $\nu(\rho((S_a))$ with **KT**, while $\nu(\rho(\nu((S_a)))$ ($= \nu(\rho((S_c)))$ coincides with **S4**. "Coincides with" can be read as saying that the consequences of the empty set are precisely the theorems of the logics-as-sets-of-formulas. (Alternatively, as *identity* of consequence relations in the case of the normal formula-logics here, taking their associated local consequence relations in the sense explained in note 115, p.146, below.) We do not go further into these matters here because the present discussion is downplaying the role of rules, consequence relations

---

[40]Below, the term *extensional* is used for a much stronger property: see Exc. 2.7.2(*i*), p. 129. That exercise concerns normal modal logics, so we pause to observe here that the L-modal system described at the end of Example 2.1.18(*iii*) is extensional; see Humberstone [555], or subsection 3.24 of [594], as well as Remark 5.1.23(*ii*) at p. 363.

and the Deduction Theorem in modal logic. See the references at the end of Example 2.1.18(*ii*) above, as well as Humberstone [592], Buvač [141], and Hakli and Negri [424]. **End of Digression.**

Lukasiewicz's L-modal system, mentioned in Example 2.1.18, is not the only proposed modal logic from the relatively early days of the subject to suffer from the defects already mentioned, such as extensionality (note 40), as well as the following further defect: it renders $\Diamond$ aggregative in the sense of containing as theorems all formulas of the form $(\Diamond A \wedge \Diamond B) \to \Diamond(A \wedge B)$. (Not to be confused with dual aggregativity for $\Diamond$, as in Example 2.1.4, p. 37.) This is not surprising given the fact that all contexts in classical propositional logic are aggregative – see Proposition 1.2.7, p. 13 as well as p. 21 – and the L-modal operators are simply hybrids of 1-ary truth-functional connectives (see Remark 5.1.23(*ii*) at p. 363). Here we pause to recall two further examples of similarly afflicted early work, the explanation perhaps being that in those days the idea of having $\Box p$ provably imply p without the converse, and p provably imply $\Diamond p$ without the converse, whether or not $\Box$ and $\Diamond$ were related as duals, overrode all other concerns. Some of the details require a familiarity not presumed elsewhere in the present text with non-modal logics weaker than classical logic:

EXAMPLES 2.1.20 (*i*) Curry [224], p. 119*f.*, taking $\Diamond$ as a further primitive alongside $\Box$ (there written as "#") proposed some rules for $\Diamond$ to be added to Johansson's 'Minimal Logic' (a proper sublogic of intuitionistic logic) whose effect he summarized by saying that they amounts to having a sentential constant M and defining $\Diamond A$ as $M \to A$. (A similar proposal, though without the subclassical background assumptions, is attributed to D. P. Henry at p. 189 of Prior [918].) Evidently this renders $\Diamond$ aggregative (as well as extensional, understanding this in terms of the "↔" of Minimal Logic). Curry also mentions that three principles his proof-theoretic proposal delivers include $A \to \Diamond A$, $\Diamond\Diamond A \to A$ and $\Diamond(A \to B) \to (\Diamond A \to \Diamond B)$, which he says are all satisfied if $\Diamond A$ is understood as $\neg\Box\neg A$, given his (roughly **S4**-like) treatment of $\Box$: something which is certainly not the case for the third of these schemata, evidently a close (and equally unwanted) relative of aggregation-for-$\Diamond$. (By the time he came to write [225], Curry had repudiated the treatment of $\Diamond$ proposed in [224], the latter originally circulated in lithographed form in 1950, seven years before the publication date given in our bibliography and five years before the appearance of [224].) When it comes to treating possibility conditionally, though we do not go into details here, mention should be made of the suggestion that for dynatic possibility – agents' abilities, that is – in which the key conditional is a subjunctive (or counterfactual) conditional and the antecedent is not just a sentential constant exhibits further complexity. In particular the suggestion is that "*a* is able to φ" (or "*a* is free to φ") amounts to "if *a* were to choose/decide/intend/want to φ then it *a* would φ". Near enough then, this amounts to: $W(\varphi a) \,\Box\!\!\to\, \varphi a$. Here "$\Box\!\!\to$" (as in Lewis [717]) is used for the conditional construction in question, along with the sentence operator "W" for "*a* wants that ___" – selected by way of example from among the various more or less boulomaic alternatives listed (on which terminology, see p. 203). Because the antecedent is not just an unstructured constant but now embeds the consequent, we no longer have the unwanted aggregation inference here. Proposals along these lines have been associated particularly with defences of compatibilism, whether directly as when being freedom to φ is identified with being able φ and the latter conditionally analysed, or less directly, as when φ-ing freely is identified with φ-ing when one was able not to φ – "could have done otherwise" formulations – and the conditional analysis is then applied at this point. One classic account is provided by Ayer [46], earlier version of the idea appear in the work of David Hume and G. E. Moore, the latter's own formulation being famously subjected to scrutiny in Austin [42]. Pointers to subsequent developments on this aspect of the compatibilism debate can be found in §3.3 of McKenna and Coates [788].

(*ii*) Moisil [814], esp. §§3, 4, proposes – as we shall put it here – to use intuitionistic logic with the standard intuitionistic negation as well as dual intuitionistic negation as the underlying logic. As in Humberstone [594], especially subsection 8.22 (and the references given on p. 1250 – to work other than Moisil's), q.v. for further information, we write these two connectives as $\neg$ and $\neg_d$. Then Moisil's idea is that $\Diamond A$ can be understood as $\neg\neg A$ and $\Box A$ as $\neg_d\neg_d A$. (Moisil writes $\neg$ and $\neg_d$ as η and γ, glossing them respectively as expressing impossibility and contingency, with $\Box$ and $\Diamond$ written as ν ($=\gamma\gamma$) and μ ($=\eta\eta$)

respectively. (The linking of intuitionistic negation with impossibility was remarked on at p. 34. There is no plausible connection between contingency – assuming this to be the appropriate English translation of the French *la contingence* use in [814] – and dual intuitionistic negation, for which, for example, $\neg_d \bot$ is provable.) While this gives p unilaterally implied by $\Box p$ and unilaterally implying $\Diamond p$, as desired, not only does $\Box$ (as $\neg_d \neg_d$) satisfy the aggregation condition, but so does $\Diamond$ (as $\neg\neg$). (The $\Diamond$-as-$\neg\neg$ idea in approximately the same subclassical setting is also mentioned at p. 120 of Curry [224]; on the following page Curry also has a further suggestion using two different negation operators together – a 'minimal' negation applied to an intuitionistic negation – to represent possibility. This amounts to saying that we take the positive fragment of intuitionistic logic and add sentential constants E and F, say, the former with no special logical behaviour and the latter provably implying every formula (so we could equally well write "$\bot$" for "F"), and then define $\neg_m$ and $\neg_i$ – for 'Minimal' Johansson-style and intuitionistic negation respectively – by $\neg_m A =_{Df} A \to E$ and $\neg_i A =_{Df} A \to F$. Curry then proposes that we introduce $\Diamond$ by $\Diamond A = \neg_m \neg_i A$, a proposal that will not be evaluated here.) Indeed, Moisil is well aware of this, remarking ([814], p. 35) that his treatment excludes the distinction between compatibility (or compossibility) on the one hand and the conjunction of possibilities on the other ($\Diamond(p \wedge q)$ and $\Diamond p \wedge \Diamond q$, respectively). For most applications of modal logic, this contrast is essential, and more fundamental than the specifically alethic obsession with the provability of **T** (without its converse – which is what the above reference to unilateral implication is meant to convey), suited only to what in Table 4.1 (p. 204) we call veridical interpretations of $\Box$. The Kripke semantics for modal logic explained in Section 2.3 below will exhibit the question of the validity of **T** as a relatively superficial aspect of how the semantic apparatus works (namely whether the accessibility relations in the models are required to be reflexive). For such reasons, in later discussions, intuitionistic double negation ($\neg\neg$) is more likely to be seen as a candidate for the interpretation of $\Box$ than of $\Diamond$ – see Došen [248]. ◂

We return to the fold of normal modal logics. For application in Section 2.8, as well as for its own interest, here we include a generalization of the aggregation property of ($\Box$ in) such logics. From amongst the ordered triples $\langle O_1, O_2, O_3 \rangle$ of modal operators $\Diamond$, $\Box$, (i.e., for each i ($1 \leqslant i \leqslant 3$), O_i is either "$\Diamond$" or "$\Box$"), we define the *distributive triples* for a normal modal logic S to be those triples $\langle O_1, O_2, O_3 \rangle$ for which we have $\vdash_S (O_1 p \wedge O_2 q) \to O_3(p \wedge q)$. (*Warning*: If you have previously encountered the phrase 'distributive triple' you have probably encountered it with an entirely different meaning – see note 41.) Since this is a point on which authors differ, a reminder is in order to the effect that the definition given in Section 1.3 builds into the notion of a normal modal logic (indeed any modal logic) the requirement of closure under uniform substitution, so the above definition is equivalent to saying that the distributive triples for S are those $\langle O_1, O_2, O_3 \rangle$ for which we have $\vdash_S (O_1 A \wedge O_2 B) \to O_3(A \wedge B)$, for all formulas A, B, C: this is the aggregation condition (from Section 1.3) which makes monotone modal logics regular. Slightly less obvious alternative formulations we collect together here:

PROPOSITION 2.1.21 *For any normal modal logic S and any ordered triple of modal operators $\langle O_1, O_2, O_3 \rangle$ the following are equivalent:*

(i) $\langle O_1, O_2, O_3 \rangle$ *is a distributive triple for S;*

(ii) $\vdash_S O_1(p \to q) \to (O_2 p \to O_3 q);$

(iii) S *is closed under the rule: From* $\vdash_S A \to (B \to C)$ *to* $\vdash_S O_1 A \to (O_2 B \to O_3 C)$.

Proof. (i) $\Rightarrow$ (ii): By (i), we have $\vdash_S O_1(p \to q) \to (O_2 p \to O_3((p \to q) \wedge p))$, and $O_3((p \to q) \wedge p)$ provably implies $O_3(q)$ in any normal modal logic, whether O_3 is $\Box$ or $\Diamond$.

(ii) $\Rightarrow$ (iii): Suppose $\vdash_S A \to (B \to C)$. Then, whether O_1 is $\Box$ or $\Diamond$, we have $\vdash_S O_1 A \to O_1(B \to C)$, and by (ii) we have $\vdash_S O_1(B \to C) \to (O_2 B \to O_3 C)$, and the conclusion of the rule mentioned in (iii) follows by truth-functional reasoning from these two.

$(iii) \Rightarrow (i)$: Take A, B, C, in the rule under (iii) as $p, q, p \wedge q$, respectively. ∎

The terminology of distributive triples is taken from the equivalence of the condition in the original definition with that provided by (ii) here, which, when $O_1 = O_2 = O_3$, says that the operator concerned distributes over (or 'across') implication in S. (Of course when this operator is $\Box$, what we have is the schema **K**.)[41]

EXERCISE 2.1.22 Can "normal" by replaced by "monotone" in Prop. 2.1.21 (and the resulting claim still be correct)? Can it be replaced by "congruential"? (Give reasons, in each case.) ✠

Note that an immediate consequence of the definition of 'distributive triple' that $\langle O_1, O_2, O_3 \rangle$ is a distributive triple for a given modal logic if and only if $\langle O_2, O_1, O_3 \rangle$ is, so by Proposition 2.1.21(ii) in the case of normal S, we have

$$\vdash_S O_1(p \to q) \to (O_2 p \to O_3 q) \text{ iff } \vdash_S O_2(p \to q) \to (O_1 p \to O_3 q),$$

something which would not have been obvious had the alternative definition of distributive triples suggested by Prop. 2.1.21(ii) been adopted – an alternative that result reports would be equivalent in its application to normal modal logics.

By an examination of cases, one may establish that the distributive triples for **K**, the smallest normal modal logic, are $\langle \Box, \Box, \Box \rangle, \langle \Box, \Diamond, \Diamond \rangle, \langle \Diamond, \Box, \Diamond \rangle$. The fact that these are distributive triples for **K** is easily seen (and indeed familiar); with only slightly more work, inspecting the remaining cases, one may verify that those listed are the only distributive triples for **K**. (See the proof of Proposition 2.9.2 below.) For stronger normal modal logics, more triples become distributive. By way of illustration: for the logic **KD** one must add $\langle \Box, \Box, \Diamond \rangle$, and for $\mathbf{KD}_c$ (here and with the "!", we use Chellas's notation, explained in Section 2.1), one must add to the **K**-distributive triples above, $\langle \Box, \Diamond, \Box \rangle, \langle \Diamond, \Box, \Box \rangle, \langle \Diamond, \Diamond, \Diamond \rangle$ and $\langle \Diamond, \Diamond, \Box \rangle$, while for **KD!** one pools these sets of triples: in other words each of the 8 combinatorially possible sequences $\langle O_1, O_2, O_3 \rangle$ with $\{O_1, O_2, O_3\} \subseteq \{\Box, \Diamond\}$ is a distributive triple for this logic.

EXERCISE 2.1.23 Prove the claim (about **KD!**) just made. ✠

We can extend the above terminology so as to apply to triples of affirmative modalities. That is, if X, Y, and Z are affirmative modalities, then $\langle X, Y, Z \rangle$ is a distributive triple ("of affirmative modalities") for S just in case $\vdash_S (Xp \wedge Yq) \to Z(p \wedge q)$, which again can be reformulated in the other ways indicated for distributive triples of modal operators $\langle O_1, O_2, O_3 \rangle$ above (these latter then being the special case of distributive triples of affirmative modalities of length 1). The definition just given does not impose any restriction to the effect that X, Y, and Z are of the same length, though this is the case with which we shall be concerned. If X is an affirmative modality of length n and $1 \leqslant i \leqslant n$, then by X_i we understand the ith element – "$\Box$" or "$\Diamond$" – of X (counting from the left).

PROPOSITION 2.1.24 *Where X, Y, and Z are affirmative modalities of length n, a sufficient condition for $\langle X, Y, Z \rangle$ to be a distributive triple of affirmative modalities for a normal modal logic S is that $\langle X_i, Y_i, Z_i \rangle$ is a distributive triple of modal operators for S, for each $i \leqslant n$.*

Proof. We use characterization (iii) of distributive triples under Prop. 2.1.21, each of the following being obtainable from its predecessor by the rule there mentioned:

---

[41]The phrase *distributive triple* has an unrelated sense in lattice theory – e.g., Birkhoff [95], p. 37 – which is connected instead to the $\wedge/\vee$ distribution law (in lattice theory). And the phrase "distributes over" is itself ambiguous, between a weaker sense in which O is said to distribute over a k-ary connective # (in S) when $O\#(A_1, \ldots, A_k) \to \#(OA_1, \ldots, OA_k)$ is provable (in S) for all $A_1, \ldots, A_k$, and a stronger sense in which what is required instead is the corresponding biconditional (with $\leftrightarrow$ in place of $\to$). This latter sense is also expressed by saying that O *commutes with* # (in S).

$\vdash_S A \to (B \to C)$
$\vdash_S X_n A \to (Y_n B \to Z_n C),$
$\vdash_S X_{n-1} X_n A \to (Y_{n-1} Y_n B \to Z_{n-1} Z_n C)$
$\vdots$
$\vdash_S X_1 \ldots X_{n-1} X_n A \to (Y_1 \ldots Y_{n-1} Y_n B \to Z_1 \ldots Z_{n-1} Z_n C)$
i.e. $\vdash_S XA \to (YB \to ZC)$, as required. ∎

For the case of $S = \mathbf{K}$, there is also a converse: the distributive triples of affirmative modalities are precisely the triples of coordinatewise distributive triples of modal operators, presented below, after we have introduced the semantics of modal logic, as Proposition 2.9.2.

We conclude this discussion of distributive triples by noting that a routine modification of the proof of Proposition 2.1.21 makes available the same equivalent formulations as were there noted for the notion of a distributive triple of operators in the case of a distributive triple of affirmative modalities:

PROPOSITION 2.1.25 *For any normal modal logic S and any triple of affirmative modalities $\langle X, Y, Z \rangle$, the following are equivalent:*

(i) $\langle X, Y, Z \rangle$ *is a distributive triple for S;*
(ii) $\vdash_S X(p \to q) \to (Yp \to Zq)$
(iii) *S is closed under the rule: From $\vdash_S A \to (B \to C)$ to $\vdash_S XA \to (YB \to ZC)$.*

As already remarked, there is no requirement here that X, Y and Z be modalities of the same length. For later use, we make an observation concerning the case in which the modalities are not just of the same length, but are the same modality:

PROPOSITION 2.1.26 *For any normal modal logic S, if $\langle X, X, X \rangle$ is a distributive triple of affirmative modalities for S, then for any formulas $A_1, \ldots, A_n$ we have:*

$$\vdash_S (XA_1 \wedge \ldots \wedge XA_n) \leftrightarrow X(A_1 \wedge \ldots \wedge A_n).$$

Proof. The $\leftarrow$ direction is available for any affirmative modality X, since such modalities are monotone in all normal modal logics (indeed in all monotone modal logics, by Prop. 1.4.2). For the $\to$ direction, use induction on n. The basis ($n = 1$) case is trivial and the induction step is proved by appeal to the fact that $\langle X, X, X \rangle$ is distributive for S, since the inductive hypothesis gives:

$$\vdash_S ((XA_1 \wedge \ldots \wedge XA_k) \wedge XA_{k+1}) \to (X(A_1 \wedge \ldots \wedge A_k) \wedge XA_{k+1}). \qquad \blacksquare$$

For an interesting application of distributive triples, see Section 2.8, which can be followed on the basis of what has been done so far (i.e., the semantics of modal logic is not required).

2.2 Models and Truth: Simplified Semantics

The formal semantics for modal logic, a simplified version of which is presented here initially, is due to Saul Kripke (and others whose names you will find in any history of the subject), and, especially in the present simplified form, can be thought of as making explicit an idea attributed – with what accuracy, we need not enquire here – to Leibniz: that necessity is truth in all possible worlds. (The name of Rudolf Carnap is often cited in connection with these simplified models, in view of [151] and subsequent work. Carnap's contributions to modal logic have been discussed in, for example, Schurz

[1014], Meadows [796], §2.4 of Williamson [1198], and Cresswell [220]. More generally, for the history of the conceptual development of the semantics, see Copeland [193], and for the history of its technical elaboration, Goldblatt [395].) A *model* will for the moment be understood to be a structure $\langle W, V \rangle$ in which W is a non-empty set, whose elements we may think of, initially at least, as 'possible worlds', and V is a mapping assigning to each propositional variable p_i a subset of W. Think of V as stipulating, in giving us $V(p_i)$, which set of worlds are those at which the various p_i are to count as true; if we identify – as is often convenient – propositions themselves with the sets of worlds at which they are true, then this can be put by saying that V assigns propositions to the propositional variables.[42] We give an inductive definition (by induction on the complexity of A) of the truth of a formula A at a point w in a model $\mathcal{M}$ (notated "$\mathcal{M} \models_w A$"). For any model $\mathcal{M} = \langle W, V \rangle$ and any $w \in W$:

$\mathcal{M} \models_w p_i$ *iff* $w \in V(p_i)$, for each variable p_i;

and for all A and B:

$\mathcal{M} \models_w A \wedge B$ *iff* $\mathcal{M} \models_w A$ *and* $\mathcal{M} \models_w B$;
$\mathcal{M} \models_w A \vee B$ *iff* $\mathcal{M} \models_w A$ *or* $\mathcal{M} \models_w B$ (*or both*);
$\mathcal{M} \models_w A \to B$ *iff* $\mathcal{M} \not\models_w A$ *or* $\mathcal{M} \models_w B$ (*or both*);
$\mathcal{M} \models_w \neg A$ *iff* $\mathcal{M} \not\models_w A$.

Nothing unexpected there, then. Defining (for arbitrary A) $v_w^{\mathcal{M}}(A) = \mathrm{T}$ iff $\mathcal{M} \models_w A$, the effect of the above clauses is that for any $\mathcal{M}$, w, the valuation $v_w^{\mathcal{M}}$ is a Boolean valuation, in the sense explained at p. 10. In fact the component V of these models (and similarly for the more general models of the following section) is itself referred to as a valuation, though evidently in a different sense of the term *valuation*. This ambiguity causes no difficulty in practice. Finally:

$\mathcal{M} \models_w \Box A$ *iff for all* $x \in W$, $\mathcal{M} \models_x A$;
$\mathcal{M} \models_w \Diamond A$ *iff for some* $x \in W$, $\mathcal{M} \models_x A$.

Digression on Notation. We have used the 'semantic' turnstile "$\models$" to indicate what is true where in the possible worlds semantics. Though common, this practice is not universal. If $\mathcal{M} = \langle W, V \rangle$ is a model and $w \in W$, then some authors write "$V(A, w) = \mathrm{T}$" rather than "$\mathcal{M} \models_w A$" to indicate that A is true at w (relative to the model $\mathcal{M}$). More accurately, we should note that something different is meant by the "V" of "$\langle W, V \rangle$" for these authors. This, in the atomic case, is a function taking a propositional variable and an element of W as arguments and delivers as its value a truth-value (T or F). That is an equivalent way of thinking of things. Everywhere the one notation has "$x \in V(p_i)$" the other would have "$V(p_i, x) = \mathrm{T}$". You will find this in, for example, Hughes and Cresswell [535], [537]. A further variation sometimes encountered uses something like "V" in the atomic case, as part of the identity of the models, and "V^+" for the extension to arbitrary formulas. Thus they would say that on the basis of a model $\langle W, V \rangle$ we defined $V^+(p_i, w) = V(p_i, w)$ for all $w \in W$, $V^+(A \wedge B, w) = \mathrm{T}$ iff $V^+(A, w) = \mathrm{T}$ and $V^+(B, w) = \mathrm{T}$, etc. It is certainly useful to have *some* notation for the set of points at which a formula A is true in a model; one convention (used in Chellas [171], for instance) is to denote this by $\|A\|$, taking it as clear from the context which model is involved.) **End of Digression.**

---

[42]While the treatment of propositions as sets of worlds is indeed frequently convenient, many will favour a different account – especially for the role of 'object of a propositional attitude' (what is believed, for example) a proposition should be taken as internally structured representation rather than just a set of worlds. See King [657] for elaboration and further references; for opposition to views of propositions as structured, see for example Cresswell [218] and Tsohampidis [1116]. As Aho ([5], p. 4) writes, with propositional attitudes mainly in mind: "There have been numerous attempts to give stricter criteria that logical equivalence for the identity of propositions, but none of them suits all purposes." Something weaker would be necessary equivalence in a sense including *a posteriori* metaphysical necessities, something we do not go into here. Chalmers [168] provides a good discussion of what he calls referentialism about propositions, which subsumes both the 'structured proposition' and the 'set of worlds' accounts of propositions, on one version of each at least, revamping Fregean objections (as in [331]) to purely reference-based accounts of propositional attitude attributions.

2.2. MODELS AND TRUTH: SIMPLIFIED SEMANTICS

We are now ready for a key (albeit provisional) semantic concept. Call a formula A *valid* if for every model $\mathcal{M} = \langle W, V \rangle$ and every $w \in W$, we have $\mathcal{M} \models_w A$. (If we were working with sequents, the corresponding definition, coinciding when $\Gamma = \varnothing$ is: for every model $\mathcal{M} = \langle W, V \rangle$ and every $w \in W$, if $\mathcal{M} \models_w C$ for each $C \in \Gamma$, then $\mathcal{M} \models_w A$. More informally stated: a sequent is valid when truth is passed from the left hand formulas to the right hand formula at any world, no matter how truth-values have been assigned to propositional variables relative to those worlds.)

The definition is provisional in more than one respect. First, we are working with simplified models, and the much more versatile Kripke models have an extra ingredient – an "accessibility relation" – which complicates the definition of truth but makes available a semantic treatment of a vast array of modal logics instead of just one, as the present treatment does (namely the modal logic whose theorems are exactly the formulas which are 'valid' in the sense here under consideration). Secondly, for various reasons it is preferable to isolate one aspect of the 'de-simplified' models, called the *frame* of the model, and speak of validity only w.r.t. frames and classes thereof. (See Section 2.5.) But for the moment we proceed with the provisional terminology. Whether we are thinking of the present models or the de-simplified Kripke models of the following section, one question we shall not be considering is the extent to which possible worlds semantics of these kinds should be regarded as some kind of *analysis* of the modal concepts of necessity and possibility; the interested reader is referred to Stalnaker [1083] and Divers [246] for discussion and references. (Such a reader is also advised not to venture into these philosophical waters without first becoming familiar with the concrete details of the semantics, as developed in the remainder of the present chapter, and preferably also the first two sections of the following chapter, where a 'local' notion of validity – validity at a point in a frame – will be explained.)

EXAMPLES 2.2.1 *(i)* $\Box p \to p$ counts as valid on the present semantics, since if for $\mathcal{M} = \langle W, V \rangle$ with $x \in W$ we have $\mathcal{M} \not\models_x \Box p \to p$ then $\mathcal{M} \models_x \Box p$ while $\mathcal{M} \not\models_x p$. But since $\mathcal{M} \models_x \Box p$, we must have $\mathcal{M} \models_w p$ for all $w \in W$, and one such w is x itself: a contradiction.
(ii) The intuitively invalid formula $p \to \Box p$, the converse of the implication considered in *(i)*, is indeed invalid on the present semantics, as we can see with a simple two-element model: putting $W = \{w_1, w_2\}$ and $V(p) = \{w_1\}$, gives a model with p true and $\Box p$ false at the point (or 'world') w_1. ◀

EXERCISE 2.2.2 Which of the following formulas are valid?
(i) $\Diamond p \to p$; *(ii)* $\Diamond(p \to \Box p)$; *(iii)* $(\Box p \land \Diamond q) \to \Diamond(p \land q)$.
(Note: *(ii)* will require careful thought.) ✠

Mention was made parenthetically above to the modal logic whose theorems are exactly the formulas which are valid on the current simplified semantics. It is about time we identified that logic axiomatically: it is **S5**. This label summarises at once either of two axiomatizations: the one which proceeds by adding to any axiomatic basis for non-modal (classical) propositional logic the axiom-schemata **K**, **T** and **5** – in other words has all formulas instantiating these schemata as axioms – and uses as rules Modus Ponens and Necessitation; and another which adds to the same non-modal basis, not the infinitely many axioms summarised by these schemata, but individual representative axioms

$$\mathbf{K}: \Box(p \to q) \to (\Box p \to \Box q) \qquad \mathbf{T}: \Box p \to p \qquad \mathbf{5}: \Diamond p \to \Box \Diamond p,$$

and adds the rule of Uniform Substitution to Modus Ponens and Necessitation in order to recover the other instances of the schemata. (Note the terminological contrast: *instances* of a schema vs. *substitution instances* of a formula.) The claim that **S5** comprises precisely the formulas valid on the present semantics falls into two halves, a soundness half – all the formulas provable in **S5** are valid – and a completeness half – all the valid formulas are provable in **S5**. The latter is an easy by-product of a completeness proof for **S5** in terms of the Kripke semantics proper – see Coro. 2.5.4 below – but let us pause for a moment on the soundness issue.

A formula A is *modally invariant* if for any model $\mathcal{M} = \langle W, V \rangle$ and any $w, x \in W$, $\mathcal{M} \models_w A$ iff $\mathcal{M} \models_x A$. A modally invariant formula – and we shall often just say "invariant", for short – in other words, is one which never changes in truth-value as we go from point to point in a model. Observe that any formula of the form $\Box B$ is modally invariant in this sense, because any point at which it is false in a model requires a point in the model at which B is false, and the latter will make $\Box B$ false at every point: so such a formula cannot be true at some but not all points in a model.

More generally, we have an observation, encapsulated as Lemma 2.2.4 below, which is formulated in terms of the notion of a *fully modalized* formula, by which is meant a formula in which every occurrence of any propositional variable lies within the scope of some occurrence of $\Box$ or $\Diamond$ (not that we need to mention the latter, $\Diamond$ not being part of our primitive vocabulary), something we exploit in the proof of Lemma 2.2.4, after some comprehension practice. Examples: the formula $\Box p \vee (\Diamond q \wedge \neg(p \to \Box\Box r))$ is not fully modalized, since the second occurrence of "p" is not in the scope of a modal operator; by contrast, the formula $\Box(p \to q) \vee \neg\Diamond(r \to p)$ *is* fully modalized.

EXERCISE 2.2.3 Say for each of the following formulas, whether or not it is fully modalized.
(i) $\Box p \to p$, (ii) $\Diamond\neg(q \wedge \Box r) \vee (\Diamond q \wedge r)$, (iii) $\Box(\Box q \to q)$. ✠

LEMMA 2.2.4 *All fully modalized formulas are invariant.*

Proof. A fully modalized formula is a truth-functional compound of formulas of the form $\Box A$. But truth-functional compounds of modally invariant formulas (such as these $\Box A$) are modally invariant, as is most easily seen by taking a functionally complete set such as $\{\neg, \wedge\}$ and checking explicitly that the negation of an invariant formula is invariant, and that the conjunction of two invariant formulas is again invariant. ■

We have called this observation a Lemma because it can be used, though the degree of generality is hardly required for this purpose, in the envisaged soundness proof for **S5** as an axiomatically presented system. On the other hand, it is just right as a lemma for a corresponding proof for the case of a natural deduction system for **S5** presented in the Appendix on Natural Deduction below (Lemma 7.4.1).

THEOREM 2.2.5 *Every formula provable in* **S5** *is valid in the sense of the present semantics.*

Proof. Suppose $\vdash_{\mathbf{S5}} A$. By induction on the length of a shortest proof of A, we check that A is valid.

Basis Case. The shortest possible proofs are of length 1, which is the case presented by the axioms. Here let us take the axiomatization of **S5** with schemata **K**, **T** and **5**, so that we avoid the use of uniform substitution. If A is an axiom it is either a substitution instance of a truth-functional tautology, or an instance of one of these three schemata. In each of these cases it is not hard to check that the formula in question cannot be false at a point in a model; in the case of **5** it will help to note that the antecedent is fully modalized, and appeal to Lemma 2.2.4.

Inductive Step. One verifies that each of the rules Modus Ponens, Necessitation, preserves the property of being valid. In fact, Modus Ponens preserves the property of being true at a point in a model, and therefore the property of being true throughout any given model, and therefore the property ('validity') of being true throughout every model. For Necessitation the story is slightly different in that we don't have preservation of truth at an arbitrary point in a model, but we do have preservation of the property of being true throughout any given model and so, as before, preservation of validity. ■

The converse result, asserting the completeness of **S5** with respect to the current notion of validity, is most easily derived by considerations from the following section, where it appears as Coro. 2.5.4. As for the soundness half, we should pause, for the record, to consider how the proof would be affected if we

2.2. MODELS AND TRUTH: SIMPLIFIED SEMANTICS

had used not axiom-schemata but individual axioms, together with the rule of uniform substitution. In this case, we have an extra case in the inductive part of the proof, to show that applications of this rule preserve validity as currently conceived. For this one uses the following observation.

PROPOSITION 2.2.6 *Suppose that $\mathcal{M} = \langle W, V \rangle$ is a model of the kind in play in our discussion. Let A be any formula and $\|A\|$ be the set of points at which A is true in $\mathcal{M}$. Suppose further that $B(p_i)$ is any formula in which the propositional variable p_i occurs and that $B(A)$ is the result of substituting the formula A for all occurrences of p_i in $B(p_i)$. Then, where $V'(p_j) = V(p_j)$ for all $j \neq i$ while $V'(p_i) = \|A\|$, and $\mathcal{M}' = \langle W, V' \rangle$, we have $\mathcal{M} \models_x B(A)$ if and only if $\mathcal{M}' \models_x B(p_i)$, for all $x \in W$.*

EXERCISE 2.2.7 (*i*) Prove Proposition 2.2.6, by induction on the complexity of $B(p)$.
(*ii*) Use Prop. 2.2.6 to show that the rule of uniform substitution preserves validity. ✠

It is noteworthy how the models we have been working with resemble the 'interpretations' (also called structures or models) $\langle D, V \rangle$ in terms of which the semantics of classical predicate logic is sometimes presented. D is a non-empty set (the domain or universe) elements of which are assigned by V to individual constants, which also assigns n-ary functions on D to n-place function symbols, and n-ary relations on D to n-ary predicate symbols. In our models $\langle W, V \rangle$ the 'W' part is functioning very much like the domain D of those interpretations $\langle D, V \rangle$: modal operators 'range' over W just as quantifiers range – no shudder quotes this time – over the domain. "□" is like "∀", "◇" is like "∃". Think about the distribution of the modal operators over connectives "∧" and "∨" and about the equivalence of □p with ¬◇¬p, with this analogy in mind. (Later we will frequently omit the quotation marks when mentioning symbols and just say, for example: connectives ∧ and ∨. Similarly with other references to the formal language of modal logic.) For the analogy to be exact, one should think specifically of *monadic* predicate logic, so that predicate symbols are assigned subsets of the domain ('1-ary relations') as their extensions, just as the V of the current modal models assigns to propositional variables subsets of W. (Much recent philosophy has been concerned with how seriously to take the analogy: do we really need to believe in a set of possible worlds, all but one of which are non-actual, in order to make sense of modal talk? Or can we use the semantics to throw light on modal language while at the same time not taking its ontological suggestions seriously? The view that there really are non-actual possible worlds, more or less like the actual world in respect of concreteness, but not spatially or temporally related to the actual world or to each other, has been called 'modal realism'; it was illustriously defended by David Lewis – no relation to C. I. Lewis, last heard of on p. 46 above – in [726].)

Naturally the existence of the analogy between modal operators and quantifiers should not blind us to the possible interest of the topic of modal *predicate* logic – also called: quantified modal logic – in which both modal operators and also ordinary quantifiers over individuals appear. Several metaphysical doctrines are naturally expressed in just such terms, and there are several delicate questions of validity to decide – e.g., over the validity of the Barcan Formula, namely the schema $\forall x \Box A \rightarrow \Box \forall x A$, which hangs on issues concerning the domains of (= sets of individuals taken to exist in) the worlds in a model and the accessibility relation; but we are simply not looking into this topic here. (Hughes and Cresswell [535] and [537] contain chapters on modal predicate logic, as does Garson [351]; for detailed information, recent works dedicated to the topic include Gabbay *et al.* [344], and Goldblatt [397], concentrating on technical matters, and Williamson [1198], concentrating on the conceptual and philosophical aspects of the subject. Williamson's discussion includes second order modal predicate logic, a topic treated also in Cocchiarella and Freund [183]. Initially surprising failures of completeness results of the kind featuring in Section 2.5 to carry over to the case of modal predicate logic – with or without the Barcan Formula – are discussed in Cresswell [214], [215] and [216]; among the systems for which completeness becomes problematic in the transition – at least in the presence of the Barcan Formula – from modal propositional to modal predicate logic is the familiar **S4.2**, as was foreshadowed in a remark in the second last paragraph of Kripke [671].) (Goldblatt and Mares [396] and – again – Goldblatt [397] are also relevant here.)

EXERCISE 2.2.8 Call a modal formula A *n-valid* (where $n \geqslant 1$) when A is true at every point in any model $\langle W, V \rangle$ with $|W| \leqslant n$. For example, the formula $\Diamond p \to \Box p$ is 1-valid but not 2-valid. Now give an example of a formula which is 2-valid but not 3-valid. ✠

Note that while every fully modalized formula is modally invariant (Lemma 2.2.4), there are obvious counterexamples to the converse: any valid formula or any formula whose negation is valid is true either everywhere or nowhere in any model and thus is modally invariant, whether or not the formula concerned is fully modalized. (But see Proposition 2.2.10 below.)

EXERCISE 2.2.9 Is the following true or false? For any formula A such that neither A nor $\neg A$ is **S5**-provable, if A is modally invariant then A is fully modalized. (Justify your answer with an informal proof or with a counterexample, as appropriate.) ✠

PROPOSITION 2.2.10 *For every formula A, A is modally invariant if and only if A is **S5**-equivalent to a fully modalized formula.*

Proof. Since the 'if' direction is clear, we address the 'only if' direction. Suppose A is modally invariant. Then $A \to \Box A$ is valid on the simplified semantics, so A is provably equivalent in **S5** to the fully modalized formula $\Box A$. ■

In the above proof, we have assumed the completeness of **S5** w.r.t. the notion of validity given by the simplified semantics. (This is proved as Coro. 2.5.4 below, p. 72.) Let us note, incidentally, that instead of $\Box A$, we could have chosen $\Diamond A$ as a fully modalized formula **S5**-equivalent to A.

EXERCISE 2.2.11 (*i*) Justify the claim just made.

(*ii*) Prove or refute the variant of Proposition 2.2.10 which replaces "**S5**-equivalent" by "**S4**-equivalent", leaving the rest of the formulation intact. ✠

2.3 Models and Truth: Kripke Semantics

In our introductory section on modal logic, the idea that "$\Box$" might be interpreted as "it ought to be the case that" was mentioned. This deontic reading is not well served by our stipulation that in a model $\mathcal{M} = \langle W, V \rangle$, we have for any $w \in W$ and any formula A, $\mathcal{M} \models_w \Box A$ iff for all $x \in W$, $\mathcal{M} \models_x A$. What ought to be the case is not what is the case in all possible worlds, but perhaps what is the case throughout some less inclusive set – for example the morally perfect (or 'ideal') worlds. This suggests that we allow a more general notion of model, in which a distinguished subset of W is selected, truth throughout which is required for the truth of a $\Box$-formula. (We will return to this idea more than once in what follows, calling it the *semi-simplified* Kripke semantics; see especially p. 206.) Once we have gone this far, though, it is natural to allow a further liberalization and let the relevant subset depend on the world at which we are evaluating the $\Box$-formula. In terms of the deontic reading of "$\Box$", this would be to allow that the morally perfect worlds *from the point of view of w* need not be the same as the morally perfect worlds *from the point of view of w'* (where $w \neq w'$).

Similarly, if we want to read $\Box A$ as "it is physically necessary that A" and understand by this that A is true at all physically possible worlds, then it seems we should allow that which set of worlds comprises the set of physically possible worlds should itself be allowed to vary from world to world: the worlds physically possible relative to w are those in which no physical laws of w are violated, those physically possible relative to w' are those in which no physical laws of w' are violated. Since the laws may be different in w and w', these sets may be different.

2.3. MODELS AND TRUTH: KRIPKE SEMANTICS

REMARKS 2.3.1 (*i*) If we want a similar notion but allowing for the possibility of laws of nature which are not physical laws – for example thinking that there might be or could have been irreducibly psychological laws – then we can say all this replacing the word "physical(ly)" but the more neutral word "nomic(ally)".

(*ii*) On the above line of thought whatever is true in all possible worlds will end up being physically (or more generally: nomically) necessary, which may cause discomfort: it would at the very least be odd to *assert* that it is physically necessary that $2 + 2 = 4$, since whatever the physical laws had been, it would have been the case that $2 + 2 = 4$. One reaction is retain the current broad conception and then separately investigate the idea of being *purely physically* necessary, meaning physically but not logically necessary. We will look into this in Section 4.3 after looking at (one take on) the broader conception in Section 4.2.

(*iii*) Note the contrast with the present line of thought and that expressed in the following remark made in passing at p. xxiii of Priest [912]: "a physically impossible world is a world where the laws of physics are different from those of the actual world." This diverges in more than one respect from the present (and indeed standard) conception. According to the present conception a world w' is physically impossible relative to w not simply if the physical laws of w differ from those of w' but specifically if the physical laws of w are violated in w'. For example if w is the actual world, assumed to have numerous physical laws, and w' is a world in which there are no laws whatever but the course of individual events exactly matches that of e – a 'Hume duplicate' of w according the material under discussion in Section 4.2 – then w' is physically possible relative to w (on our understanding) despite having a very different set (namely the empty set) of laws from w and therefore being physically impossible relative to w according to Priest's way of understanding the terminology. Similarly if a worlds w'' had all of the physical laws of w and more besides, w'' would be physically possible relative to w on the usual conception though not on Priest's. (This remark assumes that the laws of a world always hold in that world – i.e., that the relation of nomic possibility is reflexive. See in this connection Remark 4.2.1, p. 217, and note 177, p. 219.) ◀

Considerations like those given before Remarks 2.3.1 suggest that we embellish our models somewhat, and take them instead to be triples $\langle W, R, V \rangle$ in which R is a binary relation on W, intuitively to be thought of as relating any $w \in W$ to those elements of W which are 'possible relative to' w (physically possible relative to w, for the case just mentioned, morally satisfactory by the lights of w, for the deontic case, and so on).[43] The relation R is called the *accessibility relation* of the model. Recall from the discussion of notating binary relations on p. 4, that we do not have a once-and-for-all fixed policy as to whether to write "Rxy" or "xRy" to indicate the holding of this relation between x and y, tending to use the latter for 'everyday' purposes and the former when explicit first-order representation is at issue. However it is written, we read it as "y is accessible to x". Other terminology includes "y is possible relative to x", "y is a successor (or more explicitly, "is an R-successor") of x", "y is an alternative to x", "x is a predecessor of y".[44] It is models of this form that are used in what is generally called the Kripke semantics for modal logic, rather than the simplified R-less versions of the preceding section. Predictably enough, in view of this discussion, the treatment of $\Box$ in terms of these models goes: for any model $\mathcal{M} = \langle W, R, V \rangle$, and any formula A:

---

[43] Even more clearly, for a (future) tense-logical reading of $\Box$ as "It will always be that __", the points in the models are most naturally thought of as moments of time rather than possible worlds (and one may wish to use a different letter for the set of them, say thinking of the models as $\langle T, R, V \rangle$) and the set of $\Box$-pertinent points comprises those which are *later than* a given point, which will be a different set as we pass from point to point. $\Box$ is often written as "G" for this application. See further Section 3.1.

[44] The terminology of alternativeness, favoured by Hintikka – see any of his entries in the bibliography – has the disadvantage of suggesting that the relation needs to be symmetric, and moreover as being that relation holding, from x's perspective, between the various y such that xRy: these are the worlds that are alternatives to each other *qua* worlds each of which is possible relative to x. Compare the idea of alternative futures in the branching time semantics for tense logic (note 156, p. 195 below), which involve many pairs of points (moments) later than the present with neither later the other. (See the preceding note.)

$\mathcal{M} \models_x \Box A$ iff for all $y \in W$ such that xRy, $\mathcal{M} \models_y A$.

For the case of $\Diamond$, replace "all y" by "some y"; this is the result of applying the definition of truth (at a point) to $\neg\Box\neg A$, which is what our $\Diamond A$ is officially short for. Note that we can only make the comment about simply replacing "all y" by "some y" because we have used the "such that" construction in the above formulation above. What the right-hand side really means is "(iff) for all $y \in W$, if xRy, then $\mathcal{M} \models_y A$." In this more explicit formulation, one cannot simply change "all" to "some" to get the corresponding clause for "$\Diamond$", because the conditional construction has to be replaced by a conjunctive construction:

$\mathcal{M} \models_x \Diamond A$ iff for some $y \in W$, xRy and $\mathcal{M} \models_y A$.

Another reformulation involves the useful abbreviation "$R(x)$", for the set of points (in the model under consideration) accessible to x; that is, for $x \in W$:

$$R(x) = \{y \in W \mid xRy\}.$$

This highlights the respect in which the set of points truth at each of which is necessary and sufficient for the truth of a $\Box$-formula at a given point is now allowed to vary from point to point. The contrast is with this set's being fixed once and for all – as on the simplified semantics – as being the set of *all* points (the set W, that is). Using this notation, and exploiting the fact that $R(x) \subseteq W$ the above clause for $\Box$ in the definition of truth at a point becomes:

$\mathcal{M} \models_x \Box A$ iff for all $y \in R(x)$, $\mathcal{M} \models_y A$.

Obvious though it may be, it does no harm to emphasize that any way of assigning to each point a set of points, the truth of A at each of which is, as it was put above, necessary and sufficient for the truth of $\Box A$ at the given point, amounts to introducing an accessibility relation. Suppose we have a family of such '$\Box$-pertinent' sets W_x (all of them subsets of W), one for each $x \in W$. Thus we are envisaging a semantic account according to which for any formula A, $\Box A$ is true at $x \in W$ just in case A is true at each $y \in W_x$. Then we can trade in the family of subsets W_x of W for a single binary relation R, defining xRy to hold iff $y \in W_x$, and from this recover the original sets W_x as $R(x)$. There is accordingly nothing to choose between models which associate with each point x, a $\Box$-pertinent set of points (the W_x idea) and models which come equipped with a binary relation (the usual Kripke models).

For a model $\mathcal{M} = \langle W, R, V \rangle$, and a formula A, we write $\mathcal{M} \models A$ to abbreviate the claim that for all $x \in W$, $\mathcal{M} \models_x A$. One can read "$\mathcal{M} \models A$" as "$\mathcal{M}$ verifies A", or "A is true in $\mathcal{M}$", or, in an attempt to be completely explicit and to ruffle no feathers,[45] as "A is true throughout $\mathcal{M}$". A modal logic S is *sound* w.r.t. a class $\mathbb{C}$ of models if for each $A \in S$, we have $\mathcal{M} \models A$ for all $\mathcal{M} \in \mathbb{C}$, *complete* w.r.t. $\mathbb{C}$ if for every formula A, if $\mathcal{M} \models A$ for all $\mathcal{M} \in \mathbb{C}$, then $A \in S$, and S is said to be *determined* by $\mathbb{C}$ if S is both sound and complete w.r.t. $\mathbb{C}$. If we temporarily, and not entirely happily, call formulas true throughout every model in $\mathbb{C}$, $\mathbb{C}$-valid, then the soundness of S w.r.t. $\mathbb{C}$ means that every theorem of S is $\mathbb{C}$-valid, while the completeness of S w.r.t. $\mathbb{C}$ means that every $\mathbb{C}$-valid formula is a theorem of S, and determination by $\mathbb{C}$ is a matter of the S-provability coinciding with $\mathbb{C}$-validity. Further, $\mathcal{M}$ is said to be a *model for* S when $\mathcal{M} \models A$ for every $A \in S$; thus S's being sound w.r.t. $\mathbb{C}$ is equivalent to every model in $\mathbb{C}$ being a model for S.

---

[45]Zalta [1214], for example, objects to using the phrase "true in a model" for anything other than truth at a distinguished point in a model, conceived of as having the form $\langle W, R, w, V \rangle$, where $w \in W$ and truth in the model is truth at w therein. Models of this form, though differently notated, were the models originally used by Kripke, and they certainly have advantages for the semantics of certain non-normal modal logics. See Hanson [440] for a critical discussion and further references to the literature, and Nelson and Zalta [842] for a response to Hanson's criticisms and Hanson [441] for a further reply to this; Wehmeier [1158] is also relevant here, as is French [333]. The issue becomes particularly serious in the setting of modal logic with an "actually" operator: see Remark 4.7.1, p. 305 below, as well as note 177, p. 219, where additional pertinent references will be found.

If we take $\mathbb{C}$ in the foregoing definitions as the class of all models, which formulas are "$\mathbb{C}$-valid"? Certainly not any of the representative instances of any schemata appearing in Table 2.1 (p. 33). In fact, as we shall see in Coro. 2.4.3, the formulas in question are precisely the theorems of **K**, the smallest normal modal logic. It is from this fact that the latter logic gets its name – as the least logic treatable in terms of the semantics of Kripke models.

2.4 Canonical Models and Generated Submodels

To prove such facts as that just alluded to, we use the method of canonical models, as explained in this section. It would be enough, for purposes of showing that S is complete with respect to a class of models, to show for every S-unprovable formula there is a model in the class containing a point at which the formula is false. But the present method proves something stronger, namely the $\exists\forall$ statement corresponding to the $\forall\exists$ statement just noted to be sufficient: we find some single model such that for each S-unprovable formula there is a point in the model at which that formula is false. A model $\mathcal{M}$ for S is called a *characteristic* model for S if $\mathcal{M}$ verifies precisely the theorems of S.[46] (So not only does $\mathcal{M}$ verify all theorems of S – as is required for it to be a model for S – but the only formulas it verifies are the theorems of S.) The canonical model for a consistent normal modal logic is a certain very specific characteristic model for that logic. If S is the logic concerned, this model will be denoted by $\mathcal{M}_S = \langle W_S, R_S, V_S \rangle$, and we will begin by describing its universe, W_S.

A set x of formulas is called S-*inconsistent* if there exist $A_1, \ldots, A_n \in x$ with $\vdash_S \neg(A_1 \land \ldots \land A_n)$; otherwise x is S-*consistent*. A set of formulas x is called *maximally* S-consistent (or alternatively "maximal consistent w.r.t. to S") if x is S-consistent but no proper superset of x is S-consistent. The set W_S on which the canonical model for S is built is the set of all maximal S-consistent sets of formulas. If S is itself inconsistent, no sets of formulas are S-consistent, let alone maximally so, so the inconsistent modal logic (which is certainly normal) has no canonical model – as the universe of a model is required to be non-empty – which explains why we have to keep restricting attention to consistent normal modal logics when discussing canonical models. We will explain what the R_S and V_S of $\mathcal{M}_S$ are after noting some key facts about maximal consistent sets of formulas:

FACT 1. If x is a maximal consistent set of formulas w.r.t. a normal modal logic S then for all formulas A, exactly one of A, $\neg A$, is an element of x. (If both were in x, it would be inconsistent. Suppose next that neither is. Then since x is *maximal* consistent each of $x \cup \{A\}$ and $x \cup \{\neg A\}$ is inconsistent, i.e., there are conjunctions C, D, of elements of x with $\vdash_S \neg(C \land A)$ and $\vdash_S \neg(D \land \neg A)$; but these imply that $\vdash_S \neg(C \land D)$ (why?), contradicting the consistency of x.)

FACT 2. If x is a maximal consistent set of formulas w.r.t. a normal modal logic S and $\vdash_S B$ then $B \in x$. (Otherwise, by Fact 1, we should have $\neg B \in x$, contradicting the consistency of x.)

FACT 3. If x is a maximal consistent set of formulas w.r.t. a normal modal logic S and if $A_1, \ldots, A_n \in x$, and $\vdash_S (A_1 \land \ldots \land A_n) \to B$, then $B \in x$. (If $B \notin x$, then $\neg B \in x$, but rephrasing the fact that $\vdash_S (A_1 \land \ldots \land A_n) \to B$ as $\vdash_S \neg(A_1 \land \ldots \land A_n \land \neg B)$, we see that this would make x inconsistent on the hypothesis that $A_1, \ldots, A_n \in x$.) Note that Fact 2 is the $n = 0$ case of this.

Resuming our explanation of the rest of $\mathcal{M}_S$, we define xR_Sy for $x, y \in W_S$ (S assumed to be a consistent normal modal logic) by: for all formulas A, if $\Box A \in x$ then $A \in y$. This is often put more succinctly thus:

$$\{A \mid \Box A \in x\} \subseteq y.$$

---

[46] Etymology: instead of "determined by" as defined at the end of the preceding section, some authors say "characterized by".

An alternative characterization in terms of $\Diamond$-formulas will be given at Exercise 2.4.5.

Finally V_S, which must assign sets of points to the propositional variables, is defined like this: $V_S(p_i) = \{x \in W_S \mid p_i \in x\}$. This stipulation secures that truth at a point in the canonical model coincides with membership in that point – which is, after all, a set of formulas – for the propositional variables. In the following result, which is often called the Fundamental Theorem of Normal Modal Logic, we see this property holds for *all* formulas.

THEOREM 2.4.1 *If S is any consistent normal modal logic, then for any formula A, we have for all $w \in W_S$: $\mathcal{M}_S \models_w A$ iff $A \in w$.*

Proof. By induction on the complexity of (= number of connectives in) A.

Basis Case: Complexity of $A = 0$. Then A is a propositional variable p_i: in this case the way V_S was specified in the model $\mathcal{M}_S = \langle W_S, R_S, V_S \rangle$, guarantees that for any w, we have $w \in V(p_i)$ iff $p_i \in w$. But from the definition of truth (at a point in a model) the latter is equivalent to having $\mathcal{M}_S \models_w p_i$.

Inductive Step. Suppose the lemma holds for all formulas of less than the complexity (> 0) of our formula A. There are several cases to consider, of which we shall consider three: (*i*) A is $B \wedge C$ for some formulas B, C; (*ii*) A is $\neg B$ for some formula B; (*iii*) A is $\Box B$ for some formula B. Note that in each case, the formulas B and C are of lower complexity than A, so we are entitled to assume ("the Inductive Hypothesis") that the lemma holds for them: i.e. that for all $w \in W_S$: $\mathcal{M}_S \models_w B$ iff $B \in w$ (and likewise for C).

Case (*i*). Suppose A is $B \wedge C$. What has to be shown is that for all $w \in W_S$: $\mathcal{M}_S \models_w A$ iff $A \in w$, i.e., for an arbitrary such w: $\mathcal{M}_S \models_w B \wedge C$ iff $B \wedge C \in w$. By the definition of truth, the l.h.s. here holds iff $\mathcal{M}_S \models_w B$ and $\mathcal{M}_S \models_w C$, which by the inductive hypothesis is equivalent to its being the case that $B \in w$ and $C \in w$. So to complete the argument, we need to know that a conjunction belongs to a maximal consistent set iff each conjunct does. But this follows from Fact 3.

Case (*ii*). Suppose A is $\neg B$. What has to be shown is that for all $w \in W_S$: $\mathcal{M}_S \models_w A$ iff $A \in w$, i.e., for an arbitrary such w: $\mathcal{M}_S \models_w \neg B$ iff $\neg B \in w$. By the definition of truth, the l.h.s. here holds iff $\mathcal{M}_S \not\models_w B$, which, by the inductive hypothesis (B being of lower complexity than $A = \neg B$), is equivalent to saying that $B \notin w$. So what we have to show is that $B \notin w$ if and only if $\neg B \in w$: but this is just a reformulation of Fact 1.

Case (*iii*). Suppose A is $\Box B$. (This is the interesting case, from the point of view of modal logic.) We must show that for all $w \in W_S$: $\mathcal{M}_S \models_w A$ iff $A \in w$, i.e., for an arbitrary such w: $\mathcal{M}_S \models_w \Box B$ iff $\Box B \in w$. By the definition of truth, the l.h.s. holds iff for all $x \in W_S$ such that wR_Sx, $\mathcal{M}_S \models_x B$. But this is equivalent, by the inductive hypothesis, to the claim that for all $x \in W_S$ such that wR_Sx, we have $B \in x$. So what needs to be shown is that

(*) $\quad \Box B \in w$ if and only if for all $x \in W_S$, $wR_Sx \Rightarrow B \in x$.

(Here "$\Rightarrow$" is just implication in the metalanguage, for brevity.) The "only if" half of (*) is an immediate consequence of the way R_S was defined. For the "if" direction, we need to note that if $\Box B \notin w$ then the set $\{D \mid \Box D \in w\} \cup \{\neg B\}$ is S-consistent. For suppose otherwise. Then for some formulas $D_1, \ldots, D_n$ for each of which $\Box D_i \in w$, we have

$\vdash_S \neg(D_1 \wedge \ldots \wedge D_n \wedge \neg B)$ and thus

$\vdash_S (D_1 \wedge \ldots \wedge D_n) \to B$ from which by the rule of normality, we have

$\vdash_S (\Box D_1 \wedge \ldots \wedge \Box D_n) \to \Box B,$

Now since each of the formulas $\Box D_1, \ldots, \Box D_n$ belongs to w, so does their conjunction (recall Case (*i*) of this proof) and this would force $\Box B$ to be an element of w (by Fact 3), contradicting our hypothesis that

2.4. CANONICAL MODELS AND GENERATED SUBMODELS

$\Box B \notin w$. Thus $\{D \mid \Box D \in w\} \cup \{\neg B\}$ is S-consistent and it can be extended to a maximal consistent set x, for which we have $wR_S x$ and $B \notin x$. So what we have shown is that if $\Box B \notin w$ then there exists an $x \in W_S$ such that $wR_S x$ and $B \notin x$: but this is a contraposed form of the "if" direction of (*), completing the proof. ∎

It follows that the canonical model for a logic is amongst its characteristic models:

COROLLARY 2.4.2 *If S is any consistent normal modal logic, S is determined by $\{\mathcal{M}_S\}$.*

Proof. S is sound w.r.t. $\{\mathcal{M}_S\}$: this means every formula provable in S is true at all points of $\mathcal{M}_S$. ($\mathcal{M}_S$ is a model for S, in other words.) For suppose that A is provable in S. Then by Fact 2, A belongs to each element of W_S, whence by 2.4.1, A is true at each such element, in the model $\mathcal{M}_S$. On the other hand, suppose that A is not provable in S; then $\{\neg A\}$ is S-consistent and can be extended to a maximal S-consistent set x which is an element of the canonical model to which A does not belong (by consistency, since $\neg A$ does) and at which therefore, A is not true in S, by Theorem 2.4.1. Thus $\mathcal{M}_S \not\models A$. ∎

COROLLARY 2.4.3 *The modal logics $\mathbf{K}$, $\mathbf{K4}$, $\mathbf{KT}$, $\mathbf{S4}$, and $\mathbf{KB}$ are determined respectively by the class of all models, the class of all transitive models, the class of all reflexive models, the class of all transitive reflexive models, and the class of all symmetric models.*

Proof. The soundness halves of these results we take for granted, and appeal to the canonical models for the completeness halves. We consider explicitly the cases of $\mathbf{K}$ and $\mathbf{KB}$. For the first, we need to show that if $\not\vdash_\mathbf{K} A$ then there is some model or other – no particular demands on the accessibility relation – in which A is false at some point. By Coro. 2.4.2, $\mathcal{M}_\mathbf{K} \not\models A$, so we are done.

We turn to the case of $\mathbf{KB}$. This time we have to find not just any old model harbouring a point at which a given non-theorem of the logic is false, but specifically a model whose accessibility relation is symmetric. But again we use the canonical model for the logic, $\mathcal{M}_\mathbf{KB} = \langle W_\mathbf{KB}, R_\mathbf{KB}, V_\mathbf{KB}\rangle$, since we will be able to show that $R_\mathbf{KB}$ is indeed a symmetric relation. To show this, suppose otherwise, hoping for a contradiction. That means we have $x, y \in W_\mathbf{KB}$ with $xR_\mathbf{KB} y$ but not $yR_\mathbf{KB} x$. The latter means that for some formula A, $\Box A \in y$ while $A \notin x$. This in turn means that $\neg A \in x$, so by Facts 1 and 3 and the instance of the $\mathbf{B}$ schema $\neg A \to \Box\Diamond \neg A$, that $\Box \Diamond \neg A \in x$. So, since $xR_\mathbf{KB} y$, we have $\Diamond \neg A \in y$, and thus $\neg \Box A \in y$, contradicting the fact that $\Box A \in y$ in view of the consistency of y. By the reasoning given at the end of the completeness proof above for $\mathbf{K}$, any non-theorem of $\mathbf{KB}$ is false at some point in the model $\mathcal{M}_\mathbf{KB}$, and therefore false at some point in a model whose accessibility relation, as we have just shown is the case for $R_\mathbf{KB}$ is symmetric. ∎

We interrupt this discussion of canonical models to apply what we have already learnt to show something about a rule we shall need later (for the proof of Proposition 2.5.35, p. 86).

PROPOSITION 2.4.4 *Every normal extension of $\mathbf{K4}$ is closed under the rule:*

$$\frac{\Diamond A}{\Diamond(A \wedge (\Box B \to B))}$$

Proof. We want to use the fact that the formula

$$(\Diamond p \wedge \Box \Diamond p) \to (\Diamond(p \wedge \neg \Box q) \vee \Diamond\Diamond(p \wedge q))$$

is provable in **K**, without going to the trouble of finding an axiomatic proof in the style of, say, Example 1.3.16 (p. 27), so the reader is invited to check that this formula cannot be false at any point in any model and draw the conclusion that such a proof exists by appealing to the completeness result for **K**, Coro. 2.4.3. Weakening within each of the disjuncts in the consequent of this formula gives the following as also **K**-provable:

$$(\Diamond p \wedge \Box \Diamond p) \to [\Diamond(p \wedge (\neg \Box q \vee q)) \vee \Diamond\Diamond(p \wedge (\neg \Box q \vee q))],$$

in which we have used the square brackets to aid readability. (The weakening steps replace each of $\neg \Box q$ and q by $\neg \Box q \vee q$. In a formal proof like that already mentioned, this could be effected by suitable appeals to $\Diamond$Mono.) Let us re-write this, using $\to$ in those disjuncts:

$$(\Diamond p \wedge \Box \Diamond p) \to [\Diamond(p \wedge (\Box q \to q)) \vee \Diamond\Diamond(p \wedge (\Box q \to q))].$$

Now by **4** – most saliently here in the dual form $\Diamond\Diamond C \to \Diamond C$ – the second disjunct provably implies the first, so the first can be omitted, and we can conclude that:

$$\vdash_{\mathbf{K4}} (\Diamond p \wedge \Box \Diamond p) \to \Diamond(p \wedge (\Box q \to q)).$$

(Note that we could have started here rather than in **K**, and invited the reader to check the that this formula cannot be false at any point in a model with a transitive accessibility relation, and so must – by Coro. 2.4.3 again – be **K4**-provable. The present course of exposition seemed more informative, though. On the alternative exposition, beginning with the present formula, it would be timely to observe that the formula $\Box q \to q$ figuring here is what is called a 'strong Hughes formula' at Exercise 2.6.38 on p. 129.) Thus, by uniform substitution we have for any formulas A, B:

$$\vdash_{\mathbf{K4}} (\Diamond A \wedge \Box \Diamond A) \to \Diamond(A \wedge (\Box B \to B)).$$

Now suppose that a premiss $\Diamond A$ for an application of the rule we are considering is provable in a normal extension of **K4**. Then so is $\Box \Diamond A$ (by Necessitation), and so therefore is the truth-functional consequence $\Diamond A \wedge \Box \Diamond A$ of these two formulas. Thus by Modus Ponens and the last cited **K4** theorem schema, we infer that $\Diamond(A \wedge (\Box B \to B))$ is provable in the logic concerned: the conclusion of the rule in question. ■

Returning to the theme of canonical models, let us note that a different but equivalent characterization of the accessibility relations R_S for consistent normal modal logics S is more convenient than that provided in the definition given earlier:

EXERCISE 2.4.5 Show that for the canonical accessibility relations R_S we have for all $x, y \in W_S$ (S here, any consistent normal modal logic): xR_Sy if and only if for all $B \in y$, we have $\Diamond B \in x$. ✠

Using this alternative characterization of R_S in the case of $R_{\mathbf{KB}}$, we can vary the part of the argument at the end of the proof of Coro. 2.4.3 to the effect that this relation is symmetric to reduce the number of negation signs in view: Suppose that $xR_{\mathbf{KB}}y$ with a view to showing that $yR_{\mathbf{KB}}x$. Take an arbitrary $\Box A \in y$ to show that $A \in x$. Since $xR_{\mathbf{KB}}y$, by the $\Diamond$-based characterization of $R_{\mathbf{KB}}$ provided by Exercise 2.4.5, we must have $\Diamond\Box A \in x$, whence by the dual form of the **B** axiom, we get $A \in x$. (Note that in this argument we rely on the original $\Box$-based definition of the accessibility relation as well as on the new variant.)

The completeness proof for the **KB** case under Coro. 2.4.3 is representative of canonical model arguments in that the effect of a particular axiom (or axiom schema) on the behaviour of the accessibility relation of the canonical model is teased out: in this case what was actually shown was something stronger than what was claimed, namely, the following. The canonical accessibility relation for any normal modal logic in which every instance of the **B** schema is provable is bound to be *symmetric*. For some of the other cases mentioned there, we need the following similar results.

2.4. CANONICAL MODELS AND GENERATED SUBMODELS

EXERCISE 2.4.6 Show that the canonical accessibility relation, R_S for any consistent normal modal logic S in which all instances of **T** are provable – any normal extension of **KT**, that is – must be reflexive, and that if all instances of **4** are provable, it must be transitive. ✠

From the results here requested, those aspects of Coro. 2.4.3 bearing on **KT** and **K4** follow, and, combining them, for **S4** (= **KT4**). Similarly, bearing in mind that although "**S5**" was defined to be an abbreviation for "**KT5**", in fact also **S5** = **KTB4** (or **S4B** if you prefer), we have the following further ramifications of our observations to date. Recall that an equivalence relation is a relation which is transitive, symmetric and reflexive.

COROLLARY 2.4.7 *The logics* **KTB** *and* **S5** *are determined by the classes of models whose accessibility relations are, (for* **KTB***) both symmetric and reflexive, and (for* **S5***) equivalence relations.*

Some other axiomatizations of **S5** were mentioned in Exercise 2.1.15, including that suggested by the label itself, using **T** and **5**, and also weakening that suggested by the label **KTB4** to **KDB4**, so we should give the new axioms appearing here some consideration.

EXERCISE 2.4.8 (*i*) Show that all instances of **5** are true throughout any model whose accessibility relation R is *euclidean* in the sense that for all elements x, y, z (of the model), we have

$$(xRy \ \& \ xRz) \Rightarrow yRz.$$

That gives the soundness of **K5** w.r.t. the class of models with euclidean accessibility relations. For completeness, we need:

(*ii*) Show that the accessibility relation of the canonical model for any consistent normal modal logic in which all instances of **5** are provable is a euclidean relation.

Similarly, for the soundness and completeness of **KD**, we need (*iii*) and (*iv*):

(*iii*) Show that all instances of **D** are true at all points in a model $\langle W, R, V \rangle$ whose accessibility relation is *serial (frames and models)* in the sense that for all $x \in W$ there exists $y \in W$ with xRy.

(*iv*) Show that the canonical model of any normal extension of **KD** satisfies the seriality condition in (*iii*). (*Hint*: For any element x of the canonical model, we need to find some y accessible to x, so show that the set $\{A \mid \Box A \in x\}$ must be **KD**-consistent, so any maximal consistent superset of this will do as a suitably accessible y.) ✠

By comparison with reflexivity, symmetry and transitivity, euclideanness is a less well-known property of binary relations, which is why we have given the definition. The terminology, due to Lemmon, is inspired by Euclid's principle that things equal to a given thing are equal to each other: instead of "equal to", read "accessible to". We did not transcribe this verbatim in the inset definition above, which, had we done so, would have read:

$$(xRy \ \& \ xRz) \Rightarrow (yRz \ \& \ zRy),$$

but it does not take too much effort to see that dropping either of the conjuncts in the consequent here makes no difference (so long as the other remains). The 'seriality' terminology is also due to Lemmon, but its justification is rather more obvious: we get an endless series of points $x_1 R x_2 R x_3 \ldots$ whenever R is serial. (However, the "endless" may seem a bit of an exaggeration when we note that this is compatible with $x_1 = x_2 = x_3 = \ldots$: see further Exercise 2.4.9 below.)

Returning to the euclidean condition, it may be worth giving a diagrammatic representation of the condition which brings out the contrast with transitivity – a contrast which sometimes puzzles those encountering it for the first time – in a pleasant visual way. The convention we adopt for diagramming conditions on the accessibility relation of the form "for all points $x_1, \ldots, x_n$ if such-and-such accessibility relationships hold amongst them, then so-and-so relationship also holds" is to draw n nodes and indicate the such-and-such relationships by arrows between them (going in the direction from a point to a point accessible to it) and the so-and-so relationship by a broken or dotted arrow.[47] In Figure 2.3 transitivity is depicted on the left and the euclidean condition on the right; we have labelled the nodes with the variables in terms of which these conditions are most familiarly formulated (as above, in the case of the latter, and in the form "for all x, y, z, if xRy and yRz then xRz" for the former).

Figure 2.3: *Transitivity and the Euclidean Condition*

Some simpler (two-variable) conditions are represented in Figure 2.4. The property of being symmetric, represented on the left of this diagram, is familiar, while what we have called range-reflexivity, represented on the right, may be less so: see p. 79. We have used a dotted rather than a dashed form of the broken arrow at the loop on the second node for easier readability. More significantly, we have not labelled the nodes with individual variables (x, y,...), since while these may help in calling to mind a particular way of formulating the condition, the choice of variables is immaterial (as long as they are all distinct). A different choice will result in a condition, which, when written as a first-order formula,[48] is what is called a bound alphabetic variant of the original formula.

Figure 2.4: *Symmetry and Range-Reflexivity*

In fact it is convenient to extend the diagramming conventions to cover purely universal conditions[49] which, instead of requiring that "if such-and-such accessibility relationships hold amongst them, then so-and-so relationship also holds", require only that "if such-and-such accessibility relationships hold amongst them, then at least one of so-and-so relationships also holds", and we do this by including more than one broken or dotted arrow for the latter relationships. Figure 2.5 illustrates this device. On the left is represented the condition of piecewise connectedness introduced in Exercise 2.5.8(i) below. On the right is one of several weakenings of the euclidean condition that have been considered in the literature; frames satisfying it are called *semi-euclidean* in Voorbraak [1136] (p. 510), and *non-branching* frames in Georgacarakos [365] (p. 300f.), a paper in which **S4F** is referred to as S4.3.2; we use the former terminology for subsequent discussion (conspicuously in Proposition 5.5.3, p. 405), since "non-branching" suggests merely connectedness or weak connectedness, and the present condition is much more restrictive than that (taking in conjunction with transitivity and reflexivity requiring something more demanding than a weak linear ordering: see Prop. 5.5.3). The condition says that for all x, y, and z, *if* (solid arrows) xRy and xRz, *then* (broken arrows) either yRz or zRx. The logic determined by the class of all models satisfying the condition is **KF**, where **F** (to use the label from Segerberg [1025], p. 161) is the

---

[47]Garson [351], introduces at pp. 95–98 (and illustrates further throughout the book) another system of diagrammatic representation for such conditions. There are many differences between the two diagramming systems, best illustrated in the diagram given for the euclidean condition here (on the right of Fig. 2.3) and in [351], p. 129. Nor is the range of conditions susceptible of diagrammatic representation the same. See for instance note 55 below.

[48]See the discussion below, following Exercise 2.5.9 and also that following Exercise 2.5.15.

[49]That is, conditions which when written with all the quantifiers at the front – in 'prenex normal form'– have only universal quantifiers. ('At the front' means here that no quantifier lies in the scope of any sentence connective.)

schema $\Box(\Box A \to B) \vee (\Diamond \Box B \to A)$.[50] As in Figure 2.4, we drop the use of variables to label the nodes.[51]

Figure 2.5: *Piecewise Connectedness and the Semi-Euclidean Condition*

We could allow such diagrams in which there are no solid arrows; the condition represented is then the condition that at least one of the relationships indicated by the broken arrows must obtain. Removing the top node and the arrows emanating from it in the left diagram in Figure 2.5 would then represent the condition of connectedness (see p. 75), and removing the first node and its appended arrow from the diagram on the right of Figure 2.4 would express reflexivity. The dual of this convention would be to allow dropping all the dotted arrows as well, with the effect of meaning that not all of the solid relationships hold (since "at least one of so-and-so relationships also holds" must be false if there are no so-and-so relationships) – but this would be potentially confusing, so we shall not explicitly draw diagrams making use of it. (For example, it would mean a single point with a solid arrow from that point to itself indicated *irreflexivity*. In principle, there is no objection to this – it would simply look odd to anyone not fully versed in the convention. We can still represent the condition of reflexivity, by depicting a single point with a broken arrow going from that point to itself. This would be a case in which there are no conjuncts in the conjunctive antecedent, just as the previous case is one in which there are no disjuncts in the disjunctive consequent. With these extreme cases allowed for, we can now diagram any universally quantified condition on a binary relation, since considerations of conjunctive normal form and the behaviour of $\forall$ all us to write the condition as a conjunction of formulas

$$\forall v_1 \forall v_2 \ldots \forall v_k (\neg \alpha_1 \vee \ldots \vee \neg \alpha_m \vee \beta_1 \vee \ldots \vee \beta_n)$$

in which each α_i and β_j is Rtu in which t and u are individual variables and $v_1, \ldots, v_n$ lists precisely the variables involved in the various α_i and β_j taken altogether. Such a universally quantified disjunction can be ignored if it is vacuously satisfied (i.e., for some i, j, α_i coincides with β_j, so in making the corresponding diagram with distinct variables replaced by distinct nodes, we link nodes corresponding to some α_i with solid arrows in the relevant direction and nodes corresponding to some β_j with broken arrows, again appropriately directed: the absence of a case in which $\alpha_i = \beta_j$ means these instructions cannot clash. But since that deals with only one conjunct of the form inset above, in general several diagrams will be needed to convey a single condition on R. For example, diagrams corresponding to each of the conditions of reflexivity, symmetry and transitivity would collectively encode the condition of being an equivalence relation. (These are all 'Horn conditions': in each, there exactly one β_j. See note 51.) This is certainly not the only set of diagrams that would do the job, though: see Exercise 2.5.42(*iii*) below.

The final frill in this diagramming repertoire is a device for accommodating the identity relation. Since this has to be symmetric we can omit the arrow head, and we may think of an extended equals ("=") sign as an headless arrow representing the identity relation in the antecedent position of one of

---

[50] This would be expressed in the terminology introduced in Section 2.5 by saying that **KF** is determined be the class of frames $\langle W, R \rangle$ meeting this weakened euclidean condition. According to Segerberg [1025], the logic determined by the class of frames satisfying the condition is **S4F**, but, taken literally, the soundness half of this claim is false: **T**, for instance, is not valid on the frame with a single element and an irreflexive accessibility relation, even though the semi-euclidean condition is satisfied in this case. Segerberg seems to be (intermittently) following a convention whereby "frame" means "transitive reflexive frame" (and note: he uses the term *consistent* to mean *sound*).

[51] The first-order condition represented by a diagram in which there is only one broken arrow, or by a collection of such diagrams taken together, is called a *universal strict Horn condition* in the literature; see Hodges [504] for further information and references.

the conditions we are considering, like a solid arrow ("⟶")[52] for the accessibility relation, with a dotted version of this for the identity relation in the consequent, like the broken arrow for accessibility. In fact, it is never necessary to use the solid version of the sign for identity, since we can simply replace the two nodes between which it was drawn by a single node. So we will see only the dotted version, as in Figure 2.6, whose left-hand diagram represents antisymmetry, and whose right hand diagram the condition of piecewise weak connectedness (see Exercise 2.5.8(*ii*)). (One might metaphorically say in view of the dispensability of solid equality edges that those philosophers who have thought that there was 'really' no such relation as identity were *half* right; for background, see White [1169].[53] This point does not require the diagrams, though they dramatize it: the disjunctive conjuncts of the conjunctive normal forms recently alluded to, but now conceived of as involving "=" as well as "R", never need negated =-formulas as disjuncts.)

Figure 2.6: *Antisymmetry and Piecewise Weak Connectedness*

If the top node in the diagram on the right of Figure 2.6 is removed, together with its attendant solid arrows, the result is a depiction of the condition of weak connectedness as defined on p. 75. If all three nodes are left intact and the two broken arrows are removed (but the dotted equality link is retained) the condition expressed is of *partial functionality* for R – each point has at most one point accessible to it. We cannot convey the force of functionality *simpliciter* – each point has exactly one point accessible to it – by these diagrams because this involves an existential component, namely seriality (an $\forall\exists$ condition, in first-order terms). Many conditions on accessibility relations – or more accurately on what in Section 2.5 are called frames (which include not only the R but the also the W component of Kripke models) – of importance in the semantics of modal logic share with seriality this existential element,[54] and are not amenable to graphical representation in terms of the conventions introduced for the diagrams we have been considering.[55] The next examples we shall meet of this kind are convergence and piecewise convergence, in Exercise 2.5.8 and the discussion following it. Note also that sometimes several diagrams taken collectively will have to be used to represent certain conditions. For example, consider the condition, written here explicitly as a first-order formula:

$$\forall x \forall y (Rxy \rightarrow (Ryx \land Ryy)),$$

which combines symmetry and range-reflexivity. We cannot combine the diagrams in Figure 2.4 by having the same solid arrow and both the backward and the looped broken arrows, as this would instead convey

---

[52]Note that the use of such arrows to indicate the hypothesis that certain accessibility relationships hold has nothing to do with the use of an arrow (→) for the implicational connective within modal (or other) formulas.

[53]In more detail: Wittgenstein in the *Tractatus* had suggested that there is no relation of identity and a logically perspicuous language would accordingly be one in which identity and non-identity were expressed by the use of the same or different symbols. When the symbols concerned are individual variables, this leads to an *exclusive* interpretation of the quantifiers, explored in Hintikka [488], according to which a formula such as "$\forall x \exists y(Rxy)$", which, taken in standard predicate logic would express the seriality of (the relation represented by) R is interpreted as the standard formula "$\forall x \exists y(Rxy)$" is: its appearing in the scope of a quantifier binding the variable "x" excludes the value of "y" from coinciding with that of "x". To express seriality in this language one must instead say: "$\forall x(Rxx \lor \exists y(Rxy))$". The merits of such an "="-free language have been urged more recently in Wehmeier [1154], [1155], [1156].

[54]More precisely, when formulated as an explicit first-order condition in prenex normal form, there is at least one "$\exists$" in this initial quantifier prefix.

[55] Diagrams in Garson [351] using the conventions described there, are given for seriality, convergence and density. (For this last property, see Section 3.3.) The prenex normal form representation of these conditions involves prefixes $\forall \ldots \forall \exists$; the precise range of first-order conditions amenable to treatment on Garson's conventions is unclear. (Would it, for instance, include $\forall \exists \forall$ conditions?) It may be timely to recall the following words from p. 421 of van Benthem [66]: "It is a notorious fact that the usefulness of pictures is very limited in the field of quantificational logic. Once we have quantifier-combinations of a certain complexity a graphical representation becomes awkward or even impossible."

a condition like that above but with $\vee$ rather than $\wedge$ in the consequent. Instead we must use the set consisting of both diagrams in Figure 2.4 to represent the condition in question. In the Exercise 2.4.9 below, we pick up the earlier comment on the etymology of the term "serial", and introduce a condition which, unlike those considered thus far, is not even first order (let alone first order universal, as those represented in our diagrams have been).

Digression. The diagrams we are drawing are a variant on several generalizations of graphs in the literature – in the literature on graph theory that is (no connection with graphs with cartesian coordinates) – in which directed edges as well as undirected edges of a different kind, as well as loops, are allowed. There are three kinds of edges, at most one of any one kind being allowed between a pair of (not necessarily distinct) nodes or vertices: solid directed edges, broken directed edges, and undirected broken equality edges. There is an obvious notion of isomorphism for such graph-like objects. For example instead of drawing the diagram on the left of Figure 2.3 as it is – minus the labelling of the nodes by variables, of course – we could have had the broken directed edge where the unbroken directed edge is (i.e., going downward and to the left of the diagram) and vice versa, and the diagram would have looked different but been the same graph (to within isomorphism). This naturally raises the question of whether we ever get two non-isomorphic such graphs representing (to within first-order logical equivalence) the same condition. The answer to this is that we certainly do. The examples below of what we call "spurious alio-" conditions illustrate this possibility in an interesting way (see Figures 2.9 and 2.10), but the possibility already arises without the equality edges in those diagrams: compare for example the diagram on the right of Figure 2.4 with a variant on it having another directed edge from the node on its left extending to another node with a broken loop on it, just like the one already there. The graphs are non-isomorphic (since already they differ as to how many vertices they have) but represent equivalent conditions on a binary relation, since the new graph represents the condition $\forall x \forall y \forall z((Rxy \wedge Rxz) \rightarrow (Ryy \vee Rzz))$, which is equivalent to the shorter range-reflexivity condition represented by the two-node graph in Figure 2.4. 'Alio-' versions of the range- and domain-reflexivity conditions are the subject of Revision Exercise 2.7.34 (p. 137). **End of Digression.**

EXERCISE 2.4.9 (*i*) A slightly different terminology (from talk of "series", mentioned apropos of seriality after Exc. 2.4.8) will be used here for a sequence of points in a model each of which bears the accessibility relation R to the next: we call such a sequence an R-*chain*, and allow the case in which the sequence is finite (in which case its final element is not required to bear the relation R to anything; these are the R-chains of Hughes and Cresswell [537]). In the case described just now, with $x_1 R x_2 R x_3 \ldots x_n R x_{n+1} \ldots$ we say that x_1 starts an infinite R-chain. If we want to add the condition that $x_i \neq x_j$ whenever $i \neq j$, we will speak of a *proper* R-chain (infinite or otherwise). Show that the logic **KW** – with **W** from Table 2.1 on p. 33 – is sound w.r.t. the class of models $\langle W, R, V \rangle$ in which R is transitive and no $w \in W$ starts an infinite R-chain. As already remarked, this is not formulable as a first order condition – because the quantification over sets of points (as opposed to individual points) is not eliminable. (There is also a completeness result here w.r.t. this same class of models, but it is not obtainable by the canonical model method, since the model $\mathcal{M}_{\mathbf{KW}}$ doesn't itself satisfy the condition. Deleting the reference to transitivity here allows for a further completeness result for the logic **K** itself; see Theorem 2.5.3(*i*) below.)

(*ii*) As the reference to transitivity might make one suspect, **KW** = **K4W**. That is, $\vdash_{\mathbf{KW}}$ 4. Give a syntactic proof – sketch an axiomatic derivation, that is – of $\Box p \rightarrow \Box \Box p$ in **KW**. (*Note*: difficult exercise.) ✠

We will return to issues of soundness and completeness in the following section, in a slightly different setting. For the moment, we turn to the topic of generated submodels, which will be of particular use for some results discussed there, as well as for Section 2.6.

Given a model $\mathcal{M} = \langle W, R, V \rangle$ and a non-empty subset $X \subseteq W$, the submodel $\mathcal{M}_0$ of $\mathcal{M}$ *generated by* X is the model $\langle W_0, R_0, V_0 \rangle$ in which $W_0 = \{y \in W \mid \exists n \in Nat, x \in X . xR^n y\}$, $R_0 = R \cap \langle W_0 \times W_0 \rangle$, and $V_0(p_i) = V(p_i) \cap W_0$.

The notation "R^n" is understood as follows: xR^0y iff $x = y$; $xR^{k+1}y$ iff for some w, xR^kw and wRy.[56] (In other words xR^ny means that you can get from x to y by taking n steps of the relation R.) Instead of writing "$\exists n \in Nat, xR^ny$", one often conveniently just writes xR^*y and calls R^* the *ancestral* of the relation R: it is the smallest reflexive transitive relation $\supseteq R$. (Sometimes this is called the 'reflexive ancestral' to distinguish it from the smallest transitive relation extending R, called the *transitive closure* of R. Similarly the ancestral of a relation, as just defined, is sometimes called the reflexive transitive closure of the relation.)

If for some $x \in W$, the model $\mathcal{M}_0$ of $\mathcal{M}$ generated by $\{x\}$ in the sense of the above definition, we will say more simply that $\mathcal{M}_0$ is generated by the point x, and call any such model a *point-generated submodel* (of $\mathcal{M}$). The basic observation here, Theorem 2.4.10 is often called the Generation Theorem:

THEOREM 2.4.10 *If $\mathcal{M}_0 = \langle W_0, R_0, V_0 \rangle$ is a generated submodel of $\mathcal{M} = \langle W, R, V \rangle$, then for any formula A, we have for all $x \in W_0$: $\mathcal{M}_0 \models_x A$ iff $\mathcal{M} \models_x A$.*

Proof. By induction on the complexity of A (called $Comp(A)$ below).

Basis Case. $Comp(A) = 0$. This is immediate from the way V_0 is defined.

Inductive Step. $Comp(A) > 0$ and the lemma holds for all formulas of lower complexity than A. We do the three inductive cases (i) A is $B \wedge C$ for some formulas B, C; (ii) A is $\neg B$ for some formula B; (iii) A is $\Box B$ for some formula B.

Case (i). We have to show that $\mathcal{M}_0 \models_x A$ iff $\mathcal{M} \models_x A$, i.e., $\mathcal{M}_0 \models_x B \wedge C$ iff $\mathcal{M} \models_x B \wedge C$. By the definition of truth the l.h.s. here is equivalent to $\mathcal{M}_0 \models_x B$ and $\mathcal{M}_0 \models_x C$, which by the inductive hypothesis is equivalent to the claim that $\mathcal{M} \models_x B$ and $\mathcal{M} \models_x C$, which by the definition of truth again is equivalent to the r.h.s.

Case (ii). Actually, let's not do this case, which proceeds in the same manner as Case (i).

Case (iii). We have to show that $\mathcal{M}_0 \models_x A$ iff $\mathcal{M} \models_x A$, i.e., $\mathcal{M}_0 \models_x \Box B$ iff $\mathcal{M} \models_x \Box B$. For the "if" direction, suppose that $\mathcal{M} \models_x \Box B$, but (for a contradiction) $\mathcal{M}_0 \not\models_x \Box B$. The latter means that for some $y \in W_0$, xR_0y and $\mathcal{M}_0 \not\models_y B$. Since $R_1 = R_2 \cap (W_1 \times W_1)$, this means that xRy so all we need for a contradiction with $\mathcal{M} \models_x \Box B$ is that $\mathcal{M} \not\models_y B$: but this follows by the induction hypothesis from the fact that $\mathcal{M}_0 \not\models B$. For the "only if" direction, suppose that $\mathcal{M}_0 \models_x \Box B$ but $\mathcal{M} \not\models_x \Box B$. Thus there exists $y \in W$, with xRy and $\mathcal{M} \not\models_y B$. Now, since xR_2y, condition (iii) of the definition of generated submodels tells us that $y \in W_0$ and thus that xR_0y, so since the inductive hypothesis gives $\mathcal{M}_0 \not\models_y B$, we do have a contradiction with $\mathcal{M} \models_x \Box B$. ∎

The content of this result is obvious, in fact. If you throw away points that can't be reached in any number of R-steps from a given point, that won't make any difference to the truth-value of a formula at that point. A more refined version of Theorem 2.4.10 is available, again with a proof by induction on the complexity of A. Let the *modal degree* of a formula be the maximum depth of embedding of $\Box$ within $\Box$ in that formula; inductively, calling this $md(A)$ for the formula A: $md(p_i) = 0$, $md(B \wedge C) = max(md(B), md(C))$; $md(\neg A) = md(A)$; $md(\Box B) = md(B) + 1$. Then the truth-value at a point x in a model with accessibility relation R of a formula A with $md(A) = n$ is not affected by throwing away all model elements outside of $\bigcup_{i=0}^{n}(R^i(x))$ – those not reachable from x by fewer than $n+1$ R-steps, that is – though we shall not bother to state this 'more refined' result formally in the style of Theorem 2.4.10. (See Lemma 2.8 in van Benthem [73].)

---

[56] More generally, the relative product of two binary relations R and S in that order, often denoted $R \circ S$ (or even just RS) is defined to hold between x and y when for some w, xRw and wSz. Thus R^2 is $R \circ R$, R^3 is $R \circ R \circ R$ (parentheses being unnecessary since the operation $\circ$ is associative), and so on. The same notation is used for the closely relation operation of composition of 1-place functions, with $f \circ g$ being that function which, when applied to x, has as its value $f(g(x))$.

REMARK 2.4.11 The Kripke semantics with its models $\langle W, R, V \rangle$ can itself be reformulated so that truth within a model is relativized to sequences whose first element is some $x \in W$, whose second element is $R(x)$, whose third is $\{Y \mid \exists y \in R(x) \text{ with } X = R(y)\}$, and so on. A generalization of this was produced by Oskar Becker before the appearance of the Kripke semantics itself – see Section 8 of Copeland [193] – in which we work with sequences of the form $X_0, X_1, \ldots X_n, \ldots$ where $X_i \in \wp^i W$, where this notation is defined inductively: $\wp^0(X) = X$ and $\wp^{i+1}(X) = \wp(\wp^i(X))$. Relative to such a sequence, the formula $(p \wedge \Box q) \vee \Diamond \Box r$, for example, is true when either p is true X_0 (an individual world) and q is true at all worlds in X_1, or else there is a $Y \in X_2$ such that for all $y \in Y$, r is true at y. The semantics in terms of such Becker sequences is ingeniously developed in Tsai [1114], where it is observed that although every Kripke model gives rise to a Becker model, the converse is not the case, and also that the logic determined by the class of all Becker models is **K**. While the greater generality of the semantics does not lead to a weaker basic logic, it would be interesting to know if it there are normal modal logics which are not determined by any class of Kripke frames but are determined by a class of Becker frames – where (see the following Section) in both cases frames are the result of abstracting away from the V component of the respective models. In other words – the words in question being defined on p. 82: are there Kripke-incomplete but Becker-complete (normal modal) logics? In the terminology (due to Makinson) of Hansson and Gärdenfors [444], this asks whether the Becker semantics has the same *width* as the Kripke semantics, Tsai having already shown that it has the same *depth*. ◀

2.5 From Models to Frames

The soundness and completeness results of the preceding section, as well as the issue of generated submodels, can be presented in a cleaner way by abstracting from the 'V' part of the model which stipulated where in the model the various propositional variables are to be true and false. This directs us from a model $\mathcal{M} = \langle W, R, V \rangle$ to the structure $\langle W, R \rangle$, called the *frame* of the model $\mathcal{M}$. The main notion this allows us to isolate is that of a formula A's being *valid on* the frame $\langle W, R \rangle$, usually notated thus: $\langle W, R \rangle \models A$.[57] Also, we say a schema is valid on a frame if every instance of that schema is valid on the frame. We adapt the terminology of the preceding section to this setting thus: S is *sound* w.r.t. a class $\mathbb{C}$ of frames when every $A \in S$ is valid on each $\mathcal{F} \in \mathbb{C}$, *complete* w.r.t. $\mathbb{C}$ when for every formula A which is valid on each $\mathcal{F} \in \mathbb{C}$, we have $A \in S$, and *determined* by $\mathbb{C}$ when S is both sound and complete w.r.t. $\mathbb{C}$. If we abbreviate "A is valid on each $\mathcal{F} \in \mathbb{C}$ to "A is valid over $\mathbb{C}$", then a logic is determined by $\mathbb{C}$ when its theorems are exactly the formulas valid over $\mathbb{C}$. If this holds for $\mathbb{C} = \{\mathcal{F}\}$ for some $\mathcal{F}$ then we say that the logic is determined by $\mathcal{F}$, or alternatively, that $\mathcal{F}$ is a *characteristic frame* for the logic.

Similarly we need to adapt the generated submodel idea across to frames. To do this, we just ignore everything pertaining to V in the definitions given before Theorem 2.4.10. That is we say that the subframe of $\langle W, R \rangle$ *generated by* $X \subseteq W$ is the frame which has for its universe the set $\bigcup_{x \in X} R^*(x)$ – recalling that R^* is the ancestral of R – and for its accessibility relation the restricted of the relation R to this set. If X contains one element, the subframe is said to be generated by that element, and to be a *point-generated subframe*; a point-generated frame is any frame which is a point-generated subframe of itself. What this amounts to is that all points in the frame can be reached in some number of steps (typically different for different points) from the generating point.

EXERCISE 2.5.1 Is the set of formulas valid over $\mathbb{C}$ guaranteed to be a normal modal logic, for any class $\mathbb{C}$ of frames? (Give reasons.) ✠

We can reformulate the (soundness and) completeness results of Coros. 2.4.3, 2.4.7 and Exc. 2.4.8 in terms of frames as follows. We adopt the convention when $\mathcal{F} = \langle W, R \rangle$ and R is transitive, symmetric,

---

[57]Many authors say "valid in" rather than "valid on" a frame. The present usage follows Hughes and Cresswell, e.g., [537].

etc., of calling $\mathcal{F}$ transitive, symmetric, etc.:[58]

THEOREM 2.5.2 **K** *is determined by the class of all frames,* **K4**, *by the class of transitive frames,* **KT** *by the class of reflexive frames,* **KD** *by the class of serial frames,* **KD4** *by the class of serial transitive frames,* **S4** *by the class of transitive reflexive frames,* **KB** *by the class of symmetric frames,* **K5** *by the class of euclidean frames,* **KDB** *by the class of symmetric serial frames,* **KTB** *by the class of symmetric reflexive frames and* **S5** *by the class of equivalence relational frames.*

In many cases, it is possible to find several interesting classes of frames determining the same logic. We collect in Theorem 2.5.3, some examples from amongst the logics mentioned in Theorem 2.5.2. The reference to infinite R-chains was explained at Exc. 2.4.9; the key point is that the successive elements of such chains do not need to be distinct, so a point can start an infinite R-chain even in a finite frame $\langle W, R \rangle$.[59] If $R \subseteq W \times W$ and R is a partial ordering (as defined on p. 15, then $\langle W, R \rangle$ is called a *partially ordered set* (or 'poset'). Finally when R is the universal (binary) relation on a set W ($R = W \times W$, that is), we call the pair $\langle W, R \rangle$ a *universal* frame.

THEOREM 2.5.3 *(i)* **K** *is determined by the class of frames* $\langle W, R \rangle$ *in which no element starts an infinite R-chain;*

(ii) **S4** *is determined by the class of partially ordered sets;*

(iii) **S5** *is determined by the class of universal frames.*

(iv) **S5** *is determined by the frame (unique up to isomorphism)* $\langle W, R \rangle$ *in which W is a countably infinite set and $R = W \times W$.*

Proof. For (i) see Boolos [112], p. 146; for (ii), Segerberg [1025], pp. 80–82 or Blackburn *et al.* [98], Thm. 4.54. (iv) is an old result (in Scroggs [1020] in a matrix-semantical guise) provable by 'filtration' methods in Segerberg [1025], not covered in the present exposition. We concentrate on (iii). Soundness is clear, since every universal frame is an equivalence relational frame (with one equivalence class). For completeness, we know (Thm. 2.5.2) that **S5** is complete w.r.t. the class of equivalence relational frames, so if $\nvdash_{\mathbf{S5}} A$, there is a model $\mathcal{M} = \langle W, R, V \rangle$ on such a frame with $x \in W$ and $\mathcal{M} \nvDash_x A$. Let $\mathcal{M}_x$ be the submodel of $\mathcal{M}$ generated by x. By Theorem 2.4.10, $\mathcal{M}_x \nvDash_x A$. But the frame of $\mathcal{M}_x$ is a universal frame (with universe $R(x)$), so we have found a model on a universal frame together with an element therein at which A is false, making our arbitrary unprovable A invalid on some universal frame. ∎

COROLLARY 2.5.4 **S5** *is complete w.r.t. the simplified semantics of Section 2.2.*

Proof. Suppose $\nvdash_{\mathbf{S5}} A$. By Thm. 2.5.3(iii), A is false at some point in a model on a universal frame. But this means that A is false in some model of the simplified semantics of the previous section, since when R is universal, the truth-conditions for formulas in the current semantics coincide with those provided by the simplified semantics, so A is invalid in the sense of the latter semantics. ∎

EXERCISE 2.5.5 According to Theorem 2.5.3(iv), there is a characteristic frame for **S5** with a universal relation. Is there a characteristic model for **S5** whose accessibility relation is universal? (Justify your answer.) ✠

---

[58] A difficulty which need not detain us for the moment lurks beneath the surface here: see note 81, p. 90.

[59] Such a frame is described as finite when W is finite. Note that even if $W = \{w\}$ for some w, we can still have an infinite R-chain, if $R = \{\langle w, w \rangle\}$.

Also apropos of Theorem 2.5.3(*iv*), it should be mentioned that a similar result holds for any infinite W (with R universal), for example if W has the same cardinality as the set of real numbers. It is not hard to see that the universe of the canonical frame for any consistent normal modal logic S is uncountable, since even forgetting about modality, all sets of negated and unnegated propositional variables (each appearing in the set either negated or unnegated but not both) are S-consistent, and there are uncountably many such sets, all of which can only be extended to distinct maximally S-consistent sets. This is because we have used the notion of canonical model (and frame) due to Scott (and found in Lemmon and Scott [705]); an independent variation due to Makinson [758] produces countable (i.e., denumerably infinite) characteristic models very similar to the present canonical models, but we are not going into that here.[60] There is also a considerable interest in finite models, not least because of their connection with decidability (see note 61) – another topic not on our agenda – and while having a finite characteristic model is generally not something desirable in a modal logic with plausible applicability (because the logic will collapse distinctions the application will want to make), the next best thing, namely being determined by a class of finite models, is a property possessed by almost all modal logics that have been considered other than with the express purpose of finding a logic – typically in a certain range of logics (e.g., extensions of **S4**) without the property. The property in question is called the *finite model property*, and it has played a role, alluded to just now, in showing the decidability of finitely axiomatizable logics possessing it, which does not concern us here.[61] One can also define a corresponding finite *frame* property, possessed by those S for which S is determined by a (generally infinite) class of finite frames, and this can be shown to coincide with the finite model property. Logics with this property are also sometimes known as 'finitely approximable'. Much information on the topic can be found in Segerberg [1025], which describes a technique ("filtration") which produces finite countermodels for non-theorems of various logics by starting with (point-generated submodels of the) canonical model and taking certain equivalence classes of points therefrom, determined on the basis of the subformulas of the non-theorem in question. Segerberg proves several results of the form all extensions (or all normal extensions) of such-and-such a logic have the finite model property, the prototype is the theorem of R. A. Bull (from [126]) giving this result for **S4.3** – where the 'normal' turns out to make no difference.[62] More recently techniques have been developed for producing such finite countermodels for unprovable formulas from scratch: See Chapter 5 of Boolos [112] and Hughes and Cresswell [537], Chapter 8, where they are called mini-canonical models.

We interrupt our treatment of completeness to treat a rule we have occasion to refer back to from later discussions (such as Sections 3.1 and 4.7). Those in need of a basic introduction should skip straight to Exercise 2.5.8 on a first reading. The result we present is taken from the same discussion by Boolos as was cited in the proof of Theorem 2.5.3; the rule involved here is sometimes called Löb's Rule, for reasons explained in Boolos [112] (see esp. p. 59).

COROLLARY 2.5.6 **K** *is closed under the following rule:*

$$\frac{\Box A \to A}{A}$$

Proof. We argue that if $\vdash_{\mathbf{K}} A$ then $\vdash_{\mathbf{K}} \Box A \to A$. Supposing that $\nvdash_{\mathbf{K}} A$ gives us, by Theorem 2.5.3(*i*), a model on a frame $\langle W, R \rangle$ in which no point starts an infinite R-chain, with A false at some point x in that model. Suppose every $y \in R(x)$ verifies A. Then x itself is a point at which $\Box A$ is true while A is false, and therefore a point at which $\Box A \to A$ is false, showing that this is not **K**-provable. If not every $y \in R(x)$ verifies A, pick some $y \in R(x)$ at which A is false, and let x_1 be this y. If every $y \in R(x_1)$ verifies A, then by the previous reasoning re-applied, x_1 is a point at which $\Box A \to A$ is false, which would again suffice for the unprovability in **K** of this formula. If not, then we have $x_2 \in R(x_1)$ with A false

---

[60]While most texts on modal logic use canonical models in the style of Scott, Fitting [310] follows Makinson's path.

[61] A logic is said to be *decidable* if there is a mechanical method, or algorithm, for determining, of an arbitrary formula, whether or not it is provable in the logic, after finitely many steps.

[62]Bull worked with algebraic models but the result transfers across to the Kripke semantics.

at x_2, and we argue as before that if this point does not verify $\Box A$, and hence falsify $\Box A \to A$, there is $x_3 \in R(x_2)$ which does this job, unless ... we are forced to proceed on and on without end, which means that x starts an infinite R-chain $x, x_1, x_2, x_3, \ldots$: but this contradicts the hypothesis that no point starts an infinite R chain in $\langle W, R \rangle$. ∎

An elegant, purely syntactic proof of Coro. 2.5.6 can be found in Crabbé [198].

Rather surprisingly, **K4** is not closed under Löb's Rule, and we will now present a well-known example illustrating that fact.[63] A simple semantic check shows that the formula $\Box(\Box(\Box p \to q) \to \Box p) \to (\Box(\Box p \to q) \to \Box q)$ is valid on all transitive frames and is, therefore, **K4**-provable. For good measure, here is a proof(-sketch); we use the derived rule TF from p. 26:

(1) $\Box(\Box(\Box p \to q) \to \Box p) \to (\Box\Box(\Box p \to q) \to \Box\Box p)$ instance of **K** schema
(2) $\Box(\Box(\Box p \to q) \to \Box p) \to (\Box(\Box p \to q) \to \Box\Box p)$ From 2, using **4**
(3) $\Box(\Box p \to q) \to (\Box\Box p \to \Box q)$ instance of **K** schema
(4) $\Box(\Box(\Box p \to q) \to \Box p) \to (\Box(\Box p \to q) \to \Box q)$ 2, 3 TF

Now this doesn't look much like a premiss for Löb's Rule as it stands, not being of the required form $\Box A \to A$. To fix this, substitute p for q, giving:

$$\Box(\Box(\Box p \to p) \to \Box p) \to (\Box(\Box p \to p) \to \Box p).$$

Now that we do have something **K4**-provable of the form $\Box A \to A$, to see that **K4** is not closed under Löb's Rule, it suffices to note that the would-be conclusion of the rule, the A involved here, is itself the formula sometimes known as Löb's Axiom, which appears in schematic form as **W** (following Segerberg [1025]) in Table 2.1 on p. 33 above. This formula is of course not a theorem of the Sobociński-regular logic **K4**.

EXERCISE 2.5.7 Is **KB** closed under Löb's Rule? (*Hint*: Revision Exercise 2.7.5 below (p. 130) should help with this. The same is true of Exercise 2.7.6 – the point being made in this way in Crabbé [198].)✠

We shall next see some variations on Löb's Rule in Section 2.6.

EXERCISE 2.5.8 (*i*) Show that **K.3**, the smallest normal modal logic containing all formulas of the form

$$\Box((\Box A \wedge A) \to B) \vee \Box(\Box B \to A),$$

is determined by the class of *piecewise weakly connected* frames, which means frames $\langle W, R \rangle$ such that for all $w \in W$, for all $x, y \in R(w)$, if $x \neq y$ either xRy or yRx. (Note: for the completeness half of this claim, the canonical model method will suffice, but you may wish to try (*ii*) first, which is simpler.)

(*ii*) Show that the following modification of **.3**, which we will call **.3⁺**, gives a result like that of (*i*) except with reference to the class of *piecewise connected* frames, meaning frames $\langle W, R \rangle$ such that for all $w \in W$, for all $x, y \in R(w)$, either xRy or yRx:

$$\Box(\Box A \to B) \vee \Box(\Box B \to A),$$

(*iii*) Show that the schema **.2** from Table 2.1 (see p. 33) is valid on all *piecewise convergent* frames – meaning frames $\langle W, R \rangle$ such that for all $w \in W$, for all $x, y \in R(w)$, there exists $z \in W$ for which xRz and yRz – and that the canonical frame (the frame of the canonical model, that is) of any consistent normal modal logic containing all instances of this schema is piecewise convergent. ✠

---

[63]See Boolos [112], p. 59.

2.5. FROM MODELS TO FRAMES

Without the "piecewise", the terminology here means the following. $\langle W, R \rangle$ is *connected, weakly connected*, or *convergent* (also called 'directed') according as we have for all $x, y \in W$, (*i*) xRy or yRx (for connectedness), (*ii*) $x = y$, xRy or yRx (for weak connectedness), or (*iii*) some z with xRz and yRz.[64] A general description of the relation between conditions such-and-such and piecewise such-and-such follows shortly.

EXERCISE 2.5.9 (*i*) Is every piecewise connected frame piecewise convergent? Is every piecewise convergent frame piecewise connected? (For each case, give an argument or a counterexample.)

(*ii*) It is not hard to see that for every frame $\langle W, R \rangle$ which is *not* piecewise convergent, we can define a model on this frame in which there is a point at which the **.2** axiom $\Diamond \Box p \to \Box \Diamond p$ is false. For if wRx and wRy ($w, x, y \in W$) but $R(x) \cap R(y) = \varnothing$ then setting $V(p) = R(x)$ (and V can assign any other sets to the other propositional variables) the antecedent of this axiom will be true at w in this model, while its consequent will be false. Show similarly that we can falsify the **.3**$^+$ axiom (i.e., putting p and q for A and B in the schema of that name from 2.5.8(*ii*)) at some point in a model on any frame which is not piecewise connected, by making a suitable choice of $V(p)$ and $V(q)$.

(*iii*) What repercussions do your answers to (*i*) and (*ii*) have for the relative inclusion relations between **K.2** and **K.3**$^+$. (I.e., does it follow from those answers, given the answers to Exc. 2.5.8, that **K.2** is included in **K.3**$^+$ or does it follow that this inclusion does not hold, and likewise for the converse inclusions?)

(*iv*) Describe the inclusion relations between **K.3**, **K.3**$^+$ (in the sense of (*iii*)), and the smallest normal modal logic containing all instances of the following schema:

$$\Box((\Box A \land A) \to B) \lor \Box((\Box B \land B) \to A).$$

(Partial assistance follows.) ✠

To help with part (*iv*) here let us note that if we start with a representative instance of the schema inset there (a 'de-lopsided' version of **.3**):

$$\Box((\Box p \land p) \to q) \lor \Box((\Box q \land q) \to p),$$

then we can substitute $p \to q$ for q, getting

$$\Box((\Box p \land p) \to (p \to q)) \lor (\Box(p \to q) \land (p \to q)) \to p).$$

Non-modal propositional logic in the scope of the $\Box$s on the disjuncts – 'exporting' and, in the case of the first disjunct also 'contracting' – allows us to pass to:

$$\Box(\Box p \to (p \to q)) \lor \Box(\Box(p \to q) \to ((p \to q) \to p)),$$

from which further truth-functional reasoning ('importing' in the first disjunct and appealing to Peirce's Law in the second) gets us then to

$$\Box((\Box p \land p) \to q) \lor \Box(\Box(p \to q) \to p),$$

at which point it is perhaps safe to see how to proceed to

$$\Box((\Box p \land p) \to q) \lor \Box(\Box q \to p).$$

---

[64]Connectedness is sometimes called strong connectedness. The present usage, as well as that just commented on, is quite different from what is found in Goldblatt [393], in which weak connectedness is called connectedness and the phrase "weakly connected" is used for what we (inspired by Segerberg [1025]) are calling "piecewise weakly connected". Another usage of the phrase will come up in Section 6.5.

(Of course a more formal proof would require citing the various tautologies appealed to here and applying the derived rule □Mono appropriately.)

For the following exercise, we need to introduce the "⊕" notation. Where S is a modal logic, write $S \oplus \Gamma$ for the smallest normal modal logic including all the formulas in $S \cup \Gamma$. (We will only be using the notation when S is itself normal.[65]) So as not to have to bother with taking representative instances of schemata, we extend the notation so that where Γ is a set of schemata rather than of formulas $S \oplus \Gamma$ is the smallest normal modal logic extending the union of S with the set of all instances of the schemata in Γ. Thus notations like "**K4U**" can be thought of as abbreviations for $\mathbf{K} \oplus \{\mathbf{4}, \mathbf{U}\}$. For brevity, when $\Gamma = \{A\}$ for some formula A, we write "$S \oplus A$" rather than "$S \oplus \{A\}$" for $S \oplus \Gamma$.

EXERCISE 2.5.10 (i) Show that $\mathbf{KT.3} = \mathbf{KT.3}^+$.

(ii) Show that $\mathbf{S4.3} = \mathbf{S4} \oplus \Box(\Box p \to \Box q) \vee \Box(\Box q \to \Box p)$.

(iii) Do we have $\mathbf{K4.3} = \mathbf{K4} \oplus \Box(\Box p \to \Box q) \vee \Box(\Box q \to \Box p)$? (Give reasons for your answer, here and in the following part.)

(iv) Is it the case that $\mathbf{K4.3}^+ = \mathbf{K4} \oplus \Box(\Box p \to \Box q) \vee \Box(\Box q \to \Box p)$?

(v) Show that $\mathbf{K.3}^+ = \mathbf{K} \oplus (\Diamond p \wedge \Diamond q) \to (\Diamond(p \wedge \Diamond q) \vee \Diamond(\Diamond p \wedge q))$ and that $\mathbf{K.3} = \mathbf{K} \oplus (\Diamond p \wedge \Diamond q) \to (\Diamond(p \wedge \Diamond q) \vee \Diamond(\Diamond p \wedge q) \vee \Diamond(p \wedge q))$. ✠

Abstracting from the notions in play above (before the recent exercises), suppose we have a condition on frames which can be written as a closed first-order formula – containing no free variables, that is – in the language with R as a two-place predicate symbol.[66] (We allow "=" to figure in the formula.) Suppose further that this condition, which, when written in prenex normal form (i.e., with a string of quantifiers preceding a quantifier free formula), has no existential quantifier to the left of a universal quantifier, and has at least one universal quantifier. That is, we are considering first-order formulas of the form:

$$\forall x_1 \forall x_2 \ldots \forall x_m \exists y_1 \exists y_2 \ldots \exists y_n \Phi, \qquad (*)$$

where Φ is quantifier-free, and n may be 0.[67] Then we define the *piecewise condition* corresponding to the given formula to be that expressed by first-order formula:

$$\forall w \forall x_1 \forall x_2 \ldots \forall x_m ((Rwx_1 \wedge \ldots \wedge Rwx_m) \to \exists y_1 \exists y_2 \ldots \exists y_n \Phi). \qquad (**)$$

(In accordance with the policy on notating binary relations mentioned on p. 4, we write "Rwx_1", etc., here, rather than wRx_1; the informal notation of "&" and "⇒" has similarly been replaced by "∧" and "→".) In our informal way of formulating such conditions to date we would be more likely to write something along the following lines instead:

$$\forall w \forall x_1, x_2, \ldots, x_m \in R(w) \centerdot \exists y_1 \exists y_2 \ldots \exists y_n \Phi,$$

though this still leaves Φ intact and we would usually write "For all $w \in W$ (etc.)" rather than "$\forall w$". (We don't need the "∈ W" because we are interpreting our first-order formula in frames as relational structures, so any quantifier ranges over the whole of W, taken as the domain, or universe, of the structure.) The piecewise conditions corresponding in this sense to connectedness and convergence are piecewise connectedness and piecewise convergence as introduced above (Exc. 2.5.8).

---

[65] Γ could itself be a normal modal logic, in fact. The logic $S_1 \oplus S_2$, for normal modal logics S_1, S_2, is then the smallest normal modal logic extending each of S_1, S_2. (This usage was foreshadowed in note 20.) It is the *join*, or least upper bound, of these logics in the lattice of all modal logics, while their *meet*, or greatest lower bound, is $S_1 \cap S_2$.

[66] Purists will note a certain use-mention shift here, since a symbol, namely "R", which we were *using* to talk about the accessibility relations of our frames and models, is now being *mentioned* (as a relation symbol). They will also be able to see how to reformulate the discussion so that no such double usage is involved.

[67] We could also allow $m = 0$, though in this case the transformation which follows simply introduces a vacuous initial quantifier "$\forall w$".

2.5. FROM MODELS TO FRAMES

The main application of this idea is given by the following result concerning transitive reflexive frames. For the proof, recall that R^* is the ancestral – or reflexive transitive closure – of R, and observe that if x is a point in a reflexive transitive frame, then $R^*(x) = R(x)$, as is clear from that definition of the ancestral. (On the definition of $R^*(x)$ as the union of the sets $R^k(x)$ for $k \in$ Nat, we always have $R(x) \subseteq R^*(x)$ since $R = R^1$. For the converse inclusion, we need $R^k \subseteq R$ for $k = 0$, given by reflexivity, and for $k \geqslant 2$, given by transitivity.)

PROPOSITION 2.5.11 *Any point-generated frame which is reflexive and transitive satisfies a condition of the form* $(*)$ *above if and only if it satisfies the corresponding piecewise condition* $(**)$.

Proof. The "only if" part is trivial because the piecewise conditions follow from the originals. For the "if" part, suppose $\langle W, R \rangle$ is a reflexive transitive frame generated by $w_0 \in W$, i.e., $W = R^*(w_0)$. Since R is reflexive and transitive, $R^*(w_0) = R(w_0)$, by the above observation. Thus $W = R(w_0)$. So, on the assumption that $\langle W, R \rangle$ satisfies the piecewise condition, instantiating the universally quantified "w" (in the formulation inset above) to w_0 tells us that $\langle W, R \rangle$ satisfies the original condition. ∎

Putting this together with the results of Exc. 2.5.8, we conclude:

COROLLARY 2.5.12 **S4.2** *and* **S4.3** *are determined respectively by the class of transitive reflexive frames which are convergent and by the class of such frames which are connected.*

REMARK 2.5.13 Note that connected frames – by contrast with weakly connected frames – are automatically reflexive, so there is some redundancy in spelling out the "such" at the end of this Corollary as "transitive reflexive" here. ◂

EXERCISE 2.5.14 (*i*) Does Remark 2.5.13 mean that we can drop the **T** axiom and have the result claimed in Coro. 2.5.12 for **S4.3** holds for **K4.3**?
(*ii*) Whether or not **S4.3** = **K4.3**, as (*i*) asks, is it the case that **KU4.3** = **K4.3**? ✠

EXERCISE 2.5.15 A question along the lines of Exc. 2.5.9: is either of the classes of frames mentioned in Coro. 2.5.12 included in the other, and if so, in which direction does the inclusion hold? (Give reasons.)✠

REMARK 2.5.16 The class of frames mentioned in connection with **S4.3** in Coro. 2.5.12 is also called the class of weak linear orders, strict linear orders being the irreflexive, transitive and weakly connected frames (the class of which determines **K4.3**). It would be interesting to know what modal logics are determined by the class of all (reflexive) connected frames, and by the class of all weakly connected frames. The following formula, for example, is valid on all weakly connected frames, as the reader may check:

$$\Box\Box\Box p \to (\Diamond\Diamond p \to (p \vee \Diamond p)).$$

This example was given to the author by Krister Segerberg (personal communication, 1980), who observed that it is not provable in **K.3** – so one cannot just drop the reference to **4** in the preceding parenthetical remark on **K4.3** to match the absence of the condition of transitivity. Segerberg also remarked that the consequent of the formula inset here is the dual form of (a representative instance of the 'alio-transitivity' schema) **4**′ on p. 93 below. ◂

EXERCISE 2.5.17 In explaining what the piecewise version of a given first order condition was we took the latter to have the form $\forall x_1 \forall x_2 \ldots \forall x_m \exists y_1 \exists y_2 \ldots \exists y_n \Phi$, with Φ containing no further quantifiers (and $m, n \geq 0$). Would dropping this restriction on quantifier free Φ make any difference to the discussion, or would Proposition 2.5.11 still be correct for the piecewise conditions induced by the recipe given earlier when Φ contains additional quantifiers? (Explain your answer.) ✠

In the definition above of the piecewise condition corresponding to a given condition we introduced a new universally quantifier ("$\forall w$") and then restricted the originally present universal quantifiers (to $R(w)$), but we did not restrict the subsequent existential quantifiers in the same way. That is, the procedure described did not end up with (assuming for simplicity that Φ is quantifier-free, as in the original exposition, rather than in Exc. 2.5.17):

$$\forall w \forall x_1 \in R(w) \ldots, \forall x_m \in R(w) \,.\, \exists y_1 \in R(w) \ldots \exists y_n \in R(w)\Phi.$$

EXERCISE 2.5.18 Would Proposition 2.5.11 have been correct if the piecewise condition had been as immediately above rather than as originally defined? (Explain.) ✠

In some special cases it is easy to see that the difference between the piecewise condition obtained by the official procedure and that obtained by the procedure addressed in Exc. 2.5.18 are equivalent. Thus consider the convergence condition: $\forall x \forall y \exists z (Rxz \wedge Ryz)$. The initially given (∗)-to(∗∗) recipe yields as the condition of piecewise convergence that given on the left, while the version currently under consideration is that on the right:

$$\forall w \forall x, y \in R(w) \, \exists z \,.\, Rxz \wedge Ryz. \qquad \forall w \forall x, y \in R(w) \, \exists z \in R(w) \,.\, Rxz \wedge Ryz.$$

If we restrict attention to the case in which (the relation denoted by) R is transitive, however, we can see that these two are equivalent. Evidently even without this restriction, the condition on the left is a consequence of that on the right. And with the help of transitivity the condition on the right follows from that on the left since, supposing the condition on the left to be satisfied, (putting it informally) if Rxz (or for that matter Ryz) then as $x \in R(w)$, we must have $z \in R(w)$. We close our discussion of such variations on the definition of piecewise conditions with the following Exercise, though the theme of piecewise conditions as originally defined will continue to occupy us for a little further:

EXERCISE 2.5.19 (i) Show that the formula $(\Box q \wedge \Diamond\Box p) \to \Box\Diamond(p \wedge q)$ is valid on all frames $\langle W, R\rangle$ such that for all $w \in W$, if wRx and wRy then there exists $z \in W$ for which wRz, xRz and yRz.
(ii) Is the modal formula mentioned in (i) valid *only* on frames meeting the condition there given? (Justify your answer.)
(iii) Characterize in first-order terms the class of frames validating the formula $(\Diamond\Box p \wedge \Diamond\Box q) \to \Diamond(p \wedge q)$. ✠

Numerous classes of formulas which are particularly tractable for the model theory of modal logic – for example yielding normal modal logics valid on their canonical frames – have been isolated over the years, the most famous being descendants of the Sahlqvist formulas from [988]. (Or see Blackburn *et al.* [98], esp. §5.6.)

EXAMPLE 2.5.20 We illustrate such projects here with the simplest class of formulas already isolated in [705] and treated in many subsequent texts. (See for example Chellas [171], §3.3, and Boolos [112], pp. 88–91. §§3.6–7 of [98] provides a comprehensive contemporary discussion.) These generalize the **.2** formula by allowing arbitrary degrees of iteration for the occurrences of $\Box$ and $\Diamond$, the general form being as follows, superscripts ranging over natural numbers:

$$\Diamond^i \Box^m A \to \Box^j \Diamond^n A.$$

The schema **.2** (also called **G**, after Geach) is the case of $i = j = m = n = 1$, while **4** and **5** are respectively the cases in which $i = n = 0$, $m = 1$, $n = 2$, and in which $i = j = n = 1$, $m = 0$. The canonical frame for any logic containing all instances of this schema (for a fixed choice of i, m, j, n) satisfies the corresponding case of the generalized piecewise convergence condition: for all $w, x, y \in W$

if $wR^i x$ and $wR^j y$ then there exists $z \in W$ with $xR^m z$ and $yR^n z$,

2.5. FROM MODELS TO FRAMES

a condition on frames which is necessary and sufficient for the schema to be valid on a frame. See the references cited for further details. ◀

Let us apply the above definition of the piecewise condition corresponding to a given condition in the cases of reflexivity and universality, which have the following first-order formulations, respectively:

$$\forall x(Rxx) \qquad \text{and} \qquad \forall x \forall y(Rxy).$$

Applying the above procedure we get the piecewise versions

$$\forall w \forall x(Rwx \to Rxx) \qquad \text{and} \qquad \forall w \forall x \forall y((Rwx \land Rwy) \to Rxy),$$

respectively. The second condition here, we have already met: it is the condition that R (or more explicitly $\langle W, R \rangle$) is euclidean, and we continue to use the latter term rather than saying *piecewise universal*. Thus, by Proposition 2.5.11, the reflexive transitive frames which are euclidean have as their point-generated frames universal frames, recalling the axiomatization of **S5** by means of axioms **T**, **4** and **5**. (**5** is sometimes called **E** instead, because of its connection with the euclidean condition.) We would not normally use the label "**KT45**" for **S5** because the axiomatization thereby invoked is not independent: we can simply drop the **4**, as indicated in Exercise 2.1.15, to which we shall return presently. As to the first of the two "piecewise" conditions inset above, it is often referred to as quasi-reflexivity, though since this term has also been used for the quite different condition expressed by either of the equivalent first-order formulations

$$\forall w (\exists x(Rwx) \to Rww) \qquad \forall w \forall x(Rwx \to Rww),$$

it is safer to avoid this terminology. So instead we call the former *range-reflexivity* and the latter *domain-reflexivity*.[68]

EXERCISE 2.5.21 (*i*) Draw the corresponding diagram for domain-reflexivity.
(*ii*) Show by the canonical model method that **KU** is determined by the class of range-reflexive frames.
(*iii*) Find a formula (or schema) for which you can show that the normal extension of **K** containing that formula (or containing all instances of that schema) is determined by the class of domain-reflexive frames. (And show this for the chosen formula or schema. If in difficulty, consult the discussion preceding Exercise 3.2.18, p. 190, which will suggest a formula worth considering as an axiom to add to **K** for this purpose.) ✠

Let us look into the logical relations between range-reflexivity and some of the other conditions on frames we have been considering. Obviously every reflexive frame is range-reflexive. The soundness half of Exercise 2.5.21(*i*) implies that **U** is valid on every range-reflexive frame, and hence on every reflexive frame, so by the completeness of **KT** w.r.t. the class of such frames, we must have $\vdash_{\mathbf{KT}} \mathbf{U}$. It is clear how such a proof might proceed: simply apply the rule of Necessitation to (an arbitrary instance of) **T**. Slightly more interesting are the following cases.

EXAMPLES 2.5.22 (*i*) If we compare the conditions of piecewise connectedness and range-reflexivity, it is evident that any frame meeting the former condition meets the latter. The former required that for

---

[68] The *domain* of a binary R relation on a set X is the set of $x \in X$ for which there exists $y \in X$ with xRy, while the range of such an R is the set of $y \in X$ for which there exists $x \in X$ with xRy; thus a range- or domain-reflexive relation is one whose restriction to its own range or domain (resp.) is reflexive. Range-reflexivity is called *postponed reflexivity* in van der Hoek [508], and *shift reflexivity* at p. 109 of Garson [351]. Relations or frames (or models) whose accessibility relations have this property are called *almost reflexive* in Åqvist [30], p. 92. Note that because each of these conditions can be written with all initial quantifiers universal, they are graphically representable in accordance with the conventions introduced for Figures 2.3–2.6 in the preceding section. Indeed range-reflexivity appeared there diagrammed on the right of Figure 2.4, p. 66.

all $w \in W$, we have for all $x, y \in R(w)$, either xRy or yRx. Thus in particular, taking y as x, for all $w \in W$, for all $x \in R(w)$, xRx (simplifying the "xRx or xRx" that arises on putting x for y): but this is range-reflexivity. Reasoning as above apropos of **KT**, we can conclude that **U** should be provable in **K.3$^+$**, since **U** is valid on all range-reflexive frames (by the soundness half of Exercise 2.5.21(i)) and hence on all piecewise connected frames (by the reasoning just given), which means that **U** should be provable in **K.3$^+$**, since this logic is complete w.r.t. the class of such frames (by Exercise 2.5.8(ii)). Nor is there any difficulty in seeing how such a proof might go. Let us compare representative instance of the schema **.3$^+$** with one of **U**:

$$\mathbf{.3^+} \quad \Box(\Box p \to q) \vee \Box(\Box q \to p) \qquad \mathbf{U} \quad \Box(\Box p \to p).$$

It is easy to see how to obtain the second of these from the first by applying the rules (closure under which is) definitive of normal modal logics: just substitute p for q in the first and then reduce the resulting formula from $A \vee A$ to A.

(ii) A slightly harder case on the issue of provability is presented by **5** and **U**. Any euclidean frame is range-reflexive – go back to the diagram on the right of Figure 2.3 (p. 66) and identify points x and y – but it is not so easy to see how to prove **U** in **K5**. Yet by reasoning analogous to that given in (i) above (or the preceding discussion) it must be provable, since **K5** is complete w.r.t. the class of euclidean frames; this is left as part of the following Exercise.

(iii) Another case in the same vein: if $\langle W, R\rangle$ is symmetric, then it must be piecewise convergent, since if wRx and wRy, then w itself is a common successor of x and y (by symmetry). Here the matching syntactic proof (showing that $\vdash_{\mathbf{KB}} \mathbf{.2}$) is very simple, since if we take **B** axiom $p \to \Box\Diamond p$ and its dual form $\Diamond\Box p \to p$, we note that they have the **.2** axiom $\Diamond\Box p \to \Box\Diamond p$ as a truth-functional consequence. ◀

EXERCISE 2.5.23 (i) Give a formal proof in the style of Example 1.3.16 of a representative instance of **U** from the axioms and rules of **K** given there with the added axiom schema **5**. (From (ii) of the preceding examples we already know that there is such a syntactic proof, so the point is simply to find one.)

(ii) We have seen that **U** is provable in **K.3$^+$**; is it provable in **K.3**? (Justify your answer.) ✠

There is a more streamlined way of discussing such issues as we have been concerned with here. Define a formula C to be a *frame consequence* of a set of formulas Γ if every frame validating each $A \in \Gamma$ also validates C. (This definition was foreshadowed in note 16, p. 23.) Frame consequences of $\{A\}$ will be described as frame consequences of A, and formulas which are frame consequences of each other will be described as *frame equivalent*. While we are at it let us define the class of frames for a set of formulas Γ,[69] denoted by $Fr(\Gamma)$, to be the set of precisely those frames in which every formula in Γ is valid. Any frame in $Fr(\Gamma)$ is called a *frame for* Γ. As with the frame consequence terminology, if $\Gamma = \{A\}$ for some formula A, we write $Fr(A)$ rather than $Fr(\{A\})$ and read this as: the class of all frames for A; similarly $\mathcal{F}$ is a frame for A when it is a frame for $\{A\}$ (which just amounts to saying $\mathcal{F} \models A$). Thus A is a frame consequence of Γ just in case $Fr(\Gamma) \subseteq Fr(A)$. The following exercise employs the "$\oplus$" notation introduced before Exercise 2.5.10 above.

EXERCISE 2.5.24 Show in the case of each of (i), (ii), that for any set of formulas Γ the claim made is correct (by sketching an argument) or incorrect (by giving a counterexample):

(i) $Fr(\Gamma) = Fr(\mathbf{K} \oplus \Gamma)$;

(ii) For any formula A, A is a frame consequence of Γ if and only if A is a frame consequence of $\mathbf{K} \oplus \Gamma$. ✠

---

[69]Since we are treating modal logics (normal or otherwise) as certain sets of formulas, Γ may but need not be a modal logic.

2.5. FROM MODELS TO FRAMES

EXERCISE 2.5.25 Let us use the notations $S + \Gamma$ ($S + A$) according to the same conventions as the "$\oplus$" notation, except that they denote, not the smallest normal modal logic extending S and including Γ (containing A), respectively, but rather the smallest modal logics answering to these descriptions. (This is a standard use of the $+/\oplus$ notation, found for example in [166].) Note that if S is normal, the logics so described are automatically quasi-normal; in some cases they may even be normal (see (ii) and (iii) here).

(i) Is the following true or false? (Give reasons.) For every formula A, $\mathbf{S4} \oplus A = \mathbf{S4} + \Box A$.

(ii) For which of the following formulas A does it hold that $\mathbf{K} + A = \mathbf{K} \oplus A$?

$$p \to \Box p \qquad \Box p \to p \qquad \Box \bot.$$

(iii) For which of the following formulas A does it hold that $\mathbf{S4} + A = \mathbf{S4} \oplus A$?

$$\Diamond p \to \Box \Diamond p \qquad p \to \Box \Diamond p \qquad \Diamond \Box p \to \Box \Diamond p.$$

✠

In this exercise, many formulas are written out explicitly rather than referred to by their abbreviative names. Those appearing under 2.5.25(ii) are (in order) $\mathbf{T}_c$, $\mathbf{T}$ and $\mathbf{Ver}$, while those under 2.5.25(iii) are respectively $\mathbf{5}$, $\mathbf{B}$ and $\mathbf{.2}$ (more commonly called $\mathbf{G}$). In the Digression on p. 34 it was mentioned that for the converse, $\mathbf{M}$, of the last of these formulas, the answer to the corresponding question would be *No*: the quasi-normal logic $\mathbf{S4} + \mathbf{M}$ is weaker than the normal logic $\mathbf{S4} \oplus \mathbf{M}$ (alias $\mathbf{S4M}$): see Prop. 3.2.27 on p. 193. $\mathbf{S4} \oplus \mathbf{M}$ has come in for attention more recently in pp. 201–3 of Leuenberger [713].

EXAMPLE 2.5.26 Recalling that $\mathbf{S5}$ can be axiomatized in various ways, one of which is as $\mathbf{KDB4}$ (see Exercise 2.1.15, p. 45), which we can also write as $\mathbf{KD4} \oplus \mathbf{B}$, with the aid of the notation from Exercise 2.5.25, we can make the more refined observation that $\mathbf{S5}$ coincides also with $\mathbf{KD4} + \mathbf{B}$. This follows from the fact that already $\mathbf{K4} + \mathbf{B} = \mathbf{K4} \oplus \mathbf{B}$ (alias $\mathbf{KDB4}$), as shown in Example 3.2.23, p. 192. ◀

EXERCISE 2.5.27 In the axiomatization of $\mathbf{S5}$ as $\mathbf{KT} \oplus \mathbf{5}$ the initial reference to possibility – speaking loosely – in $\mathbf{5}$ can replaced by one to contingency. Show that this is the case, i.e., that $\mathbf{S5}$ is $\mathbf{KT} \oplus (\Diamond A \wedge \Diamond \neg A) \to \Box \Diamond A$. (This exercise comes from the discussion in Gallie [346], a paper noting a lapse in concentration on Lemmon's part in [695], as well as the fact that – as the present exercise shows – although this resulted in an infelicitous formulation, no lasting damage was done. The formulation given involves speaking loosely because if one says "It is possible that p" – "$\Diamond p$" – one is not making a *reference* to possibility, any more than in saying that that Barack Obama is not a vegetarian one is making a reference to negation, though one certainly *is* making a reference to Barack Obama.) ✠

In view of (the correct answer to) part (i) of Exercise 2.5.24, the frames for $\mathbf{T}$ are exactly the frames for $\mathbf{KT}$. Further, this is so whether we take $\mathbf{T}$ as the schema of that name or as a representative instance of that schema, and the class in question is the class of all reflexive frames. This amounts to the claim that for all frames $\langle W, R \rangle$,

$$\langle W, R \rangle \models \Box p \to p \text{ if and only if } \langle W, R \rangle \text{ is reflexive,}$$

which (it will be recalled) means that R is a reflexive relation on W. Now the "if" direction of the inset equivalence is just the soundness of $\mathbf{KT}$ w.r.t. the class of reflexive frames, but the "only if" is far from amounting to the corresponding completeness result, and takes us into a new area, that of modal definability: the inset equivalence says that the class of reflexive frames is modally defined by the formula $\Box p \to p$. (Well, not quite a wholly new area, since these considerations were anticipated at Exercise 2.5.9(ii).) More generally – this case being the case of singleton Γ – a class $\mathbb{C}$ of frames is *modally defined* by a collection Γ of formulas, just in case $\mathbb{C} = Fr(\Gamma)$. Anyway, to see that the "if" direction of the equivalence inset above is correct, we need to show that on any non-reflexive frame $\langle W, R \rangle$ we can find a model $\langle W, R, V \rangle$ such that $\langle W, R, V \rangle \not\models \Box p \to p$. But this is easy. Since the frame is not reflexive there is some $x \in W$ for which not xRx. Any V with $R(x) \subseteq V(p)$ will yield a model in which $\Box p$ is true at x, and any V with $x \notin V(p)$ will be one in which p is false at x. So as long as we can find a V satisfying both of these conditions we will have a model in which the antecedent of our conditional $\Box p \to p$ is true and the consequent is false a x, making the conditional false at x in this model, thereby demonstrating the invalidity of the formula on the frame. And since $x \notin R(x)$, we can certainly satisfy both conditions on V. For the sake of definiteness, one way to do this is to set $V(p) = R(x)$. Another would be to set $V(p) = W \smallsetminus \{x\}$. The first way makes p true at the smallest number of points compatible with $\Box p$'s being true (while p is false) at x, while the second way makes p true at the largest number of points compatible with p's being false (while $\Box p$ is) at x. As a simplified variation on saying that the class of reflexive frames is modally defined by **T** (as we have just seen), we sometimes say just that reflexivity is modally defined by **T**. (A 'local', or 'point by point', notion of modal definability contrasting with the present 'global' – or 'frame by frame' – notion, will be introduced in Section 3.2, though it could equally well be dealt with here, having nothing to do with the fact that there tense logic rather than monomodal logic is under discussion.)

EXERCISE 2.5.28 (i) Show that symmetry, transitivity, the euclidean condition, seriality, range-reflexivity and functionality are modally defined by **B**, **4**, **5**, **D**, **U**, **D!**, respectively.

(ii) Find a formula A, containing occurrences of a single propositional variable, which is frame equivalent to the two-variable formula $(\Diamond p \wedge \Diamond q) \to \Diamond(p \wedge q)$, and also describe (via a condition on R) the class $Fr(A)$.

(iii) As in part (ii), but this time for the two-variable formula $\Box(p \vee \Box q) \to (\Box p \vee \Box q)$. ✠

The phrase "frame for", introduced above, has a potential danger associated with its use: although we refer to the frame of the canonical model for S (viz. $\langle W_S, R_S \rangle$) as the canonical frame for S, there is no general guarantee that this frame is a frame for S in the sense just introduced. We are not much concerned with normal modal logics whose theorems are not all valid on their canonical frames, though one example, **KW**, was already mentioned in this connection in Exercise 2.4.9(i).[70] If they are determined by any class of frames, then this cannot be shown by the canonical model method (at least on its own), which would consist in observing that the canonical frame lies within the class in question. Thus the 'frame' version of Coro. 2.4.2, telling us that any consistent normal modal logic S is determined by $\mathcal{M}_S$ (or more pedantically by $\{\mathcal{M}_S\}$), would not be correct: the canonical frame need not be a characteristic frame in the sense of a frame *validating* precisely the theorems of the logic, by contrast with the case of the canonical model, which does indeed *verify* precisely the theorems of the logic. A normal modal logic S whose canonical frame *does* belong to $Fr(S)$ is called a *canonical* modal logic. Sometimes, non-canonical logics are nevertheless determined by a class of frames, or are *Kripke-complete*, as it is often put, but sometimes they aren't (even by the class of all frames for the logic), and no method will be available to prove both soundness and completeness w.r.t. any single class of frames.[71] The following exercise works with three alternative but equivalent ways of characterizing Kripke-completeness.

---

[70]To be more explicit: $Fr(\mathbf{KW})$ is the same class of frames by which this logic was noted in Exc. 2.4.9(i) to be determined – those which are (a) transitive and (b) such that no point in them starts an infinite R-chain. The canonical frame for **KW** does not satisfy (b), as we shall see in (the proof of) Coro. 2.6.10 below; a similar argument appears in Hughes and Cresswell [537], pp. 139–141, though without reference to the rule of disjunction (explained in Section 2.6). A different argument is given in Boolos [112], p. 90f., using extraneous considerations from Peano arithmetic. Boolos refers to **KW** as *GL*, for "Gödel–Löb" logic).

[71]For textbook discussions, see Chapter 9 of [537] or Chapter 11 of [112]; warning: Hughes and Cresswell, and also Boolos, speak of a formula's being "valid" in a model when it is – as we prefer to say – *true throughout* the model. Boolos's usage is even further from that followed here in that he (like many others) speaks of a formula's being valid "in" – rather than *on* – a frame. The "on a frame" formulation is that preferred by Hughes and Cresswell, perhaps beginning with the still invaluable Cresswell [207].

2.5. FROM MODELS TO FRAMES

EXERCISE 2.5.29 Show that the following are equivalent for a consistent normal modal logic S:

(*i*) S is determined by $Fr(S)$.

(*ii*) S is determined by some class of frames.

(*iii*) S is closed under frame consequences in the sense that any formula A which is a frame consequence of $\Gamma \subseteq S$ is provable in S. ✠

We cannot add, as a fourth supposed equivalent to (*i*)–(*iii*) in the above exercise: "S is determined by $\langle W_S, R_S \rangle$", in view of the existence of (Kripke-)complete normal modal logics S which are not canonical, in the sense that $\langle W_S, R_S \rangle \notin Fr(S)$, as in the case of $S = \mathbf{KW}$ recalled in note 70.

Digression. A *general frame* for monomodal logic is a structure $\langle W, R, \mathbb{P} \rangle$ with $\mathbb{P} \subseteq \wp(W)$ (intuitive idea: $\mathbb{P}$ collects up the "admissible propositions"), closed under operations corresponding to the Boolean connectives and □ so that when, in a model on such a substructure, propositional variables are assigned elements $\mathbb{P}$, all formulas end up with their truth-sets belonging to $\mathbb{P}$. ("Corresponding to" here is illustrated in the Boolean case by the requirement – to secure that conjunction preserves admissibility – that if $X, Y \in \mathbb{P}$, then $X \cap Y \in \mathbb{P}$. For □, we require that if $X \in \mathbb{P}$ then $\{w \in W \mid R(w) \subseteq X\} \in \mathbb{P}$.) Validity on a general frame is then truth in any model on the frame which makes such assignments to the propositional variables. In contrast to the case of Kripke frames, every normal modal logic is determined by a class of general frames. More information on the subject can be found in Chapters 4 and 6 of van Benthem [73], Blackburn *et al.* [98] (see the latter work's index), and in many other texts – where their intimate connection with modal algebras – cf. the discussion following Exercise 2.10.3, p. 155 is usually discussed in the same breath. A seminal treatment of this relationship can be found in the paper republished as Chapter 1 of Goldblatt [394]. **End of Digression.**

Many easily described classes of frames can be shown not to be modally definable – i.e., not modally defined by any set of formulas – by methods we are not going into at this stage, such as the class of irreflexive frames, the class of asymmetric frames, the class of antisymmetric frames, and the class of intransitive frames,[72] but in some other cases we already have the wherewithal to draw such conclusions. For one thing, we have already seen cases of normal modal logics determined by distinct classes of frames – for example with **S5**, determined by the class of equivalence relational frames and also by the class of universal frames. But such multiple completeness results always point to modal undefinability:

EXERCISE 2.5.30 Show that if S is determined by classes $\mathbb{C}_1$ and $\mathbb{C}_2$ then if $\mathbb{C}_1 \neq \mathbb{C}_2$, at least one of $\mathbb{C}_1$, $\mathbb{C}_2$, must be modally undefinable. ✠

In the case of **S5**, we know from Exercise 2.5.28(*i*) that the class of equivalence relational frames is modally definable, so the class of universal frames is not. We could draw that conclusion more directly by reflecting on a consequence of the Generation Theorem, Thm. 2.4.10, the same theorem that was used to derived the completeness proof for **S5** w.r.t. the class of universal frames, namely part (*i*) of the following, in which the *disjoint union* of frames $\langle W, R \rangle$ and $\langle W', R' \rangle$ is the frame we get by first replacing $\langle W', R' \rangle$ by an isomorphic copy of itself $\langle W'', R'' \rangle$ with $W \cap W'' = \varnothing$ and then forming the frame $\langle W \cup W'', R \cup R'' \rangle$. (An analogous process which we need not spell out in detail gives the disjoint union of an infinite collection of frames.)

THEOREM 2.5.31 (*i*) *If a formula is valid on each of several frames, it is valid on their disjoint union.*

(*ii*) *If a formula is valid on a frame, it is valid on any generated subframe thereof.*

Proof. Deriving (*i*) and (*ii*) from Theorem 2.4.10 is left to the reader. ∎

---

[72]See Revision Exercise 2.7.35 (*iii*), (*iv*); also Coro. 4.4.33.

Various other frame/model constructions preserve the validity/truth of formulas in the same way as taking disjoint unions and generated submodels. One, introduced in the preamble to Revision Exercise 2.7.35, and made use of later, is that of taking what are called p-morphic images of frames. Another, which we shall not be going into at all, is filtrations (briefly mentioned on p. 73 above).

THEOREM 2.5.32 *(i)* *If $\mathbb{C}$ is not closed under disjoint unions (i.e., contains some frames whose disjoint union it does not contain) then $\mathbb{C}$ is not modally definable.*

(ii) If $\mathbb{C}$ is not closed under generated subframes, then $\mathbb{C}$ is not modally definable.

(iii) If there is a consistent normal modal logic S with $\mathcal{F}_S \in \mathbb{C}$ while $Fr(S) \not\subseteq \mathbb{C}$, then $\mathbb{C}$ is not modally definable.

Proof. Parts *(i)* and *(ii)* are simple corollaries of the corresponding parts of Theorem 2.5.31. For *(iii)* suppose that some S and $\mathbb{C}$ meet the conditions stated but, for a contradiction, that $\mathbb{C}$ is modally definable. Thus for some set Γ of formulas we have

$$\mathcal{F} \in \mathbb{C} \text{ if and only if for each } A \in \Gamma, \mathcal{F} \models A.$$

Since $\mathcal{F}_S \in \mathbb{C}$, for each $A \in \Gamma$, $\mathcal{F}_S \models A$ (using the 'only if' direction of the inset 'if and only if' claim), so $\mathcal{M}_S \models A$, so $\vdash_S A$, for each $A \in \Gamma$. But we are also told that $Fr(S) \not\subseteq \mathbb{C}$, so choose $\mathcal{F} \in Fr(S) \smallsetminus \mathbb{C}$. Since $\mathcal{F}$ is a frame for S ($\mathcal{F} \in Fr(S)$) and $\vdash_S A$ for each $A \in \Gamma$, $\mathcal{F} \models A$ for each $A \in \Gamma$. But this contradicts the 'if' direction of the inset 'if and only if' claim above, given that $\mathcal{F} \notin \mathbb{C}$. ∎

Part *(iii)* of Theorem 2.5.32 can be thought of as allowing us to showing modal undefinability by means of the method of 'canonical accidents' – as it were, accidental properties of a canonical frame – in the sense of finding a logic not all of whose frames satisfy a certain condition but whose canonical frame does happen to satisfy that condition: in that case we conclude that (the class of all frames satisfying) the condition in question is not modally definable. Coro. 2.5.33*(iv)* illustrates an application of this method to the case of the following condition on frames: every point in the frame has a reflexive successor. **KD4** features as the logic whose canonical frame 'just happens' to satisfy this condition.

COROLLARY 2.5.33 *(i) The class of universal frames is not modally definable.*

(ii) The class of finite frames is not modally definable.

(iii) The class of frames $\langle W, R \rangle$ which are converse serial, *in the sense that for every $x \in W$ there exists $w \in W$ with wRx, is not modally definable.*

(iv) The class of frames $\langle W, R \rangle$ in which for every $x \in W$, there exists $y \in R(x)$ such that yRy, is not modally definable.

Proof. For *(i)* consider any two universal frames: their disjoint union is not a universal frame, so the modal undefinability result follows from Thm. 2.5.32*(i)*. For *(ii)*, we use the same result but in the infinite version: the disjoint union of (say) countably many copies of the one-element irreflexive frame, for example. *(iii)* follows from Thm. 2.5.32*(ii)*; for example, consider the integers (positive, zero and negative) as comprising W with R as the relation $<$. (In other words, we are dealing with the frame $\langle \mathbb{Z}, < \rangle$.) This frame is converse serial but the subframe generated by 0 (or indeed by any other point) is not. In the case of *(iv)*, we help ourselves to Prop. 2.5.35 below. Take an arbitrary $x \in W_{\mathbf{KD4}}$ with a view to finding $y \in W_{\mathbf{KD4}}$ with $xR_{\mathbf{KD4}}y$ and $yR_{\mathbf{KD4}}y$. Assuming the **KD4**-consistency of the union of the set of all implications of the form $\Box B \to B$ with $\{A \mid \Box A \in x\}$, we have, in any maximal consistent superset of this set, a $y \in W_{\mathbf{KD4}}$ for which $xR_{\mathbf{KD4}}y$ (since we put every A into y when $\Box A$ was in x) and $yR_{\mathbf{KD4}}y$ (since with every $\Box B \to B$ in y, any boxed formula in y will also appear there unboxed). It remains only to check that the set in question is **KD4**-consistent. Suppose otherwise. Then for some $A_1, \ldots, A_m$ with $\Box A_i \in x$ and some formulas $B_1, \ldots, B_n$, we have:

2.5. FROM MODELS TO FRAMES

$\vdash_{\mathbf{KD4}} (A_1 \wedge \ldots \wedge A_m) \to \neg((\Box B_1 \to B_1) \wedge \ldots \wedge (\Box B_n \to B_n))$.

By normality, we should then have

$\vdash_{\mathbf{KD4}} (\Box A_1 \wedge \ldots \wedge \Box A_m) \to \Box\neg((\Box B_1 \to B_1) \wedge \ldots \wedge (\Box B_n \to B_n))$,

and therefore, since each $\Box A_i \in x$, the consequent of this formula belongs to x, which is impossible as it would contradict the **KD4**-consistency of x, since according to Prop. 2.5.35, $\Diamond((\Box p_1 \to p_1) \wedge \ldots \wedge (\Box p_n \to p_n))$ and hence the result of substituting B_i for p_i in this formula, is **KD4**-provable, this particular substitution instance being the negation of the formula we saw had to belong to x.

Thus the canonical frame for **KD4** satisfies the condition that every point has a reflexive successor. But not every frame for **KD4** meets this condition – e.g., $\langle \mathbb{Z}, < \rangle$ and $\langle \mathbb{N}, < \rangle$ (i.e., $\langle Nat, < \rangle$) are frames for **KD4** since they are serial and transitive, but no every – and in fact *no* point – in these frames has a reflexive successor. Thus (iv) is established, in view of Theorem 2.5.32(iii). ∎

It is worth keeping track of the relations between the piecewise versions – as defined on p. 76 above – of various condition on frames and this issue of modal definability:

EXAMPLES 2.5.34 (i) The case of Corollary 2.5.33(i) illustrates how a first order condition describing a modally undefinable class of frames (namely $\forall x \forall y \,.\, Rxy$) has, for its piecewise version, a condition describing a modally definable class of frames (the euclidean frames); we have the same situation with convergence and connectedness: the piecewise versions but not the originals, give modally definable classes of frames.

(ii) In other cases, however, we have the reverse situation: forming the piecewise condition corresponding to a given condition will yield, from a modally definable class of frames, a modally undefinable class. The class of symmetric frames is modally defined by **B**, while the class of piecewise symmetric frames – those $\langle W, R \rangle$ such that for all $w \in W$, for all $x, y \in R(w)$, if xRy then yRx – is not modally definable, as can be gleaned from Revision Exercise 2.7.36.

(iii) The remaining combinatorial possibilities are also realised: classes of frames described by a condition and its piecewise version, *neither* of which is modally definable, and classes of frames described by a condition and its piecewise version *each* of which is modally definable. For an illustration of the latter case, consider the reflexive frames and the range-reflexive frames (modally defined by **T** and **U** respectively), noting that the latter comprise the piecewise reflexive frames. ◀

The modal undefinability reported in Corollary 2.5.33 (iv) was first shown, using other methods, by R. Goldblatt and S. K. Thomason in 1974. (Their argument uses a frame construction called *ultrafilter extensions* and it can be found conveniently in van Benthem [73], p. 33, which also gives another proof – p. 36 – using a model construction called *filtrations*; since neither construction lies within the ambit of the present text, we use instead the simple 'canonical accidents' proof above.) The class of frames involved – those in which every point has a reflexive successor – determines a weaker logic than **KD4**, described in Hughes [534], where many further points of interest are made about this class of frames.[73] The logic concerned is the least normal modal logic containing all the formulas $\mathbf{T}_n^\Diamond$ $(n \geqslant 1)$:

$\mathbf{T}_n^\Diamond$ $\Diamond((\Box p_1 \to p_1) \wedge \ldots \wedge (\Box p_n \to p_n))$.

Hughes shows that this logic cannot be axiomatized (using the rules definitive of normal modal logics) with only finitely many axioms, or indeed finitely many axiom-schemata,[74] making it the first example

---

[73]Issues raised in Hughes's paper have been more recently been taken up in Goldblatt *et al.* [399], Balbiani *et al.* [50], and my own [589]. A variation on Hughes's sequence $\mathbf{T}_n^\Diamond$, collectively equivalent to it, is provided in Venema [1128].

[74]Schemata which are like axioms except in having schematic letters in place of propositional variables, that is. The inset expression labelled $\mathbf{T}_n^\Diamond$ above could itself be described as a schema whose instances are the various formulas arising for each particular choice of n. We could equally well have written this using schematic letters in place of the propositional variables, the result of which would be *doubly* schematic.

of that phenomenon we have seen. Our concern with this logic goes only so far as showing that it is included in **KD4**, since this is what we appealed to above in the proof of Coro. 2.5.33(*iv*).

PROPOSITION 2.5.35 *All the formulas* $\mathbf{T}_n^\diamond$ *are provable in* **KD4**.

Proof. We argue by induction on n.

Basis Case. If $n = 1$ our formula is provably equivalent in **K** (by routine manipulations: see the third formula under Exercise 1.3.17(*ii*)) to $\Box\Box p_1 \to \Diamond p_1$, which is obviously **KD4**-provable.

Inductive Step. Suppose $\vdash_{\mathbf{KD4}} \Diamond((\Box p_1 \to p_1) \wedge \ldots \wedge (\Box p_n \to p_n))$, with a view to showing that $\vdash_{\mathbf{KD4}} \Diamond((\Box p_1 \to p_1) \wedge \ldots \wedge (\Box p_n \to p_n) \wedge (\Box p_{n+1} \to p_{n+1}))$. According to Prop. 2.4.4 every normal extension of **K4** – including therefore, in particular, **KD4** – is closed under the rule:

$$\frac{\Diamond A}{\Diamond(A \wedge (\Box B \to B))}$$

which we may therefore invoke here, taking A as $(\Box p_1 \to p_1) \wedge \ldots \wedge (\Box p_n \to p_n)$ and B as p_{n+1}, conveniently completing the inductive step. ∎

As well as appealing to the formulas $\mathbf{T}_n^\diamond$ to help, as above, with the undefinability example of Coro. 2.5.33(*iv*), we can press them into service to illustrate another issue, the possibility of a normal modal logic's not being closed under what in Humberstone [582], p. 578*f*., is called *internal* uniform substitution, or more explicitly, internal conjunctive uniform substitution.[75] In fact we need only the first two members of the sequence, $\mathbf{T}_1^\diamond$ and $\mathbf{T}_2^\diamond$ for this illustration. But first we must define internal uniform substitution.

Let us first consider the situation with the familiar rule of uniform substitution. Let $A(p_i)$ be a formula in which the variable p_i may occur and in which if it does, the formulas $A(B_1), \ldots, A(B_n)$ are the results of substituting respectively the formulas $B_1, \ldots, B_n$ for all occurrences of p in $A(p_i)$. Any logic closed under (*i*) uniform substitution and (*ii*) tautological consequence will then, if it contains $A(p_i)$, contain each of $A(B_1), \ldots, A(B_n)$ (by (*i*)) and so $A(B_1) \wedge \ldots \wedge A(B_n)$ (by (*ii*)). We can think of this last formula as derived from $A(p_i)$ by 'external' conjunctive uniform substitution. For the internal version, we allow the conjunction of substitution instances of a proper subformula of the starting formula. More exactly, let $A = A(p_i)$ be as before and $B = B(p_i)$, and substitute the latter uniformly for occurrences of p_i in the former, giving the formula $A(B(p_i))$; thus all occurrences of p_i in the latter formula lie in subformulas $B(p_i)$. This time we take conjunctions of substitution instances of $B(p_i)$: $B(C_1) \wedge \ldots \wedge B(C_n)$ (for formulas $C_1, \ldots, C_n$). We say that the logic is *closed under internal uniform substitution* just in case whenever it contains $A(B(p_i))$ it contains every formula $A(B(C_1) \wedge \ldots \wedge B(C_n))$.

PROPOSITION 2.5.36 *Classical propositional logic is closed under internal uniform substitution.*

Proof. Suppose $\vdash_{CL} A(B(p_i))$. Then by closure under US, we have $\vdash_{CL} A(B(C_j))$ for $j = 1, \ldots, n$, and so $\vdash_{CL} A(B(C_1)) \wedge \ldots \wedge A(B(C_n))$ (since the conjunction is a tautological consequence of its conjuncts). Finally, by Prop. 1.2.7, we have $\vdash_{CL} A(B(C_1) \wedge \ldots \wedge B(C_n))$. ∎

Thus the result depends on the fact that in classical truth-functional logic all contexts are aggregative, which is typically not the case in modal logics, even in normal modal logics, which require $\Box$ – but do not require $\Diamond$ – to satisfy the aggregation condition. In particular, we have the example foreshadowed above. For a given choice of n, we denote the logic $\mathbf{K} \oplus \mathbf{T}_n^\diamond$ by $\mathbf{KT}_n^\diamond$.

---

[75]In Humberstone [582], p. 579, this is illustrated by a sequence of formulas like $\mathbf{T}_n^\diamond$ except with **T** replaced by $\mathbf{T}_c$. Although only the first two elements of this sequence are discussed there, the full sequence appears in §4 of Balbiani *et al.* [50], with application to a theme from Hughes [534]. A similar sequence of formulas replaces **T** (or $\mathbf{T}_c$) with $\mathbf{D}_c$; its first member is the McKinsey axiom **M**. The whole sequence is considered in Lemmon and Scott [705], p. 74 apropos of **KM** and **S4M**; see Theorem 4.5.17, p. 286, below, and surrounding discussion, for more details.

2.5. FROM MODELS TO FRAMES

EXAMPLE 2.5.37 Consider the normal modal logic $\mathbf{KT}_1^\diamond$ and in particular the formula $\mathbf{T}_1^\diamond = \diamond(\Box p \to p)$. Taking this formula as the $A(B(p))$ of our definition of closure under internal uniform substitution, and $B(p)$ as $\Box p \to p$, choose C_1 and C_2 as p and q respectively to obtain the formula $\mathbf{T}_2^\diamond$, i.e.,

$$\diamond((\Box p \to p) \land (\Box q \to q)),$$

which would accordingly have to be $\mathbf{KT}_1^\diamond$-provable for this logic to be closed under internal uniform substitution. But it is not hard to see that $\mathbf{KT}_1^\diamond$ is determined by the class of frames $\langle W, R \rangle$ in which $R(x) \cap R^2(x) \neq \varnothing$ for each $x \in W$, while the model depicted in Figure 2.7, whose frame meets the condition just stated, falsifies the formula $\mathbf{T}_2^\diamond$ at its top point. Thus this formula is not provable in $\mathbf{KT}_1^\diamond$. So in fact for present purposes all we need to check is the soundness of this logic w.r.t. the class of frames mentioned (though a canonical model argument straightforwardly gives the corresponding completeness result).

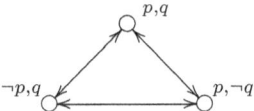

Figure 2.7: A Countermodel for $\mathbf{T}_2^\diamond$

In Figure 2.7, the appearance, for example, of $\{p, \neg q\}$ by a node means that for the point x thereby represented, we have $x \in V(p)$ and $x \notin V(q)$. Any model whose V component behaves like this will have the desired effect (of falsifying $\mathbf{T}_2^\diamond$ at the point represented by the node at the top of the diagram). The accessibility relation is indicated by the arrows: each point bears it to the other two but not to itself. (The arrows are thus not to be taken as in diagrammatic representations of universally quantified conditions on frames, as meaning "*if* such and such relationships hold, ...".) ◀

Before leaving the topic of modal definability, we enter a clarificatory warning. In the discussion after Exercise 2.5.29 various classes of frames were noted to be modally undefinable, including the class of irreflexive frames and that of antisymmetric frames. It is to emphasized that this means there is no set of formulas valid on *all and only* the frames in the classes in question. Obviously there are many formulas valid on *all* such frames – for example every $\mathbf{K}$-provable formula is, since these are valid on every frame whatever. There are also formulas valid on only such frames. Degenerate examples include the negations of $\mathbf{K}$-theorems, $\Box$-free formulas which are not tautologous, and so on, since such formulas are valid on no frames whatever.[76] But there are also formulas which are valid on some frames but only those in some modally undefinable class. For instance $\mathbf{W}$ is valid on precisely those frames with accessibility relation R such that (i) R is transitive and (ii) no point starts an infinite R-chain (see Exc. 2.4.9, p. 69); in virtue of the non-first-order condition (ii) here, it follows that every frame validating $\mathbf{W}$ is irreflexive. Antisymmetry allows for a similar example, but this time, by contrast, with a first-order definable class of frames:

EXAMPLE 2.5.38 The formula $p \lor \Box(p \to \Box p)$ modally defines the class of frames $\langle W, R \rangle$ such that for all $x, y, z \in W$, $xRyRz$ implies $x = y$ or $y = z$. (Additionally, the smallest normal modal logic containing this formula is complete w.r.t. that class of frames, since its canonical frame satisfies this condition.) Taking the case in which z is x, we conclude that all these frames are antisymmetric. (The formula under consideration here, in the variant form $p \to \Box(\diamond p \to p)$), was first considered by Sobociński, and it – or rather the corresponding schema – is discussed under the name $\mathbf{H}$ at p. 148 of Segerberg [1025], where detailed references to Sobociński's work may be found. In Chapters 4 and 5 we shall need to refer several

---

[76]Somewhat more interesting formulas also answering to this description include $\diamond\Box\bot$, which can, unlike those just listed, be what we shall (see p. 185) call valid *at a point* in a frame though not valid on any frame.

times to another principle under that name, so, to avoid ambiguity, we will refer to the current **H** as **H̲** instead. The authors just mentioned were especially interested in the logic **S4H**, as it would be called with this convention in place.) ◀

The parenthetical label just given, "**S4H̲**" has the convenience of reminding us that it is an extension of **S4** that is at issue, but the awkward feature that taking "**S4**" to abbreviate "**KT4**" the label involves redundancy: we can drop the **4**. To see this semantically, note that the condition mentioned in Example 2.5.38, namely that if $xRyRz$ then $x = y$ or $y = z$, for all $x, y, z \in W$, implies the transitivity of R, since if we have $xRyRz$ without xRz then we must have $x \neq y$ (since yRz but not xRz) and also $y \neq z$ (since xRy but not xRz). Thus we are led to expect that **KTH̲** = **S4H̲**, and more generally, since reflexivity was not appealed to in that reasoning, that $\vdash_{\mathbf{KH̲}} \mathbf{4}$, as will now be confirmed by an informal sketch of a formal derivation of **4** from **H̲**.

EXAMPLE 2.5.39 To derive **4** from $\mathbf{H̲} = p \vee \Box(p \to \Box p)$, we need to make a suitable substitution for p, and semantic considerations suggest that the appropriate formula to substitute for p is $p \wedge \neg \Box p$. (To invalidate **H̲** on a frame not meeting the condition on R given above, for a given choice of x, y, z, we may put $V(p) = \{y\}$. On a non-transitive frame with $xRyRz$ without xRz, $\Box p \to \Box\Box p$ gets to be false at x through having $p \wedge \neg \Box p$ true at y.) This substitution gives us the following formula, as provable in **KH̲**:

$$(p \wedge \neg \Box p) \vee \Box((p \wedge \neg \Box p) \to \Box(p \wedge \neg \Box p)),$$

which, with some truth-functional manipulation in the scope of the outer $\Box$ on the second disjunct gives:

$$(p \wedge \neg \Box p) \vee \Box(p \to (\Box p \vee \Box(p \wedge \neg \Box p))).$$

Weakening the first disjunct by discarding one of the conjuncts and the second disjunct by putting $\Box p$ for $\Box(p \wedge \neg \Box p)$, we get:

$$\neg \Box p \vee \Box(p \to (\Box p \vee \Box p)),$$

which again with minor reformulations becomes:

$$\Box p \to \Box(p \to \Box p).$$

Distributing the $\Box$ on the main consequent here across the implication in its scope (i.e., appealing to the **K** axiom) gives:

$$\Box p \to (\Box p \to \Box\Box p),$$

which has the desired formula **4** as a consequence by truth-functional reasoning (contracting the two antecedents to one). ◀

From here until Exercise 2.5.42 the discussion is devoted to aspects of **S5**, as illuminated by some of the concepts in play above. Exercise 2.1.15 asked for syntactic proofs that the axiomatizations suggested in these labels were all axiomatizations – and indeed independent axiomatizations – of **S5**:

KT5   KTB4   KDB4   KDB5,

and it is worth looking at this issue from a model-theoretic point of view (cf. the discussion following Exercise 2.5.21 above), given the correspondences between the axioms and various conditions on (or classes of) frames.[77] The frames for **S5** are precisely the equivalence relational frames, and equivalence relations are usually defined to be reflexive, symmetric and transitive relations. Reflexivity, symmetry

---

[77]Revision Exercise 2.7.7, (p. 130) is also relevant here.

and transitivity are respectively modally defined by **T**, **4** and **B**, so what this most obviously suggests is the axiomatization of **S5** as **KTB4** (or **KT4B**): take an axiomatization of **S4** and add **B**. However, there is a well-known alternative characterization of equivalence relations as *serial*, symmetric and transitive relations. Seriality evidently follows from reflexivity, so it remains only to show that given the other conditions, seriality implies reflexivity. Take a serial transitive symmetric frame $\langle W, R \rangle$ and a point $x \in W$. By seriality, for some $y \in W$, xRy, so by symmetry yRx, whence xRx by transitivity. Since x was arbitrary, $\langle W, R \rangle$ is a reflexive frame. This means that **T** is a frame consequence of $\{\mathbf{D}, \mathbf{B}, \mathbf{4}\}$, and the fact that the canonical model method easily shows **KDB4** to be Kripke-complete means that **T** is provable therefrom (see Exc. 2.5.29). This is a semantic gloss on the equivalence of **KTB4** and **KDB4** as axiomatizations.

There is also the issue of independence here. One can easily see that none of the conditions seriality, symmetry and transitivity, follows from the other two, and use frames illustrating this to show that none of the axioms can be dropped without loss from the **KDB4** axiomatization. Historically, there has been some tendency (outside of modal logic) to ignore seriality – or reflexivity, to which we have just seen it to be equivalent given the other conditions – however, and think one can simply define an equivalence relation as a symmetric transitive relation. Hughes and Londey [538], p. 274, explicitly say that "(e)very relation which is both symmetric and transitive is also reflexive".[78] This is, of course, false – as one sees by considering the case of a single point bearing the binary relation R to nothing. The relation is symmetric and transitive but not reflexive. Nor will setting aside the arguably degenerate case of the empty relation remove the difficulty, since we can consider a universe with, for example, three elements one of which is as before (bearing the relation to nothing) while the others bear the relation to themselves and to each other. Indeed the mistake we are now considering has something of the status of a classic 'howler', with an exercise in Birkhoff and Mac Lane [96] (Exercise 6 on p. 30) asking for a diagnosis of the mistaken reasoning to this conclusion involved in arguing "By the symmetric law, aRb implies bRa; by the transitive law, aRb and bRa imply aRa."[79] The mistake is of course that we have not been given that there *is* a b for which aRb – exactly what the condition of seriality requires, in other words. The corresponding modal-logical error would consist in thinking that **S5** can be axiomatized as **KB4**.[80]

REMARK 2.5.40 Nevertheless, the logics **KB4** and **S5** are very close: Segerberg [1025], p. 128, notes that there is no modal logic (not even just: no normal modal logic) properly between them. (See p. 417 below for a similar question asked – by Sobociński – concerning **S4.4** and **S5**; see also Exercise 2.5.41(i) below.) Note that **KB4** is determined by the class of transitive symmetric frames (the completeness half of this assertion being given by the canonical model method, *à la* Coro. 2.4.3, Thm. 2.5.2). Segerberg also observes that **KB4** = **S5** ∩ **KVer**, which will come as no surprise to anyone reflecting on the point-generated frames for **KB4**, **S5** and **KVer** – cf. Exc. 2.5.43(ii). For various normal modal logics S the logic $S \cap$ **KVer** is called in the papers Lambert *et al.* [685], Morgan [827], [828], [829], and [830], a 'liberated' version of S; in fact the papers cited the choices of S are restricted to **KT** and its extensions. The enterprise comes along with a supposed analogy with free logic, these liberated versions being analogues of the logics in need of liberation now shorn of their existential commitment to (merely) possible worlds. The semantics is not set out with standard Kripke models at all but with a new type of structure; a criticism of this motivational story can be found in Humberstone [588], p. 5, though the latter's own motivation overlaps that of Morgan [830] (final paragraph). ◀

Part (i) of the following exercise picks up a theme from early in Remark 2.5.40, while part (ii)

---

[78] Actually the authors say "symmetrical", but we have translated this into contemporary terminology, especially because it is useful to distinguish these adjectives and reserve the latter for something else: a *symmetric frame* is one whose accessibility relation is symmetric, a *symmetrical frame* is one which is isomorphic to the result of replacing its accessibility relation with the converse of that relation. Evidently neither of these classes of frames is included in the other.

[79] The same reasoning is cited as erroneous on pp. 60 (exposition) and 62 (diagnosis) of Schreider [1003], and forms the subject of Exercise 3 on p. 104 of Humphreys and Prest [604], as well as in many other texts.

[80] This oversight can be found in line 7 from the base of p. 331 of Gabbay [341].

compares the axiomatization of 'liberated **S5**' as given in Morgan [830] with that using **B** and **4**:

EXERCISE 2.5.41 (*i*) Either show that there are no normal modal logics S with the property that **KD45** $\subsetneq S \subsetneq$ **S5** or describe such a logic axiomatically.
(*ii*) Show (syntactically) that **KB4** = **K45** $\oplus \Box\bot \vee (\Box p \to p)$. ✠

Note that the formula mentioned in (*ii*) of the above exercise is a variant formulation of that more directly associated with domain-reflexivity: $\Diamond\top \to (\Box p \to p)$; for more on this, see the discussion before Exercise 3.2.18, p. 190. Either formula would have done in response to Exercise 2.5.21(*iii*).

Digression. What may look like the same mistake as was alluded to before Remark 2.5.40 appears in Pollock [890], p. 24, with a statement (3.12) which reads as follows:

> If R is transitive and symmetric then R is reflexive.

But it turns out that Pollock does not have the standard sense of "reflexive" in mind in writing this, instead defining a binary relation to be reflexive when for all x, if there exists a y such that yRx or there exists a y such that xRy, then xRx. With this specialized redefinition, transitivity and symmetry together do imply reflexivity, but the usage is confusing and not followed by Pollock in other parts of his book – even in the same chapter.[81] (For instance on p. 1 the principle that everything is identical with itself, and on p. 6 that every set is a subset of itself, are both labelled "reflexivity", even though these are not formulated in the conditional manner of the specialized redefinition.) It also results in a non-standard notion of equivalence relation in terms of which there is no longer a one-to-one correspondence between the equivalence relations on a set and the partitions of that set. (The correspondence in question is described in pp. 25–27 of [890]. 'Partition' is defined in the next paragraph below.) One may think that from Pollock's discussion we can at least see that to conclude that xRx (for a given – albeit arbitrary – x) instead of using the seriality-suggested condition that for some y, xRy, there is another hypothesis that serves equally well, namely the hypothesis, suggested by converse seriality, that for some y, yRx. But this would be confused, since this is equivalent to the original hypothesis in the setting in which the argument is cast, i.e., assuming symmetry.

While we are in critical mode, let us note that even the discussion provided by Birkhoff and Mac Lane ([96], p. 146*f*.), last cited favourably on the issue of whether reflexivity follows from transitivity and symmetry, is not without error. The standard definition of a *partition* of a set X is a collection of non-empty pairwise disjoint subsets of X whose union is X. It is the partitions of X in this sense that are in a one-to-one correspondence with the equivalence relations on X. This is what Birkhoff and Mac Lane say, however, at p. 146 of [96]:

> In general, a *partition* π of a class S is any collection of subclasses $A, B, C, \ldots$, of S such that each element of S belongs to one and only one of the subclasses of the collection.

---

[81] However, a similar discussion appears in Smullyan [1069]. This understanding of the term "reflexive" – which amounts to "both domain-reflexive and range-reflexive" – also appears in some of the literature on relation algebras: see for instance p. 98 of Maddux [757]; as with Pollock, Maddux elsewhere seems to use the term *reflexive* in its standard sense. The relation-algebraic literature also uses the phrase *equivalence relation* in the sense of transitive symmetric relation – as in Maddux [757], and Definition 2.2(*i*) of Jónsson [631]. In the usage of such authors, these equivalence relations are indeed 'reflexive' – but this only means domain-reflexive and range-reflexive ('field-reflexive', we might say, since the field of a binary relation is traditionally defined to be the union of its domain and its range), not actually reflexive in the standard sense. In correspondence, Maddux has kindly explained the reason this (non-standard) notion of reflexivity is preferred by some: if we are just given a binary relation as a set of ordered pairs, we cannot tell whether it is reflexive in the way that we can tell whether it is symmetric or transitive (for example), since we cannot tell whether there is an object among those the universal quantifier (in "for all x, xRx") ranges over which is not paired with itself. Thus the same relation R qua subset of $X \times X$ may be reflexive while qua subset of $Y \times Y$, where $Y \supsetneq X$, it is not. 'Field-reflexivity' does not suffer from this difficulty. Not that it arises for modal logic where we are concerned with reflexivity as a property of frames $\langle W, R\rangle$, W giving us the set over which the "for all x" is to range. But it does mean that the usual idea of applying a property of relations to frames derivatively on the basis of its applying to the accessibility relation of the frame does not work in this case.

2.5. FROM MODELS TO FRAMES

This coincides with the standard definition except in one respect: the authors have not required the subsets in question to be non-empty. As well as $\{\{1,2\},\{3\}\}$ as a partition of the set $\{1,2,3\}$, we also have $\{\{1,2\},\{3\},\varnothing\}$ as an equally good 'partition' of the same set, in the sense of *partition* defined in the passage quoted above. Both correspond to the same equivalence relation on that set, relating 1 and 2 to themselves and each other and 3 to itself and only itself. (The correspondence takes us from a partition to the induced equivalence relation holding between any elements both belonging to some block or 'cell' of the partition, and from an equivalence relation to the induced partition whose blocks are the equivalence classes of the elements, $R(x)$ being the equivalence class to which x belongs.) Birkhoff and Mac Lane conclude their discussion with a Theorem 28 (p. 146*f.*) which reads as follows:

> Every equivalence relation R on a set S determines a partition π of S into non-overlapping R-classes, and conversely, each partition of S yields an equivalence relation R. There is thus a one-one correspondence $R \leftrightarrow \pi$ between the equivalence relations R on S and the partitions π of S, such that elements a and b of S lie in the same subclass [= block] of the partition if and only if aRb.

The word "thus" here marks a *non sequitur*: while what precedes it is perfectly correct, the conclusion would follow only on the further condition that applying the procedure for obtaining an equivalence relation from a partition and then obtaining a partition from the equivalence relation returned us to the original partition, which it does not do as the example above shows: we have two partitions with the same induced equivalence relation.[82] The authors presumably had in mind the standard definition which excludes $\varnothing$ from being a block in a partition, inadvertently giving a definition not ruling this out. A flawless discussion of the issue will be found in many expositions of the theory of equivalence relations – e.g., to cite an example from a work criticized above (p. 45) for an unrelated reason, Green [407], pp. 15–18. Another amusing mistake in this area is made in the first of two papers reviewed in Hailperin [422], in which the author attempts to establish the correct claim that not every transitive symmetric relation is reflexive by irrelevantly citing an example of a relation which is symmetric as well as what we shall shortly call *aliotransitive* – but not actually transitive – and which fails to be reflexive. (Even serial aliotransitive symmetric relations need not be reflexive.) **End of Digression.**

Let us turn our attention to the axiomatization **KDB5**, which can be thought of as arising from the non-independent axiomatization of **S5** as **KTB5** by weakening **T** to **D** in the same way as in the case just reviewed, with a similar weakening in **KTB4**. For this new case, the key point is that **B** and **5** already have **4** as a frame consequence: every symmetric euclidean frame is transitive. For suppose that we have $xRyRz$, with a view to showing that xRz. By symmetry, since xRy, yRx. And now we see that yRx and yRz, so by the euclidean condition, xRz.

Figure 2.8: *3-Cyclicity and a Special Case of the Euclidean Condition*

Turning to **KT5**, we confine ourselves to checking that **4** is a frame consequence of **T** and **5**. The diagram on the right-hand side of Figure 2.8 depicts the relevant special case of the euclidean condition which shows that any reflexive euclidean frame is symmetric. Transitivity then follows (from the euclidean condition and symmetry) by the argument given above.

EXERCISE 2.5.42 *The diagram on the left of Figure 2.8 represents the condition that a frame (or its accessibility relation) be what we shall call 3-cyclic.*

---

[82]The same definition of a partition – omitting the requirement that the blocks be non-empty – as well as this same mistake concerning partitions in the sense defined and equivalence relations can also be found on the second page of Ore [859].

(*i*) Write down the condition in question as a first-order formula.

(*ii*) Find a modal formula (or schema if preferred) for which you can show – and indeed show this – that the smallest normal modal logic containing your formula (or all instances of your schema) is determined by the class of 3-cyclic frames.

(*iii*) Show that the reflexive 3-cyclic relations are precisely the equivalence relations.

(*iv*) What axiomatization of **S5** is yielded by the answers to previous parts of this exercise? ✠

The 3-cyclic relations have been called circular relations,[83] but one advantage of the present terminology is that it presents the property as one of a spectrum, with n-cyclicity as the condition that for all $x_1, \ldots, x_n$ if $x_1 R x_2 R x_3 \ldots x_{n-1} R x_n$ then $x_n R x_1$. (Suggestion: draw the diagram for 4-cyclicity.) Thus 2-cyclicity is symmetry and – though one must ponder the significance of a degenerate case of the "iff" clause to see this – 1-cyclicity is reflexivity. Further discussion of axiomatizations of **S5**, with a somewhat 'recreational mathematics' feel to it, can be found in Hughes [533]. It is natural to want to think about the equivalence relational frames when thinking about **S5** since this class of frames is after all $Fr(\mathbf{S5})$. But our first encounter with **S5** was via another class of frames by which this logic is determined, the class of universal frames. This class of frames deserves some attention to, but after we have had a look at some other logics in an exercise which helps call to attention a feature of the *condition* of universality on frames.

EXERCISE 2.5.43 (*i*) Using the canonical model method for the completeness half (i.e., show that the canonical frame lies in the class), find a class of frames by which the logic $\mathbf{K4}_c$ is determined. Now do the same for the logics $\mathbf{KB}_c$, **KD!**, and **KVer**. (Nomenclature as explained apropos of Table 2.1, p. 33.)

(*ii*) What do the point-generated frames for **KVer**, **KT!**, and $\mathbf{KT}_c$ look like? (Three answers required.)

(*iii*) This question concerns the logic $\mathbf{KB}_c$, mentioned on p. 42 above, and may help with that aspect of part (*i*) here concerning that logic as well as (*iv*) below. In a somewhat different use of the "cyclic" terminology from that deployed in Exercise 2.5.42, Humberstone and Williamson [603] call the 'n-cyclicity' schema $\Box^n A \leftrightarrow A$, $\mathbf{C}_n$. Thus $\mathbf{KC}_1$ is just **KT!**. Show that $\mathbf{KC}_2$ coincides with $\mathbf{KB}_c$ and that the logic with the axiomatizations these labels convey coincides with **KB!** and extends **KD!**.

(*iv*) What do the point-generated frames for $\mathbf{KB}_c$ (alias $\mathbf{KC}_2$, **KB!**) look like? ✠

The particular aspect of the condition of universality on frames to which some parts of this exercise draw attention can be isolated like this. Sometimes we have a condition on binary relations with the property that on any given set, there is exactly one binary relation with that property. Call such conditions *uniquely satisfiable*. Universality is such a condition, since given a (now let us add, non-empty) set W there is a unique universal binary relation on W, namely $W \times W$. Three obvious further such uniquely satisfiable conditions come to mind: being the empty relation of the set in question, being the identity relation on the set in question, and being the non-identity relation on the set in question.[84] We know that **S5** is determined by the class of frames in which the accessibility relation is universal, and Exercise 2.5.43 provides (along with Theorem 2.4.10, p. 70 – the Generation Theorem) the material for concluding that if we are looking for the logic determined by the class of frames $\langle W, R \rangle$ in which R is the identity relation on W, we have found it in **KT!**, while if we want the logic determined by the class of $\langle W, R \rangle$ with R the empty relation, then what we are after is **KVer**. But this leaves one case, the logic determined by the class of all frames in which accessibility coincides with non-identity. Hughes and Cresswell [537] refer to such frames for short as *NI frames*. The frame (of the model) depicted in Figure 2.7 (p. 87) is an *NI* frame, for example. Like the class of universal frames, the class of *NI* frames is not modally definable.

---

[83] E.g., in Exercise 5 at p. 30 of [96].

[84] More can be said than that these relations are uniquely satisfiable in the sense defined: they are also *logical* relations according to the permutation-invariance criterion of logicality. See van Benthem [77] for an exposition of this criterion (and some critical remarks).

2.5. FROM MODELS TO FRAMES

(By contrast the classes of frames with R the identity relation and the empty relation are modally defined respectively by $\Box p \leftrightarrow p$ and $\Box\bot$ – alias **T!** and **Ver**.) By contrast with case of R universal, there is no simple proof of completeness using generated subframes, for an axiomatization which is in fact complete, but Segerberg (with some help from Stalnaker)[85] showed that this is the case for **KB4'** where **4'** is the schema:

4' $(A \wedge \Box A) \to \Box\Box A$.

This variation on **4** modally defines the class of frames whose accessibility relations are what C. S. Peirce called *aliotransitive*,[86] where R's aliotransitivity means that for all x, y, z, if xRy and yRz and $x \neq z$ then xRz. The "alio-" (Latin *alius*, "other") reflects the $\neq$. As usual, however, we prefer to treat negated antecedents unnegated in (disjunctive) consequent position, as in this formulation:

For all x, y, z, if xRy and yRz then $x = z$ or xRz,

for which the representing diagram is that on the left of Figure 2.9.

More generally, given a condition on frames of the form $\forall v_1 \ldots \forall v_n \Phi$, in which Φ may contain further (especially, existential) quantifiers, we call any condition:

$$\forall v_1 \ldots \forall v_n (v_i \neq v_j \to \Phi)$$

with $1 \leq i \neq j \leq n$, an 'alio' version of the original condition, and we take the liberty of writing it in whatever form is convenient. (For example if Φ has the form $\Phi_0 \to \Phi_1$ we may write $\forall v_1 \ldots \forall v_n (\Phi_0 \to (\Phi_1 \vee v_i = v_j))$.) A slightly more liberal use of the phrase would also count cases in which the inset form has an antecedent which is the conjunction of several $(v_i \neq v_j)$-type conjuncts, though we do not make much use of this. ((3) under Exercise 2.5.67, p. 108 below, does involves such an example.) A given 'alio' version of a condition is *spurious* if it is equivalent to the original condition.

Figure 2.9: *Aliotransitivity and a Spurious Further Condition*

The nodes in Figure 2.9 have been labelled with individual variables to make for an easier comparison between the genuine aliotransitivity condition on its left, and a spurious alternative which might also come to mind as an 'alio-' version of transitivity – in that it adds an equality disjunct to the consequent of the transitivity condition,[87] giving us:

For all x, y, z, if xRy and yRz then $x = y$ or xRz,

but this is simply transitivity again. The new condition follows from transitivity, since we have weakened the consequent with an extra disjunct, and transitivity follows from it, since a failure of transitivity, with xRy, yRz, and not xRz, is automatically a violation of the new condition, being a case in which $x \neq y$ (since y does, while x does not, bear the relation R to z).[88]

---

[85] For references to the literature, see the notes to Section 3 of my [588].

[86] It is not obvious whether this is better spelt with a hyphen after the "alio" or not. Certainly "alio-equivalence", introduced below, seems to deserve a hyphen to prevent a vowel collision.

[87] Or equivalently: adds a negated equality conjunct to the antecedent. Note that weak connectedness (see p. 75) is an 'alio-' version of (strong) connectedness.

[88] A simpler example – mentioned in note 20 of Humberstone [588] – of such a spurious "alio-" variant is provided by the following condition of 'aliosymmetry': for all x, y, if xRy and $x \neq y$, then yRx. The aliosymmetric relations are easily seen to be precisely the symmetric relations. We return to this issue presently, after Exc. 2.5.52. What about "alio-asymmetry": xRy and $x \neq y$ imply not yRx? This is clearly equivalent to antisymmetry: Rxy and Ryx imply $x = y$. (Note: Harary [451], p. 160, wrongly – or, in case he is attempting a definition, confusingly – says *asymmetric* here. A rather more serious problem with [451] is raised in Potts [901].)

EXERCISE 2.5.44 A third condition coming to mind along the lines of those just considered – a third 'alio' version of transitivity, that is – is as follows: for all x, y, z, if xRy and yRz then $y = z$ or xRz. (Move the dotted equality link in Figure 2.9 to the base of the triangle.) Is this genuinely a weakening of transitivity or is it, like the case treated just now, only spuriously so? (Give reasons.) ✠

We return to aliotransitivity in a moment, after exhibiting another pair of spurious 'alio' conditions:

EXAMPLE 2.5.45 Recall the condition of range-reflexivity: $\forall x \forall y (Rxy \to Ryy)$, and consider the following 'alio' variant:

$$\forall x \forall y ((Rxy \land x \neq y) \to Ryy).$$

Evidently this is equivalent to range-reflexivity given the corresponding condition with "$x = y$" in place of "$x \neq y$". But we are indeed given this on purely logical grounds – by the logic of "=": if $x = y$ and Rxy, then of course Ryy. Although the new condition is accordingly a spurious 'alio' condition, we can write down a modal formula which most directly invokes the formulation inset above: either the one-variable form $p \to \Box(\Box p \to p)$ or the two-variable formula $p \to \Box(\Box q \to (p \lor q))$; here "invokes" means that either of these **K**-interdeducible formulas[89] – the latter being the more convenient – is most directly helpful in a canonical model completeness proof for any consistent normal modal logic containing them, to show that the (admittedly spurious) alio range-reflexivity condition is satisfied. This raises the question of a syntactic derivation of $\Box(\Box p \to p)$ (alias **U**) in $\mathbf{K} \oplus p \to \Box(\Box p \to p)$ or in $\mathbf{K} \oplus p \to \Box(\Box q \to (p \lor q))$. Here is a sketch of a derivation in the latter case, though we stop with **U** in the form $\Box(\Box q \to q)$, for convenience:

First re-write the new axiom, $p \to \Box(\Box q \to (p \lor q))$, reformulating what is in the scope of the $\Box$ on the consequent to a truth-functionally equivalent form:

$$p \to \Box((\Box q \to q) \lor p)).$$

Now substitute $\Box q \to q$ for p, giving:

$$(\Box q \to q) \to \Box((\Box q \to q) \lor (\Box q \to q))).$$

Simplifying the consequent, we have: $(\Box q \to q) \to \Box(\Box q \to q)$, which has the truth-functional consequence $\neg \Box q \to \Box(\Box q \to q)$. But already in **K** the formula $\Box q \to \Box(\Box q \to q)$ is provable, so from these two together, we draw the truth-functional consequence $\Box(\Box q \to q)$. ◀

EXAMPLE 2.5.46 The same considerations show that the following 'alio' variation on domain reflexivity:

$$\forall x \forall y ((Rxy \land x \neq y) \to Rxx),$$

is also spurious, and again the modal formula most directly inspired by this formulation, namely the following 'Halldén unreasonable' disjunction (to use some terminology officially introduced in note 102, p. 120 below):

$$(\Box p \to p) \lor (q \to \Box q)$$

is not quite what first comes to mind for domain-reflexivity as usually formulates, which is perhaps $\Diamond \top \to (\Box p \to p)$, or $(\Box p \to p) \lor \Box \bot$, though it is quite similarly to the following variation of the latter formula (with which the latter formula is evidently **K**-interdeducible:

$$(\Box p \to p) \lor \Box q.$$

Obviously we can derive the first formula inset above from this second one by a truth-functional weakening of its second disjunct. The converse derivation is left to the following Exercise. ◀

---

[89]*Interdeducible* as defined in note 28, p. 34.

2.5. FROM MODELS TO FRAMES

EXERCISE 2.5.47 Show syntactically (sketching an axiomatic proof) that $(\Box p \to p) \vee \Box q$ is provable in $\mathbf{K} \oplus (\Box p \to p) \vee (q \to \Box q)$. ✠

We return to the non-spurious condition of aliotransitivity. Although an aliotransitive relation need not be irreflexive, aliotransitivity is most useful as a substitute for transitivity in the case of symmetric relations which are irreflexive, such as the *sibling* relation amongst people. (What is intended is the relation of being a full sibling, i.e., not counting half-brothers or half-sisters.) If you have one older brother and one younger sister, you have two siblings rather than three: the fact that the brother is your sibling and vice versa does not imply that you are also your own sibling. Thus the relation of being a sibling is aliotransitive. Let us call a relation which is aliotransitive, irreflexive and symmetric, an *alio-equivalence relation*. The sibling relation meets all three conditions and so constitutes a simple example of an alio-equivalence relation, as does the relation of non-identity (on which more below).[90]

A one-to-one correspondence between partitions (in the standard sense of the word, as given in the previous Digression) of a set X and alio-equivalence relations on X arises as follows. Given an alio-equivalence relation $R \subseteq X \times X$, the blocks of the induced partition are the alio-equivalence classes $R(x)$ for $x \in X$; given a partition of X, define xRy iff $x \neq y$ and x, y lie in some common block of the partition.

EXERCISE 2.5.48 (*i*) Show that whether we start with an alio-equivalence relation or a partition, twice applying the procedure just described returns us to the original relation or partition.

(*ii*) How are the equivalence relation and the alio-equivalence relation, corresponding to a given partition, related to each other? ✠

REMARK 2.5.49 Before leaving Harary, mentioned in note 90, too far behind, it is worth observing that the condition of 'distinct transitivity' there mentioned is what would be expressed by the usual formulation (as on p. 14) if this formulation were to be read with the exclusive interpretation of the quantifiers described in note 53 on p. 68. One might argue that this is the only available interpretation for certain vernacular formulations taken to express transitivity, such as "If one thing bears the relation R to a second thing, which bears R to a third, then the first thing bears R to the third", on the grounds that the third thing wouldn't be a third thing if it were just the first thing over again (and similarly with the first and second, and second and third, though as observed in note 90, the non-identities here are inessential). This is of course what we have been calling aliotransitivity. A related reluctance to trade in talk of transitivity for any such alternative terminology seems to motivate Pieter Seuren, who, in §4 of [1040], as well as in [1042]. In [1042], Seuren notes that all relations which are both symmetric and transitive are field-reflexive in the sense of note 81 – for which Seuren just uses the word "reflexive".[91] When relations such as the *sibling of* and *colleague of* relations whose irreflexivity seems to present an obstacle to calling them transitive, since they are certainly symmetric, rather than resort, as here, to the Peirce-derived terminology of aliotransitivity, Seuren writes as follows ([1040], p. 132):

> We prefer to keep the attribution of transitivity to the class of predicates of equality, such as *colleague*. But in order to do that we need to formulate a general non-mathematical principle which automatically ensures that the reflexive 'exceptions' are explained.

---

[90]The siblinghood example of aliotransitivity is given in Harary [452], where aliotransitive relations are described as 'distinctly transitive' and alio-equivalence relations are called parity relations. The definition given of R's being distinctly transitive is that for any three distinct x, y, z, if xRy and yRz then xRz. Unpacking the distinctness condition reveals three non-identity conditions: $x \neq y$, $y \neq z$ and $x \neq z$, only the third of which is doing any work, for which reason this way of proceeding seems less than ideal. (The other two are, as we have been putting it, 'spurious' weakenings of the formulation obtained by omitting them.) For more on the connection between equivalence relations, alio-equivalence relations, and partitions, see Green [406]. Further variations on the theme may be found in Hazen and Humberstone [464] and Nowak [848], the latter mentioned again below in note 409, p. 462.

[91]In [1040], Seuren defines reflexivity in the usual way and so records this fact in different terms: namely, as the fact that that when the domain is the field of a relation – as he puts it (p. 130), the relation "affects" every element of the domain – of a transitive symmetric relation, that relation is reflexive.

The principle in question is: "The n terms of an n-ary predicate must have distinct denotations," which is supplemented by an insistence that a would-be predicate violating this constraint is to be replaced by a suitably reflexivized predicate of arity $< n$. (The word *denotation* in this principle has a somewhat specialized sense for Seuren but present purposes do not require an elaboration of the subtleties involved.) With such a principle assumed to be in force, and counting individual variables as being terms for applying the principle, the upshot is that the usual formulation of transitivity will then amount to a condition of distinct transitivity *à la* Harary (and thus, equivalently, to aliotransitivity), and this condition together with symmetry is indeed consistent with a relation's not being reflexive or even field-reflexive. For further motivation and repercussions, the interested reader is referred to Seuren's works. Somewhat similar ideas, also touching (as do Seuren's) on Russell's Paradox and the (so-called) Barber Paradox – already a theme in Hintikka [488] (§8) – an be found in Slater [1060], [1061], and other papers by the same author. Slater, by contrast with Seuren, explicitly denies the existence of reflexivized predicates, registering this denial in second-order terms thus ([1060], p. 28): $\neg \forall R \exists P \forall x (Rxx \leftrightarrow Px)$. ◀

Let us recall Hughes and Cresswell's notion of an *NI* frame as a frame $\langle W, R \rangle$ in which for all $x, y \in W$ we have Rxy if and only if $x \neq y$, for the sake of the following multiple completeness result of **KB4′**, due to Segerberg:

THEOREM 2.5.50 *The logic* **KB4′** *is determined* (i) *by the class of symmetric aliotransitive frames, as well as* (ii) *by the class alio-equivalence relational frames, and also* (iii) *by the class of NI frames.*

We do not give a full proof of this result here. For (i), a simple canonical model argument suffices; the transition from (i) to (ii), which adds irreflexivity, is given by a p-morphism argument (Revision Exercise 2.7.35, p. 137: see Segerberg [1029], [1030], for the present application); (iii) follows from (ii) by appeal to the Generation Theorem (Thm. 2.4.10, p. 70), since point-generated subframes of alio-equivalence relational frames are *NI* frames.

Thus we now know which modal logics are determined by all four of the classes of frames whose accessibility relations satisfy the uniquely satisfiable conditions, mentioned above, of being universal (**S5**), being empty (**KVer**), being the identity relation (**KT!**), and being the complement[92] of the identity relation (**KB4′**).

EXERCISE 2.5.51 (i) Show that (as mentioned on p. 92) the class of *NI* frames is not modally definable. (ii) Say that a normal monomodal logic S *uniquely characterizes* $\Box$, if the smallest normal bimodal logic extending S_1 and S_2 proves $\Box_1 p \leftrightarrow \Box_2 p$, where S_i is like S except in having $\Box_i$ (and $\Diamond_i$) where theorems of S have $\Box$ (and $\Diamond$), for $i = 1, 2$. Which of the logics **S5**, **KVer**, **KT!**, and **KB4′** uniquely characterize $\Box$? (Give reasons.) ✠

Most of the remainder of this section will be organized around further odds and ends falling under the "alio-" rubric.

Figure 2.10: *One Spurious and One Genuine Alio-Euclidean Condition*

---

[92]The complement relative to the class of points in the frame, that is.

2.5. FROM MODELS TO FRAMES

EXERCISE 2.5.52 (*i*) Write out the conditions directly represented by the two diagrams in Figure 2.10, and explain why they count as spurious and genuine, respectively. (The logic determined by the class of all frames satisfying the genuine condition – that on the right – is **K.4**, where the schema **.4**, listed in Table 2.1 on p. 33, which is sometimes called **R** in the literature, is best known from its presence in the axiomatization of **S4.4** suggested by that label: see Proposition 5.5.16, p. 413.)

(*ii*) Write out the conditions directly represented by the diagrams in Figure 2.11, the first of which gives yet another 'alio-' variation on the euclidean theme, and say whether the broken 'equality' edges are spurious (i.e., can be omitted to yield representations of conditions logically equivalent to the originals). ✠

Figure 2.11: Further 'Alio' Condition Diagrams

In fact the diagram on the right of Figure 2.11 depicts directly the condition of aliosymmetry, which, as mentioned in note 88, is equivalent to symmetry, so half of Exc. 2.5.52(*ii*) has already been addressed, though we did not go through the reasoning as to why aliosymmetry implies symmetry. (The converse implication is more obvious. We return to the diagram on the left of Figure 2.11 later, after Exercise 2.5.60.) The non-identity in the aliosymmetry condition – i.e., the "$x \neq y$" in "$\forall x \forall y ((Rxy \land x \neq y) \to Ryx)$" – is directly invoked in any attempt to show that either of the following formulas is valid on any frame meeting that condition:

$$\Diamond(p \land \Box p) \to p \qquad \Diamond(p \land \Box q) \to (p \lor q),$$

and either of these formulas can be derived from the other using the rules closure under which is definitive of being a normal modal logic: substitute $p \lor q$ for p to begin the derivation of the second from the first, for example.

EXERCISE 2.5.53 Follow up the above suggestion to complete the demonstration that $\mathbf{K} \oplus \Diamond(p \land \Box p) \to p$ = $\mathbf{K} \oplus \Diamond(p \land \Box q) \to (p \lor q)$. ✠

Of course this point could equally be made with the dual forms of these two formulas, which bear a close resemblance to **B** (while the formulas themselves are similar to the dual of **B**):

$$p \to \Box(p \lor \Diamond p) \qquad (p \land q) \to \Box(p \lor \Diamond q),$$

but we continue the discussion in terms of the formulas given initially. (In dual form, what follows appears in pp. 10–11 of Humberstone [588].) While it is easy to see how to derive these formulas from (the dual form of) **B**, the converse derivation is not at all obvious. We begin with the two-variable form for convenience:

(1)	$\Diamond(p \land \Box q) \to (p \lor q)$	Given
(2)	$\Diamond(\Box p \land \Box p) \to (\Box p \lor p)$	1, US
(3)	$\Box p \to (\Box p \land \Box p)$	TF
(4)	$\Diamond \Box p \to \Diamond(\Box p \land \Box p)$	3, $\Diamond$Mono
(5)	$\Diamond \Box p \to (\Box p \lor p)$	2, 4 TF
(6)	$(\Diamond p \land \Box q) \to \Diamond(p \land q)$	**K**-provable (dual of 1.3.16)
(7)	$(\Diamond \Box p \land \Box p) \to \Diamond(\Box p \land p)$	6, US
(8)	$\Diamond(\Box p \land p) \to p$	1-variable form of (1)
(9)	$(\Diamond \Box p \land \Box p) \to p$	7, 8 TF
(10)	$\Diamond \Box p \to p$	5, 9 TF

Thus we have derived the dual form of the **B** axiom from our 'aliosymmetry' principle (line (1) above). Aside from its intrinsic interest, we can use this to provide a syntactic proof of a result in Kowalski [664], there proved using the algebraic semantics for modal logic (touched on only tangentially here – for example after Exercise 2.10.3, p. 155). Kowalski is confirming a conjecture of J. Perzanowski, in a paper published in Polish, that:

$$\mathbf{K} \oplus \Box(p \to \Box p) \leftrightarrow \Box(\Diamond p \to p) = \mathbf{KB}^\Box 4,$$

where $\mathbf{B}^\Box$ is $\Box(p \to \Box \Diamond p)$ (the necessitation of the **B** axiom: see note 30, p. 37).

The first thing to notice about Perzanowski's axiom, the formula on the left here, is that instead of taking it as a biconditional, we could equally well take it simply as a one-way conditional, either in the $\to$ direction or in the $\leftarrow$ direction, since each of these added to **K** has the other deducible from it.

EXERCISE 2.5.54 Prove the claim just made, to the effect that $\mathbf{K} \oplus \Box(p \to \Box p) \to \Box(\Diamond p \to p)$ and $\mathbf{K} \oplus \Box(\Diamond p \to p) \to \Box(p \to \Box p)$ are the same logic. ✠

Digression. The equivalence of the two one-directional forms of Perzanowski's axiom involves essentially the same reason – basic duality moves (see Section 2.1) – as that given on the first page of Chellas and Segerberg [172], to the effect that for a broad range of modal logics (certainly including all the normal modal logics) closure under either of these rules implies closure under the other:

$$\frac{\Diamond A \to A}{A \to \Box A} \qquad \frac{A \to \Box A}{\Diamond A \to A}$$

Because of some examples given by J. J. MacIntosh (in [754] and elsewhere), Chellas and Segerberg call logics closed under either of these rules (and therefore under both) *MacIntosh logics*. The examples in question provide some motivation for Perzanowski's "horizontalized" version of the MacIntosh rule(s). For further discussion of the latter rule, see Williamson [1184]; the rule appeared (essentially) earlier, in Coro. 2 on p. 96 of Williamson [1181], in the equivalence between (a) and (c) there for a range of logics. Note that if we trade in the single schematic letter in these rules for a pair of letters, giving:

$$\frac{\Diamond A \to B}{A \to \Box B} \qquad \frac{A \to \Box B}{\Diamond A \to B}$$

then we again have interderivable rules, either of which is equivalent (in a normal modal logic) to the provability of **B**; from the semantic characterizations of the original MacIntosh rules supplied in the papers already referred to, one can see that these 'variegated' or two-letter rules cannot be derived from them, since symmetry does not follow from the condition that for all x, y, if xRy then y starts an R-chain which returns to x. Transitivity would enable any such chain to shrink to one of length 1, securing symmetry, so one expects that any extension of **K4** closed under the MacIntosh rule(s) will be closed under the two-letter versions, or equivalently, will contain **B**. The most straightforward syntactic confirmation of

2.5. FROM MODELS TO FRAMES

this expectation might try and begin with **4** in the form $\Box p \to \Box\Box p$, say, and then apply the MacIntosh rule on the right above: but this (taking A as $\Box p$) delivers **5**, $\Diamond\Box p \to \Box p$, rather than **B** itself. More deviously, however, if one begins with the **4**-derivable

$$(\Box p \land p) \to \Box(\Box p \land p),$$

and now applies the same rule we have as our conclusion $\Diamond(\Box p \land p) \to (\Box p \land p)$, so discarding the first conjunct of the consequent we have a proof of $\Diamond(\Box p \land p) \to p$. From we know via Exercise 2.5.53 and (1)–(10) following it, that we can derived **B**. *End of Digression.*

Kowalski [664] notes that Perzanowski had already reduced the problem of showing that the smallest normal modal logic containing Perzanowski's axiom was $\mathbf{KB}^\Box\mathbf{4}$, to the problem of showing that the former logic proves $\mathbf{B}^\Box$, leaving that as the problem [664] sets about solving. But let us look at some of the background issues first. To check that Perzanowski's logic is included in $\mathbf{KB}^\Box\mathbf{4}$, one can undertake a semantic analysis (in terms of Kripke models), though we shall not do that here, proceeding instead syntactically to show that Perzanowski's axiom, in the form $\Box(p \to \Box p) \to \Box(\Diamond p \to p)$ for definiteness, is provable in $\mathbf{KB}^\Box\mathbf{4}$. First note that already in $\mathbf{K}$ we have the following provable:

$$\Box(p \to q) \to (\Diamond p \to \Diamond q),$$

and so by Uniform Substitution and $\Box$Mono, the following is also $\mathbf{K}$-provable:

$$\Box\Box(p \to \Box p) \to \Box(\Diamond p \to \Diamond\Box p).$$

By **4** we can therefore simplify the antecedent and so in $\mathbf{K4}$ prove:

$$\Box(p \to \Box p) \to \Box(\Diamond p \to \Diamond\Box p).$$

But using $\mathbf{B}^\Box$, we can simplify the consequent in $\mathbf{KB}^\Box\mathbf{4}$ thus:

$$\Box(p \to \Box p) \to \Box(\Diamond p \to p),$$

and we are done.

EXERCISE 2.5.55 Spell out this last step in detail. ✠

It remains to show that $\mathbf{KB}^\Box\mathbf{4}$ is included in Perzanowski's logic, which means we must show that **4** and $\mathbf{B}^\Box$ are derivable from Perzanowski's axiom. The first of these is easy, especially if we take the other form of the latter axiom:

$$\Box(\Diamond p \to p) \to \Box(p \to \Box p).$$

$\Box p$ provably implies the antecedent of the above conditional in $\mathbf{K}$, by TF and $\Box$Mono, and the consequent of the above conditional provably implies $\Box p \to \Box\Box p$ in $\mathbf{K}$, so putting these together, Perzanowski's axiom yields $\Box p \to (\Box p \to \Box\Box p)$ which by TF we can simplify ("contract") to $\Box p \to \Box\Box p$.

This leaves $\mathbf{B}^\Box$ to derive from Perzanowski's axiom, a task sufficiently unstraightforward that Kowalski [664] supplies an algebraic proof that such a syntactic derivation exists rather than exhibiting one. A first thought might be to begin with the Perzanowski axiom

$$\Box(p \to \Box p) \to \Box(\Diamond p \to p),$$

and substitute $\Box p$ for p, giving an implication with a provable antecedent (since we have already seen that **4** – and therefore its necessitation – is derivable from Perzanowski's axiom), which allows us to detach its consequent. But the latter is then $\Box(\Diamond\Box p \to \Box p)$, a necessitated version of (the dual form of) **5** rather than of **B**: we want this, but with the rightmost "$\Box$" missing. Accordingly we make a more judicious substitution,[93] putting $p \land \Box p$ for p instead, as in the Digression above, which gives:

---

[93]Suggested, in fact, by the algebraic reasoning in [664].

$$\Box((p \wedge \Box p) \to \Box(p \wedge \Box p)) \to \Box(\Diamond(p \wedge \Box p) \to (p \wedge \Box p)).$$

Notice that the antecedent of this implication is also **K4**- and therefore **KB**$^{\Box}$**4**-provable, delivering the consequent, which we can further weaken (by $\Box$Mono and TF) by dropping a conjunct, getting

$$\Box(\Diamond(p \wedge \Box p) \to p).$$

Looking at the formula in the scope of the main "$\Box$" here, you will see that this is where we came in: the derivation (1)–(10) showed how to derive $\Diamond\Box p \to p$ from $\Diamond(p \wedge \Box p) \to p$, near enough: we actually started with the two-variable aliosymmetry axiom $\Diamond(p \wedge \Box q) \to (p \vee q)$, rather than with the one-variable form $\Diamond(p \wedge \Box p) \to p$ – but we already noted the interderivability of these forms in any normal modal logic.

At this point rather than writing out the desired derivation, we note that if B is derivable using the rules (and axioms) definitive of normal modal logics (Modus Ponens, Necessitation, etc.) from A, then $\Box B$ is similarly derivable from $\Box A$. This does not mean that we can simply prefix a $\Box$ to each step of (1)–(10), since, amongst other things, we have expedited matters by using derived rules without bothering to baptize the corresponding derived rules working in the scope of $\Box$: for instance, appeals to $\Box$Mono and, more to the point since this rule is applied at line (4) in that derivation, $\Diamond$Mono, would need to be replaced by appeals to the companion rules:

$$\dfrac{\Box(A \to B)}{\Box(\Box A \to \Box B)} \quad \text{and} \quad \dfrac{\Box(A \to B)}{\Box(\Diamond A \to \Diamond B)}$$

It is clear enough that any normal modal logic is closed under such rules, though. (The first rule made an appearance in Example 2.1.18 *(ii)*, p. 46. It has been called Becker's Rule – see Hughes and Cresswell [535], note 210 – after Oskar Becker, who remarked on the derivability of a similar rule with the two $\Box$s introduced in the conclusion replaced by two occurrences of any affirmative modality, thereby subsuming, *inter alia*, the second rule.) We leave the general assertion made here without proof as Proposition 2.5.56, and its application to the present case as part of the exercise below.

PROPOSITION 2.5.56 *Let S be any normal modal logic and A, B, be any formulas. If $\vdash_{S \oplus A} B$, then $\vdash_{S \oplus \Box A} \Box B$.*

EXERCISE 2.5.57 *(i) Does Proposition 2.5.56 continue to hold if the single formula A is replaced by a reference to a set of formulas Γ (and the corresponding set $\{\Box C \mid C \in \Gamma\}$? (Give reasons or a counterexample.)*
(ii) Would the following variation on Proposition 2.5.56 be correct? If $\vdash_{S \oplus A} B$, then $\vdash_{S \oplus \Diamond A} \Diamond B$. (Give reasons or else a counterexample.)
(iii) Provide an explicit syntactic derivation of $\Box(\Diamond\Box p \to p)$ from $\Box(\Diamond(p \wedge \Box p) \to p)$. ✠

Before we leave Perzanowski's axiom in the form $\Box(p \to \Box p) \to \Box(\Diamond p \to p)$ for good, some remarks on its philosophical interest and connections with other issues are in order. Its antecedent, $\Box(p \to \Box p)$ says that it cannot be contingently true that p. (We could have written it, instead, as $\neg\Diamond(p \wedge \Diamond\neg p)$.) Let us compare this with the analogous statement that it cannot be contingently false that p: $\Box(\neg p \to \Box\neg p)$. Contraposing this, we arrive at the consequent of Perzanowski's axiom. So the axiom itself says that whatever is incapable of being contingently true is incapable of being contingently false. Both antecedent and consequent are – one might say *persistence-related*[94] – notions of noncontingency differing from that provided by the standard way of defining noncontingency $\Box p \vee \Box\neg p$ (alias Δp: see Exc. 1.3.8, p. 21) in numerous modal logics intended for alethic applications, such as **S4**, though equivalent to it in **S5**. There is nothing to choose between the formulas stating that either of these new forms of noncontingency imply the traditional form:

---

[94] See Exc. 2.7.14 (p. 132) and note 426 (p. 483).

2.5. FROM MODELS TO FRAMES

EXERCISE 2.5.58 Show that $\mathbf{K} \oplus \Box(p \to \Box p) \to (\Box p \vee \Box \neg p) = \mathbf{K} \oplus \Box(\neg p \to \Box \neg p) \to (\Box p \vee \Box \neg p)$. (Here we could equally well have put "+" for "$\oplus$".) ✠

Rewriting $\Box p \vee \Box \neg p$ as $\Diamond p \to \Box p$, the first formula appearing in the above exercise is (1) here, listed alongside a variation, (2):

$$\Box(p \to \Box p) \to (\Diamond p \to \Box p) \qquad (1)$$

$$\Box(p \to \Box p) \to (\Diamond p \to p) \qquad (2)$$

Because for any formulas A and B, $\Box A \to B$ and $\Box A \to \Box B$ are **S4**-interdeducible – and indeed **S4** is precisely the weakest normal modal logic for which that is so – if we worked with **S4** as our background logic for $\Box$ there would be nothing to choose between (2) and Perzanowski's axiom,[95] and the same goes for (1) and the result of replacing its consequent with $\Box(\Diamond p \to \Box p)$. Similarly (1) and (2) are themselves **S4**-interdeducible, since each yields **5** on addition to **S4**: substitute $\Box p$ for p to see this (and make a further appeal to **T** in the case of (1)). This shows us that neither (1) nor (2) is **S4**-provable, though both are **S5**-provable. As Prior [922], p. 201, remarks, this is one of those rare occasions on which a choice between these two candidate alethic modal logics makes a difference to the validity of an argument outside of the area of modal metaphysics itself:

EXAMPLE 2.5.59 The area in question is what is sometimes called the modal ontological argument. Prior uses (1) above to illustrate the point but here (as in Humberstone [559], p. 199f. where these principles are discussed with an epistemic reading of $\Box$) we may use (2): take p as "God exists". Then the two premises for the argument are given by $\Box(p \to \Box p)$, felt to be grounded in the concept of God as the kind of thing which, if it existed would exist necessarily, and $\Diamond p$ – surely it is not an impossibility, even if it is not in fact the case, that God should exist. The conclusion, by two applications of Modus Ponens (or more accurately of the natural deduction rule ($\to$E): see Section 7.1) from (2) is that God does indeed exist. The argument is invalid if the underlying modal logic is **S4** but valid if we take the underlying logic as **S5** (whatever one might think about the plausibility of the premises). A similar line of reasoning was popularized at roughly the same time in Hartshorne [460], especially pp. 51–53 (or see [459]), though exactly what Hartshorne had in mind has been subject to some speculation and debate, of which two samples are Kapitan [638], bringing in modal predicate logic, as one may suspect is called for in view of the role played in the argument by the notion of existence, and Baker [49], which supplies additional further bibliographical references on this topic; likewise Chapter 4 of Viney [1132], as also does Chapter 3 of Sobel [1071]. The existence issue arises particularly because whereas in the version discussed by Prior we take p as "God exists", in Hartshorne's version it is "There exists some perfect being", with the premiss $\Box(p \to \Box p)$ taken to reflect an aspect of the notion of perfection involved: Hartshorne calls this 'Anselm's Principle'. See also Purtill [934], where Hartshorne's argument is cast into roughly the Prior-style format (invoking Perzanowski's axiom – (1) above with a $\Box$ on the consequent, that is), and Kane [636]. See also Rosenberg [967]. A quick guide to the broader field of ontological arguments in general can be found in Oppy [858] and Chapter 2 of Sobel [1071]. ◂

Digression. Brown [120] displays a misunderstanding of the modal ontological argument, thinking that we are to take $\Box(p \to \Box p)$ (or $p \dashv \neg\Diamond\neg p$, as he formulates it – though in fact writing "$\to$" for the strict (i.e., necessary) implication connective "$\dashv$", since he uses the notation "$\supset$" for material implication) as a kind of proposed modal axiom, in which $\neg p$ can be substituted for p in accordance with US, despite noticing that "'Anselm's Principle' is highly dubious as a postulate". But no proponent of the argument has thought of the principle as anything other than a *premiss* in a particular argument, in the formal

---

[95] In the course of presenting a simple Kripke-incomplete normal modal logic in [82], R. Benton uses (2) – or a formula truth-functionally equivalent to (2) – as one of his axioms, under the name **L** (well, in fact just L), to suggest 'loop'. See further Example 2.5.64(*i*) below.

rendering of which p is a fixed sentence letter rather than as a propositional variable open to substitution. It may help to make this point by reference to the local consequence relation associated with **S5** (see note 115, p. 146), for which we write $\vdash_{\mathbf{S5}}$: then we are not making the 'vertical' transition from $\vdash_{\mathbf{S5}} \Box(p \to \Box p)$ to $\vdash_{\mathbf{S5}} \Box(\Diamond p \to p)$ by appeal to US, but rather the horizontal transition: $\Box(p \to \Box p) \vdash_{\mathbf{S5}} \Box(\Diamond p \to p)$ by appeal to the special strength of **S5** as contrasted with **S4**. There is no suggestion of providing any plausible extension of **S5** according to which $\Box(p \to \Box p)$ is provable outright (or: extension of $\vdash_{\mathbf{S5}}$ according to which the latter formula is a consequence of $\varnothing$).

Also somewhat off track is Odegaard [854], whose opening moves are reproduced *symbolatim* here:

Let 'p' stand for 'God exists' and consider the argument:

1. $p \supset \Box p$ Prem.
2. $\sim p \supset \Box \sim p$ Prem.
3. $\sim \Box \sim p$ Prem.
4. $\sim\sim p$ 2, 3, M.T.
5. p 4, D.N.

This is meant to be a kind of natural deduction proof, with "M.T." and "D.N." indicating Modus Tollens and Double Negation; the latter is essentially the rule ($\neg\neg$E) given in Section 7.1 below. The two most conspicuous features of the above derivation are (1) the fact that there is no appeal to anything specific to **S5** – and indeed not even to much of **K**, since the argument goes through in the smallest modal logic, the conclusion being a tautological consequence of the premises, and (2) that Premise 1 is not used at all! These two facts are connected, since it was in deriving a strict implication version of (2) from a strict form of (1) that the additional strength of **S5** over **S4** came in – though as we have been seeing in the last several pages, what is special about **S5** here is simply that it extends $\mathbf{KB}^\Box \mathbf{4}$. (We will also need **T** to pass from the strict version to the material version, of course.) Thus Odegaard seems to have the right ingredients for the modal ontological argument but to have forgotten the recipe: a version of (1) is assumed to be more compelling, at first blush, than (2), and (2) is not to be taken as a *premise* at all. **End of Digression.**

The formulas (1) and (2) given after Exercise 2.5.58 above can of course be considered without the specific background of **S4**, with such questions as what classes of frames they (modally) define and whether the normal modal logics they give whether when taken as the sole additional axioms, or in conjunction with others, are Kripke-complete. Pertinent comments on both issues as they arise for (2) may be found in van Benthem [70] and [67].

It is a commonplace that in general weaker logics provide for more distinctions (though for some qualifications, see Humberstone [581]) and we have in effect just seen this illustrated in the case of **S4** and **S5**. Abbreviating our positive and negative persistence formulas to $\triangle^+ A$ and $\triangle^- A$, respectively – i.e., putting $\triangle^+ A = \Box(A \to \Box A)$ and $\triangle^- A = \Box(\neg A \to \Box \neg A)$ (or equivalently $= \Box(\Diamond A \to A)$), we have three notions of noncontingency all distinguished in **S4**, namely $\triangle p$ (or $\Box p \lor \Box \neg p$), $\triangle^+ p$ and $\triangle^- p$, all of which are equivalent in **S5**. Not that there are no logical relations among these formulas in **S4**:

EXERCISE 2.5.60 Show that $\vdash_{\mathbf{S4}} (\triangle^+ p \land \triangle^- p) \leftrightarrow \triangle p$.   ✠

REMARK 2.5.61 The upshot of our recent discussion – of Kowalski's confirmation of Perzanowski's conjecture – has been that the smallest normal modal logic in which $\triangle^+ p$ and $\triangle^- p$ are provably equivalent is the logic $\mathbf{KB}^\Box \mathbf{4}$. Writing out this equivalence in terms of $\Box$, it appears as:

$$\Box(p \to \Box p) \leftrightarrow \Box(\neg p \to \Box \neg p),$$

or alternatively (as already remarked):

$$\Box(p \to \Box p) \leftrightarrow \Box(\Diamond p \to p).$$

2.5. FROM MODELS TO FRAMES

This shows incidentally, that $\mathbf{KB}^{\square}4$ does not admit the rule of Cancellation (see Exercise 2.6.31, p. 126), which would have us remove the occurrences of $\square$ on the two immediate subformulas of this formula, thus yielding

$$(p \to \square p) \leftrightarrow (\Diamond p \to p).$$

This last formula is provably equivalent by truth-functional logic to the result of putting "$\wedge$" for its "$\leftrightarrow$" since the disjunction of the formulas flanking the "$\leftrightarrow$" is a substitution instance of the tautology $(p \to q) \vee (r \to p)$. So on no Boolean valuation are both formulas false, meaning that the only way for the biconditional to be true (on such a valuation) is for them both to be true. These two formulas themselves are just versions of $\mathbf{T}_c$, and so certainly not provable in $\mathbf{KB}^{\square}4$, a sublogic of $\mathbf{S5}$. ◄

EXERCISE 2.5.62 (*i*) Concentrating on the one-directional version of the second inset formula in the above remark as Perzanowski's axiom:

$$\square(p \to \square p) \to \square(\Diamond p \to p),$$

show that the smallest normal modal logic containing this formula – which we have seen to be none other than $\mathbf{KB}^{\square}4$ – is the same as the smallest normal modal logic containing the (apparently) more general formula:

$$\square(p \to \square q) \to \square(\Diamond p \to q).$$

(It may help to do part (*ii*) of this Exercise.) The point here could of course equally well be made with schemata $\square(A \to \square A) \to \square(\Diamond A \to A)$ and $\square(A \to \square B) \to \square(\Diamond A \to B)$: as axiom schemata, these yield the same normal extension of $\mathbf{K}$. Contrast the case of the rules $A \to \square A \,/\, \Diamond A \to A$ and $A \to \square B \,/\, \Diamond A \to B$, where the additional strength of the latter rule was noted in the Digression addressing the MacIntosh rule(s) on p. 98.

(*ii*) Show that $\mathbf{KB}^{\square}4$ is determined the class of frames $\langle W, R \rangle$ satisfying the condition that for all $x, y, z \in W$, if $xRyRz$ then $xRzRy$. (Note, incidentally, that because the consequent here is conjunctive – xRz & xRy – this condition cannot be codified in a single diagram of the kind we have been using, most recently on the left side or right side of Figure 2.11 on p. 97, but requires a pair of diagrams.) ✠

In terms of the discussion in Humberstone [559] the equivalence in Exercise 2.5.60 would be put – giving $\square$ an epistemic reading – by saying that knowing one has a positive recognitional capacity in respect of whether p (i.e., knowing that, if p, then one knows that p) combined with knowing that one has a negative recognitional capacity in this respect (knowing that if not-p, then one knows that not-p) is equivalent to knowing whether or not p. Let us pursue this a little further:

EXAMPLE 2.5.63 Most of the discussion in [559] concerns whether known possession of a positive recognitional capacity implies possession of the corresponding negative capacity; saying someone with a positive and a negative recognitional capacity w.r.t. a property has a general recognitional capacity w.r.t. it, which amounts to being in a position to know, presented with an arbitrary item, whether or not it has the property in question, we can put this by asking whether known possession of a positive recognitional capacity implies possession of the corresponding general recognitional capacity. (Put in terms of properties, having a positive recognitional capacity is a matter of knowing of anything with the property concerned, that it has that property, and having a negative recognitional capacity is a matter of knowing of anything lacking the property concerned, that it is lacks the property. Evidently this involves some delicate matters of quantified epistemic logic not touched on in Chapter 5, where our concern is exclusively with propositional logic, as here.) Naturally one must not begin by conflating positive and general recognitional capacities. Bernard Williams [1174], Appendix 3, illustrated this several times over, perhaps most memorably with the case of anoxia, which because of its deleterious effects on cognitive function was not such as to be recognisable by pilots suffering it, which does not (on the face of it, at least) mean that pilots not suffering from anoxia could not tell that they not suffering from it. This would be a matter of possessing the positive recognitional capacity w.r.t. not being anoxic, without the corresponding negative (and hence the general) recognitional capacity. Williams's point was to flag as problematic Descartes' concluding that one can't, even if awake, tell that one is not dreaming, from the premiss that one can't, when dreaming, tell that one is dreaming. A condition for which the anyone satisfying the condition has a positive recognitional capacity for satisfying it is essentially the same as what Williamson (see Example 5.2.6, p. 377, below) calls a 'luminous' condition, so the following passage from the first page of Steup [1089] shows that vigilance is still required in this area:

> Obviously, non-mental conditions are not luminous. Consider the presence of milk in my refrigerator. My refrigerator's having milk in it is not a luminous condition, since I am not always in a position to know whether it obtains.

The remark exhibits a conflation of the distinction with which we have been concerned. It argues that a condition (of having milk in one's refrigerator) is not luminous, giving as grounds that one is not always in a position to know whether it obtains. The grounds say one lacks a certain general recognitional capacity, while what is inferred from these grounds is that one lacks the positive recognitional capacity involved; of course it is obvious without the need for giving any such grounds that the positive recognitional capacity is lacking, so the point here is simply to note the incidental *non sequitur*. It is similarly surprising, in view of examples such as those above, to find in Section 6 of Greenough [408] a favourable reference to

> the assumption that if a positive mental state C is luminous, then the corresponding negative state not-C is luminous too. This assumption is surely plausible.

(There follows a footnote citing others as being of a similar view.) More on *knowing whether* can be found in Fan, Wang, and van Ditmarsch [282], in the work of Akama, Murai, and Kudo, where *knowing whether* is described as *non-ignorance* or *nonignorance* – the latter in the publication we cite here: [7], and at the end of Example 5.1.2, which begins on p. 334, as well as the Digression following it. A relatively early observation on one point of technical interest for *knowing whether* as opposed to *knowing that* appeared in Hart, Heifetz and Samet [457]. ◀

We return from the decorated triangles of Exercise 2.5.60 to the notation of $\Box$ and $\Diamond$, for a few further remarks on (1) and (2) above – the latter being Perzanowski's axiom with a $\Box$ removed from its main consequent – repeated here for convenience:

$$(1) \quad \Box(p \to \Box p) \to (\Diamond p \to \Box p) \qquad (2) \quad \Box(p \to \Box p) \to (\Diamond p \to p)$$

It is noteworthy that (1) is valid on any frame satisfying one of our earlier alio-euclidean conditions, namely that depicted on the left of Figure 2.11 (p. 97). Answering the request in part (ii) of Exercise 2.5.52 there, to spell this condition out, we have: $\forall x \forall y \forall z ((Rxy \wedge Rxz \wedge y \neq z) \to Ryz)$, taken as a condition on frames $\langle W, R \rangle$. For take any point x on such a frame in a model, falsifying the consequent of (1): this means that p is contingent at x in the traditional sense, i.e., there exist $y, z \in R(x)$ with p true at y but not z. This means that $y \neq z$ so the current alio-euclidean condition kicks in and tells us that Ryz, making the antecedent of (1) false at x since y is a point accessible to x at which $p \to \Box p$ is false (in view of the falsity of p at y's successor z). Notice, however, that even if there were an R-chain of greater length (than 1) from y to z, say $u_1, \ldots, u_n$ with $u = u_1$, $z = u_n$, then as long as each of the intervening u_i also belongs to $R(x)$, the truth of p at u_i (starting with $i = 1$) will mean that $\Box p$ is true at u_i and so p true at u_{i+1} (because of the truth of $\Box(p \to \Box p)$ at x), until we eventually reach u_n, alias z, where p is supposedly not true: a contradiction.

EXAMPLES 2.5.64 (i) A similar but more thoroughgoing discussion of conditions under which – not (1) but the simpler formula (2) is valid on a frame can be found in van Benthem [70], p. 347f. q.v. for a

2.5. FROM MODELS TO FRAMES

demonstration that the frames validating (2) are precisely those $\langle W, R \rangle$ satisfying the following condition (quantifiers ranging over W, of course):

For all x, y if xRy then for some $n \geq 0$ there exist $z_1, \ldots, z_n$ such that (a) $yRz_1Rz_2 \ldots z_nRx$ and (b) $\{z_1, \ldots, z_n\} \subseteq R(x)$.

The "for some $n \geq 0$" amounts to an infinite disjunction of the cases $n = 0, n = 1, n = 2, \ldots$. Alternatively one can construe this as quantifying over R-chains (from y to x). Either way the condition is not, and is not equivalent to, a first-order condition. Van Benthem calls is a condition of Safe Return, perhaps meaning that for y directly accessible to x, there is an R-chain – the z_i – returning us from y to x safely under the supervision of x in the sense that each of the z_i is accessible to x. This is the condition that Benton (see note 95) referred to simply as the loop condition, though that label suggests more simply just the (a) part of the formulation above – an interesting condition in its own right. (Although Benton cites other publications of van Benthem in which somewhat related considerations are raised, he seems not to be aware that the Kripke incompleteness of the normal extension of **KTM** by (2) – with $\mathbf{T_c}$ and hence **T!** unprovable though valid on every frame for the logic – was already observed in van Benthem [70], including in the original 1984 publication of [70]. Benton [82] does, however, present some additional results of interest, such as a completeness proof for the non-canonical logic **KTM** itself.)

(*ii*) For its own sake, let us look at a special case of the Safe Return condition from (*i*) – or more accurately with a fixed choice of n in that condition in place of the existential quantification "for some n". (This will be a nice technical exercise though it will not have the theoretical interest of the Safe Return condition itself, since the class of frames on which it is valid will in any given case be first-order definable, and the logic determined by the class will be canonical.) If we choose $n = 0$ part (a) of the condition just becomes yRx and part (b) is satisfied vacuously: so these are just the symmetric frames and we know that that logic determined by this class of frames is **KB**. Accordingly let us choose $n = 1$. So our condition becomes:

For all x, y if xRy then for some there exists z such that (a) $yRzRx$ and (b) xRz.

Consider the following formula, somewhat reminiscent of (2) above, and given an *ad hoc* temporary label here:

(∗) $\Box(p \to \Box q) \to (\Diamond\Box p \to q)$

It is is easy to see that (∗) is valid on any frame meeting our $n = 1$ version of the Safe Return condition, and its exact form is suggested by the following canonical model completeness argument. Here $\langle W, R \rangle$ is the canonical frame of the smallest normal modal logic containing (∗). We want to show that the frame satisfies our condition, so suppose we have $x, y \in W$ with xRy. We need to find $z \in W$ such that xRz, yRz, and zRx, so we consider the consistency (relative to this logic) of the set of formulas:

$$\{A \mid \Box A \in x\} \cup \{B \mid \Box B \in y\} \cup \{\Diamond C \mid C \in x\},$$

since if this set is consistent, any maximal consistent extension of it will serve as the desired z. We derive a contradiction from supposing that the set is not consistent. That supposition would mean that

$\vdash \hat{A} \to (\hat{B} \to \Box\neg\hat{C})$

where $\vdash$ indicates provability in the current logic, $\hat{A}$ is $A_1 \wedge \ldots \wedge A_k$ for some $A_1, \ldots, A_k$ for which $\Box A_i \in x$ ($i = 1, \ldots, k$), $\hat{B}$ is $B_1 \wedge \ldots \wedge B_\ell$ for some $B_1, \ldots, B_\ell$ for which $\Box B_i \in y$ ($i = 1, \ldots, \ell$), and $\hat{C}$ is $C_1 \wedge \ldots \wedge C_m$ for some $C_1, \ldots, C_m \in x$. But in that case by normality:

$\vdash \Box\hat{A} \to \Box(\hat{B} \to \Box\neg\hat{C})$,

and the antecedent of this formula is an element of x, so $\Box(\hat{B} \to \Box\neg\hat{C}) \in x$. Substituting in our axiom (∗) for p and q gives us

$$\vdash \Box(\hat{B} \to \Box\neg\hat{C}) \to (\Diamond\Box\hat{B} \to \neg\hat{C}),$$

and since the antecedent of this formula belongs to x so does the consequent, $\Diamond\Box\hat{B} \to \neg\hat{C}$. But this is impossible, since $\Box\hat{B} \in y$ and xRy, so $\Diamond\Box\hat{B} \in x$, which would put $\neg\hat{C}$ into x: but $\hat{C} \in x$, since x is where $\hat{C}$'s conjuncts came from. This concludes the completeness proof. Pondering the $n = 2$ version of the safe return without any such argument in mind would perhaps lead one to consider a candidate axiom a little different from (∗), and of modal degree 3 rather than 2; we use different propositional variables so as to reduce confusion in relating the two candidate axioms:

(∗∗) $\qquad \Diamond\Box(r \vee \Box s) \to (\Diamond r \vee s)$

(∗∗) might come to mind, for example, as a suitable variant on the **B**-like $\Diamond\Box\Box s \to s$ which would axiomatize the class of frames satisfying the $n = 1$ Safe Return condition *minus* the (b) part of that condition (leaving just the condition: $xRy \Rightarrow yR^2x$). However, (∗) and (∗∗) are interdeducible over **K** in the sense that $\mathbf{K} \oplus (*) = \mathbf{K} \oplus (**)$ (indeed, $\mathbf{K} + (*) = \mathbf{K} + (**)$), as we now show in one direction, leaving the other as an exercise.

First, showing that (∗∗) can be deduced from (∗), by beginning with the latter's substitution instance (putting $r \to \Box s$ for p, and s for q):

$$\Box((r \to \Box s) \to \Box s) \to (\Diamond\Box(r \to \Box s) \to s)$$

In view of the truth-functional equivalence between $(r \to \Box s) \to \Box s$ and $r \vee \Box s$, we can rewrite this as:

$$\Box(r \vee \Box s) \to (\Diamond\Box(r \to \Box s) \to s).$$

Since the antecedent here is provably implied by $\Box r$ (a point which could already have been made without the disjunctive reformulation we have effected, indeed) by $\Box$Mono we have

$$\Box r \to (\Diamond\Box(r \to \Box s) \to s).$$

Finally, substituting $\neg r$ for r and permuting antecedents gives us:

$$\Diamond\Box(\neg r \to \Box s) \to (\Box\neg r \to s),$$

which can obviously be re-written as (∗∗). ◀

EXERCISE 2.5.65 (*i*) Show that, conversely, (∗) is provable in $\mathbf{K} \oplus (**)$, using a similarly informal (but nonetheless rigorous) argument. (*Suggestion*: it is best to transform (∗∗) into the form $\Box r \to (\Diamond\Box(r \to \Box s) \to s)$ to begin with, by substituting $\neg r$ for r in (∗∗) and making some easy truth-functional manipulations. Note that this is the second-last formula inset under Example 2.5.64(*ii*) above. From here, consider substituting $p \to \Box q$ for r and q for s. The reasoning is very similar to that used to derive (∗∗) from (∗).)

(*ii*) From the fact that $\mathbf{K} \oplus (*) = \mathbf{K} \oplus (**)$, it follows that (∗) and (∗∗) modally define the same class of frames. To show that this class comprises precisely the those satisfying the $n = 1$ Safe Return condition, it suffices (taking for granted the validity, already observed at the first appearance of (∗) in Example 2.5.64(*ii*) of (∗) on all such frames) to show that on any frame not $\langle W, R \rangle$ satisfying the condition, say because we have $x, y \in W$ with xRy but no $z \in R(y)$ such that $xRzRx$, there is a model falsifying (∗) at x. Show how to define V giving $\langle W, R, V \rangle$ as such a model. (*Suggestion*: put $V(p) = W \smallsetminus \{x\}$ and $V(q) = R(y)$, and show that (∗) is false at x in the resulting model.) ✣

The form of alio-euclideanness under discussion in the paragraph before Examples 2.5.64:

$$\forall x \forall y \forall z ((Rxy \wedge Rxz \wedge y \neq z) \to Ryz)$$

2.5. FROM MODELS TO FRAMES

can be compared with the result of putting $y = z$ for $y \neq z$ here, the upshot of which is the condition of range-reflexivity, modally defined by the formula **U** ($= \Box(\Box p \to p)$). Euclideanness itself is the conjunction of these two logically independent conditions, the alio-euclidean condition itself being modally defined by the formula on the left here, or alternatively, by its two-variable variant on the right:

$$\Diamond p \to \Box(p \vee \Diamond p) \qquad\qquad \Diamond(p \wedge q) \to \Box(p \vee \Diamond q).$$

Concentrating on the left-hand formula, it is easy to see that this will be provable in **K5**, whose significance is that $Fr(\mathbf{K5})$ consists of the euclidean frames, and it is also easy to see that $\mathbf{KU} \oplus \Diamond p \to \Box(p \vee \Diamond p) = \mathbf{K5}$, since from this last formula and **U** in the form $\Box(p \to \Diamond p)$ we can simplify $\Box(p \vee \Diamond p)$ to $\Box \Diamond p$. There is an interesting contrast with the case of aliotransitivity and transitivity:

EXAMPLE 2.5.66 Parallelling the development just seen, one might ask about a condition resembling aliotransitivity:

$$\forall x \forall y \forall z ((Rxy \wedge Ryz \wedge x \neq z) \to Rxz),$$

except having "=" in place of "$\neq$". Again, we can write the condition without explicitly using "=":

$$\forall x \forall y ((Rxy \wedge Ryx) \to Rxx),$$

but this time, by contrast with the **U**-defined condition of range-reflexivity, the condition thus obtained is not modally definable. To see this, we must borrow ahead, from Theorem 4.4.32 (p. 269), which tells us that **K** is determined by the class of all asymmetric frames (as well as by numerous other classes of frames, in addition to the class of *all* frames). Since any non-theorem of **K** is false at some point in a model on an asymmetric frame, it is false at some point in a model on a frame satisfying the condition last inset above, which follows from asymmetry. Thus **K** is determined by the class of all frames meeting this condition and so if the condition were modally defined by some set of formulas Γ, all would be theorems of **K** and therefore valid on every frame, even frames not satisfying the condition, contradicting the idea that Γ modally defines the class of frames in question: if the formulas in Γ are valid on all such frames, they cannot be valid only on such frames. (An alternative version of this argument appeals directly to p-morphisms, as explained in the preamble to Exercise 2.7.35, p. 137, noting that a frame satisfying the condition inset above can have a p-morphic image not satisfying it. Here one requires Exc. 2.7.35(ii)). Note that if either of the conjuncts in the antecedent of the universally quantified first-order condition "$(Rxy \wedge Ryx) \to Rxx$" the result is modally definable, being simply domain reflexivity or range reflexivity depending on which conjunct is dropped. ◂

Digression. Understanding by the *symmetric closure* of a relation $R \subseteq W \times W$ the smallest symmetric relation which includes R, and by the *symmetric contracture* of R the largest symmetric relation included in R. Let us denote these for the moment by R^{SY} and R^{sy}. It is not hard to verify that $R^{\mathsf{sy}} = R \cap R^{-1}$ and $R^{\mathsf{SY}} = R \cup R^{-1}$, where, as usual, R^{-1} is the converse of R. To accommodate R^{-1}, one naturally reaches for the bimodal language of tense logic with an additional operator $\Box^{-1}$ with R^{-1} as its accessibility relation. (See Chapter 3, in which the more usual notation of G and H – Prior's 'tense operators' – for $\Box$ and $\Box^{-1}$ is employed, though where we shall also see that there is no inherent connection between tense logic and any specifically temporal concerns.) If we want to say that A is true at all R^{SY}-related points we can do so in this language: $\Box A \wedge \Box^{-1} A$. But to say that A is true at all R^{sy}-related points is more problematic: a special case of the problem of intersections of accessibility relations (raised in Exercise 4.4.30(i), p. 268). We could certainly add a new operator $\Box^{\mathsf{sy}}$, say, to the language and consider bimodal logics determined by various classes of frames $\langle W, R, R^{\mathsf{sy}} \rangle$, with R^{sy} being the symmetric contracture of R as the notation suggests, but the class of all such frames is not itself bimodally definable. (That is, the class of bimodal frames $\langle W, R, S \rangle$ such that $S = R^{\mathsf{sy}}$ is not modally definable. Adding $\Box^{-1}$ to the language does not help. By contrast, the tense-logical case of all bimodal frames $\langle W, R, S \rangle$ in which

$S = R^{-1}$ is modally definable – by the tense-logical axioms for the minimal normal tense logic $\mathbf{K}_t$ given on p. 179.) One can of course decide to restrict one's attention to frames in this class; with this restriction in place, those satisfying the condition of Example 2.5.66 is modally defined in the present language by $\Box^{sy}(\Box p \to p)$. Modulo the same restriction, asymmetry and antisymmetry (of R) also becomes modally definable: by $\Box^{sy}\bot$ and $p \to \Box^{sy}p$, respectively. **End of Digression.**

We proceed with some final examples in the 'alio-' vein. Recall (from p. 68) the property of partial functionality, possessed by R (or $\langle W, R \rangle$) when each point (in W) bears the relation R to at most one point. Suppose instead that we wish to consider an 'alio-' liberalization: each point bears R to at most one point other than itself. To get the ball rolling on this topic, we offer an exercise for thought, though the solution follows (which you might postpone looking at in order to do some of the thinking on this surprisingly unobvious issue).

EXERCISE 2.5.67 Consider the following four conditions on $\langle W, R \rangle$, understood as universally quantified, the quantifiers ranging over W:

(1) xRy & xRz imply $y = z$.
(2) xRy & xRz & $x \neq y$ imply $y = z$.
(2′) xRy & xRz & $x \neq z$ imply $y = z$.
(3) xRy & xRz & $x \neq y$ & $x \neq z$ imply $y = z$.

It is clear enough that (2) and (2′) are equivalent, by re-lettering and re-ordering, and also that we have the implications (1) $\Rightarrow$ (2) $\Rightarrow$ (3).[96] (1) is the condition of partial functionality. Is either of the implications just mentioned reversible? Which of the three conditions is the 'alio-' version of partial functionality alluded to above, i.e., which is equivalent to the requirement that for all $x \in W$, $R(x) \smallsetminus \{x\}$ contains at most one element? ✠

Since (3) is a fairly direct reformulation mentioned at the end of Exercise 2.5.67, the only remaining issue is the status of (2), which we can clear up by noting that the implication (1) $\Rightarrow$ (2) is reversible. (Hence (1) and (2) and equivalent, and (2) is another example of a spurious 'alio-' condition. (2) is therefore not implied by (3), since it is easy enough to see (3) does not imply (1).) The simplest way to do this is to show that a failure of (1) implies a failure of (2). So suppose that (1) is false for some $\langle W, R \rangle$. Thus there are $u, v, w \in W$ with:

$$uRv, \quad uRw, \quad v \neq w.$$

Now, we must have either $u \neq v$ or $u \neq w$, since if we had both $u = v$ and $u = w$, then we should have $v = w$, contrary to the current supposition. So suppose, first, that $u \neq v$. Then taking x as u, y as v and z as w gives a counterexample to (2). Suppose, on the other hand, that $u \neq w$. In that case, taking x as u, y as w, and z as v gives a counterexample to (2).

EXERCISE 2.5.68 (*i*) Draw diagrams in the style of Figures 2.9–2.11 representing conditions (2) and (3) from Exercise 2.5.67. (Even though condition (2) is, as we have seen, equivalent to condition (1), we are asking for a representation directly corresponding to the form of (2), and not just the modification of the diagram on the right of Figure 2.6 which is described (though not actually depicted) on p. 68.)

(*ii*) $\mathbf{KD}_c$ is easily seen to be determined by the class of frames satisfying the partial functionality condition; show that the logic determined by the class of frames satisfying the 'alio-' version of partial functionality may be axiomatized as the smallest normal modal logic containing all formulas of the form

$$(A \wedge B) \to (\Box(C \vee A) \vee \Box(C \to B)).$$

---

[96]This of course means "(1) $\Rightarrow$ (2) and (2) $\Rightarrow$ (3)". And "(1) $\Rightarrow$ (2)" in turn, for example, means that the universal quantification of what appears as (1) above implies the universal quantification of what appears as (2).

2.5. FROM MODELS TO FRAMES

(Use the canonical model method for the completeness half of the proof that the logic thus axiomatized is determined by the class of frames at issue.) ✠

The schema given in this last exercise can be simplified to:

$$A \to (\Box(C \vee A) \vee \Box(C \to A)),$$

the form in the exercise, with both "A" and "B", being given to make the completeness proof more obvious.

For our final example in this vein, we use a condition which is partly existential in force and therefore has no depiction of the kind asked for in (i) of the previous exercise: piecewise convergence. Recall that this condition (introduced in Exercise 2.5.8(iii), p. 74), was satisfied by a frame $\langle W, R \rangle$ just in case for all $w \in W$ and all $x, y \in R(w)$, there exists $z \in W$ for which xRz and yRz. An obvious 'alio-' variant would require the existence of such a $z \in R(x) \cap R(y)$ only for the case of $x, y \in R(w)$ for which $x \neq y$.

EXERCISE 2.5.69 Show that the class of frames just described determines the normal logic obtained from **K** by adding as axioms those falling under the schema: $\Diamond(A \wedge \Box B) \to \Box(A \vee \Diamond B)$. ✠

The schema figuring here should be compared with **.2**. A simpler schema,

$$\Diamond(A \wedge \Box A) \to \Box(A \vee \Diamond A),$$

gives the logic of those frames in which a $\langle W, R \rangle$ in which for all $w \in W$ and $x, y \in R(w)$ for which that $x \neq y$, not xRy, and not yRx, there exists $z \in R(x) \cap R(y)$.[97]

On p. 85 a sequence of candidate axioms $\mathbf{T}_n^\Diamond$ was considered, and we conclude the present section with some remarks on a more famous such sequence. Let **Alt**$_n$ be the formula:

$$\Box p_1 \vee \Box(p_1 \to p_2) \vee \Box((p_1 \wedge p_2) \to p_3) \vee \ldots \vee \Box((p_1 \wedge \ldots \wedge p_n) \to p_{n+1}).$$

As usual, we also use the same label for the corresponding schema (with schematic letters A_i replacing the propositional variables p_i. These schemata were introduced by Segerberg.[98] (We take **Alt**$_0$ to be $\Box p_1$, alias $\Box p$ or **Ver**.) **KAlt**$_n$ is determined by the class of frames in which each point has at most n points accessible to it ("at most n alternatives", to put it in a way that explains the label's etymology). The completeness half of this claim is readily established by the canonical model method; one can also check that **Alt**$_n$ modally defines the class of frames in question. Note that because the variable p_{n+1} appears only once in **Alt**$_n$, we can replace it with $\bot$ (by uniform substitution), and rewrite the result of doing this in **Alt**$_n$ as

$$\Box p_1 \vee \Box(p_1 \to p_2) \vee \Box((p_1 \wedge p_2) \to p_3) \vee \ldots \vee \Box((p_1 \wedge \ldots \wedge p_{n-1}) \to \neg p_n).$$

Further, the original **Alt**$_n$ can be recovered from this new form in any normal modal logic. (How?) **Alt**$_1$ and **Alt**$_2$ in this new form appear as:

$$\Box p \vee \Box \neg p \qquad \text{and} \qquad \Box p \vee \Box(p \to q) \vee \Box(p \to \neg q),$$

respectively, from which we see in the former case that **Alt**$_1$ is just a variant on $\mathbf{D}_c$.

EXERCISE 2.5.70 (i) Show that **KAlt**$_2$ coincides with the smallest normal modal logics containing the formula shown first here (or its dual form, below it):

$$(\Diamond p \wedge \Diamond q \wedge \Diamond r) \to (\Diamond(p \wedge q) \vee \Diamond(p \wedge r) \vee \Diamond(q \wedge r))$$

$$(\Box(p \vee q) \wedge \Box(p \vee r) \wedge \Box(q \vee r)) \to (\Box p \vee \Box q \vee \Box r).$$

---

[97]This is mentioned on p. 94 of Segerberg [1025].
[98][1025], p. 52

(ii) Give a formula (either with $\Diamond$s, like the first formula above, or with $\Box$s, as in the second formula, which is similarly **K**-interdeducible with **Alt**$_3$. ✠

For later reference (in Chapter 5), we need to consider a related sequence of formulas, obtained by deleting the leftmost $\Box$ from **Alt**$_n$. (In fact we could equally well delete any other single $\Box$ instead, but let us stick with this formulation.) The nth formula in this sequence for $n \geq 1$, we call **Alt**$_n^{\neq}$:

$$p_1 \vee \Box(p_1 \to p_2) \vee \Box((p_1 \wedge p_2) \to p_3) \vee \ldots \vee \Box((p_1 \wedge \ldots \wedge p_n) \to p_{n+1}).$$

KAlt$_n^{\neq}$ is determined by the class of frames in which each point has at most $n-1$ R-successors distinct from itself. The use of this terminology is meant to be compatible with but, not require, that the point in question is accessible to itself. An alternative formulation may be helpful. Given a frame $\langle W, R \rangle$, define, as in Exercise 2.7.34, the relation $R^{\neq}$ by: $R^{\neq}xy$ iff $Rxy \wedge x \neq y$ (or $Rxy \,\&\, x \neq y$, if you prefer to reserve "$\wedge$" for conjunction in the modal object-language); this is sometimes called the irreflexivization of R. (If R is already irreflexive, then $R^{\neq}$ is just R.) Then parallelling the notation "$R(x)$", we denote by "$R^{\neq}(x)$" the set $\{y \in W \mid R^{\neq}xy\}$. To put the preceding claim in this notation, we have: **KAlt**$_n^{\neq}$ is determined by the class of frames $\langle W, R \rangle$ in which for all $x \in W$, $|R^{\neq}(x)| \leq n-1$, by contrast with **KAlt**$_n$, which is determined by the class of $\langle W, R \rangle$ with $|R(x)| \leq n$ (for all $x \in W$). As in the case of the formulas **Alt**$_n$, this class of frames is modally defined by **Alt**$_n^{\neq}$ and the canonical frame for any consistent normal modal logic containing it lies in the class, so the completeness result just cited is "portable": **KTAlt**$_n^{\neq}$, for example, is determined by the class of reflexive frames in which no point has accessible to it more than $n-1$ points distinct from itself. In fact in this case, note that whenever $x \in R(x)$, $|R^{\neq}(x)| = |R(x)| - 1$, so $|R(x)| \leq n$ if and only if $|R^{\neq}(x)| \leq n-1$, and accordingly **KTAlt**$_n^{\neq}$ = **KTAlt**$_n$. Naturally, to derive **Alt**$_n^{\neq}$ from **Alt**$_n$ in this setting, we simply remove the leftmost "$\Box$", as licensed by **T**. In the other direction, we do not need assistance from **T**, since if a point has at most $n-1$ successors distinct from itself, then it can only have at most n successors in all (whether or not it is in fact one of its own successors). Let us look at the case of $n = 2$ to see how this works out syntactically; it turns out to be trickier than one might have expected.

EXAMPLE 2.5.71 Proving **Alt**$_2$ in **KAlt**$_2^{\neq}$. The above observation about eliminating the variable p_{n+1} from Segerberg's **Alt**$_n$ applies here too, so we may take **Alt**$_2^{\neq}$ to be $p \vee \Box(p \to q) \vee \Box(p \to \neg q)$ and **Alt**$_2$ to be $\Box p \vee \Box(p \to q) \vee \Box(p \to \neg q)$. What is far from obvious is how to derive the latter from the former using the resources of **K**. There is no difficulty in showing that the formula **Alt**$_2$ is provable in **KAlt**$_2^{\neq}$, since there is a straightforward canonical model argument to the effect that the canonical frame contains no point with more than one successor distinct from itself. (For suppose that x has successors y, z distinct from each other and from x. Since $x \neq y$ there is a formula, A, say, with $A \in x$, $A \notin y$, and since $x \neq z$ there is a formula B, say, with $B \in x$, $B \notin z$. Since $y \neq z$ we have C, say, with $C \in y$, $C \notin z$. Now invoke the fact that, by **Alt**$_2^{\neq}$, $\neg(A \wedge B) \vee \Box(\neg(A \wedge B) \to C) \vee \Box(\neg(A \wedge B) \to \neg C) \in x$, to get a contradiction.) But there seems in the present instance to be a greater distance than one is usually accustomed to between having a proof – via the completeness theorem – that a certain formula is provable from given axioms, and actually having a formal proof of the formula from those axioms. In fact we give only an informal description of such a proof here, and will take the starting point as the schematic form of **Alt**$_2^{\neq}$ in the following form: $A \to (\Box(A \vee B) \vee \Box(A \to \neg B))$, obtained from that given above by putting $\neg p$ for p in $p \vee \Box(p \to q) \vee \Box(p \to \neg q)$ and then making appropriate $\to$/$\vee$ interchanges to hide the resulting negations, before replacing distinct propositional variables by distinct schematic letters. In fact, it is even more convenient to work with the dual form of the above schema, namely:

$$((\Diamond(A \wedge B) \wedge \Diamond(A \wedge \neg B)) \to A.$$

A corresponding form of the formula we are trying to prove from the instances of this schema, which will accordingly do for present purposes as **Alt**$_2$, is $((\Diamond(p \wedge q) \wedge \Diamond(p \wedge \neg q)) \to \Box p$, and in this informal

2.5. FROM MODELS TO FRAMES

natural deduction style sketch we proceed by assuming the negation of this last and trying to obtain a contradiction. Spelling out the assumption, we have (1), (2) and (3):

$$(1) \quad \Diamond(p \wedge q) \qquad (2) \quad \Diamond(p \wedge \neg q) \qquad (3) \quad \Diamond \neg p.$$

Using the $\mathbf{Alt}_2^{\neq}$ schema just given, taking A, B as p, q, respectively, the antecedent of the resulting instance of the schema is the conjunction of (1) and (2), so the consequent p follows. There are now two cases to consider, depending as we have (a) $p \wedge q$ or (b) $p \wedge \neg q$. To dispose of case (a), take A as $\neg(p \wedge q)$ and B as p in $\mathbf{Alt}_2^{\neq}$ as inset above, giving:

$$(\Diamond(\neg(p \wedge q) \wedge p) \wedge \Diamond(\neg(p \wedge q) \wedge \neg p)) \to \neg(p \wedge q).$$

Now the first conjunct of the antecedent follows from (2) by truth-functional reasoning and $\Diamond$Mono, and the second conjunct of the antecedent follows similarly from (3), giving us the whole antecedent, from which we infer the consequent, $\neg(p \wedge q)$ – which is inconsistent with (a). This leaves only case (b). For this, take A and B from $\mathbf{Alt}_2^{\neq}$ as $\neg(p \wedge \neg q)$ and B as $\neg p$, giving the following:

$$(\Diamond(\neg(p \wedge \neg q) \wedge \neg p) \wedge \Diamond(\neg(p \wedge \neg q) \wedge \neg\neg p)) \to \neg(p \wedge \neg q).$$

From (3), reasoning as before, we get the first conjunct of the antecedent, and from (1), the second conjunct, so again the consequent follows, which contradicts (b). (Note that "$\neg\neg p$" appears here, rather than simply "p", just to make it clear that we are dealing with an instance of $\mathbf{Alt}_2^{\neq}$; similarly, we have the cumbersome "$\neg(p \wedge \neg q)$" rather than "$p \to q$" so as to have the negation of (b) explicitly before us.) ◀

EXERCISE 2.5.72 Convert the above argument into an axiomatic proof in the style of Example 1.3.16 etc., of the formula $\mathbf{Alt}_2$ in the logic $\mathbf{KAlt}_2^{\neq}$ (axiomatized as the label suggests). Note that the rule of Necessitation (or any rule such as $\Diamond$Mono, derived with its aid) is never actually applied to the axiom $\mathbf{Alt}_2^{\neq}$. This means that $\mathbf{Alt}_2$ is provable in $\mathbf{K} + \mathbf{Alt}_2^{\neq}$ and not just in $\mathbf{K} \oplus \mathbf{Alt}_2^{\neq}$ (alias $\mathbf{KAlt}_2^{\neq}$). This reflects the semantic fact that $\mathbf{Alt}_2$ is not just valid on every frame on which $\mathbf{Alt}_2^{\neq}$ is valid, but to use terminology to be introduced in Section 3.2 below, $\mathbf{Alt}_2$ is valid *at any given point* in a frame whenever $\mathbf{Alt}_2^{\neq}$ is valid *at that point*. ✠

Let us return to the logics $\mathbf{KAlt}_n$. The finite model property was mentioned on p. 73 above and it was claimed, though not proved, that the logics under discussion in our survey typically possess this property. In particular, this is so for $\mathbf{K}$, $\mathbf{S4}$ and $\mathbf{S5}$. Let us say that a modal logic S has the *descending chain property* if there is an infinite[99] strictly descending chain of logics $S_0 \supsetneq S_1 \supsetneq \ldots \supsetneq S_n \supsetneq S_{n+1} \ldots$ such that:

$$S = \bigcap_{i \in Nat} S_i$$

In this case S can be regarded as a lower limit of the descending chain in question. The fact that the logics just mentioned have the finite model property will be used in the proof of the following, as is the fact that the logics $\mathbf{KAlt}_0, \mathbf{KAlt}_1, \ldots, \mathbf{KAlt}_n, \ldots$ are all distinct, which is evident from the fact that one can always find a frame validating $\mathbf{Alt}_{i+1}$ but not $\mathbf{Alt}_i$, e.g., by giving every point exactly $i + 1$ successors.

PROPOSITION 2.5.73 *The normal modal logics* $\mathbf{K}$, $\mathbf{S4}$, *and* $\mathbf{S5}$ *have the descending chain property.*

---

[99] The current definition makes it clear that "infinite" here means *countably infinite*; a more general definition could be given, but this will do for our purposes. The same goes for the restriction to consideration of sequences of order type ω. (Similarly in the case of the ascending chain property, defined below, with the reverse order type.) The phrase "descending chain property" is from Schumm [1007]. *Warning*: similar terminology has been used to (what one might describe loosely as) opposite effect. For example Kurosh [681] defines a partially ordered set to satisfy the *descending chain condition* when "for every descending chain of elements $a_1 \geq a_2 \geq \ldots \geq a_n \geq \ldots$ there is an index n at which the chain becomes stationary, i.e., $a_n = a_{n+1} = \ldots$".

Proof. In the case of **K**, the finite model property implies that every non-theorem is invalid on some finite frame and so on a frame validating **Alt**$_n$ for some n. (Take n as the largest number of points accessible to any point in the frame. Since the frame is finite, such an n exists.) It follows that **K** is the intersection of the logics in the descending chain **KAlt**$_0 \supsetneq$ **KAlt**$_1 \supsetneq \ldots \supsetneq$ **KAlt**$_n \ldots$: clearly **K** is included in each of these logics, and conversely, if $A \notin$ **K** then by the above finite model property considerations, for some n, $A \notin$ **KAlt**$_n$, which keeps A out of the intersection. For **S4** and **S5** the argument is similar, using the logics **S4Alt**$_n$ and **S5Alt**$_n$, having verified that the inclusions in question are strict for these cases also.∎

The case of **S5** is somewhat different from the others, in that logics **S5Alt**$_n$ comprise its *only* consistent proper extensions, a result first established in Scroggs [1020]. For each of them there is a finite characteristic frame (with universal accessibility relation), so it can be regarded as a many-valued logic with a finite characteristic matrix, or generalized truth-table, for which reason such logics are often called *tabular*. (For this kind of matrix treatment, see Section 2.10, 'A Quick History', beginning on p. 152. A matrix is *characteristic* for a logic if validity in the matrix coincides with provability in the logic. The phrase "locally tabular" was mentioned at p. 44.) Logics which, like **S5**, are not tabular but all of whose consistent extensions are tabular, are similarly called *pretabular*; see Maksimova [766], or Esakia and Meskhi [274], for a look at the range of normal pretabular extensions of **S4**. (The latter should be consulted rather than the former, which introduces numerous extraneous considerations.)

The sequence of formulas **Alt**$_n$ is by no means the only sequence of formulas that could have been used to show a logic with the finite model property has the descending chain property. The *bounded width* formulas and *bounded length* formulas from p. 80f. provide such sequences, limiting the number of (distinct) mutually inaccessible successors – as opposed to successor in general, as with **Alt**$_n$ – for a given point, in the former case, and the length of chains of mutually accessible points (in transitive frames) in the latter. These two sequences appeared originally in Fine [290] and Hughes [532], respectively.

To round out the discussion we should briefly record the notion dual to that of the descending chain property. Say that a modal logic S has the *ascending chain property* if there is an infinite strictly ascending chain of logics $S_0 \subsetneq S_1 \subsetneq \ldots \subsetneq S_n \subsetneq S_{n+1} \ldots$ such that:

$$S = \bigcup_{i \in Nat} S_i$$

In this case S can be regarded as an *upper* limit of the ascending chain in question. In the literature this property has been associated with not being finitely axiomatizable (see Lemmon [700]), but we do not go into this here – except to note that different properties have gone by this name, the concept really needing to be relativized to a family of logics (or to a set of rules embodying the closure conditions for the family in question). The logic mentioned on p. 85 which adds each of the formulas $\mathbf{T}_n^\Diamond$ ($= \Diamond((\Box p_1 \to p_1) \land \ldots \land (\Box p_n \to p_n))$ to **K** is an example of a logic with the ascending chain property.

2.6 The Rule of Disjunction

We take up a theme from Lemmon and Scott [705], the terminology of which is slightly adapted here, despite a respect in which it is inappropriate. A (mono)modal logic S has (or 'provides', or 'enjoys') the *rule of n-ary disjunction* just in case for any formulas $A_1, \ldots, A_n$ if $\vdash_S \Box A_1 \lor \ldots \lor \Box A_n$ then $\vdash_S A_i$ for some $i \in \{1, \ldots, n\}$. (We will often say, instead, that S has the n-ary rule of disjunction in this case.) S has the rule of disjunction *tout court* if S has the rule of n-ary disjunction for all $n \in Nat$. The reason the terminology is not quite appropriate is that a rule should have a conclusion, as happens in the present case only for $n = 1$, the rule $\Box A / A$, generally called *Denecessitation*. If we have any $\Box A$ provable we can apply this rule and conclude that A is provable. But there is nothing similar in the case of the binary rule of disjunction (say): if we know that $\Box A \lor \Box B$ is provable and wish to invoke the "rule", all we know

2.6. THE RULE OF DISJUNCTION

is that either A is provable or else B is (and perhaps both are). In other words, for something to count as a rule, there should always be such a thing as the smallest logic containing such-and-such formulas and closed under the rule – a condition not satisfied here. But, having explained how it is less than ideal, we continue to use the terminology here.

THEOREM 2.6.1 **K**, **KT**, **K4**, **S4** and **KW** *provide the rule of disjunction.*

Proof. Let S be any one of the logics mentioned and suppose that for some formulas $A_1, \ldots, A_n$, we have $\nvdash_S A_1, \ldots, \nvdash_S A_n$. We must show that in that case $\nvdash_S \Box A_1 \vee \ldots \vee \Box A_n$. Let us work through the case in which S is **K**. Since no A_i is **K**-provable each of these formulas is false at some point x_i in some model $\mathcal{M}_i$, as **K** is complete w.r.t. the class of all models (equivalently: of all frames); further, we can assume without loss of generality that the universes W_i of these models are pairwise disjoint. Let $\mathcal{M}_{x_i}$ be the submodel of $\mathcal{M}_i$ generated by x_i ($i = 1, \ldots, n$). By Theorem 2.4.10, since $\mathcal{M}_i \nvDash_{x_i} A_i$, we have $\mathcal{M}_{x_i} \nvDash_{x_i} A_i$. For simplicity we think of the model $\mathcal{M}_{x_i}$ as being $\langle W_i, R_i, V_i \rangle$, to avoid further double subscripting ("W_{x_i}" etc.). We make a new model from all of these by taking a new point lying outside $\bigcup_{i=1}^n W_i$, w, say, and make the frame $\langle W, R \rangle$ of our new model by taking $W = \bigcup_{i=1}^n W_i \cup \{w\}$, and $R = \bigcup_{i=1}^n R_i \cup \{\langle w, x_1 \rangle, \ldots, \langle w, x_n \rangle\}$. Let $V(p_i) = \bigcup_{i=1}^n V_i(p_i)$. Note that the submodel of $\mathcal{M} = \langle W, R, V \rangle$ generated by x_i is simply $\mathcal{M}_{x_i}$, so by Theorem 2.4.10 again, $\mathcal{M} \nvDash_{x_i} A_i$: but therefore $\mathcal{M} \nvDash_w \Box A_1 \vee \ldots \vee \Box A_n$, as the accessible A_i-falsifying point x_i prevents the disjunct $\Box A_i$ of this disjunction from being true at w (in $\mathcal{M}$). So by the soundness of **K** w.r.t. the class of all models (alternatively put: all frames), we conclude that $\nvdash_{\mathbf{K}} \Box A_1 \vee \ldots \vee \Box A_n$. For **KT**, we need the following variation, to make sure we end up with a reflexive frame $\langle W, R \rangle$ given that the various frames $\langle W_i, R_i \rangle$ are reflexive: $R = \bigcup_{i=1}^n R_i \cup \{\langle w, w \rangle, \langle w, x_1 \rangle, \ldots, \langle w, x_n \rangle\}$. The rest of the argument is as above, putting in appropriate references to reflexive as opposed to arbitrary frames and using the fact that **KT** is determined by the class of such frames. For **K4** we need to make sure the argument goes through with transitive frames under consideration and under construction throughout, so this time from supposedly transitive frames $\langle W_i, R_i \rangle$ we take $\langle W, R \rangle$ with $R = \bigcup_{i=1}^n R_i \cup \{\langle w, y \rangle \mid y \in \bigcup_{i=1}^n W_i\}$. W in all cases just has the single additional point w over and above those in the various W_i. For **S4** we use $R = \bigcup_{i=1}^n R_i \cup \{\langle w, y \rangle \mid y \in W\}$. For **KW**, R is defined as for **K4** and as well as being transitive, the frames we start with have no point starting an infinite R_i-chain; since the new model adds one point and one R link to each point in the frames $\langle W_i, R_i \rangle$, the construction yields a transitive frame in which again no point starts an infinite R-chain. ∎

The important thing about the logics described in Theorem 2.6.1 is that they are determined by classes of frames whose point-generated members permit the adjunction of a new point bearing the accessibility relation to each of the original generating points in a frame which remains in the class in question and for which the frames thus linked by the newly adjoined point (w in the above proof) are generated subframes of the new frame. The most obvious omission from Theorem 2.6.1 is any reference to **S5**, which lacks the rule of binary disjunction (the n-ary rule of disjunction for $n = 2$, that is), as we see from the fact that $\vdash_{\mathbf{S5}} \Box p \vee \Box \neg \Box p$ while $\nvdash_{\mathbf{S5}} p$ and $\nvdash_{\mathbf{S5}} \neg \Box p$. It is not hard to see how an argument along the lines of the proof of Thm. 2.6.1 would fail if we used the fact that **S5** is determined by the class of equivalence-relational frames, since given two such frames generated by points x_1 and x_2, adding a new point w which is accessible to itself (to make the final frame reflexive) and to which all of x_1's and x_2's accessible points are accessible (for transitivity) will still not be enough to end up with an equivalence-relational frame. The relation will have to be symmetric as well: but now w will have to be accessible to x_1 and x_2, which is already enough to stop the original frames from being generated subframes of the new frame, even before we go on to observe that for transitivity we will now have x_1 and x_2 mutually accessible, and so on. The fact that we have the above counterexample to the rule of binary disjunction for **S5** means that **S5** is not determined by any class of frames that escapes this difficulty – it wasn't just bad luck that the class of equivalence-relational frames gave trouble.

REMARK 2.6.2 Recall the notation ΔA for "it is noncontingent whether A" from Exercise 1.3.8, p. 21 (as well as more recently in the discussion leading up to Execise 2.5.60, p. 102), where it was defined to be $\Box A \vee \Box \neg A$. Let us observe that in logics such as **S4** which provide the rule of disjunction (and are, like all normal modal logics, closed under Necessitation), whenever a formula A is provably noncontingent, the logic tells us 'which way' – by being necessary or by being impossible – it is noncontingent, proving one of $\Box A$, $\Box \neg A$. This does not mean that any normal modal logic which fails to provide the rule of disjunction lacks this feature of 'resolving provable noncontingencies'. For instance the counterexample above to the rule of (binary) disjunction for **S5** – using the fact that $\vdash_{\mathbf{S5}} \Box p \vee \Box \neg \Box p$ – does not immediately present us with an unresolved noncontingency, since we are not dealing here with an **S5** theorem of the form $\Box A \vee \Box \neg A$. In this case we can easily adjust the example so that it is the provability of $\Box\Box p \vee \Box \neg \Box p$ ($= \Delta \Box p$), without either $\Box\Box p$ or $\Box\neg\Box p$ being provable that gives us an unresolved provable noncontingency. The issues raised here deserve further consideration – for example the question of a normal modal logics lacking the rule of binary disjunction but resolving all provable noncontingencies, and of a variation on the latter property in the direction of the rule of disjunction: $\vdash_S \Box A \vee \Box \neg A$ implying $\vdash_S A$ or $\vdash_S \neg A$. (Since, while normality takes us from "$\vdash_S A$ or $\vdash_S A$" to "$\vdash_S \Box A$ or $\vdash_S \Box\neg A$", the converse transition uses Denecessitation and so a special case of the standard rule of disjunction. The original formulation with the all occurrences of $\Box$ intact is more closely related to what we call the *modified* rule of disjunction at p. 122 below.) ◀

EXERCISE 2.6.3 For each of the following normal modal logics either prove or refute the claim that the logic in question provides the rule of disjunction:

KD   **KB**   **K5**   **KD45**   **S4.2**   **KU**.

(Partial assistance: for **S4.2**, concentrate on the ".2" part of the label; to bring this to bear on the case of **KB**, consider Example 2.5.22(iii), p. 79) ✠

From the definitions given, it is clear that if a logic has the rule of n-ary disjunction then it has the rule of m-ary disjunction whenever $m < n$ (just repeat a disjunct), let us pause for a moment over the case in which $m > n$. As we have seen, **S5** does not provide the rule of binary disjunction. On the other hand, the singular case – Denecessitation – is certainly satisfied here, since $\vdash_{\mathbf{S5}} \Box A \to A$, so Modus Ponens delivers the denecessitated form of any $\Box$-theorem.[100] So here we have the rule for 1-ary but not 2-ary disjunctions. (Note that since closure under the rule of Denecessitation follows from enjoying the rule of disjunction, all of the logics mentioned in Theorem 2.6.1 and any for which this property was correctly claimed in response to Exercise 2.6.3, including in particular those such as **K** which do not prove $\Box A \to A$ for arbitrary A, are closed under Denecessitation.)

EXERCISE 2.6.4 (i) Describe a normal modal logic for which you can show that it provides the 2-ary but not the 3-ary rule of disjunction.

(ii) Supplying either a proof or a counterexample, say whether the following is true or false. For any formula A, if $\vdash_{\mathbf{K}} \Box A \to \Box\Box A$, then $\vdash_{\mathbf{K}} A$.

(iii) As for (ii) but this time for the following claim: $\vdash_{\mathbf{K}} \Box\Box A \to \Box A$, then $\vdash_{\mathbf{K}} A$.

Suggestions: For (ii), see the discussion after Exc. 2.6.33, p. 127 below; for (iii), consider the rule of Denecessitation and also Löb's Rule (see Coro. 2.5.6, p. 73). ✠

We can use the rule of disjunction to furnish structural information about the frames of the canonical models for logics providing that rule, via the 'only if' direction of the following:

---

[100]The admissibility of Denecessitation does not of course require the provability of the **T** axiom, Theorem 2.6.1 securing it along with all the rest of the full rule of disjunction for **K** and **K4**, which are not extensions of **KT**. Note also that since **S5** = **KT5** and $\vdash_{\mathbf{K5}} \mathbf{U}$ (see Example 2.5.22(ii), p. 79, and Exercise 2.5.23), **S5** can be axiomatized by adding the rule Denecessitation, rather than the generally stronger principle **T**, to the axiomatic basis of **K5** suggested by the latter label.

2.6. THE RULE OF DISJUNCTION

THEOREM 2.6.5 *For any consistent normal modal logic S, S provides the n-ary rule of disjunction if and only if the canonical frame $\langle W_S, R_S \rangle$ satisfies the condition that for all $x_1, \ldots, x_n \in W_S$ there exists $w \in W_S$ such that $\{x_1, \ldots, x_n\} \subseteq R_S(w)$.*

Proof. 'Only if': Suppose S is as described, and take $x_1, \ldots, x_n \in W_S$ with a view to finding a common R_S-predecessor w for all these points. Let w be a maximal S-consistent superset of the following set, whose S-consistency we shall establish presently: $\{\Diamond A \mid A \in \bigcup_{i=1}^n x_i\}$. Any such w bears the relation R_S to each of the x_i, by Exercise 2.4.5. It remains to check the S-consistency of the above set. Suppose otherwise. Then for some $A_1^1, A_2^1, \ldots, A_{k_1}^1 \in x_1$, and some $A_1^2, A_2^2, \ldots, A_{k_2}^2 \in x_2$, and ... and some $A_1^n, A_2^n, \ldots, A_{k_n}^n \in x_n$, we have, writing A^i for the conjunction of the $A_1^i, A_2^i, \ldots A_{k_i}^i$:

$\vdash_S \neg(\Diamond A^1 \wedge \Diamond A^2 \wedge \ldots \wedge \Diamond A^n).$

(Here we have used the fact that $\neg(\Diamond A_1^i \wedge \ldots \wedge \Diamond A_{k_i}^i)$ provably implies, in any normal modal logic, $\neg \Diamond(A_1^i \wedge \ldots \wedge A_{k_i}^i)$, i.e., $\neg \Diamond A^i$.) Rewriting this, we get:

$\vdash_S \Box \neg A^1 \vee \Box \neg A^2 \vee \ldots \vee \Box \neg A^n,$

so, invoking the supposition that S enjoys the rule of n-ary disjunction, we infer that $\vdash_S \neg A^i$ for some $i \in \{1, \ldots, n\}$. But this is impossible since $A^i \in x_i$, as A^i is a conjunction all of whose conjuncts are drawn from x_i. ∎

We interrupt the 'Rule of Disjunction' theme of this section to illustrate the frequently encountered weakening step parenthetically indicated in the above proof with its allusion to the fact that $\neg(\Diamond A_1^i \wedge \ldots \wedge \Diamond A_{k_i}^i)$ provably implies, in any normal modal logic, $\neg \Diamond(A_1^i \wedge \ldots \wedge A_{k_i}^i)$. Such a 'weakening' step is very common in arguments concerning canonical models, and instead of involving, as here, the passage from the provability of $\neg(\Diamond C_1 \wedge \ldots \wedge \Diamond C_m)$ to that of $\neg \Diamond(C_1 \wedge \ldots \wedge C_m)$, or alternatively in the unnegated form ('de-contraposing'), from the provability of $\Diamond(C_1 \wedge \ldots \wedge C_m)$ to the provability of $\Diamond C_1 \wedge \ldots \wedge \Diamond C_m$, or again, using $\Box$, from $\Box C_1 \vee \ldots \vee \Box C_m$ to $\Box(C_1 \vee \ldots \vee C_m)$ – or the latter's contrapositive form. Here is another example in which this kind of move appears in a less cluttered setting – free, any rate, of the separate superscripts featuring here.

EXAMPLE 2.6.6 Just as a consistent normal modal logic is closed under Denecessitation if and only if its canonical frame is converse serial – the 'only if' half of this being mentioned below, after Exercise 2.6.7, the 'if' half being obvious, so we can show that the a consistent normal modal logic is closed under the following *Brouwerian rule*: from $\Diamond \Box A$ to A if and only if the logic's canonical frame, $\langle W, R \rangle$ satisfies the following condition:

$$\forall x \exists y \forall z (Ryz \rightarrow Rzx)$$

Again it is straightforward to see that if for a consistent normal modal logic S, S's canonical frame satisfies this condition, then S must be closed under the rule, since if the conclusion A is not provable, A is false at some point x in the canonical model, in which case $\Diamond \Box A$ will be false at a y the condition promises for any such x, since every successor of y has x accessible to it and will therefore not verify $\Box A$. For the converse implication, suppose that we have $x \in W$ with $\langle W, R \rangle$ is the canonical frame for S. To find y as the condition requires, let y be any extension of the set, for the moment just presumed to be S-consistent, $\{\Box \Diamond A \mid A \in x\}$. Any R-successor of such a y will bear R to x because it contains each $\Diamond A$ for $A \in x$. So it remains only to check that the presumption of consistency holds up. If this is set is not S-consistent, then for some $A_1, \ldots A_n \in x$, we have:

(1) $\vdash_S \neg(\Box \Diamond A_1 \wedge \ldots \wedge \Box \Diamond A_n)$; reformulating this, we have:

(2) $\vdash_S \neg \Box(\Diamond A_1 \wedge \ldots \wedge \Diamond A_n)$; and so, weakening this:

(3) $\vdash_S \neg\Box\Diamond(A_1 \wedge \ldots \wedge_n)$; reformulating again:

(4) $\vdash_S \Diamond\Box\neg(A_1 \wedge \ldots \wedge_n)$, whence by the Brouwerian rule:

(5) $\vdash_S \neg(A_1 \wedge \ldots \wedge A_n)$.

But this is impossible becuase each A_i, and therefore their conjunction, belonged to x. This completes the proof of the "only if" half of the claim. Note that it is the (2) $\Rightarrow$ (3) step which embodies the weakening move this example is designed to highlight, as we pass from $\neg\Box(\Diamond A_1 \wedge \ldots \wedge \Diamond A_n)$ to the (typically) weaker $\neg\Box\Diamond(A_1 \wedge \ldots \wedge_n)$, in view of and the fact that $\Diamond(A_1 \wedge \ldots \wedge_n)$ is (typically) stronger than – i.e., implies but is not implied by – $\Diamond A_1 \wedge \ldots \wedge \Diamond A_n$, in view of the fact that $\Box$ preserves while $\neg$ reverses relative strength of antecedent and consequent.

The present example is of some interest in its own right (as is another example connecting closure under a rule to structural features of the canonical frame, to be found in Exercise 2.6.31, p. 126). Note that **K** is closed under the Brouwerian rule vacuously, since it has no theorems of the form $\Diamond B$, and *a fortiori* none of the form $\Diamond\Box A$. **KD** is non-vacuously closed under the rule, as the reader may verify by adjoining points suitably to a point-generated submodel of a serial model falsifying its conclusion. **KD4** on the other hand, is not closed under the rule, since $\vdash_{\mathbf{KD4}} \Diamond\Box(\Diamond\Box p \to p)$ while $\nvdash_{\mathbf{KD4}} \Diamond\Box p \to p$. This shows that instead of adding the Brouwerian axiom **B** to **KD4** it would suffice to add the Brouwerian rule to any axiomatization including the rule of Necessitation, since the rule immmediately delivers the axiom from the **KD4**-theorem just mentioned. (The reference to Necessitation is there to secure that the logic thus axiomatized is normal. The situation is analogous to that noted on p. 114, where we found that **T** could be replaced by its 'rule form' – Denecessitation – without loss in the axiomatization of **S5** as **KT5**.)) Note that although the **B** axiom could alternatively be given in its dual form $p \to \Box\Diamond p$ the Brouwerian rule could not be replaced by the 'rule form' $A / \Box\Diamond A$ of the latter. In any congruential modal logic, the closure under the latter rule is just equivalent to having $\Box\Diamond\top$ as a theorem.) ◀

We return to the 'rule of disjunction' theme:

EXERCISE 2.6.7 Prove the 'if' direction of Theorem 2.6.5. ✠

In particular, then, taking the $n = 1$ case of this result, we conclude that the canonical frame for any normal modal logic closed under Denecessitation is converse serial in the sense of Coro. 2.5.33(*iii*) (i.e., the frame $\langle W, R^{-1}\rangle$ is serial, R^{-1} being the converse of R). Thus we can give the following completeness result for **K**, for instance: this logic is determined by the class of all converse serial frames. (The observation just made about the canonical frame gives the completeness half, and the soundness half is trivial, since **K** is sound w.r.t. the class of all frames, and therefore w.r.t. any class of frames you care to name.) At the other end of the spectrum, given the n-ary rule of disjunction for all n, we have common predecessors not only for all finite sets of points, but all sets of points:

THEOREM 2.6.8 *If S provides the rule of disjunction then, where $\langle W_S, R_S\rangle$ is the canonical frame for S, for every $X \subseteq W_S$ there is some $w \in W_S$ with $X \subseteq R_S(w)$.*

Proof. Take $X \subseteq W_S$, and supply $w \in W_S$ with $X \subseteq R_S(w)$ by maximally extending the S-consistent set $\{\Diamond A \mid A \in \bigcup X\}$. As in the proof of Theorem 2.6.5, this gives the result that $X \subseteq R(w)$, but we need check for the set really is S-consistency as claimed. If it is not S-consistent, then we have formulas $\Diamond A_1, \ldots, \Diamond A_n$ for which $A_i \in \bigcup X$ and $\vdash_S \neg(\Diamond A_1 \wedge \ldots \wedge \Diamond A_n)$. So, as in the previous proof, we have $\vdash_S \Box\neg A_1 \vee \ldots \vee \Box\neg A_n$, and by the rule of n-ary disjunction, which S is supposed to provide for every n, we conclude that for some i, $\vdash_S \neg A_i$. But that is impossible, since $A_i \in \bigcup X$ and so $A_i \in x$ for some $x \in X$, which would contradict the S-consistency of the x in question. ∎

In particular, we may take X as W_S itself, giving:

2.6. THE RULE OF DISJUNCTION

COROLLARY 2.6.9 *For any consistent normal modal logic S, with canonical frame $\langle W_S, R_S \rangle$, if S provides the rule of disjunction then there exists some $w \in W_S$ such that for all $x \in W_S$, wR_Sx.*

COROLLARY 2.6.10 $\langle W_{\mathbf{KW}}, R_{\mathbf{KW}} \rangle \notin Fr(\mathbf{KW})$.

Proof. In a frame for **KW** no point starts an infinite R-chain, but by Theorem 2.6.1 this logic enjoys the rule of disjunction, so by Coro. 2.6.9 there is some $w \in W_S$ such that for all $x \in W_{\mathbf{KW}}$, $wR_{\mathbf{KW}}x$; for any such w, therefore, we have $wR_{\mathbf{KW}}w$, so w starts an infinite $R_{\mathbf{KW}}$-chain. Thus $\langle W_{\mathbf{KW}}, R_{\mathbf{KW}} \rangle$ is not a frame for **KW**. ■

REMARK 2.6.11 The proof of Theorem 2.6.8 makes a frequently transition in canonical model argument, which we may call the 'finite-to-arbitrary' move. In effect, we use the fact that the result which is claimed for arbitrary subsets $X \subseteq W$ follows once we have have established it for all *finite* subsets of W, because – in the present instance – if the set $\{\Diamond A \mid A \in \bigcup X\}$ is S-inconsistent, this implies the S-inconsistency of one of its finite subsets. ◀

EXERCISE 2.6.12 Prove the following converse of Coro. 2.6.9: For any consistent normal modal logic S, with canonical frame $\langle W_S, R_S \rangle$, satisfying the condition that for some $w \in W_S$, we have wR_Sx for all $x \in W_S$, then S provides the rule of disjunction. ✠

A word of warning: the use of the canonical frames for various logics is essential in the above results. In particular, apropos of the $n = 2$ case of Theorem 2.6.5, the hypothesis that S is determined by some frame in which every pair of elements have a common predecessor does *not* guarantee that S has the rule of binary disjunction. The following (obvious enough) example was given in Humberstone [596], making this point:

EXAMPLE 2.6.13 The frame $\langle \mathbb{Q}, \leq \rangle$ determines the logic **S4.3** and evidently satisfies the condition that any two elements have a common predecessor (take the minimum), but **S4.3** evidently lacks the binary rule of disjunction, in view of the formula **.3** itself. ◀

Matters are different when it comes to the $n = 1$ form of the rule, however:

EXERCISE 2.6.14 Show that if S is determined by some converse serial frame, then S has the rule of Denecessitation. ✠

Thus there is quite a difference between the n-ary rule of disjunction for $n = 1$ and for $n > 1$, not only because it is only in the former case that the term *rule* is really apposite. What allows us to show in the former case, but not the latter, that a characteristic frame in which any n points have a common predecessor guarantees the rule, is that in the latter case the unprovability of A_1 and A_2 – to illustrate with $n = 2$ – in a logic determined by $\langle W, R \rangle$, gives us models $\langle W, R, V_1 \rangle$ and $\langle W, R, V_2 \rangle$ on this frame, and points $x_1, x_2 \in W$ with $\langle W, R, V_1 \rangle \not\models_{x_1} A_1$ and $\langle W, R, V_2 \rangle \not\models_{x_2} A_2$, but even if x_1 and x_2 have a common predecessor w, say, in $\langle W, R \rangle$, we have (in general) no way of obtaining V from V_1, V_2, for which $\langle W, R, V \rangle \not\models_w \Box A_1 \vee \Box A_2$, because of interactions between the truth-value at A_1 at x_1 and that of A_2 at x_2. For the 1-ary rule (i.e., Denecessitation), however, we do not have to construct a *new* model on the frame in question, given a model with a point falsifying the conclusion of the rule: the same model will do, falsifying $\Box A$ at the promised predecessor of a point falsifying A.

An interesting application of the rule of disjunction pertains to the extensions of normal modal logics providing the rule:

THEOREM 2.6.15 *If a normal modal logic S enjoys the n-ary rule of disjunction, then S is not the intersection of n or fewer normal modal logics properly extending S.*

Proof. Suppose, for a contradiction, that S provides the rule of n-ary disjunction and that $S = S_1 \cap \ldots \cap S_n$, where all logics involved are normal and each $S_i \supsetneq S$. (It is not excluded that for some $i \neq j$, $S_i = S_j$.) Since the inclusions here are strict, for each S_i ($i = 1, \ldots, n$) there is a formula $A_i \in S_i \smallsetminus S$. By normality $\Box A_i \in S_i$ and so $\Box A_1 \vee \ldots \vee \Box A_n \in S_i$ for each i. Thus this disjunction belongs to S, by our supposition that S is the intersection of the S_i. But this is impossible since S provides the rule of n-ary disjunction, so $\Box A_1 \vee \ldots \vee \Box A_n \in S$ would imply some $A_i \in S$, whereas A_i was chosen as a formula in S_i but not in S. ∎

In Proposition 2.5.73, we found each of various logics to be the intersection of infinitely many of their proper extensions; we are now in a position to infer that they are not the intersections of any finite set of extensions:

COROLLARY 2.6.16 *None of the logics* **K**, **KT**, **K4**, **S4**, **KW**, *is the intersection of finitely many of its normal proper extensions; equivalently: none of these logics is the intersection of two of its normal extensions.*

Proof. By Theorems 2.6.1 and 2.6.15. ∎

What follows the "equivalently" in the formulation of this corollary follows from Theorem 2.6.15 because of the "or fewer", but what was intended by its insertion was the result asked for in the following exercise.

EXERCISE 2.6.17 Suppose that S, a normal modal logic, is the intersection of finitely many normal proper extensions $S_1, \ldots, S_n$. Show that S is the intersection of *two* of its normal proper extensions, and that the same holds if the word "normal" is deleted throughout. (Suggestion: since the supposition can never hold for $n = 1$, it suffices to show that whenever S is the intersection of $n \geq 3$ proper extensions, S is the intersection of $n - 1$ of its proper extensions.) ✠

The crucial question, then, is whether a logic is the intersection of two of its proper extensions. In the lattice-theoretic terminology of meets and joins, as introduced on p. 39, this is usually put in terms of meet-(ir)reducibility: an element a in a lattice is said to be *meet-reducible* (*meet-irreducible*) in that lattice if there are (resp., are not) lattice elements b, c, both distinct from a, with a being the meet of b and c. In the present case, for a logic S, this is the question of S is there are S_1 and S_2, both distinct from S, with $S = S_1 \cap S_2$. Now $S = S_1 \cap S_2$ implies that $S \subseteq S_i$) ($i = 1, 2$), so the distinctness of S from each of S_1, S_2, amounts to these inclusions being proper – as in our discussion up to this paragraph. Let us illustrate the sensitivity of these concepts to the particular lattice under consideration:

EXAMPLE 2.6.18 In the earlier Digression introducing the terminology of meets and joins in lattices of modal logics (see p. 39), mention was made of the well-known fact that – to put it in terms of the vocabulary just introduced – the Halldén-complete modal logics are precisely the meet-irreducible modal logics, along with the fact that **K** is not Halldén-complete. But aren't these facts in conflict with what Coro. 2.6.16 says about **K**, namely that it is *not* the intersection of any pair of its proper normal extensions, i.e., that it is meet-irreducible? The answer to this is that we have to be careful about whether we are considering the lattice of all modal logics or the lattice of all normal modal logics. **K** is meet-reducible in the former lattice (for example **K** = (**K** + $\Diamond\top$) ∩ (**K** + $\Box\bot$)) but meet-irreducible in the latter – so we certainly do not have, for example **K** = (**K** ⊕ $\Diamond\top$) ∩ (**K** ⊕ $\Box\bot$). (The notion of Halldén normality from Exercise 5.5.17, p. 415 below, would be relevant to a fuller discussion of these issues: see in particular Exc. 5.5.17(*iii*).) ◀

2.6. THE RULE OF DISJUNCTION

Note that by (the correct answer to) Exercise 2.5.25(*ii*) as it pertains to the last formula given there, **K** $\oplus \Box\bot$ (alias **KVer**) is in fact the same logic as **K** $+ \Box\bot$; there is no similar identity for **K** $\oplus \Diamond\top$ (alias **KD**) and **K** $+ \Diamond\top$, however. Note also that discussion of the lattice of *all* modal logics is problematic, as explained in the Digression on p. 39, because the structure of the lattice depends on the chosen Boolean primitives. This complication does not affect our example, however, as we are working above **K**. (Quasi-normal logics are *a fortiori* quasi-congruential, so note 34 on p. 39 applies.)

EXERCISE 2.6.19 Substantiating the last parenthetical remark in Example 2.6.18, provide a formula provable in (**K** $\oplus \Diamond\top$) $\cap$ (**K** $\oplus \Box\bot$), i.e., in **KD** $\cap$ **KVer**, but not in **K**. (*Hint:* make use of the rule of Necessitation.) ✠

Digression. The formula $\Box A_1 \vee \ldots \vee \Box A_n \in S$ in the proof of Theorem 2.6.15 figures as a special case of something more general. (In this discussion we revert to the formulation in terms of general n rather than the $n = 2$ form appearing in the second half of Coro. 2.6.16.) For a set Λ of modal logics, and a given modal logic S, say that a function f which maps any n formulas to a formula is an *n-ary intersection function S* w.r.t. Λ just in case for any n (or fewer) logics $S_1, \ldots, S_n \in \Lambda$, all of them extensions (not necessarily proper extensions) of S, for any $A_1, \ldots A_n$ with $A_i \in S_i$, we have:

- $f(A_1, \ldots, A_n) \in \bigcap_{i=1}^{n} S_i$, and
- $f(A_1, \ldots, A_n) \in S$ implies that for some A_i $(i = 1, \ldots, n)$, $A_i \in S$.

Then the heart of the proof of Theorem 2.6.15 establishes this: if there is an n-ary intersection function, f, for S w.r.t. Λ, then S is not the intersection of n (or fewer) of its proper extensions lying in Λ. (Abstracting from the proof, we take $A_1 \in S_1 \smallsetminus S, \ldots, A_n \in S_n \smallsetminus S$ as witnesses to the extensions' being proper, and get $f(A_1, \ldots, A_n) \in \bigcap_{i=1}^{n} S_i$ by the first of the two bulleted conditions, which implies that S is properly included in this intersection by appeal to the second condition.) For the case of Theorem 2.6.15, Λ is the class of normal modal logics, and $f(A_1, \ldots, A_n)$ is $\Box A_1 \vee \ldots \vee \Box A_n$. If S is a Halldén-complete modal logic in the sense explained on p. 39, then we can make another application of this general scheme, to show that S is not the intersection of finitely many modal logics properly extending S. (So here Λ can be taken to comprise all modal logics, not just the normal ones.) If a logic is Halldén-complete in the sense defined earlier, then the provability of $B_1 \vee \ldots \vee B_n$ where no two B_i share a common propositional variable, implies the provability of some B_i in the logic (by repeated appeal to the $n = 2$ version of this condition given in the definition). The n-place f defined as follows is an n-ary intersection function for any Halldén-complete S (w.r.t. the class of modal logics):

$$f(A_1, \ldots, A_n) = s_1(A_1) \vee s_2(A_2) \vee \ldots \vee s_n(A_n),$$

where the s_i are re-lettering (or 'invertible') substitutions so chosen that $s_i(A_i)$ and $s_j(A_j)$ are variable-disjoint when $i \neq j$;[101] you may care to verify that this f satisfies the two defining conditions on intersection functions. (Hughes [532] uses Halldén completeness in proof that **S4** has no minimal proper extensions; Schumm [1007], which settles some questions raised in [532], makes use of a combination of the rule of disjunction with Halldén completeness, in the form of a condition that either A or B should be provable in a logic proving $\Box A \vee \Box B$ when A and B have no common propositional variables.)

For the general scheme, the reference to an n-place function f cannot be replaced by a reference to a ('context') formula $C(p_1, \ldots, p_n)$, with $f(A_1, \ldots, A_n)$ replaced by $C(A_1, \ldots, A_n)$, as this second example shows: $f(A_1, \ldots, A_n)$ is not the result of substituting A_i for p_i in any formula $C(p_1, \ldots, p_n)$. **End of Digression.**

The Digression referred to in Example 2.6.18 also introduced the terminology of *covering* in a lattice. (See note 33, p. 39.) In alternative terminology adapted from Hughes [532], a modal logic S^+ is a *minimal*

---

[101]For example, s_1 can be the identity substitution, and s_{i+1} can replace distinct variables in A_{i+1} one-to-one by the first variables (in the enumeration $p_1, \ldots, p_n, \ldots$) not appearing in $s_1(A_1) \vee \ldots \vee s_{i-1}(A_{i-1})$.

proper extension of S in the lattice of Λ-logics, for some family Λ of modal logics, when S^+ covers S in that lattice. That is, $S, S^+ \in \Lambda$, $S \subsetneq S^+$ and there is no logic in Λ strictly (or 'properly') between S and S^+. In fact Hughes explicitly considers just the case in which Λ comprises the normal modal logics, and gives and argument (broadly) similar to that of the following proof, to the conclusion that **S4** has no cover amongst such logics (no minimal normal proper extension, that is).

THEOREM 2.6.20 *Let Λ be a family of modal logics, with $S \in \Lambda$. If S has the descending chain property in Λ and is meet-irreducible in the lattice of logics in Λ, then S has no cover in Λ.*

Proof. For a contradiction, suppose that $S = \bigcap_{i \in Nat} S_i$ where $S_0 \supsetneq S_1 \supsetneq \ldots \supsetneq S_n \supsetneq S_{n+1} \ldots$, with each $S_i \in \Lambda$, and that S^+ covers S in (the lattice of logics in) Λ, and that S is meet-irreducible in this lattice. Since S^+ is a proper extension of S, we may choose $A \in S^+ \smallsetminus S$. Since $A \notin S$, there is some S_i in the descending chain just described, with $A \notin S_i$. (Otherwise S could not be the intersection of the logics in the chain.) Now consider the logic (in Λ) $S^+ \cap S_i$. In contradiction to the supposition that S^+ covers S (in Λ), we claim:

$$S \subsetneq S^+ \cap S_i \subsetneq S^+.$$

The "$\subseteq$" parts of this claim being clear, it remains to verify that each of the two inclusions is proper. For the first: if, instead, $S = S^+ \cap S_i$, this would contradict the meet-irreducibility of S. For the second: if, instead, $S^+ \cap S_i = S^+$, this would conflict with the fact that $A \in S^+$ while $A \notin S^+ \cap S_i$ (since $A \notin S_i$). ∎

Thus, by Proposition 2.5.73 (p. 111) and Coro. 2.6.16, we conclude further:

COROLLARY 2.6.21 *None of the logics **K**, **KT**, **K4**, **S4**, **KW**, has a cover in the lattice of normal modal logics.*

In the case of **S4** and **KT**, because these logics are Halldén-complete, we can conclude from Theorem 2.6.20 that these logics have no covers in the lattice of arbitrary modal logics. What about in the remaining (Halldén-incomplete) cases listed here?[102] We will look in detail at the case of **K**, finding in Coro. 2.6.24 below at least one cover for this logic in the more extensive lattice. To put it another way: we exhibit an atom in the lattice of quasi-normal logics. (See note 33, p. 39.) In the case of **S5**, not listed in Coro. 2.6.21 since our rule-of-disjunction route is not available in this case, the distinction between normal and arbitrary extensions does not arise. This was already noted in the Digression on p. 34 to be a finding of Scroggs [1020]. From another result – pretabularity of **S5** – in [1020], noted in the discussion after Proposition 2.5.73, we can infer that this logic has no cover in either lattice – no minimal proper extension, normal or otherwise. But let us return to the example of finding a cover for **K** in the lattice of all modal logics. Those wanting to skip this example should pass to the paragraph leading up to Proposition 2.6.26 below; others might like to look briefly at one of the Revision Exercises – 2.7.22, p. 134, which will be drawn on here – before proceeding.

LEMMA 2.6.22 *If all substitution instances of a formula A are true at a point $x \in W$ in a model $\langle W, R, V \rangle$ with $R(x) = \varnothing$ then $\vdash_{\mathbf{K}} \Box\bot \to A$.*

Proof. Suppose that for $\mathcal{M} = \langle W, R, V \rangle$ as in the formulation of the Lemma, and so in particular with $R(x) = \varnothing$, we have $\mathcal{M} \models_x s(A)$ for every substitution s, but (for a contradiction) that $\nvdash_{\mathbf{K}} \Box\bot \to A$. Thus there is a model $\mathcal{M}' = \langle W', R', V' \rangle$ with $y \in W'$ and $\mathcal{M}' \models_y \Box\bot$ (so $R'(y) = \varnothing$) while $\mathcal{M}' \nvDash_y A$.

---

[102]A more interesting example of a *Halldén-unreasonable* disjunction for **KW** – a disjunction witnessing the Halldén incompleteness of this logic, that is – than just the usual **K** example, comes by way of Exercise 2.1.5(*i*), p. 38. This terminology echoes that of in the classic discussion of McKinsey [790], *q.v.* for references to Halldén's own work explaining why he thought of them as unreasonable.

2.6. THE RULE OF DISJUNCTION

Let v_x and v_y be the Boolean valuations uniquely fixed by the specification that for each propositional variable p_i, $v_x(p_i) = \mathrm{T}$ iff $x \in V(p_i)$ and $v_y(p_i) = \mathrm{T}$ iff $y \in V'(p_i)$. (We will not bother to superscript a reference to the two models here, as was done on p. 54.) Calling on the translation $(\cdot)^{\mathbf{Ver}}$ from Exercise 2.7.22 (p. 134 below), we infer from part (ii) of that exercise that $v_y(A^{\mathbf{Ver}}) = \mathrm{F}$. Now consider the substitution s_0 defined by $s_0(p_i) = \top$ if $v_y(p_i) = \mathrm{T}$ and $s_0(p_i) = \bot$ if $v_y(p_i) = \mathrm{F}$, noting that in this case $v_y(s_0(A^{\mathbf{Ver}})) = \mathrm{F}$. Since no propositional variables occur in $s_0(A^{\mathbf{Ver}})$, this value does not depend on v_y, and so in particular we also have $v_x(s_0(A^{\mathbf{Ver}})) = \mathrm{F}$. For any substitution s and formula B, $s(B^{\mathbf{Ver}})$ is the same formula as $(s(B))^{\mathbf{Ver}}$ (as may be checked by induction on the complexity of B), so, in particular, we conclude that $v_x((s_0(A))^{\mathbf{Ver}}) = \mathrm{F}$, which implies that $\mathcal{M} \not\models_x (s_0(A))^{\mathbf{Ver}}$. Since all substitution instances of A were supposedly true at x in $\mathcal{M}$, we should have $\mathcal{M} \models_x s_0(A)$. But this contradicts Exc. 2.7.22(ii), since that implies $\mathcal{M} \models_x s_0(A)$ iff $\mathcal{M} \models_x (s_0(A))^{\mathbf{Ver}}$. ∎

Note that in the following, $\mathbf{K} + \Diamond\top$ is the smallest quasi-normal modal logic containing $\Diamond\top$, and not the normal modal logic $\mathbf{K} \oplus \Diamond\top$, alias $\mathbf{KD}$. (Note that $\mathbf{K} + \Diamond\top = \mathbf{K} + \mathbf{D}$.)

PROPOSITION 2.6.23 *For any formula A, if $\not\vdash_{\mathbf{K}} A$ and $\vdash_{\mathbf{K}+\Diamond\top} A$, then $\vdash_{\mathbf{K}+A} \Diamond\top$.*

Proof. Suppose, for a contradiction: (1) $\not\vdash_{\mathbf{K}} A$, (2) $\vdash_{\mathbf{K}+\Diamond\top} A$, and (3) $\not\vdash_{\mathbf{K}+A} \Diamond\top$. (3) implies that there is a point with nothing accessible to it (since $\Diamond\top$ is false there) in some model, at which all substitution instances A are true. By Lemma 2.6.22, then $\vdash_{\mathbf{K}} \Box\bot \to A$. But (2) implies that $\vdash_{\mathbf{K}} \Diamond\top \to A$ (since there are no propositional variables in $\Diamond\top$ to make substitutions for). So, since $\vdash_{\mathbf{K}} \Box\bot \vee \Diamond\top$, we conclude that $\vdash_{\mathbf{K}} A$, contradicting (1). ∎

COROLLARY 2.6.24 $\mathbf{K} + \Diamond\top$ *is a cover of* $\mathbf{K}$ *in the lattice of all modal logics.*

Proof. Clearly $\mathbf{K} \subsetneq \mathbf{K} + \Diamond\top$, so it remains to rule out the existence of a modal logic S with $\mathbf{K} \subsetneq S \subsetneq \mathbf{K} + \Diamond\top$. Suppose S satisfies this condition. Then in particular, since $\mathbf{K} \subsetneq S$ there is at least one formula A for which $\vdash_S A$, $\not\vdash_{\mathbf{K}} A$. Since $S \subseteq \mathbf{K} + \Diamond\top$, $\vdash_{\mathbf{K}+\Diamond\top} A$. Thus by Prop. 2.6.23, $\vdash_{\mathbf{K}+A} \Diamond\top$, contradicting $S \subsetneq \mathbf{K} + \Diamond\top$. ∎

EXERCISE 2.6.25 Either prove or (with a counterexample) refute the following generalization of Proposition 2.6.23: For any formulas A and B, if $\not\vdash_{\mathbf{K}} A$ and $\vdash_{\mathbf{K}+B} A$, then $\vdash_{\mathbf{K}+A} B$. ✽

So much for the rule of disjunction and some repercussions of a logic's providing it. We turn to variations on the rule, beginning with a (genuine) rule bearing a passing resemblance to the special case, Denecessitation, of the rule of disjunction. The rule involved here will show up again in Section 4.7 (Proposition 4.7.24 and Exercise 4.7.26), as well as in a 'hint' for Exercise 2.6.38(i).

PROPOSITION 2.6.26 *A normal modal logic S is closed under the rule:*

$$\frac{\Box A \vee A}{A}$$

if and only if $S \supseteq \mathbf{KT}$.

Proof. 'If': If $S \supseteq \mathbf{KT}$ then $\Box A \to A$ is S-provable for any A, and for a given A this implication together with a premiss for the application of the above rule have the conclusion of the rule as a truth-functional consequence.

'Only if': Let us first observe that $\vdash_{\mathbf{K}} \Box(\Box p \to p) \vee (\Box p \to p)$, since $\Box p \vee (\Box p \to p)$ is a substitution instance of the truth-functional tautology $q \vee (q \to p)$. Since $p \to (\Box p \to p)$ is another substitution instance of a tautology, it too is **K**-provable and so therefore by the fact that $\Box$ is monotone in any normal modal logic we get

$$\Box p \to \Box(\Box p \to p)$$

provable in **K**. But this and the earlier disjunction have as a truth-functional consequence the formula, $\Box(\Box p \to p) \vee (\Box p \to p)$, we were to show **K**-provable. Since this formula is accordingly provable in any normal modal logic, any such logic closed under the rule we are considering, applied taking A as the **T** axiom $\Box p \to p$, yields the latter as conclusion and we are accordingly in some $S \supseteq \mathbf{KT}$. ∎

For later reference (in Section 5.6) mention needs to be made here of what we may call the *modified rule of disjunction*, a variation on the rule of disjunction which appears from time to time in the literature. (For example: van der Hoek et al. [510], p. 27, and [511], p. 397.) Again, we may regard S's enjoying this rule as S's satisfying a condition for each n, namely the following:

for all $A_1, \ldots, A_n$, if $\vdash_S \Box A_1 \vee \ldots \vee \Box A_n$ then $\vdash_S \Box A_i$ for some $i \in \{1, \ldots, n\}$.

Thus, unlike the rule of disjunction itself, for the modified rule we do not peel off the occurrences of $\Box$ from the disjuncts; for those familiar with intuitionistic and intermediate logics, one might say that what we have here is the Disjunction Property for $\Box$-formulas.[103] Note that, by contrast with the rule of disjunction itself where the 1-ary rule was Denecessitation, the $n = 1$ case of the above condition is automatically satisfied.

EXERCISE 2.6.27 (*i*) Is the following claim correct? (Justify your answer.) "If a normal modal logic has the rule of disjunction, then it has the modified rule of disjunction."

(*ii*) Is the following claim correct. (Again, justify.) "If a normal modal logic has the modified rule of disjunction, then it has the rule of disjunction."

(*iii*) Prove that if S is a normal modal logic closed under Denecessitation then S has the rule of disjunction if and only if S has the modified rule of disjunction. ✣

To establish (directly) the modified rule of disjunction for a normal modal logic, we make a slight variation on the proof that this or that logic enjoys the standard rule of disjunction (the proof of Thm. 2.6.1, p. 113, for instance). Instead of letting $\mathcal{M}_{x_i}$, or $\mathcal{M}_i$ for short, be the submodel of $\mathcal{M}_i$ generated by x_i ($i = 1, \ldots, n$), as in the proof of Theorem 2.6.1, we discard the generating point x_i except when $x_i R_i x_i$, and adjoin the new point w so that – mimicking the earlier proof for **K**, $R(w) = \bigcup_{i=1}^{n} R_i(x_i)$, with corresponding alterations to the **T** and **4** variants of the earlier proof (of Thm. 2.6.1, that is).

Digression. Several further variations on the rule of disjunction are given an interesting discussion in Williamson [1181]. What follow are a few remarks on a subsequent description, *inter alia*, of that paper, in §7 of Kracht [666]. Kracht claims to follow Hughes and Cresswell [536], p. 96, in saying that a logic S provides the rule of disjunction when for all n, the condition above for the *modified* rule of disjunction is satisfied,[104] and cites Williamson [1181] for a 'strong rule of disjunction' which is the (standard) form we have been discussing, i.e., like that just given but requiring the provability of one of the A_i, rather

---

[103] A logic with the disjunction property is one which proves at least one disjunct of any disjunction it proves. See 6.41 of Humberstone [594] for further information and references.

[104] In fact, there is also a typographical error in Kracht's discussion, in that where he refers to the admissibility for each n of the rule $\langle \{\Box \bigvee_{i<n} p_i\}, \{\Box p_i : i < n\} \rangle$, what he actually means is $\langle \{\bigvee_{i<n} \Box p_i\}, \{\Box p_i : i < n\} \rangle$; it is this which is being criticized in the text above, in that the second set in the pair should be $\{p_i : i < n\}$ rather than $\{\Box p_i : i < n\}$. (The idea of this 'pairs of sets' notation is that the rule is admissible – cf. note 14, p. 19 – if any substitution mapping all formulas in the first set to provable formulas maps at least one in the second set to a provable theorem.)

2.6. THE RULE OF DISJUNCTION

than one of the $\Box A_i$, whenever $\Box A_1 \vee \ldots \vee \Box A_n$ is. In fact in the passage cited, Hughes and Cresswell just give the standard form and not what Kracht calls the rule of disjunction, and there is no reference to any 'strong rule of disjunction' in Williamson [1181]; what Kracht ([666], p. 543) describes under this terminology is just the standard rule of disjunction, and it is simply called the rule of disjunction by Williamson, who does also describe ([1181], p. 89) what it takes for S to provide the *weak* rule of disjunction, meaning by this that whenever we have $\vdash_S \Box^{k_1} A_1 \vee \ldots \vee \Box^{k_n} A_n$ for all $k_1, \ldots, k_n$, then $\vdash_S A_i$ for some i $(1 \leq i \leq n)$.[105]

Theorem 6.5 of Hughes and Cresswell [536] is our Coro. 2.6.9 above, p. 117, though differently proved, since we wanted to take a route that included Theorem 2.6.5. Hughes and Cresswell's argument is more direct, generalizing the proof of Theorem 2.6.1 above: Just take the set $\{\neg\Box A \mid \nvdash_S A\}$ and this will, if S-consistent, have a maximal (S)-consistent extension, in which the only $\Box$-ed formulas are S's theorems, to which all points in the canonical frame are accessible since those theorems belong to all maximal consistent sets. The rule of disjunction serves to establish the consistency of the set $\{\neg\Box A \mid \nvdash_S A\}$, because the inconsistency of any finite subset of size n contradicts the hypothesis that S enjoys the rule of n-ary disjunction. Hughes and Cresswell's formulation of this result says that for S providing the rule of disjunction, the canonical frame[106] for S is strongly generated, by this last phrase meaning not just that it is point-generated, but that there is some point which bears the accessibility relation itself (rather than its reflexive transitive closure) to every point – as in our formulation of Coro. 2.6.9. Summarising Hughes and Cresswell's discussion, Kracht [666], p. 542, says that what they show is that if a logic "provides the rule of disjunction then the canonical frame is generated by a single point, which is the set $\{\neg\Box A \mid \nvdash_S A\}$".[107] Well, first, as just remarked, what is shown is not just that the canonical frame is point-generated, but that there is a point to which all elements are accessible in one step, and secondly, *all* maximal consistent extensions of the set $\{\neg\Box A \mid \nvdash_S A\}$ are points to which all points are accessible. This set itself is of course not maximal consistent w.r.t. any normal modal logic – all it has in it are negated $\Box$-formulas! – so Kracht's comment about a single point "which is the set $\{\neg\Box A \mid \nvdash_S A\}$" seems to be an oversight. There is a danger of ambiguity in talking about a frame being generated, or strongly generated, as Hughes and Cresswell say for the 'one R-step' version, by a single point. What the word 'single' does in cases such as the present one is to indicate an $\exists\forall$ reading rather than a $\forall\exists$ reading, even though in other contexts what the word does instead is to strengthen an $\exists$ to an $\exists!$ ("there exists exactly one"). The latter meaning may have affected Kracht's formulation.

In fact there is a connection – or at least an apparent connection – between Kracht's weakening of the result attributed to Hughes and Cresswell (though in fact as the following quotation show, it is already in Lemmon and Scott [705]) by dropping the "strongly" from their formulation using "strongly generated", and Williamson's weak rule of disjunction. The following is from Williamson [1181], p. 95:[108]

> **KD!** also constitutes a counterexample to the natural extension of Lemmon and Scott's semantic characterization of their rule to the weak rule. They show ([705], p. 45) that a consistent normal system provides the rule of disjunction iff its canonical frame $\langle W, R \rangle$ (...) is *left-directed*, in the sense that for any $x_1, \ldots, x_n \in W$ there is a $y \in W$ such that $yRx_1, \ldots, yRx_n$. One might correspondingly suppose that a normal system provides the weak rule iff its canonical frame is *ancestrally left-directed*, in the sense that for any $x_1, \ldots, x_n \in W$ there is a $y \in W$ such that $yR^{j_1}x_1, \ldots, yR^{j_n}x_n$ for some $j_1, \ldots, j_n \geq 0$. This condition seems to stand to the Lemmon–Scott condition just as the weak rule stands to the Lemmon–Scott rule. One can indeed show that if the canonical frame of a normal

---

[105] Personally, I don't like the invisible quantification over n here, and would rather say that when this condition is satisfied (i.e., for satisfied all $k_1, \ldots, k_n$) for a *given* choice of n, S provides the n-ary weak rule of disjunction – or the weak n-ary rule of disjunction, if you prefer – and then say that having the weak rule of disjunction *tout court* is a matter of providing the n-ary rule for each n.

[106] In fact Theorem 6.5 of [536] is worded in terms of the canonical *model*

[107] Notation adjusted here; in fact Kracht uses the schematic letter φ rather than A. It should be noted that the result we are considering here was already in Lemmon and Scott [705], though proved in a slightly different way from that given by Hughes and Cresswell.

[108] Minor notational adjustments have been made for conformity with our discussion.

system is ancestrally left-directed, then the system provides the weak rule of disjunction. However, the converse fails.

As you might expect from the opening sentence of this quotation, Williamson then proceeds to show how the canonical frame for **KD!** fails to satisfy the condition in question, while this logic all the same does provide the weak rule of disjunction. If the canonical frame for a normal modal logic satisfied Williamson's ancestral left-directedness condition as stated, i.e., in terms of finite sets of points having a common R^*-predecessor (R^* being the ancestral – or reflexive transitive closure – of R), then, reasoning as in the proof of Theorem 2.6.8 we could lift this to arbitrary sets of points and derive from what he describes as the expected result, the following variant of that theorem: If S enjoys the weak rule of disjunction then, where $\langle W_S, R_S \rangle$ is the canonical frame for S, for every $X \subseteq W_S$ there exists some $w \in W_S$ with $X \subseteq R_S^*(w)$. Then, à la Coro. 2.6.9, we should have that for S providing the weak rule of disjunction then there exists some $w \in W_S$ such that for all $x \in W_S$, wR_S^*x, which is precisely to say that $\langle W_S, R_S \rangle$ is generated by such a w. However, as Williamson indicates, the hypothesis on which this line of reasoning is premised – that providing the weak rule secures ancestral left-directedness – is in fact false. A residual question remains (in addition to the various questions explicitly posed in [1181]): is there a simple syntactic characterization of the property of having a point-generated canonical frame? **End of Digression.**

There was something more than could have been extracted from the proof given above of Theorem 2.6.1, specifically in respect of **K**. For this case, recall, we started with $A_1, \ldots, A_n$, for which $\nvdash_\mathbf{K} A_1, \ldots, \nvdash_\mathbf{K} A_n$, and showed that $\nvdash_\mathbf{K} \Box A_1 \vee \ldots \vee \Box A_n$, by adjoining a new point w to models generated by points x_i (with A_i false at x_i) and adding the pairs $\langle w, x_1 \rangle, \ldots, \langle w, x_n \rangle$ to the union of the original accessibility relations to obtain the accessibility relation of the new model. Now all we used this for was to conclude that no $\Box A_i$ was true at w in the new model, since this sufficed for the disjunction of these formulas to be **K**-unprovable. For this we needed only that, where R is the accessibility relation of the new model, $\{x_1, \ldots, x_n\} \subseteq R(w)$. We did not exploit, in other words, the full force of what this particular construction – unlike the variants given in the proof of Thm. 2.6.1 to handle the cases of **KT** and **K4** – supplies, namely that $\{x_1, \ldots, x_n\} = R(w)$. That is more than enough by way of help for the following:

EXERCISE 2.6.28 Say that S provides the *conditional rule of disjunction* when for any n and any formulas A and $B_1, \ldots, B_n$, if we have $\vdash_S \Box A \to (\Box B_1 \vee \ldots \vee \Box B_n)$ then for some i ($1 \leq i \leq n$), we have $\vdash_S A \to B_i$. Show that **K** provides the conditional rule of disjunction. ✠

Note that the conditional rule of disjunction as just defined,[109] could equally well be formulated with $\vee$ throughout, in the following terms:

$$\vdash_S \Diamond A \vee \Box B_1 \vee \ldots \vee \Box B_n \Rightarrow \vdash_S A \vee B_i \text{ for some } i \ (1 \leq i \leq n).$$

Note also that for normal modal logics, this property implies satisfaction of the rule of disjunction; indeed for any given n, the n-specific case of the condition in Exc. 2.6.28 implies the n-ary rule of disjunction, since we can take A as $\Box\top$. (In the thoroughly disjunctive formulation just given, with $\Diamond A$, take $A = \bot$.)

We could combine elements of the modified rule of disjunction, which leaves occurrences of $\Box$ intact, with the conditional rule of disjunction, and say that S provides the *conditional modified rule of disjunction* when for any n and any formulas A and $B_1, \ldots, B_n$, if we have $\vdash_S \Box A \to (\Box B_1 \vee \ldots \vee \Box B_n)$ then for some i ($1 \leq i \leq n$), we have $\vdash_S \Box A \to \Box B_i$. (We return to the conditional rule proper, after Exercise 2.6.31.)

EXERCISE 2.6.29 Of the logics mentioned in Theorem 2.6.1, which provide the conditional modified rule of disjunction? ✠

---

[109] In Humberstone [596] S's providing the conditional rule of disjunction (mentioned also at p. 40 of [603] under a different description) is put – for reasons there explained – in terms of $\Box$'s being 'minimally normal' in S, and it is left as an open question whether $\Box$ is minimally normal in any proper extension of **K**.

2.6. THE RULE OF DISJUNCTION

Digression. Van der Hoek *et al.* [510] and [511] take an interest in which particular formulas A satisfy the condition, relative to S, that for all $B_1, \ldots, B_n$, if $\vdash_S \Box A \to (\Box B_1 \vee \ldots \vee \Box B_n)$, then for some i ($1 \leq i \leq n$), we have $\vdash_S \Box A \to \Box B_i$, considering this as one possible explication of a notion of *honesty* with some currency in AI-oriented treatments of epistemic logic (originating in Halpern and Moses [428]), and roughly amounting the idea that an honest statement (represented here by a formula) is one that could be used to convey the whole of someone's knowledge. (So $\Box$ is given an epistemic reading here, with S being a logic supporting such a reading, say **KT** or **S4**.) On the current explication, this is identified with its not entailing that one knows something unless it is specific about which is entailed. $\Box p \vee \Box q$ is a paradigm case of a failure of honesty in the present highly technicalized sense, since to know this disjunction to be true it would have to *be* true, so one would have to know that p or else know that q, but $\Box(\Box p \vee \Box q)$ does not provably imply (in either of the candidate epistemic logics just parenthetically mentioned) either of $\Box p$, $\Box q$: so it could be true and also represent *all* one knew. (See also Van der Hoek *et al.* [509], other aspects of which will occupy us in Section 5.3; the issues involved were perhaps first raised in Halpern and Moses [428].) These considerations are closely connected with non-monotonic applications of epistemic logic, as will be clear from the references cited in this Digression, to which should be added Stalnaker [1078] and Moore [824]. Stalnaker defines a set of formulas to be *stable* if it is deductively closed (so this is really a logic-relative notion, since we are requiring that if the logic in question proves $(A_1 \wedge \ldots \wedge A_n) \to B$ and the set contains each of the A_i, it must contain B) and further contains $\Box A$ whenever it contains A, and contains $\neg \Box A$ whenever it does not contain A. These last two features inspire the terminology of 'autoepistemic logic' (found in [824][110]) since if "$\Box A$" is glossed as "I know that A", they amount to recording the contents of one's own state of mind. The second of these two features means that a consistent modal logic (considered as a set of formulas S) which does not satisfy the rule of disjunction has no chance of being stable in this sense. For suppose the failure of the rule of disjunction is witnessed by: (i) $\Box A \vee \Box B \in S$, (ii) $A \notin S$, (iii) $B \notin S$. The condition in question requires that $\neg \Box A \in S$ and $\neg \Box B \in S$ in view of (ii) and (iii), implying that S is inconsistent in view of (i). Normal modal logics to which such considerations are applied in Stalnaker [1078] include **S5** and **K45**, sometimes considered as candidate epistemic and doxastic logics respectively (particularly implausibly in the former case, as we shall see in Section 5.2). Even without such autoepistemic considerations, there may seem to be something odd about a normal doxastic logic whose theorems of the form $\Box A$ are taken as saying that a rational subject believes that A, lacking the rule of disjunction, since it seems to be telling one that there is no irrationality in failing to believe that A when $\Box A$ is unprovable (as it will be if A is unprovable) and no likewise in the case of B, but anyone taking up both options falls foul of the instruction to believe A or else believe B implicit in the provability of $\Box A \vee \Box B$. On the other hand, perhaps this is no more odd than the familiar deontic situation in which we have – to speak loosely – an obligatory disjunction with neither disjunct obligatory. ***End of Digression.***

With the various adaptations of the 'adjunction of a new point' construction given in the proof of Theorem 2.6.1 which aimed to produce models on reflexive and transitive frames and thereby establish the Rule of Disjunction for **KT** and **K4** (and, in combination, for **S4**), the above argument, with $\{x_1, \ldots, x_n\} = R(w)$, is not available. It is not hard to see that the would-be conclusion of this argument is false for these logics – that is, that the property Exercise 2.6.28 demands a proof of is not one that these logics possess. This is very clear in the case of logics containing **4** but not $\mathbf{T}_c$ – such as **K4**, **S4** and **KW** mentioned in Thm. 2.6.1 – since the $n = 1$ case of the property leads from the former to the latter. It is slightly less obvious in the case of **KT**:

EXAMPLE 2.6.30 Note that $\vdash_{\mathbf{KT}} \Diamond(p \to \Box p)$. Thus the negation of this formula provably implies all formulas in **KT**, and so in particular, taking $\Box \bot$ as the formula to be implied, we have

$$\vdash_{\mathbf{KT}} \Box(p \wedge \neg \Box p) \to \Box \bot.$$

---

[110]This is reproduced as Chapter 6 of Moore [825], Chapters 7 and 8 of which also address this theme.

Taking $n = 1$ in the condition mentioned under Exc. 2.6.28, we should be able to remove the $\Box$s from the antecedent and consequent here, telling us that, if **KT** satisfied that condition, we should have $\vdash_{\mathbf{KT}} \neg(p \wedge \neg\Box p)$, which is of course not the case (the formula in question being again $\mathbf{T}_c$, rewritten slightly). ◀

EXERCISE 2.6.31 Concerning the $n = 1$ case of the conditional rule of disjunction, which amounts, as with Denecessitation, to the admissibility of a genuine rule ("Inverse Monotony", one might say):

$$\frac{\Box A \to \Box B}{A \to B}$$

(i) Show that a consistent normal modal logic S is closed under this rule if and only if its canonical frame $\langle W_S, R_S \rangle$ satisfies the following strengthening of the condition of converse seriality – see the discussion immediately after Exercise 2.6.7 2.6.7 – that for all $x \in W_S$ there exists $w \in W_S$ with $R_S(w) = \{x\}$. (Suggestion for the 'only if' direction: consider the set $\{\Diamond A \,|\, A \in x\} \cup \{\Box\neg B \,|\, B \notin x\}$. Check its S-consistency and that any maximal S-consistent extension of this set will serve as the promised w.)

(ii) Is the above inverse monotony rule admissible for **K**? (Justify your answer.)

(iii) Show that a normal modal logic is closed under the inverse monotony rule if and only if it is closed under the rule (called Cancellation in Humberstone and Williamson [603]; see further the references there cited, as well as Williamson [1187] and Humberstone [599] for subsequent work):

$$\frac{\Box A \leftrightarrow \Box B}{A \leftrightarrow B}$$

We encountered this rule briefly above in Remark 2.5.61, p. 102. ✠

It is worth seeing how much of the development that we have seen in the case of the rule of disjunction would go through for the conditional form of the rule. (Exercise 2.6.31 makes a start on this parallel development.) Specifically, let us recall Theorem 2.6.8, which told us (on the assumption that S is a consistent normal modal logic):

"If S provides the rule of disjunction then, where $\langle W_S, R_S \rangle$ is the canonical frame for S, for every $X \subseteq W_S$ there is some $w \in W_S$ with $X \subseteq R_S(w)$."

In particular, for simplicity, invoking the *conditional* rule of disjunction (from Exc. 2.6.28), would a parallel development lead to the following conclusion (with the same assumption concerning S as above)?

"If S provides the conditional rule of disjunction then, where $\langle W_S, R_S \rangle$ is the canonical frame for S, for every $X \subseteq W_S$ there is some $w \in W_S$ with $X = R_S(w)$."

Why is this of any interest? Because we have an *a priori* assurance from general set theory that for no set U does there exist a surjective map $f : U \longrightarrow \wp(U)$. (Recall Cantor's proof of this – the inspiration behind Russell's Paradox – which considers apropos of any mapping f from U to $\wp(U)$, the subset $\{x \in U \,|\, x \notin f(x)\}$, and concludes that this set cannot itself be $f(u)$ for any $u \in U$; thus f cannot be a surjection.) With this in mind, forget about canonical frames (for S as above) and just consider any frame $\langle W, R \rangle$: there can't be a surjective mapping from W to $\wp(W)$. But this means that the following condition cannot be satisfied: for all $X \subseteq W$ there is some $w \in W$ with $X = R(w)$. For if it were, the function $R(\cdot)$ would be a surjection of the kind ruled out by Cantor's Theorem.[111]

EXERCISE 2.6.32 By the reasoning just given, Theorem 2.6.8 cannot be adapted by replacing references to the rule of disjunction with references to the conditional rule of disjunction and changing "$\subseteq$" to "$=$".

---

[111] Section 3 of Humberstone [596] discusses these matters, in the terminology mentioned in note 109 above.

2.6. THE RULE OF DISJUNCTION

But where does the reasoning break down? What happens to the analogue, with the above changes, of Theorem 2.6.5? (Exercise 2.6.31 addressed the $n = 1$ case.) Is it here, or is it in appealing to that result in the proof of Theorem 2.6.8 that the argument breaks down, and how?   ✠

In Coro. 2.5.6, we saw that **K** was closed under Löb's Rule: from $\Box A \to A$ to A. Let us consider what happens to this rule if the implicational premiss is replaced by its converse. Are any normal modal logics closed under the following rule?

$$\frac{A \to \Box A}{A}$$

Of course one thing that would – by Necessitation – secure the provability of the premiss in a normal modal logic would be the provability of the conclusion. But something else that would do the same would be the refutability of (i.e., provability of the negation of) the conclusion, since this conclusion is the antecedent of the premiss ($B \to C$ being a truth-functional consequence of $\neg B$). In particular, then, applied to the **K**-provable premiss $\bot \to \Box \bot$ this rule would deliver the conclusion $\bot$. Thus the only normal modal logic (and in fact, the only modal logics) closed under the rule would be the inconsistent logic, containing every formula. One is accordingly led to consider a metalinguistically disjunctive variant on the rule, called by Williamson, in [1181] and elsewhere, the *rule of margins*, which a normal modal logic S is said to provide when for all formulas A, $\vdash_S A \to \Box A$ implies that either $\vdash_S \neg A$ or $\vdash_S A$. As with the rule of disjunction, this definition does not strictly speaking address the admissibility of a rule (because of the "or"), and this condition is described in such terms only by courtesy. In fact, in [1181], this 'rule of margins' property is shown to follow from a condition not detailed here, called by Williamson the *alternative rule of disjunction*, which is a variant on his weak rule of disjunction, as described in the Digression beginning on p. 122 above. More information on all these phenomena, as well as motivation for the rule of margins, and the reason for the label, may be found in [1181]; see also Example 5.2.6 below (p. 377). One question which it is instructive to ponder before reading on (where an answer is provided) is as follows.

EXERCISE 2.6.33 Does **K** provide the rule of margins? (Justify your answer.)   ✠

To make some space between this question and its answer, we ask a few more (which are rather easier):

EXERCISE 2.6.34 (*i*) What does closure under Löb's Rule imply (for a normal modal logic) about the provability or refutability of A, given that $A \to \Diamond A$ is provable (in the logic)?
(*ii*) What does providing the rule of margins imply (for a normal modal logic) about the provability or refutability of A, given that $\Diamond A \to A$ is provable (in the logic)?
(*iii*) In the Digression on p. 98 the MacIntosh rule: from $A \to \Box A$ to $\Diamond A \to A$ was mentioned. Show that any normal modal logic providing the rule of margins is closed under this rule.   ✠

Let us return to Exc. 2.6.33. One might expect that if neither $\neg A$ nor A is provable in **K**, so that we can have an A-verifying point in a model and also have an A-falsifying point in a model, the absence of any conditions on the accessibility relation for the class of models (or of their underlying frames) should make it possible to set things up in such a way that a point of the former kind has accessible to it a point of the latter kind, and at such a point $A \to \Box A$ would be false, and therefore not **K**-provable. This hand-waving argument sketch, if successfully fleshed out, would show by contraposition that **K** provides the rule of margins. But a hitch in the fleshing-out procedure arises immediately: we assumed that there would be nothing to stop an A-verifying point from having an A-falsifying point accessible to it, but what if the only points verifying A in any model lack successors altogether, as in the case in which A is (or indeed **K**-implies) $\Box \bot$? Indeed, this choice of A offers a counterexample to the hypothesis that **K** provides the rule of margins since we have:

$$\vdash_{\mathbf{K}} \Box\bot \to \Box\Box\bot \quad\quad\text{while}\quad\quad \nvdash_{\mathbf{K}} \neg\Box\bot \text{ and } \nvdash_{\mathbf{K}} \Box\bot.$$

So **K** does not provide the rule of margins.

Williamson [1181], p. 97, writes as follows; the quotation begins with a reference to his 'alternative rule of disjunction', alluded to (though not spelt out) above:[112]

> Nevertheless, the rule of margins has many of the consequences of the alternative rule. If S is normal and provides the marginal rule, it extends **KD**, and if it is consistent, it does not extend **K4** or **K5**.

Let us verify the first assertion in the second sentence of this passage. We have already seen that if a normal modal logic S provides the rule of margins, then because it contains $\Box\bot \to \Box\Box\bot$ it must contain either $\neg\Box\bot$ or else $\Box\bot$. In the former case, we have $\vdash_S \Diamond\top$, so $S \supseteq \mathbf{KD}$; in the latter case $S \supseteq \mathbf{KVer}$ and S contains every formula beginning with a $\Box$. Thus $\vdash_S p \to \Box p$, so, supposing S to provide the rule of margins, we have $\vdash_S p$ or $\vdash_S \neg p$, in either of which cases, S is inconsistent. (Why?) Thus, again, $S \supseteq \mathbf{KD}$.

EXERCISE 2.6.35 Prove the second assertion from the passage quoted above, namely, that if S is consistent and provides the rule of margins, then S extends neither **K4** nor **K5**. ✠

Another interesting rule is a variation on Löb's Rule – now passing away from the rule of disjunction altogether – we obtain, not by replacing the $\Box A \to A$ premiss by its converse, as above, but by negating its consequent, and having this be the conclusion, too: from $\Box A \to \neg A$ to $\neg A$. Considered in connection with arbitrary modal logics, one would have to note that another way to have the premiss provable would be to have the negation of its antecedent provable, but – as indeed with Löb's Rule itself – this consideration does not arise for **K**, since no formulas of the form $\neg\Box A$ are provable in **K**. (Why not?) This motivates the following question, which will be answered below but which is well worth pondering before reading on:

EXERCISE 2.6.36 Is **K** closed under the rule just described (i.e., the rule taking us from $\Box A \to \neg A$ to $\neg A$)? ✠

Before proceeding to answer this question, let us reformulate what the premiss of the rule says in more suggestive terms. If the premiss is **K**-provable for a given formula A, then for another formula $B = \neg A$, we have $\Box\neg B \to B$, and thus $\Diamond B \vee B$ provable in **K**, and thus valid on every frame. This tells us that the formula B has the following significant property: for any point x in any model, either B is true at x in that model or else B is true at some successor of x. Such formulas are called *weak Hughes formulas* in Humberstone [589], as a special case of the more general property of being true at some point in every model (and the reason for naming weak Hughes formulas after G. E. Hughes is explained). The boring case would be in which this is so for B because B itself is true at every point in every model – which for B as $\neg A$ in the above rule, amounts to the case in which the conclusion of the rule is **K**-provable; so the question is whether there are any 'non-boring' cases.

EXAMPLE 2.6.37 Take B as the formula $p \to \Box p$ (alias $\mathbf{T}_c$). One easily checks (using the Kripke semantics) that $B \vee \Diamond B$ is valid on every frame, and hence that this B is a weak Hughes formula. (Note further that since in **KT** itself the first disjunct provably implies the second, as a special case of the dual form of **T**, so for the current B, $\vdash_{\mathbf{KT}} \Diamond B$. In other words, $\Diamond(p \to \Box p)$ is valid on all reflexive frames. We do not need as strong a condition as reflexivity – or, syntactically, that **T** should be provable – for this conclusion to hold, since the formula just cited is **K**-equivalent simply to $\Box p \to \Diamond\Box p$, i.e., $\mathbf{5}_c$, in view of the third equivalence mentioned under Exc. 1.3.17(*ii*). The formula we have been considering here,

---

[112] In fact Williamson uses 'E' for '5' and so the passage ends "or **KE**"; the nomenclature has been adjusted to match our conventions.

$\Diamond(p \to \Box p)$, is the first formula, called $\mathbf{T}_1^\Diamond$, of a sequence of formulas mentioned on p. 85, all of which were observed to be **KD4**-provable in Prop. 2.5.35.) ◀

Returning to Exercise 2.6.36, we need to recall that we have been operating with the negation of a candidate A, as this schematic letter figures in the rule figuring there. So to pass back to a suitable A, we take (a formula equivalent to) the negation of the B of Example 2.6.37: $p \land \neg \Box p$. This settles Exc. 2.6.36 negatively, since for this choice of A we have $\vdash_\mathbf{K} \Box A \to \neg A$, while $\nvdash_\mathbf{K} \neg A$. (Compare Example 2.6.30.)

EXERCISE 2.6.38 *(i)* Define a formula A to be a *strong Hughes formula* if for any point in any model, either A is true at that point or else A is true at all that point's successors. Give an example of a strong Hughes formula which is not valid on every frame. (*Hint:* See Proposition 2.6.26.)

(ii) Is every strong Hughes formula a weak Hughes formula? (Justify your answer with a proof or a counterexample.)

(iii) Is the set of all weak Hughes formulas a modal logic? (Justify your answer. Note that the question does not ask whether this set of formulas is a *normal* modal logic.)

(iv) Is the set of all strong Hughes formulas a modal logic? (Justification again required.) ✠

2.7 Revision Exercises

Here we collect some exercises on material presented up to this point, which could be used for purposes of revision or assessment. Some of exercises, however, involve new concepts or material – especially those with a 'Preamble' to them – and are included here because placing the material earlier would have interrupted the flow of exposition. In a few cases the answers are explicit or all but explicit in other parts of this work, but it would be less onerous to work them out than to chase them down. (Naturally, we have not given cross-references in these cases, since the intention is that these exercises should be usable as assessment tasks.) The final three exercises relate mainly to material in the Appendix (Chapter 7, especially Sections 7.2, 7.3) on natural deduction, so those who have chosen to begin with that material should skip to p. 141, at the end of the exercises bearing on the axiomatic presentation (and semantics) of Chapters 1 and 2.

EXERCISE 2.7.1 *(i)* Consider the formula $\Box(p \to q)$; answer with one or more of (a)–(e), concerning other formulas to which this formula is provably equivalent in **K** (A, B, being said to be provably equivalent in S when $\vdash_S A \leftrightarrow B$):

(a) $\Diamond p \to \Diamond q$; (b) $\Diamond p \to \Box q$; (c) $\Box p \to \Diamond q$; (d) $\Box p \to \Box q$; (e) none of (a)–(d).

(ii) Answer with one of more of (a)–(e) above but this time addressing the question of which formulas are provably equivalent in **K** to the formula $\Diamond(p \to q)$.

(iii) Do the answers to *(i)* and *(ii)* change if we replace the reference to **K** by a reference to **KD**?

(iv) Do the answers to *(i)* and *(ii)* change if we replace the reference to **K** by a reference to **KD!**? ✠

EXERCISE 2.7.2 *(i)* Which of the following is the smallest normal modal logic containing the 'extensionality' principle $(p \leftrightarrow q) \to (\Box p \leftrightarrow \Box q)$? (No justification for your answer need be given; similarly with *(ii)* and *(iii)* below.)

K KT! KT$_c$ KD$_c$ KD! KVer.

(*ii*) Does the answer change (and if so, how) if we replace the occurrences of "↔" in the above formula with "→", and consider instead the formula $(p \to q) \to (\Box p \to \Box q)$?

(*iii*) What would the answer to (*i*) be if in place of the formula there given, we considered the formula $(p \leftrightarrow q) \leftrightarrow (\Box p \leftrightarrow \Box q)$? ✠

EXERCISE 2.7.3 This exercise concerns (mono)modal logics in general rather than specifically normal – or even congruential – modal logics. (The difference pointed out by Makinson [760], noted in the Digression on p. 39 above, arising for non-congruential modal logics depending on the choice of Boolean primitives will not affect the question. Another example of reasoning in a modal logic not even assumed to be congruential, though this time also making use of propositional quantification, may be found in Remark 4.4.17, p. 257 below.) Read the suggestion below only if in difficulties.

Show that the smallest modal logic containing all instances of the schema $A \lor \Box A$, contains some disjunction of (finitely many) formulas of the form $B \land \Box\Box B$.

Suggestion: One set of three formulas whose disjunction will be provable in the logic in question is the following (with superscripts for iteration): $\{p \land \Box^2 p, \Box p \land \Box^3 p, \Box^2 p \land \Box^4 p\}$. The disjunction of these formulas is a truth-functional consequence of the set of instances of the schema $A \lor \Box A$ we get by taking A successively as $p, \Box p, \Box\Box p, \Box\Box\Box p$. (This exercise was inspired by an example in Letz [712], which involves the unobvious validity of the inference in first order logic with function symbols from $\forall x(Rx \lor Rf(x))$ to $\exists x(Rx \land Rf(f(x)))$, except that here the individual variable is replaced by a propositional variable, the roles played by both the one-place predicate letter R and the one-place function symbol f are played by the operator $\Box$, the universal quantification is replaced by the use of a schema, and the existential quantification by a disjunctive formula. Thinking of the domain as comprising people, Letz describes the first-order inference above as proceeding from a premiss to the effect that everyone is either rich ("Rx") or has a rich father ("$Rf(x)$"), to the conclusion that someone who is rich has a rich paternal grandfather.) ✠

The next three exercises concern the normal modal logic **KB**.

EXERCISE 2.7.4 Say for each of the following formulas whether or not it is provable in **KB**.

(*i*) $(p \leftrightarrow q) \to (\Box\Diamond p \leftrightarrow \Box\Diamond q)$ (*ii*) $(p \to q) \to (\Box p \to \Box\Diamond q)$
(*iii*) $(p \to q) \to (\Box p \to \Diamond q)$ (*iv*) $(p \to q) \to \Box(\Box p \to \Diamond q)$
(*v*) $\Diamond\Box\Box p \to p$ (*vi*) $\Diamond\Box p \to \Box\Diamond p$
(*vii*) $\Box\Diamond p \to \Diamond\Box p$ (*viii*) $p \to \Box\Diamond\Box\Diamond p$ ✠

EXERCISE 2.7.5 For which (if any) of the following choices of A is the formula $\Box A \to A$ provable in **KB**: (*i*) $A = \Box p$, (*ii*) $A = p$, (*iii*) $A = \Box p \lor p$, (*iv*) $A = \Box p \to p$, (*v*) $A = \Box p \land p$, (*vi*) $A = \Diamond p$. (Reasons need not be given.) ✠

EXERCISE 2.7.6 In which (if any) case or cases does one of the formulas (*i*)–(*iv*) listed here differ in respect of provability in **KB** from the formula written immediately below it ((*v*)–(*viii*) respectively)?

(*i*) $\Box\Box p \to p$ (*ii*) $\Box\Box p \to \Box p$ (*iii*) $\Box\Box\Box p \to \Box p$ (*iv*) $\Box\Box\Box p \to p$
(*v*) $\Box\Box\bot \to \bot$ (*vi*) $\Box\Box\bot \to \Box\bot$ (*vii*) $\Box\Box\Box\bot \to \Box\bot$ (*viii*) $\Box\Box\Box\bot \to \bot$ ✠

EXERCISE 2.7.7 How are the logics **KB4** and **KB5** related? That is, do we have (*i*) **KB4** = **KB5**, or do we have (*ii*) **KB4** $\subsetneq$ **KB5**, or do we have (*iii*) **KB5** $\subsetneq$ **KB4**, or (*iv*): none of (*i*)–(*iii*)? (Justify your answer with semantic or syntactic considerations.) ✠

2.7. REVISION EXERCISES

EXERCISE 2.7.8 True or false? (Give reasons in each case.)
(i) For any normal modal logic S, we have $\mathbf{KD} \subseteq S$ if and only if some formula of the form $\Diamond A$ is provable in S.
(ii) For any normal modal logic S, we have $\mathbf{KD} \subseteq S$ if and only if some formula of the form $\Box B \to \Diamond C$ is provable in S.
(iii) Is every normal modal logic closed under the following rule?

$$\frac{\Box B \to \Diamond C}{\Box A \to \Diamond A}$$

✠

EXERCISE 2.7.9 This exercise concerns the logics $\mathbf{K}$, $\mathbf{KD}$ and $\mathbf{KD}_c$. For each of the four formulas below, say whether the formula is provable or otherwise in these logics. Thus if you thought formula (ii) was provable in $\mathbf{K}$ and $\mathbf{KD}$ but not $\mathbf{KD}_c$, you would answer "Yes, Yes, No" by (ii). (Of course this would be a silly answer, as you can see even without looking at what the formula is, since provability in $\mathbf{K}$ implies provability in $\mathbf{KD}_c$.)
(i) $((p \land \Box q) \to \Box r) \to ((p \land \Diamond q) \to \Box r)$
(ii) $((p \land \Diamond q) \to \Box r) \to ((p \land \Box q) \to \Box r)$
(iii) $((p \land \Box q) \to r) \to ((p \land \Diamond q) \to r)$
(iv) $((p \land \Diamond q) \to r) \to ((p \land \Box q) \to r)$.

✠

EXERCISE 2.7.10 (i) Does $\mathbf{K}$ have any theorems of the form $(\Box A \land \neg A) \lor (\Box B \land \neg B)$?
(ii) Does $\mathbf{S5}$ have any theorems of the form $(\Box A \land \neg A) \lor (\Box B \land \neg B)$?
(iii) Does $\mathbf{K}$ have any theorems of the form $\Diamond A$?
(iv) Does $\mathbf{K}$ have any theorems of the form $\Diamond A \land \Diamond \neg A$?
(v) Does $\mathbf{S5}$ have any theorems of the form $\Diamond A$?
(vi) Does $\mathbf{S5}$ have any theorems of the form $\Diamond A \land \Diamond \neg A$?

✠

EXERCISE 2.7.11 We use the notation introduced in Example 2.1.2, p. 36.
(i) Show that $Fm_{\mathbf{K}}^{\Box} \cap Fm_{\mathbf{K}}^{\Diamond} = \varnothing$. (I.e., no $\Box$-formula is $\mathbf{K}$-equivalent to a $\Diamond$-formula. *Suggestion*: note that every $\Box$ formula is provably implied by $\Box\bot$ in $\mathbf{K}$ and that every $\Diamond$-formula provably implies $\Diamond\top$ in $\mathbf{K}$.)
(ii) Show that $Fm_{\mathbf{S5}}^{\Box} = Fm_{\mathbf{S5}}^{\Diamond}$. (Though we cite the more famous $\mathbf{S5}$, this works for any extension of $\mathbf{K5}$!, and indeed many other logics too. Note that by the observation in Example 2.1.2, it suffices to show $Fm_{\mathbf{S5}}^{\Box} \subseteq Fm_{\mathbf{S5}}^{\Diamond}$.)
(iii) Either give an example of a normal modal logic S, showing it to be such an example, for which $Fm_S^{\Box} \cap Fm_S^{\Diamond} \neq \varnothing$ but $Fm_S^{\Box} \neq Fm_S^{\Diamond}$, or else show that no such example can be given.

✠

EXERCISE 2.7.12 Evaluate the following claims as true or false, giving proofs or counterexamples, as appropriate. (On the 'proof' side, anything from the discussion which follows these exercises – Sections 2.8–2.9 – may be used as support.)
(i) For all formulas A, we have $\vdash_{\mathbf{S5}} A$ if and only if $\vdash_{\mathbf{S4}} \Box A$.
(ii) For all formulas A, we have $\vdash_{\mathbf{S5}} A$ if and only if $\vdash_{\mathbf{S4}} \Diamond A$.
(iii) For all formulas A, we have $\vdash_{\mathbf{S5}} A$ if and only if $\vdash_{\mathbf{S4}} \Diamond\Box A$.
(iv) For all formulas A, we have $\vdash_{\mathbf{S5}} A$ if and only if $\vdash_{\mathbf{S4}} \Box\Diamond A$.

✠

EXERCISE 2.7.13 Here we write "$\Gamma \vdash_{\mathbf{K}} A$" to mean that any point in any model at which all formulas in the set Γ are true, also has A true at it. Which of the following claims are correct? (Reasons need not be given.) Note that in (ii)–(iv), formulas from the left of the "$\vdash$" in (i) are removed in turn.

(i) $\Box(p \to q), \Diamond p, \Diamond q \vdash_{\mathbf{K}} \Diamond(p \wedge q)$;

(ii) $\Diamond p, \Diamond q \vdash_{\mathbf{K}} \Diamond(p \wedge q)$;

(iii) $\Box(p \to q), \Diamond q \vdash_{\mathbf{K}} \Diamond(p \wedge q)$;

(iv) $\Box(p \to q), \Diamond p \vdash_{\mathbf{K}} \Diamond(p \wedge q)$.

(v) Which of the (true) answers to (ii)–(iv) would be different, if instead of "$\vdash_{\mathbf{K}}$" we had "$\vdash_{\mathbf{KD!}}$", where "$\Gamma \vdash_{\mathbf{KD!}} A$" means that any point in a model on a functional frame, at which all formulas in the set Γ are true, also has A true at it? (A functional frame is one in which each point has exactly one point accessible to it; this definition was given at p. 68. Note that although here defined semantically, $C_1, \ldots, C_n \vdash_{\mathbf{K}} A$ and $C_1, \ldots, C_n \vdash_{\mathbf{KD!}} A$ amount to the claim that $(C_1 \wedge \ldots \wedge C_n) \to A$ is provable in $\mathbf{K}$ or $\mathbf{KD!}$, respectively.) ✠

EXERCISE 2.7.14 Call a formula A *persistent* over a class of frames when for any $\langle W, R \rangle$ in the class, for all $x, y \in W$ with xRy, if A is true at x in a model on $\langle W, R \rangle$, then A is true at y in that model.

(i) Find formulas A, B, such that A and B are both persistent over the class of all frames and yet $\not\vdash_{\mathbf{K}} A \leftrightarrow B$.

(ii) Is it possible – give an example or explain why there isn't one – to have another formula C which is $\mathbf{K}$-equivalent to neither of the A, B, provided in answer to (i) (i.e. $\not\vdash_{\mathbf{K}} A \leftrightarrow C$ and $\not\vdash_{\mathbf{K}} B \leftrightarrow C$ with C also being persistent w.r.t. the class of all frames)?

(iii) If a formula is persistent over the class of reflexive transitive frames, does it follow that the formula is $\mathbf{S4}$-equivalent to a $\Box$-formula (i.e., where A is such a formula, must there exist a formula B for which $\vdash_{\mathbf{S4}} A \leftrightarrow \Box B$)? (Justify your answer.) ✠

EXERCISE 2.7.15 Answer the following seven questions, providing reasons:

(i) If for a frame $\langle W, R \rangle$, W contains exactly one element, does it follow that the formula $\Box p \to \Box\Box p$ is valid on $\langle W, R \rangle$?

(ii) If for a frame $\langle W, R \rangle$, W contains exactly two elements, does it follow that the formula $\Box p \to \Box\Box p$ is valid on $\langle W, R \rangle$?

(iii) If for a frame $\langle W, R \rangle$, W contains exactly three elements, does it follow that the formula $\Box p \to \Box\Box p$ is valid on $\langle W, R \rangle$?

(iv) If for a frame $\langle W, R \rangle$, W contains exactly one element, does it follow that the formula $p \to \Box\Diamond p$ is valid on $\langle W, R \rangle$?

(v) If for a frame $\langle W, R \rangle$, W contains exactly two elements, does it follow that the formula $p \to \Box\Diamond p$ is valid on $\langle W, R \rangle$?

(vi) If, for a frame $\langle W, R \rangle$, W contains exactly one element, does it follow that the formula $p \to \Box p$ is valid on $\langle W, R \rangle$?

(vii) If, for a frame $\langle W, R \rangle$, W contains exactly one element, does it follow that the formula $\Box p \to p$ is valid on $\langle W, R \rangle$? ✠

EXERCISE 2.7.16 (i) Which of the four modal formulas mentioned in the preceding question has the property that the smallest normal modal logic containing the formula also contains the formula (**.2**, often also called **G**): $\Diamond\Box p \to \Box\Diamond p$? Justify your answer.

(ii) Which of the four modal formulas mentioned in Exercise (Exc. 2.7.15, that is) has the property that the smallest normal modal logic containing the formula also contains the formula $\Diamond\Diamond\Diamond p \to \Diamond p$? Justify your answer. ✠

2.7. REVISION EXERCISES

EXERCISE 2.7.17 In Section 2.1 the following was said: The label "S4.1" (...) was also used historically for another extension of **S4**, namely **S4M**, but, as has often been recognised since, this is confusing since this logic is not a sublogic of **S4.2**." On p. 146 of Chellas [171] there is a similar suggestion in the case of the label "**S4.4**", since **S4.4** is said there not to be an extension of **S4.3** or **S4.2**. Recall that **S4.4** is the normal extension of **S4** by (all instances of) the schema $(A \land \Diamond \Box A) \to \Box A$, and that the effect of .4 was detailed in Exc. 2.5.52(i) (p. 97; the schema just cited was the dual form from Table 2.1 on p. 33).

(i) Show, either syntactically (by a suitable modal deduction, that is) or semantically (by a consideration of the point-generated frames for **S4.4** and **S4.3**) that, contrary to Chellas's claim, in fact **S4.3** $\subseteq$ **S4.4**. (Indeed, more specifically, **S4.3** $\subsetneq$ **S4.4**.)

(ii) Would it be correct to say that **K.3** $\subseteq$ **K.4**? (Give reasons.)

Apropos of part (i) here: although some of the errors in the original printing of [171] were corrected in reprintings, this is one of several that were not (at least by the appearance of 1999 reprint). The following exercise concerns another such error. ✠

EXERCISE 2.7.18 Chellas writes at p. 144 (Exc. 4.56) of [171], concerning a list of 12 axiom schemes labelled with variously decorated versions of the letters "H" and "L" that they all yield the same logics as normal extensions of **S4**. The first of these two is labelled "H$^{++\Diamond}$" and the second "L$^+$" but we rename them here for convenience in the former case, and to match the label given in Exc. 2.5.8(ii) (p. 74) in the latter:

H $(\Diamond \Box A \land \Diamond \Box B) \to \Diamond \Box (A \land B)$ .3$^+$ $\Box(\Box A \to B) \lor \Box(\Box B \to A)$

Show that, despite Chellas's claim that **S4H** = **S4.3**$^+$, in fact **S4H** = **S4.2**, **S4.3**$^+$ = **S4.3**, and thus **S4H** $\subsetneq$ **S4.3**$^+$. (The present mistake was mentioned in note 15, p. 286, of Humberstone [584], having been first communicated in a letter to Chellas in 1982; a corresponding letter concerning the issue raised in Exercise 2.7.17 was sent in 1983. Chellas acknowledged the point in both cases but evidently was not as free to change the text for re-printing purposes as might have been desirable. The story of **H** will be continued in Section 4.5, beginning with p. 280, though in fact there we work with representative instances of the present schema – propositional variables replacing schematic letters, that is. ✠

EXERCISE 2.7.19 (i) Is the following true or false? (Give supporting argumentation or a counterexample, as appropriate.)

For every normal modal logic S and all formulas A, A', B, B', if $\vdash_S \Box A \to \Box A'$ and $\vdash_S \Box B \to \Box B'$, then $\vdash_S \Box(A \lor B) \to \Box(A' \lor B')$?

(ii) Same question as (i), but this time replace all occurrences of "$\lor$" by "$\land$". ✠

EXERCISE 2.7.20 (i) If $\langle W, R, V \rangle \models \Box A \to A$ for all formulas A, does it follow that R is reflexive (i.e. does this hold for every model $\langle W, R, V \rangle$)?

(ii) If $\langle W, R \rangle \models \Box A \to A$ for all formulas A, does it follow that R is reflexive (i.e. does this hold for every frame $\langle W, R \rangle$)?

(iii) If $\langle W, R, V \rangle \models \Box A \to \Diamond A$ for all formulas A, does it follow that R is serial?

(iv) If $\langle W, R \rangle \models \Box A \to \Diamond A$ for all formulas A, does it follow that R is serial?

Justify your answers. (Note that what follows the "If" in (i) is the hypothesis that all formulas of the form $\Box A \to A$ are true throughout a model, where in (ii) it is the hypothesis that all such formulas are valid on a frame; likewise with (iii) and (iv), except that here we are concerned with the schema $\Box A \to \Diamond A$.) ✠

EXERCISE 2.7.21 For this question we assume that the constants $\top$ and $\bot$ are amongst our primitive vocabulary. With their aid, it is possible to construct formulas containing no propositional variables – indeed each of them already *is* such a formula. These variable-free formulas we call *pure*. (E.g., the formula **D**, $\Box p \to \Diamond p$ is provably equivalent in **K** to the pure formula $\Diamond\top$. Saying A and B are provably equivalent in **K** just means that $\vdash_{\mathbf{K}} A \leftrightarrow B$.)

(*i*) Prove or refute the claim that (a) and (b) below are equivalent:

(a) For all models $\langle W, R, V\rangle$, all $w \in W$, $\langle W, R, V\rangle \models_w A$ implies $\langle W, R\rangle \models_w A$;

(b) A is provably equivalent in **K** to a pure formula.

(*ii*) Prove or refute the claim that (a) and (b) below are equivalent:

(a) For every model $\langle W, R, V\rangle$, $\langle W, R, V\rangle \models A$ implies $\langle W, R\rangle \models A$;

(b) A is provably equivalent in **K** to a pure formula. ✠

EXERCISE 2.7.22 Here we follow up the discussion on p. 38 about the truth-functional interpretation of $\Box$ under certain extreme conditions. Consider the two translations $(\cdot)^{\mathbf{T!}}$ and $(\cdot)^{\mathbf{Ver}}$ from modal to non-modal formulas defined as below, where the cases of $\neg$ and $\wedge$ are representative for any chosen Boolean primitives:

$$
\begin{array}{ll}
(p_i)^{\mathbf{T!}} = p_i & (p_i)^{\mathbf{Ver}} = p_i \\
(\neg A)^{\mathbf{T!}} = \neg(A^{\mathbf{T!}}) & (\neg A)^{\mathbf{Ver}} = \neg(A^{\mathbf{Ver}}) \\
(A \wedge B)^{\mathbf{T!}} = A^{\mathbf{T!}} \wedge B^{\mathbf{T!}} & (A \wedge B)^{\mathbf{Ver}} = A^{\mathbf{Ver}} \wedge B^{\mathbf{Ver}} \\
(\Box A)^{\mathbf{T!}} = A^{\mathbf{T!}} & (\Box A)^{\mathbf{Ver}} = \top.
\end{array}
$$

Show that in any model $\mathcal{M} = \langle W, R, V\rangle$ with $x \in W$:

(*i*) if $R(x) = \{x\}$ then for all formulas A, $\mathcal{M} \models_x A \leftrightarrow A^{\mathbf{T!}}$; and

(*ii*) if $R(x) = \varnothing$ then for all formulas A, $\mathcal{M} \models_x A \leftrightarrow A^{\mathbf{Ver}}$; and

(*iii*) that from (*i*) and (*ii*) we may conclude that for all A, $\vdash_{\mathbf{KT!}} A \leftrightarrow A^{\mathbf{T!}}$ and $\vdash_{\mathbf{KVer}} A \leftrightarrow A^{\mathbf{Ver}}$. ✠

EXERCISE 2.7.23 (*i*) Consider the smallest normal modal logic containing the formula

$$(\Box p \to p) \vee (\Box q \to \neg q).$$

Is this logic Halldén-incomplete? (*Reminder* – from the Digression on p. 39: a Halldén-incomplete logic is one in which some disjunction $A \vee B$ is provable, while neither A nor B is, for some formulas A and B which do not have a propositional variable in common. Justify your answer, helping yourself without further justification to any of the following facts: **K** is Halldén-incomplete, while **KD**, **KT**, **S4** and **S5** are Halldén-complete; in the case of **K**, in case you are wondering, we have the following 'Halldén unreasonable' disjunction provable: $\Box p \vee \Diamond(q \vee \neg q)$. We could equally well use as an example $\Box\bot \vee \Diamond\top$, since here there are no variables at all, let alone any shared between the two **K**-unprovable disjuncts.)

(*ii*) Answer the same question in (*i*), except this time for the logic $\mathbf{KT}_c$ (i.e., $\mathbf{K} \oplus p \to \Box p$). (Hint: consider Exercise 2.5.43(*ii*), p. 92.) ✠

EXERCISE 2.7.24 Like its predecessor, this exercise concerns Halldén completeness, but now in the context of a bimodal logic whose operators $\Box_1$ and $\Box_2$ need only be assumed to be monotone. Consider the smallest modal logic answering to this description and containing the formula $p \to (\Box_1 p \vee \Box_2 p)$. Show that this logic is Halldén-incomplete.

Suggestion: show that the logic contains $(p \to \Box_1 p) \vee (q \to \Box_2 q)$ but contains neither of the disjuncts of this (variable-disjoint) disjunction. (If $\Box_i$ is given the epistemic reading "Agent i knows that ___" this in effect says that if agents 1 and 2 are 'collectively omniscient' then at least one of them is omniscient *tout court* – or indeed *tout seul*, one might say. For a detailed presentation together with variations, see Humberstone [554]; further repercussions are touched on in [560], [583], [595]. These concern the occasional nonconservativity of adding logical principles governing ∧ to various logics, the topic of (non-)conservative extension occupying us very little here – though a definition and some discussion may be found at p. 316*ff.* below. The restriction to two knowers is incidental here, since the same reasoning can be done for any finite number of monotone knowers to deliver the conclusion that if they are collectively omniscient, one of them is omniscient *simpliciter*. One may even relax the finiteness restriction here, if propositions are identified with sets of worlds, as Roy Sorensen pointed out to the author in correspondence some years ago: conducting the discussion from the perspective of world w, if our knowers are collectively omniscient, one of them must know the truth of the proposition $\{w\}$, from which all other truths-in-w follow, and this knower will be omniscient. However, unlike that for the finite cases, this reasoning cannot be captured in the modal object language as it stands; one enrichment of that language which would help matters in this respect would be the add the 'contingent constant' of Meredith and Prior [802], perhaps the first publication to put a toe into the waters of two-dimensional modal logic – or so it is suggested in Humberstone [578].) ✠

EXERCISE 2.7.25 (*i*) What is the relation between **KT** and the smallest normal extension of **K** by the formula $\Box(\Box p \to p) \to (\Box p \to p)$? (That is: is **KT** properly included in this logic or vice versa; are they the same logic; is neither included in the other?)

(*ii*) What is the relation between **S5** and the smallest normal extension of **K** by the formula $(\Box p \wedge \Diamond q) \to (p \wedge \Box \Diamond q)$? (Again, understand this question as indicated in the parenthetical comment after (*i*).) ✠

EXERCISE 2.7.26 For the purposes of this exercise, let **5**′ be the following weakening of **5**:

$$\Diamond p \to \Box(\Diamond \top \to \Diamond p).$$

(*i*) Show that **K5**′ is determined by the class of frames $\langle W, R \rangle$ such that for any $x, y, z \in W$ if xRy, xRz and $R(y) \neq \varnothing$, then yRz.

(*ii*) Show that $\mathbf{K45}' = \mathbf{K} \oplus (\Box p \leftrightarrow \Box q) \to \Box(\Box p \leftrightarrow \Box q)$. ✠

PREAMBLE TO EXC. 2.7.27. For the exercise below, we offer some heuristic advice. If asked to find a class of frames determining a particular logic, an often helpful policy is to see what conditions would, if imposed on the frames under consideration, secure the soundness of the logic w.r.t. that class of frames, making the conditions in question no stronger than are needed for that purpose. Thus if we are dealing with the special case in which the logic under consideration is $\mathbf{K} \oplus A$ for some formula A, we look for a condition which would validate A which is no stronger than is called for to that end. Then try to show that this condition is satisfied by the canonical frame for the logic concerned, in which case we have the conclusion that the logic is determined by the class of all frames meeting the condition. Fingers need to remain firmly crossed during this latter part of the procedure, since the logic may after all not be determined by any class of frames at all (a Kripke-incomplete logic, as mentioned on p. 82), and even if it is determined by a class of frames, there is no guarantee that the canonical frame lies in that class (as was also mentioned on p. 82). The same element of luck is also required in employing the following strategy – which will, however, often prove successful in practice – in approaching a question of the form: what logic is determined by the class of frames meeting such-and-such a condition? Try to see what the logic would need to prove, without going beyond this bare minimum, in order to show that its canonical frame meets the condition in question. Then check that the logic is sound w.r.t. the class of frames meeting that condition.

EXERCISE 2.7.27 (*i*) Find a class of frames (described in terms of a first order condition) for which you can show that $\mathbf{K} \oplus \Box \Box p \to p$ is determined by that class. Sketch the proof of soundness and completeness.

(*ii*) Now do the same for the logic $\mathbf{K} \oplus p \rightarrow \Box\Box p$.

(*iii*) Same again, now for the logic $\mathbf{K} \oplus p \rightarrow \Diamond\Box p$ (alias $\mathbf{KB}_c$).

(*iv*) Same again, now for the logic $\mathbf{K} \oplus p \leftrightarrow \Diamond\Box p$ (alias $\mathbf{KB!}$).

(*v*) What can you say about the inclusion relations among the logics mentioned under (*ii*), (*iii*), and (*iv*) (in the sense explained under Exercise 2.7.25)?

Note: some of the work for the above questions was already done in addressing several parts of Exercise 2.5.43, p. 92.

(*vi*) Do the logics described in parts (*i*)–(*iv*) here provide the Rule of Disjunction? (Justify your answers.) Are the four logics described in (*i*)–(*iv*) all distinct? (Justify your answer.) ✠

EXERCISE 2.7.28 Show that $\mathbf{K5}_c$ is determined by the class of frames $\langle W, R \rangle$ satisfying: for all $x \in W$, there is some $y \in R(x)$ such that $R(y) \subseteq R(x)$. (Hint: for completeness use the canonical model method. To secure $R(y) \subseteq R(x)$ for a given x, make sure $y \supseteq \{\Box A \mid \Box A \in x\}$.) ✠

EXERCISE 2.7.29 Does there exist a formula A with the property that A is valid on at least one frame and $\neg A$ is valid on at least one frame? (Note that we ask for A to be valid on a frame, and for its negation to be valid on a frame, not just for A to be true throughout some model and its negation to be true throughout some model.) Justify you answer, with an example if there is such an A and with a proof that there isn't, otherwise. ✠

EXERCISE 2.7.30 Find formulas A and B for which you can show that $\mathbf{KT} \oplus A \rightarrow B = \mathbf{S4}$ and $\mathbf{KT} \oplus B \rightarrow A = \mathbf{S5}$, or else show that no such formulas exist. (This is a recreational exercise, purely for practice in modal manipulations.) ✠

EXERCISE 2.7.31 Is $\Box(\Box p \vee \Box q) \rightarrow (\Box p \vee \Box q)$ provable in $\mathbf{K4}_c$? (Justify your answer.) *Suggestion:* Let $\mathbb{Q}_a$ and $\mathbb{Q}_b$ be two disjoint copies of the positive rational numbers, the usual *less than* relations on which we denote by $<_a$ and $<_b$ respectively, and consider the frame $\langle W, R \rangle$ with $W = \{0\} \cup \mathbb{Q}_a \cup \mathbb{Q}_b$ and $R = \{\langle x, y \rangle \mid x, y \in \mathbb{Q}_a \text{ and } x <_a y\} \cup \{\langle x, y \rangle \mid x, y \in \mathbb{Q}_b \text{ and } x <_b y\} \cup \{\langle x, y \rangle \mid x = 0, y \neq 0\}$. Try falsifying the formula at the point 0 in a model on this (dense, transitive) frame. ✠

EXERCISE 2.7.32 (*i*) Axiomatize the (normal modal) logic determined by the class of frames $\langle W, R \rangle$ satisfying the following first-order condition (quantifiers ranging over W):

$$\forall x \exists y (Rxy \wedge \forall z(Rxz \rightarrow Ryz)).$$

(Just find a formula A, or a set of formulas Γ, concerning which you claim that $\mathbf{K} \oplus A$, or $\mathbf{K} \oplus \Gamma$, is the logic concerned. There is no need to supply a proof of soundness or completeness here or in (*ii*) below; see the remarks before 2.7.27 for advice as to how to proceed; however, this is not an easy question.)

(*ii*) Answer a similar question but this time for a condition like that inset above, only with the "$Rxz \rightarrow Ryz$" part replaced by "$Ryz \rightarrow Rxz$". ✠

EXERCISE 2.7.33 Take the canonical model $\langle W_S, R_S, V_S \rangle$ for a consistent normal modal logic S, and define a new ternary relation R'_S (on W_S) – in connection with which we write "xR'_Syz" for "R'_Sxyz" – by putting (for $x, y, z \in W_S$): xR'_Syz iff for all formulas $\Box A \in x$, we have $A \in y$ or $A \in z$. Is the following true or false? (Justify your answer.)

For all $x, y, z \in W_S$, xR'_Syz if and only if either xR_Sy or xR_Sz.

(For the interest of R'_S in modal logics based on functionally incomplete sets of Boolean connectives, see Humberstone [560].) ✠

2.7. REVISION EXERCISES

EXERCISE 2.7.34 Abbreviating "$Rxy \land x \neq y$" to "$R^{\neq}xy$", consider the following 'alio-' variations on domain-reflexivity and range-reflexivity respectively:

$$\forall x \forall y (R^{\neq}xy \to Rxx) \qquad \text{and} \qquad \forall x \forall y (R^{\neq}xy \to Ryy).$$

Example 2.5.45 (p. 94) considered the latter condition.

(*i*) Draw the diagrams which depict these two conditions.

(*ii*) In Example 2.5.45 it was noted that the second condition is a spurious 'alio' version of range-reflexivity. Is the first condition here similarly a spurious 'alio' version of domain-reflexivity, equivalent to what we get from it by replacing $R^{\neq}$ by R?

(*iii*) Here is a condition related to symmetry as range-reflexivity is related to reflexivity: $\forall x \forall y \forall z(Rxy \to (Ryz \to Rzy))$. Consider the 'alio' version of this condition obtained by replacing the leftmost occurrence of "R" with "$R^{\neq}$". This gives a condition called *remote symmetry* in Georgacarakos [365], where it is used in characterizing a logic **S4.04** introduced by Zeman and mentioned briefly in the Digression beginning on p. 431 below. The terminology was recalled more recently under the same name in Halpern *et al.* [431]; Is this a spurious or a genuine 'alio' condition (where the former, as usual, means that the condition is equivalent to the version with R rather than $R^{\neq}$)? ✠

PREAMBLE TO EXC. 2.7.35. A *p-morphism* from one frame $\langle W, R \rangle$ to another $\langle W', R' \rangle$ is a function f from W onto W' meeting these the two conditions for all $x, y \in W$:

- $xRy \Rightarrow f(x)R'f(y)$;

- $f(x)R'f(y) \Rightarrow xRz$ for some $z \in W$ with $f(z) = f(y)$.

(In this case $\langle W', R' \rangle$ is called a p-morphic image of $\langle W, R \rangle$. Note the "onto W'" part of the above definition, indicating that f is surjective; we sometimes call $\langle W, R \rangle$ a p-morphic pre-image of $\langle W', R' \rangle$.) The function f may satisfy the following further condition on models $\langle W, R, V \rangle$, $\langle W', R', V' \rangle$, based on frames so related:

$$V(p_i) = \{x \in W \mid f(x) \in V'(p_i)\},$$

in which case f is said to be a p-morphism from the first model to the second. (P-morphisms, introduced into modal logic by Krister Segerberg under the name *pseudo-epimorphism*, were much exploited in the 1970s by Kit Fine and Segerberg and others – see especially the many applications in Segerberg [1025] – whose definition is used here; they are called 'strongly isotone maps' in Esakia and Meshki [274]. A slightly different definition is common also: for this variant we drop the surjectivity condition on f and change the second bulleted condition above to: $f(x)R'y \Rightarrow xRz$ for some $z \in W$ with $f(z) = y$. Part (*ii*) of the following exercise now still goes through without the surjectivity condition, but the use of the term "image" in part (*ii*) restricts attention to the case in which the p-morphism – in the newly defined sense – is surjective. P-morphisms in this second sense are called 'zigzag morphisms' in van Benthem [70], 'bounded morphisms' in Blackburn *et al.* [98], and 'reductions' in Chagrov and Zakharyaschev [166]. P-morphisms had been used earlier under still other names in the semantics of intuitionistic and intermediate logics; this and other details may be gleaned from the history of the subject provided in Goldblatt [395]. They have since been largely replaced in theoretical work by a generalization of them called bisimulations – not defined here – and due to van Benthem.)

EXERCISE 2.7.35 (*i*) Show by induction on the complexity of formulas that if f is a p-morphism from $\mathcal{M} = \langle W, R, V \rangle$ to $\mathcal{M}' = \langle W', R', V' \rangle$, then for all formulas A:

For all $x \in W$: $\mathcal{M} \models_x A$ if and only if $\mathcal{M}' \models_{f(x)} A$.

(ii) Explain how it follows from the result in (i) that if a formula is valid on a frame $\langle W, R \rangle$, then it is valid on any p-morphic image of $\langle W, R \rangle$.

(iii) As an illustration of (ii) note that there is only one mapping from the universe of the irreflexive frame containing two mutually accessible points onto the single-point reflexive frame, but that this mapping is a p-morphism, and therefore there can be no modal formula valid on all and only the irreflexive frames, i.e., this class of frames is not modally definable. Can we draw, from this example (and the point under (ii)), the same conclusion for the class of asymmetric frames? Can we draw, from the example, the same conclusion for the class of intransitive frames?

(iv) If the answer to either of the questions in (iii) is *no*, can you modify the example and still use p-morphism considerations to show modal undefinability? ✠

The next exercise, which clarifies Example 2.5.34(ii) (p. 85), gives further practice with p-morphisms and involves the two point-generated frames depicted in Figure 2.12. Note that this is not the diagrammatic representation of a condition on frames, but a particular pair of frames. The exercise after that is based on an observation from van Benthem [73], p. 31.

Figure 2.12: *Two Frames for Piecewise Symmetry*

EXERCISE 2.7.36 Recall that the piecewise version of the condition of symmetry would be the following condition on frames $\langle W, R \rangle$: for all $w \in W$, for all $u, v \in R(w)$, if uRv then vRu. We have avoided the letters x, y, z, as variables in this formulation so as not to cause confusion with their use in Figure 2.12 as temporary labels for particular points. Consider the mapping f from the set of points in the frame on the left to those on the right, defined by: $f(x_0) = x$, $f(y_0) = y$, $f(z_0) = f(z_1) = z$.

(i) Is f, so defined, a p-morphism from the first frame to the second?

(ii) What conclusions, if any, can be drawn about the modal definability of the condition of piecewise symmetry from the answer to (i) here, taken in conjunction with Exc. 2.7.35(ii), (iii)? ✠

EXERCISE 2.7.37 Show that if a modal formula A is valid on some frame, then A valid either on the one-point reflexive frame or on the one-point irreflexive frame (or both). (*Hint*: Argue by cases from the hypothesis that A is valid on $\langle W, R \rangle$, the two cases being that $\langle W, R \rangle$ is serial and that $\langle W, R \rangle$ is not serial. For the first case, use a p-morphism argument, and for the second, appeal to generated subframes – i.e., to Theorem 2.5.31(ii).) ✠

EXERCISE 2.7.38 The class of frames satisfying the condition that every frame element has *at most one* successor is modally defined by $\mathbf{D}_c$. The class of frames in which each element has *at most two* successors is modally defined by either of the formulas mentioned in Exercise 2.5.70(i) on p. 109. Let us change "at most" to "at least". The class of frames in which each element has *at least one* successor is modally defined by $\mathbf{D}$. To complete the pattern: what about the class of frames in which each element has *at least two* successors? Is this class of frames modally definable? (*Hint*: it is no accident that this exercise is placed where it is.) ✠

2.7. REVISION EXERCISES

EXERCISE 2.7.39 We continue the theme of the preceding exercise, but now bringing in the notion of a pure formula from Exc. 2.7.21 as well us picking up a thread from Remark 2.1.3 on p. 36. We know that the class of serial frames, those in which each point has at least one successor is defined not only by **D** in its standard form, but also by a pure variant: $\Diamond\top$. It is natural to wonder whether there is a pure formula modally defining the class of ('partial functional': see p. 68) frames in which each point has at most one successor, as well as the impure formula $\mathbf{D}_c$ ($\Diamond p \to \Box p$ – or with any other sentence letter in place of p). To see that there is no such pure formula, consider the two-element frame with universe $W = \{0, 1\}$ and $R = W \times W$. Each point has two successors here and so if there is a pure formula A defining the class of partial functional frames, A is invalid on this frame. Thus there is a modal $\mathcal{M} = \langle W, R, V \rangle$ which does not verify A. Without loss of generality we may assume that A is false at 0 in $\mathcal{M}$. Now, since A is a pure formula, this fact does not depend on the particular V involved and we may trade in V for V' where, for all i, $V'(p_i) = \{0, 1\}$ whenever $0 \in V(p_i)$ and $V'(p_i) = \varnothing$ otherwise. In the model $\mathcal{M}' = \langle W, R, V' \rangle$, in which 0 and 1 verify exactly the same formulas (as each other), we still have $\mathcal{M}' \not\models_0 A$. Let $\mathcal{M}_0 = \langle W_0, R_0, V_0 \rangle$, where $W_0 = \{u\}$ for some u, $R_0 = \{\langle u, u \rangle\}$ and $V_0(p_i) = \{u\}$ if $V'(p_i) = W$ and $V(p_i) = \varnothing$ otherwise, bringing us to the first part of the present exercise:

(*i*) Show that f, defined by: $f(0) = f(1) = u$, is a p-morphism from $\mathcal{M}'$ to $\mathcal{M}_0$.

Just to complete the reasoning, for the record: the result in Exc. 2.7.35(*i*) then tells us that since $\mathcal{M}' \not\models_0 A$, $\mathcal{M}_0 \not\models_u A$. But since the frame of $\mathcal{M}_0$ is a partial functional frame, this contradicts A's claim to being valid on exactly these frames. Now, perhaps similarly:

(*ii*) Show that there is no pure formula valid on precisely the reflexive frames?

In other words, calling formulas *frame-equivalent* when they are valid on the same frames, whereas there is a pure formula frame-equivalent $\Box p \to \Diamond p$ (alias **D**), there is no pure formula frame-equivalent to $\Box p \to p$ (alias **T**). ✠

PREAMBLE TO EXC. 2.7.40 Figure 2.13 is a Hasse diagram depicting the relations of inclusion amongst normal modal logics that can be axiomatized by a selection from candidate axioms **T**, **B**, **4**. It accurately reflects *joins*, but not *meets*, in the lattice of normal modal logics (between **K** and **S5**; the partial order relation – "$\leq$" – underlying this lattice is of course the relation $\subseteq$, restricted to this range). That is: the least upper bound of two logics on the diagram really is the smallest normal modal logic in which both are included, but the greatest lower bound of two logics on the diagram is not guaranteed to be the largest normal modal logic included in both of them. In fact in no case is the meet of two distinct logics in the lattice in question given by their meet on the diagram. The meet of two normal modal logics is just their intersection, while the join is the closure of their union under the defining conditions for being a normal modal logic. (See note 20 for a more general description.) Thus if, for example **KT** really were the meet/intersection of **KTB** and **S4** it would be Halldén-incomplete, by the result cited from Lemmon [701] and Kripke [670] in the Digression on p. 39 (see also Example 2.6.18), contradicting the remark about **KT** in Exercise 2.7.23 (*i*).

In labelling the nodes of Figure 2.13, "S4" and "S5" have been used for **KT4** and **KT4B** (or **KTB4**, if you prefer), but we should also recall that officially the label "S5" abbreviates "**KT5**", and the diagram does not explicitly chart the fortunes of **5** in its own right at all. (The following discussion assumes familiarity with the notion of a Boolean algebra.) Since Figure 2.13 is a picture of the Boolean algebra with 8 (= 2^3) elements, you might at first expect that with four axioms in play, we would end up with the 16 (= 2^4) element Boolean algebra, but we already know that this is not so, because although there are 16 ways of selecting a subset from the set $\{\mathbf{T}, \mathbf{B}, \mathbf{4}, \mathbf{5}\}$, unlike the case of $\{\mathbf{T}, \mathbf{B}, \mathbf{4}\}$, in which each of the 8 subsets gives the label for a distinct normal modal logic (**K** for the subset $\varnothing$, **KTB** for the subset $\{\mathbf{T}, \mathbf{B}\}$, etc.), here distinct subsets can give different labels for the same logics – in the case just recalled, for instance, $\{\mathbf{T}, \mathbf{5}\}$ and $\{\mathbf{T}, \mathbf{B}, \mathbf{4}\}$ simply correspond to two axiomatizations of the one logic (namely

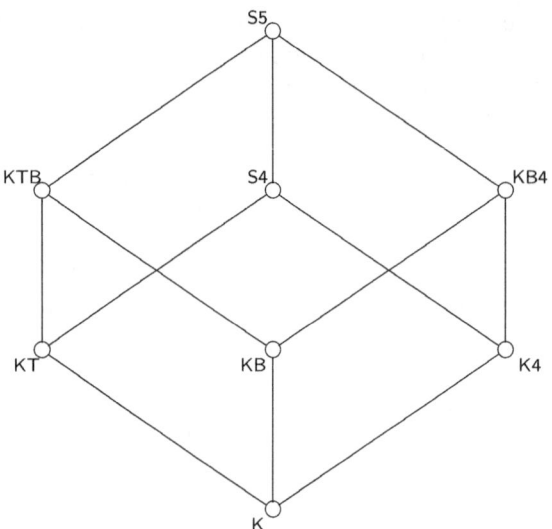

Figure 2.13: Inclusions Among Logics with Axioms **T**, **B**, **4**

S5). So there will certainly be fewer than 16 elements in the lattice in question, which leads us to ask the following, as a preliminary to making a diagram like Figure 2.13 for this new case:

QUESTION How many distinct normal modal logics can be axiomatized by choosing axioms from amongst $\{\mathbf{T}, \mathbf{B}, \mathbf{4}, \mathbf{5}\}$?

(To think about the above question – answered in what follows – it will help immeasurably to bear in mind the semantic characterizations of the logics concerned and in particular of the correspondence between the formulas **T**, **B**, **4**, **5** and the classes of reflexive, symmetric, transitive, and euclidean frames, respectively. See Coro. 2.4.3, Exc. 2.4.6, Coro. 2.4.7 – and the reformulation of such results in terms of frames, in Theorem 2.5.2.)

Behind the fact that we do not in this case have the 16-element Boolean algebra, with four atoms – elements immediately covering the bottom element – is a failure of a certain kind of independence: that no set of the axioms represented by the atoms should have an axiom not in the set as a consequence. Clearly $\{\mathbf{T}, \mathbf{5}\}$ has **B** as a consequence, for example. (Less obvious example: **4** is also a consequence of this set.) An even simpler example of this non-independence possibility would arise if we decided to throw **D** into the mix, as one might in view of the fact that **S5** = **KDB4**. In this case **T** all by itself – which is to say, the set $\{\mathbf{T}\}$ – has **D** as a consequence. But for the above question itself, Figure 2.13 provides the answer, displaying the inclusion relations amongst the logics concerned: it remains only to count the nodes on the diagram. Note that the vertical line in Figure 2.13 going up from **K4** to **KB4** (= **KB5**) has been omitted in Figure 2.14 (in accordance with the conventions for drawing such diagrams) because the inclusion in question now follows from the path going from **K4** to **KB4** via **K45**.

QUESTIONS (*i*) How many distinct normal modal logics can be axiomatized by choosing axioms from amongst $\{\mathbf{T}, \mathbf{D}, \mathbf{B}, \mathbf{4}, \mathbf{5}\}$?

(*ii*) What is the appropriate notion of *consequence*, for the remarks before the preceding question to understood in terms of?

These are, however, unofficial questions to ponder; our official question will be simpler than (*i*) here.

2.7. REVISION EXERCISES

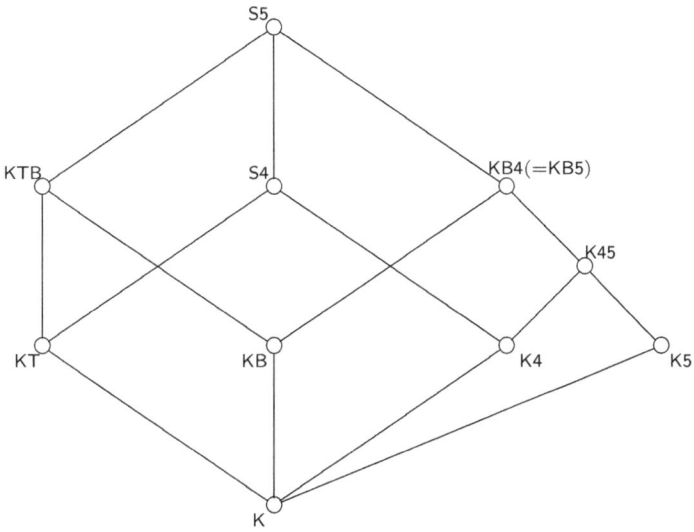

Figure 2.14: *Inclusions Amongst Logics with Axioms* **T**, **B**, **4**, **5**

EXERCISE 2.7.40 How many distinct normal modal logics are there with axioms drawn from the range $\{\mathbf{T}, \mathbf{D}, \mathbf{B}, \mathbf{4}\}$? Please draw a diagram (like those in Figures 2.13, 2.14) indicating their relations of inclusion. ✠

EXERCISE 2.7.41 Say, for each of the following sequents whether or not it is valid, according to the semantics for **S5** given in Section 2.2.

(i) $\quad p, \Diamond q \succ \Diamond (p \wedge q)$
(ii) $\quad p, \Diamond q \succ \Diamond (\Diamond p \wedge q)$
(iii) $\quad \Diamond (p \rightarrow q) \succ \Diamond p \rightarrow \Diamond q$
(iv) $\quad \Box (p \rightarrow q) \succ \Diamond p \rightarrow \Diamond q$
(v) $\quad \Box (p \rightarrow q) \succ \Diamond q \rightarrow \Diamond p$
(vi) $\quad \Box (p \vee \Diamond q) \succ \Box p \vee \Diamond q$
(vii) $\quad \Box (p \wedge (q \vee r)) \succ \Box (p \wedge q) \vee \Box (p \wedge r)$
$(viii)$ $\quad \Box (p \vee (q \wedge r)) \succ \Box (p \vee q) \wedge \Box (p \vee r)$
(ix) $\quad \Box (p \rightarrow \Box (q \rightarrow r)) \succ \Box (q \rightarrow \Box (p \rightarrow r))$
(x) $\quad \Box (p \rightarrow \Box (q \rightarrow r)) \succ \Box (\Box q \rightarrow \Box (p \rightarrow r))$
(xi) $\quad \Diamond \Box p, \Diamond \Box q \succ \Diamond \Box (p \wedge q)$
(xii) $\quad \Box \Diamond \Box p, \Box \Diamond \Box q \succ \Box \Diamond \Box (p \wedge q)$
$(xiii)$ $\quad \Box (p \vee \Box q) \succ \Box p \vee \Box q$
(xiv) $\quad \Box p, \Diamond q \succ \Diamond (\Diamond p \wedge q)$
(xv) $\quad \succ \Diamond (\Diamond p \rightarrow p)$.

✠

EXERCISE 2.7.42 Of those sequents listed under Exercise 2.7.41 which are valid according to the semantics for **S5**, pick any *four* – but not including both of (xi), (xii), since they are so similar – and provide proofs of them using the natural deduction system for **S5** in the Appendix (Sections 7.2, 7.3), using the liberalized form of ($\Box$I) if desired. ✠

EXERCISE 2.7.43 For each of the sequents listed under Exercise 2.7.41, say whether or not is valid according to the semantics given for **S4** (i.e., in terms of Kripke models with reflexive transitive accessibility relations – though, as for 2.7.41, your working need not be shown). ✠

2.8 Supplement: Matsumoto's Embedding

This section and the following section constitute something of an optional appendix to the present chapter, and both deal with the same topic. They have been separated in such a way that this section can be read straight after Section 2.1: it does not presuppose familiarity with the semantic concepts and methods introduced in Sections 2.2 and 2.3. Without appealing to that material we can still include a sample of syntactical manipulations of some interest in their own right. The aim of the exercise will be to give a syntactic proof of a 1955 result from Matsumoto (in [782]). No novelty is claimed for the discussion except in the choice of aspects to emphasize and in the explicit deployment of the concept of a distributive triple from Section 2.1 (p. 51). The following section includes materials which do, by contrast, assume familiarity with the semantic apparatus.

Concerning the normal modal logics **S4** and **S5**, Matsumoto [782] showed that for any formula A,

$$\vdash_{\mathbf{S5}} A \text{ if and only if } \vdash_{\mathbf{S4}} \Box\Diamond\Box A.$$

This appears as Theorem 2.8.4 below. As usual, we mostly consider normal modal logics as (certain) collections of formulas, and feel free, when convenient, to write this as: $A \in \mathbf{S5}$ if and only if $\Box\Diamond\Box A \in \mathbf{S4}$. (Consequence relation formulations will be employed at some points, however.) Since in general $\vdash_{\mathbf{S4}} \Box B$ if and only if $\vdash_{\mathbf{S4}} B$, we can equally well consider the following variant as giving the content of Matsumoto's embedding of **S5** into **S4**: $\vdash_{\mathbf{S5}} A$ if and only if $\vdash_{\mathbf{S4}} \Diamond\Box A$. The latter version of the result is proved by (Kripke-)model-theoretic methods in Fitting [310], pp. 222–225; we will not follow that route in the present case (for reasons explained below) but do include elsewhere a similar style of proof for the simpler embedding of **S5** into **K45**: $\vdash_{\mathbf{S5}} A$ if and only if $\vdash_{\mathbf{K45}} \Box A$. (See Coro. 4.1.6, p. 207.) We are dealing in these cases with a (faithful) *translational embedding* from one logic into another, in the sense of a mapping (or 'translation') τ of the formulas of the language of the first, or 'source' logic, to the language of the second, 'target' logic – these languages coinciding in the present instance – with the property that A is provable in the source logic if and only if $\tau(A)$ is provable in the target logic.

REMARK 2.8.1 More generally, τ *embeds* S_0 in S_1 when for all $A \in S_0$, $\tau(A) \in S_1$; τ embeds S_0 in S_1 *faithfully* when in addition we have the converse: $\tau(A) \in S_1$ implies $A \in S_0$ for A in the language of S_0. (These definitions do not require that the languages of S_0 and S_1 to coincide, though of course for monomodal logics as in the present section and that following, the languages do coincide.) There is a straightforward extension of this vocabulary to the case of logics considered as consequence relations rather than as sets of formulas, given below at p. 147. ◀

In Matsumoto's original formulation, τ is the operation of prefixing "$\Box\Diamond\Box$" to the formula, while in Fitting's version, it prefixes just "$\Diamond\Box$". (Fitting's version of Matsumoto's embedding theorem is treated as a paradigm of such results in Inoué [611]. The character of the embedding is rather different from the other translational embeddings which we shall have occasion to consider in later chapters, which are by contrast what we call *compositional*, indeed *definitional*: see Example 4.4.25, beginning on p. 264, for the explanation of this terminology.) In spite of the apparent indifference expressed above on the choice between these translations, with a slight change of context, as we shall see below – see Theorem 2.8.12 and the discussion following it – the two translations exhibit strikingly different behaviour. (The first hint of this is mentioned in Remark 2.8.5 below.)

Digression. The inspiration for Matsumoto's embedding was the "¬¬" translational embedding of classical into intuitionistic propositional logic, associated with Gödel and Glivenko and, together with the

2.8. SUPPLEMENT: MATSUMOTO'S EMBEDDING

translation(s) of intuitionistic propositional logic into **S4** associated with (Gödel and) Tarski and McKinsey, explains the presence of the initial "$\Box$" in Matsumoto's prefix (absent from Fitting's). Matsumoto [782], in which the references just alluded to may be found, gives a syntactic argument for his embedding result which draws on these connections with intuitionistic logic. (A different line of Glivenko-inspired investigation in modal logic – relating **K4** and **S5** – may be found in the 1992 paper Rybakov [981].)
End of Digression.

We shall provide a self-contained (i.e., purely modal) version of Matsumoto's syntactic proof below: see Theorem 2.8.4. (Matsumoto's original discussion draws on the material mentioned in the above Digression, though §6 of Ohnishi and Matsumoto [855] gives essentially the proof to be presented here.) As already mentioned, Fitting gives a semantic argument, but we favour the present line of thought for two reasons, one being that the key ingredient in our proof is the simple notion of a distributive triple, introduced mainly for this purpose at the end of Section 2.1, of considerably broader application. The second reason is that it leads us on to consider a special feature of Matsumoto's prefix as it appears in **S4**, namely that it behaves like a normal $\Box$ operator in its own right. (See Section 1.4.) This feature – we shall later call it Matsumoto normality – is not essential to a logic's being embeddable into **S5** by Matsumoto's translation, but it is at work, behind the scenes, in Matsumoto's proof (as reconstructed here: Theorem 2.8.4) that **S4** is so embeddable. More precisely, what is exploited in that proof is the fact that Matsumoto's prefix taken three times constitutes a distributive triple for **S4**, which amounts to saying that this prefix behaves as a *regular* modal operator – something we noted in Section 1.3 falls short of normality in not guaranteeing to obey a Necessitation-like rule. The immediate background on distributive triples is provided by Propositions 2.1.24–2.1.26. The crucial observation for our version of the proof of Matsumoto's Embedding Theorem follows.[113]

LEMMA 2.8.2 $\langle \Box\Diamond\Box, \Box\Diamond\Box, \Box\Diamond\Box \rangle$ *is a distributive triple of modalities for* **K4** *(and therefore for all extensions of* **K4***).*

Proof. By Prop. 2.1.24 on p. 52, $\langle \Box\Box\Diamond\Box, \Box\Diamond\Box\Box, \Box\Diamond\Diamond\Box \rangle$ is a distributive triple of modalities for any normal modal logic, because the component-by-component triples of operators $\langle \Box, \Box, \Box \rangle$ (first coordinate), $\langle \Box, \Diamond, \Diamond \rangle$ (second coordinate), $\langle \Diamond, \Box, \Diamond \rangle$ (third coordinate) and $\langle \Box, \Box, \Box \rangle$ (fourth coordinate) are distributive triples of operators for **K**, and hence for any normal modal logic. Now, of the 'premiss' modalities in our triple (the first two), $\Box\Box\Diamond\Box$ and $\Box\Diamond\Box\Box$, each is implied by virtue of **4** by the corresponding modality with a single $\Box$ in place of the pair, and in the 'conclusion' modality, $\Box\Diamond\Diamond\Box$, the doubled $\Diamond$ implies the single $\Diamond$ in virtue of that same principle (in its dual form), so we do not need the *equivalence* of doubled and single operators and the argument goes through for **K4**. ∎

REMARK 2.8.3 At first sight collapsing adjacent pairs of $\Box$'s or $\Diamond$'s as above may seem to demand the strength of **K4!**, in which any such pair is equivalent to the corresponding single occurrence, but, as the attention to premiss/conclusion occurrences shows, we do not actually need the equivalences: one-way implications suffice, and are provided by **K4**. ◂

For the proof of Matsumoto's embedding theorem, we take **S4** and **S5** to be axiomatized as indicated in the 'anatomical' labels for these logics, **KT4** and **KT4B**, respectively. The reason we favour these axiomatizations as against taking **S5** with axiomatization indicated in the label **KT5**, for instance (see Exc. 2.1.15, p. 45), is so that we have one additional axiom (here **B**) over and above a set of independent axioms for **S4**.

---
[113]Its content can be found in Åqvist [27], mid-page on p. 16.

THEOREM 2.8.4 (Matsumoto) *For any formula A, we have $\vdash_{\mathbf{S5}} A$ if and only if $\vdash_{\mathbf{S4}} \Box\Diamond\Box A$ (equivalently: if and only if $\vdash_{\mathbf{S4}} \Diamond\Box A$).*

Proof. The 'if' direction is clear, since if $\vdash_{\mathbf{S4}} \Box\Diamond\Box A$ then as $\mathbf{S5} \supseteq \mathbf{S4}$, $\vdash_{\mathbf{S5}} \Box\Diamond\Box A$, and it is easily seen that $\vdash_{\mathbf{S5}} \Box\Diamond\Box A \to A$. The 'only if' direction is proved by induction on the length of a shortest proof of A on the basis of the axiomatization of $\mathbf{S5}$ described above. For the basis case, we check the axioms. Since we can apply Necessitation three times to any $\mathbf{S4}$ theorem, weakening the middle $\Box$ to a $\Diamond$ by appeal to the fact that $\mathbf{S4} \supseteq \mathbf{KD}$, we have $\vdash_{\mathbf{S4}} \Box\Diamond\Box A$ for $A = \mathbf{K}, \mathbf{T}$, or $\mathbf{4}$, leaving only $\mathbf{B}$, for which a separate check is required that $\vdash_{\mathbf{S4}} \Box\Diamond\Box(p \to \Box\Diamond p)$. For the inductive step, we need to check that the three rules have the property that if their premises are such that the results of prefixing "$\Box\Diamond\Box$" to them are $\mathbf{S4}$-provable, then so are their conclusions. This is clear for Uniform Substitution and for Necessitation. For Modus Ponens, the hypothesis means that we have

$$\vdash_{\mathbf{S4}} \Box\Diamond\Box(A \to B) \quad \text{and} \quad \vdash_{\mathbf{S4}} \Box\Diamond\Box A,$$

and what we want is that $\vdash_{\mathbf{S4}} \Box\Diamond\Box B$. But by Lemma 2.8.2, $\langle \Box\Diamond\Box, \Box\Diamond\Box, \Box\Diamond\Box \rangle$ is a distributive triple for $\mathbf{S4}$, which under formulation *(ii)* of Proposition 2.1.25 (p. 53) emerges as the claim that $\vdash_{\mathbf{S4}} \Box\Diamond\Box(A \to B) \to (\Box\Diamond\Box A \to \Box\Diamond\Box B)$, so the desired conclusion follows by two applications of Modus Ponens. ■

REMARK 2.8.5 Note that this last step of the argument would not have worked with the simplified modality "$\Diamond\Box$" in place of the full Matsumoto prefix "$\Box\Diamond\Box$". ◀

Clearly the reference to $\mathbf{S4}$ in Theorem 2.8.4 could be replaced by a reference to any of several other normal modal logics. Our own proof, via Lemma 2.8.2, for example, works if in place of $\mathbf{S4}$ we have $\mathbf{KD4}$. The "4" part is needed because of Lemma 2.8.2 itself, and the "D" part because of the appeal in the proof of Theorem 2.8.4 to the fact that $\mathbf{S4} \supseteq \mathbf{KD}$, used in showing that $\vdash_{\mathbf{S4}} \Box\Diamond\Box A$ for $A = \mathbf{K}$, $\mathbf{T}$, $\mathbf{4}$; for $\mathbf{B}$ we note that $\vdash_{\mathbf{KD4}} \Box\Diamond\Box(p \to \Box\Diamond p)$. (This is actually most easily seen with the aid of the semantics introduced in Section 2.3.)

EXERCISE 2.8.6 Give a syntactic proof (i.e., a deduction using the relevant axioms and rules) establishing that $\vdash_{\mathbf{KD4}} \Box\Diamond\Box(\Box p \to p)$. ✣

EXERCISE 2.8.7 Either sketch a proof or give a counterexample to the following Matsumoto-like claim: For any formula A, we have $\vdash_{\mathbf{S5}} A$ if and only if $\vdash_{\mathbf{S4}} \Box\Diamond A$. Is the claim correct in either direction (i.e., in the 'if' direction or the 'only if' direction)? ✣

We will make one application of Matsumoto's Theorem here, to throw light on a suggestion from Parry [866]. As Dummett and Lemmon [263] report, Parry had proposed a certain logic S4.5 (or $\mathbf{S4.5}$, as we might have called it, since it is a normal extension of $\mathbf{S4}$, and indeed of $\mathbf{S4.4}$) as the smallest extension of $\mathbf{S4}$ with the axiom

$$\Box(\Box\Diamond\Box p \to \Box p).$$

Dummett and Lemmon point out, using some considerations about translations between intermediate[114] and modal logics between $\mathbf{S4}$ and $\mathbf{S5}$, that Parry's proposed new logic is nothing other than $\mathbf{S5}$ itself. We can draw this conclusion without the detour through translations by an appeal to Theorem 2.8.4 (even though, as mentioned in the Digression on p. 142, Matsumoto was led to this result through a

---

[114]*Intermediate logics* are non-modal (for present purposes, sentential) logics between intuitionistic and classical logic, including the end-points here.

2.8. SUPPLEMENT: MATSUMOTO'S EMBEDDING

consideration of such translations). First, using the **T** axiom, we can remove the initial "$\Box$" from Parry's axiom, inset above. Now take any provable formula A of **S5**. By Theorem 2.8.4, we have $\Box\Diamond\Box A$ provable in **S4**, and hence also in Parry's 'S4.5'. And by Uniform Substitution into the denecessitated version of the new axiom just remarked on, in this system $\Box\Diamond\Box A \to A$ is also provable. By Modus Ponens, then, so also is A: but A was an arbitrary theorem of **S5**. So Parry's extension of **S4** in fact takes us, as Dummett and Lemmon had remarked, all the way to **S5**. (Of course to check that we have precisely **S5** here, we need also to check that Parry's axiom is provable in **S5**, but this is readily seen with the aid of Coro. 2.1.12.)

That application of Theorem 2.8.4 concludes the main body of our discussion of Matsumoto's embedding. Another appeal to the result will occur in the proof of Theorem 5.5.24 on p. 419 below. The remainder of this section addresses some specialized issues in the vicinity.

Let us say that a normal modal logic S *satisfies Matsumoto's Embedding Condition* if Theorem 2.8.4 holds for S in place of **S4**. Some *ad hoc* terminology for the present discussion: let S_0 be the smallest normal modal logic $\supseteq \{\Box\Diamond\Box A \mid A \in \textbf{S5}\}$.

LEMMA 2.8.8 (*i*) **S5** *satisfies Matsumoto's Embedding Condition.*
(*ii*) $S_0 \subseteq \textbf{S5}$.

Proof. (*i*): It is easy to check that $\vdash_{\textbf{S5}} A$ if and only if $\vdash_{\textbf{S5}} \Box\Diamond\Box A$; (*ii*): from the "if" half of the preceding claim. ∎

LEMMA 2.8.9 *For any normal modal logic S, if S satisfies Matsumoto's Embedding Condition then $S_0 \subseteq S$.*

Proof. Suppose that S satisfies Matsumoto's Embedding Condition and $A \in S_0$. Then $A \in \textbf{S5}$ by Lemma 2.8.8(ii), so $\Box\Diamond\Box A \in \textbf{S5}$, so since S satisfies Matsumoto's Embedding Condition, $A \in S$. ∎

LEMMA 2.8.10 *For any normal modal logic S, if S satisfies Matsumoto's Embedding Condition then $S \subseteq \textbf{S5}$.*

Proof. Suppose that S satisfies Matsumoto's Embedding Condition and $A \in S$. Then since $A \to A \in \textbf{S5}$, $\Box\Diamond\Box(A \to A) \in S$ by Matsumoto's Embedding Condition, and $\Box\Box\Box A \in S$ by three applications of Necessitation. Since $\langle \Box\Diamond\Box, \Box\Box\Box, \Box\Diamond\Box \rangle$ is a distributive triple for any normal modal logic (by Prop. 2.1.24), we get $\Box\Diamond\Box A \in S$. By Matsumoto's Embedding Condition, $A \in \textbf{S5}$. ∎

We can now sum up our findings for the range of normal modal logics satisfying Matsumoto's Embedding Condition:

PROPOSITION 2.8.11 *For any normal modal logic S, S satisfies Matsumoto's Embedding Condition if and only if $S_0 \subseteq S \subseteq \textbf{S5}$.*

Proof. 'If': We must show, given $S_0 \subseteq S \subseteq \textbf{S5}$, that $\Box\Diamond\Box A \in S$ if and only if $A \in \textbf{S5}$. For the "if" direction, suppose that $A \in \textbf{S5}$. Then $\Box\Diamond\Box A \in S_0$, so since $S_0 \subseteq S$, $\Box\Diamond\Box A \in S$. For the "only if" direction, suppose that $\Box\Diamond\Box A \in S$. Then since $S \subseteq \textbf{S5}$, $\Box\Diamond\Box A \in \textbf{S5}$, and so $A \in \textbf{S5}$.
'Only if': Combining Lemmas 2.8.9, 2.8.10. ∎

We have not given an explicit description – as the smallest normal modal logic containing some finite set of formulas, for instance (or all instances of some finite set of schemata) – of the lower bound S_0 of normal modal logics satisfying Matsumoto's Condition. Indeed, we are not familiar with any description along those lines, though it is easy enough to provide a finite axiomatization of S_0 if special-purpose rules are allowed. Pick any axiomatization of **S5** with the rules Uniform Substitution, Modus Ponens and Necessitation and finitely many axioms. Let C be the conjunction of these axioms. Then S_0 can be axiomatized as an extension of **K** with those same rules and $\Box\Diamond\Box C$ as an additional axiom, together with the further rules

(MP)$^{\Box\Diamond\Box}$ From $\vdash \Box\Diamond\Box(A \to B)$ and $\vdash \Box\Diamond\Box A$ to $\vdash \Box\Diamond\Box B$.

(Nec)$^{\Box\Diamond\Box}$ From $\vdash \Box\Diamond\Box A$ to $\vdash \Box\Diamond\Box\Box A$.

These rules have been chosen so that applications of Modus Ponens and Necessitation in **S5** are mirrored in the proof-system for S_0, making it clear that this is indeed an axiomatization of that logic (defined as the smallest normal modal logic in which $\Box\Diamond\Box A$ is provable whenever A is provable in **S5**). We could avoid the second rule by beginning with an axiomatization of **S5** in with only Uniform Substitution and Modus Ponens as primitive rules. (Alternatively, we could simply replace this rule with an axiom $\Box\Diamond\top$.) Whether the first rule is similarly avoidable is unclear. We emphasize that it is a rule for passing from provable formulas to provable formula and so is much weaker than the claim that $\langle \Box\Diamond\Box, \Box\Diamond\Box, \Box\Diamond\Box\rangle$ is a distributive triple for the logic concerned (see Proposition 2.9.4 below), though, as in the case of the latter claim, it could equally well be replaced by a rule with premisses $\Box\Diamond\Box A$ and $\Box\Diamond\Box B$ and conclusion $\Box\Diamond\Box(A \wedge B)$. A more manageable axiomatization of S_0 would be desirable, though even with just the initial description of S_0 as the smallest normal modal logic containing every formula $\Box\Diamond\Box A$ for which $A \in$ **S5** – which we can think of as a lazy axiomatization with infinitely many axioms – we can provide a semantic description and will do so in Theorem 2.9.3 in a later section. (Of course any axiomatic presentation using axiom-schemata involves infinitely many axioms – the instances of the schemata – but the present point is that such a presentation in this case would involve infinitely many schemata.)

As in the Digression before Proposition 1.3.10 (see p. 22), let us call a consequence relation a *normal (modal) consequence relation* if it extends the consequence relation of classical propositional logic, formulated in some language with a functionally complete set of Boolean connectives (amongst which we presume to be the 0-ary truth constant $\top$) as well as the additional 1-ary connective $\Box$, and is substitution-invariant and satisfies the condition that

$$A_1, \ldots, A_n \vdash B \text{ implies } \Box A_1, \ldots, \Box A_n \vdash \Box B,$$

for all $A_1, \ldots, A_n, B$. An equivalent characterization (cf. Exc. 1.3.5) can make do with the $n = 2$ case of this implication alongside the $n = 0$ case, and that the latter can be replaced by the unconditional requirement that $\vdash \Box\top$ (i.e., $\varnothing \vdash \Box\top$). Given any modality X, we say that $\vdash$ is an X-*normal* consequence relation if the implication we obtain from the inset condition above by replacing $\Box$ with X. (Thus the normal consequence relations are the $\Box$-normal consequence relations.) Again, the preceding simplification of requiring only the $n = 2$ case of the implication, alongside the requirement that $\vdash X\top$, is available. The consequence relation $\vdash_{\mathbf{S4}}$ we take to be the relation holding between any Γ and A just in case for some $C_1, \ldots, C_m \in \Gamma$ when $(C_1 \wedge \ldots \wedge C_m) \to A \in$ **S4** – understanding this as just A itself when $m = 0$ – is easily seen to be a normal modal consequence relation – susceptible in fact of a more elegant description in terms of natural deduction (cf. Prawitz [905], p. 74, or Chapter 7 below) than the lazy definition just given. The lazy definition is equivalent to saying that we can obtain A from the formulas in Γ by repeated appeals to Modus Ponens and the theorems of **S4**. We understand by $\vdash_{\mathbf{S5}}$ the similarly defined consequence relation associated with **S5**.[115] Clearly the apparently double usage of "$\vdash$" here, in saying, as in the preceding sections, "$\vdash_S A$" to mean "$A \in S$" and also to mean "$\varnothing \vdash_S A$"

---

[115] In the literature, these are often called the *local* consequence relations associated with **S4** and **S5**, respectively, for semantic reasons – they preserve truth at a point in any model. Compare note 117 below.

2.8. SUPPLEMENT: MATSUMOTO'S EMBEDDING

causes no trouble since the two things come to the same. Let us put on record a form of the Matsumoto embedding result for this conception of what a modal logic is:

THEOREM 2.8.12 *For any formulas $A_1, \ldots, A_n, B$:*
if $A_1, \ldots, A_n \vdash_{\mathbf{S5}} B$ then $\Box\Diamond\Box A_1, \ldots, \Box\Diamond\Box A_n \vdash_{\mathbf{S4}} \Box\Diamond\Box B$.

Proof. To reduce work, we exploit Theorem 2.8.4, transforming the antecedent into the hypothesis that $\vdash_{\mathbf{S5}} (A_1 \wedge \ldots \wedge A_n) \to B$, and inferring by appeal to that result that
$\vdash_{\mathbf{S4}} \Box\Diamond\Box((A_1 \wedge \ldots \wedge A_n) \to B)$.
Since $\langle \Box\Diamond\Box, \Box\Diamond\Box, \Box\Diamond\Box \rangle$ is a distributive triple for **S4** (Lemma 2.8.2), we get
$\vdash_{\mathbf{S4}} \Box\Diamond\Box(A_1 \wedge \ldots \wedge A_n) \to \Box\Diamond\Box B$.
Then by Proposition 2.1.26
$\vdash_{\mathbf{S4}} (\Box\Diamond\Box A_1 \wedge \ldots \wedge \Box\Diamond\Box A_n) \to \Box\Diamond\Box B$, so
$\Box\Diamond\Box A_1, \ldots, \Box\Diamond\Box A_n \vdash_{\mathbf{S4}} \Box\Diamond\Box B$. ∎

Disappointingly, by contrast with Theorem 2.8.4, all we have here is the "if" result as stated, and not also the "only if" part – the embedding is no longer 'faithful'. The notions in play here can be explained as follows. A function τ from the language of one consequence relation $\vdash_{\mathsf{source}}$ to that of another, $\vdash_{\mathsf{target}}$ is said to *embed* the former in the latter when for all formulas $A_1, \ldots, A_n, B$ of the former language, we have the "only if" direction of the following condition satisfied, and to embed $\vdash_{\mathsf{source}}$ *faithfully* – cf. Remark 2.8.1, p. 142 – into $\vdash_{\mathsf{target}}$ if we have both the "only if" and also the "if" directions satisfied (for all formulas $A_1, \ldots, A_n, B$:

$A_1, \ldots, A_n \vdash_{\mathsf{source}} B$ if and only if $\tau(A_1), \ldots, \tau(A_n) \vdash_{\mathsf{target}} \tau(B)$.

In the case of interest here, the language of $\vdash_{\mathsf{source}}$ coincides with that of $\vdash_{\mathsf{target}}$, because it is simply the standard language of monomodal logic, though the consequence relations themselves differ, being respectively $\vdash_{\mathbf{S5}}$ and $\vdash_{\mathbf{S4}}$. Now Matsumoto's Theorem (our Thm. 2.8.4) gives us the faithful – i.e., two-way – form of the condition inset above for the special case of $n = 0$, and Thm. 2.8.12 provides the "only if" direction of the general (i.e., n arbitrary) inset condition above (for the current source and target logics), but the "if" ('faithfulness') direction – which because of a shift in formulation corresponds to the "only if" version of Thm. 2.8.12 – fails, as remarked above and as is shown by the following example, which was mentioned already in Ohnishi and Matsumoto [855], p. 125:

EXAMPLE 2.8.13 For a counterexample to the missing converse, consider the fact that while

$$\Box\Diamond\Box p \vdash_{\mathbf{S4}} \Box\Diamond\Box\Box p$$

we do not have $p \vdash_{\mathbf{S5}} \Box p$. ◀

Nevertheless, Theorem 2.8.12 reveals a difference from the case of Fitting's version of Matsumoto's embedding: replacing the "$\Box\Diamond\Box$"s by "$\Diamond\Box$" would turn it from true to false:

EXAMPLE 2.8.14 While $p, q \vdash_{\mathbf{S5}} p \wedge q$, we do not have $\Diamond\Box p, \Diamond\Box q \vdash_{\mathbf{S4}} \Diamond\Box(p \wedge q)$. The easiest way to see this is substitute $\neg p$ for q. If we had
$\vdash_{\mathbf{S4}} (\Diamond\Box p \wedge \Diamond\Box q) \to \Diamond\Box(p \wedge q)$,
then by that substitution, we should have
$\vdash_{\mathbf{S4}} (\Diamond\Box p \wedge \Diamond\Box \neg p) \to \Diamond\Box(p \wedge \neg p)$,

the negation of whose consequent is easily seen to be **S4**-provable, which would imply that the negation of its antecedent also be **S4**-provable: which it is not. The negation of the antecedent is a representative instance of the schema **.2** in Table 2.1 on p. 33, but **S4.2** is a proper extension of **S4**, and this formula is not **S4**-provable. (One can check without difficulty that **S4.2** is the weakest extension of **S4** in which $\Diamond\Box$ is normal. We return to this in connection with deontic and doxastic logic in Sections 4.5 and 5.5, respectively, as well as in the discussion of Theorem 5.2.1 on p. 372.) Certainly, this argument assumes without justification that we have here a proper extension of **S4**, but the correctness of hat assumption will be clear from the semantics given for these logics in Section 2.3; see in particular Coro. 2.5.12.[116] ◂

One might, in view of the counterexample (Example 2.8.13) to the missing converse above, entertain the thought of exchanging the present consequence relation for the consequence relation which preserves truth throughout a model rather than truth at a point – or, put syntactically, allows not only Modus Ponens but also Necessitation to be applied in passing from the formulas on the left of the "⊢" to that on the right.[117] However, this would destroy the contrast just made with the case of Fitting's variant.

The heart of that contrast can be stated without reference to **S5**: Example 2.8.14, showing that we cannot trade in Matsumoto's prefix for Fitting's, reveals that unlike $\langle\Box\Diamond\Box, \Box\Diamond\Box, \Box\Diamond\Box\rangle$, $\langle\Diamond\Box, \Diamond\Box, \Diamond\Box\rangle$ is not a distributive triple for **S4**. (We note that $\langle\Box\Diamond\Box, \Diamond\Box, \Diamond\Box\rangle$, however, is distributive for **S4**.[118]) By Proposition 2.1.25(iii), this means that while for all A, B, C, if $\vdash_{\mathbf{S4}} A \to (B \to C)$ then $\vdash_{\mathbf{S4}} XA \to (XB \to XC)$ when X is taken as $\Box\Diamond\Box$, but not when X is taken as $\Diamond\Box$. In terms of the **S4** consequence relation with which we have been working, this means that for the first but not for the second choice of X, whenever $A, B \vdash_{\mathbf{S4}} C$, we have $XA, XB \vdash_{\mathbf{S4}} XC$. Since this is the $n = 2$ case of the condition of X-normality for consequence relations, and we also have the $n = 0$ case, as $\vdash_{\mathbf{S4}} \Box\Diamond\Box\top$, this consequence relation is $\Box\Diamond\Box$-normal, in the terminology introduced for consequence relations before Theorem 2.8.12. (Of course we also have $\vdash_{\mathbf{S4}} \Diamond\Box\top$, but since the $n = 2$ case of the condition for $\vdash$ to be $\Diamond\Box$-normal – i.e., regularity for $\Diamond\Box$ – is not satisfied, this consequence relation is not $\Diamond\Box$-normal.) Because $\Box\top$ and $\top$ are provably equivalent – and accordingly everywhere interreplaceable – in any normal modal logic, we can simplify the $n = 0$ case for the modality $\Box\Diamond\Box$ to the condition that $\vdash_S \Box\Diamond\top$. Let us call a normal (i.e. $\Box$-normal) modal logic which is also $\Box\Diamond\Box$-normal a *Matsumoto normal modal logic*. We can think of this as applying to modal logics as consequence relations, or to modal logics as sets of formulas, and for the remainder of our discussion we revert to the latter framework. Thus S is a Matsumoto normal modal logic if S is a normal modal logic for which $\langle\Box\Diamond\Box, \Box\Diamond\Box, \Box\Diamond\Box\rangle$ is a distributive triple and for which we have $\vdash_S \Box\Diamond\top$. A by-product, then, of looking at what underlies Matsumoto's embedding of **S5** into **S4**, turns out to be the discovery that, quite independently of any considerations concerning **S5**, **S4** is a Matsumoto normal modal logic, something worth putting on record, using the terminology of Section 1.4:

PROPOSITION 2.8.15 *The modality* $\Box\Diamond\Box$ *is normal in* **S4**.

This does not mean that Matsumoto normality is essential for a logic to satisfy Matsumoto's Condition, but that Matsumoto's proof that **S4** satisfies that condition – reconstructed here as Theorem 2.8.4 – goes

---

[116] More directly, taking for granted the semantics of Section 2.3 – without which this footnote will not be intelligible – consider the model used for case (4) in the proof of Proposition 2.9.2 in our 'post-semantic' discussion (p. 149) of Matsumoto-related issues, changing the frame by adding $\langle 0, 0\rangle$ to R to secure reflexivity; $\Diamond\Box p, \Diamond\Box q$ are true at 0, though not $\Diamond\Box(p \wedge q)$.

[117] This would be to consider the *global* rather than the *local* consequence relation – see note 115 – determined by a class of frames, preserving the property of being true throughout any model on a frame in that class, rather than preserving truth at a point in any such model.

[118] This means that, since we may necessitate its premiss, the rule licensing the transition from $\vdash \Diamond\Box(A \to B)$ to $\vdash \Diamond\Box A \to \Diamond\Box B$ is a derivable rule of **S4**, under the usual axiomatization. (This rule can be used to give a direct syntactic proof of Fitting's version of Matsumoto's embedding theorem: it enables us to reproduce applications of Modus Ponens in **S5** within the image of **S5** in **S4** under the embedding. Of course the same could be said of the weaker rule which passes from $\vdash \Diamond\Box(A \to B)$ and $\vdash \Diamond\Box A$ to $\vdash \Diamond\Box B$, but the first rule is closer to what is missing, namely that $\Diamond\Box A \to \Diamond\Box B$ is provably implied by $\Diamond\Box(A \to B)$ – alternatively put, that we have $\Diamond\Box(A \to B) \vdash_{\mathbf{S4}} \Diamond\Box A \to \Diamond\Box B$ – which is equivalent, by Proposition 2.1.21, to the incorrect claim that $\langle\Diamond\Box, \Diamond\Box, \Diamond\Box\rangle$ is a distributive triple for **S4**.)

via a recognition that this logic is what we are calling Matsumoto normal. This is a slight overstatement in that Lemma 2.8.2 – appealed to in the proof of Theorem 2.8.4 – emphasizes only the point that $\langle \Box \Diamond \Box,\ \Box \Diamond \Box,\ \Box \Diamond \Box \rangle$ is a distributive triple for **S4**, and so leaves out the 'Necessitation' part of the story. As noted after Theorem 2.8.4, since the distributive triples idea works for **K4** and we can recover Necessitation by $\Box \Diamond \Box$ from **D**, Matsumoto's proof works for **KD4**, and we have the corresponding version of Proposition 2.8.15, from which the latter follows as a corollary:

PROPOSITION 2.8.16 *The modality $\Box \Diamond \Box$ is normal in* **KD4**.

Indeed we already know that Matsumoto normality is not necessary for Matsumoto's Embedding Condition, from the case of S_0 from Propositions 2.8.11, 2.9.4 – the latter coming in our 'post-semantic' discussion in Section 2.9 of Matsumoto-related matters. Nor is it is sufficient, as is illustrated by the case of **KT!** and – to cite a sublogic of **S5** – the normal extension of logic **KD** by $\Box p \to \Box \Box \Box p$.

2.9 Supplement: Matsumoto's Embedding (Concluded)

In this section, we deal with some issues left over from the discussion in Section 2.8 concerning distributive triples, Matsumoto's embedding, and the notion of a Matsumoto normal modal logic. While these issues were all introduced there, we promised to hive off from that discussion anything requiring the Kripke semantics for normal modal logics, which meant that we were not in a position to present certain proofs or raise certain questions, to which we can turn here. One such issue arises over a condition noted in Proposition 2.1.24 (p. 52) for a triple of affirmative modalities of the same length to a distributive triple for an arbitrary normal modal logic, namely that the operators concerned constitute coordinatewise distributive triples. As remarked there, in the case of the smallest normal modal logic, **K**, there is a converse to this, given here as Proposition 2.9.2, for which we need the initial observation:

LEMMA 2.9.1 *For any $\{O_1, O_2, O_3\} \subseteq \{\Box, \Diamond\}$, we have: for all formulas A, B, C if $\vdash_{\mathbf{K}} (O_1 A \wedge O_2 B) \to O_3 C$ then $\vdash_{\mathbf{K}} (A \wedge B) \to C$.*

Proof. By a simple adjunction-of-points argument (as in the proof of Theorem 2.6.1): if $\nvdash_K (A \wedge B) \to C$, then by the completeness of **K** w.r.t. the class of all frames, there is a point y in some model, at which A, B, and C are respectively true, true, and false, which truth-values they therefore retain in the submodel generated by y. Adjoin a new point x and extend the accessibility relation – R say – of this generated submodel by adding the pair $\langle x, y \rangle$, and in the extended model (however truth-values are allocated to propositional variables at x), $O_1 A, O_2 B, O_3 C$ are respectively true, true, and false at x. Thus by the soundness of **K** w.r.t. the class of all frames, $\nvdash_K (O_1 A \wedge O_2 B) \to O_3 C$. ■

In the proof that follows, we use the convenient notation "O^n" for a string of n occurrences of the modal operator O (either $\Box$ or $\Diamond$); when $n = 0$, $O^n A$ is just the formula A.

PROPOSITION 2.9.2 *Where X, Y, and Z are affirmative modalities of length n, a necessary and sufficient condition for $\langle X, Y, Z \rangle$ to be a distributive triple of affirmative modalities for **K** is that $\langle X_i, Y_i, Z_i \rangle$ is a distributive triple of modal operators for **K** for each $i \leqslant n$.*

Proof. In view of Prop. 2.1.24 (p. 52), we need only show the necessity of the stated condition. Accordingly, suppose – for a contradiction – that $\langle X, Y, Z \rangle$ is a distributive triple of affirmative modalities for **K** but that for some k $(1 \leqslant k \leqslant n)$, $\langle X_k, Y_k, Z_k \rangle$ is not a distributive triple (of operators) for **K**. Denote by X' the modality $X_{k+1} X_{k+2} \ldots X_n$, understanding Y', Z' correspondingly. We examine the possibilities for $\langle X_k, Y_k, Z_k \rangle$. Since this is not a distributive triple for **K**, there are five possibilities, as enumerated in our earlier discussion leading up to Exercise 2.1.23 (p. 52 above) concerning **KD** and $\mathbf{KD}_c$:

(1) $\langle\Box,\Box,\Diamond\rangle$, (2) $\langle\Box,\Diamond,\Box\rangle$, (3) $\langle\Diamond,\Box,\Box\rangle$, (4) $\langle\Diamond,\Diamond,\Diamond\rangle$ and (5) $\langle\Diamond,\Diamond,\Box\rangle$.

We shall show that in each case $\langle X_k X', Y_k Y', Z_k Z'\rangle$ is not a distributive triple of modalities for **K**. The existence of frames containing points without R-successors shows that no theorem of **K** has the form $(\Box A \wedge \Box B) \to \Diamond C$, disposing of case (1). Consider the frame $\langle W, R\rangle$ with $W = \{0, 1, 2\}$ and $R = \{\langle 0,1\rangle, \langle 0,2\rangle, \langle 1,1\rangle, \langle 2,2\rangle\}$. A model on this frame which has p true at (just) 1 and q true at (just) 2 makes $(\Diamond X'p \wedge \Diamond Y'q) \to \Diamond Z'(p \wedge q)$ false at 0, thereby disposing of case (4). (Because $\{1\}$ is the set of successors of 1, $X'p$ and p take the same truth-value at 1 in any model on this frame, whatever affirmative modality X' may be, and likewise, *mutatis mutandis*, for 2 and $Y'q$.) The same model works for the case of (5). Changing the model so that now p is true at 2 as well as 1, while q is still true just at 2 falsifies $(\Box X'p \wedge \Diamond Y'q) \to \Diamond Z'(p \wedge q)$ at 0, dealing with case (2), while reversing these assignments to p and q disposes of case (3).

Having showing that in none of cases (1)–(5) is $\langle X_k X', Y_k Y', Z_k Z'\rangle$ a distributive triple of modalities for **K**, we conclude that where X'', Y'', Z'' are any affirmative modalities of the same length, $\langle X''X_k X', Y''Y_k Y', Z''Z_k Z'\rangle$ is not a distributive triple for **K**, by as many appeals to Lemma 2.9.1 as there are modal operators in X'' (which is the same number as for Y'', Z''), taking the A, B, C of that lemma as p, q and $p \wedge q$. ∎

Putting "**K4**" for "**K**" in Proposition 2.9.2 converts it into something false, as we know from Lemma 2.8.2, concerning Matsumoto's prefix: $\langle\Box\Diamond\Box, \Box\Diamond\Box, \Box\Diamond\Box\rangle$ is a distributive triple for **K4**, but not so the triple extracted from the second coordinates, $\langle\Diamond,\Diamond,\Diamond\rangle$.

We turn our attention now to the logic S_0, defined before Lemma 2.8.8 (p. 145) as the smallest normal modal logic satisfying Matsumoto's Embedding Condition, i.e., containing every formula $\Box\Diamond\Box A$ for **S5**-provable A. Here we supply a (soundness and) completeness result for this logic. Then, in Proposition 2.9.4, we proceed to show that S_0 is not what (at the end of Section 2.8) we called a Matsumoto normal logic. Recall (from the discussion leading up to Theorem 2.4.10, p. 70) that R^* is the reflexive transitive closure of R.

THEOREM 2.9.3 S_0 *is determined by the class of frames* $\langle W, R\rangle$ *satisfying*

$$\forall x \forall y \in R(x) \centerdot \exists z \in R(y) \forall u \in R(z)(\forall v, w \in R^*(u) \centerdot Rvw).$$

Proof. Soundness: We have only to check that $\Box\Diamond\Box A$ is valid on any frame satisfying the above condition, whenever $A \in \mathbf{S5}$. The supposition that $\Box\Diamond\Box A$ is false at a point x in a model on such a frame gives us some $y \in R(x)$ at which $\Diamond\Box A$ is false, so by the condition we have $z \in R(y)$ such that for all successors u of z, any $v, w \in R^*(u)$ stand in the relation R, and $\Box A$ is false at z, so at some such $u \in R(z)$, A is false. But the condition involving R^* just described implies that the submodel generated by u has a frame on which R is the universal relation, and every theorem of **S5** is valid on all such frames, so the fact that A is false at u contradicts the hypothesis that $A \in \mathbf{S5}$.

Completeness: We show that the canonical model for S_0 is a model on a frame – which we shall call $\langle W, R\rangle$ – satisfying the above condition. Given $x, y \in W$ with Rxy, we must show how to find z as promised in the condition. In fact, z can be any maximal S_0-consistent superset of the following set, whose S_0-consistency will be shown presently:

$$\{A \mid \Box A \in y\} \cup \{\Box B \mid B \in \mathbf{S5}\}.$$

Because of the first term of the union, we shall have $z \in R(y)$, and because of the second term, we have, for any $u \in R(z)$ that for all $v, w \in R^*(u)$, Rvw: for suppose that $u \in R(z)$, and for some m, n, we have $R^m uv$, $R^n uw$ and yet not Rvw. Thus for some formula C, $\Box C \in v$ while $C \notin w$; but then also $\Diamond^m \Box C \in u$ and $\Diamond^n \neg C \in u$. This is impossible because $\Diamond^m \Box C \to \neg\Diamond^n \neg C \in \mathbf{S5}$, so $\Box(\Diamond^m \Box C \to \neg\Diamond^n \neg C) \in z$, so $\Diamond^m \Box C \to \neg\Diamond^n \neg C \in u$. We return to the issue of the consistency of the above union. If it is not

2.9. SUPPLEMENT: MATSUMOTO'S EMBEDDING (CONCLUDED)

S_0-consistent, then for some conjunction A of formulas A_i for which $\Box A_i \in y$, and some conjunction B of formulas $B_j \in \mathbf{S5}$, we have $A \to \neg \Box B \in S_0$, and so by normality, $\Box A \to \Box \neg \Box B \in S_0$; the antecedent belongs to y, so the consequent does, so since $y \in R(x)$, $\Diamond \Box \neg \Box B \in x$, or alternatively put, $\neg \Box \Diamond \Box B \in x$. But since B is a conjunction whose conjuncts are $\mathbf{S5}$-provable, we have $B \in \mathbf{S5}$, and hence by the definition of S_0, $\Box \Diamond \Box B \in S_0$, contradicting the S_0-consistency of x. ∎

Because of the "R^*" the condition on frames figuring in this result is not a first-order condition (in the language of R); the author does not know whether there is a first-order condition for which a completeness result can be given, or whether the class of frames mentioned in Theorem 2.9.3 is the class of all frames for S_0. Fortunately, we do not need to know this in order to substantiate the earlier claim about how weak this logic is:

PROPOSITION 2.9.4 $\langle \Box \Diamond \Box, \Box \Diamond \Box, \Box \Diamond \Box \rangle$ *is not a distributive triple for* S_0.

Proof. Consider the frame $\langle W, R \rangle$ with W containing seven distinct elements x, y_1, y_2, z_1, z_2, u_1, u_2, and $R = \{\langle x, y_1 \rangle, \langle x, y_2 \rangle, \langle y_1, z_1 \rangle, \langle y_1, z_2 \rangle, \langle y_2, z_1 \rangle, \langle y_2, z_2 \rangle, \langle z_1, u_1 \rangle, \langle z_2, u_2 \rangle, \langle u_1, u_1 \rangle, \langle u_2, u_2 \rangle\}$. The labelling of the elements is intended to make it easier to check that the condition on frames mentioned in Theorem 2.9.3 is satisfied. Any model on this frame which has p true at u_1 and nowhere else, and q true at u_2 and nowhere else, has the formula

$$(\Box \Diamond \Box p \land \Box \Diamond \Box q) \to \Box \Diamond \Box (p \land q)$$

false at x, establishing the present result (given the soundness half of Thm. 2.9.3). ∎

Thus S_0 is not itself a Matsumoto normal modal logic. In such a logic, the prefix $\Box \Diamond \Box$ turns out to behave just as a $\Box$-operator is required to behave in a normal modal logic, and can be thought of as a $\Box$-operator in its own right – though of course we must not use "$\Box$" to represent it, since we are not saying that prefixing the modality $\Box \Diamond \Box$ to a formula is equivalent to prefixing its first element to that formula. In some special cases, such an equivalence will hold, with $\mathbf{S5}$, for example, as a trivial case of a Matsumoto normal logic in view of the equivalence in $\mathbf{S5}$ of $\Box \Diamond \Box$ with $\Box$ (Coro. 2.1.12, p. 44). The same applies for $\mathbf{KT!}$ and for $\mathbf{KVer}$, the two Post complete normal modal logics. With some other choices of modality as X, non-trivial (i.e., X not equivalent to $\Box$) cases of a logics being X-normal are not at all puzzling: for example with X as a sequence of occurrences of $\Box$, there is nothing surprising in the fact that any normal modal logic is X-normal, either syntactically – this being a consequence of Proposition 2.1.24 (p. 52) and the fact that $\langle \Box, \Box, \Box \rangle$ is distributive for any normal modal logic – or semantically, since we can easily see that, where R is the accessibility relation of a Kripke model, $\Box^n$ (the n-termed sequence of "$\Box$"s) can be treated as a necessity operator in its own right, with associated accessibility relation R^n, the n-fold relative product of R with itself. (See note 56, p. 70, and the text to which it is appended.) But because of the "$\Diamond$" in the middle position, it is far from obvious how, starting from a model whose accessibility relation is reflexive and transitive – to take the case the $\Box \Diamond \Box$-normal (i.e., Matsumoto normal) modal logic $\mathbf{S4}$, to construe $\Box \Diamond \Box A$ as saying that A is true at all of a range of suitably accessible points. What is the suitable accessibility relation, in terms of the given R? We will not pursue this here, but a few remarks on the case of the simpler Fitting-style prefix $\Diamond \Box$ may be found in my [584], apropos of the smallest normal modal logic in which this modality is itself normal. In the discussion after Theorem 2.8.12 the general idea of faithfully embedding one logic into another via a translation τ was mentioned, in its 'consequence relation' incarnation (though here, as there, we can just restrict attention to the $n = 0$ case):

$$A_1, \ldots, A_n \vdash_{\text{source}} B \text{ if and only if } \tau(A_1), \ldots, \tau(A_n) \vdash_{\text{target}} \tau(B).$$

Whereas in the case of Matsumoto's embedding $\tau(A)$ was obtained from A simply by prefixing $\Box\Diamond\Box$, what we are concerned with in [584] and numerous other papers there mentioned are translations which effect a much more thoroughgoing transformation of the formulas on which they act, replacing each occurrence of $\Box$ in the original formula A with $\Diamond\Box$ (or the Matsumoto-related case, with $\Box\Diamond\Box$) to obtain $\tau(A)$. (See for example Remark 2.1.6, p. 38, and Example 4.4.25, p. 264 below.)

2.10 Semantical Postscript

In this section we discuss three topics from the semantics of modal logic: (1) a way of thinking of the development of the Kripke semantics for modal logic which is more a rational reconstruction of the history than a chronologically faithful record, and (2) some variations on the notion of a Kripke frame in which functions or sets of functions figure prominently, and (3) a question about the extensions of consistent congruential modal logics. (1) and (2) are headed 'A Quick History' and 'Frames with Functions', respectively. (3) does not itself directly pertain to semantics but is included here under the title 'An Application of Neighbourhood Semantics' because it makes use of material from (2) to address the question described.

A Quick History. Recalling the difficulties depicted by Figure 1.1 on p. 1 for giving a semantic account of modal notions by means of truth-tables – even just for the alethic case ($\Box$ and $\Diamond$ as necessity and possibility of a logical or metaphysical kind), one may conclude that the problem is not so much a failure of truth-functionality for these notions as a failure to consider the right set of values for them to be functional in.[119] Given that p is true, we don't know whether $\Box p$ is true – but what if what we were given was that p was necessary? In that case we might be happy (favouring a modal logic at least as strong as **S4**) to say that $\Box p$ itself had this status. So perhaps we could exchange our two values, *true* and *false*, for three values, *necessary*, *contingent*, and *impossible*, say, and, abbreviating these to 1, 2 and 3, attempt a three-valued truth-functional treatment. (This idea occurred to Łukasiewicz; references are given in the closing paragraph of the present discussion, on p. 156.) Negation presents no problems: it should be interpreted by the function mapping 1 to 3, 2 to 2, and 3 to 1. What about conjunction? We can begin easily enough. We should have $1 \wedge 1 = 1$, $1 \wedge 2 = 2$, $1 \wedge 3 = 3$, and similarly if the conjuncts are reversed. Any conjunction with one conjunct having the value 3 should evidently take the value, since if it is impossible for a given statement to be true it is also impossible for its conjunction with anything else. This leaves one case to consider: $2 \wedge 2$. But – a standard objection[120] – this case proves fatal, as there is no satisfactory answer as to what the status is, with respect to the trichotomy *necessary*, *contingent*, *impossible*, of a conjunction each of whose conjuncts has the status *contingent*. For example, if p is contingent, then $p \wedge p$ should presumably be contingent (suggesting $2 \wedge 2 = 2$) while $p \wedge \neg p$ should still be impossible (requiring, on the contrary, that $2 \wedge 2 = 3$). Evidently we need to know more about the conjuncts than their status as necessary, contingent or impossible, in order to come to a conclusion on that score as to the status of the conjunction. So the three-valued idea fails.

The case of $p \wedge \neg p$ just touched on suggests a line of revision, though: if either conjunct is contingently true, the other is contingently false. Whereas in the case of $p \wedge p$, we can harmlessly say that if the repeated conjunct is contingently true (contingently false) then the conjunction has that same status. So a *four*-valued approach may fare considerably better: we have the necessary, the contingently true, the

---

[119] A different response, not considered here, involves using many truth-tables, including partial ones, at once, and allowing the an in one truth-table to depend on entries for subformulas in other tables. This gives the notion of a truth-tabular (but in the interesting cases non-truth-functional) connective explored in Massey [777]. Massey devotes Part 3 of his logic text [780] to presenting modal logic in this way, and credits the gist of the idea to (the somewhat impenetrable) Leonard [711].

[120] See Humberstone [594], p. 270 for references. A corresponding problem is raised by this consideration for attempts such as that of Fisher [304] to treat deontic logic as three-valued (see also Prior [916]); the second last paragraph on p. 111 of [304] suggests the issue is not fully appreciated. An appeal even more far-fetched to three-valued logic in this connection may be found in Cohen [186]. (But those were early days.)

2.10. SEMANTICAL POSTSCRIPT

contingently false, and the impossible – abbreviated for convenience as 1, 2, 3, and 4, respectively,[121] and we can now resolve the earlier problem by putting $\neg 2 = 3$, $\neg 3 = 2$, $2 \wedge 2 = 2$, while $2 \wedge 3 = 4$. More generally, we have the tables of Figure 2.15 for $\wedge$ and $\neg$, with a simple-minded table for $\Box$ thrown in:

$\wedge$	1	2	3	4		$\neg$			$\Box$	
1	1	2	3	4		4	1		1	1
2	2	2	4	4		3	2		4	2
3	3	4	3	4		2	3		4	3
4	4	4	4	4		1	4		4	4

Figure 2.15: Tables for $\wedge$, $\neg$ and $\Box$

While this 4-valued account is a considerable improvement on its 3-valued predecessor, it is not hard to see that it, too, has its shortcomings. Suppose we have two cups, a and b, each of which, as with any ordinary cups, can be full or empty independently of whether the other is full or empty, and that as things stand, both are full (of sugar, milk, or whatever). Then the statement:

Cup a is full and cup b is empty,

is the conjunction of a contingent truth with a contingent falsehood, and our tables, with the entry $2 \wedge 3 = 4$ declare it to be impossible (taking value 4), whereas it should in fact be ruled to be contingently false, since although it is not true that a is full and b empty, this *could* have been true. In other words, not enough possibilities are taken into account in the four-valued story. With the standard interdefinabilities, we may suppose the tables supplemented by the tables they yield with the aid of those definitions, of $\vee$, $\rightarrow$, $\leftrightarrow$, and $\Diamond$. The resulting table for $\leftrightarrow$ then has the entry 1 along its (falling) diagonal, so $2 \leftrightarrow 2 = 1$ and $3 \leftrightarrow 3 = 1$, and thus $\Box(2 \leftrightarrow 2) = 1$ and $\Box(3 \leftrightarrow 3) = 1$, embodying the implausible claim that all contingent truths are strictly equivalent,[122] and similarly for all contingent falsehoods. Again, and as Restall [952], p. 385, puts it in a similar connection, "finitely valued logics being the cramped places that they are," we do not seem to have taken into account the full range of possibilities. (What if we disallowed the Boolean connectives which allow for the expression of these anomalous features, and pursued the pure $\{\Box, \Diamond\}$-fragment of the language – now thinking of $\Diamond$ as an independent primitive alongside $\Box$ – and thought of the desired logic as embodied in a consequence relations rather than a set of formulas? In that case the prospects for a four-valued treatment are by no means as slim: see Béziau [88].)

More precisely, the four-valued story takes into account exactly two possibilities – or what we might in the interests of familiarity describe as two possible worlds. Call them w_1 and w_2 and let us describe by means of an ordered pair of traditional (T and F) truth-values what we may call the *truth-profile* of an arbitrary statement, putting a T or an F in the first slot of such a pair according as the given statement is true or false in w_1, and putting a T or an F in to record its (ordinary bivalent) truth-value at w_2 in the second position in the pair. Now re-baptize the four combinatorially available profiles, $\langle T, T \rangle$, $\langle T, F \rangle$, $\langle F, T \rangle$, and $\langle F, F \rangle$ as 1, 2, 3, and 4, respectively. Then the tables of Figure 2.15 can be seen as computing the truth-profiles across w_1 and w_2 of compounds from the profiles of their components. For example, if A has the profile $\langle T, T \rangle$ while B has the profile $\langle T, F \rangle$, then the conjunction $A \wedge B$ has the profile $\langle T \wedge T, T \wedge F \rangle = \langle T, F \rangle$, underlying the table entry $1 \wedge 2 = 2$. In the case of $\Box$ the profile is not computed coordinatewise but instead returns a T in each position when there is a T in both positions and an F in

---
[121]Note that, since the values should represent mutually exclusive and jointly exhaustive options, it would not do to say "possibly true" and "possibly false" for "contingently true" and "contingently false" here (as Béziau does at p. 112 of [88]).
[122]Strict equivalence is the two-way version of strict implication, for which various symbols have been introduced in the literature, though not here. Talk of "the" two-way version of strict implication here presumes that we have aggregation and its converse ("$\Box$ commutes with $\wedge$"), so that there is no need to distinguish the necessitated biconditional from the conjunction of two oppositely directed strict conditionals.

both positions otherwise. From this perspective, emphasized in Prior [915], what we have been thinking of as four special modal truth-values are really just four ways of distributing the ordinary two truth-values, T and F, across two worlds; more generally, if we have n worlds to start with, then the number of such distributions of T and F – truth-profiles, as we have been calling them – is 2^n, since that is how many ordered n-tuples of Ts and Fs there are. Note that the truth-profile has a certain impartiality about it, whereas in giving $\langle T, F \rangle$, alias the value 2, the informal gloss "contingently true", we are describing matters from the point of view, specifically, of w_1: from w_2's perspective, statements with this profile would count as contingently false. Suitably generalizing the notion of n-tuple will extend this to cover the case in which there are infinitely (even uncountably) many worlds, though here for expository simplicity, we stick with the finite case. For the case of n worlds, $w_1, \ldots, w_n$ these truth-profiles $\langle \xi_1, \ldots, \xi_n \rangle$, where each ξ_i is T or F, are in a one-to-one correspondence with the subsets of $W = \{w_1, \ldots, w_n\}$, alias propositions or candidate truth-sets: given $\langle \xi_1, \ldots, \xi_n \rangle$ we obtain $\{w_i \in W \mid \xi_i = T\}$, while from $X \subseteq W$ we recover the profile $\langle \xi_1, \ldots, \xi_n \rangle$ with $\xi_i = T$ when $w_i \in X$ and $\xi_i = F$ otherwise.

The description of the behavior of $\Box$ is most easily given in terms of truth-sets rather than truth-profiles. If we ask where $\Box A$ is true in a Kripke model we find that it is at those points all of whose R-successors have A true at them. So, abstracting from the particular formulas concerned and just thinking of the propositions the model makes available (candidate truth-sets), if X is one of them, then $\Box X$ should be the set of points all of whose R-successors lie in X. In our two-world model above, if we make the accessibility relation universal, then we get the $\Box$ table of Figure 2.15 in this way.[123] By way of further illustration, we present another example in which the accessibility relation is not universal (Example 2.10.2) but first, a comment on the present case is in order.

REMARK 2.10.1 As the above discussion shows, the logic determined by the two-element universal frame, or, to put it algebraically, the logic determined by the four-element Henle matrix, is not a very good approximation to **S5**, or to any plausible modal logic. (The logic *determined by* a matrix is the set of formulas valid in the matrix in the sense of always taking a designated value however values are assigned to its sentence letters. This definition is admittedly not quite intelligible yet, however, since the notion of designated values is actually introduced below, after Exercise 2.10.3.) As the example about cups a and b may suggest, however, the approximation is exact if attention is restricted to the 1-variable fragment of **S5**: a formula, of whatever complexity, in which only a single propositional variable appears (any number of times) is **S5**-provable if and only if it is valid on the two-element universal frame. The 'only if' part here is obvious, and one way to check the 'if' part is by brute force, working through each of the 16 pairwise non-equivalent formulas of **S5** alluded to in note 39 (p. 44) in which the only variable to appear is (say) p, checking that each of them other than the single **S5**-provable case (with representative formula $\top$ or $p \to p$) can be falsified in a two-element universal model. This establishes the claim that the two-element frame is characteristic for the set of one-variable **S5** theorems, or, to put it another way that if attention is restricted to one-variable formulas, **S5** and the Scroggs logic **S5Alt**$_2$ coincide. (See the discussion after the proof of Proposition 2.5.73, p. 111.) An interesting addendum: the 16 **S5**-equivalence classes of formulas in a given propositional variable can all be obtained by varying # across all binary Boolean connectives in $p \# C(p)$ for each of several choices of $C(p)$, one of them being the noncontingency formula $\triangle p$ from Exercise 1.3.8 (p. 21). This phenomenon is discussed in Canty and Scharle [145], Massey

---

[123]Given the notion of modal algebra defined below (after Exc. 2.10.3) what we have here is what is often called a *Henle algebra*, meaning that the operation associated with $\Box$ maps the top element (of the underlying Boolean algebra) to itself and every other element to the bottom element; in the terminology of universal algebra, these are subdirectly irreducible **S5**-algebras, an **S5**-algebra being a modal algebra in which every **S5** theorem evaluates to the top element, when its propositional variables are assigned arbitrary elements of the algebra and values of compounds are computed using the corresponding operations. Henle algebras are the algebraic version of frames with universal accessibility relations and thus, essentially, of the frames of models provided by the simplified Kripke semantics in which no accessibility relation explicitly appears. For the semi-simplified case introduced below at p. 206, in which the frames have the form $\langle W, R \rangle$ with $R(\cdot)$ a constant function – or equivalently with the set of values of this function replacing R – the corresponding algebraic description is provided by the *filter algebras* of Tokarz [1109].

2.10. SEMANTICAL POSTSCRIPT

[778] (correcting an erroneous assertion in [145]), and Colonna [190] (apparently unaware of the earlier literature), as well as, more recently, in Falcão [281]. ◂

We turn now to the promised variation, by way of illustration:

$$w_1 \longrightarrow w_2$$

Figure 2.16: *A 2-Element Frame for Turning into a Modal Algebra*

EXAMPLE 2.10.2 We have the same two-element set as before, with its four subsets $1 = \{w_1, w_2\}$, $2 = \{w_1\}$, $3 = \{w_2\}$, and $4 = \varnothing$, so let us calculate the behaviour of $\Box$ as an operation on these subsets, given R as depicted in Figure 2.16: $\Box 1 = 1$ because the set of elements all of whose R-successors lie within 1 is (obviously) the set of all – or rather, both – elements. Similarly, as in the case of Fig. 2.15 we have $\Box 2 = \Box 4 = 4$. When it comes to $\Box 3$, we get a different answer, though. What is the set of points all of whose R-successors lie in 3, alias $\{w_2\}$? Unlike the case in which each point is accessible to the other (the Fig. 2.15 case), here we find that this set is not empty, but is rather the set $\{w_2\}$ itself. Thus we have, alongside the Boolean tables of Fig. 2.15, the following table for $\Box$:

$\Box$	
1	1
4	2
3	3
4	4

◂

EXERCISE 2.10.3 (*i*) What would the table for $\Diamond$ look like in the case of Example 2.10.2?
(*ii*) How would the $\Box$ table of Example 2.10.2 change if we changed the accessibility relation of the underlying two-element frame by removing $\langle w_1, w_1 \rangle$ from the R depicted in Fig. 2.16? ✠

Each set of tables for $\wedge$, $\neg$ and $\Box$ of the kind we have been considering constitutes the description of a *modal algebra*, or more explicitly, a *normal* modal algebra, which consists of a Boolean algebra with top element 1 together with a 1-ary operation $\Box$ satisfying $\Box 1 = 1$, and, for all algebra elements a, b: $\Box(a \wedge b) = \Box a \wedge \Box b$.[124] This explanation takes for granted the notion of a Boolean algebra and the associated partial ordering w.r.t. which 1 is a greatest element. (When, as in our examples, the Boolean algebra is the algebra of all subsets of a given set, the partial ordering is: $\subseteq$.) The use of 1 in the definition coincides with the deployment of '1' in the above examples (in terms of the primitives used above, it would be the element $\neg(a \wedge \neg a)$, which does not depend on the choice of a); usually the bottom element of a Boolean algebra (or any bounded lattice) is called 0. In our examples, this has been the role played by the algebra element 4. There are various ways of using an algebra or a class of algebras to give a logic, say in the form of a consequence relation. The simplest is to expand the algebra to what is called a *matrix* by distinguishing a subset of its set of elements, called the set of *designated* elements, and adding this set as a component to the structure. A is then taken to be a consequence of Γ, according to such a matrix, whenever every assignment of algebra elements to formulas which respects the tables in question and assigns a designated element to every $C \in \Gamma$ assigns a designated element to A. If we

---

[124]If we want to consider algebraic semantics for arbitrary congruential modal logics without the restriction to normal ones, we need a notion of modal algebra which does not impose either of the above two conditions on the $\Box$ operation. See Makinson [759] and Došen [249].

are interested in just the consequences of the empty set – which is to say, we are conceiving of modal logics as sets of formulas – then various choices of matrix, as we may illustrate in the present case, lead to the same set of formulas as those which are consequences of ∅ according to the matrix in question, or, expressed more briefly, the same set of formulas as those which are valid in the matrix (always assume a designated value). In the case of the matrix of Example 2.10.2, this amounts to: the set of formulas valid on the frame depicted in Figure 2.16. An axiomatic description of this set of formulas is as **S4.4M**, with various alternative labels suggestive of other axiomatizations being given on p. 418 below.

EXAMPLE 2.10.4 In the case of the matrix resulting from the algebra depicted in Figure 2.15 by taking as the set of designated elements the set $\{1,2\}$, we obtain the same set of formulas as valid as we would if we had chosen as the set of designated elements $\{1,3\}$, or again simply $\{1\}$. To see this, recall that having the value 2 amounts to being true at w_1 and false at w_2, whereas having the value 1 amounts to being true at both points. So having a value in $\{1,2\}$ amounts to being true at w_1, since the formulas true at w_1 are precisely those true at w_1 but not w_2 or else true at both w_1 and w_2. Thus, when we take $\{1,2\}$ as our set of designated values we are considering the formulas which are true at w_1 in every model on the two-element universal frame, which is evidently the same as the set of formulas which are true at w_2 in every such model, there being nothing to choose between w_1 and w_2. Thus taking instead $\{1,3\}$ as our set of designated elements makes no difference to the formulas valid in the matrix, and indeed we see from this that these are precisely the formulas that always take the value 1, corresponding to "true everywhere" in a model on the corresponding Kripke frame. So the matrix with $\{1\}$ as set of designated elements returns the same formulas as valid. When we turn to the consequence relations involved, however, we can see that there is a large difference between that determined by the matrix with $\{1,2\}$ as its set of designated elements, on the one hand, and that determined by the matrix with $\{1\}$ as its set of designated elements on the other. The first consequence relation amounts to truth-preservation at the point w_1 (and accordingly coincides, by the previous considerations, with the consequence relation determined by the matrix with $\{1,2\}$ as designated) while the latter coincides with preservation of being true throughout the model. These are respectively the local and global consequence relations distinguished in note 115, p. 146, and their most conspicuous contrast arises in the case of $\Box p$ and p, the former being a global (or here, $\{1\}$-preserving) consequence but not a local ($\{1,2\}$-preserving) consequence of the latter. (Of course it is still the case that for any formula A *valid* in the matrix with 1 and 2 both designated, $\Box A$ is valid in that matrix.) Bibliographical information: the conversion of a frame to a corresponding matrix is explained, for example, in Kripke [669], §5.2, and – for some non-normal modal logics – in Kripke [670], p. 219*f*. ◀

EXERCISE 2.10.5 (*i*) Turn now to the matrices on the algebra of Example 2.10.2 with sets of designated elements $\{1,2\}$, $\{1,3\}$, and $\{1\}$. Are the same formulas valid in these three matrices, and if not, what are the inclusion relations among the different sets of formulas involved? Do the consequence relations determined by the first two of these three matrices coincide?

(*ii*) Sticking with the same algebra as in part (*i*), what would happen if we chose as our set of designated elements (to get a matrix and its associated consequence relation) $\{2\}$ or $\{2,3\}$? ✠

Our hasty recapitulation above has been something of a just-so story rather than the literal historical truth, so what follow are some corrective comments. The passage from two-valued logic to three-valued logic described here for the purpose of accommodating modality is approximately that found in Łukasiewicz [745] (though a table for $\to$ was given by him which does not coincide with that implicit in the description above); a move to four-valued logic by Łukasiewicz about thirty years later (in the 1950s), for this same purpose, is made in his [746] but the treatment given of $\Box$ in this – as it is called – L-*modal system* is very different from that embodied in Figure 2.15. It is the latter which gives rise to the description in terms of truth-profiles which we have associated with Prior [915]; subsequent work in

2.10. SEMANTICAL POSTSCRIPT

this vein includes Rescher [946] and Massey [781]. The L-modal system itself was described syntactically under Example 2.1.18(*iii*) (p. 48) and will come up again on p. 261, at the end of Example 4.4.21.

Frames with Functions. We consider various kinds of functions in frames. A Kripke frame $\langle W, R \rangle$ with R a binary relation on W can already be regarded, instead, as having R be a function from W to $\wp(W)$. That is, instead of regarding $R(x)$ as a defined notation, abbreviating "$\{y \mid xRy\}$", we could have taken the "$R(\cdot)$" notation as primitive and defined the binary relation of interest by saying that xRy iff $y \in R(x)$. But here we are interested in functions from W to W rather than from W to $\wp W$: point-to-point functions rather than point-to-set functions. Below, we will also consider ('neighbourhood') frames with functions from W to $\wp(\wp(W))$.

A simple variation on frames $\langle W, R \rangle$ replaces the 2-ary relation $R \subseteq W \times W$ with a 1-ary function $f : W \longrightarrow W$, and defines truth in a model on such a frame using the following clause for $\Box$:

$$\langle W, f, V \rangle \models_x \Box A \text{ if and only if } \langle W, f, V \rangle \models_{f(x)} A.$$

This gives nothing particularly new, though, since its effect is already available by restricting attention to traditional frames $\langle W, R \rangle$ with R a (total) functional relation – as defined on p. 68. The upshot is that we are restricted to considering extensions of **KD!**, though certainly the functional notation is more convenient when this restriction is in force.

A more extensive use of functions in frames – both in the sense of using more functions and of being applicable to more logics – comes when we equip a frame with a whole collection of them (again, from W to W) rather than just one, by taking frames to have the form $\langle W, \mathbb{F} \rangle$ where $\mathbb{F}$ is a collection of functions $f : W \longrightarrow W$, and for the definition of truth at a point in a model on such a frame we say, for $x \in W$:

$$\langle W, \mathbb{F}, V \rangle \models_x \Box A \text{ if and only if for all } f \in \mathbb{F}, \langle W, \mathbb{F}, V \rangle \models_{f(x)} A.$$

With $\Diamond$ defined as usual ($= \neg\Box\neg$) we get:

$$\langle W, \mathbb{F}, V \rangle \models_x \Diamond A \text{ if and only if for some } f \in \mathbb{F}, \langle W, \mathbb{F}, V \rangle \models_{f(x)} A.$$

Any class of such frames determines a normal modal logic (i.e., as the set of formulas valid on every frame in the class), raising the question of what the minimal logic for this semantic framework is, i.e., the logic determined by the class of all frames (no special conditions on $\mathbb{F}$, that is). One might conjecture – or even claim[125] – that this is the minimal normal logic **K**. As we have set things up, the answer to that question turns out to be a logic slightly stronger than **K**, containing the **K**-unprovable formula called $\mathbf{D}^{\Box}$ below, though re-drafting the semantics slightly by allowing *partial* functions in $\mathbb{F}$ and altering the definition of truth slightly, counting $\Box A$ true at x just in case for every $f \in \mathbb{F}$ either f is undefined at x or A is true at $f(x)$, would allow us to bring **K** into the fold. We stick with the simple version of the semantics as above, however. Exercise 2.10.6 below introduces the minimal logic in question, but first a few historical and motivational remarks are in order. The current semantics, its bearing on various normal modal logics, and its relation to the Kripke semantics may be found in van Fraassen [327], pp. 151–153, and in §2 of Garson [350]. We are there encouraged to think of the universe W of these frames $\langle W, \mathbb{F} \rangle$ as possible worlds under some particular conceptualization (e.g., from the point of view of a particular coordinate scheme), and as the various $f \in \mathbb{F}$ as admissible transformations of that conceptualization (e.g., giving various alternative coordinatizations). Thus $\Box B$ records the truth of B as invariant under such re-conceptualizations, registering a kind of perspectival independence – absolute as opposed to scheme-relative truth, one might say. Of course, it is not necessary to buy into any (preferably, more precise) version of this motivational story to take an interest in the formal behaviour of the semantic apparatus.

EXERCISE 2.10.6 Show that $\mathbf{K} \oplus \Box\Diamond\top = \mathbf{K} \oplus \{\Box\bot \vee \Box^n\Diamond\top \mid n \geq 0\}$. ✠

---
[125] As Garson does in [350], note 5.

Using **D**$^\Box$ as a label for the formula $\Box\Diamond\top$ (in accordance with the convention of note 30, p. 37), the description of **KD**$^\Box$ on the right in Exc. 2.10.6 tells us that point-generated frames for this logic are either of the form $\langle\{w\},\varnothing\rangle$ (where w is the generating point concerned) or else serial, giving us:

PROPOSITION 2.10.7 **KD**$^\Box$ *is determined by the class of frames* $\langle W, R\rangle$ *satisfying* (i) *for all* $x \in W$, $R(x) = \varnothing$ *or* (ii) *for all* $x \in W$, $R(x) \neq \varnothing$.

Returning to our frames $\langle W, \mathbb{F}\rangle$ and the semantics given in terms of them, we have:

THEOREM 2.10.8 **KD**$^\Box$ *is determined by the class of all frames* $\langle W, \mathbb{F}\rangle$.

Proof. Soundness: left to the reader.
Completeness: Suppose $\not\vdash_{\mathbf{KD}^\Box} A$, with a view to invalidating A on some frame $\langle W, \mathbb{F}\rangle$. By Prop. 2.10.7 (i) for some V, $\langle W, R, V\rangle \not\models_x A$ for $x \in W$ and $R(x) = \varnothing$ or (ii) for some V, $\langle W, R, V\rangle \not\models_x A$ for $x \in W$ and $\langle W, R\rangle$ is serial. In case (i) put $\mathbb{F} = \varnothing$ and check that $\langle W, \mathbb{F}, V\rangle \not\models_x A$; in case (ii) let $\mathbb{F}$ contain all functions from W to W which, conceived as sets of ordered pairs, are subrelations of R, and again verify that $\langle W, \mathbb{F}, V\rangle \not\models_x A$, since for all B and all $w \in W$, $\langle W, \mathbb{F}, V\rangle \models_w B$ iff $\langle W, R, V\rangle \models_w B$. ∎

By virtue of the proof here, we have:

COROLLARY 2.10.9 **KD** *is determined by the class of all frames* $\langle W, \mathbb{F}\rangle$ *with* $\mathbb{F} \neq \varnothing$.

Among the observations recorded in [327] and [350] are those given in the following; note that the logics considered are extensions of **KD**:

EXERCISE 2.10.10 (i) Show that **KT** is determined by the class of frames $\langle W, \mathbb{F}\rangle$ in which $\mathbb{F}$ contains the identity map (the function f with $f(w) = w$ for all $w \in W$, that is).
(ii) Show that **S4** is determined by the class of frames $\langle W, \mathbb{F}\rangle$ in which $\mathbb{F}$ contains the identity map and is closed under composition of functions.
(iii) Describe a class of frames $\langle W, \mathbb{F}\rangle$ determining **S5**. ✠

Little work has been done on the current 'transformations' semantics, by comparison with that on the relational Kripke semantics. Note that the results given or asked for above all pertain to completeness rather than modal definability. It would not be correct, for instance, to claim that the *only* frames $\langle W, \mathbb{F}\rangle$ validating **T** are those in which $\mathbb{F}$ contains the identity map. A weaker condition would evidently be sufficient, namely that for every $x \in W$ there is some $f \in \mathbb{F}$ with $f(x) = x$. (That $\mathbb{F}$ should contain the identity function is the '$\exists\forall$' version of this '$\forall\exists$' condition.) A similar weakening is possible in the case of **4**. A discussion of these issues, begun in Garson [350], is continued in van Benthem [76]. (The 'transformational semantics' of Gerla [368] concerned modal predicate logic and uses functions mapping individuals – in the domains of the worlds – to individuals, rather than worlds to worlds, as here.)

A natural variation on the above semantics, though somewhat further removed from the Kripke semantics, would involve consideration of frames $\langle W, \cdot\rangle$ in which $\cdot$ is a binary operation on W – a *groupoid*, as such structures are often called – and define truth at a point in a model thus:

$$\langle W, \cdot, V\rangle \models_x \Box B \text{ if and only if for all } y \in W, \langle W, \cdot, V\rangle \models_{x \cdot y} B.$$

These groupoid frames were introduced in §3 of Garson [350], in the same year (1972) as Urquhart [1121] appeared, in which essentially structures of this form were deployed to provide a semantics for relevant implication, as part of the enterprise described in Anderson and Belnap [20]. (In Urquhart's treatment, such an implication, $A \Rightarrow B$, say, is true at a point x when for all elements y, if A is true at y, then B is true at $x \cdot y$, so the above clause makes $\Box B$ like a conditional $A \Rightarrow B$ with an antecedent A which

2.10. SEMANTICAL POSTSCRIPT

is true at all points. Urquhart suggested thinking of the elements as pieces of information and of the binary operation as representing a particular way of combining information; thus the conditional holds relative to a piece of information iff any information relative to which its antecedent holds combines with the given piece to yield information relative to which the consequent holds.) The logic determined by the class of all groupoids is **KD**. (A *groupoid* is an algebra comprising a non-empty set and a binary operation under which that set is closed.) The formula **T** is valid on precisely those frames $\langle W, \cdot \rangle$ in which for all $x \in W$ there is some $y \in W$ such that $x \cdot y = x$. As with the 'transformations' semantics, a completeness argument goes through with a stronger ($\exists \forall$ rather than $\forall \exists$) condition, and **KT** is determined by the class of all groupoids containing an identity element,[126] and **S4** similarly by the class of all semigroups containing an identity element (semigroups were defined in the Digression on p. 43); while **S5** is determined by the class of all (groupoid reducts of) groups. (See [350]. Urquhart's groupoid frames in [1121] were semilattice frames – i.e., commutative idempotent semigroups – with a distinguished neutral or identity element, though many variations have since been considered in the literature on relevant and substructural logics.)

It is worth remarking that although the frames we have just been considering are in fact algebras (sets together with selected operations under which they are closed), we have been doing model-theoretic semantics with them, just as with the Kripke semantics – i.e., defining truth at a point and obtaining the notion of validity by suitable universal quantification – rather than algebraic semantics (a topic not dwelt on in the present text, though occasionally mentioned in passing, such as in Example 4.4.21 on p. 261, and, more recently, in the discussion after Exercise 2.10.3 on p. 155).

Returning from functions of two arguments (binary operations) to 1-ary functions, we pass to the best known model-theoretic semantics involving functions as frame-ingredients: neighbourhood semantics. Recall that the functions $f \in \mathcal{F}$ of the 'transformations' semantics took us from points to points, and that the standard relational semantics can be construed as involving functions ('$R(x)$') from points to sets, for which reason we leapfrog over them and onward to the next candidate to consider in this direction: frames with functions from points to *sets of* sets of points – sets of propositions, on one understanding of the latter term (p. 54). A *neighbourhood frame* is a pair $\langle W, \mathcal{N} \rangle$ in which alongside our (as ever, non-empty) set W we have $\mathcal{N}: W \longrightarrow \wp(\wp(W))$. If $X \in \mathcal{N}(x)$, we call the set $X \subseteq W$ a 'neighbourhood' of the point x, and put (for suitable V):

$$\langle W, \mathcal{N}, V \rangle \models_x \Box B \text{ if and only if } \|B\| \in \mathcal{N}(x),$$

where $\|B\|$ is $\{w \in W \mid \langle W, \mathcal{N}, V \rangle \models_w B\}$, the truth-set of – or proposition expressed by – the formula B relative to the given model. As with R, which we can think of interchangeably as a point-to-set function or as a binary relation between points, so we could re-construe $\mathcal{N}$ here as a binary relation between points and sets (propositions), the relation holding between x and X when X is a neighbourhood of x.[127] The term *neighbourhood* is intended to recall the use of that term in connection with topological (and metric) spaces, but note that we do not require that a point x should be an element of every $X \in \mathcal{N}(x)$ – or even that it should belong to *any* $X \in \mathcal{N}(x)$, or that the intersection of two neighbourhoods of a point should itself be a neighbourhood of that point. Neighbourhood semantics is also called Scott–Montague semantics in the literature, and, as it happens, a motivating example in Scott's case (see [1019], p. 160) involves neighbourhoods in the more familiar mathematical sense:

EXAMPLE 2.10.11 The case in question concerns progressive aspect – or 'continuous' tenses – with metric neighbourhoods on the real line (W as the set of real numbers, with the standard metric). The idea was that we think of the elements as temporal instants and with B as, say, "John runs", we may then read

---

[126] An *identity element* is an element e such that for all elements x of the groupoid, we have $e \cdot x = x \cdots e = x$. If, more generally, there are several binary operations in play in an algebra, the condition just given can be taken as defining what it is for e to be an identity element – more specifically a two-sided identity element – *for the operation* $\cdot$.

[127] Cf. Gabbay [338], p. 23, and p. 36 for the contrast between point-to-point and point-to-set relations. An early appearance of the point-to-set relation version of neighbourhood semantics can be found in Cresswell [205], p. 349, where the relation concerned is notated as R.

□A as "John is running": for this to be true at an instant, that instant should have as a neighbourhood the set of instants at which John runs. The French version, in which there is no morphologically marked present simple/present continuous distinction, *Jean est en train de courir*, is particularly suggestive here. Kenny [647], p. 185, famously drew attention to the lack of any treatment of aspectual distinctions in Priorean tense logic of the kind we present in the following chapter. Such distinctions include not only the contrast ± continuous/progressive but also the simple past *vs.* present perfect distinction, concerning which some proposals of Hans Reichenbach, in [944], had been made well before Prior launched his own 'modal' approach to tense logic. Reichenbach's proposals are taken up in, for example, Åqvist [28] and [29], Comrie [191], Galton [347], Hornstein [524], Smith [1066], Taylor [1098], Verkuyl and Le Loux-Schuringa [1130], Vikner [1131], Kibort [651], Areces and Blackburn [34], and Meyer-Viol and Jones [808]. For other discussions of perfect aspect, see the references on p. 377, just before Example 5.2.6. For subsequent development of the treatment suggested by Scott [1019] for the progressive, see Shehtman [1045] and Hodkinson [506], which supplies many further references. (Note that the neighbourhood idea works only for the present progressive form of certain verbs – or, more accurately, certain verb phrases. By contrast with someone who is running at t, and so has to be in the midst of an interval of running, a person writing a book at t need not at any earlier moment *have written* a book – and certainly not the book now under construction – at a time earlier than t, and there may be no later such moment either, if the planned volume is abandoned. See the references just given, as well as the discussion and references in Portner [898], and, for the twentieth century revival of related Aristotelian themes, also the discussions in Chapter 8 of Kenny [647] already referred to, Chapter 4 of Vendler [1123], and the papers Potts [902] and Mourelatos [835], the latter supplying copious references to the literature as of the date of its publication – 1978. Galton [347], should also be mentioned again in this connection.) ◀

A better known application of neighbourhood frames is in broadening the range of model-theoretic semantics so as to encompass various non-normal modal logics which are nonetheless congruential.[128] In particular, recalling (from note 32, p. 39) that **E** is the smallest congruential modal logic, we have:

THEOREM 2.10.12 **E** *is determined by the class of all neighbourhood frames.*

A proof of this result and much else of interest concerning the neighbourhood semantics is pleasantly set out in Chapter 7 of Chellas [171], where the terminology of 'minimal semantics' is employed (with neighbourhood models called 'minimal models', etc.).[129] This further interest concerns in particular various extensions of **E**, including **EM**, where this denoted the smallest congruential extension of **E** by all formulas of the form $\Box(A \wedge B) \to \Box A$, the use of "**M**" for which of course clashes with its use here for McKinsey's axiom (the **M** of the table in Table 2.1, p. 33). **EM** could alternatively be described as the smallest monotone modal logic, and in view of Theorem 2.10.12 it will come as no surprise to learn that it is determined by the class of all frames $\langle W, \mathcal{N} \rangle$ satisfying the further condition that for all $x \in W$, if $X \in \mathcal{N}(x)$ and $X \subseteq Y \subseteq W$, then $Y \in \mathcal{N}(x)$. Alternatively, we could build the monotonicity into the truth-definition itself, following Jennings and Schotch [625],[130] and say:

$$\langle W, \mathcal{N}, V \rangle \models_x \Box B \text{ if and only if for some } X \in \mathcal{N}(x), X \subseteq \|B\|,$$

---

[128]This increase in breadth of application raises the question of whether every congruential modal logic is determined by some class of neighbourhood frames, a question settled negatively in Gerson [369]. A simpler – albeit bimodal – example can be found at the end of Hansson and Gärdenfors [444], But to avoid such incompleteness results one can use a device like that used to produce general frames, as in the Digression on p. 83 – which here we might usefully call general Kripke frames – and consider *general neighbourhod frames*, which come equipped with a set of admissible propositions: see Došen [249].

[129]Chellas denotes the neighbourhood-assigning function by N rather than $\mathcal{N}$ (with N_x for our $\mathcal{N}(x)$), but as we later use N for the set of normal points in the modified Kripke frames – also due to Kripke – introduced on p. 360, here we change fonts.

[130]Also of interest in this connection: Brown [121], mentioned above in Example 2.1.4, p. 37.

2.10. SEMANTICAL POSTSCRIPT

understanding $\|B\|$ as before. With truth defined using this clause for $\Box$, one has **EM** as the logic determined by the class of all frames – "by all locales", Jennings and Schotch say, to emphasize that the standard neighbourhood semantics is being modified (though the term *locale* has other logic-related uses which make this potentially confusing).[131] We return to this topic in Exercise 4.4.10 (p. 246), Example 5.1.6(*iii*), and p. 340. The following exercise is adapted from Chellas [171], p. 211 (Exercise 7.8):

EXERCISE 2.10.13 (*i*) For a formula A, let A^* be the result of interchanging $\Box$ and $\Diamond$ in A. Show, making use of any of the information provided above, that for all formulas A, $\vdash_\mathbf{E} A$ iff $\vdash_\mathbf{E} A^*$.

(*ii*) Understanding the notation as in (*i*), prove or refute the claim that for all formulas A, $\vdash_\mathbf{EM} A$ iff $\vdash_\mathbf{EM} A^*$. ✠

Chapters 7–9 of Chellas [171] show how to add further conditions on neighbourhood frames so as to obtain a class of frames which determines the smallest normal modal logic **K**; the interested reader is referred to that discussion for details, though these are not hard to guess at, since we have already noted how to capture **EM** in these terms, leaving only aggregation and $\Box\top$ to get us to normality.[132] Of course, there will also be many congruential logics $\subseteq$-incomparable with **K**: for example, the smallest such logic containing all instances of the schema **T** ($\Box A \to A$), which is determined by the class of all neighbourhood frames in which every point belongs to all of its neighbourhoods. It is natural to call this logic **ET**, though one must be alert to the possible danger that formulas or schemata equivalent in the setting of normal modal logic and regarded there as alternative versions of the same labelled principle – such as $\Box A \to \Diamond A$ and $\Diamond\top$, interchangeably taken (see Table 2.1 on p. 33) as versions of **D**, lead to distinct candidate congruential logics for a label such as **ED** to refer to.

A suggestive reformulation of the neighbourhood semantics is noted in Hansson and Gärdenfors [444], p. 157, especially convenient, as the authors observe, for modulating in the direction of the algebraic semantics for modal logic touched above (see the discussion following Exercise 2.10.3 on p. 155).[133] While such frames might naturally be called 'functional frames', note that they are not to be confused with the functional Kripke frames described on p. 157 above, with functions $f : W \longrightarrow W$. Now we are dealing instead with functions $f : \wp(W) \longrightarrow \wp(W)$, and where $\|\cdot\|$ understood, as above, relative to a model $\mathcal{M} = \langle W, f, V \rangle$ on such a frame, we say:

$$\|\Box A\| = f(\|A\|),$$

or, more laboriously: $\mathcal{M} \models_x \Box A$ iff $x \in f(\|A\|)$, i.e., iff $x \in f(\{y \in W \mid \mathcal{M} \models_y A\})$. As Hansson and Gärdenfors remark, given a neighbourhood model $\langle W, \mathcal{N}, V \rangle$, we get a pointwise equivalent model of this kind by defining, for $X \subseteq W$, $f(X) = \{y \mid X \in \mathcal{N}(y)\}$, and similarly if we start with $\langle W, f, V \rangle$, we can define, for $x \in W$, $N(x) = \{X \mid x \in f(X)\}$, giving a pointwise equivalent neighbourhood model, and moreover, applying either of these transformations and then the other gets us back to where we started from.

There is an unedifying variation on the neighbourhood semantics worth attending to briefly for the sake of non-congruential modal logics. We may denote a typical model $\mathcal{M} = \langle W, \underline{\mathcal{N}}, V \rangle$ in which to distinguish

---

[131]The locale semantics is also presented, though not under that name, in Exercises 7.9 (p. 211), 7.24 (p. 219), and 9.27 (p. 256) of Chellas [171].

[132]The joint imposition of all the relevant conditions will lead to neighbourhood frames which allow for the definition of a binary accessibility relation R (by: xRy iff $y \in \bigcap \mathcal{N}(x)$) which yields pointwise equivalent Kripke models when supplied with the V of a given neighbourhood model. (Models $\mathcal{M}$, $\mathcal{M}'$, with a common universe, W, say, are said to be *pointwise equivalent* when for all formulas A, and all $x \in W$, we have $\mathcal{M} \models_x A$ iff $\mathcal{M}' \models_x A$; the result just cited can be found in Chellas [171], p. 222.) An earlier survey of the relations between the various kinds of semantics in play here (and others), Hansson and Gärdenfors [444] is still well worth reading. One of their observations will given in the following paragraph.

[133]These ideas were rediscovered 20 years later in Thijsse [1101], p. 335. (Thijsse refers to the sets $\mathcal{N}(x)$ as neighbourhoods of x, but it is the elements – themselves sets of points – of $\mathcal{N}(x)$ that constitute x's neighbourhoods.) The reformulation of the neighbourhood semantics about to be described is used as the preferred presentation in numerous subsequent publications on the subject, such as Shehtman [1046].

it from the $\mathcal{N}$ recently in play, assigning sets of propositions to each $w \in W$ (w's neighbourhoods), this function $\underline{\mathcal{N}}$ simply assigns sets of formulas to each $w \in W$. W and V are as always and the definition of truth on this 'pseudo-neighbourhood' semantics features the following clause for the case of $\Box$:

$$\langle W, \underline{\mathcal{N}}, V \rangle \models_x \Box B \text{ if and only if } B \in \underline{\mathcal{N}}(x).$$

In other words, where the proposition expressed by a linguistic expression ($\|B\|$ for the formula B) in the neighbourhood semantics, in the pseudo-neighbourhood semantics the linguistic expression occurs, perhaps prompting the suggestion that we are dealing with pseudo-neighbourhood pseudo-semantics – or at any rate, semantics with a very syntactic flavour. (The utility sometimes thought to be possessed by this is its escape from congruentiality and related conditions felt to impugn – "the problem of logical omniscience" – the plausibility of possible worlds semantics for propositional attitude ascriptions. See Example 5.6.3, p. 425.) Further, although the models have a non-empty universe W, suggesting a treatment in terms of possible worlds, in fact in evaluating a formula at one $w \in W$ we are never required to attend to any other $w' \in W$: for the non-modal aspects of our formula at w, V and the Boolean clauses decide matters, while for the modal subformulas, we just look inside the local bag of favoured formulas $\underline{\mathcal{N}}(w)$. Thus a less misleading way of presenting the semantics would be in terms of Boolean valuations. This terminology was introduced on p. 10 to refer to valuations for a language whose only connectives were Boolean, to mean that the various truth-functions were associated on the, valuation on question, with these connectives in the conventional way, but we can use the same terminology when additional connectives, such as $\Box$, are present also. In this case $\Box$-formulas are subjected to no constraints at all – or, if you prefer, Boolean valuations are subject to no constraints in respect of such formulas, which accordingly may be regarded as further Boolean atoms (like the propositional variables) meaning, as Segerberg [1032] (p. 51), spells out this phrase, atoms 'from the Boolean point of view'. (*Warning*: do not confuse these with atoms in a Boolean algebra.)

Any such valuation can be regarded as $v_w^{\mathcal{M}}$ – to use the notation of p. 54 – for a pseudo-neighbourhood model $\mathcal{M}$ and a point w therein, and any $v_w^{\mathcal{M}}$ is a Boolean valuation, so the logic determined by the class of all Boolean valuations (for the language with $\Box$), or equivalently, the logic determined by the class of all pseudo-neighbourhood models, is simply the smallest modal logic. (It has been some while since modal logics in general were under discussion, without a restriction to normal, monotone, congruential, etc., modal logics. The general definition appeared on p. 18.) Let us recall ('Makinson's warning': see the Digression on p. 39) that the phrase *the smallest modal logic* has an interpretation which is more than usually dependent on the question of what the Boolean primitives are, being Halldén complete on one way of making this choice and not so on another; the semantic characterization just given is not sensitive to this choice, however. By adding further stipulations, one could characterize extensions of this smallest logic. For example, the smallest modal logic containing **4** would be determined by the class of pseudo-neighbourhood models in which $\Box B \in \underline{\mathcal{N}}(x)$ whenever $B \in \underline{\mathcal{N}}(x)$; for the analogous extension by **T**, we cannot allow $B \in \underline{\mathcal{N}}(x)$ when $\mathcal{M} \not\models_x B$, so really in this case we are dealing with a condition on models rather than the underlying frames. (Contrast the case of the neighbourhood semantics proper, where **T** can be handled with the condition on $\langle W, \mathcal{N} \rangle$ that for all $x \in W$, for all $Y \in \mathcal{N}(x)$, $x \in Y$.)

Another semantic approach to arbitrary modal logics is proposed by Williamson in [1194]. This approach evaluates $\Box$ by means of a universal quantifier over accessible worlds and thereby builds in a kind of malleability in the direction of the normal modal logics. As in Theorem 2.10.8 for the Garson–van Fraassen semantics a class of functions, which we will again call $\mathbb{F}$ is involved, but here the functions take us from formulas to formulas, rather than from points to points in the frames. Further, since the set of all such functions will be pertinent – and this is what we take $\mathbb{F}$ to denote from now on – we need not include this as an ingredient in the frames at all (for the moment, at least), which can instead to be taken to have the form $\langle W, R \rangle$ in which $R \subseteq W \times W \times \mathbb{F}$. For convenience, we write $R_f xy$ in place of $Rxyf$. A model – a *Williamson model*, let us say[134] – on such a frame equips it with a V exactly as for the Kripke

---

[134]Williamson calls these refined models, but the term "refined" is already used for two other things in the semantics of

2.10. SEMANTICAL POSTSCRIPT

semantics, but in the Williamson semantics the definition of truth brings in quantification not only over W but over $\mathbb{F}$, the inductive part for $\Box$ in the definition of truth being as follows, where $\mathcal{M} = \langle W, R, V \rangle$:

$$\mathcal{M} \models_x \Box B \text{ if and only if for all } f \in \mathbb{F}, \text{ and all } y \in W, \text{ if } R_f xy \text{ then } \mathcal{M} \models_y f(B).$$

Williamson's main interest is in the semantic description of modal logics intended as serious candidate epistemic and doxastic logics, where normality – and even mere congruentiality – may seem to involve implausible 'logical omniscience' assumptions. (See 5.1 below.) Williamson makes room for cognitive agents to violate these high ideals of rationality by having them, roughly speaking,[135] confused about the objects of their knowledge and belief: whichever of these we think of $\Box$ as representing, $\Box B$'s truth at x requires not that B itself is true at every epistemic or doxastic alternative y to x, but that $f(B)$ should be, where, if $R_f xy$, $f(B)$ is a kind of 'counterpart' of B in y as far as the agent is concerned in x. Since he is especially concerned with knowledge rather than belief in [1194], he is interested in validating the principle **T** and so takes a special interest in Williamson models $\langle W, R, V \rangle$ satisfying the condition that $R_\iota xx$ for all $x \in W$, where ι is the identity function from formulas to formulas. Restricting attention to such models, if we have $\mathcal{M} = \langle W, R, V \rangle$ and $x \in W$ with $\mathcal{M} \models_x \Box A$, since this implies that for all $f \in \mathbb{F}$, all $y \in W$, if $R_f xy$ then $\mathcal{M} \models_y f(A)$, it implies in particular that for $\iota \in \mathbb{F}$ we have: for all $y \in W$, if $R_\iota xy$ then $\mathcal{M} \models_y \iota(A)$. Taking y as x itself, and remembering that $R_\iota xx$, as well as that $\iota(A) = A$, we conclude that $\mathcal{M} \models_x A$. Williamson also shows (see the Appendix of [1194]) that the modal logic determined by the class of Williamson models with R_ι reflexive does not contain any formulas not provable in the smallest modal logic containing **T**. Here we give an analogous completeness proof for the smallest normal logic itself, and adapt it in Coro. 2.10.15 to cover the ι-reflexivity case.

THEOREM 2.10.14 *For all formulas A, A is true at every point in all Williamson models if and only if A is provable in the smallest modal logic.*

Proof. 'If' (soundness): straightforward.

'Only if' (completeness): Use the canonical model $\langle W, R, V \rangle$ in which W is the set of all sets of formulas which are maximal consistent w.r.t. the smallest modal logic, $V(p_i) = \{x \in W \mid p_i \in x\}$ and $R_f xy$ iff for all A, $\Box A \in x$ implies $f(A) \in W$. Then we show membership and truth coincide (as in Theorem 2.4.1, p. 62, without the "normal"), the inductive step for $\Box B$ requiring:

$$\Box B \in x \text{ iff for all } y \in W, \text{ all } f \in \mathbb{F}, \text{ if } R_f xy \text{ then } f(B) \in y.$$

The 'only if' direction is given by the way R_f was defined. For the 'if' direction, suppose $\Box B \notin x$. We need $f \in \mathbb{F}$, $y \in W$ with $R_f xy$ and $f(B) \notin y$. For the required function, consider $f \in \mathbb{F}$ defined as follows, for all formulas C:

$$f(C) = \begin{cases} \bot & \text{if } C = B \\ \top & \text{if } C \neq B \end{cases}$$

Select any $y \in W$. We have $R_f xy$ with f as defined here. Suppose otherwise: then for some C, $\Box C \in x$ while $f(C) \notin y$. Since $\Box C \in x$ whereas $\Box B \notin x$, $C \neq B$, so by the definition of f above, $f(C) = \top$, contradicting $f(C) \notin y$. Thus $R_f xy$ and we have only to check that $f(B) \notin y$. But by the way f was defined, $f(B) = \bot$, so certainly $f(B) \notin y$. Thus truth and membership coincide in the canonical model, so (the unit set of) any non-theorem can be extended to a maximal consistent set at which, considered as an element of this model, this formula is false. ■

As Williamson ([1194], Appendix) shows, by a slightly different argument, we have the following corollary (of the proof, rather than the content of Theorem 2.10.14 itself):

---

modal logic (see note 303, p. 361), so we change to the present terminology.

[135]For more detail, see [1194], p. 23, as well as the further references cited in note 17 on that page.

COROLLARY 2.10.15 *The smallest modal logic containing* **T** *is determined by the class of Williamson models* $\langle W, R, V \rangle$ *such that for all* $x \in W$, $R_\iota xx$, *where* ι *is the identity map on formulas.*

Proof. Again we consider only the completeness half, using the canonical model constructed as above, but now using maximal consistency w.r.t. the logic with **T**. The fact that truth and membership coincide is unaffected, leaving us only to verify that the model satisfies the condition that for all $x \in W$, $R_\iota xx$. Take $\Box A \in x$ (x any element of W) with a view to showing that if $\Box A \in x$, then $\iota(A) \in x$. Since $\iota(A)$ is A itself we get the desired result from the **T** instance $\Box A \to A$, and Fact 3 from p. 61 (which did not depend on normality). ■

What was described above as the malleability of the semantics in terms of Williamson models can be brought out, as Williamson does in [1194], by imposing various restrictions on the functions in $\mathbb{F}$, something it is perhaps most convenient to describe by re-conceiving the models as having the form $\langle W, R, \mathbb{F}, V \rangle$ in which $\mathbb{F}$ is *some* set of functions from formulas to formulas, and with the clause for $\Box$ in the definition of truth exactly as above, though now understood with reference to this model-varying $\mathbb{F}$. Then we can bring out, as Williamson does, what might be called the 'latent normality' of the semantics by imposing various restrictions on $\mathbb{F}$. Example (i) here is from [1194]:

EXAMPLES 2.10.16 (i) If we impose the condition that for all $f \in \mathbb{F}$ and all formulas A, B, we have $f(A \land B) = f(A) \land f(B)$ – rough gloss: the envisaged cognitive agent is not confused about conjunction – then we find that even with no special conditions on R, every formula of the form:

$$(\Box A \land \Box B) \leftrightarrow \Box(A \land B)$$

emerges as true throughout every model $\langle W, R, \mathbb{F}, V \rangle$, recovering a property (Aggregation and its converse) familiar from normal modal logics.

(ii) On the other hand, if we impose the condition on our models that $f \in \mathbb{F}$ and all formulas A, B, we have $f(A \lor B) = f(A) \lor f(B)$, then we recover only the forward direction (again familiar from the normal modal setting) and not the backward direction (not satisfied by all normal modal logics and not appropriate to any of them intended for doxastic or epistemic application) of:

$$(\Box A \lor \Box B) \leftrightarrow \Box(A \lor B).$$

The conditions on $\mathbb{F}$ here and in (i) nullify the de-normalizing effect of the part of the truth definition which involves the application of f to the formulas after the occurrence of $\Box$, and allows the universal quantification over accessible points to shine through, activating the latent normality of the semantics. ◀

The conditions considered in these examples involve requiring the functions in $\mathbb{F}$ to commute with conjunction (in the first case) and with disjunction (in the second), in a sense of "commute with" analogous to that sketched in note 41.[136] Using the phrase in precisely the sense given there we can summarise the point of the examples by saying that these conditions respectively force and fail to force $\Box$ to commute with the connectives concerned. For the following exercise, the phrase *Williamson frame* is used for the $\langle W, R, \mathbb{F} \rangle$-reducts of Williamson models of this kind; the exercise concerns only soundness, and we are not concerned with modal definability or completeness (concerning which, note that the kind of f used in the proof of Theorem 2.10.14 above would certainly not commute with the connectives under discussion in the previous Example, or here):

---

[136]By way of motivation: Williamson ([1194], p. 24) invites us to "consider a different agent who is credulous and undiscriminating with atomic formulas but scrupulously respects logical relations. Suppose that her counterparts for atomic q in worlds w^* and w^{**} are atomic q^* and q^{**} respectively; then her counterparts for $\neg q$ in w^* and w^{**} are $\neg q^*$ and $\neg q^{**}$ respectively. She would never treat $\neg q^{**}$ as the negation of q^* in w^* or $\neg q^*$ as the negation of q^{**} in w^{**}."

2.10. SEMANTICAL POSTSCRIPT

EXERCISE 2.10.17 (*i*) Show that every instance of the **D**-schema $\Box A \to \Diamond A$ (recalling that $\Diamond A$ is $\neg\Box\neg A$) is valid on a Williamson frame $\langle W, R, \mathbb{F}\rangle$ if (1) each $f \in \mathbb{F}$, f commutes with $\neg$ (i.e., $f(\neg A) = \neg(f(A))$, for all A), and (2) for all $x \in W$, $f \in \mathbb{F}$ there exists $y \in W$ such that $R_f xy$ (i.e., each relation R_f is serial).

(*ii*) Show that every instance of the 4-schema $\Box A \to \Box\Box A$ is valid on a Williamson frame $\langle W, R, \mathbb{F}\rangle$ if (1) each $f \in \mathbb{F}$, f commutes with $\Box$, and (2) for all $f, g \in \mathbb{F}$, their composition $g \circ f$ belongs to $\mathbb{F}$ and satisfies: for all $x, y, z \in W$, if $R_f xy$ and $R_g yz$ then $R_{g \circ f} xz$. ✠

The conditions on Williamson frames given in the above exercise are stronger than is required to validate (all instances of) the schemata mentioned, as we illustrate in the case of Exc. 2.10.17(*i*):

EXAMPLE 2.10.18 The following weaker condition (i.e., weaker than that given in 2.10.17(*i*)) on Williamson frames $\langle W, R, \mathbb{F}\rangle$:

$$\forall A \forall x \in W \exists f \in \mathbb{F} \exists y \in W \,.\, R_f xy \,\&\, f(\neg A) = \neg(f(A)),$$

would suffice to validate the **D**-schema mentioned under Exc. 2.10.17(*i*). Despite the universal quantification over formulas, this remains a condition on Williamson frames as here conceived, rather than on Williamson models, because it does not draw on any particular distribution of truth-values of those formulas in such a model. (In fact, the condition could be weakened further since all we need require of $f(\neg A)$ is that it should be incompatible with – never true at a point alongside with – the formula $f(A)$, not also that one or other of $f(\neg A)$, $f(A)$, should be true at an arbitrary point. We do not go into further complications of this kind.) Note also that the schema under discussion here, or the corresponding formula (with p for A), is not the only thing going under the name **D** in the context of normal modal logics, where we also use this label for the formula $\neg\Box\bot$, which in the present weaker setting would need to be distinguished from that schema. ($\bot$ is taken as primitive here; we choose $\neg\Box\bot$ rather than $\Diamond\top$ for convenience.) Again we find a strong condition like that figuring in Exc. 2.10.17:

(1) $\forall f \in \mathbb{F}$: $f(\bot) = \bot$ and   (2) $\forall x \in W \forall f \in \mathbb{F} \exists y \in W \,.\, R_f xy$,

and a weaker 'localized' form:

$$\forall x \in W \exists f \in \mathbb{F} \,.\, f(\bot) = \bot \,\&\, R_f xy.$$

Compare the stronger and weaker conditions alluded to for the Garson–van Fraassen semantics in the paragraph following Exercise 2.10.10 above (p. 158). ◂

If we were to constrain our formula-to-formula functions by requiring that they commute with all connectives ($\Box$ as well as the Boolean connectives), then we would be dealing with *substitutions*, and instead of interpreting $\Box$ by means of a double universal quantification over substitutions and worlds, the option arises of dropping the worlds altogether and doing all the work with the substitutions. Roughly and somewhat anachronistically speaking, this is what the 'syntactical construction' – or 'syntactical characterization' as the title of Drake [250] puts it – of modality explored in McKinsey [789] set out to do – "roughly" because McKinsey was not concerned with substitutions within a language, but from one language to another,[137] and "anachronistically" because this work appeared over ten years before

---

[137] In fact McKinsey has altogether three languages in play in [789], this being reduced to two in a streamlined version presented in Drake [250]. McKinsey considered various conditions on classes of substitutions, such as being closed under composition, some of which – like that just cited – were satisfied in the case of the class of all substitutions, and others of which were not, such as a condition to the effect that every substitution s has (oversimplifying somewhat) a left inverse s' which 'undoes' it in the sense that $s'(s(A))$ is A for all A. He proved the soundness of **S4** and **S5** w.r.t. this semantics and sets of substitutions meeting the various conditions, for which Drake [250] proved corresponding completeness results using a McKinsey-inspired variation on the algebraic semantics for modal logic; Baxter [55] relates this to the Kripke model-theoretic semantics. Compare Exercise 2.10.10, p. 158 above. Further discussion of McKinsey's project can be found in Sections 6 and 7 of Cresswell [219]. An early philosophical reaction to McKinsey can be found in Section V of Bennett [65].

anything explicitly resembling possible worlds semantics for modal logic had appeared in the literature. We leave Williamson's semantics behind here, returning in Section 5.1 to some of his motivation for developing it, and devote the remainder of the discussion to a simplified version of McKinsey's treatment. Highly simplified, in fact, since we shall have only one language in play (compare note 137), and it will be that of modal propositional logic, with all substitutions under considerations being what we may call *constantizing*, meaning by substitutions s such that for every propositional variable p_i, either $s(p_i) \in \{\top, \bot\}$ or else $s(p_i) = p_i$. If the latter possibility is not realized for any p_i, we call s, *fully constantizing*. The set of all substitutions, we denote by $\mathbb{S}$, and the set of all constantizing substitutions, by $\mathbb{S}_{\text{cst}}$. Since applying a fully constantizing substitution to a formula leaves no variables behind to be affected by any further substitutions, we have the following:

LEMMA 2.10.19 *Suppose s is a fully constantizing substitution. Then for all $s' \in \mathbb{S}$ and all formulas A, $s'(s(A)) = s(A)$.*

Let v be a Boolean valuation for the language with $\Box$ as well as a functionally complete set of Boolean connectives (among which $\top$ and $\bot$ are taken as primitive for convenience). Call such a v a *McKinsey valuation* when it satisfies the following condition for all formulas A:

$$v(\Box A) = \text{T iff for all } s \in \mathbb{S}_{\text{cst}}, v(s(A)) = \text{T},$$

and call a formula A *McKinsey valid* when for all $s \in \mathbb{S}$, and all McKinsey valuations v, $v(s(A)) = \text{T}$. We quantify over all $s \in \mathbb{S}$ in this definition to make sure the class of McKinsey valid formulas has a chance of being a modal logic, which requires closure under Uniform Substitution; recall, for example, note 22 – p. 28 above – and the text to which it is appended. (If we defined a notion of validity as truth on all McKinsey valuations, then $\Diamond p$ and $\Diamond \neg p$ would be valid while $\Diamond(p \wedge \neg p)$ was not, for instance. The "$\Diamond \neg p$" example, reformulated, gives rise to what Cresswell calls "the $\neg \Box p$ problem" in [219]; see also p. 86*f*. in Burgess [137].[138]

REMARK 2.10.20 Burgess also discusses this "$\neg \Box p$" problem in §6 of [139], but may not be right to think that this is exactly the problem being raised in a passage from p. 90 of Hintikka [499] that he quotes is raising. Here is the quoted passage:

> What is needed for the logical necessity of a sentence p in a world w_0 is more than its truth in each one of some arbitrarily selected alternatives to w_0. What is needed is its truth in each *logically possible* world. However in Kripke semantics it is not required that all such worlds are among the alterantives to a given one.

Burgess goes on ([139], p. 127) to write – in terms some of which will need to be explained after the quotation – as follows:

> Now it is certainly true, as the complaint alleges, that the valuations to be represented in the model may be "arbitrarily selected." For any set of valuations, there is a Kripke model $M = (X, a, R, V)$ where just those valuations and no others turn up as the valuations $V(x)$ attached to indices x in X

---

[138] Burgess compares in [137] the merits of various alethic modal logics as formal systematizations of logical necessity and possibility – as they had been in Lemmon [695], though with a focus on different candidate logics, Burgess's conspicuously including, of current relevance, McKinsey's. Note that the abstract at the head of [137] contains a typo, and that the underlined occurrence of the word *argument* in this passage should read *logic*: "The arguments of Halldén and others that the right validity <u>argument</u> is **S5**, and the right demonstrability logic includes **S4**, are reviewed,...". This issue – semantically *vs.* proof-theoretically oriented conception of logical necessity, usefully distinguished by Burgess from modal provability logic *à la* Boolos [112] – is also touched on at p. 563 of Humberstone [582] and will not occupy us further here. A potentially more confusing typographical error for readers of [137] occurs in the second new paragraph of p. 83 in which the "distinctive axioms" of **S4.2** and **S4M** – the latter called by Burgess "**McK**" – are "equivalent respectively to $\neg(\Box \neg \Box p \wedge \Box \neg \Box \neg p)$ and $\Box \neg \Box p \vee \Box \neg \Box \neg p$", where the modal formulas concerned appear the wrong way round, those just given being versions of **M** and **.2** (alias **G**), respectively.

2.10. SEMANTICAL POSTSCRIPT

by the function V. For instance there are Kripke models where no index is assigned a valuation that assigns the atom p the value T, and there are Kripke models where no index is assigned a valuation that assigns the atom p the value F.

Evidently the notation and terminology are a little different from what we are used to. X is the universe of the model in question – whose elements Burgess calls indices rather than points or worlds – and on which R is as usual a binary relation, and here in addition a is a distinguished element (we will meet such models in the discussion starting on p. 190) and V is a function assigning to each element $x \in X$ another function $V(x)$ which in turn assigns a truth-value to each propositional variable. (So where we would say "$x \in V(p_i)$", Burgess says "$V(x)(p_i) = $ T.") Burgess then goes on to reply to Hintikka's complaint by saying that we want our propositional variables to be thought of as stand-ins for arbitrary statements, including those that may be necessary or impossible, so it is actually desirable that they should exhibit the behaviour described at the end of the passage just quoted from him.

While that is a perfectly good response to those that might say that we want "$\neg\Box p$" to come out as valid – in fact a mild variation on the line that we want our logics to be closed under uniform substitution – it does quite take up Hintikka's complaint, as the reference to "the atom p" indicates. It is a very sensible policy (followed in this book for instance) to reserve $p, q, r, \ldots$ to be propositional variables (or sentence letters), and thus to be atomic formulas (or 'atoms') of the language of propositional modal logic, but this is not policy Hintikka follows. Rather than using, in formal work, $p, q, r, \ldots$ in the way just described, Hintikka uses these letters (as remarked above in the discussion headed 'Notation for schematic letters' beginning on p. 3) as schematic letters (alias meta-linguistic variables) for arbitrary formulas. In the passage from Hintikka above, he speaks of the "logical necessity of a sentence p in a world w_0", and there is no "atomic" before the word "sentence", so Burgess's emphasis on what happens to atomic formulas is not to the point. But the talk of sentences (rather than formulas) and of being true in a world w_0 without any explicit relativity to a model, suggests that Hintikka is not thinking after all of Kripke's model-theoretic semantics for modal logic but of a derived truth-theoretic semantics for an interpreted (rather than merely formal) modal language – getting a definition of truth-at-a-world by relativizing the model-theoretic notion of truth-at-a-world-in-a-model to some one intended model (cf. Evans [275]). Certainly anyone interested in this venture would need to attend to Hintikka's advice about getting all and only the genuinely possible worlds into the picture, but even the reference to elements of the universe of a Kripke model as possible worlds is merely heuristic, as a reminder of the motivation – the universe itself is required for model-theoretic purposes only to be some non-empty set or other. (Incidentally, the first two pages of Hintikka [499] show him very much resenting talk of the model theory of modal logic as 'Kripke' semantics, since it underplays the role of various earlier contributors including Hintikka himself. Some non-partisan historical accounts of the development of modal logic and its semantics were listed at the start of Section 2.2, p. 53. §7 of Burgess [139] itself provides some useful comparative remarks on the difference between the approaches of Kanger, Hintikka and Kripke.) ◄

Let us return to the KcKinsey semantics now. As usual, in the discussion preceding Remark 2.10.20, we are taking $\Diamond$ as defined by $\neg\Box\neg$, which has the effect that a McKinsey valuation verifies $\Diamond A$ just in case it verifies some substitution instance via a constantizing valuation, of A.[139] With the present definition, we have not only a modal logic but a normal modal logic, and, furthermore, one including the famous McKinsey axiom, **M**: $\Diamond(\Box p \vee \Box \neg p)$. To see that this is McKinsey valid (essentially as noted at p. 92 of [789]), take any substitution instance $\Diamond(\Box A \vee \Box \neg A)$ and any McKinsey valuation v, and note that there is $s \in \mathbb{S}_{\text{cst}}$ such that $v(s(\Box A \vee \Box \neg A)) = $ T, i.e., $v(\Box s(A) \vee \Box \neg s(A)) = $ T, since the latter means that $v(\Box s(A)) = $ T or $v(\Box \neg s(A)) = $ T. This is equivalent to saying that either for all $s' \in \mathbb{S}$, $v(s'(s(A))) = $ T,

---

[139]McKinsey himself takes $\Diamond$ as primitive in [789], and simply surrounds this with negation signs rather than using any special purpose notation for $\Box$. It is clear from his discussion that he is exploring a particularly formalistic notion of logical possibility, for which "it would be said that the sentence, 'Lions are indigenous to Alaska,' is possible, because of the fact that the sentence, 'Lions are indigenous to Africa,' has the same form and is true" ([789], p. 83). See further Example 2.10.23 below.

or else for all $s' \in \mathbb{S}$, $v(s'(s(A))) = \text{F}$. By Lemma 2.10.19, we may choose as the desired s any fully constantizing substitution, since under such a choice $s'(s(A)) = s(A)$, for any $s' \in \mathbb{S}$, and certainly we have $v(s(A)) = \text{T}$ or else $v(s(A)) = \text{F}$. This establishes a key part of:

PROPOSITION 2.10.21 *Every theorem of* **S4M** *is McKinsey valid.*

EXERCISE 2.10.22 Prove the rest of Prop. 2.10.21: those parts not pertaining to **M**; i.e., show that the set of McKinsey valid formulas includes all truth-functional tautologies, the axioms **K**, **T** and **4**, and is closed under Modus Ponens, Uniform Substitution and Necessitation. ✠

McKinsey's ideas, at least at the motivating level, concerned the substitution of some items of non-logical vocabulary for others in languages with greater internal complexity than that provided by propositional logic, as the following informal example shows (as does that quoted in note 139); it is based on a 1952 letter written by McKinsey, quoted in Anderson and Belnap [20], p. 123.

EXAMPLE 2.10.23 In the letter concerned, McKinsey argued that the following is false (in the actual world), taking the English words involved to have their usual meanings:

$$\Box\Diamond(\text{Sugar is sweet and vinegar is not sweet}).$$

on the grounds that from the sentence in the scope of the "$\Box$" there results by substitution (of *sugar* for *vinegar*) a certain false sentence:

$$\Diamond(\text{Sugar is sweet and sugar is not sweet}).$$

(This sentence is false because there is no substitution which will turn the sentence in the scope of the "$\Diamond$" into something true.) Thus, assuming the substitutional semantics for $\Box$ and $\Diamond$, we can have A true (in this case: "Sugar is sweet and vinegar is not sweet") without $\Box\Diamond A$" being true, so the correct modal logic must reject the **B** axiom, $p \to \Box\Diamond p$.[140] ◀

It is worth considering the case of **B** in the simplified setting of McKinsey validity as defined earlier, as we do in part (*i*) of the following Exercise; part (*ii*) concerns a variation on **M**, obtained from the formulation of this axiom as $\Box\Diamond p \to \Diamond\Box p$ (as in Table 2.1 on p. 33, though there it was the corresponding schema that was listed as **M**) by omitting the $\Box$ on the antecedent:

EXERCISE 2.10.24 (*i*) Is the formula $p \to \Box\Diamond p$ McKinsey valid? (Justify your answer.)

(*ii*) Is the formula $\Diamond p \to \Diamond\Box p$ McKinsey valid? (Again, justification required.) ✠

Some questions it would be good to have answers to:

- Are the theorems of **S4M** precisely the McKinsey valid formulas, as defined here? (That is, do we have the completeness half of the result for which Prop. 2.10.21 provides the soundness half? For a – soundness and – completeness result in terms of Kripke frames, see Theorem 4.5.17(*iii*), p. 286. The relevant class of frames comprises the transitive reflexive frames in which every point has accessible to it a reflexive end-point, the latter points being analogous to fully constantizing substitutions in that, *à la* Lemma 2.10.19, no further substitution takes us anywhere else.)

---

[140]McKinsey actually addresses his criticism to **5** rather than **B**. Likewise with the example of arithmetical equations and truth-functional combinations thereof in §2 of McKinsey [789].

2.10. SEMANTICAL POSTSCRIPT

- How sensitive is the class of McKinsey valid formulas to the exact formulation of the condition concerning $\Box$ in the definition of a McKinsey valuation? If this were changed to include arbitrary substitutions, or to the intermediate class of substitutions which map propositional variables to propositional variables or else to $\{\top, \bot\}$, rather than just what we have called constantizing substitutions, would this affect the class of formulas *absolutely all* of whose substitution instances were true on every McKinsey valuation?

Concerning the second of these questions, let us observe that if we replaced the class ("$\mathbb{S}_{cst}$") of constantizing substitutions – which perhaps should be called 'at most constantizing' substitutions, since they either substitute one of $\top$, $\bot$, for p_i or else leave p_i alone – with the class of fully constantizing substitutions, this would certainly affect the class of formulas validated:

EXAMPLE 2.10.25 $\Box(p \to \Box p)$ would count as McKinsey valid if we interpreted $\Box$ by quantifying only over fully constantizing substitutions. To see this, take an arbitrary substitution instance of the formula, $\Box(A \to \Box A)$, say, and suppose v – a McKinsey valuation in the currently modified sense – assigns the value F to this formula. Then for some fully constantizing substitution s, $v(s(A \to \Box A)) = F$, which means that $v(s(A)) = T$ while $v(\Box(s(A))) = F$, so for some further such substitution s' we have $v(s'(s(A))) = F$. This is a contradiction, since by Lemma 2.10.19, $s(A)$ and $s'(s(A))$ are the same formula. Thus the change to fully constantizing substitutions would add new formulas as valid. But the same example shows that it would also invalidate some formerly ('McKinsey') valid formulas, because we now have the following instance of the **T** axiom:

$$\Box(p \to \Box p) \to (p \to \Box p),$$

whose antecedent we have just seen to be valid on the semantics with only fully constantizing substitutions in the condition on $\Box$, but whose consequent certainly is not valid on that semantics. ◀

Still on the subject of the second of the two questions above, the reason for working mainly with what we have been calling constantizing valuations is that they do not increase the complexity of formulas, on a suitable measure of complexity (we have to count $\bot$ and $\top$ as having complexity 0, like the p_i, for this purpose, rather than as having complexity 1 on the grounds that each uses a connective – albeit nullary – once in its construction). This enables us to sidestep some apparent worries about circularity that might otherwise intrude. For example, if we allowed arbitrary substitutions at this stage and wondered about whether $v(\Box A) = T$, we are directed to the question of whether or not $v(s(A)) = T$ for the various substitution instances $s(A)$ of A, which could be of any complexity, and in particular could themselves contain the original formula $\Box A$ as a subformula. This issue, we set to one side here. (The issue is discussed in a closely related connection – the substitutional interpretation of the quantifiers – in Kripke [673], p. 331 and esp. p. 332.)

In the Kripke semantics we interpret $\Box$ as universally quantifying over points in models suitably related to the point of evaluation, while in the transformations semantics of Garson (and van Fraassen) discussed above (see p. 157), the interpretation involves instead universal quantification over point-to-point functions. McKinsey's substitutional semantics – if we may call it that – uses universal quantification over (certain) formula-to-formula functions, raising the question of whether one might provide a semantic account which uses instead universal quantification over suitably related formulas. The binary accessible relation between worlds would be replaced, on the semantics envisaged by one between formulas. Concisely put, we are asking whether it is possible to fill the blank in the following proportionality:

GARSON *is to* KRIPKE *as* McKINSEY *is to* ___.

To conclude our discussion, we have a brief look at the most obvious way of filling the blank. We need a binary relation between formulas, $\mathscr{R}$, say, to be analogous to the binary relation between points in the Kripke semantics, and we will parcel it up together with a Boolean valuation v to get what we think of as a *generalized McKinsey model*, $\langle v, \mathscr{R} \rangle$ provided the following condition is satisfied for all formulas A:

$$v(\Box A) = \mathrm{T} \text{ if and only if for all } B \text{ such that } A\mathcal{R}B, \text{ we have } v(B) = \mathrm{T},$$

and instead of writing $v(A) = \mathrm{T}$ we write $\langle v, \mathcal{R}\rangle \models A$. As with McKinsey validity, the question arises as to whether we should define a formula to be valid on the present semantics if all its substitution instances are true in all generalized McKinsey models or instead to require simply that the formula itself be true in all such models. We choose the latter, since it follows from Theorem 2.10.26 below that the set of valid formulas, so defined, is closed under Uniform Substitution anyway. Thus, we call A *valid* (on the present semantics) if and only if for all generalized McKinsey models $\langle v, \mathcal{R}\rangle$, we have $\langle v, \mathcal{R}\rangle \models A$.

THEOREM 2.10.26 *For all formulas A, A is valid (on the present semantics) if and only if A belongs to the smallest modal logic.*

Proof. 'If' (Soundness):

'Only if' (Completeness): Suppose that A is not provable in the smallest modal logic, so that $\{\neg A\}$ is consistent w.r.t. that logic and can be extended to a maximal consistent set x, say, and we define the (evidently, Boolean) valuation v to be the characteristic function of this set. Next, we need to find a suitable $\mathcal{R}$ so that A is false in the model $\langle v, \mathcal{R}\rangle$, showing our unprovable formula to be invalid as claimed. We use the following definition:

$$A\mathcal{R}B \text{ if and only if } \Box A \to B \in x \text{ (for all } A, B\text{)}.$$

From what we have already noted, that v is a Boolean valuation with $v(A) = \mathrm{F}$, we can conclude that A is invalid according to the present semantics provided that $\langle v, \mathcal{R}\rangle$ is indeed a generalized McKinsey model, once we have ascertained that the condition relating to $\Box$ is satisfied, which is here put in terms of membership in x rather than truth on v:

for all formulas B, $\Box B \in x$ if and only if for every formula C, if $B\mathcal{R}C$ then $C \in x$.

Unpacking the definition of $\mathcal{R}$, and making abbreviative use of $\forall$ and $\Rightarrow$ (universal quantification and material implication in the metalanguage), this means:

$$\begin{array}{lll} \Box B \in x & \textit{iff} & \forall C((\Box B \in x \Rightarrow C \in x) \Rightarrow C \in x) \\ & \textit{iff} & \forall C(\Box B \in x \text{ or } C \in x) \\ & \textit{iff} & \Box B \in x \text{ or } \forall C(C \in x). \end{array}$$

The right-hand sides here – that parts after the "iff", that is – are evidently equivalent and the second disjunct on the right of the last line would contradict the consistency of x, so the left and right-hand sides of the top line are equivalent, as required. ■

As with Corollary 2.10.15 to a result (Theorem 2.10.14, p. 163) for Williamson's semantics like that just given for the current semantics, there is a straightforward adaptation for the presence of **T**:

EXERCISE 2.10.27 Show that the smallest modal logic containing **T** is sound and complete for the adaptation of the above semantics in which attention is restricted to generalized McKinsey models $\langle v, \mathcal{R}\rangle$ with $\mathcal{R}$ reflexive. ✠

Other such variations, we leave to the interested reader. As with Williamson's semantics, we have a certain 'latent normality', in that structure-respecting conditions may be imposed on $\mathcal{R}$ to get the effect of having $A\mathcal{R}B$ as: $s(A) = B$ for some suitable substitution s (e.g., $s \in \mathbb{S}_{\mathsf{cst}}$), in which case we could drop the "generalized" from "generalized McKinsey models". (Thus we should have conditions as: $A \wedge B\mathcal{R}C$ iff for some formulas C_0, C_1, we have $C = C_0 \wedge C_1$ and $A\mathcal{R}C_0$, $B\mathcal{R}C_1$.) This would be analogous to seeing a binary accessibility relation between points xRy as of the form $f(x) = y$ for a suitable function

2.10. SEMANTICAL POSTSCRIPT

f (an $f \in \mathbb{F}$ from a Garson–van Fraassen frame $\langle W, \mathbb{F} \rangle$ satisfying this or that set of conditions). But with the present McKinsey-inspired discussion, we are already wandering some distance off the 'frames with functions' track, and call an end to this tour of alternative semantic approaches. There will be more on semantics for non-normal modal logics in Section 5.1, at p. 360 and the pages following.

An Application of Neighbourhood Semantics. Here we make an application of the neighbourhood semantics introduced in the preceding discussion (specifically, at p. 159). But it will take a while before we get there. We take up a question from Remark 2.1.6 (p. 38): does every consistent congruential modal logic admit of a truth-functional interpretation for $\Box$ in the sense that its theorems become tautologies when $\Box$ is interpreted as expressing one of the four 1-ary truth-functions (the identity function, the constant true truth-function, negation, the constant false truth-function)? Putting this another way, this question can be put as the question of whether every congruential modal logic is a sublogic of our of the following four modal logics: the smallest modal logic containing $\Box p \leftrightarrow p$, the smallest modal logic containing $\Box p \leftrightarrow \top$ (equivalently $\Box p$), the smallest modal logic containing $\Box \leftrightarrow \neg p$, and finally, the smallest modal logic containing $\Box p \leftrightarrow \bot$. Makinson [759] showed that every consistent monotone or antitone modal logic is a sublogic of one of these four, and we will accordingly refer to them as the four *Makinson logics*. Note that these four logics are all congruential, and indeed two of them are normal – the identity logic and the constant true logic being the Post complete normal modal logics **KT!** and **KVer**.[141] Our current question is whether the phrase "monotone or antitone" can be replaced with "congruential", thus strengthening Makinson's observation. The following discussion overlaps with that in Humberstone [601], though some supplementary information will be found in the discussion which follows Exercise 2.10.34 (p. 174).

For a modal logic S and formula A, we say that A is S-*decidable* if either A or its negation is a theorem of S;[142] alternatively we may say in this case that S *decides* A. (This is sometimes put by saying that A is S-provable or S-refutable.) We use the notion of a pure formula – one containing no propositional variables – from Exercise 2.7.21 (p. 134); this presumes we take $\top$ or $\bot$ or both – and for convenience here we take both – as primitive. For definiteness, let the remaining primitive connectives be $\wedge$, $\neg$ and $\Box$.

LEMMA 2.10.28 *Let S be a congruential modal logic for which $\Box\top$ and $\Box\bot$ are both S-decidable. Then every pure formula is S-decidable.*

Proof. Supposing that $\Box\top$ and $\Box\bot$ are both S-decidable, we argue by induction on the complexity of (the pure formula) A, understood as the number of non-nullary connectives in the construction of A, that A is S-decidable.

Basis Case. The complexity of A is 0. Then A is $\top$ or $\bot$, so evidently A is S-decidable, being respectively provable or refutable in S.

Inductive Step. Suppose A is $B \wedge C$. B and C are pure formulas of lower complexity than A, so each is assumed to be S-decidable. If both are S-provable, then so is their conjunction, A, and if not, then at least one is S-refutable, in which case A is refutable, so either way, A is S-decidable. A second case is that in which A is $\neg B$, which the reader can be left to take care of. Finally, suppose that A is $\Box B$. Here the inductive hypothesis tells us that B is a pure formula which is S-provable, in which case $\vdash_S B \leftrightarrow \top$, or S-refutable, in which case $\vdash_S B \leftrightarrow \bot$. By congruentiality we have, respectively, $\vdash_S \Box B \leftrightarrow \Box \top$ or $\vdash_S \Box B \leftrightarrow \Box \bot$. But now the S-decidablity of $\Box\top$ and $\Box\bot$ implies the S-decidability of $\Box B$, alias A, either way. ∎

---

[141] The latter is sometimes called the Verum logic; the constant false case is sometimes called the Falsum logic. Though not normal, this logic is at least regular, and will receive occasional brief mention below, for example at pp. 303 and 362.

[142] This notion of decidability is of course not the same as the notion of decidability of a logic explained in note 61 (p. 73).

Note that since normal modal logics are congruential, this already delivers a well-known result we pause to record:

COROLLARY 2.10.29 *The normal modal logics* **KD** *and* **KVer** *decide all pure formulas.*

Proof. By normality, $\Box\top$ is provable in both logics, while they decide $\Box\bot$ negatively and positively, respectively. ∎

Resuming our discussion, we have:

LEMMA 2.10.30 *Let S be a consistent congruential logic. Then*
(i) *if $\vdash_S \Box\top$ and $\vdash_S \Box\bot$, then for no formula A do we have $\vdash_S \neg\Box A$; and*
(ii) *if $\vdash_S \neg\Box\top$ and $\vdash_S \neg\Box\bot$, then for no formula A do we have $\vdash_S \Box A$.*

Proof. We do the proof for (i), that for (ii) being similar. Suppose that, for a consistent congruential S, we have $\vdash_S \Box\top$ and $\vdash_S \Box\bot$, and, for a contradiction, that for some formula A, $\vdash_S \neg\Box A$. Let B be any pure substitution instance of A (A itself if A is a pure formula). By Lemma 2.10.28, S decides B, so either $\vdash_S B \leftrightarrow \top$ or $\vdash_S B \leftrightarrow \bot$, in which case by congruentiality either $\vdash_S \Box B \leftrightarrow \Box\top$ or $\vdash_S \Box B \leftrightarrow \Box\bot$. But since $\vdash_S \Box\top$ and $\vdash_S \Box\bot$, we have $\vdash_S \Box B$ either way. Recalling that $\vdash_S \neg\Box A$, by closure under Uniform Substitution, we should also have $\vdash_S \neg\Box A$, contradicting the consistency of S. ∎

We continue with a claim whose status, as its label suggests is dubious, going on below to explain how the proof offered is fallacious:

WOULD-BE THEOREM 2.10.31 *In any consistent congruential logic, $\Box$ has a truth-functional interpretation, in the sense that S has a consistent congruential extension adding (i) $\Box A \leftrightarrow A$ for all A, or (ii) $\Box A \leftrightarrow \neg A$ for all A, or (iii) $\Box A \leftrightarrow \top$ for all A, or (iv) $\Box A \leftrightarrow \bot$ for all A.*

Proof. If $\Box$ cannot be interpreted (in S) as the identity truth-function, because the extension mentioned under (i) is not consistent, then there must be some formula A, such that either (α_1): $\vdash_S A \wedge \neg\Box A$ or (α_2): $\vdash_S \Box A \wedge \neg A$. Similarly, if $\Box$ cannot be interpreted as negation, there must be some formula B for which either (β_1): $\vdash_S B \wedge \Box B$ or (β_2): $\vdash_S \Box B \wedge B$. And if $\Box$ cannot be interpreted as the constant true truth-function, we have (γ): $\vdash_S \neg\Box C$ for some formula C. Finally, if $\Box$ cannot be interpreted as the constant false truth-function, then (δ): $\vdash_S \Box D$, for some formula D. Now, it is not hard to see that (α_1) and (α_2) are equivalent, given the congruentiality of S, to $(\alpha_1)'$ and $(\alpha_2)'$ respectively:

$$(\alpha_1)' \quad \vdash_S \neg\Box\top \qquad\qquad (\alpha_2)' \quad \vdash_S \Box\bot.$$

Similarly, (β_1) and (β_2) are equivalent respectively to $(\beta_1)'$ and $(\beta_2)'$:

$$(\beta_1)' \quad \vdash_S \Box\top \qquad\qquad (\beta_2)' \quad \vdash_S \neg\Box\bot.$$

So if $\Box$ cannot interpreted as the identity or the negation truth-function, one of conditions $(\alpha_1)'$, $(\alpha_2)'$, must be satisfied alongside one of $(\beta_1)'$, $(\beta_2)'$. Since S is consistent, we cannot have $(\alpha_1)'$ and $(\beta_1)'$ satisfied, and we cannot have $(\alpha_2)'$ and $(\beta_2)'$ satisfied. So it must be either $(\alpha_1)'$ together with $(\beta_2)'$ or else $(\alpha_2)'$ together with $(\beta_1)'$. But the first combination contradicts (δ), in view of Lemma 2.10.30(ii), while the second contradicts (γ), in view of Lemma 2.10.30(i). Thus at least one of the extensions listed under (i)–(iv) of the Theorem must be consistent, as claimed. ∎

The mistake in the proof just given does not lie in the appeals to Lemma 2.10.30 or in the claimed equivalence of (α_i) with $(\alpha_i)'$ or of (β_i) with $(\beta_i)'$, $i=1,2$. The mistake lies instead in the assumption

2.10. SEMANTICAL POSTSCRIPT

that if S is not consistently extended by $\Box A \leftrightarrow A$, to take the cases of (i), then there must be some A for which we have (α_1) or (α_2). To see that this is not so, take the case of **KW**, in which $\Box$ cannot be interpreted as the identity truth-function, or to put it another way, which cannot be consistently extended by adding $\Box B \leftrightarrow B$ for all B. (These are equivalent ways of spelling out the claim that **KW** is not Sobociński-regular.) Suppose (α_1) were correct. Then we should have A for which $\vdash_S A \wedge \neg \Box A$, and thus $\vdash_{\mathbf{KW}} A$ and $\vdash_{\mathbf{KW}} \neg \Box A$, from the first of which Necessitation (**KW** being a normal modal logic) delivers a formula whose negation the second says is **KW**-provable: since **KW** is consistent, there is no such A. Turning instead to (α_2) we get $\vdash_{\mathbf{KW}} \Box A \wedge \neg A$ and this time necessitating the second rather than the first conjunct we get $\vdash_{\mathbf{KW}} \Box(A \wedge \neg A)$, again appealing to the normality of **KW**, which is not the case. (In terms of the $(\alpha_i)'$ and $(\beta_i)'$ formulations: **KW**, since neither $(\alpha_2)'$ nor $(\beta_2)'$ is satisfied, taking S as **KW**, neither of the combinations $(\alpha_1)'$-with-$(\beta_2)'$, $(\alpha_2)'$-with-$(\beta_1)'$ can be satisfied.)

As a first approximation to correcting the reasoning in the proof of Theorem 2.10.31, and still sticking with the case of (i) there – uninterpretability of $\Box$ as the identity truth-function – the example of **KW** just considered shows that while the existence of a formula A for which either $\Box A \wedge \neg A$ or $A \wedge \neg \Box A$ is provable is *sufficient* to rule out an identity interpretation for $\Box$ in the logic concerned, it is not *necessary*. Each of these formulas provable implies their disjunction, of course, but S could prove their disjunction without proving either disjunct, and that would already show S not to be Sobociński-regular, as is especially clear if the disjunction is rewritten as a negated biconditional (as we can do in the present case because the disjuncts are mutually inconsistent):

$$\vdash_S \neg(\Box A \leftrightarrow A).$$

If we have this for some formula A, then evidently $\Box$ is not interpretable as the identity truth-function. Note that we can move the negation sign to get a formulation which is sometimes more convenient: $\vdash_S \Box A \leftrightarrow \neg A$. (We have not shifted to considering interpreting $\Box$ as negation here, since the condition in question says that we have this for at least one A, not that we have it for all A.) Sticking with the inset formulation, we can think of it as a negated conjunction (of an implication and its converse), and then De Morgan this negation across, getting:

$$\vdash_S \neg(\Box A \to A) \vee \neg(A \to \Box A).$$

But of course this does not imply, returning to the diagnostic remark already made, that S proves the first disjunct (i.e., the condition (a_1) is satisfied) or that S proves the second disjunction (meaning that (a_2) is satisfied). However, this is still only a first approximation to correcting the reasoning of our erroneous proof, since the condition inset here – or, more accurately, the condition that we have this for some A – is still only sufficient and not necessary, for $\Box$ to fail to be interpretable as the identity truth-function.

After all, suppose that there are formulas $A_1, \ldots, A_n$, for which

$$\vdash_S \neg((\Box A_1 \leftrightarrow A_1) \wedge \ldots \wedge (\Box A_n \leftrightarrow A_n)).$$

Then S would be rendered inconsistent with the addition of all instances of the **T!** schema $\Box A \leftrightarrow A$. The condition that there should be no formulas $A_1, \ldots, A_n$, is not only necessary but also sufficient for the extension of S by **T!** to be inconsistent:

EXERCISE 2.10.32 Show that if the smallest modal logic extending a consistent congruential logic S by all instances of the schema **T!** is inconsistent, then there must be formulas $A_1, \ldots, A_n$ for which we have $\vdash_S \neg((\Box A_1 \leftrightarrow A_1) \wedge \ldots \wedge (\Box A_n \leftrightarrow A_n))$. ✠

As it turns out, for the case of **KW**, reviewed above, we can choose $n = 1$ in this necessary and sufficient condition for the uninterpretability of $\Box$ as the identity truth-function (alias failure of Sobociński-regularity). We want to find a formula A for which $\vdash_{\mathbf{KW}} \neg(\Box A \leftrightarrow A)$. To this end, choose A as $\neg \Box \bot$. Rewriting $\neg(\Box A \leftrightarrow A)$ as $\Box A \leftrightarrow \neg A$ and cancelling the two negations, what we are claiming to be **KW**-provable is:

$$\Box\neg\Box\bot \leftrightarrow \Box\bot.$$

The $\leftarrow$-direction here is already **K**-provable; the $\rightarrow$-direction, rewritten as a disjunction becomes $\Diamond\Box\bot \vee \Box\bot$, which is – with the disjuncts switched around – the 'ending time' axiom **End** whose **KW**-provability was remarked on in Exercise 2.1.5(i) (p. 38).

Note that Exercise 2.10.32 asks about the smallest modal logic extending S and containing all instances of **T!**, rather than the smallest such *congruential* extension of S, but of course the logic in question (which for any consistent modal logics will be the same, namely **KT!**) is automatically congruential, as here we have all instances of the extensionality schema – from Exercise 2.7.2(i) (p. 129) and elsewhere – provable: $(A \leftrightarrow B) \rightarrow (\Box A \leftrightarrow \Box B)$, giving us congruentiality by appeal to Modus Ponens. The same goes for the other three logics mentioned in Would-Be Theorem 2.10.31, one of which three is also normal (which one?). These four logics are Post complete in the lattice of modal logics, since they are all versions of non-modal classical propositional logic with a redundant $\Box$-connective (interchangeable with traditional Boolean resources), and sufficient conditions for a consistent congruential logic to be included in one or other of them are detailed in Makinson [759]. But our present question is whether *every* congruential modal logic is included in one of the four, since this coincides with the amenability to interpretation of $\Box$ as a truth-functional connective (see Remark 2.1.6, p. 38). And it is with a view to answering that question negatively that we appeal to the neighbourhood semantics of our earlier discussion (p. 159). We use the fact, evident from that discussion, that the set of formulas valid on a neighbourhood frame is always a consistent congruential logic.

EXAMPLE 2.10.33 Consider, then, the neighbourhood frame $\langle W, \mathcal{N}\rangle$ with $W = \{a, b\}$, $\mathcal{N}(a) = \{\varnothing\}$, $\mathcal{N}(b) = \{\varnothing, \{a,b\}\}$. The formula $\Box\bot \wedge \neg\Box\Box\Box\bot$ is valid on this frame (an exercise to follow), a formula we could equally well write as $\neg(\Box\bot \rightarrow \Box\Box\Box\bot)$: but on any of the four 1-ary truth-functional interpretations of $\Box$, all formulas of the form $\Box A \rightarrow \Box\Box\Box A$ are valid. (See the discussion after Thm. 3.22.5 at p. 451 of Humberstone [594] for further information and references.) This includes the case of $A = \bot$, of course. So the consistent congruential logic determined by the present frame has no consistent extension to any of the four Post complete logics mentioned in Would-Be Theorem 2.10.31, or to put it another way, is not a sublogic of any of those four. Or, finally, to put it yet another way, this is a contra-classical consistent congruential modal logic. ◀

Example 2.10.33 is presented and discussed further in Humberstone [601]; but the presentation above was not quite complete, and we have a gap to fill:

EXERCISE 2.10.34 Show that, as claimed in Example 2.10.33, the formula $\Box\bot \wedge \neg\Box\Box\Box\bot$ is valid on the two-element neighbourhood frame there described. ✠

The remainder of this section can be treated as an optional digression, giving some additional information about the conclusion Example 2.10.33 allows us to draw – namely that there are consistent congruential modal logics which are not sublogics of the any of the four 'Makinson logics' (in which $\Box$ is interpreted as one of the four 1-ary truth-functions) of Makinson [759].[143]

The following, including Figure 2.17 (differently numbered), is quoted from Humberstone [601], in which Example 2.10.33 is under discussion, and there is a footnote at the end of the paragraph to the effect that that such an algebra's validating a formula A is a matter of having $h(A) = \mathbf{1}$ for each assignment h mapping formulas homomorphically into the algebra concerned:

> We can reconstrue the neighbourhood frame of the above example as a Boolean algebra expanded by an operation interpreting $\Box$, in the sense that for any model $\mathcal{M}$ on the frame, $\|\Box A\|^{\mathcal{M}}$ is the result of applying this operation to $\|A\|^{\mathcal{M}}$. The algebra in question is depicted in Figure 2.17, with a solid

---

[143] In fact, there are uncountably many such consistent congruential modal logics, as is observed in Fritz [336], where much further information on the present topic is also to be found.

2.10. SEMANTICAL POSTSCRIPT

line Hasse diagram for its Boolean reduct and dashed arrows indicating the action of the □-operation. **1**, **a**, **b**, and **0** represent the subsets $\{a,b\}$ ($= W$), $\{a\}$, $\{b\}$, and $\varnothing$ respectively. Such algebras – *modal algebras* in a suitably general sense, not building in normality[144] – are called "Boolean frames" in Hansson and Gärdenfors [444], where it is observed that in the finite case they correspond one-to-one with neighbourhood frames, corresponding structures validating the same formulas.

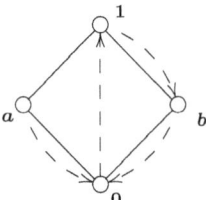

Figure 2.17: Our Neighbourhod Frame as an Expanded Boolean Algebra

In correspondence, Makinson drew attention to a respect in this modal algebra is less elegant than it might be for the role it plays, in that if we replace it with the algebra of Figure 2.18.

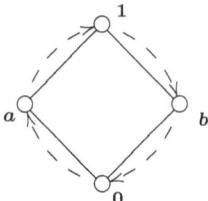

Figure 2.18: A Nicer Algebra

This is an improvement, Makinson argued, not just because of the greater symmetry we see in the behaviour of the □-operation, but because it allows us to cite as examples of formulas valid in the algebra but not susceptible of any two-valued interpretation of □ which are of modal degree 2 rather than (as in Example 2.10.33) degree 3. To quote from this correspondence, using □ not just for the 1-ary connective but for the associated operation in the algebra:

> We have $\Box\Box\mathbf{1} = \mathbf{0}$, so we cannot consistently add as axioms $\Box p \leftrightarrow p$, nor $\Box p \leftrightarrow \top$, nor $\Box p \leftrightarrow \neg p$, thus covering three of the four truth-functions. And we also have $\Box\Box\mathbf{0} = \mathbf{1}$, so we cannot consistently add $\Box p \leftrightarrow \bot$, covering the remaining one.

Reading this brought to mind the use of this very algebra in Humberstone [568], where it was used to show that one could consistently consider Boolean negation as the iteration of another operator, the latter with a congruential logic. This was treated as an exercise in the question of the existence of connectives with prescribed logical properties (as in §4.2 of [594]) rather than from the perspective of modal logics, for which reason it was not written as □ (used for another purpose in the same discussion) but instead as "§" (called 'demi-negation').[145] But of course we could write such a connective as □ and then observe

---

[144]Normality was built into the definition on p. 155 above. Here, by contrast, we are simply concerned with the algebras that result when a Boolean algebra is expanded by the addition of a new fundamental 1-ary operation (which we denote by "□").

[145]In fact the other Boolean primitives are absent from [568], their work being done by thinking of logics as generalized consequence relations (see p. 14 above), and the behaviour of the new operation is given on the right of Figure 2 on p. 5 there, with the elements of the algebra differently named (in fact named after the fashion of the discussion of truth-profiles on p. 153 above).

that the formula $\Box\Box p \leftrightarrow \neg p$, a general version of the special cases (with $\top$ and $\bot$ in place of p) in the passage from Makinson above. In fact the example is revisited in more of a modal setting, and written as $\Box$ rather than §, in Humberstone [573], which a Kripke-style semantics is described on p. 452. At any rate, the issue which Humberstone [601] set out to resolve had evidently been settled quite a few years earlier, and, rather embarrassingly, by the same (forgetful) author.

Chapter 3

Applications: Tense Logic

3.1 Axiomatizing the Basic Logic

$\mathbf{K}_t$ is a bimodal logic whose box operators $\Box_1$ and $\Box_2$ we write as G and H, with the corresponding dual operators (taken as defined) written as F and P. In particular it is the smallest normal bimodal logic in this vocabulary, in which all formulas of the forms $\varphi \to GP\varphi$ and $\varphi \to HF\varphi$ are provable.[146] Note that because we are writing "GP" etc. rather than "$\Box_1 \Diamond_2$" we have changed from using $A, B,\ldots$ as schematic letters (for formulas) to using φ, ψ, in the interests of readability. The interpretation of this language in terms of models on bi-relational frames $\langle U, R_1, R_2 \rangle$ is as follows, and note that to avoid the "worlds" connotations of using W we have used "U" for the universe of the frame, in particular because we want to suggest the possibility of thinking of the frame elements as moments of time rather than as possible worlds. (Why not use "T", then, so suggest a set of times? Because tense logic has many uses and is in no way tied to the original motivating case that gave it its name; for some of these the relevant set of entities will indeed be worlds rather than moments. So the point of using "U" is to have something neutral w.r.t. to its suggestiveness as between worlds and times.) Valuations V are as in the monomodal case, so we have models $\mathcal{M}$ of the form $\langle U, R_1, R_2, V \rangle$, relative to one of which we have, for all $u \in U$:

$\mathcal{M} \models_u G\varphi$ iff for all v such that uR_1v, $\mathcal{M} \models_v \varphi$

$\mathcal{M} \models_u H\varphi$ iff for all v such that uR_2v, $\mathcal{M} \models_v \varphi$.

(Similarly with F and P except that "all" is replaced by "some".) The significance of the schemata $\varphi \to GP\varphi$ and $\varphi \to HF\varphi$ mentioned above is that over the class of such bi-relational frames, these schemata modally define (in the sense of p. 81) the classes of frames in which xR_1y implies yR_2x for all x, y, and in which xR_2y implies yR_1x for all x, y, respectively. Thus taking them together, we have modally defined the class of frames $\langle U, R_1, R_2 \rangle$ in which R_1 and R_2 are *converses*. Restricting our attention to such frames, which is what one does in tense logic in the style of A. N. Prior,[147] we need not mention both relations, and just cite R_1, now rebaptized as R. The above clauses in the definition of truth at a point in a model now appear in the following way, where $\mathcal{M} = \langle U, R, V \rangle$:

$\mathcal{M} \models_u G\varphi$ iff for all v such that uRv, $\mathcal{M} \models_v \varphi$

$\mathcal{M} \models_u H\varphi$ iff for all v such that vRu, $\mathcal{M} \models_v \varphi$.

If we think of the elements of U as moments of time and read uRv as "u precedes (or *is earlier than*) v" then G and H emerge as future tense and past tense $\Box$-operators: mnemonically, think of "$G\varphi$" as

---

[146]The label "$\mathbf{K}_t$" is used differently in Goldblatt [393] and some other places; see note 153 below.

[147]Well, in the style inspired by Prior. Prior himself tended to merge what is usually regarded as the apparatus of the semantical metalanguage, the accessibility relation and its relata, with the object language and its tense operators.

saying "It is always Going to be the case that φ" and of $H\varphi$ as saying "It Has always been the case that φ". There will be no difficulty in remembering which of their duals is which: F is the Future-directed $\Diamond$ operator and P the Past-directed one. Sometimes one may not want all these temporal associations, however, for which reason it might be safer, as is sometimes done, to write G and H (F and P) simply as $\Box$ and $\Box^{-1}$ ($\Diamond$ and $\Diamond^{-1}$), where the superscript is intended to be reminiscent of the notation R^{-1} for the converse of the relation R. (This notation appeared in the Digression on p. 107.)

The subscript on "$\mathbf{K}_t$" can be thought of as abbreviating "tense-logical" or "temporal". Although, as already intimated, we are not particularly interested in the temporal interpretation of these operators – and they certainly do not do very well at capturing natural language tense constructions (see Partee [871] and Bäuerle [54] and references therein, not least because of the interplay between tense and aspect: see Example 2.10.11, p. 159) – we continue to use Prior's G, F, H, P notation for them and to refer to the future and the past in connection with the semantic apparatus. Their real interest is much more general: the use of binary relations in connection with modal notions in general (when the latter are treated as normal modal operators) and accessibility relations forces the "$\Box^{-1}$" notions to our attention since one thing every binary relation has is a converse. Thus even if temporal considerations had not led Prior to consider tense logic, it would have been invented eventually by modal logicians, as implicit in their subject matter. This point will recur several times in the sequel, appearing next in Example 4.1.11.

The canonical model method works as readily for $\mathbf{K}_t$ as it does in the case of its monomodal cousin $\mathbf{K}$, except that there are now four rather than two ways[148] of defining what the canonical accessibility relation should be.

EXERCISE 3.1.1 Show that the following are equivalent for maximally $\mathbf{K}_t$-consistent sets of formulas x and y:

(i) For all φ: $G\varphi \in x \Rightarrow \varphi \in y$

(ii) For all φ: $\varphi \in y \Rightarrow F\varphi \in x$

(iii) For all φ: $H\varphi \in y \Rightarrow \varphi \in x$

(iv) For all φ: $\varphi \in x \Rightarrow P\varphi \in y$.

(Suggestion: it may help to do the following exercise first.) ✠

EXERCISE 3.1.2 Show from the description of $\mathbf{K}_t$ above, or using the axiomatization below, that $\vdash_{\mathbf{K}_t} FHp \to p$ and $\vdash_{\mathbf{K}_t} PGp \to p$. ✠

Enough has been said by way of preparation for the following.

THEOREM 3.1.3 *The logic $\mathbf{K}_t$ is determined by the class of all frames.*

Because of the interest in thinking of the accessibility relation as one of temporal precedence, there is a tendency to regard Thm. 3.1.3 as of considerably less interest than the corresponding result, also available (as for monomodal $\mathbf{K}$: the completeness half following from Thm. 2.5.3(i)) with the word "irreflexive" inserted before "frames". There is also a considerable interest in the strict linear orderings (irreflexive, transitive and weakly connected frames) and further specializations thereof, since these embody natural hypotheses about the structure of time. The further specializations involve such issues as order type: are moments ordered like integers, like the rational numbers, or like the real numbers?

Lazily taking all tautologies as axioms and also the rules of Modus Ponens and Uniform Substitution, we can axiomatize $\mathbf{K}_t$ by adding the following further axioms and rules to this basis; we write "$\mathbf{K}_G$" meaning $\mathbf{K}$-for-G, i.e., the $\mathbf{K}$ axiom with $\Box$ written as G, and similarly in the case of H.[149]

---

[148]We have in mind the original definition as well as that given by Exc. 2.4.5.

[149]This handy convention can be found, for example, in Halpern *et al.* [431], where it is also – as will sometimes happen here – used for logics themselves: thus $\mathbf{KD!}_G$ would be $\mathbf{KD!}$ with $\Box$ written as G.

3.1. AXIOMATIZING THE BASIC LOGIC

$\mathbf{K}_G$	$G(p \to q) \to (Gp \to Gq)$	$\mathbf{K}_H$	$H(p \to q) \to (Hp \to Hq)$
GP-ax.	$p \to GPp$	HF-ax.	$p \to HFp$
G-Nec.	From φ to $G\varphi$	H-Nec.	From φ to $H\varphi$

As an alternative, we could replace the individual axioms by schemata with φ and ψ for p and q, meaning that any instance of these schemata counts as an axiom, thereby avoiding the need for the rule of Uniform Substitution, provided we also allow our truth-functional basis to include not only all tautologies (in the narrow sense: only Boolean connectives permitted) but also arbitrary substitution instances of tautologies in the present language. This is all familiar from the monomodal case. For something distinctively tense-logical, consider what is often called the *mirror image* of a formula φ, by which what is meant is the result of interchanging future and past tense operators in φ (putting H for G and vice versa, and P for F and vice versa). Note that the axiom called $\mathbf{K}_G$ and $\mathbf{K}_H$ are each other's mirror images, as are the two "bridging axioms", the GP and HF axioms (which enforce the demand, in terms of the bi-relational formulation with frames $\langle W, R_1, R_2 \rangle$, that R_1 and R_2 are each other's converses). With that observation made, we are well on the way to a proof by induction on the length of proofs establishing that $\mathbf{K}_t$ has what is called the *mirror image property*:

PROPOSITION 3.1.4 *A formula is provable in* $\mathbf{K}_t$ *if and only if its mirror image is.*

Alternatively, given Theorem 3.1.3, we could prove Proposition 3.1.4 using the easy observation that φ is valid on every frame just its mirror image – $\varphi^{\bowtie}$, let's call it – is, since $\langle U, R \rangle \models \varphi$ if and only if $\langle U, R^{-1} \rangle \models \varphi^{\bowtie}$, where (as before) R^{-1} is the converse of R. (To show this, verify by induction on the complexity of φ that for any model $\langle U, R, V \rangle$ we have: for all $x \in U$, $\langle U, R, V \rangle \models_x \varphi$ iff $\langle U, R^{-1}, V \rangle \models_x \varphi^{\bowtie}$.) For the moment, however, we continue with syntactic considerations, and in particular with the theme of alternative axiomatizations of $\mathbf{K}_t$.

Wansing [1147] offers one such axiomatization which differs from the more familiar axiomatization(s) above in that, as he writes ([1147], p. 60), it "underlines the interaction between the forward and backward oriented modalities [= future and past tense operators], but puts less emphasis on normality as a salient feature of the presentation".

Wans 1	$G(Pp \to q) \to (p \to Gq)$	Wans 2	$H(Fp \to q) \to (p \to Hq)$
Wans 3	$H(p \to Gq) \to (Pp \to q)$	Wans 4	$G(p \to Hq) \to (Fp \to q)$
G-Nec.	From φ to $G\varphi$	H-Nec.	From φ to $H\varphi$
GMono	From $\varphi \to \psi$ to $G\varphi \to G\psi$	HMono	From $\varphi \to \psi$ to $H\varphi \to H\psi$

These are to be subjoined to any axiomatization of the truth-functional tautologies with the aid of Modus Ponens and Uniform Substitution as rule. In fact the last rule is not used by Wansing, who instead gives what we have called Wans 1–Wans 4 as the corresponding schemata,[150] but we have chosen this format for ease of comparison with that on which our earlier description focussed. The comparison makes Wansing's axiomatization appear to come off unfavourably in respect of economy, in that, concentrating on the tense-logical part of the two bases, we have traded in two rules for four (adding GMono and HMono to GNec. and HNec.) while not reducing the number of axioms (which remains at four). We return to this below. It is easy to prove Wansing's axioms and derive his rules from those in the above axiomatization of $\mathbf{K}_t$, so the equivalence of the two axiomatizations requires us to derive the above basis for $\mathbf{K}_t$ from Wansing's axiomatization. The annotations are as in Example 1.3.16, with the rule TF from p. 26.

---

[150]These appear as A7–A10 in [1147].

(1) $Gp \to Gp$	TF
(2) $H(Gp \to Gp)$	1, HNec.
(3) $H(Gp \to Gp) \to (PGp \to p)$	Wans 3
(4) $PGp \to p$	2, 3 Modus Ponens
(5) $(p \to q) \to (PGp \to q)$	4, TF
(6) $G(p \to q) \to G(PGp \to q)$	5, GMono
(7) $G(PGp \to q)) \to (Gp \to Gq)$	Wans 1
(8) $G(p \to q) \to (Gp \to Gq)$	6, 7 TF

Thus we have a proof from Wansing's axiomatization of $\mathbf{K}_G$, by line (4) of which we have already proved the dual of the H axiom.[151] By GMono and HMono, we are in a congruential logic, so formulas and their duals are interdeducible (as in Section 2.1). To prove $\mathbf{K}_H$ and the GP axiom, thereby recovering the rest of the standard axiomatization of $\mathbf{K}_t$ from Wansing's, we can give the mirror image of the above proof – the proof, that is, which has $\varphi^{\bowtie}$ wherever φ appears in (1)–(8). This means, *inter alia*, that the remaining axioms, Wans 2 and Wans 4, will be invoked in place of Wans 1 and Wans 2. But we can do better, meeting the earlier complaint about economy:

PROPOSITION 3.1.5 *Wansing's axioms* Wans 2 *and* Wans 4 *are redundant,* Wans 1, Wans 3, GMono, GNec, HMono *and* HNec *already sufficing as the non-truth-functional part of an axiomatization of* $\mathbf{K}_t$.

Proof. Given the foregoing discussion, it suffices to sketch proofs of Wans 2 and Wans 4. We illustrate with the case of the latter, which invokes Wans 1 (where a proof of the former would invoke Wans 3); for clarity we have written out the "P" here in line (1) in primitive notation, and used "Cong." to abbreviate transitions justified by the already noted congruentiality provided by the Monotone rules for G and H; otherwise annotations are as above (including "US" for Uniform Substitution):

(1) $G(\neg H \neg q \to \neg p) \to (q \to G \neg p)$	Wans 1
(2) $G(\neg H \neg \neg q \to \neg p) \to (\neg q \to G \neg p)$	1, US ($\neg q$ for q)
(3) $G(\neg Hq \to \neg p) \to (\neg q \to G \neg p)$	2, Cong.
(4) $G(p \to Hq) \to (\neg q \to G \neg p)$	3, Cong.
(5) $G(p \to Hq) \to (\neg G \neg p \to q)$	4, TF
(6) $G(p \to Hq) \to (Fp \to q)$	5, Def. F

∎

Now, if what we want is an economical axiomatization which, moreover, in Wansing's words, "underlines the interaction between the forward and backward oriented modalities, but puts less emphasis on normality as a salient feature of the presentation," then we can do even better (as in my [558], Appendix), with the following (two-way) 'GH' rule handling all the tense-logical details:

$$\frac{\varphi \vee G\psi}{H\varphi \vee \psi} \qquad GH\ \textit{rule}$$

The double line indicates that the rule applies downwards, in which direction we shall call it G-to-H, and also upwards, in which direction we call it H-to-G. Notice that in view of the commutativity of disjunction, we could equally well have formulated the GH rule as stipulating the interderivability of $\varphi \vee H\psi$ and $G\varphi \vee \psi$. Let us recall, from Sections 1.3 and 1.4, that an operator O is normal (in a logic) when $O\top$ and $(Op \wedge Oq) \to O(p \wedge q)$ (the aggregation principle) are provable and the logic is closed under the monotony rule: from $\varphi \to \psi$ to $O\varphi \to O\psi$. Let us begin by showing that $G\top$ and $H\top$ are provable in any modal logic closed under the GH rule. Begin with the fact that ("by TF") $H\bot \vee \top$ is provable,

---

[151]Duality is as explained in Section 2.1, except that we are now in a bimodal setting: the dual form of an affirmative modality $O_1 O_2 \ldots O_k$ is $\widetilde{O}_1 \widetilde{O}_2 \ldots \widetilde{O}_k$, where $\widetilde{O}_i$ is F if O_i is G, G if O_i is F, P if O_i is H and H if O_i is P.

3.1. AXIOMATIZING THE BASIC LOGIC

so by the *G*-to-*H* rule, $\bot \vee G\top$ is provable, and so therefore (by TF again) is $G\top$. A similar argument works for $H\top$.

Next we turn our attention to *G*Mono. We need to begin with a mixed principle, however, a disjunctive version of the dual $PGp \to p$ of the HF axiom from our axiomatization of $\mathbf{K}_t$, namely: $H\neg Gp \vee p$. To prove this we simply apply *G*-to-*H* to the excluded middle premiss $\neg Gp \vee Gp$. Note that the mirror image is similarly provable. (Thus we have both the bridging axioms.) Now for *G*Mono suppose $\varphi \to \psi$ is provable, and consider the substitution instance $H\neg G\varphi \vee \varphi$ of the formula (namely, $H\neg Gp \vee p$) whose provability we just assured ourselves of. A truth-functional consequence of these two is $H\neg G\varphi \vee \psi$, so by *H*-to-*G*, $\neg G\varphi \vee G\psi$, and hence $G\varphi \to G\psi$, is provable. Again the same goes for *H*Mono. This leaves only the aggregation formula $(Gp \wedge Gq) \to G(p \wedge q)$, for which we have the following derivation:

(1) $H\neg Gp \vee p$ Proved above.
(2) $H\neg Gq \vee q$ 1, US (q for p)
(3) $\neg Gp \to (\neg Gp \vee \neg Gq)$ TF
(4) $H\neg Gp \to H(\neg Gp \vee \neg Gq)$ 3, *H*Mono
(5) $\neg Gq \to (\neg Gp \vee \neg Gq)$ TF
(6) $H\neg Gq \to H(\neg Gp \vee \neg Gq)$ *H*Mono
(7) $H(\neg Gp \vee \neg Gq) \vee p$ 1, 4 TF
(8) $H(\neg Gp \vee \neg Gq) \vee q$ 2, 5 TF
(9) $H(\neg Gp \vee \neg Gq) \vee (p \wedge q)$ 7, 8 TF
(10) $(\neg Gp \vee \neg Gq) \vee G(p \wedge q)$ 9, *H*-to-*G*
(11) $(Gp \wedge Gq) \to G(p \wedge q)$ 10, TF

Since again, we can provide a proof *mutatis mutandis* of the mirror image of this last formula, we summarise our findings as:

PROPOSITION 3.1.6 $\mathbf{K}_t$ *can be axiomatized as the smallest bimodal logic closed under the GH rule.*

Let us notice also a monomodal corollary of the above result (or, more accurately, of the proof given):

COROLLARY 3.1.7 *The normal monomodal logic* **KB** *can be axiomatized as the smallest monomodal logic closed under the following rule:*

$$\frac{\varphi \vee \Box\psi}{\Box\varphi \vee \psi}$$

EXERCISE 3.1.8 How does the reasoning which culminated in Prop. 3.1.6 justify Coro. 3.1.7? And why does the latter feature only a one-way rule whereas the former involved a two-way rule (i.e., the *GH* rule)? ✠

Jennings [622] also notes the simplifiability of typical axiomatizations of **KB** which would simply add **B** to an axiomatic basis for **K**.[152] The axiomatic basis for **K** that Jennings has in mind is one using the rule of monotony, □Mono, the aggregation schema, and the Necessitation rule Nec, and the simplification in question pertains omission of Aggregation and Necessitation from this description. Recall that aggregation is the schema $(\Box\varphi \wedge \Box\psi) \to \Box(\varphi \wedge \psi)$. Thus, Jennings's observation is the following:

---

[152]Actually, Jennings considers **KTB**, called simply B in [622]; but **T** plays no role in the simplification so we keep it out of the discussion here. The proof in [622] that the simplification works relies on the neighbourhood semantics – see Thm. 2.10.12, p. 160, and the discussion preceding it above – for modal logic, whereas we give direct syntactic derivations. (A final footnote in [622] remarks that such derivations "have since been produced by J. F. A. K. van Benthem and a referee of this paper.")

PROPOSITION 3.1.9 **KB** *is the smallest monotone modal logic containing* **B**.

Proof. It suffices to show that **B** and □Mono render the rule of Necessitation and the aggregation schema derivable. Given □Mono, to derive Necessitation all we need is some schema of the form $p \to \Box\psi$, since we can this secures the provability (on substituting ⊤ for p) of the formula $\Box\psi'$, where ψ' results from making this substitution for any occurrences of p in ψ. And once we have a □-formula provable, such as $\Box\psi'$ here, we can derive Necessitation by passing from a supposedly provable φ

$$\psi' \to \varphi$$

by TF and then by □Mono to

$$\Box\psi' \to \Box\varphi,$$

whence we get the desired conclusion of the rule Nec, $\Box\varphi$, by Modus Ponens, since $\Box\psi'$ was provable. In the present case, thanks to **B**, a formula of the form $p \to \Box\psi$ whose provability we are assured of, securing the above derivation, can be taken as having $\psi = \Box\Diamond p$.

Next, we prove a representative instance of the aggregation schema on the above basis. Recall (Prop. 1.4.1) that ◇Mono is derivable from □Mono, so on the cited basis we can prove:

$$\Diamond(\Box p \wedge \Box q) \to \Diamond\Box p \quad \text{and} \quad \Diamond(\Box p \wedge \Box q) \to \Diamond\Box q,$$

which by appealing to the $\varphi = p$, $\varphi = q$ instances of the dual form of the **B** schema $\Diamond\Box\varphi \to \varphi$, we can drop the two "◇□" prefixes in the above consequents. As a truth-functional consequence of the result of doing this, we get:

$$\Diamond(\Box p \wedge \Box q) \to (p \wedge q).$$

By □Mono, we derive:

$$\Box\Diamond(\Box p \wedge \Box q) \to \Box(p \wedge q).$$

Finally, the **B** schema allows us to remove the initial "□◇" prefix from the antecedent, giving us the desired aggregation principle, $(\Box p \wedge \Box q) \to \Box(p \wedge q)$. ∎

In view of the similarity between the **B** axiom and the bridging axioms of $\mathbf{K}_t$, HF-ax. and GP-ax. (from p. 179), it is worth seeing what can be made of such reasoning in a tense-logical context. Suppose, accordingly, that H satisfies HMono (and therefore PMono) and GMono, and that we have both the above bridging axioms. Then we can certainly derive aggregation for G:

(1)	$(Gp \wedge Gq) \to Gp$		TF
(2)	$P(Gp \wedge Gq) \to PGp$		1, PMono
(3)	$(Gp \wedge Gq) \to Gq$		TF
(4)	$P(Gp \wedge Gq) \to PGq$		3, PMono
(5)	$P(Gp \wedge Gq) \to (PGp \wedge PGq)$		2, 4 TF
(6)	$PGp \to p$		HF-ax. (dual form)
(7)	$PGq \to q$		HF-ax. (dual form)
(8)	$(PGp \wedge PGq) \to (p \wedge q)$		6, 7 TF
(9)	$GP(Gp \wedge Gq) \to G(p \wedge q)$		8, GMono
(10)	$(Gp \wedge Gq) \to GP(Gp \wedge Gq)$		GP-ax.
(11)	$(Gp \wedge Gq) \to G(p \wedge q)$		9, 10 TF

In view of the discussion of Jennings's reaxiomatization of **KB**, the following exercise will present no difficulty.

3.2. EXTENSIONS OF $\mathbf{K}_t$

EXERCISE 3.1.10 Show that the smallest bimodal logic containing HF-ax and GP-ax and closed under HMono and GMono is closed under the rules HNec. and GNec. (from p. 179). ✠

From (1)–(10) and a corresponding derivation in which each line is replaced by its mirror image (with annotations suitably adjusted), and the result of Exercise 3.1.10, we draw the following conclusion:

PROPOSITION 3.1.11 $\mathbf{K}_t$ *is the smallest bimodal logic, with primitive operators G, H, in which each of G and H is monotone and in which the axioms HF-ax. and GP-ax. are provable.*

In Coro. 2.5.6 (p. 73) it was noted that $\mathbf{K}$ was closed under the rule: *from $\Box \varphi \to \varphi$ to φ*. Exercise 2.7.5(*iii*) (p. 130) asks if $\mathbf{KB}$ proves $\Box \varphi \to \varphi$ for $\varphi = \Box p \vee p$, to which the answer is affirmative, a fact which may, given the analogies between $\mathbf{K}_t$ and monomodal $\mathbf{KB}$ we have seen recently, prompt the thought that $G(Hp \vee p) \to (Hp \vee p)$ would be $\mathbf{K}_t$-provable, and thus, since $(Hp \vee p)$ is not, serve as a counterexample to the admissibility of the above rule for $\mathbf{K}_t$, rewriting $\Box$ as G. (See also Exercise 2.5.7, p. 74.) But $G(Hp \vee p) \to (Hp \vee p)$ is not $\mathbf{K}_t$-provable, blocking this strategy. Carefully deployed, however, the $\mathbf{K}_t/\mathbf{KB}$ analogy gives us instead the $\mathbf{K}_t$-theorem $G(Hp \vee p) \to (Gp \vee p)$, which is not of the required $G\varphi \to \varphi$ form. But using it will help with the following exercise:

EXERCISE 3.1.12 Defining $\Box' \varphi$ as $G\varphi \wedge H\varphi$, show that $\mathbf{K}_t$ is not closed under the rule: from $\Box' \varphi \to \varphi$ to φ, with $\Box'$ so understood. ✠

EXERCISE 3.1.13 Is the defined operator $\Box'$, figuring in preceding exercise *normal* in $\mathbf{K}_t$? Is there an associated accessibility relation (i.e., a relation S such that in every model $\langle U, R, V \rangle$, the truth for any formula φ of $\Box' \varphi$ at a point stands or falls with the truth of φ at each S-related point, and if so, what is this relation S? ✠

3.2 Extensions of $\mathbf{K}_t$

In the case of bimodal logics such as $\mathbf{K}_t$, the simple system of nomenclature that worked for the monomodal case becomes awkward, first because one would naturally wish to consider mixed principles involving both G (or F) and H (or P), so that fresh labels need to be sought to refer to new candidate axioms or axiom-schemata, and secondly because even in the case of unmixed principles, we would need to distinguish the future directed version from the past directed version of a given monomodal principle. Take the seriality schema $\mathbf{D}$, for example, which now comes in two flavours; again we subscript a "G" or "H" to the name of a familiar monomodal principle to mean the result of writing $\Box$ as G or H respectively, though we now add also, and writing $\Diamond$ as F or P respectively:

$$\mathbf{D}_G\colon G\varphi \to F\varphi \text{ (or just } F\top\text{)} \qquad \mathbf{D}_H\colon H\varphi \to P\varphi \text{ (or just } P\top\text{).}$$

Evidently the first principle modally defines the class of frames which are serial and the second the class of frames which are converse serial, and a canonical argument shows the normal extensions of $\mathbf{K}_t$ by these principles are complete w.r.t. these classes of frames, respectively. The two logics are therefore quite different, neither being included in the other, and each properly extending $\mathbf{K}_t$. As a consequence of this, since – to take the simpler form of the axioms in question – $F\top$ and $P\top$ are each other's mirror images, neither of these extensions of $\mathbf{K}_t$ has the mirror image property, although the extension of $\mathbf{K}_t$ by *both* axioms does have this property.

Since $\mathbf{K}_t \vdash GP\top$ (by GP-ax., putting $\top$ for p) while $\mathbf{K}_t \nvdash P\top$, unlike monomodal $\mathbf{K}$, $\mathbf{K}_t$ is not closed under the rule of Denecessitation for G; by a similar consideration, nor is it closed under Denecessitation for H. Recall that these closure properties are special cases of the more general property of providing the rule of disjunction, which again bifurcates into two properties in the present setting: (G version) $\vdash_S G\varphi_1 \vee \ldots \vee G\varphi_n \Rightarrow \vdash_S \varphi_i$ for some $i \in \{1, \ldots, n\}$ and (H version) $\vdash_S H\varphi_1 \vee \ldots \vee H\varphi_n \Rightarrow \vdash_S \varphi_i$ for some $i \in \{1, \ldots, n\}$.

EXAMPLE 3.2.1 While the logic $\mathbf{K}_t \oplus \{P\top, F\top\}$ is closed under the Denecessitation rules just mentioned, this logic still fails to provide the n-ary rule of disjunction for all n, failing already for $n = 2$: This logic contains (as does $\mathbf{K}_t$ itself) $G\neg p \lor GPFp$ without containing $\neg p$ or PFp, and similarly for the case of the mirror image of this disjunction. ◀

Theorem 2.6.1 stated that $\mathbf{K}$, amongst certain other normal modal logics, provided the rule of disjunction, and the key to its proof was an 'adjunction of points' construction together with an appeal to the Generation Theorem (Thm. 2.4.10). The latter ensured that the newly adjoined points did not affect the truth-values of formulas within the submodels to which they were adjoined, because although the new points bore the accessibility relation to points generating the old submodels (and for the cases other than that of $\mathbf{K}$, to other points within those submodels also), the old points did not in turn bear the accessibility relation to the new points. Now we are in a bimodal setting we have to be careful about what a generated submodel looks like, and in particular for the case of tense logic, in which one accessibility relation is the converse of the other, it is clear that adding a point to which others are accessible by the one relation amounts to adding a point accessible to them by the other accessibility relation. The original monomodal definition ran like this (except that we have re-lettered "W" to "U" here: Given a model $\mathcal{M} = \langle U, R, V \rangle$ and a non-empty subset $X \subseteq U$, the submodel $\mathcal{M}_0$ of $\mathcal{M}$ generated by X is the model $\langle U_0, R_0, V_0 \rangle$ in which $U_0 = \{y \in U \mid \exists n \in \mathit{Nat}, x \in X \cdot xR^n y\}$, $R_0 = R \cap \langle U_0 \times U_0 \rangle$, and $V_0(p_i) = V(p_i) \cap U_0$. For the case of bimodal models $\mathcal{M} = \langle U, R_1, R_2, V \rangle$ we must say instead that the submodel generated by $X \subseteq U$ has for its universe, U_0 again (say) $\{y \in U \mid \exists n \in \mathit{Nat}, x \in X \cdot x(R_1 \cup R_2)^n y\}$: we must collect together all points that can be reached by any $(R_1 \cup R_2)$-chain from the generating points. As before, the accessibility relations and the valuation part of the generated submodel are just the restrictions of the original relations and valuation to the points thus collected (and generated subframes are what we get if we ignore the valuation components). We will not bother to reformulate the Generation Theorem for this or the more general case with n accessibility relations, but if you care to do so and then to prove it, it will be clear why for the inductive step of the proof the $\Box_i$-cases require the above definition. Of course, we are not writing our models in the explicit form $\langle U, R_1, R_2, V \rangle$ but simply as $\langle U, R, V \rangle$, with R_1 understood as R and R_2 as R^{-1}. In terms of this way of writing things, the universe of the submodel generated by a point (the special case in which X above is a singleton) comprises all those points reachable by any number of steps of R with each step going either forward or backward. (It is because of the backward steps that the monomodal Coro. 2.5.33(iii), asserting the modal undefinability of the class of converse serial frames fails in the tense-logical setting, this class being modally defined, as already noted, by $P\top$.)

EXERCISE 3.2.2 Let us consider a tense-logical analogue of the inverse monotony rule considered in Exercise 2.6.31 (p. 126):

$$\frac{G\varphi \to G\psi}{\varphi \to \psi}$$

Establish or refute the claim that $\mathbf{K}_t$ is closed under this rule. ✠

We have seen that $F\top$ and $P\top$ are independent candidate axioms to add to $\mathbf{K}_t$, so that there is nothing we could unambiguously called $\mathbf{K}_t\mathbf{D}$. In the presence of further conditions on frames – most particularly transitivity and irreflexivity – we could naturally think of these axioms as capturing the idea that time was forward-infinite and backward-infinite, respectively. When it comes to transitivity itself, however, there is an obvious contrast with seriality: there is no distinction to be made between frames which are forward-transitive and frames which are backward-transitive (or those frames which are transitive and those which are 'converse transitive'), because evidently a relation is transitive if and only if its converse is. Thus we do not expect a similar future-vs.-past bifurcation as to what might be meant by the label "$\mathbf{K}_t 4$", which should be axiomatized indifferently by adding any one of 4_G, its dual, its mirror image, or the latter's dual:

3.2. EXTENSIONS OF $\mathbf{K}_t$

$$Gp \to GGp \qquad FFp \to Fp \qquad Hp \to HHp \qquad PPp \to Pp.$$

And so it is (see Exc. 3.2.3).[153] But a more refined analysis of what is going on here is called for – rather than the mere observation that a relation is transitive just in case its converse is. For this, we shall need the notion of local (modal) definability, to be provided after a quick exercise:

EXERCISE 3.2.3 Using the canonical model method, show that the normal extension of $\mathbf{K}_t$ by any one of the four axioms inset above is determined by the class of transitive frames. ✠

To explain the local version of definability we need to back up and isolate a local notion of validity. Here (as in van Benthem [73] and Fitting [310]) "local" – as opposed to "global" – means that we are concerned with points in a frame one by one, whereas the notion of validity on a frame was global in the sense that it required the unfalsifiability of a formula at each of the points in any model on that frame. We could have introduced the corresponding local notion, which we call the *validity of a formula at a point* in a frame, in our discussion of monomodal logic. If $\langle U, R \rangle$ is the frame in question and φ is the formula, then we write $\langle U, R \rangle \models_x \varphi$ to indicate that φ is valid at $x \in U$. The definition is as follows:

$\langle U, R \rangle \models_x \varphi$ if and only if for every model $\langle U, R, V \rangle$, we have $\langle U, R, V \rangle \models_x \varphi$.

This notion deserves to be spoken of using the word "valid" because of its independence of any particular V, as in the case of validity on a frame (notation: $\langle U, R \rangle \models \varphi$) and unlike the fundamental notion of truth at a point in a model (notation: $\langle U, R, V \rangle \models_x \varphi$) or the other derivative notion of truth throughout a model (notation: $\langle U, R, V \rangle \models \varphi$). Not all authors agree, however, and it not uncommon to hear the latter spoken of as validity in a model. (See note 71, p. 82.)

EXAMPLE 3.2.4 The discussion before Exercise 2.5.28, showing that $\Box p \to p$ was valid on a frame if and only if that frame was reflexive (a global definability result), in fact shows something stronger (a corresponding local definability result): this formula is valid at a point in a frame if and only if that point is reflexive (i.e., bears the accessibility relation to itself). ◀

The following exercise returns briefly to the purely global level.

EXERCISE 3.2.5 On precisely which frames – described by a condition on the accessibility relation – is the tense-logical formula $q \to GHq$ valid? Is there a formula involving only H (or also the defined operator P) valid on exactly the same frames? Is there a formula involving only G (or also the defined operator F) valid on the exactly these frames? (Briefly justify your answers.) ✠

EXERCISE 3.2.6 (*i*) Give an example of a monomodal formula which is valid at some point in some frame but not valid on any frame.
(*ii*) Which of the following rules is guaranteed to preserve validity at a point (in a frame)? Uniform Substitution, Necessitation (or GNec and HNec, for tense logic), Modus Ponens.
(*iii*) This question uses the concept of Halldén completeness, introduced on p. 39 (in the Digression there). Show that for any frame and point therein, the set of formulas valid at that point is a Halldén-complete modal logic. ✠

To think about transitivity and the multiple axiomatizability (noted above) of $\mathbf{K}_t$, we need some terminology applying to points within frames. Given a frame $\mathcal{F} = \langle U, R \rangle$ and a point $x \in U$, say that x is 1-*transitive* in $\mathcal{F}$ just in case for all $y, z \in U$, if xRy and yRz, then xRz, that $y \in U$ is 2-*transitive* (in $\mathcal{F}$) when for all $x, z \in U$, if xRy and yRz, then xRz, and, finally, that $z \in U$ is 3-*transitive* when for all

---
[153] Confusingly, Goldblatt [393] refers to $\mathbf{K}_t\mathbf{4}$ as $\mathbf{K}_t$. He has been followed in this practice by some others, as in [505].

$x, y \in U$, if xRy and yRz, then xRz. Note that these are properties of points (relative to frames), unlike transitivity itself, which is a property of frames (or alternatively, of their accessibility relations). But there is an obvious connection between these properties and that of transitivity, in that the following are all equivalent, for a frame $\langle U, R \rangle$, to the claim that $\langle U, R \rangle$ is transitive:

- Every element of U is 1-transitive
- Every element of U is 2-transitive
- Every element of U is 3-transitive

One can obtain a pictorial representation of such 'point properties' from the diagrams used for the (universal) frame properties by circling or otherwise singling out a one of the nodes in such diagrams depicting frame properties. Thus in the left-hand diagram of Figure 2.3 (on p. 66), circling the point labelled "x" gives a representation of 1-transitivity, circling the y-node, 2-transitivity, and the z-node, 3-transitivity. The convention here would be that the condition represented by the diagram is obtained by considering the circled node as a free variable, with the remaining nodes representing (as usual) universally quantified variables. Since we have a free variable in the induced condition, the condition can be satisfied or otherwise by this or that point in a frame, as required. Figure 3.1 is not such a diagram; it just presents particular frame for discussion in Example 3.2.7 and the Exercise following it.

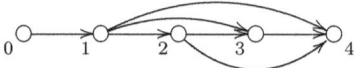

Figure 3.1: *A Frame Illustrating Various Local Transitivity Conditions*

EXAMPLE 3.2.7 Figure 3.1 depicts a particular frame, rather than representing a universally quantified condition on frames (or relations) as with the diagrams on pp. 66–68, for example. (In this respect it resembles Figure 2.7 on p. 87, except that here we are not also depicting a particular model on the frame in question.) We have labelled the points 0, 1, 2, 3, 4 so as to be able to talk about them conveniently. The point 1 is 1-transitive, but not 2-transitive, since it has a predecessor (namely 0) which is not a predecessor of at least one of its successors (in fact not a predecessor of any successor of 1). For this same reason, 0 is not 1-transitive. The remaining points are all 1-transitive, as well as 2-transitive and 3-transitive. ◀

EXERCISE 3.2.8 Remove the arrow from 2 to 4 in the frame depicted in Figure 3.1. In the resulting frame, which points are 1-transitive? Which are 2-transitive? Which are 3-transitive? ✠

Taking the initial axiomatization of $\mathbf{K}_t$ from Section 3.1 – the version with uniform substitution as a primitive rule – we show how to deduce $Hp \to HHp$ from $Gp \to GGp$, and later comment on the proof from the point of view of the three localized versions of transitivity just distinguished. To keep the key steps of the proof in prominent relief, let us in advance list some easily proved $\mathbf{K}_t$-theorems:

(i) $Fp \to HFFp$ (ii) $HFHp \to Hp$ (iii) $Hp \to HFHp$.

Further, to make life easier, we begin not with $Gp \to GGp$ but with the dual form, $FFp \to Fp$, since the straightforward interdeducibility of formulas and their duals in normal modal logics was explained (with emphasis on the monomodal case) in Section 2.1.

3.2. EXTENSIONS OF $\mathbf{K}_t$

(1)	$FFp \to Fp$	Given
(2)	$HFFp \to HFp$	1, HMono
(3)	$Fp \to HFp$	2, (i), TF
(4)	$FHp \to HFHp$	3, US
(5)	$FHp \to Hp$	4, (ii), TF
(6)	$HFHp \to HHp$	5, HMono
(7)	$Hp \to HHp$	6, (iii), TF

By way of commentary let us begin with a simple observation, whose verification is left to the reader:

PROPOSITION 3.2.9 *Each of the future-oriented* **4**-*principles* $Gp \to GGp$ *and its dual* $FFp \to Fp$ *(locally) defines 1-transitivity, while their mirror images define 3-transitivity. The formulas appearing on lines (3) and (5) of the above proof each locally define 2-transitivity.*

In connection with the first sentence in Prop. 3.2.9, let us note that our terminology could be refined, in case several relations are under discussion, relativizing these properties to the relation in question. 1-transitivity with respect to R would then be equivalent to 3-transitivity w.r.t. R^{-1}. And apropos of the second sentence: a point is 2-transitive w.r.t. R if and only if it is 2-transitive (again) w.r.t. R^{-1}. In what follows when we use the terminology without such "w.r.t." qualifications, we mean as before the understanding of this terminology w.r.t. the accessibility relation R of the frame. (Of course in terms of the explicitly bimodal description of the frames as $\langle U, R_1, R_2 \rangle$, this is the relation R_1, with R_2 going unmentioned and coinciding with $R^{-1} = R_1^{-1}$.)

Returning to the deduction (1)–(7) above, we see that it begins with a formula valid at all (and, as noted, at *only* – though we do not need this here) the 1-transitive points in a frame. To expedite the deduction, we have made use of the derived rule HMono, which using only the primitive rules (of the axiomatization in question) would be to invoke HNec and a theorem of $\mathbf{K}_t$ (a substitution instance of one of the axioms), the latter being therefore valid at every point (in every frame). The interest lies accordingly with the effect of applying HNec, and in particular with the "if" half of the following observation, again stated without (the straightforward) proof:

PROPOSITION 3.2.10 (i) *For any formula φ, $G\varphi$ is valid at a point in a frame if and only if φ is valid at every successor of that point.*
(ii) *For any formula φ, $H\varphi$ is valid at a point in a frame if and only if φ is valid at every predecessor of that point.*

Thus (2) and its truth-functional consequence (3) are valid at any point all of whose predecessors validate $FFp \to Fp$, i.e., at all points all of whose predecessors are 1-transitive. But evidently if all predecessors of a point are 1-transitive, the point itself is 2-transitive, explaining the appearance here of the 2-transitivity defining formulas (mentioned in Prop. 3.2.9) at lines (3) and (5). We move on again with another tacit appeal to HNec in line (6) and its truth-functional consequence at line (7), so we re-invoke Prop. 3.2.10(ii), observing that if every predecessor of a point is 2-transitive, the point itself is 3-transitive, explaining the appearance of the 3-transitivity defining formula at line (7).

EXERCISE 3.2.11 The above commentary made two conditional claims:

(i) If every predecessor of a point is 1-transitive, the point itself is 2-transitive.

(ii) If every predecessor of a point is 2-transitive, the point itself is 3-transitive.

Are the converse conditionals also correct? (Either explain why or give a counterexample, for each of them.) ✠

Although this completes our commentary on the deduction (1)–(7), a commentary you may care to parallel for the case of the mirror image (1)$^{\bowtie}$–(7)$^{\bowtie}$ of this deduction, it has raised another issue that cannot be allowed to pass without notice. Proposition 3.2.10 said that prefixing a G to a formula φ gives a formula valid at a point in a frame if and only if φ is valid at every successor of that point, and likewise in the case of H, putting "predecessor" for "successor". One might naively think that we could similarly assert:

> Prefixing an F to a formula φ, gives a formula valid at a point in a frame if and only if φ is valid at *some* successor of that point, and similarly in the case of prefixing a P (replacing 'successor' by 'predecessor').

While the "if" direction here is correct, in both cases (i.e., for F and for P), the "only if" direction is not. We already have the materials for a simple (and well known) counterexample, in Coro. 2.5.33(iv) (see also the discussion following that result), according to which the class of frames in which every point has a reflexive successor is not modally definable. But whether we write it as "$\Diamond(\Box p \to p)$" or as "$F(Gp \to p)$", this would be a formula defining that class of frames if it locally defined the class of points with reflexive successors, as it would have to if the above claim were correct, since $Gp \to p$ locally defines reflexivity of points. (Prop. 2.5.35 noted that all the formulas $\mathbf{T}_n^\Diamond$ – and $\Diamond(\Box p \to p)$ is $\mathbf{T}_1^\Diamond$ – are provable in **KD4** and thus are valid on all (forward-)serial transitive frames, even in the absence of any reflexive points; see also the end of Example 2.6.37.)

EXERCISE 3.2.12 Is it legitimate to appeal to Corollary 2.5.33(iv), which was established in the context of monomodal logic, to show undefinability in tense logic? Explain why, or why not. (After all, 2.5.33(iii), asserting the undefinability of the class of converse serial frames in that earlier context, makes a claim we have already noted to be false for the case of tense logic, in which we have the formula $P\top$ defining this class.) ✠

To see how the argument one might attempt in defence of the false claims we have been considering breaks down, let us start with the hypothesis that $F\varphi$ is valid at $x \in U$ in the frame $\langle U, R \rangle$:

$\langle U, R \rangle \models_x F\varphi$,

and so, unpacking via the definition of local validity:

for all V, $\langle U, R, V \rangle \models_x F\varphi$,

so by the definition of truth (at a point in a model):

for some $y \in R(x)$, $\langle U, R, V \rangle \models_y \varphi$.

Thus what we have, using formal quantifier notation to make the contrast vivid, is:

$\forall V \exists y \in R(x), \langle U, R, V \rangle \models_y \varphi$,

whereas what we needed (for the claim that φ was valid at some successor of x) was rather:

$\exists y \in R(x), \forall V \langle U, R, V \rangle \models_y \varphi$,

which is one (illicit) quantifier shift away.

EXERCISE 3.2.13 Define a point $x \in U$ to be *1-symmetric* in the frame $\langle U, R \rangle$ if for all $y \in U$, if xRy then yRx, and to be *2-symmetric* if for all $y \in U$, if yRx then xRy. (This terminology was introduced in my [588], Example 4.3.) Show that a point can satisfy either of these conditions without satisfying the other and, for each of them, find a tense-logical formula locally defining the condition in question. Note that as with 1, 2- and 3-transitivity, the global condition on frames that all their points are 1-symmetric is equivalent to the condition that all their points are 2-symmetric, amounting in each case to the frame's being symmetric. ✠

3.2. EXTENSIONS OF $\mathbf{K}_t$

EXAMPLE 3.2.14 The condition of 2-symmetry, unlike 1-symmetry (locally defined by **B**) from the preceding exercise is not locally definable in monomodal logic. As noted there the frame property of having all points 1-symmetric coincides with that of the frame's being symmetric, and similarly for having all points 2-symmetric. But what about the property of having all points being either 1-symmetric or 2-symmetric? For a point y to be neither 1- nor 2-symmetric (in a frame $\langle U, R \rangle$, $y \in U$) it is necessary and sufficient that there exist $x \in U$, $z \in U$ such that $xRyRz$ but neither yRx nor zRy: x provides a counterexample to y's being 2-symmetric and z a counterexample to y's being 1-symmetric. Thus for a frame to satisfy this condition is for it to satisfy, for all $x, y, z \in U$: if $xRyRz$, yRx or zRy. This class of frames is monomodally definable, being defined by the formula $\mathbf{B}_2$ (in which we write $\Box$ and $\Diamond$ rather than G and F:

$$p \to \Box(\Diamond p \vee (q \to \Box \Diamond q)).$$

The name $\mathbf{B}_2$ is taken from Segerberg [1025] (p. 132f.), where a whole sequence $\mathbf{B}_1, \ldots, \mathbf{B}_n, \ldots$ is introduced, the nth formula being valid on exactly those frames $\langle U, R \rangle$ satisfying: for any R-chain $x_1, \ldots, x_n$ in $\langle U, R \rangle$, there is $k \in \{2, \ldots, n\}$ with $x_k R x_{k-1}$. $\mathbf{B}_1$ is just the formula $\mathbf{B}$, while $\mathbf{B}_3$, for example, is:

$$p \to \Box(\Diamond p \vee (q \to \Box(\Diamond q \vee (r \to \Box \Diamond r)))).$$

(In fact, Segeberg uses schemata rather than formulas, and adopts a dual formulation.) We will re-encounter $\mathbf{B}_2$ in Chapter 5 (Proposition 5.6.14, p. 435 and the discussion following it), as well as in the exercise below. ◀

EXERCISE 3.2.15 Show that $\mathbf{KB}_2$ is determined by the class of frames in which every point is 1-symmetric or 2-symmetric. ✠

REMARK 3.2.16 Notice incidentally that while a point in a frame can be 1-symmetric without being 2-symmetric, and vice versa, if we defined a point $x \in U$ to be 1-*asymmetric* in the frame $\langle U, R \rangle$ just in case for all $y \in U$ with xRy, we have: not yRx, and similarly defined y to be 2-*asymmetric* just in case for all $x \in U$ with xRy we have: not yRx, then we would quickly observe that this is a distinction without a difference – the 1-asymmetric points are precisely the 2-asymmetric points. ◀

It would be a mistake to conclude from Exercise 3.2.13 on symmetry and from our discussion of the local transitivity conditions that there are only as many local conditions available, which coincide with the global condition when universally quantified, as there are universal quantifiers in the first-order formulation of the global condition. Indeed, the earlier derivation (1)–(7) already turns up further conditions – or rather, formulas locally defining such conditions – as we see from line (2), or, more simply from the formula $H(FFp \to Fp)$ in the invisible step between lines (1) and (2). As remarked above, this formula defines the property of being a point all of whose predecessors are 1-transitive, as will also have already been noted by anyone answering Exercise 3.2.11(i), this does not coincide with the property of 2-transitivity – or, we may add here, with 1- or 3-transitivity either. In fact the additional multiplicity just illustrated arises already at the monomodal level. For this example we revert to the "$\Box$"/"$\Diamond$" notation (rather than "G"/"F"):

EXAMPLE 3.2.17 The property of a being a reflexive point is not the only one possession of which by all points in a frame is equivalent to the frame's being reflexive. Consider instead the property of being a reflexive point all of whose successors are reflexive, (locally) defined by $(\Box p \to p) \wedge \Box(\Box p \to p)$. (We could of course re-letter one of the conjuncts here, writing instead: $(\Box p \to p) \wedge \Box(\Box q \to q)$.) Not all reflexive points have this property (and accordingly not all validate the formula just given), but it is again a property universally possessed in precisely the reflexive frames. It suggests the following question, to which I do not know the answer: Is there a property of points whose universal possession is equivalent to frame reflexivity but which does not imply reflexivity of an individual point with the property? ◀

Let us recall the frame properties of domain-reflexivity and range-reflexivity, respectively:

$$\forall x \forall y (Rxy \to Rxx) \quad \text{and} \quad \forall x \forall y (Rxy \to Ryy),$$

which were monomodally definable by the formulas $\Diamond\top \to (\Box p \to p)$ and $\Box(\Box p \to p)$, of modal degree 1 and 2 respectively. In the tense-logical setting we have a more equitable arrangement: domain-reflexivity is defined by $F\top \to (Gp \to p)$, a rewriting of the monomodal formula, and also by the formula $H(Gp \to p)$, while range-reflexivity is defined by $G(Gp \to p)$, again a transcription of the monomodal formula (the axiom **U**), of degree 2, and also – by contrast with the monomodal case – by the degree 1 formula $P\top \to (Gp \to p)$.

EXERCISE 3.2.18 *(i)* Which, if any, of the four tense-logical formulas just mentioned is $\mathbf{K}_t$-interdeducible with $G(Hp \to p)$, and which, if any, with $H(Hp \to p)$? (In calling φ and ψ $\mathbf{K}_t$-interdeducible, we mean of course that $\mathbf{K}_t \oplus \varphi = \mathbf{K}_t \oplus \psi$.)

(ii) Calling a point 1-domain-reflexive or 2-domain-reflexive according as, when taken as x for the former and y for the latter, it satisfies the condition

$$\forall y(Rxy \to Rxx) \quad \text{or} \quad \forall x(Rxy \to Rxx),$$

respectively, find a tense-logical formula which locally defines 1-domain-reflexivity and find one which locally defines 2-domain-reflexivity.

(iii) Calling a point 1-range-reflexive or 2-range-reflexive according as, when taken as x for the former and y for the latter, it satisfies the condition

$$\forall y(Rxy \to Ryy) \quad \text{or} \quad \forall x(Rxy \to Ryy),$$

respectively, find a tense-logical formula which locally defines 1-range-reflexivity and find one which locally defines 2-range-reflexivity. ✠

None of these properties are likely to be of much interest for the specifically temporal applications of tense logic, since the accessibility relation there is typically expected to be irreflexive, since we are thinking of it as the relation of precedence between moments of time (the relation *is earlier than*). Thus one is usually interested in tense logics determined by classes of transitive irreflexive frames, meeting further conditions that can be thought of as embodying plausible or at least investigationworthy hypotheses about the structure of time. We touch briefly on this kind of venture in Section 3.3. Here we have been concerned, by contrast, to illustrate what happens when we can "look both ways" down R-chains for an accessibility relation R: not just forward, as with G (or $\Box$), but backward too, with the aid of H (or $\Box^{-1}$). But before leaving the present considerations, there is a bit more to be said about the local notion of validity.

The first is that instead of writing "$\langle U, R \rangle \models_x \varphi$" to say that the formula φ is valid at the point x (in the frame $\langle U, R \rangle$), we may prefer to promote the point we are interested in to the status of a distinguished element of the frame, and arrive at triples $\langle U, R, x \rangle$ as structures in their own right, which might go by such names as "pointed frames", "local frames", or "frames with distinguished elements", as in Segerberg [1025] – though in fact on first introducing this phrase Segerberg (who abbreviates it to "frames w.d.e.") allows a *set* of distinguished elements in the corresponding structures, rather than, as envisaged here, one such element per structure. Validity on a local frame of this kind is just validity at the distinguished point in the ordinary frame concerned, but the alternative terminology is worth bearing in mind in case it renders certain questions more easily asked. One kind of question we can ask easily enough either way is what the logic determined by a class of local frames is in which the distinguished element is reflexive, say, or for another example, 1-transitive. (See further Remark 3.2.21 below.) In the other way of speaking, this question asks how to axiomatize the class of formulas valid at all reflexive, or again, at all 1-transitive points. To link up with our earlier monomodal discussion, let us drop H altogether and write $\Box$ rather

3.2. EXTENSIONS OF $\mathbf{K}_t$

than G. (We also revert to A, B, ... as schematic letters, rather than φ, ψ,...) Then, picking the first of the two examples just given (and using the frames w.d.e. formulation), we have the following result, used in Segerberg [1025] to illustrate the general situation. Recall (from Exc. 2.5.25, p. 81) that $\mathbf{K} + \mathbf{T}$ is not the same as $\mathbf{K} \oplus \mathbf{T}$, alias $\mathbf{KT}$. The latter is the smallest normal modal logic containing $\mathbf{T}$, while the former is the smallest modal logic extending $\mathbf{K}$ and containing $\mathbf{T}$, or equivalently (since we are dealing with extensions of $\mathbf{K}$), the smallest quasi-normal modal logic containing $\mathbf{T}$.

PROPOSITION 3.2.19 *The logic determined by the class of all $\langle U, R, x \rangle$ with xRx is $\mathbf{K} + \mathbf{T}$.*

It follows that $\mathbf{K} + \mathbf{T}$ is properly quasi-normal (quasi-normal but not normal, that is):

COROLLARY 3.2.20 $\mathbf{K} + \mathbf{T} \neq \mathbf{KT}$.

Proof. Use the fact that the $\mathbf{KT}$-theorem (e.g.) $\Box(\Box p \to p)$ – alias $\mathbf{U}$ – is not provable in $\mathbf{K} + \mathbf{T}$, since this formula can be falsified at the distinguished element x of a local frame $\langle U, R, x \rangle$ with xRx as in Prop. 3.2.19, as long as that point has at least one successor y such that not yRy: just put $V(p) = R(y)$ to get a model falsifying $\Box(\Box p \to p)$ at x. ∎

REMARK 3.2.21 Porte [897] described a non-normal modal logic (which he called T^+) between the normal modal logics $\mathbf{KT}$ and $\mathbf{S4}$, namely $\mathbf{KT} + \mathbf{4}$, which is, as he notes, though apparently without knowledge of the discussion in Segerberg [1025], determined by the class of reflexive frames w.d.e. in which the distinguished element is 1-transitive, which amounts, when taking the subframe generated by the distinguished element (a notion sometimes built into the idea of frames w.d.e., alias pointed frames, from the start) to every element being accessible to the distinguished point. (This includes the element itself, by reflexivity.) ◀

Exercise 2.5.25(*iii*) (see p. 81) asked of several formulas whether or not when taken as A they satisfied the equation $\mathbf{S4} + A = \mathbf{S4} \oplus A$, and here we will look in more detail at one of the formulas listed, namely $p \to \Box \Diamond p$, which is to say, the formula $\mathbf{B}$. Of course the only question here arises over the inclusion $\mathbf{S4} \oplus (p \to \Box \Diamond p) \subseteq \mathbf{S4} + (p \to \Box \Diamond p)$, which amounts to asking whether $\Box(p \to \Box \Diamond p)$ is provable in $\mathbf{S4} + (p \to \Box \Diamond p)$, since any further effects of Necessitation – of the form $\Box^n(p \to \Box \Diamond p)$ with $n > 1$ can be obtained from this by $\mathbf{4}$. Now, much as with Prop. 3.2.19, the latter logic is the quasi-normal modal logic determined by the class of local frames $\langle U, R, x \rangle$ with $\langle U, R \rangle$ a transitive reflexive frame and x a 1-symmetric distinguished element. So the question becomes: must $\langle U, R \rangle$ then be a symmetric frame (and so an equivalence-relational frame)? In other words: does the 1-symmetry of the distinguished element x imply the 1-symmetry of all $y \in U$? In fact, an affirmative answer can be returned to this question without even exploiting the reflexivity of R, since the following observation implies that, when attention is restricted to transitive $\langle U, R, x \rangle$ generated by x (as it may as well be, by the Generation Theorem, 2.4.10, on p. 70), the 1-symmetry of x suffices for the symmetry of R:

PROPOSITION 3.2.22 *All successors of a 1-symmetric point in a transitive frame are themselves 1-symmetric.*

Proof. Suppose we have $\langle U, R \rangle$ transitive, $x \in U$ 1-symmetric, and $xRyRz$, with a view to showing that zRy, which will mean that y is 1-symmetric. (It may help to draw a little picture.) Since $xRyRz$, by transitivity we get xRz, so by the 1-symmetry of x, zRx. Then since xRy, by transitivity we get zRy, as required. ∎

It is not immediately obvious how to transform these considerations into a syntactic proof that $\Box(p \to \Box\Diamond p)$ is provable in $\mathbf{K4} + (p \to \Box\Diamond p)$. Note that the formula after this last "+" is the axiom **B**, and that we can equally well do our reasoning with the dual form of this axiom, namely $\Diamond\Box p \to p$, as indeed we shall. The formal proof below is the result of transforming an informal argument from the hypothesis that we have $\Diamond\Box A \to A$ for all formulas A but not $\Box(\Diamond\Box p \to p)$, and so $\Diamond(\Diamond\Box p \wedge \neg p)$, and thus (i) $\Diamond\Diamond\Box p$ and (ii) $\Diamond\neg p$. By **4** in its dual form, (i) gives $\Diamond\Box p$ and thus by **4** in its original form $\Diamond\Box\Box p$. (This exploits $\Diamond$Mono, since we are replacing one $\Box$ by two in the scope of a $\Diamond$.) Since we have $\Diamond\Box A \to A$ for all A, taking A as $\Box p$ we conclude $\Box p$, which contradicts (ii). Let us set this out as an axiomatic proof, albeit a rather lazy one, which helps itself to the provability of various things in **K** and **K4**:

EXAMPLE 3.2.23 Putting a $\Box$ onto $\Diamond\Box p \to p$ with a background logic of **K4**:

(1)	$(\Box\Box\Diamond\neg p \vee \Box p) \to \Box(\Box\Diamond\neg p \vee p)$	provable in **K**
(2)	$\Box\Diamond\neg p \to \Box\Box\Diamond\neg p$	provable in **K4**
(3)	$(\Box\Diamond\neg p \vee \Box p) \to \Box(\Box\Diamond\neg p \vee p)$	1, 2 TF
(4)	$\Diamond\Diamond\neg p \to \Diamond\neg p$	**K4**-provable
(5)	$\Box\Diamond\Diamond\neg p \to \Box\Diamond\neg p$	4, $\Box$Mono
(6)	$(\Box\Diamond\Diamond\neg p \vee \Box p) \to \Box(\Box\Diamond\neg p \vee p)$	3, 5 TF
(7)	$\Diamond\neg p \to \Box\Diamond\Diamond\neg p$	Instance of **B**
(8)	$(\Diamond\neg p \vee \Box p) \to \Box(\Box\Diamond\neg p \vee p)$	6, 7 TF
(9)	$\Diamond\neg p \vee \Box p$	**K**-provable
(10)	$\Box(\Box\Diamond\neg p \vee p)$	8, 9 Modus Ponens
(11)	$\Box(\Box\Diamond\neg p \vee p) \to \Box(\Diamond\Box p \to p)$	**K**-provable
(12)	$\Box(\Diamond\Box p \to p)$	10, 11 Modus Ponens

Of course, things could have been formulated at a coarser or a finer degree of justification here. For instance, we could have spelt out (11) more fully by beginning with the **K**-provability of $(\Box\Diamond\neg p \vee p) \to (\Diamond\Box p \to p)$ and then applying $\Box$Mono to this (or even starting this justification further back). The key thing to watch is that while we applied $\Box$Mono several times, whether explicitly, as in line (4), or implicitly, as here, this is only done to **K4**-provable formulas (sometimes already even **K**-provable), and not to the **B** schema, any <u>unnecessitated</u> instance of which schema can be used in the proof. (Recall that using only the primitive rules of our axiomatization of **K**, the derivation of the rule $\Box$Mono involves Necessitation – not allowed in the present context since we are looking at **K4** + **B**: "+", not "$\oplus$".) ◀

The above discussion shows that $\mathbf{K4} + \mathbf{B} = \mathbf{K4} \oplus \mathbf{B}$ (alias **KB4** or **K4B**). What about reversing the roles of **B** and **4** here? Let us see:

EXERCISE 3.2.24 Do we have $\mathbf{KB} + \mathbf{4} = \mathbf{KB} \oplus \mathbf{4}(\ = \mathbf{KB4})$? Justify your answer. ✠

The phenomenon of Kripke-incompleteness – failure of Kripke-completeness in the sense of Exc. 2.5.29 (p. 83) – arises in a 'local' form too: just as we can have formulas A valid on every frame validating all formulas in a set Γ without A's being provable in $\mathbf{K} \oplus \Gamma$, a formula A may be valid at any point in a frame (or: on any local frame) validating all formulas in Γ, without our having A provable in $\mathbf{K} + \Gamma$. We take the following simple example, as well as Exercise 3.2.26, from van Benthem [68]:

EXAMPLE 3.2.25 Here we take A and Γ of the preceding paragraph as $\bot$ and $\{\Box(\Box p \to p) \to p\}$. The latter formula is a variation on (the formula form of) **W** from Table 2.1, p. 33, obtained by dropping the $\Box$ from the consequent (of the whole formula). It is easier to consider this formula in the alternative form (as for **W** in the final column of Table 2.1): $p \to \Diamond(p \wedge \Box\neg p)$. We want to show that any point in a frame at which this formula is valid, $\bot$ is valid – which, since $\bot$ is never valid at a point, amounts to showing that

3.2. EXTENSIONS OF $\mathbf{K}_t$

$p \to \Diamond(p \wedge \Box \neg p)$ is never valid at a point. Suppose otherwise, i.e., that we have $\langle W, R \rangle \models_x p \to \Diamond(p \wedge \Box \neg p)$, where $x \in W$. (Formulation using 'local frames': $\langle W, R, x \rangle \models p \to \Diamond(p \wedge \Box \neg p)$.) Consider V with $V(p) = \{x\}$. Since its antecedent is true at x in $\langle W, R, V \rangle$, if our formula is to be valid at x, the consequent must be true at x in this model, so for some $y \in W$, xRy and $\langle W, R, V \rangle \models_y p \wedge \Box \neg p$, so (1) $\langle W, R, V \rangle \models_y p$ and (2) $\langle W, R, V \rangle \models_y \Box \neg p$. By the choice of V, (1) implies (1'): $y = x$; by (2), no R-successor of y verifies p, and thus (2'): not yRx. But (1') and (2') together contradict the fact that xRy. So our original implicational formula cannot be valid at a point, as required. The part of the argument in van Benthem [68], showing that nonetheless $\mathbf{K} + \Box(\Box p \to p) \to p$ is *consistent* – does not have $\bot$ as a theorem – we do not reproduce here, as it involves the apparatus of *general frames* which we are not going into. (See the Digression on p. 83.) ◀

EXERCISE 3.2.26 Show that by contrast with the quasi-normal logic $\mathbf{K} + \Box(\Box p \to p) \to p$, the normal logic $\mathbf{K} \oplus \Box(\Box p \to p) \to p$ is inconsistent. ✠

We return to the case of potentially non-normal extensions of $\mathbf{K4}$.

Are there indeed any candidate axioms for modal logics one might add to $\mathbf{K4}$ which do not yield their own necessitations without the help of the Necessitation rule itself? Obviously candidate $\mathbf{K4}$-equivalent to a formula of the form $\Box A$ will not require an appeal to (Nec.) to apply a further $\Box$ to it, since we can use Modus Ponens and the appropriate instance of **4** (i.e., $\Box A \to \Box\Box A$); as Example 3.2.23 shows, the phenomenon is more widespread that this, so the question is whether it is completely ubiquitous. A similar question was raised and answered negatively by McKinsey and Tarski [791], p. 7, for the stronger logic $\mathbf{S4}$, thereby settling the question negatively for $\mathbf{K4}$. The authors' example of a non-normal extension of $\mathbf{S4}$ was $\mathbf{S4} + \mathbf{M}$, which is why $\mathbf{M}$ (namely $\Diamond(\Box p \vee \Box \neg p)$ or alternatively $\Box \Diamond p \to \Diamond \Box p$) is named after McKinsey (see the discussion of $\mathbf{M}$ in McKinsey's substitutional semantics for modal logic in Section 2.10 above, especially that immediately preceding Proposition 2.10.21 on p. 168); for the proof, it will be convenient to refer to the frame w.d.e. depicted in Figure 3.2 (the distinguished element being that labelled "x"):

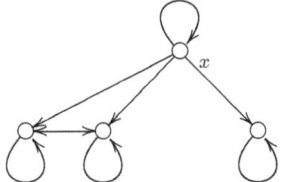

Figure 3.2: *A Local Frame Illustrating the McKinsey–Tarski Observation*

PROPOSITION 3.2.27 $\mathbf{S4} + \mathbf{M} \neq \mathbf{S4M}$.

Proof. Consider the pointed frame $\langle U, R, x \rangle$ diagrammed in Figure 3.2. As R is transitive and reflexive $\langle U, R \rangle$, and hence $\langle U, R, x \rangle$, is a frame for $\mathbf{S4}$, and since the distinguished element has accessible to it some element which is its own sole successor (on the lower right of the diagram), $\mathbf{M}$ cannot be falsified at x in any model, so this is a local frame for $\mathbf{S4} + \mathbf{M}$. But the necessitation of $\mathbf{M}$ can be falsified at x, since $\mathbf{M}$ can be falsified at either of the two other successors of x, neither of them having itself as sole successor. So the necessitation of $\mathbf{M}$ is not provable in $\mathbf{S4} + \mathbf{M}$, which is accordingly non-normal and is a proper sublogic of $\mathbf{S4M}$. ■

Since $\nvdash_{\mathbf{S5}}$ **M**, the case of **S4** + **M** does not answer another question one might naturally raise, as to whether there are any non-normal modal logics between **S4** and **S5**. As was mentioned in the Digression on p. 34, every extension of **S4.3** is normal, but this leaves plenty of room, and various papers of Goldblatt in the 1970s – especially [391] – suggest where one might look for an example:

EXERCISE 3.2.28 Show that **S4** + $(\Diamond \Box p \to p) \lor \Diamond(\Box q \lor \Box \neg q)$ is a non-normal modal logic between **S4** and **S5**. (*Hint*: to show non-normality, follow the general strategy of the proof of Prop. 3.2.27, slightly complicating the frame of Figure 3.2 by suitably interpolating one additional point.) ✠

Here we have used the dual form of **B**, $\Diamond \Box p \to p$ rather than the original form, simply because that is the form in which we most recently encountered this principle, in Example 3.2.23. Note also that the second disjunct of the formula cited is a form of **M** re-lettered so as not to overlap in variables with the first disjunct. (Why?)

EXERCISE 3.2.29 Does the logic mentioned in Exercise 3.2.28 have the following formula among its theorems?

$$\Box(\Diamond \Box p \to p) \lor \Diamond(\Box q \lor \Box \neg q)$$

(Justify your answer.) ✠

We conclude this section by returning to the traditional tense-logical language with G and H, in order to illustrate as simply as possible, how a logic can fail to have the finite model property, as this was defined on p. 73 above. The example which follows is based on one given for first-order logic by Peirce (known also to Dedekind), for a clear presentation of which, see the opening section of Legg [690]. Further information on the history of what is involved here, especially the roles of Dedekind and Peirce, is provided in §3 of Dipert [242]. Other examples of failures of the finite model property in first-order logic – (sets of) formulas true in some model but in no model with a finite domain – can be found in many classic sources and introductory presentations, such as Church [179], Exercise 43.5, p. 233, and Wang [1145], pp. 64, 66.

EXAMPLE 3.2.30 Consider the normal tense logic $\mathbf{K}_t \oplus \{F\top, Pq \to Hq\}$, evidently determined by the class of frames in which every point has at least one R-successor and at most one R-predecessor. From this description we see that $P\top$ is not a theorem of the logic, since it is invalid on the frame whose universe comprises the natural numbers with $xRy \Leftrightarrow x+1 = y$, being false at 0 in any model on this frame. But every *finite* model on a frame in this determining class verifies the non-theorem $P\top$. For suppose we have a model $\mathcal{M} = \langle U, R, V \rangle$ with $\langle U, R \rangle$ as described with $\mathcal{M} \nvDash_{u_0} P\top$. Since u_0 needs an R-successor, which in turn needs a successor, etc., we have an infinite R-chain $u_0, u_1, u_2, \ldots$, as this notion was explained in Exc. 2.4.9(i), where it was emphasized that this terminology does not imply that the elements u_i, u_j, are distinct for distinct i and j, and does not imply therefore that we are dealing with infinitely many points (or even that we dealing with more than one point). However, in the present case the points are indeed all distinct, showing that U cannot be finite: for consider the least i for which there exists $j > i$ with $u_i = u_j$; i cannot be 0, as $u_{j-1}Ru_j$ and u_0, supposedly equal to u_j, has no R-predecessor. Suppose instead that $i > 0$, since also $j > i$, from the hypothesis that $u_i = u_j$ we can conclude that $u_{i-1} = u_{j-1}$, contradicting the choice of i as providing the least such case. ◀

3.3 Temporally Motivated Concerns: Density and Discreteness

One interesting question than naturally arises when we think of moments of time is whether they are densely or discretely ordered by the relation "earlier than"; the question is often thought of against a background assumption that the ordering in a strict linear ordering (transitive, irreflexive, and weakly

3.3. TEMPORALLY MOTIVATED CONCERNS: DENSITY AND DISCRETENESS

connected, that is), and probably infinite in both directions, in which case the question is whether moments of time share with the integers the 'discreteness' property that for any given one there is an immediate predecessor and an immediate successor, or with the real numbers and the rational numbers the 'density' property that between any two there is a third (and hence infinitely many, by reinvoking the property). Of course the real numbers have the further (second-order) property of *continuity*, not shared with the rationals, and this too has been the subject of tense-logical investigation.[154] But here we will spend some time just thinking about discreteness and density.

In this chapter we will not consider ways of showing completeness w.r.t. classes of irreflexive frames here – techniques to be found in Segerberg [1025], [1024], or less indirectly, in Burgess [134]; these are straightforward bimodal adaptations of the techniques in giving monomodal completeness results for classes of such frames that appear below. See for example the proof of Theorem 4.4.32 (p. 269), which shows $\mathbf{K}$ to be determined by the class of irreflexive frames. (The same applies to $\mathbf{K}_t$.) Those considerations will not concern us here as we ponder the conditions of density and discreteness themselves.[155] And we will for the most part think about these issues without assuming the connectedness aspect of linearity, which in a tense-logical setting comes in two flavours, since we could disallow branching into the future, by imposing the G form of the piecewise weak connectedness axiom **.3**, and we could also disallow branching into the past, by imposing the H form.[156]

Anyone successfully attempting Exercise 2.5.43(i) will have already encountered the class of *dense* frames, i.e., frames $\langle U, R \rangle$ such that for all $x, z \in U$ there exists $y \in U$ with $xRyRz$. The monomodal logic determined by this class of frames is $\mathbf{K4}_c$, and the frames for this logic are precisely those which are dense. Evidently, if $\langle U, R \rangle$ is dense then so is $\langle U, R^{-1} \rangle$, so (analogously with the case of **4** itself) we can take our tense-logical density schema to be either

$$GG\varphi \to G\varphi \qquad \text{or} \qquad HH\varphi \to H\varphi.$$

EXERCISE 3.3.1 Deduce a representative instance, $HHq \to Hq$, say, of the second of these schemata using any instances of the first of them together with the rules and axioms of $\mathbf{K}_t$. (Here and frequently in what follows, we tend to prefer using 'q' over using "p" in single-variable formulas, so that they are easier to read out without a homophony clash over the tense operator "P".) ✠

Evidently one could equally well take either of the duals of these principles. In fact the formula $Fq \to FFq$ is perhaps more obviously (than the G formulation) suggestive of the idea of a dense ordering

---

[154]For historical information, see §5 of Chapter IV in Prior [924]. There is also some information on this and much else in tense logic in Rescher and Urquhart [951] (though one should also have Fine [288] to hand for the correction of some misstatements). See van Benthem [74], p. 162, on this topic, and the whole of the chapter containing that page for a study of tense-logical completeness and definability. Rescher and Urquhart use the phrase 'temporal logic' to cover various logical studies motivated (at least initially) by a concern with time and 'tense logic' for the specifically Prior–style bimodal enterprise, a usage followed here, though many subsequent writers have ignored their suggestion and just use 'temporal logic' to mean tense logic.

[155]In fact, along with irreflexivity, we will avoid, roughly speaking, any other (bi)modally undefinable frame properties, which in the case of discreteness includes one obvious way of construing that property: define a future discrete frame to be a frame $\langle U, R \rangle$ in which for all $x \in U$ there is a $z \in R(x)$ for which there exists no y such that $xRyRz$. A variation on this would restrict the condition on x here to those x for which $R(x) \neq \varnothing$. This latter condition coupled with a Past-directed version is called DISC in van Benthem [74], where the tense-logical undefinability of such conditions is also shown. (See [74], p. 18, p. 160f.) "Roughly speaking" because what we are actually most concerned to avoid is not so much classes of frames which are modally undefinable so much as conditions on frames which are not 'canonically enforceable', in the sense that there are no formulas whose provability in a logic will guarantee that the logic's canonical frame meets the condition. Not having developed the required tools here, we prefer not to get involved with completeness results that cannot be obtained directly from the canonical frame or from point-generated subframes thereof.

[156]Historically there has been much greater interest in allowing future branching without past branching, in order to think of the future – but not the past – as 'open', alternative branches representing alternative possible future courses of events. See Prior's 'Time and Determinism' (= Chapter VII of [924]) for a seminal discussion, Thomason [1104] and Burgess [133] for philosophical and logical discussion, and Zanardo [1216] and references therein for more recent – mostly technical – developments.

(though we are abstracting from all 'ordering' aspects of this description here). We encountered a variant on $\mathbf{4}_c$ in Exercise 2.7.31, p. 136, which asked whether

$$\Box(\Box p \vee \Box q) \rightarrow (\Box p \vee \Box q)$$

was provable in $\mathbf{K4}_c$. Following up the suggestion there given immediately gives a countermodel on the dense frame described there (put $V(p) = \mathbb{Q}_a$, $V(q) = \mathbb{Q}_b$, to falsify the formula at 0). Curiously enough, despite its connection with density (or what we are about to call *generalized* density), this question bears on the issue of discreteness as well, as we shall see. In fact, the contrast between these two properties – the second of which we have yet to define precisely – which one associates with linear orderings, is not really present in the current more general setting, as is clear from the fact that the above density axioms, just like the following principle we can call Hamblin's (forward, or future-oriented) discreteness axiom:

$$(q \wedge Hq) \rightarrow FHq,$$

are both Sobociński-regular and so, far from being in any sense mutually inconsistent, are both valid on the one-element reflexive frame.[157] However, as it turns out, of the two ways of precisifying the idea of discreteness, it is the non-Hamblin way that leads us to consider the formula above from 2.7.31 which generalizes density. We get to these "two ways" in a moment, first pausing to record the generalization involved. Let us do so with schemata rather than particular axioms. By the n-th Generalized Density schema we mean the following:

$$\text{GD}_n: \qquad G(G\varphi_1 \vee \ldots \vee G\varphi_n) \rightarrow (G\varphi_1 \vee \ldots \vee G\varphi_n).$$

Thus GD_1 is (the schema) $\mathbf{4}_c$, while the formula from Exercise 2.7.31 is a representative instance of GD_2. (The 'generalized density' terminology is from Kuhn [679].) We write "GD" rather than "**GD**" to avoid any suggestion of the Geach convergence axiom **G** (alias **.2**) or the seriality axiom **D**.

THEOREM 3.3.2 *For $n \geq 1$, the logic $\mathbf{K}_t \oplus \text{GD}_n$ is determined by the class of frames $\langle U, R \rangle$ such that for all $x \in U$ and all $z_1, \ldots, z_n \in R(x)$, there exists $y \in R(x)$ such that $\{z_1, \ldots, z_n\} \subseteq R(y)$.*

Proof. Soundness is clear, so only the completeness half of the claim will be addressed. To reduce clutter, we give the proof for the case of $n = 2$, which is representative of the general case. Let $\langle U, R \rangle$ be the canonical frame for $\mathbf{K}_t \oplus \text{GD}_2$, and take $x \in U$ with xRz_1 and xRz_2. We must find $y \in R(x)$ with yRz_1 and yRz_2.

We can do so by letting y be a set of formulas maximal consistent w.r.t. the present logic such that $y \supseteq \{\varphi \mid G\varphi \in x\} \cup \{F\psi \mid \psi \in z_1\} \cup \{F\chi \mid \chi \in z_2\}$. Since such a y evidently stands in the desired R-relationships with x, z_1 and z_2, it remains only to check that this set of formulas is consistent (w.r.t. the present logic), and hence that such a y exists. Suppose the set is not consistent, i.e., that for some k, m, n, with $G\varphi_1, \ldots, G\varphi_k \in x$, $\psi_1, \ldots \psi_m \in z_1$, and $\chi_1, \ldots \chi_n \in z_2$, the following is provable:

$$(\varphi_1 \wedge \ldots \wedge \varphi_k) \rightarrow \neg(F\psi_1 \wedge \ldots \wedge F\psi_m \wedge F\chi_1 \wedge \ldots \wedge F\chi_n),$$

and thus so also is

$$(\varphi_1 \wedge \ldots \wedge \varphi_k) \rightarrow \neg(F(\psi_1 \wedge \ldots \wedge \psi_m) \wedge F(\chi_1 \wedge \ldots \wedge \chi_n)),$$

and by further processing:

$$(\varphi_1 \wedge \ldots \wedge \varphi_k) \rightarrow (G\neg(\psi_1 \wedge \ldots \wedge \psi_m) \vee G\neg(\chi_1 \wedge \ldots \wedge \chi_n)).$$

---

[157] Here we extend the notion of Sobociński-regularity defined for the monomodal case at p. 37 to the general case to mean that the result of erasing all primitive $\Box$ (and thus also $\Diamond$) operators in all theorems leaves truth-functional tautologies.

3.3. TEMPORALLY MOTIVATED CONCERNS: DENSITY AND DISCRETENESS   197

Since G is normal, we get
$$(G\varphi_1 \wedge \ldots \wedge G\varphi_k) \rightarrow G(G\neg(\psi_1 \wedge \ldots \wedge \psi_m) \vee G\neg(\chi_1 \wedge \ldots \wedge \chi_n)),$$
whence, given the provenance of the φ_i, the antecedent and hence the consequent of the above implication belongs to x. Appealing now to the following instance of the schema GD_2:
$$G(G\neg(\psi_1 \wedge \ldots \wedge \psi_m) \vee G\neg(\chi_1 \wedge \ldots \wedge \chi_n)) \rightarrow (G\neg(\psi_1 \wedge \ldots \wedge \psi_m) \vee G\neg(\chi_1 \wedge \ldots \wedge \chi_n)),$$
we conclude that the latter disjunction, hence at least one of its disjuncts, belongs to x. But this is impossible, since the first disjunct's belonging to x contradicts the fact xRz_1 (as the $\psi_i \in z_1$, and so likewise for their conjunction) while the second disjunct gives a similar contradiction with the fact that xRz_2 (this time because of the χ_i). This contradiction establishes the consistency of the set of formulas we needed for producing a y as promised. ■

Notice that this is really an entirely monomodal result, making no use of H (and P), and it would not have been out of place in our earlier □-notated discussion.[158] We can now make another move familiar from the monomodal setting – as in the proof of Theorem 2.6.8 – in its shift from something about finite sets to something about arbitrary sets, highlighted in Remark 2.6.11, p. 117). Here we want to consider the possibility of an S in which all the principles GD_n are provable, for $n \in Nat$, and the reason we want to do this is that as long as a given GD_n is provable, then we have completeness w.r.t. the following condition – reformulating Theorem 3.3.2 – on frames $\langle U, R \rangle$: that for all $x \in U$ and all $Z \subseteq R(x)$ with $|Z| \leq n$, there exists $y \in R(x)$ with $Z \subseteq R(y)$. (We say "$|Z| \leq n$" rather than "$|Z| = n$", because the Thm. 3.3.2 formulation in terms of $z_1, \ldots, z_n \in R(x)$ did not require the these $z_1, \ldots, z_n$ to be distinct.) But we did not quite consider all cases of GD_n for $n \in Nat$ in Theorem 3.3.2, as we started with $n = 1$ rather than $n = 0$. In terms of cardinality, this would confront us with the condition that for all $x \in U$ and all $Z \subseteq R(x)$ with $|Z| \leq 0$, there exists $y \in R(x)$ with $Z \subseteq R(y)$. (Here the "$\leq$" really does just amount to "$=$".) The only Z meeting the cardinality restriction here is $\varnothing$, and this Z vacuously satisfies the further condition that $Z \subseteq R(u)$ for any u – including most conspicuously the cases of $u = x$, for the given point x, and $u = y$ where y is a promised R-successor of x. Thus the condition boils down to nothing but seriality: each point must have an R-successor. When we look at the schema GD_n for the case of $n = 0$, then on the usual understanding that the empty disjunction is $\bot$, this special case is simply $G\bot \rightarrow \bot$, another way of writing the seriality axiom **D** (for G). So in fact Theorem 3.3.2 does continue to hold if we allow in the $n = 0$ case, since we know that **D** secures seriality for the canonical accessibility relation. (Exercise 2.4.8(iv), p. 65.) Despite all this, there is a certain discontinuity as we drop down to the 0 case of GD_n, in that whereas for all $n \geq 1$, the provability of GD_{n+1} implies that of GD_n, as given GD_{n+1}:
$$G(G\varphi_1 \vee \ldots \vee G\varphi_n \vee G\varphi_{n+1}) \rightarrow (G\varphi_1 \vee \ldots \vee G\varphi_n \vee G\varphi_{n+1}),$$
we can always consider the special case in which φ_{n+1} is just φ_n again:
$$G(G\varphi_1 \vee \ldots \vee G\varphi_n \vee G\varphi_n) \rightarrow (G\varphi_1 \vee \ldots \vee G\varphi_n \vee G\varphi_n),$$
which is in the present normal (and hence congruential) setting, equivalent to having GD_n:
$$G(G\varphi_1 \vee \ldots \vee G\varphi_n) \rightarrow (G\varphi_1 \vee \ldots \vee G\varphi_n).$$
But we cannot pass from GD_1 to GD_0 (alias **D** for G) in this way. This discontinuity does not spoil our ("finite-to-arbitrary") analogue of Theorem 2.6.8, however; the proof, as there, consists in observing that the unavailability of a point y behaving as desired can always be traced to the inconsistency of the union of *finitely* many sets of formulas, which itself can only be due to *finite* subsets of the sets whose union is taken.

---

[158]Sometimes, it can happen that a monomodal logic is Kripke-complete while its tense-logical analogue is complete. This possibility, discovered by Frank Wolter (see Litak [737], p. 163) is evidently not to the point here.

THEOREM 3.3.3 *Let S be any consistent normal modal logic. If $\vdash_S \text{GD}_n$ for all $n \in \text{Nat}$, then where $\langle U, R \rangle$ is the canonical frame for S, we have: for every $x \in U$ and every $Z \subseteq R(x)$ there exists $y \in R(x)$ with $Z \subseteq R(y)$.*

Much as in the case of Coro. 2.6.9:

COROLLARY 3.3.4 *With S and $\langle U, R \rangle$ as above, for any $x \in U$, there exists $y \in R(x)$ with $R(x) \subseteq R(y)$.*

Proof. Take Z in Theorem 3.3.3 as $R(x)$. ■

Thus we have the following, which, as it would hold (writing G as $\Box$) for $\mathbf{K}$ no less than for $\mathbf{K}_t$, incidentally answers Exercise 2.7.32(i) (p. 136); as there, we have decided to write the condition out in an explicit first-order form:

COROLLARY 3.3.5 $\mathbf{K}_t \oplus \{\text{GD}_n \mid n \in \text{Nat}\}$ *is determined by the class of frames $\langle U, R \rangle$ satisfying:*

$$\forall x \exists y (Rxy \land \forall z(Rxz \to Ryz)).$$

If we are interested specifically in transitive frames, as we most likely would be for temporal applications, then we can produce a finite axiomatization, reported in Coro. 3.3.8 below; the relevant observation is recorded in Prop. 3.3.6:

PROPOSITION 3.3.6 *For any $n \geq 2$, GD_{n+1} is valid on any transitive frame validating GD_n and is also $\mathbf{K}_t 4$-deducible from GD_n.*

EXERCISE 3.3.7 Prove both claims in Prop. 3.3.6. (Suggestion: it is easier to think about GD_n, for present purposes, in the dual form:

$$(F\varphi_1 \land \ldots \land F\varphi_n) \to F(F\varphi_1 \land \ldots \land F\varphi_n).)$$

✠

Putting these ingredients – the syntactic (second) half of Prop. 3.3.6 – together and adjusting somewhat (we need to add back $\mathbf{D}$, here appearing as $F\top$), we have a further corollary to Theorem 3.3.3:[159]

COROLLARY 3.3.8 $\mathbf{K}_t 4 \oplus \{F\top, \text{GD}_2\}$ *is determined by the class of transitive frames meeting the condition specified in Coro. 3.3.5.*

We should check, while we are at it, that without transitivity, these collapses do not occur, and no GD_n has GD_{n+1} as a frame consequence.

EXAMPLE 3.3.9 For $n \in \text{Nat}$, let $\mathcal{F}_n = \langle U, R \rangle$ be as follows: for some $(n+1)$-element set x (e.g., of positive integers), $U = \{x\} \cup \{y \subseteq x : |y| = n\} \cup \{z \subseteq x : |z| = 1\}$ and

$$uRv \Leftrightarrow u \supsetneq v \text{ or } |u| = |v| = n.$$

Thus $\mathcal{F}_n$ has (its universe has, that is) $2n+3$ elements: 1 (x itself) $+\, n+1$ (the n-element subsets of x) $+\, n+1$ (the unit subsets of x). The claim is then that $\mathcal{F}_n \models \text{GD}_n$ while $\mathcal{F}_n \not\models \text{GD}_{n+1}$. In particular, using "$x$" as in the definition of $\mathcal{F}_n$ above, $\mathcal{F}_n \not\models_x \text{GD}_{n+1}$, since xRz for each of the $n+1$ unit subsets of x, while there is no $u \in R(x)$ with all of these unit subsets in $R(u)$. To construct a model on $\mathcal{F}_n$ falsifying an instance of – for convenience we choose the – *dual* form of GD_{n+1}, given in Exc. 3.3.7, at

---

[159]The result mentions $\mathbf{K}_t 4$, in which "4" can be taken indifferently as **4** for H ("4_H") or **4** for G ("4_G") – see Exercise 3.2.3, p. 185 – though the latter reading is more instructive for keeping an eye on what can be done monomodally here.

3.3. TEMPORALLY MOTIVATED CONCERNS: DENSITY AND DISCRETENESS

x put $V(p_i) = \{z_i\}$ for each of these unit subsets z_i. In the case where x is, say $\{1,2,3\}$, then we have $Fp \wedge Fq \wedge Fr$ true at x, because p, q, and r are true at precisely $\{1\}$, $\{2\}$ and $\{3\}$ respectively, while $F(Fp \wedge Fq \wedge Fr)$ is false at x, since no successor of x has all of $\{1\}$, $\{2\}$, $\{3\}$ amongst its successors. ◀

EXERCISE 3.3.10 (i) Complete the above argument by showing that $\mathcal{F}_n \models \mathrm{GD}_n$.

(ii) Find a counterexample (a triple of frame elements, that is) to the claim that $\mathcal{F}_2$, the seven-element frame defined as in Example 3.3.9, is transitive. ✠

We now return to the theme of discreteness, in the more general setting in which transitivity is not assumed.

Kuhn [679] calls a y as promised by Coro. 3.3.4 (or 3.3.5) above for a given x, an 'immediate successor' of x. Thus we can see that although we have been investigating a generalization of density, we are close to what might reasonably called a condition of discreteness: each point has an immediate successor. Not quite there, though, since although we are not worrying about proving completeness w.r.t. classes of irreflexive frames, we certainly don't want to impose conditions which are incompatible with irreflexivity, as the condition that every point has an immediate successor *in this sense* is. (Incompatible for the following reason: take $x \in U$, and we get $y \in R(x)$ bearing R to everything to which x bears R – and therefore to y itself, since xRy.) Thus we are led to the following Future Discreteness condition on frames $\langle U, R \rangle$:

(FDisc)$_1$ For all $x \in U$, there exists $y \in R(x)$ such that for all $z \in R(x)$, either $z = y$ or yRz.

We promised two separate notions of (future) discreteness – setting aside such modally undefinable discreteness notions as those mentioned in note 155 – and the one relevant to Hamblin's (future) discreteness axiom mentioned above, $(q \wedge Hq) \rightarrow FHq$, is not (FDisc)$_1$, but rather:

(FDisc)$_2$ For all $x \in U$, there exists $y \in R(x)$ such that for all $z \in R^{-1}(y)$, either $z = x$ or zRx.

Thus, reading uRv as "u is earlier than v" or "v is later than u", and adding the adverb "strictly" when in addition $u \neq v$, (FDisc)$_1$ demands for each point x an immediate successor y in the sense that anything later than x is either y or a point strictly later than y, while (FDisc)$_2$ demands for each point x an immediate successor y in the rather different sense that anything earlier than y is either x or a point strictly earlier than x. In Figure 3.3 the frame on the left has the point labelled y as an immediate successor to that labelled x in the sense of (FDisc)$_2$ but not in the sense of (FDisc)$_1$, while in the frame depicted on the right y is an immediate successor of x in the sense of (FDisc)$_1$ but not in the sense of (FDisc)$_2$.

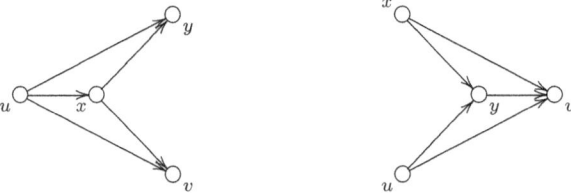

Figure 3.3: *Two Small Frames: Immediate Successors*

EXERCISE 3.3.11 Show that in a strict linear ordering $\langle U, R \rangle$, i.e., a transitive weakly connected irreflexive frame, for $x, y \in U$, y is an immediate successor of x in the sense of (FDisc)$_1$ if and only if y is an immediate successor of x in the sense of (FDisc)$_2$. (In fact we have this result already for the broader class of frames which are asymmetric – and therefore also irreflexive – and weakly connected, whether or not they are transitive.) ☖

With (FDisc)$_2$ the problem of finding the logic determined is straightforward, with the aid of Hamblin's axiom.

EXERCISE 3.3.12 Show that $\mathbf{K}_t \oplus (\varphi \wedge H\varphi) \to FH\varphi$ is determined by the class of all frames satisfying the condition (FDisc)$_2$ above. (Suggestion: for the completeness half, use the canonical model method.) ☖

For (FDisc)$_1$, we need something more elaborate, in the form of the following variation on the generalized density schemas GD$_n$ above:

GD$_n^=$: $\quad G((G\varphi_1 \wedge \varphi_1) \vee \ldots \vee (G\varphi_n \wedge \varphi_n)) \to (G\varphi_1 \vee \ldots \vee G\varphi_n)$.

We need a corresponding variation on Theorem 3.3.2:

THEOREM 3.3.13 *For $n \geq 1$, the logic $\mathbf{K}_t \oplus \mathrm{GD}_n^=$ is determined by the class of frames $\langle U, R \rangle$ such that for all $x \in U$ and all $z_1, \ldots, z_n \in R(x)$, there exists $y \in R(x)$ such that for all $z \in \{z_1, \ldots, z_n\}$, we have either yRz or $y = z$.*

Proof. As with the proof of Thm. 3.3.2, we take the soundness half of the result for granted, and for the completeness half, confine ourselves to showing that the canonical frame $\langle U, R \rangle$ for the representative case of the logic $\mathbf{K}_t \oplus \mathrm{GD}_2^=$ satisfies the required condition. (Also as with Thm. 3.3.2, the argument is entirely monomodal and could equally well be read as applying to $\mathbf{K} \oplus \mathrm{GD}_2^=$, reading G as $\Box$.) The condition for the current $n = 2$ case is that for all $x \in U$ and all $z_1, z_2 \in R(x)$, there exists $y \in R(x)$ with both (1) yRz_1 or $y = z_1$ and (2) yRz_2 or $y = z_2$. Given x, z_1, z_2 as here, let $y \supseteq \{\varphi \mid G\varphi \in x\} \cup \{F\psi \vee \psi \mid \psi \in z_1\} \cup \{F\chi \vee \chi \mid \chi \in z_2\}$, assuming the latter set is consistent (w.r.t. the current logic). On that assumption, clearly $y \in R(x)$. Suppose (1) fails and we have neither yRz_1 nor $y = z_1$. Since not yRz_1, there is a formula α with $\alpha \in z_1$ but $F\alpha \notin y$, and since $y \neq z_1$, there is a formula β with $\beta \in z_1$ but $\beta \notin y$. Now, $\alpha \wedge \beta \in z_1$, so since $y \supseteq \{F\psi \vee \psi \mid \psi \in z_1\}$ we have $F(\alpha \wedge \beta) \vee (\alpha \wedge \beta) \in y$. Thus we should have to have either $F(\alpha \wedge \beta) \in y$ or $\alpha \wedge \beta \in y$. But the first contradicts the fact that $F\alpha \notin y$ and the second the fact that $\beta \notin y$. A similar argument works for (2) and z_2. So as long as the set of formulas y was specified as some maximal consistent extension of is itself consistent – the proof of which involves a simple appeal to the axiom schema GD$_2^=$ – we are done. ■

EXERCISE 3.3.14 Complete the above proof by showing that

$$\{\varphi \mid G\varphi \in x\} \cup \{F\psi \vee \psi \mid \psi \in z_1\} \cup \{F\chi \vee \chi \mid \chi \in z_2\}$$

is indeed consistent w.r.t. $\mathbf{K}_t \oplus \mathrm{GD}_2^=$, with x, z_1 and z_2 as in the proof. ☖

Here is a simple variation (inspired by the way α and β work in the proof of Theorem 3.3.13), on GD$_n^=$:

(∗) $\quad G((G\varphi_1 \wedge \psi_1) \vee \ldots \vee (G\varphi_n \wedge \psi_n)) \to (G(\varphi_1 \vee \psi_1) \ldots \vee G(\varphi_n \vee \psi_n))$.

It is worth considering what happens if we identify the different ψ_i here, as in the schema:

(∗∗) $\quad G((G\varphi_1 \wedge \psi) \vee \ldots \vee (G\varphi_n \wedge \psi)) \to (G(\varphi_1 \vee \psi) \ldots \vee G(\varphi_n \vee \psi))$.

3.3. TEMPORALLY MOTIVATED CONCERNS: DENSITY AND DISCRETENESS 201

EXERCISE 3.3.15 *(i)* Show that for any given n, a modal logic in which G is normal contains all instances of the original schema $GD_n^=$ if and only if contains all instances of $(*)$ for that choice of n.

(ii) Does a similar relationship to that indicated under *(i)* hold between $GD_n^=$ and the corresponding schema $(**)$? (Justify your answer.) ✠

Returning to a development parallelling that of GD_n above, we have the "finite-to-arbitrary" move (compare Theorem 3.3.3); as with GD_0, take $GD_0^=$ to be $F\top$:

THEOREM 3.3.16 *If $\vdash_S GD_n^=$ for all $n \in$ Nat, then where $\langle U, R \rangle$ is the canonical frame for S, we have: for every $x \in U$ and every $Z \subseteq R(x)$ there exists $y \in R(x)$ such that for all $z \in Z$, either $y = z$ or yRz.*

And corresponding to Coro. 3.3.4, we have:

COROLLARY 3.3.17 *With S and $\langle U, R \rangle$ as above, for any $x \in U$, there exists $y \in R(x)$ such that for all $z \in R(x)$, either $y = z$ or yRz.*

COROLLARY 3.3.18 $\mathbf{K}_t \oplus \{GD_n^= : n \in \text{Nat}\}$ *is determined by the class of all frames satisfying the condition* (FDisc)$_1$.

We should pause to notice a difference in the $n = 1$ cases of our two sequences of schemata with nth members GD_n and $GD_n^=$ respectively. While GD_1 is the original density axiom $\mathbf{4}_c$ (for G), $GD_1^=$ is the triviality

$$G(G\varphi \wedge \varphi) \to G\varphi,$$

whose instances are already $\mathbf{K}_t$ provable (this schema being a special case of the still more general $G(\psi \wedge \varphi) \to G\varphi$). This is as Theorem 3.3.13 would lead one to expect, because the corresponding semantic condition given there for this case is that for all x and $z \in R(x)$, there exists $y \in R(x)$ such that for all yRz or $y = z$, a condition automatically satisfied, since we can take y as z itself (for any given x).

One naturally expects some further simplification to the axiomatization in the presence of a transitivity axiom, along the lines of Coro 3.3.8 above; we will leave this an exercise:

EXERCISE 3.3.19 Prove or refute the claim that $\mathbf{K}_t\mathbf{4} \oplus \{F\top, GD_2^=\}$ is determined by the class of transitive frames satisfying (FDisc)$_1$. ✠

Having seen (Exc. 3.3.11) that there is nothing to choose between being an immediate successor in the (FDisc)$_1$ sense and being an immediate successor in the (FDisc)$_2$ sense on strict linear orderings despite their general independence (as in Figure 3.3, p. 199), one might form certain expectations about the interdeducibility of Hamblin's axiom and the $GD^=$ axioms – expectations which are not fulfilled.

EXAMPLE 3.3.20 Let us recall the following weak connectedness axiom, a variant on **.3** given in Exercise 2.5.10(v), though here written with F in place of $\diamond$:

$$(Fp \wedge Fq) \to (F(p \wedge Fq) \vee F(Fp \wedge q) \vee F(p \wedge q)).$$

Let us compare this with the dual form of $GD_n^=$ (cf. Exc. 3.3.7 for the corresponding form of GD_n):

$$(F\varphi_1 \wedge \ldots \wedge F\varphi_n) \to F((F\varphi_1 \vee \varphi_1) \wedge \ldots \wedge (F\varphi_n \vee \varphi_n)),$$

and in particular with a representative instance of the all-important $n = 2$ case:

$$(Fp \wedge Fq) \to (F(Fp \vee p) \wedge F(Fq \vee q)).$$

We give an informal natural deduction style argument to show how this follows from the weak connectedness principle above. Assume the antecedent $Fp \wedge Fq$ of our representative $\text{GD}_2^=$ instance. By the principle in question, we infer that one of the following three cases obtains (*i*) $F(p \wedge Fq)$, (*ii*) $F(Fp \wedge q)$, (*iii*) $F(p \wedge q)$. Now each of (*i*)–(*iii*) implies $F(Fp \vee p) \wedge F(Fq \vee q)$, the consequent of the $\text{GD}_2^=$ formula. Thus $\mathbf{K}_t\mathbf{D4.3}$, where the ".D" and ".3" refer to the principles governing G (as $\Box$),[160] has all the $\text{GD}_n^=$ formulas amongst its theorems. Even throwing in **.3** for H for good measure, we still cannot prove in the resulting logic, Hamblin's formula $(q \wedge Hq) \to FHq$, because the latter is not valid on the 'rationals frame' $\langle \mathbb{Q}, < \rangle$, which is a frame for the logic in question. This may seem paradoxical, since the frame in question is a strict linear ordering, and did we not see that the contrast between every point's having an immediate successor in the $\text{GD}_n^=$-related (FDisc)$_1$ sense, and every point's having an immediate successor in the Hamblin-related (FDisc)$_2$ sense, evaporates in the setting of such structures? Yes but – resolving the puzzle – it was never claimed that every frame validating all the $\text{GD}_n^=$ axioms was itself forward discrete in the sense of (FDisc)$_1$, only that helping ourselves to all these axioms secured completeness w.r.t. this class of frames. ◀

EXERCISE 3.3.21 (*i*) Is the class of frames satisfying (FDisc)$_1$ modally (whether monomodally or tense-logically) definable? (Justify your answer, as also for part (*ii*).)

(*ii*) Let S be the logic $\mathbf{K}_t\mathbf{D4.3}$ considered in Example 3.3.20, with ".3" meaning we take the G form and also the H form of monomodal **.3**. Are the formulas $\text{GD}_n^=$ theorems of $S \oplus (q \wedge Hq) \to FHq$? ✠

A fuller discussion of these matters would of course include consideration of the mirror images of the formulas $\text{GD}_n^=$ and Hamblin's axiom, and also details of how the discrete linear orderings considered here fall short of capturing a key feature of the integer frame $\langle \mathbb{Z}, < \rangle$, namely its (non-first-order) property that no point starts an infinite R-chain ($<$-chain) all of whose elements precede a given point. (That is, there are no two points with infinitely many points strictly between them.) A discrete strict linear ordering need not have this property (for example, a frame consisting of two copies of $\mathbb{Z}$, the first preceding the second). The property is reminiscent of a condition on frames ("no point starts an infinite R-chain") associated – see Exercise 2.4.9 – with the axiom **W**, and it secures the validity of a formula similarly reminiscent of that axiom, namely $G(Gp \to p) \to (FGp \to Gp)$ (not valid on the $\mathbb{Z} + \mathbb{Z}$ frame just mentioned[161]).

Indeed, a fuller discussion of specifically temporally motivated explorations in tense logic would also cover the use of the binary *Since* and *Until* connectives introduced by Hans Kamp (in his 1968 PhD thesis, details of which can be found in the references given later in this paragraph), the idea being that $Since(\varphi, \psi)$ is true at x in a model with accessibility ("earlier than") relation R if there is y such that yRx with φ true at y and ψ true every z such that $yRzRx$. (Understand *Until* similarly, replacing R with its converse throughout this explanation.) Kamp paid special attention to to models based on the frames $\langle \mathbb{R}, < \rangle$ and $\langle \mathbb{N}, < \rangle$, over which the language with *Since* and *Until* as its non-Boolean primitives enjoys (he showed) a certain expressive completeness rendering the Prior tense-logical primitives and any other vocabulary plausibly describable as a tense operator (this being made precise by Kamp) definable; in addition, as Gabbay observed in [340], in such models every formula has an equivalent which is a Boolean compound past-directed, future-directed, and present-directed formulas, where these notions again have natural definitions. For detailed presentation of these results of Kamp and Gabbay, including full references, see see Rabinovich [937] and Hodkinson and Reynolds [507].

---

[160]So we should really write something along the lines of "$\mathbf{K}_t(\mathbf{D4.3})_G$".

[161]The "+" here is for ordinal rather than cardinal addition, so it would more accurate to say we are dealing with the frame whose order type is $\omega^* + \omega + \omega^* + \omega$ – except that then all this notation would need to be explained: see for example pp. 14–20 of Rosenstein [968].

Chapter 4

Applications: Alethic, Nomic, Deontic

4.1 Introduction

First – the terms in the title are not all on a par: there is no such phrase (in common currency) as "nomic logic". One has (alethic) modal logic with a nomic or nomological interpretation or application. That is, the kind of necessity with which one is concerned is natural or nomic necessity (of which physical necessity is perhaps to be thought of as a special case, and perhaps as the general case, depending on one's views). We will look at that topic in more detail in Sections 4.2 and 4.3. The topics collected by our title are linked by subject matter as being applications of modal logic other than temporal applications and other than applications which are epistemic in the broad sense which subsumes doxastic logic.

We could make the latter (epistemic/doxastic) category even broader by speaking of propositional attitudes in general. This would then include such lesser known areas as *boulomaic* logic, to use the terminology of Kenny [647], – also often called bouletic logic – in which the intended reading of $\Box A$ is that some fixed subject *wants it to be the case that* A. The possibility of taking a modal perspective in this case is mentioned already in von Wright [1206]. Such a perspective need not attend too much to the particular choice of verb – *want, desire, wish* – but would be at its most attractive if desire-attributions are amenable to expression with a sentential complement whichever verb is chosen, with wanting a car, for instance, being taken as amounting to desiring that one own(s) a new car. (See Sinhababu [1057], as well as Section II of Humberstone [561] for some complications. Even then, note that another complication built into the use of "one" here is a *de se* element we steadfastly ignore: see Remark 5.2.7, p. 379.)

The association of propositional attitudes in this way with the interpretation of the $\Box$ operator of normal modal logics does not extend to the dual $\Diamond$ operator. For example in doxastic and epistemic logic in which the $\Box$ is usually written B and K respectively (for: so-and-so $\underline{B}$elieves or $\underline{K}$nows that such and such) are generally spoken of as attributing propositional attitudes, give or take philosophical qualms about the existence of the would-be propositions concerned (see note 372, p. 421, for example), but the corresponding $\Diamond$ operators – which we follow Hintikka [490] in writing as C and P – do not attribute propositional attitudes. That is, the absence of belief and the absence of knowledge are not propositional attitudes toward what is not believed or known, in the way that belief and knowledge are attitudes toward what is believed or known. (*Not believing* is relevant here because C is to represent "the subject does not have the belief that not such-and-such". *Disbelieving*, on the other hand, is another matter, a perfectly straightforward propositional attitude, represented by $B\neg(\cdot)$.) This will be an occasionally recurrent theme in Chapter 5, when the logic and knowledge and belief are the principle objects of investigation.

So: one way of classifying applications of modal logic is in terms of whether the $\Box$ operator is primarily intended for the attribution of propositional attitudes or not. There is another equally natural way of classifying these areas which looks at the plausibility or otherwise of the **T** schema for the intended application. On this taxonomy, doxastic and deontic logic belong together – where for each of which weakenings of **T** such as **D** and **U** have been proposed – while epistemic logic (now in the narrow sense) and alethic modal logic belong together, in not calling for any such weakening. Of course there is no question of either classification being better than the other. The resulting cross-classifications are depicted in Table 4.1.

	VERIDICAL	NON-VERIDICAL
NON-ATTITUDINAL	alethic (inc. nomic)	deontic
ATTITUDINAL	epistemic	doxastic

Table 4.1: Two Ways of Classifying Applications

In Table 4.1 the term *veridical* is used to indicate the appropriateness of the **T** axiom for the application in question (i.e., for the intended informal interpretation of □); other terms used for this in the literature include *truth-entailing* and *factive*, though the latter term had originally a somewhat more specialized meaning (see Kiparsky and Kiparsky [658]), as is mentioned in the discussion on p. 352. (While that "factive"/"veridical" contrast concerns the contested distinction between entailment and presupposition, there are several other such terminological contrasts in the literature also. For example, Sharvit [1044] uses the factive/non-factive contrast specifically for contexts embedding declarative sentences, freeing up veridical/non-veridical for other contexts, writing in note 1: "So, for example, *tell* is non-factive when it takes a declarative complement, and veridical when it takes an interrogative complement."[162]) Still another use of the 'veridicality' terminology can be found in Giannakidou [372], esp. p. 114*f*., in which *believe* is said, like *know*, to create a veridical context, while *want* is said to create a nonveridical context, even though false beliefs are just as much a fact of life as unsatisfied desires. (More salient than anything deserving to be called veridicality here, is the contrast in respect of 'direction of fit': beliefs are at least putative representations of how the world is and when they fail to fit the way it is end up as misrepresentations, while desires are in the business of getting the world to fit *them*, rather than the other way around. But this is all highly metaphorical. See further Remark 5.1.17 on p. 356 below.) The description, appearing parenthetically in Table 4.1, of nomic necessity – physical necessity, if all natural laws are physical laws – as veridical has been contested: see note 177 (p. 219).

REMARK 4.1.1 Indeed, even the veridicality of knowledge ascriptions has been contested Hazlett [466], on the basis of what others would be regard as highly marginal 'inverted commas' uses of the word *know* – "I *knew* I was going to die in that collision" – and the like. (See the index of Hare [453] under 'inverted commas' as to the use in question. This terminology was also used in the present connection by Hintikka: see note 274 on p. 333 below.) A somewhat more precise description is given by Richard Holton's talk of *protagonist projection* in [519] (see also Hazlett [466], [467]); in the collision example just given the protagonist a prior temporal stage (to put the matter contentiously, for the sake of brevity) of the speaker. For further discussion, see Buckwalter [124]. Zangwill [1218] asks "Does knowledge depend on truth?" and returns a negative answer – but this turns out not to be a denial that only what is in fact the case can be known to be the case, and so does not bear on the issue of veridicality. This could equally well be put by saying that there is not denial here of the following: the claim that only true propositions can be known. Quite a fuss has been made of the explicit presence of the word "true" in such formulations.

---

[162]Thus "Joan told Toby that she had been at the cinema", unlike "Joan informed Toby that she had been at the cinema", does not require for its truth that Joan had been at the cinema, whereas, "Joan told Toby where she had been" requires for its truth that wherever she told Toby she had been, she really had been there. For the record, there has been some disagreement about this issue: see Holton [519] and Tsohatzidis [1115].

4.1. INTRODUCTION

See Anderson [23] and Casullo [164], both of which discuss this issue as it was raised for epistemology by Keith Lehrer. However the considerations involved here arise in theorizing about truth in general – 'redundancy' vs. other theories of truth – rather than being of any distinctively epistemological interest. ◄

By 'attitudinal' in Table 4.1 is meant: intended for the ascription of propositional attitudes. One could also subsume under this description also mechanisms simulating those in natural language suited to the expression rather than the ascription (even the self-ascription) of such attitudes; in the epistemic case, such mechanisms in English include adverbs (such as *perhaps*) and modal auxiliaries (such as *may, might, must*); except for occasional asides, as in Example 4.6.15 (p. 300), our attention will be on attitude-ascription rather than on such attitude-expressing devices. Only in the non-attitudinal, deontic case is the intended gloss for the $\Box$-operator (often written O in this case) given by a modal auxiliary-like element, namely *ought*[163] (or *should*). As well as this genuinely normative interpretation of the operator, deontic logic is generally regarded as subsuming norm-*reporting* constructions Op interpreted as saying that according to this or that moral (or indeed legal) code, it is obligatory that p.

In the present chapter we are concerned with the non-attitudinal cases, the first row of Table 4.1, passing to the second row in Chapter 5. The reason for this division of labour, as opposed to one treating the first *column* of the table in one chapter and the second in another, is that the involvement with propositional attitudes exemplified by the entries in the second row raises a number of distinctive issues.[164] Prominent among them is what we shall be calling the concept-possession issue (see Section 5.6): it would seem, at least *prima facie*, that propositions – speaking now informally (see note 372, p. 421) – involve concepts and no attitude can be taken to a proposition by one not possessing the concepts that proposition involves.[165] No corresponding issue arises for the non-attitudinal interpretations of $\Box$. Other issues arising only in the attitudinal case involve logical omniscience and introspective capacities, also discussed in Chapter 5.

We begin by addressing a simplification of the Kripke semantics which is, the above remarks notwithstanding, relevant to both the non-veridical case of obligation – our principal official business here – and to non-veridical case of belief (in connection with which it will be revisited in Section 5.4 and elsewhere in Chapter 5). We will call this: *semi-simplified* Kripke semantics.

In Section 2.2 we presented what was spoken of as – and can indeed be regarded in retrospect (from the perspective of Section 2.3) as being – a simplification of the Kripke semantics with its models $\langle W, R, V \rangle$. With these models the truth of $\Box A$ at a point $w \in W$ depends on the truth of A throughout the set $R(w)$, which can of course vary from one w to another. With the simplified models $\langle W, V \rangle$ the truth of $\Box A$ at w depended on the truth of A over the whole of W, so the $\Box$-pertinent set of points (as it was put on p. 59) is now, by contrast, constant: it does not depend on the x in question, the point at which we are considering the truth of $\Box A$. The simplification, then, does two things at once: (1) it make the $\Box$-pertinent set constant, and (2) it does so by making this set universal. Here we will consider doing the first things without doing the second. (The reverse combination is not available.) A repercussion of (2) is that instances of the **T** schema become unfalsifiable on the simplified semantics. Accordingly, retaining

---

[163] In subsequent discussion we subsume *ought* among the modal auxiliaries although it is not a full-blooded auxiliary verb, in that it takes a following *to* rather than the bare infinitival form.

[164] As the boulomaic case mentioned above shows, these are indeed only examples – the most thoroughly investigated examples in the literature, to be sure. There are many further cases involving, by contrast, the same 'direction of fit', to use the terminology explained informally before Remark 4.1.1, as that of knowledge and belief: *seeing that p* and *remembering that p*, for instance, treated respectively in Hintikka [491] and Aho and Niiniluoto [6]. Many non-attitudinal applications are also absent from Table 4.1, such as the case of tense logic (see Section 3.1); for the intended readings of G and H we have here two non-veridical $\Box$-operators, in that the **T** axioms would not be wanted. Some may doubt the appropriateness of the term *non-veridical* in this case, in its conflation of truth with present truth, while others will think it highly appropriate on these same grounds, perhaps reformulated with "identification" for "conflation": e.g., Prior [927].

[165] This will have to be modified if normality is to be retained, since from propositions not involving a given concept others will follow which do involve that concept. The latter may be deemed to be objects of propositional attitudes (such as knowledge and belief) held implicitly, in virtue of those explicitly held, with the concept-possession condition proper taken to apply to the explicit attitudes; see Remark 5.6.8, p. 428.

(1) while jettisoning (2), what we have in mind as *semi-simplified Kripke models* will be models in which the set of □-pertinent worlds is constant, but need not comprise the set of *all* worlds in the model. Accordingly such models may provisionally be taken to have the form $\langle W, W_0, V \rangle$ in which $W_0 \subseteq W$ is the constant set in question. (We require as usual that $W \neq \varnothing$, but have not imposed this requirement on W_0.) If $\mathcal{M}$ is such a model, we modify the clause in the definition of truth given in Section 2.2 in the following way, which is to be understood as prefaced by "for any formula A, and any $w \in W$":

$$\mathcal{M} \models_w \Box A \text{ iff for all } x \in W_0, \mathcal{M} \models_x A.$$

(In Section 2.2, we had, instead, "for all $x \in W$" rather that "for all $x \in W_0$".) As in the Kripke semantics and in the simplified Kripke semantics, a logic is *sound w.r.t.*, *complete w.r.t.*, or *determined by* a class of these models when its theorems comprise respectively, *only*, *all*, or *precisely* those formulas true throughout W for each model $\langle W, W_0, V \rangle$ in the class. Note that just as there is a one-to-one correspondence between simplified Kripke models and pointwise equivalent Kripke models with universal accessibility relations, there is also a one-to-one correspondence between semi-simplified Kripke models $\langle W, W_0, V \rangle$ and pointwise equivalent Kripke models $\langle W, R, V \rangle$ with $R(\cdot)$ constant: given $\langle W, W_0, V \rangle$, put $xRy \Leftrightarrow y \in W_0$ (all $x, y \in W$); conversely given $\langle W, R, V \rangle$ with $R(\cdot)$ constant, put $W_0 = R(x)$ for an arbitrarily selected $x \in W$. (Recall the notion of pointwise equivalence of models as defined in note 132, p. 161.)

The semi-simplified models may be thought of in the following way. W represents the set of all possible worlds, while W_0 represents the set of, in the deontic case, ideal worlds, in the sense of worlds in which whatever ought to be the case is the case, and in the doxastic case, belief-compatible worlds, in the sense of worlds in which whatever our cognitive agent believes to be the case is the case. But this phrasing will seem very loose: is the deontic gloss intended to mean "worlds x in which whatever ought, in w, to be the case, is the case" or instead "worlds x in which whatever ought, in x, to be the case, is the case", and, in the former case, what is w? The answer to this question – or the corresponding doxastic question – is that the semi-simplified semantics nullifies the distinction on which clarification is being asked: a formula of the form $\Box A$ is true at one point in one of these models just in case it is true at all points. Minimally adapting the notion of modal invariance from the simplified Kripke semantics (see p. 56) to the present setting gives us the idea of a formula A as modally invariant (w.r.t. the semi-simplified Kripke semantics) just in case for every model $\mathcal{M} = \langle W, W_0, V \rangle$ and any $w, x \in W$, $\mathcal{M} \models_w A$ iff $\mathcal{M} \models_x A$. (Note that here for simplicity of exposition we ignore the fact that the relevant set of worlds may depend on a temporal parameter, too, as is mentioned in Remark 4.4.6 – p. 243 below.) Further, such formulas display the key feature they had in the simplified semantics, given in Lemma 2.2.4: that of including all the fully modalized formulas. Thus this feature is revealed as due to the constancy rather than the universality of the accessibility relation (to put it in terms of the Kripke semantics proper):

EXERCISE 4.1.2 Show that all fully modalized formulas are modally invariant in the semi-simplified Kripke semantics. ✠

The apparently sloppy formulation above, then, in terms of worlds at which whatever ought to be the case is the case, or whatever is believed to be the case is the case, is perfectly in order in the context of the present semantics. This could, of course, itself be grounds for rejecting the semantics in favour of the more general Kripke semantics with accessibility relations and thus non-constant □-pertinent subsets. But for the moment, let us see what logic the semi-simplified version of the semantics leads to:

PROPOSITION 4.1.3 (*i*) *The modal logic determined by the class of all semi-simplified models is* **K45**.
(*ii*) *The modal logic determined by the class of all semi-simplified models* $\langle W, W_0, V \rangle$ *in which* $W_0 \neq \varnothing$ *is* **KD45**.

Proof. Soundness is easily checked, and completeness can be obtained from the familiar completeness results by use of generated submodels. Coro. 2.4.3 and Exercises 2.4.6 and 2.4.8 give us that **K45** is complete w.r.t. the classes of Kripke models whose accessibility relations are transitive and euclidean. So any non-theorem A is false at a point w in such a model. The universe of the submodel of that model generated by w, at which A is still false in this submodel, is $R(w) \cup \{w\}$, or rebaptizing the restriction of R to this submodel R_0, $R_0(w) \cup \{w\}$. In view of the remarks preceding the statement of the present result, it remains only to check that $R_0(\cdot)$ is constant: for all x, y, in the generated submodel, $R_0(x) = R_0(y)$. It suffices to show the $\subseteq$ direction of the equality, as x and y are on a par here. So suppose $z \in R_0(x)$, with a view to showing $z \in R_0(y)$ (equivalently, given that $z \in R_0(x)$: yRz). Since $z \in R_0(x)$, xRz. Since $x, y \in R(w) \cup \{w\}$ there are four possibilities:

$$(1)\ x, y \in R(w); \quad (2)\ x \in R(w), y = w; \quad (3)\ x = w, y \in R(w); \quad (4)\ x = w = y.$$

In case (1), since wRx and (from before) xRz, by transitivity we have wRz, so since also wRy, by the euclidean condition, yRz. For case (2) we have wRz as in the previous case, so on the further assumption that $y = w$ we again get the desired conclusion: yRz. Cases (3) and (4) are equally straightforward and are left as an exercise (below), which are all that remains for establishing part (i) of the present result; to show (ii) note that if R is serial, then $R_0(\cdot)$ is not only a constant function on the universe of the point-generated models, but also has for its value a non-empty subset of W_0. ∎

EXERCISE 4.1.4 *Complete the above proof by covering the remaining two cases (i.e., (3) and (4)) of (1)–(4).* ✠

A mild variation on the proof of the Generation Theorem (Thm. 2.4.10, p. 70) shows that Proposition 4.1.3 remains correct (both (i) and (ii)) if we restrict attention to models $\langle W, W_0, V \rangle$ with $W \smallsetminus W_0$ contains at most one element (and $W_0 \neq \varnothing$ for part (ii)), since in evaluating a formula $\Box A$ at $u \in W \smallsetminus W_0$ we a directed at once to points in W_0 rather than to other points in $W \smallsetminus W_0$. It does no harm to require that there is exactly one point in $W \smallsetminus W_0$ (see Example 4.6.5, p. 295), and we can then treat this point w, say, as something like the distinguished element of in the models mentioned in note 45, except that here we discard the accessibility relation of the model, and more importantly, if we denote the resulting model as $\langle W_0, w, V \rangle$, we do not require that $w \in W_0$.[166] Bringing accessibility back in, note that the submodels of transitive euclidean model, generated by an element w thereof has for its universe $W_0 \cup \{w\}$ where $W_0 = R(w) = R^n(w)$ for all $n \geq 1$, W_0 being non-empty for serial R. For the following reformulation of Prop. 4.1.3, we can understand A's being true in the model $\mathcal{M} = \langle W_0, w, V \rangle$, either as meaning that $\mathcal{M} \models_w A$ or as meaning that for all $x \in W_0 \cup \{w\}$, $\mathcal{M} \models_x A$. This makes no difference to the class of formulas true in all models (all serial models):[167]

COROLLARY 4.1.5 *The logics determined by the class of all models $\langle W_0, w, V \rangle$ as just defined, and by the class of all such models with $W_0 \neq \varnothing$ are respectively **K45** and **KD45**.*

Note that if, from a Kripke frame for **KD45** with accessibility relation R, we select a point w and consider the subframe generated thereby we get, not, as in the case of **S5** a frame on which the restriction of R is the universal relation – and thus essentially the 'frame' part of the models of the simplified semantics – but w together with a non-empty set of points accessible to w and all accessible to themselves and each other. The latter set is what is parcelled up into the W_0 part of the models just considered (with w as the distinguished point). If we drop **D** we no longer have the assurance that this set is non-empty.

COROLLARY 4.1.6 *For any formula A, $\vdash_{\mathbf{K45}} \Box A$ if and only if $\vdash_{\mathbf{S5}} A$.*

---

[166] Note the contrast with the usual models (on frames) with distinguished elements, as described on p. 190.
[167] Again, we refer to Example 4.6.5.

Proof. Only if: The axiom **T** will take us from any **K45** theorem $\Box A$ to an **S5** proof of A.

If: Suppose $\Box A$ is not **K45**-provable. Then $\Box A$ is false and the generating point of a semi-simplified model and so A is false at some point accessible to that point in W_0 if the original model is thought of as $\langle W, W_0, V \rangle$ in the style of the discussion after Exc. 4.1.4. But this is an **S5**-model, so A is not **S5**-provable. ∎

As noted before Remark 2.8.1 on p. 142, this can be thought of as a faithful translational embedding of **S5** into **K45**. The same argument establishes the same result for **KD45**; from this we see – though it would be evident enough on reflection without consideration of these embeddings – that **K45** and **KD45** have the same theorems of the form $\Box A$.

Both logics have some claim on our attention as candidate doxastic or deontic logics. **KD45**, in particular, is often thought of as a simple-minded such candidate: the set W_0 of the resulting models being capable of being construed as the set of worlds compatible with the agent's beliefs in w (i.e., in which all those beliefs are true) or the set of worlds permissible relative to w (i.e., in which every obligation obtaining in w is fulfilled). A word about the label **KD45**: no part of this label is redundant. In Exercise 2.1.15, and at p. 88, we noted that **KT45** = **KT5** = **KTB4** = **KDB4**, these labels evoking various axiomatizations of **S5**), the last equation reflecting the fact that we can weaken **T** to **D** without loss in the context of (**K** and) **B** and **4**. (Semantically: amongst symmetric relations, those which are reflexive are precisely those which are serial – namely equivalence relations, under either description.) But, weakening **T** to **D** in the **KT45** axiomatization gives the strictly weaker logic **KD45** with which we have been concerned here. Because of its importance for later developments, we include the first of the following two exercises, the second taking us back to another variation on the semi-simplified theme.

EXERCISE 4.1.7 (*i*) Show that $\Box \Diamond p \to \Diamond p$ (= $\mathbf{5}_c$) is provable in **KD4** (and hence in **KD45**).

(*ii*) Show that $\Box \neg (\Box p \wedge p) \to \neg \Box p$ is provable in **KD45**. Is this formula provable in **KD4**? ✠

EXERCISE 4.1.8 Let us consider a double-barrelled version of the semi-simplified semantics, with models $\langle W, W_0, W_1, V \rangle$, with $W_0, W_1 \subseteq W$. (More generally we could consider models $\langle W, W_0, W_1, \ldots, W_n, V \rangle$, but this will do.) We have a bimodal language with operators $\Box_0, \Box_1$, semantically interpreted in these models $\mathcal{M}$ by means of these clauses in the definition of truth:

$$\mathcal{M} \models_w \Box_i A \text{ iff for all } x \in W_i, \mathcal{M} \models_x A \quad (i = 0, 1).$$

Have a go at axiomatizing the set of formulas determined by the class of all such models with this definition of truth in force. ✠

Here is a reformulation of Coro. 4.1.5 in the notation for models just used:

COROLLARY 4.1.9 *The logic* **K45** *is determined by the class of semi-simplified Kripke models* $\langle W, W_0, V \rangle$ *with* $|W \smallsetminus W_0| \leq 1$, *and similarly for* **KD45** *with the additional condition that* $W_0 \neq \varnothing$.

As already intimated, we can, for both cases (**K45** and **KD45**) put "$|W \smallsetminus W_0| = 1$" here in place of $|W \smallsetminus W_0| \leq 1$: see Example 4.6.5.

Variations on this theme will arise in Sections 5.3 and 5.4 below, during the course of our discussion of specifically epistemic and doxastic matters. We resume the deontic theme in Section 4.4. This section concludes instead with an illustration of the application of tense-logical considerations to non-temporal subject matter, and our illustration concerns alethic modality – in particular, metaphysical necessity and possibility (notions which will attract less attention in what follows, though they do come up in Sections 4.2, 6.5). Before giving that illustration a general remark on necessity and possibility is in order.

4.1. INTRODUCTION

REMARK 4.1.10 We are assuming in what follows, as elsewhere, that it is reasonable to read $\Box$ and $\Diamond$ as "It is necessary that" and "It is possible that" and to take them as subject to the usual interdefinability-underlying equivalences (of $\Box A$ with $\neg\Diamond\neg A$, and of $\Diamond A$ with $\neg\Box\neg A$. Prior [919] (also [924] p. 154ff.) developed a system of modal predicate logic ("Q") which rejected these equivalences, thinking of Fa's being necessary as being true in all worlds and of that as requiring the existence in all worlds of (the denotation of) a. So if in some world a fails to exist, $\neg\Box Fa$ will be true at any world – for simplicity taking accessibility as the universal relation – which falls short of what we require for the truth of $\Diamond\neg Fa$, namely the existence of a world at which a exists and fails to be F. Early technical papers on this logic include Bull [125] and Segerberg [1023]; for further discussion, see §7 of Fine[292], Menzel [800], and for some brief introductions, Hughes and Cresswell [535], p. 303ff. and 4.2.1 in Menzel [801]. A rather different kind of opposition to taking $\Box$ and $\Diamond$ as dual is urged in Goswick [403] (also [404]). Goswick would like to leave as an option a view according to which some (or maybe even all) objects are sufficiently 'thin' as to lack non-trivial modal properties altogether, having neither powers nor essences, so that if a is such an object, both $\Diamond Fa$ and $\Box\neg Fa$ would be false (supposing in the former case that Fa is not itself true, which would make $\Diamond Fa$ trivial rather than distinctively modal). She urges that modal logic should not itself rule out such a metaphysical position, as it would if it claimed that $\Box\neg Fa$ followed logically from $\neg\Diamond Fa$. ◂

EXAMPLE 4.1.11 Consider the composite view, to be found in Karmo [642], that (1) identity statements, when true, are true of metaphysical necessity, but (2) there can be non-identity statements which are only contingently true. The justification for part (a) of the view comes from the idea that for a to be identical with b, whatever is true of a must be true of b, so since being something necessarily identical with a is true of a, this must also be true of b. In other words from

$$a = b \to (\Box(a = a) \to \Box(a = b))$$

we may detach the middle antecedent and conclude that $a = b \to \Box(a = b)$, i.e. (1), on an appropriate reading of $\Box$. (This argument was much discussed in twentieth century philosophy, and was defended by many, including perhaps most famously in Kripke [672], and attacked by others.) In fact, on any reading of $\Box$ according to which for some binary relation R between worlds $\Box A$ is true at a world just in case A is true at all worlds related by that relation to that world,[168] the corresponding version of (1) will be correct. For a will be identical with a at all suitably related worlds and so if a and b share all their properties, b will be identical with a at all such worlds. (You may worry about worlds in which a or b does not exist, in which case restrict all such quantifications over worlds to worlds in which a and b do exist.[169]) Part (2) of the view is based on examples, such as the following (from [642]): Suppose a is a tunnel started from one side of a mountain and b a tunnel started from the opposite side (at the same altitude), but the money runs out, work stops, and the tunnels never join up. It seems that $a \neq b$ while $\Diamond(a = b)$, since it would have ended up being true that $a = b$ if work had continued to completion, which it perfectly well could have. Thus unlike true identities, which must be necessarily true (à la (1)), here we have a true non-identity statement ($a \neq b$) which is not necessarily true, establishing (2).

Now, if we accept **S5** as the correct logic of the modalities in play here, then the composite position just described is untenable, since from $a = b \to \Box(a = b)$ (part (1) of the view) we get $\Diamond(a = b) \to \Diamond\Box(a = b)$ by the monotone property for $\Diamond$ (which we continue to regard as abbreviating $\neg\Box\neg$), which, by the following instance of the **S5**-endorsed **B**-schema (in its dual form here):

$$\Diamond\Box(a = b) \to a = b,$$

---

[168] R is an accessibility relation for $\Box$, as it would be put in expounding the Kripke semantics for a formal modal language.
[169] The status of the names "a" and "b" here would also come under discussion in a full treatment of these issues: are they (weakly) rigid designators, or not? All these details are somewhat beside the point. Our concern is not with the plausibility of the view that identity holds necessarily or of the view that some non-identities are contingent, but with the project of trying to defend the cotenability of these views by weakening one's modal logic.

yields the conclusion: $\Diamond(a = b) \to a = b$, contradicting part (2) of the position. In terms of the tunnels example, if it is possible for tunnels a and b to be identical, they are identical. The situation as described above is not a possible case.

Karmo's reaction in [642] to this objection is to take it as showing that we should not endorse **S5** as a logic of metaphysical necessity and possibility, and in particular that we should reject the crucial **B** schema. (We could thus embrace a weaker logic, **S4** for example, free of this objectionable feature.) Similarly, thinking in terms of the ("applied") Kripke semantics, we should reject the demand that the accessibility relation for a metaphysically interpreted "□" should be symmetric: even if a world in which tunnels a and b are one is possible relative to a world in which they are two, we should not infer that a world in which they are two tunnels is possible relative to a world in which they are one.

The *alethic* tense-logical version of this argument – not a *temporal* tense-logical analogue, already discussed and (according to what follows, erroneously) claimed not to be parallel in Karmo [642] – shows that this reply cannot work. Just take the proposed accessibility relation R of relative metaphysical possibility and consider its converse R^{-1}. The reasons for endorsing $a = b \to \Box(a = b)$, basically that $\Box(a = a)$ is a correct principle, spread across to endorsing the following (since $\Box^{-1}(a = a)$ is equally undeniable):

$$a = b \to \Box^{-1}(a = b).$$

As before, we can proceed to

$$\Diamond(a = b) \to \Diamond\Box^{-1}(a = b)$$

and hence using what is now a tense-logical bridging principle justified not by the symmetry of R but by the fact that R^{-1} is the converse of R, namely $\Diamond\Box^{-1}(a = b) \to a = b$:

$$\Diamond(a = b) \to a = b.$$

In other words, the tension between (1) and (2) in the composite position needs more than a binary relation failing to coincide with its converse (a symmetric relation), but the far more dubiously intelligible idea of such a relation's lacking a converse altogether. (This reaction to Karmo [642] is that of Humberstone [551].) ◀

The discussion here will ring a few bells. The crucial step using **B** in the original argument is essentially an application of the MacIntosh rule (see the Digression following Exercise 2.5.54 on p. 98):

$$\frac{A \to \Box A}{\Diamond A \to A}$$

Indeed the necessity of identity and non-identity was one of the motivating cases behind the general idea of MacIntosh logics. (See the sources cited in our earlier discussion.) This rule is intimately associated with the **B** schema; in fact in Coro. 3.1.7 (p. 181), we noted that **KB** is precisely the least (mono)modal logic closed under the rule:

$$\frac{A \vee \Box B}{\Box A \vee B}$$

which has the MacIntosh rule above as a special case, putting $\neg A$ for A and A for B (as well as rewriting a disjunction as the corresponding implication with first disjunct as negated antecedent; in the earlier discussion we were using $\varphi, \psi, \ldots$ as schematic letters because of the proliferation of tense operators in the vicinity). Before we met this latest rule in Section 3.1, we had already made the acquaintance of its tense-logical analogue, re-notated here and separated into its the two directions of the (two-way) GH rule given there (p. 180); there has also been a switch in the order of disjuncts in the second rule here:

4.1. INTRODUCTION

$$\frac{A \vee \Box B}{\Box^{-1} A \vee B} \qquad \frac{A \vee \Box^{-1} B}{\Box A \vee B}$$

Rephrasing these rules in terms of implication:

$$\frac{A \to \Box B}{\Diamond^{-1} A \to B} \qquad \frac{A \to \Box^{-1} B}{\Diamond A \to B}$$

we can see that it is specifically the second of these rules that takes us from

$$a = b \to \Box^{-1}(a = b).$$

to:

$$\Diamond(a = b) \to a = b.$$

The first of the two rules inset above is not needed. In terms of the original presentation of this derivation, without recourse to either of these rules, we appealed to the bridging axiom $\Diamond\Box^{-1}(a = b) \to a = b$, or, in Prior's tense-logical notation, an instance of the schema $FH\varphi \to \varphi$ (the dual of the GP-axiom, or "GP-ax." from Section 3.1), but not to its mirror image $PG\varphi \to \varphi$. The moral of this, in terms of the reply to Karmo given in Example 4.1.11 above, is that for two $\Box$-operators with $\Box_1$ and $\Box_2$ interpreted by accessibility relations R_1 and R_2 respectively, the passage from $a = b \to \Box_2(a = b)$ to $\Diamond_1(a = b) \to a = b$ (alternatively: to $a \neq b \to \Box_1(a \neq b)$) requires only that R_1 be *included in* the converse of R_2, not that it should coincide with the converse of R_2. Thus, where W is the set of all worlds, in between which we are thinking of R_1 as the relation of relative metaphysical possibility (with R_1 representing genuine metaphysical necessity), the moves in Example 4.1.11 can be made with any relation R_2 such that $R_1^{-1} \subseteq R_2 \subseteq W \times W$.[170] The discussion in 4.1.11 took R_2 from the lower end of this range; Williamson [1193] runs a similar argument taking R_2 from the top end (the universal relation). We will be shifting attention presently from logically interesting aspects of metaphysical necessity to logically interesting aspects of nomological necessity, but close with a brief look at two topics before doing so. The first (Example 4.1.12) illustrates the availability of alternatives to semantics in terms of possible worlds with one such alternative which applies in the case of nomological as well as for metaphysical necessity, and indeed even more pressingly perhaps for doxastic, epistemic and perhaps also deontic applications of modal logic. The second topic picks up a issue from Remark 1.3.9(*i*): the concept of necessitativity.

EXAMPLE 4.1.12 Instead of thinking of the entities w.r.t. which formulas are evaluated for truth (in a model) as possible worlds, fully specifying for each formula whether it is to count as true or false, we could work with something less determinate. In Humberstone [545] this is treated as somewhat analogous to a transition from considering spatial points to considering spatial regions, and what replace the worlds are there called *possibilities*, thought of as finitely specifiable regions of logical space, and, for the sake of those sceptical about the intelligibility of fully specific non-actual worlds, taken as primitive in their own right rather than as sets of worlds ('points' in logical space). Such regions come with a mereological structure, a given possibility having more and more specific subregions ('refinements'), and suitable clauses in the inductive definition of truth relative to a possibility given so that classical propositional logic is preserved. For example, with negation, one would not want to have $\neg A$ true w.r.t. a possibility just in case A is not true w.r.t. it: rather one would want to require that A is not true w.r.t. any refinement of that possibility. This has the flavour of negation as treated in the Kripke semantics for intuitionistic logic, so conditions on the models are needed to recover the desired classical behaviour. (What if we don't want to recover classicality, though? See Rumfitt [979].) The 'partial specification' idea is that propositional variables (sentence letters) are allowed to be true, false or neither relative to a possibility, and to do justice to the motivation sketched above, one could impose the additional requirement that relative to

---

[170]Here we use the fact that $R \subseteq S^{-1} \Leftrightarrow R^{-1} \subseteq S$.

any possibility, at most finitely propositional variables are in the true or false category. (A somewhat different presentation of the semantics, without such 'truth-value gaps' is given in subsection 6.44 of Humberstone [594].) While [545] gives soundness and completeness proofs in terms of the broader class of possibilities models without this additional 'finite specification' requirement in force, Holliday [514] has more recently shown how to obtain such stronger results. (See also Holliday [516], which reveals *inter alia* a special role for the 'converse modalities' $\Box^{-1}$ in play above in obtaining such stronger results, via a property of logics called by Holliday internal adjointness: see Definition 7.3 pf [516].) Incidentally, the way the transition from worlds to possibilities is described in [545] is not, as above, in terms of a spatial analogy so much as a temporal one, with the discussion presenting itself as the alethic modal analogue of the tense-logical move in Humberstone [544] from evaluation relative to temporal instants to evaluation relative to temporal intervals, with special emphasis on the motivation of those not prepared to believe in the existence of unextended points of time. It should be acknowledged that there is a widespread intuition that the modal and temporal cases are not on a par because we do not think of a possible world as a mere vehicle for things to be the case relative to, susceptible of having had things being otherwise w.r.t. it, in the way that we think of a moment of time (or a spatial point, for that matter). One could retain this intuition for the non-punctiform semantics, holding there to be a similar contrast between possibilities and intervals; on the other hand – see p. 588 of Humberstone [582] for discussion and references – it turns out not to be as easy as one might think to spell out the intuitive disanalogy here gestured toward. (Warning: the contrast between possible worlds and possibilities drawn by David Lewis in order to accommodate *de se* propositional attitudes in, for example, [719] and [730] – in the case of the latter, especially 'possibilities *de se et nunc*', p. 552 – is not related to the current worlds/possibilities contrast.) ◀

Note that although Example 4.1.11 involved identity as a relation between individuals and hence transcended our usual policy of restricting the discussion to modal *propositional* logic, little has been (or will be) said about modal predicate logic with quantifiers over individuals, except in passing – for instance in the course of Example 2.5.59 (p. 101). Several metaphysical issues for which there was until recently something of a consensus as to how they should be handled in quantified modal logic have found themselves re-opened for examination. For example, the general presumption that the domains over which quantifiers are taken to range should be allowed to vary from world to world in a model in order to accommodate the phenomenon of contingent existence has been challenged in the work of Timothy Williamson ([1198] and elsewhere, a similar view having been defended by B. Linsky and E. Zalta[171] on the grounds that this phenomenon is entirely illusory.[172] A second such example is concerns the notion of an essential property of an individual, widely taken to be a property possessed by that individual in every world in which that individual exists (or – a modification for the 'no contingent existence' position – every world in which that individual is 'concrete' – or, as it has also been put – 'has physical presencee'[173]). Kit Fine has been challenging this consensus position by suggesting that more is required for being an essential property of an individual than this, with such examples as this, which presumes that a set with given elements exists in precisely the worlds in which all those elements exist: the set whose sole element is the planet Earth exists in precisely the same worlds as the plant Earth itself exists, but while it is an essential property of the set that it has the Earth as an element, it is not an essential property of the

---

[171] For references to (and discussion of) this work see Hayaki [463] – and also the still earlier defence, Cresswell [212], of the Barcan Formula, mentioned above at p. 57.

[172] Not that there was ever quite so robust a consensus as this might suggest, since David Lewis always held that the domain should not only be allowed but in fact required to vary from world to world, as the different worlds' domains should be disjoint, and what the alternative view thinks of as the same individual in two worlds is best thought of an individual and a counterpart of that individual. See papers 3 (with Postscripts) and 4 in Lewis [724], as well as [726] on modal metaphysics more generally. Some useful historically orienting remarks on these matters can be found near the start of Williamson [1200].

[173] Levesque and Lakemeyer [715], p. 56: "The approach we take here is that properties of objects may indeed change from world to world, including perhaps having a physical presence of some sort, but that there is only one fixed universal set of objects (possibly without physical presence) to begin with."

4.1. INTRODUCTION

Earth that it belong to this set. See Fine [295], [296], Correia [195], Giordani [376].

We conclude the present discussion by looking at one way of cashing out the idea of necessitativity, mentioned in Remark 1.3.9(*i*), together with a discussion of a similar idea in an early publication by Jonathan Bennett.

EXAMPLE 4.1.13 One suggested way of cashing out necessitativity is as being strictly equivalent to something. We are interested in doing this for modal formulas and so in the first instance this seems to commit us to using propositional quantifiers, something we prefer to avoid in the interests of simplicity. Suppose we think (perhaps somewhat artificially) of necessitativity in terms of a 1-place connective, here to be written as $\mathcal{N}$, with $\mathcal{N}A$ defined to be: $\exists q \Box(A \leftrightarrow \Box q)$. This says that A is necessarily (or 'strictly') equivalent to some necessity claim or other; the variable q should be chosen to be one not occurring A to avoid spoiling this intended reading. ($\mathcal{N}A$ can be thought of as something like an attempt to bring the metalinguistic claim – see Example 2.1.2, p. 36 – that $A \in Fm_S^\Box$ into the object language of a modal logic S.) Now it turns out to possible to replace this proposed definition, on minimal assumptions about how the existential propositional quantifier "$\exists q$" behaves, for a wide range of modal logics, including many for which the reading of $\Box$ as "it is necessary that" would not be appropriate. In particular, we do not need to restrict attention to veridical modal logics – those extending **KT**. For logics S in the range in question, extended by a quantifier "$\exists$" binding propositional variables and satisfying the conditions ($\exists$I) and ($\exists$E); we use the notation "$C(p_i)$" and "$C(A)$" as introduced before Proposition 1.2.7 (and appearing within Prop. 1.2.7(*ii*)) on p. 13. An occurrence of p_i not in a quantifier prefix $\exists p_i$ and not bound by any such quantifier within a formula is said to be a *free* occurrence in that formula. Closure under uniform substitution is still assumed but with substitutions replacing only the free occurrences of a given propositional variable in the formula to which it is applied, and also not in such a way that no occurrence of a variable free before its application becomes bound after it.

($\exists$I) $\vdash_S C(A) \rightarrow \exists p_i C(p_i)$

($\exists$E)   If $\vdash_S C(p_i) \rightarrow B$ and p_i has no free occurrence in B, then $\vdash_S \exists p_i C(p_i) \rightarrow B$.

The "I" and "E" are meant to recall the Introduction and Elimination rules of natural deduction systems (as in Chapter 7). The scope of the "$\exists p_i$" in ($\exists$E) is just the antecedent of the conditional.

Taking $C(q)$ as $\Box(A \leftrightarrow \Box q)$, we have as a case of ($\exists$I):

$$\vdash_S \Box(A \leftrightarrow \Box A) \rightarrow \exists q(\Box(A \leftrightarrow \Box q)).$$

The logics S we are interested in are the normal extensions of **K4!**, since in these logics the converse of the above implication is also provable. In these logics, there will accordingly be, for any formula A, a provable equivalence between $\mathcal{N}A$, alias the propositionally quantified formula $\exists q \Box(A \leftrightarrow \Box q)$, and a particular quantifier-free instance thereof: $\Box(A \leftrightarrow \Box A)$. So it remains only to show that the converse of the implication above is **K4!**-provable. But according to Proposition 2.1.9, p. 40, the formula $\Box(p \leftrightarrow \Box q) \rightarrow \Box(p \leftrightarrow \Box p)$ is a theorem of **K4!** – something we could (having in the meantime met the Kripke semantics) have checked by noting that this formula is valid on all frames for **K4!**, i.e., on all transitive dense frames (a class of frames w.r.t. which **K4!** is complete, since these frames include **K4!**'s canonical frame). Thus by ($\exists$E), we have:

$$\vdash_{\mathbf{K4!}} \exists q(\Box(p \leftrightarrow \Box q)) \rightarrow \Box(p \leftrightarrow \Box p),$$

and by uniform substitution, where q does not have a free occurrence in A:

$$\vdash_{\mathbf{K4!}} \exists q(\Box(A \leftrightarrow \Box q)) \rightarrow \Box(A \leftrightarrow \Box A).$$

(If A contains q free, just do this picking a different variable as that found by the quantifier.) Thus for logics extending **K4!**, we can give a quantifier-free definition of $\mathcal{N}A$ as $\Box(A \leftrightarrow \Box A)$. The most famous candidate alethic modal logics extending **K4!** are **S4** and **S5**, so it is worth noting that we have in effect already met our necessitativity connective $\mathcal{N}$ for these logics before, when we were considering noncontingency. Let us recall, from Exercise 2.5.60 (p. 102), and the discussion leading up to it, that in **S4** there are (at least) three reasonably plausible candidates for something like noncontingency. We have the immediately obvious formalization $\triangle$ with $\triangle A = \Box A \vee \Box \neg A$, alongside two weaker notions there baptized $\triangle^+$ and $\triangle^-$, where $\triangle^+ A = \Box(A \to \Box A)$ and $\triangle^- A = \Box(\Diamond A \to A)$. In **S4** these are all in general distinct (not provably equivalent for $A = p$), and are related by the equivalence:

$$(\triangle^+ A \wedge \triangle^- A) \leftrightarrow \triangle A,$$

while in **S5** all three notions of noncontingency are equivalent. In **S4** we have the **T** axiom which means that $\Box(A \leftrightarrow \Box A)$ and $\Box(A \to \Box A)$ are provably equivalent, **T** supplying the $\leftarrow$-half of the (strict) biconditional version 'for free' (and in fact **U** would have sufficed for this), so in **S4** $\mathcal{N}A$ amounts to $\triangle^+ A$. $\triangle^- A$ by contrast, amounts in **S4** to $\mathcal{N}\neg A$, which is a way of saying that A is strictly equivalent to a $\Diamond$-formula. ◀

The discussion in Example 4.1.13 is partly inspired by a somewhat unsatisfactory passage from p. 55 of in Bennett [65] (briefly mentioned in note 137, p. 165), a paper concentrating on the acceptability of the principles **4** and **5** for alethic modal logic. In the passage in question, Bennett has been suggesting that there is something not quite right about thinking that **4** manages to express the idea that what is necessarily the case is *necessarily* necessarily the case, because the italicized occurrence of necessarily ends up being vacuous, when all occurrences are interpreted as $\Box$. A more robust notion of necessity, it is suggested, can be defined with the merit that it applies only to non-modal statements. The merits of this suggestion will not occupy us here, so much as the proposed solution, which is to introduce a necessity operator T that builds in the idea that what it applies to does not express a modal proposition, which is done by defining Tp as:

$$\Box p \wedge \neg\Box(p \leftrightarrow \Diamond q) \wedge \neg\Box(p \leftrightarrow \Box q) \wedge \neg\Box(p \leftrightarrow \Diamond \neg q) \wedge \neg\Box(p \leftrightarrow \Box \neg q).$$

The last four conjuncts here are intended to say that p is non-modal. (In fact, rather than a conjunction of negations as here, Bennett writes the negation of the disjunction of the four formulas $\Box(p \leftrightarrow \Diamond q)$, $\Box(p \leftrightarrow \Box q)$, etc.) Really for this to be a definition it should be written with schematic letters rather than propositional variables, defining TA for arbitrary A, but there is a more serious breach of conceptual hygiene to worry than that: what Bennett has really defined is not a one-place connective, as the Tp notation suggests, but a two-place connective, since we have not only p but also q in the *definiens*. (We stick with Bennett's use of propositional variables for the present discussion.) It is possible to treat such would-de definitions, in a variables appears free in the definiens but not in the *definiendum*, as legitimate under certain circumstances – see Humberstone [565] for more on this – namely when the additional variable appears inessentially in that the *definiens* is provably equivalent to the result of replacing that variable by another. But as this reference to provable equivalence makes clear, the propriety of a such a would-be definition now depends in part on what logic we are envisaging extended with the definition in question. For example, supposed someone proposed that we extend **K** by introducing a new sentence operator \$, defining \$$p$ as $\neg p \wedge \Box q$. We should protest immediatley that since this will render \$$p \leftrightarrow (\neg p \wedge \Box q)$, by uniform substitution we also have \$$p \leftrightarrow (\neg p \wedge \Box r)$ provable and thus also

$$(\neg p \wedge \Box q) \leftrightarrow (\neg p \wedge \Box r)$$

provable. But this is something not provable in the original system **K** without the definition and which does not involve the defined symbol \$, whereas definitions are supposed to expedite the expression of what was already expressible less concisely, not to inflate the stock of theorems we had in the original vocabulary. (The usual way of saying this is that extensions by definitions should be *conservative* extensions, or that definitions should be 'non-creative'.) Yet the proposed definition were introduced in the context of **KVer** instead of **K**, there would be no such objection, since all $\Box$-formulas are equivalent in

4.1. INTRODUCTION

this logic, so in the proposed *definiens*, $\neg p \wedge \Box q$ variable q does not appear essentially – not that there would be much point in introducing $ in the first place because $p, as so defined, is just equivalent in **KVer** to $\neg p$. (A more interesting environment into which to introduce $ by thsi definition would be one in which all $\Box$-formulas are equivalent, without either being provable or refutable, such as the smallest modal logic containing all formulas of the form $\Box A \to \Box B$. (This is a regular modal logic, the intersection of **KVer** with the 'Falsum' logic first mentioned in note 141. Since these two are truth-functional or, as its was put it on p. 171, Makinson logics, their intersection, which not treating $\Box$ as a truth-function, is extensional.) $ there behaves as a hybrid of negation and the 1-ary constant false truth-function. (See p. 261 for this 'hybrid' terminology.) But again the feature of having an extraneous variable in the *definiens* is not really earning its keep, since we could just as well have used an orthodox definition, defining $p as $\neg p \wedge \Box p$, or indeed as $\neg p \wedge \Box \top$ or $\neg p \wedge \Box \bot$. (In fact it is only in passing from the 1-place to the 0-place case, as we would be if we started with tried to define either $\top$ or $\bot$ in terms of the $\neg$ and the binary connectives, that we really meet the need to liberalize from orthodox definitions to those with extraneous but inessential variables. For some history and other discussion, see again [565].)

If Bennett intended the proposed definition of Tp by means of the 5-conjunct conjunction above with the liberalized conception of definition in mind, we face the question of what the background logic is relative to which the occurrence of q in this formula is inessential. It is certainly inessential if we take the background logic to be one in which **.2**, since then the first, second and fourth conjuncts are already inconsistent taken together: $\vdash_{\mathbf{K.2}} \neg(\Box p \wedge \neg\Box(p \leftrightarrow \Diamond q) \wedge \neg\Box(p \leftrightarrow \Diamond\neg q))$. To see this, we exploit the fact that **K.2** is determined by the class of piecewise convergent frames (see Exercise 2.5.8(*iii*), p. 74) and note that if w is a point in a such a frame and in some model on that frame we have both $\Box(p \leftrightarrow \Diamond q)$ and $\Box(p \leftrightarrow \Diamond\neg q)$ false at w, w must have a successor x at which $p \leftrightarrow \Diamond q$ is false and a successor y at which $p \leftrightarrow \Diamond\neg q$ is false. But if $\Box p$ is also true at x, p must be true at both these successors, so $\Diamond q$ will be false at x and $\Diamond q$ and false at y, meaning that x and y cannot have a common successor – what could q's truth-value be there? – violating the convergence condition. Thus while the definition of Tp would be legitimate despite the extraneous variable because that variable occurs inessentially, this good news is tempered by the bad news that if the logic to which we are adding the definition proves **.2** – for example if it is **S4.2** or **S5** – then the resulting logic proves $\neg TA$ for all formulas A, not what Bennett had in mind. In fact we can also see that several other subconjunctions of the five-conjunct conjunction would pose the same problem in the setting of **S5**, the first, second and fifth conjuncts, for example giving us points accessible to one where they are all true, with $\Diamond q$ true at one and false at the other, contradicting the modal invariance of fully modalized formulas in point-generated models for **S5**. (There is a similar problem with the first, third and fourth conjuncts taken together.) Rather than continuing to speculate about what background logic Bennett might have had in mind with his liberalized definition, let us instead consider the perhaps more likely possibility that what Bennett had in mind was not the five-conjunct conjunctive definition above but a version with propositional quantifiers, which he may have been trying to avoid because of the complications their presence raises. The suggestion is that the intended definiens for Tp was instead this:

$$\Box p \wedge \neg \exists q \Box(p \leftrightarrow \Diamond q) \wedge \neg \exists q \Box(p \leftrightarrow \Box q) \wedge \neg \exists q \Box(p \leftrightarrow \Diamond\neg q) \wedge \neg \exists q \Box(p \leftrightarrow \Box\neg q).$$

We can immediately shorten this by dropping the now redundant final two conjucts:

$$\Box p \wedge \neg \exists q \Box(p \leftrightarrow \Diamond q) \wedge \neg \exists q \Box(p \leftrightarrow \Box q).$$

The part that says "the proposition that p is non-modal" consists now of just the second and third conjuncts, which say that this proposition is not 'possibilitative' and that it is not necesssitative, the latter being the idea pursued in Example 4.1.13 above.

REMARK 4.1.14 The current suggestion amounts to saying that 'being modal' is a matter of being necessitative or the negation of a necessitative, so we might well wonder why the disjunction of two necessitatives or the implication with a necessitative antecedent and consequent should not also count. More generally this would amount to an object language analogue of being fully modalized in the sense given on p. 56. ◀

Motivated by the examples just mentioned, we interrupt this discussion of Bennett [65] with an exercise on a metalinguistic version of the issue raised in Remark 4.1.14, in which strict equivalence is replaced by provable (material) equivalence.

EXERCISE 4.1.15 (*i*) Show that in **S4** every disjunction of $\Box$-formulas is provably equivalent to a $\Box$ formula, but that this is not so in **KD4**.

(*ii*) Show that in **S5** every implication of the form $\Box A \to \Box B$ is provably equivalent to a $\Box$-formula, this is not so in **S4**.

(*iii*) Show that **S5** is the smallest normal modal logic extending **S4** and satisfying the condition that every implication of the form $\Box A \to \Box B$ is provably equivalent to a $\Box$-formula.

Suggestions: For (*i*) Recall from Exercise 2.7.31 (p. 136 and the subsequent discussion on p. 196 that $\nvdash_{\mathbf{K4}_c} \Box(\Box p \vee \Box q) \to (\Box p \vee \Box q)$. The countermodel there described has a frame which was not only dense but serial and transitive, so we can replace the subscript here with "**KD4!**". If $\Box p \vee \Box q$ were provably equivalent in **KD4** to some formula $\Box C$, then this would be so for the even stronger logic **KD4!** and we could exploit this equivalence along with the $\mathbf{4}_c$ instance $\Box\Box C \to \Box C$ – how? – to show that $\Box(\Box p \vee \Box q)$ provably implied $\Box p \vee \Box q$ in **KD4!**, something we know not to be the case. (A variation on this argument appears as Example 4.22.13 in Humberstone [594]; see also the surrounding discussion on pp. 559–561 there.)

For (*ii*) and (*iii*): Note that in **S4** (in fact already in **KD**) that for any formula A, $\neg \Box A$ is provably equivalent to $\Box A \to \Box \bot$, and if the latter were provably equivalent to a $\Box$-formula, it would provably imply its own necessitation (by **4**), and so ... ✣

Let us return now to Bennett [65]. Propositional quantifiers were used as long ago as the 1930s in modal logic – conspicuously in Lewis and Langford [716] with its "Existence Postulate" (reproduced here with Lewis's strict implication symbol $\prec$:

$$\exists p \exists q (\neg(p \prec q) \wedge \neg(p \prec \neg q)).^{174}$$

But we learn from especially Fine [287], §2.2, that other than with **S5** as the logic to be augmented by their addition, the weaker normal modal logics with natural semantics in terms of Kripke models are prone to fail to be axiomatizable, which is why we avoid them if possible (as in Example 4.1.13). (Additional references on propositionally quantified modal logic are given in note 284, 340 below.) Nevertheless, their use is strongly suggested – as we saw before Remark 4.1.14 above – for spelling out an idea of Bennett's, and a remedy for the problematic definition of T in [65]: the use of such quantifiers means that we do have an orthodox definition on our hands since the additional variable q in the *definiens* is no longer a free variable, guaranteeing that the definition gives a conservative extension. (It is 'universally non-creative' in the terminology of Humberstone [594], p. 723 – though it was definitions in first order theories rather than in logics that were under discussion there.) But along with the good news there is again bad news (from Bennett's point of view): the discussion in Example 4.1.13 showed that in propositionally quantified **S4**, $\exists q \Box(A \leftrightarrow \Box q)$ is provably equivalent to $\Box(A \leftrightarrow \Box A)$, or again, to $\Box(A \to \Box A)$ (alias $\triangle^+ A$). This is provably implied by $\Box A$ in **S4**. So the first and third conjuncts of this last attempt at a *definiens* for

---

[174] This is C. I. Lewis, of course, rather than David Lewis. An interesting motivation is offered in Lewis and Langford [716], p. 178, for adding such postulates as this: the Sobociński-regular modal logics do not exclude a vacuous interpretation of the modal vocabulary, with "◇" and "□" (not that the latter symbol is employed [716]) as both standing for the identity truth-function and "$\prec$" interpreted as "$\to$"; with such things as the Existence Postulate, modal logic can exclude these intuitively non-modal interpretations, since a propositionally quantified version of classical non-modal propositional logic contains their negations. On the latter logic, see note 318 (p. 291) of Hughes and Cresswell [535].

Bennett's T are inconsistent and $\neg TA$ is provable for all A. This is awkward because Bennett had no objections to **4** for $\Box$ (or L as he wrote it), except that it was liable not to be read as informally intended, which intentions T was introduced to help us express. (We were now meant to be able to express them by rejecting the principle $\Box p \to T\Box p$. Unfortunately Bennett then erred in suggesting we should accept instead the negation of this conditional, which would make $\Box p \wedge \neg T\Box p$, and hence $A \wedge \neg T\Box A$ for all A, provable in the – as we now see, inconsistent – logic expressing our intentions. Again, behind the error is an oversight concerning quantifiers. The requirement that modal logics be closed under uniform substitution is a way of spelling out the fact that if propositional quantifiers are made explicit, each theorem is understood as universally quantified with respect to all the propositional variables it contains. Bennett is trying to deny $\forall p(\Box p \to T\Box p)$ so what he needs is not $\forall p\neg(\Box p \to T\Box p)$, which is what in omitting the quantifier he in effect offers, but rather $\neg\forall p(\Box p \to T\Box p)$, or equivalently $\exists p(\Box p \wedge \neg T\Box p)$. The mistake involved here is very clearly described and warned against in the top half of p. 180 of Lewis and Langford [716].)

4.2 Nomic Necessity I: Pargetter

Robert Pargetter [864] defended a particular normal modal logic, an extension of **KT**, as the correct logic for nomic necessity, at least under certain assumptions – toward which [864] is favourably disposed. This was given as an example in §6 of my [582] as a counterexample to the claim sometimes made that (temporally motivated) tense logic distinguishes itself from (alethic, etc.) modal logic in starting with a class of frames felt to embody a hypothesis about the subject matter – the structure of time, typically – and looking for the logic determined thereby, as opposed to starting with an independently favoured logic and taking a derivative interest in classes of frames that might determine it. Here we treat the topic as affording a pleasant opportunity for further practice with modal techniques. Pargetter's motivation still needs to be explained, though.

The background idea is that what is nomically necessary in a world w is whatever is true in every world w' to which w bears a certain nomic accessibility relation; denoting this relation by R, we think of wRw' as meaning that all the laws of w hold as truths (not necessarily as laws) of w'. (See Remark 2.3.1(i), as well as the surrounding discussion, from p. 59.) Not unreasonably perhaps, [864] takes it that the laws of w are *at least* truths of w: so the relation R is reflexive.

REMARK 4.2.1 Given our gloss on nomic accessibility, the reflexivity issue amounts to saying that nomic necessity obeys the **T** axiom. Although in the present section we shall see John Bigelow making this assumption in his collaboration with Pargetter, it should be noted that in subsequent work, Bigelow has, under the influence of Martin Leckey, come to question this, thinking instead of what is nomically necessary as what will happen 'in the natural course of events' – that is: unless prevented by something from outside the system of laws. (See Leckey and Bigelow [689], and Leckey [688]; the gist of the present summary owes much to personal communication from Bigelow.) As a result we have something that looks more like a deontic logic than a traditional alethic modal logic – some extension of **KD** which is not an extension of **KT**, perhaps suggesting something of a *rapprochement* between laws in the nomic sense and laws in the legal sense. This would then make room for the idea, for example, of a *miracle* as a violations of the laws of nature without making it true *a priori* that miracles do not occur. If miracles in the sense of violations-at-w of w's own laws of nature are even conceivable, then whether or not they occur, they should not be ruled out by the logic of nomic necessity. There is also a "two-dimensional" notion of miracle to be found in the writings of David Lewis (such as Chapter 17 of [725]), according to which from the perspective of w, an event in w' is said to be a miracle if it violates the laws – not of w' but – of w. This means that the claim that a miracle has occurred would have the status of a 'di-propositional constant' like the "***R***" of p. 275 below. In fact an objection to **T** for nomic (or physical) necessity which is unrelated to the Bigelow–Leckey worries but is (albeit indirectly) connected with this two-dimensional issue was once raised by Richard Montague, and is described at note 177 below. We proceed here without further attention to these qualms about **T**. ◂

If, Pargetter observes, we now make the metaphysical assumption that to every world there corresponds at least one 'Hume copy' ('Hume duplicate'), mentioned in Remark 2.3.1(iii) (p. 59), in which there are *no* laws but in which the course of events is exactly as in the original world, then we can draw some conclusions about the nomic accessibility relation which have direct repercussions as to what the appropriate logic will be.[175]

In this paragraph – not required for the formal developments which follow – we touch on some respects in which Pargetter's ideas have undergone modification since the appearance of [864]. The theme is taken up in Chapter 5 of Bigelow and Pargetter [91], where the authors are keen to avoid the conflict mentioned above with Humean supervenience (though not of course defending a Humean theory of laws). They assert (p. 239) that "(t)he accessibility relations among worlds supervene on the contents of those worlds", with the following clarificatory remark on the notion of content here employed (p. 242):

> We do not believe that accessibility relations supervene on the *first-order* properties and relations of individuals in those worlds. On the contrary, we believe that two worlds may exactly match one another at the level of first-order universals and anything which supervenes on these, and yet may differ in their accessibility relations. Their differences in accessibility relations will indicate a difference in higher-order universals, at some level, but it need not indicate any first-order difference between them.

Bigelow and Pargetter believe, in particular, that physical forces are amongst these higher-order universals and that forces are absent from what they call Hume worlds (presumably meaning candidate Hume copies for various worlds), with the result that the needed Hume copy (or copies) for a world governed by force laws are not accessible to that world (the laws not being truths there). They continue (pp. 248–9):

> The reduction of logical to nomic modalities does not, however, require the accessibility of a Hume world. What is required is something weaker. The regularities which are laws in a world need not all be first-order regularities. But whatever their level, they must be regularities which hold in every accessible world. (...) Consider, then, any world with laws and compare it with a world where there are regularities that are not laws, yet which otherwise exactly match the regularities that are laws in the lawful world. Call this lawless world the *Heimson* world corresponding to the lawful world. It may not be a Hume world, because it may have lawless yet regular correlations among higher-order things like forces. Yet it can play the same role in reducing logical modalities to nomic modalities.

We take this question of reducing the logical to the nomic briefly at the end of this discussion.

Returning to the setting of our initial summary of [864], let us denote by $h(w)$ the Hume copy of an arbitrary world w (the 'Heimson' copy of w, in the preferred formulation of [91]); in case there are several such copies—the prospect of which raises matters of world-individuation which we need not go into here—let $h(w)$ be one of them. (At least one such is secured by the metaphysical assumption of the opening paragraph above.) If w itself has no laws, then we may take $h(w) = w$. Pargetter then argues as follows. For any world w, since the course of events in $h(w)$ is as in w, w bears the relation R to $h(w)$. And since $h(w)$ itself has no laws, $h(w)$ bears the relation R to every world w'. Although the idea is clear enough, Pargetter's summary of the bearing of these considerations on how the accessibility relations (in the models for our logic) are to be constrained is a little hazy:

---

[175]This assumption itself may be found implausible, since in allowing that worlds may be alike in respect of matters of particular fact – share the same 'course of history' – and yet differ in respect of which generalizations about such matters are laws, the position is a violation of what has been called (Lewis [726]) the doctrine of Humean supervenience. But the interest of Pargetter's project, as already mentioned, is in its role as a case study in the exercise of teasing out modal principles for nomic necessity from metaphysical hypotheses such as the present assumption, rather than in the comparative merits of such hypotheses themselves.

4.2. NOMIC NECESSITY I: PARGETTER

Besides accessibility being reflexive, we have accessibility being "two-step" transitive, and all worlds having access to a world from which all worlds are accessible. So to the axioms of T we add

$$\Diamond\Box\Box A \to \Box\Box A \tag{1}$$
$$(\Diamond\Diamond A_1 \wedge \ldots \wedge \Diamond\Diamond A_n) \to \Diamond(\Diamond A_1 \wedge \ldots \wedge \Diamond A_n) \text{ for every } n \geqslant 2. \tag{2}$$

[864], p. 339.[176]

In the terminology of, e.g., Chellas [171], the logic being proposed by Pargetter is the smallest normal modal logic extending **KT** and including all instances of the schemata (1) and (2).[177] Note the close resemblance of (2) to the generalized density schema GD_n discussed in Section 3.3; the difference is only that the occurrences of $\Diamond\Diamond$ in the antecedent here appeared as occurrences of $\Diamond$ in the dual form (which appeared in Exc. 3.3.7) of the latter schema there.

On the basis of remarks on the previous page of [864], Pargetter may be taken to mean, by "two-step" transitivity in the above quotation, that for all worlds w, w', we have wR^2w' (that is, for some world x, wRx and xRw'). The haziness alluded to earlier arises with respect to two features of the description quoted. In the first place, it is odd to list this (itself oddly named) two-step transitivity condition alongside the condition that all worlds have "access to a world from which all worlds are accessible", since the former condition follows from the latter. The oddity of the terminology lies in the suggestion in the word 'transitivity' that we are dealing with a *conditional* requirement; below, we shall refer to this condition simply as the *two-step condition*.[178] It is of course this latter condition – definitive of the class of what we shall call Pargetter frames below – which recapitulates the considerations above concerning Hume (or Heimson) duplicates. Secondly, the "So to the axioms of T we add" formulation suggests that schemas (1) and (2) correspond respectively to the two conditions listed, just as reflexivity corresponds to the **T** schema ($\Box A \to A$, that is). Now while there are two senses of 'correspond' to be distinguished here – associated in the one case with modal definability theory and in the other with modal completeness theory (see van Benthem [70], which tends to associate the term with 'correspondence' with the former area) – in neither of them would this suggestion be correct. (More on the dangers of the term *correspondence* can be found in Humberstone [602].)

In the modal definability sense, as we recall from p. 81, a set Γ of modal formulas – such as the set of instances of a schema, as above – corresponds to ('modally defines') a class $\mathbb{C}$ of frames when $\mathbb{C}$ comprises all and only those frames on which every formula in Γ is valid. In the completeness-theoretic setting, by contrast, the correspondences of interest between Γ and $\mathbb{C}$ arise when the smallest normal modal logic containing all formulas in Γ ($\mathbf{K} \oplus \Gamma$, that is) comprises all ('completeness w.r.t. $\mathbb{C}$') and only ('soundness w.r.t. $\mathbb{C}$') the formulas valid on every frame in $\mathbb{C}$, in which case – as we continue to recall from Section

---

[176]In the above quotation the line numbering has been added and the formal notation inconsequentially altered. In particular, Pargetter inserts an "N" into the boxes and diamonds as a reminder of the intended interpretation in terms of nomic necessity and possibility. Likewise in [91], Chapter 6.

[177] The assumption here is that **T** is a correct principle for nomic necessity; dissent on this score was mentioned in Remark 4.2.1 but a rather different objection is voiced in Section 6 of Montague [821]. The reason is that Montague thinks of the nomic necessity operator – in fact he says 'physical necessity' – as recording physical necessity *in the actual world*, taken as a distinguished point in the models, while requiring for the validity of a formula its truth at all points in all models rather than just at the distinguished point in each model – general validity as opposed to real-world validity as it is put in, e.g., Humberstone [578]. (In Rabinowicz and Segerberg [939] these notions go by the names 'strong validity' and 'weak validity', respectively, and in Gochet and Gillet [390] 'first class' and 'second class validity', while in Predelli and Stojanovic [906] these notions appear respectively as classical validity and SR-validity (the "SR" abbreviating "semantic relativism"). Morato [826] calls general validity *textbook validity* and real-world validity *Kripkean validity*, and provides a general discussion of the issue.) This is the distinction mentioned, though not with this terminology in note 45, p. 60; for the semantics of the "actually" part of the composite operator, "actually physically necessary" invoked here, see Remark 4.7.1, p. 305. A diagnosis of bad modal metaphysics putting the blame on confusing what is necessary – this time *metaphysically* necessary – with what is actually necessary can be found in Salmon [992]. That discussion is taken up in Williamson [1186], pp. 97–99, and in Gregory [411].

[178]This same condition is called 'two-connectedness' in [679], p. 177.

2.5 – that logic is said to be *determined by* $\mathbb{C}$.[179] We recall, further, that the latter correspondence does not provide a unique $\mathbb{C}$ for any arbitrarily given Γ, since the same logic ($\mathbf{K} \oplus \Gamma$) may be determined by several distinct classes of frames (even identifying isomorphic frames) or by no class of frames (the so-called Kripke incomplete logics).

Naturally we say that a schema stands in one or other of the correspondences distinguished above with a class of frames when the set of instances of the schema does so in accord with the official definitions. We can then say that Pargetter's schema (1) above, repeated here for convenience:

$$\Diamond\Box\Box A \to \Box\Box A \tag{1}$$

modally defines the class of frames $\langle W, R \rangle$ satisfying the following euclidean-like condition:

$$\text{If } xRy \text{ and } xR^2z \text{ then } yR^2z, \text{ for all } x,y,z \in W. \tag{1°}$$

Further, since the canonical frame of any consistent normal modal logic containing all instances of (1) is easily seen to satisfy (1°), we also have the completeness-theoretic correspondence in this case: $\mathbf{K} \oplus (1)$ (i.e., the smallest normal modal logic containing all instances of (1)), is determined by the class of frames satisfying (1°). Not much connection here, then, between the modal schema—taken by itself, at least— and the two-step condition (i.e., the condition that for all $x, y \in W$, xR^2y). Note incidentally, that the class of frames satisfying the latter condition is clearly not modally definable – since the disjoint union of two frames satisfying it will not itself satisfy it (see Theorem 2.5.32(i)). Of course, this does not prevent us from asking for an axiomatization of the normal modal logic determined by the class of all frames – or the class of all reflexive frames (not to forget that Pargetter's intended accessibility relation is reflexive) satisfying the two-step condition. The situation is analogous to that alluded to in §2 of [864], in which **S5** (= **KT5**) is the modal logic determined by the class of all frames $\langle W, R \rangle$ in which for all $x, y \in W$, xRy (the "one-step condition"), even though this class of frames is not modally definable for the same reason as in the two-step case.

Pargetter's second schema (2) is best thought of initially in terms of its several 'subschemes' (2_n), one for each natural number n. (There is no point in adding the restriction '$n \geqslant 2$', which is dropped in [91], p. 260.)

$$(\Diamond\Diamond A_1 \wedge \ldots \wedge \Diamond\Diamond A_n) \to \Diamond(\Diamond A_1 \wedge \ldots \wedge \Diamond A_n). \tag{2_n}$$

Here we find again that the class of frames modally defined by the schema—which comprises precisely those $\langle W, R \rangle$ such that

$$\text{For all } x, z_1, \ldots, z_n \in W, \text{ if } xR^2z_1 \text{ and } \ldots \text{ and } xR^2z_n, \text{ then there exists } y \in W \text{ such that } xRy \text{ and } yRz_1 \text{ and } \ldots \text{ and } yRz_n. \tag{$2_n°$}$$

itself contains the canonical frame for any consistent normal modal logic containing all instances of the schema. If we consider original schema (2) itself, however, we find that whereas its set of instances modally defines the class $\mathbb{C}_1 = \{\langle W, R \rangle \mid \text{for all } x \in W \text{ and all finite } Z \subseteq R^2(x), \text{ there exists } y \in W \text{ such that } xRy \text{ and } Z \subseteq R(y)\}$, the normal modal logic $\mathbf{K} + (2)$ is revealed by the canonical model technique to be determined not only by $\mathbb{C}_1$ but also by the smaller class of frames $\mathbb{C}_0 = \{\langle W, R \rangle \mid \text{for all } x \in W \text{ and all } Z \subseteq R^2(x), \text{ there exists } y \in W \text{ such that } xRy \text{ and } Z \subseteq R(y)\}$. We have specified $\mathbb{C}_0$ as above to make visible the contrast with that of $\mathbb{C}_1$, from which it differs only in the dropping of the word 'finite'. A simpler but equivalent specification would run: $\mathbb{C}_0 = \{\langle W, R \rangle \mid \text{for all } x \in W \text{there exists } y \in R(x) \text{ with } R^2(x) \subseteq R(y)\}$.[180].

---

[179] A somewhat narrower completeness-related notion of correspondence is occasionally found, combining (i) $\mathbf{K} \oplus \Gamma$ is sound w.r.t. $\mathbb{C}$ and (ii) the canonical frame for this logic belongs to $\mathbb{C}$.

[180] Compare the way the reference to finiteness disappeared in the formulation of Theorem 2.6.8, p. 116

4.2. NOMIC NECESSITY I: PARGETTER

In neither sense of the word 'correspond', then, do (1) and (2) correspond to the conditions on accessibility Pargetter cites in the passage quoted above (the two-step condition and the condition of being what we shall call a Pargetter frame, below). In their re-working of this material in [91], Bigelow and Pargetter formulate matters somewhat differently. In this quotation from p. 260 of [91], (1) and (2) are called Axiom A29 and Axiom A30:

> Axiom A29 requires, in effect, that for any world there be an accessible Heimson world. Axioms A30 requires, in effect, that any world be accessible from a Heimson world.

Although these conditions are different from those offered in Pargetter [864], they are still wide of the mark, as is evident from our discussion above, and there is again a suspicious failure of independence – though this time more subtly so than in the passage quoted from [864]. The point is that in terms of accessibility, all there is to being a Heimson (or Hume) world accessible to w is: being a world accessible to w and to which every world is in turn accessible. So the first condition given here, apropos of A29 (= (1)), in fact incorporates the second (given for A30 (= (2))). The truth is that it is only in virtue of some rather subtle interaction with the **T** schema that (1) and (2) manage to have the combined effect of securing completeness with respect to the class of frames in which every world has access to a world which has access to every world. This interaction will be evident from the proof of Theorem 4.2.5 below.

In order to prove the completeness of Pargetter's axiomatization, we proceed a little indirectly and introduce a different axiomatization of what will turn out to be the same logic. We formulate a modal schema which is 'trebly' schematic: it not only contains schematic letters for formulas (as (1) does), and a variable numerical subscript to subsume Boolean compounds of arbitrary length (as (2) does), but it involves also the use of (superscripted numerical) variables to indicate arbitrarily great iteration of modal operators. Here we use $s, r_1, \ldots, r_n$ in this last capacity. Thus **KPar** (= **K** ⊕ **Par**) contains the formulas resulting from the schema below by making any selection of $n, s, r_1, \ldots, r_n$ from *Nat*.

Par $\quad\quad\quad \Diamond^s \Box (\Box A_1 \vee \ldots \vee \Box A_n) \to (\Box^{r_1} A_1 \vee \ldots \vee \Box^{r_n} A_n)$

We can show without much trouble (see Theorem 4.2.2 below) that **KPar** is determined by the class of all frames satisfying Pargetter's condition above – excerpting from our earlier quotation – "all worlds having access to a world from which all worlds are accessible", which is to say the class of all frames $\langle W, R \rangle$, we shall call them *Pargetter frames*, in which:

> For all $x \in W$ there exists $y \in R(x)$ such that for all $z \in W$, yRz.

But before presenting that, we pause to note the way **Par** subsumes (1) and (2), and hence that for any normal modal logic S, $S \oplus \{(1), (2)\} \subseteq S \oplus \mathbf{Par}$. We obtain (1) from **Par** by setting $n = 1$, $s = 1$, $r_1 = 2$. For (2), we should consider the dual form of this schema:

$$\Box(\Box A_1 \vee \ldots \vee \Box A_n) \to (\Box\Box A_1 \vee \ldots \vee \Box\Box A_n).$$

This we get from **Par** by putting $s = 0$ and $r_1 = \ldots = r_n = 2$. Thus for any normal modal logic S, $S \oplus \{(1), (2)\} \subseteq S \oplus \mathbf{Par}$. If in addition $S \supseteq \mathbf{KT}$, then we also have, as (the proof of) Theorem 4.2.5 below shows, $S \oplus \mathbf{Par} \subseteq S \oplus \{(1), (2)\}$.

The schema inset above, for any given choice of n, is called $\mathbf{Seg}_n$ (after Segerberg) in numerous publications of Hughes and Cresswell, such as Cresswell [209] and references there cited. (In the title of that paper, "B Seg" refers to what we would call **KTBSeg** or more explicitly $\mathbf{KTB} \oplus \{\mathbf{Seg}_n \mid n \in \mathit{Nat}\}$.)

THEOREM 4.2.2 **KPar** *is determined by the class of all Pargetter frames.*

Proof. Soundness: Suppose that for some model $\mathcal{M} = \langle W, R, V \rangle$ with $\langle W, R \rangle$ a Pargetter frame, we have $\mathcal{M} \not\models_w \mathbf{Par}$, for some $w \in W$ and some $s, r_1, \ldots, r_n \in Nat$. Thus

(i) $\mathcal{M} \models_w \Diamond^s \Box(\Box A_1 \vee \ldots \vee \Box A_n)$, while

(ii) $\mathcal{M} \not\models_w \Box^{r_1} A_1 \vee \ldots \vee \Box^{r_n} A_n$.

By (i), there exists $x \in W$ with $wR^s x$ and $\mathcal{M} \models_x \Box(\Box A_1 \vee \ldots \vee \Box A_n)$. Since $\langle W, R \rangle$ is a Pargetter frame, there is some $y \in R(x)$ with $R(y) = W$; accordingly for such a y we have $\mathcal{M} \models_y \Box A_1 \vee \ldots \vee \Box A_n$, and so for some i ($1 \leqslant i \leqslant n$) we must have A_i true throughout $\mathcal{M}$. Since in particular this means that A_i is true at all points at a distance of r_i steps of the relation R from w, we get a contradiction with (ii).

Completeness: Let $\mathcal{M}_{\mathbf{KPar}}$ be the canonical model for **KPar**, and suppose that we have some formula C not provable in this logic. We have to find a model on a Pargetter frame in which there is a point at which C is false. By standard arguments, we know that for some point w in $\mathcal{M}_{\mathbf{KPar}}$, we have $\mathcal{M}_{\mathbf{KPar}} \not\models_w C$. Let $\mathcal{M}$ be the submodel of $\mathcal{M}_{\mathbf{KPar}}$ generated by w. Since $\mathcal{M} \not\models_w C$ (by Theorem 2.4.10, p. 70), we need only show that the frame $\langle W, R \rangle$ of $\mathcal{M}$ is a Pargetter frame.

We must show, that is, how to find, for any $x \in W$, some $y \in R(x)$ such that for all $z \in W$, yRz. The recipe is as follows. Given $x \in W$, we let y be any maximal consistent superset (w.r.t. the present logic) of the set

$$y_0 = \{A \mid \Box A \in x\} \cup \{\Diamond B \mid \Diamond^r B \in w \text{ for some } r \in Nat\}$$

on the assumption that this set is itself **KPar**-consistent. For such a y we clearly have xRy, as required, and we also have that yRz for all $z \in W$: otherwise there is $z \in W$, such that not yRz. Hence there exists a formula B with $\Box B \in y$ and $B \notin z$. Since $z \in W$, $wR^r z$ for some $r \in Nat$. Thus $\Diamond^r \neg B \in w$, so since $y \supseteq y_0$, $\Diamond \neg B \in y$, contradicting the fact that $\Box B \in y$. It remains, then, only to check that y_0 is consistent. Suppose otherwise: that for some $A_1, \ldots, A_m$ with $\Box A_1, \ldots, \Box A_m \in x$, and $B_1, \ldots, B_m$ with $r_1, \ldots, r_n \in Nat$ for which $\Diamond^{r_i} B_i \in w$, we have:

$\vdash_{\mathbf{KPar}} (A_1 \wedge \ldots \wedge A_m) \to \neg(\Diamond B_1 \wedge \ldots \wedge \Diamond B_n)$. Thus, by familiar manipulations (available in any normal modal logic)

$\vdash_{\mathbf{KPar}} (A_1 \wedge \ldots \wedge A_m) \to (\Box \neg B_1 \vee \ldots \vee \Box \neg B_n)$

and so by further such manipulations:

$\vdash_{\mathbf{KPar}} (\Box A_1 \wedge \ldots \wedge \Box A_m) \to \Box(\Box \neg B_1 \vee \ldots \vee \Box \neg B_n)$.

Since the antecedent here belongs to x, so does the consequent. As $x \in W$, x lies at some number of R steps from w. Let that number be s. In view of **Par**, we have:

$\vdash_{\mathbf{KPar}} \Diamond^s \Box(\Box \neg B_1 \vee \ldots \vee \Box \neg B_n) \to (\Box^{r_1} \neg B_1 \vee \ldots \vee \Box^{r_n} \neg B_n)$,

where $r_1, \ldots, r_n$ are as above. Since $\Box(\Box \neg B_1 \vee \ldots \vee \Box \neg B_n) \in x$, the antecedent of this implication belongs to w, and so therefore its consequent does also. We have a contradiction now, because since $\Diamond^{r_i} B_i \in w$, none of the disjuncts $\Box^{r_i} \neg B_i$ can belong to w. ∎

By routine modification of this argument, we can incorporate reflexivity (Part (i)), and also treat (Part (ii)) the case of the weaker "two-step" condition (weaker than the condition of being a Pargetter frame, that is):

4.2. NOMIC NECESSITY I: PARGETTER

THEOREM 4.2.3 *(i)* **KTPar** *is determined by the class of all reflexive Pargetter frames.*

(ii) $\mathbf{K} \oplus \{\Diamond^s \Box\Box A \to \Box^r A \mid r, s \in Nat\}$ *is determined by the class of all frames $\langle W, R \rangle$ in which xR^2y for all $x, y \in W$.*

We now show, as is claimed in [864], where the proof (not there sketched) of this fact is credited to a personal communication from G. E. Hughes, Pargetter's axiomatization (including the **T** schema) is (sound and) complete w.r.t. the class of reflexive Pargetter frames. Since Part *(i)* of the above Theorem already gives such a result for the case of **KTPar**, it remains only to reduce the present case to that one.

LEMMA 4.2.4 $\vdash_{\mathbf{KT}\oplus(1)} \Box\Box\Box A \leftrightarrow \Box\Box A$, *for all formulas A.*

Proof. Since the $\to$ direction is just a special case of **T**, it remains only to show that every formula of the form $\Box\Box A \to \Box\Box\Box A$ is provable in $\mathbf{KT} \oplus (1)$. Here we use our standard conventions for displaying a proof from axioms with the exception that, to avoid a clash with the labels "(1)" and "(2)", line numbers are given in roman numerals:

(i)	$\Diamond\Box\Box A \to \Box\Box A$	= (1)
(ii)	$\Diamond\Diamond\Box\Box A \to \Diamond\Box\Box A$	(i), $\Diamond$Mono
(iii)	$\Box\Diamond\Diamond\Box\Box A \to \Box\Diamond\Box\Box A$	(ii), $\Box$Mono
(iv)	$\Diamond\Diamond\Box\Box A \to \Box\Diamond\Box\Box A$	(1), dual form
(v)	$\Box\Box A \to \Diamond\Diamond\Box\Box A$	Two appeals to **T**, dual form
(vi)	$\Box\Box A \to \Box\Diamond\Diamond\Box\Box A$	(iv), (v) TF
(vii)	$\Box\Box A \to \Box\Diamond\Box\Box A$	(iii), (vi) TF
(viii)	$\Box\Diamond\Box\Box A \to \Box\Box\Box A$	(i), $\Box$Mono
(ix)	$\Box\Box A \to \Box\Box\Box A$	(vii), (viii) TF

∎

THEOREM 4.2.5 $\mathbf{KT} \oplus \{(1), (2)\} = \mathbf{KTPar}$.

Proof. The discussion preceding Theorem 4.2.2 shows that

$$\mathbf{KT} \oplus \{(1), (2)\} \subseteq \mathbf{KTPar}.$$

For the converse inclusion, we must show that every instance of **Par** is provable in $\mathbf{KT} \oplus \{(1), (2)\}$. For convenience we repeat **Par** here:

$$\Diamond^s \Box(\Box A_1 \vee \ldots \vee \Box A_n) \to (\Box^{r_1} A_1 \vee \ldots \Box^{r_n} A_n).$$

We first show that this is derivable in $\mathbf{KT} \oplus \{(1), (2)\}$ for any n and s when $r_1 = \ldots = r_n = 2$, arguing by induction on s that this is so for arbitrary n.

<u>Basis.</u> When $s = 0$, the desired schema is just the dual form of (2).

<u>Induction.</u> Suppose the result holds for a given s; that is, that $\vdash_{\mathbf{KT}\oplus\{((1),(2))\}} \Diamond^s \Box(\Box A_1 \vee \ldots \vee \Box A_n) \to (\Box\Box A_1 \vee \ldots \vee \Box\Box A_n)$. Then by $\Diamond$Mono, we have:

$$\vdash_{\mathbf{KT}\oplus\{((1),(2))\}} \Diamond^{s+1}\Box(\Box A_1 \vee \ldots \vee \Box A_n) \to \Diamond(\Box\Box A_1 \vee \ldots \vee \Box\Box A_n)$$

and therefore, distributing the $\Diamond$ in the consequent across the disjunction and appealing to (1) for $A = A_1, \ldots, A_n$, we get

$$\vdash_{\mathbf{KT}\oplus\{(1),(2)\}} \Diamond^{s+1}\Box(\Box A_1 \vee \ldots \vee \Box A_n) \to (\Box\Box A_1 \vee \ldots \vee \Box\Box A_n),$$

completing the induction.

Having shown that we have **Par** from Pargetter's axioms for the case in which $r_1 = \ldots = r_n = 2$, we must show that the schema is derivable in full generality. But if $r_i < 2$, we may appeal to **T**, and if $r_i > 2$, Lemma 4.2.4 gives the result. ∎

COROLLARY 4.2.6 (G. E. Hughes) **KT** $\oplus$ $\{(1),(2)\}$ *determined by the class of all reflexive Pargetter frames.*

Proof. By Theorems 4.2.3(i) and 4.2.5. ∎

Although the axiomatization in terms of **KT** with **Par** is more economical than that in terms of **KT** with (1) and (2) if we just count the schemata, there is a good sense in which the latter axiomatization (Pargetter's) is simpler – a sense indicated by our earlier remark that **Par** is, as we put it, 'trebly schematic', whereas (1) and (2) are only 'doubly schematic'.

We conclude with some discussion of these results, and an exercise. A quotation from [91] was given earlier (p. 218) which spoke of the "reduction of logical to nomic modalities", and similar remarks appear in [864]. The point is that if we take logical necessity to be truth at all worlds, then with the assumption that every world has some world (nomically) accessible to it, to which all worlds are in turn accessible, then an iteration of nomic necessity (or of nomic possibility) will amount to a statement of logical necessity (or logical possibility, respectively). We have to be careful about how we express this in terms of the formal semantics for modal logic (as in the preceding section). The point is *not* that in all models on frames validating **KTPar**, the truth of $\Box\Box A$ at any point is equivalent to the truth of A at every point. Rather, this holds for all models on point-generated frames: so the logic (*alias* Pargetter's logic, by Theorem 4.2.5) is indeed determined by a class of frames in which this is so (i.e., frames for which it is true that the truth of $\Box\Box A$ at any point is equivalent to the truth of A at every point in the model). This is just another way of saying that the frames mentioned in Coro. 4.2.6 (or in Theorem 4.2.3(ii)) satisfy the two-step condition. Returning to the informal mode, let us quote [91]'s explanation of the 'reduction' idea (p. 249):

> This concept of nomic possibility is closer to the pre-philosophical concept of possibility than is abstract logical possibility. According to this pre-philosophical concept of possibility, it is no more possible for a cow to jump over the moon than it is to construct a five-sided square. Consistency may be a necessary condition for possibility, but it is not a sufficient one. If nomic possibility were to be taken as providing an analysis of this possibility concept, consistency or logical possibility could in turn be provided with a reductive analysis in terms of nomic possibility. Logical possibility becomes what is possibly possible, a result which itself does not lack intuitive pre-analytic appeal.

The last three sentences here are also the final three sentences of [864], (except that 'pre-analytic' appears there as 'pre-analysis'), though [91] goes on to qualify—or perhaps rather to clarify—the position:

> We have just argued that there is an equivalence between logical necessity and iterated nomic necessity. From a formal point of view, it is tempting to construe this equivalence as a definition and to define logical necessity in terms of nomic necessity. For the reasons given Chapter 3, however, this would be a mistake. Logical and nomic necessity are semantically quite distinct. The equivalence between them is not a purely logical or definitional matter. It requires a metaphysical lemma, namely, the accessibility of a lawless Hume or Heimson world corresponding to each lawful world.

Those interested in "the reasons given in Chapter 3" can read Bigelow and Pargetter [91] for themselves. As for the remark that "(l)ogical and nomic necessity are semantically quite distinct", this is

4.2. NOMIC NECESSITY I: PARGETTER

hardly relevant to the claim under discussion, since the purported definition of 'it is logically necessary that A' is not 'it is nomically necessary that A', but 'it is nomically necessary it is nomically necessary that A'. To bear on this claim, the remark would have to have read: "logical necessity and *iterated* nomic necessity are semantically quite distinct". Taking this correction to have been made, we can see Bigelow and Pargetter as justifying the claim (thus adjusted) that this is not a matter of definition on the grounds that it holds only in the presence of a substantive metaphysical assumption (a "metaphysical lemma", as they put it). Perhaps the point is that even someone prepared to make this assumption need not treat its correctness as an *a priori* matter – not, then, a matter to be enshrined in a definition. If so, Bigelow and Pargetter are using the term 'reduction' in the way it is sometimes used in connection with scientific reductions, as when someone says that temperature is reducible to mean kinetic energy, intending an *a posteriori* identity claim which is consistent with the claim that the property-denoting terms express distinct concepts (have different senses, that is).

If that were correct, however, then a puzzle would emerge elsewhere – quite apart from the implausibility of treating the current metaphysical hypothesis as capable of *a posteriori* support – in another idea taken over from [864], where it is expressed at the base of p. 347; in [91], the relevant passage is on p. 248:

> Thus, although we have argued that it is not possible to give a reduction of the nomic concepts in terms of logical ones, it may be possible to give a reduction the other way around. Hence the theory is not committed to a multiplicity of independent, primitive modal concepts, only to the acceptance of a single core modality.

But how is the passage quoted previously to be read as doing anything (with its objections to treating the equivalence as definitional) other than acknowledging a pair of "independent, primitive modal concepts"? Perhaps the point is terminological, depending on how primitivity is in turn spelt out. One thing that becomes clear is that Bigelow and Pargetter are opposed to what they see as an orthodoxy of 'relative necessity' (the title of §5.2 of [91]; on p. 224 we have: "The orthodox theory of the necessity of the laws of nature, of natural or nomic necessity, has been that natural necessity is merely one salient sort of relative necessity". In particular this 'orthodox theory' would claim that nomic necessity is necessity relative to (= conditional upon) the natural laws. (In [91] it seems to be suggested that the 'orthodox theory' must also hold this claim to be explanatory, conceptually illuminating, non-circular, or some such thing.) Bigelow and Pargetter continue (pp. 224–5):

> Suppose, then, that we single out a class A of sentences and designate them (for one reason or another) as laws of nature. This class of sentences demarcates a class C of worlds, the worlds in which all these laws are true. Then we may say that any sentence α is *nomically necessary* just when it is true in all worlds in class C. Nomic necessity is then explained as being simply entailment by the laws. The question arises, how are we to single out the set A of sentences which count as laws? Clearly, which sentences count as laws in a world will depend on what that world is like. A sentence may be a law in one possible world, but not a law in another. (...) Corresponding to each possible world w, then, there will be a set A_w of the sentences which are laws in w. This will determine a set C_w of worlds whose events fit all the laws of w. (...) We can define a notion of accessibility as follows. Any world u is accessible from a given world w just when $u \in C_w$, that is, when u is compatible with all the laws of w. Then we can say that a sentence α is nomically necessary in a world w if and only if α is true in all worlds accessible from w. (...) Note that, in thus construing nomic necessity as a kind of relativized necessity, we define nomic necessity by appeal to a prior designation of a class of sentences which count as *laws* for a given world. Laws of nature are used to define natural necessity. Hence we cannot then define a law of nature to be a generalization which is nomically necessary.

The point of the last two sentences here is that there would be a circularity involved if we both (*a*) defined the concept of nomic (or natural) necessity in terms of the concept of a natural law, and (*b*) defined the concept of a natural law in terms of nomic necessity. [91] continues (p. 225):

> It has been widely believed that the notion of natural laws cannot be explained in terms of a modal concept of natural necessity. It has been assumed that natural necessity must be explained as relative necessity – necessity relative to laws. If that were so, laws would have to be explained nonmodally.

The last sentence presumably means that the idea of *being a law* has to be explained (that is, explicated or defined) in non-modal terms.[181]

On p. 238 of Bigelow and Pargetter [91], the authors remark that according to their usage logical necessities count as a special case of natural (*sc.* nomic) necessities, "since whatever is true in all worlds is true in all accessible worlds". Quite right. They remark parenthetically:

> Often the term 'natural necessity' is used to mean *merely* natural necessity, a natural necessity which is not logical necessity. But it is tidier if we allow that, strictly, natural necessities include logical necessities as a proper subclass.

Quite right again: it is *much* tidier to do this. What would life be like if we considered 'merely' nomic necessity in its own right? What would its logic be like, considered in isolation from nomic necessity in the broader sense? This question we will investigate in Section 4.3: although it is untidy, the technical challenge has some interest in its own right. But this issue is connected with another reason, apart from any circularity problems of the kind raised above, why we (re-quoting from above) "cannot then define a law of nature to be a generalization which is nomically necessary" – namely because there would then be no worlds without laws of nature, such as the Hume (or Heimson) duplicates that motivated the present enterprise: in these worlds it would still be nomically necessary, for instance, that *if p then p*. If the point is to be put in terms of any kind of necessity, it would seem to be that in the lawless worlds, nothing is *merely* nomically necessary.

EXERCISE 4.2.7 (*i*) If we define $\Box' A$ as $\Box\Box A$ in **KTPar**, does $\Box'$ satisfy the rules (governing $\Box$) and axioms of any selected axiomatization of **S5**? (Justify your answer, here and for (*ii*).)

(*ii*) Suppose we had a normal bimodal logic with $\Box'$, now taken as primitive and satisfying the axioms of **S5**, and $\Box$ satisfying the axioms of **KTPar**. Does it follow that the resulting logic has amongst its theorems $\Box' p \leftrightarrow \Box\Box p$? ✠

4.3 Nomic Necessity II: Bacon

John Bacon [48] addresses the question of the logic of what is there called *purely* physical necessity ('purely nomic necessity' would be the more neutral term for what he has in mind), meaning by this: the property of being physically (better: 'nomically') necessary but not logically necessary. Bacon writes "N" for the modal operator for which this is the intended reading. In a bimodal logic with operators for both types of necessity, say nomic necessity represented by $\Box$ and logical (or broadly logical, or metaphysical) necessity represented by $\Box'$ (echoing Exercise 4.2.7) for which we here use the notation introduced at the end of Section 4.2, one can define pure nomic necessity, of course: $N\varphi =_{\mathsf{Df}} \Box\varphi \wedge \neg\Box'\varphi$. (We revert to our usual practice of not writing the "Df" in such definitions for the rest of this section, and change to $\varphi, \psi, \ldots$ as schematic letters – see p. 3 – rather than $A, B, \ldots$) But Bacon is interested in what the logic of N looks like when it is considered on its own – that is, when we add to the usual expressive resources of truth-functional logic just one modal operator, N, wanting it to behave as it would if it had been defined into existence in accordance with the above definition. Thus, although only one modal operator is present (one non-Boolean primitive, that is), its semantics is given in terms of frames with two accessibility relations.[182] Unlike the material reviewed in Section 4.2, no philosophical significance

---

[181]One may wonder *by whom* it had "been assumed that natural necessity must be explained as relative necessity" – who says "must be", as opposed to "may be"? According to John Bigelow (personal communication), he had in mind Ernest Nagel, Carl Hempel, and David Lewis here.

[182]One could say "bimodal frames" for such bi-relational frames, though that label would be confusing here since the two relations are used together to give the semantics of a monomodal logic.

4.3. NOMIC NECESSITY II: BACON

is claimed for this development, recounted in what follows for its technical interest; trying to axiomatize the restricted notion of purely nomic necessity in the sense described but without the general notions of nomic and logical necessity available as separate primitives has something of the feel of working with one's hands behind one's back. One can learn from such challenges, all the same. But before getting into that development, a historical note is in order, presented here as an example. (An example of what? An example of how to make mistakes in hoping to apply one's favourite modal logic.)

EXAMPLE 4.3.1 Lemmon [694], p. 183, had suggested some modal logics he had taken a special interest in, E2 and E3 (mentioned under Example 2.1.18(i), p. 46), as providing in their $\Box$ operators something suitable for reading as "it is nomically but not logically necessary that". (In fact Lemmon writes "scientifically" rather than "nomically" here.) The favoured systems were not closed under the rule of Necessitation – which is perhaps what prompted this suggestion – but they are monotone, and this, as Cresswell [201], p. 199, observed (or see Hughes and Cresswell [535], note 355 on p. 302) undermines the proposed interpretation. For example, suppose it really is nomically but not logically necessary that p, and consider the provability in any monotone modal logic of $\Box p \to \Box(p \lor \neg p)$, or, again, that of $\Box p \to \Box(q \to q)$. The consequents here are certainly false, whatever sentences of an interpreted language one imagines being plugged in for the propositional variables, on the suggested reading of $\Box$. The present concept does not lend itself to a monotone treatment, then. ◀

Let us return to Bacon [48], and denote the accessibility associated with (the *absent* operator for) nomic necessity by S, and that for logical necessity (also absent!) by R.[183] Of course, we could confine our attention, if we did not like the idea of restricting the class of worlds which should be considered for logical necessity, to those bi-relational frames $\langle W, R, S \rangle$ in which $R = W \times W$. W, as usual, should be taken to be some non-empty set.[184] However, Bacon's restrictions are somewhat different. He requires that:

(1) $S \subseteq R$;   (2) *S is reflexive*;   (3) *R is reflexive*.

Note that (3) follows from (1) and (2). The definition of truth at a point in a model $\mathcal{M} = \langle W, R, S, V \rangle$, based on one of these frames, is as usual for the atomic formulas and compounding by Boolean connectives, with the clause for N following straightforwardly the motivation for its introduction; for all formulas φ:[185]

[N] $\mathcal{M} \models_x N\varphi$ iff (i) $\forall y \in S(x)$, $\mathcal{M} \models_y \varphi$, and (ii) $\exists z \in R(x)$, $\mathcal{M} \not\models_z \varphi$.

(The change of quantified variable from y in (i) to z in (ii) is of course purely cosmetic.) Below, we shall make use of a further abbreviation, suppressing the reference to $\mathcal{M}$ when which model is at issue is clear from the context, writing, where $T \in \{R, S\}$, "$T(x) \models \varphi$" to mean "$\forall y \in T(x), \mathcal{M} \models_y \varphi$", and writing "$T(x) \not\models \varphi$" for the negation of this claim. (Note that this does not mean that for all $y \in T(x)$, $\mathcal{M} \not\models_y \varphi$.) In this notation, the r.h.s. of [N] appears thus: (i) $S(x) \models \varphi$ and (ii) $R(x) \not\models \varphi$.

The task is, then, to find an axiomatization of the logic determined by the class of all frames $\langle W, R, S \rangle$ satisfying (1), (2) and (therefore) (3) above. No doubt, this is methodologically not the best way to

---

[183] These are the operators written above as $\Box$ and $\Box'$, respectively.

[184] When Bacon sets up the semantics (p. 137), he defines models to have, for their sets of worlds, sets of *sets of formulas*. This is ill advised, and seems to involve gratuitously imposing a special feature of so-called canonical models on models in general, or even to involve pushing some philosophical barrow – a form of 'linguistic ersatzism' about possible worlds, as it is called in Lewis [726] – on what is hardly an appropriate occasion.

[185] Greek letters "φ", "ψ",... (as well as α, β, occasionally below), as schematic letters here since we are using N rather than $\Box$ – see the remarks on p. 3 above; Bacon uses p, q,... for this purpose but as always we take these to be, instead, propositional variables. Bacon also uses "$\supset$" rather than "$\to$", for material implication. When Bacon sets up the semantics ([48], p. 137), he defines models to have, for their sets of worlds, sets of *sets of formulas*. This is misguided, and seems to involve gratuitously imposing a special feature of so-called canonical models on models in general, or even to involve pushing some philosophical barrow – a form of 'linguistic ersatzism' about possible worlds, as it is called in Lewis [726] – on what is really not the right occasion.

approach the subject: the first thing to do should be to axiomatize the basic logic in this semantic framework: the logic (for N) determined by the class of all frames. One should worry about imposing specific conditions only later. However, let us proceed to Bacon's solution of the problem he has set himself. The intermingling of positive and negative elements – i.e., (i) and (ii) respectively – in the clause [N] makes it far from obvious what the solution should be.

The axiom schemata Bacon [48] proposes to subjoin to any basis for truth-functional logic (with $\neg$, $\to$, primitive, and presumably Modus Ponens amongst the primitive or derived rules) are the following, the first two of which are the **T** (or "$\mathbf{T}_N$" to use the kind of notation suggested on p. 178) and Aggregation principles (for N); note that anyone objecting to **T** for nomic necessity (see Remark 4.2.1, p. 217 and note 177, p. 219), would also reject this principle for 'purely nomic' necessity, though this is not a question we are going into here:

(N1) $\quad N\varphi \to \varphi$

(N2) $\quad (N\varphi \land N\psi) \to N(\varphi \land \psi)$

(N3) $\quad N((\varphi \to \psi) \to \psi) \to N((\psi \to \varphi) \to \varphi)$

(N4) $\quad (\neg N(\varphi \to \psi) \land N\psi) \to (N\chi \to N((\varphi \to \psi) \to \chi))$,

together with the two rules:

$$(\text{N5}) \quad \frac{\varphi}{\neg N\varphi} \qquad\qquad (\text{N6}) \quad \frac{\varphi \to \psi}{N(\chi \to \psi) \to (N\varphi \to N\psi)}$$

In a footnote appended to (N3), Bacon says: "I suspect that N3 may be redundant". In more familiar notation, what (N3) says is:

$$N(\varphi \lor \psi) \to N(\psi \lor \varphi).$$

So it would be redundant if we had a 'rule of congruentiality' for N, derivable without recourse to (N3):

$$(N\text{Cong}) \quad \frac{\varphi \leftrightarrow \psi}{N\varphi \leftrightarrow N\psi}$$

As it happens, Bacon's derivation of this rule does make use of (N3).[186] In normal modal logics (N taken as a $\square$ operator), we have the stronger *monotone* rule for N ("NMono") – from which of course (NCong) follows – allowing passage from $\vdash \varphi \to \psi$ to $\vdash N\varphi \to N\psi$. But this rule fails for the present understanding of "N" (i.e., if fails to preserve validity on the above bi-relational frames, when interpreted in accordance with [N]), as we saw already in Example 4.3.1, with various counterinstances. For a somewhat different counterexample, take $\varphi = p \land q$, $\psi = p$; as Bacon remarks (p. 137), this formula is not valid: $N(p \land q) \to Np$; nor of course, is this: $N(p \land q) \to Nq$.

EXAMPLE 4.3.2 By contrast, the following formula is valid: $N(p \land q) \to (Np \lor Nq)$, as the reader is invited to check. (This is called T5 in Bacon [48]. Actually, what he remarks on is the *provability* of this as against the unprovability of the versions which drop one of the disjuncts, but it is in any case not hard to see that the contrast holds in respect of validity on the frames we are concerned with – indeed on any frames, provided the current truth-definition is in force.)

Instead of applying the semantics directly, one may prefer to look at the bimodal translation in which $N\varphi$ is replaced by $\square_1\varphi \land \neg\square_2\varphi$, with each of $\square_1$, $\square_2$, presumed normal. (As with what were written as $\square$ and $\square'$ in the opening paragraph of this section.) The above formula then translates into:

---

[186] The rule appears on p. 136 of Bacon [48], under the name T3.

4.3. NOMIC NECESSITY II: BACON

$$(\Box_1(p \wedge q) \wedge \neg(\Box_2(p \wedge q))) \to ((\Box_1 p \wedge \neg\Box_2 p) \vee (\Box_1 q \wedge \neg\Box_2 q)).$$

The provability of this in the smallest normal bimodal logic is clear enough; the same goes for the aggregativity of the defined N. (That is for satisfying the condition on $\Box$ called Aggregation on p. 21.) This way of obtaining aggregative non-monotone operators often arises in the literature on modal logic. For example, Jones and Pörn [628] present a bimodal version of deontic logic with operators $\Box_1$ and $\Box_2$ (as we shall write them here), each with the logic **KD**, semantically interpreted by universal quantification over worlds thought of as respectively *ideal* and *sub-ideal* w.r.t. a given world,[187] and then they take Ought φ, the primary intended formalization for a statement as to what should be the case, not to amount – as one might expect – to $\Box_1 \varphi$, but rather to $\Box_1 \varphi \wedge \neg\Box_2 \varphi$, in order to block various monoton(icit)y-related paradoxes of deontic logic as well as to implement the idea that genuine obligations are inherently susceptible of being violated; for more detail see Example 4.4.47 on p. 277 below.

Another application of the $\Box_1 \varphi \wedge \neg\Box_2 \varphi$ idea to deontic logic is suggested in Åqvist [32]. Think of this as saying that all the best worlds verify φ though it is not true that all the morally permissible worlds verify φ, to get a feel for Åqvist's proposal that this might be a good way of saying that it would be supererogatory for it to be the case that φ; see further the Digression on p. 255, and the text preceding it.[188]

Further variations on this 'partly positive, partly negative' theme are worth mentioning. Instead of N defined as above, we could consider N' defined by:

$$N'\varphi = \Box_1 \varphi \wedge \Box_2 \neg\varphi;$$

that is, using impossibility rather than non-necessity in the second conjunct. A $\Box$-operator is essentially defined along the lines of N' in vander Nat [840], though the $\Box_1$ and $\Box_2$ are not made explicit in the object language. (Rather the semantics for the single $\Box$ requires for $\Box\varphi$'s truth at w the truth of φ at all points R-related to w and the falsity of φ at all points S-related to w, where R and S are the two accessibility relations each frame comes equipped with. In the special case of in which S is the complement (relative to $W \times W$) of R, we get the modal logic of what is true at *all and only* accessible worlds, studied in Humberstone [556].) Further, one might replace conjunction with disjunction, giving:

$$N''\varphi = \Box_1\varphi \vee \neg\Box_2\varphi \quad \text{or again} \quad N'''\varphi = \Box_1\varphi \vee \Box_2\neg\varphi.$$

In all cases the operator defined is neither monotone nor antitone (as defined on p 20, though here applied to a particular operator rather than to the logic). The definition of N' would come to the same thing as that of N if the logic of $\Box_2$ were (at least as strong as) **KD!**. This is what happens in the logic discussed in Chapter 6 below, where in addition $\Box_1 \varphi$ is taken as φ itself – and N' is written as D. (Alternatively put: where $\Box_1$ enjoys the logic **KT!**; see also Steinsvold [1088].) If $\Box_2 = \Box_1$, then N''' is an operator for noncontingency (more commonly notated as Δ, mentioned, for example, in Exercise 1.3.8 on p. 21, and usually notated as "Δ"); a related case is studied in Marcos [770] and Steinsvold [1087]. ◂

More generally, Bacon shows that a formula is provable on the basis of the above axioms and rules if and only if the formula is valid on every frame satisfying conditions (1), (2) (and therefore also (3)) from the previous section. The 'only if' direction here is the *soundness* of the present proof system with respect to the class of frames in question, and the 'if' direction is the *completeness* of the system w.r.t.

---

[187]That is, frames take the form $\langle W, R_1, R_2 \rangle$ with R_1 and R_2 serial; in addition these relations are subjected to the conditions (i) $R_1 \cap R_2 = \emptyset$ and (ii) for all $x \in W$, either xR_1x or xR_2x. It is clear from their informal discussion, however, that they are thinking of R_2 as the complement of R_1; see Humberstone [550].

[188]Yet another deontic variation, favoured at one time by Anderson (see Anderson and Moore [21], p. 17), was to build in a contingency conjunct into the definition of obligation; thus $O\varphi$ – though Anderson wrote O' to avoid confusion with the usual *normal* deontic box operator – is defined as $(\Diamond\varphi \wedge \Diamond\neg\varphi) \wedge \Diamond(\neg\varphi \wedge \neg S)$ in which $\Diamond$ represents broadly logical possibility and S is Anderson's "sanction" constant, on which see the discussion beginning at p. 253 below. Note that we don't actually need the first "$\wedge \Diamond\neg\varphi$" in of the later conjunct "$\Diamond(\neg\varphi \wedge \neg S)$". Thus where O is the familiar obligation operator, Anderson's $O'\varphi$ amounts to $O\varphi \wedge \neg\Box\varphi$.

that class. The soundness claim involved is easily verified: one just checks that the axioms are all valid on the frames in question, and that the rules preserve validity on such frames. Now notice that the only axiom/rule which requires appeal to any of the conditions (1)–(3) is (N1). In particular, to check that every instance of (N1) is valid on all these frames, suppose that we have a model $\mathcal{M} = \langle W, R, S, V \rangle$ on such a frame, and that for $x \in W$ we have (for a given choice of φ) $\mathcal{M} \models_x N\varphi$. Thus, by [N], $S(x) \models \varphi$ and $R(x) \not\models \varphi$. Since $S(x) \models \varphi$ and $x \in S(x)$ – as S is reflexive – we have $\mathcal{M} \models_x \varphi$. So at any point in a model at which $N\varphi$ is true, φ is true, meaning that $N\varphi \to \varphi$ is true at every point in every model (on one of the frames under consideration). What we had to appeal to here was simply the reflexivity of S: condition (2) above. The reflexivity of R (condition (3)) was not needed, and nor was the inclusion $S \subseteq R$ (condition (1)). And, as remarked, the remaining axioms (and rules) require none of these conditions: they are valid (resp., preserve validity) on all frames, and so can be regarded as consequences of the form of the clause [N] in the truth-definition.

Consider, by way of example, the case of (N2). If the antecedent (of some instance of this schema) is true at a point x in some model, then we have (1a) $S(x) \models \varphi$ and (1b) $R(x) \not\models \varphi$ (for $\models_x N\varphi$) and also (2a) $S(x) \models \psi$ and (2b) $R(x) \not\models \psi$ (for $\models_x N\psi$). We want to show that the consequent of (N2) is also true at x. From (1a) and (2a) we conclude that $S(x) \models \varphi \wedge \psi$, while from either (1b) or (2b) alone – let alone both of them together – we conclude that $R(x) \not\models \varphi \wedge \psi$. Since $S(x) \models \varphi \wedge \psi$ and $R(x) \not\models \varphi \wedge \psi$, by [N], we have $\models_x N(\varphi \wedge \psi)$; since $N(\varphi \wedge \psi)$ is the consequent of (N2), we are done.

When it comes to proving the completeness of his axiomatization (w.r.t. the class of frames satisfying (1)–(3)), Bacon makes heavy use of a couple of derived rules, the first of which is:

(R1) $$\frac{(\chi_1 \wedge \ldots \wedge \chi_n) \to \varphi}{(N\chi_1 \wedge \ldots \wedge N\chi_n) \to (N(\varphi \vee \psi) \to N\varphi)}$$

The second rule is as follows:[189]

(R2) $$\frac{((\psi_1 \vee \chi_1) \wedge \ldots \wedge (\psi_n \vee \chi_n)) \to \varphi}{[(N\chi_1 \wedge \neg N(\psi_1 \vee \chi_1)) \wedge \ldots \wedge (N\chi_n \wedge \neg N(\psi_n \vee \chi_n))] \to \neg N\varphi}$$

Both the above schematic formulations are to be understood as allowing $n \geqslant 0$ (rather than requiring $n \geqslant 1$). Note that when $n = 0$, (R2) gives Bacon's 'non-Necessitation' rule (N5) (another respect – in addition to the failure, noted above, of monotony – in which the logic of N is not a normal modal logic). When $n = 0$ in (R1), we conclude from the provability of φ that $N(\varphi \vee \psi) \to N\varphi$ is provable. This may seem to be in tension with non-Necessitation (i.e., (N5)): we know on the same hypothesis (namely, that φ is provable) that $\neg N\varphi$ is provable. But together, these tell us (by Modus Tollens) only that $\neg N(\varphi \vee \psi)$ is provable, and this is something we had already, since from the provability of φ, we get the provability of $\varphi \vee \psi$, from which this follows by non-Necessitation.

The interest of these ingenious rules of Bacon's lies in their role in the completeness proof below. They do almost all of the work there, in fact, so that one can imagine working with the system defined by (R1) and (R2); the only other ingredients we need are the **T**-axiom (N1), and the congruentiality of N. In fact, as we shall see, the congruentiality rule for N is derivable from (R1) and (R2), so that it does not require the separate billing that the formulation just given suggests. To work with just these rules (on top of the usual truth-functional basis), one would need to check that (R1) and (R2) preserve validity—not having derived them from some basis whose soundness had already been established. The key to understanding (R1) and (R2), which provides the check just alluded to, comes from the following lemma, in which we have used the notation "$\varphi \Vdash \psi$" to mean this: in every model $\mathcal{M} = \langle W, R, S, V \rangle$,

---

[189]In fact Bacon does not put things quite this way; but his ML1 and ML2 from [48], p. 138, can be regarded as arguments that these rules are derivable. I have also replaced some implicational subformulas by disjunctions to increase intelligibility.

4.3. NOMIC NECESSITY II: BACON

regardless of whether conditions (1)–(3) are satisfied, but using the clause [N] in the definition of truth, *if* $\mathcal{M} \models_x \varphi$ *then* $\mathcal{M} \models_x \psi$, for any $x \in W$. (We reach back to the start of the Greek alphabet for schematic letters here, to avoid confusion with particular choices of formulas for "φ", "ψ", etc., below.)

LEMMA 4.3.3 *Suppose that any point in any model (not necessarily satisfying conditions (1)–(3)) verifying α verifies β. Then for any $\mathcal{M} = \langle W, R, S, V \rangle$, $x \in W$:*

(i) $\mathcal{M} \models_x N\alpha \wedge \neg N\beta$ *implies* $R(x) \models \beta$;

(ii) $\mathcal{M} \models_x N\beta \wedge \neg N\alpha$ *implies* $S(x) \not\models \alpha$.

Proof. (i): Suppose any α-verifying point verifies β, and $\mathcal{M} \models_x N\alpha \wedge \neg N\beta$. Thus (1) $\mathcal{M} \models_x N\alpha$ and (2) $\mathcal{M} \not\models_x N\beta$. By [N], (1) implies (1a): $S(x) \models \alpha$ and also (1b): $R(x) \not\models \alpha$, and (2) implies that *Either* (2a): $S(x) \not\models \beta$, or (2b): $R(x) \models \beta$. Since β is true wherever α is, (1a) implies $S(x) \models \beta$, so of the alternatives (2a) and (2b) it must be (2b) which holds, as required.

(ii): By a similar argument. ■

We are now ready to check that (R1) and (R2) preserve validity. In fact, we shall find that they have the stronger property of preserving truth throughout any model on a given frame. For let $\mathcal{M} = \langle W, R, S, V \rangle$ be such a model, and suppose that a premiss $(\chi_1 \wedge \ldots \wedge \chi_n) \to \varphi$ for the application of (R1) is true at all $w \in W$. We wish to show that the corresponding conclusion:

$$(N\chi_1 \wedge \cdots \wedge N\chi_n) \to (N(\varphi \vee \psi) \to N\varphi)$$

shares this property. So suppose, for a contradiction, that we have $x \in W$ with $\mathcal{M} \models_x N\chi_i (i = 1, \ldots, n)$ and yet $\mathcal{M} \not\models_x N(\varphi \vee \psi) \to N\varphi$. Then:

$$\mathcal{M} \models_x N(\varphi \vee \psi) \wedge \neg N\varphi.$$

By part (ii) of Lemma 4.3.3 above, with φ as φ and ψ as $\varphi \vee \psi$, we conclude that $S(x) \not\models \varphi$. Since the implication $(\chi_1 \wedge \ldots \wedge \chi_n) \to \varphi$ is true throughout the model, this means that for at least one i ($1 \leqslant i \leqslant n$) we have $S(x) \not\models \chi_i$; however this is incompatible (in view of [N]) with the fact that $\mathcal{M} \models_x N\chi_i$. So much for (R1).

Turning to (R2), again we begin by supposing that a premiss for this rule, $((\psi_1 \vee \chi_1) \wedge \ldots \wedge (\psi_n \vee \chi_n)) \to \varphi$ is true throughout $\mathcal{M}$ ($= \langle W, R, S, V \rangle$) but that the conclusion is not. So we have $x \in W$ with $\mathcal{M} \models_x N\chi_i \wedge \neg N(\psi_i \vee \chi_i)$ for $i = 1, \ldots, n$, while also $\mathcal{M} \not\models_x N\varphi$. Taking χ_i and $\psi_i \vee \chi_i$ as the φ and ψ of Lemma 4.3.3, we may apply part (i) thereof to infer that $R(x) \models \psi_i \vee \chi_i$ for each i ($1 \leqslant i \leqslant n$). Thus since the premiss for this application of (R2) is true throughout $\mathcal{M}$, we conclude that $R(x) \models \varphi$. This conflicts with our earlier finding that $\mathcal{M} \not\models_x N\varphi$, and completes the proof that (R2) preserves truth throughout a model.

Conspicuously, for both these applications of Lemma 4.3.3, the required relation between φ and ψ (namely, that for all models, ψ is true at any point at which φ is) is secured syntactically in the same way: by taking φ as one disjunct of the disjunction ψ. We should recall that this is only one possibility, though. For example, we could equally well take φ as a conjunction of which ψ is one conjunct. In fact, it is as well to give some thought to putting matters more neutrally in the formulation of (R1) and (R2). In particular, if we make a modification along these lines to (R1), so that instead of using $\varphi \vee \psi$ we use anything provably implied by φ (for which role we use 'ψ' here), we get a rule I shall call (R1)* which delivers congruentiality as a pleasant by-product, as well as yielding (R1) as derivable:

$$\text{(R1)*} \quad \frac{(\chi_1 \wedge \ldots \wedge \chi_n) \to \varphi \qquad \varphi \to \psi}{(N\chi_1 \wedge \ldots \wedge N\chi_n) \to (N\psi \to N\varphi)}$$

To obtain congruentiality for N (i.e., to derive the rule (NCong) above), let $n = 1$ and put ψ for χ_1. Then, from the provable equivalence of φ with ψ, (R1)* gives the conclusion $N\psi \to (N\psi \to N\varphi)$, which is truth-functionally equivalent to $N\psi \to N\varphi$; by interchanging the roles of φ and ψ, we get the converse implication. (In fact, as already mentioned, (NCong) is derivable from (R1) itself, taken alongside (R2), as we shall see below.)

These findings suggest that (in addition to the truth-functional basis) the pair of rules (R1)*, (R2), by themselves axiomatize the logic determined by the class of all frames, without the imposition of special requirements such as (1), (2) and (3) above. Bacon's completeness proof (essentially for this system with the addition of $N\varphi \to \varphi$) adapts to give us this conclusion. In fact, as we shall see, this result can be obtained for the rules (R1) and (R2) themselves. We turn to that proof.

Actually, rather than go through Bacon's own argument, one minor modification will allows us to conduct a canonical model completeness proof in the same general style as those given in Sections 2.4 and 2.5.[190] We need to define the relations R and S in the canonical model, and, following Bacon, put (for all maximal consistent sets x, y):

(Def. S) xSy iff for all φ: $N\varphi \in x \Rightarrow \varphi \in y$

(Def. R) xRy iff for all φ, ψ: $\neg N(\varphi \vee \psi) \wedge N\psi \in x \Rightarrow \varphi \vee \psi \in y$.

V is defined as usual, i.e., as V_S was defined for normal S just before Theorem 2.4.1 (p. 62), except that here we have not assigned a label to Bacon's logic and so have nothing to put in the subscript position. ("Normal S" here means "consistent normal modal logic S": no connection with the use of "S" for a binary relation introduced in (Def. S) above.)

We have next to show, as in the proof of Theorem 2.4.1, that a formula belongs to a point in this model if and only if it is true at that point in that model; V having settled this for the propositional variables, and the notion of maximal consistency taking care of the cases of the Boolean connectives in the inductive part of the proof, we need to deal with the case in which our formula is of the form $N\varphi$. Since φ is of lower complexity than $N\varphi$, we are entitled to assume (the 'inductive hypothesis') that φ itself is true at a point if and only if φ belongs to that point, which allows us to rephrase what has to shown in the case of $N\varphi$ thus:

(*) $N\varphi \in x \Leftrightarrow (i) \forall y \in S(x), \varphi \in y \ \& \ (ii) \exists z \in R(x), \varphi \notin z$.

The $\Rightarrow$-direction of (*) is easily established: Suppose $N\varphi \in x$. Then if xSy, it follows immediately from (Def. S) that $\varphi \in y$, giving (i). For (ii), we consider, the set $\{\psi \vee \chi \mid N\chi \wedge \neg N(\psi \vee \chi) \in x\} \cup \{\neg\varphi\}$. This set is consistent, since otherwise, by appeal to (R2), we get a contradiction with the supposition that $N\varphi \in x$, and so has a maximal consistent superset, to which x bears the relation R as defined by (Def. R), and further, completing our verification that (ii) holds, to which φ does not belong.

The argument (adapted from [48]) for the $\Leftarrow$-direction of (*) requires a lemma:

LEMMA 4.3.4 *For the canonical model $\langle W, R, S, V \rangle$ we have the following: for any $x \in W$, and any formula φ, if $\varphi \in y$ for all y such that xSy, then there are formulas $\chi_1, \ldots, \chi_n$ with each $N\chi_i \in x$ $(i = 1, \ldots, n)$, and $\vdash (\chi_1 \wedge \ldots \wedge \chi_n) \to \varphi$.*

Proof. Take any φ such that $\varphi \in y$ for all y such that xSy. Let Γ be the set all the formulas χ for which $N\chi \in x$. We want to show that some finite subset $\{\chi_1, \ldots, \chi_n\}$ of Γ has a conjunction provably implying φ. If not, then $\Gamma \cup \{\neg\varphi\}$ is consistent, in which case it has a maximal consistent superset, which by the

---

[190]This will mean by-passing the inductive definition given in the top part of Bacon [48], p. 139, and just considering, as the points in our ('canonical') model, *all* maximal consistent sets of formulas (w.r.t. Bacon's logic, as axiomatized above). Also, the proof (that truth at a point in this canonical model and membership in that point, considered as a set of formulas, coincide) below is slightly reorganized.

4.3. NOMIC NECESSITY II: BACON

way Γ was defined, is a $y \in W$ such that xSy while $\varphi \notin y$ – contradicting the choice of φ and a formula belonging to every element of the canonical model S-accessible to x. ∎

Returning to the $\Leftarrow$-direction of (*): Suppose (i) $\forall y \in S(x), \varphi \in y$ and (ii) $\exists z \in R(x), \varphi \notin z$. We want to show that $N\varphi \in x$. By (ii) we have some $z \in R(x)$ to which φ does not belong, even though, by (i), φ belongs to each point S-related to x. Therefore *not* xSz (for this choice of z). Thus, by (Def. S), there exists a formula ψ_1 such that $N\psi_1 \in x$, $\psi_1 \notin z$. Further, given (i), by Lemma 4.3.4, there must exist formulas $\chi_1, \ldots, \chi_n$ such that in each case $N\chi_i \in x$, and $\vdash (\chi_1 \wedge \ldots \wedge \chi_n) \to \varphi$, so by (R1):

$$\vdash (N\chi_1 \wedge \ldots \wedge N\chi_n) \to (N(\varphi \vee \psi) \to N\varphi),$$

for all formulas ψ, and hence the consequent $N(\varphi \vee \psi) \to N\varphi$ belongs to x for all formulas ψ, and so, in particular, for the formula ψ_1 mentioned above. If we can show that $N(\varphi \vee \psi_1) \in x$, then we shall be able to conclude, as desired, that $N\varphi \in x$. So suppose, for a contradiction, that $N(\varphi \vee \psi_1) \notin x$. Then $\neg N(\varphi \vee \psi_1) \wedge N\psi_1 \in x$, since we already knew that $N\psi_1 \in x$. Now recall the point z introduced above: since $z \in R(x)$ and $\neg N(\varphi \vee \psi_1) \wedge N\psi_1 \in x$, by (Def. R) we have $\varphi \vee \psi_1 \in z$. This is impossible, as we already had, above, that $\varphi \notin z$ and $\psi_1 \notin z$.

This is not the end of (our version of) Bacon's completeness proof, however, since although we now have, by standard arguments, that any non-theorem of the logic is false at some point in the canonical model $\langle W, R, S, V \rangle$ with which we have been working – which gives part (i) of Theorem 4.3.5 below – we have yet to check that Bacon's frame conditions (1) and (2) (and therefore (3)) are satisfied.[191] For (1), $S \subseteq R$, we argue: suppose xSy (for $x, y \in W$), and that for an arbitrarily selected pair, φ, ψ, of formulas, we have $\neg N(\varphi \vee \psi) \wedge N\psi \in x$; to show, as desired, that $\varphi \vee \psi \in y$ (and hence, since x, y are themselves arbitrary, that xRy), we note that since $\neg N(\varphi \vee \psi) \wedge N\psi$ is an element of x, so is $N\psi$, and accordingly, since xSy, $\psi \in y$, whence $\varphi \vee \psi \in y$. For (2), the condition that S is reflexive, we need to appeal to (Def. S) and to (N1) ($= N\varphi \to \varphi$), from which the reflexivity of S follows by considerations (see Exercise 2.4.6) familiar from the canonical model completeness proof for **KT**. These considerations give parts (ii) and (iii) of:

THEOREM 4.3.5 (i) *The smallest modal logic closed under* (R1) *and* (R2) *is determined by the class of all frames* $\langle W, R, S \rangle$.

(ii) *The smallest modal logic closed under* (R1) *and* (R2) *is determined by the class of all frames* $\langle W, R, S \rangle$ *in which* $S \subseteq R$.

(iii) *The smallest modal logic closed under* (R1) *and* (R2) *and containing all instances of schema* (N1) *(= Bacon's logic) is determined by the class of all frames* $\langle W, R, S \rangle$ *in which* S *is reflexive, as well as by the class of such frames with* S *reflexive and* $S \subseteq R$ *(and hence with* R *reflexive also).*

A point of incidental interest here is that since, in view of part (i) of the above result, (R1) and (R2) give—when added to a suitable truth-functional basis—the complete logic of all frames, and (NCong) preserves validity on any frame (indeed, preserves truth in any model), it follows that we can dispense with the latter rule as part of our basis. We may derive it from (R1) and (R2) – a fact most easily demonstrated if we first show another rule, (R3) let's call it, to be derivable from them.

(R3) $\dfrac{\psi \to \varphi}{N\varphi \to N(\psi \vee \varphi)}$

---

[191] Bacon himself appears to have overlooked the need for this verification, in [48], as well as the need – see the end of this paragraph – to check that S is reflexive.

The relevant derivation follows:

(1) $\psi \to \varphi$ (Provable by hypothesis)
(2) $(\psi \vee \varphi) \to \varphi$ 1, TF
(3) $(N\varphi \wedge \neg N(\psi \vee \varphi)) \to \neg N\varphi$ 2, by (R2), taking $n = 1$
(4) $N\varphi \to N(\psi \vee \varphi)$ 3, TF.

Using (R1) and (R3), we obtain (NCong), re-lettered here so as to reduce confusion with the lettering in (R1):

(NCong) $\quad \dfrac{\varphi \leftrightarrow \psi}{N\varphi \leftrightarrow N\psi}$

This we show to be derivable as follows:

(1) $\varphi \leftrightarrow \psi$ (By hypothesis.)
(2) $\varphi \to \psi$ 1, TF
(3) $\psi \to \varphi$ 1, TF
(4) $N\varphi \to N(\psi \vee \varphi)$ 3, by (R3)
(5) $N\varphi \to (N(\varphi \vee \psi) \to N\psi)$ 2, by (R1), with $n = 1$
(6) $N\varphi \to N\psi$ 4, 5, TF

By symmetry, line (1) similarly gives the converse of (6), so, appealing to TF, we can draw the desired conclusion: $N\varphi \leftrightarrow N\psi$.

4.4 Deontic Logic: Main Themes

Deontic logic was described in general terms in Section 1.1, and mentioned many times in the interim (for example as the reason the **D** axiom bears that name, as well as extensively in Section 4.1), but let us recall here that the label 'deontic logic' has been used collectively for applications of modal logic to normative matters. The subject has historically been the site of gross logical blunders (described in Føllesdal and Hilpinen [316] and Powers [903], for instance), as well as of complaints that the formalism of modal logic with its 1-ary non-Boolean connectives (or 'sentence operators') is ill-suited to its subject matter. Such complaints are often made by drawing attention to anomalies resulting from the use in the representation of normative concepts of this traditional modal language and its familiar semantics, so we have the Paradox of Commitment, the Good Samaritan Paradox, Chisholm's Paradox, Ross's Paradox, and several more, the details of which can be gathered from such surveys as Føllesdal and Hilpinen [316], Castañeda [160], Chapter II of Åqvist [30], or §4 of McNamara [793]; see also Bryant [123], Sinnott-Armstrong [1058]; these anomalies will receive less attention here, though some of them – such as Ross's Paradox – are discussed from time to time (see the appropriate index entries), in one case for a whole section (Section 4.8). In fact even the pioneer of the subject decided that the alethic modal analogy failed and deontic operators – often written as O and P rather than $\Box$ and $\Diamond$ – should be replaced with predicates of act-types (von Wright [1204]), so that, for instance, the two occurrences of "¬" in "¬O¬q" are of quite different syntactic categories, the outer occurrence being something that takes formulas to formulas (or sentences to sentences) and the inner occurrence being something that takes names of act-types to names of act-types. (Further elaboration of this point will be found in note 194.) In the main body of this section (and the next) we pursue deontic logic with sentence operators, since we are considering applications of modal logic, rather than predicates (of act-types, or individual actions, to be introduced shortly), though we continue in this introduction of the topic to touch on these non-modal alternatives. (That is a somewhat tendentious description, since even alethic modal logic is sometimes pursued by treating necessity and possibility as predicates of propositions or even sentences, rather than as 1-ary connectives – as in Zalta [1215] or Priest [907]. The latter, metalinguistic, approach

4.4. DEONTIC LOGIC: MAIN THEMES

shades off into Solovay-style completeness proofs for modal provability logic, in which the connection with a metalinguistic predicate – here provability in a system of number theory – with what happens with the $\Box$-operator of a modal logic is not motivated by way of any kind of apology for the latter.[192]

Act-tokens (i.e., individual actions) as well as act-types have found their way into the literature too, with explicit quantification over them urged in Hintikka [489]; further discussion can be found in Makinson [762] and Hilpinen [485]. See also Voorbraak [1135], Meyer [805], van der Meyden [804]. Hintikka's discussion ([489], p. 7) is suggestive in relation to the discomfort people have identifying requirements/obligations with prohibitions – the requirement to φ with the prohibition on not φ-ing ("φ" here a place-holder for a verb phrase). The idea is that what are naturally regarded as obligations are requirements that *at least one* of one's actions should be of the type required, while when prohibitions are re-packaged as obligations they emerge as requirements that *all of* one's actions be of the relevant kind (acts of non-killing, for example). A detailed view of the issues would require entering into the contested area of the identity conditions and essential properties of acts and events; related problems are aired by Makinson in [762].

REMARK 4.4.1 The transcript of a radio interview ([239]) about Judge Henry Hudson on the US Constitution and mandatory health insurance, Brannon Denning as saying "Well, Judge Hudson says that this is particularly unprecedented because what Congress is attempting to do is to force individuals to purchase a product. So in other words, merely by doing nothing, you can't avoid regulation the way you could if, say, it were some kind of prohibition." We are evidently concerned here with legal rather than moral versions of the deontic concepts, and in this connection one cannot help wondering what would be said about the legal obligation to pay income tax: is this a prohibition? One does not escape it by 'doing nothing' – a phrase that would itself clearly repay closer attention. In fact the taxation example surfaces in this connection already in the lively and still interesting interchange between Marcus Singer and Bernard Mayo on the status of the distinction between positive and negative duties ([1056] and [784]), not to mention Isaiah Berlin's negative *vs.* positive distinction. Relatedly, here is Quentin Skinner at note 54 (p. 18) of [1059]:

> Although the writers I am considering generally speak of absence of restraint (rather than constraint), they assume that your liberty is undermined when you are coerced into acting as well as when you are coercively prevented.

The ascription of such an assumption to others in these terms itself appears to involve quite an assumption of its own: that there is a distinction between being ('coercively') prevented from φ-ing, on the one hand, and being coerced not to φ, on the other. Perhaps some version of an act/omission contrast is involved, though in practice many cases do not lend themselves to such a contrast ("forced to stay in" vs. "prevented from going out", for example, sounds like a distinction without a difference). Returning to the specifically deontic realm, further discussion and references, more specific in respect of normative detail, can be found in Belliotti [60] and Lichtenberg [732]. Another very specific but this time logically focussed discussion of the distinction between φ-ing's being forbidden (or wrong) with not-φ-ing's being obligatory (or morally required) appears in the 'Talmudic deontic logic' of Abraham, Gabbay and Schild [3], p. 168; see the Digression on p. 255 below.) ◂

The discussions just remarked on are evidently connected (as there noted) to the status of the act–omission distinction, sometimes handled modally in terms of a sentence operator, D, say, for: "the agent brings it about that", with a contrast between $D\neg p$ and $\neg Dp$. See the final paragraphs of Section 6.1 for pertinent references.

---

[192]To whom might one be apologizing, anyway? To the Quine of [935], perhaps. In this paper, as elsewhere, Quine did his best to discourage work in modal logic, especially quantified modal logic. Since the present concern is with modal logic and philosophical applications thereof, rather than with the philosophy *of* modal logic, these efforts are not on the agenda. Interested readers will find in Burgess [136] a sympathetic account of them.

Digression. If one has both act types ('generic actions') and act tokens ('individual actions') in play, both with the deontic properties of being obligatory and permitted/permissible, and let us throw in also for the present aside, the property of being forbidden (or wrong, to take the normative rather than the reportive option), the question comes up – and is explicitly raised in Makinson [763], some highlights of which are presented here – as to whether the properties applying at the type level are definable in terms of those applying at the token level or vice versa. Let us use a subscript "1" for the deontic predicates as they apply to act tokens: $O_1 a, P_1 a$, and $F_1 a$ saying accordingly that the particular act a is obligatory, permitted, forbidden, respectively, and a subscript "2" for the corresponding predications at the type level, so that $P_2 X$ for instance says that the kind of action X (of telling a lie, for instance), as opposed to some particular action of this kind is forbidden below, we will use "a" as a bindable variable ranging over act tokens. Without subscripts, O and P are, as usual, not predicates but the familiar deontic sentence operators. (One could throw in F for $P\neg$ but why bother?) Makinson [763] considers the possibility of defining the properties of types in terms of those of tokens but for various reasons passes to the option, favoured by von Wright, of effecting a definition in the reverse direction, noting the plausibility of the following definitions of P_1 and F_1 (not that he uses exactly this notation, with the subscripts):[193]

$$\forall a(F_1 a \leftrightarrow \exists X(F_2 X \wedge Xa))$$
$$\forall a(P_1 a \leftrightarrow \neg\exists X(F_2 X \wedge Xa))$$

Thus an action is forbidden if it falls under some forbidden types, and is permitted otherwise. If, for types, being permitted is a matter of not being forbidden, then the r.h.s. of the second definition here is equivalent to $\forall X(Xa \to P_2 X)$: an individual act is permitted if every type it falls under is permitted. But what about obligation – defining O_1 in terms of O_2? Von Wright favoured a definition like those inset above, using (what we call) F_2, and in which $\overline{X}$ is the generic action of not performing any action of kind X:

$$\forall a(O_1 a \leftrightarrow \exists X(F_2 \overline{X} \wedge Xa)).$$

If, for types, being obligatory is matter of the complementary type being forbidden, then the r.h.s. here is equivalent to $\exists X(O_2 X \wedge Xa)$: an individual act is obligatory if it falls under some obligatory type.[194] But, as Makinson points out, this seems initially problematic in view of examples like the following (suggested by Raúl Orayen), concerning the legal obligations of a driver reaching a T-junction. As Makinson writes ([763], p. 9[195]):

> That is, the generic action $X = (X_1$ or $X_2)$ of turning right or left is obligatory, though neither of the generic actions X_1, X_2 is. Now imagine that a certain person at a certain time makes a specific act a of turning left at the junction. Then of course a is of the kind X_1, but it is also of the more general kind X. Thus according to von Wright's reduction scheme, the individual act a would itself be obligatory, although we would ordinarily regard it, in the conditions of the example, as optional.

(The word "optional" is here used for the deontic analogue of contingency, now taken as a property of individual acts: not being either obligatory or forbidden/wrong.) Makinson goes on to point observe that to say of such cases, in response, that a is obligatory insofar as it is an act of turning left or right, while it is not obligatory insofar as it is an act of turning left, is to "refuse to characterize the deontic status

---

[193] However, the second-order style of notation in which $F_2 X$ is a second level predication is taken from Makinson. If preferred, a first order notation with a new relational predicate relating act types to act tokens falling under them could be used instead.

[194] Elaborating a point from the opening paragraph of this section with the current notation to hand: when von Wright writes "$\neg O \neg p$", "p" is occupying the role of the X here (except that it cannot be bound by a quantifier) and the O amounts to O_2, while, awkwardly enough, the outer "$\neg$" is the usual negation connective while the inner occurrence of "$\neg$" is playing the role of the overlining in "$\overline{X}$". All the Boolean connectives lead such double lives on his proposed approach to deontic logic.

[195] In the quotation I have changed "right or left" to "left or right" so as to avoid an unintended 'respectively' reading which would intterfere with the later identification of X_1 with turning left.

of the individual act at all", and further that, on reflection the idea of an individual act as obligatory is itself highly problematic: how could an individual act a be such that ([763], p. 10) "failure to do *that specific action a is forbidden*. In other words, the doing of any other actions instead of a (that is, without also doing a) would still have left the agent in infraction, even if the other actions are similar to a in many interesting characteristics: it had to be that specific individual act a."

Setting aside the issues alluded to above, just before Remark 4.4.1, about individuation and essential properties of actions which arise here (in the form: to what extent could a given *individual action* have been performed in a different way?) and the question of when an individual act belongs to a kind of action (it being essential to the example that one and the same act is of several kinds – something less evidently possible on one understanding of the token/type style of formulation), it does seem that Makinson is right to find the idea of an individual act's being obligatory. In that case, why is there no corresponding problem about an individual act's being forbidden, given that at the level of sentential operators, its being forbidden that A *is* just its being obligatory that $\neg A$? A step in the direction of an answer here may be provided by Davidson's proposal, in [231] and elsewhere, that sentences reporting actions are best understood as existential in force: to say that Paul resigns is to say that *there is* an action a which is an act of resigning and has Paul as agent.[196] The details of how what follows the "which" here do not matter for present purposes, so let us just take them as parcelled up into a predicate φ of actions (or more generally, as Davidson would have it, of events). "Paul resigns" thus emerges as $\exists a(\varphi a)$. This suggests – setting aside the complications of agent-implicating vs. situational obligation-statements mentioned in Example 4.4.14 below – that "Paul ought to resign" be represented as $O\exists a(\varphi a)$. This does not even look as though it says of any particular action that that action ought to be performed. (This is just as well, since the obligation in question may not be fulfilled in which case there will be no such action – here assuming that there are no unperformed individual actions, or more generally, that there are no unoccurring particular events.) On the 'forbidden' side, for "Paul ought not to resign" we have $O\neg \exists a(\varphi a)$. We could equally well formulated this as $O\forall a\neg(\varphi a)$, emphasizing that, by contrast, implies that every φ action should not exist (= be performed).

Some explanation along the above sides of the asymmetry between forbidden and obligatory individual actions may prove successful though there are numerous issues that would need to be settled. One involves an apparent return to the approach of Hintikka [489] and the difficulties raised for that approach in Makinson [762]. However, the most serious of those difficulties pertain to obligation statements which are *de re* – or specific, as one might equally well say here – with respect to individual acts, in the sense of quantifying (over actions) across the scope of a deontic operator, a feature not evident in the examples of the preceding paragraph.[197] Another issue arises over the fact that many obligations concern matters other than the performance of actions, even in the case of what are called agent-implication obligations in Example 4.4.14, as in "Children ought to respect their parents". Whether Davidson's arguments – which have not been touched on here – for the existentially quantified representation of action sentences extend to such cases has been a frequent topic of discussion (see for example, Borowski [114]). **End of Digression.**

REMARKS 4.4.2 (*i*) Another line of exploration applies dynamic logic – the modal logic of programs – to deontic considerations by thinking of act types as analogous to programs and the performance of an action as analogous to the execution of a program. Segerberg [1031] (especially the final section) and

---

[196] In this paper Davidson comments adversely on the adequacy of representations of the "Dp" style – for "the agents brings it about that p" – mentioned just before the present Digression.

[197] In general, a formula containing one occurrence of a modal operator and one of a quantifier – typically interpreted as ranging over the domain of individuals existing at the world w.r.t. which subformula beginning with that quantifier is being evaluated – is said to be *de re* if the quantifier binds a variable in the scope of the modal operator, and otherwise to be *de dicto*. Since this topic pertains to modal predicate logic, nothing more will be said about it here, but further information and references are supplied in §19.1 of Garson [351], in Hughes and Cresswell [537], pp. 250–255, and in Chapters 4 and 5 of Boër and Lycan [104], and we touch on it again at p. 442.

[1033] represent early ventures in this area, with Segerberg [1037] being much more recent; Meyer [805], already mentioned above, falls under this same description.

(*ii*) Although we will concern ourselves very little from this point on with acts – types or tokens – and are not going into the details of the Good Samaritan Paradox, brief mention may be made herd of possible assistance from the former in thinking about the latter. Although helping an assault victim can is something that can only be done if there has been an assault, *O*Mono seems to imply that if one ought to help such a victim then there ought to have been such an assault, whereas common sense seems to say that the obligation to help does not contradict the claim that there ought never to have an assault. (This is the nub of the Good Samaritan Paradox.) However, while the performance of an act of helping an assault victim implies the existence of such an assault, it is not itself an act of assaulting anyone. Confining the effects of monotonicity to the act-typing does not seem to present the same difficulties: for instance any case of helping the victim of an assault by a gang is a case of helping the victim of an assault, and if you ought to help some victim of an assault by a gang, then you ought indeed to help some victim of an assault. (A suggestion along these just given, with the role here played by identity of actions being played by something the authors call inclusion of actions, can be found in Nozick and Routley [850].)

(*iii*) A response like that under (*ii*) here will not impress those claiming that other examples manifest the same phenomenon as the Good Samaritan Paradox – such as the 'Gentle Murder' example in Forrester [325] – but resist solution along these lines. We do not go into the merits of this claim here, since the discussion would require deployment of the conditional obligation construction described in Example 4.4.4 below but not on our main agenda, as well as the contrast between actions and circumstances, or more explicitly between actions 'prescriptively considered' and background circumstances which may themselves include its being given that this or that action is performed – cf. §7.9 of Castañeda [163], and Humberstone [549]. The idea, in the latter discussion at least, is that the unconditional obligations for a given agent are those obligations conditional on that agent's fixed circumstances, where these vary not only from one time to another but from one agent to another;[198] the trying to rely just on the temporal dimension here does not work, as one sees from a referee's objection discussed under (*iii*) at p. 343 of Åqvist [33] in connection with a related ('epistemic obligation') puzzle. But this again takes us further afield, involving, as it does, the idea of agent-relative obligation statements. See Example 4.4.14, p. 251 on these.

(*iv*) After the quoted passage from Makinson in the Digression above, optionality was described as the deontic analogue of contingency. While the context there was of this as a property of actions, the analogy more literally taken returns us to the setting of O and P as deontic operators in the style of our all-purpose $\Box$ and $\Diamond$, and here it should be noted explicitly that what is at issue in the analogy is *contingency whether* rather than *contingency that*. In other words, in the notation of Exercise 1.3.1, p. 17, we are concerned with ∇A understood as $PA \land P \neg A$ (or equivalently $\neg(OA \lor O \neg A)$). In the alethic case its being contingent that A (= its being contingently true that A) could equally well be understood as $A \land \Diamond \neg A$ or as $A \land \nabla A$, since these are equivalent in the presence of the **T** axiom. But for deontic applications of modal logic where we don't expect the latter axiom to be in place these would be non-equivalent analogues of its being contingent that A: $A \land P \neg A$ on the one hand, and the stronger $A \land (PA \land P \neg A)$, only the latter implying the optionality of A. ◀

Aside from quantification over act-tokens, there is the simpler prospect of quantifying over agents, and though deontic predicate logic is not on the agenda here, let us note its special interest for issues of

---

[198]Danielsson ([230] p. 42) writes, similarly: "Now concerning prescriptions, it seems reasonable to distinguish between *circumstances*, on the one hand, and *alternatives* on the other, the former already being the case and the latter still being open possibilities." This does not seem exactly the right formulation, though: nobody is going to confuse the circumstances in which an action is performed with the alternative actions that could have been performed instead. The potential confusion lies in classifying among the circumstances of an action the *non*-performance of those alternatives.

4.4. DEONTIC LOGIC: MAIN THEMES

perennial ethical and meta-ethical interest: the idea of universalizability, and also of what has been called the generalization argument. See respectively Hare [453] (as well as many of Hare's later writings – also very briefly at Example 4.6.15(i), p. 299) and Singer [1055] for these topics, and for some steps in the direction of treating them formally, Kroy [677] and Rabinowicz [938]. (See also the review of the latter, Kuhn [678].) We are sticking with propositional deontic logic – and though quantification over acts or agents (or other individuals) will not arise again, propositional quantification puts in a brief appearance (in note 224, p. 257; see the index under "propositional quantifiers" for additional passages outside of the discussion of deontic logic). We return to von Wright's treatment of the topic, in which what would for parity with other applications of modal logic be sentence operators, end up instead being a kind of predicate (of act types).

Von Wright's approach, as just described, has the awkward feature (over and above – though not unrelated to – that mentioned in note 194) that while we can still formulate and, if desired, endorse such principles as the (two-way) distribution of the obligation or *Ought* operator over conjunction and the principle $\mathbf{D}_O$ for this operator[199] – though "operator" is no longer an apt term on this approach – we can no longer consider either of:

$$\mathbf{T}_O \quad Op \to p \qquad \text{or} \qquad \mathbf{U}_O \colon \ O(Op \to p),$$

since these are no longer well formed. A similar line is taken in Nozick and Routley [850] in criticism of Prior [920], a paper which certainly contains some infelicitous formulations "p is obligatory", presumably intended as a kind of shorthand, though these are readily correctable (to "it is obligatory that p" or "it ought to be the case that p" in the present instance; for Prior's own reply, see [923]). Recall – for example from the start of Section 4.1 – that the usual line in deontic logic has been that while the first of these principles is to be rejected, the second, along with $\mathbf{D}_O$, is a plausible substitute.[200] But to make these judgments – or indeed to repudiate them – one needs the formulas themselves to be well formed.[201] For this reason we shall be sticking with the standard syntax of modal logic, notwithstanding the fact that *directly* iterated deontic operators are hard to make sense of ("OOq", "$OPPOq$", etc.). Such iteration has been the subject of considerable discussion.[202] Here let us simply sound a warning not to change the subject mid-formula, a tendency visible in some of the papers cited in note 202, and think of "OPq", for instance, as saying that it ought (morally) to be (legally) permissible that q: one needs to make up one's mind as to whether one is using the vocabulary to talk about moral obligation or about legal obligation and then stick to that interpretation. (Of course, if you want to talk about both in the same breath, you can always do so with the aid of a bimodal language. A similar sentiment is expressed in the middle paragraph of p. 237 of Hintikka's reply, [493], to Tranøy [1113].) This last example also serves to remind us that the deontic operators can be understood not only in terms of different sets of norms (legal and moral in the present case) but also as either merely reportive or as genuinely normative. So, sticking to the case of moral norms, "Oq" could be a report on its being required by a certain moral code that q, or as actually endorsing this requirement and so amounting to something along the lines of "It ought to be the case that q."[203] It would be a mistake for a non-cognitivist or 'expressivist' (someone holding the meta-ethical view that moral judgments are neither true nor false) to dismiss deontic logic as confused on the

---

[199]As in the case of tense logic we write $\mathbf{X}_O$ as the label for an axiom or schema in which the $\Box$ and $\Diamond$ appearing in a modal principle labelled $\mathbf{X}$ have been replaced by O and P.

[200]Prior [922], p. 229, mentions a suggestion to the effect that $\mathbf{U}_O$ represents a "synthetic *a priori*" truth, and the principle is the focus of attention in Prior [917]; according to Kanger [637], p. 54, it "may be regarded as an expression of what might be called moralism". (Naturally, it is far from clear what either of these assertions amounts to.) There is a criticism of this principle in Chellas [171], rebutted in Vorobej [1137].

[201]In some quarters even the alethic versions of these principles have been questioned as not literally unintelligible, it being urged in Massey [776] for example, in connection with $\mathbf{T} - \mathbf{T}_\Box$, we might say for emphasis, with an alethic reading of $\Box$ in mind – that there is no single way of interpreting "p" in its first and second occurrences in '$\Box p \to p$" such as to render this principal acceptable. (In fact on p. 348 of [776] this is said in the case of the dual form of $\mathbf{T}$: $p \to \Diamond p$.)

[202]Early entrants include Barcan Marcus [52] and Goble [381]; for later thoughts, see Belnap and Bartha [61] and Wansing [1149].

[203]There are many differences between the various modal auxiliary and periphrastic constructions in English in roughly

grounds that moral judgments do not enter into logical relations – even if one thought that this was what non-cognitivism entailed – since it ignores the reportive interpretation: judgments about what this or that moral code requires or permits are not themselves moral judgments. (If these meta-ethical positions are not familiar, explanations can be found in van Roojen [964]. One way of denying the entailment in question can be found in Gibbard [373], esp. Chapter 5; see also Dreier [252] for refinements, variations and qualms.)

In fact, it is probably a sensible policy to keep one's deontic logic independent of any particular meta-ethical or indeed one's normative ethical position, hoping for instance, that even a non-cognitivist will be able to cope with a semantics involving a truth-relation or truth-like relation $\models$.[204] On the normative neutrality front, the last thing one would want would be a 'utilitarian semantics for deontic logic' (the title of Jennings [620]) or a special 'semantics for a utilitarian deontic logic' (the title of Kielkopf [652]) or just plain 'utilitarian deontic logic' (Goble [383]) – not that these authors are in fact doing anything specifically utilitarian (or even, more generally, anything specifically consequentialist).[205] Of course, the recommended neutrality here requires that as well as being compatible with non-utilitarian ethics – a condition Hansson [443] implausibly suggests standard deontic logic fails to satisfy – any efforts in the name of deontic logic should also be compatible with utilitarianism itself.[206]

REMARKS 4.4.3 (*i*) The Hansson referred to just now was Bengt Hansson. There is a convention that aspiring writers on deontic logic adopt similar surnames. Our bibliography contains work in this area by (aside from Bengt Hansson) Sven Ole Hansson, William H. Hanson, and Jörg Hansen. In fact, before we leave the discussion above of von Wright's decision to opt for construing the deontic vocabulary as predicates of act-types too far behind us, let us take a moment to criticise one of those just named on this matter. While one could pursue von Wright's policy consistently – the disadvantages mentioned above notwithstanding – it would definitely *not* do to write, as S. O. Hansson does in the Abstract at the start of [447]: "This system provides a logic for a predicate 'O' for moral prescription, commonly read 'it is obligatory that'." If that is how you intend to read "O" then you are dealing with a sentence operator (one-place connective), and not a predicate.

(*ii*) Despite the claim made above about normative neutrality requiring that one avoid any specifically (say) *utilitarian* deontic principles – should there be any such at the level of deontic logic itself – one may adopt a more sophisticated line, and following Leuenberger's deployment in [713] for alethic modal

---

the "O" vicinity – *ought to/should, must, has to*, not discussed here. A discussion of some of the issues, as well as pointers to the literature – especially the study White [1168] – may be found at p. 538 of my [582]; an interesting recent discussion appears in Ninan [843]. See further: Examples 4.6.15, p. 299, as well as various places listed in the index under "modal auxiliaries". The reference in [582] to the contrast between *must* and *has to* illustrated by an example of Robin Lakoff's should be supplemented by the contrast between *ought to* and *has to* in §6 of Cameron [144]. For more recent – and more highly theoretical – discussions see Wedgwood [1153] and Chapter 2 of Broome [119].

[204] This issue is taken much more seriously in some quarters. See Makinson [764] and references there cited, as well as the second sentence of the following note. It was perhaps first raised, in the 1930s, by Jørge Jørgensen: see Jørgensen [632] and Weinberger [1160]. Those interested can pick up (one strand of) the debate at a later stage by looking at Volpe [1134].

[205] For a similar failure-of-neutrality criticism, see van Fraassen [329], esp. the opening paragraph of Section III. As to the status as utilitarian (or consequentialist) of a normative position, even deciding what ought to be done on the basis of the expected value of outcomes of actions does not secure such a status when "outcome" is construed so broadly that having kept a promise counts as an outcome of keeping a promise: the "consequent-" part of the label "consequentialism" is an allusion specifically to the *causal* consequences of actions. Sometimes theorists also include a condition of agent-neutrality in the application of this label, though that seems to be independent of the 'causal consequences' idea: for example an ethical egoism that judged the rightness of actions by the utility they (causally!) result in for the agent would in the central sense deserve to be counted as consequentialist. The blunt statement that these non-neutral proposals are 'the last thing one would want' will be qualified in Remark 4.4.3(*ii*) below. The interchange between Hare and Geach mentioned in Kuhn [678] is also of interest in connection with the neutrality issue.

[206] An early worry that this may not be the case for any monotone – let alone normal – treatment of the obligation operator was aired in Castañeda [159], and replied to in Bergström [84]; a subsequent review of this and related literature is provided by Carlson [150]. A discussion (taking a rather different view from that urged here) of the desirability of insisting that deontic logic be neutral between various normative and indeed various meta-ethical theories, with many references to earlier literature on this topic, can be found in Innala [610].

4.4. DEONTIC LOGIC: MAIN THEMES

logic and tense logic of a distinction between *de jure* and *de facto* logics, make such a distinction in the deontic case too. The *de jure* logic gives fundamental formal principles for the area, while this or that *de facto* logic would explore the ramifications of particular substantive positions. Whether such a distinction can be maintained under critical pressure is not clear. Some points of interest on Leuenberger's own presentation can be found in Melia [799]. For the general issue of the metaphysical neutrality (or otherwise) of logic, see Williamson [1199]. A negative position is there taken on the possibility of any such neutrality, as is done specifically in the deontic case in Sayre-McCord [996] – though the latter's note 9 (p. 196) draws a contrast between a 'core logic' and various 'special-interest logics' which suggests the deontic version just alluded to of Leuenberger's *de jure* and *de facto* logics. (Perhaps the 'Talmudic deontic logic' of [3] mentioned in Remark 4.4.1 is openly intended as belonging to this special interest genre.) ◀

Two other departures from the standard syntax of modal logic which have been motivated by deontic concerns are worth mentioning. We give them as Examples 4.4.4 and 4.4.14. In between presenting them, we include some examples of (and exercises on) non-normal ventures in deontic logic.

EXAMPLE 4.4.4 In connection, in particular, with some of the 'paradoxes' mentioned in the opening paragraph of this section, some have responded by exploring a dyadic obligation operator $O(\cdot/\cdot)$ with the intended reading of $O(B/A)$ (respectively $P(B/A)$) being "it ought to be that B, given that A" (resp. "it is permissible that B, given that A"). Several early suggestions for the treatment of such a notion of conditional obligation appear in Hilpinen's 1971 anthology [481] (see also the editorial introduction [316]), with classic treatments from the 1970s appearing in: van Fraassen [328], Lewis [718] and Chellas [170]. A textbook treatment discussing these and other proposals is conveniently available in Åqvist [30], Chapter 5 onward; there have been many subsequent ventures into this area, including, from among the sources cited later in this section for other reasons, Makinson [764], Horty [527] and Hansen [437]. A typical approach involves some variation on the following simple idea: equip the models with a betterness or preference ordering amongst the worlds, and take $O(B/A)$ to be true at a world if all the best worlds verifying A verify B. (Among the candidate complications is to relativize the betterness relations to worlds, making it ternary, and deem $O(B/A)$ true at w iff all the w-best A-verifying worlds verify B.) Such betterness relations have also found their way into monadic deontic logic of the kind we are considering here – as in Jennings [620], Jackson [614], Forrester [326] (e.g., pp. 45, 61), Goble [382], Hansson [447], [448], and van der Torre and Tan [1112]; frequently there is also an explicit binary preferability connective in the language, an idea originally pursued in the 1960s by von Wright and by Rescher ([1209], [947], respectively) – *prohairetic* logic, von Wright called it – followed by Chisholm and Sosa [176], [177]. Further references and discussion can be found in Mullen [836], though be on guard for typographical mistakes in the names of authors cited.[207] Note the contrast with studies of preference relations in the theory of decision and social choice (see for example Fishburn [303]), in which a dyadic predicate rather than a binary connective is the focus of attention. One reason for not going into this subject here, aside from the space it would consume, is that it would not constitute an application of the material in Chapter 2. A typical version of the semantics in such discussions (though there are many variations) uses models $\langle W, <, g, V \rangle$ in which $<$ is a strict linear ordering of W (think of $x < y$ as saying that y is morally better than x), and for $x \in W$, $X \subseteq W$ $g(w, X) \in W$ and g should satisfy some reasonable conditions given that we intend to think of $g(w, X)$ as picking out the most similar world to w among those in X – such as the conditions that if $w \in X$, then $g(w, X) = w$ and if $X \neq \varnothing$ then $g(w, X) \in X$ for all w. In the definition of truth at a point in a model, one has the following clause for O, where $\mathcal{M} = \langle W, <, g, V \rangle$ and $\|\cdot\|$ is understood as giving truth-sets in $\mathcal{M}$:

$$\mathcal{M} \models_w OA \text{ iff } g(w, \|A\|) > g(w, W \smallsetminus \|A\|).$$

---

[207]Castañeda, Fodor, Swinburne and Rescher appear as *Castañada, Foder, Swineburne* and – on one occasion – *Reacher*.

Obviously this semantic treatment will not render O normal, or indeed even monotone. (One could of course use such betterness rankings to define other one-place deontic operators of an 'axiological' flavour: 'it is or would be a good thing that A', and the like, though such ventures – which include Chisholm and Sosa [176], [177], mentioned already above – fall outside our purview here, though see Exercise 4.4.5 below.) The suggestion that $g(w, X)$ be thought of in terms of similarity links this failure of monotony to the invalidity of 'strengthening the antecedent' for counterfactuals, as treated in Lewis [717]. In fact one suggestion for treating conditional obligation (Mott [834], see also Niiniluoto [844]) was that rather than employ some special dyadic deontic construction, we simply use the resources of the monadic O alongside the counterfactual conditional, and blame the apparent inadequacy of monadic deontic logic on the early reliance on material implication as the only available conditional construction. (Also relevant here: Bonevac [109], Decew [238], and Arregui [37] and [38]; for the related 'logic of conditional intention', see Niiniluoto [845].) More recently, it has again been suggested – in Kolodny and MacFarlane [659] – that what conditional *ought*-statements require for their understanding is a careful account of *if-then* conditionality and its interplay with the semantics of *ought*. (The account of *if-then* in [659] has nothing to do with counterfactual or subjunctive conditionals, and the envisaged interplay reveals some surprises for the distinction sometimes put as that between what one should do subjectively speaking – meaning, given one's evidence – and what 'objectively' one should do. This distinction is an epistemic one and has nothing to do with the issue of realism/objectivism as a meta-ethical doctrine.) Much of this literature concerns the availability of Modus Ponens in the form '$O(B/A), B \therefore OA$' ('dyadic deontic detachment'), or more generally how to derive unconditional obligation statements from conditional obligation statements. Discussions concentrating specifically on the latter perspective include Greenspan [410], Goldman [400], Humberstone [549], Jackson and Pargetter [617], and Carlson [150]; Goble [382] and [383] are also relevant on the contrast between what Jackson and Pargetter call actualism and possibilism – no relation to positions going by these names in the metaphysics of modality. Contributions dialectically downstream from Kolodny and MacFarlane [659] bearing on the natural language version (with indicative conditionals) of dyadic deontic detachment may be found in the papers by Finlay and by Silk, both mentioned at the end of the Digression on p. 244 below. See also Carr [154]. ◀

EXERCISE 4.4.5 Typanska [1117] uses algebraic methods to explore the Post complete extensions of a logic of 'values' (good and bad) extracted from publications in Russian and in German by Aleksandr Ivin in a language extending that of truth-functional logic with additional operators G and H which are to read "it is good that" and "it is bad that". (No connection with the similarly notated tense operators of Prior, then.) The proposed axioms are all substitution instances of truth-functional tautologies in this bimodal language together with (in Typanska's numbering, from [1117], p. 162) all instances of the schemes, in which we revert to φ, ψ, rather than A, B, as schematic letters:

A1. $G(\varphi \wedge \psi) \leftrightarrow (G\varphi \wedge G\psi)$

A2. $H(\varphi \wedge \psi) \leftrightarrow (H\varphi \wedge H\psi)$

A3. $G\varphi \to \neg H\varphi$

and the primitive rules are Modus Ponens and the congruentiality rule for each operator (from $\varphi \leftrightarrow \psi$ to $G\varphi \leftrightarrow G\psi$, and from this same premiss to $H\varphi \leftrightarrow H\psi$). As Typanska notes, this gives us the monotonicity rules for these operators (u.e, replacing $\leftrightarrow$ with to in the congruentiality rules). Show – what does not seem especially promising for the intended application of this logic – that for all formulas φ, ψ, the formula

$$G\varphi \to \neg H\psi$$

is provable in this logic: so according to the logic if anything is good, then nothing is bad. (*Hint*: consider the disjunction of φ with ψ. Note that the postulated aggregativity of G and H, given by the "←" directions of A1 and A2 is not even needed for this.) ✠

4.4. DEONTIC LOGIC: MAIN THEMES

REMARK 4.4.6 The references to worlds in the discussion in Example 4.4.4 are perhaps more plausibly interpreted as world–time pairs. This comment applies across the board in the semantics of deontic logic, since even if fundamental moral principles are not plausibly regarded as changing truth-value with the passage of time, everyday *Ought*-judgments certainly do, it being natural to regard their correctness as depending on the practically available options and currently obtaining circumstances, both of which vary with time. (For more on circumstance dependence, see the present author's [549]; for some interesting historical remarks bearing on the issue, see p. 40*f.* of Prior [913].) Thus in the case of the traditional binary accessibility relation too, one should think of the relevant alternative worlds – the morally ideal ones to be considered depend not only on a world but on a time. This was mentioned parenthetically in the 'semi-simplified Kripke semantics' for deontic logic when this topic was raised earlier (see p. 206). For fuller discussion, see, for example, Thomason [1105], §6.3 of Chellas [171], and van Eck [270]. Of course a similar consideration applies also in the case of doxastic logic, from that earlier discussion, or for that matter epistemic logic, since what a cogniser believes or knows varies from time to time within a world – though Thomason thinks that in the deontic case the connection with time is more intimate (because of a theory relating truth to moments of time in a branching time semantics for tense logic, expounded in Thomason [1104]). We continue to ignore such complications for expository convenience, as well as because their significance can be overestimated, as was mentioned at the end of Remark 4.4.2(*iii*) above. But recalling them helps perhaps with objections such as the following, from Hansson [449], p. 329: "The ideal world semantics of standard deontic logic identifies our obligations with how we would act in an ideal world. However, to act as if one lived in an ideal world is bad moral advice, associated with wishful thinking rather than well-considered moral deliberation." We return to a similar issue – not so much *time* variability as *world* variability (alias contingency) for moral claims – in Remark 4.5.1, p. 279. ◀

Readers wanting to omit the discussion of further non-normal approaches to deontic logic inspired by the behaviour of (monadic) O commented on above should skip to Example 4.4.14, p. 251.

The monotone implication $O(A \wedge B) \to OA$ is opposed with various more or less plausible examples in various publications of Frank Jackson (Jackson [614], Jackson and Pargetter [617] – though see also Jackson's contribution to the symposium [616], where it is claimed that most concrete instances of this schema are seriously ambiguous, with only one disambiguation being unacceptable, and Altham's reply as co-symposiast there. Sometimes one sees a less plausible assault on the principle:

EXERCISE 4.4.7 The following passage is from p. 346 of Hansson [445]:

> Sinnott-Armstrong [1058] put forward the following example: "If I both mow and water your grass, I mow your grass, so, of it is obligatory for me to mow and water your grass, it is obligatory for me to mow your grass." However, the plausibility of this example depends on the contingent fact that mowing and watering are independently valuable. Suppose the grass was of some type that would be destroyed if it were mowed without being watered. Then an obligation to mow and water the grass would not imply an independent obligation to mow it.

Evaluate Hansson's response here as an objection to the schema $O(A \wedge B) \to OA$. ✠

The above discussion concerns the rule OMono – □Mono with O for $\Box$, that is – as applied to conjunction-to-conjunct inferences. Much attention – often in publications with 'Ross's Paradox' in their titles – has been given to its application to disjunct-to-disjunction inferences. If James should post the letter, does it really follow that Jones should post the letter or burn it? It may seem that a pragmatic response is possible – it does follow, but the disjunctive formulation would be a misleadingly weak (i.e., underinformative) thing to come out with if one thought that James should post the letter. This would be to say that the 'take-home message' of the disjunctive formulation ("$O(A \vee B)$",) to the effect that it is permissible to burn the letter ("PB"), results from a conversational *implicature à la* Grice of the utterance rather than a logical *implication* of its content. (See Remark 1.3.2, p. 18 and Example 5.1.11,

p. 349.) However, recent moves with Jackson-inspired semantics suggest that such a response may not be the best available (Cariani [148]). Like the problem of disjunctive permission statements ('free choice permission') mentioned after Example 4.4.14 below, this is an interesting area we do not pursue in any detail here, beyond a brief second mention at Remark 5.1.12 on p. 350 below. We will, however, devote a section below, beginning on p. 324, to one particular suggested response (proposed by Sven Danielsson) to Ross's Paradox, because of its inherent logical interest. (Danielsson offers a restricted version of the schema $O(A \wedge B) \to OA$, which will stop OMono, and with it $OA \to O(A \vee B)$, from being derivable.)

Digression. An unrelated puzzle (in the style of Zeno's Paradoxes) is discussed under the name "Ross's Paradox" in Allis and Keutsier[11] (as well as subsequent articles by others in the same journal), after the name of the author, Sheldon Ross, of a probability textbook; the Ross of Ross's Paradox in deontic logic is the Danish legal theorist Alf Ross, who is in turn to be distinguished from the Scottish moral philosopher (and Aristotle scholar) W. D. Ross. This last Ross's emphasis on the distinction between *prima facie* and all-thing-considered obligations was taken up from the perspective of deontic logic in Hintikka [494] and elsewhere. Horty [527] also uses the "prima facie" vocabulary in what he presumably takes to be a related sense. Many philosophers have replaced talk of *prima facie* obligations, reasons for action, etc., with talk of *pro tanto* obligations, in part to avoid distracting epistemic connotations. (A leading influence here was Hurley [605], though the relevant point had been made before, perhaps first at p. 45 of Lemmon [697]. Whether Horty would regard these connotations as a distraction is less clear, given the background provided by his [525].) Outside of the present Digression, we will have nothing to say about this distinction between *prima facie* and all-things-considered obligations. The same goes for the contrast between objective and subjective *ought*-judgments, the latter being understood as filtered through the evidential (and perhaps even moral) perspective of a particular agent:[208] should Sarah, reliably assured that a certain medicine will give her son relief from a temporary but painful condition, administer the medicine? Subjectively (one might say) she should – even if, though nobody knows it, her son is one of the very few in whom this medicine would trigger a fatal reaction, so that, 'objectively', she should not. (To repeat a point made in Example 4.4.4: note that the contrast is available even to those with meta-ethical positions that might be described as denying the objectivity – in a suitable sense – of all moral judgments.) The parenthetical reference just made to the agent's *moral* perspective is meant as a reminder of the vexed question of the extent to which one ought to do whatever it is that one believes one ought to do, discussed from time to time in the journal *Mind*, for example in the 1930s with Osborne [861] (repeated as part of a later book, [862], beginning at p. 105) and then again in the late 1960s with Cohen [184] (following up which, we have: Kordig [661], Sturch [1093], Cohen [185]), the later discussion making no reference to the earlier. In between, we have the 1945 of H. D. Lewis [731], and then in the 1950s, the subject occupied A. C. Ewing in [277], a reference taken up more recently in Section 2 of Finlay [300] and then in Silk [1052] p. 692. (These last two papers are ventures in compositional natural language semantics for modal auxiliaries in the general tradition of Kratzer [667].) Also relevant is the discussion of 'pseudo-subjectivism' in Jennings [621], mentioned at the end of Example 4.6.14, p. 298 below. **End of Digression.**

Less radically departing from the spirit of normal modal logic for deontic applications is the suggestion that while we retain the idea that O is monotone, we drop the idea that it should be aggregative, thereby defending a non-regular modal logic. This would allow us to retain $\neg O(p \wedge \neg p)$ and $\neg(O\bot)$ (or $P\top$) in our deontic logic – surely nothing contradictory could be morally required – while jettisoning $\neg(Op \wedge O\neg p)$ (or $Op \to Pp$), to allow that each of two mutually incompatible moral requirements may well obtain. (These situations are variously described as involving conflicting obligations, moral conflicts, or moral dilemmas.) This was suggested in §6.5 of Chellas [171], with a semantic treatment ([171], p. 208f.) in terms of neighbourhood models in the sense of p. 159 above. The theme was also addressed in later work

---

[208]This means that the question of a subjective/objective construal arises naturally only for what in Example 4.4.14 (p. 251) are called agent-implicating as opposed to situational *ought*-judgements, a point (essentially) noted in Gensler [364], which contains further discussion of this issue and relevant references to the literature.

4.4. DEONTIC LOGIC: MAIN THEMES

in the 1980s by Schotch and Jennings (see their [1002], [1000], [625]), presented here because of the more appealing way they package the semantics:

EXAMPLE 4.4.8 Schotch and Jennings motivate the idea just sketched by suggesting that moral conflicts could arise in which it ought to be that p and it ought to be that not-p, from which we don't want to infer that it ought to be that both p and not-p, the latter conjunction being impossible and so perhaps not the kind of thing that ought to be the case.[209] Thus we want to distinguish between the more plausible principle $\neg O\bot$ – or $\neg O(A \wedge \neg A)$ – and the less plausible $OA \to \neg O\neg A$ (or, to bring out the contrast: $\neg(OA \wedge O\neg A)$). These are conflated (as $\mathbf{D}_O$) when we are committed to normality, or more specifically, to aggregation (see note 15, p. 21), creating what this line of thought sees as the mistaken impression that $\mathbf{KD}$ is a plausible basic deontic logic. One way of presenting the semantic suggestions of [1002] is to pass to multi-relational frames $\langle W, \mathscr{R} \rangle$ in which $\mathscr{R}$ is a set of binary relations on W. For a model $\mathcal{M} = \langle W, \mathscr{R}, V \rangle$ on such a frame, the novel clause for O, which we write here as $\Box$ for purposes of comparison:

$$\mathcal{M} \models_x \Box A \text{ if and only if for some } R \in \mathscr{R}, \text{ for all } y \in R(x), \mathcal{M} \models_y A.$$

Thus this semantics reduces to the usual Kripke semantics in the case in which $\mathscr{R}$ is a singleton. If $\mathscr{R}$ contains at least two elements R_1 and R_2, say, then the aggregation formula $(\Box p \wedge \Box q) \to \Box(p \wedge q)$ can be falsified at a point by having all that point's R_1-successors verify p but not q, all its R_2-successors verify q but not p. If we have no further relations in $\mathscr{R}$, then the frame will, however, validate the following variation on the theme of aggregation:

$$(\Box p \wedge \Box q \wedge \Box r) \to (\Box(p \wedge q) \vee \Box(p \wedge r) \vee \Box(q \wedge r)),$$

and the reader will have no difficulty in supplying further formulas depending for their validity on a frame $\langle W, \mathscr{R} \rangle$ on the condition that $|\mathscr{R}| \leq n$ for values of n greater than the value (i.e. 2) involved in this illustration of the general idea. By way of motivation from deontic logic, Schotch and Jennings are making room for the position that obligations (or more generally, moral considerations) may spring from various different sources – for example, different and perhaps incommensurable types of value – and the various relations in $\mathscr{R}$ correspond to the ideal realization of the various values concerned. One may be the source of the obligation that it be the case that p, and another lie behind the requirement that q, with no value whatever dictating that it should be that $p \wedge q$. (A similar idea in epistemic and doxastic logic might lead us from normal logics for individual knowers and believers to a monotone non-aggregative logic for the impersonal constructions "it is believed/known that..." where this means that *at least one* of the cognitive agents believes that...: here the different relations in $\mathscr{R}$ would correspond to the agents' various different accessibility relations (doxastic or epistemic). A similar approach has been used at the intra-personal level to model the uncombinability of, e.g., compartmentalized belief systems, as in Lewis [722]: see note 377 on p. 423.) ◀

The multi-relational models described in Example 4.4.8 do not constitute the only semantic novelty in Schotch and Jennings [1002]; they also take an interest in frames $\langle W, R \rangle$ in which R is an n-ary relation on W for some (temporarily fixed) n, a model on such a frame verifying $\Box A$ at $x \in W$ just in case for all $y_1, \ldots, y_{n-1}$ such that $Rxy_1 \ldots y_{n-1}$, A is true at y_1 or ... or at y_{n-1}.[210] For more information on with

---

[209] A different reaction at this juncture would be to claim that because of the underlying use of classical non-modal logic this in turn implies that everything ought to be the case, since $p \wedge \neg p$ provably implies every formula. This reaction leads to paraconsistent deontic logic, as elaborated in da Costa and Carnielli [197], Coniglio and Peron [192] or Routley and Plumwood [973], and the 'semi-paraconsistent' deontic logic developed in McGinnis [787], neither of which will concern us here.

[210] See note 23, p. 30. These models were used by Goldblatt to provide a semantic account for the generalization alluded to there for a normal $(n-1)$-ary $\Box$ operator, with $\Box(A_1, \ldots, A_{n-1})$ true at x iff for all $y_1, \ldots, y_{n-1}$ such that $Rxy_1 \ldots y_{n-1}$, there is some i ($1 \leq i \leq n-1$) for which A_i is true at y_i. The Schotch–Jennings application is thus to the case in which $A_1 = \ldots = A_{n-1} = A$.

these frames, see the Schotch and Jennings publications already mentioned. It should also be mentioned that these authors do not present the multi-relational frames quite as above, but rather, again fixing temporarily on some n, take the frames to be of the form $\langle W, R_1, \ldots, R_n \rangle$.[211] It is not clear whether the option of taking $n = 0$ is envisaged here, whereas it is clear that for our frames $\langle W, \mathscr{R} \rangle$, we allow $\mathscr{R} = \varnothing$, in which case all $\Box$-formulas are false at all points (in any model on a frame exploiting this possibility). Thus as well as failing to determine a normal modal logic, by allowing $|\mathscr{R}| > 1$, blocking aggregativity, we get a further failure of normality by allowing $|\mathscr{R}| < 1$, blocking Necessitation.

REMARK 4.4.9 Closure under Necessitation has often been regarded as a defect of treating deontic logic under the banner of normal modal logic. Here is Charles Pigden on p. 139 of the feisty [881]; Necessitation for O is in this passage called (O), and its imposition is just one of the things Pigden complains about:

> Deontic logic is a dubious enterprise; its leading principles are false, bordering on the nonsensical. The principle (O) not only obliges us to keep tautologies going but, by iterating deontic modalities, to keep these obligations going too. (If OA is a theorem, so is OOA!)

Similarly, in [454], p. 23, Jonathan Harrison had referred to "$O(p \vee \neg p)$" as a "quite absurd formula". Here we must bear in mind the appropriate adaptation of Cresswell's point about nomic necessity (see Example 4.3.1, p. 227): anyone objecting to the idea that it ought to be that __, when a logical truth fills the blank, will have to object not only to Necessitation but also to Monotony, since this will deliver $Op \rightarrow O(p \vee \neg p)$. (See further Example 4.4.47 at p. 277 below.) In any case, all these objections are not objections to deontic logic *per se* but at best motivations for pursuing the subject other than under the umbrella of normal modal logics. Indeed, von Wright [1204] explicitly refrains from endorsing Necessitation; see the opening paragraph of Section 4.8, p. 324 below. Compare also note 2, added in the reprinting of the original article in [1207], p. 69, in which von Wright goes further – though in ways which are specific to his own idea that what might be thought of as deontic operators on sentences are to be taken, instead, as predicates of action-types. (The position – with or without the added footnote – is not really coherent, however, in view of Cresswell's point. The footnote, in which von Wright suggests that there are no contradictory or tautologous act types anyway, creates its own difficulties: we can no longer regard the class of act types as closed under the Boolean operations.) For further examples of opposition to Necessitation-for-O, see also Remark 4.4.15(ii), p. 253 below, and p. 11f. in Mares [771] and the discussions there cited. ◀

Let us return to Schotch and Jennings' explorations of the non-normal developments alluded to above. It would at first sight seem that the logic determined by all multi-relational frames is accordingly none other than the smallest monotone modal logic ("**EM**"), a suspicion bolstered by the closeness of the present semantics to Schotch and Jennings' locale semantics, described in the discussion after Theorem 2.10.12 (p. 161 above).

EXERCISE 4.4.10 Recall from p. 161 that locale frames and models are just neighbourhood frames and models, but that we use this terminology to indicate that a different truth-definition is in force. Consider the following line of reasoning:

> Suppose that $\langle W, \mathscr{R}, V \rangle$ is a multi-relational model. Define a locale model $\langle W, \mathcal{N}, V \rangle$ by putting, for $x \in W$: $\mathcal{N}(x) = \{X \subseteq W \mid \exists R \in \mathscr{R} . X = R(x)\}$, and verify by induction on the complexity of A that for all A, for all $x \in W$, we have $\langle W, \mathscr{R}, V \rangle \models_x A$ if and only if $\langle W, \mathcal{N}, V \rangle \models_x A$. (The two models are 'pointwise equivalent'.) Thus it follows that if a

---

[211]Frames of a similar kind and the logic determined by them are also considered in Fitting [310], pp. 329–331, except that in addition they feature a set of normal worlds – though Fitting actually isolated the complement of this set relative to W – which enter into the definition of truth much as in Kripke's version of the semantics for such systems as S2: see p. 360 below. Fitting credits the idea to a 1979 publication in German by P. Steinacker.

4.4. DEONTIC LOGIC: MAIN THEMES

formula is valid on every locale frame, it is valid on every multi-relational frame since if it is false at some point in a model on a multi-relational frame the construction just described gives us a model on a locale frame with the formula false at the same point in the latter model.

Similarly, given a locale model, $\langle W, \mathcal{N}, V \rangle$, construct a multi-relational model $\langle W, \mathcal{R}, V \rangle$ by putting, for all $R \subseteq W \times W$: $R \in \mathcal{R}$ if and only if for some $x \in W$ with $X \in \mathcal{N}(x)$, $R = \{\langle x, y \rangle \,|\, y \in X\}$, and again verifying by induction that the multi-relational model thus derived is pointwise equivalent to the original locale model. Now it follows that if a formula is valid on every multi-relational frame it is valid on every locale frame since a falsifying point in a model on a locale frame gives us a falsifying point in a multi-relational frame.

Putting these conclusions together, we infer that the formulas valid on every multi-relational frame and those valid on every locale frame coincide.

Does the above argument establish that the logic **EM** determined by the class of all locale frames coincides with the logic determined by the class of all multi-relational frames? ✠

Room for doubt about the success of the above argument is suggested by the following Exercise.

EXERCISE 4.4.11 (*i*) Is the formula $\Box\top \rightarrow \Box\Box\top$ valid on all locale frames (equivalently: **EM**-provable)?

(*ii*) Is the formula $\Box\top \rightarrow \Box\Box\top$ valid on all multi-relational frames?

Correct answers to (*i*) and (*ii*) here lead to a more focussed version of Exercise 4.4.10:

(*iii*) Where precisely does the argument given there go wrong? ✠

The caution needed over the possibility of having $\mathcal{R} = \varnothing$ illustrated by these last Exercises is reminiscent of the case of $\mathbb{F} = \varnothing$ in Garson frames $\langle W, \mathbb{F} \rangle$, discussed earlier: see p. 157 as well as Theorem 2.10.8 and Corollary 2.10.9. We will conclude this intermission on deontically motivated non-normal modal logics on a more positive note. We use the nomenclature of Chellas [171], who calls the smallest monotone modal logic containing $\Box\top$ – or equivalently, the smallest monotone modal logic closed under Necessitation – **EMN**.

PROPOSITION 4.4.12 **EMN** *is determined by the class of all multirelational frames* $\langle W, \mathcal{R} \rangle$ *with* $\mathcal{R} \neq \varnothing$.

Proof. Soundness is left to the reader. For completeness one uses a canonical model argument, with the canonical model for **EMN** or indeed any monotone extension thereof – call it $\langle W, \mathcal{R}, V \rangle$ having its W and V components defined as usual (maximal consistent sets of formulas and assigning p_i the set of such sets with p_i as an element), and $\mathcal{R}$ defined as the set of relations R_A (for each formula A) defined by: xR_Ay iff xR_Ay iff $\Box A \in x \Rightarrow A \in y$. (This "$\Rightarrow$" is just material implication in the metalanguage, so what follows the "iff" means: either $\Box A \notin x$ or $A \in y$.) Note that, as required, $\mathcal{R}$ is non-empty, since the set of formulas is non-empty. We need to check that truth (in the canonical model) at $x \in W$ and membership (of A in x) coincide, for which the inductive case of $A = \Box B$ is the only case of interest. For this, the inductive hypothesis assures us that for all $x \in W$, $B \in x$ iff $\langle W, \mathcal{R}, V \rangle \models_x B$, so our task is reduced to showing:

$$\Box B \in x \Leftrightarrow \text{for some } R \in \mathcal{R}, \text{ we have: for all } y \in R(x), B \in y.$$

We show the $\Rightarrow$-direction first. Suppose $\Box B \in x$. Then we offer $R = R_B$ as the promised $R \in \mathcal{R}$ for which all R-accessible sets contain B. For suppose xR_By. This just means that $\Box B \in x$ implies $B \in y$ and we already have that $\Box B \in x$, so $B \in y$, as desired.

Turning to the $\Leftarrow$-direction, suppose that for some formula C for all $y \in W$: $xR_Cy \Rightarrow B \in y$. Spelling out the latter more explicitly: for all $y \in W((\Box C \in x \Rightarrow C \in y) \Rightarrow B \in y)$. We can unpack this as the conjunction of (*i*) with (*ii*):

(*i*)   For all $y \in W$, $\Box C \notin x \Rightarrow B \in y$, and
(*ii*) For all $y \in W$, $C \in y \Rightarrow B \in y$.

Now (*ii*) implies $\vdash C \to B$, where "$\vdash$" indicates provability in the current logic, while (*i*) implies that for all $y \in W$, $\Box C \notin x \Rightarrow B \in y$, and hence that $\Box C \notin x \Rightarrow$ for all $y \in W$, $B \in y$. and thus that either $\Box C \in x$ or else $\vdash B$. If the first disjunct holds, $\Box C \in x$, we get the desired conclusion that $\Box B \in x$ from the fact, recently derived from the provability of $\Box C \to \Box B$, which follows from the fact, noted above, that $\vdash C \to B$, by □Mono. If the second disjunct holds, then $\vdash B$, and so by Necessitation $\vdash \Box B$, in which case certainly $\Box B \in x$.

Thus truth and membership coincide for all formulas and so any non-theorem of **EMN** is false at some point in the canonical model, so every formula valid on all multi-relational frames with non-empty $\mathscr{R}$ is **EMN**-provable. ∎

The locale semantics was introduced for deontic logic, in effect, in van Fraassen [329], out of a desire to avoid aggregativity while retaining monotony for O; van Fraassen's motivating discussion overlaps considerably with the earlier paper of Williams [1170], mentioned in note 15 (p. 21) apropos of the term "aggregation" and variations. (Ideas broadly similar to those of [329] in a more syntactic presentation can be found in earlier work by Erik Stenius and later joint work by Alchourrón and Bulygin, referred to in Makinson [764].) Recall from p. 161 that the models involved are just neighbourhood models, but that the truth-definition uses "⊆" where the standard neighbourhood semantics would have "=":

$$\langle W, \mathcal{N}, V \rangle \models_x \Box B \text{ if and only if for some } X \in \mathcal{N}(x),\ X \subseteq \|B\|.$$

A special case of the standard neighbourhood semantics or of the locale semantics arises if we insist that for all $y \in W$, $\mathcal{N}(x) = \mathcal{N}(y)$. This gives the analogue in the neighbourhood case of the semi-simplified Kripke semantics in which $R(x) = R(y)$ for all $x, y \in W$. (See p. 206.) For that case, we noted that one could re-draft the semantics in terms of models with a distinguished subset of points, to be taken as the accessible points for any point in the model. Similarly here, we could re-draft the neighbourhood models to have in the model a collection of sets of points given once and for all. For familiarity, let is retain the letter $\mathcal{N}$ for this purpose. Then the above clause would read:

$$\langle W, \mathcal{N}, V \rangle \models_x \Box B \text{ if and only if for some } X \in \mathcal{N},\ X \subseteq \|B\|,$$

and likewise for the standard neighbourhood semantics, except with "⊆" replaced by "=" (or more simply, writing the right-hand side – after the "if and only if" – as $\|B\| \in \mathcal{N}$). This is essentially what van Fraassen suggests (provisionally) in [329], with a gloss along the following lines: the elements of $\mathcal{N}$ are to be thought of as the basic moral imperatives in force (or their semantic values, if you are thinking of the imperatives as linguistically expressed; note that this is a loose use of the term *imperative*: see Example 4.6.15(*i*), p. 299). The reason for the parenthetical "provisionally" is that while this avoids aggregativity, making room for everyday moral conflicts (cases in which – reverting now to the specifically deontic notation with O for $\Box$ – OA and $O\neg A$ are both true for a given A, while not every O-statement is), it precludes some desirable inferences from non-clashing imperatives. While we are not following up the proposed amendment here, there is a considerable literature on it and further variations, including Horty [525] (see also [527]), Hansen [435] (see also [436], [437]), and Goble [387], which corrects some misformulations in [386], as well as Goble [388], which gives an extended discursive presentation of his position.

In the papers just cited, Goble considers a different reaction to cases in which it is conceded that OA and $O\neg A$ from that we saw from Schotch and Jennings in Example 4.4.8. This different reaction involves conceding the aggregation inference to $O(A \wedge \neg A)$ that Schotch and Jennings wanted to resist, and, retaining congruentiality so allowing the further inference to arbitrary $O(B \wedge \neg B)$ (or to $O\bot$, as in Example 4.4.8). But it constrains monotony in such a way as to disallow the further inference from $O(A \wedge \neg A)$ to OA (or, via $O(B \wedge \neg B)$, to OB). OMono would be replaced by a weaker rule:

4.4. DEONTIC LOGIC: MAIN THEMES

$$\frac{A \to B}{(OA \land PA) \to OB}$$

in which, as usual, PA is simply $\neg O \neg A$.[212] As Goble notes, we can replace this rule, given the Congruentiality rule for O (i.e., $A \leftrightarrow B / OA \leftrightarrow OB$), with an axiom schema, such as $PA \to (OA \to O(A \lor B))$. This together with Aggregation and Necessitation for O (the latter being formulable as the axiom $O\top$), together with the usual truth-functional apparatus, provides a basis for a logic Goble calls **DPM.1**. Goble presents this system – the letters standing for Deontic Logic with Permitted Monotony (which he calls 'Inheritance', despite the abbreviation to 'M') – as one of three candidates for a plausible conflict-allowing deontic logic; note that the present reason for dropping unrestricted monotony is rather different from those in play in Exercise 4.4.7 and the text surrounding it above. (The other candidates, **DPM.2** and **DPM.3**, involve a similar "P" restriction on aggregation for O, in addition to the restriction on monotony. There is no connection with the modal principles .2, .3, here, and indeed little reason to use a decimal point in the labels in the first place.) A neighbourhood semantical discussion of these logics can be found in [386]; note that because we are restricting monotony, we cannot use the locale semantics. (Another monotony-restricting proposal will occupy us in Section 4.8.)

EXERCISE 4.4.13 Suppose we define O^+A as $OA \land PA$. (Goble speaks in such cases of there being an *unconflicted obligation* for it to be the case that A.) Is O^+ normal in the logic **DPM.1** just described? (Justify your answer.) ✠

We conclude this highly selective survey of non-normal manoeuvres with an explanation as to why simply placing the envisaged restriction (for **DPM.2** and **DPM.3**) on Aggregation:

$$P(A \land B) \to ((OA \land OB) \to O(A \land B)),$$

would not suffice to block unwanted consequences for a deontic logic accommodating moral conflicts, if monotony itself remains unconstrained.[213] Goble remarks, plausibly enough, that we should not want to be able to prove in such a logic $(Op \land O\neg p) \to (Pq \to Oq)$, whose consequent could equally well be written $Oq \lor O\neg q$; we do not want to say that the existence of one such conflict renders everything which is permissible obligatory. Yet in a proof on p. 472 of Goble [387], the deductive machinery just described is shown to suffice for a proof of just such an unwanted result. Figure 4.1 depicts the reasoning informally.

The transitions in Figure 4.1 from the three assumptions at the top to their unwanted consequence at the bottom are justified as follows. (1) and (2) make use of unrestricted monotony, while step (3) is justified by congruentiality. (This follows from OMono of course, but with the restricted version

---

[212] An unrelated weakening of monotony for an obligation or "ought" operator appears also in Example 4.4.47, p. 277 below.

[213] Brink [117], p. 229, had also (in effect) urged the merits of this restricted form of Aggregation. Care is needed in consulting this source because "→" is used for something – it is not made clear what – other than material implication. This explains the sympathetic treatment of what look like notoriously unacceptable deontic principles such as a variant ([117], p. 233) of Ernst Mally's axiom scheme $O(A \to B) \to (A \to OB)$, whose disastrous consequences are explained in Føllesdal and Hilpinen [316], pp. 2–4; ad hoc appeal to the principle is made in the course of Forrester [325], with an acknowledgment of its not being part and parcel of standard deontic logic, coupled with the claim that it is defended as a general deontic principle by Castañeda – something for readers of the latter's [163] to assess for themselves. (In fact Mally also accepted the converse of the principle under discussion here, with equally obvious undesirable results – such as $\neg A \to O(A \to B)$ – in its own right. But these were early days – 1926 – and as any history of the subject will remind one, although it was von Wright's work that drew serious attention to deontic logic, the word "deontic" itself, or its German analogue, to be accurate, was introduced by Mally. For further historical information, see Lokhorst [740]. An interesting variation on Mally's system, recast as intuitionistically rather than classically based, can be found in Lokhorst [741] and [742].) Ernst Mally is not to be confused with the fictitious Australian poet of the 1940s, Ern Malley, though David Lewis [729] notes a connection: the former had a metaphysical *penchant* for such non-existent objects as the latter. (Similarly, to settle a question which will have been bothering many readers, the title, "Against Cresswell: A Lampoon" – of Potocki [900] – should not be taken to suggest an attack on the well-liked modal logician Max Cresswell, the target instead being the non-fictitious New Zealand poet D'Arcy Cresswell, an unrelated earlier compatriot).

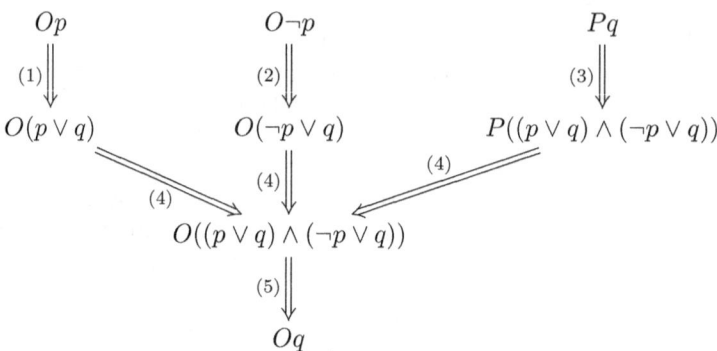

Figure 4.1: A Picture of Goble's Argument

of OMono above, we still retain the congruentiality rule unrestricted.) The links marked (4) takes us from the premisses of an inference appealing to the restricted aggregation rule to the conclusion of that rule. (Here we are thinking of a natural-deduction like rule corresponding to the implicational schema written above, with premisses OA, OB, and $P(A \wedge B)$, to the conclusion $O(A \wedge B)$.) Finally (5) is again a matter of congruentiality. Thus Goble justifies restricting OMono, blocking the argument by disallowing steps (1) and (2). Those with other theoretical predilections may wish to point the finger elsewhere: for example the relevant and semi-paraconsistent approaches of Plumwood–Routley and McGinnis (respectively) mentioned in note 209, happy with unrestricted aggregation anyway and thus not particularly concerned with the right-hand branch in Figure 4.1, would object to step (5).[214] And if we wanted to attend to (3) from this perspective, it would no longer be a case of congruentiality but would be fine as another monotone inference (like (1) and (2)) since it is in the reverse direction that there is a failure (of first-degree entailment between the formulas embedded under "P").

Having completed our brief tour of a few non-normal developments, we return to the topic of changing the *language* – rather than the logic in that language – for deontically motivated modal logic. We cannot set the (first) topic of Example 4.4.4 aside for good without at least mentioning that von Wright, as well as pursuing deontic logic with a dyadic obligation operator – in his paper in the collection [481] – explored in some detail the idea of an analogous treatment for alethic modal logic ([1207], pp. 89–126), in which he wrote $N(B/A)$ and $M(B/A)$ to with the intended readings "it is necessary that B, given A" and "It is possible that B, given A", and these were not treated as equivalent to what we would write as $\Box(A \to B)$ (or, as in Exercise 2.1.17 and the discussion preceding it, "$A \strictif B$") and $\Diamond(A \wedge B)$ respectively. The analogy is with conditional probability, the relative possibility or necessity of a statement not being identified with the outright or absolute possibility or necessity of some other statement. (See also Bryant [123].) Von Wright does also define an 'absolute' necessity – the kind of thing $\Box$ might be taken to represent – in terms of his relative necessity by abbreviating $\Box A$ (though in fact he writes "NA") as $N(A/\top)$. This venture no doubt deserves more attention than it has received in the literature or will receive here.

Example 4.4.4 began with the special conditional obligation connective $O(A/B)$, providing us with the first of two departures from the standard language of deontic logic that were to be mentioned here.

---

[214]Goble himself has gone on to take a rather different tack with these issues, moving from tinkering with conventional deductive systems and into the realm of 'adaptive logics': see [389]. This involves a transition from the non-monotone to the non-monotonic, as these terms were contrasted in Remark 1.3.6, p. 20; see also Horty [525], and indeed the whole anthology in which it occurs (Nute [852]). Further variations on these approaches to reasoning in the face of conflicting obligations can be found in Nair [838].

4.4. DEONTIC LOGIC: MAIN THEMES

What follows is the second.

EXAMPLE 4.4.14 The second kind of departure from the standard syntax of monomodal logic in need of illustration is suggested by natural language sentences such as:

"Mary ought to return the money to Norma,"

in which typically what is being said is not that something should be the case – namely that Mary returns the money in question to Norma – but specifically that it is *up to Mary* to see that this the case, rather than, for example, Norma or anyone else: it is Mary that is under the obligation in question. There are various remedies as to how to handle such 'agential' or *agent-implicating* obligation statements and the topic will be touched on below on in passing. Contributions to the debate include the following (in order of publication): Castañeda [162], McNaughton and Rawling [795], Forrester [326] (Chapters 4 and 6), Hansson [447], Horty [526], McNamara [792]. I had my own say on the matter in [563], itself synthesizing some earlier ventures – especially [539] and [549]. The former distinguishes *situational* from – the already mentioned – *agent-implicating* "ought"-statements; the commonly encountered formulation of the contrast as between ought-to-be and ought-to-do is misleading, since it can be specifically up to someone *to be* a certain way no less than it can be up to them *to act* a certain way. Indeed the proposed alternative terminology suffers somewhat from the same fault, in that "agent" brings with it connotations of action. Another paper published in the same year as [539], namely Antinucci and Parisi [26], p. 33 makes a similar point about deontic *must*, and also notes that it does not arise for epistemic *must* – something applying equally well in the case of *ought*; also relevant in this connection from the linguistic literature of the 1970s is Mittwoch [811]. (This last paper emphasizes the corresponding distinction in statements of permission and permissibility, where "agent-focussed" would perhaps be a better term than "agent-implicating": under suitable circumstances "Rob may walk with Martha" for example, permits Rob to walk with Martha rather than permitting Martha to walk with Rob, even though the situation resulting from the exercise of these two permissions – Rob and Martha's walking together – is one and the same. (Confusingly in this respect, van der Auwera and Ammann [45] refer to the agent-focussed case as involving situational possibility, contrasting the present *may* of deontic possibility with *may* expressing epistemic possibility.) Bibliographical references to earlier literature can be found in all these publications; see especially notes 1 and 7 of Humberstone [563] for further references, including to work by Potts and Geach. (Geach does not mention the earlier work of his then colleague at Leeds, Timothy Potts, in asking 'Whatever happened to deontic logic?' and complaining about the customary use of an impersonal sentence operator to render *ought*-statements.) Gilabert [374], p. 662, looks as though he is aiming for a similar distinction with the contrast there drawn between 'the "ought" of obligation' and 'the "ought" of moral desirability' but things are then rather spoilt in note 6 of the following page when formal renderings are suggested of both as simply sentence operators – so we have lost the specification of the individual to whom the obligation is attributed. One reference not supplied in [563]: St. Anselm was already on top of the distinction involved here, and the corresponding contrast with *can* statements: see Henry [474], pp. 155 and 200. It plays a significant role in Chapter 2 of Broome [119], mentioned already in note 203 (p. 239). When one attends to the needs of a syntactically plausible compositional semantics for English, it becomes harder to maintain the agent-implicating/situational contrast as corresponding to an actual ambiguity in sentences of the form "*a* ought to φ", as Finlay and Snedegar [301] observe. Bengt Hansson once suggested an explicit relativization not only to the agent under the obligation, but also to the party to whom that agent owed the obligation in question: [442], p. 243; this has a rather different flavour: the inset sentence above is actually ambiguous according as it is taken situationally or agent-implicatingly, with Mary as agent (by far the more likely reading, no doubt), but there is no further ambiguity – at most a failure of specificity – as to whether the obligation is one owed to Norma. A catalogue of normative relations involving such matters and more (rights, immunities. etc.), with a flavour avowedly more legal than ethical, is associated with W. N. Hohfeld and has occasionally attracted logical attention: Anderson [19], Kanger [637]; see also Saunders [995], which supplies numerous further references. ◀

Not so much on the subject of what the syntax should look like as on what the logic should do with it, are a host of questions that have been raised about the permissibility operator, and in particular whether, if that's how we want to read P, the definition of PA as $\neg O\neg A$ is really appropriate. One worry is the possibility of a stronger sense of (perhaps especially legal) permissibility, in which an explicit act of permitting (on the part of the legislature, say) is required, and for which the mere absence of a prohibition is not sufficient. (Here we are assuming that its being prohibited – or forbidden – that A is suitably represented by $O\neg A$.[215] The classic discussion of this issue and of the pertinent maxim *nullum crimen sine lege* is provided in §14 of Chapter 5 in von Wright [1208].[216] Another ground for dissatisfaction on this score (§9 of Pigden [882], also Goswick [404] – though the latter's reasons are somewhat different) is that one might, as a moral nihilist or error-theorist about ethics, think that moral judgments are all misguided, and so be inclined to deny all permissibility statements as well as all obligation statements. Such a person is committed to $\neg Oq$ and $\neg P\neg q$ and so needs to reject the implication from $\neg Oq$ to $P\neg q$. (In the case of Goswick, the reason is analogous to her opposition to the duality of $\Box$ and $\Diamond$ in the logic of metaphysical necessity, noted in Remark 4.1.10, p. 209, except that instead of modal statements about objects lacking non-trivial powers or essences, the issue concerns entities which are not moral agents, the attributions of obligations and permissibilities alike would both be false. Thus we are in the agent-implicating territory marked out in Example 4.4.14.)

REMARK 4.4.15 (*i*) Pigden [882], mentioned above, addresses the following difficulty for moral nihilism, understood as the claim that moral judgments are all false: the position is incoherent because the negation of a given moral judgment – which should be true if the given judgment is false – is another moral judgment: so not all moral judgments can be false. Pigden's response is as follows.

> Thus meta-ethical nihilism needs to be reformulated. I suggest the following: All nonnegative atomic moral judgments are false. This requires elucidation. First we specify a range of primitive 'thin' moral predicates 'good' (morally good), 'bad', 'right', 'wrong' 'ought to' etc. (there may be a problem about this as some of them are interdefinable). We then define an atomic moral judgment as a proposition ascribing an *n*-place moral predicate to *n* specific items. As defined these are non-negative, i.e. *not* governed by the negation operator, but we redundantly specify that they are non-negative for the sake of clarity. Nihilism now amounts to the claim that all non-negative atomic moral propositions are false. And the argument is the standard nihilistic argument that there are no moral properties

---

[215]The following remark from Lindström and Segerberg [734], p. 1205, represents a lapse of concentration: "In Standard Deontic Logic, φ is obligatory if and only if the negation of φ is not forbidden". On the standard account, this defines being permissible rather than being obligatory, for which the "not" should be deleted. (In the same vein, we have, on p. 80 of Hansson [447]: "Thus, 'allowed' may be defined as 'not wrong that not'.") For the record, we quote a dissenting voice: that of Hart [456], on p. 82 of which one reads the following: "Something more, I hope, than a blind wish to adhere to our common speech prompts the protest that it is absurd to speak of having a moral *duty* not to kill another human being, or an *obligation* not to torture a child." In a similar vein, van Fraassen, ([329] p. 6) asks us to distinguish "between what is obligatory, and hence ought to be, and what ought to be for some reason or other," and Lemmon [696] wants us to attend to the distinction between *duties* and *obligations*; for yet more in this vein, see Tranøy [1113] as well as Beran [83] and references there cited. These issues, like some raised in (and around) Remark 4.4.1, are too subtle to occupy us here; note in any case that talk of *one's duty* has now lost its normative force, serving mostly to evoke the moralism of a bygone era. (Cf. Schiff [998]: "the quaint concept of duty".) In [697], p. 45, Lemmon draws our attention to the distinction between being obliged to do something and being under an obligation to do it; one can be under obligation which one does not fulfill – perhaps because of competing obligations – whereas describing someone has having been obliged to leave the room seems to convey the claim that did in fact leave the room (and perhaps did so because of being under an obligation to). This is somewhat similar to Lemmon's claim in this same paper and elsewhere that *must*, even a moral (rather than a 'logical') *must*, distinguishes itself from *ought* in obeying the **T** axiom (to speak rather loosely). See Example 4.6.15(*ii*). Finally, while von Wright took up C. D. Broad's suggested use of the adjective *deontic* from a classical Greek verb widely translated as "ought" (the suitability of which – or any other all-purpose – translation of that verb is called into question in Collingwood [187], p. 63.)

[216]Those interested in this issue may also be interested in the discussion in Hippler and Schwartz [502].

or relations corresponding to the moral predicates and thus no moral facts. Although this new formulation of nihilism is much more restricted than the original doctrine, it captures the spirit, though not the letter of the original thesis.

The formulation "all non-negative atomic moral propositions are false" is not idea since the atomic/non-atomic contrast is one between linguistic expression (sentences of a natural language or formulas of a formal language) rather than propositions, understood as what suitably equivalent such expressions have in common as the proposition expressed by them. This is what the reference to favouring as primitive a select range of predicates is all about, but Pigden is quite right to worry that interdefinability raises a difficulty, since by a change of primitive predicates we change what is atomic and what is not. And further, the 'atomic or otherwise' decision is not well suited to traditional deontic vocabulary in any case since whether we take 'O' as primitive and define P as $\neg O \neg$ or instead take P as primitive and define O as $\neg P \neg$, neither Oq nor Pq is atomic, and there is similarly no intrinsic reason to favour "ought" over "may" if either of these is to be explained in terms of the other. (A poor example, admittedly, since *may* is used to give permission rather than to register permissibility.) Thus the suggested reformulation of moral nihilism will need further thought.

(*ii*) What about the view, however it may be related to moral nihilism, that nothing whatever is morally wrong – which, even if it is an oversimplification, we may take for present purposes to coincide with the view that nothing is morally obligatory? If deontic logic needs to make room for such a view, the normal modal logics appear at first sight not to offer a favourable environment for deontic developments, in view of (deontic) Necessitation. (This was the theme of Remark 4.4.9, p. 246.) One possibility would be to opt for the smallest regular modal logic as a basic deontic logic, rather that **K** or **KD**, in which case the current view would be embodied in its consistent extension by the (all formulas instantiating the) schema $\neg OA$: the so-called Falsum logic. (See the index.) Another option would be to remain normal but go bimodal, with an alethic $\square$ alongside O, the latter thought of as expressing perhaps narrowly logical necessity, and perhaps something broader, along the lines of the similarly supplementary alethic operator in play in the discussion of Section 4.8 below, where it is written as L (see note 270 on p. 326). The idea would be to tolerate the provability of such things as $O(p \to p)$ but explicitly record their degenerate status with a bridging axiom saying that they are the only things that are obligatory: $OA \to \square A$. Alternative put: whatever is possible is permissible. ◀

Another topic which has attracted much attention is that of free choice permission, by which is meant the phenomenon that "it is permissible that A or B" is usually construed as meaning that it is both permissible that A and permissible that B. For a P representing any notion for which this equivalence held, we cannot have P as the dual of a reasonably behaving O, since the dual form of the equivalence just noted would then be an equivalence between $O(A \wedge B)$ and $OA \vee OB$, which no-one has found tempting. There has been extensive discussion of free choice permission in the literature, a sample of which is provided by Hilpinen [484], Jennings [623], [624], and Simons [1053], [1054], van der Meyden [804]. From these references it will be clear that the issue has more to do with the interaction between "or" and various other operators rather than just permissibility. Indeed it may have little to do specifically with permissibility at all; as well as some of the above references, see §6.14 of Humberstone [594]. Further problems arising over (notions of) permissibility in deontic logic are detailed in Merin [803], Hansen [438], the references supplied in these papers, and Lewis [720], Yablo [1210] and Fine [298].

A. R. Anderson, in [17] and other works there cited, opened up a fruitful line of investigation in deontic logic by suggesting a reduction of deontic logic to alethic modal logic supplemented by a special sentential constant S with a heuristic reading "the sanction is applied". In fact, he considered two options, the first of which – call it Anderson's reduction Mark I – involves subjecting this constant to a special axiom to the effect that the sanction is escapable, namely $\diamond \neg S$ (or $\neg \square S$),[217] and a second option – Anderson's

---

[217]Prior [919], p. 140, remarks on Anderson's deployment of a variation on this Mark I treatment, in which the special axiom is stronger – $\diamond S \wedge \diamond \neg S$ – and observes that this extra strength does not pull any weight in the deontic reduction

reduction, Mark II – in which no such axiom is required, though there is a more complicated translation of the deontic language (with O and P) into the supplemented alethic language. The Mark II version we shall defer for a moment. (What we are calling the Mark I version is a simplification – see note 217 – of the presentation in [21], while the Mark II version appeared, in the following year, in [16].)

On the Mark I version, the translation of OA is $\Box(\neg A \to S)$. Early philosophical reaction overexcitedly alleged that G. E. Moore's 'naturalistic fallacy' was on display, at least for the principal – which is to say, the *moral* – interpretation of O, since there is obviously no equivalence between the claim that something ought to be the case and the claim that if it isn't, a sanction is applied, indeed, that if it isn't anything describable in non-moral terms is the case. But Anderson clearly thought (at least in [16]) of the "sanction" reading of S as a picturesque turn of phrase, explicitly mentioning the preferability of a more accurate reading along the lines of: something that ought not to be the case is the case.[218] It would not be an objection to the enterprise, though it might call into question the description of the enterprise as a *reduction*, to say that this gloss on S re-introduces the vocabulary S was supposed to define, but that doesn't mean that, having explained its significance, we can't take it as primitive and with its aid (as well as $\Box$ and Boolean vocabulary) and 'reduce' deontic logic to alethic modal logic together with a – shall we say? – *deontic* sentential constant.[219]

For a range of plausible alethic modal logics, of which **KT** is perhaps the weakest, the special axiom mentioned above brings with it $\mathbf{D}_O$ once the above definition of O is in place, since, reasoning informally, from the assumptions

$$\Box(A \to S) \qquad \text{and} \qquad \Box(\neg A \to S)$$

(alias $O\neg A$ and OA, respectively) then it would follow that $\Box S$, contradicting our special axiom $\Diamond \neg S$. In addition to rendering $\mathbf{D}_O$ (and indeed $\mathbf{K}_O$) provable, various unwanted principles remained unprovable in this setting, such as $Op \to p$ and $Pp \to Op$, making this a promising approach to deontic logic.[220] (We will be more precise in Theorem 4.4.19 below, after re-casting the reduction somewhat.)

Let us pause to note some further moves suggested by Anderson's Reduction (the Mark I version, anyway). First, recall that one topic set aside above (Example 4.4.14, p. 251) was that of agent-relative or agent-implicating obligation statements. Let us think of b as the name of one agent and of $O_b A$ as saying that there is an obligation on b to have it be the case that A. Then, as noted in Nowell-Smith and Lemmon

---

(see Prior [919], p. 141), for which reason we ignore it here. (The stronger axiom does not survive in any published work of Anderson's beyond the early paper [21] (see p. 16) he co-authored with Omar Khayyam Moore; Prior [921] refers it as 'the fuller axiom'. – which may accordingly puzzle the reader. Anderson [17], n. 40, informs us that it figured also in an early oral presentation, whose text he had sent to Prior. Certainly the simplified version (i.e., $\Diamond \neg S$) appears already in Anderson [19], where it is remarked that it can be dispensed with as special axiom – an allusion to what we call the Mark II version of Anderson's reduction. Note that the contingency issue raised in connection with Anderson in note 188, p. 229, and ignored here, is a different one: having to do with the contingency of p as a prerequisite for the truth of Op rather than the contingency of S.) Interestingly, a constant with this behaviour can be used to do modal logic with $\triangle$ – for noncontingency, as in Exc. 1.3.8, p. 21 – since it allows for the definition of $\Box$ in any extension of **KD**, as observed in Pizzi [884]. As we are not taking $\Box$ as primitive in this context (or help ourselves to $\Diamond$), think of the new axiom as having the form $\neg \triangle S$ (alias ∇S). Pizzi shows also that a weakening of this axiom governing our sentential constant – to $\triangle S \to \triangle A$ – even allows for the definition of $\Box$ in terms of this sentential constant and $\triangle$ in the setting of ($\triangle$-based) **K**. Pizzi's definition for $\Box A$, in either case, is as $\triangle A \wedge \triangle(S \to A)$. For further discussion, see my [598]. Section 2 of that paper has a fatal flaw – which does not affect the discussion of Pizzi – pointed out to me by Evgeny Zolin: Lemma 2.1 there is false, as are the results based on it (Theorem 2.2 and Corollary 2.3). Zolin has concrete counterexamples, which, one hopes, will appear in print at some stage, to the claimed results, as well as a diagnosis of the error in the proof of Lemma 2.1 (an oversight in the inductive organization of the argument). Instead of using Pizzi's definition of $\Box A$ as $\triangle S \to \triangle A$ with a special constant axiomatically stipulated to be contingent, one can make considerable mileage – see Fan, Wang and van Ditmarsch [282] – out of a sort of conditional definition using a contingency condition on a new schematic letter, thus: $\nabla B \to (\Box A \leftrightarrow (\triangle B \wedge \triangle(B \to A)))$.

[218]This is made very clear in the second new paragraph of p. 86 of Anderson [15].
[219]Prior [920] provides a thoughtful early discussion of this venture.
[220]The word "promising" is used in place of Anderson's talk of normal deontic logics, since that clashes with the contemporary meaning of "normal". Anderson required, under this heading, that certain things should be provable, in particular what amounts to the equivalence of $O(p \wedge q)$ with $Op \wedge Oq$ (though this is put in terms of P and $\vee$ by Anderson) and also $\mathbf{D}_O$ ($= Op \to Pp$), and that certain other things should *not* be provable, such as $Op \to p$ and $Pp \to Op$.

4.4. DEONTIC LOGIC: MAIN THEMES

[849], we could given an Andersonian representation of this if we traded in our sentential constant S for a (monadic) predicate constant written the same way, with the picturesque reading for Sx that x is sanctioned (or is sanctionworthy,[221] or ...). The representation of $O_b A$ would then be $\Box(\neg A \to Sb)$, and to get (agent-implicating) obligation to imply (agent-implicating) permissibility, we need axioms along the lines of $\Diamond \neg Sb$ for the agents concerned (as here in the case of b). A second variation on Anderson's theme would be to follow his lead but looking on the brighter side, with a sentential constant R with the (again, purely heuristic) reading "the reward is given", thinking of this as what happens when agents act beyond the call of duty. The resulting operator, which we shall not bother to introduce any notation for, would be defined by $\Box(\_ \to R)$ would amount to its being *supererogatory* that A; in fact this is even more naturally thought of as agent-specific, so one might follow the Nowell-Smith–Lemmon lead and replace R with Ra, Rb, Rc,...

Digression. (On supererogation.) The version with plain sentential R was explored in Humberstone [540]. The idea of a 'reward' constant is mentioned in the final sentence of Anderson [16], but not in connection with supererogation. (Likewise at p. 117 of Fisher [304]. At p. 168 of Abraham, Gabbay and Schild [3], mentioned in Remark 4.4.1 above, the idea of a – divine – reward is associated with the obligatory rather than the supererogatory, the corresponding Andersonian sanction being associated with a non-interdefinable notion of the forbidden.) Various other approaches to the logic of supererogation can be found in Chisholm and Sosa [177] (with Chisholm [174] of some interest), Mares and McNamara [774], McNamara [794] and Åqvist [32], this last mentioned in passing under Example 4.3.2, p. 228. For the philosophical background, see Urmson [1120], Heyd [477] and Flescher [313]. The main lesson is that even if some normative ethical theories are so organized as not to make room for the possibility of supererogatory acts – as is arguably the case with utilitarianism, for instance – it would be unwise to preclude this possibility in what one says about deontic logic, for instance by calling the deontic analogue of contingency ($PA \land P\neg A$) *moral indifference*. **End of Digression.**

In the Mark II variation ([16], as well as [17]) Anderson avoided the need to have the special axiom $\Diamond \neg S$ while still securing its intended effect (namely $\mathbf{D}_O$) by complicating the translation of O,

$$\begin{aligned}\text{replacing:} \quad & OA = \Box(\neg A \to S) & \text{(Mark I)}\\ \text{with:} \quad & OA = \Box(\neg A \to (S \land \Diamond \neg S)) & \text{(Mark II)}.\end{aligned}$$

At first sight, this second proposal looks like a confusion, as though we have confused strictly implying S, which is avoidable,[222] on the one hand, with strictly implying both: S and it is avoidable that S. But while there may be a difficulty (see Prior [921]) in knowing, in the Mark II reduction, whether it is S or

---

[221] Rickman [957], under (2) on p. 273, seems to offer a jaded acknowledgment of Nowell-Smith and Lemmon's suggestion, summing up with: "We are forced to the translation (of "p is wrong" as) "p necessarily implies that the perpetrator of p ought to suffer the sanction". This is of course completely garbled, adding to the familiar confusion between using a letter like "p" where a sentence might stand and where the name of a proposition might stand the additional conflation with holding a place for the name of an action (capable of being 'perpetrated'), with the related mistake that a unique perpetrator or set of perpetrators can be read off from an *ought*-statement, as well as the idea that there is any kind of *a priori* connection between the wrongness of a's action and the permissibility or otherwise of punishing a for that action. (Rickman says he is commenting on 'Escapism: the Logical Basis of Ethics' without giving bibliographical details. From the frequent mention of Nowell-Smith and Lemmon and the fact that it appeared in the journal *Mind*, it is clear that it is [849] that is under discussion, despite the additional occasional mention of A. N. Prior, who also wrote an earlier paper with this same title, namely [920], in which this suggestion of Rickman's is endorsed as a solution to the Good Samaritan Paradox, which first appeared in this very paper of Prior's. (Note that just specifying the agent won't completely help, in view of a variation in which the assailant is the Samaritan, who has had a change of heart – or indeed in any ordinary case of an obligation to make amends or apologize for what one has done; see also Remark 4.4.2(*ii*), p. 237.) Rickman's own discussion note [957] bears the same title, making it one of the four entries in our bibliography to be called 'Escapism: The Logical Basis of Ethics' – the fourth being Smiley [1065]. The subtitle here was perhaps originally intended as an echo of the somewhat similar title of Prior's 1949 book [913], mentioned in Remark 4.4.6.) The sanctionworthiness theme was more recently taken up in Chapter 14 of Forrester [326].

[222] We use the phrase "S is avoidable" to abbreviate the "$\Diamond \neg S$"; this is not quite right since S should be appearing in the English glosses where a sentence could appear rather than where a name could appear.

$S \wedge \Diamond \neg S$ that should be described as the sanction (or the 'bad thing'), the idea works out satisfactorily on a technical level, allowing us to prove $\mathbf{D}_O$ in, for example, **KT**, when O is treated *à la* Anderson Mark II. First, note that the following is a theorem of **K**:

$$(\Box(p \to q) \wedge \Box(\neg p \to q)) \to \Box q,$$

and now substitute $S \wedge \Diamond \neg S$ for q, which gives:

$$(\Box(p \to (S \wedge \Diamond \neg S)) \wedge (\Box(\neg p \to (S \wedge \Diamond \neg S)))) \to \Box(S \wedge \Diamond \neg S).$$

But the consequent here is **KT**-refutable (i.e., its negation is **KT**-provable), since distributing the $\Box$ across the conjunction and then appealing to **T** (though the much weaker $\mathbf{5}_c$ will do for this purpose) we get $\Box S \wedge \Diamond \neg S$. Thus by truth-functional reasoning, the first conjunct of the last inset formula's antecedent provably implies (in **KT** and above, supplemented by the presence of S) the negation of its second conjunct, which is to say, we have $Op \to Pp$ ($\mathbf{D}_O$, that is). The **KT**-provability of all formulas of the form $\neg \Box (A \wedge \Diamond \neg A)$, exploited here with S as A, came up in our discussion at the end of Example 2.6.37 (p. 128) in the variant form $\Diamond(A \to \Box A)$. In the **K**-equivalent form $\mathbf{5}_c$ ($\Box A \to \Diamond \Box A$ or $\Box \Diamond A \to \Diamond A$), it will appear again at several points below (such as in Proposition 5.1.18, on p. 357.) The technical demands on the consequent $S \wedge \Diamond \neg S$ for Anderson's purposes are that it should serve as a formula B with the following two properties relative to a monomodal logic (with perhaps an additional constant such as S in its language), provability in which is indicated by the $\vdash$:

(1) $\vdash \neg \Box B$ (2) $\nvdash \neg B$.

The discussion above stresses the role of (1), since it is this that makes O satisfy the **D** axiom when O is defined as $\Box(\neg A \to \neg B)$. Now as long as the $\Box$ of our unspecified monomodal logic itself satisfies **D**, any refutable formula, for example $\bot$, would serve to fulfill the condition (1). Condition (2) rules out such choices for B, and does so because if, as is plausible given that we are thinking of our logic for $\Box$ as suited to an alethic interpretation, not only **D** but **T** is satisfied by this operator, then OA would amount to $\Box A$ by the $\Box(\neg A \to \neg B)$ definition – already, worryingly, making O just another notation for $\Box$, but now in virtue of **T** making OA provably imply A. Choosing B as $S \wedge \Diamond \neg S$ gives us something meeting conditions (1) and (2). But why not use an existing formula C from the monomodal language, rather than introducing the new constant S – especially, on might think, since on the Mark II reduction S is governed by no special axioms – so why not just use a propositional variable (or sentence letter)? Certainly this last suggestion would not have the desired effect because it would not make the defined O a 1-ary connective: if the formula C has propositional variables in it and OA is defined as $\Box(\neg A \to (C \wedge \Diamond \neg C))$, then those variables occur in OA: but the only variables occurring in OA if O is a one-place connective (or sentence operator) are those occurring in A: so trouble threatens.[223] One might relax things to allow for variables occurring *inessentially* in the sense that C is equivalent (according to the logic in question) to a formula not containing the variable in question, but if C is a formula in which p_i occurs inessentially in this sense, so we have $\vdash C \leftrightarrow C'$ for C' not containing p_i, then by uniformly substituting $\top$ for p_i throughout, we get $\vdash C^* \leftrightarrow C'$, where C^* is the result of substituting $\top$ for p_i in C, and thus $\vdash C \leftrightarrow C^*$. Successive applications of this to each variable in C, if C contains no variable essentially, gives a pure formula provably equivalent to C, so let us consider directly the option of defining OA to be $\Box(\neg A \to (C \wedge \Diamond \neg C))$ for C a pure formula ('pure' as defined in Exercise 2.7.21, p. 134). That means that $C \wedge \Diamond \neg C$ is itself a pure formula, and since we are taking our candidate alethic modal logic to extend **KT**, and hence to extend **KD**, Coro. 2.10.29 (p. 172) tells us that this conjunction is provably equivalent to $\top$ or else to $\bot$, which is bad news:

EXERCISE 4.4.16 Show that the two possibilities just mentioned for pure B of the form $C \wedge \Diamond \neg C$, namely $\vdash B$ and $\vdash \neg B$, each conflict with the joint satisfaction of conditions (1) and (2) above. ✠

---

[223]Numerous methodological issues arise here, discussed in more detail in my [565] and in subsection 1.11 of [594].

4.4. DEONTIC LOGIC: MAIN THEMES

Thus is is no accident that Anderson chose to introduce a new propositional constant S for his purposes. For the moment we leave his ingenious Mark II and return to the Mark I form.

Or rather: to the Mark I form of Anderson's reduction as it was more suggestively transformed in work by Kanger and Smiley ([637] and [1064], respectively), with the present exposition picking up especially on the latter. The transformation required is nothing but contraposition. Instead of thinking of OA as $\Box(\neg A \to S)$, rewrite this as $\Box(\neg S \to A)$, and instead of using S, let us take what would have been $\neg S$ as a new primitive. Following Kanger we write this constant as Q. Thus we now take our favoured alethic modal logic and define OA as $\Box(Q \to A)$. Evidently, if O is to be read as "it is legally (resp. morally) obligatory that" then Q can be thought of as saying that the legal (resp. moral) code in question is complied with; in semantic terms Q is to be true at the ideal, perfect or permissible worlds (in which nothing that ought not to be the case is the case).[224] But as Smiley emphasized, such a treatment may be reasonable in other settings too. For example, Q can be thought of as a statement of the laws of nature, whatever they may be, in which case O, as so defined, would amount to a kind of nomological necessity operator, or again, at least abstracting from worries about logical omniscience and the like (see Sections 5.1, 5.2, 5.6), Q could be taken to represent a subject's corpus of belief and O would emerge as the belief operator for that subject: a reduction of doxastic logic to alethic modal logic with an added sentential constant. (This suggestion had also appeared in Fitch [306], 11.19 at the base of p. 69. Notice the contrast in the case of nomological necessity, with the proposal of Pargetter and Bigelow, described in Section 4.2, in which metaphysical or logical necessity is explicated in terms of nomic necessity, the former being the nomic necessity of the nomic necessity of the statement in question, rather than here, with nomic necessity explicated in terms of the broader necessity coupled with a suitable constant.) Smiley [1064] thinks of all these as different notions of *relative* necessity, with $\Box$ as providing the contrasting background notion of *absolute* necessity: O represents necessity relative to Q. Now, this is not the exotic relative necessity of von Wright's dyadic alethic modal logic from [1207], mentioned in passing above (p. 250). Rather, it is simply strict implication: necessity relative to the moral norms, the physical laws (alias nomic or physical necessity), etc., is simply a matter of being strictly implied by them. (For further issues and variations, see Smiley [1065]; one interesting suggestion to be found there – p. 240f. – is that one take the biconditional schema $OA \leftrightarrow \Box(Q \to A)$ as delivering a deontic logic when taken as giving the quasi-normal rather than the normal extension of one's favoured alethic modal logic. This amounts to treating O and $\Box(Q \to \_)$ as giving coextensive rather than synonymous operators, to put it in the combined terminology of Smiley and the more recent exploration, Williamson [1186], of similar issues.)

REMARK 4.4.17 Before note 224 is too far behind us, a few words on the negation of the propositionally quantified formula at the end of it are in order. Let us abbreviate that formula to α. Thus α is the formula $\exists p(Op \land \neg p)$. Prior [928], Chapter 6, gives an interesting argument from the assumption $O\alpha$ to the conclusion α; see also MacIntosh [752]. We present it here as an argument to the conditional conclusion $O\alpha \to \alpha$ (depending on no assumptions.) All one needs by way of logical behaviour on the part of $\exists$ here is an obvious introduction rule, appearing in implicational form as $(\exists I)$ in Example 4.1.13 on p. 213 above; and concerning "O" – which Prior did not write that way, the topic of deontic logic not being under discussion at all – no assumptions whatever need to be made. For the argument, given here in an informal natural deduction style, begin by assuming

$$O\alpha \land \neg\alpha.$$

Now by the $\exists$-introduction rule alluded to, taking p as α:

$$\exists p(Op \land \neg p).$$

---

[224] Prior [920] also considers a move along these lines, writing his constant as E – for "Escape (from the sanction)" – and acknowledging a suggestion to that effect from Lemmon. While this informal reading lacks the philosophical appeal of Kanger's and Smiley's, Prior is perfectly aware that all talk of sanctions and escape is merely heuristic, and observes that if we had propositional quantifiers in our language, the constant in question, however it is notated, would amount to: $\forall p(Op \to p)$. See, further, Remark 4.4.17.

But recall that this is none other than the formula α, so since our initial assumption entails not only (by $\wedge$-elimination – or ($\wedge$E), as it is called in Section 7.1 below) $\neg\alpha$, but also, as we have seen, α itself, by Reductio (or ($\wedge$I) and (RAA), *à la* §7.1) we have the negation of our initial assumption, depending now on no assumptions. But this is evidently equivalent to the implication $O\alpha \to \alpha$. One might here be tempted to think along the following lines: $\exists p(Op \wedge \neg p)$ and $O\exists p(Op \wedge \neg p)$ respectively modally define the classes of irreflexive frames and of frames in which all points accessible to any point are irreflexive. (The famous undefinability of the class of irreflexive frames – Exc. 2.7.35(*iii*), p. 137 – pertains to undefinability in the language without propositional quantifiers. Not that we have specified a precise semantic treatment for such quantifiers for the sake of the current discussion.) However, brief reflection reveals – this tempting line of thought continues – that any frame in which all points have only irreflexive successors is itself an irreflexive frame, so there is nothing untoward about the deducibility of $\exists p(Op \wedge \neg p)$ from $O\exists p(Op \wedge \neg p)$. To correct any such complacency, note that the above deduction made no use of the normality of O at all, and would apply equally well to, for example $\neg O$ taken as a 1-ary operator in its own right. Not even congruentiality of O was assumed. To show that all is far from well here, see Prior's original discussion in [928], which sails us straight into Liar-Paradoxical waters. How to react? Perhaps propositional quantifiers need a dose of ramified type theory, as is sometimes suggested for a problem not discussed here – a diagonalization difficulty noted by Kaplan for the sets-of-worlds conception of propositions (see Kaplan [640], Oksanen [857], Lindström [733], Anderson [22], and note 22 in Cresswell [210], as well as the text to which that note is appended, for further historical information, and also von Kutschera [682], in which Kaplan's problem provokes some interesting (modal) logical moves. (Gettier's work, mentioned by Cresswell, appears not to have been published. Other early treatments include a presentation of the problem in Appendix 9 of Davies [233], and a reaction to that presentation in Moore [822].) This would involve complicating the semantic suggestions for modal logic with propositional quantifiers in the works cited in Example 5.1.6(*ii*) and note 284, p. 340 (or the corresponding general frames: see the Digression on p. 83); see also the discussion in Example 4.1.13, p. 213 – already alluded to in connection with $\exists$-introduction – and the remainder of the section after that. Finally, let it be noted that the derivation above is connected in ways we do not go into here – in which Linsky [735], and, for a fully comprehensive discussion, Sorensen [1073] – with two other topics touched on below: Moore's Paradox (see p. 355), and the Fitch derivation (Example 5.4.11, p. 397), concerning which Linsky also cites MacIntosh [753]. ◀

To see exactly what deontic (or doxastic or whatever – though we concentrate on the deontic application in our description) logic O ends up having on the basis of a definition of the above kind, the contraposed Anderson Mark I treatment, given this or that alethic modal logic, it is convenient to think of translating deontic formulas into the language of alethic modal logic, by a translation we can "read off" from the above definition of O in terms of $\Box$ and Q. In fact, now that we have segregated the language there is no particular need to write O after all and for purposes of describing the translation we will think of each language as having a single 1-ary non-Boolean operator, written in each case as $\Box$. This follows the procedure for describing such translations as we have encountered them already, for example on pp. 142, 147, in the discussion of Matsumoto's embedding (to which we shall return in Section 4.5).

Getting down to details, then: add to the language of monomodal logic a new sentential constant Q and consider the logic we shall call **KT** $\oplus \Diamond Q$ meaning by this the smallest modal logic in this expanded language in which $\Box$ is normal and which contains $\Diamond Q$ (and therefore also $\Box^n \Diamond Q$ for all n). By **KDU** we mean the normal monomodal logic standardly meant by that label: thus the new constant Q is absent from the language of **KDU**. Define a translation τ from this language to the language of **KT** $\oplus \Diamond Q$ thus:

- $\tau(p_i) = p_i$
- $\tau(\neg A) = \neg(\tau(A))$
- $\tau(A \to B) = \tau(A) \to \tau(B)$ (and similarly for other Boolean connectives)
- $\tau(\Box A) = \Box(Q \to \tau(A))$.

4.4. DEONTIC LOGIC: MAIN THEMES

Thus τ simply gives the Kanger–Smiley translation into alethic terms (plus Q) of what we may think of as a deontic formula; you may prefer to rewrite things so that for the source and target[225] modal operators of this translation we have O and $\Box$ respectively. The present notation is chosen simply to minimize discontinuity with such findings from earlier chapters as the fact – established by combining the proofs for Exercises 2.4.8(iv) and 2.5.21(ii) on pp. 65 and 79, respectively – that **KDU** is determined by the class of serial range-reflexive frames. For a semantic description of $\mathbf{KT} \oplus \Diamond Q$ we need frames to incorporate a set of points for Q to be true at in any model on the frame, though here we will pass straight to a description of the models, superscripting Q by a name of the model in question to denote the required truth-set. Thus we are dealing with models $\mathcal{M} = \langle W, R, Q^{\mathcal{M}}, V \rangle$ in which R is reflexive (in view of the axiom **T**) and for all $x \in W$, $R(x) \cap Q^{\mathcal{M}} \neq \varnothing$ (in view of the axiom $\Diamond Q$), and we formulate the required completeness result in terms of models (as in Sections 2.3 and 2.4) – rather than frames – for convenience:

EXERCISE 4.4.18 Show that $\mathbf{KT} \oplus \Diamond Q$ is determined by the class of all models as just described. (For the completeness half, produce a canonical model $\mathcal{M} = \langle W, R, Q^{\mathcal{M}}, V \rangle$ in which $Q^{\mathcal{M}}$ is defined to be $\{x \in W \mid Q \in x\}$ and W, R and V are defined as usual. Verify that truth and membership coincide and that the model satisfies the conditions that R is reflexive – no novelties there – and that $R(x) \cap Q^{\mathcal{M}} \neq \varnothing$ for all $x \in W$.) ✠

We are now in a position to show that the translation τ defined above embeds **KDU** (faithfully) into $\mathbf{KT} \oplus \Diamond Q$, in the usual sense (as on pp. 142, 147) given by the following result, in which we adopt (as with the translations in Exercise 2.7.22) the superscript notation "A^τ" in place of "$\tau(A)$".

THEOREM 4.4.19 *For all formulas A of the language of* **KDU**, *we have*

$$\vdash_{\mathbf{KDU}} A \text{ if and only if } \vdash_{\mathbf{KT} \oplus \Diamond Q} A^\tau.$$

Proof. 'Only if': Pick any convenient axiomatization of **KDU** – for example that with all substitution instances of truth-functional tautologies and all instances of **K**, **D**, **U**, understanding these as labels for schemata, as axioms, and rules Modus Ponens and Necessitation, and argue by induction on the length of proofs that if A has a proof from these axioms and rules, then A^τ is provable in $\mathbf{KT} \oplus \Diamond Q$.

'If': Suppose $\nvdash_{\mathbf{KDU}} A$, with a view to showing that $\nvdash_{\mathbf{KT} \oplus \Diamond Q} A^\tau$. The supposition gives us a serial range-reflexive model $\mathcal{M} = \langle W, R, V \rangle$ with $w \in W$ for which $\mathcal{M} \nvDash_w A$, and we describe a procedure for getting a new model $\mathcal{M}'$, for $\mathbf{KT} \oplus \Diamond Q$, concerning which we shall have $\mathcal{M}' \nvDash_w A^\tau$, showing in view of Exc. 4.4.18, that $\nvdash_{\mathbf{KT} \oplus \Diamond Q} A^\tau$. To this end, define $\mathcal{M}' = \langle W', R', Q^{\mathcal{M}'}, V' \rangle$ thus: $W' = W$, $V' = V$, R' is the reflexive closure of R, i.e., the union of R with the set of pairs $\langle u, u \rangle$ with $u \in W'$ (alias W), and $Q^{\mathcal{M}'} = \{v \in W' \mid \exists u \in W \centerdot uRv\}$. (Note that while the alternation between "W" and "W'" has no significance, this last reference to R is definitely to R and not R'.) First, we must check that $\mathcal{M}'$, so defined, is a model in the class described before Exc. 4.4.18; in particular, one must verify that the condition is satisfied which requires that – as it would appear for the present instance – $R'(x) \cap Q^{\mathcal{M}'} \neq \varnothing$ for all $x \in W'$. This is left to the reader. Next we need to establish, by induction on the complexity of B that for all formulas B (of the language of **KDU**) we have, for all $x \in W$:

$$\mathcal{M} \vDash_x B \Leftrightarrow \mathcal{M}' \vDash_x B^\tau.$$

The basis case, in which B is p_i, is given by the fact that $V' = V$. The inductive cases for the Boolean connectives take care of themselves straightforwardly, leaving us to ponder the case of $B = \Box C$.

$\Rightarrow$-direction: for a contradiction, suppose that $\mathcal{M} \vDash_x \Box C$ while $\mathcal{M}' \nvDash_x (\Box C)^\tau$, i.e., $\mathcal{M}' \nvDash_x \Box(Q \to C^\tau)$. This last means we have $y \in R'(x)$ for which $\mathcal{M}' \vDash_y Q$ and $\mathcal{M}' \nvDash_y C^\tau$. We want to argue that xRy

---

[225] Here we recall the terminology used on p. 147 above, though there the source and target logics were consequence relations rather than simply sets of formulas.

and by the inductive hypothesis (C being of lower complexity that $B = \Box C$) $\mathcal{M} \not\models_y C$, which would give the desired contradiction since supposedly $\mathcal{M} \models_x \Box C$. So it remains to justify the claim that xRy. We know that $xR'y$, so let us distinguish two cases: (1) $x \neq y$, (2) $x = y$. In case (1), since R' differs from R only in adding in the pairs whose first and second elements coincide, we get xRy from the fact that $xR'y$. In case (2) we use the fact that $\mathcal{M}' \models_y Q$, which, from the way that $Q^{\mathcal{M}'}$ was defined, that for some $u \in W'$, yRu. Since $W = W'$, uRy for any such u, and since R is range-reflexive, yRy. As case (2) is the case in which $x = y$, this means that xRy, as desired.

$\Leftarrow$-direction: Suppose that $\mathcal{M} \not\models_x \Box C$ (aiming to show that $\mathcal{M}' \not\models_x \Box(Q \to C^\tau)$). Thus there is some $y \in R(x)$ with $\mathcal{M} \not\models_y C$. By the inductive hypothesis, then, we have (1): $\mathcal{M}' \not\models_y C^\tau$. Since xRy, $y \in Q^{\mathcal{M}'}$, so we have (2): $\mathcal{M}' \models_y Q$. (1) and (2) together imply that $\mathcal{M}' \not\models_y Q \to C^\tau$. But $R \subseteq R'$, so $xR'y$, so $\mathcal{M}' \not\models_x \Box(Q \to C^\tau)$.

At the start of this proof of the 'if' direction of the present result, we had $\mathcal{M} \not\models_w A$, for the **KDU**-unprovable formula A, so by appealing to the fact, just established, that for all B, $\mathcal{M} \models_x B$ if and only if $\mathcal{M}' \models_x B^\tau$, we conclude that $\mathcal{M}' \not\models_w A^\tau$, and thus that $\not\vdash_{\mathbf{KT} \oplus \Diamond Q} A^\tau$. ∎

As already indicated, our use of the notation "$\mathbf{KT} \oplus \Diamond Q$" is not strictly legitimate with the first 'summand' **KT** is not quite the **KT** of old but has an additional piece of logical vocabulary, the 0-place connective or sentential constant Q, though subjected to no particular logical demands. Let us be very explicit about this for a moment and write "$\mathbf{KT}^Q$" for this logic. (Åqvist [30] subscripts the "Q" for this purpose, but since we already have the convention, e.g. from p. 178, of subscripting operator notations for a different purpose – writing "$\mathbf{KT}_O$", for instance for the logic **KT** with $\Box$ rewritten as "O" – we follow the superscripting option instead.)

EXERCISE 4.4.20 (*i*) Show that the translation τ under discussion above provides a faithful embedding of **KU** into $\mathbf{KT}^Q$, in the sense that:

$$\vdash_{\mathbf{KU}} A \text{ if and only if } \vdash_{\mathbf{KT}^Q} A^\tau,$$

for all formulas A (of the language of **KU**).

(*ii*) In the proof of the 'if' part of Theorem 4.4.19 we began with the existence of a serial range-reflexive model containing a point at which a **KDU**-unprovable formula A was false; but where, if anywhere, in the proof was the *seriality* of (the frame of) this model exploited?

(*iii*) What if we have no additional axioms at all? Prove or refute the claim that, again for τ as above:

$$\vdash_{\mathbf{K}} A \text{ if and only if } \vdash_{\mathbf{K}^Q} A^\tau,$$

for all formulas A (of the language without Q).

(*iv*) Same question as (*iii*) but with **K4** on the left and $\mathbf{K4}^Q$ on the right. ✠

The result of Exc. 4.4.20(*i*) was proved by a similar method (to that of the proof of Thm. 4.4.19 here), but with matrix reasoning replacing the Kripke semantical construction (of $\mathcal{M}'$ from $\mathcal{M}$), on p. 124 of Smiley [1064]. Because Smiley calls **KT** "M" and writes $\Box$ in deontic dress as O in the source logics for these translations,[226] he registers the fact that we have an embedding by calling **KU** (or $\mathbf{KU}_O$) "OM", and gives similar results for OS4, OS5 and – before all of these – OS2 (S2 the non-normal logic of that name considered by C. I. Lewis: see Example 2.1.18(*ii*) above, p. 46, for an axiomatic description, and p. 362 below, for a semantic characterizatin). Using a similar nomenclature, Åqvist [30] (Chapters 3 and 4) presents altogether 10 embedding results along these lines for the translation τ. (Our small sample – Theorem 4.4.19 – uses the same method of proof as Chapter 4 of [30], to which the interested reader is

---

[226] And, incidentally, L for $\Box$ in the target logics.

4.4. DEONTIC LOGIC: MAIN THEMES

referred for other cases.) In Åqvist's terminology – but using our nomenclature for the logics concerned – Theorem 4.4.19 and Exercise 4.4.20(*i*) show **KDU**$_O$ and **KU**$_O$ to be the *deontic fragments* of **KT** $\oplus \Diamond Q$ (or **KT**$^Q \oplus \Diamond Q$, as we might say) and **KT**^{Q}, respectively. The term "fragment" is intended to recall the fact that the only occurrences of $\Box$ and Q in the logics so described are the combination $\Box(Q \to \_)$, written as $O\_$ (including here $\Diamond(Q \wedge \_)$ as abbreviating $\neg\Box(Q \to \neg\_)$, written as $P\_$).

EXAMPLE 4.4.21 As well as the logics mentioned above, Smiley [1064], p. 127, considers something he calls "OPC", for its general theoretical interest rather than its deontic suitability. The label has to do with thinking of **KT!** as PC in the sense of '(non-modal) propositional calculus with a redundant $\Box$ operator present', but the point of the present example will be evident from the axiomatization Smiley provides of it as **KT**$_c$**U**$_O$, but the point of interest has nothing to do with "O": rather, it concerns the joint presence of axioms **T**$_c$ and **U** and the absence of any acknowledgment that the latter is redundant. This illustrates the usefulness of the Kripke semantics for modal logic over the matrix and algebraic methods used by Smiley for seeing the deductive relations between modal formulas. No-one acquainted with the Kripke semantics would write down as separate axioms (for a normal modal logic) both **T**$_c$ and **U**, since **T**$_c$ has **U** as a frame consequence, and there is no issue of Kripke incompleteness here (a phenomenon which has indeed resulted in a revival of algebraic semantics). For suppose **T**$_c$ is valid on a frame $\langle W, R \rangle$. Then each $x \in W$ bears R to at most itself. Whether $R(x) = \{x\}$ or $R(x) = \varnothing$, we see that all points in $R(x)$ are reflexive, so the frame is range-reflexive and **U** is valid on $\langle W, R \rangle$. Exercise 4.4.22 below asks for a syntactic derivation reflecting the semantic situation in this regard. (Compare also Restall and Russell [956], p. 253 in which a deontic logic is described with a proposed axiomatization as **KD45U**, but since, as remarked in Example 2.5.22(*ii*), p. 79, all euclidean frames are range-reflexive the "**U**" is redundant in any label containing "**5**". A syntactic proof in this case was requested in the immediately following Exercise 2.5.23.)

The point of theoretical interest about this "deontic fragment of **KT!**" to put it in Åqvist's terminology is the connection Smiley makes with the Ł-modal logic of Łukasiewicz. The OA defined as $\Box(Q \to A)$ when this strict implication – frequently written, it will be recalled, as "$Q \dashv A$" – is simply equivalent to the corresponding material implication turns out to behave, not like the necessity operator of the Ł-modal logic, but like its possibility operator. Łukasiewicz actually wrote "Γ" and "Δ" for $\Box$ and $\Diamond$, and most implausibly defended against obvious criticisms – e.g., from von Wright [1207], p. 124 – what he already realised was an extensional modal logic (in the sense of Revision Exercise 2.7.2(*i*)). Following up the references in Smiley [1064] on this topic will explain that the extensionality is (as on p. 322) a by-product of treating necessity as a *hybrid* of truth-functional connectives (in this case of connectives expressing the identity truth-function and the constant false truth-function); on the notions of extensionality in play in these discussions, see further the Appendix (= §5) of Humberstone [600]. Alternatively, one can think of Łukasiewicz-style possibility as being represented as hybridizing the constant true and identity truth-functions, which is why it emerges as "$Q \dashv \_$" when "$\dashv$" behaves like "$\to$". When Q is true this amounts to whatever goes into blank (identity) and when Q is false this amounts to the constant true, regardless of how the blank is filled. For further references and a more recent discussion, see Font and Hájek [317]. Another relatively early venture into modal logic, Törnebohm [1111], seems to have wandered into extensional territory with even less of a recognition that this may not be the most suitable environment for modal reasoning. ◀

EXERCISE 4.4.22 Give a formal axiomatic proof showing that $\vdash_{\mathbf{KT}_c} \Box(\Box p \to p)$. (For motivation, see the first paragraph of Example 4.4.21.) ✠

It is worth keeping track of which principles governing O (understood as $\Box \to \_$) are inherited from $\Box$'s satisfying certain principles (without additional assistance from a $\Diamond Q$ axiom). The 'only if' part of the proof of Theorem 4.4.19, which left details to the reader, shows that – as we may put it in the current terminology – O satisfies **U** (or "**U**$_O$") when $\Box$ satisfies **T**. To spell this out syntactically, **T** gives us:

$$\Box(Q \to A) \to (Q \to A),$$

so, permuting antecedents (by 'TF') we obtain:

$$Q \to (\Box(Q \to A) \to (Q \to A)),$$

which, when necessitated, gives us $\mathbf{U}_O$. To think about the matter semantically, it helps to have the concept of the range-restriction of a (binary) relation. Suppose $R_1 \subseteq W \times W$ and $R_2 \subseteq W \times W$. Then we call R_2 a *range-restriction* of R_1 if there exists $X \subseteq W$ with:

$$xR_2y \text{ if and only if } (xR_1y \text{ and } y \in X), \text{ for all } x, y \in W.$$

More succinctly put: if $R_2 = R_1 \cap (W \times X)$. In the case of such an X, we say that R_2 is the relation R_1 range-restricted to X. (Recall from note 68 on p. 79 that the *range* of $R \subseteq W \times W$ is $\bigcup_{x \in W}(R(x))$.) The bearing of this concept on the results in play above is that O as currently defined ends up having as its accessibility relation some range-restriction of the accessibility relation for $\Box$, with $Q^{\mathcal{M}}$ playing the role of X. And one easily sees that the range-restriction of any reflexive relation must be range-reflexive, whence $\mathbf{U}$ for O given $\mathbf{T}$ for $\Box$. In the case of $\mathbf{4}$ for $\Box$, we get $\mathbf{4}$ itself for O, since any range-restriction of a transitive relation is transitive. An interesting further case covered in Åqvist's discussion is that of $\mathbf{B}_\Box$ ($\mathbf{B}$ for $\Box$, that is). What modal behaviour does this induce for O? Since $\mathbf{T}_\Box$ induced $\mathbf{U}_O$, these principles modally defining respectively the classes of reflexive and range-reflexive frames, and range-reflexivity coincides with piecewise reflexivity, one might expect to see piecewise symmetry putting in an appearance here (corresponding to symmetry, plain and simple, defined by $\mathbf{B}$: see the discussion beginning on p. 76). Recall from Example 2.5.34 that this is the condition that for all frame elements w in a frame with accessibility relation R, we have for all $x, y \in R(w)$, if xRy then yRx. That would be quite a complication, since as Revision Exercise 2.7.36 showed, this condition is not modally definable at all. All that is required, however, is the condition that for all w, and all $x \in R(w)$, if xRy then yRx, described by Åqvist [30] in terms of "almost symmetric" frames; in the terminology introduced in Exercise 3.2.13 above, the condition amounts to saying that every accessible point (every point in the range of R, that is) should be 1-symmetric. Note that by Proposition 3.2.10 and (the answer to) Exercise 3.2.13, the class of frames meeting this condition is modally defined by the formula $\Box(p \to \Box\Diamond p)$. (In our earlier discussion – Example 3.2.23 – we saw that this could be derived from $p \to \Box\Diamond p$ without using Necessitation, against the background of $\mathbf{K4}$.) As mentioned in note 30, p. 37, we call the formula $\Box(p \to \Box\Diamond p)$ – or the corresponding schema – simply $\mathbf{B}^\Box$ (not to be confused with $\mathbf{B}_\Box$, above, of course). The last time we encountered this principle was in the discussion of Perzanowski (p. 98); its first appearance in deontic logic was perhaps as (A7) on p. 178 of Hanson [439].

EXERCISE 4.4.23 Show that for the translation τ we have been considering, we have the following: For all formulas A of the language of $\mathbf{KB}$, we have

$$\vdash_{\mathbf{KB}} A \text{ if and only if } \vdash_{\mathbf{K} \oplus \mathbf{B}^\Box} A^\tau,$$

where for simplicity we have written "$\mathbf{K} \oplus \mathbf{B}^\Box$", rather than what it strictly meant, namely "$(\mathbf{K} \oplus \mathbf{B}^\Box)^Q$" – to register the presence of Q in the language, despite its not being subjected to any particular axioms. *Suggestion*: follow the pattern of the proof of Theorem 4.4.19, defining $Q^{\mathcal{M}'}$ in exactly the same way but defining R' to be the symmetric closure – rather than the reflexive closure – of R (the accessibility relation of the model on the basis of which $\mathcal{M}'$ is constructed), where the *symmetric closure* of a binary relation is the union of that relation with its converse. (This is the smallest symmetric relation including a given relation.) ✠

Let us see what becomes of the Kanger–Smiley version of Anderson's reduction Mark II, in which OA is identified with $\Box(\neg A \to (S \land \Diamond \neg S))$. As with the Mark I reduction, we contrapose the strict implication here, which, together with some truth-functional reformulation (available since we are not

4.4. DEONTIC LOGIC: MAIN THEMES

contemplating non-congruential logics), turns this □-formula into $\Box((S \to \Box S) \to A)$. To mark its passage into antecedent position, we want to notate this with the aid of "Q", and we could simply replace S with Q to effect such a change, but for continuity with our treatment of the Mark I form, it is actually more convenient to think of Q as representing $\neg S$ rather than S itself. This gives, with the help of some further contraposition, the following as the alethic (plus Q) rendering of OA:

$$\Box((\Diamond Q \to Q) \to A).$$

Note that in view of the truth-functional equivalence of $(p \to q) \to r$ with $(p \vee r) \wedge (q \to r)$, and assuming □ is normal, the inset formula above is equivalent to:

$$\Box(\Diamond Q \vee A) \wedge \Box(Q \to A),$$

of which the second conjunct represents the Mark I treatment of OA. Thus the current treatment makes OA into a stronger claim, adding in the first conjunct. Recall that the point of this is to avoid having to adding a special axiom ($\Diamond Q$) governing Q, while still recovering $\mathbf{D}_O$. Indeed a simpler formula, dropping the second disjunct of the first conjunct:

$$\Box \Diamond Q \wedge \Box(Q \to A),$$

would also deliver $\mathbf{D}_O$, but would not deliver O as normal (because of the first conjunct, since we do not have $O\top$ provable, recalling that no special axioms governing Q are envisaged).

In fact, such a decomposition of the originally inset version of OA is not the most perspicuous path to understanding how $\mathbf{D}_O$ arises. Instead, consider the **K**-provable formula:

$$\Diamond p \to (\Box(p \to q) \to \Diamond(p \wedge q)).$$

Substituting $\Diamond Q \to Q$ for p and (an arbitrary formula) A for q, we obtain a formula whose antecedent is **KT**-provable, or more accurately $\mathbf{KT}^Q$-provable, since $\vdash_{\mathbf{KT}} \Diamond(\Diamond p \to p)$, and Modus Ponens then gives us $\mathbf{D}_O$. The **KT**-provability of the antecedent dropped in this Modus Ponens is a matter of some interest. It arises for exactly the same reasons as we have $\vdash_{\mathbf{KT}} \Diamond(p \to \Box p)$, as explained on p. 256, where Example 2.6.37 was mentioned too. Elaborating somewhat more for this discussion: the point is that both $\Diamond p \to p$ and $p \to \Box p$ are what were called, in the discussion leading up Example 2.6.37 (p. 128), weak Hughes formulas, meaning formulas with the property that for any point in any model, they are either true at that point or at some successor. As we noted there, though putting the point slightly differently, if attention is restricted to reflexive frames (equivalently: frames for **KT**), then this can be simplified to: for any point, they are true at some successor of that point. Thus the results of prefixing a $\Diamond$ to them are formulas true at every point, explaining – if that is not too strong a word for it – their **KT**-provability.

We are now ready for the variant of Theorem 4.4.19 for this Mark II translation, which we shall call τ'. As for the prototype, we write □ for O in the source logic.

- $\tau'(p_i) = p_i$
- $\tau'(\neg A) = \neg(\tau(A))$
- $\tau'(A \to B) = \tau(A) \to \tau(B)$ (and similarly for other Boolean connectives)
- $\tau'(\Box A) = \Box((\Diamond Q \to Q) \to \tau'(A))$.

Having given the definition, we revert to the superscript notation used for τ, writing $A^{\tau'}$ rather than $\tau'(A)$. The proof below follows that of our earlier embedding of **KDU** (Theorem 4.4.19), with changes as needed.

THEOREM 4.4.24 *For all formulas A of the language of **KDU**, we have*

264   CHAPTER 4. APPLICATIONS: ALETHIC, NOMIC, DEONTIC

$\vdash_{\mathbf{KDU}} A$ if and only if $\vdash_{\mathbf{KT}^Q} A^{\tau'}$.

Proof. 'Only if'. Use the same strategy as with Theorem 4.4.19, arguing by induction on the length of a proof from **K**, **D** and **U**. Evidently the context $B \to \_$ is normal for any formula B (taking care of Necessitation as well as the axiom **K**). The case of **D** was covered (under the guise $\mathbf{D}_O$) in the discussion before the introduction of τ' above. The case of **U** was explicitly treated on p. 262, though now we put $\Diamond Q \to Q$ for Q. The non-modal parts of the proof go through as before.

'If': Suppose $\nvdash_{\mathbf{KDU}} A$, in the hope of showing that $\nvdash_{\mathbf{KT}^Q} A^{\tau'}$. From this supposition we have a serial range-reflexive model $\mathcal{M} = \langle W, R, V \rangle$ with $w \in W$ with $\mathcal{M} \nvDash_w A$, and we want to obtain a new model $\mathcal{M}'$, for $\mathbf{KT}^Q$, for which $\mathcal{M}' \nvDash_w A^{\tau'}$, to conclude that $\nvdash_{\mathbf{KT}^Q} A^{\tau'}$. As before, define $\mathcal{M}' = \langle W', R', Q^{\mathcal{M}'}, V' \rangle$ thus: $W' = W$, $V' = V$, R' is the reflexive closure of R, and $Q^{\mathcal{M}'}$ is the range of R. (N.B., R, not R'.) Since $\mathcal{M}$ is serial and $R \subseteq R'$, this means that $\Diamond Q$ is true throughout $\mathcal{M}'$, which in turn implies that for all $y \in W'$:

$$\mathcal{M}' \vDash_y \Diamond Q \to Q \text{ iff } \mathcal{M}' \vDash_y Q. \qquad (*)$$

This will help us to show, by induction on the complexity of B, that for all formulas B (of the language of **KDU**) we have, for all $x \in W$:

$$\mathcal{M} \vDash_x B \Leftrightarrow \mathcal{M}' \vDash_x B^\tau,$$

essentially because $(*)$ reduces the present case to that of (the proof of) Theorem 4.4.19. We need only worry about the inductive case of $B = \Box C$.

$\Rightarrow$-direction: for a contradiction, suppose that $\mathcal{M} \vDash_x \Box C$ while $\mathcal{M}' \nvDash_x (\Box C)^{\tau'}$, i.e., $\mathcal{M}' \nvDash_x \Box((\Diamond Q \to Q) \to C^{\tau'})$, which by $(*)$ amounts to: $\mathcal{M}' \nvDash_x \Box(Q \to C^{\tau'})$. Thus we have $y \in R'(x)$ for which $\mathcal{M}' \vDash_y Q$ and $\mathcal{M}' \nvDash_y C^{\tau'}$. As in the proof of Thm. 4.4.19, we want to argue that xRy and by the inductive hypothesis (C being of lower complexity that $B = \Box C$) $\mathcal{M} \nvDash_y C$, giving a contradiction since $\mathcal{M} \vDash_x \Box C$. We justify the claim that xRy, exactly as in the proof of Thm. 4.4.19: we have $xR'y$, and distinguish two cases: (1) $x \neq y$, (2) $x = y$. In case (1), since R' is the reflexive closure of R, we get xRy from the fact that $xR'y$. In case (2) we use the fact that $\mathcal{M}' \vDash_y Q$, and the definition of $Q^{\mathcal{M}'}$ as the range of the range-reflexive relation R, to conclude that yRy, and thus, that xRy, as required.

$\Leftarrow$-direction: Suppose that $\mathcal{M} \nvDash_x \Box C$ (to show that $\mathcal{M}' \nvDash_x \Box((\Diamond Q \to Q) \to C^{\tau'})$). Thus there is some $y \in R(x)$ with $\mathcal{M} \nvDash_y C$. By the inductive hypothesis, then, we have (1): $\mathcal{M}' \nvDash_y C^{\tau'}$. Since xRy, $y \in Q^{\mathcal{M}'}$, we have (2): $\mathcal{M}' \vDash_y Q$. (1) and (2) together imply that $\mathcal{M}' \nvDash_y Q \to C^\tau$. Since $R \subseteq R'$, $xR'y$, so $\mathcal{M}' \nvDash_x \Box(Q \to C^\tau)$, so by $(*)$, $\mathcal{M}' \nvDash_x \Box((\Diamond Q \to Q) \to C^{\tau'})$.

So, for the 'if' direction we reason that if $\nvdash_{\mathbf{KDU}} A$ then we transform a countermodel $\mathcal{M}$ to A into the countermodel $\mathcal{M}'$, as above, to $A^{\tau'}$, and conclude that $\nvdash_{\mathbf{KT}^Q} A^{\tau'}$ on the grounds that $\mathbf{KT}^Q$ is obviously sound w.r.t. the class of models with a reflexive accessibility relation interpreting $\Box$ and set $Q^{\mathcal{M}}$ for interpreting Q. ∎

It may occur to the reader that it is a bit odd to be adding a special sentential constant Q to the language when nothing is said about it on the syntactic side or on the semantic side, as here, where $\mathbf{KT}^Q$ has no axioms specifically addressing Q and the models just described place no constraints on the subset this constant is deemed to be true over. Similarly in the case of Exercise 4.4.20(*i*), claiming that the earlier translation τ embedded **KU** into $\mathbf{KT}^Q$. This suggests translations of a rather different kind:

EXAMPLE 4.4.25 Throughout this example, A is to be a monomodal formula ($\Box$ as primitive). Let $A^{\oplus}$ be the result of replacing all subformulas $\Box B$ of A by $\Box(p_i \to B)$, where p_i is the first propositional variable in the official enumeration of variables $p_1, \ldots, p_n, \ldots$ which does not occur in A, and let $A^{\oplus'}$ be the result of replacing all $\Box B$ subformulas of A instead by $\Box((\Diamond p_i \to p_i) \to B)$ for this same choice of p_i. Then we could drop the superscripted Q from the reference to $\mathbf{KT}^Q$ in Exc. 4.4.20(*i*) and simply say:

4.4. DEONTIC LOGIC: MAIN THEMES

$$\vdash_{\mathbf{KU}} A \text{ if and only if } \vdash_{\mathbf{KT}} A^{①},$$

for all formulas A. Likewise in the case of Thm. 4.4.24, we have:

$$\vdash_{\mathbf{KDU}} A \text{ if and only if } \vdash_{\mathbf{KT}} A^{①'},$$

for all formulas A. Note, however, that $(\cdot)^{①}$ and $(\cdot)^{①'}$ have the awkward feature that the translation, under either of them, of – for example – $A \wedge B$ is not in general the conjunction of the translations of A and B (since B may contain the first propositional variable not occurring in A, for instance).

That can be changed in the following way: Let A^+, for a formula A, be the result of replacing each p_i throughout A with p_{i+1}. Now let $A^{②}$ be the result of replacing all subformulas $\Box B$ of A' by $\Box(p_1 \to B)$, and $A^{②'}$ be result of replacing all subformulas $\Box B$ of A' by $\Box((\Diamond p_1 \to p_1) \to B)$. (Normally we would write just "p" rather than "p_1" – see p. 6 – but this makes it clearer that we are shunting all the variables along so there is a gap left for the first of them to play the Q role.) Again we have (for all A):

$$\vdash_{\mathbf{KU}} A \text{ if and only if } \vdash_{\mathbf{KT}} A^{②}, \quad \text{and} \quad \vdash_{\mathbf{KDU}} A \text{ if and only if } \vdash_{\mathbf{KT}} A^{②'}.$$

Unlike τ, τ', ① and ①$'$, these translations disturb the propositional variables. Let us call a translation *variable-fixed* if it translates each p_i by itself. This is one half of what it takes for a translation to be what is called a *definitional* translation (see Wójcicki [1202], p. 70). The other half is the condition that the translation be *compositional*, which means that for each primitive n-ary connective # of the source language there is an n-variable formula $C(p_1, \ldots, p_n)$ of the target language, for which the translation of $\#(A_1, \ldots, A_n)$ is $C(A_1^*, \ldots, A_n^*)$, where A_i^* is the translation of A_i. When one connective is defined in terms of others, the definition can be viewed as effecting a translation from the language with the defined symbol as a new primitive to the original language. For example, the definition of the 1-ary connective ∇ (a contingency operator: see Exercise 1.3.8):

$$\nabla A = \Diamond A \wedge \Diamond \neg A$$

gives the following way of translating out occurrences of "∇" – call it τ^∇, with associated context $C(p)$ as above being $\Diamond p \wedge \Diamond \neg p$:

- $\tau^\nabla(p_i) = p_i$;
- $\tau^\nabla(\neg A) = \neg(\tau^\nabla(A))$;
- $\tau^\nabla(\#(A_1, \ldots, A_n)) = \#(\tau^\nabla(A_1), \ldots, \tau^\nabla(A_n))$ for any n-ary connectives # other than ∇;
- $\tau^\nabla(\nabla A) = \Diamond \tau^\nabla(A) \wedge \Diamond \neg \tau^\nabla A)$.

(The second condition here is just a special case of the third, with $n = 1$, and is included only for familiarity. We could also list such special cases of the latter condition for binary connectives, for which purpose the usual infix notation would be used; thus: $\tau^\nabla(A \vee B) = \tau^\nabla(A) \vee \tau^\nabla(B)$, etc.)

Observe that the translations thus eliminating defined vocabulary are always compositional and variable-fixed, which is why definitions with these properties are called definitional translations. The translations ① and ①$'$, above, fail to be definitional by not being compositional, while the translations ② and ②$'$ fail to be definitional by not being variable-fixed. (The Anderson–Kanger–Smiley translations are definitional, of course.) ◀

EXERCISE 4.4.26 *(i)* Is Matsumoto's translation, from Section 2.8, which maps a formula A to the formula $\Box \Diamond \Box A$, variable-fixed? Is it compositional?

(ii) Answer the same two questions from *(i)* for the translations $(\cdot)^{\mathbf{T!}}$ and $(\cdot)^{\mathbf{Ver}}$ of Revision Exercise 2.7.22 (p. 134).

A definitional translation need not be associated with new vocabulary being defined, however, and in particular many such translations of interest in modal logic are what might be called $\Box$-*definitional*, in proceeding like that of τ^∇ in Example 4.4.25 – translated the propositional variables and Boolean connectives by themselves, that is – except that the last condition would be replaced by one for $\Box$ itself.

EXAMPLE 4.4.27 If a $\Box$-definitional translation τ_X which replaces occurrences of $\Box$ with occurrences of the modality X faithfully embeds one modal logic S into another (not necessarily distinct) modal logic S' in the sense (as on p. 142 above) that for all formulas A we have:

$$\vdash_S A \text{ if and only if } \vdash_{S'} \tau_X(A),$$

then Zolin [1226] indicates this by writing $S'(X) = S$, which we may read as saying that S''s treatment of X coincides with the treatment of $\Box$ in S. For example, **S4.2**($\Diamond\Box$) = **KD45**, as observed in Dawson [237] and by Lenzen [706], the authors here taking a special interest in **KD45** as a candidate deontic logic and as a candidate doxastic logic respectively; our proof of this result comes with its appearance below (p. 282) as Theorem 4.5.9, and the translation involved will be referred to in accordance with the present convention as $\tau_{\Diamond\Box}$. As already remarked (e.g. p. 31), Zolin has a more general understanding of what a modality is and in terms of the discussion under Example 4.4.25 of n-variable formulas $C(p_1, \ldots, p_n)$, this understanding amounts to the $n = 1$ case, with $\Box A$ translated to $C(A^*)$, A^* being the translation of A. For instance, if $C(p)$ is $\Box p \wedge p$, commonly abbreviated to $\boxdot p$, then the translation $\tau_\boxdot$ leaves everything except $\Box$ as it was, while replacing subformulas $\Box\_\_$ by $\Box\_\_ \wedge \_\_$. It is most famous for embedding **KT** into **K**:

$$\vdash_{\mathbf{KT}} A \text{ if and only if } \vdash_{\mathbf{K}} \tau_\boxdot(A),$$

or, in Zolin's handy embedding notation: $\mathbf{K}(\boxdot) = \mathbf{KT}$. We return to $\tau_\boxdot$ in Theorem 5.5.18 (p. 416) and the discussion preceding it. ◂

The sets $Q^{\mathcal{M}}$ in our models – informally the set of worlds where all goes well from the point of view of whatever normative system is in play – gained considerable prominence in the discussion before the above excursus on non-definitional translations. (We also set aside the Kanger–Smiley approach based on Anderson's reduction Mark II, and concentrate on the Mark I version.) One thing we could now do is simply interpret O by universal quantification over this set. This amounts to the semantics of the semi-simplified models of Section 4.1, and because this approach was dealt with there and will re-emerge in connection with doxastic logic below, it would be wasteful to pursue this line further here. In our current terminology, this would amount to taking the accessibility relation for O in a model $\mathcal{M}$ with universe W to be the range-restriction to $Q^{\mathcal{M}}$ of the universal relation $W \times W$, rather than concentrating, as the Anderson–Kanger–Smiley reduction suggests on the interaction between $\Box$ and O. Instead, let us consider that reduction from a different point of view, taking a normal bimodal logic with both of these as primitive and asking under what conditions we could have introduced the sentential constant Q into this bimodal language and dropped O from the primitive vocabulary on the grounds that OA was provably equivalent to $\Box(Q \to A)$. Evidently this will be the case precisely when the models $\mathcal{M} = \langle W, R_\Box, R_O, V \rangle$ have some $X \subseteq W$ with R_O being $R_\Box$ range-restricted to X, in which case we can take X as $Q^{\mathcal{M}}$ – and discard R_O if desired. To avoid issues specifically suggested by the alethic–deontic context, let us for the moment just ask under what conditions for binary relations R_1, R_2, on a set, R_2 is a range restriction of R_1. Although this is a second-order characterization (because it existentially quantifies over subsets X of the domain and not just over elements), it is not hard to see that we can find a first-order equivalent:

PROPOSITION 4.4.28 *Given $W \neq \varnothing$ and $R_1, R_2 \subseteq W \times W$, R_2 is a range-restriction of R_1 if and only if (i) for all $x, y \in W$, if xR_2y then xR_1y and (ii) for all $x, y, u \in W$ if xR_1y and uR_2y then xR_2y.*

4.4. DEONTIC LOGIC: MAIN THEMES

Proof. The fact that (i) and (ii) are necessary conditions for R_2 to be a range-restriction of R_1 is easily seen. To check their sufficiency, suppose they are satisfied for a given W, R_1, R_2. Define $X \subseteq W$ by putting $X = \{w \in W \mid \exists u \in W \,.\, uR_2w\}$ and verify, using (i) and (ii), that

$$xR_2y \text{ if and only if } xR_1y \text{ and } y \in X,$$

for all $x, y \in W$. ∎

Condition (i) here, $R_2 \subseteq R_1$, is straightforward, being modally defined and canonically secured by the schema $\Box_1 A \to \Box_2 A$, but condition (ii) of Proposition 4.4.28 is another matter entirely. We can see that the class of frames satisfying the conditions (i) and (ii) is not modally definable – and therefore, that this goes for the class of frames satisfying (ii) by itself (or else we could just add $\Box_1 p \to \Box_2 p$ to the 'defining' formulas) – by a p-morphism argument. For that we have to extend the notion of p-morphism (given in the Preamble to Revision Exercise 2.7.35) to bimodal logic; this is just a routine reduplication of the two conditions given there for one accessibility relation to cover the case of the other as well. The results asked for in Exc. 2.7.35 go through for the multimodal case, by the same arguments. Figure 4.2 depicts two frames with that on the left being a p-morphic pre-image of that on the right, the p-morphism in question mapping w_0, x_0, y_0 to w, x, y respectively, and mapping both z_0 and z_1 to z. The points w_0 and w have been included to make both frames point-generated (and avoid irrelevant worries about modal undefinability arising from disjoint unions of frames via the Generation Theorem – more specifically, Theorem 2.5.32). A "1" on an arrow indicates that R_1 holds and a "2" that R_2 holds. Because we want to satisfy the p-morphism-preserved condition that $R_2 \subseteq R_1$, the only occurrence of "2" is accompanied by "1". That was condition (i) in Proposition 4.4.28. To see that condition (ii) is not similarly preserved, observe that it is satisfied by the frame on the left of Figure 4.2, but violated by its p-morphic image on the right (since we do not have xR_2z even though xR_1z and yR_2z).

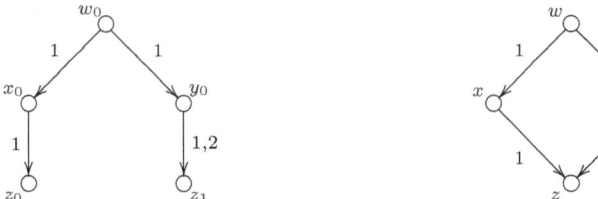

Figure 4.2: Two Frames for Range-Restriction

We pause to illustrate the utility of the tense-logical perspective on matters having nothing to do with time but rather with the general fact that every binary relation has a converse (as in Example 4.1.11). This example presumes at least a fleeting acquaintance with the material in Section 3.1.

EXAMPLE 4.4.29 Suppose that instead of just the modal primitives O and $\Box$ we had at our disposal, we filled out the picture with a 'past tense' companion O^{-1} for O, with accessibility relation R_O^{-1}. (*Warning*: if we think of O as like Prior's G, O^{-1} would correspond to H, but Prior's P is not to be confused with the current – dual of O – "P", corresponding instead to P^{-1}, if that is how we choose to write the dual of O^{-1} for the sake of the 'bridging axioms' $A \to OP^{-1}A$ and $A \to O^{-1}PA$.) Then the class of frames meeting the range restriction condition (ii) from Proposition 4.4.28 would be modally defined by:

$$OA \to \Box(A \vee O^{-1}\bot),$$

which enforces satisfaction of that condition on the canonical frames of the (now trimodal) logics containing all instances of this schema. We will not entertain this complication of the language further here, except to note one aspect of the interest of the bridging axiom scheme $A \to OP^{-1}A$. If we take the instance with A as p, say, then we seem here to have, on the normative as opposed to reportive interpretation of our deontic vocabulary, a provable implication with a non-normative antecedent and a normative consequent, where what is in the scope of the O cannot just be replaced (as in examples like $(p \land \neg p) \to Oq$) by absolutely anything, in apparent violation of one style of regimentation of 'Hume's Thesis' (or 'Hume's Law') to the effect that no valid argument with non-normative premises has a normative conclusion. Several versions of this principle are studied in Schurz [1013] – or, for edited highlights, Schurz [1012]; for further discussion see my [570] and numerous papers in the more recent collection, Pigden [883]. ◀

In something of a minor digression from the problem of dealing with range-restriction condition, we present an exercise most appropriately placed in close proximity to Figure 4.2 and with some interest for deontic logic. The exercise concerns frames $\langle W, R_1, R_2, R_3 \rangle$ for trimodal logic. The basic normal logic here has primitive operators $\Box_i$ (for $i = 1, 2, 3$) interpreted with accessibility relation R_i, in which each $\Box_i$ is normal. Suppose we ask whether the class of frames $\langle W, R_1, R_2, R_3 \rangle$ in which $R_3 = R_1 \cup R_2$ is modally definable. The answer is evidently *yes*, the class in question being defined by the formula $\Box_3 p \leftrightarrow (\Box_1 p \land \Box_2 p)$. Note that this formula can also be regarded (once p is replaced by a schematic letter) as defining (in another sense of "define") $\Box_3$ as the desired 'union' operator.

EXERCISE 4.4.30 (*i*) Is the class of frames $\langle W, R_1, R_2, R_3 \rangle$ in which $R_3 = R_1 \cap R_2$ modally definable ('trimodally definable' to be more precise)?

(*ii*) What about the class of frames $\langle W, R_1, R_2, R_3 \rangle$ with $R_3 = R_1 \circ R_2$? ($\circ$ is for relative product here, or composition of relations, as in note 56.)

(Hint: Variations on the theme of Figure 4.2 may be helpful for part (*i*) here.) ✠

To illustrate the potential interest of such 'intersective' operators as Exercise 4.4.30(*i*) asks about, we have the following case, quoted from my [554], p. 413*f*.:

> Think of $\Box_1 A$ as saying that according to a certain moral code it ought to be that A, and of $\Box_2 A$ as saying by contrast that a certain legal code requires that A. Now, if the moral code is your own, and the legal code is that under the threat of whose sanctions you live, you may take a special interest in those things which are both legally and morally permissible, and it is my impression that this must be taken in a sense in which to say:
>
> (α)   It is morally permissible that A and it is legally permissible that A,
>
> is to say something much weaker than to say:
>
> (β)   It is both morally and legally permissible that A.
>
> (I am prepared to admit an ambiguity in (β), one sense being that of (α),[227] the other being the stronger sense here at issue.) What I just termed your special interest is in facts of the type (β) rather than of the type (α). For an example, which will admittedly be somewhat farfetched, suppose that the only person it would be morally permissible for you to marry belongs to a category of persons it is legally forbidden that you marry. Then although it is legally permissible for you to get married, and also morally permissible for you to get married, it is (...) not legally and morally permissible that you should marry, since there is no way you can get married while fulfilling both your moral and your legal obligations. Putting it in terms of possible worlds: although there are morally permissible worlds in which you marry, and there are legally permissible worlds in which you marry (...), there are no worlds which are legally as well as morally permissible in which you marry).

---

[227] Which there is no problem in representing in the current modal language: $\Diamond_1 A \land \Diamond_2 A$.

4.4. DEONTIC LOGIC: MAIN THEMES

This intersection-of-accessibility relations theme is taken up briefly in Example 4.4.46 at the end of the present section. In the meantime, we resume the main 'range-restriction' theme. Having used p-morphisms to draw conclusions about modal undefinability, we can go on to use them to obtain completeness results, in particular answering the questions: Which bimodal logic is determined by the class of all frames $\langle W, R_1, R_2 \rangle$ with R_2 a range-restriction of R_1, and which such logic is determined by the class of all such frames in which R_1 is reflexive and R_2 a range-restriction of R_1? These questions are to be understood as asking for a syntactic characterization of the logics concerned. We will construct p-morphic pre-images, using a technique called *unravelling*, of arbitrary frames from within classes of frames already known to determine the logics in question, in such a way as to secure satisfaction of the range-restriction condition. While for these applications we shall need to work in a bimodal setting, the technique is best introduced by beginning with its incarnation for monomodal logics.

The idea of unravelling of frames we are about to explain was used in Sahlqvist [988], (and before that, without the terminology, in Dummett and Lemmon [263]). Let $\langle W, R \rangle$ be a frame generated by the point $w \in W$, the unravelling ("from w") of $\langle W, R \rangle$ from w is a frame $\langle \vec{W}, \vec{R} \rangle$ in which $\vec{W}$ is the set of finite R-chains w starts in $w \in W$ – sequences of elements $x_1 \ldots x_n$ for some n, that is, in which $x_1 = w$ and $x_i R x_{i+1}$, that is, and where s is one such sequence and t is another, we define

$$s \vec{R} t \Leftrightarrow t = s\hat{\,}x \text{ for some } x \in W \text{ for which } \dot{s}Rx.$$

To explain this notation: $\dot{s}$ denotes the last element of the sequence s and $s\hat{\,}x$ is the sequence s extended by adding the element x in its final position.[228] Note that because of how $\vec{W}$ was defined, we could equally well just state the r.h.s. inset above in the form: "$t = s\hat{\,}x$" for some $x \in W$, since unless $\dot{s}Rx$, we do not have $s\hat{\,}x \in \vec{W}$. But the above formulation is more convenient for extending the idea to bimodal frames as we are about to. Further note that strictly the notation $\langle \vec{W}, \vec{R} \rangle$ should register the dependence on the generating point w, since having $\langle W, R \rangle$ be a frame generated by the point $w \in W$ does not rule out that $\langle W, R \rangle$ is also generated by $v \in W$ with $v \neq w$, and as unravelled from v, the sequences in $\vec{W}$ would all begin with v rather than w. This will cause no confusion in practice, however.

PROPOSITION 4.4.31 *Let $\langle \vec{W}, \vec{R} \rangle$ be the unravelling of a frame $\langle W, R \rangle$ from some generating point $w \in W$. Then defining f from $\vec{W}$ to W by $f(s) = \dot{s}$ makes f a p-morphism from $\langle \vec{W}, \vec{R} \rangle$ to $\langle W, R \rangle$.*

Proof. This is simply a matter of checking the defining conditions for p-morphisms given in the preamble to Exercise 2.7.35. ∎

Thus in particular if, with a particular model $\mathcal{M} = \langle W, R, V \rangle$ on $\langle W, R \rangle$ in mind, we define $\vec{V}$ by setting $\vec{V}(p_i) = \{ s \in \vec{W} \mid \dot{s} \in V(p_i) \}$, then for the resulting 'unravelled model' $\vec{\mathcal{M}} = \langle \vec{W}, \vec{R}, \vec{V} \rangle$ we have for all formulas A and all $s \in \vec{W}$:

$$\vec{\mathcal{M}} \models_s A \text{ if and only if } \mathcal{M} \models_{\dot{s}} A,$$

in view of Exercise 2.7.35.

THEOREM 4.4.32 **K** *is determined by the class of irreflexive frames, by the class of asymmetric frames, by the class of intransitive frames, by the class of irreflexive intransitive frames, and the class of asymmetric intransitive frames.*

---

[228]These finite sequences can be thought of as ordered n-tuples for varying n, with $s = x_1 \ldots x_n$ being $\langle x_1, \ldots, x_n \rangle$, identifying $\langle x_1 \rangle$ with x_1; then we have $\dot{s} = x_n$ and $s\hat{\,}y = \langle x_1, \ldots, x_n, y \rangle$, which we will also denote by $x_1\hat{\,}\ldots\hat{\,}x_n\hat{\,}y$. Ranging over these sequences we use the variables "s", t", and later when we need a third variable "r", to avoid bumping into the variables from the end of the alphabet that range over elements of the frames to be unravelled. There will be no confusion with the use of "r" as a propositional variable (for what is officially p_3).

Proof. The soundness half of the claim (for each class of frames mentioned) follows from the soundness of **K** w.r.t. the class of all frames. For the completeness half, take any non-theorem A of **K**. By the completeness of **K** w.r.t. the class of all frames, and the Generation Theorem (Thms. 2.5.2 and 2.4.10), A is false at some point in w some model $\mathcal{M}$ on a frame generated by w. Letting $\overrightarrow{\mathcal{M}}$ be the unravelling of $\mathcal{M}$ from w, by the observation just made, since $\mathcal{M} \not\models_w A$, we have $\overrightarrow{\mathcal{M}} \not\models_w A$ (w being the final element of the one element sequence whose only element is w). Note that the frame of $\overrightarrow{\mathcal{M}}$ is irreflexive, asymmetric, and intransitive. Thus A is false at a point in a model whose frame belongs to each of the classes w.r.t. which completeness was to be shown. ∎

Variations on the theme of Theorem 4.4.32 are particularly useful for specifically temporal applications of tense logic, since one usually wants to think of the accessibility relation as the *earlier than* relation and accordingly as irreflexive (and therefore also asymmetric) – though transitivity, rather than intransitivity, will also typically be wanted here too. A monomodal version of this kind of combination arises in part (*i*) of Exercise 4.4.36, after which exercise we turn to bimodal logic (though not to tense logic).

By Exercise 2.5.30, since **K** is determined by the class of all frames, a class which is modally defined by (for example) the formula $\top$ (alternatively, by the set of all theorems of **K**) we get the following, which is admittedly something we already knew (from Revision Exercise 2.7.35):

COROLLARY 4.4.33 *None of the classes of frames described in Theorem 4.4.32 is modally definable.*

Another interesting property of unravelled frames which is related to something we shall exploit in our application of these to the deontic–alethic question is given in the following:

EXAMPLE 4.4.34 Say that a frame $\langle W, R \rangle$ has the *unique predecessor property* just in case for all $x, y, z \in W$ if xRz and yRz, then $x = y$. Whether or not a point-generated frame has this property, its unravelling from a generating point, $\langle \overrightarrow{W}, \overrightarrow{R} \rangle$, has this property, since if $s\overrightarrow{R}t$, then s is uniquely determined as the sub-sequence of t obtained by deleting $\dot{t}$ from the end of t. Thus **K** is determined by the class of frames with the unique predecessor property, and this property is itself not modally definable. ◀

Again for its relevance to the deontic–alethic question, but also of interest in its own right, is the fact that we can modify the unravelled frames so as to obtain further completeness results:

EXAMPLE 4.4.35 Suppose we want to show that **KT** is determined by the class of reflexive antisymmetric frames. We cannot get this result just using the canonical frame, since it is not antisymmetric. We do know, however, that **KT** is determined by the class of point-generated reflexive frames. And if we unravel such a frame (from a generating point), the result will be antisymmetric (because asymmetric) – though it will not be reflexive: so further work is required. We can take the reflexive closure of $\overrightarrow{R}$ to obtain a reflexive frame which is antisymmetric. (Not asymmetric, of course, since asymmetry implies irreflexivity. Note also that we have lost the unique predecessor property of Example 4.4.34, since $\hat{x}\hat{y}z$, for instance, has both $\hat{x}y$ and $\hat{x}\hat{y}z$ itself as predecessors.) Having made this change, we need to check that the $s \mapsto \dot{s}$ map is still a p-morphism. We have added the new pairs $\langle s, s \rangle$ into $\overrightarrow{R}$, so we need to re-check the first of the two conditions given in the definition of p-morphism in the Preamble to Exc. 2.7.35: If $s\overrightarrow{R}t$ then $\dot{s}R\dot{t}$. (Here we have used the notation "$\overrightarrow{R}$", even though it is the reflexive closure of this relation, as originally defined, that is intended; note also that in the discussion just referred to, R and R' were the accessibility relations of the p-morphic pre-image and the p-morphic image, respectively, whereas here it is $\overrightarrow{R}$ and R that play these respective roles.) The new cases are those for which $s\overrightarrow{R}s$, so what we need is that $\dot{s}R\dot{s}$: but since R was reflexive, we have what we need, and can conclude that **KT** is indeed determined by the class of reflexive antisymmetric frames. (Here we have just sketched the completeness half of this assertion, but evidently **KT** is sound w.r.t. any class of reflexive frames.) ◀

4.4. DEONTIC LOGIC: MAIN THEMES

EXERCISE 4.4.36 (i) Show that **K4** is determined by the class of transitive irreflexive frames by using unravellings.

(ii) Can you show that **KB** is determined by the class of symmetric irreflexive frames in a similar manner? ✠

We are now in a position to apply all this to a specific combined alethic–deontic logic. The first thing to do is to describe unravellings for point-generated bimodal frames $\langle W, R_1, R_2 \rangle$, but rather than do so generally, we do so for the case of interest, in which $R_2 \subseteq R_1$ and we want to arrange things so that (some version of) R_2 ends up as a range-restriction of (some version of) R_2. Since the numerical subscripts are rather unmemorable, let us revert to writing $R_\Box$ and R_O rather than R_1 and R_2. This will serve as a reminder that it is R_O that is to be the range-restriction of $R_\Box$ rather than the other way round. So we begin with a frame $\langle W, R_\Box, R_O \rangle$, generated by some $w \in W$, and for which $R_O \subseteq R_\Box$. In view of this inclusion we can take the universe of the unravelling of $\langle W, R_\Box, R_O \rangle$ from w, namely $\vec{W}$, to be just the $R_\Box$-chains from $\langle W, R_\Box, R_O \rangle$ started by w. Since $R_O \subseteq R_\Box$, we will get to every $x \in W$ this way without paying separate attention to R_O-links. To complete the definition of the unravelled frame $\langle \vec{W}, \vec{R}_\Box, \vec{R}_O \rangle$, we need to define the two relations. Unsurprisingly, we stipulate that, for all $s, t \in \vec{W}$:

$$s\vec{R}_\Box t \Leftrightarrow t = s\hat{\ }x \text{ for some } x \in W \text{ for which } \dot{s}R_\Box x;$$
$$s\vec{R}_O t \Leftrightarrow t = s\hat{\ }x \text{ for some } x \in W \text{ for which } \dot{s}R_O x.$$

EXERCISE 4.4.37 Check that with f defined from $\vec{W}$ to W, by $f(s) = \dot{s}$, where $\vec{\mathcal{F}} = \langle \vec{W}, \vec{R}_\Box, \vec{R}_O \rangle$ and $\mathcal{F} = \langle W, R_\Box, R_O \rangle$ are frames as in the preceding discussion, f is a p-morphism from $\vec{\mathcal{F}}$ to $\mathcal{F}$. ✠

We need only one further observation before making use of these unravellings. Let us recall the content of Proposition 4.4.28 but in the $R_\Box/R_O$ notation: for a frame $\langle W, R_\Box, R_O \rangle$, R_O is a range-restriction of R_1 if and only if (i) for all $R_O \subseteq R_\Box$ and (ii) for all $x, y, u \in W$ if $xR_\Box y$ and $uR_O y$ then $xR_O y$. We can show the tricky part (namely (ii)) for the current unravelled frames essentially as in Example 4.4.34 (on the unique predecessor property).

LEMMA 4.4.38 *The unravelling $\langle \vec{W}, \vec{R}_\Box, \vec{R}_O \rangle$ of a frame $\langle W, R_\Box, R_O \rangle$ (from a generating point), as above, satisfies the conditions (i) and (ii) just given, and therefore have $\vec{R}_O$ as a range-restriction of $\vec{R}_\Box$.*

Proof. For (i): since $R_O \subseteq R_\Box$, we get $\vec{R}_O \subseteq \vec{R}_\Box$ from the definition of the relations $\vec{R}_O$, $\vec{R}_\Box$. For (ii), suppose that for $s, t, r \in \vec{W}$, we have $s\vec{R}_\Box t$ and $r\vec{R}_O t$, with a view to showing that $s\vec{R}_O t$. Since $s\vec{R}_\Box t$, $t = s\hat{\ }x$ for some $x \in R_\Box(\dot{s})$, and since $r\vec{R}_O t$, $t = r\hat{\ }y$ for some $y \in R_O(\dot{r})$. But then we must have $s = r$ and $x = y$, from which it follows that $s\vec{R}_O t$, since $t = s\hat{\ }x$ for some $x \in R_O(\dot{s}) = R_O(\dot{r})$. ■

THEOREM 4.4.39 *The smallest normal bimodal logic in $\Box$ and O and in which all formulas $\Box A \to OA$ are provable is determined not only by (1) the class of all frames $\langle W, R_\Box, R_O \rangle$ with $R_O \subseteq R_\Box$, but also by (2) the class of all such frames in which R_O is a range-restriction of $R_\Box$.*

Proof. Result (1) is immediate by the canonical model method in its completeness half, the soundness half of (1) and (2) being clear. For the completeness half of (2) we take A as a non-theorem of the logic described and exploiting (1), assume that A is false at the generating point w of a model $\mathcal{M} = \langle W, R_\Box, R_O, V \rangle$ with $R_O \subseteq R_\Box$. Unravel this frame from w to get $\langle \vec{W}, \vec{R}_\Box, \vec{R}_O \rangle$ and convert it into a p-morphic pre-image $\vec{\mathcal{M}}$ of the original model by putting $\vec{V}(p_i) = \{s \in \vec{W} \mid \dot{s} \in V(p_i)\}$ for all p_i. By

Exercises 2.7.35(i) and 4.4.37 we conclude that $\vec{\mathcal{M}} \not\models_w A$. (Recall that we identify w with the $\langle w \rangle$, so w belongs to the universe of the original frame and to that of its unravelling.) But by Lemma 4.4.38 this is a model on a frame in which $\vec{R}_O$ is a range-restriction of $\vec{R}_\square$. ∎

The following example can be skipped by anyone interested only in the deontic–alethic theme, which is resumed after Exercise 4.4.42. It is included here because the proof technique used for Theorem 4.4.39 finds application here.

EXAMPLE 4.4.40 Suppose we are interested in the bimodal logic determined by the class of frames $\langle W, R_\square, R_O \rangle$ satisfying the following condition:

$$\text{For all } x, y \in W, \ x R_O y \text{ iff } x R_\square y \text{ and } x \neq y.$$

Heuristically, we retain the subscript "O", which may now be read as suggesting "Other" rather than "Ought" (or "Obligatory"): OA requires for its truth at a point x in a model that A is true at all ($R_\square$-)accessible points other than x itself. Note that this leaves it open whether A is true at x, as well as whether $xR_\square(x)$. If $R_\square$ is the universal relation on W (the relation $W \times W$, that is) then the reduct $\langle W, R_O \rangle$ of a frame satisfying the condition inset above is an *NI* frame ('non-identity frame') in the sense introduced on p. 92; Theorem 2.5.50 recorded the fact that the class of such frames determined the monomodal logic **KB4'** ('the logic of *elsewhere*'), which in the current context we might call **KB4'$_O$**. But the present question relaxes this restriction to $R_\square$ as universal, and asks about the general situation; moreover, it asks about the general situation when both operators, $\square$ and O are present.

Consider the smallest normal bimodal logic in $\square$ and O containing all formulas of the forms $\square A \to OA$ and $(OA \wedge A) \to \square A$. The canonical model method shows that this logic is determined – the soundness half of this claim being straightforward – by the class of frames $\langle W, R_\square, R_O \rangle$ for which $R_O \subseteq R_\square$ and $xR_\square y$ and $x \neq y$ imply $xR_O y$ (all $x, y \in W$), since the two schemata given secure these two properties (respectively) for the canonical frame. All we need to get completeness w.r.t. the class of frames meeting the conditions above is the further condition that $xR_O y$ implies $x \neq y$ (all $x, y \in W$), which is to say that R_O is irreflexive. Recalling from Theorem 4.4.32 that unravelling produces irreflexive accessibility relations, we might, as a first thought, take any point-generated frame $\langle W, R_\square, R_O \rangle$, with the two properties mentioned and construct its unravelling $\langle \vec{W}, \vec{R}_\square, \vec{R}_O \rangle$ exactly as above, in which $\vec{R}_O$ is irreflexive, and we will still have the first of the other two properties: $\vec{R}_O \subseteq \vec{R}_\square$. But the trouble is that we will not have the second: $s\vec{R}_\square t$ and $s \neq t$ imply $s\vec{R}_O t$ (for arbitrary $s, t \in \vec{W}$), since $s\vec{R}_\square t$ by itself already implies $s \neq t$, as $\vec{R}_\square$ is irreflexive, which would mean we had $s\vec{R}_\square t$ implying $s\vec{R}_O t$, i.e., $\vec{R}_\square \subseteq \vec{R}_O$. (Since we have already secured the converse inclusion, we would then have $\vec{R}_\square = \vec{R}_O$.) This would make all formulas $OA \to \square A$ valid on the unravelled frame and hence on its p-morphic image, the original frame $\langle W, R_\square, R_O \rangle$, which in general they are not. So we need to have a second thought, keeping the unravelled version of R_O irreflexive but preventing some suitable variation of the unravelled $R_\square$ from being irreflexive. We cannot simply take its reflexive closure, as in Example 4.4.35, or again the desired p-morphism will not be forthcoming. A rather delicate admixture of reflexivity and irreflexivity is called for.

To indicate that we are varying the usual definition of $\vec{R}_\square$, we denoted the varied relation (still a binary relation on $\vec{W}$) by $\widetilde{R}_\square$, rather than using the same notation. (The definition of $\vec{R}_O$ remains as given before Exercise 4.4.37, likewise for $\vec{W}$, since again we start with $R_O \subseteq R_\square$ and so only need $R_\square$-chains as elements of $\vec{W}$.) The definition will run as follows, assuming given a point-generated frame $\langle W, R_\square, R_O \rangle$ such that $R_O \subseteq R_\square$, and also $xR_\square y$ and $x \neq y$ imply $xR_O y$ (all $x, y \in W$):

$s\widetilde{R}_\square t$ iff (1) $\dot{s}\hat{x} = t$ for some $x \in W$ with $\dot{s}R_\square x$ and $\dot{s} \neq x$, or
 (2) $s = t$ and $\dot{s}R_\square \dot{s}$.

4.4. DEONTIC LOGIC: MAIN THEMES

We need to check that this variant unravelling $\langle \overrightarrow{W}, \widetilde{R}_\Box, \overrightarrow{R}_O \rangle$ is a frame in the class w.r.t. which we are trying to show completeness for the logic described, and also that the map $s \mapsto \dot{s}$ is a p-morphism from this frame to the originally given frame $\langle W, R_\Box, R_O \rangle$. For the first of these tasks, as well as observing that $\overrightarrow{R}_O$ is irreflexive, we need to check that $\overrightarrow{R}_O \subseteq \widetilde{R}_\Box$ and $s\widetilde{R}_\Box t$ and $s \neq t$ imply $s\overrightarrow{R}_O t$ (all $s, t \in \overrightarrow{W}$). That $\overrightarrow{R}_O \subseteq \widetilde{R}_\Box$ follows from the fact that $R_O \subseteq R_\Box$ and case (1) of the definition above of $\widetilde{R}_\Box$. Next, suppose that $s\widetilde{R}_\Box t$ and $s \neq t$. Since $s \neq t$ we are under case (1) of the definition of $\widetilde{R}_\Box$ again, so for some $x \in W$, $\dot{s}R_\Box x$ and $\dot{s} \neq x$. Now the original frame $\langle W, R_\Box, R_O \rangle$ satisfies the condition: $uR_\Box x$ and $u \neq x$ imply $uR_O x$ (all $u, x \in W$: we re-letter the condition here from the earlier form so as to avoid a clash of "x"es). Thus we have $\dot{s}R_O x$, and therefore $s\overrightarrow{R}_O t$.

It remains to check that $s \mapsto \dot{s}$ is a p-morphism. We attend only to $\widetilde{R}_\Box$, since the case of $\overrightarrow{R}_O$ presents no novelties. Checking the 'forward' condition first, suppose that $s\widetilde{R}_\Box t$ ($s, t \in \overrightarrow{W}$), with a view to showing that $\dot{s}R_\Box \dot{t}$. If we have $s\widetilde{R}_\Box t$ because of case (1) in the definition, then $\hat{s}x = t$ for some $x \in W$ with $\dot{s}R_\Box x$, so for this x, $\dot{t} = x$ and therefore $\dot{s}R_\Box \dot{t}$. If we have $s\widetilde{R}_\Box t$ because of case (2) in the definition, $s = t$ and $\dot{s}R_\Box \dot{s}$, so again $\dot{s}R_\Box \dot{t}$. We turn to the 'backward' condition. Suppose $\dot{s}R_\Box \dot{t}$, with a view to showing that $s\widetilde{R}_\Box r$ for some $r \in \overrightarrow{W}$ with $\dot{r} = \dot{t}$. If $\dot{s} = \dot{t}$, then we can take r as s, in view of case (2) of the definition of $\widetilde{R}_\Box$. If $\dot{s} \neq \dot{t}$ then we have $s\widetilde{R}_\Box \hat{s}\dot{t}$ under case (1) of that definition, and can take r as $\hat{s}\dot{t}$. ◂

Summarising the findings of Example 4.4.40, we have:

THEOREM 4.4.41 *The smallest bimodal logic (with $\Box$ and O as primitive) containing all formulas of the forms $\Box A \to OA$ and $(OA \land A) \to \Box A$ is determined by the class of all frames $\langle W, R_\Box, R_O \rangle$ such that for all $x, y \in W$, $xR_O y$ iff $xR_\Box y$ and $x \neq y$.*

EXERCISE 4.4.42 Give an explicit proof of Theorem 4.4.41, appealing as appropriate to the discussion in Example 4.4.40.

Returning to the main theme, we give one more result along the lines of Theorem 4.4.39, this time with reflexivity imposed on $R_\Box$, in accordance with the motivation for these considerations which takes $\Box$ as some reasonable alethic necessity operator and asks about the consequential impact on O when the latter is defined using Q and $\Box$ – or in the current incarnation, when $R_\Box$ is a range restriction of R_O in a class of frames which determines the logic, so R_O *could have been* defined in terms of such a Q. Example 4.4.35 was a preparation for the proof which follows, in its deployment of the idea of taking reflexive closures of unravelled relations. But this proof also uses the terminology of range-reflexive closure, which we accordingly define here, with the aid of the abbreviation ran(R) for the range of R: The *range-reflexive closure* of a binary relation R is $R \cup \{\langle u, u \rangle \mid u \in \text{ran}(R)\}$; note that this is the smallest range-reflexive relation to include R. Note that in the proof below, as in Example 4.4.40, we use the "$\widetilde{R}$" notation to indicate a modification of the standard unravelled $\overrightarrow{R}$ ($R = R_O, R_\Box$), but the modification in question here is not the same as that in Example 4.4.40.

THEOREM 4.4.43 *The smallest normal bimodal logic in $\Box$ and O extending $\mathbf{KT}_\Box$ and $\mathbf{KU}_O$ and containing all formulas $\Box A \to OA$ is determined not only by (1) the class of all frames $\langle W, R_\Box, R_O \rangle$ with $R_\Box$ reflexive, R_O range-reflexive and $R_O \subseteq R_\Box$, but also by (2) the class of all such frames in which R_O is a range-restriction of $R_\Box$.*

Proof. In outline: As with Theorem 4.4.39, the only novelty lies in the completeness half under (2), and we proceed as in the proof of that theorem by using a modification of the unravelled frames $\langle \overrightarrow{W}, \overrightarrow{R}_\Box, \overrightarrow{R}_O \rangle$ there. Given $\overrightarrow{R}_\Box$ as defined before, we use as our modified relation on $\overrightarrow{W}$ the reflexive closure of $\overrightarrow{R}_\Box$. Call this relation $\widetilde{R}_\Box$. Given $\overrightarrow{R}_O$ as previously defined, let $\widetilde{R}_O$ be its range-reflexive closure. The model playing the role played by $\overrightarrow{\mathcal{M}}$ in the proof of Theorem 4.4.39 is here played by the model $\widetilde{\mathcal{M}} = \langle \overrightarrow{W}, \widetilde{R}_\Box, \widetilde{R}_O, \overrightarrow{V} \rangle$, so we need to check that the $s \mapsto \dot{s}$ map is a p-morphism from the frame of this model onto the frame $\langle W, R_\Box, R_O\rangle$ that was unravelled in the above-modified manner and that $\widetilde{R}_O$ is a range-restriction of $\widetilde{R}_\Box$. The crucial part of the range-restriction condition ((ii) from Proposition 4.4.28, p. 266), namely:

$$s\widetilde{R}_\Box t \text{ and } r\widetilde{R}_O t \text{ imply that } s\widetilde{R}_O t,$$

for all $r, s, t \in \overrightarrow{W}$, is verified as in Lemma 4.4.38 for the simpler case of $\overrightarrow{R}_\Box$ and $\overrightarrow{R}_O$. Suppose, then, that we have $r, s, t \in \overrightarrow{W}$ with (α): $s\widetilde{R}_\Box t$, and (β): $r\widetilde{R}_O t$, in order to show that we must then have $s\widetilde{R}_O t$. Each of (α), (β) divides into two subcases, different ways for the condition in question to be satisfied:

α1: $\hat{s}x = t$ for some $x \in R_\Box(\dot{s})$ $\quad\mid\quad$ β1: $\hat{r}y = t$ for some $y \in R_O(\dot{r})$
α2: $s = t$ $\quad\quad\quad\quad\quad\quad\quad\quad\quad\quad\quad\;\mid\;$ β2: $r'\widetilde{R}_O r$ for some $r' \in \overrightarrow{W}$ and $r = t$

The subcases numbered 1 are just the cases in which the original unravelled relations $\overrightarrow{R}_\Box$ and $\overrightarrow{R}_O$ hold, while those numbered 2 are the extra cases arising from taking the reflexive and range-reflexive closures of these relations respectively. Thus the combination α1-and-β1 is the case already considered in Lemma 4.4.38, where we saw this implied $s\overrightarrow{R}_O t$; thus we conclude that $s\widetilde{R}_O t$ for this case. Of the three remaining possibilities – α2-and-β1, α1-and-β2, α2-and-β2 – we show that the first allows us also to conclude that $s\widetilde{R}_O t$, leaving the remaining cases (and the fact that we have the desired p-morphism for this construction) as an exercise below. In the α2-and-β1 case, the β part means that $r\widetilde{R}_O t$; but $\widetilde{R}_O$ is range-reflexive, so $t\widetilde{R}_O t$. According to α2, $s = t$, so we conclude that $s\widetilde{R}_O t$, as desired.

With the details of the remaining cases taken as read, we argue as in the proof of Theorem 4.4.39 that for A not provable in the logic described we have a model $\mathcal{M}$ on a frame as described under (1) of the present theorem, for which $\mathcal{M} \not\models A$ and of which $\widetilde{\mathcal{M}}$ is a p-morphic pre-image with a frame as described under (2) and for which $\widetilde{\mathcal{M}} \not\models A$. ■

EXERCISE 4.4.44 (i) Show that in the cases labelled α1-and-β2, α2-and-β2 in the above proof, it also follows that $s\widetilde{R}_O t$.

(ii) Show that the function f from $\overrightarrow{W}$ to W defined by: $f(s) = \dot{s}$ (for $s \in \overrightarrow{W}$) is a p-morphism from $\langle\overrightarrow{W}, \widetilde{R}_\Box, \widetilde{R}_O\rangle$, where the accessibility relations involved are as in the above proof. (Suggestion: see Example 4.4.35 for guidance, or Example 4.4.40.) ✡

Theorem 4.4.43 talks about the normal extension of $\mathbf{KT}_\Box$ and $\mathbf{KU}_O$ by means of a bridging axiom $\Box A \to OA$ and tell us that this is determined by an appropriate class of frames with the accessibility relation for O a range restriction of that for $\Box$, a condition that would have been satisfied had OA been taken as abbreviating $\Box(Q \to A)$. $\mathbf{U}$ and $\mathbf{T}$ also figure prominently in part (i) of Exercise 4.4.20, where the translation τ unpacking the idea of this abbreviation is said to embed $\mathbf{KU}$ faithfully into $\mathbf{KT}^Q$. Similarly, compare Theorem 4.4.39 with part (iii) of Exercise 4.4.20, where we have $\mathbf{K}$ on the deontic side as well as only the alethic side. Observe, however, that in the case of these earlier embedding results, as in the case of Theorem 4.4.19 and so on, we never consider the deontic $\Box$ (the O of $\mathbf{KU}_O$ in Thm. 4.4.43) and the alethic $\Box$ (the $\Box$ of $\mathbf{KT}_\Box$ in Thm. 4.4.43) together in the same formulas: the translation replaces the deontic operator with the alethic one, so their interactions are never addressed. The simplest way to do that is to employ the bimodal language, as in Theorems 4.4.39 and 4.4.43. Will the whole truth about their interaction be told in general, as in those two results, by the schema $\Box A \to OA$? No. While we do not attempt a general account here, any speculation to that effect is easily scotched by a counterexample.

4.4. DEONTIC LOGIC: MAIN THEMES

EXAMPLE 4.4.45 Suppose we replace the reference to the reflexivity of $R_\Box$ in Theorem 4.4.43 with transitivity, and concomitantly replace the reference to the range-reflexivity of R_O with a reference to transitivity (again), range-restrictions of transitive relations always being, as we have observed, themselves transitive. (Cf. also Exc. 4.4.20(iv).) Then we would obtain the following:

> The smallest normal bimodal logic in $\Box$ and O extending $\mathbf{K4}_\Box$ and $\mathbf{K4}_O$ and containing all formulas $\Box A \to OA$ is determined not only by (1) the class of all frames $\langle W, R_\Box, R_O\rangle$ with $R_\Box$ and R_O transitive $R_O \subseteq R_\Box$, but also by (2) by the class of all such frames in which R_O is a range-restriction of $R_\Box$.

The "but also by (2)" part here is obviously incorrect, though, since the frames there mentioned all validate $Op \to \Box Op$, which is not provable in the logic described. (For example, let W be a 3-element set $\{x, y, z\}$, say, with $R_\Box = \{\langle x, y\rangle, \langle y, z\rangle, \langle x, z\rangle\}$, $R_O = \{\langle y, z\rangle\}$. Then the frame $\langle W, R_\Box, R_O\rangle$ meets the conditions under (1) but we falsify $Op \to \Box Op$ at x in any model on the frame with p false at z. And of course R_O is not a range-restriction of $R_\Box$ here.) ◀

What, then, if one's preference happened to be for the combined alethic–deontic logic described syntactically in the inset passage under Example 4.4.45 – without, in particular, $Op \to \Box Op$? Would this mean that one could not avail oneself of the Kanger–Smiley approach, with OA understood as A's being strictly implied by some sentential constant? (The undesirable aspects of an affirmative answer here for a traditional alethic reading of "$\Box$" are stressed in Remark 4.5.1, p. 279 below.) To address this question (as in Humberstone [546]) we have to note both (1) that a sentential constant need not be interpreted as a propositional constant, and (2) that strict implication need not be understood as *alethically* necessitated material implication. Observation (1) here means that instead of assigning, in each model, to a sentential constant some set of worlds (or 'proposition'), in the way that, when we had Q explicitly in the language it was assigned, relative to a model $\mathcal{M}$, the set $Q^{\mathcal{M}}$, we could assign to it a set of ordered pairs of worlds (a 'di-proposition') instead, and thereby encode any accessibility relation that might have been used to interpret the operator O. (We will write "$\boldsymbol{R}$" for this sentential constant, as in Humberstone [546], [578]; Venema [1125], p. 70 writes "λ" instead.)

The idea will be that our new sentential constants, as well as the background notion of absolute necessity symbolized by $\Box$, make essential use of two world-parameters, so that for a smooth inductive truth-definition both must be in play throughout, though one of them will be idle for the case of the other vocabulary. In particular, we write "$\mathcal{M} \models_y^x A$" with the sometimes idle parameter in the superscript position and the familiar parameter in its accustomed subscript position. The new sentential constants will be written in boldface italic. Suppose there is just one of these to interpret, allowing each model to come equipped with just one accessibility relation. (We will get to $\Box$ later.) Thus, a model $\mathcal{M}$ now has the form $\langle W, R, V\rangle$. The relation R is playing the role of R_O in the bimodal language we have most recently been attending to, it will be matched in the object language by the sentential constant $\boldsymbol{R}$. We give this constant a two-dimensional semantics (assigning to it a 'di-proposition'), while the propositional variables continue to deserve that name and get only a traditional one-dimensional treatment (being assigned a proposition, that is). So while $V(p_i) \subseteq W$, with $y \in V(p_i)$ being necessary and sufficient for $\mathcal{M} \models_y^x p_i$, regardless of x ($x, y \in W$), we have:

$$\mathcal{M} \models_y^x \boldsymbol{R} \text{ if and only if } xRy.$$

The Boolean connectives are given the usual treatment, with no variation of parameters: $\mathcal{M} \models_y^x A \wedge B$ iff $\mathcal{M} \models_y^x A$ and $\mathcal{M} \models_y^x B$, for all A, B, and so on. The remaining novelty comes in the treatment of $\Box$, which will enable it to amount to absolute necessity in the sense of the necessity to which, when OA is written as $\Box(\boldsymbol{R} \to A)$, O is relative:

$$\mathcal{M} \models_y^x \Box A \text{ if and only if for all } z \in W, \mathcal{M} \models_z^y A.$$

Now if we consider the conditions under which $\mathcal{M} \models_y^x Op$, where the formula involved is regarded as abbreviating $\Box(\boldsymbol{R} \to p)$, we find that:

$$\begin{aligned}
\mathcal{M} \models^x_y \Box(\boldsymbol{R} \to p) \text{ iff } &\text{ for all } z \in W \ \mathcal{M} \models^y_z \boldsymbol{R} \to p \\
\text{iff } &\text{ for all } z \in W \text{ iff } \mathcal{M} \models^y_z \boldsymbol{R} \Rightarrow \mathcal{M} \models^y_z p \\
\text{iff } &\text{ for all } z \in W, yRz \Rightarrow z \in V(p),
\end{aligned}$$

which is exactly what we should have ended up with had we stuck with writing the formula concerned as Op and processed it through the standard semantics with R as R_O. If we want several such operators, we can equip the frames with corresponding relations: for example R_1 and R_2 for constants $\boldsymbol{R}_1$ and $\boldsymbol{R}_2$, where, to illustrate a point of interest, we might be thinking of the operators O_1 and O_2 as representing some agent's knowledge in the former case, and the notion of nomic necessity in the latter. Naturally, we are envisaging the definition of $O_i A$ as $\Box(\boldsymbol{R}_i \to A)$, for $i = 1, 2$. The point of interest arises over the fact that both of these are veridical notions in the sense that we should want our logic to contain $\mathbf{T}_{O_1}$ and $\mathbf{T}_{O_2}$, which in primitive notation become:

$$\Box(\boldsymbol{R}_1 \to p) \to p \qquad \text{and} \qquad \Box(\boldsymbol{R}_2 \to p) \to p.$$

Now this looks like trouble because if we substitute $\boldsymbol{R}_1$ for p (uniformly) in the first and $\boldsymbol{R}_2$ for p in the second, we get conditionals with tautologous antecedents and can thus detach their consequents, so both $\boldsymbol{R}_1$ and $\boldsymbol{R}_2$ become provable. This suffices for the provability of such things as $\Box(\boldsymbol{R}_1 \to p) \to \Box(\boldsymbol{R}_2 \to p)$ and its converse. But we didn't want knowledge and nomic necessity to be equivalent – we just wanted them each to imply the truth of what was known or necessary.

We can see what has gone wrong if we think about the semantics. We wanted two distinct accessibility relations, each reflexive – not just one relation. But on the usual treatment of relative necessity, each of these relations would be the range restriction of a single underlying accessibility relation (for $\Box$), R, say, so we should have for all $x, y \in W$ (W the universe of some arbitrary frame):

$$xR_1 y \Leftrightarrow (xRy \ \& \ y \in X_1) \qquad \text{and} \qquad xR_2 y \Leftrightarrow (xRy \ \& \ y \in X_2),$$

for some $X_1, X_2 \subseteq W$. For R_1 thus defined to be reflexive, we need $X_1 = W$; and likewise, with R_2 and X_2, so $R_1 = R_2 = W \times W$ and naturally things go wrong. The whole idea was to get away from range restrictions. We can see that a false step was taken in passing from $\Box(\boldsymbol{R}_i \to p) \to p$ to its substitution instance with $\boldsymbol{R}_i$ replacing p, by recapitulating as above, though expanding somewhat (in distinguishing the third and fourth lines below):

$$\begin{aligned}
\mathcal{M} \models^x_y \Box(\boldsymbol{R}_i \to p) \to p \text{ iff } &\text{ if for all } z \in W \ \mathcal{M} \models^y_z \boldsymbol{R}_i \to p, \text{ then } \mathcal{M} \models^x_y p \\
\text{iff } &\text{ if for all } z \in W, \mathcal{M} \models^y_z \boldsymbol{R}_i \Rightarrow \mathcal{M} \models^y_z p, \text{ then } \mathcal{M} \models^x_y p \\
\text{iff } &\text{ if for all } z \in W, yR_i z \Rightarrow \mathcal{M} \models^y_z p, \text{ then } \mathcal{M} \models^x_y p \\
\text{iff } &\text{ if for all } z \in W, yR_i z \Rightarrow z \in V(p), \text{ then } y \in V(p).
\end{aligned}$$

What follows the "iff" on this last line is correct for reflexive R_i, since given that for all $z \in W$, we have $yR_i z \Rightarrow z \in V(p)$, we must have, taking y as z, $yR_i y$: so $y \in V(p)$. But what happens if in place of p, we have, as on the derivation imagined above, the sentential constant $\boldsymbol{R}_i$ again? Then, corresponding to the third line, we would have:

$$\ldots \text{iff: if for all } z \in W, yR_i z \Rightarrow \mathcal{M} \models^y_z \boldsymbol{R}_i, \text{ then } \mathcal{M} \models^x_y \boldsymbol{R}_i,$$

and thus, corresponding to the fourth line:

$$\ldots \text{iff: if for all } z \in W, yR_i z \Rightarrow yR_i z, \text{ then } xR_i y.$$

Now the condition on z is vacuously satisfied and the right-hand side is just equivalent to $xR_i y$, echoing our earlier finding that the reflexive R_i end up being universal. The uniform substitution steps need to be blocked, then, which is not surprising since the propositional variables are treated as one-dimensional formulas, depending only on the lower index of "$\models$" (that is, relative to any model $\mathcal{M}$, we have $\mathcal{M} \models^x_z p_j$ iff $\mathcal{M} \models^y_z p_j$) while the constants $\boldsymbol{R}_i$ are two-dimensional, sensitive to variation in the upper index as

4.4. DEONTIC LOGIC: MAIN THEMES

well. One response to this – Kuhn [679], p. 193 – might be to introduce special two-dimensional sentence letters ("di-propositional variables") either replacing or supplementing the usual range of propositional variables, for which there would be no such restriction on substitution.

EXAMPLE 4.4.46 Although these matters are not further taken up here it let us recall the special sense noted on p. 268 for – to repeat the labelling from that discussion – β:

(β)   It is both morally and legally permissible that A.

For this special sense what is required for its truth is the existence of some world both morally and legally accessible at which it is true that A; we saw that this is inexpressible in the standard modal language. But it could be captured with the current di-propositional constants, say with $\boldsymbol{R}_1$ and $\boldsymbol{R}_2$ for moral and legal accessibility, simply enough, by: $\Diamond(\boldsymbol{R}_1 \wedge \boldsymbol{R}_2 \wedge A)$. Similarly, with a single accessibility relation R in mind now, we can directly express the idea that A is true at all *inaccessible* worlds by $\Box(\neg \boldsymbol{R} \to A)$ (or alternatively $\Box(\boldsymbol{R} \vee A)$), as well as the idea that A is true at *precisely* (= all and only) the accessible worlds, by $\Box(\boldsymbol{R} \leftrightarrow A)$. (More on these ideas can be found in my [550] and [556], respectively, the latter mentioned already under Example 4.3.2, p. 228. Our discussion here simplifies away temporal considerations – cf. Remark 4.4.6, p. 243 – in the interests of brevity.) ◀

As well as treating deontic, nomic, etc., modalities by means of such sentential constants, we may wish to do this for logical or metaphysical necessity itself, so as to avoid untoward repercussions (some of them described in Humberstone [546]) of casting $\Box$ in this role. The latter continues to deserve to be called absolute necessity, since these other operators are presented as relative necessities in terms of it. It has an accessibility relation of its own (see Segerberg [1027], Kuhn [679]), if we think of the universes of our models as $W \times W$ rather than W: $\langle w, x \rangle$ bears this relation to $\langle y, z \rangle$ just in case $x = y$. (Here we are writing the 'upper' index as the first element of the ordered pair and the lower index as the second.) For further information, see my [546], or the summary thereof in §5 of [578]; for subsequent developments – in particular the axiomatic description of the pure logic of $\Box$ as absolute necessity in the present sense – see Kuhn [679], and also Venema [1126]. (Some qualms about this approach, and suggestions as to how – using propositional quantification – to improve on it, can be found in Hale and Leech [426].) De Boer, Gabbay, Parent and Slavkovic, in [103] independently rediscover the utility of such two-dimensional treatments of sentential constants for deontic logic. Though their motivations are somewhat different, as it happens, they involve themselves also with the 'inaccessible worlds' idea of Example 4.4.46 in order to develop some ideas from Jones and Pörn [629] and [628], taking the latters' 'subideal worlds' as those which are not accessible by the deontic accessibility relation. We illustrate with one such idea here, mentioned already (differently notated) in Example 4.3.2, p. 228 (see also note 187 there):

EXAMPLE 4.4.47 In the papers just cited, Jones and Pörn actually work with two primitive accessibility relations, one to take us to a world's ideal alternatives and the other to its subideal alternatives, perhaps because the idea of taking the second as the complement of the first did not occur to them, and where O is interpreted by universal quantification over the ideal worlds and O' by universal quantification over the subideal worlds, the suggested response to some claimed anomalies of deontic logic arising from the monotone – indeed the normal – behaviour of O is as follows. We consider, not O itself, but a more complex defined operator as the proper obligation operator, to be written as Ought. The definition, making use of O' runs thus:

$$\mathsf{Ought}\, A = OA \wedge \neg O'A.$$

So for example, if one regards, as we noted in Remark 4.4.9 (p. 246) Jonathan Harrison [454], p. 23, does, $O(p \vee \neg p)$ as "quite absurd", presumably on the grounds that what is inevitable can hardly be a matter of obligation, then one will be relieved to see that $\mathsf{Ought}(p \vee \neg p)$ is not valid according to the Jones–Pörn treatment, since $p \vee \neg p$ does not fail to be true at some non-ideal world. There is a restricted form of normality, given here only in the special case of monotony:

$$\frac{A \to B}{(\text{Ought } A \wedge \neg O'B) \to \text{Ought } B}.$$

For some adverse reactions to even the weaker form – have we really avoided Ross's Paradox (p. 243 above), for example – along with replies and further commentary, see Hansson [446], Jones and Pörn [630], and de Boer, Gabbay et al. [103]. (Some of these issues have since been revisited by Gabbay with the aid of the somewhat exotic apparatus of 'reactive Kripke semantics': see Chapter 11 of Gabbay [342].)   ◀

4.5 Deontic Logic: More Translations, More Issues

In this section we continue the theme of translationally embedding suitable deontic logics in modal logics that might be thought plausible for an alethic interpretation. The translations of interest here do not involve additional linguistic resources such as the constants S and Q of Anderson's reduction(s) and the Kanger–Smiley variation thereon reviewed in the preceding section. Instead they treat the O operator as definable in terms of the Boolean connectives and the given alethic operator; in the particular cases we consider, O is treated as some affirmative alethic modality (in the sense introduced on p. 31). This idea was pioneered by Dawson in [237]. (Later we will look at variations on this theme, from Kielkopf [653] and, more briefly, Åqvist [27]; see Example 4.5.25 on the latter, and the discussion there promised for the following chapter.) Dawson suggested for this purpose the use of **S4.2** as an alethic modal logic with O interpreted as $\Diamond\Box$ (and accordingly with P as $\Box\Diamond$), pointing out that this gave an acceptable – or, as it was then called (see note 220), 'normal' deontic logic. So the affirmative modality in question has length 2; it is 'alethic' in the sense that the $\Box$ part of this affirmative modality (i.e., $\Diamond\Box$) is governed by the axiom **T**. In the contemporary sense of *normal*, we noted in Example 2.8.14 that $\Diamond\Box$ is normal in **S4.2** (though not in **S4**); and **.2**, with O as $\Diamond\Box$, gives $\mathbf{D}_O$.[229] From work in the following decade by Wolfgang Lenzen, who had epistemic applications in mind (as explained presently), it became evident that the 'deontic fragment' of **S4.2** under the rendering of O as $\Diamond\Box$, was precisely **KD45** – one of several systems[230] to have been called "deontic **S5**" and last encountered in and (in the discussion following, corollaries to, etc.) Proposition 4.1.3(*ii*), p. 206, as the logic of those semi-simplified frames in which the $\Box$-pertinent – or rather O-pertinent – subset was non-empty. We can put this in terms of a definitional translation (as explained in Example 4.4.25, beginning on p. 264), for which we will use the label "$\tau_{\Diamond\Box}$", thus:

- $\tau_{\Diamond\Box}(p_i) = p_i$
- $\tau_{\Diamond\Box}(\neg A) = \neg(\tau_{\Diamond\Box}(A))$
- $\tau_{\Diamond\Box}(A \to B) = \tau_{\Diamond\Box}(A) \to \tau_{\Diamond\Box}(B)$ (etc., for other Boolean connectives)
- $\tau_{\Diamond\Box}(\Box A) = \Diamond\Box(\tau_{\Diamond\Box}(A))$.

Then Lenzen's observation would be formulated as: for all formulas A,

$$\vdash_{\mathbf{KD45}} A \text{ if and only if } \vdash_{\mathbf{S4.2}} \tau_{\Diamond\Box}(A).$$

This will be proved as Theorem 4.5.9 below. In the meantime, after some philosophical remarks, we examine from a semantic point of view how $\Diamond\Box$ manages to behave like a normal modal operator in

---

[229]The presumption here is that **S4.2** is a plausible candidate alethic modal logic, and though in general the focus in such debates are the endpoints of **S4**–**S5** spectrum, in §6 of [137], Burgess mounts an argument specifically against **S4.2** as an alethic modal logic for an epistemically flavoured construal of necessity as *demonstrability*. Since this is an attack on **.2**, it would, if successful, undermine **S5**'s claim to being a plausible logic for necessity understood in these terms – not that the **S5** theorems **B** or **5** have any appeal for this reading of $\Box$ anyway. (See also note 138, p. 166.)

[230]See Humberstone [584], note 13.

4.5. DEONTIC LOGIC: MORE TRANSLATIONS, MORE ISSUES

S4.2, and indeed more generally in all extensions – **S4.2** being one of these – of a logic we shall be calling **KDH**.

In the discussion leading up to Example 4.4.45 from the preceding section, the point was made that such embedding results reflect in only a weak way the idea of reducing the logic of deontic operators to that of alethic modal operators, in that they do not address the question of the interaction of the deontic operator (the "□" of the source logic in the above formulation) and the alethic operator (the "□" of the target logic). On this weaker way of interpreting the broadly Anderson-inspired program of such reductions, all we want to do is to find an intuitively plausible deontic logic embedded inside some intuitively plausible alethic modal logic. On the stronger interpretation, however, what is wanted is a plausible combined alethic–deontic modal logic. In the case of Anderson, this stronger interpretation is entirely plausible. As Prior ([919], p. 142) puts it, having said that the reduction to alethic modal logic (called ordinary modal logic in the following quotation) "makes possible a quite remarkable simplification of the postulates of deontic logic," continues:

> But it does more than this. It also gives us new means of investigating the relations of deontic and ordinary modal logic.

Dawson's reduction cannot plausibly viewed in this stronger light, since there is evidently no logical equivalence between "It ought to be that A" and "It is possible that necessarily A" – whatever one's views about whether the latter is equivalent to "Necessarily A".[231] Dawson, however, did not see this, and wrote ([237], p. 77) that "(t)he present alethic interpretation of the deontic operators also provides a straightforward means of resolving the status of combined deontic-alethic formulae".[232] The situation in this respect is quite different with Lenzen's corresponding 'reduction of doxastic logic to epistemic logic', where the corresponding equivalence, between "a believes that A" and "For all a knows, a knows that A" has considerable plausibility, favouring the strong interpretation in this case. (See Theorem 5.2.1 on p. 372, much of the discussion in Section 5.5, and – notwithstanding all this, for an objection to the equivalence – Section 5.6.)

REMARK 4.5.1 Recall from the preceding section (in particular, the discussion after Example 4.4.45, itself on p. 275) that even for Anderson–Kanger–Smiley style treatments, there is a potential problem about what has just been called the stronger interpretation, if we think of the □ (alongside the sentential constant Q, as we wrote it) in terms of which O is defined, because if one thinks of this is as representing logical or metaphysical necessity, and accepts **4** as a correct principle for □ so interpreted, then one will be committed to accepting $Op \to \Box Op$ as correct. But to echo some wording from Remark 4.4.6 on p. 243, since even if fundamental moral principles are held not to be contingent, everyday $Ought$-judgments ("She ought to thank him") surely are. ◀

Before returning to **S4.2** itself, in which it is clear from the above remarks (and was already mentioned in Example 2.8.14, p. 147) that the modality $\Diamond\Box$ is normal, we can usefully consider the *smallest* normal modal logic in which $\Diamond\Box$ is normal, which is evidently the normal extension of **KD** by what we call **H** – the etymology behind which label is explained in Exercise 2.7.18 at p. 133 above – but note that there is no connection between this **H** and the principle **H** figuring in Examples 2.5.38 and 2.5.39 (see p. 87),

---

[231]The latter equivalence is endorsed by **S5**, of course, but not by **S4.2**, which is what blocks the provability of $Op \to p$ when O is read as $\Diamond\Box$.

[232]The present criticism of Dawson was made in my [584], on which several aspects of the discussion of **S4.2** and **KDH** in this section are based; presumably Bacon had the same point in mind when he described Dawson's treatment as being "of purely formal interest" ([47], p. 208). A similar criticism applies to Kielkopf, whose [653] suggests a deontic ("O") reading of $\Box\Diamond$; we return to this after Theorem 4.5.9 below. Dawson writes □ and ◊ as L and M, and uses parentheses even when no ambiguity would result from their omission – "$L(p)$" for "Lp, etc. He writes (p. 408): "(T)he fact that $O(p)$, viz. $LM(p)$, is not equivalent to $L(p)$ shows that ought-statements are not assertions of logical necessity if that is what $L(\ )$ is to be used for. The ought-operator contains $L(\ )$ but it is not $L(\ )$." The remarks are confused. If L is being used for logical necessity, then of course LMp – if we imagine this asserted – *is* an assertion of logical necessity – the logical necessity of Mp.

renamed from 'H' to avoid this ambiguity. The present **H** makes this modality aggregative, **D** provides Necessitation for it – note that $\Diamond\Box\top$ and $\Diamond\top$ are **K**-equivalent – and like any other affirmative modality (see Proposition 1.4.2), $\Diamond\Box$ will be monotone; the following candidate axioms all yield the same normal modal logic when added ("⊕"-added) to **K**; the first (or top left) amongst them says directly that the modality $\Diamond\Box$ is aggregative in the logic thereby axiomatized:

H $\quad(\Diamond\Box p \wedge \Diamond\Box q) \to \Diamond\Box(p \wedge q) \qquad (\Diamond\Box p \wedge \Diamond\Box q) \to \Diamond(\Box p \wedge \Box q)$

$\quad\quad\Box\Diamond(p \vee q) \to (\Box\Diamond p \vee \Box\Diamond q) \qquad \Box(\Diamond p \vee \Diamond q) \to (\Box\Diamond p \vee \Box\Diamond q).$

The four forms above are obviously similarly equivalent (as candidate axioms) by taking duals and distributing $\Box$ over $\wedge$ or $\Diamond$ over $\vee$. We address a generalized form:

H$_n \quad (\Diamond\Box p_1 \wedge \ldots \wedge \Diamond\Box p_n) \to \Diamond\Box(p_1 \wedge \ldots \wedge p_n),$

in the next exercise, at least for $n \geq 2$. Note that as with **H** above, which is the special case of **H**$_2$, we could also supply another three versions of this principle. Note also that while **H**$_1$ is vacuous as a candidate axiom, **H**$_0$ amounts to **D** (in the form $\Diamond\top$, taking $\top$ as the conjunction of 0 conjuncts), which is not provable in **KH**.

EXERCISE 4.5.2 (*i*) Show that $\vdash_{\mathbf{KH}}$ **H**$_n$ for all $n \geq 2$.

(*ii*) Show, by checking soundness, and then using a canonical model argument for completeness, that **KH** is determined by the class of all frames $\langle W, R\rangle$ satisfying the condition:

$$\forall w \forall x \forall y ((Rwx \wedge Rwy) \to \exists u (Rwu \wedge \forall v (Ruv \to (Rxv \wedge Ryv)))).$$

(Suggestion: where $\langle W, R\rangle$ is the canonical frame of any consistent normal modal logic in which **H** is provable, it suffices to show how, given $w, x, y \in W$, some $u \in W$ can be found satisfying: $u \in R(w)$ and $R(u) \subseteq R(x) \cap R(y)$. Show that

$$\{A \mid \Box A \in w\} \cup \{\Box B \mid \Box B \in x\} \cup \{\Box C \mid \Box C \in y\}$$

is **KH**-consistent and taking u as any maximal consistent superset thereof will exhibit the promised behaviour.) ✠

Of greater interest is a second completeness result we can use the axiom **H** for. (This can be found in Humberstone [584] and – essentially – in Stalnaker [1082]; see further Proposition 4.5.8 and the discussion in Section 5.5, esp. p. 410.)

EXERCISE 4.5.3 Show that the canonical frame for any consistent extension of **KH** satisfies the following condition:

$$\forall w (R(w) \neq \varnothing \to \exists u \in R(w) \forall x \in R(w) \centerdot R(u) \subseteq R(x)).$$

(Suggestion: given that $R(w) \neq \varnothing$, obtain the promised $u \in W$, as any maximal consistent extension of the set $\{A \mid \Box A \in w\} \cup \{\Box B \mid \Diamond\Box B \in w\}$, verifying that this set is indeed consistent (using an appropriate choice of **H**$_n$) and guarantees that u satisfies the required conditions.) ✠

Thus, from Exercises 4.5.2(*ii*) and 4.5.3, we have the following two completeness results for **KDH**:

PROPOSITION 4.5.4 **KDH** *is determined*

(*i*) *by the class of serial frames* $\langle W, R\rangle$ *satisfying the condition:*

$$\forall w \forall x \forall y ((Rwx \wedge Rwy) \to \exists u (Rwu \wedge \forall v (Ruv \to (Rxv \wedge Ryv))));$$

4.5. DEONTIC LOGIC: MORE TRANSLATIONS, MORE ISSUES

and

(*ii*) *by the class of frames* $\langle W, R \rangle$ *satisfying the condition:*

$$\forall w \exists u \in R(w) \forall x \in R(w) \,.\, R(u) \subseteq R(x)).$$

Part (*ii*) here is the simplification we get from Exercise 4.5.3 by dropping the "$R(w) \neq \varnothing$" antecedent (as we may since **D** gives us seriality), so what we really get is something more general on the completeness side: the canonical frame of any consistent normal extension of **KDH** satisfies the condition given. This in turn provides a semantic perspective on the normality of $\Diamond\Box$ in these logics, since we can supply a binary relation which will serve as an accessibility relation for this composite operator, with $\Diamond\Box A$ true at a point iff A is true at all points to which the given point bears this relation. We call the relation in question S. To define it on the universe of any frame $\langle W, R \rangle$ satisfying the condition in Proposition 4.5.4(*ii*), let s be a function assigning to each $w \in W$ a 'special' successor u as described there, and now define $S \subseteq W \times W$ by:

$$xSy \text{ if and only if } s(x)Ry \text{ (all } x, y \in W).$$

Now, for any model $\mathcal{M} = \langle W, R, V \rangle$ on the given frame, let $\mathcal{M}^*$ be the model $\langle W, S, V \rangle$, with S as just defined. The relation between these models is the subject of the following exercise.

EXERCISE 4.5.5 Show by induction on the complexity of A that for any formula A we have: for all $w \in W$, $\mathcal{M}^* \models_w A$ if and only if $\mathcal{M} \models_w \tau_{\Diamond\Box}(A)$. ✠

This provides a semantic gloss on the normality of the prefix $\Diamond\Box$ in normal extensions of **KDH**, since we are able to interpret it as universally quantifying over S-accessible points. But this must not be overstated, as it would be in the following claim: In any model on a frame for **KDH**, there is a model on that frame with the same universe and a binary relation for which the truth of a formula $\Diamond\Box A$ at a point is equivalent to the truth of A at all points in the model which are accessible by that relation. This cannot be concluded on the basis of the foregoing considerations, since they concerned specifically a phenomenon arising for the *canonical* frames of the (consistent normal) extensions of **KDH**, since it is only these that were shown to satisfy the condition in Proposition 4.5.4(*ii*), and hence to supply the special successors $s(w)$ in terms of which the relations S were defined. (Not that we need to introduce "S" explicitly to make this observation: $\Diamond\Box$ is interpreted at x as universally quantifying over $R(s(x))$.) To see that this conclusion cannot be obtained in any other way either, we offer a counterexample essentially derived from van Benthem [72], p. 386.

EXAMPLE 4.5.6 Consider the model $\mathcal{M} = \langle W, R, V \rangle$ in which W is the set of natural numbers, R the relation $<$, and $V(p_i) = \{n \in W \mid n \geqslant i\}$. In this model, $\Diamond\Box p_1$ is true at 0 since there is an accessible point at which $\Box p_1$ is true (namely 1 – and indeed all greater numbers), and $\Diamond\Box p_2$ is true at 0, since there is an accessible point at which $\Box p_2$ is true (namely 2, and again every $n \geq 2$), and so on for each formula $\Diamond\Box p_i$. Now suppose (for a contradiction) that there is a relation $S \subseteq W \times W$ serving as an accessibility relation for $\Diamond\Box$ in the sense that for all formulas A and all $x \in W$,

$$\mathcal{M} \models_x \Diamond\Box A \text{ iff for all } y \in W, \text{ if } xSy \text{ then } \mathcal{M} \models_y A.$$

Since $\mathcal{M} \models_0 \Diamond\Box p_i$ for each i ($i = 1, 2, \ldots$), it follows from the "only if" direction of the biconditional inset above that every $y \in W$ such that $0Sy$, $\mathcal{M} \models_y p_i$ for all i. But there are no $y \in W$ for which $\mathcal{M} \models_y p_i$ for all i; thus, vacuously, for every $y \in W$ such that $0Sy$, we have $\mathcal{M} \models_y \bot$, which should imply, by the "if" half of the above biconditional, that $\mathcal{M} \models_0 \Diamond\Box\bot$ – which is not the case, since 0 has no R-successor all of whose R-successors verify $\bot$ (i.e., 0 has no R-successors without R-successors). ◀

The above example could equally well be presented with R as $\leq$ rather than $<$, in which case the frame concerned would be a frame for **S4.2** – indeed, a frame for **S4.3**, but it is **S4.2** that we are especially concerned with here as the target of the embedding $\tau_{\Diamond\Box}$. We shall get to this soon (Thm. 4.5.9), after an observation in the style of Proposition 4.5.4(*ii*). For this, we need to assure ourselves that **S4.2** extends **KDH**:

EXERCISE 4.5.7 Show, either by a semantic argument using the completeness of **S4.2** w.r.t. the class of transitive, reflexive, convergent frames (Coro. 2.5.12), or else by an axiomatic deduction, that $\vdash_{\mathbf{S4.2}}$ **H** (and therefore, since obviously $\vdash_{\mathbf{S4.2}}$ **D**, that **KDH** $\subseteq$ **S4.2**). ✠

Now we already know that **S4.2** is determined by the class of all transitive reflexive frames which are piecewise convergent (combining Thm. 2.5.2, as proved by the canonical model method, with Exc. 2.5.8(*iii*)), and also that we can drop the "piecewise" here (Coro. 2.5.12), by taking point-generated subframes. A still narrower class of frames also determines this logic, as we are now in a position to observe.

PROPOSITION 4.5.8 **S4.2** *is determined by the class of transitive reflexive frames $\langle W, R \rangle$ for which there is a non-empty $X \subseteq W$ such that for all $x \in W$, $X \subseteq R(x)$.*

Proof. Soundness is clear enough since the condition in question guarantees convergence – any element of X serving as a common successor for any other pair of elements. For the completeness half, by Exercise 4.5.7 and the proof of Proposition 4.5.4(*ii*), we know that the canonical frame for **S4.2** satisfies the condition: $\forall w \exists u \in R(w) \forall x \in R(w) \centerdot R(u) \subseteq R(x))$. Now take the subframe of this frame generated by an arbitrarily selected element w, calling this $\langle W, R \rangle$, and take some $u \in R(w)$ as promised by the condition cited (a 'special successor' $s(w)$ of w as we have been putting it). Thus for all $x \in R(w)$ we have $R(u) \subseteq R(x)$. But since **S4.2** $\supseteq$ **S4**, the relation R is transitive and reflexive, so $R(w) = W$; taking $R(u)$ as X, then, we have, as required (non-empty) $X \subseteq W$ satisfying: for all $x \in W$, $X \subseteq R(x)$. ∎

The above result is mentioned in Stalnaker [1082], and we shall return to it in our discussion of epistemic (and doxastic) logic in the following chapter, when it makes a slightly transformed appearance as Theorem 5.5.14 on p. 413; the relevant passage from Stalnaker is quoted on p. 410.

THEOREM 4.5.9 *For all formulas A:* $\vdash_{\mathbf{KD45}} A$ *if and only if* $\vdash_{\mathbf{S4.2}} \tau_{\Diamond\Box}(A)$.

Proof. For the "only if" direction we proceed by induction on the length of proof of A from the axiomatization of **KD45** suggested by that label. For the "if" direction we suppose that $\nvdash_{\mathbf{KD45}} A$ with a view to showing that $\nvdash_{\mathbf{S4.2}} \tau_{\Diamond\Box}(A)$. The supposition, taken in conjunction with Coro. 4.1.5, gives us a model $\langle W_0, w, V \rangle$ with universe $W = W_0 \cup \{w\}$ in which truth anywhere for $\Box$-formulas amounting to truth throughout $W_0 \neq \emptyset$, and moreover with A false at w. The implicit accessibility relation – call it R – is $W \times W_0$. Call this model $\mathcal{M}$. Let $\mathcal{M}^*$ have W and V as for $\mathcal{M}$ but with accessibility relation $R^* \cup \{\langle w, w \rangle\}$. (Note that this is the reflexive closure of R, since the only potentially non-reflexive point in W is w.) One sees by induction on the complexity of a formula B that for all $x \in W$, $\mathcal{M} \models_x B$ if and only if $\mathcal{M}^* \models_x \tau_{\Diamond\Box}(B)$. So $\mathcal{M}^* \not\models A$, and hence $\nvdash_{\mathbf{S4.2}} \tau_{\Diamond\Box}(A)$, since the frame of $\mathcal{M}^*$ is reflexive, transitive and convergent (Coro. 2.5.12, p.77). ∎

We turn to the option, canvassed in Kielkopf [653] and touched on in note 231 above, of carving out a deontic logic from **S4M** with $\Box\Diamond$ as O, in approximate analogy to Dawson's doing so from **S4.2** with $\Diamond\Box$ as O. As in the case of Dawson's embedding (just proved), we begin with the smallest normal logic in which the new modality is itself normal. In the previous case, this logic was **KDH**, while in the present

4.5. DEONTIC LOGIC: MORE TRANSLATIONS, MORE ISSUES

case it is something we shall call $\mathbf{KFi} \oplus \Box\Diamond\top$, and we begin with $\mathbf{Fi}$ (which plays a role analogous to that played above by $\mathbf{KH}$. Corresponding to the four forms given under $\mathbf{H}$ on p. 280, we have the following:

$\mathbf{Fi}$ $\quad(\Box\Diamond p \wedge \Box\Diamond q) \to \Box\Diamond(p \wedge q) \qquad \Box(\Diamond p \wedge \Diamond q) \to \Box\Diamond(p \wedge q)$
$\qquad \Diamond\Box(p \vee q) \to (\Diamond\Box p \vee \Diamond\Box q) \qquad \Diamond\Box(p \vee q) \to \Diamond(\Box p \vee \Box q).$

The label here is chosen to recall Fine [289] where an equivalent axiom (given normality) is described, given in Example 4.5.15 below.[233] A simple generalization, of which the above represent the $n = 2$ case, is the following:

$\mathbf{Fi}_n \quad (\Box\Diamond p_1 \wedge \ldots \wedge \Box\Diamond p_n) \to \Box\Diamond(p_1 \wedge \ldots \wedge p_n),$

with alternative forms parallelling those given under the special case of $\mathbf{Fi} = \mathbf{Fi}_2$ before. Note that while $\mathbf{Fi}_1$ is vacuous as a candidate axiom, $\mathbf{Fi}_0$ amounts to $\Box\Diamond\top$, and this is not provable in $\mathbf{KFi}$.

EXERCISE 4.5.10 Show that $\vdash_{\mathbf{KFi}} \mathbf{Fi}_n$ for all $n \geq 2$, and that $\mathbf{KFi} \oplus \Box\Diamond\top$ is the smallest normal modal logic in which the modality $\Box\Diamond$ is normal. ✠

A canonical model completeness proof for $\mathbf{KFi}$ will require a preliminary observation.

LEMMA 4.5.11 *Suppose that for elements $x, y \in W$ for the canonical frame $\langle W, R\rangle$ of some consistent normal modal logic we have $y \supseteq \{\Box A \vee \Box B \mid \Box(A \vee B) \in x\}$. Then $|R(y)| \leq 1$ and $R(y) \subseteq R(x)$.*

Proof. Make the supposition. Since $\Box(A \vee \neg A) \in x$ for all formulas A, the supposition gives $\Box A \in y$ or $\Box\neg A \in y$ for each formula A, from which it follows that $|R(y)| \leq 1$. To show that $R(y) \subseteq R(x)$ it suffices to show that $\Box A \in x$ implies $\Box A \in y$ for all formulas A. (Why?) Now if $\Box A \in x$ we have $\Box A \vee \Box A \in x$, so by the supposition $\Box(A \vee A) \in y$ and therefore $\Box A \in y$ and we are done. ∎

A second result we need for our completeness proof for $\mathbf{KFi}_2$ was already given as Proposition 1.2.6 (p. 12), recalled here for convenience:

> For a formula $D = (A_1 \wedge B_1) \vee \ldots \vee (A_n \wedge B_n)$, let Φ_D be the set of formulas $C_1 \vee \ldots \vee C_n$, where each C_i is either A_i or B_i $(1 \leq i \leq n)$, and let D' be the conjunction of the formulas in Φ_D. Then D' is truth-functionally equivalent to D.

We also make use of the fact (Exercise 4.5.10) that $\mathbf{Fi}_n$ is provable for all $n \geq 2$ in $\mathbf{KFi}_2$.

THEOREM 4.5.12 $\mathbf{KFi}$ *is determined by the class of frames $\langle W, R\rangle$ satisfying: for all $w \in W$, for all $x \in R(w)$ there exists $y \in R(w)$ such that for all $z_1, z_2 \in R(y)$, $z_1 = z_2$ and xRz_1.*

Proof. The soundness half of the result presents no difficulties. For the completeness half we show that the canonical frame – call it $\langle W, R\rangle$ – for this logic satisfies the condition described. To avoid a clash of lettering with that used in Proposition 1.2.6, as recalled above, we reach for "E" as a schematic letter in the argument where otherwise an "A" might be expected (and in fact, below, the conjunctions $\neg A_i \wedge \neg B_i$ correspond to the conjunctions $A_i \wedge B_i$ in the formulation of Prop. 1.2.6). Taking $w \in W$ with $x \in R(w)$, we shall find y as promised by taking y as a maximal consistent superset of the following set, whose consistency will be shown presently:

$$\{E \mid \Box E \in w\} \cup \{\Box A \vee \Box B \mid \Box(A \vee B) \in x\}.$$

---

[233] We cannot use just "$\mathbf{F}$", since this label is reserved for something else: see note 26 (p. 34) and Section 5.5 below.

Any $y \in W$ extending this set has the required properties: at most one successor and any successor it may have being accessible to x, by Lemma 4.5.11, and also being accessible to w in view of including $\{E \mid \Box E \in w\}$. So we need only check the **KFi**-consistency of the above set of formulas. If the set is not consistent we have, for some $m \geq 0, n \geq 1$:

$$\vdash_{\mathbf{KFi}} (E_1 \wedge \ldots \wedge E_m) \to (\neg(\Box A_1 \vee \Box B_1) \vee \ldots \vee \neg(\Box A_n \vee \Box B_n)),$$

where for each E_i, $\Box E_i \in w$ and for each $\Box A_j \vee \Box B_j$, $\Box(A_j \vee B_j) \in x$. (We can require $n \geq 1$ here, and hence avoid the need to appeal below to **Fi**$_0$ – alias $\Box \Diamond \top$ – since, as recalled in the proof of Lemma 4.5.11, there are plenty of formulas $\Box(A_j \vee B_j) \in x$.) Thus:

$$\vdash_{\mathbf{KFi}} (E_1 \wedge \ldots \wedge E_m) \to ((\Diamond \neg A_1 \wedge \Diamond \neg B_1) \vee \ldots \vee (\Diamond \neg A_n \wedge \Diamond \neg B_n)),$$

and so, appealing to Proposition 1.2.6, $(E_1 \wedge \ldots \wedge E_m)$ provably implies (in **KFi**) the conjunction of the various formulas

$$\Diamond C_1 \vee \ldots \vee \Diamond C_n$$

for all choices of C_i as either $\neg A_i$ or $\neg B_i$, whose conjuncts we can rewrite as

$$\Diamond(C_1 \vee \ldots \vee C_n).$$

By normality, $(\Box E_1 \wedge \ldots \wedge \Box E_m)$ then provably implies the necessitation of this conjunction, and therefore, distributing the $\Box$ over $\wedge$, provably implying the necessitation,

$$\Box \Diamond (C_1 \vee \ldots \vee C_n),$$

of each of its conjuncts. For a suitable choice of **Fi**$_m$ ($m \geq n \geq 1$), we have this conjunction implying the result of prefixing "$\Box \Diamond$" to the conjunction of these formulas $C_1 \vee \ldots \vee C_n$. By Proposition 1.2.6 again, however, this conjunction is truth-functionally equivalent to $(\neg A \wedge \neg B) \vee \ldots \vee (\neg A_n \wedge \neg B_n)$, so we have:

$$\vdash_{\mathbf{KFi}} (\Box E_1 \wedge \ldots \wedge \Box E_m) \to \Box \Diamond ((\neg A \wedge \neg B) \vee \ldots \vee (\neg A_n \wedge \neg B_n)), \text{ and thus:}$$

$$\vdash_{\mathbf{KFi}} (\Box E_1 \wedge \ldots \wedge \Box E_m) \to \Box(\Diamond(\neg A \wedge \neg B) \vee \ldots \vee \Diamond(\neg A_n \wedge \neg B_n))$$

Since each of $\Box E_1, \ldots, \Box E_m \in w$, we have the consequent of the above implication in x, and therefore at least one of its disjuncts. But we cannot have $\Diamond(\neg A_i \wedge \neg B_i) \in x$, as, by the choice of these A_i, B_i, we have $\Box(A_i \vee B_i) \in x$. ■

The most obvious difference to this result that would be made by adding $\Box \Diamond \top$, as we shall from now on spell out this formula (rather than referring to it as **Fi**$_0$) as a further axiom to **KFi** is that its proof delivers us a class of frames satisfying the condition mentioned in Theorem 4.5.12 and the further condition that for all $w \in W$ and $x \in R(w)$, $R(x) \neq \varnothing$. But a little thought shows that we are entitled also to the following reformulation, more convenient for extracting a tacit accessibility relation for $\Box \Diamond$ considered as a $\Box$ operator (which of course, given present concerns, we might choose to write as "O") as we shall below:

COROLLARY 4.5.13 **KFi** $\oplus \Box \Diamond \top$ *is determined by the class of frames* $\langle W, R \rangle$ *satisfying: for all* $w \in W$, *for all* $x \in R(w)$ *there exists* $y \in R(w)$ *such that* $|R(y)| = 1$ *and* $R(y) \subseteq R(x)$.

We are now in a position to discern our hidden accessibility relation, which again we shall call S, defined as follows, for all $w, z \in W$:

$$wSz \text{ if and only if for some } y, wRy \text{ and } R(y) = \{z\}.$$

4.5. DEONTIC LOGIC: MORE TRANSLATIONS, MORE ISSUES 285

EXERCISE 4.5.14 Show that if $\mathcal{M} = \langle W, R, V \rangle$ is any model on a frame meeting the condition described in Coro. 4.5.13, then with S as just defined, we have for all formulas A and all $w \in W$: $\mathcal{M} \models_w \Box\Diamond A$ iff for all z such that wSz, $\mathcal{M} \models_z A$. (Note: the frame condition is only needed for the 'if' direction here.)◮

We shall move on presently in the direction of **S4M**, which extends **KFi** $\oplus$ $\Box\Diamond\top$ and in which ($= \Box\Diamond p \to \Diamond\Box p$) is $\mathbf{D}_O$ when O is taken as $\Box\Diamond$. But first let us pause here to remark on the role of **KFi** in Fine [289].

EXAMPLE 4.5.15 **KFi** figures in [289] as an example, or more precisely, as a counterexample to the converse of a general result there proved to the effect that every complete normal modal logic with a first-order definable class of frames is canonical (i.e., recalling the definition on p. 82, all of its theorems are valid on its canonical frame). In particular, Fine shows ([289], pp. 28–30) that while the canonical frame for **KFi** validates **Fi** and is therefore a frame for this logic (Thm. 4.5.12 here), the class of all frames for **KFi** is not first-order definable. (The proof of the latter claim involves concepts and results beyond the scope of the present work. We have also oversimplified the result itself here, since Fine considers canonical frames for versions of a different logic in which the set of propositional variables has different cardinalities, and also works with a generalization of first-order definability.) Rather than working with **Fi** as formulated above – a formulation directly capturing the aggregativity of $\Box\Diamond$ in this logic, Fine works with the axiom:

$$\Diamond\Box p \to ((\Diamond\Box(p \land q) \lor \Diamond\Box(p \land \neg q)),$$

which it will be more convenient to consider in its dual form (highlighting the role of $\Box\Diamond$ as in our discussion):

$$((\Box\Diamond(p \lor q) \land \Box\Diamond(p \lor \neg q)) \to \Box\Diamond p.$$

To derive this from **Fi** above, $(\Box\Diamond p \land \Box\Diamond q) \to \Box\Diamond(p \land q)$, substitute $p \lor q$ and $p \lor \neg q$ for p and q respectively, and simplify the resulting consequent, $\Box\Diamond((p \lor q) \land (p \lor \neg q))$ to the equivalent $\Box\Diamond p$. To derive **Fi**, as just recalled, from the new axiom, substitute $p \land q$ for p, and $p \land \neg q$ for q in the latter, giving:

$$\Box\Diamond((p \land q) \lor (p \land \neg q)) \land \Box\Diamond((p \land q) \lor \neg(p \land \neg q)) \to \Box\Diamond(p \land q).$$

Then note that $\Box\Diamond p$ provably implies (indeed, is provably equivalent to) the first conjunct of the antecedent of this formula, while $\Box\Diamond q$ provably implies the second conjunct, so the conjunction of $\Box\Diamond p$ with $\Box\Diamond q$ provably implies the consequent of the above implicational formula. (Here we make tacit use of the fact – a consequence of Prop. 1.4.2 on p. 31 – that affirmative modalities such as $\Box\Diamond$ are monotone in all normal modal logics.) There is more of interest concerning **KFi** in Venema [1128], in which it **Fi** is taken in the form:

$$\Diamond\Box(p \lor q) \leftrightarrow (\Diamond\Box p \lor \Diamond\Box q),$$

the backward half direction for any normal modal logic in view of the monotonicity considerations just parenthetically recalled, and arbitrary candidate axioms replacing $\Diamond\Box(\cdot)$ with other contexts for which the axiom lays down a similar 'distribution across disjunction' (or 'additivity') property are shown to yield canonical logics. ◭

First, let us note that **KFi** $\oplus$ $\Box\Diamond\top$ (indeed **KFi** itself) has amongst its theorems the following disjunction: $\Box\bot \lor (\Box\Diamond p \to \Diamond\Box p)$. For an informal sketch of the proof, begin by assuming the negation of this distinction, which means that we have (1) $\Diamond\top$, (2) $\Box\Diamond p$, and (3) $\Box\Diamond\neg p$. From (2) and (3) by **Fi** we have $\Box\Diamond(p \land \neg p)$, or more concisely: $\Box\Diamond\bot$. Since in any normal modal logic this last formula is equivalent to $\Box\bot$ (since $\Diamond\bot$ is equivalent to $\bot$), which contradicts (1). So if we strengthen $\Box\Diamond\top$ to $\Diamond\top$

(alias D) we have the negation of the first disjunct of our disjunction $\Box\bot \vee (\Box\Diamond p \to \Diamond\Box p)$ as a theorem, showing that the second disjunct is provable in the extension in question, **KDFi**.[234] Thus $\vdash_{\mathbf{KDFi}}$ **M**. In fact we can strengthen this observation, to, as in part (i) of the following, with part (ii) recording providing a semantic description for **KDFi** and **KTFi**.

PROPOSITION 4.5.16 (i) **KDFi** *is the smallest normal modal logic in which* $\Box\Diamond$ *is normal and* **M** *is provable.*

(ii) **KDFi** *is determined by the class of serial frames satisfying the condition mentioned in Coro. 4.5.13, and* **KTFi** *is determined by the class of reflexive such frames, for which the condition can be reformulated as: for all* $w \in W$, *for all* $x \in R(w)$ *there exists* $y \in R(w)$ *such that* $R(y) = \{y\}$ *and* $y \in R(x)$.

Proof. (i): We know already that $\Box\Diamond$ is normal in any normal extension of $\mathbf{KFi} \oplus \Box\Diamond\top$, such as **KDFi**, and have just seen that **M** is provable in this logic. To justify the "smallest" in the above formulation, we add to the result of Exc. 4.5.10 the observation that $\vdash_{\mathbf{KM}}$ **D**, something which is especially clear when **M** is rewritten in the **K**-equivalent form $\Diamond(\Box p \vee \Box\neg p)$, since $\vdash_{\mathbf{K}} \Diamond A \to \Diamond\top$ for any formula A. (Here we take A as $\Box p \vee \Box\neg p$.)

(ii) Just using the canonical frame in the case of each logic mentioned, and for the case of **KTFi**, simplifying the condition that $|R(y)| = 1$, i.e., that for some $z \in W$, $R(y) = \{z\}$, to "$R(y) = \{y\}$", since reflexivity already demands that $y \in R(y)$, leaving y as the only candidate z. ∎

It is now time to see how McKinsey's axiom **M** gets into the discussion. According to Prop. 4.5.16(ii), **KDFi** is determined by the class of serial frames $\langle W, R \rangle$ such that: for all $w \in W$, for all $x \in R(w)$ there exists $y \in R(w)$ such that $|R(y)| = 1$ and $R(y) \subseteq R(x)$. Taking any $w \in W$ in such a frame, by seriality there exists at least one $x \in R(w)$, and so at least one such x for which $y \in R(w)$ with $|R(y)| = 1$ and $R(y) \subseteq R(x)$. Discarding the parts about x here, note that this implies that every point w has some successor y which in turn has exactly one successor (which in the reflexive case is accordingly y itself) – a condition on frames interestingly connected with the axiom **M**, which is **K**-equivalent to $\mathbf{M}_1$ in the sequence of formulas[235] whose n^{th} member is:

$\mathbf{M}_n \quad \Diamond((\Diamond p_1 \to \Box p_1) \wedge \ldots (\Diamond p_n \to \Box p_n)).$

Lemmon and Scott (pp. 74–76) establish the following:

THEOREM 4.5.17 (i) *The smallest normal modal logic containing all the* $\mathbf{M}_n$ *is determined by the class of frames* $\langle W, R \rangle$ *in which for each* $w \in W$ *there exists* $y \in R(w)$ *with* $|R(y)| = 1$.

(ii) $\vdash_{\mathbf{K4M}_i} \mathbf{M}_{i+1}$ *for all* i ($i \geq 1$) *and* **K4M** ($= \mathbf{K4M}_1$) *is determined by the class of transitive frames satisfying the condition in* (i).

(iii) **S4M** *is determined by the class of reflexive transitive frames satisfying the condition in* (i), *or equivalently – in virtue of reflexivity – by the class of reflexive transitive frames* $\langle W, R \rangle$ *such that for each* $w \in W$ *there exists* $y \in R(w)$ *with* $R(y) = \{y\}$.

The completeness claims here are all established by verifying that the canonical frame of the logic in question satisfies the cited condition, using the first claim under (ii) (as well as part (i)) to establish the second claim (and to establish (iii)). The situation with **KM** itself is considerably less straightforward, since **M** is not valid on **KM**'s canonical frame (Goldblatt [392]), – this is not a canonical modal logic –

---

[234]We no longer need to add "$\oplus \Box\Diamond\top$", since we have $\Box\Diamond\top$, by Necessitation, from **D**. Indeed by analogy with the use of the label "$\mathbf{B}^\Box$" in pp. 98–104, we could have decided to call this formula – formerly $\mathbf{Fi}_0$ – "$\mathbf{D}^\Box$", and indeed called it precisely this in the discussion including Exc. 2.10.6 on p. 157 above.

[235]Compare the sequence of formulas $\mathbf{T}_n^\Diamond$ from p. 85; the present sequence is from Lemmon and Scott [705], p. 74.

4.5. DEONTIC LOGIC: MORE TRANSLATIONS, MORE ISSUES

but the logic is nonetheless determined by the class of all its frames[236] – and even has the finite model property[237] – though that class itself is not first-order definable. (The Kripke completeness of **KM** was shown in Fine [291]. Further details and references may be found in Goldblatt and Hodkinkson [398].)

EXERCISE 4.5.18 (*i*) Show by sketching the deductions involved (from **M** to **Fi** and back, against the background of **S4**) that **S4M = S4Fi**.
(*ii*) Is it the case that **KD4M = KD4Fi**? (Justify your answer.) ✠

The route to **S4M** which goes via **Fi** is useful in assuring us that $\Box\Diamond$ is normal in this logic (since plainly $\Box\Diamond\top$ is available in **S4**). Recall that the interest of **S4M** in the present context is that this logic gives the modality $\Box\Diamond$ a somewhat similar status *qua* candidate deontic O as **S4.2** gives to $\Diamond\Box$, as was observed in Kielkopf [653],[238] where the interest is not in normality in the present sense, but in the Anderson 'normal deontic logic' sense – see note 220, p. 254. But satisfying the conditions there cited does not guarantee that one has a plausible deontic logic on one's hands, as we proceed to show with the present candidate.

To that end, let us ask about the behaviour of a would-be O extracted from **S4M** by defining O as $\Box\Diamond$. Recall the S defined in terms of a frame $\langle W, R \rangle$ as for Exercise 4.5.14 above, by (for all $w, z \in W$):

$$wSz \text{ if and only if for some } y, wRy \text{ and } R(y) = \{z\}.$$

Given that R is reflexive and transitive, as we may assume for a class of frames determining any extension of **S4**, this can be simplified in the **S4M** setting to:

$$wSz \text{ if and only if for some } wRz \text{ and } R(z) = \{z\}.$$

Thus the S-accessible points to w are just the reflexive end-point R-successors of w, so the point-generated subframes are either as in Figure 4.3, with the generating point at the top, or else consist of a single reflexive point (like those at the bottom). Thus in particular these frames are all serial and the logic determined by them is an extension of **KD**. The two sets of ellipsis dots in Fig. 4.3 are meant to indicate that there is no fixed cardinality to the set of reflexive end-point successors of the top point, and the relation here is of course S, not R, which is why the top point is not reflexive (since although for $w \in W$ in a frame $\langle W, R \rangle$ for **S4M** we have wRw, only if $R(w) = \{w\}$ do we have wSw).

Letting $\mathbf{T!}^\Box$ be the formula $\Box(\Box p \leftrightarrow p)$, the logic determined by the class of frames just described is evidently $\mathbf{KDT!}^\Box$ (i.e., $\mathbf{KD} \oplus \mathbf{T!}^\Box$). This accordingly gives the $\Box\Diamond$-fragment of **S4M**:

THEOREM 4.5.19 *Where $\tau_{\Box\Diamond}$ is the definitional translation which replaces $\Box$ with $\Box\Diamond$, we have, for all formulas A:*

$$\vdash_{\mathbf{KDT!}^\Box} A \text{ if and only if } \vdash_{\mathbf{S4M}} \tau_{\Box\Diamond}(A).$$

Now, while one half of the strict equivalence, $\mathbf{T!}^\Box$, namely $\mathbf{U}$: $\Box(\Box p \to p)$, is a well-regarded deontic principle, the other half $\Box(p \to \Box p)$ has no plausibility when $\Box$ (= the $\Box\Diamond$ of **S4M**) is interpreted as O: why ought it to be the case that whatever is the case *ought to* be the case?[239] Not only does the proposed deontic logic contain $O(p \to Op)$, it also contains $O(Pp \to Op)$: it ought to be that either $O\neg p$ or Op, in other words.

---
[236]It is thus Kripke-complete, in the sense explained on p. 82.
[237]See p. 73.
[238]In [653], **S4M** is referred to as $K1$. This is part of an older nomenclature used by Sobociński and his students, which has no connection with that taken from Lemmon and Scott [705], Segerberg [1025], Chellas [171], etc., which is used here.
[239]Prior [917], p. 57, objects to this principle, using the phrase "excessive rigorism", and observes that it implies that it implies that it ought to be that no acts are morally indifferent. (A typographical error has "no acts" appearing as "not acts".) By the phrase "modally indifferent" Prior means: permissible but not obligatory; deontic $\mathbf{U}$ is being assumed here. One danger of using the phrase in this sense – as the deontic analogue of contingency – was mentioned in the Digression on p. 255.

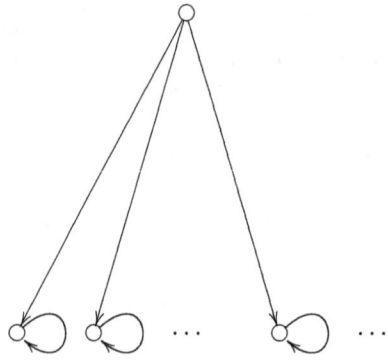

Figure 4.3: Point-generated Frames for Kielkopf

Just because **M**, which on Kielkopf's project here provided **D**$_O$, delivered the untoward results just noted for □◇ when added to **S4**, there is no reason to employ an off-the-shelf candidate such as **S4M**. Perhaps a tailor-made product will have better prospects. We want □◇ (alias O with the above translation in mind) to be normal, so we need at least **KFi** ⊕ □◇⊤, but since we also want **M**, Proposition 4.5.16(i) suggests the following line of enquiry:

EXERCISE 4.5.20 Investigate the credentials of □◇ in **KDFi** as a suitable deontic O, providing an axiomatic description, if possible, of **KDFi**(□◇), to use Zolin's embedding notation, explained in Example 4.4.27 (p. 266). ✠

REMARK 4.5.21 As a matter of fact, we have already met with a close relation of the logic, **KDT!**$^□$, figuring in Theorem 4.5.19, namely **KT!**$^□$, determined by the class of all frames of the kind depicted in Figure 4.3 understood, instead, as permitting no reflexive end-points at all: in other words, containing also the one-point irreflexive frame. We encountered this logic under another description in Exercise 2.1.13 (p. 44), which asked for an axiomatization of the smallest normal modal logic with the Outermost Reduction Property, i.e., in which every affirmative modality of length ≥ 1 was equivalent to its outermost (= leftmost) modal operator (□ or ◇). The expected answer – by analogy with the case of the Innermost Reduction Property, the normal modal logics possessing which were seen in Prop. 2.1.11 to be precisely the extensions of **K5!** – was as the logic **K** ⊕ □◇p ↔ □p, or with the dual axiom ◇□p ↔ ◇p instead, if preferred. (This is most easily shown by reasoning as in the Digression on p. 43, except that instead of beginning with (1) and (2) as there, we begin with (1)′: $ab = a$, and (2)′: $ba = b$, to get to the conclusion that $a^2 = a$ and $b^2 = b$, completing the argument inductively as in the proof of Prop. 2.1.11.) ◀

Digression. Fitting [309] describes an translation from the language of classical logic (in fact classical predicate logic rather than classical propositional logic) to that of monomodal logic which proceeds by prefixing □◇ to every subformula (including atomic subsformulas) of the formula to be translated, and shows that this embeds the source logic faithfully in **S4** (a predicate-logical version thereof, in fact), by an ingenious model-theoretic argument. (Note that, unlike $\tau_{□◇}$, this is not a definitional translation.) But unless one is particularly interested in this translation for its own sake – and Fitting does mention some connections with 'forcing' in set theory – it is not easy to see the point of this, since the same could be said of the identity translation which simply translates every formula as itself. Numerous variations on this theme of Fitting are explored in Czermak [227], [228]. **End of Digression.**

4.5. DEONTIC LOGIC: MORE TRANSLATIONS, MORE ISSUES

EXERCISE 4.5.22 Show syntactically (i.e., by giving suitable derivations) that the logic mentioned in Remark 4.5.21, $\mathbf{K} \oplus \Box\Diamond p \leftrightarrow \Box p$, is in fact none other than $\mathbf{KT!}^{\Box}$. (The easy direction is proving $\Box\Diamond p \leftrightarrow \Box p$ in $\mathbf{KT!}^{\Box}$. A hint for the harder – converse – direction, which we may think of as requiring the derivation of $\Box(\Box p \to p)$, alias $\mathbf{U}$, and of $\Box(p \to \Box p)$, pertaining to the former derivation: informally, we may derive a contradiction from the hypothesis that $\Diamond(\Box p \land \neg p)$ using $\Box\Diamond p \leftrightarrow \Box p$ and the dual form by reasoning successively: $\Diamond(\Box p \land \neg p)$, $\Diamond\Box(\Box p \land \neg p)$, $\Diamond(\Box\Box p \land \Box\neg p)$, $\Diamond(\Box p \land \Box\neg p)$, $\Diamond\Box(p \land \neg p)$, $\Diamond(p \land \neg p)$. The negation of the last formula in this list is $\mathbf{K}$-provable, and the earlier transitions are either routine $\mathbf{K}$ moves or justified by the equivalence in the present logic of the modalities $\Diamond\Box$ with $\Diamond$, or else that of $\Box\Box$ with $\Box$.) ✠

The semantic side of the picture is rather interesting, as we illustrate with the following explanation as to why any frame validating $\Diamond\Box p \leftrightarrow \Diamond p$ validates $\mathbf{T!}^{\Box}$:

EXAMPLE 4.5.23 First, note that a frame $\langle W, R \rangle$ validates the $\to$ direction of the formula $\Diamond\Box p \leftrightarrow \Diamond p$ iff $\langle W, R \rangle$ satisfies the condition (1), validates the $\leftarrow$ direction of this formula iff the frame satisfies the condition (2), and validates $\mathbf{T!}^{\Box}$ iff it satisfies the condition (3); in each case understand the conditions as prefixed by "for all $x, y \in W$":

$$xRy \Rightarrow R(x) \cap R(y) \neq \varnothing \tag{1}$$

$$xRy \Rightarrow \text{ for some } z \in W : xRz \,\&\, R(z) \subseteq \{y\} \tag{2}$$

$$xRy \Rightarrow R(y) = \{y\}. \tag{3}$$

Thus it remains only to show that (1) and (2) have (3) as a consequence. So suppose that $\langle W, R \rangle$ satisfies (1) and (2) and that we have $x, y \in W$ with xRy with a view to concluding that $R(y) = \{y\}$, thereby showing that (3) is satisfied. By (2), since xRy, x has an R-successor z which has at most y as a successor. Since xRz, x and z have a common successor, by (1) (instantiating y as z), so $R(z) \neq \varnothing$ and thus $R(z) = \{y\}$. Since zRy, by (1) again z and y have a common successor, so since the only successor of z is y, this common successor must be y itself, and therefore yRy. Also, since zRy, there must by (2) be some z' accessible to z with $R(z') \subseteq \{y\}$. But $R(z) = \{y\}$, so this z' must be y itself and accordingly $R(y) \subseteq \{y\}$. As we have already established that yRy, we conclude, as required to establish (3), that $R(y) = \{y\}$, completing the argument. (Note further, however, that since this shows an arbitrary $y \in R(x)$ in $\langle W, R \rangle$ satisfying (1) and (2), $R(y) = \{y\}$, we also have, for the z just under discussion, $R(z) = \{z\}$; but since, for the particular y under discussion, $R(z) = \{y\}$, we can in fact conclude that $z = y$.) ◀

REMARK 4.5.24 The logic featuring in Theorem 4.5.19, $\mathbf{KDT!}^{\Box}$, appears under a different name and in a very different light in Fine [297] as an alethic modal logic there argued to be implicit in a passage in Aristotle, as well as being motivated by certain considerations of natural language semantics. Among the latter is the fact that, according to Fine [297], p. 1011,

> "Possibly, Pam might be drunk and Quentin might be drunk"

taken as having the form $\Diamond(\Diamond p \land \Diamond q)$, is naturally interpreted as requiring the existence (accessibility) of a world in which both Pam and Quention are drunk, and thus as equivalent to the corresponding $\Diamond(p \land q)$ rather than as equivalent (as in $\mathbf{S4}$, $\mathbf{S5}$) to the quite different first-degree formula $\Diamond p \land \Diamond q$. A glance at Figure 4.3 reveals why the equivalence obtains for the present logic: once we are under the scope of a $\Diamond$, we are directed to the accessible points, at which $\Diamond$ is equivalent to the null modality. But, by way of objection to this line of defence, they are also equivalent to $\Box$, and there is no tendency to express the $\Diamond(p \land q)$ thought with a sentence like that inset above except with a modal auxiliary dual to "might" – "must", perhaps – replacing it, the equivalence of $\Diamond(p \land q)$ with $\Diamond(\Box p \land \Box q)$ in the present logic notwithstanding. (There is also the fact that the presumably desirable – for alethic purposes – principle $\mathbf{T}$ is not forthcoming in $\mathbf{KDT!}^{\Box}$; Fine [297] does address this, however.) ◀

We conclude with a return to deontically motivated concerns, though the embedding suggested here will occupy us more fully in the discussion of doxastic logic (especially in Section 5.6). This is because it offers a response to certain worries about concept-possession which arise in epistemic and doxastic logic but not in deontic logic, Op not being intended to represent the ascription of a propositional attitude; see the discussion after Table 4.1 (p. 204). As that table indicates, there is also a conspicuous respect in which the doxastic and deontic cases resemble each other, presented there in the division into rows rather than that into columns (the latter guiding our distribution of the material across this chapter and the next): the fact that the $\Box$ operator in question is not veridical: beliefs can be mistaken and obligations unfulfilled. In fact the latter issue can be put more broadly: it is not just that there is no implication from $\Box p$ to p in a plausible doxastic or deontic logic, but a complete lack of *any* logical relations between the two. In the following section we will make this more precise, describing it in terms of the phenomenon of fully modalized logics. But we conclude the present section with the promised alternative embedding.

EXAMPLE 4.5.25 Dawson extracted a reasonable deontic logic from **S4.2** by interpreting O as $\Diamond\Box$, and as we saw the resulting logic was **KD45**. (Much was later made of this embedding, with a doxastic interpretation of $\Diamond\Box$ and an epistemic interpretation of $\Box$ in Lenzen [706], as we shall see in the following chapter: Theorem 5.2.1 on p. 372, and Section 5.5, for instance.) Kielkopf, as we have been noting, considered another extension of **S4**, namely **S4M**, and deemed it, too, to conceal a reasonable deontic logic – though, as we saw (after Theorem 4.5.19), his enthusiasm on this score seems to have been misplaced – with O disguised, this time, as $\Box\Diamond$. In the case of Dawson's embedding, another proper extension of **S4** had been pertinent, namely, as just recalled, **S4.2**. But Åqvist [27] noticed that we do not need to go beyond **S4** itself in order to simulate a deontic O. Just glance at Figure 2.1 on p. 41, to see how $\Box\Diamond\Box$ appears there, neither implying nor being implied by the null modality, implying (but not being implied by) its own dual, $\Diamond\Box\Diamond$, giving us **D** (without $\mathbf{D}_c$), and most importantly of all, making O normal.[240] (The fact that $\Box\Diamond\Box$ is normal in **S4** was noted in our discussion of Matsumoto's embedding: Proposition 2.8.15, p. 148.) We defer the precise identification of what in the notation of Zolin [1226] – see Example 4.4.27 (p. 266) – would be called **S4($\Box\Diamond\Box$)** – the logic of $\Box\Diamond\Box$ (treated as an unstructured operator) as it behaves according to **S4**, to the following chapter, where it will come up in connection with doxastic and epistemic logic: see Theorem 5.5.24 (p. 419), as well as the further discussion in Section 5.6. ◀

4.6 Logics Which Are Fully Modalized

> In the systems K and D [= **K** and **KD**] we have no connexion between modalised and ordinary unmodalised formulae.
> Lucas [744], p. 24

The notion of a fully modalized formula was defined on p. 56, but here, following Zolin [1226], we adapt it to apply it to modal logics themselves. In fact Zolin just uses the term *modalized* for both applications, and more generally says that a formula A is *modalized in* the variable p_i if every occurrence of p_i in A lies within the scope of some occurrence of $\Box$; thus a (fully) modalized formula A is a formula modalized in every propositional variable occurring in A. Since a modal logic is a set of formulas – and note that here we are considering arbitrary modal logics and not just (for instance) normal modal logics – one might think to extend the definition by counting a modal logic, or any other set of formulas, as fully modalized just in case every formula in the set is fully modalized. But no modal logic would on this definition be fully modalized, or f.m. as we shall say for short here, since $p \to p$ is not f.m. but belongs to every modal logic; likewise, whenever A is provable in such a logic, even if A is f.m., then also provable is $A \vee p$, which

---

[240]Of course a modality, not being a formula, does not literally speaking (provably) imply anything, and when we say that a modality X implies a modality Y in a modal logic S, when is meant is that Xp provably implies Yp in S – or equivalently, that for all formulas A, $\vdash_S XA \to YA$.

4.6. LOGICS WHICH ARE FULLY MODALIZED

is certainly not f.m. The intuitive idea behind calling a modal logic fully modalized (or modalized, in Zolin's terminology) is that such derivatively non-f.m. theorems are the only theorems which are not f.m. Thus we want, for example, **K** to count as an f.m. logic but not **KT**, since the latter contains $\Box p \to p$, forging a characteristic logical link between a modal and a non-modal formula which cannot be regarded as derivative (by contrast with the case of the **K** theorem $p \to \Box\top$, for instance, in which the identity of the particular non-modal formula is irrelevant).

It seems that one would not want **T** in a modal logic intended for doxastic or deontic interpretation – $\mathbf{T}_B$ or $\mathbf{T}_O$, as we have been putting it (though see Examples 4.6.15, p. 299 for some qualms in the latter case) – and indeed seriously proposed doxastic and deontic logics tend to be f.m., in the sense just explained: they do not postulate *any* substantive logical relations between how things are believed to be and how things are, in the one case, or between how things ought to be and how things are, in the other. That would of course be a very hazy formulation, since one aspect of how things are is how they ought to be or are believed to be. What we don't expect, rather, is the forging of any such logical connections between $\Box A$ and A for any given A – other than those which hold, as it was put above, derivatively. Let us look at trying to make this idea more precise, with Zolin's definition of an f.m. logic. The definition is fairly complex and simplification will follow in due course.[241]

Before quoting the passage from Zolin ([1226], Definition 5.4, p. 875), defining the property of current interest, an explanation of some of the notation (introduced at p. 864 of [1226]) is called for. Zolin calls the truth-values T and F, $\top$ and $\bot$ respectively, writing **2** for $\{\top, \bot\}$, as well as using these symbols (as here) as nullary connectives in the object language, and uses them also in superscript position in accordance with the convention that $p_i^\top$ is p_i and $p_i^\bot$ is $\neg p_i$. σ_i is either $\top$ or $\bot$, with $\vec{p}$ being the sequence $(p_1, \ldots, p_n)$, $\vec{\sigma}$ for a sequence $(\sigma_1, \ldots, \sigma_n)$, and $\vec{p}^{\vec{\sigma}}$ for the formula $p_1^{\sigma_1} \wedge \ldots \wedge p_n^{\sigma_n}$. Finally, in the following quotation, **Fm** denotes the set of formulas of the language:[242]

> A formula A is modalized in p if every occurrence of the variable p is in the scope of $\Box$. In particular, if p does not occur in A then A is modalized in p. A is called *modalized* if it is modalized in every variable; in other words, if A is a truth-functional compound of formulas of the form $\Box F$. If $\vec{p} = (p_1, \ldots, p_n)$ is the list of all variables in which A is not modalized then A is truth-functionally equivalent to a decomposition w.r.t. $\vec{p}$ of the form [on the right here]:
>
> $$A \leftrightarrow \bigvee_{\vec{\sigma} \in \mathbf{2}^n} (\vec{p}^{\vec{\sigma}} \wedge B_{\vec{\sigma}}), \qquad (\star)$$
>
> where $B_{\vec{\sigma}}$ are modalized formulas.
>
> A logic S is *modalized* if, for all $A \in \mathbf{Fm}$, $\vdash_S A$ implies $\vdash_S B_{\vec{\sigma}}$, for all $\vec{\sigma} \in \mathbf{2}^n$, where $B_{\vec{\sigma}}$ are taken from the decomposition $(\star)$ of A. To put it in another way, S is modalized if it does not prove any nontrivial truth-functional combination of variables and modalized formulas.

To give a simple example, suppose that A is $\Box p \vee (p \wedge \Box \Diamond (p \vee q))$ (where, $\Diamond$, as usual, abbreviates $\neg \Box \neg$); $\Box p$ and $\Box \Diamond(p \vee q)$ are $\Box$-formulas not themselves within the scope of an occurrence of $\Box$ (the formulas $\Box F$ in the above passage from Zolin), and we construct a truth-table – Table 4.2 – treating these maximal $\Box$-subformulas as though they were simply new propositional variables (sentence letters) alongside p itself (like q and r: the table reflects this alphabetic ordering, later variables oscillating more rapidly in truth-value).

We now read off a decomposition (à la $(\star)$) w.r.t. p of the sample formula A – in which p is the only variable occurring unmodalized – by looking at the lines in which A is true when p is true (lines 1, 2, and 3) and then the lines in which the formula A is true when p is false (lines 5 and 6). For brevity,

---

[241]This is not to say Zolin's definition was not appropriate in the setting of [1226], since he there goes on to make other uses of the Boolean decompositions figuring in the definition.

[242]This quotation has been adjusted to match our notation and spelling in some respects (e.g., putting "modalized" for "modalised"), though Zolin's use of "p" as a variable over propositional variables (rather than for a specific such variable) has been retained, and likewise with the use of $p_1, \ldots, p_n$. Zolin uses "M" where this has been replaced here by "S", and writes "$M \vdash A$", etc., rather than "$\vdash_S A$".

$\Box p$	$\vee$	$(p$	$\wedge$	$\Box\Diamond(p \vee q))$
T	T	T	T	T
T	T	T	F	F
F	T	T	T	T
F	F	T	F	F
T	T	F	F	T
T	T	F	F	F
F	F	F	F	T
F	F	F	F	F

Table 4.2: A Simple Zolin Truth-table

we denote the maximal $\Box$-subformulas of A,[243] $\Box p$ and $\Box\Diamond(p \vee q)$, by M_1 and M_2 respectively ("M" suggesting *modalized*):

$$A \leftrightarrow [p \wedge ((M_1 \wedge M_2) \vee (M_1 \wedge \neg M_2) \vee (\neg M_1 \wedge M_2))] \vee [\neg p \wedge ((M_1 \wedge M_2) \vee (M_1 \wedge \neg M_2))]$$

The right-hand side here could be greatly simplified, but the point of setting it out in full like this is to illustrate the general situation, even though we have done so in the special case in which the sequence $\vec{\sigma}$ only has a length of 1, so the second conjunct $B_{\vec{\sigma}}$ of Zolin's $(\star)$ is simply $(M_1 \wedge M_2) \vee (M_1 \wedge \neg M_2) \vee (\neg M_1 \wedge M_2)$ for $\vec{\sigma} = \langle \text{T} \rangle$ (or $\langle \top \rangle$, in Zolin's notation) and in the $\vec{\sigma} = \langle \text{F} \rangle$ case is simply $(M_1 \wedge M_2) \vee (M_1 \wedge \neg M_2)$. Thus for S to be f.m. (or 'modalized') where $\vdash_S A$, we must have:

$$\vdash_S (M_1 \wedge M_2) \vee (M_1 \wedge \neg M_2) \vee (\neg M_1 \wedge M_2) \text{ and } \vdash_S (M_1 \wedge M_2) \vee (M_1 \wedge \neg M_2).$$

Of course, there is the analogous demand for every other S-provable formula which must be satisfied if S is to count as f.m. under the present definition. It is worth noting that instead of formulating the r.h.s. of $(\star)$ as a disjunction, we could equivalently use a conjunctive formulation:

$$A \leftrightarrow \bigwedge_{\vec{\sigma} \in 2^n} (\vec{p}^{\vec{\sigma}} \to B_{\vec{\sigma}}), \qquad (\star\star)$$

In this case S's being f.m. requires for each S-provable A that the various $B_{\vec{\sigma}}$ should be provable outright rather than simply conditionally on $\vec{p}^{\vec{\sigma}}$: any case in which the conditionality was unavoidable would amount to logical dependencies between modal and non-modal formulas of the kind a logic's being fully modalized is intended to rule out.

EXAMPLE 4.6.1 To illustrate the equivalence of $(\star)$ and $(\star\star)$ for the case of a formula A which is not modalized in the variables p and q, let us list the right-hand sides of these equivalences to see that they are in turn equivalent (in any modal logic). The disjuncts of the r.h.s. of $(\star)$ are in the first column and the corresponding conjuncts of the r.h.s. of $(\star\star)$ in the second:

$$(p \wedge q) \wedge B_{\langle \text{T},\text{T} \rangle} \qquad (p \wedge q) \to B_{\langle \text{T},\text{T} \rangle}$$
$$(p \wedge \neg q) \wedge B_{\langle \text{T},\text{F} \rangle} \qquad (p \wedge \neg q) \to B_{\langle \text{T},\text{F} \rangle}$$
$$(\neg p \wedge q) \wedge B_{\langle \text{F},\text{T} \rangle} \qquad (\neg p \wedge q) \to B_{\langle \text{F},\text{T} \rangle}$$
$$(\neg p \wedge \neg q) \wedge B_{\langle \text{F},\text{F} \rangle} \qquad (\neg p \wedge \neg q) \to B_{\langle \text{F},\text{F} \rangle}$$

---

[243]That is, $\Box$-formulas which are not proper subformulas of other $\Box$-subformulas of A. These are called (possibly complex) 'Boolean atoms' in [1032], p. 51, where Segerberg notes that to reduce the appearance of a contradiction in terms here (*complex* with *atom*), one could instead say 'atom from the Boolean point of view' – a formulation quoted already at p. 162 above.

4.6. LOGICS WHICH ARE FULLY MODALIZED

That the disjunction of the formulas in the first column have the conjunction of those in the second column as a tautological consequence follows from the fact that the four 'state descriptions' $\vec{p}^{\vec{\sigma}}$ here are mutually exclusive (in the sense that the negation of the conjunction of any two is a substitution instance of a truth-functional tautology), while the converse implication follows from the fact that they are jointly exhaustive (in the sense that their disjunction is a substitution instance of a truth-functional tautology).

◀

At the end of the passage quoted from Zolin [1226] above, there is a remark worthy of comment, to the effect that "S is modalized if it does not prove any nontrivial truth-functional combination of variables and modalized formulas." If one thought of triviality and non-triviality as absolute properties of formulas, this would invite the following mistaken reaction. The gloss just given implies that if S is f.m. and $S_0 \subseteq S$, then S_0 must be f.m. too. For suppose that S_0 is not f.m., thus proving some "nontrivial combination of variables and modalized formulas"; since $S_0 \subseteq S$, S too will prove this combination of variables and modalized formulas. Since we can have an f.m. extension of a non-f.m. logic – for example **KVer** is an f.m. extension of the non-f.m. **KT**$_c$, something has gone wrong here. And it is not hard to see what: a formula does not constitute a nontrivial combination of propositional variables and modalized formulas *absolutely*, but only *relative to* a logic: $p \to \Box p$ would be a trivial combination relative to **KVer**, for instance, but not relative to **KT**$_c$. The same goes for the wording in terms of merely derivative provability in the opening paragraph of this section. Note that the inconsistent modal logic counts as f.m. because the property of being f.m. requires that certain formulas be provable whenever certain others are, a demand in its case is automatically satisfied since every formula whatever is provable. So in fact we could have used that logic rather than **KVer** as an example of an f.m. extension of a non-f.m. logic.

Let us turn to the promised simplification of the definition of a logic's being f.m., which will have a structural resemblance to the definitions of Halldén completeness[244] and of having the (binary) rule of disjunction.[245] On the new definition, we say that a modal logic S is *fully modalized* (f.m. for short, as usual) if and only if whenever $\vdash_S M \vee N$, with M a fully modalized formula and N a non-modal (i.e., $\Box$-free) formula, we have $\vdash_S M$ or else $\vdash_S N$.

PROPOSITION 4.6.2 *A modal logic is f.m. according to the definition just given if and only if it is f.m. (or 'modalized') according to Zolin's definition, given earlier.*

Proof. Since the inconsistent modal logic is f.m. by both definitions, we may restrict attention in the proof to consistent modal logics.

'If': For a contradiction, suppose that a consistent modal logic S is f.m. according to Zolin's definition but we have M, N, as in the new definition, with $\vdash_S M \vee N$, $\nvdash_S M$, $\nvdash_S N$. The general situation can be treated by means of a sufficiently representative example, so let us suppose that the variables occurring in N are p, q, and that M is a truth-functional compound of maximal $\Box$-formulas M_1, M_2, M_3. This means that a full truth-table in the style of Table 4.2 would have $2^5 = 32$ lines in it, on some of which N is false, since if N is true on every line it is a truth-functional tautology and we should have $\vdash_S N$. By way of example, suppose that $p \wedge \neg q$ describes the assignment to p and q on one such line. (I.e., this conjunction is $\vec{p}^{\vec{\sigma}}$ for $\vec{\sigma} = \langle T, F\rangle$, where as usual we write p_1, p_2, and p, q.) Similarly there are lines on which M is true, because otherwise $\neg M$ is a substitution instance of a tautology and would be provable in S, which since $\vdash_S M \vee N$ would imply $\vdash_S N$; again, for illustration, let us suppose that these lines are as summarised in the disjuncts of:

$$(M_1 \wedge M_2 \wedge \neg M_3) \vee (\neg M_1 \wedge M_2 \wedge M_3),$$

making this disjunction a representation – call it M' – in disjunctive normal form, of the formula M, treating the M_i as though they were new variables. As new variables they are independent of p and q, so

---
[244]See p. 39.
[245]See Section 2.6, beginning on p. 112.

one disjunct of the r.h.s. of Zolin's $(\star)$ will be $(p \wedge \neg q) \wedge M'$, when A is taken as $M \vee N$, because in this line $M \vee N$ is true (since M is). This contradicts the hypothesis that S is f.m. in Zolin's sense, since that hypothesis then requires us to have $\vdash_S M'$ – impossible as M and M' are truth-functionally equivalent and $\nvdash_S M$.

'Only if': Suppose S is not f.m. according to Zolin's definition. Thus for some A, $\vdash_S A$, but for some $B_{\vec{\sigma}}$ as in $(\star\star)$, $\nvdash_S B_{\vec{\sigma}}$, even though $\vdash_S \vec{p}^{\vec{\sigma}} \to B_{\vec{\sigma}}$ (since $\vdash_S A$ and every modal logic contains $(\star\star)$, whose r.h.s. is a conjunction with this implication as a conjunct). We can rewrite the implication concerned in the form $M \vee N$ as $B_{\vec{\sigma}} \vee \neg \vec{p}^{\vec{\sigma}}$, provable in S but for which *ex hypothesi* $\nvdash_S M$, and also $\nvdash_S N$, since N ($= \neg \vec{p}^{\vec{\sigma}}$) is not tautologous and the only non-modal formulas any consistent modal logic proves are the truth-functional tautologies. Thus S is not f.m. by the new definition. ∎

REMARK 4.6.3 Still another characterization of f.m. logics is given (though this terminology is not used) in §9 of Tokarz [1109], which begins by defining the substitutions (we shall call) s_k as those satisfying $s_k(p_i) = p_{k+i}$ for all sentence letters p_i, and then defines mappings w_k from formulas to formulas by setting:

- $w_k(p_i) = p_i$ (all $i \geq 1$);
- $w_k(\#(A_1, \ldots, A_n)) = \#(w_k(A_1), \ldots, w_k(A_n))$, for any n-ary Boolean connective $\#$;
- $w_k(\Box A) = s_k(\Box A)$.

Finally, where $max(A)$ is the largest value of i for which p_i occurs in the formula A, one puts $w(A) = w_{max(A)}(A)$, and then defines a modal logic S to be fully modalized just in case for all formulas A, $\vdash_S A$ if and only if $\vdash_S w(A)$. The work here is being done by the "only if" direction, the converse just being a matter of closure under Uniform Substitution (and so part of the definition of *modal logic*). To illustrate these concepts, suppose A is the formula $\Box(p \wedge r) \to (p \vee \Box\Box r)$. Then $w(A) = w_{max(A)}(A)$ and in the present case $max(A)$ here is 3 (as r is p_3), and we calculate:

$$w_3(\Box(p_1 \wedge p_3) \to (p_1 \vee \Box\Box p_3)) = w_3(\Box(p_1 \wedge p_3)) \to w_3((p_1 \vee \Box\Box p_3))$$
$$= w_3(\Box(p_1 \wedge p_3)) \to w_3((p_1)) \vee w_3(\Box\Box p_3))$$
$$= (\Box(p_4 \wedge p_6)) \to (p_1 \vee \Box\Box p_6)$$

Here we see that the effect of w on a formula is to re-letter all occurrences of propositional variables in the scope of occurrences of "$\Box$" with different occurrences of the same variable re-lettered in the same way, and in such a way (since $w(A)$ was defined as $w_{max(A)}(A)$) that they do not coincide with any of the remaining non-$\Box$-embedded occurrences. If doing this to an S-theorem still yields an S theorem, that the S-provability of the formula concerned cannot be due to any interaction between embedded and unembedded occurrences of the same subformula, and for this to be the case for all theorems of S is for S to be f.m. For example the above $w(A)$ is **K4**-provable because of the appearances of p_6 (and the axioms **4**), just as the original A was, while if we change the $\vee$ in the consequent to $\wedge$ we get an **S4**-provable A with $w(A)$ not **S4**-provable, because by re-lettering p_1 to p_4 we have lost the former effect of the **T** axiom, $\Box p_4 \to p_1$ no longer being of the form $\Box B \to B$. (In Tokarz [1109] **KD45**, there called LB and axiomatized as **KD4!5!**, has this full modalization property. One aspect of the redundancy in this axiomatization was mentioned in Lemma 2.1.10, p. 42: even in the absence of **D** we can derive **4!** from **5!**. The presence of **D** causes a further complication, detailed in the Digression below.) ◀

Digression. Tokarz [1109] considers also a strengthening of his LB – which letters are meant to stand for "Logic of Belief" – boosting the **D** of **KD45** to a **D!**, and points out that the resulting logic, called by him LCB, is tabular, having a 4-element characteristic matrix, or as we would put it in Kripke-semantical

4.6. LOGICS WHICH ARE FULLY MODALIZED

terms, a 2-element characteristic frame consisting of an irreflexive point to which a reflexive point is accessible. Compare the two-element frame of Figure 2.16, in which both points were reflexive, discussed in Example 2.10.2, p. 155. (Note that **KD!45** = **KD!4** = **KD!5**.) Tokarz also attempts to consider also a weakening of *LB*, which he calls *LIB* (for: logic of potentially inconsistent belief), by removing **D** from the axiomatization of *LB* as **KD4!5!**. But by (the correct, affirmative answer to) Exercise 2.7.8(*ii*), p. 131, **D** is already derivable from the 5_c half of **5!**, since any implication of the form $\Box B \to \Diamond C$ provably implies $\Diamond \top$ even in **K** – and 5_c is an implication of this form. (This special case is also the subject of Exercise 5.1.21, p. 360.) So **D** is redundant in the axiomatzation suggested by the label **KD4!5!** and in dropping it, Tokarz has not weakened the logic axiomatized. It is clear from the claims made about it, however, that we should take *LIB* to be **KD45**. *End of Digression.*

In view of Proposition 4.6.2, we can follow some of Zolin's steps in §5 of [1226][246] using the simpler definition of an f.m. modal logic (though we do not touch on his main concern there, which lies with the application of this concept to the question of translational embeddability of one logic in another). Two of these are a simple sufficient semantic condition for a normal logic to be f.m., and the fact that extending an f.m. modal logic by f.m. axioms yields an f.m. logic. After that, food for thought will be provided in the shape of another condition on modal logics that might be an alternative explication of what is typically wanted from modal logics with doxastic or deontic (as opposed to epistemic or alethic) applications – as suggested in the second paragraph of this section.

THEOREM 4.6.4 *Suppose S is a normal modal logic determined by a class of models* $\mathcal{M} = \langle W, R, V \rangle$ *with the property that whenever* $\nvdash_S A$ *there is a point x in such a model such that* $\mathcal{M} \nvDash_x A$ *and for no* $n \geq 1$ *do we have* $xR^n x$. *Then S is f.m.*

Proof. Suppose with S as described, and M fully modalized and N $\Box$-free, $\nvdash_S M$ and $\nvdash_S N$. The latter implies that N is not a truth-functional tautology. So there is a Boolean valuation v with $v(N) =$ F. By the former, there is a model $\mathcal{M} = \langle W, R, V \rangle$ with $\mathcal{M} \nvDash_x M$. Without loss of generality we may take M to be generated by the point x. Now adjust V to V' thus: $V'(p_i) = V(p_i) \cup \{x\}$ if $v(p_i) =$ T and $V'(p_i) = V(p_i) \smallsetminus \{x\}$ if $v(p_i) =$ F. Where $\mathcal{M}' = \langle W, R, V' \rangle$, this makes $v_x^{\mathcal{M}'}$ (as defined on p. 54) coincide with the valuation v, so $\mathcal{M}' \nvDash_x N$. But we still have $\mathcal{M}' \nvDash_x M$ because, since M is fully modalized, its truth-value at x depends only on points R^n-related to x for various $n \geq 1$, which by hypothesis do not include x itself, and on which $\mathcal{M}'$ does not differ from $\mathcal{M}$. ∎

As a corollary we can deduce that **K**, **KD**, **K4**, **K5**, **KD45**, **KVer** and many other normal modal logics are f.m., in the last case, for instance because the class of models on the one-point irreflexive frame evidently satisfies the condition in Theorem 4.6.4. Once we have this result for **K**, though, as in Coro. 4.6.6, we can draw such conclusions for normal and quasi-normal modal logics by appeal to a simple syntactic condition (Prop. 4.6.7). However, it is worth pausing to discuss the case of **KD45** as mentioned affording an instructive illustration:

EXAMPLE 4.6.5 In Coro. 4.1.9 on p. 208, **KD45** was described as being determined by the class of 'semi-simplified' (as we put it) Kripke models $\langle W, W_0, V \rangle$ with $|W \smallsetminus W_0| \leq 1$, with that $W_0 \neq \varnothing$, and it was remarked that we could change "$\leq$" to "$=$" and still have a correct claim. Recall that W_0 is the set of points accessible to any point in W, this being the same for all points in W. In terms of the usual notion of a model, then, we need to show that **KD45** is determined by the class of models $\langle W, R, V \rangle$ such that for some $w \in W$, $W = \{w\} \cup W_0$ for some non-empty W_0 with $w \notin W_0$, and $R = (\{w\} \times W_0) \cup (W_0 \times W_0)$, with any non-theorem of **KD45** being false at the point w in some such model. The only part of the completeness half of such a result that we do not get by considering point-generated submodels of the canonical model for **KD45** is the $w \notin W_0$ part, when w is taken as a point at which some non-theorem

---

[246]In particular, Zolin's Lemmas 5.5 and 5.6 correspond fairly closely to Theorem 4.6.4 and Proposition 4.6.7 below.

is false in that model and $\mathcal{M} = \langle W, R, V \rangle$ is the submodel generated by w: it may happen that R is the universal relation on W rather than as required above. But in that case, we duplicate w, adding a new point w', say, and consider the model $\mathcal{M}' = \langle W', R', V' \rangle$ with $W' = W \cup \{w'\}$, $V'(p_i) = V(p_i) \cup \{w'\}$ if $w \in V(p_i)$, $V'(p_i) = V(p_i)$ otherwise, and $R' = R \cup \{w'\} \times W$. The mapping f from W' to W defined by: $f(x) = x$ for all $x \in W$ and $f(w') = f(w) = w$ for the new point w', is then a p-morphism from $\mathcal{M}'$ to $\mathcal{M}$, allowing us to conclude (by Exercise 2.7.35(i), p. 137) that our non-theorem, false at w in $\mathcal{M}$, is false at w' in $\mathcal{M}'$. So this w' will do as the required w in the formulation "for some $w \in W$, $W = \{w\} \cup W_0$ for some non-empty W_0 with $w \notin W_0$" (with the roles of W and W_0 in this formulation played by W' and W). Since w' is not accessible to any point, the condition of Theorem 4.6.4 is satisfied and we may conclude that **KD45** is f.m. ◀

For the reason given before the above Example, we touch only on the case of **K** in the following corollary to Theorem 4.6.4:

COROLLARY 4.6.6 **K** *is f.m.*

Proof. The unravellings from their generating points of point-generated models (see p. 269) give us the class of models required for Thm. 4.6.4 to apply in the case of **K**. ■

PROPOSITION 4.6.7 *If S is a fully modalized modal logic and Γ is a set of fully modalized formulas, then $S + \Gamma$ and $S \oplus \Gamma$ are fully modalized logics.*

Proof. Suppose S and Γ are as described. Then for all B, $\vdash_{S+\Gamma} B$ iff B is provably implied in S by some conjunction of substitution instances of formulas from Γ; in the case in which B is $M \vee N$, M f.m. and N non-modal, where we need to show that either M or N is $(S + \Gamma)$-provable, we have, for some $A_1, \ldots A_n \in \Gamma$ and some substitutions $s_1, \ldots, s_n$:

$$\vdash_S (s_1(A_1) \wedge \ldots \wedge s_n(A_n)) \to (M \vee N),$$

and thus (by truth-functional reasoning):

$$\vdash_S ((s_1(A_1) \wedge \ldots \wedge s_n(A_n)) \to M) \vee N,$$

in which case we have now an S-provable disjunction whose first disjunct is f.m., as all of the $s_i(A_i))$ are (substitution instances of f.m. formulas being themselves f.m.) and so is M, and whose second disjunct ($= N$) is non-modal. So by the hypothesis that S is f.m., we conclude that one or other of these disjuncts is provable in S, and thus that either M or N is provable in $S + \Gamma$ (in the former case because the $s_i(A_i)$) are provable in $S + \Gamma$, in the latter because N is already provable in S). The argument is similar in the case of $S \oplus \Gamma$, except that here the provability of $M \vee N$ gives us instead that for some $A_1, \ldots A_n \in \Gamma$, substitutions $s_1, \ldots, s_n$, and some choice of $k_1, \ldots, k_n$:

$$\vdash_S (s_1(\Box^{k_1} A_1) \wedge \ldots \wedge s_n(\Box^{k_n} A_n)) \to (M \vee N),$$

with the argument from this point proceeding as before. ■

Evidently, the fact that all the logics mentioned in the paragraph before Example 4.6.5 (as well as many others) are f.m. can be derived from Coro. 4.6.6 and Prop. 4.6.7. The latter result can sometimes be hard to apply because a candidate axiom may not be f.m., but be interdeducible with one which is:

4.6. LOGICS WHICH ARE FULLY MODALIZED

EXAMPLE 4.6.8 If we are wondering whether $\mathbf{K} \oplus q \to \Box(\Box p \to p)$ is a fully modalized logic, we see immediately that this is the same logic as $\mathbf{K} \oplus \Box(\Box p \to p)$ (i.e., $\mathbf{KU}$), since the new – shorter – axiom can be weakened to the original by truth-functional reasoning, and from the original we obtain this new version by substituting, for example, $\top$ for q. It is harder to conclude, using Proposition 4.6.7, that $\mathbf{K} \oplus p \to \Box(\Box p \to p)$ is fully modalized, because it is less obvious that this is just $\mathbf{KU}$ again: see Example 2.5.45, p. 94. ◀

The description of $\mathbf{KD45}$ as 'deontic $\mathbf{S5}$' was mentioned in the opening paragraph of Section 4.5; a similar description with *doxastic* in place of *deontic* would be no less appropriate (though in Section 5.6 we will urge the merits of a weaker logic, $\mathbf{KDB}^\Box 4$, where $\mathbf{B}^\Box$ is the necessitation of the $\mathbf{B}$ axiom). In view of the in-practice consensus that plausible deontic and doxastic logics should be fully modalized, we can make such descriptions precise by introducing the idea of the 'f.m. core' of a modal logic, which will coincide with that logic in case the latter is itself fully modalized, and otherwise be a distinguished f.m. proper sublogic of it. For a modal logic S, we define the *f.m. core* of S, denoted by $fm(S)$, to be the smallest modal logic containing all f.m. theorems of S. It follows from this definition that $fm(S) \subseteq S$ and thus also that $fm(S)$ not only contains every f.m. formula in S, but contains exactly the same such formulas.

EXERCISE 4.6.9 Show that if S is a normal modal logic $fm(S)$ is also a normal modal logic. ✠

We are now in a position to make the promised connection between $\mathbf{S5}$ and $\mathbf{KD45}$:

PROPOSITION 4.6.10 $fm(\mathbf{S5}) = \mathbf{KD45}$.

Proof. Since $\mathbf{KD45} \subseteq \mathbf{S5}$, we need only show that for any f.m. formula A such that $\nvdash_{\mathbf{KD45}} A$, we have $\nvdash_{\mathbf{S5}} A$. But we recall from Example 4.6.5 that this implies we have $\mathcal{M} = \langle W, R, V \rangle$ such that for some $w \in W$, $W = \{w\} \cup W_0$ for some non-empty W_0 with $w \notin W_0$, and $R = (\{w\} \times W_0) \cup (W_0 \times W_0)$, with $\mathcal{M} \nvDash_w A$. Since $\Box$-formulas are modally invariant in such ('semi-simplified') models and A is a truth-functional compound of $\Box$ formulas, we can select any $w' \in W_0$ (recalling that $W_0 \neq \varnothing$) and A will be false at w' in $\mathcal{M}$, and hence at w' in the submodel generated by w'. But in this model, the accessibility relation is universal, so $\nvdash_{\mathbf{S5}} A$, as required. ∎

In view of the frequency with which (the f.m. axioms) $\mathbf{D}$ and $\mathbf{U}$ are cited as weakenings of $\mathbf{T}$ suitable for doxastic and deontic logic, one might reasonably conjecture that the f.m. core of $\mathbf{KT}$ is $\mathbf{KDU}$. This is not quite right, however. In the first place, notice that $\mathbf{5}_c$ ($= \Box p \to \Diamond \Box p$, or $\Box \Diamond p \to \Diamond p$) is a fully modalized $\mathbf{KT}$-theorem which is not $\mathbf{KDU}$-provable (as one may check semantically). Since $\vdash_{\mathbf{K5}_c} \mathbf{D}$, one might revise the conjecture to: $fm(\mathbf{KT}) = \mathbf{KU5}_c$, but this is still not strong enough. What we need is not $\mathbf{5}_c$, but a considerable generalization of this – namely an infinite sequence of axioms, which one can find in Kuhn [679], p. 184, whose n^{th} term is:

$$(\Box A \wedge \Diamond B_1 \wedge \ldots \wedge \Diamond B_n) \to \Diamond(\Box A \wedge \Diamond B_1 \wedge \ldots \wedge \Diamond B_n),$$

or, in the dual form:

$$\Box(\Diamond A \vee \Box B_1 \vee \ldots \vee \Box B_n) \to (\Diamond A \vee \Box B_1 \vee \ldots \vee \Box B_n).$$

Note here the blend of $\mathbf{5}_c$ (the $n = 0$ case) with the generalized density principles GD_n from p. 196. These respectively secure that the canonical frame $\langle W, R \rangle$ of any consistent normal modal logic containing them (all of the GD_n taken together in the latter case) satisfies respectively:

$$\forall x \in W \,.\, \exists y \in R(x),\, R(y) \subseteq R(x) \quad \text{and} \quad \forall x \in W \,.\, \exists y \in R(x),\, R(x) \subseteq R(y),$$

by Exercise 2.7.28 (p. 136) and Coro. 3.3.5 (p. 198), respectively. The reader is left to check that the placement of the main $\Diamond$ on the consequent of Kuhn's schemata (or of the $\Box$ on the consequent in the second formulation) secure that the canonical frame for a (consistent normal modal) logic containing all instances of these schemata satisfies:

$$\text{For all } x \in W, \text{ there exists } y \in R(x),\ R(y) = R(x).$$

Specialising at once to the logic of current interest, this, gives the completeness half of the following, which also incorporates the range-reflexivity axiom **U**:

PROPOSITION 4.6.11 *The modal logic determined by the class of all range-reflexive frames $\langle W, R \rangle$ satisfying the inset condition immediately above is*

$$\mathbf{KU} \oplus \{(\Box A \wedge \Diamond B_1 \wedge \ldots \wedge \Diamond B_n) \rightarrow \Diamond(\Box A \wedge \Diamond B_1 \wedge \ldots \wedge \Diamond B_n) \,|\, n \in \mathit{Nat}\}.$$

The reason for the current interest in this logic is given by Proposition 4.6.13, but to tease out some consequences of the condition on x and y that $R(x) = R(y)$, we pause for an Exercise:

EXERCISE 4.6.12 (*i*) Show that for any frame $\langle W, R \rangle$ with $x, y \in W$, if $R(x) = R(y)$, then $R^k(x) = R^k(y)$, for all $k \geq 1$.
(*ii*) Show that with $\langle W, R \rangle$ with $x, y \in W$ with $R(x) = R(y)$, for any f.m. formula A and any model $\mathcal{M}$ on that frame, $\mathcal{M} \models_x A$ if and only if $\mathcal{M} \models_y A$.
(*iii*) Show that the axiomatic description of the logic mentioned in Prop. 4.6.11 is equivalent to the following: the normal extension of **KU** by all fully modalized instances of the **T** schema – or equivalently, by all formulas $\Box A \rightarrow A$ for f.m. A. ✠

PROPOSITION 4.6.13 *The modal logic described in Prop. 4.6.11 is fm(**KT**)*.

Proof. Since the logic in question is a fully modalized sublogic of **KT**, it remains only to show that for any f.m. formula A unprovable in the present logic is unprovable in **KT**. Such an A is false at some point x in a model on a frame of the kind described in Prop. 4.6.11, so take $y \in R(x)$ with $R(x) = R(y)$ and consider the submodel generated by y. This is a model on a reflexive frame, since the original frame was range-reflexive (and $y \in R(x)$), and any $\Box$-formula true at x in the original model is true at y in this model while any $\Box$-formula false at x in the original model is false at y in this model, by Exc. 4.6.12(*ii*) and the Generation Theorem (Thm. 2.4.10, p. 70). Thus A is false at y in this reflexive model and so $\nvdash_{\mathbf{KT}} A$. ■

It is now time to raise the question of whether the property of being f.m. is as pertinent to the distinctive character of modal logics suited to deontic and doxastic applications as we have been supposing. From time to time a proposal surfaces in the literature to the effect that this requirement is too strong:

EXAMPLE 4.6.14 At p. 358f. of [18], Anderson suggested that $\mathbf{B}_O$ ($= p \rightarrow OPp$) had some plausibility, on the grounds that since only what is permissible should be the case, if it *is* the case that p then it ought to be that it is permissible that p.[247] Now any extension of **KB** in which $\Box \Diamond p$ is not provable – to revert to this notation rather than that of O and P – will not be f.m., so if one accepted Anderson's proposal, one would be rejecting the restriction to f.m. logics for deontic purposes. This particular proposal seems, as Powers [903] pointed out, to be based on a confusion between $p \rightarrow OPp$ and $O(p \rightarrow Pp)$, but perhaps

---

[247]In fact in Anderson's discussion the "$\rightarrow$" is interpreted not as material implication but as implicational connective from *relevant* logic (representing entailment, understood as a modalized cousin of 'relevant implication': see [20]); this aspect of the situation is independent of the criticism about to be made. For further discussion of this theme from Anderson [18], see Goble [384], Mares [771] and Lokhorst [739].

4.6. LOGICS WHICH ARE FULLY MODALIZED

better reasons against the restriction could be found, so it is of some interest to notice another possible restriction in the vicinity – to what (below) we call modal logics with the 'compositional independence property' – especially since **KB**, as it happens, possesses the latter property despite not being f.m. A similar though perhaps subtler scope ambiguity with O is remarked on by Jennings [621], mentioned above at p. 244 (in the Digression), in connection with what he calls *pseudo-subjectivism*, the idea that one ought to do what one thinks one ought to do; but which idea is that? As Jennings notes, one might have in mind, making things explicit with a bimodal deontic-doxastic logic, the principle $BOp \to Op$, and one might instead have in mind: $O(BOp \to p)$. (Jennings in fact subscripts the O and the B with an "a" for the believer in the former case and – perhaps – as the agent implicated in an agent-implicating *ought*-judgment à la Example 4.4.14, p. 251, in the latter.) One subsequent commentator – Bickenbach [89] – offered as an alternative regimentation $Op \to BOp$, though this would seem to stand a better chance at representing "One ought only to do what one thinks one ought to do" than it does in the case of the original sentence without *only*. There has been considerable subsequent discussion of the mixed deontic–doxastic principle $O(BOp \to p)$ with O given the special reading "rationality requires that" under the name the Enkratic Principle; for discussion and further references see Titlenbaum [1107], where thie "O" is written as "R". ◀

The most obvious violation of being f.m. which has been taken to be something for deontic logic to avoid is $\mathbf{T}_O$, but even here dissent is frequently voiced:

EXAMPLES 4.6.15 (*i*) Fisher [305] (especially §4) may well be the earliest attempt to defend the schema $OA \to A$ as a deontic principle. But Fisher does not think that whatever ought to be the case is in fact the case. A representative instance of the schema he takes to be something written – abstracting from the Polish notation used in the paper – as $O\dot{p} \to \dot{p}$, where the dots indicate that the propositional variable they appear above is to be taken imperatively rather than declaratively. The meta-ethical background is R. M. Hare's universal prescriptivism (see [453]), according to which *Ought*-principles represent universally addressed imperatives or prescriptions, and the justification from this perspective for this version of $\mathbf{T}_O$ is the inconsistency of assenting to the claim that something ought to be the case while not assenting to the imperative that it be the case – the imperative 'Make it the case that such-and-such', where assenting to an imperative is being resolved to act on it. Particular moral judgments are supposed to be susceptible of being derived from such universal principles together with the non-modal details of the particular case at issue, and in this sense to be universalizable; for the benefit of those not familiar with the literature on meta-ethics, it should be emphasized that talk of universality in this connection means universal application and has nothing to do with universal acceptance. (The idea of treating *ought*-judgments as imperatives goes back to Carnap in the 1930s, as one may glean from any historical survey – such as van Roojen [964] – of meta-ethics, so what is new here is the admixture of the arguably Kantian ingredient of universali[zabili]ty. Note that the word 'imperative' in Kantian formulations such as 'categorical vs. hypothetical imperative' does not itself force the narrow imperative-as-opposed-to-declarative interpretation that is relevant here. Similarly Chellas [169] is called gives what the author calls a logic of imperatives but which is essentially just a system of deontic logic with an additional 'now-unpreventability' style necessity operator thrown in, to use the terminology of Chapter 7 of Prior [924]; see further note 270 on p. 326 below. This is sometimes called historical necessity, a phrase unfortunately redolent of dialectical materialism.) A similar imperative-based idea is to be found in Castañeda [162]. (This is connected with the appearance of a reference to the same work in Example 4.4.14: the individual to whom the imperative is addressed is the agent to whom the corresponding *ought*-judgment is relativized.) In a review of the latter work ([541], p. 217 and n. 14) a suggestion by Anthony Price is recalled to the effect that these imperatival treatments of moral language provide more plausible treatments of *must* than of *ought*; a somewhat similar remark can be found in Tranøy [1113], p. 224, to the effect that "ought" is suited to *prima facie* rather than absolute obligations. Indeed in his reply to Tranøy's paper, Hintikka [493], p. 234, agrees that "the operator 'O' is best thought of as formalizing a weakish 'ought to' that has to be disitnguished sharply from the typical 'must' of a moral imperative." The deontic version of **T** takes us from a declarative antecedent to an imperatival consequent (O turning imperatives into declaratives), so again we are not considering – what the rejection of $\mathbf{T}_O$ has traditionally had in mind – an implication from (e.g.) "Harry should/must leave now" to "Harry will leave now," with the latter understood as a prediction rather than a third-person command. Many valuable aspects of Castañeda's work in deontic logic are not touched on here; see further the rest of [541], and §4 of Humberstone [578]; whether or not one agrees that the subjunctive turn – or the propositions/practitions distinction, as Castañeda came to formulate matters – resolves such things as the Good Samaritan Paradox or other OMono-induced anomalies (on which question Tomberlin [1110] might be consulted), the early recognition by Castañeda of expressive difficulties for the standard modal language of deontic logic is striking. (As [578] explains, this is related to the increase in expressive power that results from adding an "actually" operator to the language, as in Davies and Humberstone [235], and enjoys a more recent incarnation the work of Kai Wehmeier – see the bibliographical entries for him, and work cited in those sources. These OMono-induced anomalies, which include alongside the Good Samaritan Paradox, Åqvist's 'Paradox of the Knower' – in fact the focus of [1110] – and Forrester's 'Gentle Murder Paradox', from [325], are among the 'paradoxes of deontic logic' explained in many popular surveys of deontic logic: see the opening paragraph of Section 4.4). We also ignore the occasional discrepancies that have been noted between deontic *should* and *ought to*: see for example Gailor [345], p. 348, right-hand column.

(*ii*) That there is a valid implication from the antecedent to the consequent in the ('Harry') example just mentioned is defended – with *must* rather than *ought* in the antecedent – on Werner [1165], largely on the grounds of the oddity of saying such things as "We must but we won't" (by contrast with "We should but we won't"). This is thus a defence of *one* deontic version of **T**, as traditionally understood (as opposed to its reconstrual along the lines seen in (*i*) above); this idea also appears in Lemmon [696], p. 149. (See also Lemmon [697], mentioned in this connection in note 216, p. 252.) There may, however, be alternative explanations of the oddity involved which do not require the inference from antecedent to consequent to be valid: see Ninan [843] for example. (Note also the preponderance of first-person – whether singular or plural – examples used to illustrate the oddity in question in [1165]. See also note 9 in Humberstone [582].)

(*iii*) There are several references to the literature (such as Jones and Pörn [629]) on the semantics of modal auxiliaries in [582], p. 538*ff*., and these will not be repeated here,[248] though an exception should be made in the case of Wertheimer [1166] and Kratzer [667], two (from among several) publications attesting to the move in the 1970s away from the view that these auxiliary verbs simply exhibited a brute ambiguity in having deontic, epistemic, etc., uses. (See also the considerably later Sweetser [1096] for an interesting treatment along these lines, as well as Bybee et al. [142], esp. Chapter 6, and also Ziegeler [1225], for examples of the historical processes involved as various auxiliaries acquire and lose corresponding dynatic, epistemic and deontic uses.) Kratzer's account of modality has become standard fare for textbooks of natural language semantics – such as the useful Portner [899] (in which see §3.1 for this material). Wertheimer's proposals were less clear, partly because of some methodological weaknesses (such as not knowing what a definition should look like), but we should at least illustrate the *must*/*ought* contrast as it surfaces for epistemic uses. Wertheimer ([1166], p. 111)[249] notes the contrast between the perfectly intelligible:

> "That's funny. The roast ought to be done by now. I wonder why it isn't."

---

[248]Since [582] was written there has, however, been an explosion of work on the semantics of, in particular, epistemic modality and the associated auxiliaries (*might*, *must*), in the wake of contextualism and relativism as previously under-exploited strategies in the philosophy of language. For these developments, which we do not treat here, see the papers in Egan and Weatherson [273] and the further references there given in the editors' introduction; see also Braun [116] as well as Remark 5.4.8, p. 394, below.

[249]See also Rivière [960] for many telling examples, concentrating on the *must*/*should* contrast, and Rubinstein [977] for further discussion of the relative strength of deontic modals, with cross-linguistic comparisons.

4.6. LOGICS WHICH ARE FULLY MODALIZED

and the deeply bizarre:

> "That's funny. The roast must be done by now. I wonder why it isn't."

Roughly speaking, Wertheimer's account is that *ought* indicates that something would follow from some general principles and the obtaining of various circumstances, without implying that those circumstances in fact obtain, while *must* adds the implication that those circumstances do indeed obtain. This account is neutral between what Wertheimer calls Systems of Actuality ("SAs") and Systems of Ideality ("SIs"): a distinction, to put it vaguely, between what follows tha modal auxiliary pertains to how the world *in fact is*, on the one hand, or to how the world *is to be*, on the other. The latter subsumes the traditional ethical subject matter that underlies deontic logic, so the *must/ought* distinction can still be drawn without a commitment to **T** for deontic *must*; presumably this distinction coincides essentially with the veridical/non-veridical distinction in Table 4.1, p. 204, and surrounding discussion. Of course, in speaking as though *ought*, for instance, can be treated as a sentence operator, we here oversimplify away the issue of agent-relativity touched on in Example 4.4.14, p. 251, which also arises for certain other modal auxiliaries, such as the *can* of ability, discussed in Kenny [649], Walton [1143] and Brown [121], generally suppressing this feature by using a bare sentence operator. As well as *can* and its dynatic dual (*cannot but*, perhaps), a modal treatment has been suggested of the degree-admitting intermediate notions of difficulty and ease, given a modal treatment with explicit attention agent-relativity in Denyer [240], with such principles as: if it is easy for a to φ then it is possible for a to φ. (A further, less logically oriented discussion of some of the distinctions in play here – ability *vs.* opportunity, etc. – can be found in Nordenfelt [847].) One response to such considerations is to treat these operators as predicate-modifiers; another, as with the "O_b" notation mentioned on p. 254 in view of Example 4.4.14, treats the operators as expressions which take a name to yield a sentence operator. For the formal study of the logic of ability, the present author suggested in [585] that the name *dynatic logic* – used above in Example 2.1.4, p. 37 – would be preferable to the sometimes encountered label "dynamic logic", to avoid confusion with dynamic logic as the 'logic of programs' enterprise mentioned in Remark 4.4.2(i), p. 237, to say nothing of the rather different *dynamic epistemic* logic, mentioned in the Digression before Example 5.1.2, p. 334.[250]

(iv) For a more recent discussion of how the deontic and epistemic *should* (or *ought to*) are related to each other, see Cariani [149]. Here is an erroneous claim concerning the interaction of aspect and modality, from p. 83 of Frajzyngier *et al.* [330]:

> The progressive can be used with auxiliaries that otherwise code root modalities. With the progressive, however, these auxiliaries have an epistemic rather than a deontic meaning...

The claim is evidently incorrect (reflecting the limitations of 'corpus linguistics', in fact), since *ought* and *should* are clearly deontic in "You should be talking to your wife about this, rather than to me," and "I ought to be fixing the fence now but decided to spend an extra hour in bed." A more promising observation concerning the epistemic/deontic contrast concerns the examples (13a) and (13b) in Moltmann, [819], reproduced here (with her numbering) alongside another pair; the phrase 'presentation pronoun' applies to the special use of the word *this* in the examples, which is not the usual demonstrative pronoun. (The latter is not available for reference to a person. Thus, while "I bought this spoon in Victoria Markest", the "spoon" can be omitted, one cannot omit "woman" from "I saw this woman in Victoria Market".) From [819], p. 47:

---

[250] An overview subsuming the logic of programs and the epistemic case can be found in van Benthem [80]. Humberstone [585], just mentioned, looks at the modal aspects of claims of sufficiency and excess, for example (working the cases with excess rather than sufficiency, by way of illustration) "He is too late to have caught the train (since if he had caught it he would be here by now)" – epistemic; "He was too late to attend the meeting (since it started half an hour after he arrived)" – dynatic; "She is too young to be reading that book (which contains material that will confuse or upset her)" – deontic. [585] contains several typos, one significant one being that in the bottom line of the main text of p. 309, "was committed" should read "wasn't committed".

The interpretation of modals as well shows the nonreferential pronouns. Presentational pronouns appear to allow only for an epistemic interpretation of a modal, as seen in the contrast beween (13a) and (13b) and between (14a) and (14b):

(13)   a.   John must be a student.
        b.   This must be a student.

(14)   a   Mary could be a gymnast.
        b.   This could be a gymnast.

Whereas *must* in (13a) allows for a deontic interpretation of the modal. (13b) allows only for an epistemic interpretation; and whereas *could* in (14a) can express physical possibility, *could* in (14b) can express only epistemic possibility.

As Moltmann urges with a large range of other considerations, such asymmetries suggest that identificational sentences ("This is Mary") are not identity sentences ("This person = Mary"), or (14b), for example, would admit of a *de re* reading with the same non-epistemic interpretation as (14a). However, these matters are much contested; see the discussion in Moltmann [819] as well as the references there to joint work by Daphna Heller and Lynsey Wolter, and the pioneering explorations of F. R. Higgins. ◀

Digression. Wertheimer's "roast ought to be done by now" case from Example 4.6.15(*iii*) prompts mention of the close-call survivor's standard form of words:

"I shouldn't be alive."

Much as in the case of the with the roast meat example, this cannot be interpreted as indicating the high probability, given the speaker's current evidence, of the proposition that he or she is not alive – by contrast with "Since they called to say there were on the way, they should be here within the hour". There is perhaps an allusion to the high probability of this proposition given salient earlier evidence. But the tone in which such remarks as that inset above are made suggests more that more than evidential and epistemic considerations are in play: something along the lines of "I had no right to expect to survive the incident". This intermingling of the moral – or at least normative – with the epistemic is evident in many of the ways we speak, and would merit further investigation. Consider the use of the word *expect* both in talk of expecting it to rain and in expecting Harriet to return a favour, the latter taken as expecting *of* Harriet that she return the favour. (Compare the agent-implicating *ought*-judgments of Example 4.4.14, p. 251.) Or again: the notoriously slippery talk of what is *normal*. There is probably more going on here than brute ambiguity. Some pertinent diachronic considerations can be found in Ziegeler [1225], mentioned above in Example 4.6.15(*iii*). **End of Digression.**

We introduce the compositional independence property alluded to in Example 4.6.14 in a rather abstract and general fashion. Given a class $\mathcal{V}$ of valuations for some language L, let us say that the formulas in a set $\Gamma \subseteq L$ are *independent* – or 'completely independent', as it is sometimes put – w.r.t. $\mathcal{V}$, when for any $\Gamma_0 \subseteq \Gamma$ there is some $v \in \mathcal{V}$ with $v(A) = T$ for all $A \in \Gamma_0$ and $v(A) = F$ for all $A \in \Gamma \smallsetminus \Gamma_0$. This means that $\mathcal{V}$ does not rule out any combinatorially possible allocation of truth-values to the formulas in Γ. Next we specialise this to the case in which the formulas involved are a (representative) compound and its components using some connective. If $\#$ is an n-ary connective of the language L, we say that $\#$ has the *compositional independence property* w.r.t. $\mathcal{V}$ if the formulas $\#(p_1, p_2, \ldots, p_n)$, $p_1, p_2, \ldots,$ and p_n are independent w.r.t. $\mathcal{V}$ (in the sense just defined). For present purposes we take a special interest in the case in which $n = 1$, and we write $\#$ as $\Box$.[251] In this setting, if we fix attention on the cases in which $\mathcal{V}$ contains only Boolean valuations and in addition the set of all propositional variables is independent

---

[251]In [571], I suggested that with a suitably chosen $\mathcal{V}$ to hand, a kind of non-vicious circularity for candidate definitions might be provided by allowing the item of vocabulary being defined to appear on the right-hand side, provided it did so only in the scope of operators with the compositional independence property w.r.t. $\mathcal{V}$; Keefe [643] provides some critical discussion (see also Burgess [131]).

4.6. LOGICS WHICH ARE FULLY MODALIZED

in the sense just defined, then the set of all formulas true on each $v \in \mathcal{V}$ is a modal logic, and we can give the following more syntactically formulated version of the notion of compositional independence for $\Box$: a (mono)modal logic S has the *compositional independence property* just in case:

$$\nvdash_S \Box p \to p \qquad \nvdash_S p \to \Box p \qquad \nvdash_S \Box p \vee p \qquad \text{and} \qquad \nvdash_S \neg(\Box p \wedge p).$$

In the current setting, these conditions amount to saying respectively that the logic does not rule out the combinations {$\Box p$ true, p false}, {$\Box p$ false, p true}, {$\Box p$ false, p false}, {$\Box p$ true, p true}, respectively. (If we are not in a multimodal setting, but $\Box$ is any one of several modal operators, the above can be taken as defining "$\Box$ has the compositional independence property according to S.") Note the use of propositional variables here and the attendant presumption of closure under uniform substitution. We could not require that *no* formula of the form $\Box A \to A$, for example, be provable – at least if we want a requirement that can be satisfied (consider $A = p \to p$) – so we say instead that *not every* such formula is provable, which is equivalent given US to saying that $\Box p \to p$ is unprovable.

It is clear from the form of the definition of the compositional independence property, which simply requires the unprovability of certain formulas, that it is very different in nature from the property of being f.m., for example, in view of the feature just remarked on, not applying to the inconsistent logic – which by contrast possesses the latter property. Among the consistent modal logics, too, we have the case of **KVer**, for example, which is f.m., but evidently lacks the compositional independence property, since we have both $\vdash_S p \to \Box p$ and $\vdash_S \Box p \vee p$ for this choice of S. (Venturing into the non-normal area, we could consider also the 'Falsum' logic, mentioned in note 141, p. 171, with $\neg \Box p$ provable – see Segerberg [1025], p. 201*f.* – and similar repercussions for the compositional independence property. Such modal logics are not just non-normal but as one might say, anti-normal in that they have no consistent normal extensions.) Thus being f.m. does not imply having the compositional independence property, even among consistent normal modal logics. Conversely, having the compositional independence property does not imply being f.m., as the example of **KB**, mentioned above, shows – though of course in **KB** the derived connective $\Box \Diamond$ is not, in the obvious sense, compositionally independent according to this logic (and the same goes for the dual modality, $\Diamond \Box$).

REMARK 4.6.16 Still another approach to the independence issue here, stressed for doxastic applications in von Kutschera [682], e.g., p. 105*f.*, actually formulates independence conditions in the object language using an alethic modal operator, alongside the doxatic operator B and von Kutschera's objectivity operator O (see Example 5.4.11). Such formulations might include as an 'object language' version of compositional independence for B, the conditional with antecedent $Op \wedge \Diamond p \wedge \Diamond \neg p$ and consequent $\Diamond(p \wedge Bp) \wedge \Diamond(p \wedge \neg Bp) \wedge \Diamond(\neg p \wedge Bp) \wedge \Diamond(\neg p \wedge \neg Bp)$. (Compare also the informal use of alethic formulations of compositional independence in Keefe [643]. Note that the current "$\Diamond$" is not the dual of the "$\Box$" in the definition of the compositional independence property, the latter being the role played by B here.) Von Kutschera envisages **KD45** as the background logic for B here, so the antecedent has for example $\Diamond p$ as a condition, since the envisaged alethic extension will not allow it as possible that $B\bot$ (because of the "D" part of **KD45**. We still need the "Op" part – it is an objective (= non-doxastic) matter whether p – because of things mandates as believed even thought they could be false, such as, $Bp \to p$ in the present case (as **KD45** contains **U**). While we need both $\Diamond p$ and $\Diamond \neg p$) in the antecedent, we do not need to add $O \neg p$ alongside Op, since, as the informal reading just given suggests, this follows from Op: see Example 5.6.7 (p.427) for further details on von Kutschera's O. ◀

We return to compositional independence as defined above:

EXERCISE 4.6.17 (*i*) Say, giving reasons, which of the following normal modal logics have the compositional independence property: $\mathbf{KT}_c$, **K5**, $\mathbf{K} \oplus (p \leftrightarrow q) \to (\Box p \leftrightarrow \Box q)$ – this last being the smallest extensional modal logic (as defined in Exc. 2.7.2(i) on p. 129).

(*ii*) If an affirmative modality, considered as a single derived 1-ary connective, has the compositional independence property according to a normal modal logic, does it follow that the same goes for the dual modality? (Give a proof or a counterexample.)

(*iii*) If two modal logics have the compositional independence property, does it follow – and if so, *how* – that their intersection has this property? ✠

Digression. Since so much reference has been made in this section to Zolin [1226], it is worth remarking that the extensional modal logics (those containing the formula after the "⊕" in part (*i*) of the above exercise – whether normal or not) are discussed in [1226] under the description *prime* modal logics, though this concept is introduced rather differently in [1226]. **End of Digression.**

Having raised the question of which out of the compositional independence property and the property of being f.m. is the better generalization of what distinguishes doxastic and deontic logic from alethic and epistemic logic, we will not attempt to answer it. The opening quotation from Lucas [744] may suggest the latter property since it employs a distinction between modalized and unmodalized formulas, and entertains the idea that there may be no logical connection between the two. In the chapter of [744] from which this quotation comes, however, the main examples of modalization come from sentence modifiers which would most naturally be treated as single modal operators rather than (for instance) modalities of length greater than 1 (as with the recent **B** example) – such as 'Pravda reports that ...' (p. 8) and 'Lucas believes that ...' (p. 11).

4.7 "Nothing in Between": A Remark by A. N. Prior

This section takes as its point of departure a remark made by A. N. Prior on deontic logic, and while our discussion contains occasional asides which are specific to the deontic interpretation of the modal operators, the real interest is in a general issue in logical theory. However, before pursuing that theme fully, we get to an observation (Proposition 4.7.10) which will be of use to us in discussing one reaction to Ross's Paradox, in Section 4.8.

In the 1951 publication quoted below, Prior is commenting on the analogies between alethic (called 'ordinary' in the passage quoted) modality, deontic modality, and the quantifiers: the relation between $\Box$, alethically or deontically interpreted, and $\Diamond$ being like that between $\forall$ – or some Aristotelian ancestor thereof – and $\exists$, that had been observed in von Wright [1204] (also from 1951).[252] In discussing alethic modality, Prior uses the old adjectives "apodictic" and "problematic" to describe statements of necessity and possibility, respectively. When he wants to distinguish *p simpliciter* from $\Box p$ and $\Diamond p$, he uses the adjective "assertoric" – not the ideal word in this connection, really, because $\Box p$ and $\Diamond p$ are no less capable of being asserted (and neither of them, nor the unmodalized form, needs to be considered in an assertoric context).

> There is one point, however, where the analogy between ordinary modality and quantity is kept up, but the analogy between "modal modes" and both the others breaks down. In respect of their quantity, propositions do not just divide into universals and particulars, but into universals, particulars and singulars. Quite similarly, in respect of their modality, propositions do not just divide into apodictic and problematic, but into apodictic, problematic and assertoric. In between "S must be P" and "S may be P", stands the simple "S is in fact P", just as "This S is P" stands in between "Every S is P" and "Some S is P". And just as "This S is P" is implied by "Every S is P" and implies "Some S is P", so "S is in fact P" is implied by "S must be P" and implies "S may be P". But so far as I can see, there is nothing among the moral or 'deontic' modalities that corresponds to these intermediary 'existential' or 'alethic' modalities. (...) There is no moral word, then, which is related to "ought" as "is in fact" is related to the non-moral "must be", or "this" is related to "every".
>
> Prior [914], p. 145*f*.

---

[252]The issue was taken up soon after the advent of the Kripke semantics for modal logic in Montague's 1960 paper, [821].

4.7. "NOTHING IN BETWEEN": A REMARK BY A. N. PRIOR

The passage ellipsed here considers and rejects the obviously flawed suggestion that the deontic modality corresponding to contingency ("It is indifferent whether A does X or not", as Prior puts it) is the desired intermediary. We will use the above passage to launch an exploration (of the "nothing in between" issue it raises) which will take us quite a long way from deontic logic. But first, four qualms about the content of the passage deserve to be aired.

A first point is that the "in fact" of Prior's discussion here suggests something different from the way p is lodged inferentially between $\Box p$ and $\Diamond p$ – being provably implied by the former and provably implying the latter – in any extension of **KT**.[253] The phrase "in fact" may suggest not so much p, but rather, $\mathcal{A}p$ where "$\mathcal{A}$" is the distinctive operator of modal actuality logic. This is not something to go into here, beyond that supplied by Remark 4.7.1; anyone puzzled at the very idea of a distinction between p and "Actually p" (which is the intended reading of $\mathcal{A}p$) should consult §2 of [578], as well as the references there cited.

REMARK 4.7.1 For the case of adding $\mathcal{A}$ to **S5**, we can take the accessibility relation for $\Box$ to be the universal binary relation on the frames of our models and leave it out of account, getting approximately the simplified Kripke models of Section 2.2, but we will also need a distinguished element (of W) to play the role of the actual world. Calling the latter w^* the idea is that we count $\mathcal{A}A$ true at an arbitrary point x in such a model just in case A itself is true at w^* in the model. When it comes to defining validity, we have to decide between whether we want to count being true at the distinguished point in every model as sufficient, or whether, more stringently, we require true at every point in every model. This is the contrast between, respectively, real-world and general validity mentioned in note 177, p. 219. ◀

Digression. For the case(s) in which the underlying *Actually*-free modal logic is not **S5**, see Hazen, Rin and Wehmeier [465]; for an interesting later treatment, compare Restall [955]; for a treatment of phenomena which provoked the introduction of an actuality operator using, instead, a simulation of natural language subjunctivity, see Humberstone [547] (summarised and typographically corrected in Section 4 of [578]) or – in a more streamlined form – Wehmeier [1157]. **End of Digression.**

Secondly, we might be worried about Prior's thinking that we get a disanalogy in the deontic case because "there is no moral word" standing in the relation of interest to "ought". Why should we require that the missing word should count as a *moral* word? If we could find anything at all to fill the gap, that would perhaps suffice. And why should the gap-filler be a *word*, anyway? Any linguistic expression, such as a multi-word phrase, would surely do. But what seems more pertinent is whether such an expression, even if not extant (in English, say), could coherently be introduced (if not already present) to play this role.[254] Thirdly, the "S/P" format – mnemonic in traditional syllogistic or Aristotelian logic for S̲ubject/P̲redicate (though both are positions for predicates in the contemporary Frege-derived understanding of that term) – is not really right for the deontic and alethic modalities. "Every student is patriotic" (or "All students are patriotic") and "Some student is (or "students are") patriotic" make clear what someone saying "Students are patriotic" might have in mind. But the unclarity thereby resolved would remain in "Students must be/may be/are patriotic", and explicit quantification would be required to remove it, whereas there is presumably not supposed to be any such unclarity.[255]

Finally, we might be worried about what the relation in question is supposed to be. One salient point about p as it stands in relation to $\Box p$ is that noted above: it lies inferentially between $\Box p$ and $\Diamond p$, in any modal logic plausibly intended to be of alethic application – which we may take here to at least

---

[253] Not that **KT** was of current concern in 1951, but certainly modal logics containing **T** were, to which the "inferential betweenness" point still applies, even though not all of them were even quasi-normal.

[254] Would such an expression, inferentially sandwiched between ethical or moral expressions automatically count as ethical (or moral)? Presumably not: see the discussion of an analogous issue in Williamson [1189], pp. 27–28.

[255] The quantificational case cited by Prior – "This S is P" between "Every S is P" and "Some S is P" – is in any case obscure (because of the status of the supposed intermediary) and is better replaced by using Fregean singulary quantifiers, and taking Fa as an intermediary between $\forall x(Fx)$ and $\exists x(Fx)$.

require the provability of **T** and the unprovability of $\mathbf{T}_c$, its converse. Not just between, but *strictly* (or 'properly') between: i.e., we do not have the converse of either of the implications (*i*) from p to $\Box p$ or (*ii*) from $\Diamond p$ to p. Without this restriction, then "between" one thing and something it provably implies there will always be at least these two intermediaries: the first thing, and the second thing. (Of course there will still be *some* formulas A for which (*i*) holds (with A for p) and some for which (*ii*) holds; indeed if we take A as provable or refutable – i.e., such that its negation is provable, then (*i*) and (*ii*) will both hold for the same choice of A.) But in general, when someone asks for an item standing in the same relation to X as Y stands in to Z, you know that no question has been clearly put. What number stands in the same relation to 3 as the number 2 stands in to the number 4? Well *one* relation 2 bears to 4 is that the latter is twice the former. The number that stands to 3 in this relation is 6. But another relation that 2 bears to 4 is that the latter is the square of the former, and the number that stands in *that* relation to 3 is 9. And a third relation that 2 stands in to 4 is that we can get from the former to the latter by adding 2; the number that 3 bears this relation to is 5. We could go on – especially as all the relations just considered are functional relations (relations that something bears to exactly one thing, that is). There is, then, in general no such thing as the item – or even (to cover the non-functional case) the set of items – bearing to X the same relation that Y bears to Z, and it becomes a matter of looking for a relation that seems especially salient in the context.[256] Here what is selected is the following relation between (alethic) $\Box p$ and p: the latter is strictly between the former and its dual – or more accurately, the formula we get by attaching the dual of $\Box$ to p.

Well, here is one way we could obtain such an intermediary in the deontic case, on the assumption that our favoured logic is at least as strong as **KD**, and the further assumption that it is determined by some class of serial frames. (For definiteness, imagine we are working with **KD** itself so that it is the class of all serial frames that is relevant.) The idea is that we work with expansions of these frames $\langle W, R \rangle$ which have an extra ingredient: a function $f : W \longrightarrow W$ satisfying the condition that for all $x \in W$, $f(x) \in R(x)$. (Recall our convention: $R(x) = \{y \in W \mid xRy\}$.) Since for x in a serial frame, $R(x)$ is never empty, there is always at least one way of expanding $\langle W, R \rangle$ to a structure $\langle W, R, f \rangle$) meeting this condition. Now let us add to the language a new operator, to be written as "·", and, since we are thinking of our $\Box$ deontically we will write the latter as O (with P for the corresponding $\Diamond$). O is treated in the usual way (in terms of R, that is) in models $\mathcal{M} = \langle W, R, f, V \rangle$ on such expanded frames, while for the new operator we stipulate:

$$\mathcal{M} \models_w \cdot A \text{ if and only if } \mathcal{M} \models_{f(w)} A.$$

Now, we have obtained something in between O and P in the sense that the following formulas are true throughout every one of these models:

$$Op \to \cdot p \qquad \cdot p \to Pp,$$

while the same cannot be said for either of their converses.

Let us further observe that if we took our monomodal deontic starting point to be **KD45** then we could throw away the accessibility relation R from the frames to be expanded, working with frames $\langle W, X \rangle$ in which $\varnothing \neq X \supseteq W$: X comprises the ideal worlds, once and for all, and what ought to be the case anywhere is what is true at all of them. (See the discussion of semi-simplified Kripke models, beginning at p. 206 above.) In that case the expanded frames can just be frames with a distinguished element $\langle W, X, u \rangle$, with $u \in X$, and we would write

---

[256]Thus in Exercise 2.7.34(*iii*) (p. 137), the following words appear: "Here is a condition related to symmetry as range-reflexivity is related to reflexivity: $\forall x \forall y \forall z (Rxy \to (Ryz \to Rzy))$". Here we are saying that all successor points are 1-symmetric, as in the case of range-reflexivity we say that all successor points are reflexive. But range-reflexivity is also the piecewise condition corresponding to reflexivity, as defined on p. 76, whereas the piecewise condition corresponding to symmetry would be: $\forall x, y, z((Rxy \wedge Rxz) \to (Ryz \to Rzy))$, which is weaker than the earlier condition. This illustrates the fact that the phrase "the condition related to symmetry as range-reflexivity is related to reflexivity" does not pick out a unique condition (not that this phrase appeared in the material just quoted from Exercise 2.7.34).

4.7. "NOTHING IN BETWEEN": A REMARK BY A. N. PRIOR

$$\mathcal{M} \models_w \cdot A \text{ if and only if } \mathcal{M} \models_u A.$$

(In effect, we are thinking of u as $f(x)$ for all $x \in W$, chosen from $R(w) = X = R(y)$ for all $y \in W$.) The notation "·" for a sentence operator (1-ary connective) is somewhat eccentric, and has been adopted here for the sake of a connection with a notation for what are called 'transitional sequents' in the natural deduction system(s) for modal logic of Section 7.5 below (where, however, "·" is not a connective but part of an elaborated turnstile which, like the present connective, looks forward to successor points). Having already used O and P we could use $\Box$ instead for this purpose. We also have $\Diamond$ but the monomodal logic determined by the class of functional frames, frames that is, satisfying the condition that every point has exactly one point accessible to it, is easily seen to be **KD!** which essentially identifies $\Box$ and $\Diamond$. (See Segerberg [1022], [1034]; see also the discussion at p. 157 above, where for **KD!** we threw R out of the frames and took them to have the form $\langle W, f \rangle$.) In fact, this provokes a variation on the above example:

EXAMPLE 4.7.2 We can interpret the bimodal language with operators O and $\Box$ in frames in which each has a serial accessibility relation and that for $\Box$ is included in that for O. We do not impose the stronger requirement that the subrelation here should be functional, as we did before. Now we have, as valid on all the frames under consideration $Op \to \Box p$, $\Box p \to \Diamond p$, and $\Diamond p \to Pp$, and none of their converses, so we have not just one but two intermediaries between O and P. If we want to consider what all this might mean, suppose that O and $\Box$ are read with the aid of the arguably stronger and weaker deontic auxiliary verbs "must" and "ought" respectively (despite the 'O' notation for the former). ◀

We now turn to some more general considerations raised by the quotation from Prior, which will lead us outside the area of deontic logic, indeed for the moment away from modal logic altogether (though we shall return to the latter, if not the former). In Section 1.2, in the course of introducing the idea of consequence relations in general and of the classical ('tautological') consequence relation $\vdash_{CL}$ in particular, it was pointed out that Exercise 1.2.2(v) (from p. 8) admitted of the following reformulation:

> Does there exist a formula A such that $p \wedge q \vdash_{CL} A$ and $A \vdash_{CL} p$ while $A \nvdash_{CL} p \wedge q$ and $p \nvdash_{CL} A$?

This is a question in the same line of territory as the passage quoted from Prior [914] has led us to explore: it asks if there is something strictly between one formula (here $p \wedge q$, there, Op) and another, unilaterally implied formula (here p, there Pp), the logics in question being classical propositional logic in the one case and some unspecified deontic logic (perhaps an extension of **KD**) in the other. If one draws a diagram of all the equivalence classes of formulas constructed from the two propositional variables p, q, by means of any functionally complete set of Boolean connectives, taking the equivalence of A with B to amount to its being the case that $A \dashv\vdash_{CL} B$, one obtains the 16-element Boolean algebra, and representing this in a Hasse diagram in the usual manner with elements representing the equivalence classes $[C]$ (say) of formulas C, and $[A] \leq [B]$ just in case $A \vdash_{CL} B$: see Humberstone [594], Figure 2.13a on p. 225, for this diagram), one sees by the most casual inspection that the equivalence class of $p \wedge q$ covers that of p in the sense given by the definition (already given in note 33, p. 39 above): x *covers* y iff $y \lneq x$ and there is no z with $y \lneq z \lneq x$. So there is nothing in this two-variable fragment that will do as the intermediate A between $p \wedge q$ and p. But instead of relying on such considerations there is a relatively short and attractive argument to be given for this conclusion, presented in the proof which follows. The argument relies on Proposition 1.2.5 (p. 12), according to which two Boolean valuations agreeing on the propositional variables in a formula must agree on the formula itself.

PROPOSITION 4.7.3 *There is no formula A constructed from p, q, by Boolean connectives for which (i) $p \wedge q \vdash_{CL} A$, (ii) $A \vdash_{CL} p$, (iii) $A \nvdash_{CL} p \wedge q$, and (iv) $p \nvdash_{CL} A$.*

Proof. Suppose, for a contradiction, we have A constructed from p, q, and satisfying (i)–(iv). By (iii), there exists a Boolean valuation v with $v(A) = T$ while $v(p \wedge q) = F$. By (iv) there is a Boolean valuation

v' with $v'(p) = $ T while $v'(A) = $ F. Since $v(A) = $ T, (ii) implies that $v(p) = $ T, so since $v(p \wedge q) = $ F we have $v(q) = $ F. Since $v'(A) = $ F, (i) implies that $v'(p \wedge q) = $ F, but $v'(p) = $ T, so $v'(q) = $ F. Thus $v(p) = v'(p)$ and $v(q) = v'(q)$, so since A is supposedly constructed from no variables other than these, by Prop. 1.2.5 we must have $v(A) = v'(A)$, contradicting the above findings that $v(A) = $ T while $v'(A) = $ F. ∎

So if we are to find an intermediary between $p \wedge q$ and p, within the expressive resources of classical propositional logic, that formula will have to involve a new propositional variable. We are being careful to use the word 'intermediary' here and to avoid talk of interpolating a formula between $p \wedge q$ and p, since such talk has a technical sense which tends in exactly the opposite direction from that we have just seen to be required. We describe that technical sense in the following Digression.

Digression. When A provably implies C, according to some logic, an *interpolant* between A and C is a formula B in the common non-logical vocabulary of A and C – which in the case of propositional logic means that the only propositional variables allowed to occur in B are those occurring *both* in A and in B, and for which we have A provably implying B and B provably implying C. Such an interpolant trivially exists when either the variables of A all occur in B or vice versa, as we can take either A or C as the desired B, not having required B to be strictly between A and C in this case. If such a 'common vocabulary' interpolant can always be found for A and C whenever the former provably implies the latter in a given logic, then that logic is said to have the Interpolation Property or satisfy the Interpolation Theorem (asserting the existence of such interpolants), though if neither of the constants $\top$, $\bot$, is taken as primitive, it is usual to exempt the case in which A is refutable or C is provable. If we are thinking of logics as sets of sequents or as consequence relations, the reference to the provability of $A \to C$ here can be replaced by the hypothesis that $A \succ C$ is provable or that $A \vdash C$. The earlier remark that an "interpolant trivially exists when either the variables of A all occur in B or vice versa, as we can take either A or C as the desired B" is strictly speaking only trivial for the last, consequence relation, formulation, since we *could* have a 'formula' logic in which not all – or even not any – instances of $A \to A$ were provable, and similarly for sequents $A \succ A$, in the latter case omitting a basic structural rule corresponding to the condition on consequence relations (see p. 8) that every formula should count as a consequence of itself. (A strengthened version of the result that such interpolants can already be found, which claims that for any A and consequence C of A a consequence B of A can be found whose variables are those shared by A and C which serves as an interpolant between A and any formula C' whose shared variables with A coincide with the common variables of A and C can also be shown for classical and intuituitionistic propositional logic but not, for example for **S4**. The envisaged result is one version of what is called a 'uniform' interpolation theorem. For its vicissitudes across a range of normal modal logics, see Ghilardi and Zawadowski [371] and Bílková [93]. A book-length treatment of interpolation results in modal and other logics is provided by Gabbay and Maksimova [343].) In summary, our intermediaries are meant to be strict intermediaries whereas interpolants are not required to be, and we have just seen from Prop. 4.7.3 that in at least the case raised by Exc. 1.2.2(v), our intermediary, if it is to be found, will have to involve non-logical vocabulary in neither $p \wedge q$ nor p, while an interpolant has to drawn only on vocabulary appearing in both the implying and the implied formula. As well as this difference in respect of vocabulary, the starting point in the demand for an interpolant is the hypothesis that A has C as a consequence – or otherwise put, that A provably implies C – in either formulation understood relative to a particular logic, whereas the starting point in the question for an intermediary is the rather different hypothesis that A has C as a consequence *and not conversely* – or that A *unilaterally* implies B – according to the logic in question. (We have used the phrase "unilaterally implies" from time to time, beginning with Example 2.1.20(ii), p.50.) One might think that the other question, the interpolation question for formulas each of which provably implies the other, is trivial – and arguably it is (see Section 4 of [572]). But the present point is simply that this *is* a question which needs an affirmative answer no less than the unilateral version for the interpolation property to hold, whereas the question about

4.7. "NOTHING IN BETWEEN": A REMARK BY A. N. PRIOR

(strict) intermediaries arises only in the unilateral case (as well as not involving the 'common vocabulary' condition). **End of Digression.**

EXAMPLE 4.7.4 Since we are going to need a new variable in any strict intermediary for $p \wedge q$ and p, we may as well choose r. Here is an example of a formula which is a tautological consequence of $p \wedge q$ and which has p as a tautological consequence:

$$p \wedge (q \vee r),$$

and for which neither of these consequence claims can be reversed – our formula does not have $p \wedge q$ as a consequence, and nor is it a consequence of p. So this will do as an answer to Exc. 1.2.2(v). It meets all the demands on A in Prop. 4.7.3 except the demand that A be constructed exclusively from the variables p, q. ◀

We might do well to follow up the use of conjunction and disjunction in this example for a more general account of when there is a strict intermediary B between A and C with $A \vdash C$ and $C \nvdash A$, taking $\vdash$ as $\vdash_{CL}$. To suggest "intermediary", let us write "I" for a candidate, which means that we want, for a given A, C, as just described:

(i) $A \vdash I$;   (ii) $I \vdash C$;   (iii) $C \nvdash I$;   (iv) $I \nvdash A$.

The way conjunction and disjunction enter the picture is through the following observation.

PROPOSITION 4.7.5 *Suppose that $A \nvdash_{CL} C$ and there is some formula B such that (*) $B, C \nvdash_{CL} A$ and (**) $\neg B, C \nvdash_{CL} A$. Then putting $I = A \vee (B \wedge C)$ gives an I satisfying the four conditions (i)–(iv) above.*

Proof. We have (i) by the properties of $\vee$ (according to $\vdash_{CL}$), and likewise for (ii), where we also need the properties of $\vee$ and the hypothesis that $A \vdash_{CL} C$. (From now on, we omit the "CL" subscript.) To show (iii), $C \nvdash A \vee (B \wedge C)$, it suffices to show that $C \nvdash A \vee B$; but this is equivalent to the special assumption (**) on A, B, C. For (iv) $A \vee (B \wedge C) \nvdash A$ we must show $B \wedge C \nvdash A$, which is equivalent to assumption (*). ■

Digression. A more symmetrical way (emerging in the above proof) of formulating the conditions (*) and (**) here would be as:

(*) $B \wedge C \nvdash_{CL} A$ (**) $C \nvdash_{CL} A \vee B$,

but if we were working with the generalized ("multiple conclusion") consequence relation of classical logic – denoted as on p. 14 (*q.v.* for further explanation) by $\Vdash_{CL}$ – we could write more elegantly still:

(*) $B, C \nVdash_{CL} A$ (**) $C \nVdash_{CL} A, B$.

End of Digression.

EXERCISE 4.7.6 (i) Show that as well as the candidate I described in Prop. 4.7.5, the following would also do as an I satisfying (i)–(iv): $(A \vee B) \wedge C$.

(ii) How are the candidates for the role of I here and in Prop. 4.7.5 related to the choice of intermediate formula in Example 4.7.4? ✠

Does the existence of candidate intermediaries as provided by Prop. 4.7.5 and Exc. 4.7.6(i) mean we can expect to turn up a mass of candidates? Well, not exactly, because given the background assumption that $A \vdash_{CL} C$, the formulas $A \vee (B \wedge C)$ and $(A \vee B) \wedge C$, from 4.7.5 and 4.7.6, respectively, are in fact equivalent. (This is a form of the what is called in lattice theory the *modular law*, which would normally find expression in the form: for all lattice elements a, b, c, if $a \leqslant c$, then $a \vee (b \wedge c) = (a \vee b) \wedge c$.[257]) Nothing said here rules out the choice of many alternatives to play the role of B, however.

EXERCISE 4.7.7 *Verify the claim just made, showing that for any formulas A, B, C, if $A \vdash_{CL} C$ then $A \vee (B \wedge C) \dashv\vdash_{CL} (A \vee B) \wedge C$.* ✠

Further, to show we have not just stumbled onto a special case, let us make a further observation.

PROPOSITION 4.7.8 *For formulas A, C, such that $A \vdash C$ while $C \nvdash A$, there exists a formula I, satisfying the conditions (i)–(iv) if and only if there exists a formula B satisfying (*) and (**) in Prop. 4.7.5.*

Proof. The "if" part is given by Prop. 4.7.5 itself. For the "only if" part, suppose that we have I satisfying (i)–(iv). Then I itself can be taken as such a B, since $I \dashv\vdash (A \vee I) \wedge C$ (left as an exercise, below) and I satisfies the conditions imposed on B by (*), (**). First, suppose that (*) fails, and we have $I \wedge C \vdash A$. By (ii), we get $I \vdash A$, contradicting (iv). Similarly, if (**) fails and $C \vdash A \vee I$, then by (i) we get $C \vdash I$, contradicting (iii). ■

EXERCISE 4.7.9 *Show that, on the supposition that I satisfies for a given A, C, the conditions (i)–(iv), that $I \dashv\vdash (A \vee I) \wedge C$.* ✠

Now while this has all been a pleasant logical excursus, you might be forgiven for feeling that the conditionality ("*if* there exists an I or a B satisfying such and such conditions for a given A, C,...") is somewhat disingenuous, since we can just follow the lead of Example 4.7.4 and reach for a propositional variable not occurring in the given A, C. Suppose the first variable (in the ordering $p_1, \ldots, p_n, \ldots$) not so occurring is p_k. Then we can just take p_k for the B in (*) and (**) and those conditions will automatically be satisfied. For if $A \nvdash_{CL} C$, a failure of (*) for this choice of B would mean $p_k, C \vdash A$. Since p_k does not occur in C or A, by uniform substitution (or more accurately the substitution-invariance of $\vdash = \vdash_{CL}$), we can put C for p_k, getting: $C, C \vdash A$, i.e., $C \vdash A$, which we are given is *not* the case. Similarly a failure of (**) would mean we had $\neg p_k, C \vdash A$ and we could similarly substitute $\neg A$ for p_k and again conclude, contrary to hypothesis, that $C \vdash A$. (The situation is clearer in the formulations provided by the Digression after Prop. 4.7.5, since here we substitute A for p_k in the case of (*) and C (itself) for p_k in the case of (**).) To record this concisely let us say a logic *provides strict intermediaries* if whenever one formula unilaterally implies a second according to that logic, there is a formula unilaterally implied by the first and unilaterally implying the second.[258] The above argument goes through in the presence of any further non-Boolean connectives, so we record its verdict in the following form:

PROPOSITION 4.7.10 *Every modal logic provides strict intermediaries.*

Proof. Summarizing the foregoing discussion, where A unilaterally implies C in any modal logic, one strict intermediary we can find between A and C is $(A \vee p_k) \wedge C$, or equivalently $A \vee (p_k) \wedge C$), where p_k is a propositional variable not occurring in A, C. ■

It is Proposition 4.7.10 that we shall need to appeal to in the following section, so readers whose main concern is with deontic logic can safely omit the remainder of the present section, which addresses of general logical (rather than specifically deontic) interest.

---

[257]Or equivalently: For all a, b, c, if $a \geqslant c$, then $a \wedge (b \vee c) = (a \wedge b) \vee c$.

[258]In other words, the relation unilateral implication for the logic in question is a *dense* strict partial ordering. (The last three words here just mean we have an irreflexive transitive relation.)

4.7. "NOTHING IN BETWEEN": A REMARK BY A. N. PRIOR

The idea to be explored from this point on is the possibility of constructing our desired intermediary formula as some kind of compound of the very formulas between which it is to be an intermediary, rather than one involving new propositional variables.[259] This will involve a new connective and a rule with the property that any substitution-instance of an application of the rule is itself an application of the rule. Uniform substitution itself lacks this property – sometimes called substitution-invariance (though this is not to be confused with the substitution-invariance of a consequence relation) – whereas the familiar rules involving various connectives (such as Modus Ponens) possess it. As a first step in this direction, we can convert the conditions (*) and (**) into rules, replacing "B" by a compound of A and C and a new binary connective we shall write as $\circ$ from those components. (Later, in the paragraph preceding Proposition 4.7.27, p. 318, we make a more direct approach, with a binary connective $\star$ for which $A \star C$ provides a candidate I between A and unilaterally implied C, rather than detouring via a candidate B, as here with $\circ$. Readers in a hurry may care to pass straight to that discussion.) We revert to the 'logics as sets of formulas' style here.

(*) Rule $\quad \dfrac{(C \wedge (C \circ A)) \to A}{C \to A} \qquad$ (**) Rule $\quad \dfrac{(C \wedge \neg(C \circ A)) \to A}{C \to A}$

These rules could be reformulated in several ways, for example as

$$\dfrac{(C \circ A) \to (C \to A)}{C \to A} \qquad \dfrac{\neg(C \circ A) \to (C \to A)}{C \to A}$$

though, again, the clearest reformulations would involve multiplicity on the left and the right, so here we use sequent-to-sequent rules with such sequents in mind. (One of these sequents $\Gamma \succ \Delta$ holds on a valuation v just in case v does not assign T to all formulas in Γ and F to all formulas in Δ. A logic taken as a set of such sequents is determined by a class $\mathcal{V}$, of valuations, if its sequents are precisely the $\mathcal{V}$-valid sequents in the sense of: sequents holding on every $v \in \mathcal{V}$.)

For the case of (*): $\quad \dfrac{C \circ A, C \succ A}{C \succ A} \qquad$ For (**): $\quad \dfrac{C \succ A, C \circ A}{C \succ A}$

The last formulation brings out very clearly what is demanded, for a given A and C, of the formula $C \circ A$ in terms of the valuations we are interested in (the collection of which determines the logic in the sense just parenthetically explained): if there is a valuation v in the class, with $v(C) = \mathrm{T}$ and $v(A) = \mathrm{F}$, there are valuations v' and v'' in the class, each of which is like v in respect of C and A, with $v'(C \circ A) = \mathrm{T}$ (for the first rule) and $v''(C \circ A) = \mathrm{F}$ (for the second rule). Of course, since v itself either verifies or falsifies $C \circ A$, v can itself be taken as one or other of v', v''.

Now it would be simpler if we could consider a singulary rather than a binary connective in thinking about these matters, and it turns out that we can do just this, assuming we have $\to$ at our disposal. To avoid confusion with the "A", "C" (and "B", "I") of the discussion to this point, it will help to change to another range of schematic letters and we will use "φ" (and "ψ", "χ", if necessary) in this capacity. Let $\#$ be a 1-ary connective which we wish to subject to the following pair of (formula-to-formula) rules:

(#1) $\quad \dfrac{\#\varphi \to \varphi}{\varphi} \qquad$ (#2) $\quad \dfrac{\#\varphi \vee \varphi}{\varphi}$

If we wanted to keep the $\to$-style formulation from (#1) we could instead have written the premiss for (#2) as $\neg\#\varphi \to \varphi$, and again the symmetries are more evident with such formulations as

---

[259]This means that as far as shared vocabulary is concerned, we are at something like the opposite end of the spectrum for intermediaries from the position with interpolants: the intermediary will be constructed from the union of the sets of propositional variables used for the antecedent and consequent, rather than the intersection of these sets.

$$\frac{\#\varphi \succ \varphi}{\succ \varphi} \quad \text{and} \quad \frac{\succ \varphi, \#\varphi}{\succ \varphi}.$$

(Those familiar with the sequent calculus approach to logic will note that a *single* rule with the two premisses $\#\varphi \succ \varphi$ and $\succ \varphi, \#\varphi$ and conclusion φ would be a special case of the Cut rule, with $\#\varphi$ as the cut-formula.)

Given a functionally complete set of Boolean connectives, the existence of a 1-ary connective $\#$ satisfying (#1) and (#2) guarantees the existence of a 2-ary connective $\circ$ satisfying the (*) and (**) rules stated earlier, since we can take $C \circ A$ as $\#(C \to A)$. Conversely, given $\circ$ satisfying the (*) and (**) rules, putting $\#\varphi = \top \circ \varphi$ gives a $\#$ satisfying (#1) and (#2).

EXERCISE 4.7.11 Provide a detailed verification of the claims of the preceding paragraph. ✠

PROPOSITION 4.7.12 *There is no 1-ary connective $\#$ definable in classical propositional logic for which the rules (#1) and (#2) both preserve tautologousness.*

Proof. Every substitution-invariant formula-to-formula rule which preserves $\mathcal{V}$-validity for $\mathcal{V}$ = the class of Boolean valuations, preserves the property of holding on an arbitrarily selected $v \in \mathcal{V}$ (see, e.g., Setlur [1038]), which implies that for $\vdash = \vdash_{CL}$ and all formulas φ we have:

$$(1) \quad \#\varphi \to \varphi \vdash \varphi \quad \text{and} \quad (2) \quad \#\varphi \vee \varphi \vdash \varphi.$$

Unfortunately (1) is equivalent to $\vdash \#\varphi \vee \varphi$, which together with (2) implies that $\vdash \varphi$ – for all φ! Since that is not the case – recall that $\vdash$ is $\vdash_{CL}$ – we conclude that there is no candidate $\#$ within the language of this consequence relation for which (#1) and (#2) both preserve tautologousness. ∎

It follows from this that the addition of rules (#1) and (#2) in a more general form, allowing side-formulas, can be expected to be quite disruptive. We give the more general forms here as sequent-to-sequent rules, but if you wish to have formula-to-formula versions make the premiss for the first be schematically represented as $\psi \to (\#\varphi \to \varphi)$, with similar prefixing by "$\psi \to$" for the conclusion, and likewise for the premiss and conclusion of (#2).

$$(\#1)^+ \quad \frac{\Gamma \succ \#\varphi \to \varphi}{\Gamma \succ \varphi} \qquad (\#2)^+ \quad \frac{\Gamma \succ \#\varphi \vee \varphi}{\Gamma \succ \varphi}$$

If, to the natural deduction system for classical propositional logic presented in Section 7.1, $(\#1)^+$ and $(\#2)^+$ are added as rules governing a *new* 1-ary connective $\#$, then the sequent $\succ \varphi$ becomes provable for every formula φ. To see how, consult Example 7.1.4. (The derivation given there is rather different from what one would by following the sketch given above in the proof of Prop. 4.7.12, which would be considerably more cumbersome.)

EXERCISE 4.7.13 Show that the above rules, $(\#1)^+$, $(\#2)^+$ are interderivable, respectively, with the simplified versions $(\#1)_0^+$, $(\#2)_0^+$, below, given the natural deduction rules (in Section 7) for $\to$, $\neg$ and $\vee$:

$$(\#1)_0^+ \quad \frac{\Gamma \succ \neg\#\varphi}{\Gamma \succ \varphi} \qquad (\#2)_0^+ \quad \frac{\Gamma \succ \#\varphi}{\Gamma \succ \varphi} \qquad \text{✠}$$

There is no such equivalence between the original theorem-to-theorem rules (#1) and (#2) – which are in effect the special $\Gamma = \varnothing$ cases of $(\#1)^+$, $(\#2)^+$ – and correspondingly simplified rules, $(\#1)_0$: *from $\neg\#\varphi$ to φ*, and $(\#2)_0$: *from $\#\varphi$ to φ*.

4.7. "NOTHING IN BETWEEN": A REMARK BY A. N. PRIOR

EXERCISE 4.7.14 Show (i) that (#1) and (#2) are respectively derivable from (#1)$_0$ and (#2)$_0$, in the sense that every modal logic (as defined on p. 18, regarding "#" as a notation for $\Box$) closed under the latter rules is closed under the former, and (ii) that the converse is not the case. ✠

We are dealing, as the last exercise points out, with a modal logic, once we accept "#" as an alternative notation for "$\Box$" – the least modal logic closed under (#1) and (#2), which is of course not to say that such rules would be appropriate for any traditionally modal (e.g., alethic) reading of the 1-ary operator concerned, however it may be notated. (We will stick with the "#" notation.) That is not yet to say we have a consistent modal logic on our hands, however, and we can press the device of alien propositional variables (from the discussion following Exercise 4.7.9) into service to establish consistency, as a consequence of the following, which says that our rules are admissible for the smallest modal logic:

PROPOSITION 4.7.15 *The smallest modal logic closed under* (#1) *and* (#2) *is consistent, its theorems in fact being simply those of the smallest modal logic.*

Proof. The envisaged logic can be presented axiomatically as having all substitution instances of truth-functional tautologies as axioms, and as rules, Modus Ponens, (#1) and (#2). We claim that every formula provable from these axioms by these rules is a substitution instance of a truth-functional tautology, and so is in fact already one of the axioms. It is clear enough that Modus Ponens preserves the property of being such a formula. As for (#1), suppose that the premiss $\#\varphi \to \varphi$ for an application of this rule is a substitution instance of a tautology. Then the tautology in question must have the form $p_i \to \varphi_0$, with φ substituted for p_i and with p_i not occurring in φ_0. The latter is guaranteed because if p_i occurred in φ_0, then $\#\varphi$ would be a subformula of the result, φ, of applying the substitution to φ_0: but since # is a (1-ary) connective, $\#\varphi$ cannot be a subformula of φ. Since $p_i \to \varphi_0$ is a tautology and p_i does not occur in φ_0, φ_0 must itself be a tautology. Otherwise any Boolean valuation v with $v(\varphi_0) = $ F could be adjusted (if necessary) to a Boolean valuation v', agreeing with v on variables occurring in φ_0 with $v'(p_i) = $ T and $v'(\varphi_0) = $ F (by Prop. 1.2.5). Thus φ is itself a substitution instance of the tautology φ_0, as was claimed. A similar argument applies in the case of (#2). ∎

All that is used in the above proof is the fact that the smallest normal modal logic is closed under uniform substitution, so what the proof really establishes is something more general:

THEOREM 4.7.16 *Suppose S is a modal logic with any number of modal operators, not necessarily of the same arity, and that the 1-ary connective # is not one of them. Let S^+ be the smallest modal logic extending S in the language with # as an additional connective. Then S^+ is closed under the rules* (#1) *and* (#2).

In a temporary return to the theme suggested more directly by the quotation from Prior with which we began, we have the following.

COROLLARY 4.7.17 *If* **KD** *is extended to a bimodal logic with the addition of new 1-ary #, the resulting logic is closed under* (#1) *and* (#2) *and so, writing the $\Box$ and $\Diamond$ of* **KD** *as O and P, in this logic we have a formula I with the properties that $Op \to I$ and $I \to Pp$ are provable in it, but neither of their converses is.*

Proof. The part before the "and so" is an appeal to Thm. 4.7.16, for which we rely on the proof of Prop. 4.7.15. The part after the "and so" is a matter of retracing the steps of our discussion, in which Op and Pp play the roles marked out by the earlier schematic letters "A", "C", for which we wanted a formula B satisfying the two conditions (*) and (**) of Prop. 4.7.5. In the current logic the formula $\#(C \to A) = \#(Pp \to Op)$ satisfies these conditions, because the logic is closed under the rules (#1) and (#2). But this still doesn't give the desired I of the current formulation, since although we have neither $\#(Pp \to Op)$

provably implying Op nor being provably implied by Pp, we have not secured the other two conditions demanded of I, namely that it should be provably implied by Op and provably imply Pp. To secure all four properties, we invoke Prop. 4.7.5, to trade in $\#(C \to A)$, the current B, for I as $A \wedge (B \vee C)$, which is to say, for $A \wedge (\#(C \to A) \vee C)$, in other words for the formula $Op \wedge (\#(Pp \to Op) \vee Pp)$. ∎

Let us note, apropos of Prior's idea, in the passage with which we began, that a suitable strict intermediary in the deontic case should itself be deontic is satisfied by the intermediary given at the end of this proof at least in the following sense: every occurrence of p lies within the scope of some occurrence of a deontic operator. (Compare the discussion of deontic logic as f.m. – fully modalized – in Section 4.6.)

Nevertheless, the last part of the proof is rather awkward – modulating from B to I via conjunction and disjunction – and we should give some attention to the possibility of obtaining the desired end-product, a strictly intermediary formula between A and C, from the very beginning, by suitably enhanced rules. But before proceeding to explore that possibility, which will take us back to the binary connective ("∘") formulation in play earlier, there are some interesting issues to raise about the current 1-ary #. Those keen to pass straight to the promised development and skip these issues are invited to pick up the discussion following Exercise 4.7.26 on p. 317 below (though some of that discussion will refer back to what follows here).

For convenience we repeat the rules governing #:

$$(\#1) \quad \dfrac{\#\varphi \to \varphi}{\varphi} \qquad (\#2) \quad \dfrac{\#\varphi \vee \varphi}{\varphi}$$

We use boldface capital letters **V**, **F**, **I** and **N** for the four one-place truth-functions: constant-true ("V" for *Verum*), constant-false, the identity truth-function, and the negation truth-function. Recall that a Boolean valuation is one associating the conventional truth-functions with $\wedge, \neg, \to \perp$ etc. This leaves open the question of how the (currently) new connective # is to be treated. So let $\mathcal{V}_\mathbf{V}, \mathcal{V}_\mathbf{F}, \mathcal{V}_\mathbf{I}, \mathcal{V}_\mathbf{N}$, be the classes of Boolean valuations which respectively associate with # (interpret # as expressing, that is) the truth-functions **V**, **F**, **I** and **N**. For a class of valuations $\mathcal{V}$, we call a formula A $\mathcal{V}$-*valid* if $v(A) = T$ for each $v \in \mathcal{V}$.[260] Now we apply these concepts in connection with the above rules.

EXAMPLE 4.7.18 Consider the following alternative argument to that given in the proof of Prop. 4.7.15. All substitution-instances in the present language (i.e., with additional connective #) of truth-functional tautologies are $(\mathcal{V}_\mathbf{I} \cup \mathcal{V}_\mathbf{N})$-valid. Modus Ponens preserves this property. And so do the above rules (#1) and (#2), for the following reason. Suppose a premiss $\#\varphi \to \varphi$ for (#1) is $(\mathcal{V}_\mathbf{I} \cup \mathcal{V}_\mathbf{N})$-valid. That means it comes out true (on any Boolean valuation) whether # is interpreted as the identity truth-function or as negation. In particular, then, it comes out true when # is interpreted as negation (i.e., it is $\mathcal{V}_\mathbf{N}$-valid). But that means that the formula $\neg\varphi \to \varphi$ is true on every Boolean valuation, in which case so is the (classically equivalent) formula φ, the conclusion for the application of (#1) under consideration. Now take (#2), and suppose the premiss, $\#\varphi \vee \varphi$ for an application for this rule is $(\mathcal{V}_\mathbf{I} \cup \mathcal{V}_\mathbf{N})$-valid. Then *a fortiori* it is $\mathcal{V}_\mathbf{I}$-valid. But when # is interpreted as the identity truth-function, as on the valuations in $\mathcal{V}_\mathbf{I}$, the premiss $\#\varphi \vee \varphi$ is equivalent again to the conclusion of this application of the rule. Therefore every theorem of the smallest modal logic closed under (#1) and (#2) is $(\mathcal{V}_\mathbf{I} \cup \mathcal{V}_\mathbf{N})$-valid and is therefore already provable in the smallest modal logic. ◀

Note that the above reasoning is being put forward as an example to ponder, rather than being claimed as a successful proof (which, as we shall see at Example 4.7.22, it is not). It involves treating # as a *hybrid* of truth-functional connectives, a treatment which is often of considerable utility and interest. (See §6 of

---

[260]There is a similar notion for sequents of course, $\mathcal{V}$-validity amounting to holding on each $v \in \mathcal{V}$.

4.7. "NOTHING IN BETWEEN": A REMARK BY A. N. PRIOR

my [586] or §3 of [587] for some examples, and below at Prop. 4.7.32.) Before coming to a verdict on the reasoning of Example 4.7.18, let us make the provisional observation that everything said about **I** and **N** could equally well have been said in terms of **F** and **V** instead, respectively. Thus if instead of looking at $(\mathcal{V}_\mathbf{I} \cup \mathcal{V}_\mathbf{N})$-validity, we have considered $(\mathcal{V}_\mathbf{F} \cup \mathcal{V}_\mathbf{V})$-validity, we should have been able to argue with no less plausibility that a (#1) premiss $\#\varphi \to \varphi$ with the latter property is *a fortiori* $\mathcal{V}_\mathbf{I}$-valid, and when # is interpreted as the constant true truth-function, this implicational premiss collapses into its consequent, the conclusion of the application of (#1) in question; a similar argument applies in connection with the rule (#2), this time stressing the fact that $(\mathcal{V}_\mathbf{F} \cup \mathcal{V}_\mathbf{V})$-validity implies $\mathcal{V}_\mathbf{F}$-validity. But in that case, why are we bothering with a 1-ary connective? These truth-functional interpretations are constant and so we can run a version of Example 4.7.18 with a 0-ary connective (sentential constant):

EXAMPLE 4.7.19 Let Ω be a 0-ary connective and consider the following argument (again, simply put forward as a putative proof) concerning the smallest extension of classical propositional logic in the language with this as an additional primitive, closed under Modus Ponens and Uniform Substitution and also under the rules:

$$(\Omega 1) \quad \frac{\Omega \to \varphi}{\varphi} \qquad (\Omega 2) \quad \frac{\Omega \vee \varphi}{\varphi}$$

The argument we have in mind would purport to show that only substitution instances of tautologies are provable in the logic just described, by reference to the observation that the rules preserve the property of being what we shall just call (in an ad hoc way) *hybrid valid*, by which is meant true on every Boolean valuation on which Ω is associated with the nullary truth-function (i.e., truth-value) T and also true on every Boolean valuation on which it is associated instead with F. Because we include the former valuations, $\Omega \to \varphi$'s being hybrid valid implies that φ is, so rule (Ω1) preserves hybrid validity, and because we include the latter, $\Omega \vee \varphi$'s being hybrid valid implies again that φ is, so again hybrid validity is preserved. Thus all provable formulas are hybrid valid. But this implies that every provable formula is a substitution instance of a tautology, since if a formula φ is provable, consider the formula φ' resulting from φ on replacing every occurrence of Ω in φ by some propositional variable p_k not occurring in φ. This formula is a tautology, because if it is not, in view of some Boolean valuation v with $v(\varphi') = $ F, let v' be the valuation like v on all variables and with $v(\Omega) = v(p_k)$. We should then have $v(\varphi) = $ F, making φ not hybrid valid. This contradicts the provability of φ, so the formula φ' just described must be tautologous after all. And φ is a substitution instance (substituting Ω for p_k) of φ'. ◀

The appearance of an argument here notwithstanding, adding the rules (Ω1, 2) would in fact render classical propositional logic inconsistent; the use of the term 'logic' in the following statement is intended to indicate closure under uniform substitution:[261]

PROPOSITION 4.7.20 *The smallest logic extending classical propositional logic and closed under the rules Modus Ponens, (Ω1), and (Ω2) contains every formula.*

Proof. Let S be the logic described. Since $p \to p$ is a tautology, $\Omega \to \Omega \in S$. Thus by (Ω1), $\Omega \in S$. Since $\Omega \to (\Omega \vee \varphi)$ is also a substitution instance of a tautology, this implication is S-provable for any formula φ, so by Modus Ponens $\Omega \vee \varphi \in S$ for any φ, whence by (Ω2), $\varphi \in S$ for any formula φ. ■

The bad news delivered by Prop. 4.7.20 is only to expected, on closer inspection of the concept of hybrid validity deployed in Example 4.7.19. For, despite the fancy appearance ("Ω") and its description as a sentential constant (or nullary connective), since hybrid validity in this degenerate case amounted merely to truth whether Ω was interpreted as though it was $\top$ (associated with T on a Boolean valuation)

---

[261]Thus we are dealing with a monomodal logic in the generalized sense of Section 1.4, with a single *nullary* (i.e. 0-ary) □ operator, written as "Ω".

or whether it was interpreted as $\bot$ (associated instead with F), this amounts to treating Ω as just another propositional variable. And *neither* of the rules governing it would preserve tautologousness if we formulated them in such a way as to make this explicit. Take (Ω1) thus reformulated:

$$\frac{p_i \to \varphi}{\varphi}$$

Instantiating the schematic letter "φ" to "p_i" itself, we have a provable premiss, and so the rule delivers p_i as conclusion, and hence by uniform substitution, every formula is provable. (Note that we are taking φ as p_i, rather than substituting φ for p_i here; the latter substitution could disrupt our consequent, since the possibility has not been excluded that p_i occurs in φ.)

EXERCISE 4.7.21 Make similar trouble for the rule (Ω2) reformulated with an arbitrary propositional variable in place of "Ω". ✠

It is not immediately clear whether the original rules (#1, 2) are in similar trouble, because a key feature of the proof of Prop. 4.7.20 is absent from the case of 1-ary #: that proof instantiates the schematic "φ" to "Ω", giving $\Omega \to \Omega$ as a premiss for the application of (Ω1). The corresponding move is not available in the case of (#1), with premiss $\#\varphi \to \varphi$, since no formula of the form $\psi \to \psi$ is also of the form $\#\varphi \to \varphi$.[262] However, φ can still be instantiated to a formula containing #, which is enough to cause trouble for the argument of Example 4.7.18:

EXAMPLE 4.7.22 (Example 4.7.18 revisited.) Consider a premiss for the supposedly $(\mathcal{V}_\mathbf{I} \cup \mathcal{V}_\mathbf{N})$-validity-preserving rule (#1):

$$\#\#\bot \to \#\bot.$$

This formula is indeed $(\mathcal{V}_\mathbf{I} \cup \mathcal{V}_\mathbf{N})$-valid since every valuation in $\mathcal{V}_\mathbf{I}$ and every $v \in \mathcal{V}_\mathbf{N}$ assigns the value F to its antecedent. But the conclusion of an application of (#1), namely $\#\bot$, is not $(\mathcal{V}_\mathbf{I} \cup \mathcal{V}_\mathbf{N})$-valid, as this formula is not $\mathcal{V}_\mathbf{I}$-valid. ◀

EXERCISE 4.7.23 (*i*) Find an example to show that the rule (#2) also fails to preserve $(\mathcal{V}_\mathbf{I} \cup \mathcal{V}_\mathbf{N})$-validity.
(*ii*) Where exactly does the argument given in Example 4.7.18 go wrong? ✠

When the language of a logic is extended by the addition of a new connective and the logic is extended by new principles (rules or axioms) in which the new connective (perhaps amongst others) figures, the extension is said to be *conservative* if the extended logic proves no formulas in the unextended language that were not provable in the original logic. We have seen, in effect, that # as governed by (#1) and (#2) does provide a conservative extension of non-modal classical propositional logic as well as of any modal logic, since this follows from Prop. 4.7.15 and Thm. 4.7.16. But does this give us a grasp as to what # might be taken to *mean*? Normal modal logics, especially when Kripke-complete, allow us some way of understanding their modal operators, since we know at least that we are concerned (in any model) with an accessibility relation, with $\Box$ or $\Diamond$ interpretable as universally or existentially quantifying over accessible points. The behaviour of $\Box$ (or $\Diamond$) in an arbitrary modal logic, however, affords us no comparable grasp of its significance.[263] Proposition 4.7.15 told us that the present # could be construed as the $\Box$ operator

---

[262]This is because $\#\varphi$ is the result of applying a one-place connective to φ, here represented by #; the latter is not simply a metalinguistic notation for any function from formulas to formulas – such as – what might be especially relevant here – the identity function. See the index entries in Humberstone [594] under "non-connectival operations on formulas".

[263]On exposure to the above material at a Melbourne logic seminar in 2007, Greg Restall noted the existence of a semantic treatment of # satisfying not only the independence principles (#1) and (#2) but also the dual principles – rules *From* $\#\varphi \to \neg\varphi$ *to* $\neg\varphi$ and *From* $\varphi \to \#\varphi$ *to* $\neg\varphi$ (also satisfied by the 'non-connectival' construal of $\#\varphi$ as denoting the first propositional variable not occurring in φ) – in terms of *general* frames (see the Digression on p. 83), and used the resulting

4.7. "NOTHING IN BETWEEN": A REMARK BY A. N. PRIOR

of the smallest modal logic, so one question that comes to mind is whether this result can be improved: can we insert before "modal" here, some more restrictive adjective, such as "congruential", "monotone". "regular", or – picking up the claim just made about intelligibility, best of all – "normal"? We address the last question. The answer here is negative, even waiving the reference to the *smallest* such logic (for which – i.e., for **K** – we saw that (#1), rewritten as on the left in the proof below, was indeed admissible, in Coro. 2.5.6, p. 73):

PROPOSITION 4.7.24 *No consistent normal modal logic is closed under the rules* (#1) *and* (#2), *construing* # *as* $\Box$.

Proof. In the $\Box$ notation the rules (#1) and (#2) become:

$$\frac{\Box A \to A}{A} \quad \text{and} \quad \frac{\Box A \vee A}{A}$$

Any normal modal logic closed under the second of these rules extends **KT** (contains **T**, that is), as we saw in Prop. 2.6.26, thereby rendering provable every premiss for an application of the first rule. Thus any normal modal logic closed under both rules contains every conclusion of the first rule, i.e., contains every formula. ∎

Note that we have here reverted to using $A, B, \ldots$ as schematic letters, since the reasons for switching over to $\varphi, \psi, \ldots$ have now lapsed.

EXERCISE 4.7.25 Is every monotone modal logic closed under the above rules also inconsistent? (Justify your answer.) ✠

The fact that taking # (obeying (#1, 2)) as $\Box$ in a normal modal logic leads to trouble, as reported in Proposition 4.7.24, does not rule out the possibility of some alternative modal interpretation of # faring more successfully. For example, we could canvas the options of taking # to be some (affirmative or negative) modality. A first option along these lines might be to take # as $\neg\Box$, but this can be quickly dismissed since it merely interchanges (#1) and (#2), and so leads to inconsistency for closure under both rules as in Prop. 4.7.24. Another simple option would be to choose instead $\Box\neg$, which is equivalent, by the interchange considerations just deployed, to taking the following pair of rules:

$$\frac{\Diamond A \to A}{A} \quad \text{and} \quad \frac{\Diamond A \vee A}{A}$$

EXERCISE 4.7.26 Show that no consistent normal modal logic is closed under both of these last two rules. (*Suggestion*: in place of $\Box(\Box p \to p) \vee (\Box p \to p)$, used in the proof of Proposition 2.6.26, p. 121, consider the formula $\Diamond(\Diamond p \to p) \vee (\Diamond p \to p)$. We could equally well use the formula $\Diamond(p \to \Box p) \vee (p \to \Box p)$.) ✠

We have hardly begun to explore the options here, and the interested reader is invited to pursue them further. I have not attempted to show that no modality (or more generally, no modal formula in one variable) can be used to provide a consistent interpretation of # in a normal modal logic, and am not in a position to rule this out. But it is high time we returned to the prospect of replacing # by a binary connective, as promised in the discussion after the proof of Coro. 4.7.17. Recall that the idea was to

---

interpretation to raise philosophical questions about worlds and propositions. Roughly: if there is a strict intermediary between any two formulas one of which unilaterally implies the other, then there is a strict intermediary between $\bot$ and any consistent formula, depriving the associated Lindenbaum algebra of atoms – which are supposed to be the unit sets of the possible worlds of the Kripke semantics. A published version of Restall's considerations appeared as [954]; perhaps more would need to be said about the significance of atomlessness in the present case, since this already arises for the case of classical propositional logic (with its infinite supply of sentence letters).

create, in one fell swoop, an intermediary B between A and C for which, as we can most conveniently put it, using the consequence relation notation ("$\vdash$" here for $\vdash_{CL}$, though very little depends on the choice specifically of *classical* logic), $A \not\Vdash C$. This is to be a 'strict intermediary' in that we want $A \not\Vdash B$ and $B \not\Vdash C$.

To avoid confusion with the earlier use of "$\circ$", we use the notation "$\star$" for a binary connective, though now we choose rules specifically aimed at having $A \star C$ behave as the desired formula B just described. Accordingly, we propose the following four rules (in which "A", "C" are just schematic letters, as usual, though chosen with the above formulations of our desiderata in mind:

$$\text{I} \quad \dfrac{A \to C}{A \to (A \star C)} \qquad \text{II} \quad \dfrac{A \to C}{(A \star C) \to C}$$

$$\text{III} \quad \dfrac{(A \star C) \to A}{C \to A} \qquad \text{IV} \quad \dfrac{C \to (A \star C)}{C \to A}$$

Rules I and II deliver the first 'positive' desiderata: that if A provably implies C then $A \star C$ is provably implied by A and in turn provably implies C. Rules III and IV deliver the 'negative' aspects (the crossed out "$\dashv$"s of our recent formulation), if C does not provably imply A then $A \star C$ does not provably imply A – by III – and nor is it provably implied by C – by IV. So taken together, the rules give us a binary connective that manufactures strict intermediaries (for unilateral implications) on demand. In summary:

PROPOSITION 4.7.27 *Any logic closed under rules I–IV for some connective $\star$ in its vocabulary provides strict intermediaries.*

Now since all modal logics provide strict intermediaries, as reported in Prop. 4.7.10, thanks to such devices as Example 4.7.4 illustrates (resorting to extraneous variables, that is), the interest of rules I–IV lies not in what is said about them in Proposition 4.7.27, but in the fact that the intermediaries provided via $\star$ are constructed from the very formulas they are intermediaries between. By iterated appeal to the existence of intermediaries, any logic providing strict intermediaries provides infinitely many non-equivalent intermediaries between any formula and any formula it unilaterally implies. But when these intermediaries are constructed from the formulas they mediate between we have infinitely many pairwise non-equivalent formulas constructed from the propositional variables occurring in A (say) and C, when the latter is unilaterally implied by A in the logic, since we have the unilateral implications (for convenience here written with consequence relation notation, with each "$\vdash$" understood unilaterally – i.e., as being a case of "$\not\Vdash$"):

$$A \vdash A \star C \vdash (A \star C) \star C \vdash ((A \star C) \star C) \star C \ldots \vdash C$$

(The notation here is of course inexcusable. We are attempting to evoke the corresponding claim, in similarly lazy notation, about the corresponding Lindenbaum algebra elements: $[A] \lneq [A \star C] \lneq [(A \star C) \star C)] \ldots$.[264]) Thus recalling from p. 44, where the notion of local finiteness was defined and a classic early finding of work in modal logic was reported – to the effect that **S5** is, while **S4** is not, locally finite – we can add to Proposition 4.7.27 the following observation:

PROPOSITION 4.7.28 *No consistent modal logic closed under rules I–IV for some connective $\star$ in its vocabulary is locally finite.*

To return to the point at which Proposition 4.7.27 was formulated: we wanted from rules I–IV the result there recorded. We must check they don't deliver too much more than that, however, and in particular that they provide us with a conservative extension of any modal logic to which we might

---

[264] Here we write "$\lneq$" in place of "$<$" for emphasis.

4.7. "NOTHING IN BETWEEN": A REMARK BY A. N. PRIOR

append them. (We were trying to provide intermediaries between formulas with the aid of a special purpose connective, not to disrupt logical relations between formulas free of that connective.) "Modal logic" here means modal logic with any number (including 0) of primitive $\Box$ operators, but otherwise as defined on p. 18. (In fact it would not matter if the $\Box$ operators concerned were restricted to the 1-ary case, a restriction we considered lifting in Section 1.4.)

First we define a translation from the language of a modal logic in the broad sense just recalled, expanded by the addition of a binary connective $\star$ (presumed not to be amongst the primitives of the language in question), to the set of $\star$-free formulas of that language. The translation, which we will call τ, is very much inspired by our earlier discussion (Example 4.7.4 etc.); note the resemblance to the non-definitional translations considered in Example 4.4.25 (p. 264).

- $\tau(p_i) = p_i$, for each propositional variable p_i;
- $\tau(\#(A_1,\ldots,A_n)) = \#(\tau(A_1),\ldots,\tau(A_n))$ for any connective $\#$ (of arity n) other than $\star$, including modal operators;
- $\tau(A \star C) = (\tau(A) \vee p_k) \wedge \tau(C)$ where p_k is the first propositional variable not occurring in $\tau(A)$ or $\tau(C)$. ("First" here means: earliest in the official enumeration $p_1, p_2, \ldots, p_n, \ldots$ of all propositional variables.)

THEOREM 4.7.29 *Let S be any modal logic and S^+ be the smallest modal logic extending S with the additional connective $\star$ and closed under the rules I–IV. Then $\vdash_{S^+} A$ implies $\vdash_S \tau(A)$.*

Proof. We argue by an induction on length of proof (of A) in an axiomatization of S^+ which takes all substitution-instances (in the present language, with additional connective $\star$) of theorems of S as axioms and Modus Ponens and I–IV as rules. For the basis case, note first that any axiom which is S-provable does not contain $\star$ at all, on which formulas τ is the identity map, so the claim holds in this subcase. Secondly, for the subcase in which we have an axiom which is a substitution instance of an S-theorem, $A(B_1,\ldots,B_n)$, say, resulting by substituting formulas, B_i uniformly for p_{k_i} in the S-theorem $A(p_{k_1},\ldots,p_{k_n})$, by uniform substitution (under which S is closed) now of $\tau(B_i)$ for p_{k_i}, the formula $A(\tau(B_1),\ldots,\tau(B_n))$ is S-provable. But this formula is just $\tau(A(B_1,\ldots,B_n))$, so the claim holds here too. The inductive step requires us to check that the five primitive rules in the axiomatization of S^+ preserve the property of having an S-provable τ-translation. This is clear for Modus Ponens, in view of the fact that $\tau(A \to B)$ is $\tau(A) \to \tau(B)$. It remains only to check that rules I–IV preserve the property that if their premisses have S-provable τ-translations, then so do their conclusions:

Rule I takes us from $A \to C$ to $A \to (A \star C)$, so the inductive hypothesis is that $\tau(A \to C)$, i.e., $\tau(A) \to \tau(C)$ is S-provable. We need to establish this for the conclusion, which means showing that $\tau(A) \to ((\tau(A) \vee p_k) \wedge \tau(C))$ is S-provable: so the result follows from the fact that this last formula is a tautological consequence of $\tau(A) \to \tau(C)$.

We pass to the case of rule II. The inductive hypothesis here is again that $\tau(A) \to \tau(C)$ is S-provable, and we want to be able to infer that $((\tau(A) \vee p_k) \wedge \tau(C)) \to \tau(C)$ is. The latter is evident even without appeal to the inductive hypothesis.

For rule III, the inductive hypothesis is that $((\tau(A) \vee p_k) \wedge \tau(C)) \to \tau(A)$ is S-provable, where p_k occurs only as displayed, and the desired conclusion is that $\tau(C) \to \tau(A)$ is S-provable. By uniformly substituting $\tau(C)$ for p_k we have that $((\tau(A) \vee \tau(C)) \wedge \tau(C)) \to \tau(A)$ is provable in S. But the antecedent here is CL-equivalent to $\tau(C)$, so this implication is similarly equivalent to the one that had to be shown to be S-provable.

In the case of rule IV, the inductive hypothesis is that $C \to ((\tau(A) \vee p_k) \wedge \tau(C))$ is S-provable, which implies that $C \to (\tau(A) \vee p_k)$ is, in which case we may substitute $\tau(A)$ for p_k and end up with something obviously equivalent to the desired conclusion that $\tau(C) \to \tau(A)$ is S-provable. ∎

Note that we could have defined τ as in the above proof but setting $\tau(A \star C) = \tau(A) \vee (p_k \wedge \tau(C))$ instead – cf. the earlier discussion of the modular law (p. 309) – and in that case the argument proceeds exactly as above with one difference: now it is the inductive hypothesis in the case of rule I rather than rule II that is unnecessary.

COROLLARY 4.7.30 *Rules I–IV conservatively extend any modal logic.*

Proof. We must show that with S, S^+, as in Theorem 4.7.29, any $\star$-free formula provable in S^+ is already provable in S. Let A be such a formula. Since $\vdash_{S^+} A$, by Thm. 4.7.29, $\vdash_S \tau(A)$, for the translation τ under consideration there. But since A is not constructed with the aid of $\star$, $\tau(A) = A$, so $\vdash_S A$. ∎

Thus we reach a point with $\star$ like that reached for $\circ$ in Coro. 4.7.17 on p. 313 (and the paragraph following it): take any deontic logic in which Op provably implies Pp and not conversely, and we have $Op \star Pp$ as a strict intermediary of the kind that the passage from Prior at the start of this section queried the existence of.

A word of clarification may be in order on the subject of Theorem 4.7.29. We are currently taking $\star$ as a binary connective in its own right, which means that the only propositional variables occurring in the formula $A \star C$ for any given A, C, are the variables occurring in A or C (or both). The proof of Theorem 4.7.29 does not reconstrue $\star$ as a mapping which takes formulas A and C to the formula $(A \vee p_k) \wedge C$, but provides a mapping (namely, τ) which maps the formula $A \star C$, in which the variable p_k, being foreign to A and C, does not occur, to a formula in which it does occur,[265] and then Coro. 4.7.30 uses a key fact about this mapping (provided by Thm. 4.7.29) to conclude that the rules governing the *bona fide* connective $\star$ provide a conservative extension of any modal logic.

Now all of this would be unnecessary if we could show that any normal modal logic already provided the materials in terms of which our binary $\star$, required only to satisfy rules I–IV, could be defined. Since this includes the modal logic with no modal operators but only Boolean connectives, this amounts to asking if a truth-functional interpretation could be found for $\star$. That would be analogous to the question answered negatively by Prop. 4.7.12 for the singulary connective # under discussion there. We can return a quick negative answer to the present question too, via the following considerations. Because of rules I and II, the truth-function concerned would have to be idempotent. Now, all binary truth-functions (in two-valued logic!) which are idempotent are associative, as was observed in Exercise 1.2.4 on p. 11 (in the preamble to which, this terminology was defined). Taking idempotence and associativity together, we see that the formula

$$((p \star q) \star q) \to (p \star q)$$

must be tautologous. But then, applying rule III, we get $q \to (p \star q)$, whence by rule IV, $q \to p$. Since this last formula is not tautologous, there is no such truth-functional interpretation available for $\star$. The argument just given is conclusive but underinformative in making use of all four of our $\star$ rules, since by appeal to the same considerations as were used in the proof in Proposition 4.7.12, we can easily see that even with just rules III and IV, the prospects for a truth-functional interpretation are doomed:

PROPOSITION 4.7.31 *There is no binary connective $\star$ definable in classical propositional logic for which the rules III and IV both preserve tautologousness.*

---

[265] Roughly speaking, the formula just mentioned, $(A \vee p_k) \wedge C$, where p_k is the first variable not occurring in A or C. More exactly, since A and C might themselves be constructed with the aid of $\star$, the formula $(\tau(A) \vee p_k) \wedge \tau(C)$.

4.7. "NOTHING IN BETWEEN": A REMARK BY A. N. PRIOR

Proof. As in the proof of Prop. 4.7.12 (p. 312), we appeal to the fact that every substitution-invariant formula-to-formula rule which preserves tautologousness preserves truth on any given Boolean valuation. So suppose that the substitution-invariant rules III and IV preserve tautologousness, for some truth-functional interpretation of $\star$, and therefore by the fact just cited, also truth an arbitrary Boolean valuation. For any such valuation v, and any formulas A, C, either $v((A \star C) \to A) = T$ or else $v(C \to (A \star C)) = T$. Since these are premises respectively for rules III and IV, with (in each case) conclusion $C \to A$, we conclude that $v(C \to A) = T$. Since v was arbitrary, this would mean that every implicational formula $C \to A$ is a tautology, contrary to fact. We conclude that the hypothesis that there was a truth-functional interpretation of $\star$ on which both III and IV preserved tautologousness is false. ■

The possibility of a hybrid-of-Boolean-connectives interpretation arises here as earlier. For convenience, here are the four rules again:

$$\text{I} \quad \frac{A \to C}{A \to (A \star C)} \qquad \text{II} \quad \frac{A \to C}{(A \star C) \to C}$$

$$\text{III} \quad \frac{(A \star C) \to A}{C \to A} \qquad \text{IV} \quad \frac{C \to (A \star C)}{C \to A}$$

Let $\mathcal{V}_\wedge$, $\mathcal{V}_\vee$, $\mathcal{V}_1$ and $\mathcal{V}_2$ be the classes of Boolean valuations on which $\star$ is associated respectively with the conjunction truth-function, the disjunction truth-function, the (binary) projection-to-the-first-coordinate and the projection-to-the-second coordinate functions.

PROPOSITION 4.7.32 *Rules I–IV preserve* $(\mathcal{V}_\wedge \cup \mathcal{V}_\vee)$-*validity and also* $(\mathcal{V}_1 \cup \mathcal{V}_2)$-*validity.*

EXERCISE 4.7.33 (*i*) For each of the rules I–IV, say whether it preserves the property of being true on an arbitrarily selected $v \in \mathcal{V}_\wedge \cup \mathcal{V}_\vee$, and likewise for $v \in \mathcal{V}_1 \cup \mathcal{V}_2$.
(*ii*) Would Prop. 4.7.32 be correct for $(\mathcal{V}_\wedge \cup \mathcal{V}_2)$-validity? ✠

One can also imagine a semantic apparatus like the *general* frames which have been used in modal logic – though touched on only tangentially here (see the Digression on p. 83) – and which here we can think of schematically as having the form $\langle W, \sim\sim, \star, \mathbb{P} \rangle$ in which the "$\sim\sim$" holds a place – or several places – for various unspecified pieces of semantic apparatus (accessibility relations and the like) which may be present depending on the nature of the logic whose extension with the addition of the intermediary-providing $\star$ connective we are considering. We have simply used the same symbol for an operation in the structures which will be used to interpret that connective. $\mathbb{P}$ is a collection of subsets of W (intuitively: the set of available or 'admissible' propositions) closed under intersection and complementation (relative to W) and appropriate operations – a phrase explained presently – corresponding to the apparatus schematically indicated by the "$\sim\sim$", as well under the binary operation $\star$, on which we impose conditions corresponding to rules I–IV, each of them to be understood as holding for all $X, Y \in \mathbb{P}$:

(*i*) If $X \subseteq Y$ then $X \subseteq X \star Y$.   (*ii*) If $X \subseteq Y$ then $X \star Y \subseteq Y$.
(*iii*) If $X \star Y \subseteq X$ then $Y \subseteq X$.   (*iv*) If $Y \subseteq X \star Y$ then $Y \subseteq X$.

Observe (for future reference) that (*i*) and (*ii*) have the consequence that the operation $\star$ is idempotent. (Put $X = Y$ in those conditions.)

The above reference to $\mathbb{P}$'s being closed under "appropriate operations" simply means that we need to make sure that for every formula A, $\|A\|$, the truth-set of A in a given model $\mathcal{M}$ – see below – on such a general frame, is an element of $\mathbb{P}$. For $\star$ itself the intention is that $\|A \star B\|$ should be $\|A\| \star \|B\|$, or in other words, that

$$\mathcal{M} \models_x A \star B \text{ if and only if } x \in \|A\| \star \|B\|.$$

Note that because we required that $\mathbb{P}$ was closed under the operation $\star$, $\|A\| \star \|B\|$ will be one of its elements as long as $\|A\|$ and $\|B\|$ are. It is the presence of the collection of admissible propositions $\mathbb{P}$ that makes the underlying structure a "general" frame – not the fact that we have not fussed about what goes on in the part labelled "$\sim\!\sim$". (See the Digression on p. 83.) The traditional frames emerge as the special case in which $\mathbb{P} = \wp(W)$, but for a reason which will be clear presently, we cannot restrict our attention to that special case. The Boolean connectives are given the usual treatment, the effect of which in the above notation is that, for $\wedge$, for instance, $\|A \wedge B\| = \|A\| \cap \|B\|$.[266] A formula is *valid* on a general frame if it is true at all points (elements of W) in any model $\mathcal{M} = \langle W, \sim\!\sim, \star, \mathbb{P}, V \rangle$, where V is constrained by the condition that for each p_i, $V(p_i) \in \mathbb{P}$. (Satisfying this constraint is what makes $\mathcal{M}$ a *model on* the general frame in question; it supplies the basis clause for a proof by induction on the complexity of A that for any formula A, $\|A\| \in \mathbb{P}$. More explicitly, we should write something like "$\|A\|^{\mathcal{M}}$" for $\|A\|$ here, but as usual – e.g., in the Digression on p. 54 – we omit this.)

In view of the uninformative way we have simply mimicked the rules I–IV with the conditions (i)–(iv) the following will come as no surprise:

PROPOSITION 4.7.34 *All substitution instances of tautologies are true throughout any model $\mathcal{M}$ on a general frame satisfying conditions (i)–(iv) and rules I–IV preserve the property of being true throughout any such $\mathcal{M}$.*

An additional rule with the preservation behaviour attributed to rules I–IV here is the congruentiality rule for $\star$:

$$\frac{A \leftrightarrow A' \qquad B \leftrightarrow B'}{(A \star B) \leftrightarrow (A' \star B')}$$

The earlier 'hybrid of binary truth-functions' treatment of $\star$ exhibits an even stronger replacement property, sometimes called extensionality,[267] namely that all instances of the following schema are valid (whether validity is understood as $(\mathcal{V}_\wedge \cup \mathcal{V}_\vee)$-validity or as $(\mathcal{V}_1 \cup \mathcal{V}_2)$-validity):

$$((A \leftrightarrow A') \wedge (B \leftrightarrow B')) \to ((A \star B) \leftrightarrow (A' \star B')),$$

as is easily verified.

EXERCISE 4.7.35 Say that a monomodal logic is *extensional* when its non-Boolean operator $\Box$ satisfies an extensionality condition analogous to that just mentioned for $\star$, i.e. if it contains all formulas of the form $(A \leftrightarrow B) \to (\Box A \leftrightarrow \Box B)$. Identify, by an axiomatic description such as "**KD4**", "**K** $\oplus \Box\Box p \leftrightarrow p$", etc. (not that either of the logics just cited is extensional), the three normal modal logics which are extensional, and say for which of them $\Box$ is truth-functional in the sense that some formula of the form $\Box p \leftrightarrow C(p)$ is provable in it, with $C(p)$ a formula constructed from p with the aid of the Boolean connectives. (Cf. Exercise 2.7.2.) ✠

After Proposition 4.7.28, the remark was made that, apropos of Prop. 4.7.27, while what we wanted from rules I–IV was the latter result, it needed to be checked that "they don't deliver too much more than that", and satisfied ourselves that at least they were conservative in their effect. But for our final observation, let us note that these rules do in fact do more than they were designed to do: provide strict intermediaries $A \star C$ between formulas A and unilaterally implied formulas C.

---

[266] The treatment of the non-Boolean connectives (other than $\star$) will of course depend on what they are and on the unspecified "$\sim\!\sim$" aspects of the general frames. They need not concern us here as they do not figure essentially in the rules I–IV.

[267] See Williamson [1192], §3.2 of Humberstone [594] or §2 of [600], for example, as well as Revision Exercise 2.7.2(i) on p. 129 above, where the 1-ary version of extensionality was considered, as it was also in Exercise 4.6.17(i).

4.7. "NOTHING IN BETWEEN": A REMARK BY A. N. PRIOR

To make this point it is convenient to continue working with the general frames we have at our disposal and to consider tailoring the conditions (i)–(iv) more closely to the purpose at hand. In fact to mirror the strict intermediaries idea directly, we should forget those conditions and replace them with the following condition on our the general frames $\langle W, \leadsto, \star, \mathbb{P}\rangle$, for all $X, Y \in \mathbb{P}$ (for which as before we require $X \star Y \in \mathbb{P}$):

$$\text{If } X \subsetneq Y \text{ then } X \subsetneq X \star Y \subsetneq Y.$$

The remainder of our earlier presentation remains intact: the constraint that V should assign elements of $\mathbb{P}$ to the propositional variables to obtain a model, the definition of truth, etc. Note that the above condition follows from the earlier (i)–(iv), which makes evident the need to work, there and here, with *general* frames: if we are considering arbitrary subsets of W, then we could not guarantee to be able to find a set properly between X and an arbitrarily selected proper superset of X – consider for example the proper superset(s) $X \cup \{u\}$ for $u \in W \smallsetminus X$.

It is not hard to see that our new condition, though implied by (i)–(iv), does not imply them, and to that extent there has been an element of excess strength in the earlier discussion. The following example is reminiscent of aspects of the main construction given in Restall [954] (see note 263, p. 316 above).

EXAMPLE 4.7.36 Consider the set of all rational numbers (the real numbers would do equally well for this purpose), and for the duration of this example, understand 'p', 'q', 'r' as ranging over this set, rather than playing their usual role (as propositional variables). Let X_p be $\{r \mid p \leqslant r\}$, so that

$$X_p \subseteq X_q \text{ if and only if } p \leqslant q,$$

and, more to the point, since this will allow us to exploit the density of the order $<$ to obtain a dense $\subsetneq$ for the sets concerned:

$$X_p \subsetneq X_q \text{ if and only if } p < q.$$

From this last observation we conclude that the condition of current interest is satisfied, since $X_p \subsetneq X_q$ implies that $X_p \subsetneq X_{(p+q)/2} \subsetneq X_q$. Accordingly define $\star$ thus:

$$X_p \star X_q = \begin{cases} X_{(p+q)/2} & \text{if } p < q \\ 2.41 & \text{otherwise.} \end{cases}$$

The reference to 2.41 in the second ('don't care') case is to a rational number chosen at random, to keep the $\star$ operation defined everywhere, and to do so in such a way that the operation defined is clearly not idempotent – a quick way to see that conditions (i) and (ii) from our earlier list are not both satisfied. Thus the earlier fourfold condition is strictly stronger than the current condition.

However, let us not hide the fact that if the $\mathbb{P}$ of the present example is supposed to comprise the various X_p for rational p (the latter themselves making up W), then it is patently *not* closed under the set-theoretic operations – in particular complementation (relative to W) – corresponding to the Boolean connectives. We will simply leave this lacuna unfilled here, noting that the construction appearing in Restall [954] fixes the problem. ◀

EXERCISE 4.7.37 Are *any* of the earlier conditions (i)–(iv) satisfied by $\star$ in the above example? ✣

Let us return to what we have been calling the current condition, i.e. the condition that strictly between any X and any proper superset Y of X, lies the set $X \star Y$, and inquire as to what changes this condition above might suggest for rules I–IV. To that end, let us decompose the condition into four components, formulated with "$\subseteq$" rather than "$\subsetneq$" (since we want rules in which it is the former that corresponds to provable implication):

$$(X \subseteq Y \ \& \ Y \not\subseteq X) \Rightarrow X \subseteq X \star Y \tag{1}$$
$$(X \subseteq Y \ \& \ Y \not\subseteq X) \Rightarrow X \star Y \not\subseteq X \tag{2}$$
$$(X \subseteq Y \ \& \ Y \not\subseteq X) \Rightarrow X \star Y \subseteq Y \tag{3}$$
$$(X \subseteq Y \ \& \ Y \not\subseteq X) \Rightarrow Y \not\subseteq X \star Y \tag{4}$$

Now, there are negated occurrences of "$\subseteq$" here (appearing as "$\not\subseteq$"), so we get rid of them by moving them across to the other side of the "$\Rightarrow$":

$$X \subseteq Y \Rightarrow (X \subseteq X \star Y \ or \ Y \subseteq X) \tag{1'}$$
$$(X \subseteq Y \ \& \ X \star Y \subseteq X) \Rightarrow Y \subseteq X \tag{2'}$$
$$X \subseteq Y \Rightarrow (X \star Y \subseteq Y \ or \ Y \subseteq X) \tag{3'}$$
$$(X \subseteq Y \ \& \ Y \subseteq X \star Y) \Rightarrow Y \subseteq X \tag{4'}$$

Rules III and IV can easily be altered, giving III* and IV* below, to pick up an extra premiss, so that the two premisses correspond to the two conjuncts of the antecedents of (2') and (4') respectively:

$$\text{III*} \quad \frac{A \to C \quad (A \star C) \to A}{C \to A} \qquad \text{IV*} \quad \frac{A \to C \quad C \to (A \star C)}{C \to A}$$

But (1') and (2'), with their disjunctive consequents, do not straightforwardly lend themselves to a corresponding rule formulation (for the same reason that we observed in Section 2.6 the 'rule' of n-ary disjunction for, $n \geqslant 2$, is not really any kind of rule at all). We close our discussion at this *impasse*, leaving the interested reader to ponder the question of how to axiomatize the complete logic of, say, just the connective $\star$ and the Boolean connectives, with validity over the class of general frames satisfying (1)–(4) (equivalently, (1')–(4')),[268] our main point having been made: that this logic is weaker than the corresponding logic with rules I–IV governing $\star$.

4.8 The Fatal Disjunction: Danielsson on Ross's Paradox

We can usefully bring some of the findings of the previous section to bear on one relatively recent reaction to Ross's Paradox, namely that found in Danielsson [230], who extracts from von Wright's 1951 paper [1204] as a basic deontic logic, the smallest congruential modal logic containing all formulas instantiating the following schemata (here numbered as in [230]):

(A1) $OA \to \neg O \neg A$
(A2) $(OA \land OB) \to O(A \land B)$
(A3) $O(A \land B) \to OA$

Extracting this from von Wright, as Danielsson observes, involves overlooking some things, most conspicuously the fact that since von Wright takes "O" as a predicate of act types rather than as a sentence operator (1-ary connective), he is not actually offering us a modal logic at all; and congruentiality is not mentioned along with the axioms in Danielsson on the opening page of Danielsson [230] but introduced via the rule $A \leftrightarrow B \ / \ OA \leftrightarrow OB$ two pages later. Because of (A3), this congruentiality (or 'replacement') rule, which we shall call $OCong$, secures that the current logic, though not normal, is monotone, and that we could equivalently replace the congruentiality rule with an $OMono$ rule and drop (A3) altogether. ((A1) and (A2) are respectively **D** and Aggregation from our discussion in Chapters 1 and 2.)

---

[268] Without supplying any filling for the schematically indicated "$\sim\sim$" part of these general frames, that is.

4.8. THE FATAL DISJUNCTION: DANIELSSON ON ROSS'S PARADOX

EXERCISE 4.8.1 This exercise is prompted by the comment just made to the effect that the present logic is not normal. Is the following true or false? (Give reasons for your answer.) "The smallest modal logic extending the logic axiomatized here, by the addition of $O\top$ is **KD**." ✠

Let us now recall (from p. 243) that while Ross's Paradox originating in the logic of imperatives as involving the apparently problematic inference from 'Post the letter' to 'Post the letter or burn it', in deontic logic it is taken to be the corresponding inference from, for example, (*) to (**), the latter being the kind of thing in the passage quoted from Danielsson below is called 'the fatal disjunction':

(*)   James ought to post the letter.
(**)   James ought to post the letter or burn it.

The problem, then, is with the provability in the basic logic described above – or indeed in any monotone modal logic – of all formulas of the form $OA \to O(A \vee B)$, and while many some responses to Ross's Paradox involve saving monotony while softening the blow of (2)'s following from (1) by invoking considerations of what it would be appropriate to say rather than what would be the case – Gricean implicature and all that – Danielsson's own proposal is to block the unrestricted derivation of formulas $OA \to O(A \vee B)$, by restricting (A3) itself.[269] An initial version of Danielsson's ingenious response runs as follows. (The eventual proposal will emerge with (A3″) below.)

First, we decide to pursue the matter bimodally and will help ourselves to a possibility operator Danielsson writes as M, as in the use of L and M for $\Box$ and $\Diamond$ in some discussions of alethic modal logic, though the exact logic which is assumed to govern them is not specified. (It will be enough here to assume that the operator L which is dual to M is governed by some normal modal logic. More detail on Danielsson's favoured interpretation appears in note 270 below.) Next we replace (A3) in the above axiomatization with a restricted version:

(A3′) $\big(O(A \wedge B) \wedge M(B \wedge \neg A)\big) \to OA.$

To see how this blocks a proof of the Ross-paradoxical $Op \to O(p \vee q)$, let us consider what such a proof would have looked like using (A3) and the congruentiality rule, rather than helping ourselves to the derivable rule OMono:

(1) $O((p \vee q) \wedge p) \to O(p \vee q)$ (A3)
(2) $((p \vee q) \wedge p) \leftrightarrow p$ TF
(3) $O((p \vee q) \wedge p) \leftrightarrow Op$ 2, OCong
(4) $Op \to O(p \vee q)$ 1, 3 TF

In terms of the schematic letters used to formulate (A3), A here is the formula $p \vee q$ and B the formula p. So when we consider an appeal to the weaker (A3′) in place of (A3) at line 1 here, what we would be considering is the formula:

$$\big(O((p \vee q) \wedge p) \wedge M(p \wedge \neg(p \vee q))\big) \to O(p \vee q),$$

and this newly added (conjunct of the) antecedent imposes a condition we know can never be satisfied: since $p \vee q$ follows from p, it is *not* possible to have p true together with $\neg(p \vee q)$. Putting it another way, in the present instance the centered formula above is provable but this is no form of Ross's Paradox, since on the minimal assumptions about M that we have made, for any formula A whatever

$$(C \wedge M(p \wedge \neg(p \vee q))) \to O(p \vee q)$$

is provable, as indeed is, for any formulas C and D, the formula

$$(C \wedge M(p \wedge \neg(p \vee q))) \to D.$$

---

[269]See our earlier discussion at p. 243 and also Danielsson [230] for descriptions of further responses that have been made.

This is because the logic already tells us that the negation of the second conjunct of the antecedent is provable. Danielsson's restriction will sometimes not have this dramatic form, which results from the formal incompatibility between p and $\neg(p \vee q)$. If in a particular application we were thinking of a case in which we agreed that it ought to be that A and B (now using these letters as schematic for English sentences), where A followed from B but not as a matter of the current logic, or perhaps not on the basis of any kind of formal logic at all. (In fact it may not be a purely logical matter at all, but of A's following from B's truth and the prevailing historical circumstances: see note 270 below.) For example – taking the case where the present logical resources are not up to the task since there is no epistemic operator – we thought of A as "James's children are safe" and B as "James knows that his children are safe", since it *not* possible for James to know that his children are safe without its being the case that his children are safe", we will not have the required 'second premise' to take us from the premise that it ought to be that James's children are safe and he knows they are, to the conclusion that it ought to be that James's knows that his children are safe. While this does not seem to be an especially attractive outcome, it is being urged here not as an objection to Danielsson's proposal – of which we have so far only seen a preliminary version in any case, and which faces a much more interesting difficulty we shall get to in due course – but simply as illustrating the way in which the additional "$M(B \wedge \neg A)$" restriction works. We could put it less negatively if we preferred, writing (A3′) in the form:

$$O(A \wedge B) \to (OA \vee L(B \to A));$$

or again, still equivalently:

$$L(B \to A) \vee \big(O(A \wedge B) \to OA\big).$$

The choice of A and B still follows Danielsson's; personally, I would have preferred some re-lettering throughout, along with commuting some conjuncts – as we may, since congruentiality is not being restricted – and written (A3′) as $\big(O(A \wedge B) \wedge M(A \wedge \neg B)\big) \to OB$ to start with, in which case this latest incarnation of the schema would have appeared as:

$$L(A \to B) \vee \big(O(A \wedge B) \to OB\big).$$

However, to follow Danielsson's discussion in the passage to be quoted below, we revert to his preferred formulation of (A3′) above and turn now to the improved formulation he offers, with an additional possibility condition over and above that in (A3′). This further condition is credited by Danielsson to Erik Carlson. Although Danielsson does not use this labelling, let us call it (A3″); we follow Danielsson in not cluttering up the conjunctive antecedent with further parentheses, but omit some time-representing subscripts on the operators he uses but which play no role in the discussion:

(A3″) $\quad \big(O(A \wedge B) \wedge M(B \wedge \neg A) \wedge M(\neg B \wedge \neg A)\big) \to OA.$

Quite a mouthful, then. But it is worth quoting Danielsson at some length to have his way of motivating the first restriction on (A3), already embodied in (A3′), as well as newly added M-conjunct. In this passage, Danielsson's own reference to (A3″) – which he calls (A6), having in the meantime mentioned some further candidate axiom schemes (A4) and (A5) which are not part of the present proposal. Also the initial references to its being an open question whether such-and-such as glosses on M-formulas should really be to its being an open possibility that such-and-such:[270]

---

[270] This is because on p. 42 Danielsson introduces M as the dual of a now-unpreventability operator of the kind mentioned above on p. 299. Writing this, as have been doing, as L (in fact Danielsson writes N instead), LA means that there is nothing anyone can do now to prevent its being the case that A, or: it is inevitable that A. So MA amounts to the possibility's still being open that it be the case that A, and this does not imply $M\neg A$, and so talk of it being an 'open question' whether such-and-such not only introduces an irrelevant epistemic element, but switches from possibility to contingency.

4.8. THE FATAL DISJUNCTION: DANIELSSON ON ROSS'S PARADOX

> Since it is clearly not an open question whether you post-the-letter-but-neither-post-the-letter-nor-burn-it or not (it is simply impossible to post the letter but neither post it nor burn it), we cannot use (A3″) to derive "you ought to post the letter or burn it" from "you ought to post the letter and post-it-or-burn-it". (The second clause in the antecedent of (A3″) is the operative one.) Nor can we derive the fatal disjunction from "you ought to post-the-lettter-or-burn-it and not burn it", since you cannot not-post-the-letter and not-burn-it and at the same time burn it (the very last clause in the antecedent is now the operative one). (...) (A3″) says that we can detach the obligation that A and the obligation that B from the obligation that A-and-B provided that A and B are alternatives which are independent of each other, in the sense that each can be realized without the other, and neither of them is realized just because the other is not, in the situation under consideration. Why this proviso? Well, the fact that (A3″) solves Ross's Paradox while still permitting the inference from "A and B ought to be" to "A ought to be" in the great majority of cases may be reason enough. Perhaps we could also say that the proviso explains why, for instance, *you close the door* seems to be a proper part of *you close the door and turn on the light* in a sense in which *you close the door or turn on the light* is not a proper part of *you close the door and either close the door or turn on the light*. This notion of being a part fits the basic intuitions behind (A3″). The proper parts of a complex obligation are themselves obligations.[271]

Now that we have the proposal before us, we can begin to assess it. Before looking at whether (A3″) is up to the task demanded of it, however, a remark is in order on final part of the above quotation from Danielsson.

REMARK 4.8.2 While italics are often used in place of quotation marks, it is clear that the talk of one thing being or not being a part of another toward the end of the passage above is not meant to concern one sentence's being a part of another, but more something like a relation between the states of affairs involved. Making this explicit, perhaps one might say:

> Your closing the door seems to be a proper part of your closing the door and turning on the light, in a sense in which your closing the door or turning on the light is not a proper part of your closing the door and either closing the door or turning on the light.

It remains unclear whether the "is a proper part of" is intended in its usual technical sense (meaning "is a part of but not the whole of") or instead in its informal sense "is properly speaking, a part of" (or perhaps "is a genuine part of"), but either way, since this is supposed to underpin a proposal in a congruential deontic logic whose underlying non-modal logic is classical propositional logic, "your closing the door and either closing the door or turning on the light" should be interchangeable with the shorter description "your closing the door". This is an example of the *absorption law* of that logic: the equivalence figuring in line (2) of the derivation (1)–(4) above. With this simplification in place we can see that the parthood intuitions essentially underlie the oft-noted asymmetry between conjunction-to-conjunct inferences and disjunct-to-disjunction inferences, an intuitive asymmetry which the absorption equivalences (that mentioned above and the dual form) threaten to collapse together. Some discussion of the asymmetry may be found on pp. 680 and 961 (top paragraph and Remark 7.12.4(*ii*), respectively) of Humberstone [594]; the former reference is to a discussion of the problematic topic of logical subtraction, and it is these very problems that suggest the line of investigation pursued below. (Essentially: the second and third conjuncts of the conjunctive antecedent of (A3″) try to select and negate the conjuncts of the $A \wedge B$ in the scope of the "O" on the left, but these cannot be uniquely recovered from the equivalence class of $A \wedge B$ itself, just as in the subtraction case, one cannot recover the equivalence class of A from those of $A \wedge B$ and B to obtain the result of subtracting B from $A \wedge B$, since there will be many A' not

---

[271]This quotation spans pages 43 and 44 of [230], occupying two paragraphs split at the ellipsis point here (where a remark about an earlier part of the discussion has been removed). I have also corrected "none of them is realized" to "neither of them is realized", but left in place the double use of A and B (in fact in roman rather than italic in the original), sometimes standing where a sentence might stand and sometimes where a name might stand.

equivalent to A but such that $A' \wedge B$ is equivalent to $A \wedge B$. The talk of equivalence here is a matter of provable equivalence in any plausible logic that might be under consideration.) One reaction which would allow congruentiality to be retained would be to use an underlying logic in which the absorption equivalences were not forthcoming; examples of this include Humberstone [577] and Fine [299]. But for Danielsson's current project, neither rejecting of these equivalences nor suspending the congruentiality condition is on the table. ◀

What made for trouble, as those who think of Ross's Paradox as worth avoiding, in the (1)–(4) derivation above was the absorption equivalence in line (2): $((p \vee q) \wedge p) \leftrightarrow p$. This was blocked by the first M-condition in (A3″), which is here rewritten but with the schematic letters appearing (for easier comprehension) in the same order in each of the three conjuncts of the antecedents:

(A3″) $\bigl(O(A \wedge B) \wedge M(\neg A \wedge B) \wedge M(\neg A \wedge \neg B)\bigr) \to OA$.

For the blockage just alluded to, we envisage taking AB as $(p \vee q) \wedge p$, something we know is equivalent to p itself and find that because the second conjunct of this conjunction implies its first conjunct, we can never have $M(\neg A \wedge B)$ true. But what if we could find, for the same choice of A, another choice of B with the effect that $A \wedge B$, i.e., $(p \vee q) \wedge B$, was (logically) equivalent to p? Then perhaps there would be a chance for the two M-conditions to be satisfiable in such a way that we would in practice have Ross's Paradox on our hands again. The idea here can be brought out by noting apropos of (A3″) that although if we are given a formula $A \wedge B$ we have uniquely determined for us the formulas in the M-conditions, if all we have is a 'formula to within provable equivalence' (a *proposition*, on one use of that term), there are no such uniquely given conjunctions – even to within equivalence – prefixing M to which would yield corresponding M-conditions. One and the same formula will be equivalent to many conjunctions $A_1 \wedge B_1, A_2 \wedge B_2, A_3 \wedge B_3, \ldots$ for which the formulas figuring in the first M-condition, $\neg A_i \wedge B_i$ will not be equivalent to each other, and likewise wit the second M-condition and $\neg A_i \wedge \neg B_i$. All we need to find is one such decomposition that works in a given case to have Ross's Paradox on our hands. We do not need every such decomposition to pass the test set by the M-conditions. For example, even if decomposing $C \wedge D$ into conjuncts C and D passes the test in the sense that in the circumstances envisaged, C's truth is compatible with D's falsity and so is C's falsity, decomposing $C \wedge D$ into the conjuncts $C \leftrightarrow D$ and $C \vee D$ and of an equivalent conjunction – pause to check the equivalence in question if you are not familiar with it – can never pass the first M condition, since $C \leftrightarrow D$'s falsity is not never compatible with $C \vee D$'s truth.

We return to the problem of finding B for which $(p \vee q) \wedge B$, is classically equivalent to p:

EXERCISE 4.8.3 (*i*) Show a formula B satisfies the above condition if and only if (1) $\vdash p \to B$, and also (2) $\vdash B \to (q \to p)$ (where $\vdash$ may be taken to be $\vdash_{CL}$; this is equivalent to saying that for all A, $\vdash A$ iff every modal logic contains A).

(*ii*) Show that if attention is restricted to formulas constructed out of the two propositional variables p, q, those satisfying (1) and (2) of part (*i*) are precisely (to within equivalence) the formulas p and $q \to p$. (For a similar issue, see Prop. 4.7.3, p. 307, and its proof.) ✠

Given Exercise 4.8.3(*i*), the news in Exercise 4.8.3(*ii*) is not good, if one wants to remain within the two-variable fragment, since if we check the two candidates for a B for which $\vdash ((p \vee q) \wedge B) \leftrightarrow p$, we find that in each case one of the M-conditions is unsatisfiable, since with the former choice of B we have $\vdash \neg(\neg(p \vee q) \wedge B)$ and in the latter case $\vdash \neg((p \vee q) \wedge \neg B)$. (Thus in any normal modal logic we have the results of replacing the intial $\neg$ in these cases by $\neg M$ – a version of Necessitation for L – rendering the M-condition unsatisfiable.) This is not really news: certainly the first choice is just that already considerd before and within the quotation from Danielsson above. (The second choice $q \to p$, or $p \vee \neg q$ if you prefer, does not quite coincide with the second of the two worries mentioned: you ought to post the letter or burn it, and not burn it, since this would be $O((p \vee q) \wedge \neg q)$, and what lies in the scope of O here is not equivalent to p.)

4.8. THE FATAL DISJUNCTION: DANIELSSON ON ROSS'S PARADOX

There is, however, no reason to confine ourselves to the two-variable fragment. Exercise 4.8.3(i) tells us that anything between q and its consequence $p \to q$ will serve as a B whose conjunction with $p \lor q$ is equivalent to p, and the bad news just recalled means that we will need to consider something strictly between these formulas. The preceding section told us how to find such strict intermediaries y stepping out of the fragment concerned, in Proposition 4.7.10 (p. 310), so let us help ourselves to that free advice and consider now B as the following intermediary between p and $q \to p$:

$$(p \lor r) \land (q \to p).$$

(Here r is the p_k of the proof of Prop. 4.7.10.) Since this is provably implied by p and provably implies $q \to p$, by the 'if' direction of the result mentioned in Exercise 4.8.3(i), when conjoined with $p \lor q$, this formula yields something equivalent to p.[272] So it remains only to check that the M-conditions can be satisfied for the O-formula concerned:

$$O\big((p \lor q) \land ((p \lor r) \land (q \to p))\big).$$

The M-conditions to conjoin with this in the relevant instance of the schema (A3″) to arrive at its antecedent (with $O(p \lor q)$ as consequent) are then, first for the $M(\neg A \land B)$ case:

$$M(\neg(p \lor q) \land ((p \lor r) \land (q \to p))) \tag{M$_1$}$$

and next, for the $M(\neg A \land \neg B)$ case:

$$M(\neg(p \lor q) \land \neg((p \lor r) \land (q \to p))) \tag{M$_2$}$$

By contrast with our earlier candidates for B (namely $B = p, q \to p$) the negations of (M$_1$) and M$_2$ are unprovable (in any consistent normal modal logic other than **KVer**, in which all negated M-formulas are provable), since the formula after the "M" in each case is capable is true on some Boolean valuation. More informatively, the two formulas involved, the A ($= p \lor q$) and the B ($= (p \lor r) \land (q \to p)$), are truth-functionally independent (cf. the reference to independence in the earlier quotation from Danielsson) in the sense that there are Boolean valuations assigning to these formulas the values T and T, respectively, T and F, respectively, F and T respectively, and F and F respectively. This claim takes the propositional variables p, q, r at face-value as being similarly independent, and for the famous example in Ross's Paradox, this may not be so: there p and q say that the subject posts the letter and that the subject burns the letter, and these may be held to be incompatible (so that under the circumstances $\neg M(p \land q)$ is true), perhaps on the grounds that putting a burning letter into the postal system hardly counts as posting it. But excluding these cases this makes no difference to the independence of $p \lor q$ and $(p \lor r) \land (q \to p)$, as we can see by looking below the single horizontal line running across Table 4.3. The cases in which p and q are true together lie above this line and below it we see all combinatorially possible $\{T, F\}$-assignments for $p \lor q$ and $(p \lor r) \land (q \to p)$.

Putting this all together, then, if we begin in a informal natural deduction style with the assumption that James ought to post such-and-such particular letter and represent this as Op, we now proceed to reformulate this using q and r which we think of representing respectively, "James burns the letter" and "James whistles the Marseillaise" – the latter selected as something which can be true or false independently of any truth-values the current p and q may have:

$$O((p \lor r) \land (q \to p)),$$

---

[272] Any residual doubts can be settled by putting an $\land$ between the two compound formulas on the right of Table 4.3 below and completing the truth-table, comparing the resulting column with that under p on the left. Note that while r occurs essentially in the second conjunct of this conjunction, i.e., in the formula $(p \lor r) \land (q \to p)$, in the sense of occurring in every formula which is logically equivalent to this one, it does not occur essentially in the whole conjunction (and neither does q), since this formula is equivalent to plain p.

p	q	r	$p \vee q$	$(p \vee r)$	$\wedge$	$(q \to p)$
T	T	T	T	T	T	T
T	T	F	T	T	T	T
T	F	T	T	T	T	T
T	F	F	T	T	T	T
F	T	T	T	T	F	F
F	T	F	T	F	F	F
F	F	T	F	T	T	T
F	F	F	F	T	F	T

Table 4.3: Truth-Table Calculations

since here we have simply replaced p in our original assumption by something equivalent to it, and O is being supposed to be congruential. We conjoin this new formula with the M-conditions (M_1) and (M_2) which are, by the choice of the current p, q, r, both satisfied. This gives us the antecedent of (A3″) whose consequent is $O(p \vee q)$. *Voilà* – Ross's Paradox reinstated.

Chapter 5

Applications: Doxastic and Epistemic Logic

5.1 The Logical Omniscience Issue

In this section the issue of logical omniscience will be used to introduce epistemic and then doxastic logic, as well as to air a range of qualms those enterprises have raised. Later sections raise and discuss various further topics that have arisen in the pursuit of these applications of modal logic.

There is a big difference between describing someone, Alf, say, on the one hand, as knowing that either p or q (imagine English sentences put in for these variables), and on the other hand, as either knowing that p or knowing that q. If only two candidates, b and c, are standing for election to a certain office, Alf may well know that b will win or c will win, without knowing which of them will win – where the latter amounts to: without knowing (more specifically) that b will win or knowing that c will win.[273] The latter can equally well be expressed by: knowing that p or that q, without the repetition of *know* but with repetition of *that*. This suffices to force the $Kp \vee Kq$ reading rather than the $K(p \vee q)$ reading. Here we trust the reader to recall – e.g., from p. 3 under 'Schematic Letters', that our □ operators for epistemic

---

[273]This is something of an oversimplification. In general, knowing who will φ is not a matter of knowing, concerning the individuals who will φ, that they will φ, but is something closer to knowing of each of contextually salient set of individuals, whether or not that individual will φ. (See Sharvit [1044], esp. §2, for discussion and references, as well as George [366] for a more recent application and illustration of the point; similar examples were also given in Bogusławski [105]. This complication would render the claim that $\mathbf{4}_K$ ($K\varphi \to KK\varphi$) captures the idea that we are dealing with "agents who know what they know" (Jago [618], p. 349) problematic since such an agent would also need to know, concerning what they do not know, that they do not know it (i.e., $\mathbf{5}_K$; naturally we are setting aside from the start here the reading of "Sally knows what she knows" that parallels "Sally eats what she eats".) A general discussion of other aspects of what it takes to know who or what satisfies a given condition can be found in Boër and Lycan [104];) see further Hintikka [500]. (A proposed extension of epistemic logic which takes "a knows who/what t is" for a term t, as an additional primitive construction, can be found in Plaza [885].) The issue is further complicated by the contrast between non-exhaustive and exhaustive interpretations of questions – "Where can I get coffee?" as "Tell me somewhere I can get coffee" vs. "Tell me everywhere I can get coffee" – a contrast which persists into the realm of embedded interrogatives, as with the *knowing wh-* constructions, *knowing whether* aside (or maybe not: see Barker [53]). An informative discussion of the exhaustivity issue can be found in Guerzoni [416]. In view of the theme of the present chapter, it should be noted that Åqvist and Hintikka both attempted to use epistemic logic coupled with an imperative-forming construction in order to throw light on direct (or unembedded) questions (consider the "tell me" paraphrases – the version eventually favoured by Åqvist – just given, and for the epistemic version put "make it the case that I know"). References to the relevant papers can be found in the survey article Cross and Roelofsen [222]. *Warning*: what looks like a *knowing wh-* construction may not be quite what is seems. In "Mary knows what an incompetent plumber Paul is, and what number she has to call to hire a better one," the second *what*-clause is an embedded interrogative, but the first is an embedded exclamative; their painless conjoinability here constrains the appropriate treatment of the two constructions, which had better allow for it. See Huddleston and Pullum [530] pp. 918–924 and esp. pp. 991–993.

and doxastic logic are respectively K and B. (In that discussion it was remarked that in when these notations are on the stage, in order to avoid a mass of overlapping italic capitals, the role of schematic letters for formulas would be played in this discussion by $\varphi, \psi, \ldots$ rather that $A, B, \ldots$.)

REMARKS 5.1.1 (*i*) Dutton [268], p. 364, concerning *whether*, writes:

> One interpretation of 'whether' would be in the form of an alternative 'that ... or that not ...' wherein the expanded form [of "I know whether p"] would be "I know that p or that not p". The result of such a reformulation remains odd, for what is claimed as known is expressed tautologously.

This last remark confuses "I know that p or that not p" with "I know that p or not p". (In the quotation, I have put "p" in place of Dutton's choice of a variable.)

(*ii*) Because in English the complementizer (clause-introducer) *that* can often be omitted, if someone writes (say) "Tom knows P" it is not clear whether they intend "P" to stand where a sentence could stand (behaving, accordingly, like the p or the φ of our discussion, or instead where the name of a proposition (or some such abstract object) could stand. They may of course be confused and mean now the one thing and now this other by this or similar notation; an example will be found at Remark 5.1.17, p. 356. For example, Williamson [1189] and elsewhere systematically uses lower case p and q in name position rather than in sentence position; see p. 422.

(*iii*) Although this would too detailed be a topic in natural language semantics to pursue at any length here, it should not be supposed from (*ii*) that the presence or absence of *that* with verbs propositional attitude and indirect speech makes no difference. Many such verbs do not permit deletion of the *that*, and for many that do, the sentences with and without the *that* are not quite synonymous. The *locus classicus* for this topic is Bolinger [107], from which the following examples come (p. 58*f*.). (Also relevant are Thompson and Mulac [1106], and, with a historical emphasis, Watts [1150].) In the first, you notice someone struggling to change a tyre on his car, and, as Bolinger continues,

> Feeling charitable you go over to him and say *I thought you might need some help*. Under these circumstances *I thought that* would be inappropriate. But if the other person looks at you wondering why you cam over, you might explain by saying *I thought that you might need some help*.

Similarly, and still concerned with car tyres, Bolinger offers:

> A passes a slow-moving car driven by B, and calls out to him: *Did you know you had a flat?* Alternatively, A passes B, notices the flat and says nothing, but B observes A's curiosity and calls *What are you staring at?* B replies *Did you know that you had a flat?*

Clearly the contrasts here are fairly subtle, as they are in Bolinger's verdicts on the acceptability or grammaticality of the *that*-free constructions triggered by various verbs and adjectives; thus he judges ([107], p. 39*f*.) "It's clear he did it", "It's possible he did it", and "It's unlikely he can" to be acceptable, but "It's unclear he did it", "It's impossible he did it", and "It's improbable he can" all to be unacceptable. For further discussion, though not in (to use J. L. Austin's phrase) the 'linguistic phenomenology' style of Bolinger, see pp. 952–967 of Huddleston and Pullum [530]. Differences in the grammatical options for *possible* and *impossible* are not unheard of; aside from the obvious issue of negative polarity – "It's impossible to ever get an appointment" (or with to and ever reversed) vs. *"It's possible to ever get an appointment" – there is an interesting contrast in respect of (what was once widely called) *tough-movement*: "Alfred is impossible to deal with" vs. *"Alfred is possible to deal with". (See Akatsuka [8].)

◀

> By contrast with the disjunctive case, there is less of a fuss to be made, if any, of the corresponding conjunctive distinction – the distinction between, on the one hand, knowing that both p and q ("$K(p \wedge q)$"),

5.1. THE LOGICAL OMNISCIENCE ISSUE

and both knowing that p and knowing that q ("$Kp \wedge Kq$", on the other. This is very much the pattern of distribution behaviour we are accustomed to from various notions of necessity, and found displayed by $\Box$ in, for example, all the normal modal logics between **K** and **S5** – and, more generally, any normal modal logic not extending $\mathbf{KD}_c$. (We find the same situation with belief: believing a conjunction is at least very close to believing its conjuncts, whereas believing a disjunction is much weaker than believing one of its disjuncts – though some have attempted to give belief a $\Diamond$-like treatment: see the Digression on p. 421.) If we add in the consideration that only what is true can literally be described as known,[274] we can narrow down the area of interest (for application to epistemic logic) to the range of normal modal logics between **KT** and **S5**. Hintikka, who pioneered the study of knowledge along such modal lines in [490], wrote the relevant $\Box$-operator as K_a where a is the name of the knowing subject (or cognitive agent) concerned, with P_a for the dual $\Diamond$-like notion.[275] Mostly in what follows these appear simply as K and P, since we are not considering epistemic logic with several agents at once (with the exception of Example 5.1.2 where the implication from $K_a K_b \varphi$ to $K_a \varphi$ is touched on), or the logic of common knowledge, impersonal knowledge, collective knowledge and distributed knowledge relative to a set of agents.[276]

Digression. Continuing the preceding aside about what we are not covering: numerous other recent developments in epistemic (and doxastic) logic will not be receiving attention here. In particular, nothing will said – except in an aside in the course of Example 5.4.11 (p. 397) on the logic of 'public announcements' or, more generally (considering changes in epistemic state, however induced), *dynamic* epistemic logic: (see Gerbrandy [367], van Ditmarsch and Kooi [243], Pacuit [865], Plaza [885], and references in these papers, or, for a full textbook treatment – also with extensive references – van Ditmarsch, van der Hoek and Kooi [244]), The same goes for enriching epistemic or doxastic logic with a mechanism for discussing the *evidence* or justification on which beliefs are based (see van Benthem [75], van Benthem, Fernández-Duque and Pacuit [81], Artemov and Nogina [40], Artemov [39], Fitting [312]). Nor, somewhat further afield, will anything be said on the subject of belief revision, on which see Hansson [450] and references there, or – for connections with doxastic logic – Segerberg [1036], Bonnano [110]. Further excluded from the present coverage is *autoepistemic logic*, a first-person incarnation of epistemic logic (mentioned above in the Digression beginning on p. 125) with various interesting technical features (cf. Levesque [714], Moore [824] – N.B.: this is R. C. Moore, not G. E. Moore) and philosophical applications, the latter often going by the name *autoepistemology* (see Humberstone [575] and references there, as well as Hendricks [471], pp. 89–96). And in view of the focus throughout on modal propositional rather than modal predicate logic, quantified epistemic logic will receive only incidental mention, despite the emphasis placed on it – for example as throwing light on the subject of questions and answers – by Hintikka in post-[490] work, some of it summarised and referred to in Hintikka [501]. (The phrase "second generation epistemic logic" used in the title of that paper refers to the use of Hintikka's 'independence friendly' as opposed to standard first order predicate logic, as the logic to be extended by adding epistemic operators.)
End of Digression.

---

[274] The word 'literally' is inserted here to ward off examples of what is sometimes (e.g., Hintikka [490], p. 23) called an 'inverted commas' use of the word – a terminology recalled already in Remark 4.1.1 at p. 204 above – here indicated by italics: "As the cabin pressure dropped and the lights went off, I just *knew* we were all going to die," where what is literally meant is something like "I felt as though I knew we were...", "I felt sure we were ...", etc.

[275] Mnemonically, think of $P_a \varphi$ as saying that it is epistemically <u>P</u>ossible from a's perspective that φ – or: possible for all a knows that φ – but note that these are simply colloquial ways of saying that a does not know that *not*-φ: it does *not* mean that a considers it to be in any sense possible that φ, since it does not, for example, require a to be capable of entertaining the thought that φ. (See further p. 420, and before that, p. 383.) Some authors, such as Schwarz and Truszczyński (in their publications listed in the bibliography, as well as Stalnaker [1082] and others) have taken to writing "M" – reminiscent of the M/L notation for $\Diamond/\Box$ – rather than "P". Hintikka's notation is followed here, despite the potential for a confusion (which will not arise in what follows) with the similarly notated deontic or tense-logical operators.

[276] For some discussion of these topics, see Hilpinen [482], Humberstone [554] – already mentioned apropos of 'collective omniscience' in Exercise 2.7.24 – as well as p. 134, Chapters 2, 6, and 11 of Fagin *et al.* [280], §2.3 of van Ditmarsch et al. [244], and Allo [12]. [280] gives a thorough overview of work in epistemic logic with an Artificial Intelligence and theoretical computer science orientation (mostly) up to the mid-1990s.

EXAMPLE 5.1.2 For the case in which we are considering at least two cognitive agents a and b, Hintikka [490] (p. 60) raised the question of the acceptability of the implicational formula $K_a K_b p \to K_a p$, mentioned above in a more schematic form (φ for p), and returned a favourable verdict. If our epistemic operators here have a logic at least as strong as **KT** this is obviously the right verdict, delivered by an application of KMono for K_a to the instance (for K_b) of **T**: $K_b p \to p$. So any qualms about this implication are likely to hang on worries about logical omniscience held to be implicit in the appeal to KMono – whether a might know that b knows that p without realising that b can only know that p if it is in fact the case that p, so that a fails to know that p. While this worry may seem far-fetched, it (or something like it) appears to have motivated P. Weingartner to suggest that we need to add a second – somehow weaker – epistemic operator for each agent, written as K'_a, K'_b, and while $K_a K_b p \to K_a p$ should be valid, Weingartner regards $K'_a K_b p \to K_a p$ as invalid ([1162], p. 255), though one would have thought the relevant question on introduction of the new K' operators would have been over the implication $K'_a K_b p \to K'_a p$ instead. Be that as it may, Weingartner ([1162], p. 257) offers the following in support of his introduction of the weaker K' operators:

> A case for the weaker version is for example if a child (or in general a person a whose knowledge is on a lower level as compared to another person b) says that he knows that his father (b) knows that something is the case. Although it should then follow (from a normal interpretation of the statement) that his father knows that p but one does not assume (and does not want) in general the consequence that the child knows that p. He may have a much weaker understanding of the case such that one wouldn't like to use 'know' here although we accept that the child knows that his father knows that p. For such a kind of "weaker knowledge" the system presented provides the weaker notions K'_a and K'_b.[277]

This does not seem particularly persuasive. Certainly there are plenty of cases in which the inference from "a knows that b knows __" to "a knows __" is faulty: the problem is that these cases do not arise from putting an ordinary declarative *that*-clause into the blank. A noun phrase representing what has been called (e.g. in Heim [469]) a 'concealed question', on the other hand, readily yields a *non sequitur*: from the hypothesis that a knows that b knows the location of the treasure, it does not follow that a knows the location of the treasure. And a traditional indirect (now often called *embedded*) question will do just as well: replace "the location of the treasure" with "where the treasure is" in the preceding example. In fact Weingartner thinks that there is a K/K'-like distinction in the case of *knowing whether* no less than for *knowing that*. Recall that a common account of knowing whether p, for purposes of epistemic logic, is as an epistemic incarnation of noncontingency: $Kp \vee K\neg p$.[278] (This would not do for the purposes of natural language semantics with the latter's need for a compositional and uniform account – consider "Harriet was wondering, while her brother already knew, whether their parents were still alive" – for which project a 'two-dimensional' account along the lines pioneered in Lewis [723] and Groenendijk and Stokhof [414] seems more promising; see the index to the present work under "two-dimensional (modal semantics)" for our brief forays into that area here. (There have also been some who deny that there is even an *equivalence* between "a knows whether p" and "Either a knows that p or a knows that not-p". According to Bogusławski [106], for instance, while we might be happy to say that a cat knows that there is a mouse present, we would not be as happy to say in such a situation that the cat knows whether – or *if*, to use Bogusławski's preferred complementizer – there is a mouse present, since this would involve "a kind of anthropomorphism: there is a distinct suggestion that the cat is pondering, as it were, two contradictory possibilities, or at least, that it is capable of pondering them." Readers may decide for

---

[277]In fact Weingartner writes "$aK'bKp$" rather than "$K'_a K_b p$", etc., and below, again to avoid inelegance, we reproduce his "$aK^O p$" as "$K^o_a p$" instead. Otherwise, the passage has been reproduced verbatim here, including the "but" in the second sentence, which should presumably just be a comma.

[278]Similarly the epistemic analogue of contingency would be a matter of something's being neither known to be the case nor not known to be the case. The phrase 'epistemically contingent' has itself sometimes – e.g., Lebar [687] – been used for something rather different, namely for something neither knowable *a priori* to be the case nor knowable *a priori* not to be the case.

5.1. THE LOGICAL OMNISCIENCE ISSUE

themselves whether they share the sentiment expressed in this passage, from p. 38 of [106]. A somewhat similar contrast in the case of direct questions is a concern of Bolinger [108], which also points out – p. 93 – the degree to which *whether* and *if* are not interchangeable as question embedding complementizers.) These issues are not specific to the present case but arise for *wh*-questions generally, using this term to include rather than, as is often done, to exclude the case of *whether*. For further information, some of the references given in note 273 will be of use, which also explains why none of the suggestions canvassed here can be quite correct.) Weingartner writes K_a° for "*a* knows whether", and correctly points out ([1162], p. 258) that we have the following equivalence:

$$K_a^\circ K_b p \leftrightarrow (K_a K_b p \vee K_a \neg K_b p),$$

labelling this as 'D8' in the mistaken belief that it might constitute some kind of definition. (Only unstructured expressions, accompanied by appropriate variables, are candidate *definienda*.) On the preceding page we have another equivalence, again proclaimed to be a definition (D6, p. 257), this time governing K° in the scope of K:

$$K_a K_b^\circ p \leftrightarrow (K_a K_b p \vee K_a K_b \neg p),$$

which he says records "the strong proposition '*a* knows that *b* knows whether *p* is the case' " prompting Weingartner to introduce a weaker notion of knowing whether ("$K^{\circ\circ}$") for the "weak proposition which is usually expressed by the same sentence in everyday language". But we can see easily that the right-hand side of the equivalence last inset above is wrong. Since we are treating knowing whether φ as either knowing that φ or knowing that $\neg\varphi$, the right-hand formula should instead have been $K_a(K_b p \vee K_b \neg p)$, which does not (in any plausible epistemic logic) have as a consequence $K_a K_b p \vee K_a K_b \neg p$ and there is no need for a mysterious second knowing-whether operator. Weingartner illustrates ([1162], p. 258) what he thinks is an ambiguity in talk of knowing whether something is the case, which the distinction between K° and the weaker $K^{\circ\circ}$ is supposed to resolve:

> If one says: The professor of mathematics (*a*) knows that his student (*b*) knows whether a certain sentence *p* is a theorem of mathematics then one supposes usually that the professor of mathematics will also know whether *p* is a theorem (or not). Thus in this case $K_a K_b^\circ p$ is applied. On the other hand if one says: The student (*a*) knows that his professor (*b*) knows whether *p* is a theorem of mathematics then one doesn't presuppose that the student will also know whether *p* is a theorem; i.e., $K_a K_b^{\circ\circ} p$, which is weaker since $K_a^\circ p$ is not derivable from it.

The case for the $K^\circ / K^{\circ\circ}$ distinction is thus even weaker than that for the K / K' distinction.[279] At least the latter was motivated by the desire to react to the provability of the implication from $K_a K_b p$ to $K_a p$ in Hintikka-style epistemic logic. There is no corresponding implication from $K_a K_b^\circ p$ to $K_a^\circ p$, and so no weakening on the above lines is called for; Weingartner is confusing his having different expectations about who might know what in different situations with the expression "knows whether" having different meanings in our descriptions of those situations. ◂

(A more sympathetic discussion of Weingartner's views on strong and weak knowledge ascriptions, aiming to invalidate – for the 'weak' case – the implication just mentioned, without getting into the issue of confusing knowing that *p* with knowing whether *p*, may be found in Gochet and Gillet [390]. The authors provide a semantic description using a variation on the 'non-normal worlds' theme – p. 360 – which invalidates the formula $K_a K_b p \to K_a p$, while validating the **K** and **T** axioms for each of K_a, K_b – and any other agent-specific knowledge operators there may be. They do not seem embarrassed by the fact – an inevitable consequence of those just cited – that $K_a(K_b p \to p)$ is invalid according to the semantics, further details of which can be found in Example 5.1.27, p. 366, below. It is hard to see *a* as possessing

---

[279] Weingartner provides a matrix semantics for his favoured logic in §3 of [1162], which gives different tables for K and K', but the significance of the ten numerical values involved is not clarified – and neither is his conception of how such a semantical description functions. (See in particular, footnote 15 on p. 260.)

the concept of knowledge if a doesn't know that whatever b knows to be the case is indeed the case. But the objection cannot be that, since those that lack the concept of knowledge can still know things, epistemic logic should not tell us that φ is known for φ essentially containing epistemic vocabulary, since the Gochet–Gillet semantics validates $K_a(K_a p \to p)$. Such concept-possession considerations will come to the fore, briefly on p. 343 below and then more fully in Section 5.6. One may respond that even if a is capable of knowing things about what is known, a may never have heard of b and accordingly not be in a position even to entertain the proposition that b knows that p, whereas perhaps the "a" embedded under "K_a" in "$K_a(K_a p \to p)$" can be given a *de se* interpretation – see Remark 5.2.7, p. 379 below – but such an objection would again apply to things that are validated by Gochet and Gillet's semantics, such as $K_a(K_b p \vee \neg K_b p)$.)

There is an intertemporal analogue of the above interpersonal example, for which one could think of the subscripted "a" and "b" as denoting temporal stages of a person – though in the following there is a shift of time as well as person:

EXAMPLE 5.1.3 This passages is excerpted from an newspaper article from 1994 (specifically from p. 1 of *The Australian*, November 3 of that year, byline: Richard McGregor).

> Australia's ambassador to Phnom Penh said last night the remains of three men including one Australian had been found near the village in southern Cambodia where they had been held by the Khmer Rouge. (...) The ambassador, Mr Tony Kevin, said there "will be an examination by a pathologist which will establish beyond doubt, as now seems to be the case, that these are three young men (who were taken hostage)".

Of course the point of interest is the (confused) claim that some later examination will establish beyond doubt – not simply *whether*, *that* something is the case; if the ambassador did say this he would have represented himself as already established, at least to his own satisfaction, that things are as he is predicting they are yet to be established. You can only know now that tomorrow you will know that p if you already know that p (setting aside temporal indexicality in any sentence substituted for p – the bread-and-butter motivation for tense logic – as not relevant here). ◂

Digression. A reader noting the title ('Logic for Reasoning About Knowledge') of Orłowska [860], and jumping in at p. 571 there, might be surprised to see the axiom schema $K\varphi \leftrightarrow K\neg\varphi$ (though actually Orłowska uses F, G, as schematic letters rather than φ, ψ), given the association since at least Hintikka [490] between "K" and "the subject knows that __". As you may have guessed, Orłowska is using this notation for "the subject knows whether __".

In Lloyd [738] we read (p. 266) the following, though I have changed the negation notation to match that used elsewhere:

> In Hintikka's system (...) 'it may be that p' is interpreted as 'I do not know whether p' and this as 'I do not know that p and I do not know that $\neg p$,...

As well as conflating the use of epistemic modal auxiliaries to express ignorance, on the one hand, with self-ascriptions of ignorance on the other, we have here a conflation of epistemic possibility with epistemic contingency – or, more strictly speaking, with the epistemic analogue of alethic modal contingency. **End of Digression.**

At least if we are restricting our attention to normal modal logics, it is the range of logics in between **KT** and **S5** that is likely to harbour any plausible candidate epistemic logics. (Hintikka [490] settled on **S4** as the favoured epistemic logic, in fact. In Section 5.5, we shall see some considerations from Wolfgang Lenzen in favour of the stronger logic **S4.2** as a minimal candidate – already evident from Theorem 5.2.1, in the following section, in fact.) Then we would have available all the advantages of the Kripke semantics, its models deeming "a knows that φ" to be true at a world w just in case φ is true at all those worlds epistemically possible (for a) relative to w, these being the worlds in which whatever a

5.1. THE LOGICAL OMNISCIENCE ISSUE

knows (in w) to be the case is the case, but otherwise differing arbitrarily from each other and from w. (Is this talk of worlds in connection with epistemic and doxastic logic talk of the same range of possible worlds pertinent to discussion of metaphysical possibility? Fortunately, for our purposes here, we can leave this question open. For two opposing answers to it, see Jackson [615] and Chalmers [168]; the point is somewhat incidental to the latter paper, which addresses a wide range of views as to how to construe propositions so that they can be the objects of the propositional attitudes.)

The restriction to normal modal logics implicit in the availability of a semantic gloss on K along these lines, brings with it an aspect of much work in epistemic logic that has provoked criticism: the phenomenon of *logical omniscience*. Since in any normal modal $\Box\varphi$ is provable whenever φ is, the use such logics for application to the epistemic arena seems to credit the envisaged knower with something like knowledge of all the (relevant) logical truths. Each of the conditions of being regular, of being monotone, and of being congruential – all of them strictly weaker than normality (as we recall from Section 1.3) – seems to involve a similarly implausible assumption, sometimes labelled deductive omniscience rather than logical omniscience, though we will not distinguish these labels here, for example in the case of monotony to the effect that the knower knows all the logical consequences of anything known. (This last assumption played a role in the 'collective omniscience' example featuring in Exercise 2.7.24, p. 134.)

REMARK 5.1.4 Gochet and Gillet [390] introduce the present theme thus (p. 97):

> Epistemic logic construed in this way is plagued by the well known paradox of logical omniscience which takes three main forms:
>
> (5) All tautologies are known.
> (6) We know whatever is materially implied by what we know.
> (7) We know whatever is logically implied by what we know.
>
> A realistic epistemic logic should deny the human knower logical omnisience. At the same time, it should recognise that rational agents are endowed with limited deductive power.

(5) is a reference to the effects of Necessitation in the 'normal modal logic' approach to the subject (or a special case thereof, if *tautology* is intended in its strict sense –that of *truth-functional tautology*), and (7) to the monotone rule. (6), on the other hand, has probably never been seriously held by anyone. One would, that is, have to search the literature very thoroughly indeed to find a proposed epistemic logic in which

$$(p \to q) \to K(p \to q)$$

was provable, since in any normal (even any congruential) modal logic this would have $q \to Kq$ as a consequence (put $\top$ for p). Formulas corresponding to that inset here would be equally implausible for any of the other applications of modal logic with which we have been concerned, though one cannot say in their case that they have never actually been espoused: Mally was committed to the deontic analogue of this formula (putting O for K, that is), for example, as one sees from p. 4 of Føllesdal and Hilpinen [316] (where it appears as formula (9)). See also note 213, p. 249 above. What looks like the formula inset above (though with "K" together with a variable ranging over potential knowers) receives favourable attention at p. 483 of Rescher [950], but since is gloss on this is that "(k)nowers automatically know things that follow from what they know", the intended reading of "$\to$" (which is not made explicit) is presumably something along the lines of strict implication (perhaps with **S5** as the background alethic modal logic, since this is the only alethic modal logic mentioned). Indirect support for such a reading comes from other writings of Rescher's, such as [949], on p. 100 of which "$\to$" is said to symbolize implication and a rather strangely worded parenthetical comment is appended to the effect that we "need not assume that the implication at issue material implication but could assume it represents a stronger relationship, such as entailment". ◀

The current problem (or problem-cluster) was already familiar to Hintikka by the time he came to write the first book on epistemic logic, [490], and rather than rely on the idea that in this respect the logics on offer are descriptions of idealized fully rational cognitive agents, he had a different angle, reported with convenient conciseness in the following passage from a review ([699]) by Lemmon of [490]; before giving it, let us quote a motivating remark from Hintikka himself ([490], p. 37): "Logical truths are not truths which logic forces upon us; they are not necessary truths in the sense of being unavoidable. They are not truths we *must* know but truths which we *can* know without making use of any factual information." Now for Lemmon's summary. In order to reduce discomfort, "p" and "q", which we have been using as propositional variables (sentence letters), have been replaced by φ and ψ; $K_a\varphi$ can be read as "a knows that φ":

> An illustration may help: suppose φ logically implies ψ; then "$K_a\varphi$" and "$\neg K_a\psi$" are jointly indefensible; they are not of course inconsistent, since a man may well know φ yet not have drawn the consequence ψ; but if a persists in claiming "$\neg K_a\psi$" after ψ's entailment by φ is shown to him, then he is unreasonable, and it is this notion that Hintikka aims to capture by the term "indefensible." Defensibility is immunity to a certain *kind* of criticism: namely, the criticism that you can be shown wrong on the basis of the consequences of what you already know. Thus Hintikka evades the paradox, common in interpretations of epistemic logic, that we all know all logical truths or all the logical consequences of what we know; but he remains committed to the subtler paradox that to claim not to know a truth of classical logic is indefensible in his sense. (...) Of course, we need to remember that the 'theorems' of Hintikka's system are not logical truths but *self-sustaining* sentences, sentences the negation of which is indefensible rather than inconsistent.[280]

Another response is also mentioned by Hintikka and subsequent writers on this topic, namely, to read "$K_a\varphi$" not as "a knows that φ" but rather as "it follows from what a knows that φ".[281] This sounds like a suitably conciliatory suggestion, though an explanation for that fact itself would be desirable. For, by contrast, suppose someone were to propose an epistemic logic closed under the 'antitone' rather than monotone rule: from $\psi \to \varphi$ to $K_a\varphi \to K_a\psi$, and to be presented with the objection: but surely one can know that $p \vee q$ without knowing that p, contrary to the dictates of this rule, which takes us from the provability of $p \to (p \vee q)$ to that of $K_a(p \vee q) \to K_a p$. The proponent of the rule then says, "Well, never mind about that: if you feel uncomfortable with these consequences, just read '$K_a\varphi$' as 'something that a knows follows from φ'."

Such a suggestion would be presumably greeted with a response along the following lines:

> Thanks, but no thanks – we were interested in the logical properties of "a knows that φ", not of this new notion you propose to investigate, the concept of φ's *logically implying* something that a knows.

The fact that Hintikka's suggested reading does not immediately evoke a similar "don't change the subject" reaction is perhaps connected with the initial naive reaction to discrepancies between what a

---

[280] This passage is spread across two paragraphs, the break between which occurs in the ellipsis indicated, on pp. 381–382 of Lemmon [699]. As well as the change in schematic letters, the notation for negation has been altered to match our usual conventions. It is something of a routine response by reviewers of works in epistemic logic to raise the spectre of logical omniscience and complain of implausibility. See also, for example, Castañeda [155], reviewing Hintikka [490], and Sadegh-Zadeh [987], reviewing Lenzen [710]. Apropos of the passage from Lemmon quoted here: anyone reading a summary of Hintikka's treatment in Girle [379] should note that in line 19 on p. 124 thereof, "self-satisfying" is a typo for "self-sustaining", with the same slip on pp. 114 and 115. (More serious in the summary of Hintikka's views is Girle's remark, on p. 107, that "'not $P_a p$' reads as *It does not follow from what a knows that p*," which should rather be: It follows from what a knows that not-p; and also Girle's formulations of Hintikka's conditions (A.PK*) and (A.PKK*) at p. 113f., where the conditionals given – if such-and-such is consistent then so-and-so is consistent – need to be replaced by the converse conditionals. These conditional constraints of Hintikka's do not appear in our discussion; Lemmon [703] criticizes Hintikka's appeal to (A.PKK*) as a blatantly circular or question-begging attempt to justify $\mathbf{4}_K$.)

[281] Hintikka [490], p. 38; Humberstone [554], p. 405; Williamson [1189], p. 228. Hintikka later had some further thoughts on the logical omniscience issue, reported in Remark 5.6.2, p. 424 below. Aho [5] makes a similar move to justify congruentiality, immediately after the passage quoted in note 286 below in which the congruential versions of these logics are described as privileged – though such a policy is described as "deliberately unrealistic psychologically".

5.1. THE LOGICAL OMNISCIENCE ISSUE

normal epistemic logic says and the behaviour of actual knowing subjects: the former is an idealized version of the latter. There is, by contrast, no tendency to think of someone not actually knowing something from which something else that they know follows, as falling short of any such ideal – or, to put it in terms of Hintikka's own way of thinking, as being vulnerable to a certain line of rational criticism. This is perhaps also the reason we find it natural to think of a's *implicitly* knowing something when it follows from what a can be uncontentiously said to know ("explicitly").

REMARK 5.1.5 The same could be said for implicit *belief*, though a corresponding move in connection with *desire* would be more problematic. In the case of boulomaic logic ($\Box$ as "a wants it to be the case that") those who resist the $\Box$Mono rule – see Remark 5.1.19, p. 358 below – would be correspondingly unhappy to acknowledge that φ's following from something that a explicitly desires renders it legitimately describable as something a implicitly desires. ◄

A further variation – suggested by various passages in Williamson [1189], where it is used more in connection with introspection issues (see the following sections) than with the current topic – is that we read "$K_a\varphi$" as something more like "a is in a position to know that φ". The suggestion seems most welcome – pending clarification of the distinction between being in a position to know and being in a position to find out.[282] Several suggestions for combining in a single language something like implicit and explicit knowledge (or belief) operators have arisen in the literature, as we pause to note:

EXAMPLES 5.1.6 (*i*) One suggestion is that a separate modal operator, A, say, for "awareness" is added, subject to perhaps no special axioms (or subject to some suitably chosen for the particular application to hand – see the discussion which begins with Example 5.6.4, p. 425 below). The idea is that $A\varphi$ is read along the lines of "the subject is consciously aware of the proposition that φ", and we then regard K (as the $\Box$ of a normal modal logic) as marking implicit knowledge, in terms of which the subject's knowing explicitly that φ might be defined as: $K\varphi \wedge A\varphi$. For details and variations of this line of thought, see §9.5 in Fagin *et al.* [280] – or the original source, Fagin and Halpern [278], which also contains references to work by others (H. Levesque, G. Lakemeyer: see further Remark 5.1.7 below) on related projects – as well as Thijsse [1100], [1101] and Thijsse and Wansing [1102], and Dalla Chiara [229]. Some such proposals are presented in (slightly) greater detail, though with special reference to implicit belief rather than implicit knowledge, in Section 5.6: see especially pp. 425–426.

---

[282] At p. 95 of [1189], Williamson makes the following clarificatory remarks: "To be in a position to know p, it is neither necessary to know p nor sufficient to be physically and psychologically capable of knowing p. No obstacle must block one's path to knowing p. If one is in a position to know p, and one has done what one is in a position to do to decide whether p is true, then one does know p. The fact is open to one's view, unhidden, even if one does not yet see it. Thus being in a position to know, like knowing and unlike being physically and psychologically capable of knowing, is factive: if one is in a position to know p, then p is true." A similar notion – of the *feasibly knowable* – is deployed in Wright [1203], p. 109. Also in this area is the notion of the availability of evidence, used in characterizing epistemic possibility as consistency with the available evidence and in various principles relating objective chance to subjective probability given the total evidence available: see the discussion in pp. 23–30 of Handfield [433], which concentrates on the probabilistic motivation but also recalls some of the epistemic possibility literature (Ian Hacking, Keith DeRose), which can also be found in our discussion toward the end of the Digression on p. 384. A formal proposal for the treatment of being in a position to know that __ can be found in Iacona [606], esp. Section 3, but little is said to connect the suggested semantic treatment with the informal idea, and some of what *is* said looks distinctly unpromising, as with the remark (p. 9) that "[t]o say that it is definitely the case that p is to say that one is in a position to know that p". There is also an interesting schematic suggestion at p. 119 of Hilpinen [480], according to which given any proposed analysis of knowledge in terms of belief and other conditions (such as truth and justification, in the most famous case) not implying belief (for which reason in the case just cited justification is not a matter of having a justification but having reasons which *would* justify the belief if one had it), one has to hand a corresponding notion of being in a position to know something when these other conditions are satisfied (justification and truth in the case cited, for instance). A more detailed discussion than can be ventured here would note that versions of the a knower's belief must be justified need not insist on possession of something called 'a justification', since being justified may simply be – e.g., in the case of directly perceptual beliefs – a matter of the absence of undermining counterevidence.

(*ii*) Williamson [1189], Appendix 6 (a reworking of material in the appendix to Williamson [1183]), takes up – as it happens, apropos of the Fitch derivation reviewed below in Example 5.4.11 (p. 397) – the issue of a knowledge operator K for which characteristically monotone inferences such as that from $K(\varphi \wedge \psi)$ to $K\varphi \wedge K\psi$ do not go through, or in his terminology "knowledge does not distribute over conjunction", but couches his discussion more generally:

> There are two obvious ways of trying to approximate an operator O which lacks a feature F with an operator with F. One is to seek the weakest operator O^+ stronger than O which has F; the other is to seek the strongest operator O^- weaker than O which has F. Of course, there is no general guarantee that either O^+ or O^- exists. However when F is the feature of distributing over conjunction, we can define both O^+ and O^- in terms of O by quantifying into sentence position.

Williamson proceeds to introduce a new binary connective Con governed by some reasonable axioms for what is given as its intuitive interpretation by the remark that "$\mathrm{Con}(p,q)$ is true if and only if p is semantically a conjunct of q". The definition of O^+ does not concern us here; for O^- it runs:

$$O^-\varphi = \exists q(\mathrm{Con}(\varphi, q) \wedge Oq).^{283}$$

Williamson shows that O entails O^- (in the sense that $O^-\varphi$ follows from $O\varphi$ for any φ), that O distributes over conjunction (in the sense just explained), and that any other operator entailed by O and distributing over conjunction is entailed by O^-. Thus if we think of O as K, a non-monotone knowledge operator, then even if $K(p \wedge q)$ does not entail (provably imply) Kp, it still entails K^-p: p being a conjunct of something known. Dealing with propositional quantifiers (the "$\exists q$" here) is no straightforward matter, however – see Fine [287][284] – and the more one can avoid them, the better. Nor is it clear what being "semantically a conjunct of" amounts to, and the axioms given do not narrow this down very tightly. For these reasons – on the second of which, see the (brief) discussion of analytic implication starting at p. 423 below – we turn to another approach in (*iii*) below. It should be mentioned that Williamson's concerns in the passage quoted above are not exactly those under discussion here, but pertain instead to variations on the theme of the Fitch derivation presented as Example 5.4.11 below (p. 397).

(*iii*) Humberstone [567], again with K in mind but writing "O" and making no special assumptions about its behaviour other than congruentiality, uses a neighbourhood semantics of the kind figuring in Thm. 2.10.12 (p. 160), and the discussion leading up to it, with the following clause in the definition of truth at a point ($x \in W$):

$$\langle W, \mathcal{N}, V \rangle \models_x O\varphi \text{ iff } \|\varphi\| \in \mathcal{N}(x),$$

where as usual $\|\varphi\|$ is the set of points in the model at which φ is true. This is then supplemented by the following for an additional 1-ary operator $\Box_O$ which when O is K or B says that something follows from something known or believed:

$$\langle W, \mathcal{N}, V \rangle \models_x \Box_O\varphi \text{ iff } \exists Y \in \mathcal{N}(x) \,.\, Y \subseteq \|\varphi\|.$$

Thus we combine the standard neighbourhood semantics, here supplied for O, with the Jennings–Schotch 'locale' semantics (from p. 161), used for the interpretation of $\Box_O$. This bimodal logic is described as *congruential-to-monotone* in [567], where it is noted that one could instead regard the 'implicitly cognised' operator $\Box_O$ (analogous to the 'O^-' of (*ii*) above) to take us also to consequences of *sets* of propositions (here thought of as subsets of W) by saying instead:

$$\langle W, \mathcal{N}, V \rangle \models_x \Box_O\varphi \text{ iff } \exists \mathscr{Y} \subseteq \mathcal{N}(x) \,.\, \bigcap \mathscr{Y} \subseteq \|\varphi\|,$$

---

[283] This definition, like the passages quoted above, appears on p. 318 of [1189]; I have changed "p" to "φ".

[284] Fine's paper was already mentioned in this connection on p. 216. In addition: Bull [127] and Kaplan [639], in which papers earlier work of Kripke on this topic is also cited.

5.1. THE LOGICAL OMNISCIENCE ISSUE

in which case we have a congruential-to-normal bimodal logic.[285] (The 'implicitly cognised' talk here covers more than one thing: further detail in Remark 5.6.8, p. 428.) Returning to the monotone case, [567] shows that with O satisfying the congruentiality rule for the basic logic, a complete axiomatization of the extension with $\Box_O$ is obtained by adding the monotone rule for the new operator ('$\Box_O$Mono') and the obvious bridging axiom $O\varphi \to \Box_O\varphi$, as well as the somewhat less obvious $\Box_O\bot \to O\bot$, we get a complete axiomatization of the valid formulas (those true throughout all models $\langle W, \mathcal{N}, V \rangle$, that is), though incompleteness lurks among the simple extensions of this system. An obvious objection, however, is that the starting assumption of congruentiality for O goes against the motivation of locating all deductive omniscience features in the 'implicit' $\Box_O$ rather than the initially given O. ◀

The question of the best philosophical response to the omniscience issue is not something to go into detail here, but a few general few remarks addressed specifically in defence of congruentiality are in order.[286] The objection that I cannot infer that the subject knows (or believes) that ψ from the assumption that the subject knows (or believes) that φ, given the logical equivalence of φ and ψ (recorded by their provable equivalence in the logic in use), because the subject may be ignorant of this equivalence is in danger of confusing direct and indirect speech – or rather the cognitive analogue of this distinction, cast in Routley and Routley [976], p. 189, as the distinction between sentential and propositional assent, respectively. When *I* say that the subject knows/believes that ψ, I am using *my* language and cognitive resources to make this attribution, not the subject's, and am entitled to use what Routley and Routley, in [976], call the principle of fair redescription (of belief), though might perhaps better be called fair reformulation. (See also note 377, p. 423, for more on the theme of the following Remark.)

REMARK 5.1.7 The Routleys, however, take a much narrower view of the logic that underlies such reformulations than that provided by the specifically classical background of congruential modal logics. For the moment, what matters is the general idea. Very close to their own favoured (*relevant*) logic for this purpose is endorsed in the same capacity in Chapter 12 of Levesque and Lakemeyer, [715], who adopt a similar philosophical perspective as to how it functions, when it comes to the attribution of explicit as opposed to implicit belief. Belief attributions for the latter notion are subjected to axioms saying *inter alia* that when φ and ψ entail each other according to Anderson and Belnap's logic of first-degree entailment, and on p. 201 we read:

> Another way to understand the axiomatization is as constraints on the individuation of beliefs. For example, $(\varphi \vee \psi)$ is believed iff $(\psi \vee \varphi)$ is because these are two lexical notations for the *same* belief. In this sense, it is not that there is an automatic inference from one belief to another, but rather two ways of describing a single belief.

A position like that taken by Routley and Routley [976] and Levesque and Lakemeyer [715] has also been suggested for iterative principles such as 4_B, though not quite for this principle itself, in Nelkin [841], on p. 226 of which one reads:

---

[285] By adding the proviso to the right-hand side here that $\mathscr{Y} \neq \varnothing$, we would obtain a congruential-to-regular variant. (Note that $\bigcap \varnothing = W$, putting making $\Box\top$ true everywhere and thus making the logic closed under Necessitation.) Closely related considerations are aired in §2.3 of Hansen [435], with deontic applications in mind.

[286] A thoughtful discussion going well beyond the widely encountered and certainly understandable implausibility verdicts will be found in Stalnaker [1076] and [1077]. Instead, we briefly consider some different sources for being worried about logical omniscience and a few technical responses. For more on the latter front, see Chapter 9 of Fagin *et al.* [280], Halpern and Pucella [429] and Jago [618], all of which also supply numerous further references. (See note 377, p 423 below, for some terminological problems with Jago's discussion.) Note that when propositions are taken as sets of worlds and also as the objects of the propositional attitudes, congruentiality is inescapable, as logically equivalent sentences are true in the same worlds and so determine a single proposition to be known or believed or whatever; this has been one motive for introducing 'structured propositions' instead – see note 42, p. 54, including in particular the remark quoted there from Aho [5] – a paper providing a thoughtful discussion of the area and in which one reads (p. 3) "I think that the first and privileged type of attitude logic should accept unrestricted logical omniscience."

> Consider for the moment beliefs. (i) 'A believes p' and (ii) 'A believes that A believes p' seem pretty clearly different. But is is so obvious that (iii) 'A believes that A believes that A believes p' expresses a different proposition from (ii)? Not to me.

It is natural to interpret this as the suggestion that the correctness of the principle $BB\varphi \to BBB\varphi$ (or its converse) is does result from an idealization of rationality or self-awareness but from the fact that the belief attributed by the consequent *just is* that attributed by its antecedent. Let us not go into the plausibility of this view, or of the suggested implausibility of a corresponding view in the case of $B\varphi \to BB\varphi$ itself, pausing only to notice that the suggestion is an interesting one, with a somewhat different flavour from those taken from [976] and [715].

We return to Levesque and Lakemeyer and the quotation from [715] above; the italic boldface notation ($\boldsymbol{B}, \boldsymbol{K}$) to be used here will be explained presently. The way the quoted passage articulates the position would need some emendation in view of the fact that some of the transitions legitimated are not reversible. For example according to p. 200 of φ and ψ [715] $\boldsymbol{B}\varphi \vee \boldsymbol{B}\psi$ provably implies $\boldsymbol{B}(\varphi \vee \psi)$, where $\boldsymbol{B}$ is Levesque and Lakemeyer's operator for explicit belief. This tells us that $\boldsymbol{B}(\varphi \vee \psi)$ follows from $\boldsymbol{B}\varphi$ and also from $\boldsymbol{B}\psi$, reflecting the fact that disjunctions follow from their disjuncts in relevant logic. (The reference to provable implication is to provable material implication, however: there is no relevant implication connective in the language.) So on the basis of someone's (explicitly) believing p, say, we may attribute to them the (explicit) belief that $p \vee q$, presumably again not on the basis of a presumed automatic inference from p to $p \vee q$ by the believer, but instead on the part of the belief-ascriber's right of redescription: but here there is no question of an equivalence of content, even by the lights of the weaker-than-classical logic employed, so talk of adopting a particular set of constraints on the individuation of beliefs is not quite to the point. A potentially more serious difficulty – raised below on p. 369 – for such accounts is the use of any particular logic, classical or otherwise, for arbitrating the legitimacy of such redescriptions, since the issue of logical omniscience threatens to arise anew w.r.t. the chosen logic. One other thing: as well as using "$\boldsymbol{B}$" as a sentence operator, "$\boldsymbol{K}$" is used by Levesque and Lakemeyer in [715]. Anyone familiar with Hintikka's notation (i.e., "B" and "K", employed in the present exposition also) might expect that these represent belief and knowledge respectively, but alas no: they stand for explicit belief and implicit belief! Neither $\boldsymbol{B}\varphi$ nor $\boldsymbol{K}\varphi$ provably implies φ, though there is an axiom schema telling us that $\boldsymbol{B}\varphi$ provably implies $\boldsymbol{K}\varphi$, which is reasonable since what is explicitly believed is (at least) implicitly believed, but only encourages confusion when it is described (as on p. 202 of [715]) as "an axiom that every belief is known". For example, the modal principle **5** is much more plausible for belief than it is for knowledge, a point next mentioned on p. 371 (including note 310). Evidently somewhat worried by the charge of misusing the word *know*, Levesque and Lakemeyer ([715], p. 61) write:

> If "know" is not the appropriate term here (since what is known in our sense is not required to be true), neither is "believe," at least in the sense of allowing for the fact that you might be mistaken. Perhaps a more accurate term would be "is absolutely sure of" which would not require truth, but would preclude doubts.

This doesn't quite work as a reply to the charge, however, since fully confident belief – the subject matter of doxastic logic, at least as standardly conceived – is indeed a kind of belief, whereas 'false knowledge' is no kind of knowledge. This criticism aside, however, Levesque and Lakemeyer do have an interesting proposal here: While the explicit version of belief (or again of knowledge) is closed under first-degree (relevant) entailment, the corresponding closure condition for implicit knowledge or belief is the converse of the classical consequence; thus the latter presents us with a normal modal logic for knowledge or belief. The treatment in Chapter 12 of [715] goes beyond propositional logic in adding quantifiers, but does not consider formulas in which there is an occurrence of the explicit belief operator in the scope of another such operator. We return to other approaches to the explicit/implicit distinction – already touched on (for knowledge) in Example 5.1.6(i) above – in more detail on p. 425ff. below. ◀

However the details are spelt out, such considerations provide a counterweight to the 'logical nihilism'

5.1. THE LOGICAL OMNISCIENCE ISSUE

of Cresswell ([206], p. 11) that all there is to (a plausible) doxastic logic is substitution instances in the language of doxastic logic of truth-functional tautologies, and of Lemmon ([703] in particular) and Williamson (e.g., [1194]) to the effect that the same holds for epistemic logic, except that here we also throw in the implication from knowledge to truth – $\mathbf{T}_K$ (the $\mathbf{T}$-axiom for K, that is). On such a view if I have every reason to think someone is sincere in saying (something of the form) "Neither a nor b is F," I have to choose carefully between attributing to that subject the belief that $\neg(Fa \lor Fb)$ on the one hand, and the belief, on the other hand, that $\neg Fa \land \neg Fb$, since despite the equivalence (let us take it) of these two formulations of the content of the belief, the corresponding belief ascriptions are – apparently – far from equivalent. Something like the 'fair redescription' point from [976] is surely pertinent here. Whether one can extend this rebuttal of extreme logical pessimism *à la* Cresswell–Lemmon–Williamson all the way up from congruentiality to normality (via monotony and regularity) remains to be seen. Some moves in this direction are attempted in Section 5.6 below. (It should, however, be conceded that our practices in attributing propositional attitudes and in using indirect speech constructions may be based on principles which are not altogether coherent, or at least that these practices threaten to defy any compositional semantic account. See Kripke [674] and Partee [872] for considerations tending in this direction; opposing considerations and pointers to the literature spawned by [674] may be found in Kemmerling [644].)

Let us for the moment focus on the fact that K is monotone in any normal epistemic logic: $K\varphi \to K\psi$ will be provable whenever $\varphi \to \psi$ is. The informal hypothesis this seems to embody is that whenever our rational subject a knows that φ, then a knows that ψ, if ψ itself follows from φ. Three types of worry can be distinguished for this hypothesis:

(1) Deductive Competence Worries. The degree of idealization involved in the term "rational" here means that no actual people will count as rational, since none are so intelligent as to be able to realize, for an arbitrary consequence of something they know, that it *is* such a consequence. A variation on this worry is that no kind of rationality should be assumed at all, and so, since it is possible for some particularly slow-witted knower to know that either q or *not-not-q* without knowing that q, even though q follows from (reverting to formal notation now) $q \lor \neg\neg q$, no epistemic logic should have as a theorem $K(q \lor \neg\neg q) \to Kq$. (Another aspect of rationality, not immediately relevant here, is doxastic *consistency*, represented by the $\mathbf{D}_B$ schema $B\varphi \to \neg B\neg\varphi$ – in Hintikka's notation $B\varphi \to C\varphi$; of course, one will only take this to represent an ideal of rationality if one takes the joint truth of φ and $\neg\varphi$ to be ruled out *a priori*. For an example of someone not thus persuaded, see Priest [908].) We will have occasion to allude to $\mathbf{D}_B$ as a consistency idealization in what follows (for example in Remark 5.4.7, p. 392).

(2) Concept Possession Worries. Quite aside from the question of the subject a's deductive prowess at seeing that various consequences $\psi_1, \psi_2, \ldots$ of something, φ, known to a are indeed consequences of φ, there is the further issue that a may not even be in a position to, as it is traditionally put, 'grasp the propositions' expressed by $\psi_1, \psi_2, \ldots$ at all, through not possessing all the concepts required. (This theme has arisen already in note 275 on p. 333 and at the end of Example 5.1.2, p. 335; see also note 372 on p. 421 below on not taking talk of grasping – or being capable of entertaining – this or that proposition too seriously.) This itself may not reflect adversely on a's rationality at all; for example, on the view that in the absence of causal contact with Madagascar (even indirect contact, such as having encountered people who have ... encountered people who have seen or visited Madagascar), a cannot have the thought that Madagascar is an island, or the thought that if Madagascar is an island then it has greater annual precipitation on one side than on the other.[287] Thus even if a knows that every island has greater annual precipitation on one side than on the other, a does not know that if Madagascar is an island then it has greater annual precipitation on one side than on the other, even though this follows from what a knows. Unlike the worry under (1) above, there is no problem about a's logical acumen in this kind of case.

---

[287] This example is adapted from p. 246 of Mackie [755].

(3) Closure Principle Worries. In epistemological discussions since the 1970s, the idea that knowledge is closed under known entailment – in the sense that if a knows that φ, and, knowing ψ to be a logical consequence of φ, infers ψ from φ, a thereby comes to know that ψ – has come under attack from various sources. The attack was first launched in Dretske in [254], as an attack on the simpler formulation that if φ entails ψ then $K_a\varphi$ entails $K_a\psi$, which approximates to the more complicated version just given if K_a is read along Williamson's lines as "a (either knows or at least) is in a position to know that", and was later popularized in Chapter 3 of Nozick [851].[288] In the case of both these (and numerous subsequent) authors is associated with an analysis of the concept of knowledge with a subjunctive (or counterfactual) conditional component, such as "If it weren't the case that φ, then a would not believe that φ" or "a believes that φ for a reason which a would not have had, if it hadn't been he case that φ." (Here the analysandum is of course "a knows that φ", and there will be further components – further necessary conditions – in the analysis as well, perhaps even beyond the conditions that it should be true that φ and that a should believe that φ.[289] See Kripke [676] for critical discussion.) Part of the interest of such proposals lies in their potential as replacements for the "justified true belief" analysis widely agreed to have been refuted in Gettier [370] (see Example 5.1.10 below), and another part of the interest lies in their deployment in response to scepticism – roughly the view that in this or that domain we in fact know far less than we customarily take ourselves to know. This is because sceptical arguments frequently take the form of getting one to agree than one does not know one thing, for example that one is not a brain in a vat whose sensory inputs are all of them controlled by a skilled neurologist making it now appear to one that one is sitting on a sofa, and that one does not therefore know that one is sitting on a sofa. The argument attempts to persuade by a version of the monotone/closure principle with an admixture of contraposition: φ entails ψ (where φ is the statement that one is on the sofa, and ψ the statement that one is not being made by the neurologist to believe falsely that one is on the sofa), so $\neg K\psi$ entails $\neg K\varphi$. The anti-closure theorist will be able to maintain that, the entailment notwithstanding, one can know that one is sitting on a sofa, if indeed one is and believes that one is, because had one not been sitting on a sofa one would not – unless the situation is very unusual – have believed that one was. So at least the counterfactual requirement mentioned above would be satisfied. On the other hand, one cannot know that one is not a brain in a vat with experiences (induced by a neuroscientist) as of sitting on a sofa because if that had been false, and one was having the induced experiences, one would still believe that one was sitting on a sofa. This shows how something can be known by someone even though something else that follows what is known is not itself known. (Note that a defender of this line of thought would be justifiably resistant to the suggestion that the everyday notion of knowledge is only idealized somewhat in normal epistemic logic when $K\chi$ is taken to mean that a knows something from which it follows that χ.)

REMARK 5.1.8 Before note 289 is too far behind us, it should be remarked that the notion of safety there mentioned – if the subject had the belief in question, it would have been true – is a counterfactual explication of a more informal notion of safety which can be thought of as an outgrowth of the idea that "if you know, you can't be wrong". (In chronological order, a sampling of discussions in which this

---

[288]Dretske [254] puts the denial of closure by saying that the knowledge operator is not a "fully penetrating operator", and accompanies this with what seems like a more positive thesis – that it *is* a 'semi-penetrating' operator: but no clear account of this latter property is provided. Some further problems with [254] are detailed in Stine [1091].

[289] Using the "$\square\!\!\rightarrow$" notation for subjunctive or counterfactual conditionals from Lewis [717], and making the subject concerned explicit with a subscript, we may write "If it weren't the case that φ, then a would not believe that φ" as "$\neg\varphi \square\!\!\rightarrow \neg B_a\varphi$". The provisional formulation given by Nozick of the counterfactual analysis of knowledge is that this taken in conjunction with $\varphi \wedge B_a\varphi \wedge (\varphi \square\!\!\rightarrow B_a\varphi)$ is necessary and sufficient for the truth of $K_a\varphi$. Thus there are altogether there are two separate counterfactual conjuncts. These "$\varphi \square\!\!\rightarrow B_a\varphi$" and "$\neg\varphi \square\!\!\rightarrow \neg B_a\varphi$" conditions are called 'adherence' and 'variation' by Nozick, though the latter is more commonly referred to as a condition of *sensitivity*; a different approach – different because subjunctive conditionals do not contrapose (see [717]) – replaces it with a condition of *safety*: $B_a\varphi \square\!\!\rightarrow \varphi$. (See Sosa [1075] and Greco [405], and the references in Remark 5.1.8.) The provisional formulation recalled here is then complicated (to meet certain objections) in Nozick [851] when it it turned into an analysis of "a knows that φ by method M".

5.1. THE LOGICAL OMNISCIENCE ISSUE

idea is prominent: Austin [43], Broyles [122], Harrison [455], Kaplan [641].) This idea in turn is usually introduced with a warning, as in the second paragraph of note 17 in Hughes and Cresswell [535], not to commit what is often called the *modal fallacy* of confusedly passing from the premiss $\Box(p \to q)$ to the conclusion "$p \to \Box q$" (with $\Box$ given an alethic interpretation), aided and abetted in this transition by the ambiguity in such constructions as "If p then, necessarily, q" in which with the comma the former is an appropriate formalization and without it, the latter. (*Necessitas consequentiae* and *necessitatis consequentis*, respectively, to go all medieval for a moment – "the necessity of the consequence" vs. "the necessity of the consequent".) Neither option here conveys the intended idea: taking p as "a knows that q" we have the trivial claim that knowledge is veridical (or factive – i.e., the principle $\mathbf{T}_K$, here $\Box$-necessitated) on the one hand and the unfortunate claim that only necessary truths are known, on the other. Perhaps it will be possible to spell out something of interest and with a chance of being correct, to mean by "If you know, you can't be wrong", but in the meantime an option worth exploring is a weakened – one might say, more *fallibilist*, version: "If you know, you can't easily be wrong", and here we arrive at the intuitive idea of safety: knowledge as safety from error on the matter in hand. Although this idea is present in Williamson [1189], it is closer to the front of the stage in Williamson [1194]. A good discussion of the idea can be found in Sainsbury [989], published before either those and taking as its stimulus some remarks by Williamson on knowledge made in the course of earlier work on vagueness (not included in our bibliography). Williamson makes this idea of its not being easy for the believer's belief to false in the case of knowledge using a binary accessibility relation (roughly, relating a world to 'close' alternative possibilities) rather than using counterfactuals (and their associated semantic apparatus); this was part of the apparatus in play in our discussion of Williamson models (beginning on p. 162 – see also the end of Remark 5.3.4, p. 384) but for the full philosophical motivations, [1194] and other work by Williamson should be consulted. A comparative critical discussion of different incarnations of safety including those of Sosa and Williamson is provided in Dutant [267], especially §§3.4.5, 3.4.6 and Appendix D. There is also a most helpful orienting discussion, with applications, in Manley [767]; see also Pritchard [931] and references there supplied, as well as Collins [188] and Williamson [1194]. ◀

Let us return to the logic omniscience issue. The worry under (1) is the main concern of most criticisms of epistemic logic *à la* Hintikka as unrealistically imposing an ideal of logical omniscience, on which the discussion above bears, and on which Stalnaker [1076] and [1077] bear particularly; there are also interesting discussions in §3.2 of Lenzen [707] and in Dubucs [258].[290] Lenzen [707] touches also on worry (3), and ably disposes (p. 58) of a spurious argument of Dretske's against the closure principle, though it treats Dretske's main examples (having the pattern of the brain-in-a-vat case just described), with the rather weak response (p. 57) that the entailments concerned are not purely formal or logical, instead relying on non-logical analyticities. A similar informal element is present in one of the examples used in a powerful reply, Hawthorne [462], to one of Dretske's more recent defences of the anti-closure position (in [255]). First, let us quote (from [462], p. 29) Hawthorne's formulation of what is at issue:

> If one knows P and competently deduces Q from P, thereby coming to believe Q, while retaining one's knowledge that P, one comes to know that Q.[291]

---

[290]The literature on logical omniscience as a potential difficulty for epistemic logic is voluminous and no pretense is made here of surveying it. Failures of omniscience are problematic in their own way, since fully understanding a claim is in some tension with failing to grasp (already) what follows from it, creating a need to explain how knowledge can be gained by deduction. Here, we restrict ourselves to citing two classic discussions from the 1970s and one from thirty years later: Dummett [260], Powers [904], Rumfitt [978].

[291]Strictly speaking there is some equivocation here over the roles of the variables P, Q. The first occurrence of P in the passage here quoted could be taken to stand in place of the name of a specific proposition (or whatever else the objects of knowledge are taken to be), and it could also be taken to stand in place of a sentence (expressing the proposition in question), since the word "that" can be omitted in the latter case. But in talking of deducing Q from P, the variables must be intended in the former way, while in the subsequent phrase "one's knowledge that P", the latter interpretation is required. To be able to echo these formulations in our discussion, a similar blurring of categories will be required.

This is evidently not quite the mere condition – the monotone condition – that if Q follows from P (or ψ from φ, as we might say) then anyone knowing the latter knows the former. The subject in Hawthorne's condition comes to believe that Q on deducing it from P, and so we are not worrying here about the case in which the subject does not notice that Q follows from P (as it must for the deduction to count as competent) or does not have the conceptual capacities to entertain the proposition that Q. This formulation is specifically aimed to test the crucial ingredient in views like those of Dretske's and Nozick's according to which despite all this, the subject may not know that Q because Q fails the counterfactual sensitivity requirement (despite P's passing it).[292]

There is also a congruentiality-related closure condition in play in Hawthorne's defence of the above closure condition:

> *The Equivalence Principle*: If one knows *a priori* (with certainty) that P is equivalent to Q and knows P, and competently deduces Q from P (retaining one's knowledge that P), one knows Q. [462], p. 31.

He continues:

> Interestingly, Dretske's reasons for denying closure have no force against the Equivalence Principle. His argument against closure relies on the following idea: following recent usage, let us say that R is "sensitive" to P just in case were P not the case, R would not be the case. Suppose one believes P on the basis of R, and that P entails Q. R may be sensitive to P and still not to Q. But notice that where P and Q are equivalent, there can be no such basis for claiming that while R can underwrite knowledge that P, it cannot underwrite knowledge that Q. We may thus safely assume that Dretske will accept the Equivalence principle. Here is a second, equally compelling, principle:
>
> *Distribution*: If one knows P and Q, then so long as one is able to deduce P, one is in a position to know that P, and so long as one is able to deduce Q, one is in a position to know that Q.

With these ingredients to hand, Hawthorne points out that anyone denying the closure principle while accepting the condition labelled Equivalence – as Dretske and company must – is forced to reject, most implausibly, the appealing condition labelled Distribution. (This observation can be found already in Lemmon [703], p. 77. Nozick [851], defending a Dretske-like account of knowledge – see note 289 above – also conceded the point ([851], p. 228 and note 63 on p. 692), for further discussion of which in this setting see Kripke [676], esp. note 60.) He continues as follows:[293]

> Suppose one knows some glass g is full of wine on the basis of perception (coupled, perhaps, with various background beliefs). The proposition that g is full of wine is *a priori* equivalent to the proposition
>
> g is full of wine and ¬(g is full of non-wine that is colored like wine).
>
> So by equivalence one knows that conjunction. Supposing distribution, one is in a position to know that
>
> ¬(g is full of non-wine that is colored like wine).
>
> But the whole point of Dretske's position is to deny that one can know the latter in the type of situation that we have in mind.

---

[292]On p. 29 of [462], Hawthorne notes the existence of a closure condition related to regularity as the above condition is to monotony: "If one knows some premisses and competently deduces Q from those, thereby coming to believe Q, while retaining one's knowledge of those premises throughout, one comes to know that Q.". In Hawthorne [461] the monotone-related and regularity-related principles are called single-premise and multi-premise closure respectively. It should be added that the present summary downplays Dretske's deployment of the notion of a *reason* for belief, which plays a considerable role in (one version of) his case against closure.

[293]This passage is from [462], p. 33f.; Hawthorne's own tilde notation for negation has been replaced with "¬" here for continuity with our general discussion, and parentheses placed around what is negated, to ease readability.

5.1. THE LOGICAL OMNISCIENCE ISSUE

Modulating to the idiom of conditions on modal logics (and writing $\Box$ rather than K): for a congruential modal logic S, if we have $\vdash_S \Box(p \wedge q) \to \Box p$, then S is monotone.[294]

EXERCISE 5.1.9 Prove the claim just made. ✠

Digression. To put what is going on here into a more general theoretical setting, it is best to abstract from the behaviour of $\to$ in the above claim and think of the inference from conjunctions to their conjuncts – say, for definiteness, to their second conjuncts, as recorded in the fact that, where $\vdash_{CL}$ is the consequence relation of classical propositional logic ('truth-functional consequence', 'tautological consequence'), we have $\varphi \wedge \psi \vdash_{CL} \varphi$. Now take any one-premiss inference, say from χ_1 to χ_2, which is sanctioned by classical logic in the sense that $\chi_1 \vdash_{CL} \chi_2$. This can be regarded as implicitly subsumed under the conjunction-to-second-conjunct form in the sense that we can find φ, ψ, for which $\chi_1 \dashv\vdash_{CL} \varphi \wedge \psi$ and $\chi_2 \dashv\vdash_{CL} \psi$, in view of which we may describe the rule of inference which takes us from a conjunction to its second conjunct as *archetypal* for classical logic. It turns out that for classical logic every such one-premiss rule of inference is archetypal unless its premiss is always inconsistent or its conclusion is always provable outright, or else its premiss is equivalent to its conclusion. (Thus Exercise 5.1.9 could have been asked apropos of congruential S proving $\Box p \to \Box(q \to p)$, for example, in place of $\Box(p \wedge q) \to \Box p$.) The situation is very different for intuitionistic logic – and indeed for any intermediate logic short of classical logic. See Humberstone [579] and Połacik [888] and [889] for elaboration and justification of these claims. ***End of Digression.***

Formulations of closure-related conditions on knowledge such as Hawthorne's, quoted earlier, can of course be provided for belief and indeed specifically for justified belief, the latter being the focus of a particularly well-known such formulation:

EXAMPLE 5.1.10 Gettier [370] refutes the traditional analysis of knowledge as justified true belief – specifically, the claim that these conditions are *sufficient* (rather than that the claim that they are necessary) for knowledge with a style of example of the following form. A subject acquires evidence sufficient to justify the belief that p and infers with justification from this that $p \vee q$. (The second disjunct here is unrelated to the first, and in Gettier's presentation, is chosen at random. The concrete instantiations given in [370] for p and q are respectively by the sentences "Smith owns a Ford" and "Brown is in Barcelona".) Though the evidence was strong enough to justify the belief that p, and so also this inferred belief, it was in fact misleading, and it is not true that p. As luck would have it, however, it is true that q, and therefore true that $p \vee q$. Thus the subject has a belief that $p \vee q$ which is true and also justified. But intuitively, one does not regard such a subject under these circumstances as *knowing* that p. So justified true belief is not sufficient for knowledge. So Gettier argued, explicitly noting that the argument required two plausible enough background assumptions to be granted, one of which was that the notion of justification employed by the justified-true-belief account would have to be sufficiently lenient that one could be in possession of a justification for believing something which was in fact false (in the above case, the proposition that p): if by contrast one required justifying evidence to be logically conclusive then (a) hardly any of our beliefs would be justified and (b) there would have been no need to add the condition that the belief be not only justified *but also* true. The second background assumption is the closure-like principle which prompts mention of the example in the present setting, here quoted from the opening page of [370], in which "S" is a variable for an arbitrary subject (or cognitive agent), with the addition of some disambiguating commas:

---

[294] Just after the passage excerpted here: "Often a stronger principle is defended: that if one knows P and Q, one knows P and one knows Q. I do not need that here, though I do not wish to question it here either." ([462], p. 31.) Schotch and Jennings, aware of the role of KMono in arguments for scepticism, defended in [1001] a logic according to which K was not monotone – or even congruential – though it did have the aggregation property. Some of these ideas are revised in Schotch [999].

(F)or any proposition P, if S is justified in believing P, and P entails Q, and S deduces Q from P and accepts Q as a result of this deduction, then S is justified in believing Q.

We call this a closure-related condition rather than a closure condition since it does not say that the class of propositions justifiably believed by a subject is closed under entailment – i.e., whatever follows from something the subject believes is in turn believed by the subject. The preconditions listed also include the hypothesis that the subject deduces the entailed proposition and accepts it as a result. Thus the principle would not be open to a concept-possession objection to the effect that the entailed proposition might not be graspable by the subject and therefore would not be believed rationally or otherwise by that subject. The objection does not apply to the principle as quoted, since the proposition in question could not in that case have been deduced ("and accepted") by the subject in the first place.

The "P" and "Q" in the general formulation above are instantiated to "the proposition that p" and "the proposition that $p \vee q$" in our preliminary discussion before presenting this quotation. That, is Gettier's capital letters in that formulation stand in for names of propositions whereas our lower case letters stand in for sentences expressing the propositions concerned (and "$\vee$" is written in place of "or" simply to avoid a misunderstanding that somehow exclusive disjunction is involved – in which case the required entailment from P to Q would not hold anyway). Often people are less fastidious about whether we are using variables in name position or in sentence position, and this will probably be found to be the case in some of our own formulations. Certainly, this is so for Gettier, since he presents the justified-true-belief analysis which such examples are intended to refute, in the following terms:

S knows that P if and only if
(i)   P is true;
(ii) S believes that P; and
(iii) S is justified in believing that P.

Here "P" in "S knows that P" is standing in sentence position (since it follows the word *that*) while in (i) it is name position, and in (ii) and (iii) it is back in sentence position. So there is no single coherent way of construing the "P" – not that this worries anyone, since one makes the obvious adjustments to restore intelligibility either way (so that all occurrences can be construed nominally or else all occurrences can be construed sententially). Williamson (in [1189] and all other publications) is scrupulously careful to retain a nominal interpretation throughout for the (lower case p, q, etc.) variables he uses: they are to stand in place of names of propositions rather than for the sentences that might express those propositions. This does occasionally lead to awkwardness, as in the case of the following formulation, from [1189], p. 24: "For example, one is sometimes in no position to know whether one is in the mental state of hoping p." Such wording makes sense in the case of belief and arguably also knowledge, since one can believe a proposition, but there is no such thing as hoping a proposition, so the formulation just quoted is not quite right – one would strictly need to introduce a new verb, *hope'*, say, explaining that to hope' a proposition p is to hope that p is true. Perhaps the use of "know" in talk of knowing a proposition is similarly a technical usage understood along parallel *know'* lines, i.e., as short for knowing the proposition in question to be true. Such a suggestion amounts to saying that *know* itself patterns like *think* rather than *believe*:

(a)   Alex believes that p.
(b)   Alex believes the proposition that p.
(c)   Alex thinks that p.
(d)   *Alex thinks the proposition that p.

(The asterisk on (d) indicates ungrammaticality. Further examples in the same vein, along with a discussion of the issues they raise, can be conveniently found in King [656], and references there given, especially to the work of K. Bach, M. McKinsey, and T. Parsons; see also Moltmann [815] or [818], esp. Chapter 4.) The subtle difference in meaning between (a) and (c) here is not something to worry about teasing out in a brief overview of doxastic logic, and is in any case not germane to the contrast

5.1. THE LOGICAL OMNISCIENCE ISSUE

of current interest; of greater relevance would be, for example, Alan White's animadversions on the phrase "probable proposition" ([1168], p. 62*f.*). Incidentally, one finds in Williamson [1189] the view that no explicit analysis of the concept of knowledge, whether along 'justified true belief' lines or others, is likely to be forthcoming; a similar view is defended in Craig [199]. Both writers maintain that there is nevertheless much to be said about the concept – perhaps even enough to characterize it uniquely ([1189], §1.4): so there are evidently some methodological subtleties involved here. ◀

Digression. What are usually called propositional variables – alias sentence letters – are things like the p and q of our discussion of Gettier in Example 5.1.10 rather than things like P and Q in that discussion (or p and q as used by Williamson), so the terminology is slightly misleading in that whereas individual variables range over individuals, propositional variables do not similarly 'range over' propositions – at least in such informal discussions as we are considering here, where they plainly stand where a complete declarative sentence could stand. Yet MacIntosh [750], p. 177, writes that in such expressions as "Bring it about that p", "'p' is not a *propositional* variable," apparently on the grounds that one cannot be said to bring about a proposition ([750], p. 178). But the example is "Bring it about that p", not "Bring about P", so the point – which seems to be essentially that of Geach [358], [359], mentioned in note 298 below – is not quite cleanly made. **End of Digression.**

Reactions to Gettier's examples and others making the same point less artificially have been highly varied and form the bread and butter of many a university course in epistemology. Going into them in our overview of basic epistemic logic would hardly be appropriate – not that there has been no contact between epistemic logic and Gettier cases, construed broadly enough to subsume any counterexamples to the justified-true-belief account: see Williamson [1197]; for a thorough and informative survey of early reactions, see Shope [1051] and for pointers to the more recent literature, Hetherington [476]. However, one reaction raising interesting methodological issues is worth citing here:

EXAMPLE 5.1.11 Jonathan Harrison [454] digresses in a discussion of some views of Stephan Körner to inject the following reaction to the case from Gettier [370] mentioned in Example 5.1.10. The references to Jones and Brown are as there, with Smith being an individual acquainted with both of these people, it being Smith whose belief that $p \vee q$ is justified and true but fails to count as knowledge (according to [370]):

> That Jones owns a Ford certainly entails that either Jones owns a Ford or Brown is in Barcelona, and that Smith believes that Jones owns a Ford certainly entails that either Smith believes that Jones owns a Ford or Smith believes that Brown is in Barcelona. Equally certainly, that Smith believes that Jones owns a Ford does not *entail* that Smith believes (that Jones owns a Ford or Brown is in Barcelona). But does Smith in fact believe that either Jones owns a Ford or Brown is in Barcelona? There is a case for saying that he does not. If someone asks me where my wife is, and I, believing that she is at the pictures, say 'I believe that she is either at the pictures or at the hairdressers', I may well be saying what is false. If I believe she is at the pictures, then it cannot be that I believe she is either at the pictures or at the hairdressers; the two statements are incompatible with one other. But in this case, Gettier has failed to produce a case of justified true belief which is not knowledge, for Smith does not believe that either Jones owns a Ford or Brown is in Barcelona, precisely because he does believe that Jones owns a Ford. It may be, however, that the implication that Smith does not believe either of these two disjuncts severally is a merely contextual one.

Note that the answer to Harrison's question, "But does Smith in fact believe that either Jones owns a Ford or Brown is in Barcelona?" was given already in Gettier's original description of the example, in which it is stipulated that Smith deduces this belief from the (as it happens, well-justified though false) belief that Jones owns a Ford. Harrison continues in such a way as to suggest that any such stipulation is inconsistent: "If I believe she is at the pictures, then it cannot be that I believe she is either at the pictures or at the hairdressers; the two statements are incompatible with one other." Not only is this simply a rejection of Gettier's closure-related condition inset under Example 5.1.10, but it is rather clearly absurd: if required to answer, with a *yes* or a *no*, the question whether his wife is either at the pictures or at the hairdresser – in which the "either" simply disambiguates the question as a yes/no question (with disjunctive content) rather than a disjunctive question – would Harrison's answering *yes* mean that he was answering insincerely, i.e., that he did not believe the disjunction in question? Presumably not. Of course it would be misleading to say, our of the blue, "I believe that my wife is either at the cinema or at the hairdresser's" because that would not be being conversationally cooperative, and in particular being gratuitously less informative that one might be, as Grice emphasized (e.g., [412], pp. 44–50) was the case with a similar utterance minus the initial "I believe", whose truth is already conceded by Harrrison. The final sentence of the above quotation, in which the reference to contextual implication is presumably intended to call Gricean conversational implicature (touched on in Remark 1.3.2, p. 18) to mind, acknowledges this – for some reason as a mere possibility ("it may be...") – but it undermines the whole response to Gettier. (The same goes for some of Harrison's criticisms of Körner, as he notes in [663], the reference to informal appropriateness in the title of which is to the avoidance of such implicature violations.) It should of course be acknowledged that such semantics-vs.-pragmatics moves were less broadly familiar in 1979, the year in which the above remarks of Harrison's were published, with [412] yet to appear in print; for a quick guide to issues of implicature, see Horn [522]. Incidentally, the initial reference to Körner above arises because of the latter's longstanding project of using logical tools to diagnose pragmatic inappropriateness in terms of inessential occurrences: thus for instance – Ross's Paradox on the assumption that that Mary ought to post the letter, it is inappropriate to assert that Mary ought to post the letter or burn it, not so much because the latter is a less informative thing to say (violating Grice's 'maxim of quantity', because though less informative it is more long-winded), as because on that same assumption one could equally infer that Mary ought to post the letter or *not* burn it. Here Körner's mark of inessential occurrence is in evidence: the occurrence of q in $O(p \vee q)$ is replaceable by $\neg q$, given the background assumption Op. (See Chapter 2 of [662] for a more accurate presentation, here oversimplified, of Körner's position.) ◀

REMARK 5.1.12 Harrison's objection above not only to any proposed entailment from $B\varphi$ to $B(\varphi \vee \psi)$, but even to the compatibility of $B\varphi$ and $B(\varphi \vee \psi)$ for suitably chosen φ, ψ (as with the p, q, of Example 5.1.11) will perhaps call to mind the case of Ross's Paradox in deontic logic, concerning which on p. 244 we remarked that the proposed pragmatic response in terms of conversational implicature had been challenged in Cariani [148]. (In all currently relevant respects, this work is substantially similar to that cited as the unpublished conference paper "Cariani [2007]" in Humberstone [594]. Also relevant is the work cited there at p. 811, as well as in [148], by Mandy Simons.) In fact the Jackson – or Jackson–Pargetter – view and that developed by Simons just parenthetically alluded to are somewhat different from Cariani's, except in respect of rejecting the pragmatic response. Cariani's treatment, in particular, makes considerable play with a relativization of truth to, *inter alia*, a set of available options, as a kind of contextually given parameter. The acceptability of the account provided in [148] requires that this be seen to be a plausible posit, as well as that the deontic modal auxiliaries behave quite differently from what seem to be their epistemic counterparts. Cariani holds ([148], p. 19) that it would not be at all odd to come out with (a), it would be very odd to come out with (b):

(a) I doubt that Lynn ought to either wear a tie or a scarf. In fact she ought to wear a scarf.

(b) ?I doubt that Lynn must have either worn a tie or a scarf. In fact she must have worn a scarf.

Accordingly, while the permissibility of wearing a tie is among the requirements for the *truth* of "Lynn ought to either wear a tie or a scarf", its being a live epistemic possibility that Lynn wore a tie is not required for the truth, but only for the *appropriateness as an assertion* in the case of "Lynn must have either worn a tie or a scarf". This asymmetrical treatment is not part of Cariani's treatment of the deontic side of the deontic–epistemic contrast, however, since one might (unlike Cariani) attempt to take a similar line with the epistemic side. Minor issues: the "either" does not seem essential in (a) and (b),

5.1. THE LOGICAL OMNISCIENCE ISSUE

and does not seem to be correctly positioned w.r.t. the "or" in the examples as they stand. One might also have preferred to see deontic *must* contrasted with the epistemic *must* in the two cases, or else deontic *ought* with epistemic *ought* – though Cariani may have avoided the latter because of further potential complications – see Cariani [149] (where actually it is *should* rather than *ought* that is involved). ◀

With this excursus on some of the perturbations arising over the monotone/nonmonotone behaviour of modal notions at an end, we conclude the treatment of our third source of worry – doubts about (this or that form of) the closure principle. The second source – concept-possession issues – will come up for discussion below in Section 5.6. Both this and the first source of complaint – excessive optimism about deductive competence, arise just as much for *belief* as for *knowledge*, making this a suitable point to recall that Hintikka [490] introduced not only epistemic but also *doxastic* logic. Corresponding to the epistemic $\Box$ and $\Diamond$ notated as K and P (or K_a and P_a, when a reference to the knowing subject is required), we have a doxastic $\Box$ and $\Diamond$, which in Hintikka's notation appear as B and C (or B_a, C_a, to be precise).[295] Not all responses to those worries will sound as good as the corresponding moves with K. Although talk of implicitly believing is fine, for example – which is not to say that it is entirely *clear* – talk of being in a "position to believe" does not lend itself to an interpretation parallel to that of the phrase "position to know" (as in note 282). On the other hand, re-reading $B\varphi$ as "it follows from what the subject believes that φ" is available; it is natural also to describe this in terms of what the subject is *committed* to.[296]

It is often observed that the syntactic behaviour of the verbs *know* and *believe* is very different in English, for example that only the former can take embedded (or 'indirect') questions as complements, that one asks how a person knows something but only why a person believes something, and so on. (Many such observations can be found at the start of Lemmon [703].[297]) Some philosophers have, because of

---

[295] Suggested mnemonic reading for "$C\varphi$" (alias $\neg B\neg\varphi$): it is consistent with a's beliefs that φ. This reading works only on the assumption that a's beliefs are themselves consistent, i.e. with the axiom $\mathbf{D}_B = Bp \to Cp$ in force. *Warning*: in Lenzen's publications, "C" is mostly used differently, in place of the usual "B", but with the reading "a is certain that" to emphasize the strength of conviction, allowing "B" itself to be used for a weaker strength belief operator; as with K and P, some authors follow Hintikka in retaining the subscript "a" even when no contrast is being marked with other cognitive agents. As remarked in note 275 (p. 333), P in the present context can be thought of as suggesting (epistemic) possibility: what is possible given what the subject knows, though this reading is only available in the presence of a deductive omniscience idealization (to the effect that whatever is impossible *tout court* is known not to be the case), as several writers have complained (see for instance Girle [377]). Note that there is a typographical error on the top line of p. 11 of [490], in the explanation of "C_a": the word *knows* should be *believes*.

[296] Routley and Routley ([976], p. 192) say that "the notion which doxastic logics such as that of Hintikka [490] study" is "the notion of commitment-to-believing" – which, they add, "is not really a belief notion at all".

[297] For other references to similar discussions, see the top of p. 43 of Williamson [1189]. The reference to embedded questions should be understood as subsuming question-word + infinitive constructions. Famously, Gilbert Ryle urged that what he called *knowing how* was not a species of *knowing that* (such and such was the case); Lemmon [703] pointed out that what Ryle wanted to hive off from 'knowledge that' – roughly speaking, propositional knowledge – was not so much *knowing how*, where this includes, for example, a's knowing how steam engines work, but more specifically *knowing how to*, as in a's knowing how to tie a shoelace. (Hintikka [495], p. 114, made the same point later.) More recently, Stanley and Williamson [1086] have argued that even (what Lemmon calls) 'knowing how to' is still just a special case of knowing that; see also Stanley [1085]. Most parallel question words (*wh* words) admit a similar construction with *to* in English, as Lemmon notes on p. 55 of [703] – or see A. W. Moore [823], p. 292, for the same point again – though curiously not the word "why", *pace* [1085], p. 70, which lists "John knows why to hit a ball hard" as grammatical. (Of course it is *intelligible*, but that is another matter; the same mistake appears on p. 178f. of the editors' own contribution to the collection Bengson and Moffat [64]. Lemmon revealed some sensitivity to this point in note 3 of [703] when he asks "But why not 'I heard on the radio this morning why to wash lettuce before eating it'?" However, in the main body of his discussion, even Lemmon ([703], p. 55) is happy with "know why to apply bandages to a wound". Incidentally, Lemmon's lettuce question itself begins with "Why not": another illustration of the rather special syntactic behaviour of *why*; consider also "Why bother?" and so on. Huddleston and Pullum [530], p. 906, say that *why to* is "just possible in the titular use," meaning: in titles and headings – but whether by "just" they mean *barely* or *only* is not clear. I think the writers of such titles are simply confused.) An assimilation in the reverse direction – knowing that as a special case of knowing how – had been urged in Hartland-Swann [458]; McGinn [786], somewhat similarly, invites us to think of all knowledge in terms of discriminatory abilities; see also Hetherington [475]. And for a different story again, arguing that the *knowing that* construction itself already involves reference to an implicit question, consult Schaffer [1043] (for more on which see Example 5.4.9, p. 395 below). Williams [1177] and the contributions in the already mentioned [64], as well as Carr [153], Snowdon [1070], Rosefeldt [966], Glick [380]

such discrepancies or for other reasons, wanted to say that lumping together knowledge and belief as two propositional attitudes – to use the popular term – embodies a confusion, in that while the objects of belief (what is believed, that is) are indeed propositions or something very like them (see the discussion following Example 5.2.6 below), knowledge has objects of a quite different type: facts (according to Vendler [1124], Chapter 5) or 'parts of the whole truth' (Unger [1119], Chapter 7). Since it remains the case that "*a* believes that __" and "*a* knows that __" still both make sentences when the blank is filled by a sentence, even if such metaphysical distinctions between the objects of knowledge and belief existed, they would make no difference to epistemic, doxastic, and mixed epistemic–doxastic logic.[298] One can go too far in this direction, though:

EXAMPLE 5.1.13 Bacon [47] opens with the following words:

> The one obvious logical characteristic of belief, as opposed to mere belief, is that it follows from knowledge. Every knowing is a believing: knowledge is belief of some special kind.

Even if we agree that "*a* believes that p" follows from "*a* knows that p" – and not all do, as is noted in the Digression on p. 391 – it does not follow that every knowing is a believing (as is noted in passing on p. 170 of MacIntosh [750]). Adapting an example from Davies [234], let us define *gwalking* to be the activity of walking on the part of an agent who is at the same time chewing gum. Note that when a is chewing gum, a's gwalking is a's walking – it is not a's chewing gum, and nor is it the composite activity on a's part of both walking and chewing gum; thus if a, while chewing gum slowly, is walking quickly, then a is gwalking quickly. With these understandings in place, "*a* is chewing gum" follows from "*a* is gwalking", just as '*a* believes that p" follows from "*a* knows that p". But *ex hypothesi* not every (and in fact, not any) act of gwalking is an act of chewing gum, so Bacon's comment that every knowing is a believing is not a mere re-statement of what has gone before it. (The distinction between mental states themselves, and mental state *concepts*, involved here is assiduously observed in Williamson [1189], especially Chapter 1; aspects of this distinction, and of the – by Williamson, contested – type/token distinction for mental states, are taken up in Fricker [335].) ◀

Two further tricky features of the verbs *know* and *believe* which can threaten to cause logical confusions were discussed already in Hintikka [490], §1.7. The second of them will allow us to touch on an important issue in doxastic logic below, namely what is usually called Moore's Paradox. The first point concerns the tendency for embedded occurrences of "knows that __" within a sentence to appear to commit anyone using that sentence to the truth of whatever fills the blank. This includes embedding under negation and in *yes/no* questions (embedding under "Is it that case that ...?"), and it was traditionally discussed in the terminology of presupposition. "*a* knows that pearl meat is expensive" was said to *presuppose*, rather than simply *entail*, that pearl meat is expensive, since the commitment to the latter remains when the sentence is turned into "*a* does not know that pearl meat is expensive", or "Does *a* know that pearl meat

---

and Cath [165] provide further discussion of the *knowing how* issue, as does Abbott [2], which also raises some interesting methodological issues.

[298]For an elaboration of this point, see the middle paragraph of p. 43 in Williamson [1189]; similar moves were made in Peterson [880]. Williamson points out, incidentally, that the linguistic evidence apropos of *knows* and *believes* is not all one way in any case, as witness the grammaticality of: "Long before I knew those things about you I believed them." In other cases, corresponding linguistic data tell against a unified treatment of the that-clauses involved. The following example is adapted from §3 of Parsons [870]: from "Sue believed that war had broken out and Tom regretted that war had broken out" we cannot infer – and indeed this is not even grammatical – "Sue believed something that Tom regretted." (The inference would also fail for *knew* in place of both occurrences of *believed*, where the conclusion would be grammatical but a meaning that prevents it from following from the premiss, namely that Sue knew concerning something that Tom regretted that thing.) Somewhat similar examples had been given by Geach in [358], [359], where they are taken as witnessing two kinds of intentionality. (See also p. 177f. of MacIntosh [750], already mentioned in the Digression after Example 5.1.10 above. MacIntosh is sympathetic to the main Vendler–Unger point that the objects of knowledge and belief differ.) Even in the alethic arena, such moves have been made: Forbes [320], Chapter 5, argues that necessity and possibility are primarily properties of states-of-affairs and apply only derivatively to propositions (taken as sentence senses – Fregean thoughts).

5.1. THE LOGICAL OMNISCIENCE ISSUE

is expensive?" (or "If a knows that pearl meat is expensive,..."). An extensive literature grew up on this phenomenon, whose very classification as semantic or pragmatic – the latter because the commitment in question appears to be "cancellable" – "No, he *doesn't* know Tom is dead, because in fact Tom is very much alive" – has been a matter of contention. (See Beaver [56] for a survey; the cancellability point was stressed originally in §4.3 of Kempson [646] and Chapter 2 of Wilson [1201]. A review of subsequent developments in thinking about presupposition in general is provided by Beaver and Geurts [57].) We can ignore it for present purposes, noting with Hintikka that when the occasion arises on which one wants to force the "unbeknown to a, pearl meat is expensive" interpretation of "a does not know that pearl meat is expensive", one can always resort to writing explicitly "$p \land \neg K_a p$" (where p is for "pearl meat is expensive"). Verbs presupposing the truth of their sentential complements were called *factive* verbs in Kiparsky and Kiparsky [658]. The same label is given in Chapter 1 of Williamson [1189] for (propositional attitude) verbs for which "a *verbs that* φ" entails (i.e. has a logical consequence) the truth of φ. There is thus some threat of confusion, at least for those believing that there is a viable semantic notion of presupposition which is distinct from entailment, creating a case for reserving the term *veridical* (used by Williamson in earlier writings, as well as by others) for the latter notion. We won't need either term here, since we can simply say that the operator, O, say, used to represent the verb in question (with a subject term supplied) satisfies the principle **T**, or, to make it explicit – as for tense logic on p. 178 above – that O is playing the $\Box$ role, the principle $\mathbf{T}_O$ ("**T**-for-O", that is).

EXAMPLE 5.1.14 To see that Hintikka's careful treatment of these matters has not been universally heeded, consider the following passage from Heathcote [468], p. 287:

> When it is the case that *I do not know that p is false*, this may seem to be adequately represented by the $\Diamond$ of knowledge. But whenever I don't know that p is false I also don't know that it is true. Our ignorance is equally well represented, and expressed, by $\neg K \neg p$ and $\neg K p$.

Since Heathcote thinks these last two are equivalent, he thinks there is a radical disanalogy with the case of alethic modal logic. But of course they are far from equivalent. Whenever Kp is true $K\neg p$ is false, so $\neg K\neg p$ is true. If this even entailed – never mind the converse – what Heathcote suggests is equivalent to it, namely $\neg Kp$, then the truth of Kp would accordingly imply that of $\neg Kp$, which is to say: Kp could never be true. Heathcote says "But whenever I don't know that p is false I also don't know that it is true," but Heathcote does not know that it is false that $2 + 2 = 4$, since this, being true, is not something anyone knows is false; according to his own principle, he does not in that case know that it is true that $2 + 2 = 4$, either. Here the presupposition issue (from p. 352) raises its head: one would not normally say "a does not know that it is false that $2 + 2 = 4$" without special intonation and surrounding de-commitment to what would otherwise be conveyed by such an utterance: that one is reporting a's ignorance of something one takes to be a fact. The risk of confusing the appropriateness of something with the truth of what is said is further compounded by the use, as here, of first-person examples. (See also Example 5.6.11, p. 431, and Remark 5.6.12.) Yet a further complication in the wings here is one mentioned by Heathcote himself – the potential for confusion between the *knows that* construction and the *knows whether* construction: the interested reader is invited to evaluate his response to this charge ([468], note 4). The above argument involved $\mathbf{D}_K$: the implication from Kp to $\neg K\neg p$ (or Pp), and Heathcote says this principle must be rejected precisely because it is in tension with his equivalence claim. But – never mind the logic of K – what is the background logic for the non-modal (non-epistemic) connectives? Apparently, not classical propositional logic, since as we read (again on p. 287 of [468]) "$\neg K\neg p$ and $\neg Kp$ are equivalent despite the fact that their negations are distinct," – presumably meaning by 'distinct', *non-equivalent*, in which case something would need to be said by way of explanation as to what kind of non-classical negation is involved here. On the other hand, perhaps 'distinct' means: not freely interchangeable to give equivalent formulas when embedded in arbitrary linguistic contexts; in this case again more information as to precisely what kind of failure of congruentiality is being envisaged. ◀

The second feature to watch out for concerns the verb *believe*, and its ability to trigger a process known in the heyday of transformational grammar as neg-raising, by which was meant that a negation governing a sentence embedded under "believes" could move ("be raised") to a position in which syntactically it came to have "believes" in its scope, creating a scope ambiguity. Thus, "*a* doesn't believe he got the job" ends ambiguous between $B_a \neg p$ (with p for "*a* got the job") and $\neg B_a p$. The former interpretation takes neg-raising to have moved an inner (or 'lower') negation outwards ('upwards').

EXAMPLE 5.1.15 The resulting ambiguity in things like "*a* doesn't believe that *p*" between a $B\neg p$ and a $\neg Bp$ reading can cause confusion. The following example is concocted and is not exactly to be found in Gensler [363], which also purports to show that (freely) acting "commits one to ethical beliefs", on the – admittedly contentious – 'internalist' assumption that if one believes that an action is morally wrong one does not freely perform that action. Take an example of such an action, say, Susan's eating a hamburger for lunch one day. By the internalist assumption, if she had believed that it was wrong to eat meat, she would have not performed this action of eating meat, so she didn't believe that it was wrong, i.e., she believed that it was morally permissible to eat meat. So her free action implies that she had this permissibility belief.

Of course, the reasoning here falters at the "i.e." conflating

$$\neg BO\neg(\text{Susan eats meat}),$$

which *does* follow from the details of the case (and the internalist assumption), and

$$B\neg O\neg(\text{Susan eats meat}),$$

which was reformulated as BP(Susan eats meat) – since O and P are deontic $\Box$ and $\Diamond$ respectively – which is what would be required as a moral belief, but does not follow. While some may refuse to think of beliefs with negated O-content as moral *judgments*, describing only beliefs about what people should and shouldn't do as 'judgmental', here we assume that a commitment to the denial of such a belief is also a moral belief. (This shows that some care is needed in characterizing the *amoralist* – see the opening chapter of Williams [1172].) That would still be the case even if one did not agree that negated *ought*-judgments amounted to permissibility judgments, though it must be admitted that the relevant deontic vocabulary should really be of the agent-implicating rather than the situational type distinguished in Example 4.4.14, p. 251 above. Again: the fallacious conflation encouraged by neg-raising which the above reasoning illustrates is not being claimed to drive Gensler's own argument in Gensler [363], which the interested reader is invited to compare with that considered here. The internalist assumption alluded to takes various forms. We encountered one version, that provided by Hare's universal prescriptivism via the notion of assenting to an imperative, mentioned in Example 4.6.15(*i*) (p. 299). For a milder formulation, compare this, from Dreier [251], according to which a "widely held thesis, internalism, tells us that to accept (sincerely assert, believe, etc.) a moral judgment logically requires having a motivating reason." A motivating – though not necessarily overriding – reason to act as the judgment enjoins, that is: so this, incidentally, is a use of the phrase *moral judgment* which excludes negated O-judgments. ◀

The verbs *believe*, *expect*, *think*, *want*, amongst others, are possible neg-raising triggers, while the verbs *know*, *hope*, *remember*, amongst others, are not. There is a good discussion in §5.2 of Horn[299] [521] (where, *inter alia*, analogous phenomena in languages other than English are noted, and also noted to differ in respect of whether verbs with the same meaning are alike in respect of triggering neg-raising: *hoffen* – "hope" in German – is said to be a trigger, for example. Also of interest is Seuren [1039]. See Horn [523], p. 252, for a more recent discussion).

---

[299]This is Lawrence Horn, the linguist – not Alfred Horn, the mathematician – in honour of whom Horn formulas are named (as encountered in note 51, p. 67, for example).

5.1. THE LOGICAL OMNISCIENCE ISSUE

REMARK 5.1.16 Something very like the neg-raising phenomenon occurs with modal auxiliaries in English and other languages, and with modal verbs in languages in which the relevant lexical items pattern like another verbs. Thus *may not* in "She may not eat" can be interpreted as $\neg\Diamond$ or as $\Diamond\neg$, depending somewhat on the intonation, with a deontic $\Diamond$ ("P"), although, as has often been observed, with an epistemic reading "She may not have eaten" permits only the "$\Diamond\neg$" reading. (Cormack and Smith [194], pp. 136, 143, suggest that deontic/dynatic *can* behaves like *may* in this respect, though that seems wrong for standard English, in which "She cannot eat" is contrasted with "She can not eat".) Evidently the area is highly idiomatized. For example, *must not* has only a "$\Box\neg$" reading while *need not* (or *needn't*) is "$\neg\Box$" – though main verb *need* is available for a "$\Box\neg$" interpretation: *He needn't be here* vs. *He needs to not be here.* In January 2015, the present author received a call-for-papers notification (in English), sent by an Italian, for a semantics/pragmatics workshop, which contained the following cryptic instruction:

> Abstracts must be blind and do not have to exceed 3500 characters in total (including examples and references).

Presumably "do not have to" (= "$\neg\Box$") is intended meant to be "must not" (= "$\Box\neg$"). The Italian modal verb *dovere* is (at least analogous to) a neg-raiser, in that *Non devo andare* can mean not only "I must not go" but also "I do not have to go". (This comment is intended to be neutral as to the exact syntactic processes involved here, and on the semantic front should not be taken to imply that "don't have to" is really the (external) negation of "must", since *have to* and *must* typically function respectively as reportive and normative deontic "$\Box$"s. Further bibliographical references on this and related matters may be found in Humberstone [582].) ◀

This second feature, as Williams [1176] points out, gives rise to an ambiguity in an example (adapted) from G. E. Moore:

(1)   It is raining but I don't believe that it is.

While (1), with p for *it is raining*, is generally interpreted as having the force of (2)), because of the neg-raising issue, it could instead be taken to mean (3); Williams calls the patterns instantiated by (2) and (3) the *omissive* and the *commissive* forms of Moore's Paradox, respectively:

(2) $p \wedge \neg Bp$ (3) $p \wedge B\neg p$.

Both readings are discussed in the first two chapters of Sorensen [1073], but here we consider only the omissive (2)-style reading. Note that if the consistency condition $\mathbf{D}_B$ is in force, (3) has (2) as a consequence, and that, in view of the plausibility – to say the least – of $\mathbf{T}_K$, with "I know" in place of "I believe", we get an interesting epistemic analogue of (2) but not of (3), the latter being inconsistent rather than just something to which assent would be problematic. (More recently the significant variation with "You" for "I", which raises rather different questions when considered as something that might be asserted, has also been considered – see for example Holliday and Icard [517]. Public assertion is, by contrast and despite much early literature on the subject, something of a side-issue when it comes to the original first-person form, as remarked, for example, in note 14 of Humberstone [575].)

The example is usually called Moore's Paradox, (after G. E. Moore) though 'Moore's Problem', as it is called in Sorensen [1073], is perhaps a better label. (As well as [1073], the collection Green and Williams [409] provides a discussion of this topic and the introductory chapter by the editors gives full historical detail. One such historical source is reprinted in a collection in our bibliography for independent reasons: see Chapter 12 of Baldwin [51]. For some debate as to whether Moorean assertions are always anomalous, see Hájek and Stoljar [423], Rosenthal [969], and Coliva [189] and references therein.) What *is* the problem, though? It is that (1) – understood along the lines of (2) – would be an absurd thing to say, or indeed to believe, but not because it is in any way inconsistent: it could very well be true that it was raining but not believed by any envisaged utterer (or thinker) of (1) to be raining (at the place and time of the utterance). A useful line of thought here, presented in Sorensen [1073] is that (1) is a kind of doxastic blindspot: it could be true but it couldn't be both believed and true. (Related issues connected with a famous derivation associated with F. B. Fitch are discussed under Example 5.4.11, p. 397.) Since, as it is often metaphorically put, 'belief aims at truth', this sounds like bad news for (1) as a candidate object of belief. We resume discussion of this metaphor in Remark 5.1.17 below.

Digression. Here the fact that we are dealing, in (1), with the first-person pronoun is essential. The fact that "p although a does not believes that p" could be true and believed by a, not realizing him/herself to be a, and so lacking the intended *de se* belief, raises the kind of issues for the individuation of propositions as the objects of propositional attitudes that we are resolved to steer clear of. (See Remark 5.2.7, p. 379.) As is also often pointed out in connection with the consistency of the *content* of the belief expressed by (1), a past tensed version "Yesterday it was raining but I did not believe that it was" is unproblematically assentable to. Another consideration pointing to the same conclusion but not involving any indexicality is that the disjunction of two doxastic blindspots is unproblematic, showing that neither disjunct could be inconsistent. For example, in the relatively strong doxastic logic **KD45** the formula $B((p \land \neg Bp) \lor (q \land \neg Bq))$ is consistent. (I.e., its negation is not provable.) Indeed, putting q for $\neg p$, this is what is true for anyone aware of being ignorant as to whether or not p. For another application of the fact that disjunctions of blindspots need not themselves be blindspots, see Humberstone [599]. For an interesting appeal to disjunctions *one* disjunct of which is an epistemic blindspot, see the top of p. 874 of Dorr and Hawthorn [247]: the authors note a contrast between this case, involving "I don't know that p" and the corresponding epistemic modal auxiliary construction "It might not be the case that p". (Recall that our main focus here is on epistemic logic for knowledge ascriptions – to oneself or others – and their negations, etc., rather than to these 'epistemic modals'.) **End of Digression.**

REMARK 5.1.17 The question of how to unpack this 'aiming' metaphor – also put, as on p. 204, in terms of *direction of fit* – has been the subject of considerable debate in the philosophical literature. The contrast itself comes originally from Anscombe [25] (esp. p. 56). Williams[300] [1171], p. 137 takes a step in the wrong direction here:

> (T)o believe that p is to believe that p is true. To believe that so and so is one and the same as to believe that that thing is true. This is the second point under the heading of 'beliefs aim at truth'.

Correcting the double use of "p" to stand at its first occurrence here where a sentence might stand and at its second occurrence where a name (such as the name of a proposition) might stand, we have the same phenomenon desiring that your children will prosper or hoping that they have not been injured in that accident you have just heard about is the same as desiring that it is true that they will prosper or hoping that it is true that they are uninjured. (Perhaps not exactly the same – see the contrasting belief-attributions about Fido at the end of Example 5.2.9, which begins on p. 379 – but close enough for present purposes.) But hope and desire are paradigms of attitudes with the direction of fit reversed from that of belief. In his later discussion, from [1175], things are much better. From p. 67:

> Beliefs can also be said to aim at the truth, to be supposed to be true, to be subject to norms of truth. It is an objection to a belief that it is false. In fact, in the case of *belief* it is a *fatal* objection, in the sense that if the person who has the belief accepts the objection, he thereby ceases to have the belief, or at least it retreats to the subconscious: if a person recognizes that the content of his belief is false, in virtue of this alone he abandons his belief in it.

In the discussion from which this passage is excerpted, the main contrast is not with wanting (hoping, intending, etc.) but with *asserting*. For discussions of how to characterize direction of fit as it pertains to

---

[300]This is Bernard Williams, not John Williams, just referred to in connection with the omissive and commissive forms of Moore's Paradox; As it happens, Bernard Williams himself also has something to say about Moore's Paradox in the paper from which we are about to quote here, as part of pressing this same point: that belief aims at truth.

5.1. THE LOGICAL OMNISCIENCE ISSUE

the doxastic vs. boulomaic attitudes see Humberstone [564], Zangwill [1217], and references cited therein. For a sceptical perspective on the distinction between the two directions of fit, see Frost [337]. ◀

To the extent that doxastic logic is in the business of expounding formally stateable conditions on rational belief, a natural candidate suggested by the 'aiming at truth' idea might be closure under the following rule. Because of its connection to Moore's Paradox – also (as mentioned in note 300) appealed to by Williams in his discussion of belief in [1171] (on the same page as the passage quoted from this work in Remark 5.1.17) – let us name the rule after G. E. Moore, and consider the merits of requiring its admissibility as a desideratum in any proposed logic of belief:

(Moore's Rule) $$\frac{\neg(B\varphi \wedge \varphi)}{\neg B\varphi}$$

That is: if the logic tells us that φ can't both be true and (rationally) believed, then it should tell us that φ cannot be (rationally) believed.

Now, an early question to consider on being presented with a proper (i.e., non-zero-premiss) rule like this is whether or not being closed under the rule is equivalent – or having the rule admissible, to use the alternative terminology – for the range of logics of interest, to the provability of (all instances of) a schema. Taking that range for present purposes to comprise the normal modal logics, this question has an easy answer, given here with B written as $\Box$.

PROPOSITION 5.1.18 *A normal modal logic is closed under Moore's Rule if and only if it extends* $\mathbf{K5}_c$.

Proof. For convenience, we take Moore's Rule in the (truth-functionally equivalent) form $\Box\varphi \to \neg\varphi$ / $\neg\Box\varphi$, and $\mathbf{5}_c$ in the form $\Box p \to \Diamond\Box p$ (i.e., as the converse of the dual form of the schema $\mathbf{5}$ as given in Section 2.1). For the 'only if' direction of the proof, suppose a normal modal logic S is closed under this rule, with a view to showing that $\vdash_S \mathbf{5}_c$ (equivalently: that $S \supseteq \mathbf{K5}_c$). Since p is a truth-functional consequence of $p \wedge \neg\Box p$, by $\Box$Mono we have $\vdash_S \Box(p \wedge \neg\Box p) \to \Box p$ and so by further truth-functional reasoning:

$$\vdash_S \Box(p \wedge \neg\Box p) \to \neg(p \wedge \neg\Box p),$$

from which by the assumption that S is closed under Moore's Rule (taking φ as $p \wedge \neg\Box p$) we conclude that $\vdash_S \neg\Box(p \wedge \neg\Box p)$. Reformulating this (since S is normal) as $\vdash_S \neg\Box(p \wedge \Box\neg\Box p)$, and then finally as $\vdash_S \Box(p \to \neg\Box\neg\Box p)$ we have $\vdash_S \mathbf{5}_c$ (writing the "$\neg\Box\neg$" as "$\Diamond$").

For the 'if' direction, suppose $\vdash_S \mathbf{5}_c$ and that we have the premiss $\Box\varphi \to \neg\varphi$ for an application of Moore's Rule, this premiss being provable in S. By $\Diamond$Mono, the following is also provable in S:

$$\Diamond\Box\varphi \to \Diamond\neg\varphi,$$

which we can rewrite as

$$\Diamond\Box\varphi \to \neg\Box\varphi.$$

Since $\vdash_S \mathbf{5}_c$, putting φ for p we have $\vdash_S \Box\varphi \to \Diamond\Box\varphi$, so by truth-functional reasoning we have as a theorem of S

$$\Box\varphi \to \neg\Box\varphi.$$

But this last is truth-functionally equivalent to $\neg\Box\varphi$, which is accordingly S-provable, showing S to be closed under Moore's Rule. ∎

REMARK 5.1.19 In view of the different direction of fit desires and beliefs are taken to have – beliefs aiming at matching the world, desires at getting the world to match them – and the appeal to Moore's Paradox mentioned just after Remark 5.1.17, the question naturally arises as to whether there is an analogue to Moore's Paradox for the case of desire. The question, raised tangentially in Shoemaker [1047], is explored more thoroughly in Wall [1140] and Williams [1178] and since we are not concentrating on boulomaic logic here, we do not go into the details. Reading W as "a wants it to be the case that" the above reasoning would terminate in $\mathbf{5}_c$ in the form $W \neg W \neg p \to \neg W \neg p$, which is easier to think about if we substitute $\neg p$ for p and help ourselves to at the lease the congruentiality involved in replacing the resulting $\neg \neg p$ by p, giving: $W \neg W p \to \neg W p$. It does not seem on the face of it to be a demand of rationality and the other idealizations involved in giving a modal treatment of propositional attitudes – introspective success, etc. – that a desire not to want something should be a desire which is guaranteed to be satisfied. (See further note 309, p. 370.) That is, there is not the pressure to treat as an ideal the alignment between desires about one's desires and the latter desires themselves, in the same way as there is over aligning one's beliefs about one's beliefs and the latter beliefs. One may respond that this shows the lack of even treatment between theoretical and practical rationality, but for present purposes, rather than explore this further, let us note that the very idea that boulomaic logic is as plausibly treated *normally* as doxastic logic is – and note the dependence of Prop. 5.1.18 on the condition of normality – is quite controversial. As in the case of what one *should* do, concerning which see the references given before Exercise 4.4.7, p. 243, and for essentially the same reason, Frank Jackson thinks that even the monotone condition is not satisfied by what we are writing as W. Here he is from p. 111 of [613]; the reference to sloops is an invitation to recall Quine [936]:

> For our purposes, what is of especial interest is one way it can be true that I want a particular sloop, without it being true that I want relief from slooplessness—namely, when there is some particularly desirable sloop, but I'm sure I wouldn't be getting that one if I got a sloop. Again, the explanation of why I do not want a job in the army, although there is a job in the army that I want, may be that I am sure the job I want is not the one I'd get. That is, whether I want a job in the army or a sloop depends in part on which particular job or sloop I think I'd get. I gave just illustrated this point by noting that I may want a particular F without wanting relief from F-lessness, because I think relief from F-lessness would be achieved by acquiring an undesirable F.

Thus although getting a glass of red wine entails getting a glass of wine, according to Jackson you can want a glass of red wine without wanting a glass of wine – if circumstances are such that you believe that were you to get a glass of wine, it would be a glass of white wine (which you detest). On the 'normal' approach wanting a glass of red wine is *ipso facto* wanting a glass of wine, and what we have in the imagined case is a situation in which although you do indeed want a glass of wine, that would be unwise way in which to order a drink.

Without taking sides on this issue, we can see what is involved as a choice over whether the content of a desire is being treated as the antecedent or as the consequent of a suitable conditional. The 'normal modal logic' approaches place the content in the consequent of what is, in the Anderson–Kanger–Smiley 'relative necessity' approach to deontic logic a straightforward strict conditional: what you ought to do is what, if you are following the demands of morality, you do do, the "if" being understood as a necessitated material conditional, necessity being more or less logical necessity, broadly construed. (The demands of morality are represented by the constant Q on p. 258. There is a more sophisticated two-dimensional version introduced in the discussion starting on p. 275, which still has the content – what is being said to be obligatory – located in the consequent of a necessitated material conditional though with a fancier kind of necessity involved: "$\Box$ as absolute necessity".) In the case of desire, an analogous approach could be taken: What you want is what is (strictly) implied by the proposition that things are fine, by your lights. This gives rise to a normal boulomaic logic. But Jackson's approach puts the content into the antecedent, thinking of what you want as given by what would be the case if that content obtained. In addition, we see that he has moved from strict implication to a counterfactual conditional, which has further ramifications for the logic concerned. Without this further variation – which could alternatively be achieved by using conditional probabilities – one would have the unfortunate consequence that from "Harriet wants to buy a scarf" there follows "Harriet wants to buy a scarf and strangle herself with it", since if it is strictly implied by Harriet's buying a scarf that (to retain the earlier formulation) things are fine by Harriet's lights, then it will be strictly implied by Harriet's buying a scarf and doing whatever else you care to name, that things are fine by Harriet's lights. (Here we assume that Lewis [717] is right to describe strengthening the antecedent as fallacious for subjunctive or counterfactuals, and we ignore the fact that despite this, it is after all possible to treat these conditionals as strict conditionals, given suitable care of what their antecedents are: see Humberstone [543].) ◀

Digression. The preceding ruminations on the topic of desire to one side, we have been thinking of 5_c, $\Box p \to \Diamond \Box p$, in doxastic terms, which is to say: as $Bp \to CBp$ in Hintikka's notation. (A semantic characterization showing what this axiom does for the canonical frames of logics containing it is mentioned in Exercise 2.7.28, p. 136.) An interesting variant arises if we consider a mixed doxastic-epistemic variant: $Kp \to CKp$. In its contraposed form, as $B\neg Kp \to \neg Kp$, this is raised – though no commitment to it is expressed – as a plausible principle at p. 172 of Hawthorne [461]: "It may be suggested that our negative verdicts about knowledge in our own case are decisive on account of a constitutive first-person authority. If I believe of myself that I do not know, then, automatically as it were, I do not know." ***End of Digression.***

REMARK 5.1.20 The adjective "constitutive" in the passage just quoted (in the Digression) flags a cluster of possible positions on the question of whether or why one is especially privileged with regard to some of one's own current mental states ('first-person authority'), whose common content is usefully summarized in Smithies and Stoljar [1068] thus:

> According to *constitutivism*, there are limits on the possibility of limits an error about one's own mental states because there is a constitutive, internal, or necessary connection between being in a certain mental state and believing that one is in that mental state.

For example, Peacocke once suggested that it may be partially constitutive possessing the concept *pain* that one be disposed to apply that concept to oneself only when one is in fact in pain. (Similarly: see p. 157 of Peacocke [876], and pp. 159*ff.* there for discussion of another constitutivist account of Crispin Wright – on which see also Holton [518].) Smithies and Stoljar go on to discuss under the same heading some views of Sydney Shoemaker, who in [1047] Chapter 2 [1048] (see also Chapter 4 of [1048], in which Chapter 2 reprints [1047], as well as Shoemaker [1049]) advanced an interesting argument to the effect that there is no quasi-perceptual faculty of introspection which informs us of our mental states, at least of our current states of belief and non-belief. The argument is that if there were such a faculty then it would have to be a logical possibility that in an otherwise rational and conceptually unconfused person, the faculty could fail (just as there not only could be, but are, people in whom sight and hearing do not function); such people Shoemaker calls self-blind: their only knowledge of their own mental states is acquired by the same processes as they acquire knowledge of the mental states of others. He continues (at p. 35 of the reprinting in [1048]):

> Assuming that self-blindness is possible, it would not of course prevent its victim from having beliefs about his environment, such as that it is raining. And offhand it would not seem that it should prevent him from expressing such beliefs by making assertions. But since he would have plenty of information about his behavior, and can be presumed not to be cognitively or conceptually deficient in any way (his deficiency is supposed to be quasi-perceptual), it ought also to be possible for him to have, and give verbal expression to, behaviorally based beliefs about his own beliefs – first-person beliefs he acquires in a "third-person way." Now it seems possible that the total evidence available to a man at a given time should support the propositions that it is raining, while the total "third-person" evidence available to him should support the proposition that he does not believe that it is

raining. (...) So if a self-blind man were in such circumstances, it seems he might be led, on perfectly reasonable grounds, to assert the Moore-paradoxical sentence "It is raining but I do not believe that it is raining."

Shoemaker takes it that such an assertion would indicate conceptual or cognitive confusion and thus that the envisaged state of self-blindness supposedly unaccompanied by any such deficiency is not a possibility after all, and that therefore introspection conceived of as a quasi-perceptual faculty does not exist. ◀

In the next few sections we will be taking as the default logic for B a popular candidate from the literature, namely **KD45**, which certainly has 5_c as a theorem and therefore is closed under Moore's rule. Indeed, 5_c is provable not only in **KD45** but already in **KD4**, as noted in Exercise 4.1.7. (A semantic gloss on this fact can be gleaned from Exercise 2.7.28: if a frame is transitive and serial, then certainly each point x has a successor y with $R(y) \subseteq R(x)$, since by transitivity *every* successor y has this property and by seriality there is at least one such successor.) A simpler exercise:

EXERCISE 5.1.21 Show that $\vdash_{\mathbf{K5}_c} \mathbf{D}$. ✠

To resume the discussion of the logical omniscience issue, we will have a look first at an early attempt, in Lemmon [694], to respond to the simplest form of this worry: that for $\Box$ (written as L in [694]) intended for an epistemic interpretation, there should be no provable formulas of the form $\Box\varphi$. Lemmon axiomatizes five logics he calls E1, E2, ..., E5 ("E" for *epistemic*, we recall from Example 2.1.18(*i*), p. 46) in the first four of which no such formulas are provable. So they are not normal, not being closed under Necessitation: see Example 2.1.18(*i*), p. 46 for more details concerning E2–E4.[301] We take this as an opportunity to pause and for several pages we look into the semantic treatment of some of the non-normal logics mentioned under Examples 2.1.18; this survey of options will finish with Examples 5.1.28 on p. 368 below. As far as the Lemmon systems themselves are concerned, the fact that they are all monotone modal logics means that they are hopeless as responses to the more general issue of logical omniscience, for any φ for which it might be inappropriate for our logic to tells us that the subject knows that φ (for whatever reason, for example because φ might be some very unobvious truth-functional tautology), $\Box p \to \Box\varphi$ will still be provable, telling us (reading $\Box$ as K_a) that if a knows anything at all, then a knows that φ. The reason is that from the provability of φ in any modal logic there follows the provability of $p \to \varphi$, at which point we invoke $\Box$Mono.[302]

Although undermotivated philosophically (at least in respect of the current issue of logical omniscience in epistemic logic), non-normal logics like these do have interesting semantic characterizations, first worked out in Kripke [670] for Lewis's own non-normal systems (principally S2 and S3), in which although there *are* theorems of the form $\Box\varphi$, by contrast with those devised for epistemic application by Lemmon, there are none of the form $\Box\Box\varphi$, so again we have a failure of Necessitation. Kripke's idea, re-notated for convenience here, is to work with models $\langle W, R, N, V \rangle$ which are as for the standard Kripke models $\langle W, R, V \rangle$, except that we have the additional component N, concerning which we require $N \subseteq W$. Points in N are called *normal* worlds and those in $W \smallsetminus N$, *non-normal* worlds. The definition of truth takes an interesting turn, essentially to make a world's being normal a necessary condition for it to verify any $\Box$-formula. Since the clauses in the definition of truth for propositional variables and Boolean compounds

---

[301] Lemmon [694] inadvertently formulated E5 in such a way that it coincided with (C. I.) Lewis's **S5**, as Kripke [670], p. 205, pointed out. In this case Lemmon used **5** as an axiom, which in any monotone logic has $\mathbf{U} = \Box(\Box p \to p)$ as a consequence (noted with 'normal' for 'monotone' in Example 2.5.22(*ii*), p. 79), and any monotone logic with a provable $\Box$-formula, such as this one, is closed under Necessitation – by essentially the reasoning given immediately in what follows. But – a further complication – Kripke's own discussion of E5 is not quite right: See the first paragraph of §6 in Porte [896]. 'Not right' means: not matching Lemmon's revised characterization of E5, given in [702]. Porte [896] describes earlier work by M. Boll and J. Reinhart isolating this same logic.

[302] The epistemic interpretation of his E-systems was the second of two interpretations Lemmon suggested, the first being as having a $\Box$ susceptible of the reading "it is scientifically but not logically necessary that", whose inadequacy, on similar grounds, was pointed out by Cresswell: see Example 4.3.1 above, at p. 227.

5.1. THE LOGICAL OMNISCIENCE ISSUE

are as before (i.e. as in Section 2.3, p. 58*f*.), here we can content ourselves with just the case of $\Box$-formulas; where $\mathcal{M}$ is a model $\langle W, R, N, V \rangle$ of the kind just described, for any $x \in W$ and any formula φ we have the following, in which, as usual, $R(x)$ is the set $\{y \in W \mid xRy\}$ and $\|\varphi\|$ is $\{w \in W \mid \mathcal{M} \models_w \varphi\}$:

$$\mathcal{M} \models_x \Box\varphi \text{ if and only if } x \in N \text{ and } R(x) \subseteq \|\varphi\|.$$

Thus all that has been changed in the definition of truth is that an extra conjunct "$x \in N$" has been added to the right-hand side. But it has yet to be said what is wanted from a formula for it to be valid on the frame $\langle W, R, N \rangle$ of one of these models with non-normal worlds, and so we don't yet know what logic is determined by the class of all frames, all reflexive frames, etc.

Two options come to mind: (1) we require, as usual, that for all $w \in W$, the formula in question should be true at w in every model on the frame; and (2) we require, instead, only that for all $w \in N$, the formula should be true at w in every model on the frame. (Of course we could equally well put the current distinction in terms of models, defining truth in a model *à la* (1) to be truth at all points in the model, or instead defining truth in a model *à la* (2) to be truth at all normal points in the model. Validity on a frame is then truth in every model on the frame with the latter taken in one or other of these two ways.) Roughly speaking, for the non-normal Lemmon ('E-') systems option (1) is the way to go, while for Lewis's own systems, (2) is appropriate. One can see that taking route (1), no $\Box$-formula is valid on any frame in which $W \smallsetminus N \neq \varnothing$ – in which there are some non-normal worlds, that is – since all $\Box$-formulas are false at such elements in every model on the frame, thanks to the new conjunct "$x \in N$" in the clause for $\Box$. On the other hand, taking route (2) will still give us $\Box$-formulas valid on frames even in the presence of non-normal worlds, since such worlds continue to verify, for example, all (substitution-instances of) truth-functional tautologies, and indeed also various further formulas (such as: all negated $\Box$-formulas and their truth-functional consequences). So prefixing a $\Box$ to a formula of this kind will not jeopardise the truth of the "$\Box$"-ed formula at normal points with non-normal points accessible to them. (By contrast, what is accessible to a non-normal world makes no difference to the truth of a $\Box$-formula there – since it has to be false, regardless – and it is quite common to standardize matters by restricting attention to models on frames $\langle W, R, N \rangle$ with $R(x) = \varnothing$ for all $x \in W \smallsetminus N$. These are the *refined* models of Lemmon [702], p. 193.[303]) On the other hand one could equally standardize in the opposite direction, putting $R(x) = W$ for $x \in W \smallsetminus N$. When concentrating only – as, for historical reasons, early work did – on modal logics containing **T**, it is even possible to leave the set N out of the models altogether and use the set of reflexive points as a surrogate, having standardized in the "$R(x) = \varnothing$ (for $x \notin N$)" direction, so that truth is defined for $\Box$-formulas using the clause:

$$\mathcal{M} \models_x \Box\varphi \text{ if and only if } xRx \text{ and } R(x) \subseteq \|\varphi\|.$$

This is the approach taken in Hughes and Cresswell [535], Chapter 15; both it and the models-expanded-with-N approach are described in Kripke [670]. On the other hand, in the interests of generality, one could enrich rather than impoverish the structures $\langle W, R, N \rangle$, thereby subsuming various developments, as in the following elegant formulation from Chagrov and Zakharyaschev [166]. The authors separate out the idea of normal worlds as the only ones that can verify $\Box$-formulas and the idea of normal worlds – for which that terminology is not used in their discussion – as the worlds truth throughout which is required for a formula to be valid on the frame of the model on Option (2) above (allowing Option (1) to be recovered by taking this to coincide with set of all worlds in the model). The latter they call distinguished worlds, as above at p. 190, where we stressed the version in which there is only one of these per model (or frame). (The same separation appears in Segerberg [1025], p. 226.) At some points in this quotation, we have changed some of the notation of [166] to match our own though the authors' boldface

---

[303] In this work the set N in the current models is replaced by its complement $W \smallsetminus N$, denoted by Q, so that the truth-definition requires for $\Box\varphi$ at x that φ be true throughout $R(x)$ and that x *not* be an element of Q. The informative discussion in Segerberg [1025] follows suit. Note also that the word "refined" has an unrelated use in connection with general frames (see the Digression on p. 83), explained for instance at p. 306 of Blackburn *et al.* [98].

"**S2**" etc., remain intact (= "**S2**" in our notation, in which "**S**" followed by a numeral indicates a normal extension of **KT**), as does their use of "valid in a frame" (where we would say "valid on a frame" and this amounts to being true in any model – as defined in the quotation – based on the frame in question); recall that the term "quasi-ordered" here means *pre-order* (reflexive and transitive), and 2^W is $\wp(W)$:

> For example, sometimes it is useful to consider frames as quadruples $\mathcal{F} = \langle W, N, R, D \rangle$, where $\langle W, R \rangle$ is a usual Kripke frame, $N \subseteq W$ is a set of so called *normal worlds* and $D \subseteq W$ is a set of *distinguished worlds*. A valuation in such a frame is, as before, a function V from [the set of propositional variables] into 2^W, and the pair $\mathcal{M} = \langle \mathcal{F}, V \rangle$ is a model. However the truth relation for $\Box$ is defined now as follows:
>
> $$\mathcal{M} \models_x \Box\varphi \text{ iff } x \in N \text{ and } \mathcal{M} \models_y \varphi \text{ for all } y \in W \text{ such that } xRy,$$
>
> and a formula is regarded to be true in $\mathcal{M}$ if it is true at all points in D. We get usual Kripke frames if $D = N = W$. By imposing various conditions on R, N and D we can define many modal logics known in the literature. For instance:
>
> - the set of formulas that are valid in all reflexive frames with $D \subseteq N$ is known as the logic **S2**;
> - the set of formulas that are valid in all quasi-ordered frames such that $D \subseteq N$ is the logic **S3**;
> - the set of formulas that are valid in all reflexive frames such that $D \subseteq N$ and $\forall x \in D \exists y \in W \smallsetminus N \,.\, xRy$ is **S6**.
>
> Chagrov and Zakharyaschev [166], p. 99.

The logic S1 (or **S1**, to use a notation in the above style) is missing from this quotation, demanding a more specialized semantic treatment; even a syntactic description is nor a simple matter, for which reason this logic did not appear under Examples 2.1.18 on p. 46.[304] The logic S6 appearing here was last mentioned in Example 2.1.18(*iii*) (48) along with some bibliographical information. It has the unusual property that all formulas of the form $\Diamond\Diamond\varphi$ (and all formulas of the form $\neg\Box\Box\varphi$) are provable, reflecting the motivating idea (see the discussion in [535]) that, thinking of $\Box$ as expressing necessity, since what is necessary depends on (in particular, linguistic) conventions that could have been otherwise, nothing is necessarily necessary.[305] An even more extreme proposal would have all formulas of the form $\Diamond\varphi$ (and thus all formulas of the form $\neg\Box\varphi$) provable; this was mentioned as the 'Falsum' logic mentioned in note 141, p. 171 as well as on p. 303 above. The smallest regular logic satisfying this condition is discussed, amongst other places, at p. 204 of Segerberg [1025]. The canonical model method applies to regular logics as it does to normal logics, with the canonical accessibility relation R_S for a consistent regular modal logic defined exactly in the case of normal modal logics, amongst maximally S-consistent sets of formulas. Of course one needs also to supply the set N_S of normal points, which is done by taking these to be the maximal consistent sets containing the formula $\Box\top$ (or equivalently, given regularity – indeed given only monotony – containing some $\Box$-formula or other); the logic just described is easily seen to be determined, using this method (and with validity defined as truth throughout the models, rather than just at the normal points), by the class of frames $\langle W, R, N \rangle$ in which $N = \varnothing$. (Note that we are writing "$\langle W, R, N \rangle$" rather than "$\langle W, N, R \rangle$", as the above passage from Chagrov and Zakharyaschev [166] might suggest for the case in which there is no explicit D.)

---

[304]See Cresswell [213], Chellas and Segerberg [173]; there is also an unedifying semantic characterization in pp. 133–5 of Thijsse and Wansing [1102] – unedifying because of extreme reliance on special conditions on the V of the models described.

[305]This is, admittedly, a *bad* idea, based on confusing a sentence's necessarily expressing a truth with a sentence's expressing a necessary truth. For more on the idea of necessity's being a matter of convention, see p. 565*ff.* in my [582], and references there cited. An alternative motivation for accepting that nothing is necessarily necessary will be explored in Remark 5.1.24 below.

5.1. THE LOGICAL OMNISCIENCE ISSUE

EXERCISE 5.1.22 Give an explicit axiomatization (i.e., spell out axioms and rules) of the smallest regular modal logic containing all formulas of the form $\Diamond\Diamond\varphi$ and using the canonical model method with the adaptation just outlined describe a class of frames $\langle W, R, N \rangle$ w.r.t. which the axiomatization is sound and complete. (As before, validity on a frame is taken to mean truth at every element of W in every model on the frame.) ✠

REMARKS 5.1.23 (*i*) Models $\langle W, R, N, V \rangle$ as above suggest the addition of a sentential constant, $\boldsymbol{n}$ say, stipulated to be true at precisely the elements of N, so that $\Box\varphi$ as understood in the discussion above could be written as $\boldsymbol{n} \wedge \Box\varphi$, where this last $\Box$ is understood as in the Kripke semantics for normal modal logics: requiring truth at all R-related points. Using this idea and variations on it Cresswell [201] and Aanderaa [1] give several interesting translational embeddings of non-normal modal logics into normal modal logics (and vice versa, in the case of [1], which uses a propositional variable not occurring in the formula being translated rather than a new constant). Cresswell considers on the non-normal side not only S2 and S3 but also Lemmon's E2 and E3, while in addition to the Lewis non-normal logics, Aanderaa brings in Łukasiewicz's Ł-modal system (see p. 261, and (*ii*) below). Some of these translations are "$\Box$-definitional" in the sense explained before Example 4.4.27 on p. 266, but those embedding the non-normal modal logics into normal ones (e.g., S2 into **KT**) one first replaces $\Box\_$ subformulas by $\boldsymbol{n} \wedge \Box\_$ and then prefixes an initial $\boldsymbol{n} \rightarrow \_$ to this compositionally translated formula to reach the final translation, in view of the role played by the points in N over and above the part they play in the truth-conditions of $\Box$-formulas in the non-normal models but also because of their status in arbitrating matters of validity. (The $D = N$ idea above.) This last 'non-compositional' part stops the translation from being definitional. (Here we use terminology from Example 4.4.25, p. 264).

(*ii*) We can downplay the role of accessibility in the models with N by taking this as the identity relation, so that with $N = W$ we have the logic **KT!**. Concentrating on one-element models, which are entirely representative in view of 'generated submodels' considerations, we have a single point which is either normal or not, making $\Box\varphi$ equivalent to φ in the former case and to $\bot$ in the latter. The formulas true at the sole point in all such models, whether or not it is, is then the set of theorems of the Ł-modal logic mentioned, e.g., at the end of Example 2.1.18(*iii*) (p. 48). If we work in the language of (*i*) above, with its constant $\boldsymbol{n}$, so that $\Box\varphi$ emerges as $\boldsymbol{n} \wedge \varphi$, what we have is a representation of this logic as disclosed in Porte [892], in which $\boldsymbol{n}$ is written as Ω (no connection – other than in being a nullary connective – with the Ω of our Example 4.7.19 at p. 315 above). Since we allow $\boldsymbol{n}$ to be true (at the single point in these model) and also allow it to be false, the valid formulas end up including those in which $\boldsymbol{n} \wedge \varphi$, alias $\Box\varphi$ is equivalent to φ in the former cases and to $\bot$ in the latter, recalling the original Łukasiewiczian conception of $\Box$ as the matrix product of the two-valued identity and constant-false truth-functions (see Łukasiewicz [746], Smiley [1062]). This explains the extensionality of the Ł-modal system remarked on in note 40, p. 49 – one of its unattractive features as a plausible alethic modal logic. The choice of these particular 1-ary truth-functions was driven by the fact that they are the only such functions as are compatible with the validity of $\Box\varphi \rightarrow \varphi$ (though in fact Łukasiewicz concentrated on $\Diamond$ – not that he used this notation). Indeed we have already seen this point about extensionality, both in the present Ł-modal context (Example 4.4.21 on p. 261) and in another (apropos of $\star$ on p. 322). ◀

The clause for $\Box$ given above, according to which $\Box\varphi$ is true at a point just in case the point is normal and has only φ-verifying successors, can be rewritten in the form: for a normal point x, $\Box\varphi$ is true at x iff $R(x) \subseteq \|\varphi\|$, while for a non-normal point x, $\Box\varphi$ is false. Perhaps because of this, one sometimes hears the suggestion that this style of semantics treats "$\Box$" as *ambiguous*. Thus, Mares [772], p. 52f. writes:

> In Kripke's semantics the truth conditions for the modal operators at non-normal worlds are not the same as their truth conditions at normal worlds. If, as we have assumed, these truth conditions at least in part constitute the meanings of these operators, then on Kripke's semantics the meanings of the modal operators are not the same at normal worlds as they are at non-normal worlds.

There is no justification for any of this talk, given the availability of the initial clause in the truth-definition, according to which a *perfectly uniform* necessary and sufficient condition for the truth of $\Box\varphi$ at *arbitrary* x is given by: $x \in N$ and $R(x) \subseteq \|\varphi\|$.[306]

REMARK 5.1.24 A similar sentiment is expressed in Kajamies [635] as part of a criticism of the attempt in Curley [223] to render coherent Descartes's view that the necessary truths were made so by God, who could have chosen to make other truths necessary instead. This second part of the summary of the position suggests that the things which are necessary might have been false (had God chosen otherwise), which clashes with the idea that what is necessary is what could not have been false, unless it is taken to imply that nothing is necessary, going against what is expressed in the first part of the summary, in which it is assumed that there are necessary truths. If one wants to deny that anything whatever is necessary – e.g., as in Mortensen [831] – then one has the Falsum system, last mentioned before Exercise 5.1.22 above, as a modal logic embodying one's position. Curley discerns the attribution of such a position to Descartes in an earlier paper by Harry Frankfurt but argues this attribution is incorrect, urging the following solution to the apparent inconsistency in that position: hold that while some things are necessary, they could have been – not *false*, had God chosen otherwise – but simply not *necessary*. Nothing is necessarily necessary, according to this view, or, to put it in terms of possibility, everything is possibly possible. Curley calls this the 'iterated modality' interpretation of Descartes, and provisionally suggests, as a modal logic embodying the position, something along the lines of S6, in which every formula of the form $\Diamond\Diamond\varphi$ is provable. (Or again, one might consider some alternative strengthening of the logic of Exc. 5.1.22.) Later, Curley has second thoughts about this suggestion and in particular about the semantics currently under consideration with its normal and non-normal worlds, for reasons somewhat in the spirit of Geach [357], p. 11, who had earlier considered the iterated modality solution and its implementation at the hands of "shyster logicians", whose "whose technical ingenuity is mistaken for rigour" – and rejected it. Since this was Curley's own considered position, it is some odd for Kajamies [635] to treat Curley as the target for an attack on the iterated modality interpretation, though that is not to the point here, where our concern is with charges of ambiguity and equivocation. At p. 24 of [635] we read that at least the resort to S6 and the semantics currently under consideration for it "implies an equivocation on the concept of possibility" between "the sense of 'possible' that fits better with out modal intuitions and a completely different one". To this suggestion it must be replied that as with $\Box$ above, the present semantics treats $\Diamond$ in a thoroughly uniform way: for all x, $\Diamond\varphi$ true at x iff $x \notin N$ or for some $y \in R(x)$, φ is true at y. The relevant charge should not be one of equivocation but of being too distantly related to the everyday notion of possibility to be regarded as a formal representation of that notion (and this is indeed an objection Kajamies presses – as would Curley and no doubt also Geach). Incidentally, the above suggestion of the Falsum system as embodying the position of Mortensen [831] should not be taken as attributing to Mortensen the kind of semantics currently under consideration for such non-normal logics, which secures the truth of $\Diamond\Diamond\varphi$ at normal worlds by a double fiat – that $\Diamond$ formulas are true at all non-normal points and that every normal point has at least one such point accessible to it (the last condition in the quotation from Chagrov and Zakharyaschev [166] on p. 362 above). Rather it would seem that some version of the Kripke semantics with impossible worlds (at which contradictions, etc., end up evaluated as true), though 'impossible' would not be the best description of them for one claiming that a statement's being true at one of them makes its truth possible. For example, the evaluations can proceed in accordance with the semantics of (the first degree entailments of) relevant logic as detailed in Routley and Routley [975], for other applications of which in the present work, see the index entry under 'relevant logic' – or for further references and a comparative study, Wansing [1146]. (See also Berto [87] and, for philosophical reflections on the idea of impossible worlds, Stalnaker [1081].) Another caveat in connection with the view that everything is possible: taken together with an endorsement of the schema 5, $\Diamond\Box\varphi \to \varphi$, this is in danger of leading to the conclusion – sometimes called *trivialism* in the literature

---

[306] For more on this issue of uniformity, see §5 of Humberstone [593] and references there cited.

5.1. THE LOGICAL OMNISCIENCE ISSUE

– that everything is true. For further discussion of this point, as well of the question of whether one needs to distinguish the claim that anything is possible (the title of Mortensen [831]) from the claim that everything is possible, see Humberstone [593]. ◀

It is worth considering the role of the "and" in the last sentence before Remark 5.1.24, underlined in this recapitulation of the clause for $\Box$ (for $\Box\varphi$ to be true at a point x in a model, that is) in the current definition of truth:

$$x \in N \text{ \underline{and} } R(x) \subseteq \|\varphi\|.$$

with a view to asking what would happen if we changed it to another (metalinguistic) Boolean connective – for example to "or", giving the following clause in the definition of truth at a point in a model $\mathcal{M} = \langle W, R, N, V \rangle$:

$$\mathcal{M} \models_x \Box\varphi \text{ if and only if: } x \in N \text{ or } R(x) \subseteq \|\varphi\|.$$

To see that this gives nothing that the simple $\langle W, R, V \rangle$ models with the usual truth-definition in force don't already provide, let us process the right-hand side a bit. Spelling out the abbreviative notation in a more explicit first-order form, it becomes the following:

$$x \in N \vee \forall y (Rxy \to \mathcal{M} \models_y \varphi),$$

and thus, pulling the quantifier to the front:

$$\forall y (x \in N \vee (Rxy \to \mathcal{M} \models_y \varphi)).$$

Getting rid of the "$\to$":

$$\forall y (x \in N \vee (\neg Rxy \vee \mathcal{M} \models_y \varphi)).$$

Re-associating:

$$\forall y ((x \in N \vee \neg Rxy) \vee \mathcal{M} \models_y \varphi),$$

and then re-introducing "$\to$":

$$\forall y (\neg (x \in N \vee \neg Rxy) \to \mathcal{M} \models_y \varphi),$$

and so, finally (by a De Morgan transformation), this is equivalent to:

$$\forall y ((x \notin N \wedge Rxy) \to \mathcal{M} \models_y \varphi).$$

All these steps are reversible, so what we began with is just a condition of the usual (N-less) form, writing $R'xy$ for $x \notin N \wedge Rxy$:

$$\forall y (R'xy \to \mathcal{M} \models_y \varphi).$$

Thus any model $\langle W, R, N, V \rangle$ on whose points truth is defined with the "or" version of the Kripkean non-normal clause for $\Box$, yields a point-by-point equivalent model $\langle W, R', V \rangle$ of the traditional kind. Conversely, any model of this latter kind can be regarded as a model $\langle W, R, N, V \rangle$ with the "or" version of the $\Box$ clause in place by taking N as $\varnothing$. From this is follows that **K** is the logic of those formulas true throughout all models $\langle W, R, N, V \rangle$ with this definition of truth in force.

EXERCISE 5.1.25 (*i*) Justify the comment just made, showing that, with the semantics just described, **K** is sound w.r.t. the class of such models (or their underlying frames $\langle W, R, N \rangle$, if that formulation is preferred) and also that **K** is complete w.r.t. this class.
(*ii*) Would the same be true for an arbitrary class of models (or frames) defined in terms of the behaviour of the accessibility relation? For instance, would it be the case that **S4** is determined by the class of models $\langle W, R, N, V \rangle$ with R transitive and reflexive when the truth-definition just considered is in force? (Justify your answer.)

Note that it is not necessary to consider separately the possibility of replacing the original "and" with implication:

$$\mathcal{M} \models_x \Box\varphi \text{ if and only if: } x \in N \text{ implies } R(x) \subseteq \|\varphi\|,^{307}$$

since this is just the "or" version all over again – replacing the N you first thought of with $W \smallsetminus N$. For the same reasons as every monadically representable binary relation is $\wedge$-, $\vee$-, or $\leftrightarrow$-representable – see the discussion before Exercise 1.2.11 (p 15) – it is natural to pass from "and" and "or", here, to "iff":

EXERCISE 5.1.26 Consider now changing the clause for $\Box$ in the definition of truth in a model $\langle W, R, N, V \rangle$ ($N \subseteq W$) with:

$$\mathcal{M} \models_x \Box\varphi \text{ if and only if: } x \in N \text{ iff } R(x) \subseteq \|\varphi\|.$$

How does the set of formulas true at all points in all such models, with this definition of truth in force, compare with the set of theorems of **K**? (That is, is one set included properly in the other, do they coincide, or is neither included in the other?) ✠

EXAMPLE 5.1.27 Our discussions on varying the "and" to another metalinguistic connective in the definition of truth have assumed that the set N of normal points plays no special role in the definition of validity and in particular that validity is simply truth all points in all models rather than just at the points in N. (In terms of the discussion from Chagrov and Zakharyaschev at p. 362 above, $D = W$.) On p. 335, mention was made of the logical investigations reported in Gochet and Gillet [390]; it is intended for epistemic application to the case of several knowers, and what they write as "K^i" (with associated accessibility relation R^i and 'normal-like' set of points W^i) will be written here as $\Box_i$ (with accessibility relation R_i and normal set N_i; on p. 335, we took there to be two K-operators and wrote them as K_a and K_b rather than as $\Box_1$ and $\Box_2$). The models have a distinguished element $w_0 \in W$ and the kind of validity we are currently interested in is truth at this point in all the models (with universe W), which are required to satisfy the condition that $w_0 \in \bigcap_i N_i$, and some further conditions given below. Correcting and slightly reformulating the key clause in the definition ([390], p. 107) of truth at a point w in such a model $\mathcal{M}$ (again with universe W) we have:

- $\mathcal{M} \models_w \Box_i\varphi$ iff either $w \in N_i$ and $R(x) \subseteq \|\varphi\|$ or else $x \notin N_i$.

Thus $\Box$-formulas are automatically true at 'non-normal points' (points outside of N_i, for $\Box = \Box_i$, that is), rather than automatically false there as in the more familiar Kripke-derived treatments. More generally, we have the following from Porte [897], p. 216, with "Lx" replaced by "$\Box\varphi$" for continuity with our discussion: "A non-normal world of the first kind (...) is a world in which $\Box\varphi$ can be true even though φ is false in at least one accessible world. A non-normal world of the second kind (...) is a world in which $\Box\varphi$ can be false even though φ is true in every accessible world." We are now dealing with a special case of non-normality of Porte's first kind: $\Box\varphi$ not only *can be* but *must be* true at x – i.e., is true at x whatever point x is and whatever formula φ is – even if $R(x) \nsubseteq \|\varphi\|$, whenever x is non-normal (that is, lies outside N_i for $\Box = \Box_i$). On the face of it, this strong form of Porte's non-normality of the second kind would appear not to serve much of a purpose, since the same effect could be achieved without any normal/non-normal division, just by having $R(x)$ be $\varnothing$ for the would-be non-normal x. (The weaker version, according to which $\Box\varphi$ is permitted to be true but also permitted to be false at a non-normal worlds, in the former case even with accessible φ-falsifying worlds, is used in a semantic characterization of S0.5 in Hughes and Cresswell [535], a descendant of the treatment in Cresswell [200], to which this description does not quite apply directly since there is no explicit accessibility relation in

---

[307]Of course instead of saying "implies" here one should really say "only if", to avoid an unwanted change of language level. But this casual formulation is less cumbersome given that we are already in the scope of an "if and only if".

5.1. THE LOGICAL OMNISCIENCE ISSUE

that formulation. (See also Example 5.1.28(*i*) below.) These references, and others to Porte's own work, are mentioned in [897] p. 216.) This aspect of Gochet and Gillet [390] is brought out more clearly by simplifying the formulation given above to the following equivalent – in fact the formulation mentioned earlier, before Exercise 5.1.26, except with subscripts added for the current multi-modal setting:

- $\mathcal{M} \models_x \Box_i \varphi$ if and only if: $x \in N_i$ implies $R_i(x) \subseteq \|\varphi\|$,

whose r.h.s. simply amounts to: $x \notin N_i$ or $R_i(x) \subseteq \|\varphi\|$. When we considered this before, this was taken as just the "or" case that had already been treated – trading in a model $\langle W, R, N, V \rangle$ with truth defined using such a clause for a pointwise equivalent model $\langle W, R', V \rangle$, without any normal/non-normal subdivision and the familiar clause for $\Box$, where $R'xy$ was defined in terms of R and N as: $x \notin N \wedge Rxy$. Similarly, we said that the "implies" case could be subsumed under the "or" case by replacing N with $w \smallsetminus N$. Here we have several candidate Ns – the N_i – but pretend we have just one: N_1. All $\Box_1$-formulas are true at the elements in not in N_1, in any model, so replacing N_1 with $W \smallsetminus N_1$ would make $\Box_1 \bot$ valid: at least if the distinguished element w_0 belongs to (the new) N_1. In any case, by contrast with Porte, Gochet and Guillet do not even describe the subsets we have been calling N_i in terms of normality, and use instead the notation W_i for them (actually "W^i"), thinking of a point x such that $x \in N_i$ and $x \notin N_j$ for $j \neq i$ as a world 'private to' knower i. Instead they use precisely the device already alluded to above apropos of Porte: we make $\Box_i \varphi$ true at worlds x outside of N_i by putting $R_i(x) = \varnothing$. This follows from the condition called *Restriction* below. The condition mentioned above on the Gochet–Gillet models, for whose representation of which we use "I" to stand for an arbitrary index set, $\langle W, \{N_i\}_{i \in I}, \{R_i\}_{i \in I}, w_0, V \rangle$ is repeated here:

- $w_0 \in \bigcap_{i \in I} N_i$ (Intersection)

To this condition one adds the following, just alluded to:

- For all $x, y \in W$, if $xR_i y$ then $x, y \in N_i$.   (Restriction)

(Gochet and Gillet put this instead by saying that R_i is only 'defined on' N_i.) And we need further the condition – the strongest reflexivity-like condition that can be imposed consistently with Restriction:

- For all $x \in N_i$, $xR_i x$.   (Restricted Reflexivity)

(The authors also consider imposing transitivity and other conditions but here it will suffice to consider their weakest candidate multi-agent epistemic logic.)

For brevity, let us call a formula *Gochet–Gillet valid* if it is true at the element w_0 in any such model. (We recall from note 177, p. 219, that our authors call this *second class* validity on their semantics – an unusual choice of terminology, given the interest they take in this notion of validity.) Observe that any formula φ, $\Box_i \varphi \to \varphi$ is Gochet–Gillet valid, because in any model $wR_i w$, by Restricted Reflexivity, since $w_0 \in R_i$ by Intersection. Let us specialise i to 1 and φ to p to observe that not only is $\Box_1 p \to p$ Gochet–Gillet valid, but so also is $\Box_1(\Box_1 p \to p)$. The reason is that if $w_0 R_1 x$ then by Restriction, $x \in N_1$, so once again $xR_1 x$ by Restricted Reflexivity. On the other hand – the motivation behind the whole apparatus – the formula $\Box_2(\Box_1 p \to p)$ is not Gochet–Gillet valid, since w_0 can bear the relation R_2 to a point x outside of N_1 (as long as this point is in N_2) and at such a point all $\Box_1$ formulas are true (by the consequence, $R_1(x) = \varnothing$ of Restriction), so to falsify $\Box_1 p \to p$ it suffices to set $V(p)$ in such a way that $x \notin V(p)$: now $\Box_2(\Box_1 p \to p)$ is false at w_0; this example of non-closure under Necessitation shows the set of Gochet–Gillet valid formulas not to be a normal multimodal logic (although each of its monomodal fragments is). It would perhaps be more accurate to describe the desire to invalidate $\Box_2 \Box_1 p \to \Box_2 p$ – which more directly illustrates the non-monotone character of the logic – rather than $\Box_2(\Box_1 p \to p)$ itself, as the motivation behind the present enterprise; in our discussion on p. 335 we urged the invalidity of (a rewritten version of) $\Box_2(\Box_1 p \to p)$ as a mark against these ideas, which originated, as we saw there, with Weingartner's objections to $\Box_2\Box_1 p \to \Box_2 p$ (see Example 5.1.2, where variously decorated versions of "$K_a K_b p \to p$" were in play). Although there are numerous differences between Weingartner's proposals (e.g., in [1162]) and those of Gochet and Gillet [390], they say this about the relevant notion of knowledge (which they call opaque knowledge, referring to the implication just mentioned as a transparency principle):

> Opaque knowledge amounts to being confident in the reliability of the information about knowledge supplied by other agents. When agent i knows indirectly that agent j knows that α, we implicitly recognize that he is unable to view all the data which support j knowing that α, though he is confident that j knows.

If this is supposed to be a reason for rejecting $K_i K_j \alpha \to K_i \alpha$ we need i not just to *be confident* (i.e., to firmly believe) that j knows that α. But whatever one feels about the philosophical motivation for the rejection, one has to admire the ingenuity of the proposed semantics – though the attempted completeness proof offered in §9 of [390] for their axiomatization presents several mysteries, as the interested reader may discover. ◀

Before leaving the apparatus of models with normal and non-normal worlds, we give two examples from work of Cresswell's that are reminiscent of the simplified and semi-simplified Kripke models of Sections 2.2 and 4.1, in the sense that the role of an accessibility relation is downplayed:

EXAMPLES 5.1.28 *(i)* As well as in the course of Example 5.1.27, Lemmon's logic S0.5 was mentioned parenthentically under Example 2.1.18(*ii*), p. 46. Cresswell [200], cited in both places, gives models with a distinguished element and no explicit accessibility relation, which we could notate as having the form $\langle W, w^*, V \rangle$, as usual thinking of the reducts without V as the frames of these models. The easiest way to explain how truth is defined – or better, constrained – in such models makes use of the notation v_x (for $x \in W$) from the proof of Lemma 2.6.22 (p. 120) in which $v_x(\varphi)$ is T when φ is true at x in the model, and F otherwise; as usual we put $v_x(p_i) = T$ ff $x \in V(p_i)$. In the present case, to extend this specification inductively to formulas of greater complexity, we require (1) that for all $x \in W$, v_x is a Boolean valuation and (2) that $v_{w^*}(\Box \varphi) = T$ iff $v_w(\varphi) = T$ for all $w \in W$. The key idea here is that the truth-values of $\Box$ formulas at points x other than w^* are not subject to any constraints. Validity on a frame is truth at the disitnguished point w^* in each model on the frame, and validity on all frames coincides with provability in S0.5; the formulas enjoying this property include $\Box(p \to (p \vee q))$ and $\Box p \to \Box(p \vee q)$ but not $\Box(\Box p \to \Box(p \vee q))$ or $\Box\Box p \to \Box\Box(p \vee q)$, since even if all points in a model verify $\Box p$, they are not therefore required to verify $\Box(p \vee q)$, which just behaves like another unstructured expression semantically when evaluated at points other than w^*. There has been some controversy as to whether this semantic account of Cresswell's vindicates Lemmon's own original motivating idea (e.g., in [695]) that S0.2 captures a reading of $\Box\varphi$ as "it is tautologous (valid by truth-tables) that φ": see Routley [970] (esp. p. 420*f*.) and Cresswell [204] for a disagreement on this issue. (They disagreed also over S1, as one may gather from Cresswell [213], in which Routley appears under his later name, Richard Sylvan.) Cresswell slipped up in the informal discussion on the opening page of [200]. He writes "x_1" for the above w^* and says concerning these models:

> The basic assumption is that x_1 is the real world and in it necessity is evaluated as in the models of Kripke [669], while the rest are worlds in which only PC tautologies are true.

Certainly the truth-functional tautologies are all true at all the elements of these models, including those other than the distinguished element, so this includes, for example, $p \vee \neg p$; but since $\vee$ behaves classically at these points (i.e., for such points x, the valuation v_x is ∨-booolean, in the sense explained at Remark 1.2.3, p. 10). Perhaps what Cresswell meant to say was that the only formulas true at all such points in every model are the truth-functional tautologies. As noted in Example 5.1.27, the semantic treatment

5.1. THE LOGICAL OMNISCIENCE ISSUE

of S0.5 in Hughes and Cresswell [535], p. 286, is slightly different, presumably for continuity with their immediately surrounding discussion, in the models there do make use of explicit accessibility relations.

(*ii*) The logic S3.5 was described under Example 2.1.18(*iii*) (p. 48). By contrast with the retention of accessibility relations in [535] for S0.5 just remarked on, for S3.5 the authors' semantic treatment (p. 284) eschews them, and presents the models of interest as triples $\langle W, N, V \rangle$ with $N \subseteq W$, with N (the normal points) playing the double role of being the points truth at all of which is required for validity on the frame of the model – the set D of the discussion quoted from Chagrov and Zakharyaschev [166] on p. 362 – and also the N role there, of entering into the definition of truth as it concerns $\Box$-formulas, belonging to N being a necessary condition for verifying any such formula. The theorems of S3.5 are the formulas valid on all such frames when validity on a frame is understood in this "unfalsfiability at points in N" way, with $\Box\varphi$ deemed to be true at a point x just in case (1) $x \in N$, and (2) φ is true throughout W. ◄

While this is all very interesting technically, no doubt, let us leave these models with normal and non-normal worlds now. As explained in the discussion following Exercise 5.1.21, non-normal modal logics of the kind treatable using semantics along these lines do not provide a philosophically satisfactory response to the logical omniscience problem (for anyone convinced there is such a problem). In a nutshell, we may be avoiding omniscience in respect of logical truths but we are not avoiding deductive or inferential omniscience – and this (a) brings back omniscience of logical truths for anyone who knows anything, and (b) is implausible in its own right. The earlier objection was formulated in terms of (a) rather than (b), and it should be noted that one popular line of response has always been: logical omniscience may not be so implausible if you pick the right logic. In something of a *locus classicus* for this response, Routley and Routley [976], the authors defend, at least as far as inferences involving the Boolean vocabulary are concerned (say, with, ¬, ∧, and, redundantly, ∨, as primitive connectives) the account provided by the first-degree entailments of relevant logic (as we recalled in Remark 5.1.7. Roughly speaking, ∧ and ∨ have a distributive lattice logic and ¬ enjoys the double negation and the (full range of) De Morgan equivalences, but in general φ does not entail $\psi \vee \neg\psi$ (though it will if φ happens to entail one of ψ, $\neg\psi$ – e.g., because φ *is* ψ, since disjunctions are always entailed by their disjuncts), and $\varphi \wedge \neg\varphi$ does not (in general – with exceptions dual to those just mentioned) entail ψ. A similar suggestion by H. J. Levesque, with some additional details we can omit here, is reviewed in Fagin and Halpern [278], at p. 48 of which the following highly pertinent philosophical point is made, undermining this kind of strategy as a response to worries about deductive omniscience:

> Unfortunately, it seems no more clear that people can do perfect reasoning in relevance logic than that they can do perfect reasoning in classical logic!

As we saw in Remark 5.1.7, Levesque does not seem to have been persuaded by this objection, since the same moves are repeated in Chapter 12 of Levesque and Lakemeyer [715].

After reviewing the relevant-logical strategy in their more recent survey Fagin, Halpern *et al.* ([280], p. 357) adopt a more conciliatory tone, in which "weaker" means "weaker than classical implication":

> It may not be so unreasonable to be closed under logical implication if we have a weaker notion of logical implication.

Thus the authors have returned to essentially the position defended in Routley and Routley [976]. Discussing the suggestion from Cresswell [208] (and elsewhere) that there is no non-trivial doxastic logic – in the sense that logic of belief is simply the smallest modal logic – on the grounds that the belief-pertinent 'worlds' invoked in his favoured semantic treatment do not constrain the logical vocabulary in any way, Routley and Routley reply as follows:

> The argument rests on a false dichotomy – either classical semantical constraints or no constraints at all. The normal worlds of relevance logic alone reveal the falseness of the options offered, that

it is not an all-or-nothing matter.[308] In particular, the treatment of negation in this semantics – initially unveiled in Routley and Routley [975] and briefly summarised in [976] – is 'non-classical' while conjunction and disjunction are treated in the usual manner.

A variant approach would be to try to constrain classical reasoning in a relevant-logical way only as it applies within the scope of epistemic and perhaps more to the point doxastic operators, with a 'semi-paraconsistent doxastic logic' along the lines of the semi-paraconsistent deontic logic of McGinnis [787]; as Remark 5.1.7 explained, this is exactly the approach taken in Levesque and Lakemeyer [715].

5.2 Introspection Issues

Suppose that any obstacles posed by logical omniscience to treating epistemic and doxastic matters by seeing them as applications of the apparatus of normal modal logic, with □-operators K and B, have been overcome. We proceed to ponder the plausibility of some principles involving iterations of these operators, either involving occurrences of the one of these operators within the scope of an occurrence of the same operator or of the other operator. The basic normal logics for the two notions are usually taken to be **KT** for K, on the grounds that only what is in fact the case can be known to be the case, giving us $\mathbf{T}_K$ – as we may call **T** with K as □. Since it is not correspondingly true that only what is in fact the case can be believed to be the case, no similar plausibility would attach to $\mathbf{T}_B$ (this label understood analogously), and, as has been mentioned already, some weaker principles, such as $\mathbf{D}_B$ and $\mathbf{U}_B$ are often taken seriously as alternatives. The first is, perhaps like logical omniscience, usually treated as an aspect of the idealization of rationality on the subject's part (for the same φ). $\mathbf{U}_B$ is a kind of principle of self-trust. If we were working with languages allowing propositional quantifiers, we would no doubt distinguish between a more and a less rash form of self-trust ("$B\forall q(Bq \to q)$" and "$\forall q B(Bq \to q)$"), respectively,[309] but in their absence let us note that what $\mathbf{U}_B$ amounts to is the latter, less arrogant assumption, and keep our eyes open for its appearance in this or that doxastic logic. One thing we already know is that it follows (given normality) from $\mathbf{5}_B$, as we saw in Example 2.5.22(*ii*), where this fact is thought of as explained/predicted by the fact that every euclidean frame is range-reflexive, since the frames on which **5** (or $\mathbf{5}_B$) is valid are all euclidean, and any range-reflexive frame validates **U** (or $\mathbf{U}_B$). $\mathbf{5}_B$ and $\mathbf{5}_K$ have been called 'negative introspection' principles for belief and for knowledge, and they have traditionally been embraced (or at least tolerated) by philosophers and vehemently opposed, respectively. The corresponding 'positive introspection' principles are of course the doxastic and epistemic incarnations of **4**:

$$\mathbf{4}_B: \quad B\varphi \to BB\varphi \qquad \mathbf{5}_B: \quad \neg B\varphi \to B\neg B\varphi$$
$$\mathbf{4}_K: \quad K\varphi \to KK\varphi \qquad \mathbf{5}_K: \quad \neg K\varphi \to K\neg K\varphi$$

The use of the 'introspection' terminology here is contentious but has become standard. It is contentious because, for example, Hintikka in [490] accepts $\mathbf{4}_K$ – which is often called the "KK Principle" in the literature on this topic – while insisting that his grounds for doing so have nothing to do with introspection; more generally, philosophers have often defended non-introspective accounts of self-knowledge (of which this is one example), or, more neutrally expressed, accounts of self-knowledge which consist in something

---

[308][976], p. 204.

[309] This was mentioned in Humberstone [582], note 42, p. 568. The boulomaic analogue of the rashly immodest form, writing the operator as W, namely $W\forall q(Wq \to q)$, is open to the objection that one may reasonably not want all one's desires to be satisfied, perhaps because this would be a very dull state to be in, and perhaps for some other reason (as suggested in Kenny [648], p. 95*f*.). Even even the boulomaic analogue of 'less rash' propositionally quantified form of $\mathbf{U}_B$, which is what the version without the quantifiers amounts to, namely $\forall q W(Wq \to q)$, is not a plausible constraint on rational desire – since one may well reasonably want to have a *specific* unsatisfied desire – for example because one would like to know what it would be like to have that desire, already knowing that the consequences of acting on it would be disastrous. (Related issues were raised in Remark 5.1.19, which began on p. 358.) There are, admittedly, further complications coming from distinctions between kinds of desire – appetitive desires, all-things-considered desires, etc.

5.2. INTROSPECTION ISSUES

other than the exercise of an inward-directed quasi-perceptual faculty: see Remark 5.1.20, p. 359. (This is what the etymology of the word *introspection* suggests.) Here and in the immediately following sections we will be mainly concerned with $\mathbf{5}_K$, to which Hintikka objected on the obvious grounds that through no failure of rationality or self-awareness in any ordinary sense one could mistakenly take oneself to know that p when p is false (on the basis of overwhelming though misleading evidence to the contrary), in which case one would not know that p and yet be in no position to know that one did not. This is the usual position taken by philosophers who have offered opinions on the matter.[310] The point attained a certain public prominence in the early 2000s because of a famous 2002 remark by then US Defense Secretary, Donald Rumsfeld, with its category of *unknown unknowns* alongside *known unknowns*. (See Pullum [933] for some pertinent remarks and references.) The latter comprise those φ for which it is not known that φ and it is known that it is not known that φ (for instance because one knows one has no idea whether or not φ) and the former those φ for which one does not know that φ but this fact is itself not something of which one is aware, perhaps through mistakenly overestimating one's epistemic position. It is these cases that $\mathbf{5}_K$ would incorrectly declare not to arise.

The case against $\mathbf{5}_K$ can be dramatized, as Wolfgang Lenzen ([706], [707]) pointed out, by setting it in the context of a combined doxastic-epistemic logic with bridging axioms KB and Positive Certainty, or PC for short, listed here along with some further 'mixed' principles discussed in the literature; the labels KB, PI, and NI, the latter two abbreviating 'Positive Introspection' and 'Negative Introspection', are taken from Stalnaker [1082]. The phrases 'Positive Certainty' and 'Negative Certainty', abbreviated here to PC and NC, are from Halpern [427]. These are all just *ad hoc* labels for short-term convenience; in particular 'KB' (abbreviating 'Knowledge-to-Belief' implication) has nothing to do with the logic **KB**, for instance, and 'NI' has no connection with the *NI* frames of p. 92. The phrase "negative introspection" in the titles of succeeding sections refers not to NI, but to the purely epistemic negative introspection principle $\mathbf{5}_K$; further as already remarked in the preceding paragraph, even the word 'introspection' in the disabbreviations of NI and PI need not be taken as especially connected with the usual sense of that word.

$$\text{KB:} \quad K\varphi \to B\varphi \qquad \text{PC:} \quad B\varphi \to BK\varphi \qquad \text{NC:} \quad \neg B\varphi \to B\neg K\varphi$$

$$\text{PI:} \quad B\varphi \to KB\varphi \qquad \text{NI:} \quad \neg B\varphi \to K\neg B\varphi$$

PI and NI here obviously imply, given KB, the earlier monomodal doxastic introspection principles $\mathbf{4}_B$ and $\mathbf{5}_B$. Even taken together with the other principles listed above (and the presumption of normality for K and B, as well as $\mathbf{T}_K$) they do not yield $\mathbf{D}_B$ as a consequence, a point worth noting in view of the popularity of **KD45** as a doxastic logic. To see this, note that we could interpret B as representing the one-place constant truth-function with value T – or alternatively put, choose **KVer** as our doxastic logic – consistently with these principles, whereas $\mathbf{D}_B$ would not be correct for this interpretation.

Before resuming our presentation of Lenzen's argument against $\mathbf{5}_K$, let us take advantage of having the above list of principles at hand to report some further observations of Lenzen's ([706]; see also [707], [710]), as Theorem 5.2.1; they will be important later (Section 5.5). They show that a selection of the potential axioms above suffice, taken together with the missing principle $\mathbf{D}_B$, for a doxastic-epistemic logic in which B is definable in terms of K, and that **S4.2** has a special status as a candidate epistemic

---

[310] Hintikka makes the point briefly at p. 106 of [490]; there is a thorough and clear exposition of it at p. 79 of Lenzen [707], on some of whose commentary we shall be drawing in what follows. Perhaps the earliest occurrence of the objection to $\mathbf{5}_K$ occurs at p. 39 of Lemmon [695]: "one does not know all the points on which one is ignorant" (though this does not make explicit the fact much if not all confident belief, false as well as true, involves believing that one knows and so will not know that one does not know in such cases). Subsequent endorsements include Humberstone [559], p. 187, although, in a complaining footnote (note 5) about writers from the Artificial Intelligence and computer science community finding $\mathbf{5}_K$ unproblematic – cf. Remark 5.1.7 above – there is an incorrect citation (corrected in [575]), Williamson [1189], p. 23 (see also p. 167), and Stalnaker [1082] – a very informative discussion. For a discussion of **5** in a different (though still epistemically relevant) context, see Williamson [1195]; this paper also contains a fuller listing (than the previous sentence of the present note does) of passages in Williamson [1189] addressing the unacceptability of **S5** as an epistemic logic: see the discussion on p. 444 of [1195] which follows the proof of Proposition 2 there.

logic, for anyone convinced that such a logic should be at least as strong as **S4**, which itself is strong enough to supply most of the above bridging principles (when construed in the light of the definition in question); Lenzen particularly emphasizes parts (iv) and (v), which also brings in the stronger candidate epistemic logic **S4.4** – which is not of particular significance for the concerns of the present section, though we return to it on p. 413 – and a rather simple-minded possible definition of knowledge as (merely) true belief:[311]

THEOREM 5.2.1 (i) $Bp \leftrightarrow PKp$ *is provable in the normal bimodal logic with* $\mathbf{D}_B$, *and bridging principles* PC, NI *and* KB *as axioms.*

(ii) $.2_K$ *is provable in the logic described under* (i).

(iii) *All of the bridging principles above,* KB, PC, NC, PI, NI, *are provable in* **S4.2** *when B is taken as defined to be PK, and in fact all but one of them, namely* PI, *are provable already in* **S4**.

(iv) $\mathbf{KD45}_B \subseteq \mathbf{S4.2}_K \oplus Bp \leftrightarrow PKp$.

(v) $\mathbf{S4.4}_K \oplus Bp \leftrightarrow PKp = \mathbf{KD45}_B \oplus Kp \leftrightarrow (Bp \wedge p)$.

Proof. (i): We sketch the required deductions informally. Applying PMono to a representative instance of KB gives $PKp \rightarrow PBp$. Contraposing NI gives $PBp \rightarrow Bp$. From these two formulas we have $PKp \rightarrow Bp$ as a truth-functional consequence. For the converse implication, note that for any formula φ we have $B\varphi \rightarrow P\varphi$ provable, since the negation of this implication $B\varphi \wedge K\neg\varphi$ gives $B\varphi \wedge B\neg\varphi$ by KB, and this conjunction is inconsistent with $\mathbf{D}_B$. Thus in particular, taking φ as Kp, the implication $BKp \rightarrow PKp$ is provable, which taken together with the instance of PC, $Bp \rightarrow BKp$, has $Bp \rightarrow PKp$ as a truth-functional consequence.

(ii) In view of the equivalence described under (i), the provable $Bp \rightarrow \neg B\neg p$ (alias $Bp \rightarrow Cp$) can be rewritten as $PKp \rightarrow KPp$, which is $.2_K$.

(iii)–(v) Left as an exercise below. ∎

The fact that in part (iii) all the listed principles with the exception of PI here require for their provability only **S4** may seem surprising given the special status these considerations of Lenzen give to **S4.2**. The extra strength of the latter logic is involved instead in making B, when defined as PK, a normal modal operator, as well as in securing $\mathbf{D}_B$, mentioned in the proof of (ii). (See Theorem 4.5.9 on p. 282 above, originally noted in Dawson [237], with a deontic application in mind, and Section 5.5 below.) The information under (iii) is summarised in the column headed "B as PK" in Figure 5.5 (p. 433 below); for a repackaging of aspects of the pure B and pure K fragments of these logics and their relations, see also Proposition 5.6.13(i), on p. 434 below. ('Pure' in the preceding sentence is not being used in the technical sense introduced in Exercise 2.7.21; by the pure B fragment is meant, for instance, the fragment using B alongside the Boolean connectives but not K.)

In a 1982 letter responding to the author's request for information in English as to the main thrust of [706], Lenzen emphasized the following two points (here translated into the present notation and nomenclature), captured above as parts (iv) and (v) of Theorem 5.2.1:

> (a) If we accept something at least as strong as $\mathbf{S4.2}_K$ as our epistemic logic, then defining $B\varphi$ as $PK\varphi$ gives (at least) $\mathbf{KD45}_B$ as our doxastic logic.
>
> (b) If we accept (at least) $\mathbf{KD45}_B$ as our epistemic logic, then defining $K\varphi$ as $B\varphi \wedge \varphi$ yields $\mathbf{S4.4}_K$ as our epistemic logic.

---

[311] Recall that as a schema $.4$ is $(A \wedge \Diamond \Box A) \rightarrow \Box A$, so with K and P for $\Box$ and $\Diamond$ and PK taken to amount to B, this schema provides the contentious direction of the simple-minded definition; references to the defence of this definition as serious analysis of the concept of knowledge can be found in note 358, p. 415.

5.2. INTROSPECTION ISSUES

For anyone accepting both of the equivalences used as candidate definitions here ($B\varphi \leftrightarrow PK\varphi$ and $K\varphi \leftrightarrow (B\varphi \wedge \varphi)$ – stated using p rather than the schematic φ in Thm. 5.2.1$(iv),(v)$, exploiting closure under Uniform Substitution) this renders **S4.2** an unstable option as preferred epistemic logic, since combining (a) and (b) would drive a proponent of that logic to accept instead the stronger **S4.4**. However, there is no particularly compelling reason – to put it mildly – to accept the identification of knowledge with true belief. (See the discussion after Proposition 5.5.16, p. 413 below.) We return to the other equivalence, $B\varphi \leftrightarrow PK\varphi$, querying its $\leftarrow$-direction, in Example 5.6.9(iv) below.

Part (v) of Theorem 5.2.1 provides an example of the *definitional equivalence* of two logics (in different languages), which is to say: their having a common definitional extension. In our case the logics are **S4.4** and **KD45** with the extension of the first by the equivalence $B\varphi \leftrightarrow PK\varphi$ and of the second by $K\varphi \leftrightarrow (B\varphi \wedge \varphi)$ coinciding. Note that treating these equivalences as definitions, respectively of B and K, amounts to a departure from the usual policy we have been following – from Exercise 1.2.1 onward – of regarding definitions as metalinguistic stipulations to the effect that a defined expression represents *our* way of abbreviating the *definiendum*.[312]

EXERCISE 5.2.2 Give syntactic proofs of parts (iii)–(v) of Theorem 5.2.1. (For syntactic proofs of (iv) and (v) – semantic proofs for aspects of which appear in Theorems 4.5.9 above and 5.5.18 below (pp. 282, 416, respectively) – show the $\subseteq$-half of the equality:

$$\mathbf{S4.2}_K \oplus Bp \leftrightarrow PKp = \mathbf{KD45}_B \oplus Kp \leftrightarrow (Bp \wedge p),$$

by deriving, using the resources of normal modal logics, suitable axioms for $\mathbf{S4.2}_K$, such as $\mathbf{T}_K$, $\mathbf{4}_K$ and $.\mathbf{2}_K$, as well as $Bp \leftrightarrow PKp$, from the axioms suggested by the labelling of $\mathbf{KD45}_B \oplus Kp \leftrightarrow (Bp \wedge p)$; similarly in the reverse direction for the $\supseteq$-half, and likewise, *mutatis mutandis* for the equality involving $\mathbf{S4.4}_K$ and $\mathbf{KD45}_B$.) ✠

EXAMPLE 5.2.3 The table referred to above – that in Figure 5.5 – makes heuristic use of the device of underlining parts of the monomodal formula representing a bimodal formula. We write $\Box$ and $\Diamond$ rather than K and P respectively, in giving the monomodal representation. Thus we write $\Box\Diamond\Box p$ rather than $KPKp$, say, to emphasize the fact that with the $B = PK$ definition in force this same modality can be 'unpacked' in two ways, distinguishable by underling the parts corresponding to the doxastic operators: $\underline{\Box\Diamond}\Box p$, amounting to KBp, on the one hand, and $\Box\underline{\Diamond\Box} p$, amounting to CKp, on the other (since $\Box\Diamond$ is equivalent to B, the dual modality is equivalent to the dual doxastic operator C). Of course these are two parsings of the same formula, rather than two formulas, and any qualms about whether in a combined doxastic-epistemic logic the modalities KB and CK should really be equivalent are simply qualms about Lenzen's positive proposal – all of which are set aside here until Section 5.6. In Figure 5.5 (in that section) the proposed definition of B in which the use of the above underlining device is actually used is not of B as PK ("$\Diamond\Box$") but rather of B as KPK ("$\Box\Diamond\Box$"), a suggestion which we will there see receives some support from a consideration of the qualms just alluded to. ◀

Let us return to the main theme: Lenzen's argument against $\mathbf{5}_K$.[313] For this purpose, we only need to make use of KB and PC. The latter principle can be thought of as a way of spelling out the idea that $B\varphi$ is intended to be read as saying that the agent has a firm fully confident belief that φ – not just a tentative belief, hunch, suspicion, or even high subjective probability as to its being the case that φ;[314], even here one may be suspicious – but the role of this principle in what follows does not actually require that one endorse it. (We return to some of the other principles in Section 5.4.)

The interpretation of Figure 5.1 is as for Figure 4.1 on p. 250, except that here we are reasoning from the initial assumption at the top to two separate conclusions, one down each branch. The double-shafted arrows indicate that what is at the tail end of the arrow is a consequence of what is at the head end.

---

[312]This contrast is taken from Meyer [807]; see also 3.16 in Humberstone [594]. The biconditionals used here work to secure the interrepleaceability of *definiens* and *definiendum* – something we expect a definition to secure – because the

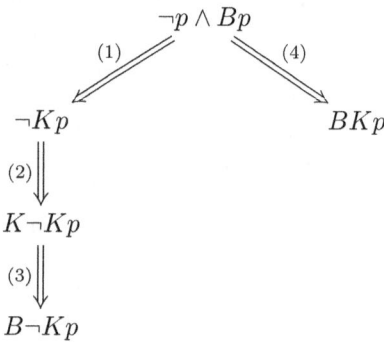

Figure 5.1: A Picture of Lenzen's Argument

Roughly speaking, this is recorded in a logic of the kind we are considering by the provability of the corresponding implication ("tail-formula $\to$ head-formula"), but we shall also consider an approach (in [509]) according to which these relations are encoded as sequents ("tail-formula $\succ$ head-formula") or metalinguistically, as consequence relation statements ("tail-formula $\vdash$ head-formula" – though in fact in Section 5.3 we will use the symbol "$\Vdash$" rather than "$\vdash$" for this purpose). Setting this last refinement to one side for the moment, arrows (1), (2), and (3) are justified by $\mathbf{T}_K$, $\mathbf{5}_K$ and KB, respectively, while arrow (4) is given by PC. In particular this part of the argument amounts to the standard proof of $\mathbf{B}$ from $\mathbf{T}$ and $\mathbf{5}$ (reflecting the fact that every euclidean reflexive frame is symmetric).[315] The formulas at the bottom of the two branches are inconsistent given $\mathbf{D}_B$, so in any logic containing all the principles just appealed to, the top formula is inconsistent, which means that the logic in question proves $Bp \to p$. While this is already an unacceptable consequence, we can transform it into the provability of $Bp \to Kp$ by substituting Kp for p and appealing to PC. The standard philosophical response, as already indicated, has been to block the argument at step (2), rejecting $\mathbf{5}_K$, but there are examples in the literature suggesting that the whole of $\mathbf{S5}$ be retained for K, and that the argument should be blocked somewhere else. Ver der Hoek *et al.* [509] suggest blocking the argument at step (1), which may seem to amount to denying that φ follows from $K\varphi$, but as we shall see in Section 5.3, is not quite this. Halpern [427] investigates a response along these lines, restricting KB in such a way that it cannot be appealed to license step (3). We will look at this response in Section 5.4; Voorbraak [1136], even suggests dropping KB rather than merely restricting it, partly for this reason. On last point at which the argument against $\mathbf{5}_K$ may seem vulnerable is the passage at (4) from Bp to BKp. But that would be a poor reaction, since we can just as well start the argument with $\neg p \land BKp$. One does not have to think that *every* case of believing

---

logics under consideration are congruential.

[313]The "against $\mathbf{5}_K$" here will often be omitted. Lenzen's *argument*, as we then call it for short, is to be distinguished sharply from Lenzen's *theorem*: Thm. 5.2.1 above.

[314]What is the relation between these subjective probabilities – or credences, or degrees of belief – that figure so prominently in, for a conspicuous example, decision theory, and the on/off notion of belief *simpliciter* that figures in standard doxastic logic? A good question, though not one we go into here. For some discussion, and pointers to the literature, see Christensen [178], Chapter 2, and several article in the special issue – No. 3 – devoted to this topic in Vol. 66 (for 2012) of the Journal *Dialectica*. Any attempt to identify the non-quantitative notion in terms of a threshold degree of the quantitative notion seems to pose a problem for the aggregativity of the former, and hence a threat to normality for doxastic logic, even with any familiar rationality assumptions in place: this is what Kyburg's 'Lottery Paradox' appeared to show. But perhaps the appearances were deceptive: see Leitgeb [692], [693].

[315]Recall that 'KB' is the name of a mixed epistemic-doxastic principle and has nothing to do with the normal modal logic $\mathbf{KB}$. As to the $\mathbf{B}$ axiom, just mentioned in the main text, in an epistemic setting, we return to this in Remark 5.3.4, p. 384 below.

5.2. INTROSPECTION ISSUES

that p *must be* a case of believing that one knows that p to see what is wrong with $\mathbf{5}_K$. It is enough to agree that in *some possible case*, a person may believe themselves to know that p even though it is not the case that p. Steps (1), (2) and (3) then lead from the first conjunct (of "$\neg p \wedge BKp$") to a formula incompatible with the second conjunct, given $\mathbf{D}_B$. This is why Lenzen's argument was described above as merely *dramatizing* Hintikka's original line of reasoning against $\mathbf{5}_K$. In fact, we do not even need $\mathbf{D}_B$. One may wish to pursue doxastic logic for believers some of whose beliefs are mutually inconsistent – though if so, one would probably reject the setting of normal modal logic for the investigation, since any such believer would, given that setting, believe everything. It still seems wrong to say that if a believes that a knows that p, then a must ipso facto have a pair of inconsistent beliefs – to the effect that a knows that p, on the one hand, and to the effect that a does not know that p, on the other: the two terminal nodes of the branches in Figure 5.1. As already stressed in describing Lenzen's argument as merely dramatizing the basic point, however, the core of the argument does not require explicit mention of B, and is already contained in steps (1) and (2) of the left branch in Fig. 5.1. It is this part of the reasoning that is addressed in the first of the responses – that of van der Hoek *et al.* – we shall be considering below (Section 5.3).

EXAMPLE 5.2.4 Another somewhat different way of dramatizing the shortcomings of $\mathbf{5}_K$, in the sense of illustrating its unwanted repercussions in the interplay between knowledge and belief, appears in Voorbraak [1136], who notes the following consequence of KB and $\mathbf{D}_B$ in the presence of $\mathbf{5}_K$: $BKp \to Kp$. The dual form of KB is the schema $C\varphi \to P\varphi$, where we use Hintikka's notation of P and C for the duals of K and B. Taking φ as Kp, we get

$$CKp \to PKp,$$

and so, by $\mathbf{D}_B$:

$$BKp \to PKp,$$

from which, finally, we appeal to $\mathbf{5}_K$ (in its dual form, $PK\varphi \to K\varphi$) to derive:

$$BKp \to Kp.$$

Voorbraak (p. 509), having formulated this as $BK\varphi \to K\varphi$, goes on to say:

> This principle contradicts the intuition that you might believe you know that φ, while in fact φ is not true. In a sense, the principle is self-defeating: Let ψ be the sentence "$BK\varphi \to K\varphi$ is not valid". The author of this paper believes [himself] to know that ψ. Hence $BK\psi$, where the author is the agent of K and B. But then the validity of $BK\varphi \to K\varphi$ implies its invalidity.

We resume the discussion of this issue at Example 5.4.5. ◀

Having noted that the epistemic aspects of Figure 5.1, it is worth pausing for the moment to consider what happens if, instead, we forget about knowledge altogether and just think about 'negative introspection' (i.e., an instantiation of **5**) for *true belief*. Spelling out **5**, $\neg\Box\varphi \to \Box\neg\Box\varphi$, for true belief (i.e., taking $\Box\varphi$ as $B\varphi \wedge \varphi$), we have:

$$\neg(B\varphi \wedge \varphi) \to (B\neg(B\varphi \wedge \varphi) \wedge \neg(B\varphi \wedge \varphi)).$$

Disregarding the second conjunct of the consequent, which makes no difference since it is already the antecedent, we get:

$$\neg(B\varphi \wedge \varphi) \to B\neg(B\varphi \wedge \varphi).$$

So since $\neg\varphi$ implies (even in **K**) the antecedent, and the consequent **KD45**-implies $\neg B\varphi$ – see Exc. 4.1.7(ii) – we get $\neg\varphi \to \neg B\varphi$, i.e., $B\varphi \to \varphi$: but it was supposed to be $\Box$, not B, that represented true belief!

The fact that true belief – a topic to which we return in Section 5.5 – cannot plausibly be supposed to obey **5** should give pause to those favourably disposed toward this principle for knowledge. It has its defenders nonetheless, as we shall see in the following sections. As to straight doxastic negative introspection – $\mathbf{5}_B$, that is – we include a few remarks in the discussion of concept-possession issues: see, in particular, Example 5.6.11, p. 431. Of the pure (i.e., purely doxastic or purely epistemic) positive introspection principles $\mathbf{4}_B$ and $\mathbf{4}_K$ we shall have little to say; nothing to say concerning the former, in fact, and some illustrations of counterarguments from the extensive philosophical literature cited here purely for the record. Our sample arguments come from Lemmon [703] and Williamson [1189] (cf. also Williamson [1182] and [1188]).

EXAMPLE 5.2.5 Lemmon [703] suggested that we might plausibly analyze "a knows that p" as: a has learnt that p and a has not forgotten that p. (This proposal is foreshadowed in Lemmon [695], p. 38$f\!f$.) This seems particularly unpromising in respect of conceptual priority, in that we naturally think of learning as *coming to know*, giving a strong impression of circularity. Lemmon's suggestion is a bit like saying that *being in Paris* consists in having gone to Paris and not left. (The fact that learning that p is a matter of coming to know that p, rather than just coming to believe that p, is not universally recognised in the literature: see p. 237 of Ma [747], for example.) Sometimes an account of the application conditions of a concept can be illuminating even if the account would be objectionably circular if proposed as a reductive analysis (see Burgess [131] and the other references cited in note 251, p. 302), though here there is even a problem about whether the application conditions have been correctly captured, since it does not seem *a priori* ruled out that some epistemic subject – God, perhaps – should always have known something to be the case. (Lemmon, [703], p. 60, makes the rhetorical move on this score of saying that such an example "would require some such doctrine as the reverse immortality of the soul", which sounds like a position to the effect that it is *in fact* the case that *every* cognising individual has always existed, whereas all that is need is that it is *conceivable* that *some* individual be in this situation.) Setting aside all these qualms about Lemmon's proposal, the upshot, had it been correct, would apparently be that one might know that p without knowing that one knows that p because one might have learnt and not forgotten that p while forgetting that one had indeed learnt that p. "Apparently" because, were the proposal formalized, we are making replacements of equivalents – *analysans* for *analysandum* – within the scope of the "knows" operator, and so are in effect assuming congruentiality for the sake of the discussion, as Lemmon remarks in Section 8 of [703], and one may wish to query this, whether in connection with Lemmon's own preferred analysis or of numerous others – knowledge as justified true belief, and so on – also detailed in [703]; similar points are made in Hilpinen [480]. Several further such analyses are considered in Okasha [856], which rediscovers Lemmon's point about the congruentiality assumption – to which he is evidently much less sympathetic than Lemmon – being required for the relevant anti-$\mathbf{4}_K$ argument to go through. See note 318 below for another such rediscovery. (Okasha speaks of an 'intensional fallacy' in the making of the required replacements, though 'hyperintensional' would be more to the point.[316]) Some of Lemmon's examples may even be persuasive on this score for those not attracted to the proposed analysis – which raises many more difficulties than have been touched on here (cf. Mannison [768]); Lemmon (to repeat a point already made) also raises similar objections to $\mathbf{4}_K$ for other candidate analyses of the concept of knowledge, proceeding by substituting in the scope of

---

[316]See Cresswell [208]. The exact spelling out of the relevant distinctions varies from author to author; for example, Williamson [1192] contrasts non-congruential contexts with hyperintensional contexts. A full discussion would also note the emergence of special issues, over and above deductive omniscience worries, that are raised by failures of replacement in propositional attitude contexts of *analysandum* by *analysans*, rather than just the interchange of arbitrary logical equivalents. See O'Connor [853] and Chapter 7 of King [657]; note that while what is called in this connection the 'paradox of analysis' is particularly associated with G. E. Moore, it has nothing to do with Moore's Paradox (discussed on p. 355).

5.2. INTROSPECTION ISSUES

"a knows that ___" the proposed analysis of "a knows that p"; as remarked at the end of note 280, he dismisses at least one of Hintikka's arguments for this principle as entirely question-begging. Interestingly Williamson [1189] thinks there is no such general analysis to be had, but still mounts an attack on 4_K: see Example 5.2.6. (Vendler [1124], p. 118, remarks: "X *knows that p* in most cases entails that X has learnt and has not forgotten that p", though quite what might be meant by saying that something *in most cases* entails something else is anything but clear.) ◀

While Lemmon, in the paper from which the above example is drawn, makes many astute semantic points about various propositional attitude verbs in English, one topic on which he is conspicuously silent is the use of perfect aspect (sometimes called the present perfect tense) in the key proposal:[317]

a knows that p if and only if a has learned that p and has not forgotten that p,

and he goes on to defend the entailment, embodied in this proposal, from "a knows that p" to "a has not forgotten that p". But on one reading of "a has forgotten that p" – that forced by "a has at some time in the past forgotten that p" – this is compatible with having subsequently recalled that p, and presumably therefore also compatible with "a knows that p". So the specific use of "has (not) forgotten" in Lemmon's discussion is what is sometimes called the 'present state' or 'present relevance' use of the perfect, as when one takes "I have closed the door to Room 325" to be false if the door to Room 325 is open (at the time the utterance is made), however many times the speaker may indeed have closed that door in the past. (See Huddleston and Pullum [530], pp. 139–148, and McCawley [785] for further discussion, as well as Vlach [1133], esp. §6.2.4; and Kibort [651], Iatridou, Anagnostopoulou and Izvorski [607], and other references given under Example 2.10.11 – see p. 159 – are also relevant, as is the joke by Groucho Marx to a dinner party host, exploiting the double usage of the English present perfect: "I've had a perfectly wonderful evening. But this wasn't it.")

EXAMPLE 5.2.6 Williamson's argument against 4_K (from §5.1 of [1189]). Suppose that there is a tree in view of our knowing subject a and let q_i be the statement that the tree in question is of height i centimetres to the nearest centimetre. If the height of the tree is $i+1$ centimetres, then given the distance a is standing from the tree and the difficulty of making such judgments just on the basis of looking, then a can't know that it doesn't have height i, and moreover, a can plausibly be presumed to know this. Thus we have for each $i \in Nat$:

(φ_i) $K(q_{i+1} \to \neg K \neg q_i)$,

which it will be more convenient to consider in a contraposed form:

$(\varphi_i)'$ $K(K\neg q_i \to \neg q_{i+1})$.

Distributing the K across, we get:

$(\varphi_i)''$ $KK\neg q_i \to K\neg q_{i+1}$.

One thing a can tell just by looking, however, is that the height of the tree is not 0 centimetres:

---

[317][703], p. 60; Lemmon conducts the discussion in terms of the first person, with "I know that p", rather than, as here "a knows that p" – an unwise move because the danger of distractions from things like Moore-paradoxicality in this area. See pp. 355 and 379. Incidentally, Lemmon records a debt to Austin for his account in note 14 of [703]: "J. L. Austin once suggested to me that 'I know' was a 'disguised perfect' tense." Austin is better known for a different proposal ([43], p. 144) to the effect that in telling someone that you know that p you are not ascribing a propositional attitude to yourself but rather giving the addressee your word that p; such 'speech-act theories' of concepts like the concept of knowledge are fraught with danger: see §6.2 of Searle [1021]. Note also the essentially first-person nature of the proposal: addressing "I know that p" rather than "x knows that p" in general – a point noted in Mayo [783], along with the concomitant disanalogy between "know" and a performative verb like "promise". While we are sampling this early literature, mention should be made of Wheatley [1167], who sees the need to avoid interference by changing second person to third person pronouns (p. 123) but continues to persist with first person formulations such as (p. 125) "I am ignorant of the fact that he went and so am not pleased because he went" despite the distracting Moore-paradoxicality.

$$K\neg q_0;$$

If we accepted $\mathbf{4}_K$, it would follow that $KK\neg q_0$ and hence by $(\varphi_0)''$, that

$$K\neg q_1,$$

whence by $\mathbf{4}_K$ again, that $KK\neg q_1$, and so by $(\varphi_1)''$:

$$K\neg q_2.$$

Repeated use of this same pattern of reasoning, invoking $\mathbf{4}_K$ and the appropriate $(\varphi_i)''$ at every step, allows us to conclude that $K\neg q_n$ for every $n \in \mathit{Nat}$, which is absurd, not least for the fact that this includes the n for which n is indeed the height of the tree to the nearest centimetre. (There are several published reactions – not considered here – to this argument, conveniently accessible in the bibliography of Dutant [266], which provides its own reply to Williamson; see also Bonnay and Égré [111], Ramachandran [941] and further references there cited, as well as Heathcote [468], already mentioned in another connection in Example 5.1.14, p. 353 and Zardini [1219]. Greenough [408], also mentioned already in Example 2.5.63, p. 103, looks sympathetically at variations of the argument with fewer or weaker background assumptions.) In fact, Williamson is also opposed, on the basis of somewhat similar examples and their general moral (knowledge requires a 'margin for error'), to PI ($B\varphi \to KB\varphi$), which explains his interest in the rule of margins, from p. 127 above, for epistemically interpreted $\Box$. In the current notation, S satisfies this condition when for all φ: $\vdash_S \varphi \to K\varphi$ implies that either $\vdash_S \neg\varphi$ or $\vdash_S \varphi$. In Williamson's terminology, this amounts to there being no non-trivial cases in which φ is *luminous*. (Actually Williamson [1189], esp. Chapter 4, speaks of luminous *conditions*, where in the present case the condition is that of having a given belief, and the luminosity consists in anyone satisfying that condition *being in a position to know* that he or she satisfies it. See note 282, p. 339. Compare also the notion of having a positive recognitional capacity, from Example 2.5.63, p. 103, and the preceding text; the link with luminosity was mentioned there already.) Williamson's §4.3 presents a general argument against the luminosity of all but trivial conditions, which is somewhat similar to that reviewed above, and on which the articles cited previously also bear, though for a specific focus on this argument, which by contrast does not involve knowledge of conditionals and the distribution, as in the step from $(\varphi_i)'$ to $(\varphi_i)''$, of "K" across "$\to$" – $\mathbf{K}_K$, as we might put it – Berker [85] provides a good discussion, citing numerous further references; see also Steup [1089], mentioned already in Example 2.5.63, p. 103. In other terminology, one might speak of φ's being *self-intimating* in this case, as in Hintikka [490], p. 84, or being *strongly evident*, as in Appendix 1 of Williams [1174]; Alston [14] and Roberts [961] are also helpful on the relevant distinctions, and MacIntosh [751] has some interesting logical moves. Rather surprisingly perhaps, given his endorsement of $\mathbf{4}_K$, Hintikka argues against PI in §3.7 of [490]; this reinforces the repeated claim in [490] that his support for $\mathbf{4}_K$ is not based on any considerations of introspection. (That will not stop us from calling this a principle of positive epistemic introspection, however, following the literature; as already remarked, this terminology is not to be taken literally, but simply as a convenient abbreviation.) In a more recent paper, namely [1196], Williamson has argued that one can know something to be the case compatibly with one's assigning the proposition that one knows it to be the case arbitrarily low probability. ◀

Recalling that "$K\varphi$" is generally taken as amounting to an abbreviation for something like "$K_a\varphi$" – "a knows that φ" – brings immediately to mind 'Geach–Castañeda style' (or "*de se*") worries about $\mathbf{4}_K$: even if one *were* guaranteed to know that one knew that φ, whenever one knew that φ, there is still a problem with inferring

$$a \text{ knows that } a \text{ knows that } \varphi$$

from

$$a \text{ knows that } \varphi.$$

5.2. INTROSPECTION ISSUES 379

The problem arises because a might correctly self-ascribe the knowledge that φ, but (through traumatic amnesia, say) not know that he/she is himself/herself a, and accordingly not be in position to know *that a knows that* φ. Such issues were raised – not specifically for epistemic logic (and note that 4_B is no less affected) – in Geach [353] and Castañeda [156], [157], [158], and were later given extensive discussion in Lewis [719] and Chisholm [175], as well as in many subsequent publications, from among which we mention here only: Nolan [846], Folescu and Higginbotham [315], Higginbotham [478], Peacocke [877],, the collection Feit and Capone [284], Cappelen and Dever [147] and Holton [520]. (This issue was raised briefly in the early review, Castañeda [155], of Hintikka [490]; it is explicitly brought to bear on the question of 4_K – or "KK" – in Castañeda [161]. The same issue of *Synthese* in which the last paper appears contains many others on the 4_K theme. See [492] for Hintikka's own preferred treatment of attributions of *de se* knowledge, and also the interesting treatment in Williamson [1185], esp. the first page and half for this issue. For a fuller discussion placing the issue in a broader setting, see Williamson [1186] and Edgington [272].)

REMARK 5.2.7 We studiously avoid all of the issues – raised in the literature just cited – which are connected with *de se* attitude ascriptions, having done so most recently in the discussion of Moore-paradoxicality (p. 355): this afflicts a's believing: p but I do not believe that p, whereas it does not afflict a or anyone else's believing: p but a does not believe that p. Readers are invited to select their favoured accounts of belief *de se*/first person content from the literature just alluded to. ◀

Many cases arise for considering the acceptability of principles like **4** for notions which either attribute (as with 4_K) – or seem somehow to allude to – propositional attitudes and we conclude with one such case, involving the prefix "It is obvious that":

EXAMPLE 5.2.8 An anecdote is often told, with numerous variations, about the Cambridge mathematician G. H. Hardy. (See Boas [102], p. 128, for example.) Hardy is reported as having described a certain point he was making in a lecture as obvious, and then asking aloud whether it really *was* obvious. According to the story, he then leaves the room to ponder the matter and returns a quarter of an hour later to announce "Yes, it is obvious," and continue with the lecture. This story is usually told to provoke some kind of adverse reaction: if it took Hardy 15 minutes to work it out, surely it can't have been obvious at all. But a natural alternative reaction is that, writing the initial claim as 'It is obvious that p', Hardy was not taking 15 minutes to check that it was true that p – which would indeed, given his intelligence, have refuted the claim that it was obvious that p – but rather, checking that it was *obvious* that p. That this might take such a time while still being the case would show that **4** fails for obviousness: something can be obvious without being its being obvious that it is obvious. ◀

Returning to the case of knowledge, it would appear that sometimes the opposition to 4_K goes beyond the grounds offered for it:

EXAMPLE 5.2.9 The following passage is from a discussion by F. Suppe ([1094], p. 313) of Paul Feyerabend, to whom the pronoun "he" refers; the bracketed numbering has been added here for back-reference below:

> Third, as with Berkeley, the sorts of arguments he offers in defense of his position lend support to the contention that truth via correspondence can play no significant role in an account of knowledge, only if one accepts the *K-K* Thesis ('*S* knows that *P*' entails '*S* knows that he knows that *P*'). For if this thesis is accepted, then from the fact that experience cannot show that such a correspondence holds and from the vaguely plausible assumption (tacit in his discussion) that such a correspondence could not be known except if displayed in experience, it would follow [1] that one cannot know that P is true; hence [2] one cannot know that he knows that P; hence via the K-K Thesis, [3] one cannot know that P.

It is difficult to share Suppe's conviction that $\mathbf{4}_K$ plays a central role in this line of thought, even if such a heavily encumbered notion of truth is employed that the transition from (1) to (3) (which it is represented here, via (2), as facilitating) – from knowing that P to knowing that (the proposition that) P is true – is seen as substantive. If there is an entailment from the claim that P to the claim that the proposition that P is true then the contested implication arises by KMono and does not require $\mathbf{4}_K$. Of course, one is free to object to KMono on such grounds (logical omniscience and all that) if one wants to, but this is a different matter from saying that it is "only if one accepts the K-K thesis" that such a transition is licensed.[318] (Incidentally, the parenthetical insertion of the words "the proposition that" – and in the preceding note, the words "its being the case that" – aims to adjust Suppe's formulation to something making sense, as would be needed with his remark (p. 330) that "Via the K-K thesis, to know that P one must know that P is true, which is to say..." in which the first occurrence of "P" stands where a sentence could stand but the second occurrence stands in the position that a name could stand.) There is indeed, and not just on the assumption of some arguably overblown version of the correspondence theory of truth, a good illustration of qualms, on concept-possession grounds (on which see Section 5.6), even when there is no explicit invoking of the concept of a proposition, is given with the following pair from p. 43 of Burgess and Burgess [130]:

(a) Fido believes that his master is at the door.

(b) Fido believes that it is true that his master is at the door.

Concerning these the authors remark (p. 44) that they "do not seem equivalent. For surely a dog, though it may have the belief ascribed in (a), does not have the concepts needed to have the belief ascribed in (b)."[319] ◀

5.3 Negative Introspection: Van der Hoek *et al.*

Van der Hoek, Jaspars and Thijsse are interested in the negative introspection principle for epistemic logic only in passing in [509], but it is this aspect of the paper we concentrate on here.[320] They work with a language with Boolean connectives $\neg$, $\wedge$, and $\vee$ and a single modal operator they write as $\Box$, in which respect the present exposition follows suit so that adjustments do not have to be made in quoted passages; $\Diamond \varphi$ abbreviates $\neg \Box \neg \varphi$, as usual; if you prefer to think of $\Box \varphi$ and $\Diamond \varphi$ as "$K\varphi$" and "$P\varphi$" feel free to do so, since the intended application is epistemic. (And although we use "$\Box$" rather than "K", we continue to use, as schematic letters for formulas, "φ", "ψ", ... rather than "A", "B", ..., for continuity both with the discussion of surrounding sections and with the notation of [509].) The logic is presented as a consequence relation, semantically and proof-theoretically characterized, though here we consider

---

[318] On the other hand, the objection to $\mathbf{4}_K$ at p. 396 of Suppe [1095] that this principle delivers the implausible claim that 'S knows that p' entails 'S knows that S's evidence for p is adequate' is specific to $\mathbf{4}_K$. Of course it also assumes that there is an entailment from 'S knows that p' to 'S's evidence for (its being the case that) p is adequate'. This is simply a form of Lemmon's original objection to the principle again, described in Example 5.2.5.

[319] In fact the examples are numbered (20a) and (20b) in [130] and this is how they are referred to in the remark quoted. The passage quoted continues: "any more than (to borrow Wittgenstein's example) it has those needed to believe that his master will come the day after tomorrow," but – even after the "it"/"his" mismatch is corrected – this does not seem like a salient parallel, since no equivalent though less conceptually demanding reformulation comes to mind for "his master will come the day after tomorrow". The vague talk of reformulation here is deliberate, there being some interesting cases in the literature where the two formulations are not logically equivalent but instead differ, for example, in respect of co-referential but non-synonymous descriptions, as in Bernard Williams's illustration involving the dog owned by the US President about whom we may be happier to say that it recognised that its master was coming up the drive than we would be to say that it recognised that the President of the United States was coming up the drive; Williams argues ([1171], p. 149) that this is not exactly a matter of concept-possession, construed literally, after all.

[320] For the significance of the word "honesty" in the title of this paper, see the Digression on p. 125 above.

5.3. NEGATIVE INTROSPECTION: VAN DER HOEK ET AL.

only the semantic characterization.[321] We will denote this relation by ⊩, and it will be defined to hold between Γ and ψ just in case in any model of a certain kind (explained presently) ψ is true at any point at which each $\varphi \in \Gamma$ is true. The models are a three-valued – Van der Hoek *et al.* say "partial" – variation on the models which figured in Section 4.1 (Coro. 4.1.5). Rather than writing as there, "$\langle W_0, w, V \rangle$" we will change notation and write instead "$\langle W, w_0, V \rangle$". The essential features are that $W \neq \varnothing$, that we do not require (though we allow) $w_0 \in W$, and that V is now not a function assigning sets of points (subsets of $W \cup \{w_0\}$) to propositional variables, but is rather a pair $\langle V^+, V^- \rangle$ of such sets, subject to the further stipulation that for any variable p_i, $V^+(p_i) \cap V^-(p_i) = \varnothing$. The idea here is that $V^+(p_i)$ tells us where p_i is true, while $V^-(p_i)$ tells us where p_i is false, where the terms *true* and *false* are not required to exhaust the possibilities – though by the stipulation just mentioned, they are required to be mutually exclusive. A model in this sense, meeting one further condition, Van der Hoek *et al.* [509] calls a *balloon model*, since in the case of $\langle W, w_0, V \rangle$ with $w_0 \notin W$, a picture of the model will look like a balloon (comprising the elements of W) if we think of w_0 as the other end of the piece of string to which this balloon is attached. The further condition in question is the following, to which we have given the name 'precursor condition', thinking of the $x \in W$ it requires, as a precursor of the distinguished point of the model in the sense that while the distinguished point may verify or falsify further formulas, it must at least verify all formulas verified by the precursor point and falsify all those falsified by the precursor point;[322] actually the present condition requires this only for the atomic formulas:

<u>Precursor Condition:</u>
There exists $x \in W$ such that for all p_i, $x \in V^+(p_i)$ implies $w_0 \in V^+(p_i)$, and $x \in V^-(p_i)$ implies $w_0 \in V^-(p_i)$.

It remains to define truth at a point in such balloon models. Or rather, truth and falsity at a point, since, as with the $V = \langle V^+, V^- \rangle$ ingredient in the models, we have to work with both notions rather than taking them to be complementary. For this exposition the truth of φ at x in $\mathcal{M}$, $\mathcal{M}$ being a balloon model, will be indicated by writing "$\mathcal{M} \models^+ \varphi$" and the falsity of φ at x in $\mathcal{M}$ by writing instead, "$\mathcal{M} \models^- \varphi$". The basis cases are clear enough, since we have V^+ and V^- to go by:

Where $\mathcal{M} = \langle W, w_0, V \rangle$, $V = \langle V^+, V^- \rangle$, and $x \in W \cup \{w_0\}$:

$$\mathcal{M} \models^+_x p_i \text{ iff } x \in V^+(p_i); \qquad \mathcal{M} \models^-_x p_i \text{ iff } x \in V^-(p_i).$$

We have the following inductive clauses, in which "∀", "∃" and "&" have been used as metalinguistic abbreviations:

$\mathcal{M} \models^+_x \neg\varphi$ iff $\mathcal{M} \models^-_x \varphi$ $\mathcal{M} \models^-_x \neg\varphi$ iff $\mathcal{M} \models^+_x \varphi$;
$\mathcal{M} \models^+_x \varphi \wedge \psi$ iff $\mathcal{M} \models^+_x \varphi$ & $\mathcal{M} \models^+_x \psi$ $\mathcal{M} \models^-_x \varphi \wedge \psi$ iff $\mathcal{M} \models^-_x \varphi$ or $\mathcal{M} \models^-_x \psi$;
$\mathcal{M} \models^+_x \varphi \vee \psi$ iff $\mathcal{M} \models^+_x \varphi$ or $\mathcal{M} \models^+_x \psi$ $\mathcal{M} \models^-_x \varphi \vee \psi$ iff $\mathcal{M} \models^-_x \varphi$ & $\mathcal{M} \models^-_x \psi$;
$\mathcal{M} \models^+_x \Box\varphi$ iff $\forall w \in W, \mathcal{M} \models^+_w \varphi$ $\mathcal{M} \models^-_x \Box\varphi$ iff $\exists w \in W, \mathcal{M} \models^-_x \varphi$.

The upshot for the defined symbol ◇ is a clause like that for □, but with "∀" and "∃" interchanged.

The Precursor Condition can now be shown to hold in the more general form indicated by the following exercise.

---

[321] In the interests of accuracy: in fact the authors focus on a generalized consequence relation (see p. 14 above) rather than a consequence relation. There are numerous further inessential differences between the content of [509] and the present exposition.

[322] The 'precursor' terminology introduced here may not be ideal, since it may suggest 'predecessor' as the converse of 'successor (by an accessibility relation)', which is not intended. Making the accessibility relation explicit would have precursors accessible to the distinguished point rather than conversely. The sense in which a precursor point precedes the distinguished point is that it bears the relation denoted in [509] by "⊑": $x \sqsubseteq y$ in a given model when whatever is true or false at x is respectively true or false at y.

EXERCISE 5.3.1 Show that if $\mathcal{M} = \langle W, w_0, V\rangle$ is a balloon model as above, then there exists $x \in W$ with the property that for all formulas φ, if $\mathcal{M} \models^+_x \varphi$ then $\mathcal{M} \models^+_{w_0} \varphi$ and if $\mathcal{M} \models^-_x \varphi$ then $\mathcal{M} \models^-_{w_0} \varphi$. (Argue by induction on the complexity of φ.) ✠

Define $\Gamma \Vdash \psi$ to hold just in case in every balloon model $\mathcal{M} = \langle W, w_0, V\rangle$ and every $x \in W \cup \{w_0\}$, if $\mathcal{M} \models^+_x \varphi$ for each $\varphi \in \Gamma$, then $\mathcal{M} \models^+_x \psi$.

EXERCISE 5.3.2 Show that, with $\Vdash$ as just defined, $\Gamma \Vdash \psi$ if and only if for every balloon model $\mathcal{M} = \langle W, w_0, V\rangle$, if $\mathcal{M} \models^+_{w_0} \varphi$ for each $\varphi \in \Gamma$, then $\mathcal{M} \models^+_{w_0} \psi$. (In the style, though not exactly the notation, of [509], "$\mathcal{M} \models_{w_0} \varphi$" would be written simply as "$\mathcal{M} \models^+ \varphi$".) ✠

Calling formulas φ and ψ equivalent when $\varphi \Vdash \psi$ and $\psi \Vdash \varphi$, Van der Hoek et al. [509] note that for any formulas φ, ψ, the formulas $\Box(\varphi \wedge \psi)$ and $\Box\varphi \wedge \Box\psi$ are equivalent, and that the same goes for $\Diamond(\varphi \vee \psi)$ and $\Diamond\varphi \vee \Diamond\psi$, much as one would expect from a normal modal logic; indeed we have the collapse of modalities expected from a rather strong such logic, **KD45** in which a formula prefixed by any affirmative modality is equivalent to that formula prefixed by the rightmost $\Box$ or $\Diamond$ in the modality (*rightmost* meaning the same as *innermost* in Proposition 2.1.11, p. 43). In the current $\Vdash$, however, we are not dealing with a normal modal consequence relation in the sense of p. 146 (or the Digression before Proposition 1.3.10), because the underlying non-modal consequence relation is not $\vdash_{CL}$. Most conspicuously we do not have $\Vdash p \vee \neg p$ (i.e., $\varnothing \Vdash p \vee \neg p$), since any point in a balloon model – even the distinguished point thereof – can lie outside the set $V^+(p) \cup V^-(p)$. Similarly, from the fact that $\varphi \Vdash \psi$ it does not follow that $\neg\psi \Vdash \neg\varphi$, the most significant illustration of which, emphasized by Van der Hoek et al. is provided by the following.

EXAMPLE 5.3.3 $\Box\varphi \Vdash \varphi$ for all formulas φ, but we do *not* have $\varphi \Vdash \Diamond\varphi$ for all φ. For the first, claim suppose that $\Box\varphi$ is true at a point in a balloon $\mathcal{M} = \langle W, w_0, V\rangle$. This means that φ is true throughout W. If the point in question is in W we thus conclude immediately that φ is true there. On the other hand if the point is w_0 and $w_0 \notin W$ then we appeal to the result of Exercise 5.3.1 to conclude that there is some $y \in W_0$ with the precursor property: that everything true there is true at w_0. Since by the previous reasoning, φ is true at y, we conclude that $\mathcal{M} \models_{w_0} \varphi$, as required. (This, indeed, is the rationale behind imposing the Precursor Condition on balloon models.) On the other hand we do not have $\varphi \Vdash \Diamond\varphi$ in general. For example $p \not\Vdash \Diamond p$, since we can have $w_0 \notin W$ and $V^+(p) = \{w_0\}$, $V^-(p) = \varnothing$, so although p is true at w_0, $\Diamond p$ is not, there being no point in W at which p is true. We did not even need to set $V^-(p) = \varnothing$ for this purpose, as long as we have some point $x \in W$ with $x \notin V^-(p)$, then we can arrange for x to be a precursor of w_0 (as required by the Precursor Condition) by having x 'agree' with w_0 on the other propositional variables – or alternatively by keeping x out of $V^+(p_i) \cup V^-(p_i)$ for all p_i. (Note that we could not make such countermodels if we required w_0 to be a 'precursor' of some point in W – but this is not what was required.) ◀

What does all this have to do with $\mathbf{5}_K$ (or indeed $\mathbf{5}_\Box$, i.e., $\mathbf{5}$ simpliciter, since we are using the $\Box$ notation here)? At p. 326 of van der Hoek et al. [509], before the semantics (presented above) is introduced, the authors write, explaining their own choice of primitives (slightly different on the Boolean front than ours above):

> Here the intended meaning of $\Box\varphi$ is that 'φ is known'. We write $\top$ for $\neg\bot$, $\varphi \vee \psi$ for $\neg(\neg\varphi \wedge \neg\psi)$ and $\Diamond\varphi$ for $\neg\Box\neg\varphi$. It is important to note that in our set-up, $\Diamond\varphi$ does not just mean that $\neg\varphi$ is not known, but that the agent considers some epistemic alternative to be possible, in which φ has a meaning: it is true!

Taking up the theme after presenting the definition of truth, the authors write as follows, in which their notation[323] "$\mathcal{M} \models \varphi$" means that φ is true at the distinguished point of the model $\mathcal{M}$ and $\mathcal{M}_w$ is the result of changing the distinguished point of $\mathcal{M}$ to w:

---

[323]They actually write "M" rather than "$\mathcal{M}$".

5.3. NEGATIVE INTROSPECTION: VAN DER HOEK ET AL.

Also note that the truth-definitions yield the intended effect for $\Diamond$-formulas: we have $\mathcal{M} \models \varphi \Leftrightarrow \mathcal{M}_w \models \varphi$ for some $w \in W$. In particular, our partial semantics makes $\Box \varphi \vee \neg \Box \varphi$, and hence $\Box \neg \varphi \vee \Diamond \varphi$ invalid. This reflects the idea that, in our opinion, $\Diamond$-formulas should express some positive evidence about φ, not just lack of knowledge of $\neg \varphi$.

Finally, for our purposes at least, on p. 333 of [509], we have:

> Negative introspection is now better motivated than in classical **S5**: if some fact is considered possible by the agent, it is explicitly present in his set of alternatives, so he knows that particular possibility. This should be contrasted to the classical case where merely not knowing the opposite is supposed to involve knowledge of the possibility.

Some of these formulations are evidently very murky, if not downright misleading (such as the final sentence just cited), but their bearing on the original negative introspection issue is threatened rather than enhanced by the 'partial' approach. In the first passage quoted above, we are told that while "$\Diamond \varphi$" abbreviates "$\neg \Box \neg \varphi$", "It is important to note that in our set-up, $\Diamond \varphi$ does not just mean that $\neg \varphi$ is not known,...". That is, the authors are saying that $\neg \Box \neg \varphi$ means something other than that $\neg \varphi$ is not known (by the agent in question), or more generally, that we have a case of ψ (take ψ as $\neg \varphi$) in which $\neg \Box \psi$ does not ("just") mean that it is not the case that $\Box \psi$. In other words the authors' use of $\neg$ does not reflect even their own use of "not". Its considerable ingenuity notwithstanding, it is hard to see how the resulting discussion can bear on the question of whether to accept the principle that whenever a does not know that p, a knows that a does not know that p.

The authors would seem to have been better advised, if they want to build into $\Diamond \varphi$ the idea that the subject explicitly considers it to be an open epistemic possibility that φ, *not* to have gone along the usual path of defining $\Diamond \varphi$ as $\neg \Box \neg \varphi$ (or $\neg K \neg \varphi$: see note 275, p. 333, on this very point) but to have taken it as an independent primitive notion – or to have defined it using other resources. (Related ideas are explored in Fagin and Halpern [278] and in some further references given both there and in Wansing [1146]. We do not consider this issue here.) However, even giving van der Hoek *et al.* the benefit of the doubts so far expressed, and seeing them as successfully blocking that part of the argument depicted in Figure 5.1 consisting of the steps there (i.e. on p. 374) numbered (1) and (2), which in the current notation and set out horizontally, becomes:

$$\neg p \underset{(1)}{\Rightarrow} \neg \Box p \underset{(2)}{\Rightarrow} \Box \neg \Box p,$$

there remains the following problem. As we have seen, van der Hoek *et al.* [509] block step (1) by producing a logic in which $\neg \Box p$, or equivalently $\Diamond \neg p$, is not a consequence of $\neg p$. But, as we also saw in Example 5.3.3, the authors are quite happy with the contraposed (or de-contraposed, if you prefer) inference from $\Box p$ to p. So let us contemplate running the argument in reverse. Start by assuming $\neg \Box \neg \Box p$, or more simply put, $\Diamond \Box p$: it is epistemically possible for our agent a, say, in the sense favoured by the authors – a consciously and actively contemplated possibility, rather than merely one not known not to obtain – that a knows that p. From this by their favoured logic, it follows that a does indeed know that p (since $\Diamond \Box p \Vdash \Box p$), from which it follows again by their own lights, as just recalled, that p. So even if we don't have the $\Vdash$-version of the **B** axiom, we have the $\Vdash$-version of its dual form. But why should it be thought to follow from a's actively acknowledging the epistemic possibility that a knows a's car is parked in the garage, that that is in fact where a's car is? This seems no more plausible than it would have been had the premiss involved the ordinary "not known not" kind of epistemic possibility, since what is added – explicit awareness of the possibility – has no power to guarantee that the possibility in question is realized.

Digression. Talk of its being epistemically possible (for a) that p, as in the preceding paragraph, can in the context of normal modal logic, epistemically interpreted, be taken to be a voicing of Pp (or $P_a p$, to make the knower explicit), where P is $\neg K \neg$). But in a more general setting ambiguities arise for spelling out what this means – alternatively put, for saying what it is for it to be the case that a for all

a knows. For example, in the absence of a monotony assumption ("KMono"), if a knows that φ and φ has as a logical consequence $\neg\psi$ but – exploiting the absence of monotony – a does not know that $\neg\varphi$, are we to say that it is epistemically possible for a that φ? If P is still defined as $\neg K \neg$ and we want to read P as conveying (epistemic) possibility, we will say *yes*; yet ψ is not compatible with what a knows in this case, so it hardly seems to right to say that for all a knows, it may be the case that ψ. (This point is made in Humberstone [567], and no doubt by others elsewhere too.) If we assume monotony but not regularity, a related difficulty arises if a knows that φ_1, a knows that φ_2, and φ_1 and φ_2 together have $\neg\psi$ as a consequence (though neither taken separately does), and a does not know that $\neg\psi$, is it an epistemic possibility for a that ψ? Further complications (which would arise even assume full normality), some people use the phrase "epistemically possible" to be the dual of "knowable *a priori*", which is a rather different matter from simply "not known not to be the case", and it has also been suggested that epistemic possibility is a matter of not being known *with certainty* to be false, which gives us a new notion provided that we think some things which are known are not known with certainty. See respectively Fiocco [302] and Reed [943] (and references in both works). Aside from the notion of an individual's knowing something with certainty there is the more impersonal construction "It is certain that φ", for which the dual notion is usually expressed as "It is possible that φ" and because of the popularity of cashing this out in terms of not being ruled out by the practically available evidence – for discussion see (in chronological order) the classic papers Hacking [419] and [420], and DeRose [241] – the notion figuring here is naturally enough regarded as some kind of epistemic possibility. Hacking reminds us of some grammatical markers of the notion of possibility, whatever we want to call it, that contrasts with certainty as opposed to that which contrasts with necessity ("possible that a is F" vs. "possible for a to be F", etc.) some of which go back to G. E. Moore, and all of which are also treated in [1168]. In fact, let us let Alan White have the last word on the subject here, expressing his displeasure with the above terminology in [1168], p. 86, though leaving the reader to investigate whether the discussion preceding it there supplies a justification: "The philosophical popular name of 'epistemic possibility' is a misnomer and the ideas based on it are mistaken." **End of Digression**

Talk of possibility as contrasted with certainty, as in the above Digression, will bring with it thoughts of treating "it is probable that" (or perhaps some more quantitatively specific version of it, such as its being probable to more than degree .5) as a kind of modal operator, and not surprisingly there have been several published ventures in this direction, of which the first, Hamblin [432] erred in ways corrected in the second, Burgess [132]. A more 'semantics of natural language' poriented approach can be found in Yalcin [1213]. Related work concentrating on comparative probability judgments but treated with binary probability connective includes Segerberg [1026], Gärdenfors [349], and Burgess [138]. Rather than go into these matters here, however, let us return to the main thrust of this section's discussion.

REMARK 5.3.4 What about **B** in the straightforward formula presentation of modal logic intended for epistemic application ($\mathbf{B}_K$, as we might say)? Hintikka [490], p. 54*f.*, was already clear that this, like $\mathbf{5}_K$, is not something to endorse. It is not hard to find variations on the case against $\mathbf{5}_K$ presented in Section 5.2 that apply here too. Suppose that Alice is firmly convinced that there is no gun in her apartment; occurrences of K will be taken to indicate Alice's knowledge. Thinking of p as representing "There is a gun in the apartment", suppose that her firm – and let us suppose also, well grounded – conviction that $\neg p$, is, as it happens, false: in fact it is true that p, which according to $\mathbf{B}_K$ implies that KPp ($\Box\Diamond p$ in the neutral notation), or, more explicitly: $K\neg K\neg p$. We said she was convinced that $\neg p$ – this is something she thinks she knows – but we have now reached the conclusion that she *knows* that she does *not* know this. So much the worse for $\mathbf{B}_K$, which is also evidently open to quite independent concept-possession objections (see Section 5.6): take p to be the proposition that there is deuterium – something Alice has never heard of (under any description) – in her apartment, and suppose again that p is true. Alice is not in any position to know anything about deuterium, including such things as her own states of knowledge or ignorance about its presence or absence, despite the fact that $\mathbf{B}_K$ would have us conclude by Modus Ponens that $K\neg K\neg p$. (More cautiously: she will know anything about deuterium which does follow from things she knows involving concepts she does possess. For example, she might be credited with knowledge that if deuterium is an element then hydrogen and deuterium are both elements, on the grounds that this follows from something which – we may suppose – she unproblematically knows, namely that hydrogen is an element. This qualification was foreshadowed in note 165, p. 205.) Such considerations notwithstanding, Floridi [314], esp. p. 236f., defends **KTB** as a plausible logic of a not-quite-epistemic notion he calls holding (or having) the information that such-and-such; the defence offered will not be discussed here, as it would require going into the (non-mental) notion of information possession involved. Williamson at various points discusses models for epistemic logic featuring, as an expository simplification, a (reflexive and) symmetric accessibility relation (e.g., [1194], [1197]) but this should not be interpreted as his favouring **B**$_K$. (See the index entries under 'Brouwersche principle' in [1189] for further details.) ◂

5.4 Negative Introspection: Halpern

We present Halpern's reaction to Lenzen's argument (against 5_K) in the same way as it is presented in Halpern [427]: by considering first a very straightforward logic for K and B which is vulnerable to the argument, or would be at any rate if the Positive Certainty principle (PC – which you might prefer to think of as standing for Positive Confidence) were added to its axioms. The problem of how to add this principle and retain 5_K is then ingeniously addressed by the provision of a new logic for K and B with an especially interesting semantical treatment of the latter operator. For those sharing the opinion already expressed, that Lenzen's argument is merely a dramatization of the objection to 5_K, the objection itself not depending on the interaction between K and B, the interest of this treatment is perhaps largely technical.

The strongest, most simple-minded combined epistemic–doxastic logic one could imagine that still respected the distinction between knowledge and belief would be that determined by the class of semi-simplified models (see Section 4.1[324]) $\mathcal{M} = \langle W, W', V \rangle$ with $\varnothing \neq W' \subseteq W$, with, for $x \in W$:

$$\mathcal{M} \models_x K\varphi \text{ iff for all } y \in W, \mathcal{M} \models_y \varphi; \quad \mathcal{M} \models_x B\varphi \text{ iff for all } y \in W', \mathcal{M} \models_y \varphi.$$

EXERCISE 5.4.1 Find a class of bimodal frames with accessibility relations for K and B – i.e., characterize the frames in terms of conditions on these relations – for which the point-generated subframes $\langle U, R_K, R_B \rangle$ can be taken to underlie the frames $\langle W, W' \rangle$ of the models just described, in the sense that, where $\langle U, R_K, R_B \rangle$ is generated by $w \in U$, if we put $W = U = R_K(w)$ and $W' = R_B(w)$, we have $\varnothing \neq W' \subseteq W$ and for all $x \in W$, $W = R_K(x)$ and $W' = R_B(x)$. (This guarantees that in any model $\langle U, R_K, R_B, V \rangle$ on $\langle U, R_K, R_B \rangle$, a formula is true at $x \in U$ iff it is true at x in the model $\langle W, W', V \rangle$, and so, by Theorem 2.4.10 – on generated submodels – that truth at all points in such $\langle W, W' \rangle$-based models coincides with validity over the class of bimodal frames $\langle U, R_K, R_B \rangle$ you have described.). ☩

The set of all formulas true throughout all such models can be (somewhat redundantly, as we shall see below) axiomatized by taking **KD45** for B and **S5** for K and adding suitable bridging (or 'mixed') axioms from our list in Section 5.2, repeated here for convenience:

KB:	$K\varphi \to B\varphi$	PC:	$B\varphi \to BK\varphi$	NC:	$\neg B\varphi \to B\neg K\varphi$
PI:	$B\varphi \to KB\varphi$	NI:	$\neg B\varphi \to K\neg B\varphi$		

---

[324]Notice here, however, that we are working with a bimodal rather than a monomodal logic and also, relatedly, that we cannot restrict attention to the cases in which $|W \smallsetminus W'| \leq 1$. Halpern [427], incidentally, uses structures $\langle W, W' \rangle$ as models, rather than, as in our exposition, $\langle W, W', V \rangle$. This is because he is thinking of the elements of W and W' as already coming equipped with a truth-value assignment (to the propositional variables); the present exposition sticks with the usual frames-&-models division of labour.

EXERCISE 5.4.2 *(i)* Which of the above schemata might be taken together to complete the axiomatization (added to the pure logics of K and B already described) of the set of formulas true at all points in models $\langle W, W', V \rangle$ with truth defined as above?

(ii) Show that any normal bimodal logic (in K and B) proving all instances of $\mathbf{T}_K$ and PI has $B\varphi \to \varphi$ as a theorem if and only if it has $B\varphi \to K\varphi$ as a theorem, for any formula φ. (In the terminology of Williams [1174], esp. Appendix 1, this would be put by saying that the logic tells us that φ is *incorrigible* iff it tells us that φ is *solid*.)

✠

While Williams's use of the term *solid* (as reported in Exercise 5.4.2(*ii*)) is stipulative, his understanding of *incorrigible* is standard – not with the formal characterization suggested in our parenthetical gloss but meaning "such that it could not be falsely believed". Here, saying that a subject falsely believes that p just means: it is not the case that p, but the subject believes that p.[325] The concept of incorrigibility is sometimes handled without due care and attention:

EXAMPLES 5.4.3 *(i)* The following passage from Armstrong [36], p. 417, begins with an inappropriate reference to sincere assertions rather than beliefs – one of several objections to the paper that can be found in Jackson [612], which has extensive references to other early incorrigibility literature – and goes on to make a more serious logical mistake:

> Introspective reports of current mental events are alleged to be logically incorrigible or logically indubitable. If I make the sincere statement "I seem to be seeing something green now," then, it is alleged, it is *logically impossible* for me to be mistaken in my statement. I may be lying, of course, but then I will know that my statement is untrue. For, it is argued, if mistake were a possibility then it would make sense to say "*I think* I seem to be seeing something green now, *but perhaps I am wrong.*" But this is nonsense, it is said, and so introspection is logically incorrigible or logically indubitable. (In the rest of this paper I shall simply say "incorrigible" or "indubitable," and I shall use the two words interchangeably.)
>
> Incorrigibility, or indubitability, must be distinguished from logical necessity. Whether or not the sincere statement "I seem to be seeing something green now" is incorrigible, it is certainly not logically necessary. This is most easily seen if we remember that a logically necessary truth is true in all possible worlds. Now we can certainly describe worlds where I do not seem to be seeing something green now. Contrariwise, it may be noted, a logically necessary statement need not be incorrigible. It is not true that we assent to any logically necessary statement as soon as we understand it. We may mistakenly think it is false.

The more serious mistake is in the incorrect claim that a logically necessary statement need not be incorrigible – incorrect, because if it cannot be false that p then it cannot be falsely believed that p – and in the argument offered for this erroneous conclusion.[326] If one were trying to show that necessity does not imply incorrigibility the appropriate counterexample to present would be of something which is necessary but not incorrigible: saying, as Armstrong does, that we may mistakenly think something concerning something which is logically necessary – p, say – that is false. Since we need a failure of

---

[325] The situation here is analogous to talk of truly believing that p, by which epistemologists typically mean: it is true that p and the subject believes that p – though beginning students often misinterpret "truly believes" as "genuinely (or "sincerely") believes". Note that the present discussion ignores all Geach–Castañeda *de se* considerations (see Remark 5.2.7), their relevance to a satisfactory treatment of the present issue notwithstanding.

[326] In the interests of not blunting one's linguistic tools, one should not go to the opposite extreme of *identifying* incorrigibility with necessity, as Gasking [352], esp. pp. 100–102, comes close to doing. (Gasking is a bit vague about whether he is concerned with statements that could not be false or with statements that could not be shown to be false.) Incidentally there are different conceptions of incorrigibility to be found in the literature, including one offered in Nakhnikian [839], p. 72, with the express purpose of not having necessity imply incorrigibility: see p. 125*ff*. of [839]. What Nakhnikian means by incorrigibility is (a minor variation on) Bernard Williams's notion of solidity, mentioned in Exercise 5.4.2(*ii*). but we need not go into further detail here.

5.4. NEGATIVE INTROSPECTION: HALPERN

incorrigibility we need to find a q for which Bq but not q (to make informal use of our doxastic notation), so since we are given that p is what is believed to be false, we must take q as $\neg p$. We have on our hands a failure of $B\neg p$ to entail $\neg p$, so $\neg p$ is not incorrigible – but this is no counterexample to the claim in question as it wasn't $\neg p$ but p itself that was being supposed to be logically necessary. Thus, as with luminosity (see Example 5.2.6, p. 377, and Example 2.5.63, p. 103) we must not assume that negation leaves unaffected the status of a proposition or condition in respect of its status as incorrigible.

(ii) A second, perhaps even more egregious, illustration of the cavalier treatment of incorrigibility is that of Weingartner [1163], p. 1, here called infallibility when applied to the cognitive agent in question rather than to any particular object of belief. But what Weingartner says expresses God's infallibility is the principle $Kp \to p$, understood in the logical setting in which closure under uniform substitution is assumed (and K represents specifically "God knows that __": in fact Weingartner writes "gKp"). But, as reviewers of [1163] (for instance, Hill [479], von Wachter [1138]) were quick to point out, this is simply $\mathbf{T}_K$, holding for knowledge in general and having nothing to do with divine infallibility. It is hard to know that Weingartner was thinking – perhaps he meant to ask whether whatever God believed had to be true, rather than whether whatever God knew had to be. ◀

Before leaving the subject of incorrigibility altogether, it should be remarked that such discussions of first-person authority or privileged access as are alluded to in Remark 5.1.20, p. 359, addressing the topic of introspection, do not by these phrases intend to invoke an incorrigibility thesis for the mental states they address. It's not so much that one couldn't be wrong in self-ascribing such a state so much as that some such self-ascriptions can be correct for distinctively first-person reasons – whether because there is literally an introspective quasi-perceptual faculty (fallible though it may be), or alternatively for the kind of reasons collected under the umbrella term 'constitutivism' in Remark 5.1.20. This said, let us return to the exploration of the formal principles given above.

From among the principles listed before Exercise 5.4.2 which are valid according to the present semantics, a suitable collection of bridging principles to supplement the monomodal B and K logics **KD45** and **S5** in an axiomatic description of the set of valid formulas may be found in Halpern [427]. The details do not matter for present purposes since we are interested instead in Halpern's attempt to alter the logic minimally so as to provide a response to Lenzen's argument, as diagrammed on p. 374 above. Accordingly the following discussion, down to the first new paragraph after Theorem 5.4.4 may be skipped by those not requiring these details (which answer Exercises 5.4.1 and 5.4.2).

Exercise 5.4.1 asks for conditions on relations which will guarantee that models on the point-generated subframes of frames satisfying them can be traded in for the 'semi-simplified' models; as it was put in the Exercise: for $\langle U, R_K, R_B \rangle$ generated $w \in U$, putting $W = U = R_K(w)$ and $W' = R_B(w)$, gives $\emptyset \neq W' \subseteq W$ and for all $x \in W$, $W = R_K(x)$ and $W' = R_B(x)$. If we make the accessibility relation for K be an equivalence relation, with that for B a serial subrelation of it, then the in the subframe generated by a point w, every point can be reached by a sequence of steps of the relation R_K from w, since every R_B-step is an R_K-step – which also secures that $W' \subseteq W$, as required – and thus every point can be reached in a single R_K step, since this is an equivalence relation, and W' is nonempty, since R_B is serial. But we need to make sure that the choice of W' does not depend on w; in the words of the Exc. 5.4.1, that for any $x \in W$, $W' = R_B(x)$. (We also required that $W = R_K(x)$ but that is clear since W is an equivalence class of the accessibility relation whose restriction to the generated subframe we are calling R_K.) This calls for a further condition relating R_B and R_K, namely the conjunction of (1) and (2), which are bi-relational variations on the themes of transitivity and the euclidean condition, respectively:

(1) $\quad wR_Kx \,\&\, xR_By \Rightarrow wR_By$ $\qquad$ (2) $\quad wR_By \,\&\, wR_Kx \Rightarrow xR_By$.

Here "R_K" and "R_B" have been written for the accessibility relations associated by K and B rather than their restrictions to some point-generated subframe, and the free variables are intended to be read as universally quantified, though the letters have been chosen so as to display the desired consequence that for any $x \in W$, $W' = R_B(x)$, i.e., for arbitrary $x \in R_K(w)$ (now taking w as a subframe-generating point) we have $R_B(w) = R_B(x)$. For $R_B(x) \subseteq R_B(x)$ we appeal to (1) above, while for the converse inclusion we appeal to (2). But we do not actually need both of these conditions, since R_K is symmetric. To derive either (1) or (2) from the other, interchange the variables "w" and "x", and then in the result, interchange "xR_Kw" and "wR_Kx" (as we may by the symmetry of R_K). In the literature, the condition in question is usually given as (i) rather than (ii), since this has a conveniently concise formulation in terms of relative products (see note 56), namely as: $R_K \circ R_B \subseteq R_B$.

If we were not assuming that R_K was symmetric, we should have two schemata in the modal language corresponding to (1) and (2), namely (i) and (ii) respectively:

(i) $B\varphi \to KB\varphi$ (ii) $PB\varphi \to B\varphi$

where "P" – let us add since we have not seen it in play much – abbreviates "$\neg K \neg$". The **B** axiom for K allows us to move between (i) and (ii) in a manner especially familiar from tense logic (cf. p. 179), but seen in a monomodal incarnation on p. 98 above, at the end of a digression of the MacIntosh rules. Note that (i) is just the linking principle PI, while (ii) is a contraposed version of NI.

We summarize the situation in the following:

THEOREM 5.4.4 *The logic determined by the class of all models $\langle W, W', V \rangle$, with the truth-definition associated with that conception of model in force, coincides with the logic determined by the class of all models $\langle U, R_K, R_B, V \rangle$ with R_K an equivalence relation, R_B a serial subrelation of R_K satisfying: $R_K \circ R_B \subseteq R_B$. This logic is axiomatized as the smallest normal bimodal logic in K and B with* **S5** *for K (e.g., with axioms* **T**$_K$ *and* **5**$_K$*), and* **D**$_B$ *along with mixed schemata* KB $= K\varphi \to B\varphi$ *and* PI $= B\varphi \to KB\varphi$.

This may seem a far cry from the original description of the logic as combining **S5** for K with **KD45** for B with suitable mixed axioms, since we have only specified **D** for B. In view of the informal completeness argument sketched above, however, we know that **4**$_B$ and **5**$_B$, which are valid on the frames singled out here, must be forthcoming. Indeed **4**$_B$ is an easy consequence of PI and KB. The earlier discussion of (i) and (ii) shows that NI ($= \neg B\varphi \to K\neg B\varphi$) is also derivable, and from here KB allows us to weakening the final K to a B, giving **5**$_B$. (Note that KB also means that there would be some redundancy in laying down a rule of Necessitation for B, since we can derive it from that for K.)

EXAMPLE 5.4.5 (Example 5.2.4 revisited.) Since the current logic evidently has the innermost reduction property (as defined before Proposition 2.1.11, p. 43) $K\varphi$ and $BK\varphi$ are equivalent according to it, and so certainly we have the implication $BK\varphi \to K\varphi$ complained of by Voorbraak in [1136], as we saw in Example 5.2.4 (p. 375). Let us take the discussion there a little further by imagining a conversation between Voorbraak, who believes that q (where this could be the ψ of Voorbraak's self-refutation argument in Example 5.2.4, for instance) and someone else, advocating the current logic but believing instead that not-q. Let us call the latter party 'Disbeliever'.

> VOORBRAAK: I believe that I know that q, so by your favoured logic, I know that q, from which it follows – again according to that logic – that q.
> DISBELIEVER: It is true that Kq follows from BKq and that q follows from Kq. Since q is false, however, it follows that you do not know, and therefore also that you do not even believe you know that q.
> VOORBRAAK: Well, I certainly *thought* that I believed that I knew that q, and, when it comes to that, I still do think – whatever you may say about it – that I believe that I know that q, and thinking that I believe something is the same as believing that I believe it, so I do

5.4. NEGATIVE INTROSPECTION: HALPERN

indeed believe that I believe that I know that q. According to the logic you accept, though, this implies that I believe that I know that q and so, by the previous reasoning, that q.

DISBELIEVER: I'm afraid not. You are right that BKq follows from $BBKq$, but by *my* previous reasoning, you do not believe that you know that q, so $BBKq$ is in the current instance false: it may *appear* to you that you believe that you believe that you know that q, but this appearance is deceptive.

VOORBRAAK: ???

While we have left Voorbraak speechless at the end of this dialogue, it is perhaps clear that once Disbeliever allows some such notion as that of its appearing or seeming to someone that they (believe that they believe that they ... believe that they) know that q which is capable of being deceptive – i.e., which is compatible with the subject's not knowing that q – then the parenthetical iterated "believes that"s are beside the point, since it is being conceded that it can seem (or appear) to a subject that they know something without in fact knowing it. For the reading of $BK\varphi \to K\varphi$ according to which it is objectionable, this matter of how things appear is what the B in the antecedent is there for, if it is there for anything. (Verbs like *appear*, *seem*, *look*, ... are susceptible of two interpretations, as has often been noted in the philosophical literature; for example Dretske [253], p. 20, Clark [181], p. 296. There is a purely phenomenal or experiential interpretation – how things perceptually strike one – and there is also a doxastic/epistemic interpretation – what such perceptual encounters lead one to believe; it is not clear whether this is a semantic matter (two senses) or a pragmatic matter (two options in respect of whether implicatures are cancelled), or indeed whether the binary distinction just drawn should be replaced by a ternary distinction, as is suggested by the discussion of the early chapters of Peacocke [874], with the phenomenal side subdivided into sensational features of experience on the one hand, and representational content of experience, on the other.) A hard-line defender of the spirit behind the semantics of the simple-minded system would presumably suggest that however deeply embedded under whatever kind of propositional attitude (or other) verbs, K still directs us to the same subset of W, refusing the concession just contemplated, and not going along with the last line given to the q-disbeliever in the dialogue above; this character insists, "You are simply wrong to think that it appears to you that ..." ◀

Let us return to the response to Lenzen's argument developed in Halpern [427]. Since this argument involves appeal to PC, it is important to notice that the part of the answer to Exercise 5.4.2(*i*) is that this schema is not valid on the present semantics.[327] $BK\varphi$ is just equivalent to $K\varphi$, each of these being true at any point in a model just in case φ is true throughout W (the set of K-pertinent worlds[328]), and for this it is not sufficient that φ be true throughout W' (the set of B-pertinent worlds), which is all that the antecedent of PC gives us. So Halpern's aim is to modify the semantics so as to validate PC, while not giving up on **S5** – and in particular on the controversial **5** – for K itself, and yet not ending up with the unwanted $\mathbf{T}_B$ (i.e., $B\varphi \to \varphi$) which Lenzen's argument seems to commit us to (by deriving an inconsistency from its negation). Clearly something else from the mixed logic described above has to go, and what Halpern sacrifices is KB, the principle that knowledge entails belief, which he calls the *entailment property*. Of more theoretical significance is that the particular way Halpern sacrifices KB, by accepting some though not all of its instances, costs us closure under Uniform Substitution.[329] Which is not, however, to say that his ingenious proposal is unmotivated.

---

[327] As Halpern remarks concerning the logic to be modified: "Of course, what this logic does not have is positive certainty. This does not make it bad; indeed, of all the approaches to modeling knowledge and belief, this is the one with which I have the most sympathy. My reasons are pragmatic rather than philosophical; this is the approach that I suspect will be most useful in applications, because it is relatively easy to model situations this way." [427], p. 488.

[328] This is intended to recall the phrase "□-pertinent", as last used on p. 205, for the set of points at which φ has to be true for $□\varphi$ to be true at a given world.

[329] Closure under uniform substitution has often been taken as a necessary condition for being a logic. It is built into the standard definition of what a modal logic is, for instance, as on p. 18 above, but is not always laid down as a requirement: see the discussion after Exercise 5.4.10 below.

The motivation in question is that the new semantic proposal offered by Halpern "enforces the idea that the agent believes that his knowledge coincides with his beliefs" ([427], p. 489). This formulation makes it sound as though everyone believes themselves to be infallible, so perhaps the idea is better – if less precisely – expressed by saying that from the subject's own perspective there is no contrast, in any given case, between what is (confidently) believed and what is known. The interest lies in the way Halpern changes the semantics so as to implement this idea. As before, models $\mathcal{M}$ can be taken to have the form $\langle W, W', V \rangle$ and to satisfy the additional condition(s) that $\varnothing \neq W' \subseteq W$, but there is a twist in the definition of truth (at a point in a model). Everything goes as at the start of this subsection except for the clause for B, though here to keep the contrast visible, we include also the clause for K. With $x \in W$ as above, we say, for any formula φ:

- $\langle W, W', V \rangle \models_x K\varphi$ iff for all $y \in W$, $\langle W, W', V \rangle \models_y \varphi$.
- $\langle W, W', V \rangle \models_x B\varphi$ iff for all $y \in W'$, $\langle W', W', V \rangle \models_y \varphi$.

Note the novelty in this second clause: in the course of evaluating $B\varphi$ at a point in one model, we are directed to what happens to φ at various points not in this same model, but in a *different* model.[330] To bring this out even more clearly, Halpern defines, for any model $\mathcal{M} = \langle W, W', V \rangle$, the model $\mathcal{M}^=$ to be the model $\langle W', W', V \rangle$ we get by setting the first component of the new model equal to the second component of the old one (and while retaining the second component without change), and writes the above clauses in the definition of truth as:

$$\mathcal{M} \models_x K\varphi \text{ iff for all } y \in W, \mathcal{M} \models_y \varphi; \quad \mathcal{M} \models_x B\varphi \text{ iff for all } y \in W', \mathcal{M}^= \models_y \varphi.$$

The effect of this striking suggestion is that on the new semantics, any occurrences of K in the scope of an occurrence of B in a formula will be processed as though they were themselves occurrences of B, since once we hit a B in evaluating a formula, the evaluation of whatever is in its scope proceeds in terms of the models $\mathcal{M}^=$ in which the set of K-pertinent and of B-pertinent worlds have been identified. That is, we will always have

$$\cdots B(\sim\!\!\sim\! K\!\sim\!\!\sim\! K\!\sim\!\!\sim)\cdots \quad equivalent\ to \quad \cdots B(\sim\!\!\sim\! B\!\sim\!\!\sim\! B\!\sim\!\!\sim)\cdots$$

Here the outer dots indicate any context within which the main B displayed is embedded, and the inner wavy lines any material within its scope, so that while for the sake of example two occurrences of "K" have been explicitly written on the left, there may be any number, and the replacement of all these by further "B"s on the right is to be understood.[331] 'Equivalent' in this formulation means that the formula on the left and that on the right have the same truth-value at any point in any model (with Halpern's semantics in force); see further Exercise 5.4.6(*ii*) below.

Let us extend the excerpt given in the previous section of the left-hand branch of our pictorial version of Lenzen's argument in Figure 5.1 (see p. 374) by adding to steps (1) and (2), the further step (3); note that we revert to the use of "K" rather than "$\Box$":

$$\neg p \underset{(1)}{\Rightarrow} \neg Kp \underset{(2)}{\Rightarrow} K\neg Kp \underset{(3)}{\Rightarrow} B\neg Kp.$$

(Recall that in another branch of the argument we pass from Bp to BKp, so that altogether, from the hypothesis that $\neg p \land Bp$, the subject ends up with inconsistent beliefs, on the one hand the belief that $\neg Kp$ and on the other, that Kp.) As we have seen already, the standard philosophical objection to this argument is an objection to step (2), this being an appeal to the dubious principle, $\mathbf{5}_K$, of epistemic

---

[330] A general discussion of this kind of model-shifting semantics, along with further illustrations, may be found in my [590], especially Section 4 there, where the term *model-changing* is used rather than *model-shifting*.

[331] In fact it doesn't matter whether all are replaced or not – the point is simply that B and K are freely interchangeable within the scope of B.

5.4. NEGATIVE INTROSPECTION: HALPERN

negative introspection. Van der Hoek *et al.*, as we saw in the previous section, want to retain this principle and object instead to (a version of) step (1). Halpern is exploring a response which again retains $\mathbf{5}_K$ but blocks step (3). To see how, consider the general equivalence above between a formula with occurrences of K in the scope of B and a formula which has (some or all of) them replaced by occurrences of B. The formula at the tail end of step (3) here is $B\neg Kp$, so making such a replacement, we conclude that on the present semantics this is equivalent to the formula $B\neg Bp$, which is in turn equivalent to $\neg Bp$. On the other hand the formula at the start of step (3) is $K\neg Kp$, which on the present account is equivalent to $\neg Kp$. So step (3) amounts to a passage from $\neg Kp$ to $\neg Bp$, which would only be justified if we accepted the implication $Bp \to Kp$, as of course we do not. (There is no questioning of classical logic for the underlying non-modal connectives here, by contrast with what we saw in Section 5.3, with the contraposition-dodging manoeuvres of van der Hoek *et al.* [509].) The converse $Kp \to Bp$, of this rejected implication is valid on the present semantics, and indeed more generally, so are all instances of the following restricted version of the KB schema, which is formulated in terms of what Halpern is not alone in calling *objective* formulas, by which is meant formulas containing no occurrences of K or B (or of course the dual operators defined in terms of them):

KB': $\quad K\varphi \to B\varphi \qquad$ for all objective formulas φ.

and just to complete Halpern's axiomatic characterization of the set of formulas valid according to his semantics, let us add here what else is used in [427] to supplement the bimodal combination of **KD45** for B with **S5** for K, alongside KB', namely:

$$\text{PI} \quad B\varphi \to KB\varphi \qquad \text{PC} \quad B\varphi \to BK\varphi \qquad \text{NC} \quad \neg B\varphi \to B\neg K\varphi.$$

Note that another of the mixed principles listed earlier, NI (the doxastic–epistemic form of negative introspection, namely $\neg B\varphi \to K\neg B\varphi$) is derivable given **KD45** for B, since every formula $\neg B\psi$ is equivalent to $B\neg B\psi$ so we can recover NI from PI.

EXERCISE 5.4.6 (*i*) Spell out the proof of NI just hinted at.
(*ii*) Show that if PC and NC are replaced by the axiom-schema $B(K\varphi \leftrightarrow B\varphi)$ in Halpern's axiomatization, the replaced schemata are derivable.
(*iii*) Show that any instance of KB, $K\varphi \to B\varphi$ in which φ is K-free (whether or not it is 'objective' in the sense of being also B-free) is valid according to the semantics described above. ✠

By way of commentary on the use of the specially restricted KB', which blocks the Lenzen argument in its transition from $K\neg Kp$ to $B\neg Kp$ (step (3), as we labelled it), rather than – as on the usual philosophers' reaction – from $\neg Kp$ to $K\neg Kp$, Halpern writes ([427], (p. 484)):

> Since the standard arguments in favour of the entailment property (see Lenzen [707] for an overview) are typically given only for objective formulas, this observation suggests that care must be taken in applying intuitions that seem reasonable in the case of objective formulas to arbitrary formulas.

The intuitions concerned would seem, on the contrary, to be completely general: the idea is simply that to know that something is the case is *inter alia* to believe correctly (i.e., truly) that it is the case. It could equally well be said, in the manner that Halpern is conducting the discussion here, that the standard arguments for the principle that $K\varphi$ entails φ are typically given only for objective formulas, suggesting that care must be taken in applying intuitions arising from such choices of φ to arbitrary choices of φ. As it happens, he does not make this suggestion, and it would not help with blocking the problematic derivation above; but without further backing, the remark just quoted would seem somewhat lame. The following Digression shows that occasional philosophical qualms about whether knowing that φ entails believing that φ have come from considerations quite different from those motivating Halpern.

Digression. A longstanding philosophical tradition has it that knowledge is, *inter alia*, true belief, or at least that it requires the knower to have the true belief in question (see the beginning of the quotation from Bacon in Example 5.1.13), and is thus committed to KB in its original unrestricted form. But many philosophers have dissented from the idea,[332] holding that the concept of knowledge, if indeed it is amenable to analysis at all, is not to be thought of as consisting in true belief plus something else (as in the accounts famously criticized in Gettier [370]: see Example 5.1.10, p. 347 above), because in particular, it is not required for a subject to know something to be the case that the subject should believe it to be the case – a point one could continue to insist on regardless of whether one took the concept of knowledge to admit of an informative analysis. Among those taking this line have been Colin McGinn, who has described knowledge as a 'subrational' achievement like perception, not requiring the cognitive sophistication of belief, and available to non-rational animals to which we would be reluctant to ascribe beliefs. (See esp. p. 547$f.$ of McGinn [786].) In a somewhat similar vein, Bernard Williams had earlier written of a "rather deep prejudice in philosophy, that knowledge must be at least as grand as belief" which seemed to him to be "largely mistaken", holding that – never mind McGinn's non-rational animals – we should be more willing to speak of a machine's knowing something to be the case when it was functioning properly in such a way as to register states of its environment, than to speak of its believing something to be the case under those circumstances. (See p. 146$f.$ of Williams [1171]. Williams also touches interestingly on the case of belief-attribution to non-linguistic animals, at p. 138$f.$; for a contrasting view and references to the early literature, see Routley [971].) Similarly, Lemmon [703], p. 58, had written: "it seems to me unlikely that knowing entails believing". And in a publication ([940]) from the previous year, Colin Radford had persuaded many people that the student who answers correctly but unconfidently – having forgotten ever having been taught the relevant facts – may know that such-and-such a king was born in such-and-such a year, without (firmly) believing this to be the case. (This is roughly speaking a reaction along the same lines as the discussion from Lemmon [703], as reviewed in Example 5.2.5 above, p. 376: forgetting having learnt the date of birth not implying forgetting the date of birth, and thus consistent with knowing the latter. "Roughly speaking", because of course remembering and forgetting having Φ-ed are not the same as remembering and forgetting – respectively – that one Φ-ed, and it is the latter construction that figured in Lemmon [703].) For further discussion and references, see Mannison [769] as well as the more recent Myers-Schulz and Schwitzgebel [837]. **End of Digression.**

The Radford reference in the above Digression is cited favourably by David Lewis in the course of his defence of an account of knowledge which does not imply belief. (See p. 556 of Lewis [730].) Lewis takes a subject to know a proposition just in case that proposition holds in every possibility uneliminated by the subject's evidence, where the 'every' is weakened by some permissive rules which allow certain possibilities to ignored when taken in conjunction with some prohibitive rules which (when they conflict) override them and explicitly disallow the ignoring of possibilities of certain types. Amongst the latter rules is the 'Rule of Belief' which states that "a possibility the subject believes to obtain is not properly ignored" ([730], p. 555) which, translated into epistemic-doxastic terms, gives us a bridging principle weaker (assuming the **D** schema, $B\varphi \to \neg B\neg\varphi$) than the entailment property itself, namely:

(∗) $K\varphi \to \neg B\neg\varphi$

REMARK 5.4.7 Note incidentally that if one did not accept $\mathbf{D}_B$ as a logical principle, one might do well to endorse (∗) explicitly in addition to KB. (A variation on this theme occurs in Nozick's analysis of knowledge using subjunctive conditionals – [851], p. 178 – in a move which refines the condition (one conjunct in the analysis of "S knows that p") "If it were that p then it would be that S believed that p" to "If it were that p then it would be that (S believed that p" and S did not believe that not-p).) Nakhnikian [839], p. 125$f.$, wants to appeal to (∗) but thinks he has to appeal to to $\mathbf{D}_B$ to justify this, writing in a footnote:

---

[332] Not under consideration here are those misled into thinking that knowledge and belief are mutually exclusive because one would not normally *say* "*a* believes ..." when one was in a position to say "*a* knows ...". The usual diagnosis of the latter would be pragmatic (rather than semantic), appealing to conversational implicature (cf. Example 5.1.11, p. 349); an especially interesting move in this area can be found at p. 18 (footnote included) of Unger [1119].

5.4. NEGATIVE INTROSPECTION: HALPERN

> In this step I am assuming that "S believes that p" entails "It is not the case that S believes that not p". (...) The principle I am using seems to be very basic. I am unable to think of any principle more evident from which to deduce it. I believe that the principle is logically connected with our conception of *believing that p*. As far as I have been able to determine, this principle is not deducible from the assumption that it is impossible for anyone to believe a proposition and its contradictory at the same time.

In fact of course the principle in question, $\mathbf{D}_B$, far from being something which cannot be derived from the assumption Nakhnikian formulates here, precisely *is* that assumption. The remainder of the footnote shows that what he intended the assumption to be was that it is impossible for anyone to believe the conjunction of a statement and its negation (or: the conjunction of a proposition and its contradictory, to use Nakhnikian's preferred terminology). From this one could derive $\mathbf{D}_B$ in any normal modal logic, appealing to aggregation, a principle accordingly rejected by anyone wanting to block the derivation, as in the analogous case of deontic logic, for those thinking one cannot have an obligation it is impossible to fulfill but one can have separate obligations not jointly fulfillable. (This is a theme in Schotch and Jennings [1002], mentioned in Example 4.4.8. The doxastic version can be found in, for example, Lewis [722].) ◄

Returning to Halpern, whose reasons – we might notice – for questioning the entailment property seem to have little in common with those of Lewis or the philosophers cited in the Digression, it must be said, unlike Lewis, he is committed to rejecting even the weaker (*) principle. For recall that according to Halpern's favoured logic, BKp and $K\neg Kp$ are consistent, which conflicts with (*), taking φ as $\neg Kp$. It may indeed strike us as odd that while the latter logic embodies the assumption that any two things known by the subject must be consistent (since they must be true) and also the assumption that any two things believed by the subject are consistent (by $\mathbf{D}$), it does not embody the assumption that anything known by the subject must be consistent with anything the subject believes: this is what (*) says.

Two points are raised by this tour through some of the opposition to the entailment property. One is a question as to whether we really need to exploit that property in order to question the appropriateness of $\mathbf{S5}$ as a logic for knowledge. The other is an observation that although we earlier described the standard philosophical position in relation to that logic as one of hostility, the case of David Lewis might be thought to show that not all philosophers share that hostility. Let us take up these points in turn. As for the first, it may seem unnecessary to detour through an explicit consideration of belief in order to put the case against $\mathbf{5}_K$: surely, one might think, a person could rationally fail to know that p, because in fact not-p, without knowing that they fail to know that p. How indeed, if the sole blame for their not knowing that p is to be laid at the feet of the fact that it is not true that p, could they be expected, by mere exercise of introspection and rationality, to come by knowledge of this ignorance? (This is just the familiar objection to $\mathbf{5}_K$, as in note 310 on p. 371, for example.) And what grounds would we have for attributing it to them? A person who does not know that p because, convincing though misleading evidence available to them notwithstanding, it is not in fact the case that p, would, if asked, "Do you know that p?" reply affirmatively, while we are supposed to say that despite this apparently sincere claim to know that p, they really know all along that they do not know that p. Halpern would no doubt regard this line of thought as question-begging, since the reference to sincerity and the rest is really a reference to belief, which it was our intention to keep out of the argument altogether. By bringing it back in, we have in effect smuggled the entailment principle back into the argument rather than providing a supposedly independent criticism of $\mathbf{5}_K$.

To take up the second point, Lewis's apparent commitment to $\mathbf{S5}$ for 'K': this impression is created by the above gloss on "S knows that φ" as "φ holds in every possibility uneliminated by S's evidence" (where the 'every' is allowed to exempt certain possibilities in accordance with the permissive and prohibitive rules). To see how this involves an equivalence relation as the accessibility relation for the 'K' operator (and hence determines the logic $\mathbf{S5}$, as remarked in, for example, note 16 of Williamson [1194]), we should see how Lewis spells it out in terms of possibilities – which can be thought of, for the sake of the present point, simply as possible worlds:

> There is one possibility that actually obtains (for the subject and at the time in question); call it *actuality*. Then a possibility W is uneliminated iff the subject's perceptual experience and memory in W exactly match his perceptual experience and memory in actuality. (Lewis [730] p. 553.)

If, tacitly fixing on a given subject and time, we dub the epistemic accessibility relation R_K – where $K\varphi$ is true in w just in case φ is true at every w' such that wR_Kw' – then I am taking it that Lewis is proposing generally that for any w, w', this relation R_K holds between w and w' just in case the subject's perceptual experience and memory in w' exactly match the subject's perceptual experience and memory in w (at the time in question). But this initial impression may not survive closer inspection; for more detail in this vein than the following Remark provides, consult Holliday [515]:

REMARK 5.4.8 There is a potentially fatal complication involved in distilling the above remarks on the model theory of epistemic logic from Lewis, since, as already mentioned, Lewis doesn't quite say that what is known is what is true in absolutely every uneliminated possibility (every world R_K-related to the world in which the knowledge is possessed, in our recent terminology), but is rather what is true in every – subject to a *sotto voce* proviso – such possibility. The proviso exempts possibilities which are 'properly ignored', with proper ignoring governed by the permissive and prohibitive rules already alluded to with the case of the (prohibitive) Rule of Belief; it is, incidentally, because of another such prohibition that the proviso has this '*sotto voce*' status: the rule of Attention, deeming any possibility being attended to (in the conversational context) to be one not properly ignored. On Lewis's account, it is this 'contextualist' feature which makes for the dialectical success of citing sceptical counterpossibilities: their very citation renders them not properly ignored in the setting of epistemological discussion, thus posing precisely the threat to claims to knowledge, in that setting, that the sceptic claims for them, though not, Lewis holds, posing a threat to the correctness of claims to knowledge on the part of the same subject of the same proposition, evaluated against the more everyday conversational contexts in which they are typically made. (Hanfling [434], p. 437, had put this by describing knowledge as "a concept with suicidal tendencies".) As for Lewis's rules as presenting a complication for the claim that his view makes **S5** the correct epistemic logic (and thereby stands refuted), this is most particularly so for one rule, another from the prohibitive stock, namely: the Rule of Actuality. The rule states that the possibility which is actual is never among those properly ignored. Lewis provides an extensive discussion of 'whose actuality' it is that is involved here ([730], p. 555 – esp. the paragraph beginning 'What is more'). What is particularly relevant is that it is the subject's actuality that is relevant, rather than that of someone attributing knowledge to the subject. Rules relating to the latter, such as those bearing on conversationally salient possibilities, are part what makes Lewis's account contextualist – meaning that according to the account, the truth-conditions for a knowledge attribution depend, like those for assertions containing explicit indexicals ("I", "here", etc.), on aspects of the context of utterance in which the attribution is made. For logical purposes we judge validity against a fixed such context, whereas since this is not so for all the rules – such as the rule of actuality (the subject's actuality, recall, not that of the attributor); these other rules allow for variation in the set of worlds pertinent as we pass from an occurrence of K to a further occurrence within its scope. Or, to put it in the earlier terminology, they allow R_K not to be an equivalence relation. Numerous writers besides Lewis have defended contextualism about knowledge ascriptions, including prominently Keith DeRose and Stewart Cohen. References to their pertinent publications and a critical discussion of the position can be found in Williamson [1190]. (See also Williamson [1191], Weatherson [1151] and Davis [236].) A potentially interesting variant – one of several different positions to which the label 'relativism' has been applied – was defended in MacFarlane [748], whose opening characterization of the position reads: "But where the contextualist takes the relevant standards to be those in play at the context of *use*, I take them to be those in play at the context of *assessment*: the context in which one is assessing a particular use of a sentence for truth or falsity." (See also several entries in the collection García-Carpintero and Kölbel [348]. An extended defence of the semantic strategy involved can be found

5.4. NEGATIVE INTROSPECTION: HALPERN

in Macfarlane [749], some warning flags on its precise formulation are raised in López de Sa [743] and Parsons [869], and for various counterarguments see Cappelen and Hawthorne [146].) Either version of the insistence on context-dependence in connection with knowledge attributions should be distinguished from the corresponding position in respect of 'epistemic modals' (modal auxiliaries with an epistemic interpretation): see Example 4.6.15(iii), p. 300. ◂

Many types of context-dependence for knowledge attributions have been postulated, including those in play in Remark 5.4.8 a dependence on which possibilities are salient in the context and what the conversationally appropriate standard of justification might be. One further example is worth mentioning here, before return to the discussion of Halpern.

EXAMPLE 5.4.9 John Bigelow [90] gives a contextualist twist to counterfactual accounts of knowledge (see note 289, p. 344 above), inspired by Lewis [721], which at the time [90] was written had been presented orally but had not yet appeared in print: instead of just having $\neg\varphi \,\square\!\!\rightarrow\varphi$ and $\varphi\,\square\!\!\rightarrow B_a\varphi$ as our combined counterfactual requirement for $K_a\varphi$ we require a range of mutually exclusive alternatives $\varphi_1, \varphi_2, \ldots$, and say that w.r.t. such a range and for φ_i in the range, a knows that φ_i when for each φ_j in the range, $\varphi_j\,\square\!\!\rightarrow B_a\varphi_j$. (This is reminiscent of what are called 'relevant alternatives' accounts of knowledge, alluded to in note 339 below.) Bigelow writes ([90], p. 20):

> Perhaps the best way of presenting the counterfactual theory of knowledge is by introducing a specific kind of context-dependence into the semantics for the verb "to know". Whether we are to say a person knows or not will depend on the range of alternative possible facts we have in mind. To make this explicit we could say, not simply "a knows that p", but rather, "a knows that p_1 *rather than* $p_2, p_3, \ldots$. (...) Thus (...) Descartes me be said to know that he is looking at his slippers rather than his bare feet, but he may not be said to know that he is looking at his slippers rather than dreaming he is looking at his slippers. If he were looking at his bare feet, he would have believed he was looking at his bare feet; but if her were dreaming he was looking at his slippers, he might still have believed he was actually looking at his slippers.

Bigelow [90] never did appear in print, but similar ideas have been developed elsewhere and are clearly articulated under the name 'contrastivism' in Schaffer [997], which is readily available. One point of contrast with Bigelow's is that the range of alternatives is restricted to two: knowledge is a ternary relation relating a cognitive agent to the proposition known and a contrast proposition.[333] Some discussion, from an 'epistemic logic' perspective, of Schaffer's proposal may be found in Aloni, Égreé and de Jager [13].

These suggestions would fit quite well with the theory of propositional allomorphs as the objects of knowledge, suggested in Dretske [256]. The allomorphs of the proposition expressed by the sentence "Mary kissed John" are the semantic values of sentences with differential contrastive stress – "<u>Mary</u> kissed John", "Mary <u>kissed</u> John" and "Mary kissed <u>John</u>". To know that Mary kissed <u>John</u>, rather than Harry or Kim or ... is then distinguished from knowing that Mary <u>kissed</u> John, as opposed to, e.g., merely shaking hands with him. Semantically one could think of such a propositional allomorph as a partition of the proposition concerned, with a distinguished block, though really the notion is relational: the allomorphs of a proposition $P \subseteq W$ (for W as the set of worlds) are the pairs $\langle \Pi, P\rangle$ in which P is one block of the partition Π of W. The remaining blocks of Π partition are the contrast propositions.[334] (Strictly we should be speaking of Ore-partitions rather than partitions proper, allowing ∅ as a block, in other words: see note 82, p. 91.) Thus "*John* kissed Mary" and "John kissed *Mary*" express allomorphs of the proposition that John kissed Mary, in the former case contrasting with propositions to the effect

---

[333] Unfortunately, Schaffer uses "trinicity" for this feature, which would have been fine if there had been such an adjective as "trinic".

[334] This use of partitions is not to be confused with the partition-based modelling of knowledge in the economics (and other) literature on epistemic states initiated by R. J. Aumann, and represented in our bibliography by Hart, Heifetz and Samet [457]., mentioned at the end of Example 2.5.63, which began on p. 103. This use of partitions amounts to buying into **S5** as an epistemic logic.

that this that or the other person kissed Mary, and suitable as an answer to the question "Who kissed Mary?", and in the latter contrasting with propositions to the effect that John kissed this, that, or the other person, and functioning as a candidate answer to the question "Who did John kiss?". (As we see from this example, really the whole of W is not partition, but only that subset of W in which question's presuppositions are satisfied.) The pairs $\langle \Pi, P \rangle$ are perhaps a workable approximation to the propositions-as-the-units-of-thought according to the treatment in Chapter 5 of Collingwood [187].[335] Aside from their obvious role in the semantics of interrogative constructions, partitions of the set of worlds have proved useful representations of the idea of subject matter: see Lewis [727] (and for more: Humberstone [574]). ◀

Let us return, finally, to Halpern [427]. Halpern gives a proof that the axiomatization given above (before Exc. 5.4.6) is sound and complete w.r.t. the class of all models $\langle W, W', V \rangle$ with the model-shifting ("$\mathcal{M}$-to-$\mathcal{M}^=$") truth-definition described above in force. Given the way we have defined the terms, this is not strictly speaking a (bi)modal logic, and so not a normal such logic, because as we have already noted, KB′ is laid down only for objective choices of φ, which include the case of $\varphi = p$, whereas the substitution-instance of this schema – used at step (3) of Lenzen's argument – which puts $\neg Kp$ for p, namely:

$$K\neg Kp \to B\neg Kp,$$

is not valid according to the semantics and so certainly not provable from the above axiomatization. (In fact, never mind axiomatic descriptions: the set of valid formulas is not closed under Uniform Substitution, we see.)

EXERCISE 5.4.10 What happens if the negation signs are removed from the example just given? In other words: is $KKp \to BKp$ valid (true throughout all models) on the current semantics, and if so, is this also the case for all of its substitution instances? ✠

As foreshadowed in note 329 above, there is a certain degree of variation in discussions of modal logic as to whether the requirement of closure under substitution is imposed; it plays no role in the canonical model completeness technique used for the completeness results of Section 2.4, for example – though it does when such results are reformulated (as in Section 2.5) in terms of determination by a class of *frames*. (Clearly the set of formulas valid on each frame in any class is closed under Uniform Substitution – at least when the standard semantics is in play,[336] as opposed to something like Halpern's.) That is, for the result that every consistent normal modal logic S is determined by its canonical model $\mathcal{M}_S$ (Coro. 2.4.2), it would have sufficed to define a normal modal logic as a set of formulas containing all theorems of **K** and closed under Necessitation and Modus Ponens, without actually requiring that it contain all substitution-instances of such formulas as are in the set but are not **K**-provable.

In any case there are many cases of proposed formal systems whose proponents are happy to call them logics (modal or otherwise) despite not being so closed. Cases already cited in our bibliography for other reasons include: Chapter 7 of Prior [924], Davies and Humberstone [235], and Schurz [1014], as well as references cited therein (for a discussion of Carnap's modal logic); see also the more recent publication Cresswell [219], and the references on Carnap's work given in the opening paragraph of Section 2.2 (p. 53 above). We take a slightly longer look at a further illustration in Example 5.4.11, because of its inherent interest. (Another specifically epistemic example, involving a novel binary modal connective, is remarked

---

[335]No attempt is here made to capture the strand in Collingwood's discussion concerning the significance of historical exegesis in settling which questions writers appearing to give the same or different answers to were actually addressing.

[336]This use of "standard" echoes the following remark from Halpern [427], p. 489: "Van der Hoek [508] attempts to provide a thorough analysis of how close we can get to having the entailment property, positive certainty, **S5** for knowledge, and **KD45** for belief. However, he considers only variants of the 'standard' semantics. In particular he does not consider changing the set of possible worlds inside the scope of a B operator." (In this quotation boldface has been added for the names of logics, and the reference adjusted to match the current bibliography.)

on in van Ditmarsch and Kooi [243], p. 206, and is implicit in Prop. 2.3 in Plaza [885].) A longer list of proposed logics not closed under Uniform Substitution, can be found in the discussion headed 'Uniform Substitution' on p. 191*f.* of Humberstone [594]. In that work, the objection to considering such things to be logics on the grounds that we will then not be able to distinguish propositional logics from arbitrary propositional *theories* (based on this or that logic) elicits the reply that we can easily do this by insisting that a logic is closed under variable-for-variable (uniform) substitution – a test passed by all the proposed non-US-closed logics, and eminently suitable for cases in which one is thinking of the propositional variables – or sentence letters, if you prefer that terminology – as having some property (such as objectivity, in Halpern's terminology) not shared by all formulas. In view of these many examples and the fact that Halpern has in the present section been seen proudly defending such a non-substitution-closed logic, it is surprising to read, in a paper co-authored by Halpern published four years before Halpern [427], the following: "In all cases of which we are aware, a formula is valid iff all substitution instances of the formula are valid."[337]

EXAMPLE 5.4.11 Fitch [307] drew attention to an interesting derivation in mixed alethic–epistemic logic, which has been the subject of much subsequent attention. In a normal bimodal logic for $\Box$ and K with the intended interpretations just indicated, and *prima facie* plausible principles governing these operators, suppose we consider a new candidate axiom expressing the idea that whatever is the case, whether or not it is in fact known to be the case, is at least *capable of* being known to be the case:

$$p \to \Diamond Kp.$$

Such a principle has been associated with verificationism and anti-realism: see the editor's introduction to Salerno [990] for detailed references. What Fitch's derivation then shows is, rather surprisingly, that the resulting logic proves also a formula like that inset here but without the "$\Diamond$", thereby collapsing the distinction between φ and $K\varphi$ (since we have the converse as $\mathbf{T}_K$ – presumed to be amongst the "plausible principles" just alluded to) and showing the cost of postulating the original axiom to be considerably higher than one might have thought. Evidently, it will be good enough to show that $q \to Kq$ follows from the given axiom ("the principle of knowability") since this is simply a re-lettered form of $p \to \Diamond Kp$; the re-lettering will make the derivation clearer. The derivation begins with

$$(q \land \neg Kq) \to \Diamond K(q \land \neg Kq),$$

a substitution instance of the new axiom, and then distributes the K, within the monotone "$\Diamond$" context, across the conjunction in the consequent:

$$(q \land \neg Kq) \to \Diamond(Kq \land K\neg Kq).$$

Appealing again to $\mathbf{T}_K$ and the fact that the surrounding context is monotone, we get:

$$(q \land \neg Kq) \to \Diamond(Kq \land \neg Kq).$$

But what follows the "$\Diamond$" here is a contradiction, so in fact the negation of the consequent of this last implication is provable, and therefore the negation of the antecedent, which is equivalent to the promised $q \to Kq$. One reaction of those wanting to continue endorsing some form of the principle of knowability without of course agreeing that everything is known would be to say that the principle must be restricted in its coverage to objective or non-epistemic matters. This restriction is mentioned on p. 102 (second column) of Rabinowicz and Segerberg [939]. It would block the above formal derivation at the first step, at which we appeal to uniform substitution to put $q \land \neg Kq$ for p in the axiom $p \to \Diamond Kp$. The informal justification would be something along the following lines. What people had in mind in saying that anything true could be known was that anything about the world, rather than about our epistemic

---

[337]Fagin, Halpern and Vardi [279], p. 1019.

relations to the world, could be known to be the case; as the Fitch derivation shows, without exempting matters concerning our epistemic relations to the world, we get interference effects from the K used to report them and the K in the principle of knowability itself. (The present point is not to endorse this response, but to use it to illustrate how naturally the idea of restricting Uniform Substitution arises.) Rather than restricting the propositional variables to range over objective propositions, von Kutschera [682] introduces a special 'objectivity' operator O in order for a suitably conditional version of the knowability principle to be formulated in the object language, without any need to restrict Uniform Substitution: $(p \land Op) \to \Diamond Kp$. (Von Kutschera's formulation is slightly different, in terms of true belief rather than knowledge, and Fitch is not mentioned by name, though it seems clear that this is what he has in mind.) Now the derivation is blocked because it would begin only with

$$((q \land \neg Kq) \land O(q \land \neg Kq)) \to \Diamond K(q \land \neg Kq),$$

and the refutable consequent gives allows us to conclude only to $(q \land \neg Kq) \to \neg O(q \land \neg Kq)$, something whose own consequent we expected to endorse, given the informal reading of O anyway – not that this consequent is actually provable in von Kutschera's logic.

For more information on the Fitch derivation and the extensive philosophical literature it has spawned, see Chapter 12 of Williamson [1189] and the collection Salerno [990]; Williamson's contribution to that volume also discusses Dummett [262], in which a proposal to restrict the knowability principle to 'basic' statements is aired, of which the present example is roughly the formal analogue. "Roughly" here because Dummett envisages the basic statements as atomic, so the principle of knowability does not even cover, for example, conjunctions of atomic statements, which would be 'objective' on the usage of Halpern and others. (As already noted, the precise analogue appears in [939].) Transferred to the present setting, this means endorsing $(p \land q) \to (\Diamond Kp \land \Diamond Kq)$, rather than $(p \land q) \to \Diamond K(p \land q)$. On the general theme of the Fitch derivation, Williamson [1179] should also be mentioned as should some more recent attempts to attend to the -*able* part of the word *knowable*. Van Ditmarsch, van der Hoek and Iliev [245] discuss the matter from the perspective of dynamic epistemic logic with knowability amounting to knowability on the basis of a successful public announcement; likewise the earlier paper van Benthem [78]. Fara [283] defends a version of the principle of knowability on which it claims (roughly) that whatever is true is such that one has a *capacity* to know it, the Fitch derivation showing only that some capacities are impossible to exercise. (Capacities like these, though, look suspiciously like incapacities.) The resemblance between the $q \land \neg Kq$ (figuring in the antecedent of the case of the knowability principle inset above) and Moore's Paradox – see p. 355 – is discussed extensively in Sorensen [1073], where they are described respectively as epistemic and doxastic blindspots; the issue is taken up more formally as one in (standard and also dynamic) epistemic logic in Holliday and Icard [517]. Occasionally the converse of the knowability principle is considered alongside it, so that we are dealing with the biconditional $p \leftrightarrow \Diamond Kp$: see Williamson [1179] and Brogaard and Salerno [118]. ◀

We should accordingly not regard its failure to be closed under Uniform Substitution as a decisive objection to the logic presented in Halpern [427]. Another line of discontentment might begin to be developed as follows. Let us contrast φ_1 and φ_2 here:

φ_1: Laura believes that there are parrots in Queensland.

φ_2: Laura knows that there are parrots in Queensland.

φ_2 but not φ_1 entails that there are parrots in Queensland. So if I say, "Laura believes that φ_1", I am attributing a belief to Laura, which belief *does not* require for its truth, that there be parrots in Queensland, whereas if I say instead "Laura believes that φ_2" I am attributing to Laura a belief which *does* require for its truth, that there be parrots in Queensland. But Halpern's semantics addressed to BKp (taking B, K, as 'Laura believes that', 'Laura knows that', and thinking of p as 'there are parrots in Queensland') processes this exactly as though it were instead BBp, since once we are in the scope of the outer "B" we are directed to the B-pertinent worlds rather than to the K-pertinent worlds, as we

5.4. NEGATIVE INTROSPECTION: HALPERN

should have been, had the following K not been embedded. Thus there is no way of expressing what was just represented as "Laura believes that φ_2", since as soon as we embed the K in φ_2 under the scope of "Laura believes that..." its full epistemic force is drained out of it.

Whether anything can be made of an objection along the above lines is unclear. In particular, there is a potentially worrying uniqueness assumption in thinking that when one says "Laura believes that φ_1", etc., there is a unique belief here attributed to Laura, about which we can fuss as to whether or not it entails that there are parrots in Queensland. (If $B\varphi$ and $B\psi$ are provably equivalent according to one's favoured doxastic logic, it need not follow that φ and ψ are – consider $\mathbf{4!}_B$ – but if "What – i.e., which belief – $B\varphi$ attributes to the believer" makes sense it should surely refer to the same thing as "What $B\psi$ attributes to the believer" under these circumstances. What we might try to make respectable under such a name as "inverse congruentiality" is in fact nothing but what Geach [355] calls the cancelling-out fallacy.) An alternative airing of the qualm under consideration here is provided by the following (discussion) exercise.

EXERCISE 5.4.12 Recall that we are considering a semantic treatment with the following clause for B-formulas in the definition of truth:

$$\langle W, W', V\rangle \models_x B\varphi \text{ iff for all } y \in W', \langle W', W', V\rangle \models_y \varphi.$$

Two things are happening at once on the right here. In the first place, we have universal quantification over the set W' which was in the second position on the left; but secondly, we are shifting the first component of the model so that it matches the second. Suppose that we separate out the two moves. We have two new primitive sentence operators, B_0, say, treated by the first move, and Ω, say, handled with – or signalling, perhaps – the second:

- $\langle W, W', V\rangle \models_x B_0\varphi$ iff for all $y \in W'$, $\langle W, W', V\rangle \models_y \varphi$.

- $\langle W, W', V\rangle \models_x \Omega\varphi$ iff $\langle W', W', V\rangle \models_x \varphi$.

Thus $B\varphi$ is treated by Halpern's semantics as though it were the combination $\Omega B_0\varphi$, or equivalently, $B_0\Omega\varphi$. Assess the following critical reaction to the discussion of Halpern [427]: "But it is B_0 rather than B that should be taken as representing belief, since the presence of Ω illicitly alters the content of the belief ascribed (any epistemic elements to that content being affected by lying within the scope of Ω). Halpern has changed the subject, in passing from B_0 to B_0-together-with-Ω."

✠

Let us return to the details of Halpern [427], where the following axiomatization is provided:

Axiom schemata (with, as rules, Modus Ponens and Necessitation rules for K and B) providing a basis for **KD45** for B, **S5** for K, in both cases as normal modal logics together with four further bridging principles:

KB′	$K\varphi \to B\varphi$	for all objective formulas φ,
PI	$B\varphi \to KB\varphi$,	
PC	$B\varphi \to BK\varphi$,	
NC	$\neg B\varphi \to B\neg K\varphi$.	

The schematic letter φ is of course unrestricted in the last three schemata, as well as in the axiom-schemata summarized by the labelling '**KD45**', '**S5**' (thus, e.g., we have the **T** schema $K\varphi \to \varphi$ for arbitrary φ).

The soundness half of the proof of soundness and completeness (Thm. 2.2, p. 490), is described as "straightforward and left to the reader". (Halpern gives a completeness proof via normal forms, rather than a canonical model argument of the kind seen illustrated in Sections 2.4, 2.5.) But because of the model-shifting feature in the clause for B, it is not quite so obvious that the rules preserve validity – in particular, the Necessitation rule for B. When this rule is applied to formulas containing K, the resulting occurrences of K will come to fall within the scope of B, so those that were not already in the scope of some other occurrence of B will in effect be re-interpreted afterwards (as quantifying over W' rather than W). This was an aspect of the semantics to which polemical attention is drawn by Exercise 5.4.12 and the discussion preceding it, but now what needs to be noticed about it is that while on the standard semantics rules like B-Necessitation preserve truth in (i.e. truth throughout) a model, on Halpern's, they do not (or rather, it does not).

EXAMPLE 5.4.13 For instance, take $\mathcal{M} = \langle W, W', V\rangle$ with $W' \subseteq V(p)$, $W \nsubseteq V(p)$. Then $\mathcal{M} \models \neg Kp$, while $\mathcal{M} \nvDash B\neg Kp$. ◀

However, for the soundness proof, we don't need the rules to preserve truth in a model: we just need them to preserve validity (in the sense of truth in all models).

EXERCISE 5.4.14 Show that the rule $\varphi \,/\, B\varphi$ does have the preservation characteristic just described. ✣

In fact, however, if we are interested in giving a more economical basis for the present logic, as in Theorem 5.4.4 for the simple system with which we began, then we should not be mentioning Necessitation for B anyway. We saw for the case of that system, one did not need such a rule in the presence of the schema KB: if φ is provable, so is $K\varphi$ and therefore, by KB and Modus Ponens, so is $B\varphi$. We have now lost KB, but with a little care a similar derivation can be given, drawing only on the resources of its restricted cousin KB':

(1)	φ	Given, by hypothesis.
(2)	$\top \to \varphi$	1 TF
(3)	$\top$	2 TF
(4)	$B\top \to B\varphi$	2 BMono
(5)	$K\top$	3 Nec (for K)
(6)	$K\top \to B\top$	KB'
(7)	$B\top$	5, 6 MP
(8)	$B\varphi$	4, 7 MP

Recalling Halpern's remark already quoted in note 336 above concerning van der Hoek (as represented in [508]), that "he does not consider changing the set of worlds inside the scope of a B operator", it may be of interest to consider an alternative semantics for his logic (i.e., the logic axiomatized), somewhat closer to conventional Kripke-style semantics, which does not have the 'model-shift' $\mathcal{M}$-to-$\mathcal{M}^=$ (i.e., $\langle W, W', V\rangle$-to-$\langle W', W', V\rangle$) feature that this remark might suggest is crucial. This alternative (presentation of the) semantics – the last item on the agenda for the present section – is not without interest in its own right. (Those wanting to move on to new pastures should skip to the following section now.) Of course, in view of the failure of Uniform Substitution, we know that no class of Kripke frames determines the logic. Instead, we define a class of frames and consider only models based on them which meet a special condition. This is the reason for including the relation E in the frames (see immediately below), which does not itself function as an accessibility relation for any operator in the language.

The frames will be of the form $\langle W, R_K, R_B, E\rangle$ in which R_K is an equivalence relation (on W) and R_B is a serial transitive euclidean relation (on W); like R_K, E is just another equivalence relation on W. Truth at a point in a model will be defined for K-formulas by universal quantification over R_K-related points, and for B-formulas, by universal quantification over R_B-related points. For the sake of the bridging axioms PI, PC, and NC, we need to impose the following conditions, which we will label by using lower-case versions of the names for the respective axioms:

5.4. NEGATIVE INTROSPECTION: HALPERN

(pi) for all $x, y, z \in W$ if $xR_K y R_B z$ then $xR_B z$;

(pc) for all $x, y, z \in W$ if $xR_B y R_K z$ then $xR_B z$;

(nc) for all $x, y, z \in W$ if $xR_B y$ and $xR_B z$ then $yR_K z$.

Note that the condition (pi) was imposed on the frames figuring for the simple logic of Theorem 5.4.4, though without the label "(pi)", and in the more concise notation: $R_K \circ R_B \subseteq R_B$. We could formulate (pc) similarly, as: $R_B \circ R_K \subseteq R_B$. For the sake of the remaining bridging principle, the weakened 'entailment property' schema KB', we need to impose an additional condition on the frames, and also the special conditions on models alluded to above. The condition on frames is:

(kb') for all $x, y \in W$ if $xR_B y$ then for some $z \in W$, yEz and $xR_K z$.

The models we are interested in are not *arbitrary* models $\langle W, R_K, R_B, E, V \rangle$ on such frames, but rather 'special' models, satisfying the constraint:

$$xEy \Rightarrow (x \in V(p_i) \Leftrightarrow y \in V(p_i))$$

for all $x, y \in W$ and all sentence letters p_i. So we now mean by a *special model* a model $\langle W, R_K, R_B, E, V \rangle$ meeting this condition, where the frame $\langle W, R_K, R_B, E \rangle$ of the model satisfies all our earlier conditions, including the conditions (kb'), (pi), (pc), and (nc).

LEMMA 5.4.15 *If $\mathcal{M} = \langle W, R_K, R_B, E, V \rangle$ is a special model, then for any $x, y \in W$ with xEy, we have*

$$\mathcal{M} \models_x \varphi \text{ iff } \mathcal{M} \models_y \varphi \text{ for all objective formulas } \varphi.$$

Proof. By induction on the complexity of φ, appealing to the specialness condition on models (for the basis case). ∎

THEOREM 5.4.16 *For all formulas φ, φ is provable from Halpern's axioms if and only if φ is true throughout every special model.*

Proof. 'Only if' (Soundness). It is a routine matter to check that the axioms are all true in every special model, given the frame conditions (pi), (pc), (nc), with the exception of (kb'), whose instances are true in every special model by the condition on models. Here one appeals to Lemma 5.4.15 and to (kb').

'If' (Completeness). Build the canonical model, $\langle W, R_K, R_B, E, V \rangle$, in more or less the usual fashion (though since there is enough subscripting going on with the accessibility relations, we will not label the logic and then use the label in subscript position). To be precise, the usual definitions are given for all ingredients except E, the current novelty, which is defined thus: xEy iff for all objective formulas φ, we have $\varphi \in x \Leftrightarrow \varphi \in y$; note that this guarantees that xEy implies $x \in V(p_i) \Leftrightarrow y \in V(p_i)$, as required. Then one sees by the usual arguments that R_K is an equivalence relation, R_B is a serial transitive euclidean relation, that (pi), (pc), (nc), are satisfied (courtesy of PI, PC, NC). This leaves, amongst the frame conditions, only the condition (kb'), that for $x, y \in W$ with $xR_B y$, there exists $z \in W$ such that yEz and $xR_K z$ to check. So suppose $xR_B y$; to find z with the desired properties, it suffices to take z as a maximal consistent extension of the set

$$\{\psi \,|\, \psi \text{ an objective formula} \in y\} \cup \{\varphi \,|\, K\varphi \in x\}$$

on the assumption that this set is itself consistent. To vindicate this assumption, suppose the set is not consistent. Then there are objective formulas belonging to y, $\psi_1, \ldots, \psi_n$, and formulas $\varphi_1, \ldots, \varphi_m$ with $K\varphi_i \in x$ ($1 \leqslant i \leqslant m$), for which (where '$\vdash$' indicates provability on the basis of Halpern's axiomatization):

$\vdash (\varphi_1 \wedge \ldots \wedge \varphi_m) \to \neg(\psi_1 \wedge \ldots \wedge \psi_n)$.

Thus:

$\vdash (K\varphi_1 \wedge \ldots \wedge K\varphi_m) \to K\neg(\psi_1 \wedge \ldots \psi_n)$.

Since each of the conjuncts of the antecedent belongs to x, so does the consequent. And since what follows the 'K' in the consequent is an objective formula, by KB′, $B(\psi_1 \wedge \ldots \wedge \psi_n) \in x$; but $xR_B y$, so $\psi_1 \wedge \ldots \wedge \psi_n \in y$, contradicting the fact that each of $\psi_1, \ldots, \psi_n$ belongs to y. Thus the displayed set is consistent and the desired z exists. By the usual reasoning, truth at a point in the canonical model and membership in that point coincide, every unprovable formula is false at some point in the canonical model, concluding our completeness argument. ∎

5.5 Logics Between S4 and S5

The title here refers to a range of candidate epistemic logics, evidently, rather than doxastic logics, since we do not want $\mathbf{T}_B$ (the **T** axiom for B as $\Box$, that is). Some significant contenders in this range are the following:

$$\mathbf{S4} \subsetneq \mathbf{S4.2} \subsetneq \mathbf{S4.3} \subsetneq \mathbf{S4F} \subsetneq \mathbf{S4.4} \subsetneq \mathbf{S5}$$

though **S5** is listed mainly for old time's sake, the reader with any luck by now having seen the error of the ingenious attempts to defend it in the previous sections, and **S4.3** is included only as a landmark, special merit having seldom been claimed for it as a suitable epistemic logic. (See note 339.) Hintikka's favoured candidate was **S4**; on the other hand, Lenzen's Theorem 5.2.1 on p. 372 above – (*ii*) in particular – makes a strong case for $.\mathbf{2}_K$, and **S4.4** as epistemic logics which suggests taking instead **S4.2** as a minimal candidate – setting aside for this section the worries about **4** that have exercised others (see Examples 5.2.5, 5.2.6, beginning on p. 376 above). Lenzen also suggested **S4.4** as a maximal plausible candidate: we shall get to this later. There are occasional additional logics between **S4** and **S5** that come up in passing in the discussion (such as **S4Sch** on p. 418) and several further such logics – in particular **S4.04** and **S4B**$_2$ – will receive brief attention in Section 5.6.

A useful general discussion of modal logics between **S4** and **S5**, though with some idiosyncratic terminology, can be found in Chapter 15 of Zeman [1224];[338] Aucher [41] also includes a survey of the epistemically significant logics in this range (essentially: those discussed by Lenzen), as well as of various mixed epistemic–doxastic principles. There is also a characteristically comprehensive and informative

---

[338] Zeman's use of the terminology of model structures results in some garbled passages in [1224]. The usual meaning of 'model structure' (Kripke's terminology) is: pointed frame – i.e., frame with a distinguished element. Zeman, p. 130f., instead says "The total set of models constructible for worlds with an accessibility relation having a given set of properties is called a 'model structure'." Hm... This seems to mean that a model structure is to be any collection of models comprising all those on a given class of frames. Zeman p. 138: "As we have noted, we may specify a model structure by stating the properties its accessibility relation is to have. Suppose we consider first of all the model structure whose accessibility relation has no specified properties; call this structure MT⁰. We shall develop a correlation between MT⁰ and the systems T⁰." (T⁰ is Zeman's name for **K**.) The "its" in Zeman's phrase "the properties its accessibility relation is to have" is rather confusing, since on the official definition we are not dealing with a single model or even with the class of all models on a single frame, so there is no such thing as 'its' accessibility relation if the 'it' refers back to what Zeman calls a model structure. By the time we get to "the model structure whose accessibility relation has no specified properties" we seem to be grappling with the equivalent of the generic triangle one would like to think Berkeley had disposed of some centuries ago. (Perhaps not all of us: Fine [293] insists all is well with such things; and then there is also the 'nonexistent objects' brigade – see the discussion of incomplete objects, and the references to Richard Routley and Graham Priest, in Reicher [945].)

5.5. LOGICS BETWEEN S4 AND S5

treatment in pp. 96–167 of Segerberg [1025], which includes mention of numerous normal extensions of **S4** which are not included in **S5**. An excellent discussion of some of the systems in the **S4**–**S5** interval with epistemic applications in mind, paying special attention to their connection with traditional epistemological themes – as well as considerations from game theory and belief revision – is provided by Stalnaker [1082].[339]

REMARK 5.5.1 And what of logics not included in **S5** (whether or not they extend **S4**)? Although some **S5**-unprovable formulas will be mentioned later – **Alt**$_2$ and **M** at 417, for example – they do not appear there because of any specifically epistemic or doxastic appeal. One such formula, **B**$_c$, is considered as a candidate axiom for epistemic logic in Alexander [9] and [10], so let us write it here in the form $p \to PKp$. The similarity in form to the principle of knowability $p \to \Diamond Kp$ is not a coincidence, since Alexander is essaying a variation on the theme of the Fitch derivation of $p \to Kp$ from this principle (see Example 5.4.11, p. 397). Many aspects of the discussion in [9] and [10] call out for comment, but here our concern is only with $p \to PKp$, or the schematic form $\varphi \to PK\varphi$, and its comparative implausibility by contrast with the (at least initial) appeal of the bimodal form with $\Diamond$ for P. Take something our knower, Paul, say, has no opinion at all about, which has come up for discussion, for example whether the number of 18-year-olds in Martinique (at some specified time) is even, and think of p as the statement that this number is indeed even, and further, let us suppose that p is true. Paul does not believe that p (or that not-p), and Paul knows that he does not believe that p, and so *a fortiori* knows that he does not know that p: this perfectly everyday situation, in which what is true is $p \land K\neg Kp$, contradicting the schema $\varphi \to PK\varphi$. (If p is false, take φ as $\neg p$ to the same effect. The "*a fortiori*" here presumes, incidentally, that one is not persuaded by Halpern [427], discussed in Section 5.4.) The reasoning behind such counterexamples is surprisingly similar to that showing that $\mathbf{5}_K$ and $\mathbf{B}_K$ are implausible epistemic principles. (See Remark 5.3.4, p. 384, in the latter case, and much of Section 5.2, beginning with p. 371, in the former.) Perhaps in the case of **B**, this is not such a surprise, since **KB**$_c$ has **B** as a theorem, not that Alexander is in these papers committed to working among the normal modal logics, and in [10] is keen on an avowedly non-normal epistemic logic containing $Kp \to p$ but not $K(Kp \to p)$ – (**U**$_K$ as we might naturally call the latter. However, having raised the matter of normal modal logics in this connection, we should note that concerning **Alt**$_2$ and **M**, mentioned above, both are theorems of **KB**$_c$ (alias **KB!**).. In the case of **Alt**$_2$, we have the even stronger **Alt**$_1$ (= **D!**). None of this, of course, would be wanted for even doxastic applications, and in the epistemic case the additional presence of **T**$_K$ would leave us with the 'trivial' modal logic **KT!**. ◄

Although we begin by thinking about knowledge, we have not seen the last of belief. Let us recall a theme from Halpern, whose [427], p. 489, was already quoted in the preceding section (p. 390) as presenting a semantics to embody "the idea that the agent believes that his knowledge coincides with his beliefs", a formulation, it was suggested, is only approximate for the intuition involved. Schwarz and

---

[339]For example, Stalnaker discerns a connection between **S4.2** and what is called *internalism* in epistemology ([1082], p. 180) – i.e., the view that what is required for knowledge over and above true belief must consist of considerations available to the believer, such a justification for the belief that could be produced on demand, or at least reconstructed after reflection. (Different positions go under the name of internalism in ethics – see Example 5.1.15, p. 354 – and again in the philosophy of mind – the latter notion arising for propositional attitudes generally and so also including knowledge as a special case, though still not to be confused with epistemological internalism as just characterized. This is sometimes internalism about *content*.) Stalnaker similarly connects **S4.3** and *defeasibility* analyses of knowledge (p. 190) – before going on to point out a difficulty for such analyses (which, roughly speaking, require the non-existence of evidence which, if brought to the subject's attention, would undermine or dislodge the belief whose status as something known is at issue). Such connections are of necessity somewhat informal and speculative, and we do not go into them further here. Discussions directly relating epistemic logic to issues of more general epistemological interest to the extent that Stalnaker [1082] does are quite rare; other examples include Holliday [513] and [515] on the 'Relevant Alternatives' account of knowledge and on the contextualism of Lewis [730], and Williamson [1197] on Gettier examples (counterexamples to the account of knowledge as justified, true belief: see Example 5.1.10). Williamson [1194] should also be mentioned, and the considerably earlier Lenzen [706], though unfortunately this paper is in German and much of the material it contains has not been translated or paraphrased in publications in English, by Lenzen or others.

Truszczyński [1018] make a semantic proposal which is similarly inspired by the "intuition that the agent believes that it knows everything it believes"[340] – a formulation which is, again, wildly ambiguous.[341] The semantic proposal is best explained by defining the concept of a *cluster*, which figures extensively in Segerberg [1025] (or see Bull and Segerberg [128]) – though our definition is less general than that of [1025], since we are only considering transitive reflexive frames while Segerberg had arbitrary transitive frames in mind. Given a transitive reflexive frame $\langle W, R \rangle$, we say that a non-empty $X \subseteq W$ is a *cluster* if it is $\subseteq$-maximal w.r.t. the property of being closed under the relation $\sim_R$ defined by:

$$x \sim_R y \text{ if and only if } xRy \text{ and } yRx.$$

Note that when (as presumed here) R is reflexive and transitive, $\sim_R$ is an equivalence relation on W and the clusters of $\langle W, R \rangle$ are the $\sim_R$-equivalence classes. We say that one cluster, Y, is *accessible to* another X, if for every $x \in X$, $y \in Y$, we have xRy. Later, we shall also find it convenient to speak of a cluster Y, say, being accessible to a point x when xRy for all $y \in Y$. Since every point (in the frames of current interest) lies in exactly one cluster, this amounts to saying that Y is accessible to the unique cluster containing x.

EXERCISE 5.5.2 Show that if some $x \in X$ bears the relation R to some $y \in Y$, then every $x' \in X$ bears R to every $y' \in Y$; i.e., Y is accessible to X in the sense just defined. ✠

This brings us to the class of frames of interest. Say that $\langle W, R \rangle$ is a *two-cluster frame* when $W = X \cup Y$ for clusters X, Y, with Y accessible to X. Since this includes the possibility that $X = Y$ (in which case $R = W \times W$), we are subsuming the case of 'one-cluster' frames under the rubric *two-cluster frames* as what we may call a degenerate case.[342] A more explicit description than "two-cluster frames" would be: frames consisting either of a single cluster or of two clusters, one of which is accessible to the other. Note that in a two-cluster frame with clusters X, Y, the degenerate case in which $X = Y$ is the only case in which $X \cap Y \neq \varnothing$. (Why?) In the non-degenerate cases, we call X and Y (accessible to X) the first and second cluster, respectively.

Now just why might such frames be of special interest? To see this, let us recall the start of Section 5.4, in which, following Halpern [427], we looked initially at the "most simple-minded combined epistemic–doxastic logic one could imagine," with a variation on the semi-simplified Kripke models idea which had $\mathcal{M} = \langle W, W', V \rangle$ with $\varnothing \neq W' \subseteq W$, and in the truth-definition, the following treatment of K and B. For any $x \in W$:

$$\mathcal{M} \models_x K\varphi \text{ iff for all } y \in W, \mathcal{M} \models_y \varphi; \quad \mathcal{M} \models_x B\varphi \text{ iff for all } y \in W', \mathcal{M} \models_y \varphi.$$

In terms of non-degenerate two-cluster frames, the first cluster (X above) is $W \smallsetminus W'$ and the second cluster (Y) is W'. (In the degenerate case, $W = W' = X = Y$.) By contrast with the semi-simplified models $\langle W, W_0, V \rangle$ under discussion in Section 4.1, after Exercise 4.1.4 (p. 207), where we found we could discard all but one element of $W \smallsetminus W_0$ without altering the logic,[343] here we certainly can't discard all but one element of the first cluster, since all points in it are mutually accessible and must survive into the subframe generated by any one of them.

---

[340][1018], p. 127 As with van der Hoek [509], discussed in Section 5.3 above, the authors are more concerned in this paper, as well as in their other publications on this material ([1016], [1017]) with issues arising in nonmonotonic logic (defaults, 'minimizing' knowledge, etc.) than with just getting an appropriate (doxastic–)epistemic logic, though this last is the only aspect of their work in which an interest is taken here.

[341]Using propositional quantifiers, we can distinguish three possible readings as $\forall p(Bp \to BKp)$, $B\forall p(Bp \to Kp)$ and $\forall pB(Bp \to Kp)$.

[342]Segerberg himself does not require reflexivity and calls the set consisting of a single irreflexive point a *degenerate cluster*. As we are (for the moment, at least) assuming reflexivity as well as transitivity in this discussion, there will be no ambiguity in the word *degenerate* to worry about.

[343]**K45** or **KD45**, depending on whether W_0 was allowed to be empty or not (Coro. 4.1.5).

5.5. LOGICS BETWEEN S4 AND S5

The treatment of K and B above means that the same K-formulas (and also the same B-formulas) were true at all points in any model, and it cost us the validity of the P(ositive) C(ertainty) principle $Bp \to BKp$, which is an embodiment of the "intuition that the agent believes that it knows everything it believes", to re-quote Schwartz and Truszczyński. The trouble is that in evaluating a K-formula from the perspective of any world in the model, we are always directed to the same set of K-pertinent worlds (namely the whole of W), whereas, to reflect the intuition in question, we should, when evaluating such a formula at a point in W', the set of B-pertinent worlds, only look at B-pertinent worlds. Recall from Section 5.4 that Halpern [427] achieved something like this effect, also validating PC, by a stipulation which amounted to saying that once we are evaluating subformulas of a given formula within the scope of an occurrence of B, we have again identified the B-pertinent and the K-pertinent (sets of) worlds, which had as a side-effect that the set of valid formulas was not closed under Uniform Substitution – all for the sake of saving $\mathbf{5}_K$. The present suggestion invalidates that principle (an advantage rather than a disadvantage, according to the tenor of the discussion in this chapter), since the R of the two-cluster frames is not in general euclidean, not being symmetric.[344] Not "in general" but only in the degenerate case of two-cluster frames in which there is after all only one cluster. We have not supplied an accessibility relation for B, and will return to B in due course, after investigating just the logic for K which emerges from these considerations.

To isolate the epistemic logic concerned, we need a characterization of the two-cluster frames (themselves defined in second-order terms, by quantification over subsets of the universe of the frame) in terms of first-order conditions on the accessibility relation. Evidently we are between **S4** and **S5**, and strictly weaker than the latter in view of the failure of the euclidean condition. The *semi-euclidean* condition, as it is called in Voorbraak [1136], turns out to be what is needed instead – well, not exactly needed, since we shall be encountering numerous alternative conditions which are not equivalent on arbitrary frames but are on transitive reflexive frames. This condition is depicted on the right of Figure 2.5 (p. 67); the condition itself, with all variables understood as universally quantified, reads as follows:

If xRy and xRz, then either yRz or zRx.

PROPOSITION 5.5.3 *A transitive reflexive frame $\langle W, R \rangle$ is a two-cluster frame if and only if $\langle W, R \rangle$ is a point-generated and semi-euclidean frame.*

Proof. 'Only if': Suppose $\langle W, R \rangle$ is a (transitive, reflexive) two-cluster frame. In the degenerate case with $R = W \times W$ the frame is generated by any $w \in W$ and the frame is evidently transitive, reflexive and semi-euclidean (euclidean, indeed). For the non-degenerate case, suppose X and Y are the first and second clusters. Then the frame is generated by any $w \in X$. To see that R satisfies the semi-euclidean condition, suppose otherwise. Thus we have $x, y, z \in W$ with xRy, xRz, and neither yRz nor zRx. Since not yRz, y must lie in the second cluster and z in the first (as the second cluster is accessible to the first). But since also not zRx, z must lie in the second cluster (and x in the first, though we do not need this): a contradiction, since the clusters are disjoint.

'If': Suppose $\langle W, R \rangle$ is a transitive, reflexive, semi-euclidean frame generated by the point x. Thus $R(x) = W$. If R is $W \times W$ then we have a degenerate two-cluster frame. If $R \neq W \times W$ then there must be at least one point $y \in W$ such that not yRx, since $R(x) = W$, so if yRx, then by transitivity, $R(y) = W$ also. Take $R^{-1}(x)$ as the first cluster, and $W \smallsetminus R^{-1}(x)$ as the second cluster. All points in the first cluster bear R to themselves and each other, as required (exploiting transitivity and reflexivity) and to all points in the second cluster; but we need to check that all points in the second cluster bear the relation R to each other. Let y_1 and y_2 be a pair of such points. We have xRy_1 and xRy_2, so by the semi-euclidean condition either y_1Ry_2, which is what we want, or else y_2Rx. Since the second cluster was defined to comprise the points not bearing R to x, we have what we want, y_2 being such a point. ∎

---

[344] Recall that a transitive reflexive frame is symmetric if and only if it is euclidean.

Thus we have two semantic characterizations of the same logic: as that determined by the class of two-cluster frames, and as that determined by the class of transitive, reflexive, semi-euclidean frames. Appropriate appeal to the Generation Theorem (Thm. 2.4.10) and Proposition 5.5.3 assures us that one and the same logic – which we shall see below is in fact **S4F** – is determined by these two classes of frames, the former comprising just the point-generated subframes of frames in the latter. Before presenting the relevant completeness results, we investigate another aspect of the semantic situation for its considerable intrinsic interest as well as its bearing on alternatives to **F** which give the same logic as a normal extension of **S4**.

Let us recall the binary relation on a set W defined, for subsets X, Y, of W and called $X + Y$ at p. 15 (where we wrote "U" rather than "W"):

$$X + Y = \{\langle x, y \rangle \mid x \in X \text{ or } y \in Y\}$$

We called a binary relation on W ∨-representable when some $X, Y \subseteq W$ could be found with $R = X + Y$ in this sense. As noted in Proposition 1.2.12 (p. 15), there is a first-order characterization of the ∨-representable binary relations on a set, namely as those satisfying the following (quantifiers taken to range over the set in question), which we label with an abbreviation of "∨-representable":

(∨-rep) $\qquad \forall x \forall y \forall u \forall z (Rxz \rightarrow (Rxy \vee Ruz))$.

The reason all this is of interest now is that two-cluster frames $\langle W, R \rangle$ are all of them ∨-representable (i.e., R, considered as a binary relation on W, has this property). If the first cluster is X and the second is Y, then R is precisely $X + Y$.[345] For considering its connection with modal formulas on transitive reflexive frames via the Generation Theorem (Thm. 2.4.10), the diagram on the left of Figure 5.2 depicts the piecewise version – see p. 76 – of the first-order condition (∨-rep), rather than (∨-rep) itself. (To obtain the latter, delete the topmost point and all arrows extending from it.) More accurately, that condition would be depicted by a similar diagram with an addition arrow extending from the topmost point to the point on the right at the bottom, but since we are considering only transitive frames and already have a sequence of two arrows leading to that point, we can omit it without loss in the present context. And since we are also considering only reflexive frames, we can make a further simplification and identify the two points at the bottom, leading to the diagram on the right of Figure 5.2, depicting what we call Stalnaker's condition, namely the following in which quantifiers range, as usual, over W for a frame $\langle W, R \rangle$:

$$\forall x \forall y (Rxy \rightarrow (\forall z(Rxz \rightarrow Ryz) \vee \forall z(Rxz \rightarrow Rzy))),$$

or, more informally: for any $x \in W$ and any $y \in R(x)$, we have $R(x) \subseteq R(y)$ or $R(x) \subseteq R^{-1}(y)$.[346]

The diagram on the right of Figure 5.2 represents a condition corresponding to $\Diamond(\Box\varphi \wedge \psi) \rightarrow (\Box\varphi \vee \Box\Diamond\psi)$, while that on left diagram for that corresponding to $\Diamond(\Box\varphi \wedge \Diamond\psi) \rightarrow (\Box\varphi \vee \Box\Diamond\psi)$; the two conditions are equivalent on reflexive transitive frames, and the corresponding schemata are equivalent in **S4**. The schemata may appear in somewhat different guises. For example, instead of the first, we may see $\Diamond\Box\psi \rightarrow (\Box\varphi \vee \Box(\Box\varphi \rightarrow \Box\psi))$ or indeed $(\Diamond\Box\psi \rightarrow \Box\psi) \vee \Box(\Box\varphi \rightarrow \Box\psi)$, each of which is **K**-interdeducible with the first schema. Several alternative formulations along these lines appear in Figures 5.3 and 5.4 below.

EXERCISE 5.5.4 *(i)* Show that any reflexive frame satisfying Stalnaker's condition is semi-euclidean.

(ii) Show that any transitive semi-euclidean frame satisfies Stalnaker's condition.

---

[345] This is explicitly how the accessibility relation for two-cluster frames is defined in Schwarz and Truszczyński [1018]: Definition 3.4, p. 130.

[346] The appearance of the condition in question – as '(F)' on p. 194 of [1082] is marred by typographical errors, but the condition given above – so Stalnaker informs me – is what was intended.

5.5. LOGICS BETWEEN S4 AND S5

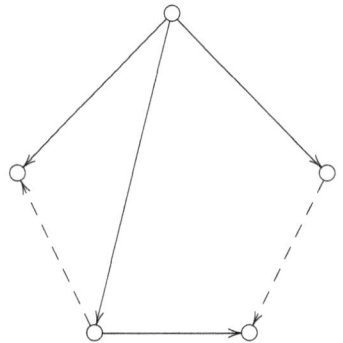
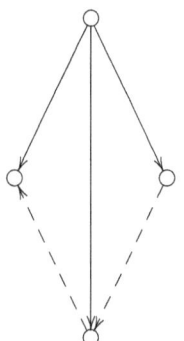

Figure 5.2: Piecewise (∨-Rep) and Stalnaker's Condition

Thus on transitive and reflexive frames, these two conditions coincide, and their point-generated subframes are precisely the two-cluster frames. Transitivity and reflexivity also bear on another aspect of the situation: ∨-representability. In a two-cluster frame with universe W and first and second clusters X and Y we don't just have $R = X + Y$, however. We have the additional information that either X and Y are complements relative to W or else $X = Y = W$ (the degenerate case). Where does this fit into the present picture?

Well, in Proposition 1.2.14 (p. 16) and Exercise 1.2.15 respectively, we saw that for $R \subseteq W \times W$ with $R = X + Y$, (i) R is transitive if and only if $X = W$ or $Y = W$ or $X \cap Y = \emptyset$, and (ii) R is reflexive if and only if $X \cup Y = W$.[347] And this supplies the additional information alluded to (given that we are considering reflexive transitive frames).

EXERCISE 5.5.5 Prove or refute (with a counterexample) the following claim:

$\langle W, R \rangle$ is a two-cluster frame if and only if R is transitive and satisfies (∨-rep).

✠

In a non-degenerate two-cluster frame the two clusters, X and Y, are each other's complements (relative to the universe of the frame), so instead of thinking of xRy as holding just in case $x \in X$ or $y \in Y$, we can reformulate the first disjunct here as $x \notin Y$. This motivates the following exercise.

EXERCISE 5.5.6 Prove or refute the following claim: conditions (i) and (ii) below are equivalent.
(i) $\langle W, R \rangle$ satisfies for some $Y \subseteq W$ the condition:

$$\forall x \forall y (Rxy \leftrightarrow (x \in Y \rightarrow y \in Y));$$

(ii) R is transitive and satisfies: $\forall x \forall y \forall z (Rxy \vee Ryz)$.

✠

Turning now to matters of axiomatics, we have a syntactic description of the logic concerned – that determined by the class of two-cluster frames – as **S4F**, as we shall see in Coro. 5.5.10. This logic lies nicely within the Lenzen-approved range **S4.2**–**S4.4**, and it was indeed mentioned already by Lenzen in the papers cited at the started of this section, as well as in Voorbraak [1136], from which we have taken the term "semi-euclidean" (though see the Digression below). What, then, is the **F** which added to **S4** to produces the normal modal logic **S4F**?

On p. 66 above the schema **F** was given as $\Box(\Box\varphi \rightarrow \psi) \vee (\Diamond\Box\psi \rightarrow \varphi)$, though actually we used "A", "B", as schematic letters there, whereas here we are sticking with "φ" and "ψ" for the reason already

---

[347]In Chapter 1, we used "U" for the underlying set, rather than "W".

given: we may want to write the schema in the form $K(K\varphi \to \psi) \vee (PK\psi \to \varphi)$ to emphasize is status as a candidate epistemic principle (in which case the adjacent capital roman letters for operators and formulas would be messy). However, for the moment let us stick with the $\Box/\Diamond$ formulation. This is the form of the schema as it appeared in Zeman [1221] and Lenzen [706] – though they refer to **S4F** by another name (see note 26), and as it is discussed in Segerberg [1025], p. 121f. But in some more recent literature reviving interest in **S4F** as an epistemic logic (Schwarz and Truszczyński [1018], Stalnaker [1082]) **F** is given in different forms; in the case of Schwarz and Truszczyński as (1) below, and by Stalnaker as:

$$(\Diamond\varphi \wedge \Diamond\Box\psi) \to \Box(\Diamond\varphi \vee \psi),$$

which certainly looks rather different on the face of it, and indeed $\mathbf{K} \oplus \mathbf{F}$ would be a different logic depending on which of these two schemata we took as **F**. In fact, we stick with **F** as the original schema, so let us call that figuring under this name in the papers by Schwarz and Truszczyński and by Stalnaker **F′** to distinguish it. However, as candidate axiom schemata for extending **S4** to get a new logic **S4F**, we would get the same result in either case: that is, $\mathbf{S4} \oplus \mathbf{F} = \mathbf{S4} \oplus \mathbf{F'}$ – or, more concisely, $\mathbf{S4F} = \mathbf{S4F'}$. To see this most easily let us compare a representative instance of **F′** (which we will call by the same name), with another similar-looking formula (from [1018]), here numbered (1):

F′ $(\Diamond p \wedge \Diamond\Box q) \to \Box(\Diamond p \vee q)$

(1) $(p \wedge \Diamond\Box q) \to \Box(\Diamond p \vee q)$.

We return to **F** itself presently. To see that **F′** and (1) are interdeducible over **S4**, note that (1) is a truth-functional consequence of **F′** and the **T** axiom (in dual form) $p \to \Diamond p$. In the other direction, we may substitute $\Diamond p$ (uniformly) for p in (1), which gives us something like **F′** except that we have a $\Diamond\Diamond p$ in the consequent where we want just $\Diamond p$, but using **4** – again in the dual form $\Diamond\Diamond p \to \Diamond p$ – it should not be hard to see how to derive **F′** itself from this.

What, then, of **F** itself? In fact (1) is just a minor variant on (a representative instance of) that schema. To see this, substitute $\neg p$ for p in (1), getting:

(2) $(\neg p \wedge \Diamond\Box q) \to \Box(\neg\Box p \vee q)$,

though here we have taken the liberty of replacing $\Diamond\neg$ by $\neg\Box$ in the consequent at the same time as making this substitution. A little further re-writing gives the following:

(3) $(\Diamond\Box q \wedge \neg p) \to \Box(\Box p \to q)$.

Further, it is not hard to see that the steps (1)–(2)–(3) are all reversible; finally note that (3) is truth-functionally equivalent to a representative instance of **F** as originally given, namely the following instance:

$$\Box(\Box p \to q) \vee (\Diamond\Box q \to p).$$

Thus we have established:

PROPOSITION 5.5.7   **S4F** = **S4F′**.

EXERCISE 5.5.8 In this exercise we consider **F** in isolation from the **T** and **4** which accompany it in **S4F**, taking **F**, as above, in the simple form:

$$\Box(\Box\varphi \to \psi) \vee (\Diamond\Box\psi \to \varphi).$$

(*i*) Show that a representative instance of this schema (distinct propositional variables replacing distinct schematic letters, that is) is valid on all and only frames which are semi-euclidean.

(*ii*) Show that the canonical frame for any consistent normal extension of **KF** is semi-euclidean.

5.5. LOGICS BETWEEN S4 AND S5

From half of part (i) together with part (ii) of this Exercise concerning **F**, together with corresponding observations for **T** and **4** we conclude:[348]

PROPOSITION 5.5.9 *S4F is determined by the class of transitive, reflexive, semi-euclidean frames.*

Thus by Proposition 5.5.3, and help (as already indicated) from Theorem 2.4.10 we have, further:

COROLLARY 5.5.10 *S4F is determined by the class of two-cluster frames.*

Since several similar-looking formulas have been in play in this discussion, it may help to reduce confusion by charting their resemblances and differences in a tabular form: see Figure 5.3; **F** as described above actually makes its appearance in a second table (Fig. 5.4) giving disjunctive formulations of these various candidate axioms. The first formula appearing under '**F**' in Fig. 5.3 is what was referred to as (1) in the discussion leading up to (and establishing) Proposition 5.5.7.

Label	Axiom	Dual Form
F	$(p \wedge \Diamond \Box q) \to \Box(\Diamond p \vee q)$	$\Diamond(\Box p \wedge q) \to (p \vee \Box \Diamond q)$
F'	$(\Diamond p \wedge \Diamond \Box q) \to \Box(\Diamond p \vee q)$	$\Diamond(\Box p \wedge q) \to (\Box p \vee \Box \Diamond q)$
F''	$(\Diamond p \wedge \Diamond \Box q) \to \Box(\Diamond p \vee \Box q)$	$\Diamond(\Box p \wedge \Diamond q) \to (\Box p \vee \Box \Diamond q)$

Figure 5.3: **F** *and Friends – Implicational Formulations*

Figure 5.3 tabulates some **F**-related candidate axioms. The formulas on any given line of the table are **K**-interdeducible (as in Table 2.1 from p. 33, where it was schemata rather than formulas that were listed). On the first lines we have versions of **F** itself, and on the second, versions of **F'**; this ad hoc labelling has been extended to dubbing the formulas on the third line **F''**. Disjunctive formulations are given in Figure 5.4, again set out to make the contrasts readily visible. The right-hand column gives a form in which the two disjuncts have been re-written in dualized form in accordance with the description of dualizing implicational formulas given in Chapter 2 (p. 35); then the propositional variables have been interchanged so as to appear in alphabetical order in the final result. Of course there are many other disjunctive forms that could have been provided instead; for example we could rewrite all the formulas $\varphi \to (\psi \vee \chi)$ in the final column of Figure 5.3 as $(\varphi \to \psi) \vee \chi$ – or, for that matter, just as $\neg \varphi \vee (\psi \vee \chi)$ – but what are given in Fig. 5.4 are the more commonly encountered disjunctive formulations.

Label	Disjunctive Axiom	Internally Dualized Form
F	$\Box(\Box p \to q) \vee (\Diamond \Box q \to p)$	$\Box(p \to \Diamond q) \vee (q \to \Box \Diamond q)$
F'	$\Box(\Box p \to q) \vee (\Diamond \Box q \to \Box p)$	$\Box(p \to \Diamond q) \vee (\Diamond q \to \Box \Diamond q)$
F''	$\Box(\Box p \to \Box q) \vee (\Diamond \Box q \to \Box p)$	$\Box(\Diamond p \to \Diamond q) \vee (\Diamond q \to \Box \Diamond q)$

Figure 5.4: **F** *and Friends – Disjunctive Formulations*

---

[348] As in Segerberg [1025], Lemma 7.10 on p. 161, though the formulation there appears to be defective, as mentioned in note 50 on p. 67 above. The semi-euclidean condition as Segerberg cites it is (the universal closure of): If xRz & *not* zRx, then $xRy \Rightarrow yRz$; this interesting re-arrangement of the parts of the version given above is clearly equivalent to that version. Coro. 5.5.10 below is Theorem 7.8 of [1025], though expressed there using Segerberg's notion of an index frame, which does not appear in our exposition.

EXERCISE 5.5.11 Show that **S4F** = **S4F′** = **S4F″**. (Note that in view of Prop. 5.5.7, it suffices to show either of **S4F** = **S4F″**, **S4F′** = **S4F″**.)

Over **K**, in place of **S4**, our axioms have rather different effects, mostly implicit if not explicit in the discussion above:

THEOREM 5.5.12 **KF**, **KF′** and **KF″** *are respectively determined by the class of frames satisfying the semi-euclidean condition, Stalnaker's condition, and the piecewise version of the (∨-rep) condition.*

Digression. Returning to the issue of additions to **S4**, as in Exc. 5.5.11, there is one candidate axiom(-schema) whose status remains less clear (to the author): namely that called **5′** in the paper from which we have taken the term *semi-euclidean*, Voorbraak [1136] (p. 510):

$$((\varphi \land \neg\Box\varphi) \land \Box(\varphi \lor \Box(\varphi \to \Box\varphi))) \to \Box\neg\Box\varphi.$$

Rewriting this by some truth-functional manipulations as

$$(\varphi \land \Box(\varphi \lor \Box(\varphi \to \Box\varphi))) \to (\neg\Box\varphi \to \Box\neg\Box\varphi),$$

makes it clear that **S4** extended by Voorbraak's schema is a sublogic of **S5**, since the consequent is now a variant on **5** itself. But do we have **S4F** = **S45′** (or perhaps even **KF** = **K5′**)? At the point in Voorbraak's discussion ([1136], p. 510) where we expect from the language used that a completeness proof for the latter system is on offer – soundness w.r.t. to the class of transitive, reflexive, semi-euclidean frames being clear enough – the discussion (Lemma 6.4) turns instead to trying to show that on any transitive frame which is not semi-euclidean a model can be constructed falsifying an instance of **5′**. This attempt itself is not successful as it stands, being marred by typographical errors in the proof of Lemma 6.4, in the description of the desired model. The discussion turns, that is, from completeness to modal definability,[349] and the ultimate status of Voorbraak's **5′** from the former perspective does not appear to have been settled. The present author was not successful in trying to derive **F** from **S4** ⊕ **5′**. ***End of Digression.***

A word was promised on the question of an accessibility relation for B as opposed to K (lately written as □) in the logics we have been considering. Taking our cue from Lenzen – Theorem 5.2.1 (p. 372) – we think of B as PK. One way of identifying an accessibility relation for B on this basis is mentioned in Stalnaker [1082], p. 181 (slightly adapted to conform to our notation):

> The usual constraint on the accessibility relation that validates **S4.2** is the following convergence principle (added to the transitivity and reflexivity conditions): if xRy and xRz, then there is a w such that yRw and zRw.[350] But **S4.2** is also sound and complete w.r.t. the following stronger convergence principle: for all x, there is a y such that for all z, if xRz then zRy. The weak convergence principle (added to reflexivity and transitivity) implies that for any *finite* set of worlds accessible to x, there is a single world accessible with respect to all of them. The strong convergence principle implies that there is a world that is accessible to all worlds that are accessible to x. The semantics for our logic of knowledge requires the stronger convergence principle.[351]
>
> Just as, within the logic, one can define belief in terms of knowledge,[352] so within the semantics one can define a doxastic accessibility relation for the derived belief operator in terms of the epistemic

---

[349]The wording of Lemma 6.4 of [1136] says that (class of) transitive, semi-euclidean frames is "characterized by" the set {**4**, **5′**}, the term "characterized by" usually being another word for "determined by", though here the direction is the wrong way round and the intended meaning seems to be "modally defined by". (Perhaps Voorbraak is tacitly appealing to a well-known corollary of Bull's Theorem on **S4.3** to the effect that all of its extensions have the finite model property, the corollary being that they are all complete.)

[350]This is piecewise convergence, in the terminology of Exercise 2.5.8(*iii*) from p. 74 above.

[351]Stalnaker appends a note at this point to the effect that while the two classes of frames here determine the same propositional modal logic, they determine different quantified modal logics.

[352]For more in this vein, see Examples 5.6.9 below, p. 429.

5.5. LOGICS BETWEEN S4 AND S5

accessibility relation. If 'R' denotes the epistemic accessibility relation and 'D' denotes the doxastic relation, then the definition is as follows: $xDy =_{df} \forall z(xRz \to zRy)$.[353] Assuming that R is transitive, reflexive and strongly convergent, it can be shown that D will be serial, transitive and euclidean – the constraints on the accessibility relation that characterize the logic **KD45**.

Defining the relation D here can be regarded as a uncovering the hidden accessibility relation which explains why the modality $\Diamond\Box$ (or PK) is normal in **S4.2**. See the discussion in pp. 280–281 in Section 4.5, where, concentrating on a sublogic, **KDH**, of **S4.3**, we went through a similar manoeuvre to explain – or at least provide a semantic gloss on – the normality of this same modality. (**KDH** is the weakest modal logic in which not only $\Box$ but also the composite operator $\Diamond\Box$ is normal.) The condition on frames mentioned under (ii) of Proposition 4.5.4 is analogous to Stalnaker's strong convergence; the accessibility relation S corresponds to Stalnaker's defined D above.[354] Rather than rework such details here, we will give a simpler route to the extraction of an accessibility relation for $PK/\Diamond\Box$ (alias B) after the following Digression.

Digression. The lengthy quotation from Stalnaker above is followed immediately in [1082] by the following:

> One can also define, in terms of D, and so in terms of R, a third binary relation on possible worlds that is relevant to describing the epistemic situation of our ideal knower: Say that two possible worlds x and y are *epistemically indistinguishable* to an agent (xEy) if and only if she has exactly the same beliefs in world x as she has in world y. That is, $xEy =_{df} \forall z(xDz \leftrightarrow yDz)$. E is obviously an equivalence relation, and so any modal operator interpreted in the usual way in terms of E would be an **S5** operator. But while this relation is definable in the semantics in terms of the epistemic accessibility relation, we cannot define, in the object language with just the knowledge operator, a modal operator whose semantics is given by this accessibility relation.

Whether or not it is definable in terms of the Boolean connectives and the other modal primitive(s), the envisaged operator concerned could be added to the language and the question of how to axiomatize the logic determined by this or that class of frames in this enriched language naturally arises. Some remarks on that question follow here, with reference to the class of all frames – or rather without imposing any conditions on the doxastic accessibility relation called D by Stalnaker. (We will of course be interested in the special condition relating this to Stalnaker's E – namely, that xEy iff $D(x) = D(y)$, though we will be writing this, instead, as: $xR_\Box y$ iff $R_B(x) = R_B(y)$.)

With the combined normal bimodal logic minimally extending **S5** for $\Box$, **K** for B, and containing all instances of the mixed schema $B\varphi \to \Box B\varphi$, we have a logic determined by the class of all frames $\langle W, R_B, R_\Box \rangle$ in which $R_\Box$ is an equivalence relation satisfying the condition that for all $x, y \in W$, $xR_\Box y$ implies $R_B(x) = R_B(y)$, as one can ascertain using the canonical model method. (Instead of $B\varphi \to \Box B\varphi$, we could equally well have taken the schema $\Diamond B\varphi \to B\varphi$.[355]) We revert to the notation R_B here in place of Stalnaker's D. It is not so obvious how to strengthen this logic so as to obtain one determined by the class of all frames of the above type satisfying instead:

$$xR_\Box y \Leftrightarrow R_B(x) = R_B(y).$$

For all one knows without further thought, the logic just described is already complete with respect to this narrower class of frames. But a little such further thought reveals that this is not so. Suppose that we had an additional operator $\dot{B}$ which we wanted to interpret semantically by the following unusual condition relative to models $\mathcal{M}$ on the frames $\langle W, R_B, R_\Box \rangle$ with which we are concerned:

---

[353]R and D here correspond to the R_K and R_B of our discussion in, for example, Section 5.4.
[354]We repeat a warning from Section 4.5: that the **H** of **KDH** – an aggregation principle for the modality $\Diamond\Box$ – has nothing to do with the **H** of Examples 2.5.38 and 2.5.39 (p. 87).
[355]In fact for any definable n-ary Boolean connective #, and any formulas $\varphi_1, \ldots, \varphi_n$, we have $\Diamond\#(B\varphi_1, \ldots, B\varphi_n) \to \#(B\varphi_1, \ldots, B\varphi_n)$ provable in the present logic.

$$\mathcal{M} \models_x \dot{B}\varphi \text{ iff for all } y \in W: \mathcal{M} \models_y \varphi \Leftrightarrow xR_By.$$

The unusual appearance of this condition notwithstanding (with its $\Leftrightarrow$ where one expects to see just $\Rightarrow$), such operators have been considered in the literature.[356] Thus $\dot{B}\varphi$ is true at a point when that point's R_B-successors comprise precisely the points verifying φ. It follows that if $\dot{B}\varphi$ and $\dot{B}\psi$ are both true at a point, then the set of points at which φ is true in the model coincides with the set of points at which ψ is true, since each is the set of R_B-alternatives to the given point. Thus where $\chi(p)$ is any formula – here regarded as a context into which (replacing p) we shall splice other formulas – we will be unable to falsify the formula

$$(\dot{B}\varphi \wedge \dot{B}\psi) \to (\chi(\varphi) \leftrightarrow \chi(\psi)),$$

and in fact we can say more. Even all formulas of the following form, in which W and S – mnemonic for "weak" and "strong" – are any sequences of occurrences of C and $\Diamond$ in the first case and of B and $\Box$ in the second, are valid:

$$\mathsf{W}(\dot{B}\varphi \wedge \dot{B}\psi) \to \mathsf{S}(\chi(\varphi) \leftrightarrow \chi(\psi)).$$

Of greater current relevance is the fact that, with a similar understanding of W and S, all instances of the following schema are valid:

$$\mathsf{W}(\dot{B}\varphi \wedge \Box\psi) \to \mathsf{S}(\dot{B}\varphi \to \Box\psi),$$

for the following reason: if we can get from a given point by m steps of the relations R_B and $R_\Box$ in some combination, supposing the prefix W is of length m, to a point x with $R(x)$ being $\|\varphi\|$ – to revive, e.g. from Prop. 2.2.6, that notation for the set of φ-verifying worlds (in a given, though unspecified, model) – then for any point y n steps of those relations away (taking n as the length of a candidate S), if $R(y)$ is also equal to $\|\varphi\|$ then $R(x) = R(y)$, which prevents $\Box\psi$ having different truth-values as x and y, which would have been necessary to falsify the formula at the original point.

But so what? We are not working with an operator like $\dot{B}$ here, only with B itself (and $\Box$). We can't say that the R_B-accessible worlds are precisely the φ-worlds, we can only say, by saying $B\varphi$ that the R_B-accessible worlds are *amongst* the φ-worlds. However, there is one choice of φ for which an arbitrary set's being a subset of $\|\varphi\|$ is equivalent to that set's being $\|\varphi\|$ itself, and that is when $\|\varphi\| = \varnothing$. In other words what we have just been understanding by $\dot{B}\varphi$ coincides with $B\varphi$ for the case of $\varphi = \bot$, and thus all instances of the following schema are valid on all frames $\langle W, R_B, R_\Box \rangle$ with $xR_\Box y$ iff $R_B(x) = R_B(y)$:

$$\mathsf{W}(B\bot \wedge \Box\psi) \to \mathsf{S}(B\bot \to \Box\psi).$$

EXERCISE 5.5.13 (*i*) Find an instance of the last schema which is not valid on all frames $\langle W, R_B, R_\Box \rangle$ with $R_\Box$ an equivalence relation satisfying the condition $xR_\Box y \Rightarrow R_B(x) = R_B(y)$. Such a formula shows that the logic determined by the class of frames $\langle W, R_B, R_\Box \rangle$ with $xR_\Box y$ iff $R_B(x) = R_B(y)$ is stronger than that determined by the class just mentioned, and thus that the bimodal logic for B and $\Box$ described syntactically above is not complete w.r.t. the latter class of frames.

(*ii*) By way of illustration of the point made in note 355, sketch a proof (from the axiomatization described) of the formula $\Diamond(Bp \wedge (\neg Bq \vee Br)) \to (Bp \wedge (\neg Bq \vee Br))$. ✠

We leave the question of axiomatizing the formulas valid on all frames $\langle W, R_B, R_\Box \rangle$ such that $xR_\Box y$ iff $R_B(x) = R_B(y)$ as an open problem. **End of Digression.**

The promised simpler way of thinking of $\Diamond\Box$ as having its own accessibility relation uses a characterization of **S4.2** provided by Segerberg ([1025], p. 77), whose proof involves a step for finitizing

---

[356] See Humberstone [556] – already mentioned in Example 4.4.46 above, as well as Levesque [714], and Chapters 8–11 of Levesque and Lakemeyer [715]. Some slightly less closely related references can be found in note 79 of Humberstone [582].

5.5. LOGICS BETWEEN S4 AND S5

countermodels to non-theorems by a technique ('filtration') we have not covered, thereby showing **S4.2** to have the finite model property (mentioned at p. 73). We offer, instead, a proof not involving any such finitization stage:

THEOREM 5.5.14 *S4.2 is determined by the class of transitive reflexive frames in which there is some cluster which is accessible to all points in the frame.*

Proof. This result is just a reformulation in terms of clusters of Prop. 4.5.8 (p. 282). One needs only to check that the X constructed in the proof there is a cluster. ∎

If $\langle W, R \rangle$ is a (transitive, reflexive) frame with a cluster Z, say, accessible to every $x \in W$ – which is of necessity unique, since one cannot have two distinct clusters accessible to each other – then we may interpret $\Diamond\Box$ as universally quantifying over Z:

EXERCISE 5.5.15 Show that given a model $\mathcal{M}$ on a frame as just described, and a point $x \in W$: $\mathcal{M} \models_x \Diamond\Box\varphi$ iff for all $z \in Z$, $\mathcal{M} \models_z \varphi$. ✠

Thus we can define an accessibility relation for B – call it D (for *doxastic*) to stick with Stalnaker's notation – by simply putting: xDy iff $y \in Z$ (where Z is as above). For the extensions of **S4.2** we have been concerned with, stronger conditions on the clusters provide a semantical characterization and we can retain this definition. For **S4.3** we can add to their being a cluster accessible to all clusters, that for any clusters X and Y, either X is accessible to Y or Y is accessible to X, while for **S4F**, as we have seen, we have as a determining class of frames those in which there is at most one other cluster other than the 'last' cluster Z (which we have been calling Y in discussing these two-cluster frames). The most extreme constraint that could be imposed on clusters in terms of their number would be to restrict it to one, which would yield **S5**. But we could constrain the size of the clusters rather than their number, and in particular we could consider the special two-cluster frames in which the first cluster either coincides with the second cluster (the degenerate case) or else contains at most one point; these are the point-generated frames for **S4.4** and are essentially the semi-simplified frames of Coro. 4.1.5 (p. 207) except with the generating point required to be reflexive here. By similar reasoning as in the case of the two-cluster frames in general and the transitive reflexive semi-euclidean frames (Prop. 5.5.3), one sees that these special two-cluster frames coincide, in the point-generated case, with the frames which are transitive, reflexive, and satisfy the *alio-euclidean* condition depicted on the right of Figure 2.10 (p. 96),

For all x, y, z, if xRy and xRz then either yRz or $x = y$,

which was the focus of Exercise 2.5.52(*i*), associating this condition with the schema .4 from Table 2.1 (p. 33). On the basis of that exercise, we conclude part (*i*) of the following result, part (*ii*) then following by considering generated submodels.

PROPOSITION 5.5.16 (*i*) *S4.4 is determined by the class of transitive, reflexive, alio-euclidean frames;* (*ii*) *S4.4 is determined by the class of two-cluster frames in which, if distinct from the second cluster, the first cluster contains at most one point.*

Digression. Note that there is no loss in generality in insisting, in a variant of Prop. 5.5.16(*ii*) that the two clusters are distinct so that the frames in question have the form $\langle W, R \rangle$ with $W = \{w\} \cup W_0$, $w \notin W$, and $R(w) = W$, $R(x) = W_0$ for $x \in W_0$, and we could accordingly package up the models in the style of the semi-simplified models of the discussion above (and from Section 4.1) but it is worth observing that none of this makes any essential play with distinguished points in 'pointed' frames (p. 190), or anything like the distinction between normal and non-normal worlds (p. 360). This is worth mentioning because Porte [897] writes (p. 209) as follows, beginning by noting that the use of models with a distinguished point originally favoured by Kripke (which were based on what he called 'model structures' – pointed frames, that is, as mentioned in note 338 – though what are called Kripke-style frames in the following passage from Porte are not these, but simply the familiar $\langle W, R \rangle$ frames) had gone out of favour, at least for work on normal modal logics:

> Then the distinguished real world all but disappeared from the literature. A notable exception was Zeman's semantics for Sobociński's system **S4.4**: see (...) Zeman [1222], [1224], (p. 256) – a "real world" is singled out and it accessible only from itself while it has access to all worlds. (...) And indeed **S4.4** has been characterized by a class of Kripke-style frames, in Georgacarakos [365], with an accessibility relation which is reflexive, transitive, "convergent" and "remotely accessible" (two new fairly complex conditions). Zeman's semantics (with distinguished real world) is noticeably simpler.

Remote symmetry was explained in Exercise 2.7.34(iii) (p. 137), and it is indeed unnecessarily complicated to conjoin this condition with convergence (or piecewise convergence), as when attention is restricted to transitive reflexive frames, the conjunction is equivalent to the single alio-euclidean condition (which moreover, is a purely universal condition, whereas convergence introduces an $\exists$). But there is no need to pass to pointed frames: the Zeman frames are just the point-generated subframes of those in play in Prop. 5.5.16(i) (or in Georgacarakos [365]). In fact the somewhat confused tone complained of here is already evident in the original discussions of Zeman and Georgacarakos themselves. Thus the latter quotes the former has that in his models, "(o)ne kind of world will have one and only one representative in any **S4.4** model (this is the 'real' world); this world has access to all worlds in the model including itself; the other kind of world may have any number of representatives in an **S4.4** model; this kind of world has access to all worlds in the model *except* for the real world." Georgacarakos continues on the same (opening) page of [365] with this "kinds of worlds" talk, promising a semantic description of **S4.4** "which does not distinguish between two kinds of worlds, but rather is characterized by the additional requirements it imposes on the accessibility relation." If the talk of two kinds of worlds just amounts to saying the universes of frames of interest are the unions of two disjoint sets of points (disjoint *clusters*, in the present case) as described, then this would be the case also for the frames for the logic **KEnd**, i.e., frames validating the formula $\Box\bot \lor \Diamond\Box\bot$ ("**End**"): there are the end-points, verifying the first disjunct in any model on the frame, and the points with an end-point as a successor, verifying the second disjunct in any model. But to say that the points call into these two classes *is* to impose a condition on the accessibility relation, and involves no suggestion that one of these classes comprises representatives of the real world (or the present moment, if we are thinking of this as the tense logic of ending time) while the other does not: validity on the frame quite safely be taken as truth at all points in all models on the frame. Similarly in the present case, validity can be taken as truth at all points in all models on the point-generated frames currently under consideration for **S4.4**. Some people have a general philosophical objection to using models without a distinguished element – see note 45, p. 60 – but they have no differential bearing on the current case. There remain commentworthy differences between the **KEnd** example and that of **S4.4**. All frames for the former logic are what is called in Humberstone [594] (p. 868) *strongly heterogeneous*, meaning that their universes are the (not necessarily disjoint) unions of pairs X, Y of sets of points with formulas A, B such that A is valid at all points in X but not at all points in Y while but not B is valid at all points in Y but not at all points in X. (As noted in Humberstone [594], this set of formulas valid on any strongly heterogeneous frame, such as the canonical frame for **KEnd**, is Halldén incomplete – as the logic **KEnd** evidently is.) The Zeman frames for **S4.4** with two clusters, one of which is a singleton, are not strongly heterogeneous since (by a consideration of p-morphisms) any formula valid at the would-be distinguished point – whose unit set is the first cluster – is valid at each point in the latter cluster (though not conversely, these points all being 1-symmetric and hence validating $p \to \Box\Diamond p$ (alias **B**).[357] This means that, by contrast with the **KEnd** case, the set of formulas valid on each frames coincides with the set of formulas valid at the generating point in all point-generated frames,

---

[357]In the terminology of [594] this is heterogeneity – or a failure of homogeneity (= all points in the frame validating the same formulas) – without being *strong* heterogeneity in the sense just explained. The term *homogeneous* has also been used

5.5. LOGICS BETWEEN S4 AND S5

thus *allowing* – but certainly not *requiring* – **S4.4** to be characterized semantically in terms of pointed frames of the kind apparently thought to be preferable in the earlier quotation from Porte. A second respect of difference between this case and the **KEnd** case, which again goes no way toward justifying the talk of 'two kinds of points' as though this binary classification had some theoretical significance, is that the passage to point-generated subframes plays no role in the **KEnd**, whereas in the **S4.4** case, it gives us a unique point having every point accessible to it. A more closely analogous situation in this respect is of the frames schematically indicated in Figure 4.3 (p. 288), in which the passage to generated subframes secures the uniqueness of the irreflexive point to which every other point is accessible. ***End of Digression.***

The following Exercise deriving from the above Digression is included for aficionados of Halldén completeness. It concerns in part homogeneous frames, as defined in note 357.

EXERCISE 5.5.17 A modal logic will be called *Halldén-normal* if it contains the axiom **K** and is closed under the rule (called 'Halldén Necessitation' in Humberstone [594], on p. 868f. of which this exercise is based):

$$\frac{\varphi \vee \psi}{\Box \varphi \vee \psi}$$

whenever φ and ψ have no propositional variables in common.

(*i*) Show that any Halldén-normal modal logic is normal.

(*ii*) Show that the set of formulas valid on any homogeneous frame is Halldén-normal.

(*iii*) Show that any Halldén-normal modal logic S and formulas φ and ψ with no propositional variables in common, if $\vdash_S \varphi \vee \psi$, then $S = (S \oplus \{\varphi\}) \cap (S \oplus \{\psi\})$. (From this, as noted in [594], p. 869, it can be seen to follow that A Halldén-normal modal logic is Halldén incomplete precisely when it is the intersection of two normal modal logics neither of which is included in the other.)

✠

Returning now to the main agenda of this section: in what was listed as the dual form of **.4** in Table 2.1 (p. 33), the schema appeared as:

$$(\varphi \wedge \Diamond \Box \varphi) \to \Box \varphi$$

though "A" rather than "φ" was our main schematic letter there. In this form, and reading $\Box$ and $\Diamond$ as K and P, this means that whatever is both true and such that it is for all one knows, known to be true, is indeed known to be true. Reading this now through the lens of Lenzen's Theorem (Thm. 5.2.1(*iv*), p. 372) and identifying PK with B, **.4** says that belief and truth suffice for knowledge (as remarked in note 311, p. 372). Since the conditions of Thm. 5.2.1 already provide the converse, what we now have is the identification of knowledge with true belief – something philosophers have typically found to be highly implausible even if they have disagreed as to what more is required for knowledge than true belief.[358] For

---

somewhat differently in the model theory of first order logic, and even in the model theory of modal propositional logic (in the form "*d*-homogeneous" in Kracht [665], at least); a simpler such purely structural notion of homogeneity (i.e., one not defined in terms of the truth or validity of formulas) appears apropos of Halldén completeness at the end of van Benthem and Humberstone [71]. A comparatively early reference to homogeneity outside of the context of standard first order modal theory appears in Section 3 of Chapter IV, in connection with a kind of temporal logic which differs from tense logic – on which contrast see note 154 p. 195 – in permitting explicit reference to and quantification over times in the object language.

[358] A notable exception, claiming that no more is required than true belief: Sartwell [993], [994]; also relevant is Weinberg, Nichols, and Stich [1159]. In the Digression on p. 391 we also encountered numerous philosophers who did not think belief was a necessary condition for knowledge, whose criticism of the identification of knowledge with true belief was that it required (for someone to count as knowing something) too much rather than too little. Chapter 3 of Williamson [1189] ('Primeness') provides a general theoretical basis for thinking that no decomposition of knowledge into independent mental and non-mental components can be given.

this reason **S4.4** is at the top end of Lenzen's range of modal logics with serious epistemic pretensions. Indeed Lenzen [706] expresses a clear preference for **S4.2** as an epistemic logic. Both he and Stalnaker regard its proper extension **S4F** – to say nothing of the even stronger **S4.4** – as suitable for application only under special conditions.[359] The idea that **S4.4** is certainly not something to contemplate going beyond – whether or not it already goes too far – can be bolstered by a more technical observation due to Rohan French, for which we must give a little background. Lenzen observed that for the translation $\tau_{\diamond\square}$ discussed in Section 4.5 and for which we there recorded, as Theorem 4.5.9 (p. 282), the following result:

$$\text{For all formulas } \varphi, \text{ we have } \vdash_{\mathbf{KD45}} \varphi \text{ if and only if } \vdash_{\mathbf{S4.2}} \tau_{\diamond\square}(\varphi),$$

we have this same result with **S4.4** in place of **S4.2**, and in this case we can invert the translation $\tau_{\diamond\square}$ – in the sense that (iii) and (iv) of Theorem 5.5.18 below are satisfied – by using $\tau_{\boxdot}$, defined, as in Example 4.4.27 (p. 266 above), to be the identity translation except in the case of $\square$-formulas, for which we stipulate that $\tau_{\boxdot}(\square\varphi) = \square\tau_{\boxdot}(\varphi) \wedge \tau_{\boxdot}(\varphi)$. (In the epistemic arena, $\tau_{\boxdot}$ spells out the 'knowledge as true belief' idea; the notation is explained by the fact that $\boxdot\varphi$ is often used, especially in modal logics not extending **KT**, to abbreviate $\square\varphi \wedge \varphi$. For more information on the translation $\tau_{\boxdot}$ see the discussion and references in French and Humberstone [334]; the main conjecture of that paper has since been settled: see Jeřábek [626].) The following result, whose proof we omit,[360] says that these translations are mutually inverse embeddings between **S4.4** and **KD45**.

THEOREM 5.5.18 *For all formulas φ:*

(i) $\vdash_{\mathbf{KD45}} \varphi$ *if and only if* $\vdash_{\mathbf{S4.4}} \tau_{\diamond\square}(\varphi)$;

(ii) $\vdash_{\mathbf{S4.4}} \varphi$ *if and only if* $\vdash_{\mathbf{KD45}} \tau_{\boxdot}(\varphi)$;

(iii) $\vdash_{\mathbf{KD45}} \varphi \leftrightarrow \tau_{\boxdot}(\tau_{\diamond\square}(\varphi))$;

(iv) $\vdash_{\mathbf{S4.4}} \varphi \leftrightarrow \tau_{\diamond\square}(\tau_{\boxdot}(\varphi))$.

There is some redundancy in putting matters this way. (In particular, the "if" directions of (i) and (ii) – which say that the embeddings concerned are faithful – follow from the remaining conditions.) But it allows for a convenient illustration of some general terminology: (i) and (ii) say that **S4.4** and **KD45** are *intertranslatable* via $\tau_{\diamond\square}$ and $\tau_{\boxdot}$, while (iii) and (iv) boost this to the claim that these translations render the logics concerned *translationally equivalent*.[361] Indeed, because these translations are definitional in the sense explained in Example 4.4.25 (beginning on p. 264), **S4.4** and **KD45** are definitionally equivalent in the sense of possessing – when suitably re-notated to have different $\square$ operators – a common definitional extension (as mentioned on p. 373).

It is worth pausing to extract from these findings a result on embedding **S4.2** in **S4.4**, originally appearing in Thomas [1103]. This involves the translation we shall call τ_{Th}, which is the identity translation everywhere except on $\square$-formulas, for which we have $\tau_{\mathsf{Th}}(\square\varphi) = \diamond\square\tau_{\mathsf{Th}}(\varphi) \wedge \tau_{\mathsf{Th}}(\varphi)$.

COROLLARY 5.5.19 *For all formulas $\vdash_{\mathbf{S4.4}} \varphi$ if and only if $\vdash_{\mathbf{S4.2}} \tau_{\mathsf{Th}}(\varphi)$.*

---

[359]Stalnaker [1082], p. 187, writes that **S4F** "might be the appropriate idealization for a certain limited context" (which he goes on to describe).

[360]For part (i), the proof of Theorem 4.5.9 (p. 282) suffices, substituting **S4.4** for **S4.2**; the proof ends with the remark that frame of $\mathcal{M}^*$ is reflexive, transitive and convergent – making it a frame for **S4.2**, but the frame in question (a two-cluster frame with first cluster containing at most one element) is also evidently a frame for the stronger **S4.4**.

[361]Or 'synonymous' in the terminology of Pelletier and Urquhart [879],[879], who leave us still waiting for an example to be given of two normal monomodal logics which are intertranslatable but not translationally equivalent. Here 'monomodal' is intended strictly, as excluding the presence of additional without additional sentential constants (nullary connectives).

5.5. LOGICS BETWEEN S4 AND S5

Proof. Combining Theorems 4.5.9 and 5.5.18(*ii*) gives $\vdash_{\mathbf{S4.4}} \varphi$ iff $\vdash_{\mathbf{S4.2}} \tau_{\Diamond\Box}(\tau_{\Box}(\varphi))$, so it remains only to note that $\tau_{\mathsf{Th}} = \tau_{\Diamond\Box} \circ \tau_{\Box}$ (proving, if necessary, by induction on the complexity of φ, that $\tau_{\mathsf{Th}}(\varphi) = \tau_{\Diamond\Box}(\tau_{\Box}(\varphi))$). ∎

The proof in [1103] runs along essentially the above lines, though it is put in somewhat different terminology. That is more than enough by way of background for the Lenzen-related maximality consideration alluded to above.

Theorem 4.2.18 of French [332] shows that $\tau_{\Diamond\Box}$ does not faithfully embed **KD45** into any proper extension of **S4.4** – giving another respect in which **S4.4** is a maximal candidate epistemic logic from Lenzen's perspective (i.e., a perspective taking **KD45** as a plausible doxastic logic and thinking of B as PK). Note that this is not to say that every logic into which $\tau_{\Diamond\Box}$ faithfully embeds **KD45** is included in **S4.4**: as French observes, $\tau_{\Diamond\Box}$ faithfully embeds **KD45** into itself and **KD45** is not a sublogic of **S4.4** (since the latter does not contain **5**), illustrating the fact that the set of normal modal logics into which $\tau_{\Diamond\Box}$ embeds **KD45** do not form an interval.[362]

With **S4.4** we have nearly reached the end of our survey of candidate epistemic logics between **S4** and **S5**, and insert here some remarks of general 'modal-logical' interest arising over a question raised about **S4.4**, as well as the answer to this question, before briefly considering an alternative candidate to Lenzen's $\Diamond\Box$ (or PK) as an epistemic rendering of belief.

Sobociński once asked – see Schumm [1004] for references – if there were any modal logics properly between **S4.4** and **S5**, or whether on the other hand, in the terminology introduced on p. 120, **S5** was a minimal proper extension or 'cover' of **S4.4** in the lattice of (normal) modal logics.[363] This is a very natural question, especially in hindsight, with the benefit of the Kripke semantics for **S4.4**, which, as already noted, provides generated models based on two-cluster frames in which the first cluster contains a single point. If one tries to think of imposing a further condition on such a frame it is hard to think of any condition weaker than one which makes the first cluster's only point in turn accessible to the second cluster, which amounts to considering degenerate two-cluster frames: but these are just universal frames, determining the logic **S5**. Now, we could obviously narrow down the class of two-cluster frames whose first clusters are singletons by restricting the cardinality of the second cluster too – for example in the most extreme case insisting that it contains at most one single point.[364] This case gives the frame depicted in Figure 2.16 (p. 155) above.

While this would yield (i.e., determine) a logic properly extending **S4.4**, it would not help with Sobociński's question because it would not produce a sublogic of **S5**, validating such **S5**-unprovable formulas as the McKinsey axiom **M** and the 'at most two alternatives' axiom and $\mathbf{Alt}_2$, which figured in Exercise 2.5.70 (p. 109). At first sight, such considerations accordingly seem irrelevant to Sobociński's question. Nevertheless, Schumm [1004] ingeniously returned an affirmative answer to that question by taking not the extension of **S4.4** in question – and in [1004] he chose **S4.4M** – but rather the intersection of that logic with **S5**. And he axiomatized **S5** ∩ **S4.4M**, as the extension[365] of **S4.4** by:

$$\Box(\Diamond p \to \Box\Diamond p) \vee \Box(\Box\Diamond q \to \Diamond\Box q).$$

Although Schumm put this forward simply as answering Sobociński's question affirmatively, coming up with some logic or other logic strictly between **S4.4** and **S5**, Zeman [1222] (see also [1224], Chapter

---

[362] By this is meant that there do not exist normal modal logics S_0 and S_1 for which the set concerned is $\{S \mid S_0 \subseteq S \subseteq S_1\}$.
[363] The presence or absence of the word *normal* here is immaterial, in view of the fact, as mentioned in the Digression on p. 34, that all extensions of **S4.3**, are normal. See also Remark 2.5.40, p. 89, for a similar question about whether there is any modal logic properly between **KB4** and **S5** – answered negatively, as reported there, by Segerberg.
[364] Of course, one could alternatively leave the first cluster unrestricted and require instead that the second cluster be a singleton. The logic determined by the resulting class of frames has no special significance for epistemic applications but is of general logical interest as one of the five pretabular extensions of **S4** (see p. 112 above) described in Esakia and Meskhi [274],— where such frames are called *tacks*; an axiomatic description of it is given there as **S4FM**.
[365] We could equivalently say *normal extension* here.

15) showed that Schumm had in fact answered the question by providing a logic which itself has **S5** as a minimal proper extension. To make this clear, Zeman pointed out ([1222], p. 353) that we could replace **M** above by **Alt**$_2$, since **S4.4M** = **S4.4Alt**$_2$, a reformulation which makes the connection with the two-point frame more evident; to simplify the labelling further, note that this logic is in fact **S4Alt**$_2$.[366] Schumm's logic is thus also axiomatized (in [1222]) by means of:

$$\Box(\Diamond p \to \Box\Diamond p) \vee \Box(\Box q \vee (\Box(q \to r) \vee \Box(q \to \neg r))).$$

After the main $\Box$ in the second disjunct here, we have a variant on **Alt**$_2$, mentioned just before Exercise 2.5.70 (p. 109) – though formulated in terms of p and q there rather than q and r.[367] Here, as in the axiomatization using **M**, we must change variables to keep away from those in the first disjunct to obtain a Halldén-unreasonable disjunction (as it was put in note 102, p. 120). The logic concerned, which Segerberg [1025] calls **S4Sch**,[368] is the intersection of two of its $\subseteq$-incomparable extensions, so we expect Halldén incompleteness here, and the behaviour of **S4** is also being exploited: putting one box on each disjunct allows us replace it with no box at all (thanks to **T**) or with any greater number of boxes (thanks to **4**). The general pattern is elicited by the following exercise:

EXERCISE 5.5.20 Show that for any extension (normal or otherwise) S of **S4**, and any formulas φ and ψ, which have no propositional variables in common, we have $(S \oplus \varphi) \cap (S \oplus \psi) = S \oplus \Box\varphi \vee \Box\psi$. ✠

Digression. As it happens, we can reduce some of the occurrences of $\Box$ in the formulas inset above by exploiting the fact that for any formulas φ and ψ, the formulas $\Box(\Box\varphi \vee \Box\psi)$ and $\Box\varphi \vee \Box\psi$ are provably equivalent in **S4**. This allows us to simplify the first formula inset above (and associated there with Schumm) to:

$$(\Diamond p \to \Box\Diamond p) \vee \Box(\Box\Diamond q \to \Diamond\Box q),$$

thinking of the "$\Diamond p \to \_$" as "$\Box\neg p \vee \_$", and to simplify the second formula inset above (and associated with Zeman) to:

$$(\Diamond p \to \Box\Diamond p) \vee (\Box q \vee (\Box(q \to r) \vee \Box(q \to \neg r))).$$

End of Digression.

Returning to the main theme of this section, Lenzen-inspired investigations of logics between **S4** and **S5**, let us pick up the suggestion – made initially with deontic rather than doxastic applications in mind – of Åqvist [27] that we consider the behaviour of $\Box\Diamond\Box$ in **S4** itself, with a view to its prospects as a B (with $\Box$ as K).[369] (See Example 4.5.25, p. 290.) But what exactly is this behaviour? One would like an axiomatic description of the logic **S4**$(\Box\Diamond\Box)$, to use Zolin's notation from Example 4.4.27 (p. 266). Spelling out the question further, for the translation $\tau_{\Box\Diamond\Box}$ which replaces $\Box$ with $\Box\Diamond\Box$, we are looking for S satisfying, for all φ:

$$\vdash_S \varphi \text{ if and only if } \vdash_{\mathbf{S4}} \tau_{\Box\Diamond\Box}(\varphi).$$

Everything §3 of Åqvist [27] tells us belongs to the desired S is provable in **KD4U**, but this logic – plausible though it may be for deontic or doxastic purposes – is too weak to be the desired S. Recall that **U** is the necessitation of the **T** axiom. We need also the necessitation of **5** – at least as a theorem

---

[366]Segerberg [1025], esp. p. 152, repeatedly uses labels such as '**S4.3MAlt**$_2$' and '**S4HAlt**$_2$' – the **H** here being what was called **H̲** above (Example 2.5.38, p. 87) – but there seems no reason to avoid the irredundant label **S4Alt**$_2$.

[367]And the formula in the scope of the $\Box$ on the first disjunct in Zeman's case is actually a version of **B** rather than **5**, as here. The change has been made to reduce complications.

[368]**Sch** is Segerberg's name (honouring Schumm) for the disjunction inset above, or rather, for a variant with the $\Box$ on the second disjunct removed (as justified by the remarks following Exc. 5.5.20 below); see [1025], p. 159, for this and many alternatives which are interdeducible with it over **S4**.

[369]This is an option raised but set aside in Lenzen [706] (p. 51): see the Digression on p. 431 below.

5.5. LOGICS BETWEEN S4 AND S5

even if we do not take it as an axiom. Indeed, we need the necessitation of every **S5** theorem. A simple way to achieve this axiomatically is to extend the logic called **KB**$^\Box$**4**, (where **B**$^\Box$ is $\Box(p \to \Box\Diamond p)$) in our discussion of a conjecture of J. Perzanowski (pp. 98–104), by adding the axiom **D**. Thus we are dealing with the logic **KDB**$^\Box$**4**. (Recall from Exercise 2.1.15, p. 45, that **KDB4** = **S5**, because the serial transitive symmetric relations are precisely the equivalence relations, as noted on p. 89.) Some familiarization exercises follow, beginning with one on the fate of **U** and necessitated **5** (**T**$^\Box$ and **5**$^\Box$, though the former abbreviation was eschewed in note 30, p. 37):

EXERCISE 5.5.21 Show that $\vdash_{\mathbf{KDB}^\Box\mathbf{4}} \Box(\Box p \to p)$ and $\vdash_{\mathbf{KDB}^\Box\mathbf{4}} \Box(\Diamond p \to \Box\Diamond p)$. ✠

For the following, giving the generalization underlying Exc. 5.5.21, it may help to recall Proposition 2.5.56: $\vdash_{S\oplus\varphi} \psi$ implies $\vdash_{S\oplus\Box\varphi} \Box\psi$, for any normal modal logic S and any formulas φ, ψ.

EXERCISE 5.5.22 Show that for all formulas φ: $\vdash_{\mathbf{KDB}^\Box\mathbf{4}} \Box\varphi$ if and only if $\vdash_{\mathbf{S5}} \varphi$. ✠

From this we see that if $\langle W, R \rangle$ is a frame for **KDB**$^\Box$**4** generated by some $w \in W$, then for any $x \in R(w)$, every theorem of **S5** is valid at x in $\langle W, R \rangle$; but we can get a more detailed picture – as a kind of 'multi-cluster version' of a point-generated **KD45** frame – of what such frames look like; here we use the term *cluster* even though the frame, which is indeed transitive, is not reflexive, but again to mean: maximal subset closed under the relation $R \cap R^{-1}$. By a *terminal* cluster is meant a cluster to which no other cluster is accessible.[370] Case (1) gives the degenerate case of one of these multi-cluster frames and case (2) the non-degenerate case. The proof is straightforward, with point-generated subframes of the canonical frame sufficing for part (*ii*):

THEOREM 5.5.23 (*i*) *If* $\langle W, R \rangle$ *is a frame for* **KDB**$^\Box$**4** *generated by* $w \in W$, *then either* (1) wRw *and* $R(w) = W$, *or* (2) *not* wRw *and* W *is the union of* $\{w\}$ *with a set of terminal clusters (restricted to each of which R is universal), all of them accessible to w.*

(*ii*) **KDB**$^\Box$**4** *is determined by the class of frames described in* (*i*); *moreover, its theorems are exactly the formulas true at the generating point w in all models on such frames.*

For the proof the promised faithful embedding result, we use the same strategy as in the case of Theorem 4.5.9 (p. 282); recall that $\tau_{\Box\Diamond\Box}$ is the translation which replaces every $\Box$ in a formula with $\Box\Diamond\Box$:

THEOREM 5.5.24 *For all formulas* φ, $\vdash_{\mathbf{KDB}^\Box\mathbf{4}} \varphi$ *if and only if* $\vdash_{\mathbf{S4}} \tau_{\Box\Diamond\Box}(\varphi)$.

Proof. 'Only if': By induction on the length of a shortest proof of φ in the axiomatization suggested by the label **KDB**$^\Box$**4**, for which purpose, since the modality $\Box\Diamond\Box$ in normal in **S4**, it suffices to show the **S4**-provability of $\tau_{\Box\Diamond\Box}(\mathbf{D})$, $\tau_{\Box\Diamond\Box}(\mathbf{B}^\Box)$ and $\tau_{\Box\Diamond\Box}(\mathbf{4})$. The case of **D** was already mentioned – along with the normality of $\Box\Diamond\Box$ – in Example 4.5.25, p. 290, so here we deal with the $\tau_{\Box\Diamond\Box}$ translations of $\mathbf{B}^\Box$ and **4**. Let us take the latter case first: $\tau_{\Box\Diamond\Box}(\mathbf{4})$ is $\Box\Diamond\Box p \to \Box\Diamond\Box\Box\Diamond\Box p$ (taking **4** as $\Box p \to \Box\Box p$). But this implication is **S4**-provable, as we see by simplifying the modality on the consequent (replacing the two adjacent occurrences of $\Box$ by one, as we may in **S4**), making it $\Box\Diamond\Box\Diamond\Box$, or, with some internal bracketing so that the **S4**-equivalence of $\Box\Diamond$ with $\Box\Diamond \circ \Box\Diamond$ recorded in Figure 2.2 (p. 42) is easily seen to apply: $(\Box\Diamond)(\Box\Diamond)\Box$. Thus we can further simplify the modality on the consequent to $\Box\Diamond\Box$, at which point it coincides with the modality in the antecedent, showing the original implication to be provable. Finally, the case of $\tau_{\Box\Diamond\Box}(\mathbf{B}^\Box)$, which is:

---

[370]These are called 'final clusters' in Segerberg [1025], but the distinction Segerberg makes between final clusters, so understood, and *last* clusters, meaning clusters accessible to every point in the frame, is a little hard to remember, so the word 'terminal' is used here instead, recalling terminal nodes in a tree, etc.

$$\Box\Diamond\Box(p \to \Box\Diamond\Box\Diamond\Box\Diamond p).$$

Again using the **S4**-equivalence $\Box\Diamond$ with $(\Box\Diamond)^2$ and thus with $(\Box\Diamond)^3$ as here, so that what has to be shown to be **S4**-provable is $\Box\Diamond\Box(p \to \Box\Diamond p)$. The easiest way to see that this is indeed provable is to note that what appears in the scope of the initial $\Box\Diamond\Box$ – the axiom **B** – is **S5**-provable, so prefixing $\Box\Diamond\Box$ to it gives something **S4**-provable by Matsumoto's Theorem (Thm. 2.8.4, p. 144).**S4**

'If': Suppose now that $\nvdash_{\mathbf{KDB}^\Box 4} \varphi$, so that φ is false at a point in a frame of the kind described in Theorem 5.5.23, with a view to showing that $\nvdash_{\mathbf{S4}} \tau_{\Box\Diamond\Box}(\varphi)$. We may assume that the point in question is the cluster-external point w as in (ii) of Thm. 5.5.23. Make a new model by reflexivizing w (as in the proof of Thm. 4.5.9, p. 282) and observe that any formula has the same truth-value at any point in the original model as its $\tau_{\Box\Diamond\Box}$ translation has at that point in the new model, so that $\tau_{\Box\Diamond\Box}(\varphi)$ is false at w in the new model, which is a model on a transitive reflexive frame, showing that $\nvdash_{\mathbf{S4}} \tau_{\Box\Diamond\Box}(\varphi)$. ∎

We will see in the following section that some interest attaches to $\mathbf{KDB}^\Box 4$ for doxastic logic: that even though it may have its faults for this application, it fares somewhat better than the traditional favourite, **KD45**. So we conclude the present section by paying this logic a little more attention.

EXERCISE 5.5.25 (i) Show by suitable formal proofs that $\mathbf{KDB}^\Box 4 = \mathbf{KD45}^\Box \mathbf{U}$ (where $\mathbf{5}^\Box$ is $\Box(\Diamond p \to \Box\Diamond p)$).

(ii) Does the label on the right in (i) here, i.e. $\mathbf{KD45}^\Box\mathbf{U}$, represent an independent axiomatization? Justify your answer by establishing or refuting the ('independence') claim that none of the label constituents **D**, **4**, $\mathbf{5}^\Box$, **U**, is redundant in this label.

(iii) According to Prop. 4.6.10 (p. 297), $fm(\mathbf{S5}) = \mathbf{KD45}$ ("**KD45** gives the f.m. core of **S5**"); how exactly would the claim that $fm(\mathbf{S5}) = \mathbf{KDB}^\Box 4$ go wrong? ✠

The following observation (related to the topic of 'Gabbay-style rules', not covered in our exposition[371]) – which would hold equally well for the case of **KD45** – will be useful at one point:

PROPOSITION 5.5.26 *Suppose that the formula φ is $\mathbf{KDB}^\Box 4$-consistent. Then so is the formula $\varphi \land (\Box p_i \land \neg p_i)$ where p_i is any propositional variable not occurring in φ.*

Proof. If φ is $\mathbf{KDB}^\Box 4$-consistent then there is a model $\langle W, R, V \rangle$ with irreflexive $w \in W$ and $R(w) = W \setminus \{w\}$, and $\langle W, R, V \rangle \models_w \varphi$. Suppose p_i does not occur in φ. Adjust V to V' by putting $V(p_j) = V(p_j)$ for all $j \neq i$ and $V(p_i) = R(w)$. Since this change to the model does not affect φ, we have $\langle W, R, V' \rangle \models_w \varphi$; but $\langle W, R, V' \rangle \models_w \varphi$, so $\langle W, R, V' \rangle \models_w \varphi \land (\Box p_i \land \neg p_i)$, showing the $\mathbf{KDB}^\Box 4$-consistency of $\varphi \land (\Box p_i \land \neg p_i)$. ∎

5.6 Concept-Possession Problems

Here we take up the issue raised under the heading 'Concept Possession Worries' at p. 343 in Section 5.1; see also note 275 there, on p. 333, Example 5.2.9, p. 379, and p. 383 in Section 5.3. As we shall see, it is not only in connection with the issue of logical omniscience, discussed in Section 5.1, that concept possession worries arise: various introspection principles (from Section 5.2) are vulnerable to objections on this score, such as $\mathbf{5}_B$ (see Example 5.6.11). Our discussion will also include, because of their interconnections with this theme, various proposals for defining belief (or B) in terms of knowledge (K).

---

[371]See Gabbay [339], Venema [1127].

5.6. CONCEPT-POSSESSION PROBLEMS

In note 275 (p. 333), a mistake was mentioned that it might be suspected no-one has ever made, so we can begin by illustrating it from the published literature, in this case from van der Hoek and Meyer [512], p. 182; as with some others mentioned in note 275, our authors write M rather than P for $\neg K \neg$, and they are seriously considering the proposal that this operator might represent some notion of belief:

> Considered as a description of belief, the operator M corresponds with a belief of a very credulous person or agent: φ is believed, as long as $\neg\varphi$ is not known for sure. M means believing in the sense of considering it possible, not excluding the possibility.

Here we are not principally concerned with the proposal being made (see the Digression below), but with this final gloss on the notation, in which $\neg K\varphi$ ("not excluding the possibility that $\neg\varphi$") is being glossed in terms of the agent's considering it possible that φ. And the problem is that if the agent lacks the conceptual resources to entertain the proposition represented by φ here, the agent is not in a position to consider it possible that φ, for the same reason that the agent cannot know that φ or know that $\neg\varphi$, or take any other propositional attitude to the proposition that φ.[372] Thus *not knowing* that φ is not itself, and does not imply, *considering it possible* – or considering it anything else – that φ.[373] Some proposition's being an epistemic possibility for a subject is not a matter of the subject's taking *any* attitude to that proposition (such as considering it to constitute a possibility) but rather of the *absence* of a certain attitude (knowledge) to its negation; see the criticism of van der Hoek *et al.* [509] on p. 383 above, for trying to mean more by the dual of K (or $\Box$, read epistemically) than this.

Note, incidentally, that very little congruentiality is needed to get us from the suggestion that $B\varphi$ is equivalent $\neg K\neg\varphi$ to the reformulation that $K\varphi$ is equivalent to $\neg B\neg\varphi$ – so little that either might be informally cashed as the claim that belief and knowledge are dual notions. In the latter form, however, the suggestion may seem less plausible, equating as it does, not having a belief with possession of knowledge. (That is, if B is equivalent to $\neg K\neg$, then K and $\neg B\neg$ are also equivalent.) To be fair to van der Hoek and Meyer [512], though, we should recall they are simply considering the epistemic–doxastic logic of a strange ('very credulous') kind of cognitive agent, rather than describing an ideal which might be recommended. (This does not undermine the above criticism, however, of the 'considers it possible' gloss used in describing the agent's situation.)

Digression. The proposal that belief should be treated as a $\Diamond$-style operator rather than a $\Box$-style operator which appears in the quotation above from van der Hoek and Meyer [512] was also toyed with in MacPherson [756], though the latter notes (p. 19) the implausible consequence that a disjunction is believed only if one of its disjuncts is believed,[374] and both [512] and [756] observe that since belief will

---

[372] No great weight should be placed on the talk of "the proposition that φ" here, since the issue of the individuation of *propositions* – entities taken as the objects of propositional attitudes – is one on which different views of the significance of concept possession considerations will take different positions on. (Indeed even the existence of propositions – or any other kind of entities – as the objects of propositional attitudes is open to serious doubt. The case against them, with ample bibliographical information on others who have pressed this case, is well put in Moltmann [815], [818]; somewhat similar sentiments can be found in Saarinen [985]. While these authors are happy with – what are called – propositional attitudes and unhappy with their traditionally postulated propositional objects, an alternative line of thought has it that problems – e.g., individuation problems – with these would-be objects mean propositional attitudes themselves do not exist: that the 'folk psychological' concepts of belief, desire, etc., have no application. Stich [1092] provides a classic defence of this *eliminativist* position, a position he later came to reject.) Instead of saying "a cannot entertain the proposition that __", where the gap makes a place for a sentence, it would be safer to employ the less ontologically committal formulation of, say: "a has no idea what it would be for it to be the case that __." Having issued this disclaimer, we continue with the more concise (if misleading) way of speaking.
[373] A similar criticism of some formulations in Lenzen [710] appears on p. 229 of Thijsse [1100].
[374] It is because a disjunction can be believed without either disjunct being believed that there is no such thing – even under the familiar idealizing assumptions (such as consistency and logical omniscience) of doxastic logic – as a person's 'belief world': a world in which exactly what that individual believes to be the case *is* the case. (It is a set of possible worlds – the doxastically accessible worlds – rather than a single such world, that conveys the content of a believer's beliefs. If a single world-like entity is sought to play this role it had better be one truth w.r.t. which does not distribute over disjunction: perhaps a *possibility* in the sense of Humberstone [545], described in Example 4.1.12, p. 211 above.)

still be monotone on such a treatment, even though not regular, there is no real escape from 'logical omniscience' objections to be had down this path. **End of Digression.**

The distinction in play above between taking something to be possible and not excluding it as a (doxastic or epistemic) possibility comes up repeatedly – as a distinction overlooked – in the literature, as we illustrate here with a further example from S. Yalcin:

EXAMPLE 5.6.1 Yalcin [1212], discussing – as we are not – epistemic modals (modal auxiliaries or auxiliary-like verbs with a doxastic or epistemic interpretation), remarks (p. 996) that the sentence:

"Vann believes that Bob might be in his office"

would (on a certain hypothesis) be "true just in case Bob's being in his office is compatible with what Vann believes. That is the intuitively correct result." Not quite: if Vann has never heard of Bob (the Bob in question, that is) or lacks the concept of an office, Bob's being in his office could be compatible with what Vann believes while the sentence inset here would definitely not be true. ◀

Williamson ([1189], §1.5) provides an interesting discussion – the concept possession issue will emerge in due course – of the possibility of analysing belief in terms of knowledge, as opposed to the common reverse direction, and in particular of a well known position – or cluster of positions – called *disjunctivism* according to which believing that p is a matter of either knowing that p or ___ , where what fills the blank is something that covers all those mental states the subject might be in but in which for some reason the subject's belief falls short of knowledge (of knowing that p, that is).[375] One most conspicuous case is that the belief in question is false. But the project would not succeed if the second disjunct directly alluded to the subject's falsely believing that p – or more generally having a belief that falls short of knowing that p in some way or other – since we are supposed to be using this disjunction to analyse the concept of belief, and the analysis would be circular if we just employed that very concept in the second disjunct. So Williamson uses as a technical term the verb "opines" to give the schematic form of the disjunctivist's proposal:

one believes p if and only if either one knows p or one opines p.

Concerning the somewhat stilted syntax of this we need to observe that Williamson uses "p" to stand in for the name of a proposition rather than for a sentence expressing the proposition – for the latter usage, one would prefer to see the word "that" before each occurrence, and especially the occurrence after *opine*, of "p" here. (See Remark 5.1.1(*ii*), p. 332.) Williamson's precedent will be followed only in quoting him and discussing what he says.) He proceeds ([1189], p. 45) to ask:

> Can we explain 'opine' in terms of 'know'? A first attempt is this: one opines the proposition p if and only if one is in a state which one cannot discriminate from knowing p, in other words, a state which is, for all one knows, knowing p. That cannot be quite right, for if one cannot grasp the proposition p then one cannot discriminate one's state from knowing p; but does not believe p, and therefore does not opine it.

Williamson goes on to revise the suggestion to this: "one opines p if and only if one has an attitude to the proposition p which one cannot discriminate from knowing" – and now the part about having an

---

[375]For historical information and philosophical reflection on disjunctivism (which is in fact a somewhat more diverse approach than the present remarks may convey), in addition to Williamson's discussion and the references there cited, see the collections Byrne and Logue [143] and Haddock and Macpherson [421], the article Martin [775] and the monograph Pritchard [932]. The disjunctivist's *disjunction* – though not an endorsement of the position itself – made an early appearance in the following passage from G. E. Moore, which can be found at p. 192 of Baldwin [51]: "But, now, even if it is not certain that I have at this moment the evidence of my senses for anything at all, it is quite certain that I *either* have the evidence of my senses that I am standing up *or* have an experience which is *very like* having the evidence of my senses that I am standing up." (Emphasis Moore's.)

5.6. CONCEPT-POSSESSION PROBLEMS

attitude (some propositional attitude or other, that is) to p blocks the possibility that the subject not grasp the proposition p – and goes on to make trouble for the revised suggestion, as least as part of the disjunctivist program. But for us the interesting aspect of the discussion is in the initial proposal and Williamson's objection to it invoking the idea that the subject must possess the concepts involved in order to take a propositional attitude to any content in which those concepts figure. A similar sentiment emerges in the following passage from p. 282f. of the same work (i.e., [1189]):

> Even so ∧-elimination has a rather special status. It may be brought out by a comparison with the equally canonical ∨-introduction inference to the disjunction $p \vee q$ from the disjunct p or from the disjunct q. Although the validity of ∨-introduction is closely tied to the meaning of ∨, a perfect logician who knows p may lack the empirical concepts to grasp (understand) the other disjunct q. Since knowing a proposition involves grasping it, and grasping a complex proposition involves grasping its constituents, such a logician is in no position to grasp $p \vee q$, and therefore does not know $p \vee q$. In contrast those who know a conjunction grasp its conjuncts, for they grasp the conjunction.

Humberstone [554], p. 407, makes a similar point about belief, though without the contestable presumption that there are such things as complex propositions:

> For while few have had occasion to entertain a conception of entailment according to which conjunctions fail to entail their conjuncts, that disjuncts should less obviously entail disjunctions in which they are constituents is evidenced by Parry's system of analytic implication.[376] As I understand it, it is with just such applications in mind as doxastic closure principles that Parry developed the idea. If I do not have the concepts involved in the formulation of a statement, is it *so* obvious that on pain of irrationality, I must believe the disjunction of that statement with, say, "Some bulls have horns"? (This problem does not arise with conjunction-to-conjunct inferences; even without assumptions of rationality, a belief to the effect that p and q seems *inter alia* to be a belief to the effect that p.[377])

Those interested in following up the option of basing epistemic and doxastic logic on Parry's logic of analytic implication or some variant thereof are encouraged to do so; pertinent references would include Parry [867], [868], Smiley [1063], Fine [294], Dunn [264], Urquhart [1122] and Ferguson [285]. (Just as propositional logics of *relevant implication* can be thought of as trying to salvage as much as possible of the classical story of how implication, negation, conjunction, and disjunction are related while stopping short of rendering provable any implications $\varphi \to \psi$ in which φ and ψ do not have a propositional variable in common, so in the case of analytic implication one wants as generously classical an account as possible without rendering provable any $\varphi \to \psi$ in which ψ contains variables not occurring in φ.) Although he makes no mention of Parry, Roberts [962], p. 376, remarks that the inference from p to $p \vee q$ is not what he calls *t*-valid, where this is defined to mean that necessarily anyone believing the premiss believes the conclusion, and it is fairly clear that at least one his reasons for this claim is a concept-possession worry. The same notion is called *conceptual entailment* in Chisholm [175] and elsewhere. Parry-like themes are pursued with rather different motivations and historical influences in Weingartner [1161], Weingartner

---

[376]There is a footnote reference here in the original article to Parry [867].

[377]A footnote here expresses agreement on this point with the central paragraph of p. 211 in Routley and Routley [976]. The Routleys here defend also the converse principle – aggregation for B – with the following remarks: "The view of conjunction as a sort of mental glueing operation which one can fail to perform is an error. It is not as if believing (A & B) is something over and above believing A and believing B; it is not a further act which our hypothetical author may have omitted." A later expression of the same view, from Levesque and Lakemeyer, [715], was quoted in Remark 5.1.7, p. 341 above. For the contrary view, see Lewis [722], or the 'society of minds' approach described in §9.6 of Fagin *et al.* [280]. See the end of Example 4.4.8, p. 245 above. Aggregation for B is naturally and widely described as the principle that the subject's set of beliefs is closed under conjunction, since it says that if φ and ψ are in this set, then so is their conjunction $\varphi \wedge \psi$. Instead of concentrating on the operation, we could focus on the rule and call this: closure (of the set of beliefs) under ∧-introduction. Jago [618], p. 327, idiosyncratically uses "closed under conjunction" to mean closure under ∧-*elimination*. And whereas calling a set closed under disjunction should mean that when both disjuncts are in the set, so is the disjunction, Jago uses it to mean that if even one of the disjuncts is in the set, so is the disjunction – closure under ∨-introduction, in other words.

and Schurz [1164], and Schurz [1011]. A related (shall we say?) 'conceptually non-ampliative' idea – under the name *analytic containment* – is explored in Angell [24] (see also Correia [196]); only slightly further afield are the "logic of meaning containment" theme in Brady [115] and the concepts of narrow consequence and 'basic content part' in Grimes [413] and Gemes [361] (also [362]) respectively. A more recent venture along similar lines is that of *logical inclusion*, guiding a principle of 'topical' or *immanent* closure in Chapter 7 of Yablo [1211]. (Cf. also the 'tautopical implication' of Makinson [761], as well as 'inclusive entailment' in Chapter 4 of Yablo [1211].)

REMARK 5.6.2 The term 'analytic' in the phrase *analytic implication* is intended to echo Kant's idea of analytic truths as those in which the predicate is contained within the subject, now reading *subject* and *predicate* as *antecedent* and *consequent*, and taking containment as measured by the presence of non-logical vocabulary (in particular, propositional variables). With the same reading of 'subject' and 'predicate' but now concentrating on quantificational languages and cashing out containment in terms of quantificational depth, Hintikka (see the papers in [496]) has suggested a different notion of analyticity for implications, called in Hintikka [497] 'surface implication' and defined there in terms of notions of the depth of a formula, as well as the depth of something called a 'surface model', for which notions the reader is referred to [497], thus (with minor alterations to fit our notation):

> A *surface implication* holds between φ_1 (of depth d_1) and φ_2 (of depth d_2) if and only if every surface model of depth d_1 (with the same predicates as $\varphi_1 \to \varphi_2$) in which φ_1 is true can be extended to a surface model of depth d_2 (with the same predicates) in which φ_2 is true. [497], p. 146.

Roughly speaking, while the Parry-style notion of (provable) analytic implication demands that the consequent import no new conceptual content beyond that present in the antecedent, Hintikka's version demands that no more thoroughgoing investigation of the quantificationally specified relations among individuals than is required to (as he argues) understand the antecedent is required to secure the truth of the consequent. In [498], Hintikka suggests that in place of the unrestricted KMono and BMono rules endorsed in his [490], these principles might plausibly be restricted to the case in which the premiss records not just an arbitrary provable implication but specifically one of these 'analytic' or 'surface' implications. Some critical discussion of Hintikka [496] may be found in van Benthem [66], though this does not include the current change of heart from the unrestrictedly monotone deductive omniscience endorsed in [490], reported in Section 5.1 above (p. 331 and onward). ◀

Here, however, we confine ourselves to standard classical propositional logic and its modal extensions – with a defeasible preference in favour of normality for the currently envisaged interpretation of those extensions. Such a preference leaves one having to explain away examples such as the provability of $Bp \to B(p \vee q)$, in the face of the objection that q may be thought of as representing a proposition whose grasp requires the possession of concepts not required for grasping, or more strongly still, for believing, that represented by p. For such a subject the antecedent of this conditional would be true but not the consequent, on a strict version of the concept possession requirement. One way of elaborating a less strict version, favoured here, would consist in distinguishing between explicit and implicit versions of the attitudes under consideration, and applying the requirement only to the explicit version.[378]

Several ways of doing this were described among Examples 5.1.6 (beginning on p. 339), two of which we briefly revisit here: those numbered as *(iii)* and as *(i)*, beginning with the former. Whereas that earlier discussion concentrated on knowledge, here we concentrate on the belief case. (A full discussion of these issues would consider also the proposed threefold distinction between explicit, implicit and inferable

---

[378]Recall Williamson's notion of *being in a position to know* – note 282 (p. 339 above) – concerning which he suggests ([1189], p. 95): "Thus people who lack the concept of pain – perhaps because their concepts carve up the space of possible sensations in an alternative way – and so never know that they are in pain, may still count as being in a position to know that they are in pain."

5.6. CONCEPT-POSSESSION PROBLEMS

knowledge or belief proposed in Duží, Jespersen and Müller [269], but that would necessitate introducing the framework of Pavel Tichý's 'transparent intensional logic', taking us well away from the kind of technical apparatus in play in the present book.)

EXAMPLE 5.6.3 Instead of considering 'congruential-to-monotone' bimodal logics (mentioned on p. 340), we could consider 'arbitrary-to-normal' such logics. The models will be those of the pseudo-neighbourhood semantics (p. 162, where this was described as rather 'syntactical' for anything one would expect from semantics[379]) for which we recall the treatment of $\Box$, now thought of as representing explicit belief:

$$\langle W, \underline{\mathcal{N}}, V \rangle \models_x \Box\varphi \text{ if and only if } \varphi \in \underline{\mathcal{N}}(x),$$

with $\Box^+$, say, being the 'normal extension' of $\Box$:

$$\langle W, \underline{\mathcal{N}}, V \rangle \models_x \Box^+\varphi \text{ iff for some } \psi_1, \ldots, \psi_n \in \underline{\mathcal{N}}(x), \|\psi_1\| \cap \ldots \cap \|\psi_n\| \subseteq \|\varphi\|,$$

where we allow the "some $\psi_1, \ldots, \psi_n$" to be $\varnothing$, with $\bigcap \varnothing$ understood as being W. Evidently the smallest bimodal logic in which $\Box^+$ is normal and $\Box\varphi \to \Box^+\varphi$ is provable (for all φ) is sound w.r.t. this semantics in the sense of having its theorems true at all points in all models of the above kind; the question of completeness causes some difficulties for a straightforward canonical model style proof, which we need not go into here. On this approach, what is implicitly believed is defined to be what follows from what is explicitly believed, where what is explicitly believed is taken to be the set – in the present exposition, a finite set, though this is not essential – of statements explicitly believed. ◀

Another bimodal – or trimodal[380] – approach is that mentioned on p. 339 under Example 5.1.6(i), $q.v.$ for references, which gives a treatment like that envisaged in Example 5.6.3 for $\Box$ but this time for an operator written as A, intended to recall the word *aware*. For the present application this is specifically a matter of being in a position to entertain this or that proposition – conceptual awareness, one might say.

EXAMPLE 5.6.4 Expand the pseudo-neighbourhood models just seen in Example 5.6.3, though now we write "$\mathcal{A}$"[381] in place of "$\underline{\mathcal{N}}$", to go with the new operator A, with an accessibility relation R (for B). Thus now we are working with two 1-ary connectives, A and B (informally: 'awareness' in the attenuated sense just explained, and implicit belief, respectively), where $\mathcal{M} = \langle W, R, \mathcal{A}, V \rangle$ is such an expanded model and $x \in W$, put:

$$\mathcal{M} \models_x A\varphi \text{ iff } \varphi \in \mathcal{A}(x); \qquad \mathcal{M} \models_x B\varphi \text{ iff for all } y \in R(x), \mathcal{M} \models_y \varphi.$$

For the current application one might naturally require a further condition: where $\varphi(p_1, \ldots, p_n)$ is any formula in which the variables displayed are precisely those occurring,

$$\varphi(p_1, \ldots, p_n) \in \mathcal{A}(x) \text{ iff } p_1 \in \mathcal{A}(x) \& \ldots \& p_n \in \mathcal{A}(x).$$

The idea is to then regard explicit belief, B^+ say, as defined by $B^+\varphi = B\varphi \wedge A\varphi$. Thus on this approach the direction of analysis is the reverse of that in Example 5.6.3, in that we start with what is implicitly believed and then filter out the beliefs which are not entertainable by the subject, to arrive at the explicit beliefs. ◀

---

[379]See Priest [911] for a rather different way of grafting a syntactical element onto standard neighbourhood semantics in response to issues such as those of present concern.

[380]If we regard the implicit belief operator, B^+, defined below as a third non-Boolean primitive. (This depends on the object-linguistic *vs.* metalinguistic conception of definition: see note 312 on p. 373.)

[381]No confusion should arise with the different use of this notation on p. 305 above, as an actuality operator in the object language (or with the use of the same letter as a schematic letter for formulas when Latin rather than Greek letters are playing that role).

For practice with the semantics of this last example:

EXERCISE 5.6.5 With the inset condition above in force on $\mathcal{A}$, but no conditions on R, which of the following formulas are valid (true throughout all models meeting the condition)?

(i) $(Ap \wedge Aq) \leftrightarrow A(p \wedge q)$ $\quad$ (ii) $(Ap \wedge Aq) \leftrightarrow A(p \vee q)$
(iii) $(Ap \vee Aq) \leftrightarrow A(p \wedge q)$ $\quad$ (iv) $(Ap \vee Aq) \leftrightarrow A(p \vee q)$
(v) $B^+(p \wedge q) \to B^+p$ $\quad$ (vi) $B^+p \to B^+(p \vee q)$

Suppose we add two further conditions: that R is transitive and that xRy implies $\mathcal{A}(x) \subseteq \mathcal{A}(y)$. Is the following formula valid in the sense of being true throughout all models meeting the earlier condition as well as these two?

(vii) $B^+p \to B^+B^+p$.

(Suggestion: first write the formula out in terms of B and A, eliminating the occurrences of B^+. For discussion of some of the issues involved with iterated explicit belief – though in a different notation from that used here – see [1102].)

✠

As Fagin et al. [280], §9.5, emphasize, there are many possible applications of models of the present kind other than to (as we would put it) issues of concept possession, and for these the special condition from Example 5.6.4 would not be appropriate. For example, on a perfectly natural view (a view we saw contested by R. and V. Routley in note 377, p. 423), one might explicitly believe that p and explicitly believe that q without explicitly believing that $p \wedge q$, not because one is incapable of grasping the proposition that $p \wedge q$ but simply because one does not give that proposition explicit consideration, for which purpose models requiring that $\mathcal{A}(x)$ contains $p \wedge q$ whenever it contains p and q would not be appropriate. Further, even restricting attention to this particular application of the 'awareness' machinery, one may doubt whether the special condition on $\mathcal{A}$ in Example 5.6.4 is exactly what is wanted. It reduces all concept possession issues to the atomic level, ignoring the possibility that these issues may arise with respect to the compounding devices used to rise about this level, i.e., roughly speaking, to the logical vocabulary. One way of responding to this consideration is as follows, with a generalization of the inset condition (constraining when $\varphi(p_1, \ldots, p_n) \in \mathcal{A}(x)$) from Example 5.6.4:

EXAMPLE 5.6.6 The condition on (the frames of) models $\mathcal{M} = \langle W, R, \mathcal{A}, V \rangle$ mentioned under Example 5.6.4 could be replaced with the following. For any primitive n-ary connective $\#$, any $x \in W$, and any formulas $\varphi_1, \ldots, \varphi_n$:

If $\#(\varphi_1, \ldots, \varphi_n) \in \mathcal{A}(x)$ then

- $\varphi_i \in \mathcal{A}(x)$ (for $i = 1, \ldots, n$)

- for all $\psi_1, \ldots, \psi_n \in \mathcal{A}(x)$, we have $\#(\psi_1, \ldots, \psi_n) \in \mathcal{A}(x)$.

Note that like the condition from Example 5.6.4, the present condition still guarantees that whenever $\varphi \in \mathcal{A}(x)$ and ψ is a proper subformula, atomic or otherwise, of φ, then $\psi \in \mathcal{A}(x)$, in virtue of the first part of this double-barrelled condition. But we are not guaranteed to be able to reassemble the components without a check that the subject has mastery over the new mode of composition, as the second part of the condition requires. For example, we could have $p \vee q$ in $\mathcal{A}(x)$ to indicate that the subject grasps (in world x) the proposition that p or q, from which it follows that the subject grasps each of p, q, but whereas according to the proposal of Example 5.6.4, this suffices for the subject to grasp the proposition that p and not q, on the present suggestion the fact that $p, q \in \mathcal{A}(x)$ does not entitle us to conclude that $p \wedge \neg q \in \mathcal{A}(x)$, since the subject may not possess the concepts – to put it rather clumsily – of negation and conjunction. But as long as for some ψ, ψ' and ψ'', we have $\neg \psi \in \mathcal{A}(x)$

5.6. CONCEPT-POSSESSION PROBLEMS

(attesting to a grasp of negation) and $\psi' \wedge \psi'' \in \mathcal{A}(x)$ (attesting to a grasp of conjunction), the second part of the double-barrelled condition does secure the result that $p \wedge \neg q \in \mathcal{A}(x)$ (from the hypothesis that $p, q \in \mathcal{A}(x)$). If we were considering the richer area of predicate logic with its internally structured atomic formulas, there would be analogous conditions to consider, such as the following, in which F and G are monadic predicate letters and a and b, individual constants: if $Fa, Gb \in \mathcal{A}(x)$, then $Fb \in \mathcal{A}(x)$ – and a more elaborate variation for the case of n-adic predicate letters for $n \geq 2$. Such conditions are versions of the 'Generality Constraint' defended in §4.3 of Evans [276] – cf. also the related 'Structural Constraint' discussed in Chapter 3 of Davies [233]. A more recent treatment with many references to the literature is provided by Beck [58]. ◀

Setting aside the 'concept of conjunction' type of worry, the condition in Exanple 5.6.6 requiring that if $\psi_1, \ldots, \psi_n \in \mathcal{A}(x)$, then $\#(\psi_1, \ldots, \psi_n) \in \mathcal{A}(x)$ lends itself to a variation giving concept possession considerations their due without compromising congruentiality:

EXAMPLE 5.6.7 The variation in question appears in Forrest [324] in which the condition just recalled comes to life in – not pseudo-neighbourhood but genuine – neighbourhood models $\langle W, \mathcal{N}, V \rangle$:

$$\langle W, \mathcal{N}, V \rangle \models_x A\varphi \text{ if and only if } \|\varphi\| \in \mathcal{N}(x).$$

Forest writes "N" rather than "A" to suggest the idea of a 'natural' proposition, but this turns out to be essentially a matter of being a humanly graspable proposition (rather than purporting to have any objective metaphysical significance, as is sometimes intended when people speak of *natural properties*), and here we use the 'A'-for-'awareness' notation because of its usage above in connection with conceptual resources. The important thing for present purposes is that these models are required to above the condition that the set of neighbourhoods $\mathcal{N}(x)$ of a point x are required to be closed under the Boolean operations of intersection and (W-relative) complementation, so that for any Boolean compound $\#(\psi_1, \ldots, \psi_n)$ of $\psi_1, \ldots, \psi_n$:

if $\|\psi_i\| \in \mathcal{N}(x)$, for $i = 1, \ldots, n$) then $\|\#(\psi_1, \ldots, \psi_n)\| \in \mathcal{N}(x)$.

(One may want to erase the dependence of $\mathcal{N}(x)$ on x, as Forrester in essence does by adding a necessity operator $\Box$ governed with an **S5** logic and a condition validating the schema $A\varphi \to \Box A\varphi$.) It is possible to obtain similar a logical effect by using, in place of the apparatus of neighbourhoods, suitably marshalled equivalence relations on W: see von Kutschera [682] and Humberstone [576], where "A" is written as "O" – in the former case to suggest "objective": see the end of the first paragraph of Example 5.4.11, which begins on p. 397. (In the case of the latter work, rather different motivating considerations are in play). The logical effect in question is of course that of validating the schema

$$(A\psi_1 \wedge \ldots \wedge A\psi_n) \to A\#(\psi_1, \ldots, \psi_n),$$

with $\#$ understood as above. (The $n = 0$ instances in which only the consequent $A\top$ and $A\bot$ survives, are to be taken as subsumed under this schema. There is a typo concerning this in Forrest [324]: Axiom $N1$ on p. 95 should read "$N\top$" rather than simply "$\top$".) Note that the converse schema is not in general valid (though some special cases, e.g., with $\# = \wedge$ and $\psi_1 = \psi_2$, will be, in virtue of congruentiality): we have a version of the traditional principle of compositionality – understanding the parts suffices for understanding the whole (though setting aside the issue of the mode of composition – here represented by $\#$) – but not the principle of *reverse* compositionality, as it has been called (see Robbins [963] and Johnson [627] for discussion and further references). Similarly, the Generality Constraint is not part of the story: nothing forces us to have $\|\varphi \vee \chi\| \in \mathcal{N}(x)$ whenever, say, $\|\varphi \vee \psi\| \in \mathcal{N}(x)$ and $\|\chi \vee \psi\| \in \mathcal{N}(x)$. The backward version of the schema inset above would spell disaster in a congruential setting, since for arbitrary φ and ψ, this validates the implication $A\varphi \to A(\varphi \vee (\varphi \wedge \psi))$ and so by 'reverse compositionality', $A\varphi \to A(\varphi \wedge \psi)$, and then, again, $A\varphi \to A\psi$: if any proposition is entertainable by the subject, so is every proposition. In fact we have the antecedent here, since $A\top$ is valid. Thus A has become a 'Verum' operator, and adding an accessibility relation for the belief operator B would make the second conjunct in $B\varphi \wedge A\varphi$ redundant rather than a restriction giving us – one version of – explicit belief. (By rewriting $A\top$ as $A(\varphi \vee \neg\varphi)$ we get even more immediately via the converse of the above schema to the conclusion that $A\varphi$.) ◀

REMARK 5.6.8 Recall from the discussion (see p. 343) of deductive competence, concept possession considerations, and 'failures of closure' arising on some accounts of knowledge from neither of those sources, that there is more than motivation for distinguishing explicit from implicit knowledge in a non-monotone (or more generally non-normal) setting; see also Example 5.1.6(*iii*), p. 340. Here we begin by recapitulating a point from the Digression on p. 383. A subject not knowing something that follows from something that subject knows might be said to know it implicitly in a narrow sense if capable of entertaining the proposition not known, and to know it implicitly in a more attenuated sense if not capable of entertaining that proposition. To the extent that one is entitled to intuitions about the application of the phrase *epistemic possibility* – not a great extent, perhaps, since this clearly a piece of technical terminology – the question as to whether in the non-monotone case for a subject who does not know that $\neg A$ while $\neg A$ follows from what the subject knows, its being the case that A counts as an epistemic possibility is something on which to test such intuitions. Or perhaps better, simply to distinguish the two notions involved (Humberstone [567] p. 602*f*.). Matters are complicated further if the proposition ruled out by what the subject knows is not graspable by the subject. Here is a case from Huemer [531], p. 122:

> Rigel 7 is the seventh planet in the Rigel star system. Sam, however, knows nothing of Rigel and consequently has no thoughts about Rigel or any of its planets. Sam looks at his couch in normal conditions and sees nothing on it. Mary (who happens to know of Rigel 7) says "For all Same knows, Rigel 7 might be on the couch".

Commenting on the example (p. 123), Huemer writes:

> To see the bizarreness of the claim in the Rigel 7 case, it is important to note that Mary's claim is not to be understood as a metalinguistic one. Mary is not asserting that for all Same knows, the *sentence* "Rigel 7 is on the couch" might express a truth. Rather, Mary is to be understood as using (not mentioning) "Rigel 7" in the way she, having a correct understanding of the term, would normally use it, and asserting that a certain proposition, one that would be true only if the seventh planet orbiting the star Rigel were on Sam's couch, is among the epistemic possibilities open to Sam.

Here, presumably the point is that we are not happy to say that for all Sam knows, Rigel 7 is on the couch, even though he does not explicitly know that this possibility does not obtain, because this is a possibility ruled out by what Sam does know, despite his inability to articulate the proposition thereby ruled out. (For example, he knows that there is not a planet on his couch, which we may take to entail that Rigel 7 is not on the couch.) ◀

Curiously, the discussion in Routley and Routley [976] of concept possession worries concentrates on the case of connectives, such as $\wedge$, rather than on the role of descriptive or non-logical concepts, and they argue that such worries are taken too seriously by opponents of their 'principle of fair redescription', one of the arguments involving the sentence "I live in a red house". They argue that if a sincerely asserts this sentence, then a can be reasonably reported as believing that they live in a house and it is red. If we were to speak a variant of English in which predicative (as opposed to attributive) adjectives cannot appear pre-nominally, this would be the only way – at least if relative clauses are also absent from the envisaged dialect – to report a's beliefs, and it would be churlish to object to the intrusion of the word "and" in our belief-attribution.

Although in Example 5.6.4, as well as in the example of the Routleys just given, we have concentrated on Boolean connectives playing the role of # there, we should also attend to the case of non-Boolean

5.6. CONCEPT-POSSESSION PROBLEMS

too, and in particular in the present context, of B and K appearing there. Such a shift of attention may cause us to question not just negative introspection but also positive introspection principles on concept possession grounds. Many have objected to 4_K, for instance as described in Examples 5.2.5 and 5.2.6 (pp. 376 and 377), though the present objection would be rather different from those. It would be that surely some cognising agent – a non-language-using animal or a young child, for example – could know various things to be the case, for example, that food is present, without having the concept of knowledge itself and thus without being in a position even to entertain the proposition (the proposition that they know that food is present) which according to 4_K they know to be true. The same would apply, *mutatis mutandis*, in the case of 4_B, as well as the mixed principle of positive certainty ('PC': $B\varphi \to BK\varphi$). To get any doxastic or epistemic logic of interest under way, then, it may be necessary to restrict attention to cognitive agents who are assumed not only to be rational but also to be in possession of the concepts of belief and knowledge themselves. The corresponding negative introspection principles will not be salvaged by making such a move, however, as we note in Example 5.6.11 below.

For the remainder of our discussion we will not employ the kind of two-tier notation of Examples 5.6.3 ($\Box/\Box^+$) and 5.6.4, 5.6.6 (B/B^+) but instead treat the pure doxastic or pure epistemic case monomodally and the mixed doxastic–epistemic case as a bimodal logic, though we continue to make informal use of the idea of explicit belief or explicit knowledge.

EXAMPLES 5.6.9 (*i*) We have proposals of the disjunctivist kind alluded to earlier, according to which we have some operator O ("opines" in the first passage from Williamson quoted above) and in terms of B and this operator – about which more would have to said, of course – we define $B\varphi$ as $K\varphi \vee O\varphi$. More would have to said not only at the informal level, to make such an identification of knowledge both plausible and non-circular, but formally too, if we expect the B that emerges to have various logical properties (such as normality, for example). An artificially crude version of this kind of disjunctivism would arise if we took O not as a 1-ary connective but as a 0-ary connective (a constant), which we accordingly write differently, as μ – mnemonic for "the subject is mistaken" – defining $B\varphi$ as $K\varphi \vee \mu$. This would lead to disaster, enabling us to prove (in any modal logic) $(B\varphi \wedge \neg K\varphi) \to B\psi$. (Evidently when one says "Unless I am mistaken, such and such is the case", this is elliptical for "Unless I am mistaken about the matter in hand,...", so the content is not well treated by a propositional constant. One could of course choose to study the logic of false belief taking this as a primitive in its own right: see Steinsvold [1088].)

(*ii*) A more sophisticated version of the artificial proposal just mentioned would introduce parentheses: define $B\varphi$ as $K(\varphi \vee \mu)$. This option was explored (without in the end being endorsed) in Bacon [47], a paper whose title, 'Belief as Relative Knowledge', is intended to recall that of Smiley's 'Relative Necessity' (= [1064]): see p. 257 above, in the notation of which (taken from Kanger) we would write the negation of μ as Q. ("μ" corresponds to Anderson's 'sanction constant' "S" from our discussion in Section 4.4.) This returns us to the familiar theme of completeness w.r.t. classes of frames specified in terms of range restrictions – linguistically, propositional constants as the antecedents of strict implications – as in Theorems 4.4.39 (p. 271) and 4.4.43 (p. 273), and indeed to the two cluster frames that figured so prominently in Section 5.5 (as at Coro. 5.5.10, Prop. 5.5.16(*ii*), from pp. 409 and 413, respectively), though Bacon's discussion is complicated by an admixture of non-normality (normal vs. non-normal worlds as on p. 360), and it is not summarized here. Dissatisfied in the end with this attempt to define belief in terms of knowledge, Bacon reverts to the traditional direction of analysis and asks ([47], p. 207f.) what has to be conjoined with $B\varphi$ to give something equivalent to $K\varphi$, and his answer is what we are writing as Q. This would, however, have the unfortunate consequence that $K\varphi \wedge B\psi$ would provably imply $K\psi$.[382]

(*iii*) The title of Moses and Shoham [833] should be compared with that of Bacon [47] given under (*ii*)

---

[382] Bacon uses a different notation, using "p" as a schematic letter – like the "φ" just seen – instead of as a propositional variable. Also, he writes ∇ and Δ for the present μ and Q. [47] is also marred by serious comprehension-threatening misprints such as various turnstiles appearing on p. 190 where negated turnstiles are intended.

above: 'Belief as Defeasible Knowledge'. Their informal remarks suggest a similar intention: "Roughly speaking, we will translate each occurrence of 'the agent believes that φ' into 'the agent knows that either φ is the case, or else some specific (perhaps unusual) circumstances obtain'." That would be Bacon's idea (from (ii)) again, with the second disjunct here being μ. In fact, instead, the authors go on to give three alternative definitions of what they write as "$B^\alpha \varphi$" indicating which definition is in force by subscripting with a 1, 2 or 3, and consider their effects against various antecedently given logics for K:

$$B_1^\alpha \varphi = K(\alpha \to \varphi) \quad B_2^\alpha \varphi = K(\alpha \to \varphi) \land (K\neg\alpha \to K\varphi) \quad B_1^\alpha \varphi = K(\alpha \to \varphi) \land P\alpha,$$

though the authors actually write $\neg K \neg \alpha$ rather than $P\alpha$ here ([833], p. 304, as well as writing $\supset$ for $\to$). It is clear that "α" is a schematic letter for formulas, just like "φ", and writing it in superscript position and in a smaller font should not deceive us into thinking that anything other than a *binary B-connective* has been supplied by any of these definitions, not the 1-ary B operator the remark just quoted from the authors – to say nothing of their title – promised to deliver. (Here we are concerned with analysing belief, or explicitly defining B with the aid of K. There is also an interesting question concerning implicit definition, raised in Halpern, Samet and Segev [431], treated below; see, in particular, Proposition 5.6.21, p. 438.)

(iv) We come, finally, to the proposed account of B in terms of K that has been the focus of most attention in this chapter (Theorem 5.2.1, p. 372, and the discussion immediately thereafter, as well as intermittently throughout Section 5.5): Wolfgang Lenzen's suggested identification of $B\varphi$ with $\neg K \neg K \varphi$, or, as we usually write it for short: with $PK\varphi$. This proposal is vulnerable to the same concept-possession objection as the otherwise considerably less plausible identification from the start of the present section, of $B\varphi$ with $P\varphi$: If the subject – a, say – is not in a position to grasp the proposition expressed by φ, then $B\varphi$ (with B for "a believes that") is false while $\neg K \neg K \varphi$ is true since a is not in a position to grasp the proposition expressed by $K \neg K \varphi$ either. ◀

Example 5.6.9(iv) returns us to the main theme of the present section, so let us consider one possible reaction to it. The general moral would appear to be that we cannot endorse any equivalence of the form $O_2\varphi \leftrightarrow \neg O_1\psi$, where $O_1, O_2 \in \{K, B\}$, or more generally where the O_i are any intended to be interpreted as propositional attitude ascribing operators. In particular, the $\leftarrow$-direction of this equivalence – $\neg O_1\psi \to O_2\varphi$ – seems vulnerable to the objection that even an ideally rational cogniser might lack the concepts required to grasp the propositions expressed by φ and ψ, thereby verifying the antecedent and falsifying the consequent. Admittedly, that can't be quite the right thing to say for someone wishing to press some concept-possession objections while still endorsing normality for K and B, though, in view of the following:

EXAMPLES 5.6.10 For instance, any normal – indeed any monotone – bimodal logic for K and B will contain as theorems all of the following:

(i) $\neg Kp \to K(q \to q)$ $\quad$ (ii) $\neg Bp \to K(q \to q)$
(iii) $\neg Bp \to B(q \to q)$ $\quad$ (iv) $\neg Kp \to B(q \to q)$.

◀

Whatever the appropriate reaction to such cases may be (see Example 5.6.16 below for a suggestion), there is a modification of Lenzen's identification that we could make, which would avoid the present objection: we replace the Lenzen's favoured epistemically constituted candidate for B, namely PK, with another: KPK – mentioned but passed over in Lenzen [706] (see the Digression below). The behaviour of this 'Matsumoto' modality $\Box \Diamond \Box$ in **S4** is something we have seen quite a bit of: Matsumoto's (non-compositional) embedding of **S5** in **S4** (Thm. 2.8.4, p. 144) as well as the Åqvist embedding (via the compositional translation $\tau_{\Box \Diamond \Box}$), of $\mathbf{KDB}^\Box \mathbf{4}$ in **S4** – Thm. 5.5.24 on p. 419 – with Example 4.5.25 (p. 290) for further background.

5.6. CONCEPT-POSSESSION PROBLEMS 431

Digression. The reasons Lenzen gives in [706] for setting this to one side are rather complex,[383] but the key concern arises from Lenzen's Theorem (Thm. 5.2.1(iv)–(v), p. 372 above), which tells us that

$$\mathbf{KD45}_B \subseteq \mathbf{S4.2}_K \oplus Bp \leftrightarrow PKp,$$

and also

$$\mathbf{S4.4}_K \oplus Bp \leftrightarrow PKp = \mathbf{KD45}_B \oplus Kp \leftrightarrow (Bp \wedge p).$$

What happens if we replace $Bp \leftrightarrow PKp$ with $Bp \leftrightarrow KPKp$ here? Lenzen's idea is that if one were to accept the equivalence of knowledge with true belief then one's preferred epistemic logic would be **S4.04**, where **.04** is like **.4** but with an extra "□": $(\varphi \wedge \Box\Diamond\Box\varphi) \to \Box\varphi$, or, in epistemic notation:

$$(\varphi \wedge KPK\varphi) \to K\varphi.$$

(**.04** is called **Zem** – after Zeman – and **S4.04**, **S4Zem**, in Segerberg [1025], p. 152; we use the traditional nomenclature for the convenience of anyone following up Lenzen's work as originally published.) But $\mathbf{S4.04}_K \oplus Bp \leftrightarrow KPKp \neq \mathbf{KD45}_B \oplus Kp \leftrightarrow (Bp \wedge p)$, since $\mathbf{5}_B$ cannot be derived in $\mathbf{S4.04}_K$ when $B\varphi$ is defined to be $KPK\varphi$; indeed, Lenzen notes, written out in terms of K and P for this definition, $\mathbf{5}_B$, $Cp \to BCp$, is

$$PKPp \to KPKPKPp,$$

whose consequent reduces in any extension of $\mathbf{S4}_K$ to KPp (see Figure 2.2 on p. 42, and in particular the entry $\Box\Diamond \circ \Box\Diamond = \Box\Diamond$), and $PKPp \to KPp$, or the dual form $PKp \to KPKp$, is $\mathbf{S4}_K$-interdeducible with $PKp \to KPp$. (Optional exercise: Prove the claim just made.) Lenzen then points out that the cost of recovering $\mathbf{5}_B$ is thus that we are committed to at least $\mathbf{S4.2}_K$, and hence to the even stronger $\mathbf{S4.4}_K$, if we also accept the $Kp \leftrightarrow (Bp \wedge p)$. (See the discussion of **S4.2** as an 'unstable option' at p. 373 above.) Even if we do not accept the latter equivalence, we are still forced to a stronger epistemic logic (i.e., **S4.2**) than the supposed candidate **S4.04**. A semantic description of **S4.04**, incidentally, is as the logic determined by the class of transitive reflexive frames which are remotely symmetric in the sense introduced in Exercise 2.7.34(iii) and recently recalled – on p. 414 – as was observed in Georgacarakos [365]. ***End of Digression.***

The worries Lenzen had with the KPK alternative to his own proposal reviewed in the above Digression arise out of a desire to retain $\mathbf{5}_B$. But this motivation illustrates very clearly a reluctance to take concept possession considerations seriously; see Remark 5.3.4 (p. 384) for a similar discussion over $\mathbf{B}_K$:

EXAMPLE 5.6.11 The candidate axiom $\mathbf{5}_B$, taken as $\neg Bp \to B\neg Bp$, is of precisely the forbidden form $\neg O_1 \psi \to O_2 \varphi$ noted above, with B as O_1 and O_2, and ψ, φ, respectively being p, Bp: if Sam is not conceptually equipped to entertain the proposition that p then Sam does not believe that p, while failing to believe that he does not believe that p.[384] So for anyone with whom these concept possession objections weigh heavily, it is a point in favour of, rather than against, a doxastic–epistemic logic that it should lack $\mathbf{5}_B$. (And corresponding considerations, quite independently of the usual objection to $\mathbf{5}_K$ – as recalled early in Section 5.2 – would similarly tell against the latter principle.) The idealization was mentioned above that for purposes of doxastic (epistemic) logic, the subject is presumed to possess the concept of belief (resp., of knowledge), which would essentially forestall (by fiat) concept possession objections to such principles as $\mathbf{4}_B$; this would not help with $\mathbf{5}_B$, though, since when $\neg B\varphi$ is true because of lack of the concepts deployed in φ even having the concept of belief will still leave the subject unable to entertain,

---

[383] I am most grateful to Lenzen for a communication in 2010 providing me with a paraphrase in English of the relevant portion of [706] (beginning with "DEF. 2*" on p. 51), no less than for his similarly informative 1982 letter, mentioned on p. 372 above.

[384] Reminder: despite appearances, as explained in note 372 (p. 421), there is no commitment to propositions as entities in their own right with their own identity conditions involved in such formulations as this.

and therefore to believe, the proposition that $\neg B\varphi$, through lack of those same concepts. Often candidate principles in doxastic and epistemic logic are thought of informally with first-person examples: supposing myself to be introspectively aware of my beliefs, I imagine that if I were not to believe that average house prices across the world will fall in the next year – then I would know, by monitoring my stock of beliefs, that this one (i.e., the belief that house prices will fall – to put it briefly) is not among them, and so I would certainly believe that it wasn't among them – that is, I would believe that I did not believe that house prices would fall. The trouble with this way of thinking is that since the person considering the example is the same as the doxastic subject in the example, it is not possible for a specific concept to be deployed in the example which is not possessed by the subject – for example the concept of (house) prices.[385] But, moving into third person mode, we can easily consider another subject, Rani, say, not possessing this concept, for example having lived entirely in a cooperative tribal culture in which nothing plays the role of money. Then *we* can see that Rani lacks the belief about house prices even though Rani herself, however great her introspective powers may be, is not in a position to believe that she lacks this belief. ◀

Since we wish to retain the idea that B has a normal modal logic, the mere fact that a belief attribution involves a concept not possessed by the subject should not make us count it as false, though the contrary appearance may be created by such discussion as is provided in Example 5.6.11. For suppose that, while the Rani of that example is not in a position to entertain the proposition that q (for example, think of this as the falling house prices proposition in that example), she is not only in a position to entertain, but in fact believes, the proposition that p. Then since any normal modal logic has $B(p \vee q)$ as a consequence of Bp, we will be agreeing that Rani believes that $p \vee q$, which seems inconsistent since she does not grasp the proposition that $p \vee q$ which we have just used to attribute to the content of a belief to her. The reply will be that she believes implicitly that $p \vee q$ in the sense that she explicitly believes something (namely p) from which this follows.[386] But in the case of Example 5.6.11, Rani not only fails to believe explicitly – through lack of the requisite concepts – that she does not believe house prices will fall: in this case one cannot point to anything she does believe explicitly, from which it follows that she does not believe that house prices will fall. (Of course this is all very vague, since what follows from what will vary with the background logic assumed in force. But the present attempt is just to present an informal case for the availability of a doxastic logic which is both normal – with all the logical omniscience that involves – and also sensitive to concept-possession considerations.)

REMARK 5.6.12 More formally, where S is a candidate normal doxastic logic extending **KD**, we want to argue on concept possession grounds that S should not prove **5**, by consideration of its instance $\neg Bq \to B\neg Bq$, by reference to the example just detailed. Rani cannot entertain the proposition that q, so the antecedent of this implication is true; for the consequent to be true we need some φ – or a set of φ_i playing this role collectively, though we simplify here to the monotone version – such that Rani explicitly believes that φ and $\vdash_S \varphi \to \neg Bq$. Since Rani explicitly believes that φ, she possesses the necessary concepts and the formula φ is not constructed with the aid of q. So the uniform substitution of $\top$ for q will not disturb φ and we have $\vdash_S \varphi \to \neg B\top$, and thus, contraposing: $\vdash_S B\top \to \neg\varphi$, so since S proves the antecedent here, $\vdash_S \neg\varphi$. Therefore $\vdash_S B\neg\varphi$ and so $\vdash_S \neg B\varphi$ (by **D**): contradicting the assumption that φ is something Rani believes. ◀

Let us pause to note one aspect of the description of **5** (for B) in Jaspars [619] (p. 142):

---

[385]Similarly, while there are doubtless many propositions I am not capable of entertaining, I cannot cite any of them by filling the blank with a sentence expressing the proposition in question, in "I cannot entertain the proposition that __". It is possible, all the same, to reason about propositions one cannot thus express; a striking example of such reasoning appears in Chalmers [167].

[386]That is a monotone rather than a fully normal formulation, which requires only that it should follow from some set of statements she explicitly believes. The present issue was touched on in Remark 5.3.4, p. 384.

5.6. CONCEPT-POSSESSION PROBLEMS

> Sometimes the axiom of negative introspection is accepted in doxastic logic (...). This axiom states that if an agent disbelieves a proposition, he believes that he does not believe it.

Jaspars goes on to give the axiom in schematic form as $\neg\Box\varphi \to \Box\neg\Box\varphi$. The needed correction arises over the word *disbelieves*: disbelieving something means believing that it is not the case, rather than failing to believe that it is the case – whence the traditional trichotomy between belief, disbelief, and suspension of judgment (this last being the doxastic analogue of contingency). As remarked on p. 203, disbelief is a propositional attitude, while absence of belief is not. So the gloss in terms of disbelief misrepresents what **5** says and at the same time makes it seem more plausible than it is. (Or at least, so one may react initially. There is room for a little more subtlety, though: one may interpret disbelief as something weaker than believing-that-not but still stronger than mere absence of belief – as rather, considered withholding of belief. This would be somewhat analogous to the way we treat *refraining from flicking a switch* as more then merely not flicking the switch, but instead as involving, at the very least, the *ability* to flick the switch – and perhaps also the conscious decision not to exercise this the ability. A similarly conscious decision to refrain from believing a proposition, even in the absence of believing its negation, would require possession of any concepts involved and deserve to count as (adopting) a propositional attitude. But this is still no help to **5** because its antecedent, $\neg B\varphi$, does not ascribe this attitude to the subject. Note that here we are replaying, in a doxastic rather than an epistemic setting, the objection to van der Hoek, Jaspars and Thijsse [509] from p. 383 above.

As we saw in the Digression above, when B is defined as KPK against the background of $\mathbf{S4}_K$, $\mathbf{5}_B$ is not provable, which is accordingly, now that we attending to concept possession considerations, a good thing. The KPK proposal should also be compared with Lenzen's favoured PK proposal in respect of their repercussions for mixed doxastic–epistemic principles. By way of reminder, those in play in this chapter are repeated here with their abbreviations, so that the latter can be used in Figure 5.5 which charts their fate under the two proposals, notation – and especially underlining – as in Example 5.2.3, p. 373.

$$\text{KB:} \quad K\varphi \to B\varphi \qquad \text{PC:} \quad B\varphi \to BK\varphi \qquad \text{NC:} \quad \neg B\varphi \to B\neg K\varphi$$

$$\text{PI:} \quad B\varphi \to KB\varphi \qquad \text{NI:} \quad \neg B\varphi \to K\neg B\varphi$$

Label	Schema	B as PK	B as KPK
KB	$K\varphi \to B\varphi$	$\Box\varphi \to \Diamond\Box\varphi$	$\Box\varphi \to \Box\Diamond\Box\varphi$ (✓)
PC	$B\varphi \to BK\varphi$	$\Diamond\Box\varphi \to \Diamond\Box\Box\varphi$	$\Diamond\Box\varphi \to \Box\Diamond\Box\varphi$ (✓)
NC	$CK\varphi \to B\varphi$	$\Box\Diamond\Box\varphi \to \Diamond\Box\varphi$	$\Diamond\Box\Diamond\Box\varphi \to \Box\Diamond\Box\varphi$ (✗)
PI	$B\varphi \to KB\varphi$	$\Diamond\Box\varphi \to \Box\Diamond\Box\varphi$ (⋆)	$\Box\Diamond\Box\varphi \to \Box\Box\Diamond\Box\varphi$ (✓)
NI	$PB\varphi \to B\varphi$	$\Diamond\Diamond\Box\varphi \to \Diamond\Box\varphi$	$\Diamond\underline{\Box\Diamond\Box}\varphi \to \Box\Diamond\Box\varphi$ (✗)

All in column 3 provable in **S4.2**, with only that marked with a star requiring the ".2" part of this label. (See Thm. 5.2.1(*iii*), p. 372.) Ticks and crosses in column 4 indicate provability and unprovability in **S4**. Underlining in column 4 to indicate doxastic component: see Example 5.2.3, p. 373.

Figure 5.5: *Bridging Principles with B defined in terms of K.*

We use the $\Box/\Diamond$ notation in the third and fourth columns, rather than the K/P notation in order to have something neutral and amenable to potentially different epistemic and doxastic glosses, as indicated by the underlining in lines three and five – where the formulas concerned, in the case of the fourth column actually coincide. One has a similar situation with the modality $\Box\Diamond\Box\Diamond$, which, taken as $\Box + \Diamond\Box\Diamond$ has a doxastic–epistemic reading as KC, and, taken as $\Box\Diamond\Box + \Diamond$, a reading as BP. Thus the "B as KPK" approach is not suitable for anyone rejecting the equivalence $KCp \leftrightarrow BPp$. Note, however, that this equivalence does not identify a B or K claim with a negated B or K claim, and so does not fall foul of the ban on such identifications. The contentious direction (for concept possession concerns), of the unwanted equivalences is the implication from negated to unnegated propositional attitude ascriptions. In Figure 5.5 we see that NC and NI disappear with the KPK treatment of belief (on which, from the final column, we see that they correspond to the same $\Box/\Diamond$ formula – with different underlinings to highlight the correspondences with the mixed doxastic–epistemic principles). Note also, when it comes to *unmixed* principles, that in using **S4** as the basic epistemic logic rather than **S4.2** we avoid the **.2** axiom $PK\varphi \to KP\varphi$ which is also of the forbidden form. (This is a mild variation on the argument of Example 5.6.9(iv) above; note that the objection does not apply to the original doxastic principle $\mathbf{D}_B$ in the form $B\varphi \to C\varphi$, which Lentzen encodes as $.2_K$, since this is an implication from a propositional attitude ascription to the negation of a propositional attitude ascription rather than the other way around. So much the worse for the encoding – which was the message of Example 5.6.9(iv).)

When it comes to unmixed principles, we can usefully compare Lenzen's proposal with the suggested revision by recording the relations between the monomodal logics involved with the aid of Zolin's notation (Example 4.4.27, p. 266, *q.v.* also for the $\boxdot$ notation used here); we set out the contrast as Proposition 5.6.13, given here without a separate proof as most of the ingredients have already been established elsewhere (Thms. 4.5.9, 5.2.1, 5.5.24), while that under (ii) pertaining to **S4.04**, mentioned in the Digression above (p. 431) is dealt with as for the **S4.4** part of (i):

PROPOSITION 5.6.13 (i) $\mathbf{S4.2}(\Diamond\Box) = \mathbf{KD45}$; $\mathbf{KD45}(\boxdot) = \mathbf{S4.4}$.
(ii) $\mathbf{S4}(\Box\Diamond\Box) = \mathbf{KDB}^{\Box}\mathbf{4}$; $\mathbf{KDB}^{\Box}\mathbf{4}(\boxdot) = \mathbf{S4.04}$.

The pure epistemic and doxastic fragments are of less interest, however, than the full system in which B and K interact, some of which interaction is highlighted in Figure 5.5. This is why it is these aspects of Lenzen's work that come first in the statement of Theorem 5.2.1 above (p. 372), and it is in respect of them that Lenzen's project distinguishes itself from such work as that of Dawson [237], as was argued on p. 279 (including note 232).

While, as we have seen, defining B as KPK in the setting of $\mathbf{S4}_K$ produces a mixed epistemic–doxastic logic with several advantages over Lenzen's original proposal (B as PK in the setting of $\mathbf{S4.2}_K$) on the concept possession front, there are some disappointing features. One is the obvious complexity of the *definiens* if it is taken seriously as giving an account of belief: surely believing that those people are your parents cannot be a matter of knowing that you don't know that you don't know that they are not – the latter is far too complicated! Even if we waive the concept possession objection to positive introspection principles, which in the present context takes the form of objecting that one can believe something to be the case without even having the concept of knowledge here deployed as part of the content of the knowledge that the belief is identified with (the $PK\varphi$ subformula of $KPK\varphi$), it is hard to swallow the depth of embedding with which it is here deployed: at least with Lenzen's proposal, Bq when unpacked in epistemic terms only had modal degree 2, while it now emerges as having degree 3.[387]

Since this section is mainly concerned with belief rather than knowledge, before Proposition 5.6.13 is too far behind us, it is worth noting that part (ii) thereof does not commit a B-as-KPK advocate of $\mathbf{KDB}^{\Box}\mathbf{4}$ as a doxastic logic to one or other of the logics $\mathbf{S4}$, $\mathbf{S4.04}$, mentioned there, as an epistemic logic, and just as we have, for example, fleshing out the first half of Prop. 5.6.13(i) in accordance with findings from Section 5.5:

$$\mathbf{S4.2}(\Diamond\Box) = \mathbf{S4F}(\Diamond\Box) = \mathbf{S4.4}(\Diamond\Box) = \mathbf{KD45},$$

so also in the present case we have the following expansion of the first half of Prop. 5.6.13(ii):

---

[387]See p. 70 for the notion of modal degree.

5.6. CONCEPT-POSSESSION PROBLEMS

PROPOSITION 5.6.14 $\mathbf{S4}(\Box\Diamond\Box) = \mathbf{S4B}_2(\Box\Diamond\Box) = \mathbf{S4.04}(\Box\Diamond\Box) = \mathbf{KDB}^\Box\mathbf{4}$.

Hardly any further work beyond that of the proof of Theorem 5.5.24 is required for the cases of $\mathbf{S4B}_2$ and $\mathbf{S4.04}$ here: just the observation at the end of the 'If' part of that proof that the model constructed has R-chains of clusters of length at most 2 in the former case, and further that the first cluster, when there are two, contains only a single (reflexive) point. Since $\mathbf{B}_2$ has not received attention since Example 3.2.14 (p. 189) and Exercise 3.2.15, let us recall that this axiom, $p \to \Box(\Diamond p \vee (q \to \Box\Diamond q))$, precludes having $xRyRz$ without either yRx or zRy, and thus on a reflexive transitive frame rules out there being three clusters C_1, C_2, C_3, with C_3 accessible to C_2 and C_2 accessible to C_1 (and therefore also C_3 accessible to C_1), since we could then choose points x, y, z in these clusters, respectively, and the condition just considered would collapse at least two of the clusters into one. We include the logic $\mathbf{S4B}_2$ here for those wanting a KPK-oriented version of the 'two clusters' logic $\mathbf{S4F}$ touched on in Section 5.5, where the (qualified) enthusiasm of Stalnaker [1082] for its epistemic credentials was mentioned.

The epistemic logics mentioned in Proposition 5.6.14 are listed there in order of increasing strength, as is seen by reflection on their semantic characterization. One aspect of the situation is worth reviewing syntactically – the inclusion of $\mathbf{S4B}_2$ in $\mathbf{S4.04}$:

EXAMPLE 5.6.15 Recalling from p. 431 – though we continue writing "$\Box$" rather than "K" here – that .04 is, in schematic form:

$$(\varphi \wedge \Box\Diamond\Box\varphi) \to \Box\varphi,$$

we sketch of a derivation of this formula in $\mathbf{S4.04}$. To make things easier we appeal to Matsumoto's Theorem (Thm. 2.8.4, p. 144), which tells us that $\vdash_{\mathbf{S4}} \Box\Diamond\Box(\Diamond p \vee (q \to \Box\Diamond p))$, since the disjunction after the "$\Box\Diamond\Box$" is $\mathbf{S5}$-provable in virtue of its second disjunct. This is the second conjunct of the antecedent in the following instance of the .04-schema just mentioned:

$$((\Diamond p \vee (q \to \Box\Diamond p)) \wedge \Box\Diamond\Box(\Diamond p \vee (q \to \Box\Diamond p))) \to \Box(\Diamond p \vee (q \to \Box\Diamond p)).$$

In view of the $\mathbf{S4}$-provability of the second conjunct of the antecedent, we infer that this simplified version is $\mathbf{S4.04}$-provable:

$$(\Diamond p \vee (q \to \Box\Diamond p)) \to \Box(\Diamond p \vee (q \to \Box\Diamond p)).$$

But in any extension of $\mathbf{KT}$, p provably implies the antecedent here (in virtue of its first disjunct), and therefore in $\mathbf{S4.04}$ provably implies its consequent. The implication in question is $\mathbf{B}_2$: $p \to \Box(\Diamond p \vee (q \to \Box\Diamond q))$. ◄

We return to the concentration on B (and to writing K rather than $\Box$, as in the preceding Example).

The examples mentioned under 5.6.10 have little to do specifically with the particular doxastic logic now under sympathetic investigation (viz. $\mathbf{KDB}^\Box\mathbf{4}$) or the combined epistemic–doxastic logic we obtain from $\mathbf{S4}_K$ by the KPK definition of B, but looking at them again will help us focus on aspects of these specific normal modal logics:

EXAMPLE 5.6.16 (Examples 5.6.10 revisited.) For definiteness, let us take Example 5.6.10(iii): $\neg Bp \to B(q \to q)$ as a theorem of every normal modal logic for B, apparently problematic since it is of the $\neg B\varphi \to B\psi$ form. Since the consequent is already provable (assuming normality) this should only count as a degenerate case of violating the restriction on theorems of the form $\neg B\varphi \to B\psi$. Writing this as a disjunction we have $\vdash_S B\varphi \vee B\psi$, and we know already that we have $\vdash_S B\psi$ for the present case, in which ψ is $q \to q$. If our cognitive subject does not possess the concepts required to entertain the proposition that q then, as long as the subject has some explicit belief or other, say the belief that φ, then implicitly and derivatively the subject believes that $q \to q$ since this follows from φ. So we note a further assumption of the present approach: the subject has at least one belief. This does not seem too extravagant an assumption for purposes of doxastic logic. There may be logical omniscience and concept possession objections to being told that a logic tells us that a believes such-and-such, but one does not usually hear it raised as an objection that a may be, for instance, a leather armchair – and so not the kind of thing to have any beliefs at all.[388] ◀

The disjunctive formulations of Example 5.6.16 are reminiscent both of the rule of disjunction (p. 112) and of the modified rule of disjunction (p.122), both here recalled for the case of $\Box = B$; having the (modified) rule of disjunction is a matter of having the (modified) n-ary rule of disjunction for all n:

S has the n-ary Rule of Disjunction: Whenever $\vdash_S B\varphi_1 \vee \ldots \vee B\varphi_n$, we have $\vdash_S \varphi_i$ for some i $(1 \leq i \leq n)$.

S has the Modified n-ary Rule of Disjunction: Whenever $\vdash_S B\varphi_1 \vee \ldots \vee B\varphi_n$, we have $\vdash_S B\varphi_i$ for some i $(1 \leq i \leq n)$.

There would be something strange about a doxastic logic which did not provide the modified rule of disjunction: for the $n = 2$ case the logic is telling us that the ideally rational cognitive agent believes that φ or else believes that ψ, but apparently that there is nothing wrong with not believing that φ (since $\nvdash_S B\varphi$) and nothing wrong with not believing that ψ (since $\nvdash_S B\psi$). How, then, does something go wrong (on grounds of rationality, consistency, etc.) when both beliefs are absent?[389] This rhetorical question is not exactly an argument, however, but *prima facie* may seem to offer support for **KDB**$^\Box$**4** over **KD45**, in the following counterexample to the thought that S's having the modified rule of disjunction implies S's having the rule of disjunction, for normal S:

EXAMPLE 5.6.17 Neither **KDB**$^\Box$**4** nor **KD45** provides the rule of disjunction. **KDB**$^\Box$**4** does – whereas **KD45** does not – provide the modified rule of disjunction. For the claim that neither of these logics has the rule of disjunction, note that both lack already the 1-ary rule ('Denecessitation') since, for example, both prove **U** (see Exc. 5.5.25(i)), but not **T**. **5** in the form $Bp \vee B\neg Bp$ provides a counterexample to the modified rule of disjunction for **KD45**$_B$. The multi-cluster semantical characterization of **KDB**$^\Box$**4** given in Thm. 5.5.23 (p. 419) makes available a simple proof that this logic enjoys the modified rule: if each $B\varphi_i$ fails at the irreflexive generating point w_i of a model $\langle W_i, R_i, V_i \rangle$ as described there, then φ_i is false at some point w_{φ_i} in a cluster accessible to w_i. So discard the points w_i and retain, for each i, the cluster containing w_{φ_i}, making a new multi-cluster model by adjoining a new point, w^+, say, to which all these clusters are accessible (and defining R and V in the obvious way): $\bigvee_{i=1}^n B\varphi_i$ will then be false at w^+ and hence not provable in **KDB**$^\Box$**4**. ◀

Denecessitation is crucial to the contrast just illustrated here between **KDB**$^\Box$**4** and **KD45** (even though these logic are alike in failing to be closed under this rule): See Exercise 2.6.27(iii) (p. 122). At the same time, the failure of Denecessitation for **KDB**$^\Box$**4** is potentially problematic, as again we illustrate with the simple case of **U** and **T** – though we could equally well have chosen, for instance, the case of **B**$^\Box$ (provable) and **B** (unprovable):

EXAMPLE 5.6.18 Can an informal justification for **U**$_B$ be provided like that in Example 5.6.16, or not (as in the case of **5**$_B$ – Remark 5.6.12, p. 432)? The latter would appear to be the case – bad news for **KDB**$^\Box$**4**. The latter logic proves $B(Bp \to p)$; let us suppose that our subject is not conceptually equipped to entertain the proposition that p. Thus there must be some φ not containing p for which (a) the subject believes φ explicitly, and (b) according to our favoured logic S, $\vdash_S \varphi \to (Bp \to p)$. To make trouble for

---

[388]There is a bit of rhetorical oversimplification here, as the reader will recognise, with an unacknowledged gap between being capable of having beliefs and in fact having at least one. Perhaps one might bridge this gap by arguing that the capacity to have beliefs required the possession of concepts, and the latter is impossible without actually having at least some beliefs involving (which is not to say beliefs *about*) the concepts in question.

[389]We could consider a mixed doxastic–epistemic version of this condition, or indeed much more generally, where the O_i are intended to be read as propositional attitude ascribing operators, the condition that if $\vdash_S O_1\varphi_1 \vee \ldots \vee O_n\varphi_n$ then $\vdash_S O_i\varphi_i$ for some i.

5.6. CONCEPT-POSSESSION PROBLEMS

KDB$^\Box$**4**, note that by $\mathbf{D}_B$, since the subject believes φ, φ must be consistent (i.e., the logic does not prove its negation). Now recall Proposition 5.5.26 (from p. 420), according to which for **KDB**$^\Box$**4**-consistent φ, the formula $\varphi \wedge (\Box p_i \wedge \neg p_i)$ is also **KDB**$^\Box$**4**-consistent, where p_i is any propositional variable – such as (taking $i = 1$) p itself not occurring in φ. This contradicts the requirement that $\vdash_S \varphi \to (Bp \to p)$ for our choice of S. ◀

With that observation we conclude our presentation of some of the points in favour of and against the suitability of **KDB**$^\Box$**4** as a doxastic logic. The remainder of this section detours through some investigations of Halpern and co-authors on implicit and explicit definition in modal logic (Halpern, Samet and Segev [431] and [430]), which will land us in a position to pose a final difficulty for a position – to which most of this section has been highly sympathetic – which takes concept-possession objections seriously. (This discussion was foreshadowed at the end of Example 5.6.9(*iii*) above.)

The twin papers [431] and [430] of Halpern, Samet and Segev make numerous observations and suggestions regarding notions of definability for normal modal operators in multi-modal logics, going well beyond anything that will be covered in the remarks which follow. One distinction they adapt from the case of first-order theories is that between such a logic S's implicitly defining an operator O, which, in its syntactic incarnation, amounts to its being the case that if we combine S with a reduplicated version, S' of S with O' rather than O in its language, which proves all the same things about O' its interaction with the remaining modal operators as S does about O, then the combined logic proves $Op \leftrightarrow O'p$.[390] On the other hand, S explicitly defines O if S itself proves something of the form $Op \leftrightarrow \varphi$, with φ not containing O. Evidently the provability of such an equivalence in S guarantees that S also implicitly defines O because S will prove $Op \leftrightarrow \varphi$ and S' will prove $O'p \leftrightarrow \varphi$, so the combined logic will prove the desired $Op \leftrightarrow O'p$.[391] But whereas explicit definability implies implicit definability the converse implication – by contrast with the first-order case ('Beth's Theorem') – does not obtain here. A simple and well-known case (mentioned, for example, in [553] and appearing as Observation 4.31.3 on p. 581 of [594]) is provided by tense logic:

EXAMPLE 5.6.19 The basic tense logic $\mathbf{K}_t$ with non-Boolean operators G, H, axiomatized on p. 179 above, implicitly defines H: in the combined logic with H and its duplicate H', $H'p \to HFH'p$ is provable (by HF-ax. and US – $H'p$ for p in the combined language, where we recall that F is $\neg G\neg$), as is $HFH'p \to Hp$, by HMono applied to the dual form of what we might call GP'-ax (P' for $\neg H\neg$); by TF we have $H'p \to Hp$, with a similar proof for the converse. But of course there is no explicit definition of H in $\mathbf{K}_t$ or the expressive power of $\mathbf{K}_t$ would coincide with that of $\mathbf{K}$ (or $\mathbf{K}_G$, if you prefer): the class of converse serial frames would not be (contra p. 183) modally definable in the language of $\mathbf{K}_t$, any more than it is in the monomodal case (Coro. 2.5.33(*iii*), p. 84). ◀

The case of tense logic may seem far from our present concerns, though any such appearance – doubtless encouraged by the tendency to think of tense-logical considerations as having something to do with *time* – will turn out to be deceptive. (For attempts to correct the tendency here alluded to, see Examples 4.1.11 and 4.4.29, on pp. 209 and 267 respectively.) But, returning to those concerns, it must first be conceded that Halpern *et al.* give their own epistemic–doxastic version of the phenomenon

---

[390]We may take the combined logic to be $S \oplus S'$, though [431] and [430] call this $S + S'$ (and call what we call $S + S'$, $S \oplus S'$). In Exercise 2.5.51(*ii*) on p. 96 above, this is expressed by saying that S uniquely characterizes O. In the material abstracted as [553], later appearing in §4.3 of Humberstone [594], this would be put by saying: S uniquely characterizes O in terms of the Boolean and other connectives in the language of S. This terminology is motivated by a desire to avoid certain complications coming from the analogy with the first-order case, though here we follow Halpern *et al.* in exposition of their work. Also, the way unique characterization is defined in [553] and [594] makes it a relation between a proof system (a collection of rules, including axioms as 0-premiss rules) and a connective, rather than a logic and a connective; on the present account the allusion to rules is repackaged into the notion of combination of logics – the rules becoming closure conditions on what counts as a logic for the relevant lattice of logics in which combinations are joins.

[391]There is no connection – other than a purely etymological one – between the terminological contrast between implicit and explicit definability on the one hand and that between implicit and explicit knowledge and belief on the other.

illustrated in Example 5.6.19 with a combined logic for K and B which incorporates $\mathbf{S5}_K$, meaning the discussion does not embody a plausible epistemic logic. On the B side of the story, we have $\mathbf{KD45}$, which, as we have lately been observing, raises some questions of its own in respect of concept-possession considerations. However, the chief interest of [431] (in the case of [430], less so) is a general theoretical one – of getting clear about implicit and explicit definition and their relation to a third notion, of what the authors call reducibility (which we will not be going into here). As well as $\mathbf{S5}$ for K and $\mathbf{KD45}$ for B, we need a couple of the bridging principles that have been in play in this chapter: KB ($= K\varphi \to B\varphi$) and PI ($= B\varphi \to KB\varphi$). It is this logic – $\mathbf{KD45}_B \oplus \mathbf{S5}_K \oplus \{\text{KB}, \text{PI}\}$ – which Halpern, Samet and Segev [431] show implicitly defines K, in their Theorem 4.4 (proof on p. 483 of [431]); the proof which follows is organized slightly differently from theirs, in order to isolate and highlight the adverse effects of $\mathbf{5}_K$ in the (evidently implausible) $\to$-direction of the schema mentioned in the following lemma:

LEMMA 5.6.20 *For* $S = \mathbf{KD45}_B \oplus \mathbf{S5}_K \oplus \{\text{KB}, \text{PI}\}$ *we have* $\vdash_S BK\varphi \leftrightarrow K\varphi$ *for all formulas* φ.

Proof. $\to$-direction: by $\mathbf{5}_K$ we have $\vdash_S \neg K\varphi \to K\neg K\varphi$, so by KB we may weaken the consequent: $\vdash_S \neg K\varphi \to B\neg K\varphi$. By $\mathbf{D}_B$ we get: $\vdash_S \neg K\varphi \to \neg BK\varphi$; contraposing gives $\vdash_S BK\varphi \to K\varphi$.
$\leftarrow$-direction: $\mathbf{4}_K$ gives $\vdash_S K\varphi \to KK\varphi$, from which KB delivers: $\vdash_S K\varphi \to BK\varphi$. ∎

Note that very little of the logic $\mathbf{KD45}_B \oplus \mathbf{S5}_K \oplus \{\text{KB}, \text{PI}\}$ is used in the above proofs, so a more general statement could have been made.

PROPOSITION 5.6.21 $\mathbf{KD45}_B \oplus \mathbf{S5}_K \oplus \{\text{KB}, \text{PI}\}$ *implicitly defines* K.

Proof. We reduplicate K to K', indicating the versions of the axioms involved by subscripting with one or other of these. (Thus, $\text{KB}_{K'}$ would be: $B\varphi \to K'B\varphi$.) It will suffice to prove $Kp \to K'p$ in the reduplicated logic, since we can interchange K and K' in such a proof to obtain a proof of the converse, giving the conclusion (in view of US) that for all φ, $K\varphi$ and $K'\varphi$ are equivalent (and indeed fully interreplaceable, since the logic is congruential). Here we sketch the (formal) proof, in which we omit the repeated antecedent from line (1) on:

(1) $Kp \to BKp$ Lemma 5.6.20, $\leftarrow$-direction
(2) $\quad\to K'BKp$ From 1 by $\text{PI}_{K'}$
(3) $\quad\to K'Kp$ see below
(4) $\quad\to K'p$ see below

The transition from (2) to (3) here uses the replacement of BKp by Kp in the scope of the monotone K', justified by the $\to$-direction of (the formula mentioned in) Lemma 5.6.20, while that from (3) to (4) replaces Kp with p again in the scope of monotone K', as justified by $\mathbf{T}_K$. ∎

For a contrast between implicit and explicit definition along the lines of Example 5.6.19, we of course need also what is proved in Halpern *et al.* [431] (p. 481) as Theorem 4.1:

PROPOSITION 5.6.22 $\mathbf{KD45}_B \oplus \mathbf{S5}_K \oplus \{\text{KB}, \text{PI}\}$ *does not define* K *explicitly*.

We do not give the (model-theoretic) proof here, since we want to proceed to a semantic reflection of implicit definability offered by the authors, but before that we should pause to note that one might think that Proposition 5.6.22 contradicts Lenzen's work (Theorem 5.2.1 etc., most recently in Prop. 5.6.13(i) at p. 434) contradicts this in supplying a definition of K using already $\mathbf{S4.4}_K$, let alone the stronger epistemic component $\mathbf{S5}$, of the epistemic–doxastic logic in play here, namely the definition: $Kp \leftrightarrow (Bp \wedge p)$. However, what the latter amounted to was the fact that if one started with a monomodal logic for B then adopting that definition would give (at least) $\mathbf{S4.4}$ for K: *not* that, if one started with

5.6. CONCEPT-POSSESSION PROBLEMS

a bimodal logic having both B and K primitive, and the K-fragment had at least the strength of **S4.4** (with the B obeying **KD45** and related to K by KB and PI), then one was guaranteed that K (the K we started with, that is) was related to B by having $Kp \leftrightarrow (Bp \wedge p)$ provable. Halpern *et al.* [431] and [430] discuss these Lenzen-related issues with the aid of their notion of reducibility, alluded to above, and not of specific concern to the theme we are pursuing here. For that, recall that the official definition of implicit definition above was described as giving the latter notion in its syntactic incarnation. Halpern *et al.* are also concerned with various semantic aspects of the situation.

In striking contrast to Proposition 5.6.22 above, Proposition 5.2 of [430] tells us that in the first-order theory of the frames $\langle W, R_K, R_B \rangle$ for the logic $\mathbf{KD45}_B \oplus \mathbf{S5}_K \oplus \{\mathrm{KB}, \mathrm{PI}\}$, R_K turns out to be explicitly definable in terms of R_B, by means of:

$$\forall x \forall y (R_K xy \leftrightarrow \exists z (R_B xz \wedge R_B yz)),$$

even though in the modal logic itself (Prop. 5.6.22), there is no explicit definition of K in terms of B (and the Boolean connectives). What the inset first-order definition calls to mind is that if the "y" and "z" in the second conjunct had been the other way round, then K would have been explicitly definable in the modal logic – by means of: $K\varphi \leftrightarrow BB\varphi$. We would have a straightforward composition of accessibility relations. As it is, the second conjunct has these variables the wrong way round, suggesting that we work with the converse of the accessibility relation for B as well as that relation itself: this is the promised tense-logical twist. Writing B^{-1} for a new operator interpreted in terms of this converse, we should then have a definition of K along the following lines: $K\varphi \leftrightarrow BB^{-1}\varphi$.[392] Of course this isn't a definition of K in terms of B and the Boolean connectives, since we have used B^{-1}, which is not explicitly definable in terms of B (though we do have implicit definition in the appropriate logic, as Example 5.6.19 indicates, taking G and H as B and B^{-1}). It seems interesting, all the same. What follows is a syntactic verification of the explicit definability of K in terms of B and B^{-1} – a proof of $Kp \leftrightarrow BB^{-1}p$ from the axioms, that is – though it begins by stepping back and making some more general moves for later application to that specific issue.definability!explicit vs. implicit—)

Suppose we have a normal bimodal logic with primitive box operators $\Box$ and $\Box_+$, and a linking axiom (schema) $\Box_+ \varphi \to \Box \varphi$. (The subscripted "+" is meant as a reminder that the operator so marked is the stronger of the two. A more natural notation would be to use the "+" in a superscript position, but we need that position to house the "-1" notation; cf. the notation "$\Box^{-1}$" of Example 4.1.11, p. 209.) Now consider its tense-logical extension with 'past tense' versions $\Box^{-1}$ and $\Box_+^{-1}$ – meaning only that we intend to interpret them using the converses of the accessibility relations interpreting $\Box$ and $\Box_+$, not that we are introducing any temporal concepts – with appropriate Lemmon–Prior style bridging axioms (à la HF-ax. and GP-ax. from p. 179, given the intended interpretation, and using diamonds for the respective duals):

$$\varphi \to \Box \Diamond^{-1} \varphi \text{ (or } \Diamond \Box^{-1} \varphi \to \varphi)$$

and

$$\varphi \to \Box_+ \Diamond_+^{-1} \varphi \text{ (or } \Diamond_+ \Box_+^{-1} \varphi \to \varphi)$$

then one can replicate the considerations showing that for their respective accessibility relations R, R_+, that $R \subseteq R_+$ implies $R^{-1} \subseteq R_+^{-1}$, with the following tense-logical derivation from the axiom $\Box_+ \varphi \to \Box \varphi$ of the 'mirror image' $\Box_+^{-1} \varphi \to \Box^{-1} \varphi$:

---

[392]The "-1" superscript is intended to recall the use of the notation R^{-1} for the converse of the relation R, as with "O^{-1}" on p. 267 above.

(1) $\Diamond p \to \Diamond_+ p$ \hspace{2em} instance of the dual of the given schema
(2) $\Box^{-1}\Diamond p \to \Box^{-1}\Diamond_+ p$ \hspace{2em} (1), $\Box^{-1}$Mono
(3) $p \to \Box^{-1}\Diamond p$ \hspace{2em} tense-logical linking axiom
(4) $p \to \Box^{-1}\Diamond_+ p$ \hspace{2em} (2), (3) TF
(5) $\Box_+^{-1} p \to \Box^{-1}\Diamond_+ \Box_+^{-1} p$ \hspace{2em} US, $\Box_+^{-1} p$ for p in (4)
(6) $\Box_+^{-1} p \to \Box^{-1} p$

where line (6) comes from (5) by replacing $\Diamond_+ \Box_+^{-1} p$ with its tense-logical consequence p in a monotone context.

For the application at hand, $\Box_+$ and $\Box$ are of course respectively K and B in the combined epistemic–doxastic logic $\mathbf{KD45}_B \oplus \mathbf{S5}_K \oplus \{\text{KB}, \text{PI}\}$, for which "+" notation on "$\Box_+$" records the effect of KB. Since for this application K satisfies the $\mathbf{B}$ axiom – or, put semantically, in any frame validating the logic the accessibility relation for K is symmetric – the distinction between K and K^{-1} collapses and there is no need to introduce the latter notation. So we need only consider the addition of B^{-1} to the language of $\mathbf{KD45}_B \oplus \mathbf{S5}_K \oplus \{\text{KB}, \text{PI}\}$, and the (normal) extension of that logic by suitable bridging axioms to link B^{-1} and B. In this setting, line (6) of the above proof becomes $Kp \to B^{-1}p$. Since B is monotone, we can pass from here to:

$$BKp \to BB^{-1}p,$$

and since we have $Kp \to KKp$ and $KKp \to BKp$ and therefore also $Kp \to BKp$ provable, we can pass from the last inset formula to:

$$Kp \to BB^{-1}p.$$

The provability of the converse implication is a simpler matter. (Recall that C is $\neg B \neg$.) By $\mathbf{D}_B$ we have $BB^{-1}p \to CB^{-1}p$. Since the consequent here provably implies (as one of the tense-logical bridging principles) p, we get

$$BB^{-1}p \to p$$

and so, by $\Box$Mono we get:

$$KBB^{-1}p \to Kp.$$

Using an instance of the schema $B\varphi \to KB\varphi$ (namely with $\varphi = B^{-1}p$), we can drop the initial K, arriving at the desired converse:

$$BB^{-1}p \to Kp.$$

Thus in the present logic we have K explicitly definable, Kp and $BB^{-1}p$ being provably equivalent. This is of course not what one expects, since the perennial possibility of being mistaken, however confident one is, makes it implausible that arbitrary attributions of knowledge can be equivalent to belief-ascriptions – notwithstanding the odd content (in view of the involvement of B^{-1}) of the beliefs ascribed in the present case. But then again, all this is premised on the use of $\mathbf{S5}_K$, so one may be inclined to take it as simply further magnification of an initial flaw. However, the recent tense-logical excursus is a threat on a broader front than this.

In particular, let us write out in the B, B^{-1}, notation (with duals C, C^{-1}) the tense-logical linking principles formulated above with $\Box$ and $\Box^{-1}$:

$$\varphi \to B^{-1}C\varphi \hspace{2em} \text{and} \hspace{2em} \varphi \to BC^{-1}\varphi.$$

The first of these is perhaps not too alarming, since it is not clear what we are being told when we are told – on the hypothesis that φ is true – that so is $B^{-1}C\varphi$. In particular the latter does not clearly ascribe a propositional attitude to the envisaged cognitive subject. In terms of the Kripke semantics it tells us, taking the assumed truth of φ to be its truth at world w, that at any world to which w is R_B-accessible, $C\varphi$ is true. We know what that means: that the subject does not (at the world in question) believe $\neg\varphi$. The second linking principle is problematic especially for the concerns of the present section, since its consequent is the attribution of a *belief*, and whatever contribution to the content of that belief may be made by the somewhat mysterious C^{-1} its content also involves whatever concepts are deployed in φ. Suppose that our subject is not conceptually equipped to entertain the proposition that q then, but that q is true. (If q is not true, run the argument with $\neg q$ instead.) This principle tells us that the subject believes that $C^{-1}q$: but whatever that amounts to, it certainly deploys some non-doxastic concepts – those involved in q – she does not possess. So she does not have an explicit belief with the content $C^{-1}q$. Now, our concession to normality allows that a belief-attribution to that effect is to count as true nonetheless if $C^{-1}q$ follows from some things that the subject does explicitly believe. But against this possibility an argument along the lines of Example 5.6.18 appears to be available: letting φ be the conjunction of some such explicit beliefs, which accordingly does not contain q, φ would provably imply $C^{-1}q$, the Uniform Substitution of $B\bot$ for q would leave φ unaltered, so that φ provably implies $C^{-1}B\bot$, and therefore $\bot$ itself, contradicting the subject's consistency.

More generally still, in the sense that the following point does not require what may seem to be excessive attention to considerations of concept-possession, the idea that a doxastic logic should be a fully modalized logic in the sense of Section 4.6, seems to need to be sacrificed when we consider adding B^{-1} to the language, for exactly the same reason that while **K** is f.m., **K**$_t$ is not: yet if the possible worlds semantics for doxastic logic is of any value at all the intelligibility of this new operator is not open to question, since, as remarked under Example 4.1.11 (p. 209), any operator whose semantics is presented by quantification over points related by an accessibility relation gives rise to an operator for whose semantics the converse of that relation plays the same role, however little one may be inclined to attend to the latter operator. As observed in connection with deontic logic (Example 4.4.29, p. 267), the presence of these mirror image operators is apt to play havoc with the expectations associated with full modalization – in the deontic case, for instance, with the issue of Hume's Thesis. Quite how to react to these surprises is a matter on which we will not speculate here.

5.7 Another Use for B^{-1}

With B^{-1} and its dual, C^{-1}, fresh in our minds, let us pause to observe that in quantified doxastic logic – not an area on our official agenda – under certain circumstances we can use the 'tense-logical' version to simulate some two-dimensionality. Two-dimensional modal logic is also not something we have gone into in any depth, but here we recall the double-indexing notation "$\models_y^x$" from the discussion of relative necessity beginning on p. 275 above, in which the upper index represents a reference point and the lower index represents the point of evaluation. The Vlach operators $\uparrow$ and $\downarrow$ can be used to store and recover a current reference point;[393] we work with models $\mathcal{M} = \langle W, R, V \rangle$, interpreting our $\Box$ operator B[394] not in the style of the 'absolute necessity' $\Box$ of our earlier discussion, but more standardly:

$$\mathcal{M} \models_v^u B\varphi \text{ if and only if for all } w \in W \text{ such that } vRw\ \mathcal{M} \models_w^u \varphi.$$

---

[393] For references and discussion, see p. 63f. in Lewis [717] and p. 142ff. in Fine [292], where the original notation of an obelisk (dagger) and inverted obelisk are used in place of $\uparrow$ and $\downarrow$. Note that the $\downarrow$ operator can be pressed into service, given suitable further conditions on the models, as a version of the actuality operator we have had occasion to mention several times in passing – see the index – as is explained in Davies and Humberstone [235], Humberstone [578], as well as the backward(s) looking operators of Esa Saarinen and Harold Hodes. Pertinent bibliographical references in the case of Hodes can be found in [578]; for Saarinen, who coined the phrase "backwards-looking operators", see [983].

[394] We no longer consider B alongside K as in the preceding section, so there is no need to write "R_B".

The Vlach operators are then interpreted like this:

$$\mathcal{M} \models^u_v \uparrow \varphi \text{ iff } \mathcal{M} \models^v_v \varphi \qquad \mathcal{M} \models^u_v \downarrow \varphi \text{ iff } \mathcal{M} \models^u_u \varphi.$$

Since we the present observation concerns quantified doxastic logic, we need to envisage our models $\mathcal{M}$ equipped not only with the above W, R and V (this last re-construed so as to interpret predicate letters relative to elements of W) but with a function assigning domains to the various $w \in W$ (the set of individuals taking as existing in w), but it is not necessary to labour over these details, except to say that we are envisaging the quantifiers evaluated as ranging over the domain of the world of evaluation (the lower index – or the sole index, in standard 'one-dimensional' Kripke semantics): this is sometimes called actualistic quantification. This gives the desired contrast between the *de re* $\exists x B \varphi(x)$ – there is something the subject believes to be φ (i.e., believes to be an x such that $\varphi(x)$) – and the *de dicto* $B \exists x \varphi(x)$ – the subject believes that something is φ, perhaps having no idea as to what that thing might be: believes that something or other, as we say to play up this nonspecificity, is φ (Quine [936] being the *locus classicus* for the present distinction – though see also the references given in note 197, p. 237 above). (See also the quotation from [936] given in Remark 5.1.19, p. 358, for the boulomaic analogue of this distinction. We return to relative specificity in desire-attributions in Example 5.7.8 below.) There is a well-known third possibility here,[395] however, in that the subject may have a specific idea as to what or who it is that is φ but there does not exist an individual believed by the subject to be φ, raising the question of how to represent such belief attributions involving *specificity without existence*. For such a belief attribution to be true in w, we need something to be φ in each world R-related to w and it needs to be the same thing in each world ('specificity') but is does not need to be in the domain of w ('without existence').[396] This issue would not arise for knowledge, because of the presumed reflexivity of the accessibility relation: to have knowledge in w that a specific individual is φ, that individual has to exist (and be φ) in w for the same reason as knowing in w that chocolate is toxic to dogs requires toxic to be toxic to dogs in w. Such specificity of belief (or indeed desire) – with or without existence – is often indicated in English with the aid of the quantifying expression "a certain": "Sam thinks a certain woman will be there tonight".[397]

We can follow the lead of Fine [292], p. 144, where a 'possibilist' existential quantifier (binding the variable x in the formula $\varphi(x)$) – meaning a quantifier ranging over the union of the domains of accessible worlds – is defined in terms of the actualist $\exists$ by taking as the *definiens*:

$$\uparrow \Diamond \exists \downarrow \varphi(x).$$

Along similar lines, we can use the Vlach operators to attribute a specific belief to the effect that something is φ compatibly with there not existing anything believed to be φ:

$$\uparrow C \exists x \downarrow B \varphi(x).$$

Evaluating this formula w.r.t. a given world u in a model $\mathcal{M} = \langle W, D, R, V \rangle$ (where D assigns a domain to each element of W), regardless of the reference world, the initial $\uparrow$ stores u as the reference world for later retrieval, leaving us to consider whether:

$\mathcal{M} \models^u_u C \exists x \downarrow B \varphi(x)$, which in turn holds iff:

---

[395] See for example Saarinen [984], esp. §4.

[396] Here we are taking it that only individuals in the domain of a world can be φ in that world. With this assumption in force, the "C" before "$\exists$" in our formulations could equally well be a "B".

[397] What about occurrences of "a certain' not embedded in such contexts? For example I might just say "A certain woman will be there tonight". At least part of the story may be (as suggested in Saarinen [986], p. 257) that this has the same truth-conditions (relative to the contextual parameters fixing where and when *there* and *tonight* are) as "A woman will be there tonight", but the conditions for the appropriate use of the former require that the speaker have a particular woman in mind. Returning to the embedded case, it even seems possible to have transpersonal specificity (without existence), where the "having in mind" crosses from one mind to another, the most famous such cases being those of Geach [354]. Subsequent discussion of this 'intentional identity' issue is provided by Saarinen [982] and Edelberg [271], the latter supplying many further references.

5.7. ANOTHER USE FOR B^{-1}

for some $v \in R(u)$, $\mathcal{M} \models_v^u \exists x \downarrow B\varphi(x)$, which holds iff:

for some $d \in D(v)$, $\mathcal{M} \models_{v\downarrow}^u B\varphi(x)$ when x is assigned the value d, which holds iff:

$\mathcal{M} \models_u^u B\varphi(x)$ when x is assigned the value d.

And now that $\downarrow$ has restored the original world u to the status of evaluation point, B quantifies over worlds accessible to u, i.e., the above holds iff:

for all $w \in R(u)$, $\mathcal{M} \models_v^u \varphi(x)$ when x is assigned the value d. Putting all this together, then, and rather sloppily writing "$\mathcal{M} \models_v^u \varphi(x)$ when x is assigned the value d" as "$\mathcal{M} \models_v^u \varphi(d)$", the formula inset above is true at evaluation point u in $\mathcal{M}$ (regardless of the reference point) just in case:

$$\text{For some } v \in R(u), \text{ for some } d \in D(v), \text{ for all } w \in R(u), \mathcal{M} \models_w^u \varphi(d).$$

Here we have the desired effect of specificity – the same promised $d \in D(v)$ (for some $v \in R(u)$) has to be φ at all $w \in R(u)$ (here our formulation assumes that φ does not contain "$\downarrow$" so being φ depends only on the lower index) – without existence: we did not require that this d was an element of $D(u)$.

REMARK 5.7.1 What happens if we leave out the Vlach operators altogether and consider as a candidate representation $C\exists x B\varphi(x)$? For this to be true at u in a model – one-dimensionalizing the semantics in the absence of those operators – we need $d \in D(v)$ (for some $v \in R(u)$) but this d then has to be φ at all $w \in R(v)$, rather than at all $w \in R(u)$. That would work for the u-relative truth-conditions of "the subject believes a certain thing to be φ" only if we were restricting attention to models on frames in which whenever uRv, we had $R(u) = R(v)$ – in other words, in transitive Euclidean frames. We will turn next to a less stringent condition allowing us to dispense with $\uparrow$ and $\downarrow$ (provided we have at our disposal not only B but B^{-1}). ◀

Having seen this apparatus in action, let us now return to the language with B^{-1} and C^{-1} alongside B and C allows us to achieve a similar effect, at least when certain conditions are met. The idea is us use C much as before to get from the initially given world u to a world v from whose domain to pluck an object and then use C^{-1} to return to u to check that the object chosen has a given property in each doxastic alternative to u itself (rather than to the v in whose domain we found it). Here we do not keep track with double indexing u as to be able to return to it (as we did with $\downarrow$) but hope to do so by just travelling back to some R-predecessor of v or other. Thus we need to know that whichever predecessor we end up at, its R-successors are precisely u's R-successors. If each point has at most one R-predecessor, then any predecessor of a successor of u would be u itself and so of course have precisely u's successors. But we not need to insist on this unique predecessor condition – encountered above at Example 4.4.34, p. 270 – which is just as well since the condition is very stringent and would not be satisfied in general by frames for popular candidate doxastic logics such as **KDU, KDU4, KD45**. (This last case is already covered by Remark 5.7.1 of course, allowing us to get by without B^{-1}.) Fortunately, we can allow a point to have more than one predecessor, as long as these in turn have precisely the same successors. Since we are using C^{-1} in place of $\downarrow$, relying on the converse of the accessibility relation for B to get to an earlier point rather than reaching for a specific stored reference point, we do not need $\uparrow$ to store it in the first place, and conduct the discussion entirely in one-dimensional semantic terms. The needed is accordingly the following, called *generalized equivalence* (or 'GE') in Humberstone [562], here given in first-order terms, and represented pictorially after the fashion of Chapter 2 (p. 66ff.) in Figure 5.6:

$$\forall u_0 \forall v_0 \forall u_1 \forall v_1 ((Ru_0 v_0 \wedge Ru_1 v_1 \wedge Ru_0 v_1) \rightarrow Ru_1 v_0).$$

Alternatively, one might prefer an abbreviated informal version, variables understood as universally quantified: *If $R(u_0) \cap R(u_1) \neq \varnothing$ then $R(u_0) = R(u_1)$*.

The generalized equivalence condition is somewhat unusual as a condition on binary relations (or frames) in that no point in Fig. 5.6 is at the head of an arrow and also at the tail of an arrow. The first order

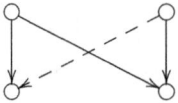

Figure 5.6: The Generalized Equivalence Condition

characterizations of ∧- and ∨-representability given in Proposition 1.2.12 on p. 15 have this same feature, and are accordingly called *separated* conditions,[398] in Humberstone [562], where some relationships among various such conditions – including all those mentioned here – are described. But for present purposes we are concerned with just this generalized equivalence condition. By way of familiarization:

PROPOSITION 5.7.2 *The class of frames satisfying the generalized equivalence relation (i) is not monomodally definable but (ii) it is bimodally (and specifically tense-logically) definable.*

Proof. (*i*) Suppose φ is a monomodal formula valid on precisely the frames satisfying the generalized equivalence condition. (This argument works equally well if we suppose that Γ is a set of formulas valid on precisely the frames in question.) If we had $\vdash_{\mathbf{K}} \varphi$ then φ would be valid on all frames, and so not only on the frames satisfying our condition. So $\nvdash_{\mathbf{K}} \varphi$, and we have a frame on which φ is not valid, being false at a point in some model on this frame. As per Example 4.4.34 (p. 270), unravel the model from that point and we have a model on a frame and we have invalidated a formula in which each point has a unique predecessor, and according on a frame satisfying the generalized equivalence condition, contradicting the supposition that φ is valid on all such frames.

(*ii*) We use the "$\Box$" notation rather than "B" notation, since the current claim is not of interest only for doxastic readings; $\Box^{-1}$ is accordingly interpreted by the converse of the accessibility relation R interpreting $\Box$ (as in Example 4.1.11, p. 209). The formula $\Box p \to \Box\Box^{-1}\Box p$ defines the class of frames in question. The reader is left to check that it is valid on all such frames. To see that is is valid on only such frames, suppose that we have a frame not satisfying the condition, i.e., we have $\langle W, R \rangle$ with $u_0, v_0, u_1, v_1 \in W$ such that $u_0 R v_0, u_1 R v_1, u_0 R v_1$ but not $u_1 R v_0$. Take any model $\langle W, R, V \rangle$ on this frame with $V(p) = R(u_1)$ and check that the formula $\Box p \to \Box\Box^{-1}\Box p$ is false at u_1 in this model. (Of course, there may be additional points in the frame variously related to each other, but we do not care about them, as they do not affect the falsity of our formula at u_1.) ■

REMARK 5.7.3 An argument like that given in the proof of Prop. 5.7.2(*i*) here can be found in Garson [351] (p. 191) apropos of irreflexivity, but the discussion goes awry somewhat: see Humberstone [602]. Whether for irreflexivity or the generalized equivalence condition, a modal undefinability argument going via the facts that (1) not every frame lies in the class while (2) every non-theorem of **K** is invalid on some frame in the class eventually needs some direct proof of (2). In the present case, the appeal to unravelled frames traces back to the role of p-morphisms – see Prop. 4.4.31, (p. 269) – as also does the case of irreflexivity (Thm. 4.4.32). ◀

Part (*ii*) of the following asks for a check of canonicality for the formula mentioned under Proposition 5.7.2(*ii*), which follows from a general result of Sahlqvist [988] (mentioned on p. 78 above) concerning a general class of potential modal axioms because this is a 'Sahlqvist formula' (in fact, a bimodal variant of the kind discussed in Example 2.5.20, p. 78), but since we have not gone into the details of this general result, here we are expecting an argument addressed at the specific formula involved.

---

[398] Because the variables can be separated into those (here u_0, u_1) occurring in the first position of the predicate letter T, and those (here v_0, v_1) occurring in the second position in a first order (and "="-free) linguistic formulation of the condition.

5.7. ANOTHER USE FOR B^{-1}

EXERCISE 5.7.4 (*i*) Is the 'piecewise' version of the generalized equivalence condition monomodally definable? (Add a node above the top two in Fig. 5.6 and arrows from it to them. A more general discussion can was provided on p. 76.)

(*ii*) Show that the frame of the canonical model of the normal extension of the basic tense logic $\mathbf{K}_t$ – though we write $\Box$ and $\Box^{-1}$ rather than the G and P of (e.g.) p. 179, since temporal considerations are the last thing we have in mind here – by the formula $\Box p \to \Box\Box^{-1}\Box p$ satisfies the generalized equivalence condition (and thus, given Proposition 5.7.2(*ii*), that this logic is determined by the class of all such frames).

(*iii*) If in the generalized equivalence condition $\forall u_0 \forall v_0 \forall u_1 \forall v_1((Ru_0v_0 \land Ru_1v_1 \land Ru_0v_1) \to Ru_1v_0)$, we consider the case in which $u_1 = v_0$ we can see that the condition implies that any point all of whose successors are reflexive is itself 1-transitive (as defined on p. 186), and thus that any frame for the logic described under (*ii*) which validates **U** (i.e., $\Box(\Box p \to p)$) validates **4** (for $\Box$ and also $\Box^{-1}$). Show syntactically – i.e., by giving an axiomatic proof – that **4** (for $\Box$) is forthcoming as a theorem when **U** is added to the above logic as a new axiom. (*Suggestion*: first show that $\Box(\Box^{-1}p \to p)$ is a theorem, and then substitute $\Box p$ for p in this and distribute the outer $\Box$ across the resulting implication, suitably invoking the generalized equivalence axiom.) ✠

At some risk, perhaps, of labouring the obvious, let us go slowly through our new candidate for the attributing of the belief that "a certain individual is φ": $C\exists x C^{-1}B\varphi(x)$, relative, as before to a point u in a model $\mathcal{M} = \langle W, D, R, V \rangle$, where $\langle W, R \rangle$ satisfies the generalized equivalence condition:

$\mathcal{M} \models_u C\exists x C^{-1}B\varphi(x)$, which in turn holds iff:

for some $v \in R(u)$, $\mathcal{M} \models_v \exists x C^{-1}B\varphi(x)$, which holds iff:

for some $d \in D(v)$, $\mathcal{M} \models_v C^{-1}B\varphi(d)$ which holds iff:

$\mathcal{M} \models_{u'} B\varphi(d)$ for some $u' \in R^{-1}(v)$, i.e., for all $w \in R(u')$, $\mathcal{M} \models_w \varphi(d)$.

Now we have $v \in R(u) \cap R(u')$, so by the generalized equivalence condition, $R(u) = R(u')$, and so $\mathcal{M} \models_u B\varphi(d)$, i.e., for all $w \in R(u)$, $\mathcal{M} \models_w \varphi(d)$. Since we did not require $d \in R(u)$, we have, as before – in the two-dimensional version – specificity without existence.

REMARK 5.7.5 It is worth emphasizing that despite appearances, "$\mathcal{M} \models_{u'} B\varphi(d)$" does not say that a certain (closed) formula, namely one in which the variable "x" in "$\varphi(x)$" is replaced by the name (or individual constant) "d". That would be a move distasteful to anyone with actualistic motivations, who would want no such truck with the non-existent as even that kind of use of a non-denoting name (from u''s perspective here). Rather, "$\mathcal{M} \models_{u'} B\varphi(d)$" just abbreviates the claim that $B\varphi(x)$ is true at u' in $\mathcal{M}$ when the variable x is assigned the value d. But, one might wonder, is this any more actualistically acceptable? And is the situation any different in this respect from the earlier ↑-with-↓ proposal? That was at least formulated in a doubly indexed semantic framework in which the actualist's scruples might focus not so much on the lower index as on the (now lost) upper index. (Note that all this discussion presumes that domains should be allowed in our semantic treatment to vary from world to world, a suggestion rejected in some quarters – at least in the case of metaphysical modality, as was mentioned on p. 212.) ◀

Let us note here that a further simplification of the B/B^{-1} approach would consist in replacing the occurrence of "C^{-1}" in "$C\exists x C^{-1}B\varphi(x)$" with "$C$" itself. A starting point here is provided by:

EXERCISE 5.7.6 Show syntactically that $\vdash_{\mathbf{K}_t \mathbf{B}} \Box^{-1}p \leftrightarrow \Box p$. (Here of course **B** is **B**-for-$\Box$, though we have the same result in the case of **B** for $\Box^{-1}$.) ✠

We could equally well have asked for a proof of $\Diamond^{-1}p \leftrightarrow \Diamond p$ in the above exercise, and since we also have the necessitations (w.r.t. $\Box$ or the equivalent $\Box^{-1}$) of these formulas, all superscripted occurrences of -1 can be deleted from formulas giving formulas equivalent to those from which they have been deleted. These are syntactic reflections of the fact that in a symmetric frame, the accessibility relation coincides with its converse. We do not need to go to such extremes, however, to get rid of the "-1" in "$C\exists x C^{-1}B\varphi(x)$", since this occurrences is embedded in the scope of an outer occurrence of "C" – which is just as well since any logic containing $\mathbf{U}$ and $\mathbf{D}$ has $\Diamond\Box(\Box p \to p)$ as a theorem, from which $\mathbf{B}$ delivers $\Box p \to p$: not something we want if $\Box p$ is taken as Bp. (In fact, the much weaker Brouwerian *rule* of Example 2.6.6, p. 115, would equally well deliver this conclusion.) Instead, we can consider the formula $\mathbf{B}^{\Box}$ ($= \Box(p \to \Box\Diamond p)$), which is provable in $\mathbf{K45}$ and (so also in) $\mathbf{KD45}$, logics which have at least been taken seriously for doxastic applications.[399]

EXERCISE 5.7.7 Show syntactically that $\vdash_{\mathbf{K}_t \mathbf{B}^{\Box}} \Box(\Box^{-1}p \leftrightarrow \Box p)$. ✠

The provability of the formula (and of all its substitution instances) mentioned in Exercise 5.7.7 then allows us within the scope of – as we now write in place of $\Box$ – "B" to replace an occurrences of B^{-1} or C^{-1} with the plain B or C, so we can write for our existentially noncommittal while still specificity belief attribution: $C\exists x CB\varphi(x)$. (In $\mathbf{KD45}$ we can go still further and simplify CB to B, since here we have **5**!.)

Finally, to conclude this discussion of specificity in belief-attributions let us observe that there appears to be an extra degree to the specific/nonspecific distinction when it is desires rather than beliefs that are at issue:

EXAMPLE 5.7.8 For example, suppose a has reason to believe that one of three people waiting to be interviewed as a witness to a crime is a genuine witness while the other two are only pretending to be witnesses, and whose versions of what happened a is not interested in hearing. Unfortunately a has no idea which one is the genuine witness. As our boulomaic operator ("a wants that") let us write W. We can distinguish between:

(1) $W\exists x(a \text{ interviews } x)$,

and

(2) $\exists x(W(a \text{ interviews } x))$,

and of these two, (1) is strongest desire-attribution that is true, since there is no particular individual a is in a position to want to interview (falsifying (2)), having no belief as to who the real witness is. This presents no contrast with the case in which as far as a is concerned, any one of the three is as suitable interviewee (perhaps because time is short, so a potential witness can be selected at random, there being no reason to believe that one of them is more likely to be reliable than the others). If one worked instead with a language in which desires are attributed using a composite prefix $B + D$, where D is an optative marker with an informal reading along some such lines as "it would be nice if". Then we can make the following threefold distinction:

(1) $BD\exists x(a \text{ interviews } x)$;
($2\frac{1}{2}$) $B\exists x(D(a \text{ interviews } x))$;
(2) $\exists x(BD(a \text{ interviews } x))$.

With this apparatus in place we can point to ($2\frac{1}{2}$) as the most specific desire-attribution that would be correct in the (original) witness interview scenario, rather than just (2). This issue does not concern specificity without existence, and the point could be made with disjunctions rather than existential quantifications. For some other advantages of such a treatment of desire-attributions, see Humberstone

---

[399]Our concept-possession worries about $\mathbf{5}_B$ from the preceding section notwithstanding. The simplest way to see that $\mathbf{B}^{\Box}$ is $\mathbf{K45}$-provable is to note that it is the necessitation of an $\mathbf{S5}$-theorem, and then invoke Coro. 4.1.6 from p. 207.

5.7. ANOTHER USE FOR B^{-1}

[557]. Note that the informal gloss "it would be nice if" suggests that the content of the desire is in antecedent rather than consequent position, to recall the contrast sketched in Remark 5.1.19 (p. 358), whereas for a normal treatment of D (and hence of W as BD) one would prefer a consequent position approach, but this is one of many details that would require consideration. (Also, in "it would be nice", "nice" has to be taken as from the subject's perspective, since we don't want to represent two individuals, one wanting it to rain and the other wanting it not to, as holding inconsistent beliefs.) A more general worry would be that since beliefs are supposed to have one direction of fit and desires another, (see p. 204) – namely attitude-to-fit-the-world in the former case and world-to-fit-attitude in the other – beliefs could not be desires. But one could reply that in the case of $BD\varphi$ this means that is is $D\varphi$ that has to fit the world since this is the propositional object of the belief attributed, whereas it is φ itself that the world has to fit, as behooves a desire, and so there is no contradiction after all. On the other hand, it remains an uphill struggle to defend the proposal against related Humean considerations: Lewis [728]. On the general idea of attitudes with both directions of fit, see also Laitinen [684]. ◂

One turn of phrase used in Example 5.7.8 deserves closer attention than it can be given here: "the propositional object of the belief attributed". Thinking of belief-attributions linguistically, one would expect that that belief attributed by a given sentence should be the same belief as is attributed by any sentence equivalent to it according to one's favoured doxastic logic – making it possible to think of belief-attributions propositionally rather than just sententially. To the extent that one finds this idea attractive, one would be drawn to doxastic logics closed under the Cancellation Rule of Exercise 2.6.31(iii), p. 126. This would exclude, to take the simplest case, **K4!**, or any otherwise (doxastically) plausible extension of it (such as **K4U**), since the rule would allow us to pass from $B\varphi \leftrightarrow BB\varphi$ to $\varphi \leftrightarrow B\varphi$ – a paradigmatically implausible doxastic principle. Or so it is urged in Humberstone [599], where the extent of failures of cancellation is the main object of study: the range of non-equivalent φ and ψ for which $B\varphi$ and $B\psi$ are provably equivalent in a given logic. (For example, does this range have a logically strongest or a weakest formula in it?) In fact $B\varphi$ is there written as $\Box\varphi$, the issue arising for modal logics in general and certainly with equal motivation for propositional attitudes other than belief.

But is this 'inverse congruentiality' (closure under Cancellation) constraint plausible? Consider for example the belief that it will at some time be the case that p and it will at some time be the case that $\neg p$. This is not the belief that p, but it *is* a belief and it is *some* propositional attitude with the proposition that p as its object. In a mixed doxastic tense logic, that is, we could introduce, if we cared to, an operator B^* by defining $B^*\varphi$ to be $B(F\varphi \wedge F\neg\varphi)$. But from $B^*\varphi$ we certainly can't recover φ uniquely, since $B^*\varphi$ would always be equivalent to $B^*\neg\varphi$ (given a modicum of congruentiality for the original B – enough at least to permit us to reverse the conjuncts in a conjunction in its scope, as well as to interchange $\neg\neg\varphi$ and φ). So as a constraint on propositional attitude attributions generally, it certainly seems too demanding to ask that the propositional object of such an attribution be recovered from the proposition expressed by the attribution.

One can imagine a further move in the case of the **4!** example above for B itself. This principle, **4!**$_B$, tells us, for example, that a subject, Margot, say, believes that there are ski resorts in Austria if and only if Margot believes that she believes that there are ski resorts in Austria. But the hypothesis that this biconditional holds as a matter of logic is not the same as saying that Margot's belief (if she does believe it) that there are ski resorts in Austria is one and the same belief – *qua* mental state – as her belief that she believes that there are such resorts, for the same reason that underlies the point about 'gwalking' in Example 5.1.13, p. 352. Recall that gwalking was a species of walking, namely walking while chewing gum. Let us now introduce a verb for a particular kind of gum-chewing. We will call an act of chewing gum performed while the agent is walking an act of chewking. It is a matter of logic in a suitably generous interpretation of that phrase that a person gwalks (at an arbitrary time) iff and only if that person chewks at that time, even though the gwalking and the chewking are not to be confused with each other, and the chewking may be slow while the gwalking is quick, to adapt the discussion from Example 5.1.13 (where this observation was credited to Davies [234]). Similarly, it might be worth entertaining the possibility that although as a matter of logic one believes that p if and only if one believes that one believes that p these believings are not to be confused with each other, each having its own (unique!) propositional object. Whether this would be a fruitful line of thought to develop remains to be seen.

Chapter 6

Coming to Stand in a Relation

6.1 Introduction

In what follows we apply some general considerations about the logical properties of operators expressing the idea that some state of affairs is brought about by an agent (or at any rate comes about), on the more specific question of bringing it about that (or more accurately: its *coming* about that) certain objects stand in a certain relation. To introduce these considerations we begin by presenting some of the content of a paper delivered in 1976 but only published as an abstract (Humberstone [542]). In contrast with the other chapters of this book, some of the discussion that follows will have a conspicuously informal – indeed, impressionistic and speculative – character. On the formal side, we pick up the theme of Example 4.3.2, which began on p. 228: loosely speaking, of operators given a mixed positive/negative interpretation.

It was at one stage common to suggest that for an agent a, the operator "a brings it about that" ("a makes it that case that") is appropriately treated as a normal modal operator. (See, e.g., Pörn [891].) If we agree to denote the operator in question by "D" (as above at p. 235), this means – it will be recalled from, e.g., Exercise 1.3.5(iii) – that the set of provable formulas in an appropriate logic is closed under the following 'rule of normality':

$$\frac{(\varphi_1 \wedge \ldots \wedge \varphi_n) \to \psi}{(D\varphi_1 \wedge \ldots \wedge D\varphi_n) \to D\psi}$$

It will also be recalled from our earlier discussion (following 1.3.2) that the $n = 1$ form of the above rule would make D monotone, and it is in connection with this special case, which gives us the consequence $D(p \wedge q) \to Dp$ when we take φ_1 as $p \wedge q$ and ψ as p, that an objection to normality for D is most easily put.[400] To see why we might wish to question whether bringing about is best treated as normal, we consider an example from [542] that suggests it should not even be regarded as monotone.

Suppose that in front of a there is a table with a notepad on it, and that a places a pen beside this previously solitary notepad. Thus a brings it about that there is a pen on the table. But a also thereby brings it about that there is both a notepad and a pen on the table, since, whether intentionally or not, a performs an action as a result of which it is the case that there is both a notepad and a pen on the table. (Think of it this way: a turns a situation of which it is false that there is notepad and a pen on

---

[400]Since we shall be casting doubt on the acceptability of $D(p \wedge q) \to Dp$, proposals for a non-normal logic of bringing about which still endorse this formula, such as those suggested by F. B. Fitch and D. Walton (see [307] – already the subject of Example 5.4.11, , p. 397 above – and [1141], respectively), are vulnerable to the criticism which follows. Also relevant in this connection, though not similarly vulnerable, is Guigon [415]. The latter is a reply to Bigelow [92], which combines the arguments of Fitch [307] and Humberstone [554], the latter summarised as Exercise 2.7.24, p. 134.

the table in question into a situation of which it is true that there is a notepad and a pen on the table.) We have here a counterexample to the principle that $D(p \wedge q) \to Dp$, taking p to represent "there is a notepad on the table" and q, "there is a pen on the table", since though a brings it about that a notepad and a pen are on the table, a does not bring it about that there is a notepad on the table: the notepad was there anyway and would, we may suppose, have remained there even if a had not put a pen on the table. For at least one reasonable interpretation of the phrase "bring it about that", then, we should look somewhere other than the arena of normal modal logic for an appropriate systematization of the logical properties of the corresponding operator.

A precisely analogous point holds, as one might expect, in the case of the universal quantifier. Just as one can bring it about as in the example just given that a conjunction is true without bringing it about that each conjunct is true, so one can bring it about that a universally quantified proposition is true without bringing about the truth of each instance thereof. For example, if some of the windows in a house are open and others are closed, and I go around and close all of the open ones, I bring it about that all the windows in the house are closed, even though it is not the case concerning each window in the house that I bring it about that *that* window is closed: I do this only in the case of the previously open ones. Similarly, by killing the last surviving O'Connell, you make it the case that all the O'Connells are dead, even if you never laid a finger on any of his relatives.

In what follows, we concentrate on the propositional rather than the quantificational logic of bringing about, for the sake of simplicity. More accurately, if it is simplicity we are interested in – this being a departure from [542] – then it is not so much *bringing* about as *coming* about that should be the focus of our attention. The point of the above examples, in other words, has nothing to do with the element of agency in the bringing about locution. We may think of "a brings it about that p" as meaning that a causes it to come about that p, and make our observations instead about the impersonal construction "it comes about that p". In the notepad and pen example, for instance, it comes about (at the time under consideration) that the table has both a notepad and a pen on it, without its then coming about that the table has a notepad on it (since that was so already). Since, except for a few informal asides, we shall have no further need to consider bringing about, we retain the symbol "D" for this new reading: $D\varphi$ is to be interpreted as saying that it comes about that φ.

The following principles for a logic of "D" were suggested in Humberstone [542]:

(D0) φ φ any substitution instance of a truth-functional tautology
(D1) $D\varphi \to \varphi$
(D2) $(\varphi \wedge D\psi) \to D(\varphi \wedge \psi)$
(D3) $D(\varphi \wedge \psi) \to (D\varphi \vee D\psi)$

Note that (D1) is **T** for D ("$\mathbf{T}_D$" – in the notation from p. 178) but we stick with the current labelling since the remaining axiom schemata have no correspondingly familiar labels.

These schemata were supplemented by some rules which need not detain us here, though there is a rule (not mentioned in [542]) which is worth singling out:

(RD)$_{0,1}$: $\dfrac{\varphi}{\neg D\varphi}$

The labelling of this rule will be explained in Section 6.2, where it will emerge as a special case of a more general rule which, together with Modus Ponens:

(MP): $\dfrac{\varphi \to \psi \quad \varphi}{\psi}$

and (D0), (D1), gives a logic for "D" we shall prove (in the final 'Postscript' – Section 6.7) to be complete for the semantics sketched in the following paragraph and spelt out more explicitly in Section 6.2. Notice

6.1. INTRODUCTION

that although not regular (because not monotone: see the discussion before 1.3.3 and part (iii), (iv), of that exercise), D does satisfy the Aggregation Principle from p. 21, i.e., that for all formulas φ and ψ we have:

$$\vdash (D\varphi \wedge D\psi) \to D(\varphi \wedge \psi),$$

where $\vdash$ indicates provability from the rules and axioms described. This is an immediate consequence of (D1) and (D2) and truth-functional reasoning (made available by (D0) and (MP)). In the following section, we shall see that the present modal logic is congruential.

The semantics alluded to involves a function f which can be thought of assigning to a state of the world at a given time the background state: the state which changes from one with respect to which what comes about is not yet the case, to a state in which it is the case. If we think of time as a discrete linear ordering (without a first element), then $f(u)$ represents the state of the world as of the instant preceding u; however we think of the states or times, the idea is that $D\varphi$ is true at u just in case φ goes from being false at $f(u)$ to being true at u. If φ is true at every instant, as is required for the validity of φ (to invoke a concept for which the precise definition appears in the following section) then φ cannot change from being false at $f(u)$ to being true at u, so $D\varphi$ cannot be true at any instant, explaining the rule $(RD)_{0,1}$. We have here, in addition to the earlier non-monotone phenomenon, another ground for rejecting the path of normal modal logic for "D", since the $n = 0$ instance of the rule of normality – alias Necessitation (for D) – would give $D\varphi$, rather than $\neg D\varphi$, as provable on the assumption that φ is provable. "In addition to" rather than "independently of", since evidently both of these departures from normality are due to the 'negative' element in the truth-conditions for a formula $D\varphi$ at u: that φ should be *false* at $f(u)$. Both of these departures from normality, we encountered in Section 4.3, in which $(RD)_{0,1}$ appeared as (N5) – see p. 228 – alias *non-Necessitation* for the operator N under consideration there, for which (D3) and the aggregation principle also held; the latter appearing as (N2) and the former noted in Example 4.3.2 (mentioned in the opening paragraph of this Chapter), which remarks on the similarities between the N of that discussion and the current D.

The justification for (D1) is that we think of u, in such glosses, as the present instant; a more accurate reading of $D\varphi$ might therefore be "it *has come* about that φ" rather than "it *comes* about that φ". (D2) is directly suggested by examples such as that of the notepad and pen: since there is (still) a notepad on the table at u (as at $f(u)$), its coming about that there is a pen on the table is its coming about that there is a notepad and a pen on the table. And for (D3), note that a conjunction can only pass from being false (at $f(u)$) to being true (at u) if at least one of its conjuncts passes from being false to being true.

The above justification of the principles under discussion works equally well for the reading of "D" in terms of bringing about rather than coming about, though there is a complication in this case because modal (or counterfactual) as well as temporal considerations may well be thought appropriate. This is so for von Wright's seminal discussion of bringing about in [1208], which requires for an agent to bring it about that φ that not only should φ be false before the agent acts but that φ should be false in the world as it would have been after the action, had the agent not acted. (This is not very explicit in von Wright's discussion, but is well brought out by the diagrams on p. 227 of [1035], in which Segerberg summarizes this and other relatively early work in the logic of action;[401] the same feature is clearly present in a more recent treatment of these matters, Belnap and Perloff [62]: see the paragraph straddling pages 181 and

---

[401]In the *Studia Logica* printing of [1035] cited in our bibliography, the relevant page is p. 353; the double issue *Studia Logica* in which it appears is devoted to the logic of action and contains many examples of then current work in the area; see also Hilpinen [486], and other references mentioned in Remark 4.4.2(i) (p. 237). A more recent survey appears in 3.1, and for present purposes especially 3.1.1, of Lindström and Segerberg [734]. Explicit attention to the logic of bringing about (and refraining from bringing about, omitting to bring about, etc.) goes back to work by Saint Anselm in the 11th Century. See Henry [473], [474], esp. pp. 119–129; also Pörn [891] itself of course, as well as Walton [1141], [1142], [1144], Talja [1097], and Uckelman [1118].. For philosophical qualms about the enterprise, see Davidson [231], as well as the reply to this paper by Lemmon in the volume, [948], in which it appeared. (This volume also contains a restatement of his position by von Wright.)

182. A convenient introduction to Belnap's logical approach – "*Stit* theory" – is provided at the start of Chapter 2 of Horty [526]. (Etymology: "Stit" is an acronym for "see(s) to it that".) If it was going to be true anyway that φ, we don't want to say that the agent brought it about that φ. The simpler course, strategically, is to isolate either the temporal aspect, which takes us from bringing about to coming about, or else the modal aspect, which gives a notion of making something the case which requires not that what is made the case was not previously the case, but simply that it would not have been the case had action not been taken: as it was put above, that φ should be false in the world as it would have been after the action, had the agent not acted. This formulation invites the objection that there is no such thing as the world as it would have been...: there are rather several things equally well qualified to count as how the world might have been... In other words, the earlier formulation involves what is called in Lewis [717] the Limit Assumption as well as 'Stalnaker's Assumption' (as does the $g(w, X)$ idea in Example 4.4.4, p. 241). The latter is also involved in the purely temporal version, in the fiction of an immediately preceding moment.

More accurately for that case – at least if moments of time are taken to have the order type of the real numbers) – one should no doubt deem it to have come about that φ at a moment if it is then true that φ but has from some earlier moment up until then been the case that not φ. (This raises the issue of perfect aspect indicating present state/relevance, mentioned in the discussion between Example 5.2.5, p. 376, and Example 5.2.6.) These complications will not be taken up here, however, because everything of interest (especially for Section 6.3 below) remains intact if we simply think of u as the present moment and $f(u)$ as some contextually salient earlier – or indeed simply some other – moment;[402] likewise for the modal case, $f(u)$ is simply some world we are concerned to contrast with the actual world for purposes of making a comparison over the truth value of a given statement.

6.2 The Logic of Coming About

In this section, we will render some of the above ideas with greater precision and present a complete axiomatization of the class of valid formulas. The language is that of monomodal logic, with the primitive non-Boolean operator written as "D". This language is interpreted with the aid of models of the form $\langle U, f, V \rangle$, in which U is a non-empty set, $f : U \longrightarrow U$, and V maps propositional variables to subsets of U. The truth of a formula φ at a point $x \in U$ in such a model (notation, as usual: "$\langle U, f, V \rangle \models_x \varphi$") is defined by induction on the complexity of φ:

$\langle U, f, V \rangle \models_x p_i$ iff $x \in V(p_i)$;

$\langle U, f, V \rangle \models_x \varphi \wedge \psi$ iff $\langle U, f, V \rangle \models_x \varphi$ and $\langle U, f, V \rangle \models_x \psi$;

and similarly for the remaining truth-functional primitives.

$\langle U, f, V \rangle \models_x D\varphi$ iff $\langle U, f, V \rangle \models_x \varphi$ and $\langle U, f, V \rangle \not\models_{f(x)} \varphi$.

Call a formula φ *valid* just in case for every model $\langle U, f, V \rangle$, and all $x \in U$, we have $\langle U, f, V \rangle \models_x \varphi$. Naturally we could put this in terms of validity over the class of all frames, validity on a frame $\langle U, f \rangle$ being understood in the obvious way; see Section 2.5. We are not pursuing this refinement here, since the logics determined by various subclasses of the class of all frames will not be receiving attention.

EXERCISE 6.2.1 (*i*) Show that every instance of the schemata (D0)–(D3) of the previous section is a valid formula.

---

[402] In the terminology of Wajszczyk [1139], where somewhat similar themes are explored, we are concerned with 'dichotomic' rather than continuous change.

6.2. THE LOGIC OF COMING ABOUT

(ii) Since in our main discussion no mention is made of iterated occurrences of D, here we take the opportunity to ask whether either of $Dp \to DDp$ or $\neg Dp \to D\neg Dp$ (respectively **4** and **5** for D) is a valid formula. ✠

To axiomatize the class of valid formulas, we take as axioms all instances of the schemata (D0) and (D1) of the previous section, the validity of which (*inter alia*) we have just asked the reader to check, along with the rule (MP) given there and the following rule:

$$(\text{RD}) \quad \frac{(\psi_1 \wedge \ldots \wedge \psi_m) \to (\varphi_1 \vee \ldots \vee \varphi_n)}{(\psi_1 \wedge \ldots \wedge \psi_m) \to ((D\varphi_1 \wedge \ldots \wedge D\varphi_n) \to (D\psi_1 \vee \ldots \vee D\psi_m))}$$

To subsume the case of $m = 0$ ($n = 0$) we stipulate that the empty conjunction (empty disjunction) is $\top$ ($\bot$). If we individuate rules in such a way that for each rule there is a single transition from formulas to formulas such that the applications of the rule comprise precisely the transitions which arise from that transition by (uniform) substitution, then (RD) is a family of rules, one rule $(\text{RD})_{m,n}$ for each choice of m and n in the above schema. The rule $(\text{RD})_{0,1}$ of the previous section then deserves its name (though we have replaced $\top \to \varphi$ by φ, and $D\varphi \to \bot$ by $\neg D\varphi$ for that formulation). Another noteworthy special case is $(\text{RD})_{1,1}$:

$$\frac{\psi \to \varphi}{\psi \to (D\varphi \to D\psi)}$$

With the aid of this rule and (D1), we can show that the operator D is congruential in the present logic – admits the substitutivity of provable equivalents within its scope, that is – by deriving the rule:

$$(D\text{Cong}) \quad \frac{\varphi \leftrightarrow \psi}{D\varphi \leftrightarrow D\varphi}$$

For, suppose that $\vdash \varphi \leftrightarrow \psi$. Applying $(\text{RD})_{1,1}$ to the $\to$ direction of this equivalence, we get $\vdash \psi \to (D\varphi \to D\psi)$, from which the $\leftarrow$ direction of the equivalence gives $\vdash \varphi \to (D\varphi \to D\psi)$. By (D1) we have $\vdash D\varphi \to \varphi$, so by truth-functional reasoning we conclude that $\vdash D\varphi \to D\psi$. Similar reasoning delivers the converse implication.

EXERCISE 6.2.2 *(i)* Derive representative instances of the previous section's (D2) and (D3), namely, $(p \wedge Dq) \to D(p \wedge q)$ and $D(p \wedge q) \to (Dp \vee Dq)$ from the present axiomatic basis (i.e., (D0), (D1), (RD) and (MP)).

(ii) Can the current rule (RD) be derived from the axioms and rules of the previous section ((D0)–(D3), MP and $(\text{RD})_{0,1}$? (Justify your answer. Do not assume a negative answer is rendered probable by the decision to work with the more general rule (RD): this is the most convenient rule to have available for the completeness proof of Section 6.7.) ✠

Very little remains to be done to show the soundness of the present logic, i.e., to show that only valid formulas are provable:

PROPOSITION 6.2.3 *For any formula φ, if $\vdash \varphi$ then φ is valid.*

Proof. In view of observations already made, we need only to check that the rule (RD) preserves validity (and also that (MP) does so: but the latter is obvious). Suppose, then, that a conclusion

$$(\psi_1 \wedge \ldots \wedge \psi_m) \to ((D\varphi_1 \wedge \ldots \wedge D\varphi_n) \to (D\psi_1 \vee \ldots \vee D\psi_m))$$

of some application of (RD) is invalid, with a view to showing that the corresponding premiss

$$(\psi_1 \wedge \ldots \wedge \psi_m) \to (\varphi_1 \vee \ldots \vee \varphi_n)$$

is also invalid. The supposition means that for some model $\mathcal{M} = \langle U, f, V \rangle$ and some $x \in U$, we have

$\mathcal{M} \models_x \psi_1, \ldots, \mathcal{M} \models_x \psi_m$;
$\mathcal{M} \models_x D\varphi_1, \ldots, \mathcal{M} \models_x D\varphi_n$;
and
$\mathcal{M} \not\models_x D\psi_1, \ldots, \mathcal{M} \not\models_x D\psi_m$.

Since these $D\psi_i$ are all false at x ($i = 1, \ldots, m$) in $\mathcal{M}$ while each ψ_i is true at x, it must be that each ψ_i is also true at $f(x)$: otherwise ψ_i would "come true" at u and $D\psi_i$ would be true there. So, if the premiss for this application of (RD) is valid, we must have $\mathcal{M} \models_{f(x)} \varphi_1 \vee \ldots \vee \varphi_n$, so for some i ($1 \leqslant i \leqslant n$): $\mathcal{M} \models_x \varphi_i$. But this contradicts the earlier conclusion that $\mathcal{M} \models_x D\varphi_i$ for all i ($1 \leqslant i \leqslant n$). The premiss formula is therefore not after all valid. We have now seen that every instance of an axiom schema is valid and that each of the rules preserves validity, completing the proof of soundness. ■

The converse is shown in Section 6.7. For the moment, we devote ourselves to noting an alternative choice of primitives giving rise to a definitionally equivalent logic, and explaining the reason for concentrating, despite its greater complexity in an obvious respect on the formulation above, on the language with "D" as primitive.

The semantic apparatus at our disposal suggests consideration of a singulary operator, which we shall write as "Y" (mnemonic for "Yesterday"), interpreted semantically thus:

$$\langle U, f, V \rangle \models_x Y\varphi \text{ iff } \langle U, f, V \rangle \models_{f(x)} \varphi$$

for all models $\langle U, f, V \rangle$, all $x \in U$, and all formulas φ. The logic of "Y", is accordingly just the modal logic determined by the class of all functional frames, i.e. **KD!**; we saw much of this logic in Section 2.7 and before (as well as since: see the index), though of course Y was notated as $\Box$ there. We can introduce the operator "D", behaving semantically as stipulated earlier, by means of the definition:

$$D\varphi = \varphi \wedge Y\neg\varphi$$

The "$Y\neg\varphi$" on the right can equivalently be replaced by "$\neg Y\varphi$". Because of the now explicit appearance of negation in the *definiens*, the striking $\wedge$ to $\vee$ transformation in (D3) [$= D(\varphi \wedge \psi) \to (D\varphi \vee D\psi)$] now emerges as a familiar De Morgan phenomenon rather than something to be taken as a brute novelty: a point which may favour the idea of taking "Y" rather than "D" as primitive, especially in view of the fact that the two are actually interdefinable. (We see the same transformation, as well as the similarly explicable subsumption of the earlier rule (RD)$_{0,1}$ of the previous section, in the case of the rule (RD) of the present section. Again, compare Example 4.3.2.) For if we begin, as above, with "D", then we can recover "Y", behaving (you may check) semantically as required, by putting:

$$Y\varphi = D\neg\varphi \vee (\varphi \wedge \neg D\varphi).$$

These interdefinabilities and the arguably greater simplicity (and certainly greater antecedent familiarity) of the Y-based language notwithstanding, it is the D-based language which is more suggestive for the material in Section 6.3. For this reason we have concentrated on proving soundness for a D-based axiomatization in above (and completeness in Section 6.7).

6.2. THE LOGIC OF COMING ABOUT

EXAMPLE 6.2.4 The above discussion illustrates the fact that if we want to abstract from logics as here conceived, to the extent of identifying definitionally equivalent logics, then considerable care needs to be exercised with such terminology as *normal, monotone*, etc., as applied to logics on the more abstract conception, since the D-based formulation of the present logic is non-normal, in that D, its sole non-Boolean primitive, is not normal in this logic, while the Y-based formulation is normal, since *its* sole non-Boolean primitive, Y, is normal. ◀

The point of departure for what follows is the schema (D3) lately touched on, which tells us that the only way for a conjunction to become true is for one or other of the conjuncts to become true. We can say in view of this that there are two ways for it to come about that $\varphi \wedge \psi$: by its coming about that φ, and by its coming about that ψ. Some will feel that a more perspicuous classification is one in which the 'ways' alluded to do not overlap. It can of course come about that $\varphi \wedge \psi$ by its coming about that φ *and* its coming about that ψ; the relevant schema (i.e., $(D\varphi \wedge D\psi) \to D(\varphi \wedge \psi)$) is an immediate consequence of (D1) and (D2) from the previous section. For those so inclined, we can provide a disjunctive consequent for a variant on (D3) whose disjuncts are mutually exclusive.[403] To do so it is convenient to introduce an operator to signal that it remains the case that φ; we write $R\varphi$ for this ("R" for "Remains"). In terms of "D", we put: $R\varphi = \varphi \wedge \neg D\varphi$. (In terms of '$Y$', we can put $R\varphi = \varphi \wedge Y\varphi$.) The variant in question is then:

(D3′) $D(\varphi \wedge \psi) \to ((D\varphi \wedge D\psi) \vee (R\varphi \wedge D\psi) \vee (D\varphi \wedge R\psi))$.

The disjuncts here are not only mutually exclusive, but together jointly exhaustive of the ways it can come about that $\varphi \wedge \psi$: which is to say that the "$\to$" in (D3′) can be strengthened to a "$\leftrightarrow$" and every formula instantiating the resulting schema is valid. With (D3′) in mind, one may say that there are exactly three ways for a conjunction to become true, but I do not think we must treat this as correcting the (D2)-inspired idea that there are only two ways for a conjunction to become true. Compare the situation of disjunction and (not *becoming*, but) *being* true. It is perfectly legitimate to say that there are only two ways for a disjunction to be true namely by its having a first disjunct which is true and by its having a second disjunct which is true and, the apparent incompatibility notwithstanding, that there are three ways for a disjunction to be true: namely by both disjuncts' being true, by the first but not the second's being true, and by the second but not the first's being true.

The talk, whichever one prefers, of two or of three ways for conjunctions to become true, is in any case really only appropriate at the schematic level. For a particular choice of φ, ψ, one might want to say something different; for example if $\varphi \wedge \psi$ is inconsistent – whether because one of φ, ψ, is already inconsistent or because even though this is not so φ still has the negation of ψ as a consequence – then there won't be two ways for the conjunction to come true: there will be *no* ways. Likewise, if φ has ψ are equivalent (for example, if they are the same formula) then there will be only one way for $\varphi \wedge \psi$ to become true. But even here, we are still thinking too schematically, since if $\varphi = \psi = p \wedge q$ then by what as already been said, there are two ways for $\varphi \wedge \psi$ to become true – but they are not the φ way and the ψ way! (Similarly, if φ and ψ are respectively p and $q \wedge r$, then there are three ways for $\varphi \wedge \psi$ to become true on the mode of counting that allows overlap, then there are *seven* ways for this $\varphi \wedge \psi$ to become true, since any one of p, q, r, could be the only one whose coming true turns the conjunction from false to true (which could happen in any of three ways), or any two making the shift (another three ways), or all three together shifting from false to true (one last way).

To round out the discussion, we should note that a principle analogous to (D3) holds for the coming about of disjunctions:

(D3-for-∨) $D(\varphi \vee \psi) \to (D\varphi \vee D\psi)$.

---

[403]Such disjunctions are called 'exclusionary disjunctions' in Humberstone [594] – not to be confused them with exclusive disjunctions: the disjunction connective involves is still the familiar (inclusive) "∨".

It is easy to check that every instance of (D3-for-$\vee$) is a valid formula; together with the original (D3) this principle provides the material for the inductive part of a proof by induction on the complexity of φ that for any formula $\varphi(p_1, \ldots, p_n)$ containing at most the propositional variables exhibited and constructed out of those it does contain exclusively by means of the connectives $\wedge$ and $\vee$ formula

(D3$^+$) $D\varphi(p_1, \ldots, p_n) \to (Dp_1 \vee \ldots \vee Dp_n)$

is a valid formula. The condition that φ be constructed without connectives other than conjunction and disjunction must be scrupulously observed. If, for example $\varphi(p,q)$ is taken as $p \leftrightarrow q$, then (D3$^+$) does not hold: such a formula can become true by p's becoming false while q remains false. We can think of any D-free formula $\varphi(p_1, \ldots, p_n)$ as standing for an n-ary truth-functional connective, and (D3$^+$) is valid for a given choice of φ precisely when the truth-function concerned is monotonic, which is to say that for any truth value assignments v_1, v_2, with $v_1(p_i) \leqslant v_2(p_i)$ for $i = 1, \ldots, n$, $v_1(\varphi(p_1, \ldots, p_n)) \leqslant v_2(\varphi(p_1, \ldots, p_n))$. Here $\leqslant$ is defined, for $x, y \in \{T, F\}$, by: $x \leqslant y$ iff not both $x = T$ and $y = F$.[404]

If we want something along the lines of (D3$^+$) for arbitrary Boolean (i.e., D-free) formulas, then the best we can do is let $C\varphi$ (read "there has been a change in respect of whether or not φ") abbreviate $D\varphi \vee D\neg\varphi$, and note the validity of:

(D3^{++}) $D\varphi(p_1, \ldots, p_n) \to (Cp_1 \vee \ldots \vee Cp_n)$.

Indeed, more generally, for such φ the D in the antecedent can itself be replaced by C: clearly the only way for a truth-functional compound built from certain sentence letters to change in truth value is for at least one of them to change, since the compound's truth value is fixed given those of its contained sentence letters. It is plain old (D3) itself that will be of greatest significance for what follows, however.

6.3 Relational Change and Coming About

We have been emphasizing that there are two ways (or, on an equally legitimate taxonomy, three ways) for a conjunction to become true. The time has now arrived to note a precisely analogous situation for the case of certain statements involving no explicit Boolean complexity. We take as our paradigm the statement(-form) that objects a and b stand in the relation R (which we represent by: Rab). Roughly, for suitable choices of R we notice that it can come about that Rab in virtue of a change in a, and it can come about that Rab in virtue of a change in b, just as, in the case with which we have been concerned, it can came about that $\varphi \wedge \psi$ in virtue of a change in the truth value of φ and it can come about that $\varphi \wedge \psi$ in virtue of a change in the truth value of ψ. In both cases, we might add, with a nod in the direction of (D3$'$) of the preceding discussion, the case in which there is a change in both relata (or in both conjuncts' truth values) is not to be forgotten.

In spite of its apparent vagueness, the phrase "in virtue of a change in a" (or "in b") is intended to bear some weight in the above description. If the relation R is the relation of *being the same shape (as)*, for example, then it can come about that Rab in virtue of a change in a's shape, or in virtue of a change in b's shape (or both); but if R is the relation of *liking*, then it is only in virtue of a change in a that it may come about that Rab, even if it is *because of* a change in b that a comes to like b. Changes of the latter sort are causally productive of the change in respect of whether or not Rab, but cannot *constitute* (in the given circumstances) a's coming to like b. This will remind the reader of the discussion, initiated by Geach, of the notion of a 'merely Cambridge' change, as opposed to a real change, in an

---

[404] The non-constant truth-functions with this property are precisely those obtainable using conjunction and disjunction; see Exercise 5.9 on p. 76 of Gindikin [375], and the solution thereto on p. 81. Note that this notion of monotonicity is not that with which we were concerned when urging that "D" not be treated as a monotone operator above. (Here we exploit the convenience of having the two adjectives *monotonic* and *monotone*.)

6.3. RELATIONAL CHANGE AND COMING ABOUT

individual: coming to be liked by a is a merely Cambridge change in b (a change only in the sense that something formerly false of b comes to be true of b), but cannot be anything less than a real change in a. (See Geach [356], pp. 72, 99, and, for the following example, Kim [655] and Helm [470].) Xanthippe's becoming a widow is similarly a merely Cambridge change in Xanthippe, even if we wanted to say it is causally productive of genuine change – grief, relief, or whatever – on her part. (Of course we might not want to say this, since we might think of these as caused by Socrates' death rather than Xanthippe's becoming a widow: this depends on issues in the individuation of events discussed in the references just given, and need not detain us here. Much more relevant to the issues of this chapter, but also not gone into because of its notorious difficulty, is the notion of an *intrinsic* property; see Weatherson [1152].)

Instead of thinking about relations R such as the *same shape* relation considered above, we can usefully begin with something simpler, in the case of which the analogy between the two ways for it to come about that Rab and the two ways for it to come about that $\varphi \wedge \psi$ is more than an analogy: the former is here nothing less than a special case of the latter. The example involves binary relations which are, in the sense explained at p. 15 above ∧-*representable*; that is – by way of reminder – relations R for which there exist properties F and G such that for any individuals x and y, x bears R to y if and only if x has F and y has G. When R, F and G are so related, its coming about that Rab is its coming about that $Fa \wedge Gb$, so by (D3) we have, treating Rab as interchangeable with the latter conjunction:

(D3-for-R) $D(Rab) \to (D(Fa) \vee D(Gb))$.

There are, we may say, two (overlapping, i.e., non-mutually-exclusive) ways for an ∧-representable relation to come to hold between a pair of objects, just as and indeed precisely because there are two ways for a conjunction to come to be true. Let us introduce a bit of informal terminology here to save space: call a binary relation *multiply inchoative* if there are in some fairly direct analogue of the case presented by (D3-for-R) two (or more) ways for a pair of individuals to come to stand in that relation. It is the vague phrasing "some fairly direct analogue" here which gives this terminology its informality: but perhaps we can tighten things up in what follows. (In Section 6.6 a case will be given in which it is natural to speak of a conjunction – the conjunction of a relational predication, Rab, and its converse, Rba – as coming to be true in two ways, one per conjunct, even though this is not a case of ∧-representability since the predicates involved are not monadic.) Since on any explication everything will count (to say the least) as a 'fairly direct analogue' of itself, we can at least say definitely that any ∧-representable relation is multiply inchoative. And one further comment is in order on the terminology. The "inchoative" in "multiply inchoative" is doing the work of stressing that it is a multiplicity (a pair, at least) of ways of *coming to stand in* the relation that is important, rather than of ways of *standing in* the relation. One might say that there are any number of ways in which a might stand in the *taller than* relation to b, since a might be taller than b by one centimetre, by two centimetres, and so on; derivatively, a might come to be taller than b by coming to exceed b in height by one centimetre, or by coming to exceed b in height by two centimetres, or … But this isn't enough to make the relation *taller than* multiply inchoative in the intended sense: the locus of multiplicity is in the range of determinates of the *taller than* relation – conceived of as a determinable – rather than in the coming-to-hold of that relation between individuals. The irrelevance of this point about specific height differences notwithstanding, the *taller than* relation does seem to deserve to be called multiply inchoative, since, as the 'Cambridge change' literature, some of which was recalled above, tells us, one way for a to become taller than b is for a to grow, while another is for b to shrink.[405] This is a genuinely inchoative multiplicity, since it is a matter of how the relation comes to hold between a and b rather than of how a and b are related after the change. And we might wish to see it as an existentially quantified form of ∧-representability: there are heights h and h' with

---

[405] The example is from Geach; if you think that the English sentence "a becomes taller than b" only allows for the former possibility, then substitute "It comes about that a is taller than b". In saying that one way for a to become taller than b is for a to grow, we should take this as allowing the possibility that b also grows, but by less; analogously in the case of b's shrinking.

$h > h'$ and it comes about that a's height is h and b's height is h'. The two ways for this to come about correspond to the coming about of the two conjuncts *à la* (D3). Somewhat similar moves – as well as qualms about them – will be considered in Section 6.5.

We would like an explication of multiple inchoativity which rules in not only the $\wedge$-representable relations but also such examples as the *same shape* and *taller than* relations, whether or not the latter are to be subsumed (as just suggested) under some generalization of the former. Perhaps the discussion of those examples has given the notion enough intuitive content for us not only to appraise any candidate explications, but also to ask about how this notion is related to another which seems at first sight to apply to much the same relations; indeed a consideration of this other notion, which emerged in philosophical discussion in the late 1970's and early 1980's, and which can be described in terms of 'cross-predication', may be of assistance in seeking such an explication. Modal cross-predication is called cross-world predication in Salmon [991], and both this and Peacocke [873], the two works from the period in question – the latter, unfortunately, not published – which provide the most extensive discussion of the notion, note the existence of what we may call temporal cross-predication, an analogue in which it is times rather than possible worlds which the predication 'crosses'.[406] By way of example, we have modal cross-predication in "This bridge is stronger than it would have been had it not been fortified" as well as in "This box is the same shape as that one might have been", and temporal cross-predication in "I am taller than I was last year", "This box is now the same shape as that box will be tomorrow".[407] Here the same relations figure in our examples as were used above to illustrate multiple inchoativity. As will be apparent the label "multiply inchoative" is too one-sidedly temporal to be ideal. Recall (from the end of Section 6.1) that the D operator does not need to be thought of as saying that things are different at one time from how they were at another: any contrast between a given point and a background reference point will do, whether these points are times or worlds (or something else), and whether or not, we noted, "in the background", for the temporal case, meant 'in the (immediate) past'. Accordingly 'multiply contrastive' would perhaps be a more neutral terminology, though we avoid it because of the suggestion of a contrast between more than two things ('points'): what we want to indicate is a contrast which arises in a multiplicity of ways between (how things are at) the two points concerned. We retain 'multiply inchoative' as the generic term and ask the reader to disregard its temporal connotations when they are not pertinent.

As we have mentioned, the *same shape* relation and the *taller than* relation serve both as examples of multiply inchoative relations and as examples of cross-predicable relations. More strikingly still, Salmon ventures into the territory of $\wedge$-representable relations, which as we noted provide the clearest or at least the most direct cases of multiple inchoativity, in giving the following example of cross-predication:

> Thus, for instance, consider the following contrived relation R that holds between a pair of individuals x and y whenever x is hungry and y is angry. We might say that a pair of individuals x and y stand in R *across* possible worlds w_1 and w_2 when x is hungry in w_1 and y is angry in w_2. (Salmon [991], p. 129.)

---

[406] There is also an extensive discussion of cross-predication from a logical point of view (with special emphasis on the temporal case) in §4 of Butterfield and Stirling [140], as well as a helpful presentation of (a simplified version of) Peacocke's own proposals in Forbes [319], pp. 92–3. For some more recent thoughts of Forbes's on a topic much at the centre of attention in the following section, namely comparatives, see [321]. Wehmeier [1157] reviews much of this earlier literature (as well as material not mentioned here) and offers his own proposal as to how to represent cross-world predication in a modal language – something which will not be under discussion here. Some further discussion and references with a view to metaphysical applications in the temporal case may be found in Crisp [221].

[407] We could have cross-predication which is both modal and temporal, as Peacocke [873] illustrates with this example: "My car x might have been in 1976 exactly the same colour as yours y actually was in 1975". (Observe that the "actually" is not redundant here, playing a significant disambiguating role.) Note apropos of the *taller than* case, we are simply assuming that inter-world height comparisons make definite sense, an assumption that could be contested. (It has even been contested that one can make objective convention-independent *intra*-world height and length comparisons; at least that seems to be the obvious interpretation of numerous works by Adolf Grünbaum, for references to and discussion of which, see Horwich [528].)

6.4. MULTIPLE INCHOATIVITY

Are we dealing here with just a coincidence, or are perhaps the classes of cross-predicable relations and of multiply inchoative relations one and the same? Let us look more closely at multiple inchoativity, and then in Section 6.5 at cross-predication. We will consider the above example further in the course of the latter discussion, returning (briefly and inconclusively) to the question of how cross-predicability and multiple inchoativity are related in Section 6.6.

6.4 Multiple Inchoativity

Let us begin by repeating the schema (D3) from Section 6.1, since it represents our paradigm for the multiplicity (or at least duality) of ways something might come about:

(D3) $\quad D(\varphi \wedge \psi) \to (D\varphi \vee D\psi).$

In Section 6.2 we noted a similar principle – we called it (D3-for-∨) – with the conjunction in the antecedent replaced by disjunction, but the discussion of Section 6.3 suggests that we should not regard this as representing multiple inchoativity: for the two 'ways' in which it may *come about* that φ or ψ are here nothing more than the two ways for it to *be the case* that φ or ψ. A side-benefit of such a ruling is that it allows us to deal with a consideration threatening to make everything which could come about, capable of coming about in more than one way, since as an instance of (D3-for-∨) we have:

(1) $\quad D((p \wedge q) \vee (p \wedge \neg q)) \to (D(p \wedge q) \vee D(p \wedge \neg q)).$

Thus since, as we have noted, equivalents may be replaced by equivalents in the scope of D, we have

(2) $\quad Dp \to (D(p \wedge q) \vee D(p \wedge \neg q)).$

But this is not to the point when we are interested in specifically inchoative multiplicity, since the two ways in question (p and q, on the one hand, and p and not q, on the other) are simply two ways for it to *be* the case that p.

Trouble may threaten, however, when we exploit the distributive law dual to that licensing the replacement just considered. That is, we should consider the following instance of (D3) itself:

(3) $\quad D((p \vee q) \wedge (p \vee \neg q)) \to (D(p \vee q) \vee D(p \vee \neg q)),$

which gives us

(4) $\quad Dp \to (D(p \vee q) \vee D(p \vee \neg q)).$[408]

---

[408] As an aside here we note that if we had taken D to have the logic of a normal modal operator, then the conjunction of $D(p \vee q)$ with $D(p \vee \neg q)$, and not just their disjunction, would follow from Dp, since the monotone property would give each of $D(p \vee q)$ and $D(p \vee \neg q)$ as a consequence of Dp. But the objection to these monotony inferences runs along the same lines as for the case of the notepad and pen (on the table) from Section 6.1: it may come about that p without its coming about that $p \vee q$ (for instance), since q may have been true all along, which would have meant that $p \vee q$ was true all along. In [542], a different line was taken (partly explaining the parenthetical part of that paper's title). We allowed prefixing of D to the antecedent and consequent of provable disjunct-to-disjunction implications, but not in the case of conjunction-to-conjunct implications; the general picture was a kind of restricted monotone condition: from $\vdash \varphi \to \psi$ to $\vdash D\varphi \to D\psi$ provided that every sentence letter occurring in φ occurred in ψ. The idea that if one brought it about that p one thereby brought it about that $p \vee q$ was defended against the objection that $p \vee q$ might have already – or 'anyway' – true, by talk of overdetermination, held not to apply in the case of the conjunction-to-conjunct inference. Whatever the merits of any such line of thought, which would require classically equivalent formulas not to be interreplaceable in the scope of "D" – might have been, they do not carry over to the interpretation of D in terms of coming about rather than bringing about.

We cannot respond here by saying that the formulas in the scope of the "D"s in the consequent represent two ways for it to be the case that p, at least if that phrase is to bear the same sense as in the previous response, since neither of the formulas in question has p as a consequence. (On the usage in question, for it to be the case that p is one way for it to be the case that $p \vee q$ – to focus on the first formula by way of example – rather than conversely.)

The idea of there being different ways for something to come to be the case may be too vague to settle the question of (4). But even if we decide that (4) does show that any state of affairs can come about in more than one way, we can consider whether this is really as troublesome as it might seem. Trouble would consist in this fact of the sentential logic of coming about spilling over and forcing us to classify every relation as multiply inchoative as a result. But is there any such threat? Let us recall our paradigm of multiple inchoativity, the $\wedge$-representable relation R for which we had:

(D3-for-R) $D(Rab) \to (D(Fa) \vee D(Gb))$.

The important point here is that b does not occur in the first disjunct of the consequent, and a does not occur in the second. So one way of its coming about that Rab is specifiable without reference to (the denotation of) b and the other, without reference to a. But any attempt to substitute an arbitrary binary relational predication Sab for p in (4) above is going to leave a and b together, rather then segregated into distinct atomic subformulas, when we consider the consequent. More generally, if p is logically equivalent to the conjunction of ψ with χ then at least one of ψ, χ, must contain p (and in fact if ψ and χ are logically independent, both must).

EXERCISE 6.4.1 Prove this last assertion, understanding "logically equivalent" to mean *classically* (alias *truth-functionally*) equivalent. ✠

6.5 Cross-Predicability

Having already introduced the topic of "predication at a modal distance" as we might think of cross-world predication, with some references to the earlier literature in Section 6.3, here we take a closer look at the topic, doing so with the aid the discussion supplied by Kemp [645], in which we read:

> In what sorts of relation may an object as it actually is stand to either a merely possible object, or to an object, not as it is, but as it might have been? ...What relations are suitable candidates for *crossworld* relations? And what makes them so? Kemp [645], p. 305.

The contrast Kemp has in mind between those relation which are cross-world predicable and those which are not he illustrates with a pair of examples we shall number as (*) and (**):

(*)   John could have been taller than Mary actually is.

(**)   John could have kissed Mary.

Without the last two words (*) is ambiguous between an intra-world height comparison and an inter- or cross-world height comparison. But there are no two corresponding readings for (**) or anything in its vicinity. As Kemp remarks, "Try as he might, it does not seem that John could actually kiss merely possible girls; whereas he is actually taller than many merely possible girls." In the following quotation from Kemp [645], p. 305, our labelling – (*) and (**) – replaces his numbering:

> What is the relevant different between (*) and (**)? For the relation x *is taller than* y, as it appears in (*), we can substitute relations such as x *is smarter than* y, x *is prettier than* y, x *is as witty as* y, and also x *is a compatriot of* y and x *is the same size as* y; but not x *kissed* y, x *is facing* y, or x *taught* y. Apparently, the difference is that the former grouping comprises only comparative relations and equivalence relations. I shall assume that this is true: only comparative and equivalence relations may sensibly be ascribed across possible worlds. My aim in this paper is to say why this is.

6.5. CROSS-PREDICABILITY

One's first reaction to this is perhaps that what we are told is being assumed is such a strong claim that a bit should be said as to *why* we are assuming it, before setting out on a project of explaining its truth. Note that the assumption is not that the modally cross-predicable relations comprise exactly the comparative and equivalence relations, but merely that the modally cross-predicable relations comprise a subset of the comparative and equivalence relations. There is still something a bit strange about the assumption, in that if two relations are cross-predicable, then presumably their union is, whereas just taking the case of equivalence relations (since the notion of comparativity has yet to be defined), the union of two of these is typically not an equivalence relation, so unless such unions all turn out to be comparative (a question we cannot settle until the definition is given), there will be trouble for the proposal. Similarly if we have a cross-predicable equivalence relation, its complement is never an equivalence relation (since it is irreflexive), so unless it turned out, on the definition to be given, to be a comparative relation (which in fact it does not), there would be trouble for the suggestion. But such technicalities aside, the proposal looks unlikely to succeed, since purely structural or formal features like the property of being an equivalence relation (or that of comparativity as defined below), do not look as though they will connect with Kemp's earlier remark about its not seeming "that John could actually kiss merely possible girls". This appears to be due not to the fact that the relation of kissing fails to satisfy some first order condition on binary relations stated just using a symbol for the relation (and the usual logical machinery of quantifiers, identity, and the Boolean sentence connectives), but rather to the fact that relations requiring spatial or causal contact, such as kissing, can only hold between world-mates. Let us shelve these worries and proceed to see what comparativity consists in, according to Kemp:

> By a *strict comparative* binary relation R I mean one that effects a strict partial ordering of its field – is asymmetrical and transitive – but which is also what we might call weakly-connected: if x and y are both members of the field of R, then either xRy or yRx, unless for all z, xRz iff yRz and zRx iff zRy (...). For example if both x and y are such as to be either taller or shorter than something or other, then either one is taller than the other or they are taller and shorter than all and only the same things (i.e., they are the same height). For each strict comparative we can speak also of its *derived comparative*: this is the relation which holds between members x and y of the field of R just in case either xRy or for all z, xRz iff yRz and zRx iff zRy. The derived comparative for *taller than* for example is *x is at least as tall as y*. The derived comparatives are transitive, reflexive and strongly connected. Together the strict and derived comparatives make up the class of comparatives. Other relations are *noncomparative*.
>
> Kemp [645] p. 307f.

At the end of the first sentence of this passage, giving the definition of comparativity for relations, there is an endnote saying that this is equivalent to a definition given by van Benthem [69] and later used in Humberstone [566], in which the defining characteristics are irreflexivity, transitivity and the following condition which van Benthem calls almost-connectedness:

> *for all x, y, z, if xRz then xRy or yRz.*

This condition is also sometimes called *negative transitivity* (see Fishburn [303]), because it amounts to requiring that the complement of the given relation is transitive (its complement relative to the domain over which our quantifiers in these conditions range, that is, on which R is supposed to be some binary relation). And indeed the conditions are equivalent – as long as we forget about the restriction to the 'field' of the relation, which does not appear in the van Benthem definition. Working through a demonstration of equivalence will help to get a feel for what is going on, but anyone not wanting to do so should skip to the end of the following intermezzo.

Detailed Logical Intermezzo. First, notice that since all asymmetric relations are irreflexive, and all transitive irreflexive relations are asymmetric, it makes no difference, once only transitive relations are under consideration, whether an additional further restriction of irreflexivity is imposed or, instead, a further condition of asymmetry. Either way, we get the notion of a strict partial ordering. It is this notion which is being further strengthened by almost-connectedness, which amounts, as already remarked, to saying that the *complement* of the given relation is transitive. This further condition is, as Kemp says in the endnote in question (note 2 of [645], p. 318*f.*) "more compact but less intuitive" than his own condition that R should be what he calls 'weakly connected'.[409] Let us proceed to verify the (far from obvious) equivalence between the two characterizations.

First, we check that van Benthem's condition implies Kemp's condition.[410] Suppose that R is a strict partial order satisfying almost-connectedness, with a view to showing that R is in Kemp's sense weakly connected. That is, we must show that for all x, y:

$$\forall x \forall y [(Rxy \lor Ryx) \lor (\forall z (Rxz \leftrightarrow Ryz) \land \forall z (Rzx \leftrightarrow Rzy))].$$

Here I have taken the liberty of going over to an explicit first-order formalization of the condition in question, so that we can reason about it with some confidence. (I have translated the "unless" in Kemp's verbal formulation as a disjunction, used the formula initial rather than infix notation for R, and also for a reason that will become clear in due course, put the "$\forall z$" onto both of the conjuncts of the conjunction rather than having it occur just once with the whole conjunction as its scope.) To derive this condition from van Benthem's, it will suffice to get a contradiction from any supposed violation of the condition. So suppose we have a, b, with (i) $\neg Rab$ and (ii) $\neg Rba$. We will first show that $\forall z(Raz \leftrightarrow Rbz)$ and then that $\forall z(Rza \leftrightarrow Rzb)$. Let us break down the first task into that of showing that $\forall z(Raz \to Rbz)$ and showing that $\forall z(Rbz \to Raz)$. For the former, suppose Rac, for arbitrarily selected c, with a view to concluding that Rbc. By van Benthem's condition, since Rac, we have either Rab or Rbc. But the first alternative contradicts (i), so we are done. For the latter, suppose that Rbc, with a view to concluding that Rac. Since Rbc, by van Benthem's condition, either Rba or Rac. The first alternative contradicts (ii), so again we are done. That leaves us to show that $\forall z(Rza \leftrightarrow Rzb)$, which again breaks down into showing that $\forall z(Rza \to Rzb)$ and $\forall z(Rzb \to Rza)$. For the former, suppose that Rca, with a view to concluding that Rcb. By van Benthem's condition, it follows that either Rcb or Rba, and the second alternative is ruled out by (ii). The argument for the converse is again similar, and we can safely omit it. Thus comparativity by van Benthem's lights implies comparativity by Kemp's.

For the converse implication, suppose that R satisfies Kemp's comparativity conditions, and we have Rac. To derive van Benthem's almost-connectedness condition, we must show that either Rab or Rbc, for an arbitrarily chosen b. For a contradiction, suppose that while Rac, we have $\neg Rab$ and $\neg Rbc$. Suppose that Rba; then by the transitivity of R and the fact that Rac, we could conclude that Rbc, contradicting our assumption that $\neg Rbc$. Thus we have $\neg Rba$. Then instantiating x and y in Kemp's weak connectedness condition to a and b respectively, we have

---

[409]Actually, as we saw, he writes "weakly-connected", but there is no call for a hyphen between an "-ly" adverb and the adjective it modifies in English. The whole terminology is ill-advised in view of the standard meaning of this phrase for another property – see p. 75 – and we will try and use it only in close proximity to the name "Kemp", so that it is clear what is meant. On the subject of ill-advised terminology, for the property of negative transitivity mentioned above, Nowak [848] uses the term "intransitivity", which is standardly used for the quite different property: for all x, y, z, if xRy and yRz then not xRz. Here things are bad enough because the word "intransitive" is already widely misused – Temkin [1099] is only one of many examples that could be given – in much of the literature on (individual and social) choice to mean simply "not transitive".

[410]The description "Kemp's condition" is not historically accurate but convenient for our purposes. See the discussion in §III at p. 254 of Milne [809] for the relevant considerations; Milne cites a 1970s paper by Max Cresswell for the definition of the indifference relation – the last disjunct of Milne's disjunctive characterization – though there is a misprint, with "Rxz" appearing as "$P\chi\zeta$". (It was not uncommon for the journal concerned sometimes to set whole lines of text inappropriately in a Greek font.) Milne there also gives a version of the van Benthem characterization which brings back in the "field of a relation" idea, as well as prefixing a "□" to each initial universal quantifier. The latter is connected to Milne's idea that the *possible* existence of heights, ages, etc., suffices for transworld comparisons in those respects, whereas, as will become clear, Kemp requires the within-a-world coexistence of these entities with each of the objects being compared. The idea of using abstract entities – in the form of *degrees* – to mediate cross-predication was advanced for the temporal case in Morton [832]. Geach pours scorn on the use of degrees in the analysis of comparatives in his interesting paper [360]; more recently another candidate – *tropes* (no doubt equally distasteful from Geach's perspective) – for a similar role was suggested in Moltmann [816].

6.5. CROSS-PREDICABILITY

$$(Rab \lor Rba) \lor (\forall z(Raz \leftrightarrow Rbz) \land \forall z(Rza \leftrightarrow Rzb)),$$

and since we have already seen that $\neg Rab$ and $\neg Rba$, we can pass to the second disjunct, instantiating the z of the first conjunct to c and discarding the second conjunct, we conclude that $Rac \leftrightarrow Rbc$. But we supposed that Rac, so we get Rbc, contradicting our assumption that $\neg Rbc$ and completing our *Reductio* argument. Kemp's and van Benthem's characterizations are thus shown to be equivalent.

It is instructive to see what was just used to derive van Benthem's condition from Kemp's. Since we discarded the second conjunct above, we have in fact derived van Benthem's condition from what might at first sight seem to be a considerable weakening of Kemp's condition, namely

$$\forall x \forall y[(Rxy \lor Ryx) \lor \forall z(Rxz \leftrightarrow Ryz)].$$

Thus this condition, without the extra "$\forall z$" turns out to be no weakening at all, since we have seen that it yields the van Benthem condition which in turn, as we saw before, yields the full Kemp condition. This shortening of Kemp's condition is equivalent to the original form. In fact we can go further, since in the argument of the preceding paragraph, all that we exploited of $Rac \leftrightarrow Rbc$ was the left-to-right direction. Thus Kemp's condition is equivalent to the following further simplification of it:

$$\forall x \forall y[(Rxy \lor Ryx) \lor \forall z(Rxz \to Ryz)].$$

These simplifications fall under the heading of 'potentially weakening', in that dropping a conjunct from a (positively occurring) conjunctive subformula typically produces something weaker – and here we subsume the "$\leftrightarrow$"-to-"$\to$" move under this description also – but there is a further simplification that could be made, this time under the heading of 'potentially strengthening', in that we can drop one of the disjuncts from the above formulation and still have something equivalent. We can drop the "Ryx" disjunct, getting:

$$\forall x \forall y(Rxy \lor \forall z(Rxz \to Ryz)).$$

The availability of this simplification is evident once we note that we have already seen the Kemp weak connectedness condition to be equivalent (given transitivity and irreflexivity, though actually only transitivity was exploited) to van Benthem's almost-connectedness condition. As remarked, the latter amounts to saying that the relation concerned has a transitive complement. But that is what the last formula inset above says, since we can remove the disjunction and (for conformity with the conventional way of expressing transitivity) pull the "$\forall x$" top the front:

$$\forall x \forall y \forall z(\neg Rxy \to (Rxz \to Ryz)),$$

and, finally contrapose the consequent:

$$\forall x \forall y \forall z(\neg Rxy \to (\neg Ryz \to \neg Rxz)),$$

which, for dramatic effect, you could of course choose to write like this:

$$\forall x \forall y \forall z(\bar{R}xy \to (\bar{R}yz \to \bar{R}xz)).$$

None of this is intended to be critical of anything that Kemp says. His original formulation of the weak connectedness condition does convey the idea very directly that for any x and y, the relation holds in one direction or the other between them, or else they are indiscernible in respect of how they are related to all other objects, as with the three options: *taller than, shorter than, the same height as*. So this is recognisably a sort of trichotomy law; the almost-connectedness condition of van Benthem is, exactly as Kemp says, not readily graspable intuitively for some reason (perhaps because the most readily intelligible implicational conditions are Horn conditions – without disjunctive (atomic) consequents, that is – though I suspect that the fact that a new variable appears in the consequent adds to the psychological difficulty of processing the condition in the present case).

The most interesting thing to come out of the above discussion is perhaps the first simplification: instead of characterizing being the same height in terms of being taller than the same things and also being shorter than the same things, it suffices to just use being taller than the same things – even though, for a *fixed pair* x, y, of individuals, this does not imply also being shorter than the same individuals. More precisely, abbreviating the claim that R is transitive and irreflexive to $\mathsf{SPO}(R)$ (the letters abbreviating Strict Partial Order, though again irreflexivity is not actually relevant here), using the notations $R(\cdot)$ and $R^{-1}(\cdot)$ as in our discussion of accessibility relations, and using $\vdash$ to indicate the consequence relation of classical predicate logic, we have:

(1) $\mathsf{SPO}(R), \forall x \forall y((Rxy \vee Ryx) \vee R(x) = R(y)) \vdash \forall x \forall y((Rxy \vee Ryx) \vee$
$$(R(x) = R(y) \wedge R^{-1}(x) = R^{-1}(y)),$$

we have a contrast at the 'local' level, in that:

(2) $\mathsf{SPO}(R), (Rab \vee Rba) \vee R(a) = R(b)) \nvdash (Rab \vee Rba) \vee$
$$(R(a) = R(b) \wedge R^{-1}(a) = R^{-1}(b)),$$

because, in particular,

(3) $\mathsf{SPO}(R), (Rab \vee Rba) \vee R(a) = R(b)) \nvdash (Rab \vee Rba) \vee R^{-1}(a) = R^{-1}(b)$.

To see how the inference from the left to the right of (3) fails, draw a diagram with points a and b, neither of which bears the relation R to the other, and add a point c which bears R to one of them but not the other – say, to a but not to b. Since they are R-incomparable, to keep the left-hand side of (3) true, we can put $R(a) = R(b) = \varnothing$. We had to have $\neg Rbc$ in any case, since transitivity would then require Rba, contrary to hypothesis. But stopping here we have a countermodel illustrating (3). If we had instead tried to verify the 'global' hypothesis on the left of (1), however, we should now have b and c incomparable, forcing $R(b) = R(c)$ which *would* give us a contradiction since Rca while $\neg Rcb$. **End of Detailed Logical Intermezzo.**

Whether characterized in van Benthem's way or in Kemp's, the important thing about comparative relations is that they give rise to structures each of which, as van Benthem ([69], p. 197) remarks, "may be pictured as a linear order of 'indifference groups' in the obvious way". The indifference groups are equivalence classes for the relation of what in the above intermezzo was called R-incomparability. Because a set with a comparative relation defined on it yields a homomorphic pre-image of a linear ordering in this way, p. 118 of [566] suggested that "pre-linear orderings" would be a good name for such relations (arguably superior to some other terminology current in the social choice literature).[411] With respect to the relation *older than*, these are age groups (sets of individual of the same age), and the homomorphism alluded to maps individuals to their respective age groups. Kemp's idea in [645] is to the think of the images under this homomorphism not so much as age groups as *ages*, and in the case of the relation *taller than*, as *heights*, and so on; he also considers equivalence relations and the introduction of corresponding entities by Fregean 'abstraction', which we needn't go into here. Kemp writes f for the homomorphism involved and "$<_R$" for the induced linear ordering, so we have an equivalence between xRy and $f(x) <_R f(y)$. We can now think of these abstract objects as existing in all possible worlds, and to take into account world-to-world variations, refine the notation somewhat so $f(x, w)$, say, represents x's height, age, or whatever, in world w. (Kemp actually writes "$[f(x)](w)$".) From here it is only a small step to bring transworld comparisons into the picture since we have (Kemp, [645], p. 312) a way of representing the idea that, as he puts it "x-at-w_i bears R to y-at-w_j", we have:

$$f(x, w_i) <_R f(y, w_j),$$

---

[411] We are not here concerned with defending the claim that all comparative adjectives give rise to these structures, whatever one chooses to call them. In particular, some comparatives (or similar constructions) seem to give rise to incommensurability (or non-transitive indifference): see Lehrer and Wagner [691] for discussion.

which "tells us to compare the value of f for x at w_i with that of f for y at w_j". In summary ([645], p. 313):

> I suggested that only comparative and equivalence relations make crossworld sense. What I want to suggest now is that the correctness of the abstraction technique for handling them would explain *why* this is so, and that fact constitutes a concise and sweeping argument for the correctness of the abstraction technique. So why would it explain it? The reason is very simple. The relations that can be predicated across worlds are precisely the equivalence relations and comparatives. But that is precisely the class of relations that support abstraction: it comprises those which support it directly because they already are equivalence relations, and those which support it via their use in defining equivalence relations as explained above [[i.e., indifference classes from comparatives]]. This is not merely coincidence: given abstraction on the equivalence relations defined in terms of a comparative, we can always replace comparisons of objects by relations amongst their associated abstracta (xRy is always equivalent to $f(x) <_R f(y)$). The position might be put, then, by saying that crossworld relations make sense if and only if they are reducible to world-bound relations amongst abstracta. If so then to advance the reduction as their analysis seems irresistible.

Kemp adds that "from a strictly logical point of view the truth of crossworld relational predications provides no reason to doubt that all modal facts must be reducible to facts-at-worlds; in fact the account explains why they should be". The sentiment about reducibility seems laudable, but in the past people have traditionally wanted modal languages to *save* them from having to postulate the abstract entities Kemp is keen to invoke, even if that means increasing the expressive power of the languages by novel devices such as Peacocke's indexed actuality operators, referred to in note 406. Note 36 (written by Davies) of [235] mentions a use of these operators in connection with the equivalence relation *is same hairstyled with* rather than be committed to an ontology of hairstyles, and still to be able to deal with the sense of "Fraser might have been same hairstyled with Whitlam" according to which the existence of a world in which both have the same hairstyle (as we may pre-theoretically put it) is neither necessary nor sufficient for the truth of sentence.[412] Likewise with the case of heights (from the preceding section). More generally, examples of attempts to reduce ontology by increasing modal sophistication can be found in Burgess [135], Lacey and Anderson [683], and numerous other places; for some worries about this enterprise (as applied in a relational theory of space), see Belot [63]. One of the most ingenious attempts to use – not cross-world but – cross-time predication to ontologically reductive effect was that of David Armstrong, who in [35] suggested that we could make sense of absolute motion without thinking of it as motion relative to (the parts of) an ethereal container ('absolute space' and its regions) once we allowed transtemporal spatial relations as primitive: something has moved absolutely since t_0 of if it is not now where it was at t_0, it has moved such and such a distance if it is now at that distance from where it was at t_0, and so on. The suggestion did not win widespread acceptance,[413] but its precise merits are less interesting than the striking idea of using cross-time predication as an alternative to ontological inflation. Well, evidently no such motivation attracts Kemp. While we are on the subject of spatial relations, we should consider an example involving cross-world rather than (as in Armstrong) cross-time spatial predication, from Salmon [991], p. 116:

> Physical spatial properties, such as being to the left of, are also naturally construable as cross world relations. (If I were on the left side of the room right now, my right hand would be to the left of where my left hand actually is, although my right hand would still be to the right of where my left hand would be.)

Clearly the point would go through with predicates not exhibiting the potentially complicating perspectival aspect of the predicate "is to the left of", such as, to recycle an example from the Armstrong

---

[412] In other words, there is an understood "as he (= Whitlam) actually is" at the end of the sentence.

[413] See Peetz [878] for an early negative reaction, and Khamara [650], pp. 49–51 for some more recent thoughts.

discussion above, precisifying it as we go: "is exactly five metres from".[414] The relation thereby signified is neither an equivalence relation nor a comparative relation, serving accordingly as a counterexample to the assumption, whose dubiousness was noted at the time, quoted above from Kemp: "I shall assume that this is true: only comparative and equivalence relations may sensibly be ascribed across possible worlds. My aim in this paper is to say why this is."

The incorrectness of Kemp's assumption was already implicit in the discussion in Section 6.2, where Salmon [991] was quoted as citing the example of x and y being hungry and angry in w_1 and w_2 respectively, an example of a relation between x and y Salmon described as contrived, and we observed amounted to venturing into the territory of $\wedge$-representable relations. On David Lewis's view of worlds and individuals (e.g. Lewis [726]), no one individual – at least of the kind we normally take an interest in – exists in more than one world, and the properties of being hungry and being angry can, suppressing the possibility of change over time in respect of them, be identified with the modally scattered classes of (possible) individuals possessing these properties.[415] In that case Salmon's relation is literally $\wedge$-representable, with x (existing only in w_1) and y (existing only in w_2) standing in the relation just in case the former is hungry and the latter angry. Matters are less straightforward for the more usual approach to the semantics of modal predicate logic which allows overlapping domains, but at least the object language itself, in concealing reference to worlds, presents the appearance of separate conjuncts in which the explicit predications are monadic, as in the case of straightforwardly $\wedge$-representable relations – with such formulations as "$Fx \wedge \Diamond Gy$." How such 'modally $\wedge$-representable relations' fare by comparison with the straightforward case dealt with in Proposition 1.2.12 is far from clear, so let us ignore this complication and ask whether $\wedge$-representable relations *tout court* pass Kemp's test for being cross-world predicable, recalling (from 1.2.12) that a relation $R \subseteq U \times U$ is $\wedge$-representable just in case it satisfies:

$$\forall x \forall y \forall u \forall z ((Rxy \wedge Ruz) \to Rxz),$$

where the quantifiers range over U. Let the monadic representing predicates be F and G (i.e., $Rxy \leftrightarrow (Fx \wedge Gy)$, all x, y). Clearly such an R is transitive, the inset condition here being a generalized ('lazy', 'sloppy', 'forgetful') version of the condition of transitivity itself (which is the special case where $y = u$). But it is not guaranteed to satisfy van Benthem's almost-connectedness condition, and therefore – by the foregoing discussion – is not guaranteed to be what Kemp calls a strict comparative. Let us write out the almost-connectedness condition for this choice of R:

$$\forall x \forall y \forall z \big((Fx \wedge Gz) \to ((Fx \wedge Gy) \vee (Fy \wedge Gy))\big).$$

Clearly, this condition need not be satisfied since there is nothing in the antecedent to prevent our making any given choice of y lie outside the extensions of both the predicates F and G. Similarly, the relation is not reflexive, so it is not what Kemp calls a derived comparative relation – or indeed any kind of equivalence relation. So his proposal does not even subsume the paradigmatically cross-world predicable relations, at least as Salmon saw them as being. Nor is there any reason to take a special interest, amongst the monadically representable relations, in those which are $\wedge$-representable. The discussion leading up to Proposition 1.2.12 mentioned also the $\vee$-representable and the $\leftrightarrow$-representable relations, and for the former we also spelt out a corresponding first-order condition on R (referring the interested reader to Humberstone [552] for such a condition in the case of the latter).

---

[414]Notice that if we vary Salmon's example to a temporal one – "After I walk to the left side of the room, my right hand will be to the left of where my left hand now is, although ..." – we naturally understand it as compatible with Salmon's room being (as most rooms tend to be) a room in a building on the surface of a planet rotating on its own axis and revolving round a star, a consideration telling in favour of the suggestion in Peetz [878] that direct cross-temporal spatial comparisons have less to do with absolute motion than Armstrong [35] had supposed.

[415]World-mates, in Lewis's parlance, are thus individuals both existing in one world, with neither existing in any other. The casual use of this term above, in the comment that "kissing(,) can only hold between world-mates" was not meant to imply all of this, but just meant that individuals can only kiss in a world in which both exist.

6.6 Cross-Predicability and Multiple Inchoativity

At the end of Section 6.3 the question was raised as to how cross-predicability and multiple inchoativity were related. It is difficult to address this question since we have both modal and temporal cross-predicability to reckon with, and multiple inchoativity was itself not precisely defined. A few words are in order nonetheless, and for simplicity we consider only temporal cross-predicability.

On the assumption that such cases as transtemporal height comparisons are representative of temporal cross-predication and multiple inchoativity, it would seem that the latter feature requires the former. In saying that we take a's increasing in height and b's decreasing in height to be two ways for it to become the case that a is taller than b – our paradigm of multiple inchoativity – we are comparing two (classes of) scenarios within each of which there is an earlier and a later time, at each of which, to draw the contrast between two ways for it to come about that a is taller than b, we compare a's height before and after the change, as well as b's height before and after the change. So we presuppose the intelligibility of temporal cross-predication in spelling out multiple inchoativity.[416] But before concluding on the basis of such considerations that multiply inchoative relations have to be cross-predicable, we should notice the following source of counterexamples.[417] If we take the intersection of a binary relation and its converse, defining S to hold, for some given R, by:

$$\forall x \forall y (Sxy \leftrightarrow (Rxy \wedge Ryx)),$$

then, just as in the monadic representability case (the case of $\wedge$-representability, to be more specific) there will be two ways for it to come about that Sab: by its coming about that Rab and by its coming about that Rba (including also the case in which both of these come about). Unlike the monadic case, we cannot on the face of it regard these as changes in respect of first one and then the other of a, b, but still we have two ways for it to come about that a and b bear R to each other. Now suppose we select for R a relation that does not admit of cross-predication: for example, think of Rxy as *x strikes y*. Now we have, in S, a relation which is multiply inchoative but not cross-predicable.[418]

What about the converse: does temporal cross-predicability (for a given relation) require that we can make sense of a multiplicity of ways of coming to stand in that relation? The relation of remembering appears to qualify as temporally cross-predicable, in that you can now (perhaps) come to remember your grandmother (as she was before she died, let's say). But it does not seem to be multiply inchoative: it is only in virtue of a change in you that it can come about that you – suddenly recollecting some incident involving her, for example – remember your grandmother, not also by a virtue of a change in her. So this is not like the cross-temporal *taller than* relation. And it differs from the latter in another – no doubt related – respect too, namely that whereas there is no natural answer to the question of *when* a is taller on Wednesday than b was on Tuesday, there is a perfectly natural answer to the question of when a on Wednesday remembers b as of Tuesday: *viz.* on Wednesday. Perhaps this shows that the category of temporally cross-predicable relations is too inappropriately heterogeneous (to employ in thinking about these issues). But if such relations as that of remembering are taken as belonging to it, then we have a counterexample to the idea that cross-predicability implies multiple inchoativity. There is also the question of whether there are cross-world analogues of remembering, on which no opinion will be ventured here; see Forbes [323] and Moltmann [817] for pertinent material.

---

[416] But aren't what are being called scenarios here simply possible worlds, indicating a breach of the promise to consider temporal rather than modal cross-predication? Yes, they are possible worlds, but no, transworld comparisons are not involved – only transtemporal comparison within worlds.

[417] For this observation I am indebted to Julian Shortt (a former student).

[418] Instead of taking the intersection (or conjunction) of R with its converse, we could equally well have taken the intersection of R with another suitably independent binary relation, with neither relation being cross-predicable.

6.7 Postscript: Completeness for the Logic of §6.2.

We use a canonical model argument in the style of Section 2.4, here taking the canonical model for the logic of Section 6.2 as the structure $\langle U, f, V \rangle$ in which U is the set of all sets of formulas which are maximal consistent w.r.t. this logic; f is defined by setting, for $u \in U$:

$$f(u) = \{\chi \mid D\neg\chi \vee (\chi \wedge \neg D\chi) \in u\};$$

and V is defined by putting $u \in V(p_i)$ iff $p_i \in u$. (Note that in terms of the operator 'Y' mentioned in Section 6.2, $f(u)$ is $\{\chi \mid Y\chi \in u\}$.) To check that the structure $\langle U, f, V \rangle$ thus defined is a model, we need to verify that $f(u) \in U$ whenever $u \in U$. The first thing to verify is that $f(u)$ is consistent, on the assumption that u is; so suppose $f(u)$ is inconsistent (w.r.t. the present logic, that is): thus for some formulas $\chi_1, \ldots, \chi_k$, where for each χ_i ($1 \leqslant i \leqslant k$), $D\neg\chi_i \vee (\chi_i \wedge \neg D\chi_i) \in u$, we have:

$$\vdash \neg(\chi_1 \wedge \ldots \wedge \chi_k). \tag{1}$$

(Since we are not giving a name to to the present logic, the provability of a formula φ in which we simply indicate by writing "$\vdash \varphi$", with no subscripted label.)

Let $\varphi_1, \ldots, \varphi_n$ be those formulas from amongst $\chi_1, \ldots, \chi_k$ for which $D\neg\chi_i \in u$, and let $\psi_1, \ldots, \psi_m$ be those of $\chi_1, \ldots, \chi_k$ for which $\chi_i \wedge \neg D\chi_i \in u$. (Because of (D1), no χ_i satisfies both of these conditions.) Thus we have

$$\vdash \neg(\psi_1 \wedge \ldots \wedge \psi_m \wedge \varphi_1 \wedge \ldots \wedge \varphi_n), \tag{2}$$

which we may rewrite as:

$$\vdash (\psi_1 \wedge \ldots \wedge \psi_m) \rightarrow (\neg\varphi_1 \vee \ldots \vee \neg\varphi_n). \tag{3}$$

By the rule (RD) then, we have

$$\vdash (\psi_1 \wedge \ldots \wedge \psi_m) \rightarrow ((D\neg\varphi_1 \wedge \ldots \wedge D\neg\varphi_n) \rightarrow (D\psi_1 \vee \ldots \vee D\psi_m)). \tag{4}$$

From the fact that for each ψ_i we have $\psi_i \wedge \neg D\psi_i \in u$, it follows that $\psi_1 \wedge \ldots \wedge \psi_m \in u$ while $D\psi_1 \vee \ldots \vee D\psi_m \notin u$; thus, given (4), we must have $D\neg\varphi_1 \wedge \ldots \wedge D\neg\varphi_n \notin u$. As each $D\neg\varphi_i \in u$, however, this is impossible. This contradiction shows that $f(u)$ as defined here, is consistent. Since we want $f(u) \in U$, we must show that $f(u)$ is actually *maximal* consistent; for this it suffices to note that every formula or its negation belongs to $f(u)$, since (5) is a substitution instance of a truth-functional tautology:

$$(D\neg\chi \vee (\chi \wedge \neg D\chi)) \vee (D\chi \vee (\neg\chi \wedge \neg D\neg\chi)). \tag{5}$$

Thus $f(u) \in U$ as long as $u \in U$, and the canonical model $\langle U, f, V \rangle$ is indeed a model.

THEOREM 6.7.1 *For the canonical model $\langle U, f, V \rangle$ and any formula φ:*

$$\langle U, f, V \rangle \models_x \varphi \text{ if and only if } \varphi \in x, \text{ for all } x \in U.$$

Proof. The proof is as usual by induction on the complexity of φ, and the only case of special interest is that in which φ is $D\psi$ for some formula ψ. So we must show that for all $x \in U$, $\langle U, f, V \rangle \models_x D\psi$ iff $D\psi \in x$, on the assumption (the inductive hypothesis) that a similar equivalence holds for ψ. Recall that $\langle U, f, V \rangle \models_x D\psi$ has been defined to hold just in case $\langle U, f, V \rangle \models_x \psi$ and $\langle U, f, V \rangle \not\models_{f(x)} \psi$, so this assumption means that it will suffice to show that:

6.7. POSTSCRIPT: COMPLETENESS FOR THE LOGIC OF §6.2.

$$[\psi \in x \text{ and } \psi \notin f(x)] \text{ iff } D\psi \in x.$$

Unpacking the "$f(x)$" part, what has to be shown is:

$$[\psi \in x \text{ and } D\neg\psi \vee (\psi \wedge \neg D\psi) \notin x] \text{ iff } D\psi \in x.$$

'If': Supposing $D\psi \in x$, we get $\psi \in x$ by (D1), and on the same supposition, we get $D\neg\psi \vee (\psi \wedge \neg D\psi) \notin x$ because clearly the second disjunct cannot belong to x, or x would be inconsistent, containing not only $D\psi$ but also $\neg D\psi$, and the first disjunct cannot belong to x either, since if $D\psi \in x$ and $D\neg\psi \in x$, we should have $\psi \in x$ and also $\neg\psi \in x$ (by (D1) again).

'Only if': Suppose that $\psi \in x$ and $D\neg\psi \vee (\psi \wedge \neg D\psi) \notin x$, with a view to showing that $D\psi \in x$. Since $D\neg\psi \vee (\psi \wedge \neg D\psi) \notin x$, $\psi \wedge \neg D\psi \notin x$, so since $\psi \in x$, we have $\neg D\psi \notin x$, and thus $D\psi \in x$. ∎

COROLLARY 6.7.2 *For any formula φ, if φ is valid then $\vdash \varphi$.*

Combining this with our earlier soundness result (Proposition 6.2.3), we have:

COROLLARY 6.7.3 *For any formula φ, $\vdash \varphi$ if and only if φ is valid.*

Chapter 7

Appendix: Natural Deduction for S4 and S5

7.1 Non-Modal Rules

In this Appendix, we give natural deduction rules for various normal modal logics, concentrating on those in the title, but touching on weaker logics in Section 7.5. We begin with some natural deduction rules for the Boolean connectives, based on Lemmon [698] and thus indirectly on Gentzen's original natural deduction system *NK* for classical (propositional) logic. (See Prawitz [905] for details and references.) Here we give the rules as sequent-to-sequent rules, with "Γ", "Δ", "Θ" ranging over sets of formulas. Lemmon's presentation has formula-to-formula rules with asides about the discharging of assumptions (formulas on the left of the "$\succ$", that is – which Lemmon himself writes as "⊢": see p. 8 for our different use of this symbol), and since he allows the same formula to be multiply assumed (with different line numbers in each case) his system would be more accurately represented by having the capital Greek letters range over multisets of formulas – though even that is not quite accurate. Those details need not affect us here. The main point (Gentzen's idea) is that the rules for $\rightarrow$, $\wedge$, $\vee$ are organized into pairs, one for introducing an occurrence of the connective in question onto the right of the "$\succ$", and one for eliminating it from that position. They are accordingly called (#I) and (#E), where # is the connective concerned. The rules for $\neg$ rather spoil the pattern, with (RAA) – *Reductio ad Absurdum*, to use the full name – and ($\neg\neg$E) – 'double negation elimination' – involving more than one occurrence of the connective concerned, and in the former case an occurrence of another connective altogether as well (namely $\wedge$). Slightly more elegant possibilities are available, but some residual awkwardness is perhaps to be expected, and is usually put down to the disparity between classical and intuitionistic logic. The last two rules, ($\bot$E) and ($\top$I) are included because these constants are often useful in modal logic (to have formulas not constructed from propositional variables). Note that there is no introduction rule for $\bot$ and no elimination rule for $\top$. Semantically, we consider only valuations assigning T to $\top$ (or: models in which $\top$ is true at every point) and F to $\bot$ (or: models in which $\bot$ is false at every point).

$$A \succ A \quad \text{Initial Sequents}$$

$$(\rightarrow \text{I}) \quad \frac{\Gamma, A \succ B}{\Gamma \succ A \rightarrow B} \qquad\qquad (\rightarrow \text{E}) \quad \frac{\Gamma \succ A \rightarrow B \qquad \Delta \succ A}{\Gamma, \Delta \succ B}$$

$$(\wedge \text{I}) \quad \frac{\Gamma \succ A \qquad \Delta \succ B}{\Gamma, \Delta \succ A \wedge B} \qquad\qquad (\wedge \text{E}) \quad \frac{\Gamma \succ A \wedge B}{\Gamma \succ A} \quad \frac{\Gamma \succ A \wedge B}{\Gamma \succ B}$$

$$(\vee\mathrm{I})\ \dfrac{\Gamma\succ A}{\Gamma\succ A\vee B}\quad \dfrac{\Gamma\succ B}{\Gamma\succ A\vee B}\qquad (\vee\mathrm{E})\ \dfrac{\Gamma\succ A\vee B\quad \Delta,A\succ C\quad \Theta,B\succ C}{\Gamma,\Delta,\Theta\succ C}$$

$$(\mathrm{RAA})\ \dfrac{\Gamma,A\succ B\wedge\neg B}{\Gamma\succ\neg A}\qquad (\neg\neg\mathrm{E})\ \dfrac{\Gamma\succ\neg\neg A}{\Gamma\succ A}$$

$$(\bot\mathrm{E})\ \dfrac{\Gamma\succ\bot}{\Gamma\succ A}\qquad\qquad (\top\mathrm{I})\ \ \Gamma\succ\top$$

Explicit sequent-to-sequent derivations look rather different from the derivations in Lemmon [698]. The latter proceed from assumptions, which are, in the case of some rules, *discharged* by the application of the rule: (RAA), (∨E), (→I); with annotations in the style of Lemmon, this means that the line numbers of these assumptions appear in the far left ('dependency record') column before the rule is applied but not after. A Lemmon-style proof of $p\to(q\to r), q\succ p\to r$, for example, would proceed as in Example 7.1.1, though the dash used in the annotation of the application of (→I) here would appear in [698] as a comma. (The dash notation is preferable since it indicates that the consequent has been derived *from* the antecedent at the preceding line.)

EXAMPLE 7.1.1 A proof of the sequent $p\to(q\to r), q\succ p\to r$ in something like Lemmon's style:

1	(1) $p\to(q\to r)$	Assumption
2	(2) q	Assumption
3	(3) p	Assumption
1,3	(4) $q\to r$	1, 3 →E
1,2,3	(5) r	2, 4 →E
1,2	(6) $p\to r$	3–5 →I

◀

Each line in the proof corresponds to the sequent we obtain by writing the formulas corresponding to the entries in the assumption dependency column on the far left (indicating which assumptions the formula written on that line depends on – themselves in the case of the assumed formulas, these being lines corresponding to the "initial sequents" of the sequent-to-sequent formulation), and then the separator ≻ ("⊢" in Lemmon's notation, as remarked above), and on the right the formula appearing on the line itself. The above proof corresponds to the following proof using the explicit sequent-to-sequent format. Note that only a single line number appears in place of the "3" and "5" in the annotation for line (6) above, because (→I) is a sequent-to-sequent rule with only one sequent-premiss.

EXAMPLE 7.1.2 An explicitly sequent-to-sequent version of the above proof:

(1) $p\to(q\to r)\succ p\to(q\to r)$	Initial Sequent
(2) $q\succ q$	Initial Sequent
(3) $p\succ p$	Initial Sequent
(4) $p\to(q\to r), p\succ q\to r$	1, 3 →E
(5) $p\to(q\to r), p, q\succ r$	2, 4 →E
(6) $p\to(q\to r), q\succ p\to r$	5 →I

◀

Since assumptions can be discharged by the rules, we can end up with the proof of a sequent having no formulas on the left; the following example illustrates several further aspects (commented on below) of Lemmon's natural deduction system, some of which may cause one to query the "natural" part of this description:

7.1. NON-MODAL RULES

EXAMPLE 7.1.3 A Lemmon-style proof of the sequent $\succ (p \to q) \vee p$:

1	(1)	$\neg((p \to q) \vee p)$	Assumption
2	(2)	p	Assumption
2	(3)	$(p \to q) \vee p$	2 $\vee$I
1, 2	(4)	$((p \to q) \vee p) \wedge \neg((p \to q) \vee p)$	3, 1 $\wedge$I
1	(5)	$\neg p$	2–4 RAA
6	(6)	$\neg q$	Assumption
1, 2	(7)	$p \wedge \neg p$	2, 5 $\wedge$I
1, 2, 6	(8)	$(p \wedge \neg p) \wedge \neg q$	7, 6 $\wedge$I
1, 2, 6	(9)	$p \wedge \neg p$	8 $\wedge$E
1, 2	(10)	$\neg\neg q$	6–9 RAA
1, 2	(11)	q	10 $\neg\neg$E
1	(12)	$p \to q$	2–11 $\to$I
1	(13)	$(p \to q) \vee p$	12 $\vee$I
1	(14)	$((p \to q) \vee p) \wedge \neg((p \to q) \vee p)$	13, 1 $\wedge$I
	(15)	$\neg\neg((p \to q) \vee p)$	1–14 RAA
	(16)	$(p \to q) \vee p$	15 $\neg\neg$E

◀

In the annotations (right hand justifications) for (RAA) in lines 5, 10, and 15, we have used Lemmon's convention of putting the line numbers of the place at which the formula (whose negation is to be inferred by applying the rule) is originally assumed, and then the line on which a contradiction has been derived from it (and perhaps other assumptions), though Lemmon writes a comma and we have used a dash to separate the lines, in order (as in the earlier examples) to suggest that it is a subderivation and not just a pair of formulas that are involved here.

The first thing that strikes one about the above proof is perhaps the conspicuous presence of $\neg$, especially given that the sequent being proved does not involve this connective. The reason for that is that the sequent in question is classically but not intuitionistically provable and all of the rules except for $(\neg\neg E)$ are intuitionistically acceptable. As Lemmon explains in the preface to [698], to convert his proof system to one for intuitionistic logic, drop this rule and replace it by the (now no longer derivable) rule allowing passage from $A \wedge \neg A$ (depending on some set of assumptions) to B (on the same assumptions).[419] (Dropping $(\neg\neg E)$ and not adding in the latter rule gives the still weaker *Minimal Logic*.) So any intuitionistically unprovable sequent which is classically provable will require the use of the rule $(\neg\neg E)$, whether or not $\neg$ is present in any of the formulas in the sequent. Lemmon was quite open about wanting intuitionistically unacceptable sequents to require longer proofs than those which were intuitionistically acceptable. ("Acceptable" here can be understood semantically, but we are not here going into the semantics of intuitionistic logic – the easiest version of which was supplied by Kripke on the basis of a translation between intuitionistic logic and **S4**, and the semantics he had already devised for normal modal logics including **S4**. Similarly, the sequents provable in Lemmon's system for classical logic are exactly those which are tautologous.)

Lemmon was by contrast much less open about the aspect of his proof system brought out by lines (5)–(12) of the proof in Example 7.1.3, which, if line (5) had been an assumption rather than a formula derived from an earlier assumption, would themselves constitute a proof for the sequent $\neg p \succ p \to q$. For this case classical and intuitionistic logic line up together, in favour of the sequent, and on the opposing side we have *relevant* logic, which finds the mere falsity of an antecedent to be insufficient grounds for a conditional, in the absence of some relevance between antecedent and consequent. A similar point holds for premises and conclusions (and here I mean premiss-formulas on the left of the "$\succ$" and conclusion

---

[419]Lemmon's discussion actually lumps together $(\neg\neg E)$ and a rule, $(\neg\neg I)$ which interchanges its premiss and conclusion, collectively as (DN) – for "double negation"; the $(\neg\neg I)$ half of this rule is intuitionistically acceptable and is derivable from (RAA).

formulas to the right, not, as above, premiss-sequents and conclusion-sequents in the application of a rule). Thus there would be a similar objection to the sequent $p, \neg p \succ q$, essentially established at line 11 in the proof above. Nor is there anything particular about negation that raises these objections, which would also apply in connection with the sequents $p, q \succ p$ and $p \succ q \to p$. If you attempt proofs for them using the rules presented here, you will again be involved in something like the bizarre procedure of lines (7)–(9) above in which ($\wedge$I) is applied and then immediately afterward, ($\wedge$E), returning us to one of the conjuncts we started from – but now as depending, somewhat magically, on a different set of assumptions. This feature is concealed in Lemmon's presentation of his rules in [698], but it becomes clearer if one considers the $\to$-subsystem, in which only the introduction and elimination rules for $\to$ are allowed (as well as the procedure for making assumptions, or, in the sequent-to-sequent version, initial sequents). Where there is no $\wedge$ (or $\vee$) available to do the assumption dependency juggling just described, there is no way left of proving such sequents as $p \succ q \to p$ and $\succ p \to (q \to p)$, and what we have is a natural deduction system for the implicational fragment of a relevant (or 'relevance') logic.[420]

We close the discussion of the non-modal rules by picking up an issue from the discussion immediately following Proposition 4.7.12, in which the possibility was mentioned of adding to the language in play here a new 1-ary connective # subject to the following two rules:

$$(\#1)^+ \quad \frac{\Gamma \succ \#A \to A}{\Gamma \succ A} \qquad (\#2)^+ \quad \frac{\Gamma \succ \#A \vee A}{\Gamma \succ A}$$

Here we have changed the schematic letter "φ" from our earlier discussion to match the current choice of schematic letters for formulas. These two rules are not respectable natural deduction rules, indeed, on various grounds, but since they are still sequent-to-sequent rules, we can throw them into the mix and see what happens. As mentioned in the discussion in Section 4.7, the results are disastrous. They also provide us with an opportunity to illustrate the rule ($\vee$E). We give a derivation in Lemmon's style and also in explicit sequent-to-sequent form.

EXAMPLE 7.1.4 To make trouble for the above # rules, we begin by selecting an arbitrary formula A and providing a proof not requiring the use of these rules, of the sequent $\succ (\#A \to A) \vee \#A$. To obtain such a proof, just go back to Example 7.1.3 and put $\#A$ in place of "p" and A in place of "q" throughout. This will be the starting point for our Lemmon-style proof of $\succ A$:

	(1)	$(\#A \to A) \vee \#A$	Given (as explained)
2	(2)	$\#A \to A$	Assumption
2	(3)	A	2 $(\#1)^+$
4	(4)	$\#A$	Assumption
4	(5)	$\#A \vee A$	4 $\vee$I
4	(6)	A	5 $(\#2)^+$
	(7)	A	1, 2–3, 4–6 $\vee$E

We have followed Lemmon's annotational practice in the case of ($\vee$E), with five line numbers cited corresponding to those on which the disjunction appears, the point at which the first disjunct is assumed, the point at which the common conclusion (to be derived from each disjunct) is reached, the point at which the second disjunct is assumed, and finally the point at which the common conclusion is derived from the second disjunct. Naturally we have replaced some of what Lemmon writes as commas, with dashes, to separate the beginnings and ends of subderivations. ([698] would just have written by line 7: "1, 2, 3, 4, 6 $\vee$E".) Rewriting this in sequent-to-sequent form:

---

[420] As presented here, the rules would then give the logic called **RM0** in Anderson and Belnap [20], and with the added condition "provided $A \notin \Gamma$" on ($\to$I), the logic there called $\mathbf{R_\to}$. At least this is so if we confine our attention to sequents with no formulas on the left of the $\succ$, and these logics are presented in [20] axiomatically anyway. Allowing formulas on the left in the comparison just made of provable sequents would raise the issue, touched on above, as to whether these constitute a sequence, a multiset or – as here – a set of formulas.

(1) $\succ (\#A \to A) \lor \#A$ Given (as before)
(2) $\#A \to A \succ \#A \to A$ Initial Sequent
(3) $\#A \to A \succ A$ 2 $(\#1)^+$
(4) $\#A \succ \#A$ Initial Sequent
(5) $\#A \succ \#A \lor A$ 4 $\lor$I
(6) $\#A \succ A$ 5 $(\#2)^+$
(7) $\succ A$ 1, 3, 6 $\lor$E

◀

Thus the result of adding the two rules (#1) and (#2) is that the system becomes inconsistent: every sequent is now provable.

EXERCISE 7.1.5 (i) Just because every sequent of the form $\succ A$ is provable in the extension of Lemmon's natural deduction system just described, how does it follow that *every* sequent is provable?

(ii) Why are there only three line numbers in the right-hand annotation column of the sequent-to-sequent proof in Example 7.1.4 above at line 7, whereas in the formula-to-formula ("Lemmon-style") proof given before it five line numbers were cited at the corresponding line 7? ✠

7.2 Natural Deduction Rules for □

Let us consider extending Lemmon's natural deduction system for truth-functional propositional logic by an introduction rule and an elimination rule for "□". In this section and the next, we are concerned with providing a natural deduction system specifically for the modal logic **S5**,[421] in the sense that we aim to have $A_1, \ldots A_n \succ B$ provable in the system just in case the formula $(A_1 \land \ldots \land A_n) \to B$ is provable in the formula logic **S5** (alias **KT5**, from Section 2.1). In the same way as Lemmon takes "↔" to be a defined symbol – with definition $A \leftrightarrow B =_{\text{Df}} (A \to B) \land (B \to A)$, though we have usually written such definitions without the "Df" – rather than a primitive symbol, so as not to have to give separate introduction and elimination rules for it, we could, if we wanted, take "◇" as given by the definition $\Diamond A =_{\text{Df}} \neg\Box\neg A$. That is one perfectly respectable option, indeed one taken at p. 16 and followed throughout our axiomatically based discussion from that point onward. However, in view of the analogies between □ and ∀ and between ◇ and ∃, it seems preferable to take both □ and ◇ as primitive, just as Lemmon [698] takes both ∀ and ∃ as primitive. So we shall need to have introduction and elimination rules not only for □, but also for ◇. We address the case of □ first. The elimination rule for "□" is easy to predict, if we concentrate on its alethic interpretation as expressive of some kind of (perhaps broadly logical) necessity. We can read such a rule off the bottom row of the partial truth-table for □ in Figure 1.1. Stated informally as a formula-to-formula rule, the rule says that given □A (depending on certain assumptions), one may infer A (depending on those same assumptions). As with the Boolean connectives of the preceding section, we prefer to make the parenthetical comments about dependencies explicit and to state the rule as a sequent-to-sequent rule:

$$(\Box E) \quad \frac{\Gamma \succ \Box A}{\Gamma \succ A}$$

---

[421] No particular originality is claimed for this proof system, which, one or two cosmetic differences aside, can be found in §4 of Chapter 9 of Forbes [322]. (Some novelties will appear in Section 7.5, though.) Fitch-style natural deduction systems – i.e., those inspired by [306] and [308] – for modal logic as opposed to those in the style of Lemmon or those in the style of Gentzen and – what is essentially the same – Prawitz (see [905]) are very common in the literature. Some corrections to Prawitz's treatment of one aspect of the system – normalization – not under discussion here can be found in Medeiros [797].

The choice of a suitable introduction rule is considerably less obvious. Since we do not want to end up with a proof of such intuitively invalid sequents as $p \succ \Box p$, we cannot write down simply an upside-down version of ($\Box$E). As it happens, a suitably restricted form of this idea will serve us well: We allow the passage from A to $\Box A$ when all the assumptions on which A depends – represented by the set of formulas on the left of the "$\succ$" – are fully modalized, in the sense introduced originally just before Exc. 2.2.3 (p. 56).[422] Here is a sequent-to-sequent formulation of the rule:

$$(\Box \text{I}) \quad \frac{\Gamma \succ A}{\Gamma \succ \Box A} \quad \textit{Provided all formulas in } \Gamma \textit{ are fully modalized.}$$

Thus the would-be two-line proof, by applying ($\Box$I) after assuming p – in sequent-to-sequent terms, using the initial sequent $p \succ p$ – of the intuitively invalid sequent $p \succ \Box p$ breaks down, since the assumption "p" is not a fully modalized formula. By contrast, the analogous proof of $\Box p \succ \Box \Box p$ (a sequent version of the modal axiom **4**) which begins with $\Box p$ as assumption and then applies ($\Box$I) is perfectly acceptable; you might think that this sequent is not a clear case of an intuitively valid sequent, and you would not be alone in that opinion. What we are developing here is a natural deduction system for the stronger **S5**, containing not only **4** but also **5**. This is indeed the strongest modal logic among those ever seriously considered as candidate formalizations of the logic of (anything deserving to be called) necessity and possibility, and is discussed extensively in Sections 2.1, 2.8, and elsewhere. A proof of the sequent corresponding to **5** in our natural deduction system is just like that for the **4** sequent, sketched above:

EXAMPLE 7.2.1 A proof of the sequent $\Diamond p \succ \Box \Diamond p$:

1	(1)	$\Diamond p$	Assumption
1	(2)	$\Box \Diamond p$	1, $\Box$I

◀

The interesting point to note here is that although $\Diamond$ features essentially in the proof, in that the corresponding sequent with both occurrences of $\Diamond$ deleted is not provable, we can give the proof – as above – without having to use any introduction or elimination rules for $\Diamond$ (which is just as well, since we have yet to formulate such rules). The reason, of course, is that we have already said that, just like $\Box$, $\Diamond$ counts as a modal operator, and it was in terms of this notion that the restriction on ($\Box$I) was formulated. That restriction is satisfied in the application of ($\Box$I) at line 2 of the above proof, since $\Diamond p$ is a fully modalized formula.

Before passing, in Section 7.3, to the question of appropriate rules for $\Diamond$, we pause to illustrate further the way our rules for $\Box$ work. Let us give proofs for three ("intuitively valid") sequents:

EXAMPLES 7.2.2
(*i*) $\Box(p \land q) \succ \Box p \land \Box q$; (*ii*) $\Box p, \Box q \succ \Box(p \land q)$; (*iii*) $\Box(p \to q) \succ \Box p \to \Box q$.

Proof of (*i*):

1	(1)	$\Box(p \land q)$	Assumption
1	(2)	$p \land q$	1, $\Box$E
1	(3)	p	2, $\land$E
1	(4)	$\Box p$	3, $\Box$I
1	(5)	q	2, $\land$E
1	(6)	$\Box q$	5, $\Box$I
1	(7)	$\Box p \land \Box q$	4, 6 $\land$I

---

[422] As formulated there, the definition covered the case in which both $\Diamond$ and $\Box$ were taken as primitive: to be fully modalized a formula needs occurrences of propositional variables to lie within the scope of an occurrence of one or other of these operators.

7.2. NATURAL DEDUCTION RULES FOR $\Box$

Notice that the applications of ($\Box$I) in lines 4 and 6 are legitimate because the assumption (1) is fully modalized. (A similar comment applies for line 6 in each of the following two examples.)

Proof of (ii)

1	(1)	$\Box p$	Assumption
2	(2)	$\Box q$	Assumption
1	(3)	p	1, $\Box$E
2	(4)	q	2, $\Box$E
1, 2	(5)	$p \wedge q$	3, 4 $\wedge$I
1, 2	(6)	$\Box(p \wedge q)$	5, $\Box$I

Proof of (iii):

1	(1)	$\Box(p \to q)$	Assumption
2	(2)	$\Box p$	Assumption
1	(3)	$p \to q$	1, $\Box$E
2	(4)	p	2, $\Box$E
1, 2	(5)	q	3, 4 $\to$E
1, 2	(6)	$\Box q$	5, $\Box$I
1	(7)	$\Box p \to \Box q$	2–7, $\to$I

◀

Some practice:

EXERCISE 7.2.3 Write out the proof just given for (*iii*) as a sequent-to-sequent proof. ✠

EXERCISE 7.2.4 Try your hand at giving proofs for:
(*i*) $\Box p \succ \Box(p \vee q)$,   (*ii*) $\succ \neg\Box(p \wedge \neg p)$,   (*iii*) $p \succ \neg\Box\neg p$. ✠

Now there is one respect in which our rules ($\Box$I) and ($\Box$E) give rise to a cumbersome natural deduction system. Consider the sequent $\Box p \wedge q \succ \Box\Box p$. Since, as we noted above, the formula $\Box\Box p$ follows from the first conjunct alone, it ought to be easy to derive it from the conjunction $\Box p \wedge q$. But the trouble is: this is no longer a fully modalized formula, so we cannot proceed by ($\wedge$E) and ($\Box$I). Not that we can find *no* proof for the sequent; we could always argue as in:

EXAMPLE 7.2.5 *A circuitous proof of* $\Box p \wedge q \succ \Box\Box p$:

1	(1)	$\Box p \wedge q$	Assumption
1	(2)	$\Box p$	1, $\wedge$E
3	(3)	$\Box p$	Assumption
3	(4)	$\Box\Box p$	3, $\Box$I
	(5)	$\Box p \to \Box\Box p$	3–4, $\to$I
1	(6)	$\Box\Box p$	2, 5 $\to$E

◀

It would be better to avoid this unseemly detour through the properties of "$\to$", which does not even appear in the sequent we are trying to prove.[423] Therefore, let us liberalize the rule ($\Box$I) above, and allow the rule to prefix a $\Box$ to a formula which is itself fully modalized, even if it has been derived from assumptions not all of which are fully modalized. Here is the liberalized form of the rule (stated in sequent-to-sequent form):

---

[423] A similar phenomenon is discussed in Prawitz [905], p. 76, in connection with **S4**; see also Forbes [322], p. 311*f*., where it is illustrated with the **S4**-unavailable **B** principle $p \succ \Box\Diamond p$.

Liberalized ($\Box$I) $\dfrac{\Gamma \succ A}{\Gamma \succ \Box A}$ *Subject to the proviso below*

provided all formulas in Γ are fully modalized, or the formula A is itself fully modalized.

Using this revised form of the rule, the short proof contemplated above, but blocked by our previous version of ($\Box$I), goes through:

1	(1)	$\Box p \wedge q$	Assumption
1	(2)	$\Box p$	1, $\wedge$E
1	(3)	$\Box\Box p$	2, $\Box$I (liberalized)

EXERCISE 7.2.6 Give a general demonstration that if a sequent can be proved using the liberalized version of ($\Box$I) (along with all the other rules, of course), then it can be proved, even if at greater length, using the original form of ($\Box$I). ✠

It should be noted that the rule ($\Box$I) could be liberalized even further, as in Prawitz [905], p. 77, to cover cases in which neither the original assumptions nor the immediate premise-formula for an application of the rule are fully modalized, so that, for example, a non-circuitous proof can be provided for $p \wedge \Box(q \wedge r) \succ \Box r$. The formulation of a further variation on the ($\Box$I) theme to avoid such difficulties is too complicated to warrant its consideration here, however.

7.3 Adding Rules for $\Diamond$

In the case of $\Box$, it was the elimination rule that was obvious and the introduction rule that was a little tricky. For $\Diamond$ it is the other way round. The introduction rule that comes first to mind is perfectly adequate, namely the rule that licenses a passage from any formula A (depending on given assumptions) to the formula $\Diamond A$ (depending on those same assumptions). To make these assumption-dependency manipulations explicit, we give a sequent-to-sequent formulation of the rule:

($\Diamond$I) $\dfrac{\Gamma \succ A}{\Gamma \succ \Diamond A}$

To illustrate this rule in action (alongside the $\Box$ rules of Section 7.2), we offer some sample proofs:

EXAMPLES 7.3.1 (*i*) $\Box p \succ \Diamond p$ and (*ii*) $\neg\Box\neg p \succ \Diamond p$.

Proof of (*i*):

1	(1)	$\Box p$	Assumption
1	(2)	p	1 $\Box$E
1	(3)	$\Diamond p$	2 $\Diamond$I

For practice, you might like to see if you can provide a proof of sequent (*ii*) above before going on to look at our solution below.

Proof of (*ii*):

7.3. ADDING RULES FOR ◇

1	(1)	¬□¬p	Assumption
2	(2)	¬◇p	Assumption
3	(3)	p	Assumption
3	(4)	◇p	3 ◇I
2,3	(5)	◇p ∧ ¬◇p	4,2 ∧I
2	(6)	¬p	3–5 RAA
2	(7)	□¬p	6 □I
1,2	(8)	□¬p ∧ ¬□¬p	7,1 ∧I
1	(9)	¬¬◇p	1–8 RAA
1	(10)	◇p	9 ¬¬E

◂

What about an elimination rule for ◇? In view of the analogy between possibility and existential quantification, it is the natural deduction rule (∃E), not covered in our purely propositional *résumé* of Lemmon [698] in Section 7.1, that should guide us in formulating an appropriate such rule. Two of the four introduction and elimination rules for the quantifiers (∀ and ∃) are hedged about with restrictions to the effect that the term figuring in a crucial instance of the quantified formula should not occur in various other places. The restricted rules are (∀I) and (∃E).[424] In the case of (∀I), the restriction was that this term should not occur in any of the assumptions on which the instance depended. Now in the rule (□I) of Section 7.2, there was a restriction to the effect that the assumptions on which the formula to be 'necessitated' depended should be *fully modalized*. (This wording is tailored to the original formulation of the rule here, rather than the liberalized version we gave later.) This suggests that what corresponds to "not containing the crucial term" in the modal case is "being fully modalized". In the case of (∃E), the restriction is that the crucial term should not occur, *inter alia*, in the formula derived from the chosen instance which is to be repeated as depending on whatever the existential formula depended on, or in any additional assumptions that were used in deriving it from that instance. Replacing this reference to not containing the crucial term by a reference to being fully modalized, what we get is the following:

(◇E) in informal 'inference figure' representation:

$$[A]$$
$$\vdots$$
$$\frac{\diamond A \quad C}{C}$$

Provided C is fully modalized, and any additional assumptions used in deriving C from A are fully modalized.

In line with a not uncommon convention, the square brackets indicate that an application of the rule discharges the assumption A on which C depends: After applying the rule, then, C depends on whatever $\diamond A$ depends on, rather than on A itself. It also depends, as in the similar cases of (∃E) and (∨E), on any additional assumptions used in the derivation schematically indicated by the column of three dots. The subderivation indicated by the dots will be represented in the annotation – on the far right – of proofs below by using a dash between the numbers of the lines on which A is assumed and on which C is derived from it. When giving a Lemmon-style formula-to-formula proof the line numbers cited are those on which A is assumed and the start and finish of the subderivation indicated by the column of dots above.

(Above, the crucial term was prohibited from occurring, *inter alia*, in the two places mentioned; "*inter alia*", because actually there was a third place it was not allowed to occur, in the case of (∃E) – namely, in the existential formula itself. This would correspond to saying that the formula $\diamond A$ is itself fully modalized: but as this condition is automatically satisfied, there is no need to mention separately in the above proviso on (◇E).)

---

[424]In fact Lemmon does not use "∀", writing "∀x(Fx)" as "(x)(Fx)", and he calls (∀I) and (∃E), "UI" and "EE".

To make clear the assumption-dependency manipulations and restrictions, we give a sequent-to-sequent formulation of the rule, as for the other rules:

$$(\Diamond E) \quad \frac{\Gamma \succ \Diamond A \quad \Delta, A \succ C}{\Gamma, \Delta \succ C} \quad \textit{If C and all formulas in } \Delta \textit{ are fully modalized.}$$

Let us see this rule in action:

EXAMPLES 7.3.2 (i) $\Box p, \Diamond q \succ \Diamond(p \wedge q)$; (ii) $\Diamond \Box p \succ \Box p$; (iii) $\Diamond p \succ \neg \Box \neg p$.

Proof of (i):

1	(1)	$\Box p$	Assumption
2	(2)	$\Diamond q$	Assumption
1	(3)	p	1 $\Box$E
4	(4)	q	Assumption
1, 4	(5)	$p \wedge q$	3, 4 $\wedge$I
1, 4	(6)	$\Diamond(p \wedge q)$	5 $\Diamond$I
1, 2	(7)	$\Diamond(p \wedge q)$	2, 4–6 $\Diamond$E

Proof of (ii):

1	(1)	$\Diamond \Box p$	Assumption
2	(2)	$\Box p$	Assumption
1	(3)	$\Box p$	1, 2–2 $\Diamond$E

Proof of (iii):

1	(1)	$\Diamond p$	Assumption
2	(2)	$\Box \neg p$	Assumption
3	(3)	p	Assumption
2	(4)	$\neg p$	2 $\Box$E
2, 3	(5)	$p \wedge \neg p$	3, 4 $\wedge$I
3	(6)	$\neg \Box \neg p$	2–3 RAA
1	(7)	$\neg \Box \neg p$	1, 2–6 $\Diamond$E

EXERCISE 7.3.3 Which formulas correspond in the applications of ($\Diamond$E) in these three proofs to the A and C of our schematic formulation of that rule, and which sets of formulas correspond to the Γ and Δ of that formulation?

Using our rules for $\Box$ and $\Diamond$ we have now proved the sequents $\Diamond p \prec \succ \neg \Box \neg p$ (to use a convenient shorthand to denote a sequent and its converse[425]) which together record the equivalence of $\Diamond$ and $\neg \Box \neg$ which is exploited by the option (followed in our earlier discussion but not – at least until Section 7.5 – in this Appendix) of defining $\Diamond A$ as $\neg \Box \neg A$. The forward direction of the equivalence is given in 7.3.2(iii), the backward direction in 7.3.1(ii). Altogether there is in the proofs of those sequents only one application of ($\Diamond$E), namely that made at line 7 of the proof of 7.3.2(iii). And in that application, for the set schematically represented by the "Δ" of our sequent-to-sequent formulation of the rule, we had $\Delta = \emptyset$. This strongly suggests that there would be no loss (when it comes to the question of what sequents are provable) in replacing the above formulation of ($\Diamond$E) by the apparently weaker rule given below, which restricts our earlier formulation by requiring that $\Delta = \emptyset$.

---

[425] Not that we wish to suggest every sequent has a converse; since here all sequents have exactly one formula on the right, the only sequents with converses are those which also have exactly one formula on the left. See note 7.

7.4. THE SEMANTICS OF S5 AND SOME RULES FOR S4

Restricted ($\Diamond$E) $\quad\dfrac{\Gamma \succ \Diamond A \qquad A \succ C}{\Gamma \succ C} \quad$ Provided C is fully modalized.

EXERCISE 7.3.4 Show how any application of ($\Diamond$E) can be replaced by an application of Restricted ($\Diamond$E), together, perhaps, with applications of some of the other rules. ✠

There are, on the other hand, good reasons for wanting to liberalize the $\Diamond$-elimination rule rather than to restrict it, closely parallelling those we saw for liberalizing ($\Box$I) in connection with Example 7.2.5. We will not be detailing a liberalized form of ($\Diamond$E), here, but simply indicating how things can get cumbersome in its absence.

EXAMPLE 7.3.5 An unsuccessful attempt at a proof of $p, \Diamond q \succ \Diamond(\Diamond p \wedge q)$.

1	(1)	p	Assumption
2	(2)	$\Diamond q$	Assumption
3	(3)	q	Assumption
1	(4)	$\Diamond p$	1 $\Diamond$I
1,3	(5)	$\Diamond p \wedge q$	4, 3 $\wedge$I
1,3	(6)	$\Diamond(\Diamond p \wedge q)$	5 $\Diamond$I
1,2	(7)	$\Diamond(\Diamond p \wedge q)$	2, 3–6 $\Diamond$E

The problem here is with the application of ($\Diamond$E) at the end. In terms of the sequent-to-sequent presentation of this rule, taking us from $\Gamma \succ \Diamond A$ and $\Delta, A \succ C$ to the conclusion $\Gamma, \Delta \succ C$ on the condition that C and all formulas in Δ are fully modalized. In the present instance, $\Gamma = \{\Diamond A\} = \{\Diamond q\}$, $\Delta = \{p\}$ and $C = \Diamond(\Diamond p \wedge q)$, so while the condition is satisfied in the case of C, in the case of Δ, it is not, as p is not a fully modalized formula. ◀

EXERCISE 7.3.6 Provide a correct proof in the present natural deduction system for **S5** of the sequent for which Example 7.3.5 displays an incorrect proof. (*Hint*: a detour through implication in the manner of Example 7.2.5 on p. 477 may help. Note incidentally that the sequent at issue here is not **S4**-provable.) ✠

7.4 The Semantics of S5 and Some Rules for S4

We adapt the simplified Kripke semantics of Section 2.2 which offered models $\langle W, V \rangle$ later seen to have the property that the theorems of the formula logic **S5** were exactly those formulas that were valid according to the semantics, in the sense of being true at every $w \in W$, for all such models $\langle W, V \rangle$. The adaptation requires no change to the definition of truth (at a point in a simplified model), but it does require that we provide a notion of validity for *sequents* as opposed to just formulas. This is easily done: $A_1, \ldots, A_n \succ B$ is *valid* (on the simplified semantics) just in case for every (simplified) model $\langle W, V \rangle$, for all $w \in W$:

if $\langle W, V \rangle \models_w A_1$ and ... and $\langle W, V \rangle \models_w A_n$, then $\langle W, V \rangle \models_w B$.

If the inset condition above is satisfied for all $w \in W$ in the case of a given $\langle W, V \rangle$, it is convenient to say that the sequent in question *holds* in $\langle W, V \rangle$. This is the analogue for sequents of the property of being true throughout a model; thus a sequent is *valid* according to the present semantics just in case it holds in every model. To check that our natural deduction system for **S5** is *sound* w.r.t. the simplified semantics, it suffices to show that all the initial sequents are valid and that the rules for the various connectives all preserve for any model, the property of holding in that model, since this implies that they preserve validity. This is very straightforward for the rules governing the Boolean connectives, and for the rules ($\Box$E) and ($\Diamond$I), but the rules ($\Box$I) and ($\Diamond$E) require some attention.

LEMMA 7.4.1 *The rule* ($\Box$I) *preserves, for any model, the property of holding in that model.*

Proof. Suppose a premiss-sequent $\Gamma \succ A$ for some application of ($\Box$I) holds in $\langle W, V \rangle$ but that the conclusion sequent $\Gamma \succ \Box A$ does not. The latter means that for some $w \in W$, $\langle W, V \rangle \models_w C$ for all $C \in \Gamma$ while $\langle W, V \rangle \not\models_w \Box A$; thus for some $x \in W$, $\langle W, V \rangle \not\models_x A$. For the sequents concerned here to be premisses and conclusions of an application of ($\Box$I), the formulas in Γ must all be fully modalized. Thus by Lemma 2.2.4 (p. 56), these formulas are modally invariant. But all of them are true at w in this model, so all of them are true at x, where A is false, contradicting the supposition that $\Gamma \succ A$ holds in $\langle W, V \rangle$. ∎

EXERCISE 7.4.2 Show that the rule ($\Diamond$E) also preserves the property of holding in any given model. ✣

THEOREM 7.4.3 *A sequent is provable in our natural deduction system for* **S5** *if and only if it is valid on the simplified semantics.*

Proof. We tackle just the "only if" ("soundness") direction; the "if" ("completeness") direction can be dealt with (via the canonical model method, taking generated submodels) as for the formula logic **S5** in Coro. 2.5.4 (p. 72). Having noted that all the initial sequents are valid and that the rules for the Boolean connectives preserve validity, as well as ($\Box$E) and ($\Diamond$I), since they preserve, for an arbitrary model, the property of holding in that model), it remains only to add the rules ($\Box$I) and ($\Diamond$E) to the list of rules with this same preservation behaviour. But this was done by Lemma 7.4.1 and Exercise 7.4.2, respectively. ∎

Alternatively, taking for granted the soundness and completeness results for logics-as-sets-of-formulas in Chapter 2, it suffices, for both the "if" and the "only if" to note that $A_1, \ldots, A_n \succ B$ is provable in the natural deduction system for **S5** just in case the formula $(A_1 \wedge \ldots \wedge A_n) \to B$ (understanding this to be simply B when $n = 0$) is provable in the formula logic **S5**, and that a similar equivalence holds in respect of the validity of the sequent and of the corresponding formula in the simplified semantics.

Retaining the notion of a sequent's holding in a model (= the right-hand formula true at any point at which all the left-hand formulas are true, in the model in question), we apply this now to the 'de-simplified' models $\langle W, R, V \rangle$ with accessibility relations (as explained in Section 2.3, and we shall also assume familiarity with the notion of a frame, from Section 2.5). We can sensibly call a sequent *valid* on a frame if it holds in every model on that frame and consider, for example, the class of sequents valid on every transitive reflexive frame: this will be a sequent version of the formula logic **S4**, and it turns out to be amenable to a simple natural deduction treatment, to which we now proceed.

The rules governing the Boolean connectives of course remain as before. Because we are considering the logic determined by a class of frames all of which are reflexive, the rules ($\Box$E) and ($\Diamond$I) can be retained intact. But we need to vary the $\Box$-Introduction rule to avoid proofs of sequents like $\neg\Box p \succ \Box\neg\Box p$ which are obviously not valid on every transitive reflexive frame. We need to replace the concept of being fully modalized, which served us well for the case of **S5**, with something else, for which we choose the concept of being a "$\Box$-formula", by which is meant: a formula of the form $\Box B$. The idea is that we can pass from A to $\Box A$ by the new rule, provided that A depends only on assumptions which are $\Box$-formulas. Writing this in sequent-to-sequent form:

($\Box$I)$_{\mathbf{S4}}$ $\qquad \dfrac{\Gamma \succ A}{\Gamma \succ \Box A}$

Provided every formula in Γ *is a* $\Box$*-formula.*

Note that we could equally well have written:

7.4. THE SEMANTICS OF S5 AND SOME RULES FOR S4

$(\Box I)_{S4}$ $\quad \dfrac{\Box C_1,\ldots,\Box C_n \succ A}{\Box C_1,\ldots,\Box C_n \succ \Box A}$

It would actually do no harm to make the restriction (on the sequent-to-sequent rule as originally formulated above) less stringent, and allow in Γ any formulas which are – shall we say? – *positively boxed*, meaning by this a formula which is constructed from $\Box$-formulas by the use of conjunction and disjunction (as in Prawitz [905], p. 77, where these formulas are called "essentially modal formulas for M_{S4}"). Thus for example $\Box p \vee q$ is not positively boxed, and nor is $\neg \Box p$, or $p \rightarrow \Box q$, or $\Box p \rightarrow \Box q$, but the formulas $\Box p \vee \Box q$ and $\Box(p \vee q) \wedge (\Box r \vee \Box(q \rightarrow \neg \Box r))$ are positively boxed – the $\neg \Box r$ subformula notwithstanding in this last case.

Corresponding to the syntactic shift from fully modalized to being a $\Box$-formula (or being positively boxed), is a semantic shift from the property of being modally invariant to that of being 'persistent'. As in Revision Exercise 2.7.14 (p. 132), we call a formula A *persistent* over a class of frames $\mathbb{C}$ provided that in any model $\mathcal{M} = \langle W, R, V \rangle$ with $\langle W, R \rangle \in \mathbb{C}$, $\mathcal{M} \models_x A$ implies $\mathcal{M} \models_y A$, whenever xRy.[426] The easy proof of the following observation is omitted.

LEMMA 7.4.4 *All $\Box$-formulas (indeed all positively boxed formulas) are persistent over any class of transitive frames.*

For the moment, let us forget about $\Diamond$ (thinking of it, if at all, as a defined symbol – $\Diamond A =_{\text{Def}} \neg \Box \neg A$ – rather than a primitive in its own right). Then we could say, concerning the natural deduction system which replaces $(\Box I)$ with $(\Box I)_{S4}$ above :

THEOREM 7.4.5 *A sequent is provable in our natural deduction system for S4 (with $\Box$ as the sole modal primitive) if and only if it is valid on every transitive reflexive frame.*

Proof. As with S5, the completeness ("if") half of this claim is best settled by the methods of Section 2.4. For soundness, in view of earlier remarks, it remains only to check that the rule $(\Box I)_{S4}$ preserves the property of holding in any model on a transitive frame. Here a simple appeal to Lemma 7.4.4 will suffice. ∎

Again, we could obtain this result by using the correspondence between sequents and formulas and the equivalences (*mutatis mutandis*) mentioned after Theorem 7.4.3.

Unfortunately the simultaneous presence of $\Diamond$ as a primitive requires us to modify the rule $(\Box I)$ above, since we now need to treat negated $\Diamond$-formulas in the same way as $\Box$-formulas; similarly in the rule for $\Diamond$-Elimination we will need to treat $\Diamond$-formulas and negated $\Box$-formulas together. To that end, let us introduce the following terminology:

A formula of the form $\Box A$ or of the form $\neg \Diamond A$ will be called a *strongly modal* formula, and a formula of the form $\Diamond A$ or $\neg \Box A$ will be called a *weakly modal* formula.

Note the following variation of Lemma 7.4.4 for this terminology; a formula A is said to be *inversely persistent* over a class of frames when for any model on a frame $\langle W, R \rangle$ in the class in which A is true at $y \in W$, we also have A true at $x \in W$ (in the model in question) whenever xRy. (Equivalently, if the *falsity* of the formula at a point implies its falsity at any R-related, or 'accessible', point, in any model on a frame in the class.)

---

[426] The term 'persistent' is chosen here because of the idea that truth at a point in a model persists as we pass to accessible points for such formulas. The same term is used in connection with a topic not covered in the present work, for formulas whose validity on a general frame (in the sense of the Digression on p. 83) is retained as we pass to the underlying Kripke frame: see, for example, 5.6 of Blackburn et al. [98].

LEMMA 7.4.6 *All strongly modal formulas are persistent, and all weakly modal formulas inversely persistent, over any class of transitive frames.*

Note that as in the case of Lemma 7.4.4 we could have been more generous here and allowed strongly modal formulas to be conjunctions and disjunctions of strongly modal formulas in the narrow sense defined, and likewise with weakly modal formulas.

Now we can reformulate ($\Box$I) and ($\Diamond$E) in a form suitable for a natural deduction system for **S4** in which both $\Box$ and $\Diamond$ are taken as primitive; the other modal rules, ($\Box$E) and ($\Diamond$I) remain as for the **S5** case (i.e., as in Sections 7.2, 7.3); as in that discussion, the issue of liberalizing the rules further naturally arises, though it will not be pursued here:

$$(\Box I)_{\mathbf{S4}} \qquad \frac{\Gamma \succ A}{\Gamma \succ \Box A}$$

Provided every formula in Γ is a strongly modal formula.

And:

$$(\Diamond E)_{\mathbf{S4}} \qquad \frac{\Gamma \succ \Diamond A \qquad \Delta, A \succ C}{\Gamma, \Delta \succ C}$$

Provided every formula in Δ is a strongly modal formula, and C is a weakly modal formula.

EXERCISE 7.4.7 Using Lemma 7.4.6, show that this last rule preserves the property of holding in a model on a reflexive transitive frame. ✠

This exercise supplies the main ingredient for the soundness half of a proof we will not give for the following (whose completeness half is obtained by the canonical model method, as in Coro. 2.4.3):

THEOREM 7.4.8 *A sequent is provable in this latest natural deduction system for **S4** (with $\Box$ and $\Diamond$ both primitive) if and only if it is valid on every transitive reflexive frame.*

EXAMPLE 7.4.9 To illustrate these **S4** rules in action, here is a proof of the sequent $\Diamond\Diamond\Box p \succ \Diamond\Box\Box p$, first in the Lemmon formula-to-formula style:

1	(1)	$\Diamond\Diamond\Box p$	Assumption
2	(2)	$\Diamond\Box p$	Assumption
3	(3)	$\Box p$	Assumption
3	(4)	$\Box\Box p$	3 $\Box$I$_{\mathbf{S4}}$
3	(5)	$\Diamond\Box\Box p$	4 $\Diamond$I
2	(6)	$\Diamond\Box\Box p$	2, 3–5 $\Diamond$E$_{\mathbf{S4}}$
3	(7)	$\Diamond\Box\Box p$	1, 2–6 $\Diamond$E$_{\mathbf{S4}}$

and now in sequent-to-sequent format:

(1)	$\Diamond\Diamond\Box p \succ \Diamond\Diamond\Box p$	Initial sequent
(2)	$\Diamond\Box p \succ \Diamond\Box p$	Initial sequent
(3)	$\Box p \succ \Box p$	Initial sequent
(4)	$\Box p \succ \Box\Box p$	3 $\Box$I$_{\mathbf{S4}}$
(5)	$\Box p \succ \Diamond\Box\Box p$	4 $\Diamond$I
(6)	$\Diamond\Box p \succ \Diamond\Box\Box p$	2, 5 $\Diamond$E$_{\mathbf{S4}}$
(7)	$\Diamond\Diamond\Box p \succ \Diamond\Box\Box p$	1, 6 $\Diamond$E$_{\mathbf{S4}}$

7.4. THE SEMANTICS OF S5 AND SOME RULES FOR S4

EXERCISE 7.4.10 (*i*) What, if anything, is wrong with the following as a putative **S4** natural deduction proof of the sequent $\Diamond\Diamond\Box p \succ \Box\Box p$? (We give a Lemmon style presentation.)

1	(1)	$\Diamond\Diamond\Box p$	Assumption
2	(2)	$\Diamond\Box p$	Assumption
3	(3)	$\Box p$	Assumption
3	(4)	$\Box\Box p$	3 $\Box I_{S4}$
2	(5)	$\Box\Box p$	2, 3–4 $\Diamond E_{S4}$
1	(6)	$\Box\Box p$	1, 2–5 $\Diamond E_{S4}$

(*ii*) Is the sequent of which the above purports to be an **S4**-proof, provable in **S4**? (Note that if you answered, under (*i*), that nothing was wrong with the proof, the present answer must be *Yes*, while if you found fault with the above would-be proof, there may be an alternative **S4** proof with no rules misapplied. You are not asked to supply such a proof however; the easiest way to decide whether such a proof exists is to apply Theorem 7.4.8.) ✠

In earlier sections we have seen proofs – or requests for proofs – of the following sequents:

7.2.2(*i*) $\Box(p \wedge q) \succ \Box p \wedge \Box q$ (*ii*) $\Box p, \Box q \succ \Box(p \wedge q)$ (*iii*) $\Box(p \to q) \succ (\Box p \to \Box q)$;
7.2.4(*i*) $\Box p \succ \Box(p \vee q)$ (*ii*) $\succ \neg \Box(p \wedge \neg p)$ (*iii*) $p \succ \neg\Box\neg p$;
7.3.1(*i*) $\Box p \succ \Diamond p$ (*ii*) $\neg\Box\neg p \succ \Diamond p$;
7.3.2(*i*) $\Box p, \Diamond q \succ \Diamond(p \wedge q)$ (*ii*) $\Diamond\Box p \succ \Box p$ (*iii*) $\Diamond p \succ \neg\Box\neg p$.

And here are some additional sequents provable in **S5**, several of them already provable in **S4**, making altogether 20 sequents when added to those listed above; these are the subject of the exercise below:

(a) $\Box p \succ \Box\Box p$
(b) $\Diamond p \succ \Box\Diamond p$
(c) $\Diamond\Diamond p \succ \Diamond p$
(d) $\Box p \succ \neg\Diamond\neg p$
(e) $\neg\Diamond\neg p \succ \Box p$
(f) $\Diamond\Diamond(p \vee \Box q) \succ \Diamond p \vee \Box q$
(g) $\Diamond\Diamond(p \vee \Box q) \succ \Diamond p \vee \Diamond\Box q$
(h) $\Box p \succ \Box(\Box p \wedge p)$
(i) $\Box(p \to q) \succ \Box(\Box p \to \Box q)$.

EXERCISE 7.4.11 (*i*) Which of these 20 sequents are valid on all frames $\langle W, R \rangle$ (with no restrictions placed on the relation R, that is)?

(*ii*) Which of the 20 sequents are valid on every transitive reflexive frame?

(*iii*) Give proofs of *six* sequents listed under (a)–(i) in the natural deduction system for **S4** (with $\Box$ and $\Diamond$ both primitive, that is); since not all are **S4**-provable, some care will have to be exercised over the choice. ✠

Further exercises on proofs and validity for sequents concentrating on the natural deduction systems for **S4** and **S5** can be found at the end of Section 2.7.

The natural deduction approach, as we have seen it deployed so far, does not lend itself to systematizing the wide range of normal modal logics, though one can manage somewhat artificially by reformulating familiar axioms as rules. The resulting rules tend to involve modal as well as Boolean connectives rather than enjoying the 'purity' of their Gentzen-inspired prototypes. (The notion of a *pure* rule as one involving only a single connective or quantifier in its schematic formulation can be found along with related concepts in p. 256*f*. of Dummett [261]. See also the index entries under 'rules, pure and simple' in Humberstone [594], though the notion of simplicity there in play does not quite coincide with Dummett's.)

The following section explores a variation on this approach which shows more promise, but here let us conclude with one example of a logic between **S4** and **S5** where a minor adaptation, due to Zeman [1220], *is* available:

EXAMPLE 7.4.12 For simplicity we revert to **S4** as treated with $\Box$ as the sole non-Boolean primitive, and $\Diamond$ taken as $\neg\Box\neg$. The elimination rule for $\Box$ ("($\Box$E)") is as above, taking us from $\Box A$ depending on certain assumptions, to A, depending on those same assumptions – or, making the dependencies explicit, from $\Gamma \succ \Box A$ to $\Gamma \succ A$ ($\Box$E). The introduction rule requires for its formulation the notion of what we shall call a *Zeman-modalized* formula, meaning a formula of the form $\Diamond^n \Box B$ for some n. The ($\Box$I) rule is that we may pass from A as depending on a set of Zeman-modalized formulas to $\Box A$ as depending on those same assumptions; since we may take $n = 0$, this subsumes the rule ($\Box$I)$_{\mathbf{S4}}$. A formulation in the sequent-to-sequent style would be:

$$(\Box\text{I})_{\mathbf{S4.2}} \qquad \frac{\Gamma \succ A}{\Gamma \succ \Box A} \qquad \text{Provided every formula in } \Gamma \text{ is Zeman-modalized.}$$

Since, as already remarked, we have the ($\Box$I) rule of **S4** already as a special case, we could restrict the n in the definition of Zeman-modalized formulas to $n = 0, 1$, since $\Diamond^m A$ and $\Diamond A$ will be equivalent whenever $m > 1$. (Here the fact that we retain the familiar elimination rule for $\Box$ is also being used.) One can show that $A_1, \ldots, A_m \succ B$ is provable in this natural deduction system just in case $\vdash_{\mathbf{S4.2}} (A_1 \wedge \ldots \wedge A_m) \to B$, though here we recall only the key ingredient of a semantically based proof – using Corollary 2.5.12 from p. 77 – of this fact: On all reflexive transitive convergent frames, Zeman-modalized formulas are persistent (i.e., true in a model on such a frame at all successors of a point at which they are true). Is ($\Box$I)$_{\mathbf{S4.2}}$ a pure rule in the sense informally explained? The explanation was simply too informal for an answer to be forthcoming. If we count what is happening in the proviso on the rule as part of the schematic formulation of the rule then the answer will be negative: taking $\Diamond$ as defined by $\neg\Box\neg$ then the proviso includes reference to (or "involves") $\neg$, and if $\Diamond$ is taken as an additional primitive then the proviso includes reference to $\Diamond$, so either way it involves more than just $\Box$. ◀

7.5 Natural Deduction for Weaker Normal Modal Logics

There is no difficulty in giving sequent-to-sequent rules for modal logics weaker than **S4**; for example we can use Scott's Rule from the Digression on Sequent Formulations (see p. 22). But this does not have the character of a natural deduction rule eliminating or introducing $\Box$. (Nor does it have the character of a left or right insertion rule for a sequent calculus/Gentzen system.) The elimination rule we have been using for $\Box$ – and for the present section $\Diamond$ will be taking a back seat in the exposition – very obviously encodes the **T** axiom(-schema), raising difficulties for a treatment on similar lines of, for example, the basic normal modal logic **K**. So something new has to be done, and in what follows the way in which we break with these similarities is by describing a natural deduction system in which we operate with not one but two kinds of sequent.[427] As well as the familiar sequents $\Gamma \succ A$ we have been using, which will now be called *plain* sequents, we will add sequents of a new kind, notated with a 'dotted' version, "$\succ\cdot$" of the sequent separator "$\succ$". These new sequents $\Gamma \succ\cdot A$ will be called *transitional* sequents, since

---

[427]Curry [225] introduced two kinds of sequent in his treatment of modal logic – though for a sequent calculus rather than a natural deduction system, and the novel sequents had a different motivation and interpretation from those appearing here. The novel – we call them 'transitional' – sequents understood along the present lines appear in Indrzejczak [608], though for the sake of, again, a sequent calculus (a cut-free sequent calculus, more specifically) rather than a natural deduction system, and for **S5** rather than **K** or other logics weaker than **S4**. (Ohnishi and Matsumoto [855] used standard sequents.) A similar device can be found in the formulation of the rules governing $\Box$ on p. 29f. in Garson [351], and a similar use of the 'transitional' terminology (for tableau systems rather than natural deduction) appears in Goré [401] (and Goré and Nguyen [402]).

7.5. NATURAL DEDUCTION FOR WEAKER NORMAL MODAL LOGICS

in evaluating them semantically the passage from the left-hand side of the separator to the right-hand side will involve a transition from one point in a model to an arbitrary successor of that point (a point accessible to the original point, that is). More precisely, let us say that a transitional sequent $\Gamma \succ\!\cdot A$ *holds at* $x \in W$ in a model $\mathcal{M} = \langle W, R, V \rangle$ just in case if $\mathcal{M} \models_x C$ for each $C \in \Gamma$, then for any $y \in R(x)$, $\mathcal{M} \models_y A$. $\Gamma \succ\!\cdot A$ *holds in* $\mathcal{M}$ when it holds at each $x \in W$, and is *valid on* the frame $\langle W, R \rangle$ when it holds in every model on that frame. The corresponding semantic notions for plain sequents remain as before – i.e., with no transition from x to $y \in R(x)$ involved. Note that, by contrast with plain sequents, transitional sequents of the form $A \succ\!\cdot A$ are not in general valid (on every frame), since that would be for the truth of an arbitrary formula to be persistent. (Thus, for A of the form $\Box B$, such sequents are valid on all transitive frames.)

Many of the rules for the Boolean connectives from Section 7.1 continue to preserve the property of holding at a point (in an arbitrary model) when the "$\succ$" is replaced by "$\succ\!\cdot$" in premisses and conclusion alike. As just remarked, for the zero-premiss rule, as we may call it, for using initial sequents[428] ($A \succ A$), this is not the case. But of the one- or more-premiss rules listed there, namely ($\wedge$I), ($\wedge$E), ($\vee$I), ($\neg\neg$E), ($\to$E), ($\to$I) and (RAA), all but the last two exhibit this preservation characteristic. (We will get to ($\vee$E) below.) The problem with the last two rules is that they move material from the left to the right of the turnstile, and when the new turnstile "$\succ\!\cdot$" is at issue, the occurrence of a formula on the left has a different significance from its occurrence on the right – truth at the given point in the former case, and truth at an arbitrary successor point in the latter. (More generally, and setting aside the zero-premiss rule just mentioned, we may offer this formulation to cover also the missing ($\vee$E): the rules which are problematic when "$\succ$" is replaced by "$\succ\!\cdot$" are those rules which *discharge assumptions*.) Thus, to take the case of (RAA), suppose that $\Gamma, A \succ\!\cdot B \wedge \neg B$ holds at a point in a model, which means that if all formulas in $\Gamma \cup \{A\}$ are true at the point, then it has no successors (since a contradiction would have to be true at any such successor). It does not follow from this supposition that $\Gamma \succ\!\cdot \neg A$ holds at the original point: for this sequent not to hold at x (in $\langle W, R, V \rangle$, with $x \in W$) we must have all formulas in Γ true at x and $\neg A$ false at some point R-related to x, and thus A true at the latter point. But this does not make A true at x itself, so we are in no position to invoke the supposition that premiss sequent $\Gamma, A \succ\!\cdot B \wedge \neg B$ holds at x to derive a contradiction. Nor does the 'dotted' version of (RAA) even have the weaker preservation characteristic of preserving, for an arbitrary model, the property of holding in that model, or even the still weaker preservation characteristic of preserving the property of holding in all models (equivalently: the property of to being valid on every frame).

EXAMPLE 7.5.1 For instance, consider the application of this rule which would pass from premiss sequent $\Box p, \Box \neg p \succ\!\cdot p \wedge \neg p$, which does indeed have this property, to the conclusion sequent $\Box p \succ\!\cdot \neg \Box \neg p$, which lacks it. This last sequent says that if a point verifies $\Box p$ then all of its successors verify what we would normally write as $\Diamond p$: but just because $\Box p$ is true at x (in a given model), and thus p is true throughout $R(x)$, there is no reason for every $y \in R(x)$ to have a successor verifying p, as would be required for $\Diamond p$ to be true throughout $R(x)$. ◀

For similar reasons, the 'dotted' version of ($\to$I) is not a rule – and the same goes for the dotted version of ($\vee$E), to which we return below (Example 7.5.8) – we can incorporate into any natural deduction system designed to prove only such transitional sequents as are valid on every frame, and so has no place in our natural deduction system of **K**, in which we aim to have provable precisely those sequents – plain and transitional – which are valid on all frames. For this purpose, only the dotted versions of the safe rules listed above, which we may call ($\wedge$I$\cdot$), ($\wedge$E$\cdot$), ($\vee$I$\cdot$), ($\neg\neg$E$\cdot$), and ($\to$E$\cdot$), should be used. But of course, we also need rules for $\Box$ itself. (Recall that we are ignoring $\Diamond$ here.)

In this last capacity, as the discussion of (RAA) may already have suggested, we offer the following introduction and elimination rules, which modulate between plain and transitional sequents, exposing the rationale behind having the latter sequents in the proof system, as well as a purely structural rule

---
[428]Recall that in Lemmon [698] this corresponds to the 'rule of assumptions'.

(not governing any particular item of logical vocabulary, that is) which we simply call *Modulation* (or "Modul" for short):

$$(\Box\text{I})\ \dfrac{\Gamma \succ\cdot A}{\Gamma \succ \Box A} \qquad (\Box\text{E})\ \dfrac{\Gamma \succ \Box A}{\Gamma \succ\cdot A} \qquad (\text{Modul})\ \dfrac{\succ A}{\succ\cdot A}$$

That concludes our list of natural deduction rules for a proof system for **K**, summarized below, so we turn to the expected soundness result (Proposition 7.5.4) and to some illustrations of the rules in action. After that and some further discussion, we get around to the corresponding completeness result (Theorem 7.5.9) and have a brief look at some extensions of the system.

It may come as something as a surprise that the dotted form of the rule (¬¬I), which would be perfectly acceptable from a semantic point of view, is not listed. The undotted form, it will be recalled from note 419, is derivable from the remaining rules with assistance, in particular, from the undotted form of (RAA). But since we do not have the dotted form of (RAA), a derivation along those lines will not be available. Another derivation is possible, however, and the same goes for many of the dotted rules that have been listed as primitive dotted rules here, as we shall see, which could accordingly have been omitted. But their presence makes for more natural proofs. For good measure, we include versions of the rules for ⊤ and ⊥ from Section 7.1, though these will not be illustrated in our sample proofs.

Plain Sequent Rules

(∧I), (∧E), (∨I), (∨E), (→I), (→E), (RAA), (¬¬E), (⊤I), (⊥E).

Transitional Sequent Rules

(∧I)·, (∧E·), (∨I·), (→E·), (¬¬E·), (⊤I·), (⊥E·).

Mixed Rules

(□I), (□E), (Modul).

The point just made about the derivability of (¬¬I·) from these rules, and the redundancy of many of the other rules, can be gleaned from the following observation and its proof.

PROPOSITION 7.5.2 *The following rule, with one plain and several transitional sequent premises (and a transitional sequent conclusion), is derivable from those given above (for any n):*

$$\dfrac{A_1,\ldots,A_n \succ B \qquad \Gamma_1 \succ\cdot A_1 \quad \ldots \quad \Gamma_n \succ\cdot A_n}{\Gamma_1,\ldots,\Gamma_n \succ\cdot B}$$

Proof. From the leftmost premiss sequent by n applications of (→I), we have the sequent $\succ A_1 \to (A_2 \to \ldots \to (A_n \to B)\ldots)$ provable, and thus the result of replacing the "$\succ$" with "$\succ\cdot$", by (Modul), so by n applications of (→E·) and the remaining premiss sequents we obtain the conclusion sequent. ∎

Thus, in view of the provability of $\neg\neg A \succ A$ for any A – think of this as the left premiss for the above rule – we can derive $\Gamma \succ\cdot A$ from $\Gamma \succ\cdot \neg\neg A$ (i.e., the rule (¬¬I·) is derivable). Similarly, we could drop all the transitional rules above except for (→I·) and they would be similarly derivable. But, as already remarked, this would make our natural deduction system very far from natural. More precisely, the conditional $A_1 \to (A_2 \to \ldots \to (A_n \to B)\ldots)$ in the proof of 7.5.2 is close to being what Prawitz

7.5. NATURAL DEDUCTION FOR WEAKER NORMAL MODAL LOGICS

[905] calls a maximum formula: at once figuring as the conclusion of an introduction rule and the premiss of an elimination rule. ([905] shows how to transform proofs with such detours in them into 'normal' proofs in which they do not appear.) "Close to", because while the introduction rule is the plain form of $\rightarrow$-introduction, the elimination uses the transitional form of $\rightarrow$-elimination.

EXERCISE 7.5.3 (i) Show that the novel rules of the present proof system ($\wedge$I·), ($\wedge$E·), ($\vee$I·), ($\neg\neg$E·), and ($\rightarrow$E·), all preserve the property of being valid on every frame. Note that this means, in the case of a rule like (Modul) with a plain sequent premiss and a transitional sequent conclusion, that this means that whenever the premiss sequent is valid (on all frames) in the sense defined for plain sequents, then so is the conclusion sequent, in the sense of validity defined for transitional sequents. Note that this last rule is the only rule which does not preserve the property of holding at a point in a model – though it does preserve the property of holding in a model (and thus *a fortiori* preserves the property of being valid on a frame and so certainly the property of being valid on every frame).

(ii) Would the result asked for in (i) go through if (Modul) had been stated with side formulas, i.e., with schematic premiss $\Gamma \succ A$ and conclusion $\Gamma \succ\cdot A$? What about allowing such side formulas (non-empty Γ) but requiring all of them to be fully modalized? ✠

Since a preservation result like that demanded in (i) of the above exercise for the novel rules obviously holds for all the (non-modal) introduction and elimination rules from Section 7.1, we have here the proof of soundness result for our system:

PROPOSITION 7.5.4 *All sequents, plain or transitional, provable using the rules of this section are all valid on every frame.*

Let us see some of the rules in action:

EXAMPLE 7.5.5 (i) A proof of $\Box p \succ \Box(p \vee q)$:

(1) $\Box p \succ \Box p$ Initial sequent
(2) $\Box p \succ\cdot p$ 1 $\Box$E
(3) $\Box p \succ\cdot p \vee q$ 2 $\vee$I·
(4) $\Box p \succ \Box(p \vee q)$ 3 $\Box$I

Of course, we could consider the first three lines of the above proof as a proof in its own right of the transitional sequent appearing there. We are mainly interested in transitional sequents, however, for their instrumental utility in providing a simple system for delivering proofs of plain sequents. Let us see the same proof (i.e., the full proof (1)–(4)) in Lemmon's style, with assumption numbers keeping track of dependencies instead of explicit formulas on the left of the successive sequents. For this we need to distinguish somehow whether a line corresponds to a plain sequent or to a transitional sequent, which we do by leaving the plain cases as they are and adding a dot after the line number for the transitional sequents. Thus the above proof becomes:

1   (1) $\Box p$ Assumption
1   (2)· p 1 $\Box$E
1   (3)· $p \vee q$ 2 $\vee$I·
1   (4) $\Box(p \vee q)$ 3 $\Box$I

Let us use the proof just given as part of the proof (in the same style) of another sequent, $\Box p \vee \Box q \succ \Box(p \vee q)$:

1	(1)	$\Box p \vee \Box q$	Assumption
2	(2)	$\Box p$	Assumption
2	(3)·	p	2 $\Box$E
2	(4)·	$p \vee q$	3 $\vee$I·
2	(5)	$\Box(p \vee q)$	4 $\Box$I
6	(6)	$\Box q$	Assumption
6	(7)·	q	6 $\Box$E
6	(8)·	$p \vee q$	7 $\vee$I·
6	(9)	$\Box(p \vee q)$	8 $\Box$I
1	(10)	$\Box(p \vee q)$	1, 2–5, 6–9 $\vee$E

◂

We turn to an example in which the assumption dependency column grows wider.

EXAMPLE 7.5.6 A proof of the sequent $\Box p, \Box q \succ \Box(p \wedge q)$:

1	(1)	$\Box p$	Assumption
2	(2)	$\Box q$	Assumption
1	(3)·	p	1, $\Box$E
2	(4)·	q	2, $\Box$E
1, 2	(5)·	$p \wedge q$	3, 4 $\wedge$I·
1, 2	(6)	$\Box(p \wedge q)$	5, $\Box$I

This is just the proof which was given for this sequent in our natural deduction system for **S5** as Example 7.2.2(*ii*), except for a judicious sprinkling of dots. ◂

EXERCISE 7.5.7 Sequents (*i*) and (*iii*) from Example 7.2.2 were $\Box(p \wedge q) \succ \Box p \wedge \Box q$ and $\Box(p \to q) \succ \Box p \to \Box q$, respectively. Give proofs of these sequents using the present proof system. ✠

To see what would go wrong if we also allowed the 'dotted' version of ($\vee$E), ($\vee$E·):

$$\frac{\Gamma \succ \cdot A \vee B \qquad \Delta, A \succ \cdot C \qquad \Theta, B \succ \cdot C}{\Gamma, \Delta, \Theta \succ \cdot C}$$

we offer the following illustration.

EXAMPLE 7.5.8 This proof, using the disallowed rule ($\vee$E·), is set out in sequent-to-sequent style; note that the sequent proved is not valid on every frame – not even on every frame for $\mathbf{K4}_c$, in answer to Exc. 2.7.31 (p. 136) though it was the corresponding implicational formula we were concerned with there. (The latter is the Generalized Density principle GD$_2$, last seen on p. 196.)

(1)	$\Box(\Box p \vee \Box q) \succ \Box(\Box p \vee \Box q)$	Initial sequent
(2)	$\Box(\Box p \vee \Box q) \succ \cdot \Box p \vee \Box q$	1 $\Box$E
(3)	$\Box p \succ \Box p$	Initial sequent
(4)	$\Box p \succ \cdot p$	3 $\Box$E
(5)	$\Box p \succ \cdot p \vee q$	4 $\vee$I
(6)	$\Box q \succ \Box q$	Initial sequent
(7)	$\Box q \succ \cdot q$	6 $\Box$E
(8)	$\Box q \succ \cdot p \vee q$	7 $\vee$I
(9)	$\Box(\Box p \vee \Box q) \succ \cdot p \vee q$	2, 5, 8 $\vee$E·
(10)	$\Box(\Box p \vee \Box q) \succ \Box(p \vee q)$	9 $\Box$I

◂

7.5. NATURAL DEDUCTION FOR WEAKER NORMAL MODAL LOGICS

Recall from p. 8 that a finitary consequence relation on a language is a relation between sets of formulas and individual formulas of that language which holds between a set Γ of formulas and a formula A only when it holds between some finite $\Gamma_0 \subseteq \Gamma$, and satisfies the following three conditions for all formulas A, B, of the language and all sets Γ, Δ, of such formulas:

- $A \vdash A$,
- $\Gamma \vdash A$ implies $\Gamma, \Delta \vdash A$,
- $\Gamma \vdash A$ and $\Delta, A \vdash B$ imply $\Gamma, \Delta \vdash B$.[429]

For the consequence relation $\vdash$ defined by: $\Gamma \vdash A$ if and only if for some finite $\Gamma_0 \subseteq \Gamma$, the sequent $\Gamma_0 \succ A$ is provable in the natural deduction system of the previous section, we have $\vdash = \vdash_{\mathbf{S4}}$, as the latter was defined on p. 146. A similar definition for the proof system of our earlier discussion in this appendix gives $\vdash = \vdash_{\mathbf{S5}}$, while for the current proof system – as we shall see in Theorem 7.5.9 – the corresponding consequence relation is $\vdash_{\mathbf{K}}$. In the current case, note that this definition involves plain sequents $\Gamma_0 \succ A$, and not the transitional sequents $\Gamma \succ\cdot A$. Suppose we made a similar definition, but this time on the basis of the transitional sequents. We use a dot as a reminder of this change, calling the relation $\vdash\cdot$. Thus, we define:

$\Gamma \vdash\cdot A$ if and only if for some finite $\Gamma_0 \subseteq \Gamma$, the sequent $\Gamma_0 \succ\cdot A$ is provable,

where provability is taken as provability in the current system. What kind of a consequence relation would this be?

We have already seen enough to return the answer: no kind of consequence relation at all. The first of the three defining conditions on consequence relations is not satisfied: the requirement that $A \vdash\cdot A$, for all A – since $A \succ\cdot A$ is not in general provable. (Actually, at p. 487, we noted that such transitional sequents were not valid on all frames; so by Proposition 7.5.4, they are not provable.) What Example 7.5.8 tells us is that the third of the defining conditions on consequence relations is not satisfied either. The sequent there involved disjunction and was used to illustrated the fact that allowing the dotted form of ∨-Elimination would jeopardise the soundness of our system (w.r.t. the class of all frames), but imagine the case in which the disjuncts coincide. This amounts to a structural rule, i.e., one whose schematic formulation does not involve any particular connective (or more generally, any particular item of logical vocabulary), namely what would be called the Cut rule (because of the way we cut out the formula represented by A on passage from premisses to conclusion) for dotted sequents:

(Dotted Cut) $\quad \dfrac{\Gamma \succ\cdot A \qquad \Delta, A \succ\cdot C}{\Gamma, \Delta \succ\cdot C}$

Evidently we can derive this rule from (∨E)· using the other (acceptable) rules, since we can pass from the left premiss-sequent to $\Gamma \succ\cdot A \vee A$ by (∨I)· and then, using the right-hand premiss sequent twice, via (∨E)·, to the conclusion sequent. And conversely, we can derive (∨E)· from Dotted Cut using the remaining rules; here we derive its conclusion-sequent from its premiss-sequents:

(1) $\Gamma \succ\cdot A \vee B$ Given
(2) $\Delta, A \succ\cdot C$ Given
(3) $\Theta, B \succ\cdot C$ Given
(4) $\Delta, A \succ \Box C$ 1 □I
(5) $\Theta, B \succ \Box C$ 2 □I
(6) $\Gamma, \Delta, \Theta \succ \Box C$ 1, 4, 5 ∨E
(7) $\Gamma, \Delta, \Theta \succ\cdot C$ 5 □E

---

[429] For this last condition the formulation differs from that at p. 8 in that we have used "Δ" in place of the second occurrence of Γ there; the formulations are equivalent given the second condition, however.

THEOREM 7.5.9 *A sequent (plain or transitional) is provable in the present system if and only if it is valid on every frame.*

Proof. The soundness, or 'only if' part of this result was given as Proposition 7.5.4. For the completeness (or 'if') part, we first note that it will suffice to show that all *plain* sequents valid on every frame are provable. The reason is that if this is established it gives the corresponding result for a transitional sequent by thus: Suppose that $\Gamma \succ \cdot A$ is valid on every frame; then so is $\Gamma \succ \Box A$, by the definition of validity for transitional sequents (and the clause for $\Box$ in the definition of truth). So on the assumption of completeness for plain sequents, $\Gamma \succ \Box A$ is provable, in which case (by ($\Box$E)), so is the original transitional sequent $\Gamma \succ \cdot A$.

To show completeness for the plain sequents, given, that every substitution instance of a tautologous sequent is provable, it suffices to show that whenever $A_1, \ldots, A_n \succ B$ is provable, so is $\Box A_1, \ldots, \Box A_n \succ \Box B$, and for this it suffices to show that this holds for $n = 0$ and $n = 1$ (versions respectively of Necessitation and $\Box$Mono), and that $\Box C, \Box D \succ \Box(C \wedge D)$ (a version of Aggregation). (Here we rely on the definition of normality given on p. 19, including the comment that follows it, with some recasting into the sequent formulation, as given by the Digression on p. 22.) The first of these three requirements involves passing from $\succ B$ to $\succ \Box B$, which can be done by applying (Modul) to get $\succ \cdot B$ and then ($\Box$I), giving $\succ \Box B$. The third is given by 7.5.6 above (putting C, D, for p, q). So it remains only to show – the monotone condition – that the provability of $A \succ B$ secures the provability of $\Box A \succ \Box B$. So suppose that $A \succ B$ is provable, and hence (*i*) that $\succ \cdot A \to B$ is too, by ($\to$I) and (Modul). From $\Box A \succ \Box A$ we derive (*ii*) $\Box A \succ \cdot A$ by ($\Box$E), and from (*i*) and (*ii*) we derive $\Box A \succ \cdot B$ by ($\to$E), and so finally $\Box A \succ \Box B$ by ($\Box$I). ∎

Let us consider extending the natural deduction system for **K** by means of rules named after the corresponding formula logics; thus, just as (formulating Theorem 7.5.9 non-semantically), it tells us that a sequent $A_1, \ldots, A_n \succ B$ is provable in the natural deduction system for **K** just in case – where the subscript now indicates the formula logic **K** – $\vdash_{\mathbf{K}} (A_1 \wedge \ldots \wedge A_n) \to B$, so adding the rule here called **4** produces a similarly correspondence with the formula logic **K4**, and so on for the other cases. The **4** rule is like ($\Box$I) except that it has a transitional sequent premiss as well as a transitional sequent conclusion. The **U** rule similarly deals exclusively in transitional sequents, being the special case of 'dotted' $\to$-introduction with Γ empty. The other two rules pass from a transitional premiss sequent to a plain conclusion sequent, in the case of **T** with the same left-hand and right-hand formulas, while with **D** we have an (RAA)-like passage from premiss to conclusion.

$$\mathbf{4} \quad \frac{\Gamma \succ \cdot A}{\Gamma \succ \cdot \Box A} \qquad\qquad \mathbf{D} \quad \frac{\Gamma, A \succ \cdot B \wedge \neg B}{\Gamma \succ \neg A}$$

$$\mathbf{T} \quad \frac{\Gamma \succ \cdot A}{\Gamma \succ A} \qquad\qquad \mathbf{U} \quad \frac{A \succ \cdot B}{\succ \cdot A \to B}$$

EXERCISE 7.5.10 (*i*) Show that these four rules give rise to natural deduction systems corresponding to the formula logics their labels suggest, the correspondence being as explained above.

(*ii*) What would be the corresponding formula logic if the above **U** rule had instead been formulated with arbitrary side formulas Γ on the left of the premiss and conclusion sequents? (This is the rule ($\to$I·) rejected for inclusion in the basic system for **K** in our earlier discussion.)

(*ii*) Same question as (*ii*), but with the addition to the basic system of the rule ($\vee$E·) which was the subject of Example 7.5.8. ✠

Of the four rules listed above, the least interesting from one perspective is **4**, in that here the formulation of the rule explicitly includes a $\Box$. The same would apply in the case of a similar rule for **B**, taking us from $\Gamma \succ A$ to $\Gamma \succ \cdot \Diamond A$ (with $\Diamond$ understood as $\neg \Box \neg$). We will not go into the extent to which avoiding this feature is possible, or indeed desirable, though there are some remarks on the subject in Blamey and Humberstone [99], where, however, the sequent-to-sequent rules on offer laid no claim to being of the natural deduction type (and sequents with more than two – left and right – positions for formulas were used, rather than more than one kind of sequent in the same system).[430] The interested reader is encouraged to experiment further with rules in the above style (minimizing occurrences of $\Box$ in their formulation) to give proof systems for other normal modal logics.

---

[430]For further variations along these and other lines (as to what sequents might usefully be taken to be for modal logic), supplementing the references in note 427, see the articles in Wansing [1148]. For more recent ventures, undertaken with Gentzen-style sequent calculi in mind rather than natural deduction systems, see Poggiolesi [887], Restall [953], Bednarska and Indrzejczak [59], as well as the survey article Indrzejczak [609].

Bibliography

[1] S. Aanderaa, 'Relations Between Different Systems of Modal Logic', *Notices of the American Math. Soc.* **13** (1966), 391.

[2] Barbara Abbott, 'Linguistic Solutions to Philosophical Problems: The Case of Knowing How', *Philosophical Perspectives* **27** (2013), 1–21.

[3] M. Abraham, D. M. Gabbay and U. Schild, 'Obligations and Prohibitions in Talmudic Deontic Logic', pp. 166–178 in G. Governatori and G. Sartor (eds.), *'DEON 2010': Deontic Logic in Computer Science*, Springer-Verlag Berlin 2010.

[4] J. L. Ackrill, *Aristotle's Categories and De Interpretatione*, (translation and notes by Ackrill) Oxford University Press, Oxford 1963.

[5] Tuomo Aho, 'On the Interpretation of Attitude Logics', pp. 1–11 in D. Prawitz and D. Westerståhl (eds.), *Logic and Philosophy of Science in Uppsala*, Kluwer, Dordrecht 1994.

[6] Tuomo Aho and Ilkka Niiniluoto, 'On the Logic of Memory', pp. 408–429 in L. Haaparanta, M. Kusch and I. Niiniluoto (eds.), *Language, Knowledge, and Intentionality: Perspectives on the Philosophy of Jaakko Hintikka*, Helsinki 1990 (= *Acta Philosophica Fennica* **49**).

[7] Seiki Akama, T. Murai, and Y. Kudo, 'Epistemic Logic Founded on Nonignorance', *International Journal of Intelligent Systems* **28** (2013), 883–891.

[8] Noriko Akatsuka, 'Why Tough-Movement is Impossible with *possible*', *(Papers from the Regional Meeting of the) Chicago Linguistic Society* **15** (1979), 1–8.

[9] Samuel Alexander, 'A Purely Epistemological Version of Fitch's Paradox', *The Reasoner* **9** (2012), 59–60.

[10] Samuel Alexander, 'An Axiomatic Version of Fitch's Paradox', *Synthese* **190** (2013), 2015–2020.

[11] V. Allis and T. Keutsier, 'On Some Paradoxes of the Infinite', *British Journal for the Philosophy of Science* **42** (1991), 187–194.

[12] Patrick Allo, 'The Many Faces of Closure and Introspection: an Interactive Perspective', *Journal of Philosophical Logic* **42** (2013), 91–124.

[13] M. Aloni, P. Égré and T. de Jager, 'Knowing Whether A or B', *Synthese* **190** (2013), 2595–2621.

[14] W. P. Alston, 'Varieties of Privileged Access', *American Philosophical Quarterly* **8** (1971), 223–241.

[15] A. R. Anderson, 'The Logic of Norms', *Logique et Analyse* **1** (1958) 84–91.

[16] A. R. Anderson, 'A Reduction of Deontic Logic to Alethic Modal Logic', *Mind* **67** (1958), 100–103.

[17] A. R. Anderson, 'The Formal Analysis of Normative Systems', pp. 147–213 in [948].

[18] A. R. Anderson, 'Some Nasty Problems in the Formal Logic of Ethics', *Noûs* **1** (1967), 345–360.

[19] A. R. Anderson, 'The Logic of Hohfeldian Propositions', *Logique et Analyse* **13** (1970) 231–242; reprinted in *University of Pittsburgh Law Review* **32** (1971), 29–38.

[20] A. R. Anderson and N. D. Belnap, *Entailment: the Logic of Relevance and Necessity, Vol. I*, Princeton University Press, Princeton, NJ 1975.

[21] A. R. Anderson and O. K. Moore, 'The Formal Analysis of Normative Concepts', *American Sociological Review* **22** (1957), 9–17.

[22] C. Anthony Anderson, 'The Lesson of Kaplan's Paradox about POssible World Semantics', pp. 85–92 in J. Almog and P. Leonardi (eds), *The Philosophy of David Kaplan*, Oxford University Press, Oxford 2009.

[23] Kent C. Anderson, 'A Note on Knowledge', *Mind* **86** (1977), 249–251.

[24] R. B. Angell 'Deducibility, Entailment and Analytic Containment', pp.119–143 in Jean Norman and Richard Sylvan (eds.), *Directions in Relevant Logic*, Kluwer, Dordrecht 1989.

[25] Elizabeth Anscombe, *Intention*, Basil Blackwell, Oxford 1957.

[26] F. Antinucci and D. Parisi, 'On English Modal Verbs', *Chicago Linguistics Society* **7** (1971), 28–39.

[27] Lennart Åqvist, 'On Dawson Models for Deontic Logic', *Logique et Analyse* **7** (1964), 14–21.

[28] Lennart Åqvist, 'Formal Semantics for verb tenses as analyzed by Reichenbach', pp. 229–236 in T. A. van Dijk (ed.), *Pragmatics of Language and Literature*, North-Holland, Amsterdam 1976.

[29] Lennart Åqvist, 'A conjectured axiomatization of two-dimensional Reichenbachian tense logic', *Journal of Philosophical Logic* **8** (1979) 1–45.

[30] Lennart Åqvist, *Introduction to Deontic Logic and the Theory of Normative Systems*, Bibliopolis, Naples 1987.

[31] Lennart Åqvist, 'Systematic Frame Constants in Defeasible Deontic Logic', pp. 59–77 in Nute [852].

[32] Lennart Åqvist, 'Three Characterizability Problems in Deontic Logic', *Nordic Journal of Philosophical Logic* **5** (2000), 65–82.

[33] Lennart Åqvist, 'Deontic Tense Logic with Historical Necessity, Frame Constants, and a Solution to the Epistemic Obligation Paradox (The "Knower")', *Theoria* **80** (2014), 319–349.

[34] Carlos Areces and Patrick Blackburn, 'Reichenbach, Prior and Montague: A Semantic Get-together', pp. 77–87 in S. Artemov, H. Barringer, A. d'Avila Garcez, L. C. Lamb and J. Woods (eds.), *We Will Show Them! Essays in Honour of Dov Gabbay on his 60th Birthday, Vol. 1*, College Publications, London, 2005.

[35] D. M. Armstrong, 'Absolute and Relative Motion', *Mind* **72** (1963), 209–223.

[36] D. M. Armstrong, 'Is Introspective Knowledge Incorrigible?', *Philosophical Review* **72**, (1963) pp. 417–432.

[37] Ana Arregui, 'Counterfactual-Style Revisions in the Semantics of Deontic Modals', *Journal of Semantics* **28** (2011), 171–210.

[38] Ana Arregui, 'Detaching *if*-clauses from *should*', *Natural Language Semantics* **18** (2010), 241–293.

[39] S. Artemov, 'Why Do We Need Justification Logic?', pp. 23–37 in J. van Benthem, A. Gupta and E. Pacuit (eds.), *Games, Norms and Reasons*, Springer-Verlag, Berlin 2011.

[40] S. Artemov and E. Nogina, 'On Epistemic Logic with Justification', pp. 279–294 in R. van der Meyden (ed.), *Proceedings of the Tenth Conference of Theoretical Aspects of Rationality and Knowledge (TARK X)*, National University of Singapore 2005.

[41] Guillaume Aucher, 'Intricate Axioms as Interaction Axioms', *Studia Logica* **103** (2015), 1035–1062.

[42] J. L. Austin, 'Ifs and Cans', Chapter 9 of [44]. (Paper first published in 1956.)

[43] J. L. Austin 'Other Minds', pp. 123–158 in A. G. N. Flew (ed.), *Logic and Language, Second Series*, Basil Blackwell, Oxford 1966; reprinted as chapter 3 in [44].

[44] J. L. Austin, *Philosophical Papers*, Second Edition, Oxford University Press 1970, Oxford. (Eds. J. O. Urmson and G. J. Warnock.)

[45] Johan van der Auwera and Andreas Ammann, 'Situational Possibility', in M. S. Dryer, and M. Haspelmath (eds.), *The World Atlas of Language Structures Online*, Leipzig: Max Planck Institute for Evolutionary Anthropology, 2013; at <http://wals.info/chapter/74>.

[46] A. J. Ayer, 'Freedom and Necessity', Chapter 12 in Ayer, *Philosophical Essays*, Macmillan, London 1954.

[47] John Bacon, 'Belief as Relative Knowledge', pp. 189–210 in A. R. Anderson, R. Barcan Marcus, and R. M. Martin (eds.), *The Logical Enterprise*, Yale University Press, New Haven CT 1975.

[48] John Bacon, 'Purely Physical Modalities', *Theoria* **47** (1981), 135–141.

[49] J. R. Baker, 'What is Not Wrong with a Hartshornean Modal Proof', *Southern Journal of Philosophy* **18** (1980), 99–106.

[50] P. Balbiani, I. Shapirovsky and V. Shehtman, 'Every World Can See a Sahlqvist World', pp. 69–85 in G. Governatori, I. Hodkinson and Y. Venema (eds.), *Advances in Modal Logic, Vol. 6*, College Publications, London 2006.

[51] Thomas Baldwin (ed.), *G. E. Moore: Selected Writings*, Routledge, London 1993.

[52] Ruth Barcan Marcus, 'Iterated Deontic Modalities', *Mind* **75** (1966), 580–582.

[53] John A. Barker, 'A Paradox of Knowing Whether', *Mind* **84** (1975), 281–283.

[54] Rainer Bäuerle, 'Tense Logics and Natural Language', *Synthese* **40** (1979), 225–230.

[55] R. J. Baxter, 'On Some Models of Modal Logics', *Notre Dame Journal of Formal Logic* **14** (1973), 121–22.

[56] David Beaver, 'Presupposition', pp. 939–1053 in J. van Benthem and A. ter Meulen (eds.), *Handbook of Logic and Language*, Elsevier, Amsterdam, 1997.

[57] David Beaver and Bart Geurts, 'Presupposition', *Stanford Encyclopedia of Philosophy* (Winter 2014 Edition), Edward N. Zalta (ed.), URL = <http://plato.stanford.edu/archives/win2014/entries/presupposition/>.

[58] Jacob Beck, 'The Generality Constraint and the Structure of Thought', *Mind* **121** (2012), 563–600.

[59] Kaja Bednarska and Andrzej Indrzejczak, 'Hypersequent Calculi for S5: The methods of cut elimination', *Logic and Logical Philosophy* **24** (2015), 277–311.

[60] Raymond A. Belliotti, 'Negative and Positive Duties', *Theoria* **47** (1981), 82–92.

[61] Nuel Belnap and Paul Bartha, 'Marcus and the Problem of Nested Deontic Modalities', pp. 174–197 in W. Sinnott-Armstrong, D. Raffman and N. Asher (eds.), *Modality, Morality and Belief*, Cambridge University Press, Cambridge 1995.

[62] Nuel Belnap and Michael Perloff, 'Seeing To it That: A Canonical Form for Agentives', *Theoria* **54** (1988), 175–99. Also at pp. 167–190 in H. E. Kyburg *et al.* (eds.), *Knowledge Representation and Defeasible Reasoning*, Kluwer, Dordrecht 1990.

[63] Gordon Belot, *Geometric Possibility*, Oxford University Press, Oxford 2011.

[64] John Bengson and Marc A. Moffett (eds.), *Knowing How: Essays on Knowledge, Mind, and Action*, Oxford University Press, Oxford 2012.

[65] Jonathan Bennett, 'Iterated Modalities', *Philosophical Quarterly* **5** (1955), 45–56.

[66] Johan van Benthem, 'Hintikka on Analyticity', *Journal of Philosophical Logic* **3** (1974) 419–431.

[67] Johan van Benthem, 'Two Simple Incomplete Modal Logics', *Theoria* **44** (1978), 25–37.

[68] Johan van Benthem, 'Syntactic Aspects of Modal Incompleteness Theorems', *Theoria* **45** (1979) 63–77.

[69] Johan van Benthem, 'Later Than Late: On the Logical Origin of the Temporal Order', *Pacific Philosophical Quarterly* **63** (1982), 193–203.

[70] Johan van Benthem, 'Correspondence Theory', pp. 167–247 in *Handbook of Philosophical Logic, Vol. II: Extensions of Classical Logic*, ed. D. M. Gabbay and F. Guenthner, Reidel, Dordrecht 1984. Reprinted with additions under the same title in pp. 325–408 of Gabbay and Guenthner (eds.), *Handbook of Philosophical Logic, Second Edition, Vol. 3*, Kluwer, Dordrecht 2001.

[71] Johan van Benthem and Lloyd Humberstone, 'Halldén-completeness by Gluing of Kripke Frames', *Notre Dame Journal of Formal Logic* **24** (1983), 426–430.

[72] Johan van Benthem, 'Possible Worlds Semantics: A Research Program that Cannot Fail?', *Studia Logica* **43** (1984), 379–393.

[73] Johan van Benthem, *Modal Logic and Classical Logic*, Bibliopolis, Naples 1985.

[74] Johan van Benthem, *The Logic of Time*, Kluwer, Dordrecht 1991. (First edn. 1983.)

[75] Johan van Benthem, 'Reflections on Epistemic Logic', *Logique et Analyse* **34** (1991), 5–14.

[76] Johan van Benthem, 'Beyond Accessibility: Functional Models for Modal Logic', pp. 1–18 in M. de Rijke (ed.), *Diamonds and Defaults*, Kluwer, Dordrecht 1993.

[77] Johan van Benthem, 'Logical Constants: The Variable Fortunes of an Elusive Notion', pp. 420–440 in W. Wieg, R. Sommer and C. Talcott (eds.), *Reflections on the Foundations of Mathematics: Essays in Honor of Solomon Feferman*, Association for Symbolic Logic, A. K. Peters, Natick, MA 2002.

[78] Johan van Benthem, 'What One May Come to Know', *Analysis* **64** (2004), 95–105.

[79] Johan van Benthem, *Modal Logic for Open Minds*, CSLI Publications, Stanford, CA 2010.

[80] Johan van Benthem, 'Dynamic Logic in Natural Language', pp. 652–666 in [980].

[81] Johan van Benthem, D. Fernández-Duque and E. Pacuit, 'Evidence Logic: A New Look at Neighborhood Structures', pp. 97–118 in T. Bolander, T. Braüner, S. Ghilardi, and L. Moss (eds.), *Advances in Modal Logic, Vol. 9*, College Publications, London 2012.

[82] R. A. Benton, 'A Simple Incomplete Extension of T Which is the Union of Two Logics With the F.M.P.', *Journal of Philosophical Logic* **31** (2002), 527–541.

[83] Harry Beran, 'Ought, Obligation and Duty', *Australasian Journal of Philosophy* **50** (1972), 207–221.

[84] Lars Bergström, 'Utilitarianism and Deontic Logic', *Analysis* **29** (1968), 43–44.

[85] Selim Berker, 'Luminosity Regained', *Philosophers' Imprint* **8** (2008), No. 2, 1–22.

[86] J. Bernert and A. Biela, 'On Two Different Modal Logics Denoted by S9', *Reports on Mathematical Logic* **13** (1981), 3–9.

[87] Francesco Berto, 'Impossible Worlds', *Stanford Encyclopedia of Philosophy* (Fall 2009 Edition), Edward N. Zalta (ed.), URL = <http://plato.stanford.edu/archives/fall2009/entries/impossible-worlds/>.

[88] Jean-Yves Béziau, 'A New Four-Valued Approach to Modal Logic', *Logique et Analyse* **54** (2011), 109–121.

[89] Jerome E. Bickenbach, 'One Ought to Do What One Thinks One Ought to Do'. *Dialogue* **14** (1975), 667–670.

[90] John Bigelow, 'Because it is True', unpublished paper, La Trobe University 1979.

[91] John Bigelow and Robert Pargetter, *Science and Necessity*, Cambridge University Press, Cambridge 1990.

[92] John Bigelow, 'Omnificence', *Analysis* **65** (2005), 187–96.

[93] Marta Bílková, 'Uniform Interpolation and Propositional Quantifiers in Modal Logics', *Studia Logica* **85** (2007), 1–31.

[94] M. Bílková, A. Palmigiano, and Y. Venema, 'Proof Systems for the Coalgebraic Cover Modality', pp. 1–21 in C. Areces and R. Goldblatt (eds.), *Advances in Modal Logic, Vol. 7*, College Publications, London 2008.

[95] Garrett Birkhoff, *Lattice Theory* (3rd Edition, 8th Printing), Colloquium Publications, American Math. Society, Providence, RI 1995 (1st Edition: 1940).

[96] Garrett Birkhoff and Saunders Mac Lane, *A Survey of Modern Algebra*, Third Edition, Macmillan, New York 1965.

[97] Patrick Blackburn, Johan van Benthem, and Frank Wolter (eds.), *Handbook of Modal Logic*, Elsevier, Amsterdam 2007.

[98] Patrick Blackburn, Martin de Rijke, and Yde Venema, *Modal Logic*, Cambridge University Press, Cambridge 2001.

[99] Stephen Blamey and Lloyd Humberstone, 'A Perspective on Modal Sequent Logic', *Publications of the Research Institute for Mathematical Sciences, Kyoto University* **27** (1991), 763–782.

[100] W. J. Blok and P. Köhler, 'Algebraic Semantics for Quasi-Classical Modal Logics', *Journal of Symbolic Logic* **48** (1983), 941–964.

[101] W. J. Blok and D. Pigozzi, 'Algebraizable Logics', *Memoirs of the American Mathematical Society* **77** (1989), #396.

[102] Ralph P. Boas, *Lion Hunting and Other Mathematical Pursuits* (eds. Gerald L. Alexanderson and Dale H. Mugler), Cambridge University Press, Cambridge 1996.

[103] M. de Boer, D. M. Gabbay, X. Parent and M. Slavkovic, 'Two-Dimensional Standard Deontic Logic', *Synthese* **187** (2012), 623–660.

[104] Steven E. Boër and William G. Lycan, *Knowing Who*, MIT Press, Cambridge MA, 1986.

[105] Andrzej Bogusławski, 'Indirect Questions: One Interpretation or More?', *Linguistica Silesiana* **3** (1978), 31–40.

[106] Andrzej Bogusławski, '*Know if* and *know that*', pp. 37–48 in Jørgen D. Johansen and Harly Sonne (eds.), *Pragmatics and Linguistics*, Odense University Press, Odense 1986.

[107] Dwight Bolinger, *That's That*, Mouton, The Hague 1972.

[108] Dwight Bolinger, 'Yes-No Questions are Not Alternative Questions', pp. 87–105 in H. Hiż (ed.), *Questions*, Reidel, Dordrecht 1978.

[109] Daniel Bonevac, 'Against Conditional Obligation', *Noûs* **32** (1998), 37–53.

[110] G. Bonnano, 'A Simple Modal Logic for Belief Revision', *Synthese* **147** (2005), 193–228.

[111] D. Bonnay and P. Égré, 'Inexact Knowledge with Introspection', *Journal of Philosophical Logic* **38** (2009), 179–227.

[112] George Boolos, *The Logic of Provability*, Cambridge University Press, Cambridge 1993.

[113] L. Borkowski (ed.), *Jan Łukasiewicz: Selected Works*, North-Holland, Amsterdam 1970.

[114] E. J. Borowski, 'Adverbials in Action Sentences', *Synthese* **28** (1974), 483–512.

[115] R. T. Brady, *Universal Logic*, CSLI Publications, Stanford University, CA 2006.

[116] David Braun, 'An Invariantist Theory of "Might" Might be Right', *Linguistics and Philosophy* **35** (2012), 461–489.

[117] David Brink, 'Moral Conflict and Its Structure', *Philosophical Review* **103** (1994), 215–247.

[118] Berit Brogaard and Joe Salerno, 'Knowability and a Modal Closure Principle', *American Philosophical Quarterly* **43** (2006), 261–270.

[119] John Broome, *Rationality Through Reasoning*, Wiley Blackwell, Oxford 2013.

[120] Charles D. Brown, 'The Ontological Theorem', *Notre Dame Journal of Formal Logic* **19** (1978), 591–592.

[121] Mark A. Brown, 'On the Logic of Ability', *Journal of Philosophical Logic* **17** (1988), 1–26.

[122] James E. Broyles, 'Knowledge and Mistake', *Mind* **78** (1969), 198–211.

[123] John Bryant. 'The Logic of Relative Modality and the Paradoxes of Deontic Logic', *Note Dame Journal of Formal Logic* **21** (1980), 78–88.

[124] Wesley Buckwalter, 'Factive Verbs and Protagonist Projection', *Episteme* **11** (2014), 391–409.

[125] R. A. Bull, 'An Axiomatization of Prior's Modal Calculus Q', *Notre Dame Journal of Formal Logic* **5** (1964), 211–214.

[126] R. A. Bull, 'That All Normal Extensions of S4.3 Have the Finite Model Property', *Zeitschr. für math. Logik und Grundlagen der Math.* **12** (1966) 341–344.

[127] R. A. Bull, 'On Modal Logic with Propositional Quantifiers', *Journal of Symbolic Logic* **34** (1969), 257–263.

[128] R. A. Bull and K. Segerberg, 'Basic Modal Logic', pp. 1–81 in D. M. Gabbay and F. Guenthner (eds.), *Handbook of Philosophical Logic, Second Edition, Vol. 3*, Kluwer, Dordrecht 2001.

[129] B. H. Bunch, *Mathematical Fallacies and Paradoxes*, Van Nostrand Reinhold, NY 1982. (Reprinted with corrections, Dover Publications, NY 1997.)

[130] Alexis G. Burgess and John P. Burgess, *Truth*, Princeton University Press, Princeton 2011.

[131] John A. Burgess, 'When is Circularity in Definitions Benign?', *Philosophical Quarterly* **58** (2008), 214–233.

[132] John P. Burgess, 'Probability Logic', *Journal of Symbolic Logic* **34** (1969), 264–274.

[133] John P. Burgess, 'The Unreal Future', *Theoria* **44** (1978), 157–179.

[134] John P. Burgess, 'Logic and Time', *Journal of Symbolic Logic* **44** (1979), 566–582.

[135] John P. Burgess, 'Non-Classical Logic and Ontological Non-Commitment, Avoiding Abstract Objects through Modal Operators', pp. 287–305 in D. Prawitz, B. Skyrms, and D. Westerståhl (eds.) *Logic, Methodology and Philosophy of Science IX*, Elsevier, Amsterdam 1994.

[136] John P. Burgess, 'Quinus ab Omni Nævo Vindicatus', *Canadian Journal of Philosophy* Supplementary Volume **23** (1997), 25–65.

[137] John P. Burgess, 'Which Modal Logic is the Right One?', *Notre Dame Journal of Formal Logic* **40** (1999), 81–93.

[138] John P. Burgess, 'Axiomatizing the Logic of Comparative Probability', *Notre Dame Journal of Formal Logic* **51** (2010), 119–126.

[139] John P. Burgess, 'Kripke Models', pp. 119–140 in A. Berger (ed.), *Kripke*, Cambridge University Press, Cambridge 2011.

[140] J. Butterfield and C. Stirling, 'Predicate Modifiers in Tense Logic', *Logique et Analyse* **30** (1987), 31–49.

[141] Saša Buvač, 'A Deduction Theorem for Normal Modal Propositional Logic', pp. 107–115 in P. Blackburn *et al.* (eds.), *CONTEXT 2003*, Lecture Notes in Artificial Intelligence #2680, Springer-Verlag, Berlin 2003.

[142] Joan Bybee, Revere Perkins and William Pagliuca, *The Evolution of Grammar: Tense, Aspect and Modality in the Languages of the World*, University of Chicago Press, Chicago 1994.

[143] Alex Byrne and Heather Logue (eds.), *Disjunctivism: Contemporary Readings*, MIT Press, Cambridge MA 2009.

[144] J. R. Cameron, ' "Ought" and Institutional Obligation', *Philosophy* **46** (1971), 309–322.

[145] J. T. Canty and T. W. Scharle, 'Note on the Singularies of S5', *Notre Dame Journal of Formal Logic* **7** (1966), 108.

[146] Herman Cappelen and John Hawthorne, *Relativism and Monadic Truth*, Oxford University Press, Oxford 2010.

[147] Herman Cappelen and Josh Dever, *The Inessential Indexical: On the Philosophical Insignificance of Perspective and the First Person*, Oxford University Press, Oxford 2013.

[148] Fabrizio Cariani, ' "Ought" and Resolution Semantics', *Noûs* **47** (2013), 534–558.

[149] Fabrizio Cariani, 'Epistemic and Deontic *Should*', *Thought* **2** (2013), 73–84.

[150] Erik Carlson, 'Consequentialism, Alternatives, and Actualism', *Philosophical Studies* **96** (1999), 253–268.

[151] Rudolf Carnap, 'Modalities and Quantification', *Journal of Symbolic Logic* **11** (1945), 33–66.

[152] Walter Carnielli and Claudio Pizzi, *Modalities and Multimodalities*, Springer-Verlag, Berlin 2008.

[153] David Carr, 'The Logic of Knowing How and Ability', *Mind* **88** (1979), 394–409.

[154] Jennifer Carr, 'Subjective *Ought*', *Ergo* **2** (2015), 678–710.

[155] H.-N. Castañeda, Review of Hintikka [490], *Journal of Symbolic Logic* **29** (1964), 132–134.

[156] H.-N. Castañeda, ' "He": A Study in the Logic of Self-Consciousness', *Ratio* **8** (1966), 130–157.

[157] H.-N. Castañeda, 'Indicators and Quasi-indicators', *American Philosophical Quarterly* **4** (1967), 85–100.

[158] H.-N. Castañeda, 'On the Logic of Self-Knowledge', *Noûs* **1** (1967), 9–21.

[159] H.-N. Castañeda, 'A Problem for Utilitarianism', *Analysis* **28** (1968), 141–142.

[160] H.-N. Castañeda, 'The Paradoxes of Deontic Logic: The Simplest Solution to All of Them in One Fell Swoop', pp. 37–85 in [483].

[161] H.-N. Castañeda, 'On Knowing (or Believing) that One Knows (or Believes)', *Synthese* **21** (1970), 187–203.

[162] H.-N. Castañeda, 'On the Semantics of the Ought-to-Do', pp. 675–694 in Davidson and Harman [232].

[163] H.-N. Castañeda, *Thinking and Doing*, Reidel, Dordrecht 1975.

[164] Albert Casullo, 'Discussion Note: Knowledge and the Elimination of Truth', *Erkenntnis* **25** (1986), 169–175.

[165] Yuri Cath, 'Knowing How Without Knowing That', pp. 113–35 in [64].

[166] A. V. Chagrov and M. Zakharyaschev, *Modal Logic*, Oxford University Press, Oxford 1997.

[167] David Chalmers, 'Actuality and Knowability', *Analysis* **71** (2011), 411–419.

[168] David Chalmers, 'Frege's Puzzle and the Objects of Credence', *Mind* **120** (2011), 587–635.

[169] Brian Chellas, 'Imperatives', *Theoria* **37** (1971), 114–129.

[170] Brian Chellas, 'Conditional Obligation', pp. 23–33 in S. Stenlund (ed.), *Logical Theory and Semantic Analysis*, Reidel, Dordrecht 1974.

[171] Brian Chellas, *Modal Logic: An Introduction*, Cambridge University Press, Cambridge 1980. Reprinted (with corrections) 1988 and subsequently.

[172] Brian Chellas and Krister Segerberg, 'Modal Logics with the MacIntosh Rule', *Journal of Philosophical Logic* **23** (1994), 67–86.

[173] Brian Chellas and Krister Segerberg, 'Modal Logics in the Vicinity of S1', *Notre Dame Journal of Formal Logic* **37** (1996), 1–24.

[174] R. M. Chisholm, 'Supererogation and Offence: A Conceptual Scheme for Ethics', *Ratio* **5** (1963), 1–14.

[175] R. M. Chisholm, *The First Person: An Essay on Reference and Intentionality* University of Minnesota Press, Minneapolis 1981.

[176] R. M. Chisholm and E. Sosa, 'On the Logic of "Intrinsically Better"', *American Philosophical Quarterly* **3** (1966), 244–249.

[177] R. M. Chisholm and E. Sosa, 'Intrinsic Preferability and the Problem of Supererogation', *Synthese* **16** (1966), 321–331.

[178] David Christensen, *Putting Logic in it Place: Formal Constraints on Rational Belief*, Oxford University Press, Oxford 2004.

[179] Alonzo Church, *Introduction to Mathematical Logic (Vol. I)*, Princeton University Press, Princeton NJ 1956.

[180] Guglielmo Cinque, *Adverbs and Functional Heads: A Cross-Linguistic Perspective*, Oxford University Press, Oxford 1999.

[181] Austen Clark, 'The Particulate Instantiation of Homogeneous Pink', *Synthese* **80** (1989), 277–304.

[182] Nino B. Cocchiarella, 'Logical Atomism and Modal Logic', *Philosophia* **4** (1974), 41–66.

[183] Nino B. Cocchiarella and Max Freund, *Modal Logic: An Introduction to its Syntax and Semantics*, Oxford University Press, Oxford 2008.

[184] Brenda Cohen, 'An Ethical Paradox', *Mind* **76** (1967), 250–259.

[185] Brenda Cohen, 'Non-Dogmatism and Ethical Paradoxes', *Mind* **81** (1972), 432–433.

[186] L. J. Cohen, 'Three-Valued Ethics', *Philosophy* **26** (1951), 208–227.

[187] R. G. Collingwood, *An Autobiography*, Oxford University Press, Oxford 1939; reprinted with an introduction by S. Toulmin, 1978.

[188] John Collins, 'Lotteries and the Close Shave Principle', pp. 83–96 in S. Hetherington (ed.), *Aspects of Knowing: Epistemological Essays*, Elsevier, Oxford 2006.

[189] Annalisa Coliva, 'How to Commit Moore's Paradox', *Journal of Philosophy* **112** (2015), 169–192.

[190] Giovanni Colonna, 'Note per un approcio vero-funzionale alla teoria delle proposizioni modali', pp. 559–571 in *Atti del Convegno Nazionale di Logica, 1–5 ottobre 1979*, Bibliopolis, Naples 1980.

[191] Bernard Comrie, 'On Reichenbach's Approach to Tense', *Papers from the Chicago Linguistics Society* **17** (1981), 24–30.

[192] M. E. Coniglio and N. M. Peron, 'A Paraconsistent Approach to Chisholm's Paradox', *Principia* **13** (2009), 299–326.

[193] B. Jack Copeland, 'The Genesis of Possible Worlds Semantics' *Journal of Philosophical Logic* **31** (2002), 99–137.

[194] Annabel Cormack and Neil Smith, 'Modals and Negation in English', pp. 133–163 in S. Barbiers, F. Beukema and W. van der Wurff (eds.), *Modality and its Interaction with the Verbal System*, Benjamins, Amsterdam 2002.

[195] F. Correia, 'Propositional Logic of Essence', *Journal of Philosophical Logic* **29** (2000), 295–313.

[196] F. Correia, 'Semantics for Analytic Containment', *Studia Logica* **77** (2004), 87–104.

[197] Newton C. A. da Costa and Walter A. Carnielli, 'On Paraconsistent Deontic Logic', *Philosophia* **16** (1986), 293–305.

[198] Marcel Crabbé, 'Deux Redondances de la Règle de Löb en Logique Modale', *Logique et Analyse* **34** (1991), 15–22.

[199] Edward Craig, *Knowledge and the State of Nature: An Essay in Conceptual Synthesis*, Oxford University Press, Oxford 1999.

[200] M. J. Cresswell, 'The Completeness of S0.5', *Logique et Analyse* **9** (1966), 263–266.

[201] M. J. Cresswell, 'Interpretation of some Lewis Systems of Modal Logic', *Australasian Journal of Philosophy* **45** (1967), 198–206.

[202] M. J. Cresswell, 'Note on a System of Åqvist', *Journal of Symbolic Logic* **32** (1967), 58–60.

[203] M. J. Cresswell, 'A Conjunctive Normal Form For S3.5', *Journal of Symbolic Logic* **34**, (1969), 253–255.

[204] M. J. Cresswell, 'Note on the Interpretation of S0.5', *Logique et Analyse* **13** (1970), 376–378.

[205] M. J. Cresswell, 'Classical Intensional Logics', *Theoria* **36** (1970), 347–372.

[206] M. J. Cresswell, 'Intensional Logics and Logical Truth', *Journal of Philosophical Logic* **1** (1972), 2–15.

[207] M. J. Cresswell, 'Frames and Models in Modal Logic', pp. 63–86 in J. N. Crossley (ed.), *Algebra and Logic* (Springer Lecture Notes in Mathematics #450), Springer-Verlag, Berlin 1975.

[208] M. J. Cresswell, 'Hyperintensional Logic', *Studia Logica* **34** (1975), 25–38.

[209] M. J. Cresswell, 'B Seg Has the Finite Model Property', *Bulletin of the Section of Logic* **8** (1979), 154–160.

[210] M. J. Cresswell, 'The Autonomy of Semantics', pp. 69–86 in S. Peters and E. Saarinen (eds.), *Processes, Beliefs, and Questions*, Reidel, Dordrecht 1982.

[211] M. J. Cresswell, 'Necessity and Contingency', *Studia Logica* **47** (1988), 145–149.

[212] M. J. Cresswell, 'In Defence of the Barcan Formula', *Logique et Analyse* **34** (1991), 271–282.

[213] M. J. Cresswell, 'S1 is Not So Simple', pp. 29–40 in W. Sinnott-Armstrong, D. Raffman and N. Asher (eds.), *Modality, Morality and Belief: Essays in Honor of Ruth Barcan Marcus*, Cambridge University Press, Cambridge 1995.

[214] M. J. Cresswell, 'Incompleteness and the Barcan Formula', *Journal of Philosophical Logic* **24** (1995), 379–403.

[215] M. J. Cresswell, 'Some Incompletable Modal Predicate Logics', *Logique et Analyse* **40** (1997), 321–334.

[216] M. J. Cresswell 'How to Complete Some Modal Predicate Logics', pp. 155–178 in M. Zakharyaschev, K. Segerberg, M. de Rijke and H. Wansing (eds.), *Advances in Modal Logic, Vol. 2*, CLSI Publications, Stanford, CA 2001.

[217] M. J. Cresswell, 'Modal Logic', Chapter 7 (pp. 136–158) in Goble [385].

[218] M. J. Cresswell,'Why Propositions Have No Structure', *Noûs* **36** (2002), 643–662.

[219] M. J. Cresswell, 'Carnap and McKinsey: Topics in the History of Possible-Worlds Semantics', pp. 53–75 in R. Downey, J. Brendle, R. Goldblatt, B. Kim (eds.), *Proceedings of the 12th Asian Logic Conference (2011)*, World Scientific, Singapore 2013.

[220] M. J. Cresswell, 'The Completeness of Carnap's Predicate Logic', *Australasian Journal of Logic* **11** (2014), 46–61.

[221] Thomas M. Crisp, 'Presentism and "Cross-Time" Relations', *American Philosophical Quarterly* **42** (2005), 5–17.

[222] Charles Cross and Floris Roelofsen, 'Questions', *Stanford Encyclopedia of Philosophy* (Fall 2014 Edition), Edward N. Zalta (ed.), URL = <http://plato.stanford.edu/archives/fall2014/entries/questions/>.

[223] E. M. Curley, 'Descartes on the Creation of the Eternal Truths', *Philosophical Review* **93** (1984), 569–597; Errata: *ibid.* **94** (1985), p. 172.

[224] H. B. Curry, *A Theory of Formal Deducibility*, Notre Dame Mathematical Lectures #6, Notre Dame, IN 1957.

[225] H. B. Curry, 'The Elimination Theorem When Modality is Present', *Journal of Symbolic Logic* **17** (1952), 249–265.

[226] H. B. Curry, Review of Church [179], *Journal of the Franklin Institute* **264** (1957), 244–246.

[227] J. Czermak, Embeddings of Classical Logic in S4', *Studia Logica* **34** (1975), 87–100.

[228] J. Czermak, Embeddings of Classical Logic in S4: Part II', *Studia Logica* **35** (1976), 257–271.

[229] Maria Luisa Dalla Chiara, 'Epistemic Logic Without Logical Omniscience', pp. 87–95 in [1015].

[230] Sven Danielsson, 'Taking Ross's Paradox Seriously: A Note on the Original Problems of Deontic Logic', pp. 39–49 in K. Segerberg and R. Sliwinski (eds.), *Logic, Law, Morality: Thirteen Essays in Practical Philosophy in Honour of Lennart Åqvist*, Uppsala Philosophical Studies #51, Department of Phiosophy, Uppsala University, Uppsala 2003.

[231] Donald Davidson, 'The Logical Form of Action Sentences', pp. 81–95 in [948].

[232] Donald Davidson and Gilbert Harman (eds.), *Semantics of Natural Language*, Reidel, Dordrecht 1972.

[233] Martin Davies, *Meaning, Quantification, Necessity: Themes in Philosophical Logic*, Routledge and Kegan Paul, London 1981.

[234] Martin Davies, 'Acts and Scenes', pp. 41–82 in N. Cooper and P. Engel (eds), *New Inquiries into Meaning and Truth*, Harvester Wheatsheaf, Hemel Hempstead 1991.

[235] Martin Davies and Lloyd Humberstone, 'Two Notions of Necessity', *Philosophical Studies* **38** (1980), 1–30.

[236] Wayne A. Davis, 'Are Knowledge Claims Indexical?', *Erkenntnis* **61** (1004), 257–281.

[237] E. E. Dawson, 'A Model for Deontic Logic', *Analysis* **19** (1959), 73–78.

[238] J. W. Decew, 'Conditional Obligation and Counterfactuals', *Journal of Philosophical Logic* **10** (1981), 55–72.

[239] Brannon Denning, National Public Radio interview transcript (from 2010) at http://www.npr.org/2010/12/14/132060874/Health-Mandate-Argument-Hinges-On-Commerce-Clause

[240] Nicholas Denyer, 'Ease and Difficulty: A Modal Logic with Deontic Applications', *Theoria* **56** (1990), 42–61.

[241] Keith DeRose, 'Epistemic Possibilities', *Philosophical Review* **100** (1991), 581–605.

[242] Randall Dipert, 'Peirce's Philosophical Conception of Sets', pp. 53–76 in N. Houser, D. Roberts, and J. Van Evra (eds.), *Studies in the Logic of Charles Sanders Peirce*, Indiana University Press, Bloomington, IN 1997.

[243] H. van Ditmarsch and B. Kooi, 'The Secret of My Success', *Synthese* **153** (2006), 201–232.

[244] H. van Ditmarsch, W. van der Hoek and B. Kooi, *Dynamic Epistemic Logic*, Springer-Verlag, Berlin 2007.

[245] H. van Ditmarsch, W. van der Hoek and P. Iliev, 'Everything is Knowable – How to Get to Know *Whether* a Proposition is True', *Theoria* 78 (2012), 93–114.

[246] John Divers, 'The Analysis of Possibility and the Extent of Possibility', *Ratio* **67** (2013), 183–200.

[247] Cian Dorr and John Hawthorne, 'Embedding Epistemic Modals', *Mind* **122** (2013), 867–913.

[248] Kosta Došen, 'Intuitionistic Double Negation as a Necessity Operator', *Publications de l'Institut Mathématique* (Belgrade) **35** (1984), 15–20.

[249] Kosta Došen, 'Duality Between Modal Algebras and Neighbourhood Frames', *Studia Logica* **48** (1989), 219–234.

[250] F. R. Drake, 'On McKinsey's Syntactical Characterization of Systems of Modal Logic', *Journal of Symbolic Logic* **27** (1962) 400–406.

[251] James Dreier, 'Internalism and Speaker Relativism', *Ethics* **101** (1990), 6–26.

[252] James Dreier, 'Transforming Expressivism', *Noûs* **33** (1999), 558–572.

[253] Fred I. Dretske, *Seeing and Knowing*, Routledge and Kegan Paul, London 1969.

[254] Fred I. Dretske, 'Epistemic Operators', *Journal of Philosophy* **67** (1970), 1007–1023.

[255] Fred I. Dretske, 'The Case Against Closure', pp. 13–26 in Steup and Sosa [1090].

[256] Fred I. Dretske, 'The Content of Knowledge', pp. 77–93 in B. Freed, A. Marras and P. Maynard (eds.), *Forms of Representation*, North-Holland, Amsterdam 1975.

[257] Lech Dubikajtis and Lafayette de Moraes, 'On Single Operator For Lewis **S5** Modal Logic', *Reports on Mathematical Logic* **11** (1981), 57–61.

[258] J. Dubucs, 'On Logical Omniscience', *Logique et Analyse* **34** (1991), 41–55.

[259] J. Dugundji, 'Note on a Property of Matrices for Lewis and Langford's Calculi of Propositions', *Journal of Symbolic Logic*, **5** (1940), 150–151.

[260] M. A. Dummett, 'The Justification of Deduction', *Procs. of the British Academy* **49** (1973), 201–232; reprinted in pp. 290–318 of Dummett, *Truth and Other Enigmas*, Duckworth, London 1978.

[261] M. A. Dummett, *The Logical Basis of Metaphysics*, Harvard University Press, Cambridge, MA 1991.

[262] M. A. Dummett, 'Victor's Error', *Analysis* **61** (2001), 1–2.

[263] M. A. Dummett and E. J. Lemmon, 'Modal Logics Between **S4** and **S5**', *Zeitschr. für math. Logik und Grundlagen der Math.* **5** (1959), 250–264.

[264] J. Michael Dunn, 'A Modification of Parry's Analytic Implication', *Note Dame Journal of Formal Logic* **13** (1972), 195–205.

[265] J. Michael Dunn, 'Positive Modal Logic', *Studia Logica* **55** (1995), 301–317.

[266] Julien Dutant, 'Inexact Knowledge, Margin for Error and Positive Introspection', pp. 118–124 in Dov Samet (ed.), *Theoretical Aspects of Rationality and Knowledge: Proceedings of the 11th Conference on Theoretical Aspects of Rationality and Knowledge (TARK 2007)*, Brussels 2007.

[267] Julien Dutant, *Knowledge, Methods and the Impossibility of Error*, doctoral thesis, University of Geneva, 2010.

[268] Johhn D. Dutton, '"Whether"', *Synthese* **13** (1961), 364–371.

[269] M. Duží, B. Jespersen and J. Müller, 'Epistemic Closure and Inferable Knowledge', pp. 125–140 in L. Běhounek and M. Bílková (eds.), *The Logica Yearbook 2004*, Filosofia, Prague 2005.

[270] J. A. van Eck, 'A System of Temporally Relative Modal and Deontic Predicate Logic and its Philosophical Applications', *Logique et Analyse* **25** (1982), 248–290 and (Part 2) 339–381.

[271] Walter Edelberg, 'Intrasubjective Intentional Identity', *Journal of Philosophy* **103** (2006), 481–502.

[272] Dorothy Edgington, 'Williamson on Iterated Attitudes', pp. 135–158 in T. J. Smiley (ed.), *Philosophical Logic* (*Procs. of the British Academy* **95**), Oxford University Press, Oxford 1998.

[273] Andy Egan and Brian Weatherson (eds.), *Epistemic Modality*, Oxford University Press, New York 2011.

[274] L. Esakia and V. Meskhi, 'Five Critical Modal Systems', *Theoria* **43** (1977), 52–60.

[275] Gareth Evans, 'Semantic Structure and Logical Form', pp. 199–222 in G. Evans and J. McDowell (eds.), *Truth and Meaning*, Oxford University Press, Oxford 1976.

[276] Gareth Evans, *The Varieties of Reference*, Oxford University Press, Oxford 1982.

[277] A. C. Ewing, *Ethics*, English Universities Press, London 1953.

[278] Ronald Fagin and Joseph Y. Halpern, 'Belief, Awareness, and Limited Reasoning', *Artificial Intelligence* **34** (1987), 39–76.

[279] Ronald Fagin, Joseph Y. Halpern, Moshe Y. Vardi, 'What is an Inference Rule?', *Journal of Symbolic Logic* **57** (1992), 1018–1045.

[280] Ronald Fagin, Joseph Y. Halpern, Yoram Moses and Moshe Y. Vardi, *Reasoning About Knowledge*, MIT Press, Cambridge, MA 1995. (Paperback Edn., slightly revised, 2003.)

[281] Pedro Alonso Amaral Falcão, 'Aspectos da Teoria de Funções Modais', Master's Thesis, University of São Paolo 2012.

[282] Jie Fan, Yanjing Wang, and Hans van Ditmarsch, 'Contingency and Knowing Whether', *Review of Symbolic Logic* **8** (2015), 75–107.

[283] Michael Fara, 'Knowability and the Capacity to Know', *Synthese* **173** (2010), 53–73.

[284] N. Feit and A. Capone (eds.) *Attitudes De Se: Linguistics, Epistemology, Metaphysics*, CSLI Publications, Stanford CA 2013.

[285] T. M. Ferguson, 'Logics of Nonsense and Parry Systems', *Journal of Philosophical Logic* **44** (2015), 65–80.

[286] Robert Feys, 'A Simplified Proof of the Reduction of All Modalities to 42 in S3', *Boletın de la Sociedad Matemática Mexicana* **10** (1953), 53–57.

[287] Kit Fine, 'Propositional Quantifiers in Modal Logic', *Theoria* **36** (1970), 336–346.

[288] Kit Fine, Review of Rescher and Urquart [951], *Philosophical Quarterly* **22** (1972), 370–371.

[289] Kit Fine, 'Some Connections Between Elementary and Modal Logic', pp. 15–31 in S. Kanger (ed.), *Proceedings of the Third Scandinavian Logic Symposium*, North-Holland, Amsterdam 1973.

[290] Kit Fine, 'Logics Containing $K4$: Part I', *Journal of Symbolic Logic* **39** (1974), 31–42.

[291] Kit Fine, 'Normal Forms in Modal Logic', *Notre Dame Journal Formal Logic* **16** (1975), 229–237.

[292] Kit Fine, 'Postscript' (= Chapter 8) in Prior and Fine [930].

[293] Kit Fine, *Reasoning With Arbitrary Objects*, Basil Blackwell, Oxford 1985.

[294] Kit Fine, 'Analytic Implication', *Notre Dame Journal of Formal Logic* **27** (1986), 169–179.

[295] Kit Fine, 'Essence and Modality', *Philosophical Perspectives* **8** (1994), 1–16.

[296] Kit Fine, 'Semantics for the Logic of Essence', *Journal of Philosophical Logic* **29** (2000), 543–584.

[297] Kit Fine, 'Aristotle's Megarian Manoeuvres', *Mind* **120** (2011), 993–1034.

[298] Kit Fine, 'Permission and Possible Worlds', *Dialectica* **68** (2014), 317–336.

[299] Kit Fine, 'Angellic Content', *Journal of Philosophical Logic*, to appear. (Online first: 2015.)

[300] Stephen Finlay, 'What *Ought* Probably Means, and Why You Can't Detach it', *Synthese* **177** (2010), 67–89.

[301] Stephen Finlay and Justin Snedegar, 'One Ought Too Many', *Philosophy and Phenomenological Research* **89** (2014), 102–124

[302] M. Oreste Fiocco 'Conceivability and Epistemic Possibility', *Erkenntnis* **67** (2007), 387–399.

[303] P. C. Fishburn, 'Transitivity', *Review of Economic Studies* **46** (1979), 163–173.

[304] Mark Fisher, 'A Three-Valued Calculus for Deontic Logic', *Theoria* **27** (1961), 107–118.

[305] Mark Fisher, 'A System of Deontic-Alethic Modal Logic', *Mind* **71** (1962), 231–236.

[306] Frederic B. Fitch, *Symbolic Logic: An Introduction*, The Ronald Press Company, NY 1952.

[307] Frederic B. Fitch, 'A Logical Analysis of Some Value Concepts', *Journal of Symbolic Logic* **28** (1963), 135–142.

[308] Frederic B. Fitch, 'Natural Deduction Rules for Obligation', *American Philosophical Quarterly* **3** (1966), 27–38.

[309] Melvin Fitting, 'An Embedding of Classical Logic in S4', *Journal of Symbolic Logic* **35** (1970), 529–534.

[310] Melvin Fitting, *Proof Methods for Modal and Intuitionistic Logics*, Reidel, Dordrecht 1983.

[311] Melvin Fitting, 'On Quantified Modal Logic', *Fundamenta Informaticae* **39** (1999), 105–121.

[312] Melvin Fitting, 'A Logic of Explicit Knowledge', pp.11–22 in L. Běhounek and M. Bílková (eds.), *The Logica Yearbook 2004*, Filosofia, Prague 2005.

[313] A. M. Flescher, *Heroes, Saints, and Ordinary Morality*, Georgetown University Press, Washington D.C. 2003.

[314] Luciano Floridi, *The Philosophy of Information*, Oxford University Press, Oxford 2011.

[315] Marina Folescu and James Higginbotham, 'Two Takes on the *De Se*', pp. 46–61 in S. Prosser and F. Recanati (eds.), *Immunity to Error Through Misindentification: New Essays*, Cambridge Universitiy Press, Cambridge 2012.

[316] D. Føllesdal and R. Hilpinen, 'Deontic Logic: An Introduction', pp. 1–35 in [481].

[317] Josep M. Font, and P. Hájek 'On Łukasiewicz's Four-Valued Modal Logic', *Studia Logica* **70** (2002), 157–182.

[318] Josep M. Font and R. Jansana, 'Leibniz Filters and the Strong Version of a Protoalgebraic Logic', *Archive for Mathematical Logic* **40** (2001), 437–465.

[319] Graeme Forbes, *The Metaphysics of Modality*, Oxford University Press, Oxford 1985.

[320] Graeme Forbes, *Languages of Possibility*, Basil Blackwell, Oxford 1989.

[321] Graeme Forbes, 'Comparatives in Counterpart Theory: Another Approach', *Analysis* **54** (1994), 37–42.

[322] Graeme Forbes, *Modern Logic*, Oxford University Press, New York 1994.

[323] Graeme Forbes, *Attitude Problems: An Essay On Linguistic Intensionality*, Oxford University Press, Oxford 2006.

[324] Peter Forrest, 'The Logic of Naturalness', *Logique et Analyse* **30** (1987), 91–102.

[325] J. W. Forrester, 'Gentle Murder, or the Adverbial Samaritan', *Journal of Philosophy* **81** (1984), 93–197.

[326] J. W. Forrester, *Being Good and Being Logical: Philosophical Groundwork for a New Deontic Logic*, M. E. Sharpe, Armonk, NY 1996.

[327] Bas van Fraassen, *Formal Semantics and Logic*, Macmillan, London 1971.

[328] Bas van Fraassen, 'The Logic of Conditional Obligation', *Journal of Philosophical Logic* **1** (1972), 417–438.

[329] Bas van Fraassen, 'Values and the Heart's Command', *Journal of Philosophy* **70** (1973), 5–19.

[330] Z. Frajzyngier with M. Bond, L. Heintzelman, D. Keller, S. Ogihra and E. Shay, 'Towards an Understanding of the Progressive Form in English', pp. 81–96 in W. Abraham and E. Leiss (eds.), *Modality-Aspect Interfaces*, John Benjamins, Amsterdam 2008.

[331] Gottlob Frege, 'On Sense and Reference', pp. 56–78 in P. T. Geach and M. Black (eds.), *Translations from the Philosophical Writings of Gottlob Frege*, Basil Blackwell, Oxford 1970. (Original German publication: 1892.)

[332] Rohan French, *Translational Embeddings in Modal Logic*, PhD Thesis, Monash University 2010.

[333] Rohan French, 'An Argument against General Validity?', *Thought* **1** (2012), 4–9.

[334] Rohan French and Lloyd Humberstone, 'Partial Confirmation of a Conjecture on the Boxdot Translation in Modal Logic', *Australasian Journal of Logic* **7** (2009), 56–61.

[335] Elizabeth Fricker, 'Is Knowing a State of Mind? The Case Against', pp. 31–59 in P. Greenough and D. Pritchard (eds.), *Williamson on Knowledge*, Oxford University Press, Oxford 2009.

[336] Peter Fritz, 'Post Completeness in Congruential Modal Logics', in preparation 2016.

[337] Kim Frost, 'On the Very Idea of Direction of Fit', *Philosophical Review* **123** (2014), 429–484.

[338] Dov M. Gabbay, *Investigations in Modal and Tense Logics with Applications to Problems in Philosophy and Linguistics*, Reidel, Dordrecht 1976.

[339] Dov M. Gabbay, 'An Irreflexivity Lemma with Applications to Axiomatizations of Conditions on Tense Frames', pp. 67–89 in [820].

[340] Dov M. Gabbay, 'Expressive Functional Completeness in Tense Logic', pp. 91–117 in [820].

[341] Dov M. Gabbay, *Fibring Logics*, Oxford University Press, Oxford 1999.

[342] Dov M. Gabbay, *Reactive Kripke Semantics*, Springer-Verlag, Berlin 2013.

[343] Dov M. Gabbay and Larisa Maksimova, *Interpolation and Definability: Modal and Intuitionistic Logics*, Oxford University Press, Oxford 2005.

[344] Dov M. Gabbay, V. B. Shehtman and D. P. Skvortsov (eds.), *Quantification in Nonclassical Logic, Vol. 1*, Eslevier, Amsterdam 2009.

[345] Denis Gailor, 'Reflections on "should", "ought to", and "must"', *ELT Journal* **37** (1983), 346–349.

[346] R. D. Gallie, 'A Correction to Lemmon on S5', *Analysis* **28** (1968), 128–130.

[347] Antony Galton, *The Logic of Aspect*, Oxford University Press, Oxford 1984.

[348] Manuel García-Carpintero and Max Kölbel (eds.), *Relative Truth*, Oxford University Press, Oxford 2008.

[349] Peter Gärdenfors, 'Qualitative Probability as an Intensional Logic', *Journal of PHilosophical Logic* **4** (1975), 171–185.

[350] James W. Garson, 'Two New Interpretations of Modality', *Logique et Analyse* **15** (1972), 443–459.

[351] James W. Garson, *Modal Logic for Philosophers* (Second edn.), Cambridge University Press, Cambridge 2014 (First edn. 2006).

[352] Douglas Gasking, 'Mathematics and the World', *Australasian Journal of (Psychology and) Philosophy* **18** (1940), 97–116; reprinted as pp. 204–221 in A. G. N. Flew (ed.), *Logic and Language, Second Series*, Basil Blackwell, Oxford 1966.

[353] P. T. Geach, 'On Beliefs About Oneself', *Analysis* **18** (1957), 23–24.

[354] P. T. Geach, 'Intentional Identity', *Journal of Philosophy* **64** (1967), 627–632.

[355] P. T. Geach, *Reference and Generality*, Emended Edn., Cornell University Press, Ithaca, NY 1968. (First Edn., 1962.)

[356] P. T. Geach, *God and the Soul*, Routledge and Kegan Paul, London 1969.

[357] P. T. Geach, 'Omnipotence', *Philosophy* **48** (1973), 7–20.

[358] P. T. Geach, 'Two Kinds of Intentionality', *Monist* **59** (1976), 306–320.

[359] P. T. Geach, 'Intentionality of Thought versus Intentionality of Desire', *Grazer Philosophische Studien* **5** (1978), 131–138.

[360] P. T. Geach, 'Comparatives', *Philosophia* **13** (1983), 235–246.

[361] Ken Gemes, 'A New Theory of Content, I: Basic Content', *Journal of Philosophical Logic* **23** (1994), 595–620.

[362] Ken Gemes, 'A New Theory of Content, II: Model Theory and Some Alternatives', *Journal of Philosophical Logic* **26** (1997), 449–476.

[363] Harry J. Gensler, 'Acting Commits One to Ethical Beliefs', *Analysis* *43* (1983), 40–43.

[364] Harry J. Gensler, 'Paradoxes of Subjective Obligation', *Metaphilosophy* **18** (1987), 208–213.

[365] G. N. Georgacarakos, 'Semantics for S4.04, S4.4, and S4.3.2', *Notre Dame Journal of Formal Logic* **17** (1976), 297–302.

[366] B. R. George, 'Knowing-'wh', Mention-Some Readings, and Non-Reducibility', *Thought* **2** (2013), 166–177.

[367] Jelle Gerbrandy 'The Surprise Examination', *Synthese* **155** (2007), 21–33.

[368] G. Gerla, 'Transformational Semantics for First Order Logic', *Logique et Analyse* **30** (1987), 69–79.

[369] M. S. Gerson, 'The Inadequacy of the Neighbourhood Semantics for Modal Logic', *Journal of Symbolic Logic* **40** (1975), 141–148.

[370] Edmund Gettier, 'Is Justified True Belief Knowledge?', *Analysis* **23** (1963), 121–123.

[371] S. Ghilardi and M. Zawadowski, 'Undefinability of Propositional Quantifiers in the Modal System S4', *Studia Logica* **55** (1995), 259–271.

[372] Anastasia Giannakidou, *Polarity Sensitivity as (Non)veridical Dependency*, John Benjamins, Amsterdam 1998.

[373] Allan Gibbard, *Wise Choices, Apt Feelings*, Harvard University Press, Cambridge, MA 1990.

[374] Pablo Gilabert, 'The Feasibility of Basic Socioeconomic Human Rights: A Conceptual Exploration', *Philosophical Quarterly* **59** (2009), 659–681.

[375] S. G. Gindikin, *Algebraic Logic* (transl. R. H. Silverman), Springer Verlag, Berlin 1985. (Russian orig.: 1972.)

[376] Alessandro Giordani, 'A New Sementics for Systems of Logic of Essence', *Studia Logica* **102** (2014), 411–440.

[377] Rod Girle, 'Epistemic Logic, Language and Concepts', *Logique et Analyse* **16** (1973), 359–373.

[378] Rod Girle, *Modal Logics and Philosophy*, Acumen, Teddington, UK 2000. Second edition 2009.

[379] Rod Girle, *Possible Worlds*, Acumen, Chesham, UK 2003.

[380] Ephraim Glick, 'Abilities and Know-How Attributions', pp. 120–139 in J. Brown and M. Gerken (eds.), *Knowledge Ascriptions*, Oxford University Press, Oxford 2012.

[381] Lou Goble, 'The Iteration of Deontic Modalities', *Logique et Analyse* **9** (1966), 197–209.

[382] Lou Goble, 'The Logic of Obligation, "Better" and "Worse"', *Philosophical Studies* **70** (1993), 134–163.

[383] Lou Goble, 'Utilitarian Deontic Logic', *Philosophical Studies* **82** (1996), 317–357.

[384] Lou Goble, 'The Andersonian Reduction and Relevant Deontic Logic', pp. 213–246 in B. Brown and J. Woods (eds.), *New Studies in Exact Philosophy: Logic, Mathematics and Science*, Hermes Science Publishers, Oxford 2001.

[385] Lou Goble (ed.), *The Blackwell Guide to Philosophical Logic*, Blackwell, Oxford 2001.

[386] Lou Goble, 'A Proposal for Dealing with Deontic Dilemmas', pp. 74–113 in A. Lomuscio and D. Nute (eds.), *DEON 2004*, Lecture Notes in Artificial Intelligence #3065, Springer-Verlag, Berlin 2004.

[387] Lou Goble, 'A Logic for Deontic Dilemmas', *Journal of Applied Logic* **3** (2005), 461–483.

[388] Lou Goble, 'Normative Conflicts and the Logic of "Ought"', *Noûs* **43** (2009), 450–489.

[389] Lou Goble, 'Deontic Logic (Adapted) for Normative Conflicts', *Logic Journal of the IGPL* **22** (2013), 206–235.

[390] Paul Gochet and Eric Gillet, 'On Professor Weingarner's Contribution to Epistemic Logic', pp. 97–115 in [1015].

[391] Robert Goldblatt, 'A Study of $\mathcal{Z}$ Modal Systems', *Notre Dame Journal of Formal Logic* **15** (1974), 289–294.

[392] Robert Goldblatt, 'The McKinsey Axiom is Not Canonical', *Journal of Symbolic Logic* **56** (1991), 554–562. Reproduced as Chapter 10 of Goldblatt [394].

[393] Robert Goldblatt, *Logics of Time and Computation*, Center for the Study of Language and Information, Stanford, Second edition 1992.

[394] Robert Goldblatt, *Mathematics of Modality*, CSLI Publications, Stanford CA 1993.

[395] Robert Goldblatt, 'Mathematical Modal Logic: A View of its Evolution', *Journal of Applied Logic* **1** (2003), 309–392.

[396] Robert Goldblatt and E. D. Mares, 'A General Semantics for Quantified Modal Logic', pp. 227–246 in G. Governatori, I. Hodkinson and Y. Venema (eds.), *Advances in Modal Logic, Vol. 6*, College Publications, London 2006.

[397] Robert Goldblatt, *Quantifiers, Propositions, and Identity: Admissible Semantics for Quantified Modal and Substructural Logics*, Cambridge University Press, Cambridge 2011.

[398] Robert Goldblatt and Ian Hodkinson, 'The McKinsey–Lemmon Logic is Barely Canonical', *Australasian Journal of Logic* **5** (2007), 1–19.

[399] Robert Goldblatt, Ian Hodkinson, and Yde Venema, 'Erdős Graphs Resolve Fine's Canonicity Problem', *Bulletin of Symbolic Logic* **10** (2004), 186–208.

[400] Holly S. Goldman, 'Dated Rightness and Moral Imperfection', *Philosophical Review* **85** (1976), 449–487.

[401] Rajeev Goré, 'Tableau Methods for Modal and Temporal Logics', pp. 297–396 in M. D' Agostino, D. M. Gabbay, R. Hahnle and J. Posegga (eds.), *Handbook of Tableau Methods*, Kluwer, Dordrecht 1999.

[402] Rajeev Goré and Linh Anh Nguyen, 'A Tableau Calculus with Automaton-Labelled Formulae for Regular Grammar Logics', pp. 138–152 in B. Beckert (ed.), *TABLEAUX 2005*, Springer-Verlag, Berlin 2005.

[403] Dana Goswick, 'Why Being Necessary Really is Not the Same as Being Not Possibly Not', *Acta Analytica* **30** (2015), 267–274.

[404] Dana Goswick, 'Does Standard Modal Logic Adequately Represent Metaphysics?', unpublished paper 2015.

[405] John Greco, 'Better Safe than Sensitive', pp. 192–206 in K. Becker and T. Black, *The Sensitivity Principle in Epistemology*, Cambridge Univesity Press, Cambridge 2012.

[406] C. Green, 'A Note on Partition-Inducing Relations', *American Mathematical Monthly* **74** (1967), p. 580.

[407] J. A. Green, *Sets and Groups*, Routledge and Kegan Paul, London 1965.

[408] Patrick Greenough, 'Discrimination and Self-Knowledge', pp. 329–349 in [1067].

[409] Mitchell Green and John N. Williams (eds.), *Moore's Paradox: New Essays on Belief, Rationality, and the First Person*, Oxford University Press, Oxford 2007.

[410] Patricia Greenspan, 'Conditional Oughts and Hypothetical Imperatives', *Journal of Philosophy* **72** (1975), 259–276.

[411] Dominic Gregory, 'Iterated Modalities, Meaning and A Priori Knowledge', *Philosophers' Imprint* **11** (2011), 1–11.

[412] H. P. Grice, 'Logic and Conversation', William James Lectures, Harvard University 1967, published as Part 1 of Grice, *Studies in the Way of Words*, Harvard University Press, Cambridge, MA 1989.

[413] Thomas R. Grimes, 'Truth, Content, and the Hypothetico-Deductive Method', *Philosophy of Science* **57** (1990), 514–522.

[414] J. Groenendijk and M. Stokhof, 'Semantic Analysis of "Wh"-Complements', *Linguistics and Philosophy* **5** (1982), 175–233.

[415] Ghislain Guigon, 'Bringing About and Conjunction: A Reply to Bigelow on Omnificence', *Analysis* **29** (2009), 452–458.

[416] Elena Guerzoni, 'Weak Exhautstivity and *Whether*: A Pragmatic Approach', pp. 112–119 in T. Friedman and M. Gibson (eds.), *SALT XVII* (i.e., proceedings of the 17th annual 'Semantics and Linguistic Theory' Conference) Cornell University Press, Ithaca, NY 2007.

[417] David S. Gunderson, *Handbook of Mathematical Induction: Theory and Applications*, CRC Press (Taylor and Francis), Boca Raton 2011.

[418] Ferdinand de Haan, 'Typological Approaches to Modality', pp. 27–69 in W. Frawley, E. Eschenroeder, S. Mills and T. Nguyen (eds.), *The Expression of Modality*, Mouton de Gruyter, Berlin 2005.

[419] Ian Hacking, 'Possibility', *Philosophical Review* **76** (1967), 143–168.

[420] Ian Hacking, 'All Kinds of Possibility', *Philosophical Review* **84** (1975), 321–337.

[421] Adrian Haddock and Fiona Macpherson (eds.), *Disjunctivism – Perception, Action, Knowledge*, Oxford University Press, Oxford 2008.

[422] T. Hailperin, Reviews of articles by Daya and Bhattacharyya, *Journal of Symbolic Logic* **24** (1959), 185–186.

[423] Alan Hájek and Daniel Stoljar, 'Crimmins, Gonzales and Moore', *Analysis* **61** (2000), 208–213.

[424] R. Hakli and S. Negri, 'Does the Deduction Theorem Fail for Modal Logic?' *Synthese* **187** (2012), 849–867.

[425] Bob Hale and Aviv Hoffmann, *Modality: Metaphysics, Logic, and Epistemology*, Oxford University Press, Oxford 2010.

[426] Bob Hale and Jessica Leech, 'Relative Necessity Reformulated', to appear, *Journal of Philosophical Logic*.

[427] Joseph Y. Halpern, 'Should Knowledge Entail Belief?', *Journal of Philosophical Logic* **25**, (1996) 483–494.

[428] Joseph Y. Halpern and Y. Moses, 'Towards a Theory of Knowledge and Ignorance', pp. 459–476 in K. Apt (ed.), *Logics and Models of Concurrent Systems*, Springer-Verlag, Berlin 1985.

[429] Joseph Y. Halpern and R. Pucella, 'Dealing with Logical Omniscience: Expressiveness and Pragmatics', *Artificial Intelligence* **175** (2011), 220–235.

[430] Joseph Y. Halpern, Dov Samet and Ella Segev, 'On Definability in Multimodal Logic', *Review of Symbolic Logic* **2** (2009), 451–468.

[431] Joseph Y. Halpern, Dov Samet and Ella Segev, 'Defining Knowledge in Terms of Belief: the Modal Logic Perspective', *Review of Symbolic Logic* **2** (2009), 469–482.

[432] Charles L. Hamblin, 'The Modal "Probably"', *Mind* **68** (1959), 234–240.

[433] Toby Handfield, *A Philosophical Guide to Chance*, Cambridge University Press, Cambridge 2012.

[434] Oswald Hanfling, 'How is Scepticism Possible?', *Philosophy* **62** (1987), 435–453.

[435] Jörg Hansen, 'Sets, Sentences, and Some Logics about Imperatives', *Fundamenta Informaticae* **48** (2001), 205–226.

[436] Jörg Hansen, 'Problems and Results for Logics about Imperatives', *Journal of Applied Logic* **2** (2004), 39–61.

[437] Jörg Hansen, 'Conflicting Imperatives and Dyadic Deontic Logic', *Journal of Applied Logic* **3** (2005), 484–511.

[438] Jörg Hansen, 'Reasoning About Permission and Obligation', pp. 287–333 in S. O. Hansson (ed.), David Makinson on Classical Methods for Non-Classical Problems, Springer-Verlag, Dordrecht 2014.

[439] William H. Hanson, 'Semantics for Deontic Logic', *Logique et Analyse* **8** (1965), 177–190.

[440] William H. Hanson, 'Actuality, Necessity, and Logical Truth', *Philosophical Studies* **130** (2006), 437–459.

[441] William H. Hanson, 'Logical Truth in Modal Languages: Reply to Nelson and Zalta', *Philosophical Studies*, **167** (2014), 327–339.

[442] B. Hansson, 'Deontic Logic and Different Levels of Generality', *Theoria* **36** (1970), 241–248.

[443] B. Hansson, 'The Dependency of Deontic Logic upon the General Theory of Decision', pp. 74–81 in A. G. Conte, G. H. von Wright and R. Hilpinen (eds.), *Deontische Logik und Semantik.* (*Linguistische Forschungen* series, #15) Athenaion, Wiesbaden 1977.

[444] B. Hansson and P. Gärdenfors, 'A Guide to Intensional Semantics', pp. 151–167 in *Modality, Morality and Other Problems of Sense and Nonsense (Essays Dedicated to Sören Halldén)*, Lund 1973.

[445] S. O. Hansson, 'Deontic Logic Without Misleading Alethic Analogies, Part 1', *Logique et Analyse* **31** (1988), 337–353.

[446] S. O. Hansson, 'A Note on the Deontic System DL of Jones and Pörn', *Synthese* **80** (1989), 427–428.

[447] S. O. Hansson, 'Preference-Based Deontic Logic', *Journal of Philosophical Logic* **19** (1990), 75–93.

[448] S. O. Hansson, 'Semantics for More Plausible Deontic Logics', *Journal of Apllied Logic* **2** (2004), 3–18.

[449] S. O. Hansson, 'Ideal Worlds – Wishful Thinking in Deontic Logic', *Studia Logica* **82** (2006), 329–326.

[450] S. O. Hansson, 'Logic of Belief Revision', *Stanford Encyclopedia of Philosophy* (Fall 2011 Edition), Edward N. Zalta (ed.), URL = <http://plato.stanford.edu/archives/fall2011/entries/logic-belief-revision/>.

[451] Frank Harary 'A Very Independent Axiom System', *American Mathematical Monthly* **68** (1961), 159–162.

[452] Frank Harary, 'A Parity Relation Partitions its Field Distinctly', *American Mathematical Monthly* **68** (1961), 215–217.

[453] R. M. Hare, *The Language of Morals*, Oxford University Press, Oxford 1952.

[454] Jonathan Harrison, 'More Deviant Logic', *Philosophy* **53** (1978), 21–32.

[455] Jonathan Harrison, 'If I Know, I Cannot Be Wrong', *Procs. of the Aristotelian Society* **79** (1979), 137–150.

[456] H. L. A. Hart, 'Legal and Moral Obligation', pp. 82–107 in Melden [798].

[457] S. Hart, A. Heifetz, and D. Samet, '"Knowing Whether," "Knowing That," and the Cardinality of State Spaces', *Journal of Economic Theory* **70** (1996), 249–256.

[458] J. Hartland-Swann, 'The Logical Status of "Knowing That"', *Analysis* **16** (1956), 111–115.

[459] Charles Hartshorne, 'The Logic of the Ontological Argument', *Journal of Philosophy* **58** (1961), 471–473.

[460] Charles Hartshorne, *The Logic of Perfection*, Open Court, La Salle, IL 1962.

[461] John Hawthorne, *Knowledge and Lotteries*, Oxford University Press, Oxford 2004.

[462] John Hawthorne, 'The Case for Closure', pp. 26–43 in Steup and Sosa [1090].

[463] Reina Hayaki, 'Contingent Objects and the Barcan Formula', *Erkenntnis* **64** (2006), 75–83.

[464] Allen P. Hazen and Lloyd Humberstone, 'Similarity Relations and the Preservation of Solidity', *Journal of Logic, Language and Information* **13** (2004), 25–46.

[465] Allen P. Hazen, Benjamin G. Rin, Kai F. Wehmeier, 'Actuality in Propositional Modal Logic', *Studia Logica* **101** (2013), 487–503.

[466] Allan Hazlett, 'The Myth of Factive Verbs', Philosophy and Phenomenological Research 80 (2010), 497–522.

[467] Allan Hazlett, 'Factive Presupposition and the Truth Condition', *Acta Analytica* **27** (2012), 461–478.

[468] Adrian Heathcote, '**KT** and the Diamond of Knowledge', *Analytic Philosophy* (formerly *Philosophical Books*) 2004, 286–295.

[469] Irene Heim, 'Concealed Questions', pp. 51–60 in Rainer Bäuerle, Urs Egli and Arnim von Stechow (eds.), *Semantics From Different Points of View*, Springer-Verlag, Berlin 1979.

[470] Paul Helm, 'Are "Cambridge" Changes Non-events?', *Analysis* **35** (1975), 140–144.

[471] Vincent F. Hendricks, *Mainstream and Formal Epistemology*, Cambridge University Press, Cambridge 2006.

[472] H. E. Hendry and G. J. Massey, 'On the Concepts of Sheffer Functions', pp. 279–293 in K. Lambert (ed.), *The Logical Way of Doing Things*, Yale University Press, New Haven and London 1969.

[473] D. P. Henry, 'St Anselm on the Varieties of "Doing"', *Theoria* **19** (1953), 178–183.

[474] D. P. Henry, *The Logic of Saint Anselm*, Oxford University Press, Oxford 1967.

[475] Stephen Hetherington, 'How to Know (that Knowledge-that is Knowledge-how)', pp. 71–94 in S. Hetherington (ed.), *Epistemology Futures*, Oxford University Press, Oxford 2006.

[476] Stephen Hetherington, 'Gettier Problems', *Internet Encyclopedia of Philosophy*, http://www.iep.utm.edu/ (as of January 1, 2015).

[477] David Heyd, *Supererogation*, Cambridge University Press, Cambridge 1982.

[478] James Higginbotham, 'Remembering, Imagining, and the First Person', pp. 496–533 in A. Barber (ed.), *Epistemology of Language*, Oxford University Press, Oxford 2003.

[479] Daniel J. Hill, Review of Weingartner [1163], *Notre Dame Philosophical Reviews* 2008 at http://ndpr.nd.edu/news/23850-omniscience-from-a-logical-point-of-view/

[480] Risto Hilpinen, 'Knowing That One Knows and the Classical Definition of Knowledge', *Synthese* **21** (1970), 109–132.

[481] Risto Hilpinen (ed.), *Deontic Logic: Introductory and Systematic Readings*, Reidel, Dordrecht 1971.

[482] Risto Hilpinen, 'Remarks on Personal and Impersonal Knowledge', *Canadian Journal of Philosophy* **7** (1977), 1–9.

[483] Risto Hilpinen (ed.), *New Studies in Deontic Logic*, Reidel, Dordrecht 1981.

[484] Risto Hilpinen, 'Disjunctive Permissions and Conditionals with Disjunctive Antecedents', pp. 175–194 in I. Niiniluoto and E. Saarinen (eds.), *Intensional Logic: Theory and Applications* (= *Acta Philosophica Fennica* **35**), Helsinki 1982.

[485] Risto Hilpinen, 'Actions in Deontic Logic', pp. 85–100 in J.-J. C. Meyer and R. J. Wieringa (eds.), *Deontic Logic in Computer Science: Normative System Specification*, Wiley, Chichester 1993.

[486] Risto Hilpinen, 'On Action and Agency', pp. 3–27 in E. Ejerhed and S. Lindström (eds.), *Logic, Action and Cognition – Essays in Philosophical Logic*, Kluwer, Dordrecht 1997.

[487] Risto Hilpinen, 'Deontic Logic', Chapter 8 (pp. 159–182) in Goble [385].

[488] Jaakko Hintikka, "Identity, Variables, and Impredicative Definitions", *Journal of Symbolic Logic* **21** (1956), pp. 225–245.

[489] Jaakko Hintikka, 'Quantifiers in Deontic Logic', *Societas Scientiarum Fennica (Commentationes Humanarum Litterarum)* **23** (1957) #4.

[490] Jaakko Hintikka, *Knowledge and Belief: an Introduction to the Logic of the Two Notions*, Cornell University Press, Ithaca, NY 1962.

[491] Jaakko Hintikka, 'On the Logic of Perception', pp. 152–183 in Hintikka, *Models For Modalities*, D. Reidel, Dordrecht 1969.

[492] Jaakko Hintikka, 'On Attributions of "Self-Knowledge" ' *Journal of Philosophy* **67** (1970), 73–87.

[493] Jaakko Hintikka, '"Prima Facie" Obligations and Iterated Modalities', *Theoria* **36** (1970), 232–240.

[494] Jaakko Hintikka, 'Some Main Problems of Deontic Logic', pp. 59–104 in [481].

[495] Jaakko Hintikka, 'Different Constructions in Terms of the Basic Epistemological Terms: A Survey of Some Problems and Proposals', pp. 105–122 in R. E. Olson and A. M. Paul (eds.), *Contemporary Philosophy in Scandinavia*, Johns Hopkins Press, Baltimore 1972.

[496] Jaakko Hintikka, *Logic, Language Games and Information*, Oxford University Press, Oxford 1973.

[497] Jaakko Hintikka, 'Surface Semantics: Definition and Its Motivation', pp. 128–147 in H. Leblanc (ed.), *Truth, Syntax and Modality*, North-Holland, Amsterdam 1973.

[498] Jaakko Hintikka, 'Knowledge, Belief, and Logical Consequence', Chapter 9 of Hintikka, *The Intensions of Intensionality and Other New Models for Modalities*, D. Reidel, Dordrecht 1975.

[499] Jaakko Hintikka, 'Is Alethic Modal Logic Possible?', pp. 89–105 in I. Niiniluoto and E. Saarinen (eds.), *Intensional Logic: Theory and Applications* (= Acta Philosophica Fennica **35**), Helsinki 1982.

[500] Jaakko Hintikka, 'Different Constructions in Terms of "Know" ', pp. 99–104 in J. Dancy and E. Sosa (eds.), *A Companion to Epistemology*, Blackwell, Oxford 1992.

[501] Jaakko Hintikka, 'A Second Generation Epistemic Logic and Its General Significance', pp. 33–55 in V. F. Hendricks, K. F. Jørgensen and S. A. Pedersen (eds.), *Knowledge Contributors*, Kluwer, Dordrecht 2003.

[502] H.-J. Hippler and N. Schwartz, 'Not Forbidding isn't Allowing: the Cognitive Basis of the Forbid–Allow Asymmetry', *Public Opinion Quarterly* **50** (1986), 87–96.

[503] H. Hiż, 'A Warning about Translating Axioms', *American Math. Monthly* **65** (1958), 613–614.

[504] Wilfrid Hodges, 'Logical Features of Horn Clauses', pp. 449–503 in D. Gabbay, C. J. Hogger and J. A. Robinson (eds.), *Handbook of Logic in Artificial Intelligence and Logic Programming, Vol. 1: Logical Foundations*, Oxford University Press, Oxford 1993.

[505] Ian Hodkinson and Mark Reynolds, 'Temporal Logic', pp. 655–720 in [97].

[506] Ian Hodkinson, 'On the Priorean Temporal Logic with "Around Now" Over the Real Line', *Journal of Logic and Computation* **24** (2014), 1071–1110.

[507] Ian Hodkinson and Mark Reynolds, 'Separation – Past, Present and Future', pp. 117–142 in S. Artemov, H. Barringer, A. d'Avila Garcez, L. C. Lamb and J. Woods (eds.), *We Will Show Them! Essays in Honour of Dov Gabbay on his 60th Birthday, Vol. 2*, College Publications, London, 2005.

[508] W. van der Hoek, 'Systems for Knowledge and Belief', *Journal of Logic and Computation* **3** (1993), 173–195.

[509] W. van der Hoek, J. Jaspars and E. Thijsse, 'Honesty in Partial Logic', *Studia Logica* **56** (1996), 323–360.

[510] W. van der Hoek, J. Jaspars and E. Thijsse, 'Persistence and Minimality in Epistemic Logic', *Annals of Mathematics and Artificial Intelligence* **27** (1999), 25–47.

[511] W. van der Hoek, J. Jaspars and E. Thijsse, 'Theories of Knowledge and Ignorance', pp. 381–418 in S. Rahman, J. Symons, D. Gabbay and J. P. van Bendegem (eds.), *Logic, Epistemology, and the Unity of Science*, Kluwer, Dordrecht 2003.

[512] W. van der Hoek and J.-J. Meyer, 'Possible Logics for Belief', *Logique et Analyse* **32** (1989), 177–194.

[513] Wesley H. Holliday, 'Epistemic Logic, Relevant Alternatives, and the Dynamics of Context', pp. 109–129 in D. Lassiter and M. Slavkovik (eds.), *ESSLLI Student Sessions, Lecture Notes in Computer Science #7415*, Springer-Verlag, Heidelberg 2012.

[514] Wesley H. Holliday, 'Partiality and Adjointness in Modal Logic', pp. 313–332 in R. Goré, B. Kooi and A. Kurucz (eds.), *Advances in Modal Logic, Vol. 10*, College Publications, London 2014.

[515] Wesley H. Holliday, 'Epistemic Closure and Epistemic Logic I: Relevant Alternatives and Subjunctivism', *Journal of Philosophical Logic*, **44** (2015), 1–62.

[516] Wesley H. Holliday, 'Possibility Frames and Forcing for Modal Logic', Working Papers of the Group in Logic and the Methodology of Science, UC Berkeley, 2015. Available at http://escholarship.org/uc/item/5462j5b6

[517] Wesley H. Holliday, and Thomas F. Icard III, 'Moorean Phenomena in Epistemic Logic', pp. 178–199 in Valentin Goranko and Valentin B. Shehtman (eds.), *Advances in Modal Logic, Vol. 8*, College Publications, London 2010.

[518] Richard Holton, 'Intention Detecting', *Philosophical Quarterly* **43** (1993), 298–318.

[519] Richard Holton, 'Some Telling Examples: A reply to Tsohatzidis', *Journal of Pragmatics* **28** (1997), 625–628.

[520] Richard Holton, 'Primitive Self-Ascription: Lewis on the *De Se*', pp. 399-410 in B. Loewer and J. Schaffer (eds.), *A Companion to David Lewis*, Wiley 2015.

[521] Laurence Horn, *A Natural History of Negation*, University of Chicago Press, Chicago 1989.

[522] Laurence Horn, 'Implicature', pp. 53–66 in [980].

[523] Laurence Horn, 'On the Contrary: Disjunctive Syllogism and Pragmatic Strengthening', pp. 241–265 in A. Koslow and A. Buchsbaum (eds.), *The Road to Universal Logic*, Birkhäuser–Springer, Cham, Switzerland 2015.

[524] N. Hornstein, 'Towards a Theory of Tense', *Linguistic Inquiry* **8** (1977), 521–557.

[525] John F. Horty, 'Moral Dilemmas as Nonmonotonic Logic', *Journal of Philosophical Logic* **23** (1994), 35–65.

[526] John F. Horty, *Agency and Deontic Logic*, Oxford University Press, Oxford 2001.

[527] John F. Horty, 'Reasoning with Moral Conflicts', *Noûs* **37** (2003), 557–605.

[528] Paul Horwich, 'Grünbaum on the Metric of Space and Time', *British Journal of Philosophy of Science* **26** (1975), 199–211.

[529] B. Hösli and G. Jäger, 'About Some Symmetries of Negation', *Journal of Symbolic Logic* **59** (1994), 473–485.

[530] Rodney Huddleston and Geoffrey K. Pullum (with further collaborators), *The Cambridge Grammar of the English Language*, Cambridge University Press, Cambridge 2002.

[531] Michael Huemer, 'Epistemic Possibility', *Synthese* **156** (2007), 119–142.

[532] G. E. Hughes, 'Modal Systems With No Minimal Proper Extensions', *Reports on Mathematical Logic* **6** (1976), 93–98.

[533] G. E. Hughes, 'Equivalence Relations and S5', *Notre Dame Journal of Formal Logic* **21** (1980), 577–584.

[534] G. E. Hughes, 'Every World Can See a Reflexive World', *Studia Logica* **49** (1990), 174–181.

[535] G. E. Hughes and M. J. Cresswell, *An Introduction to Modal Logic*, Methuen, London 1968.

[536] G. E. Hughes and M. J. Cresswell, *A Companion to Modal Logic*, Methuen, London 1984.

[537] G. E. Hughes and M. J. Cresswell, *A New Introduction to Modal Logic*, Routledge, London, 1996.

[538] G. E. Hughes and D. E. Londey, *The Elements of Formal Logic*, Methuen, London 1965.

[539] Lloyd Humberstone, 'Two Sorts of "Ought"s', *Analysis* **32** (1971), 8–11.

[540] Lloyd Humberstone, 'Logic for Saints and Heroes', *Ratio* **16** (1974), 103–114.

[541] Lloyd Humberstone, Review of Davidson and Harman [232], *York Papers in Linguistics* **5** (1975), 195–224.

[542] Lloyd Humberstone, 'An Alternative Account of Bringing About (with a Pinch of Relevance)', delivered to the annual conference of the Australasian Association for Logic, Canberra 1976. Abstracted in *Bulletin of the Section of Logic (Polish Academy of Sciences)* **6** (1977), 144–145, and in *The Relevance Logic Newsletter* **2** (1977), 107–108.

[543] Lloyd Humberstone, 'Two Merits of the Circumstantial Operator Language for Conditional Logics', *Australasian Journal of Philosophy* **56** (1978), 21–24.

[544] Lloyd Humberstone, 'Interval Semantics for Tense Logic: Some Remarks', *Journal of Philosophical Logic* **8** (1979), 171–196.

[545] Lloyd Humberstone, 'From Worlds to Possibilities', *Journal of Philosophical Logic* **10** (1981), 313–339.

[546] Lloyd Humberstone, 'Relative Necessity Revisited', *Reports on Mathematical Logic* **13** (1981), 33–42.

[547] Lloyd Humberstone, 'Scope and Subjunctivity', *Philosophia* **12** (1982), 99–126.

[548] Lloyd Humberstone, 'Necessary Conclusions', *Philosophical Studies* **41** (1982), 321–335.

[549] Lloyd Humberstone, 'The Background of Circumstances', *Pacific Philosophical Quarterly* **64** (1983), 19–34.

[550] Lloyd Humberstone, 'Inaccessible Worlds', *Notre Dame Journal of Formal Logic* **24** (1983), 346–352.

[551] Lloyd Humberstone, 'Karmo on Contingent Non-identity', *Australasian Journal of Philosophy* **61** (1983), 188–191.

[552] Lloyd Humberstone, 'Monadic Representability of Certain Binary Relations', *Bulletin of the Australian Mathematical Society* **29** (1984), 365–375.

[553] Lloyd Humberstone, 'Unique Characterization of Connectives' (Abstract) *Journal of Symbolic Logic* **49** (1984), 1426–1427. (This forms the core of §4.3 [594].)

[554] Lloyd Humberstone, 'The Formalities of Collective Omniscience', *Philosophical Studies* **48** (1985) 401–423.

[555] Lloyd Humberstone, 'Extensionality in Sentence Position', *Journal of Philosophical Logic* **15** (1986), 27–54; also *ibid.* **17**(1988), 221–223, 'The Lattice of Extensional Connectives: A Correction'.

[556] Lloyd Humberstone, 'The Modal Logic of "All and Only"', *Notre Dame Journal of Formal Logic* **28** (1987), 177–188.

[557] Lloyd Humberstone, 'Wanting as Believing', *Canadian Journal of Philosophy* **17** (1987), 49–62.

[558] Lloyd Humberstone, 'Heterogeneous Logic', *Erkenntnis* **29** (1988), 395–435.

[559] Lloyd Humberstone, 'Some Epistemic Capacities', *Dialectica* **42** (1988), 183–200.

[560] Lloyd Humberstone, 'Expressive Power and Semantic Completeness: Boolean Connectives in Modal Logic', *Studia Logica* **49** (1990), 197–214.

[561] Lloyd Humberstone, 'Wanting, Getting, Having', *Philosophical Papers* **19** (1990), 99–118.

[562] Lloyd Humberstone, 'A Study of Some "Separated" Conditions on Binary Relations', *Theoria* **57** (1991), 1–16.

[563] Lloyd Humberstone, 'Two Kinds of Agent-Relativity', *Philosophical Quarterly* **41** (1991), 144–166.

[564] Lloyd Humberstone, 'Direction of Fit', *Mind* **101** (1992), 59–83.

[565] Lloyd Humberstone, 'Zero-Place Operations and Functional Completeness (and the Definition of New Connectives)', *History and Philosophy of Logic* **14** (1993), 39–66.

[566] Lloyd Humberstone, 'Comparatives and the Reducibility of Relations', *Pacific Philosophical Quarterly* **76** (1995), 117–141.

[567] Lloyd Humberstone, 'A Basic System of Congruential-to-Monotone Bimodal Logic and Two of its Extensions', *Notre Dame Journal of Formal Logic* **37** (1996), 602–612.

[568] Lloyd Humberstone, 'Negation by Iteration', *Theoria* **61** (1995), 1–24.

[569] Lloyd Humberstone, 'The Logic of Non-Contingency', *Notre Dame Journal of Formal Logic* **36** (1995), 214–229.

[570] Lloyd Humberstone, 'A Study in Philosophical Taxonomy', *Philosophical Studies* **83** (1996), 121–169.

[571] Lloyd Humberstone, 'Two Types of Circularity', *Philosophy and Phenomenological Research* **57** (1997), 249–280.

[572] Lloyd Humberstone, 'Equivalential Interpolation', pp. 36–53 in K. Segerberg (ed.), *The Goldblatt Variations*, Department of Philosophy (Preprints Series), Uppsala University 1999.

[573] Lloyd Humberstone, 'Contra-Classical Logics', *Australasian Journal of Philosophy* **78** (2000), 437–474.

[574] Lloyd Humberstone, 'Parts and Partitions', *Theoria* **66** (2000), 41–82.

[575] Lloyd Humberstone, 'Invitation to Autoepistemology', *Theoria* **68** (2002), 13–51.

[576] Lloyd Humberstone, 'The Modal Logic of Agreement and Noncontingency', *Notre Dame Journal of Formal Logic* **43** (2002), 95-127.

[577] Lloyd Humberstone, 'False Though Partly True – An Experiment in Logic', *Journal of Philosophical Logic* **32** (2003), 613–665.

[578] Lloyd Humberstone, 'Two-Dimensional Adventures', *Philosophical Studies* **118** (2004), 17–65.

[579] Lloyd Humberstone, 'Archetypal Forms of Inference', *Synthese* **141** (2004), 45–76.

[580] Lloyd Humberstone, 'Yet Another "Choice-of-Primitives" Warning: Normal Modal Logics', *Logique et Analyse* **47** (2004), 395–407.

[581] Lloyd Humberstone, 'Logical Discrimination', pp. 207–228 in Jean-Yves Béziau (ed.), *Logica Universalis: Towards a General Theory of Logic*, Birkhäuser, Basel 2005.

[582] Lloyd Humberstone, 'Modality', pp. 534–614 (= Chapter 20) in F. C. Jackson and M. Smith (eds.), *The Oxford Handbook of Contemporary Philosophy*, Oxford University Press, Oxford and New York 2005.

[583] Lloyd Humberstone, 'For Want of an "And": a Puzzle About Non-conservative Extension', *History and Philosophy of Logic* **26** (2005), 229–266.

[584] Lloyd Humberstone, 'Weaker-to-Stronger Translational Embeddings in Modal Logic', pp. 279–297 in G. Governatori, I. Hodkinson and Y. Venema (eds.), *Advances in Modal Logic, Vol. 6*, College Publications, London 2006.

[585] Lloyd Humberstone, 'Sufficiency and Excess', *Aristotelian Society Supplementary Volume* **80** (2006), 265–320.

[586] Lloyd Humberstone, 'Identical Twins, Deduction Theorems, and Pattern Functions: Exploring the Implicative *BCSK* Fragment of **S5**', *Journal of Philosophical Logic* **35** (2006), 435–487; Erratum, *ibid.* **36** (2007), 249.

[587] Lloyd Humberstone, 'Investigations into a Left-Structural Right-Substructural Sequent Calculus', *Journal of Logic, Language and Information* **16** (2007), 141–171.

[588] Lloyd Humberstone, 'Modal Logic for Other-World Agnostics: Neutrality and Halldén Incompleteness', *Journal of Philosophical Logic* **36** (2007), 1–32.

[589] Lloyd Humberstone, 'Modal Formulas True at Some Point in Every Model', *Australasian Journal of Logic* **6** (2008), 70–82. (http://www.philosophy.unimelb.edu.au/ajl/2008).

[590] Lloyd Humberstone, 'Can Every Modifier be Treated as a Sentence Modifier?', *Philosophical Perspectives* **22** (2008), 241–275.

[591] Lloyd Humberstone, 'Collapsing Modalities', *Note Dame Journal of Formal Logic* **50** (2009), 119–132.

[592] Lloyd Humberstone, 'Smiley's Distinction Between Rules of Inference and Rules of Proof', pp. 107–126 in J. Lear and A. Oliver (eds.), *The Force of Argument: Essays in Honor of Timothy Smiley*, Routledge, NY 2010.

[593] Lloyd Humberstone, 'Variation on a Trivialist Argument of Paul Kabay', *Journal of Logic, Language and Information* **20** (2011), 115–132.

[594] Lloyd Humberstone, *The Connectives*, MIT Press, Cambridge MA 2011.

[595] Lloyd Humberstone, 'On a Conservative Extension Argument of Dana Scott', *Logic Journal of the IGPL* **19** (2011), 241–288.

[596] Lloyd Humberstone, 'Minimally Congruential Contexts: Observations and Questions on Embedding **E** in **K**', *Notre Dame Journal of Formal Logic* **53** (2012), 581–598.

[597] Lloyd Humberstone, 'Replacement in Logic', *Journal of Philosophical Logic* **42** (2013), 49–89.

[598] Lloyd Humberstone, 'Zolin and Pizzi: Defining Necessity from Noncontingency', *Erkenntnis* **78** (2013), 1275–1302.

[599] Lloyd Humberstone, 'Inverse Images of Box Formulas in Modal Logic', *Studia Logica* **101** (2013), 1031–1060.

[600] Lloyd Humberstone, 'Aggregation and Idempotence', *Review of Symbolic Logic* **6** (2013), 680–708.

[601] Lloyd Humberstone, 'Note on Extending Congruential Modal Logics', *Notre Dame Journal of Formal Logic* **57** (2016), 95–103.

[602] Lloyd Humberstone, Review of Garson [351], *Studia Logica*, **104** (2016), 365–379.

[603] Lloyd Humberstone and Timothy Williamson, 'Inverses for Normal Modal Operators', *Studia Logica* **59** (1997), 33–64.

[604] J. F. Humphreys and M. Y. Prest, *Numbers, Groups and Codes*, Cambridge University Press, Cambridge 1989.

[605] Susan Hurley, *Natural Reasons*, Oxford University Press, Oxford 1989.

[606] A. Iacona, 'T × W Epistemic Modality', *Logic & Philosophy of Science* **10** (2012), 3–14.

[607] S. Iatridou, E. Anagnostopoulou and R. Izvorski, 'Observations about the Form and Meaning of the Perfect', pp. 189–238 in M. Kenstowicz (ed.), *Ken Hale: A Life in Language*, MIT Press, Cambridge, MA 2001.

[608] A. Indrzejczak, 'Cut-free Double Sequent Calculus for **S5**', *Logic Journal of the IGPL* **6** (1998), 505–516.

[609] A. Indrzejczak, 'A Survey of Nonstandard Sequent Calculi', *Studia Logica* **102** (2014), 1295–1322.

[610] H.-P. Innala, 'On the Non-Neutrality of Deontic Logic', *Logique et Analyse* **43** (2000), 393–410.

[611] T. Inoué, 'On Compatibility of Theories and Equivalent Translations', *Bulletin of the Section of Logic* **21** (1992), 112–119.

[612] Frank Jackson, 'Is There a Good Argument Against the Incorrigibility Thesis?', *Australasian Journal of Philosophy* **51** (1973), 51–62.

[613] Frank Jackson, 'Internal Conflicts in Desires and Morals', *American Philosophical Quarterly* **22** (1985), 105–114.

[614] Frank Jackson, 'On the Semantics and Logic of Obligation', *Mind* **94** (1985), 177–195. Reprinted pp. 197–219 in Jackson, *Mind, Method and Conditionals: Selected Papers*, Routledge, NY; first publ. 1998.

[615] Frank Jackson, 'Possibilities for Representation and Credence: Two-Space-ism vs. One-Space-ism', pp. 131–143 in [273].

[616] Frank Jackson and J. E. J. Altham (symposiasts), 'Understanding the Logic of Obligation' *Aristotelian Society Supplementary Volume* **62** (1988), pp. 255–283.

[617] Frank Jackson and Robert Pargetter, 'Oughts, Options and Actualism', *Philosophical Review* **95** (1986), 233–255.

[618] Mark Jago, 'Hintikka and Cresswell on Logical Omniscience', *Logic and Logical Philosophy* **15** (2006), 325–354.

[619] J. Jaspars, 'Logical Omniscience', pp. 129–146 in M. de Rijke (ed.), *Diamonds and Defaults*, Kluwer, Dordrecht 1993.

[620] R. E. Jennings, 'A Utilitarian Semantics for Deontic Logic', *Journal of Philosophical Logic* **3** (1974), 445–456.

[621] R. E. Jennings, 'Pseudo-Subjectivism in Ethics', *Dialogue* **13** (1974), 515–518.

[622] R. E. Jennings, 'A Note on the Axiomatisation of Brouwersche Modal Logic', *Journal of Philosophical Logic* **19** (1981), 341–343.

[623] R. E. Jennings, 'Can There Be a Natural Deontic Logic?', *Synthese* **65** (1985), 257–273.

[624] R. E. Jennings, 'The *Or* of Free Choice Permission', *Topoi* **13** (1994), 3–10.

[625] R. E. Jennings and P. K. Schotch, 'Some Remarks on (Weakly) Weak Modal Logics', *Notre Dame Journal of Formal Logic* **22** (1981), 309–314.

[626] E. Jeřábek, 'Cluster Expansion and the Boxdot Conjecture', <arxiv.org/pdf/ 1308.0994.pdf> (2013).

[627] Kent Johnson, 'On the Nature of Reverse Compositionality', *Erkenntnis* **64** (2006), 37–60.

[628] Andrew J. I. Jones and Ingmar Pörn, 'Ideality, Sub-Ideality and Deontic Logic', *Synthese* **65** (1985), 275–290.

[629] Andrew J. I. Jones and Ingmar Pörn, '"Ought" and "Must"', *Synthese* **66** (1986), 89–93.

[630] Andrew J. I. Jones and Ingmar Pörn, 'A Rejoinder to Hansson', *Synthese* **80** (1989), 429–432.

[631] Bjarni Jónsson, 'Varieties of Relation Algebras', *Algebra Universalis* **15** (1982), 273–298.

[632] J. Jørgensen, 'Imperatives and Logic', *Erkenntnis* **7** (1937–1938), 288–296.

[633] Michael Jubien, *Possibility*, Oxford University Press, Oxford 2009.

[634] Reinhard Kahle, 'Modalities Without Worlds', pp. 101–118 in S. Rahman, G. Primiero and M. Marion (eds.), *The Realism-Antirealism Debate in the Age of Alternative Logics*, Springer, Dordrecht 2012.

[635] Timo Kajamies, 'Iterated Modalalities and Modal Voluntarism', *Teorema* **27** (2008), 17–28.

[636] R. Kane, 'The Modal Ontological Argument', *Mind* **93** (1984), 336–350.

[637] Stig Kanger, 'New Foundations for Ethical Theory', pp. 36–58 in [481]; earlier draft circulated 1957.

[638] Tomis Kapitan, 'Perfection and Modality: Charles Hartshorne's Ontological Proof', *International Journal for Philosophy of Religion* **7** (1976), 379–385.

[639] David Kaplan, 'S5 with Quantifiable Propositional Variables' (Abstract), *Journal of Symbolic Logic* **35** (1970), 355.

[640] David Kaplan, 'A Problem in Possible World Semantics', pp. 41–52 in W. Sinnott-Armstrong, D. Raffman and N. Asher (eds.), *Modality, Morality and Belief: Essays in Honor of Ruth Barcan Marcus*, Cambridge University Press, Cambridge.

[641] Mark Kaplan, 'If You Know You Can't Be Wrong', pp. 180–98 in S. C. Hetherington (ed.), *Epistemology Futures*, Oxford University Press, Oxford 2006.

[642] Toomas Karmo, 'Contingent Non-Identity', *Australasian Journal of Philosophy* **61** (1983), 185–187.

[643] Rosanna Keefe, 'When Does Circularity Matter?', *Proceedings of the Aristotelian Soc.* **102** (2002), 275–292.

[644] Andreas Kemmerling, 'Kripke's Principle of Disquotation and the Epistemology of Belief Ascription', *Facta Philosophica* **8** (2006), 119–143.

[645] Gary Kemp, 'The Interpretation of Crossworld Predication', *Philosophical Studies* **98** (2000), 305–320.

[646] Ruth Kempson, *Presupposition and the Delimitation of Semantics*, Cambridge University Press, Cambridge 1975

[647] Anthony Kenny, *Action, Emotion and Will*, Routledge and Kegan Paul, London 1963.

[648] Anthony Kenny, 'Happiness', *Proceedings of the Aristelian Society* **66** (1966), 93–102.

[649] Anthony Kenny, 'Human Ability and Dynamic Modalities', pp. 209–232 in J. Manninen and R. Tuomela (eds.), *Essays on Explanation and Understanding*, Reidel, Dordrecht 1976.

[650] Edward J. Khamara, *Space, Time and Theology in the Leibniz–Newton Controversy*, Ontos Verlag, Frankfurt 2006.

[651] Anna Kibort, 'Modelling "the Perfect", a Category Between Tense and Aspect', pp. 1390–1404 in *Current Issues in Unity and Diversity of Languages*, Linguistic Society of Korea, Seoul 2009.

[652] Charles F. Kielkopf, 'Semantics for a Utilitarian Deontic Logic', *Logique et Analyse* **14** (1971), 783–802.

[653] Charles F. Kielkopf, 'K1 as a Dawson Modelling of A. R. Anderson's Sense of "Ought"', *Notre Dame Journal of Formal Logic* **15** (1974), 402–410.

[654] Charles F. Kielkopf, 'A Completeness Proof for Porte's S_a^0 and S_a', *Logique et Analyse* **25** (1982), 435–441.

[655] Jaegwon Kim, 'Noncausal Connections', *Noûs* **8** (1971), 41–52.

[656] Jeffrey C. King, 'Designating Propositions', *Philosophical Review* **111** (2002), 341–371.

[657] Jeffrey C. King, *The Nature and Structure of Content*, Oxford University Press, Oxford 2007.

[658] Paul Kiparsky and Carol Kiparsky, "Fact", pp. 345–369 in D. Steinberg and L. Jakobovits (eds.), *Semantics: An Interdisciplinary Reader*, Cambridge University Press, Cambridge 1971.

[659] Niko Kolodny and John MacFarlane, 'Ifs and Oughts', *Journal of Philosophy* **107** (2010), 115–143.

[660] Michiro Kondo, 'Solutions for Porte's Conjectures', *Publications of the Research Institute for Mathamtical Sciences, Kyoto University* **23** (1897), 575–582.

[661] Carl R. Kordig, 'Another Ethical Paradox', *Mind* **78** (1969), 598–599.

[662] Stephan Körner, *Experience and Conduct*, Cambridge University Press, Cambridge 1976.

[663] Stephan Körner, 'On Logical Validity and Informal Appropriateness', *Philosophy* **54** (1979), 377–379.

[664] Tomasz Kowalski, 'Perzanowski's Hypothesis Confirmed', *Bulletin of the Section of Logic* **25** (1996), 58–59.

[665] Marcus Kracht, 'Invariant Logics', *Mathematical Logic Quarterly* **48** (2002), 29–50.

[666] Marcus Kracht, 'Modal Consequence Relations', pp. 491–545 in [97].

[667] Angelika Kratzer, 'What "Must" and "Can" Must and Can Mean', *Linguistics and Philosophy* **1** (1977), 337–355; reprinted in [668].

[668] Angelika Kratzer, *Modals and Conditionals: New and Revised Perspectives*, Oxford University Press, Oxford 2012.

[669] Saul A. Kripke, 'Semantical Analysis of Modal Logic I. Normal Modal Propositional Calculi', *Zeitschr. für math. Logik und Grundlagen der Math.* **9** (1963), 67–96.

[670] Saul A. Kripke, 'Semantical Analysis of Modal Logic II. Non-Normal Modal Propositional Calculi', pp. 206–220 in J. W. Addison, L. Henkin, and A. Tarski (eds.), *The Theory of Models*, North-Holland, Amsterdam 1965.

[671] Saul A. Kripke, Review of Lemmon [702], *Mathematical Reviews* **34** (1967), #5661. (Online version at MathSciNet.)

[672] Saul A. Kripke, 'Identity and Necessity', pp. 144–149 in M. K. Munitz (ed.), *Identity and Individuation*, New York University Press, New York 1971.

[673] Saul A. Kripke 'Is There a Problem About Substitutional Quantification?', pp. 325–419 in G. Evans and J. McDowell (eds.), *Truth and Meaning*, Oxford University Press, Oxford 1976.

[674] Saul A. Kripke, 'A Puzzle About Belief', pp. 239–283 in A. Margalit (ed.), *Meaning and Use*, Redidel, Dordrecht 1979. (Reprinted as Chapter 6 of [675].)

[675] Saul A. Kripke, *Philosophical Troubles (Collected Papers, Vol. I)*, Oxford University Press, Oxford 2011.

[676] Saul A. Kripke, 'Nozick on Knowledge', Chapter 7 of [675].

[677] Moshe Kroy, 'A Partial Formalization of Kant's Categorical Imperative', *Kant-Studien* **67** (1976), 192–209.

[678] Steven T. Kuhn, Review of Rabinowicz [938] *International Studies in Philosophy* **15** (1983), 107–109.

[679] Steven T. Kuhn, 'The Domino Relation: Flattening a Two-Dimensional Logic', *Journal of Philosophical Logic* **18** (1989), 173–195.

[680] Steven T. Kuhn, 'Minimal Non-Contingency Logic', *Note Dame Journal of Formal Logic* **36** (1995), 230–234.

[681] A. G. Kurosh, *Lectures on General Algebra*, Chelsea Publ. Co., New York, NY 1963.

[682] Franz von Kutschera, 'Global Supervenience and Belief', *Journal of Philosophical Logic* **23** (1994), 103–110.

[683] H. M. Lacey and E. Anderson, 'Spatial Ontology and Physical Modalities', *Philosophical Studies* **38** (1980), 261–285.

[684] Arto Laitinen, 'Against Representations with Two Directions of Fit', *Phenomenology and the Cognitive Sciences* **13** (2014), 179–199.

[685] K. Lambert, H. Leblanc and R. K. Meyer, 'A Liberated Version of S5', *Archiv für mathematische Logik und Grundlagenforschung* **12** (1969), 151–154.

[686] Morris Lazerowitz, 'Necessary and Contingent Truths', *Philosophical Review* **45** (1936), 268–282.

[687] Mark Lebar, 'Three Dogmas of Response-Deptendence', *Philosophical Studies* **123** (2005), 175–211.

[688] Martin Leckey, 'The Naturalness Theory of Laws', pp. 77–82 in H. Sankey (ed.), *Causation and Laws of Nature*, Kluwer, Dordrecht 1999.

[689] Martin Leckey and John Bigelow, 'The Necessitarian Perspective: Laws as Natural Entailments', pp. 92–119 in F. Weinert (ed.), *Laws of Nature*, de Gruyter, Berlin 1995.

[690] Catherine Legg, 'Argument-Forms Which Turn Invalid over Infinite Domains: Physicalism as Supertask?', *Contemporary Pragmatism* **5** (2008), 1–11.

[691] K. Lehrer and C. Wagner, 'Intransitive Indifference: the Semi-Order Problem', *Synthese* **65** (1985), 249–256.

[692] Hannes Leitgeb, 'Reducing Belief Simpliciter to Degrees of Belief', *Annals of Pure and Applied Logic* **164** (2013), 1338–1389.

[693] Hannes Leitgeb, 'The Review Paradox: On The Diachronic Costs of Not Closing Rational Belief Under Conjunction', *Noûs* **48** (2014), 781–793.

[694] E. J. Lemmon, 'New Foundations for Lewis Modal Systems', *Journal of Symbolic Logic* **22** (1957), 176–186.

[695] E. J. Lemmon, 'Is There Only One Correct System of Modal Logic?', *Aristotelian Society Supplementary Vol.* **33** (1959), 23–40.

[696] E. J. Lemmon, 'Moral Dilemmas', *Philosophical Review* **71** (1962), 139–158.

[697] E. J. Lemmon, 'Deontic Logic and the Logic of Imperatives', *Logique et Analyse* **8** (1965), 39–71.

[698] E. J. Lemmon, *Beginning Logic*, Nelson, London 1965. (This book has since gone through several subsequent printings and editions, at the hands of various publishers.)

[699] E. J. Lemmon, Review of Hintikka [490], *Philosophical Review* **74** (1965), 381–384.

[700] E. J. Lemmon, 'Some Results on Finite Axiomatizability in Modal Logic', *Notre Dame Journal of Formal Logic* **6** (1965), 301–308.

[701] E. J. Lemmon, 'A Note on Halldén-incompleteness', *Notre Dame Journal of Formal Logic* **7**, (1966) 296–300.

[702] E. J. Lemmon, 'Algebraic Semantics for Modal Logics. II' *Journal of Symbolic Logic* **31** (1966), 191–218.

[703] E. J. Lemmon, 'If I Know, Do I Know That I Know?' in A. Stroll, ed., *Epistemology: New Essays in the Theory of Knowledge*, Harper and Row, NY 1967.

[704] E. J. Lemmon, C. A. Meredith, D. Meredith, A. N. Prior and I. Thomas, 'Calculi of Pure Strict Implication', pp. 215–250 in J. W. Davis *et al.*, (eds.), *Philosophical Logic*, Reidel, Dordrecht 1969.

[705] E. J. Lemmon and D. S. Scott, *An Introduction to Modal Logic*, ed. K. Segerberg, American Philosophical Quarterly Monograph Series, Basil Blackwell, Oxford 1977. (Originally circulated 1966.)

[706] Wolfgang Lenzen, 'Epistemologische Betrachtungen zu [S4,S5]' *Erkenntnis* **14** (1979), 33–56.

[707] Wolfgang Lenzen, 'Recent Work in Epistemic Logic', Issue 1 of *Acta Philosophica Fennica* **30** (1978).

[708] Wolfgang Lenzen, 'On some Substitution Instances of **R1** and **L1**', *Notre Dame Journal of Formal Logic* **19** (1978), 159–164.

[709] Wolfgang Lenzen, 'A Rare Accident', *Notre Dame Journal of Formal Logic* **19** (1978), 249–250.

[710] Wolfgang Lenzen, *Glauben, Wissen und Wahrscheinlichkeit: Systeme der epistemischen Logik*, Springer-Verlag, Vienna 1980.

[711] Henry S. Leonard, 'Two-Valued Truth Tables for Modal Functions', pp. 42–67 in P. Henle, H. M. Kallen and S. K. Langer (eds.), *Structure, Method and Meaning*, Liberal Arts Press, NY 1951.

[712] Reinhold Letz, 'First-Order Tableau Methods', pp. 125–196 in M. D'Agostino, D. Gabbay, R. Hähnle and J. Posegga. (eds.), *Handbook of Tableau Methods*, Kluwer, Dordrecht 1999.

[713] Stephan Leuenberger, '*De Jure* and *De Facto* Validity in the Logic of Time and Modality', *Thought* **2** (2013), 196–205.

[714] Hector J. Levesque, 'All I Know: A Study in Autoepistemic Logic', *Artificial Intelligence* **42** (1990), 263–309.

[715] Hector J. Levesque and Gerhard Lakemeyer, *The Logic of Knowledge Bases*, MIT Press, Cambridge, MA 2001.

[716] C. I. Lewis and C. H. Langford, *Symbolic Logic*, Dover Publications, New York, NY 1959. (Originally publ. 1932.)

[717] David Lewis, *Counterfactuals*, Blackwell, Oxford 1973.

[718] David Lewis, 'Semantic Analyses for Dyadic Deontic Logic', pp. 1–14 in S. Stenlund (ed.), *Logical Theory and Semantic Analysis*, Reidel, Dordrecht 1974.

[719] David Lewis, 'Attitudes *De Dicto* and *De Se*', *Philosophical Review* **88** (1979), 513-43.

[720] David Lewis, 'A Problem About Permission', pp. 163–175 in E. Saarinen *et al.* (eds.), *Essays in Honour of Jaakko Hintikka*, Reidel, Dordrecht 1979.

[721] David Lewis, 'Veridical Hallucination and Prosthetic Vision', *Australasian Journal of Philosophy* **58** (1980), 239–249.

[722] David Lewis, 'Logic for Equivocators', *Noûs* **16** (1982), 431–441.

[723] David Lewis, '"Whether" Report', pp. 194–206 in T. Pauli (ed.), *<320311>: Philosophical Essays Dedicated to Lennart Åqvist on his Fiftieth Birthday*, University of Uppsala 1982.

[724] David Lewis, *Philosophical Papers, Vol. 1*, Oxford University Press, New York 1983.

[725] David Lewis, *Philosophical Papers, Vol. I1*, Oxford University Press, New York 1986.

[726] David Lewis, *On the Plurality of Worlds*, Blackwell, Oxford 1986.

[727] David Lewis, 'Statements Partly About Observation', *Philosophical Papers* **17** (1988), 1–31.

[728] David Lewis, 'Desire as Belief', *Mind* **97** (1988), 323–332, and 'Desire as Belief. II' *Mind* **105**, 303–13.

[729] David Lewis, 'Ern Malley's Namesake', *Quadrant*, March 1995, 14–15.

[730] David Lewis, 'Elusive Knowledge', *Australasian Journal of Philosophy* **74** (1996), 549–567.

[731] H. D. Lewis, 'Obedience to Conscience', *Mind* **54** (1945), 227–253.

[732] Judith Lichtenberg, 'Negative Duties, Positive Duties, and the "New Harms"' *Ethics* **120** (2010), 557–578.

[733] Sten Lindström, 'Possible Worlds Semantics and the Liar', pp. 297–314 in A. Rojszczak, J. Cachro and G. Kurczewski (eds.), *Philosophical Dimensions of Logic and Science*, Kluwer, Dordrecht 2003. Reprinted in pp. 93–108 of J. Almog and P. Leonardi (eds.), *The Philosophy of David Kaplan*, Oxford University Press, Oxford 2009.

[734] Sten Lindström and Krister Segerberg, 'Modal Logic and Philosophy', pp. 1149–1214 in [97].

[735] Bernard Linsky, 'Factives, Blindspots and Some Paradoxes', *Analysis* **46** (1986), 10–15.

[736] Christian List and Philip Pettit, 'Aggregating Sets of Judgments: an Impossibility Result', *Economics and Philosophy* **18** (2002), 89–110.

[737] Tadeusz Litak, 'On Notions of Completeness Weaker than Kripke Completeness' pp. 149–169 in R. Schmidt, I. Pratt-Hartmann, M. Reynolds, and H. Wansing (eds.), *Advances in Modal Logic, Vol. 5*, King's College Publications, London 2005.

[738] A. C. Lloyd, 'Talk About Knowing', *Mind* **89** (1980), 263–268.

[739] G.-J. C. Lokhorst, 'Andersonian Deontic Logic, Propositional Quantification, and Mally', *Notre Dame Journal of Formal Logic* **47** (2006), 385–395.

[740] G.-J. C. Lokhorst, 'Mally's Deontic Logic', *Stanford Encyclopedia of Philosophy* (Winter 2013 Edition), Edward N. Zalta (ed.), URL = <http://plato.stanford.edu/archives/win2013/entries/mally-deontic/>.

[741] G.-J. C. Lokhorst, 'An Intuitionistic Reformulation of Mally's Deontic Logic', *Journal of Philosophical Logic* **42** (2013), 635–641.

[742] G.-J. C. Lokhorst, 'Mally's Deontic Logic: Reducibility and Semantics' *Journal of Philosophical Logic* **44** (2015), 309–319.

[743] Dan López de Sa, 'Relativizing Utterance-Truth?', *Synthese* **170** (2009), 1–5.

[744] J. R. Lucas, *Moods and Tenses*, written up version of the author's lectures at Oxford University, Merton College, Oxford 1982.

[745] J. Łukasiewicz, 'Philosophical Remarks on Many-Valued Systems of Propositional Logic', pp. 153–178 in Borkowski [113]; originally delivered as a talk in 1920 and first published, in German, in 1930.

[746] J. Łukasiewicz, 'A System of Modal Logic', *Journal of Computing Systems* **1** (1953), 111–149; reprinted as pp. 352–390 of Borkowski [113].

[747] Minghui Ma, 'Dynamic Epistemic Logic of Finite Identification', pp. 227–237 in X. He, J. Horty, and E. Pacuit (eds.), *Logic, Rationality, and Interaction, Second International Workshop (LORI 2009)*, Chongqing, China, Lecture Notes in Artificial Intelligence #5834, Springer-Verlag, Berlin 2009.

[748] John MacFarlane, 'The Assessment Sensitivity of Knowledge Attributions', pp. 197–234 in T. S. Gendler and J. Hawthorne (eds.), *Oxford Studies in Epistemology, Vol. 1*, Oxford University Press, Oxford 2005.

[749] John MacFarlane, *Assessment Sensitivity: Relative Truth and its Applications*, Oxford University Press, Oxford 2014.

[750] J. J. MacIntosh, 'Knowing and Believing', *Procs. of the Aristotelian Society* **80** (1980), 169–185.

[751] J. J. MacIntosh, 'The Logic of Privileged Access', *Australasian Journal of Philosophy* **61** (1983), 142–151.

[752] J. J. MacIntosh, 'Some Propositional Attitude Paradoxes', *Pacific Philosophical Quarterly* **65** (1984), 21–25.

[753] J. J. MacIntosh, 'Fitch's Factives', *Analysis* **44** (1984), 153–158.

[754] J. J. Macintosh, 'Theological Question-Begging', *Dialogue* **30** (1991), 531–547.

[755] J. L. Mackie, 'The Possibility of Innate Knowledge', *Procs. of the Aristotelian Society* **70** (1970), 245–257.

[756] Brian MacPherson, 'Is it Possible that Belief Isn't Necessary?', *Notre Dame Journal of Formal Logic* **34** (1993), 12–28.

[757] Roger D. Maddux, *Relation Algebras*, Elsevier, Amsterdam 2006.

[758] David Makinson, 'On Some Completeness Theorems in Modal Logic', *Zeitschr. für math. Logik und Grundlagen der Math.* **12** (1966), 379–384.

[759] David Makinson, 'Some Embedding Theorems for Modal Logics', *Notre Dame Journal of Formal Logic* **12** (1971), 252–254.

[760] David Makinson, 'A Warning About the Choice of Primitive Operators in Modal Logic', *Journal of Philosophical Logic* **2** (1973), 193–196.

[761] David Makinson, *Topics in Modern Logic*, Methuen, London 1973.

[762] David Makinson, 'Quantificational Reefs in Deontic Waters', pp. 87–91 in Hilpinen [483].

[763] David Makinson, 'Individual Actions are Very Seldom Obligatory', *Journal of Non-Classical Logic* **2** (1983), 7–13.

[764] David Makinson, 'On a Fundamental Problem of Deontic Logic', pp. 29–53 in P. McNamara and H. Prakken (eds.), *Norms, Logics and Information Systems: New Studies in Deontic Logic and Computer Science*, IOS Press, Amsterdam 1999.

[765] David Makinson, *Bridges from Classical to Nonmonotonic Logic*, King's College Publications, London 2005.

[766] L. L. Maksimova, 'Pretabular Extensions of Lewis' **S4**', *Algebra and Logic* **14** (1975), 16–33.

[767] David Manley, 'Safety, Content, Apriority, Self-knowledge', *Journal of Philosophy* **104** (2007) 403–423.

[768] D. S. Mannison, 'Lemmon on Knowing', *Synthese* **26** (1974), 383–390.

[769] D. S. Mannison '"Inexplicable Knowledge" Does Not Require Belief', *Philosophical Quarterly* **26** (1976), 139–148.

[770] João Marcos, 'Logics of Essence and Accident', *Bulletin of the Section of Logic* **34** (2005), 43–56.

[771] E. D. Mares, 'Andersonian Deontic Logic', *Theoria* **58** (1992), 3–20.

[772] E. D. Mares, *Relevant Logic: A Philosophical Interpretation*, Cambridge University Press, Cambridge 2004.

[773] E. D. Mares, 'A Lewisian Semantics for S2', *History and Philosophy of Logic* **34** (2013), 53–67.

[774] E. D. Mares and Paul McNamara, 'Supererogation in Deontic Logic: Metatheory for DWE and Some Close Neighbours', *Studia Logica* **59** (1997), 397–415.

[775] M. G. F. Martin, 'The Limits of Self-Awareness', *Philosophical Studies* **120** (2004), 37–89.

[776] G. J. Massey, 'Four Simple Systems of Modal Propositional Logic', *Philosophy of Science* **32** (1965), 342–355.

[777] G. J. Massey, 'The Theory of Truth Tabular Connectives, both Truth Functional and Modal', *Journal of Symbolic Logic* **31** (1966), 593–608.

[778] G. J. Massey, 'Normal Form Generation of S5 Functions via Truth Functions', *Notre Dame Journal of Formal Logic* **9** (1968), 81–85.

[779] G. J. Massey, 'Binary Closure-Algebraic Operations that are Functionally Complete', *Notre Dame Journal of Formal Logic* **11** (1970), 340–342.

[780] G. J. Massey, *Understanding Symbolic Logic*, Harper & Row, NY 1970.

[781] G. J. Massey, 'The Modal Structure of the Prior–Rescher Family of Infinite Product Systems', *Notre Dame Journal of Formal Logic* **13** (1972), 219–233.

[782] K. Matsumoto, 'Reduction Theorem in Lewis' Sentential Calculi', *Math. Japonicae* **3** (1955), 133–135.

[783] Bernard Mayo, 'A Note on Austin's Performative Theory of Knowledge' *Philosophical Studies* **14** (1963), 28–31.

[784] Bernard Mayo, 'Negative and Positive Duties: A Reply', *Philosophical Quarterly* **16** (1966), 159–164.

[785] James D. McCawley, 'Tense and Time Reference in English', pp. 96–113 in C. J. Fillmore and D. T. Langendoen (eds.), *Studies in Linguistic Semantics*, Holt, Rinehart and Winston, NY 1971.

[786] Colin McGinn, 'The Concept of Knowledge', pp. 529–554 in P. A. French *et al.* (eds.), *Midwest Studies in Philosophy, IX, Causation and Causal Theories*, University of Minnesota Press, Minneapolis 1984.

[787] Casey McGinnis, 'Semi-Paraconsistent Deontic Logic', pp. 103–125 in J.-Y. Béziau, W. Carnielli and D. Gabbay (eds.), *Handbook of Paraconsistency*, College Publications, London 2007.

[788] M. McKenna and D. J. Coates, 'Compatibilism', *Stanford Encyclopedia of Philosophy* (Summer 2015 Edition), Edward N. Zalta (ed.), URL = ⟨ http://plato.stanford.edu/archives/sum2015/entries/compatibilism/⟩

[789] J. C. C. McKinsey, 'On the Syntactical Construction of Systems of Modal Logic', *Journal of Symbolic Logic* **10** (1945), 83–94.

[790] J. C. C. McKinsey, 'Systems of Modal Logic Which are Not Unreasonable in the Sense of Halldén', *Journal of Symbolic Logic* **18** (1953), 109–113.

[791] J. C. C. McKinsey and A. Tarski, 'Some Theorems About the Sentential Calculi of Lewis and Heyting', *Journal of Symbolic Logic* **13** (1948), 1–15.

[792] Paul McNamara, 'Agential Obligation and Non-Agential Personal Obligation Plus Agency', *Journal of Applied Logic* **2** (2004), 117–152.

[793] Paul McNamara, 'Deontic Logic', pp. 197–288 in D. M. Gabbay and J. Woods (eds.), *Handbook of the History of Logic, Vol. 7*, Elsevier, Amsterdam 2006.

[794] Paul McNamara, 'Praise, Blame, Obligation, and Beyond: Toward a Framework for Classical Supererogation and Kin', pp. 233–247 in R. van der Meyden and L. van der Torre (eds.), *Deontic Logic in Computer Science: 9th International Conference*, Springer-Verlag, Berlin 2008.

[795] David McNaughton and Piers Rawling, 'Agent-Relativity and the Doing–Happening Distinction', *Philosophical Studies* **63** (1991), 167–185.

[796] Toby Meadows, 'Revising Carnap's Semantic Conception of Modal Logic', *Studia Logica* **100** (2012), 497–515.

[797] M. da Paz N. Medeiros, 'A New Classical S4 Modal Logic in Natural Deduction', *Journal of Symbolic Logic* **71** (2006), 799–809.

[798] A. I. Melden (ed.), *Essays in Moral Philosophy*, University of Washington Press, Seattle 1958.

[799] Joseph Melia, "Comments on '*De Jure* and *De Facto* Validity in the Logic of Time and Modality'", *Thought* **2** (2013), 206–209.

[800] Christopher Menzel, 'The True Modal Logic', *Journal of Philosophical Logic* **20** (1991), 331–374.

[801] Christopher Menzel, 'Actualism' entry in *Stanford Encyclopedia of Philosophy* (Spring 2015 Edition), Edward N. Zalta (ed.), URL = <http://plato.stanford.edu/archives/spr2015/entries/actualism/>.

[802] C. A. Meredith and A. N. Prior, 'Modal Logic with Functorial Variables and a Contingent Constant', *Notre Dame Journal of Formal Logic* **6** (1965), 99–109.

[803] Arthur Merin, 'Permission Sentences Stand in the Way of Boolean and Other Lattice-Theoretic Semantics', *Journal of Semantics* **9** (1992), 95–162.

[804] van der Meyden, R., 'The Dynamic Logic of Permission', *Journal of Logic and Computation* **6** (1996), 465–479.

[805] J.-J. C. Meyer, 'A Different Approach to Deontic Logic: Deontic Logic Viewed as a Variant of Dynamic Logic', *Notre Dame Journal of Formal Logic* 29 (1988), 109–136.

[806] J.-J. C. Meyer, 'Epistemic Logic', Chapter 9 (pp. 183–202) in Goble [385].

[807] Robert K. Meyer, 'Entailment is not Strict Implication', *Australasian Journal of Philosophy* **52** (1974), 211–231.

[808] W. P. M. Meyer-Viol and H. S. Jones, 'Reference Time and the English Past Tenses', *Linguistics and Philosophy* **34** (2011), 223–256.

[809] Peter Milne, 'Modal Metaphysics and Comparatives', *Australasian Journal of Philosophy* **70** (1992), 248–262.

[810] Grigori Mints, *A Short Inroduction to Modal Logic*, CSLI Publications, Stanford, CA 1992.

[811] Anita Mittwoch, 'Equi or Raising or Both—Another Look at the Root Modals and at Permissive *Allow*', *Papers in Linguistics* **10** (1977), 55–75.

[812] Y. Miyazaki, 'Kripke Incomplete Logics Containing **KTB**', *Studia Logica* **85** (2007), 303–317.

[813] Y. Miyazaki, 'A Splitting Logic in NExt(**KTB**)', *Studia Logica* **85** (2007), 381–394.

[814] G. C. Moisil, 'Logique Modale', *Disquisitiones Mathematicae et Physicae* **2** (1942), 3–98.

[815] Friederike Moltmann, 'Propositional Attitudes Without Propositions', *Synthese* **135** (2003), 77–118.

[816] Friederike Moltmann, 'Comparatives Without Degrees', pp. 155–160 in P. Dekker and M. Franke (eds.), *Proceedings of the Fifteenth Amsterdam Colloquium*, ILLC/Department of Philosophy, University of Amsterdam, Amsterdam 2005.

[817] Friederike Moltmann, 'Intensional Verbs and their Intentional Objects', *Natural Language Semantics* **16** (2008), 239–270.

[818] Friederike Moltmann, *Abstract Objects and the Semantics of Natural Language*, Oxford University Press, Oxford 2013.

[819] Friederike Moltmann, 'Identificational Sentences', *Natural Language Semantics* **21** (2013), 43–77.

[820] Uwe Mönnich (ed.), *Aspects of Philosophical Logic*, Reidel, Dordrecht 1981.

[821] Richard Montague, 'Logical Necessity, Physical Necessity, Ethics and Quantifiers', pp. 71–83 in R. H. Thomason (ed.), *Formal Philosophy: Selected Papers of Richard Montague*. (This paper orig. publ. 1960.)

[822] A. W. Moore, 'Possible Worlds and Diagonalization', *Analysis* **44** (1984), 21–22.

[823] A. W. Moore, 'Ineffability and Reflextion: An Outline of the Concept of Knowledge', *European Journal of Philosophy* **1** (1993), 285–308.

[824] Robert C. Moore, 'Semantical Considerations on Monmonotonic Logic', *Artificial Intelligence* **25** (1985), 75–94.

[825] Robert C. Moore, *Logic and Representation*, CSLI Publications, Stanford, CA 1995.

[826] Vittorio Morato, 'Validity and Actuality', *Logique et Analyse* **57** (2014), 379–405.

[827] C. G. Morgan, 'Liberated Brouwerian Modal Logic', *Dialogue* **13** (1974), 505–514.

[828] C. G. Morgan, 'Liberated Versions of T, $S4$ and $S5$', *Archiv für mathematische Logik und Grundlagenforschung* **17** (1975), 85–90.

[829] C. G. Morgan, 'Weak Liberated Versions of T and $S4$', *Journal of Symbolic Logic* **40** (1975), 25–30.

[830] C. G. Morgan, 'Note on a Strong Liberated Modal Logic and its Relevance to Possible World Scepticism', *Notre Dame Journal of Formal Logic* **20** (1979), 718–722.

[831] Chris Mortensen, 'Anything is Possible', *Erkenntnis* **30** (1989), 319337.

[832] Adam Morton, 'Comparatives and Degrees', *Analysis* **44** (1984), 16–18.

[833] Y. Moses and Y. Shoham, 'Belief as Defeasible Knowledge', *Artificial Intelligence* **64** (1993), 299–321.

[834] Peter L. Mott, 'On Chisholm's Paradox', *Journal of Philosophical Logic* **2** (1973), 197–211.

[835] A. Mourelatos, 'Events, Processes, and States', *Linguistics and Philosophy* **2** (1978), 415–434.

[836] J. D. Mullen, 'Does the Logic of Preference Rest on a Mistake?', *Metaphilosophy* **10** (1979), 247–245.

[837] B. Myers-Schulz and E. Schwitzgebel, 'Knowing That P without Believing That P', *Noûs* **47** (2013), 371–384.

[838] Shyam Nair, 'Consequences of Reasoning with Conflicting Obligations', *Mind* **123** (2014), 753–790.

[839] George Nakhnikian, *Introduction to Philosophy*, Alfred A. Knopf, New York 1967.

[840] Arnold vander Nat, 'Beyond Non-normal Possible Worlds', *Notre Dame Journal of Formal Logic* **20** (1979), 631–635.

[841] Norton Nelkin, 'What is it Like to be a Person?', *Mind and Language* **2** (1987), 220–241.

[842] Michael Nelson and Edward N. Zalta, 'Defense of Contingent Logical Truths', *Philosophical Studies* **157** (2012), 153–162.

[843] Dilip Ninan, 'Two Puzzles About Deontic Necessity', in J. Gajewski *et al.* (eds.), *New Work on Modality* (= *MIT Working Papers in Linguistics* **51** (2005)).

[844] Ilkka Niiniluoto, 'Hypothetical Imperatives and Conditional Obligations', *Synthese* **66** (1986), 111–133.

[845] Ilkka Niiniluoto, 'Conditional Intentions', pp. 167–179 in P. I. Bystrov and V. N. Sadovsky (eds.), *Philosophical Logic and Logical Philosophy*, Kluwer, Dordrecht 1996.

[846] Daniel Nolan, 'Selfless Desires', *Philosophy and Phenomenological Research* **73** (2006), 665–679.

[847] Lennart Nordenfelt, 'On Ability, Opportunity and Competence: An Inquiry Into People's Possibility for Action', pp. 145-158 in G. Holmström-Hintikka and R. Tuomela (eds.), *Contemporary Action Theory. Vol. I*, Kluwer, Dordrecht 1997.

[848] Marek Nowak, 'On Some Generalizations of the Concept of Partition', *Studia Logica* **102** (2014), 93–116.

[849] P. H. Nowell-Smith and E. J. Lemmon, 'Escapism: The Logical Basis of Ethics', *Mind* **69** (1960), 289–300.

[850] Robert Nozick and Richard Routley, 'Escaping the Good Samaritan Paradox', *Mind* **71** (1962), 277–382.

[851] Robert Nozick, *Philosophical Explanations*, Oxford University Press, Oxford, and Harvard University Press, Harvard, MA 1981.

[852] Donald Nute (ed.), *Defeasible Deontic Logic*, Kluwer, Dordrecht 1997.

[853] David O'Connor, 'Moore and the Paradox of Analysis', *Philosophy* **57** (1982), 211–221.

[854] D. Odegaard, 'Modality and the Ontological Argument', *Logique et Analyse* **20** (1977), 134–137.

[855] M. Ohnishi and K. Matsumoto, 'Gentzen Method in Modal Calculi', *Osaka Mathematical Journal* **9** (1957), 113–130; correction *ibid.* **10** (1958), 147.

[856] S. Okasha, 'On A Flawed Argument Against the KK Principle', *Analysis* **73** (2013), 80–86.

[857] Mika Oksanen, 'The Russell-Kaplan Paradox and Other Modal Paradoxes: New Solutions', *Nordic Journal of Philosophical Logic* **4** (1999), 73–93.

[858] Graham Oppy, 'Ontological Arguments', *Stanford Encyclopedia of Philosophy* (Winter 2012 Edition), Edward N. Zalta (ed.), URL = <http://plato.stanford.edu/archives/win2012/entries/ontological-arguments/>.

[859] Oystein Ore, 'Theory of Equivalence Relations', *Duke Mathematical Journal* **9** (1942), 573–627.

[860] Ewa Orłowska, 'Logic for Reasoning About Knowledge', *Zeitschr. für math. Logik und Grundlagen der Math.* **35** (1989), 559–572.

[861] H. Osborne, 'A Contradiction in Commonsense Ethics', *Mind* **39** (1930), 332–337.

[862] H. Osborne, *Foundations of the Theory of Value*, Cambridge University Press, Cambridge 1933.

[863] F. R. Palmer, '*Can, Will* and Actuality', pp. 91–99 in S. Greenbaum and J. Svartvik (eds.), *Studies in English Linguistics for Randolph Quirk*, Longman, London 1979.

[864] Robert Pargetter, 'Laws and Modal Realism', *Philosophical Studies* **46** (1984), 335–347.

[865] Eric Pacuit, 'Dynamic Epistemic Logic I: Modeling Knowledge and Belief', *Philosophy Compass* **8** (2013), 798–814.

[866] W. T. Parry, 'Modalities in the *Survey* System of Strict Implication', *Journal of Symbolic Logic* **4** (1939), 137–154.

[867] W. T. Parry, 'The Logic of C. I. Lewis', pp. 115–154 in P. Schilpp (ed.) *The Philosophy of C. I. Lewis*, Cambridge 1968.

[868] W. T. Parry, 'Analytic Implication; its History, Justification and Varieties', pp. 101–118 in J. Norman and R. Sylvan (eds.), *Directions in Relevant Logic*, Kluwer, Dordrecht 1989.

[869] Josh Parsons, 'Assessment-Contextual Indexicals', *Australasian Journal of Philosophy* **89** (2011), 1–17.

[870] Terence Parsons, 'On Denoting Propositions and Facts', *Philosophical Perspectives* **7** (1993), 440–460.

[871] Barbara Hall Partee, 'Some Structural Analogies between Tenses and Pronouns in English', *Journal of Philosophy* **70** (1973), 601–609.

[872] Barbara Hall Partee, 'Belief-sentences and the Limits of Semantics', pp. 87–106 in S. Peters and E. Saarinen (eds.), *Processes, Beliefs, and Questions*, Reidel, Dordrecht 1982.

[873] Christopher Peacocke, 'Abstract Objects and Arithmetical Relations', unpublished typescript, circa 1976.

[874] Christopher Peacocke, *Sense and Content*, Oxford University Press, Oxford 1983.

[875] Christopher Peacocke, 'Understanding Logical Constants: A Realist's Account', *Procs. British Academy* **73** (1987), 153–199. (Reprinted in T. Baldwin and T. Smiley, eds., *Studies in the Philosophy of Logic and Knowledge: British Academy Lectures*, Oxford University Press, Oxford 2004.)

[876] Christopher Peacocke, *A Study of Concepts*, MIT Press, Cambridge MA 1992.

[877] Christopher Peacocke, 'Explaining *De Se* Phenomena', pp. 144–157 in S. Prosser and F. Recanati (eds.), *Immunity to Error Through Misindentification: New Essays*, Cambridge Universitiy Press, Cambridge 2012.

[878] Vera Peetz, 'Note on Armstrong's "Absolute and Relative Motion"', *Mind* **79** (1970), 427–430.

[879] F. J. Pelletier and A. Urquhart, 'Synonymous Logics', *Journal of Philosophical Logic* **32** (2003) 259–285; Correction: *ibid.* **37** (2008), 95–100.

[880] Philip L. Peterson, 'How to Infer Belief from Knowledge', *Philosophical Studies* **32** (1977), 203–209.

[881] Charles R. Pigden, 'Logic and the Autonomy of Ethics', *Australasian Journal of Philosophy* **67** (1989), 127–151.

[882] Charles R. Pigden, 'Nihilism, Nietzsche and the Doppleganger Problem', *Ethical Theory and Moral Practice* **10** (2007), 441–456.

[883] Charles R. Pigden (ed.), *Hume on Is and Ought*, Palgrave Macmillan, Basingstoke 2010.

[884] Claudio Pizzi, 'Necessity and Relative Contingency', *Studia Logica* **85** (2007), 395–410.

[885] Jan Plaza, 'Logics of Public Communications', *Synthese* **158** (2007), 165–179. (Orig. publ. 1989.)

[886] K. E. Pledger 'Modalities of Systems Containing S3', *Zeitschr. für math. Logik und Grundlagen der Math.* **18** (1972), 287–283.

[887] F. Poggiolesi, 'The Method of Tree-Hypersequents for Modal Propositional Logic', pp. 31–51 in D. Makinson, J. Malinowski and H. Wansing (eds.), *Towards Mathematical Philosophy*, Springer-Verlag, Berlin 2009.

[888] Tomasz Połacik, 'The Unique Intermediate Logic Whose Every Rule is Archetypal', *Logic Journal of the IGPL* **13** (2005), 269–275.

[889] Tomasz Połacik, 'Archetypal Rules and Intermediate Logics', pp. 227–237 in M. Peliš and V. Punčochář (eds.), *The Logica Yearbook 2011*, College Publications, London 2012.

[890] John L. Pollock, *Technical Methods in Philosophy*, Westview Press, Boulder, CO 1990.

[891] Ingmar Pörn, *The Logic of Power*, Blackwell, Oxford 1970.

[892] Jean Porte, 'The Ω-system and the L-system of Modal Logic', *Notre Dame Journal of Formal Logic* **20** (1979), 915–920.

[893] Jean Porte, 'A Research in Modal Logics', *Logique et Analyse* **23** (1980), 3–34.

[894] Jean Porte, 'Notes on Modal Logics', *Logique et Analyse* **24** (1981), 399–406.

[895] Jean Porte, 'The Deducibilities of **S5**', *Journal of Philosophical Logic* **10** (1981), 409–422.

[896] Jean Porte, 'Boll-Reinhart Modal Logic', *Logique et Analyse* **25** (1982), 181–190.

[897] Jean Porte, 'The Real World: Completeness and Incompleteness of a Modal Logic', *Logique et Analyse* **30** (1987), 209–220.

[898] Paul Portner, 'The Progressive in Modal Semantics', *Language* **74** (1998), 760–787.

[899] Paul Portner, *Modality*, Oxford University Press, Oxford 2009.

[900] Geoffrey, Count Potocki of Montalk, *Against Cresswell: A Lampoon*, Maidment Press, London 1930.

[901] D. H. Potts, review of Harary [451], *Journal of Symbolic Logic* **39** (1974), p. 604.

[902] Timothy Potts, 'States, Activities, and Performances', *Aristotelian Society Supplementary Volume* **39** (1965), 65–84.

[903] Lawrence Powers, 'Some Deontic Logicians', *Noûs* **1** (1967), 381–400.

[904] Lawrence Powers, 'Knowledge by Deduction', *Philosophical Review* **87** (1978), 337–371.

[905] Dag Prawitz, *Natural Deduction*, Almqvist and Wiksell, Stockholm 1965.

[906] Stefano Predelli and Isidora Stojanovic, 'Semantic Relativism and the Logic of Indexicals', pp. 63–79 in [348].

[907] Graham Priest 'Modality as a Meta-Concept', *Notre Dame Journal of Formal Logic* **17** (1976), 401–414.

[908] Graham Priest, 'Contradiction, Belief and Rationality', *Proceedings of the Aristotelian Society* **86** (1985–6), 99–116.

[909] Graham Priest, *An Introduction to Non-Classical Logic*, Cambridge University Press, Cambridge 2001.

[910] Graham Priest, *In Contradiction: A Study of the Transconsistent*, Second Edn., Oxford University Press, Oxford 2006.

[911] Graham Priest, 'Neighborhood Semantics for Intentional Operators', *Review of Symbolic Logic* **2** (2009), 360–373.

[912] Graham Priest, *One: Being an Investigation into the Unity of Reality and of its Parts, including the Singular Object which is Nothingness*, Oxford University Press, Oxford 2014.

[913] A. N. Prior, *Logic and the Basis of Ethics*, Oxford University Press, Oxford 1949.

[914] A. N. Prior, 'The Ethical Copula', *Australasian Journal of Philosophy* **39** (1951), 137–154; reprinted as Chapter 1 in Prior [929].

[915] A. N. Prior, 'Many-Valued and Modal Systems: an Intuitive Approach', *Philosophical Review* **64** (1955), 626–630.

[916] A. N. Prior, Review of papers by Jerzy Kalinowski, *Journal of Symbolic Logic* **21** (1956), p. 191*f*.

[917] A. N. Prior, 'A Note on the Logic of Obligation', *Revue Philosophique de Louvain* **54** (1956), 86–87. (With commentary, *ibid.*, pp. 88–90, by Robert Feys.)

[918] A. N. Prior, 'Logicians at Play; or Syll, Simp and Hilbert', *Australasian Journal of Philosophy* **34** (1956), 182–192.

[919] A. N. Prior, *Time and Modality*, Oxford University Press, Oxford 1957.

[920] A. N. Prior, 'Escapism: the Logical Basis of Ethics', pp. 135–146 in Melden [798].

[921] A. N. Prior, Review of Anderson [16] and other papers, *Journal of Symbolic Logic* **24** (1959), 177–178.

[922] A. N. Prior, *Formal Logic* (Second Edn.), Oxford 1962. (First edn. 1955.)

[923] A. N. Prior, 'The Done Thing', *Mind* **73** (1964), 441–442.

[924] A. N. Prior *Past, Present and Future*, Oxford University Press, Oxford 1967.

[925] A. N. Prior, 'The Logic of Ending Time' (= Chapter 10 of Prior, *Papers on Time and Tense*, Oxford 1968).

[926] A. N. Prior, 'Time and Change', *Ratio* **10** (1968), 173–177.

[927] A. N. Prior, 'The Notion of the Present', *Studium Generale* **23** (1970), 245–248.

[928] A. N. Prior, *Objects of Thought*, ed. P. T. Geach and A. J. Kenny, Oxford University Press, Oxford 1971.

[929] A. N. Prior, *Papers in Logic and Ethics*, ed. P. T. Geach and A. J. Kenny, Duckworth, London 1976.

[930] A. N. Prior and Kit Fine, *Worlds, Times and Selves*, Duckworth, London 1977.

[931] Duncan Pritchard, 'Safety-Based Epistemology: Whither Now?' *Journal of Philosophical Research* **34** (2009), 33–45.

[932] Duncan Pritchard, *Epistemological Disjunctivism*, Oxford University Press, Oxford 2012.

[933] Geoffrey K. Pullum, 'No Foot in Mouth', pp. 27–29 in Tom Sumner (ed.), *Untidy: The Blogs on Rumsfeld*, William, James & Co., Wilsonville, Oregon 2005.

[934] R. L. Purtill, 'Hartshorne's Modal Proof', *Journal of Philosophy* **63** (1966), 397–409.

[935] W. V. Quine, 'Three Grades of Modal Involvement', pp. 156–174 in Quine, *The Ways of Paradox and Other Essays*, Random House, New York 1968. (Paper first publ. 1953.)

[936] W. V. Quine, 'Quantifiers and Propositional Attitudes', pp. 183–194 in Quine, *The Ways of Paradox and Other Essays*, Random House, New York 1968. (Paper first publ. 1955.)

[937] Alexander Rabinovich 'A Proof of Kamp's Theorem', *Logical Methods in Computer Science* **10** (2014), 1–16

[938] W. Rabinowicz, *Universalizability: A Study in Morals and Metaphysics*, Reidel, Dordrecht 1979.

[939] W. Rabinowicz and K. Segerberg, 'Actual Truth, Possible Knowledge', *Topoi* **13** (1994), 101–115.

[940] Colin Radford, 'Knowledge – By Examples', *Analysis* **27** (1966), 1–11.

[941] Murali Ramachandran, 'The KK-Principle, Margins for Error, and Safety', *Erkenntnis* **76** (2012), 121–136.

[942] T. J. Ramsamujh, 'A Paradox—(1) All Positive Integers are Equal', *Mathematical Gazette* **72** (1988), 113.

[943] Baron Reed, 'Fallibilism, Epistemic Possibility, and Epistemic Agency', *Philosophical Issues* **23** (2013), 40–69.

[944] Hans Reichenbach, *Elements of Symbolic Logic*, Macmillan, London 1947.

[945] Maria Reicher, 'Nonexistent Objects', *Stanford Encyclopedia of Philosophy* (Winter 2014 Edition), Edward N. Zalta (ed.), URL = <http://plato.stanford.edu/archives/win2014/entries/nonexistent-objects/>.

[946] Nicholas Rescher, 'An Intuitive Interpretation of Systems of Four-Valued Logic', *Notre Dame Journal of Formal Logic* **6** (1965), 154–156.

[947] Nicholas Rescher, 'Semantic Foundations for the Logic of Preference', pp. 37–62 in [948].

[948] Nicholas Rescher (ed.), *The Logic of Decision and Action*, University of Pittsburgh Press, Pittsburgh, PA 1967.

[949] Nicholas Rescher, *Studies in Modality* (with the collaboration of R. Manor, A. vander Nat, Z. Parks), Basil Blackwell, Oxford 1974.

[950] Nicholas Rescher, 'Epistemic Logic', pp. 478–490 in D. Jacquette (ed.), *A Companion to Philosophical Logic*, Blackwell, Oxford 2002.

[951] Nicholas Rescher and Alasdair Urquhart, *Temporal Logic*, Springer-Verlag, Vienna 1971.

[952] Greg Restall, 'How to be *Really* Contraction-Free', *Studia Logica* **52** (1993), 381–391.

[953] Greg Restall, 'Proofnets for S5: Sequents and Circuits for Modal Logic', pp. 151–172 in C. Dimitracopoulos, L. Newelski, and D. Normann (eds.), *Logic Colloquium 2005*, Cambridge University Press, Cambridge 2007.

[954] Greg Restall, 'Always More', pp. 223–229 in Michal Peliš (ed.), *Logica Yearbook 2009*, College Publications, London 2010.

[955] Greg Restall, 'A Cut-Free Sequent System for Two-Dimensional Modal Logic, and Why it Matters', *Annals of Pure and Applied Logic* **163** (2012), 1611–1623.

[956] Greg Restall and Gillian Russell, 'Barriers to Implication', pp. 243–258 in Charles Pigden (ed.), *Hume on 'Is' and 'Ought'*, Palgrave Macmillan, NY 2010.

[957] H. P. Rickman, 'Escapism and the Logical Basis of Ethics', *Mind* **72** (1963), 273–274.

[958] M. de Rijke and H. Wansing, 'Proofs and Expressiveness in Alethic Modal Logic', pp. 422–441 in D. Jacquette (ed.), *A Companion to Philosophical Logic*, Blackwell, Oxford 2002, 2006.

[959] A. A. Rini and M. J. Cresswell, *The World-Time Parallel: Tense and Modality in Logic and Metaphysics*, Cambridge University Press, Cambridge 2013.

[960] Claude Rivière, 'Is *Should* a Weaker *Must*?', *Journal of Linguistics* **19** (1981), 179–195.

[961] George W. Roberts, 'Some Questions in Epistemology', *Procs. of the Aristotelian Society* **70** (1969), 37–60.

[962] George W. Roberts, 'Some Aspects of Knowledge (I)', pp. 348–383 in G. W. Roberts (ed.), *Bertrand Russell Memorial Volume*, Allen and Unwin, London 1979.

[963] Philip Robbins, 'The Myth of Reverse Compositionality', *Philosophical Studies* **125** (2005), 251–275.

[964] Mark van Roojen, "Moral Cognitivism vs. Non-Cognitivism", *Stanford Encyclopedia of Philosophy* (Summer 2014 Edition), Edward N. Zalta (ed.), URL = <http://plato.stanford.edu/archives/sum2014/entries/moral-cognitivism/>.

[965] Alan Rose, 'Self-Dual Primitives for Modal Logic', *Mathematische Annalen* **125** (1953), 284–286.

[966] Tobias Rosefeldt, 'Is Knowing-how Simply a Case of Knowing-that?', *Philosophical Investigations* **27** (2004), 370–379.

[967] Shalom Rosenberg, 'On the Modal Version of the Ontological Argument', *Logique et Analyse* **24** (1981), 130–133.

[968] J. G. Rosenstein, *Linear Orderings*, Academic Press, NY 1982.

[969] David M. Rosenthal, 'Moore's Paradox and Crimmins's Case', *Analysis* **62** (2002), 167–171.

[970] R. Routley, 'The Decidability and Semantical Incompleteness of Lemmon's System S0.5', *Logique et Analyse* **11** (1968), 413–421.

[971] R. Routley, 'Alleged Problems in Attributing Beliefs, and Intentionality, to Animals', *Inquiry* **24** (1981), 385–417.

[972] R. Routley and H. Montgomery, 'The Inadequacy of Kripke's Semantical Analysis of D2 and D3', *Journal of Symbolic Logic* **33** (1968), 568.

[973] R. Routley and V. Plumwood, 'Moral Dilemmas and the Logic of Deontic Notions', pp. 653–690 in G. Priest, R. Routley and J. Norman (eds.), *Paraconsistent Logic: Essays on the Inconsistent*, Philosophia Verlag, Munich 1989.

[974] R. Routley and V. Routley, 'A Fallacy of Modality', *Noûs* **3** (1969), 129–153.

[975] R. Routley and V. Routley, 'Semantics of First Degree Entailment', *Noûs* **6** (1972), 335–359.

[976] R. Routley and V. Routley, 'The Role of Inconsistent and Incomplete Theories in the Logic of Belief', *Communication and Cognition* **8** (1975) 185–235.

[977] Aynat Rubinstein, 'On Necessity and Comparison', *Pacific Philosophical Quarterly* **95** (2014), 512–554.

[978] Ian Rumfitt, 'Knowledge by Deduction', *Grazer Philosophische Studien* **77** (2008), 61–84.

[979] Ian Rumfitt, 'On a Neglected Path to Intuitionism', *Topoi* **31** (2012), 101–109.

[980] Gillian Russell and Delia Graff Fara (eds.), *The Routledge Companion to Philosophy of Language*, Routledge, New York 2012.

[981] V. V. Rybakov, 'A Modal Analog for Glivenko's Theorem and its Applications', *Notre Dame Journal of Formal Logic* **33** (1992), 244–248.

[982] Esa Saarinen 'Intentional Identity Interpreted', *Linguistics and Philosophy* **2** (1978), 151–224.

[983] Esa Saarinen, 'Backwards-Looking Operators in Tense Logic and in Natural Language', pp. 341–367 in J. Hintikka, I. Niiniluoto, and E. Saarinen (eds.), *Essays on Mathematical and Philosophical Logic*, Reidel, Dordrecht 1979. (Also at pp. 215–244 in E. Saarinen (ed.), *Game-Theoretical Semantics*, Reidel, Dordrecht 1979.)

[984] Esa Saarinen, 'Quantifier Phrases are (at Least) Five Ways Ambiguous in Intensional Contexts', pp. 1–45 in F. Heny (ed.), *Ambiguities in Intensional Contexts*, Reidel, Dordrecht 1981.

[985] Esa Saarinen, 'Propositional Attitudes are not Attitudes Towards Propositions', pp. 130–162. I. Niiniluoto and E. Saarinen (eds.), *Intensional Logic: Theory and Applications* (= *Acta Philosophica Fennica* **35**), Helsinki 1982.

[986] Esa Saarinen, 'How to Frege a Russell-Kaplan' *Noûs* **16** (1982), 253–276.

[987] Kazem Sadegh-Zadeh, Review of Lenzen [710], *Metamedicine* **3** (1982), 297–307. (Note that this journal has since been renamed as *Theoretical Medicine*.)

[988] H. Sahlqvist, 'Completeness and Correspondence in the First and Second Order Semantics for Modal Logic', pp. 110–143 in S. Kanger (ed.), *Procs. of the Third Scandinavian Logic Symposium, Uppsala 1973*, North-Holland, Amsterdam 1975.

[989] R. M. Sainsbury, 'Easy Possibilities', *Philosophy and Phenomenological Research* **57** (1997), 907–919.

[990] Joe Salerno (ed.), *New Essays on the Knowability Paradox*, Oxford University Press, Oxford 2009.

[991] Nathan Salmon, *Reference and Essence*, Princeton University Press, Princeton 1981.

[992] Nathan Salmon, 'The Logic of What Might Have Been', *Philosophical Review* **98** (1989), 3–34.

[993] Crispin Sartwell, 'Knowledge is Merely True Belief', *American Philosophical Quarterly* **28** (1991), 157–165.

[994] Crispin Sartwell, 'Why Knowledge is Merely True Belief', *Journal of Philosophy* **89** (1992), 167–180.

[995] K. W. Saunders, 'A Formal Analysis OF Hohfeldian Relations', *Akron Law Review* **23** (1989–90), 499-506.

[996] Geoffrey Sayre-McCord, 'Deontic Logic and the Priority of Moral Theory', *Noûs* **20** (1986), 179–197.

[997] Jonathan Schaffer, 'From Contextualism to Contrastivism', *Philosophical Studies* **119** (2004), 73–103.

[998] Stacy Schiff, 'All the Kings Siblings', Review of Stella Tillyard, *A Royal Affair: George III and His Scandalous Siblings*, available at <http://www.nytimes.com/2006/12/31/books/review/Schiff.t.html>.

[999] P. K. Schotch 'Skepticism and Epistemic Logic', *Studia Logica* **65** (2000), 187–198.

[1000] P. K. Schotch and R. E. Jennings, 'Modal Logic and the Theory of Modal Aggregation', *Philosophia* **9** (1975), 265–278.

[1001] P. K. Schotch and R. E. Jennings, 'Epistemic Logic, Skepticism, and Non-Normal Modal Logic', *Philosophical Studies* **40** (1981), 47–67.

[1002] P. K. Schotch and R. E. Jennings, 'Non-Kripkean Deontic Logic', pp. 149–162 in R. Hilpinen (ed.), *New Studies in Deontic Logic*, Reidel, Dordrecht 1981.

[1003] J. A. Schreider, *Equality, Resemblance, and Order*, (transl. Martin Greendlinger) Mir Publishers, Moscow 1975.

[1004] George F. Schumm, 'On Some Open Questions of B. Sobociński', *Notre Dame Journal of Formal Logic* **10** (1969), 2610–261

[1005] George F. Schumm, 'Disjunctive Extensions of **S4** and a Conjecture of Goldblatt's', *Zeitschr. für math. Logik und Grundlagen der Math.* **21** (1975), 81–86.

[1006] George F. Schumm, 'An Incomplete Nonnormal Extension of S3', *Journal of Symbolic Logic* **43** (1978), 211–212.

[1007] George F. Schumm, 'Modal Logics with no Minimal Proper Extensions', *Studia Logica* **37** (1978), 233–235.

[1008] George F. Schumm, 'The Number of $\{\Box, \to\}$-Logics', *Zeitschr. für math. Logik und Grundlagen der Math.* **36** (1990), 517–518.

[1009] George F. Schumm, 'Why Does Halldén-Completeness Matter?', *Theoria* **59** (1993), 192–206.

[1010] George F. Schumm and R. Edelstein, 'Negation-free Modal Logics', *Zeitschr. für math. Logik und Grundlagen der Math.* **25** (1979), 281–288.

[1011] Gerhard Schurz, 'Relevant Deduction', *Erkenntnis* **35** (1991), 391–437.

[1012] Gerhard Schurz, 'Hume's Is-Ought Thesis in Logics with Alethic-Deontic Bridge Principles', *Logique et Analyse* **37** (1994), 265–293.

[1013] Gerhard Schurz, *The Is-Ought Problem: An Investigation in Philosophical Logic*, Kluwer, Dordrecht 1997.

[1014] Gerhard Schurz, 'Carnap's Modal Logic', pp. 365–380 in W. Stelzner and M. Stöckler (eds.), *Zwischen traditioneller und moderner Logik*, Mentis Verlag, Paderborn 2001.

[1015] Gerhard Schurz and Georg Dorn (eds.), *Advances in Scientific Philosophy: Essays in Honour of Paul Wiengartner on the Occasion of the 60th Anniversary of his Birthday*, Rodopi, Amsterdam 1991.

[1016] G. Schwarz, 'Minimal Model Semantics for Nonmonotonic Modal Logics', pp. 34–43 in *Seventh Annual IEEE Symposium on Logic in Computer Science*, IEEE Computer Society Press, Los Alamitos, CA 1992.

[1017] G. Schwarz and M. Truszczyński, 'Modal Logic **S4F** and the Minimal Knowledge Paradigm', pp. 184–198 in *Theoretical Aspects of Rationality and Knowledge: Proceedings of the 4th conference on Theoretical Aspects of Reasoning about Knowledge*, Morgan Kaufmann, San Francisco, CA 1992.

[1018] G. Schwarz and M. Truszczyński, 'Minimal Knowledge Problem: A New Approach', *Artificial Intelligence* **67** (1994), 113–141.

[1019] Dana Scott, 'Advice on Modal Logic', pp. 143–173 in K. Lambert (ed.), *Philosophical Problems in Logic*, Reidel, Dordrecht 1970.

[1020] Schiller Joe Scroggs, 'Extensions of the Lewis System **S5**', *Journal of Symbolic Logic* **16** (1951), 112–120.

[1021] John R. Searle, *Speech Acts*, Cambridge University Press, Cambridge 1969.

[1022] Krister Segerberg, 'On the Logic of "Tomorrow"', *Theoria* **33** (1967), 45–52.

[1023] Krister Segerberg, 'Some Modal Logics based on a Three-valued Logic' *Theoria* **33** (1967), 53–71.

[1024] Krister Segerberg, 'Modal Logics with Linear Alternative Relations', *Theoria* **36** (1970), 301–322.

[1025] Krister Segerberg, *An Essay in Classical Modal Logic*, Filosofiska Studier, Uppsala 1971.

[1026] Krister Segerberg, 'Qualitative Probability in a Modal Setting', pp. 341–352 in J. E. Fenstad (ed.), *Procs. of the Second Scandinavian Logic Symposium*, North-Holland, Amsterdam 1971.

[1027] Krister Segerberg, 'Two-Dimensional Modal Logic', *Journal of Philosophical Logic* **2** (1973), 77–96.

[1028] Krister Segerberg, 'That Every Extension of **S4.3** is Normal', pp. 194–196 in S. Kanger (ed.), *Procs. of the Third Scandinavian Logic Symposium*, North-Holland, Amsterdam 1975.

[1029] Krister Segerberg, '"Somewhere else" and "Some other time"', pp. 61–64 in *Wright and Wrong: Mini-Essays in Honor of George Henrik von Wright on his Sixtieth Birthday*, Publications of the Group in Logic and Methodology of Real Finland, Vol. 3, Åbo Akademi 1976.

[1030] Krister Segerberg, 'A Note on the Logic of Elsewhere', *Theoria* **46** (1980), 183–187.

[1031] Krister Segerberg , 'Applying Modal Logic', *Studia Logica* **39** (1980), 275–295.

[1032] Krister Segerberg, *Classical Propositional Operators*, Oxford 1982.

[1033] Krister Segerberg, 'A Deontic Logic of Action', *Studia Logica* **41** (1982), 269–282.

[1034] Krister Segerberg, 'Modal Logics with Functional Alternative Relations', *Notre Dame Journal of Formal Logic* **27** (1986), 504–522.

[1035] Krister Segerberg, 'Getting Started: Beginnings in the Logic of Action', *Atti del Convegno Internazionale della Logica: Teoria delle Modalite*, CLUEB, Bologna 1989; also in *Studia Logica* **51** (1992), 347–378.

[1036] Krister Segerberg, 'Two Traditions in the Logic of Belief: Bringing Them Together', pp. 135–147 in H. J. Ohlbach and U. Reyle (eds.), *Logic, Language and Reasoning*, Kluwer, Dordrecht 1999.

[1037] Krister Segerberg, 'Blueprint for a Dynamic Deontic Logic', *Journal of Applied Logic* **7** (2009) 388–402.

[1038] R. V. Setlur, 'On the Equivalence of Strong and Weak Validity of Rule Schemes in the Two-Valued Propositional Calculus', *Notre Dame Journal of Formal Logic* **11** (1970), 249–253.

[1039] Pieter A. M. Seuren, 'Negative's Travels', Chapter 8 in [1041]. (Originally publ. 1974.)

[1040] Pieter A. M. Seuren, 'The Paradoxes and Natural Language', pp. 119–135 in [1041]. (First published in French, 1987.)

[1041] Pieter A. M. Seuren, *A View of Language*, Oxford University Press, Oxford 2001.

[1042] Pieter A. M. Seuren, 'Reflexivity and Identity in Language and Cognition', Chapter 7 in Seuren, *From Whorf to Montague*, Oxford University Press, Oxford 2013.

[1043] Jonathan Schaffer, 'Knowing the Answer'. *Philosophy and Phenomenological Research* **75** (2007), 383–403.

[1044] Yael Sharvit, 'Embedded Questions and De Dicto Readings', *Natural Language Semantics* **10** (2002), 97–123.

[1045] V. B. Shehtman, 'A Logic With Progressive Tenses', pp. 255–285 in M. de Rijke (ed.), *Diamonds and Defaults*, Kluwer, Dordrecht 1993.

[1046] V. B. Shehtman, 'On Neighbourhood Semantics Thirty Years Later', pp. 663–691 in S. Artemov, H. Barringer, A. d'Avila Garcez, L C. Lamb and J. Woods (eds.), *We Will Show Them! Essays in Honour of Dov Gabbay on his 60th Birthday, Vol. 2*, College Publications, London 2005.

[1047] Sydney Shoemaker, 'On Knowing One's Own Mind', pp. 183–209 in James E. Tomberlin, ed., *Philosophical Perspectives* **2** (Epistemology), Ridgeview Publ. Co., Atascadero, California 1988. (Also appearing as Chapter 2 of [1048].)

[1048] Sydney Shoemaker, *The First-Person Perspective and Other Essays*, Cambridge University Press, Cambridge 1996.

[1049] Sydney Shoemaker, 'Self-Intimation and Second-Order Belief', pp. 239–257 in [1067].

[1050] D. J. Shoesmith and T. J. Smiley, *Multiple-Conclusion Logic*, Cambridge University Press, Cambridge 1978.

[1051] Robert K. Shope, *The Analysis of Knowing: A Decade of Research*, Princeton University Press, Princeton, NJ 1983.

[1052] Alex Silk, 'Evidence Sensitivity in Weak Necessity Deontic Modals', *Journal of Philosophical Logic* **43** (2014), 691–723.

[1053] Mandy Simons, 'Dividing Things Up: The Semantics of *Or* and the Modal/*Or* Interaction', *Natural Language Semantics* **13** (2005), 271–316.

[1054] Mandy Simons, 'Semantics and Pragmatics in the Interpretation of *Or*', pp. 205–222 in E. Georgala and J. Howell (eds.), *Semantics and Linguistic Theory* **15** (Proceedings *SALT XV*), CLC Publications, Cornell, Ithaca, NY 2005.

[1055] Marcus G. Singer, *Generalisation in Ethics*, Alfred A. Knopf, NY 1961.

[1056] Marcus G. Singer, 'Negative and Positive Duties', *Philosophical Quarterly* **15** (1965), 97–103.

[1057] Neil Sinhababu, 'Advantages of Propositionalism', *Pacific Philosophical Quarterly* **96** (2015), 165–180.

[1058] Walter Sinnott-Armstrong, 'A Solution to Forrester's Paradox of Gentle Murder', *Journal of Philosophy* **82** (1985), 162–168.

[1059] Quentin Skinner, *Liberty Before Liberalism*, Cambridge University Press, Cambridge 1998.

[1060] Hartley Slater, 'Frege's Hidden Assumption', *Crítica: Revista Hispanoamericana de Filosofía*, **38** (2006), 27–37.

[1061] Hartley Slater, 'Consistent Truth', *Ratio* (new series) **27** (2013), 247–261.

[1062] Timothy Smiley, 'On Łukasiewicz's L-modal System', *Notre Dame Journal of Formal Logic* **2** (1961), 149–153.

[1063] Timothy Smiley, 'Analytic Implication and 3-valued Logic' (Abstract), *Journal of Symbolic Logic* **27** (1962), 378.

[1064] Timothy Smiley, 'Relative Necessity', *Journal of Symbolic Logic* **28** (1963), 113–134.

[1065] Timothy Smiley, 'The Logical Basis of Ethics', *Acta Philosophica Fennica* **16** (1963), 237–246.

[1066] Carlota S. Smith, 'The Syntax and Interpretation of Temporal Expressions in English', *Linguistics and Philosophy* **2** (1978), 43–99.

[1067] Declan Smithies and Daniel Stoljar (eds.), *Introspection and Consciousness*, Oxford University Press, Oxford 2012.

[1068] Declan Smithies and Daniel Stoljar, 'Introspection and Consciousness: An Overview', pp. 3–25 in [1067].

[1069] Raymond Smullyan, 'Equivalence Relations and Groups', pp. 261–271 in C. Anthony Anderson *et al.* (eds.), *Logic, Meaning and Computation: Essays in Memory of Alonzo Church*, Kluwer, Dordrecht 2001.

[1070] Paul Snowdon, 'Knowing How and Knowing That: A Distinction Reconsidered', *Procs. of the Aristotelian Society* **104** (2004), 1–29.

[1071] J. H. Sobel, *Logic and Theism: Arguments for and against Beliefs in God*, Cambridge University Press, Cambridge 2003.

[1072] B. Sobociński, 'Note on G. J. Massey's Closure-Algebraic Operation', *Notre Dame Journal of Formal Logic* **11** (1970), 343–346 (and errata, p. 584).

[1073] Roy Sorensen, *Blindspots*, Oxford University Press, Oxford 1988.

[1074] Roy Sorensen, 'Does Apriority Agglomerate?', Chapter 6 of Sorensen, *Vagueness and Contradiction*, Oxford University Press, Oxford 2001.

[1075] Ernest Sosa, 'How to Defeat Opposition to Moore', *Philosophical Perspectives* **13** (1999), 141–153.

[1076] Robert Stalnaker, 'The Problem of Logical Omniscience, I', *Synthese* **89** (1991), 425–440; reprinted as item 13 in the collection [1079], pp. 241–254.

[1077] Robert Stalnaker, 'The Problem of Logical Omniscience, II', pp. 255–273 in [1079].

[1078] Robert Stalnaker, 'A Note on Non-monotonic Logic', *Artificial Intelligence* **64** (1993), 183–196.

[1079] Robert Stalnaker, *Context and Content*, Oxford University Press, Oxford 1999.

[1080] Robert Stalnaker, *Ways a World Might Be: Metaphysical and Anti-Metaphysical Essays*, Oxford University Press, Ocxford 2003.

[1081] Robert Stalnaker, 'Impossibilities', pp. 55–67 in [1080]. (Expanding an article orig. publ. 1996.)

[1082] Robert Stalnaker, 'On Logics of Knowledge and Belief', *Philosophical Studies* **128** (2006), 169–199.

[1083] Robert Stalnaker, 'Possible Worlds Semantics: Philosophical Foundations', pp. 100–115 in A. Berger (ed.), *Kripke*, Cambridge University Press, Cambridge 2011.

[1084] Robert Stalnaker, *Mere Possibilities: Metaphysical Foundations of Modal Semantics*, Princeton University Press, Princeton, NJ 2012.

[1085] Jason Stanley, *Know How*, Oxford University Press, Oxford 2011.

[1086] Jason Stanley and Timothy Williamson, 'Knowing How', *Journal of Philosophy* **98** (2001), 411–444.

[1087] Christopher Steinsvold, 'Completeness for Various Logics of Essence and Accident', *Bulletin of the Section of Logic* **37** (2008), 93–101.

[1088] Christopher Steinsvold, 'Being Wrong: the Logic of False Belief', *Note Dame Journal of Formal Logic* **52** (2011), 245–253.

[1089] Matthias Steup, 'Are Mental States Luminous?', pp. 217–236 in P. Greenough and D. Pritchard (eds.), *Williamson on Knowledge*, Oxford University Press, Oxford 2009.

[1090] Matthias Steup and Ernest Sosa, *Contemporary Debates in Epistemology*, Blackwell Publishing, Oxford 2005.

[1091] Gail Stine, 'Dretske on Knowing the Logical Consequences', *Journal of Philosophy* **68** (1971), 296–299.

[1092] Stephen P. Stich, *From Folk Psychology to Cognitive Science: the Case Against Belief*, MIT Press, Cambridge, MA 1983.

[1093] R. L. Sturch, 'Moral Non-Dogmatism', *Mind* **79** (1970), 122–125.

[1094] Frederick Suppe, *The Semantic Conception of Scientific Theories*, University of Illinois Press, Urbana 1989.

[1095] Frederick Suppe, 'Science Without Induction', pp. 386–429 in J. Earman and J. D. Norton (eds.), *The Cosmos of Science: Essays of Exploration*, University of Pittsburgh Press, Pittsburgh 1997.

[1096] Eve Sweetser, *From Etymology to Pragmatics*, Cambridge University Press, Cambridge 1990.

[1097] Jari Talja, 'On the Logic of Omissions', *Synthese* **65** (1985), 235–248.

[1098] B. M. Taylor, 'Tense and Continuity', *Linguistics and Philosophy* **1** (1977), 199–220.

[1099] Larry S. Temkin, 'A Continuum Argument for Intransitivity', *Philosophy and Public Affairs* **25** (1996), 175–210.

[1100] E. Thijsse, 'Logics of Consciousness Explained and Compared: Partial Approaches to Actual Belief', *Logique et Analyse* **34** (1991), 221–250.

[1101] E. Thijsse, 'On Total Awareness Logics', pp. 309–347 in M. de Rijke (ed.), *Diamonds and Defaults*, Kluwer, Dordrecht 1993.

[1102] E. Thijsse and H. Wansing, 'A Fugue on the Themes of Awareness Logic and Correspondence', *Journal of Applied Non-Classical Logics* **6** (1996), 127–136.

[1103] Ivo Thomas, 'A Theorem on **S4.2** and **S4.4**', *Notre Dame Journal of Formal Logic* **8** (1967), 335–336.

[1104] Richmond H. Thomason 'Indeterminist Time and Truth-Value Gaps', *Theoria* **36** (1970), 264–281.

[1105] Richmond H. Thomason, 'Deontic Logic as Founded on Tense Logic', pp. 165–176 in Hilpinen [483].

[1106] Sandra A. Thompson and Anthony Mulac, 'A Quantitaive Perspectove on the Grammaticization of Epistemic Parentheticals in English', pp.313–329 in E. C. Traugott and B. Heine (eds.), *Approaches to Grammaticalization, Volume II: Focus on Tyoes of Grammatical Markers*, John Benjamins, Amsterdam 1991.

[1107] Michael Titelbaum, 'How to Derive a Narrow-Scope Requirement from Wide-Scope Requirements', *Philosophical Studies* **172** (2015), 535–542.

[1108] Marcin Tkaczyk, 'On Axiomatization of Łukasiewicz's Four-Valued Modal Logic', *Logic and Logical Philosophy* **21** (2011), 215–232.

[1109] Marek Tokarz, 'On the Logic of Conscious Belief', *Studia Logica* **49** (1990), 321–332.

[1110] James E. Tomberlin, 'Good Samaritans and Castañeda's System of Deontic Logic', pp. 255–2272 in Tomberlin (ed.), *Hector-Neri Castañeda*, Reidel, Dordrecht 1986.

[1111] Håkan Törnebohm, 'Notes on Modal Operators', *Theoria* **24** (1958), 130–135.

[1112] L. van der Torre and Y.-H. Tan, 'Prohairetic Deontic Logic (PDL)', pp.77–91 in J. Dix, L. Fariñas del Cerro and U. Furbach (eds.), *JELIA '98*, Lecture Notes in Artificial Intelligence #1489, Springer-Verlag, Berlin 1998

[1113] K. E. Tranøy, 'Deontic Logic and Deontically Perfect Worlds', *Theoria* **36** (1970), 221–231.

[1114] Cheng-Chih Tsai, 'The Genesis of Hi-Worlds: Towards a Principle-Based Possible World Semantics', *Erkenntnis* **76** (2012), 101–114.

[1115] Savas L. Tsohatzidis, 'More Telling Examples: A response to Holton', *Journal of Pragmatics* **28** (1997), 629–636.

[1116] Vasilis Tsompanidis, 'The Structure of Propositions and Crosslinguistic Syntactic Variability', *Croatian Journal of Philosophy* **13** (2013), 399–419.

[1117] Lidia Typanska, 'A Note on Post-complete Extensions of a Logic of Values of A. Ivin', *Bulletin of the Section of Logic* **42** (2013), 161–168.

[1118] Sara L. Uckelman, 'Anselm's Logic of Agency', *Logical Analysis and History of Philosophy* **12** (2009), 248–268.

[1119] Peter Unger, *Ignorance: A Case for Scepticism*, Oxford University Press, Oxford 1975.

[1120] J. O. Urmson, 'Saints and Heroes', pp. 198–216 in Melden [798].

[1121] Alasdair Urquhart, 'Semantics for Relevant Logics', *Journal of Symbolic Logic* **37** (1972), 159–169.

[1122] Alasdair Urquhart, 'A Semantical Theory of Analytic Implication', *Journal of Philosophical Logic* **2** (1973), 212–219.

[1123] Zeno Vendler, *Linguistics in Philosophy*, Cornell University Press, Ithaca, NY 1967.

[1124] Zeno Vendler, *Res Cogitans: An Essay in Rational Psychology*, Cornell University Press, Ithaca NY 1972.

[1125] Yde Venema, *Many-Dimensional Modal Logic*, Academisch Proefschrift, University of Amsterdam 1991.

[1126] Yde Venema, 'A Note on the Tense Logic of Dominoes', *Journal of Philosophical Logic* **21** (1992), 172–182.

[1127] Yde Venema, 'Derivation Rules as Anti-Axioms in Modal Logic', *Journal of Symbolic Logic*, **58** (1993), 1003–1034.

[1128] Yde Venema, 'Canonical Pseudo-Correspondence', pp. 421–430 in M. Zakharyaschev, K. Segerberg, M. de Rijke and H. Wansing (eds.), *Advances in Modal Logic, Vol. 2*, CSLI Publications, Stanford, CA 2001.

[1129] Yde Venema, 'Temporal Logic', Chapter 10 (pp. 203–223) in Goble [385].

[1130] H. J. Verkuyl and J. A. Le Loux-Schuringa 'Once Upon a Tense', *Linguistics and Philosophy* **8** (1985), 237–261.

[1131] S. Vikner, 'Reichenbach Revisited: One, Two, or Three Temporal Relations', *Acta Linguistica Hafniensia* **19** (1985), 81–95.

[1132] D. W. Viney, *Charles Hartshorne and the Existence of God*, State University of New York Press, Albany, NY 1985.

[1133] Frank Vlach, 'Temporal Adverbials, Tenses and the Perfect', *Linguistics and Philosophy* **16**, 231–283 (1993).

[1134] Giorgio Volpe 'Minimalism and Normative Reasoning: A Reply to Sean Coyle', *Ratio Juris* **15** (2002) 319-27.

[1135] Frans Voorbraak, 'The Logic of Actual Obligation, an Alternative Approach to Deontic Logic', *Philosophical Studies* **55** (1989), 173-194.

[1136] Frans Voorbraak, 'The Logic of Objective Knowledge and Rational Belief', pp. 499–515 in J. van Eijck (ed.), *Logics in AI: European Workshop JELIA '90*, Lecture Notes in Artificial Intelligence #478, Springer-Verlag, Berlin 1991.

[1137] Mark Vorobej, 'Deontic Accessibility', *Philosophical Studies* **41** (1982), 317–319.

[1138] Daniel von Wachter, 'Belief, Knowledge, and Omniscience' (review of [1163]), *Grazer Philosophische Studien* **83** (2011) 267–279.

[1139] Józef Wajszczyk, 'The Logic of Dichotomic Changes', *Bulletin of the Section of Logic* **24** (1995), 89–97.

[1140] David Wall, 'A Moorean Paradox of Desire', *Philosophical Explorations* **15**, 63–84.

[1141] Douglas Walton, 'Modal Logic and Agency', *Logique et Analyse* **18** (1975), 103–111.

[1142] Douglas Walton, 'Principles of Interpersonal Agency in the Free Will Defense', *Bijdragen* **37** (1976), 36–46.

[1143] Douglas Walton, 'Time and Modality in the "Can" of Opportunity', pp. 271–287 in M. Brand and D. Walton (eds.), *Action Theory*, Reidel, Dordrecht 1976.

[1144] Douglas Walton, 'Logical Form and Agency', *Philosophical Studies* **29** (1976) 75–89.

[1145] Hao Wang, *Popular Lectures on Mathematical Logic*, Van Nostrand, NY 1981.

[1146] Heinrich Wansing, 'A General Possible Worlds Framework for Reasoning About Knowledge and Belief', *Studia Logica* **49** (1990), 523–539.

[1147] Heinrich Wansing, 'A New Axiomatization of K_t', *Bulletin of the Section of Logic* **25** (1996), 60–62.

[1148] Heinrich Wansing (ed.), *Proof Theory of Modal Logic*, Springer-Verlag, Berlin 1996.

[1149] Heinrich Wansing, 'Nested Deontic Modalities: Another View of Parking on Highways', *Erkenntnis* **49** (1998), 185–199.

[1150] R. J. Watts, 'The Conjunction *That*: A Semantically Empty Particle?', *Studia Anglica Posnaniensia* **15** (1982), 13–37.

[1151] Brian Weatherson, 'Questioning Contextualism', pp.133–147 in S. Hetherington (ed.), *Aspects of Knowing: Epistemological Essays*, Elsevier, Oxford 2006.

[1152] Brian Weatherson, 'Intrinsic vs. Extrinsic Properties', *Stanford Encyclopedia of Philosophy* (Fall 2008 Edition), Edward N. Zalta (ed.), URL = <http://plato.stanford.edu/archives/fall2008/entries/intrinsic-extrinsic/>.

[1153] 'The Meaning of "Ought"', pp. 127–160 in R. Shafer-Landau (ed.), *Oxford Studies in Metaethics, Vol. 1*, Oxford University Press, Oxford 2006.

[1154] Kai F. Wehmeier, 'Wittgensteinian Predicate Logic', *Notre Dame Journal of Formal Logic* **45** (2004), 1-11.

[1155] Kai F. Wehmeier, 'Wittgensteinian Tableaux, Identity, and Co-Denotation', *Erkenntnis* **69** (2008), 363–376.

[1156] Kai F. Wehmeier, 'How to Live Without Identity—and Why', *Australasian Journal of Philosophy* **90** (2012), 761–777.

[1157] Kai Wehmeier, 'Subjunctivity and Cross-World Predication', *Philosophical Studies* **159** (2012) 107–122.

[1158] Kai Wehmeier, 'Nothing But d-Truth', *Analytic Philosophy* **55** (2014), 114–117.

[1159] Jonathan M. Weinberg, Shaun Nichols, and Stephen Stich, 'Normativity and Epistemic Intuitions', *Philosophical Topics* **29** (2001), 429–460.

[1160] Ota Weinberger, 'The Logic of Norms Founded on Descriptive Language', *Ratio Juris* **4** (1991) 284–307.

[1161] Paul Weingartner, 'A Simple Relevance-Criterion for Natural Language and its Semantics', pp. 563–575 in G. Dorn and P. Weingartner (eds.), *Foundations of Logic and Linguistics*, Plenum Press, New York 1985.

[1162] Paul Weingartner, 'Conditions of Rationality for the Concepts Belief, Knowledge, and Assumption', *Dialectica* **36** (1982), 243–263.

[1163] Paul Weingartner, *Omniscience: From a Logical Point of View*, Ontos Verlag, Frankfurt 2008.

[1164] Paul Weingartner and Gerhard Schurz, 'Paradoxes Solved by Simple Relevance Criteria', *Logique et Analyse* **29** (1986) 3–40

[1165] Tom Werner, 'You Do What You Gotta Do, or Why *must* Implies *will*', pp. 294–308 in E. Georgala and J. Howell (eds.), *Semantics and Linguistic Theory* **15** (Proceedings *SALT XV*), CLC Publications, Cornell, Ithaca, NY 2005.

[1166] Roger Wertheimer, *The Significance of Sense*, Cornell University Press, Ithaca, NY 1972.

[1167] J. M. O. Wheatley, 'Wishing and Hoping', *Analysis* **18** (1958), 121–131.

[1168] Alan R. White, *Modal Thinking*, Basil Blackwell, Oxford 1975.

[1169] Roger White, 'Wittgenstein on Identity', *Proceedings of the Aristotelian Society* **78** (1978), 157–174.

[1170] Bernard Williams, 'Ethical Consistency', *Aristotelian Society Supplementary Vol. 39* (1965) 103–124; reprinted in [1173], pp. 166–186.

[1171] Bernard Williams, 'Deciding to Believe', pp. 136–51 in Williams [1173].

[1172] Bernard Williams, *Morality: An Introduction to Ethics*, Cambridge University Press, Cambridge 1972.

[1173] Bernard Williams, *Problems of the Self*, Cambridge University Press, Cambridge 1973.

[1174] Bernard Williams, *Descartes: The Project of Pure Enquiry*, Penguin Books, Harmonsdworth, Middlesex, England 1978.

[1175] Bernard Williams, *Truth and Truthfulness*, Princeton University Press, Princeton, NJ 2002.

[1176] John N. Williams, 'Moore's Paradox: One or Two?', *Analysis* **39** (1979), 141–142.

[1177] John N. Williams, 'Propositional Knowledge and Know-How', *Synthese* **165** (2008), 107–125.

[1178] John N. Williams, 'Moore's Paradox in Belief and Desire', *Acta Analytica* **29** (2014), 1–23.

[1179] Timothy Williamson, 'Two Incomplete Anti-realist Modal Epistemic Logics', *Journal of Symbolic Logic* **55** (1990), 297–314.

[1180] Timothy Williamson, 'A Relation Between Namesakes in Modal Logic', *Bulletin of the Section of Logic* **20** (1991), 129–135.

[1181] Timothy Williamson, 'An Alternative Rule of Disjunction in Modal Logic', *Note Dame Journal of Formal Logic* **33** (1992), 89–100.

[1182] Timothy Williamson, 'Inexact Knowledge', *Mind* **101** (1992), 217–242.

[1183] Timothy Williamson, 'Verificationism and Non-Distributive Knowledge', *Australasian Journal of Philosophy* **71** (1993), 78–86.

[1184] Timothy Williamson, 'Non-Genuine MacIntosh Logics', *Journal of Philosophical Logic* **23** (1994), 87–101.

[1185] Timothy Williamson, 'Self-Knowledge and Embedded Operators', *Analysis* 56 (1996), 202–209.

[1186] Timothy Williamson, 'Iterated Attitudes', pp. 85–133 in T. J. Smiley (ed.), *Philosophical Logic* (*Procs. of the British Academy* **95**), Oxford University Press, Oxford 1998.

[1187] Timothy Williamson, 'Continuum Many Maximal Consistent Normal Bimodal Logics with Inverses', *Notre Dame Journal of Formal Logic* **39** (1998), 1128–1134.

[1188] Timothy Williamson, 'Rational failures of the KK principle', pp. 101–118 in C. Bicchieri, R. Jeffrey and B. Skyrms, eds., *The Logic of Strategy*, Oxford University Press, Oxford 1999.

[1189] Timothy Williamson, *Knowledge and its Limits*, Oxford University Press, Oxford 2000.

[1190] Timothy Williamson, 'Knowledge, Context, and the Agent's Point of View', pp. 91–114 in G. Preyer and G. Peter (eds.), *Contextualism in Philosophy: Knowledge, Meaning, and Truth*, Oxford University Press, Oxford 2005.

[1191] Timothy Williamson, 'Contextualism, Subject-Sensitive Invariantism and Knowledge of Knowledge', *Philosophical Quarterly* **55** (2005), 213–235.

[1192] Timothy Williamson, 'Indicative versus Subjunctive Conditionals, Congruential versus Non-Hyperintensional Contexts', *Philosophical Issues* **16** (2006), 310–333.

[1193] Timothy Williamson, 'Stalnaker on the Interaction of Modality with Quantification and Identity', pp. 123–147 in J. Thomson and A. Byrne (eds.), *Content and Modality: Themes from the Philosophy of Robert Stalnaker*, Oxford University Press, Oxford 2006.

[1194] Timothy Williamson, 'Probability and Danger', *The Amherst Lecture in Philosophy* **4** (2009), 1–35.

[1195] Timothy Williamson, 'Some Computational Constraints in Epistemic Logic', pp. 437–456 in S. Rahman, J. Symons, D. M. Gabbay, and J. P. van Bendegem (eds.), *Logic, Epistemology, and the Unity of Science, Vol. 1*, Springer, Dordrecht 2009.

[1196] Timothy Williamson, 'Improbable Knowing', pp. 147–164 in T. Dougherty (ed.), *Evidentialism and Its Discontents*, Oxford University Press, Oxford 2011.

[1197] Timothy Williamson, 'Gettier Cases in Epistemic Logic', *Inquiry* **56** (2013), 1–14.

[1198] Timothy Williamson, *Modal Logic as Metaphysics*, Oxford University Press, Oxford 2013.

[1199] Timothy Williamson, 'Logic, Metalogic and Neutrality', *Erkenntnis* **79** (2014), 211–231.

[1200] Timothy Williamson, 'How Did We Get Here from There? The Transformation of Analytic Philosophy', to appear in *Belgrade Philosophical Annual*.

[1201] Deirdre Wilson, *Presupposition and Non-Truth-Conditional Semantics*, Academic Press, London 1975.

[1202] Ryszard Wójcicki, *Theory of Logical Calculi*, Kluwer, Dordrecht 1988.

[1203] Crispin Wright, 'Relativism and Classical Logic', pp. 95–118 in A. O'Hear (ed.), *Logic, Thought and Language*, Royal Institute of Philosophy Supplement: 51, Cambridge University Press, Cambridge 2002.

[1204] G. H. von Wright, 'Deontic Logic', *Mind* **60** (1951), 1–15; reprinted in pp. 58–74 of [1207].

[1205] G. H. von Wright, *An Essay in Modal Logic*, North-Holland, Amsterdam 1951.

[1206] G. H. von Wright, 'On the Logic of some Axiological and Epistemological Concepts', *Ajatus* **17** (1952), 213–234.

[1207] G. H. von Wright, *Logical Studies*, Routledge and Kegan Paul, London 1957.

[1208] G. H. von Wright, *Norm and Action*, Routledge and Kegan Paul, London 1963.

[1209] G. H. von Wright, *The Logic of Preference: An Essay*, Edinburgh University Press, Edinburgh 1963.

[1210] Stephen Yablo, 'A Problem About Permission and Possibility', pp. 270–294 in [273].

[1211] Stephen Yablo, *Aboutness*, Princeton University Press, Princeton, NJ 2014.

[1212] Seth Yalcin, 'Epistemic Modals', *Mind* **116** (2007), 983–1026.

[1213] Seth Yalcin, 'Probability Operators', *Philosophy Compass* **5** (2010), 916–937.

[1214] Edward N. Zalta. 'Logical and Analytic Truths that are Not Necessary', *Journal of Philosophy* **85** (1988), 57–74.

[1215] Edward N. Zalta, 'A Philosophical Conception of Propositional Modal Logic', *Philosophical Topics* **21** (1993), 263–281.

[1216] Alberto Zanardo, 'Moment/History Duality in Prior's Logics of Branching Time', *Synthese* **150** (2006), 483–507.

[1217] Nick Zangwill, 'Direction of Fit and Normative Functionalism', *Philosophical Studies* **91** (1998), 173–203.

[1218] Nick Zangwill, 'Does Knowledge Depend on Truth?', *Acta Analytica* **28** (2013), 139–144.

[1219] Elia Zardini, 'Luminosity and Vagueness', *Dialectica* **66** (2012), 375–410.

[1220] J. Jay Zeman, 'Bases for S4 and S4.2 Without Added Axioms', *Notre Dame Journal of Formal Logic* **4** (1969), 227–230.

[1221] J. Jay Zeman, 'The Propositional Calculus **MC** and its Modal Analog', *Notre Dame Journal of Formal Logic* **9** (1968), 294–298.

[1222] J. Jay Zeman, 'A Study of Some Systems in the Neighborhood of S4.4', *Notre Dame Journal of Formal Logic* **12** (1971), 341–357.

[1223] J. Jay Zeman, 'S4.6 is S4.9', *Notre Dame Journal of Formal Logic* **13** (1972), 118.

[1224] J. Jay Zeman, *Modal Logic: The Lewis-Modal Systems*, Oxford University Press, Oxford 1973.

[1225] Debra Ziegeler, 'Propositional Aspect and the Development of Modal Inferences in English', pp. 43–79 in W. Abraham and E. Leiss (eds.), *Modality-Aspect Interfaces*, John Benjamins, Amsterdam 2008.

[1226] E. E. Zolin, 'Embeddings of Propositional Monomodal Logics', *Logic Journal of the IGPL* **8** (2000), 861–882.

Index

.2, 33, 74
.3, 33, 74
.3⁺, 74
.4 and **K.4**, 33, 97
4′, 77, 93
4$_c$, 36, 196
    coincides with **GD$_1$**, 201
5$_c$, 36, 44, 128, 256, 297
    a mixed doxastic-epistemic variant of, 359
    and Moore's Rule, 357
    implies **D**, 295
    provable in **KD4**, 208

4, 33
    4$_K$, 338n, 370, 378
        Hintikka supports, 378
        Lemmon and others oppose, 376
        Williamson opposes, 377
    rule for natural deduction, 492
5, 33, 370
    5$_K$, 371, 371n, 380–402
        Lenzen's argument against, 373
        the basic objection to, 371

A

    actuality operator, 305
    function in the semantics of explicit awareness, 425
Aanderaa, S., 363
Abbott, B., 352n
abbreviations (used here), 4
ability (logic of), *see* dynatic (logic)
Abraham, M., 235, 255
absorption (laws), 327
accessibility relation, 59
    absent from simplified models, 55
    between clusters, 404
Ackrill, J. L., 22
acts, actions (*see also* bringing about)
    'gwalking' example, 352
    individual and generic, tokens and types, 234–238, 246
actuality operator, 219n, 300, 305, 441, 458n, 465
admissible
    propositions, 83, 321, 322
    rules, 19n, 122, 313, 317
agent-neutral (principles), 240n
agent-relative obligation statements, 251
aggregation, 21, 24, 229
    dual aggregation (for $\Diamond$), 37
    in deontic logic, 248
    in doxastic logic, 374n, 423n
    unwanted aggregativity for $\Diamond$, 31
        in Łukasiewicz, Curry, Moisil, 31n, 50–51
aggregativity, *see* aggregation
Aho, T., 54n, 205n, 338n, 341n
Akama, S., 104
Akatsuka, N., 332
Alban, M. J., 48
Alchourrón, C., 248
alethic (modal logic), 1, 16
Alexander, S., 403
algebras, (normal) modal, 83, 155
    algebraic semantics and Kripke incompleteness, 261
    correspond to general frames, 83
    without the normality restriction, 175
'alio-' versions of conditions on relations, 93–109
'alio-asymmetry' (= antisymmetry), 93
alio-equivalence (relation), 95
'aliosymmetric' (relations), 93n
aliotransitive, 91, 93
Allis, V., 244
Allo, P., 333n
Aloni, M., 395
Alston, W. P., 378
Alt$_n$ (**Alt$_2$**, **Alt$_3$**, etc.), 109
"alternative" as a term for *accessible*, 59n
Altham, J. E. J., 243
ambiguity
    scope —, 299, 354

INDEX

wrongly claimed, 45, 335, 364
Ammann, A., 251
'the amoralist', 354
Anagnostopoulou, E., 377
analytic implication, 340, 423
analyticity à la Hintikka (surface implication), 424
ancestral (of a binary relation), 28, 70
Anderson, A. R., 46, 158, 168, 474n
    his reduction of deontic to alethic modal logic, 229n, 253–258, 262, 279, 429
    on $\mathbf{B}_O$, 298
    relevant implication in the reduction of deontic to alethic modal logic, 298n
Anderson, A. T., 23
Anderson, C. A., 258
Anderson, E., 465
Anderson, K. C., 205
Angell, R. B., 424
Anscombe, G. E. M., 356
St. Anselm, 101, 251, 451n
Antinucci, F., 251
antisymmetric, 15, 68, 93n
antitone
    epistemic logic, 338
    modal logics, 20
"apodictic", 304
Åqvist, L., 48, 79n, 143n, 160, 234, 238, 241, 255, 260–262, 278, 290, 418, 419
    on questions, 331n
    on supererogation, 229
Areces, C., 160
Aristotle, 22
    apparent iterated possibility examples, 289
    Aristotelian syllogistic, 305
    Aristotelian themes in time, aspect, action (kinesis, energeia), 160
    Aristotelian version of universal quantification, 304
    confused over possibility and contingency?, 22
Armstrong, D. M.
    confused about incorrigibility, 386
    on absolute motion w/out absolute space, 465
Arregui, A., 242
Artemov, S., 333
ascending chain property, 112
aspect

continuous (or progressive), see progressive aspect
perfect, see perfect aspect
associative, 11
asymmetric, 14
    "1, 2-asymmetric", 189
Aucher, G., 402
Aumann, R. J., 395n
Austin, J. L., 50, 332, 345, 377n
autoepistemic (logic), 125, 333
van der Auwera, J., 251
Ayer, A. J., 50

B (doxastic $\Box$), 203, 351
$\mathbf{B}$, 33
B^{-1} (operator interpreted by the converse of the doxastic accessibility relation), 439
$\mathbf{B}_2$, $\mathbf{B}_3$, etc., 189
Bach, K., 348
Bacon, J., 279n, 392
    on nomic necessity, 226–234
Baker, J. R., 101
Balbiani, P., 85, 86n
Baldwin, T., 355, 422n
Barcan Marcus, R., 239n
    the Barcan Formula, 57, 212n
Barker, J. A., 331n
Bartha, P., 239n
Bäuerle, R., 178
Baxter, R. J., 165n
$\mathbf{B}^{\Box}$, 37n, 98–100, 262
    in doxastic logic, 446
Beaver, D., 353
Beck, J., 427
Becker, O.
    Becker's Rule, 100
    his semantics for modal logic, 71
Bednarska, K., 493n
belief
    degrees of — vs. outright belief and its absence, 374n
belief (see also doxastic logic)
    "belief world", 421n
    confident, 342, 371n, 373, 390
    consistent: $\mathbf{D}_B$, 343
    defined in terms of knowledge, 371, 410, 429–430
        B as KPK, 373, 430, 433
        B as PK, 430, 433
Belliotti, R., 235

Belnap, N. D., 23, 46, 158, 168, 239n, 451, 474n
Belot, G., 465
Bengson, J., 351n
Bennett, J., 213
    on a notion of necessity building in that what it applies to is not a modal statement, 214–217
    on entailment, 46
    on McKinsey, 165n
van Benthem, J., 3, 5, 137, 138, 181, 195, 219, 333
    'local' Kripke incompleteness example, 192, 193
    definability in tense logic, 195
    dynamic epistemic logic, 398
    evidence in epistemic logic, 333
    homogeneity, 415n
    local/global contrast, 185
    modal degree sensitive version of the Generation Theorem, 70
    on $\Box(p \to \Box p) \to (\Diamond p \to p)$, 102, 104
    on almost-connected relations, 463
    on comparatives, 461, 464
    on dynamic epistemic logic, 301n
    on general frames and completeness, 83
    on Hintikka on analyticity, 424
    on logicality, 92n
    on the Garson–van Fraassen 'transformations' semantics, 158
    on the limits of diagrammatic representation, 68
    on ultrafilter extensions, 85
    variation on an example of —, 281
Benton, R. A., 101n, 105
Beran, H., 252n
Berlin, I., 235
Bernert, J., 46, 48
Berto, F., 364
Beth, E. W.
    Beth's Theorem, 437
Béziau, J.-Y., 153
Bickenbach. J. E., 299
Biela, A., 46, 48
Bigelow, J., 218, 221, 224, 225, 226n, 257, 449n
    on knowledge, 395
    on nomic necessity not entailing truth, 217
Bílková, M., 22, 308
Birkhoff, G., 90
bisimulations, 137

Blackburn, P., 22, 72, 83, 160, 483n
Blamey, S. R., 7, 493
Blanché, R., 17
blindspots, 356, 398
    disjunctions of, 356n
Blok, W., 23n, 39n
Boas, R., 379
Boër, S. E., 237n, 331n
de Boer, M., 277
Bogusławski, A., 331n, 334
boldface
    for congruential extensions of **E**, 49
    for normal extensions of **K**, 33, 362
Bolinger, D., 332, 335
Boll, M., 360n
Bonevac, D., 242
Bonnano, G., 333
Bonnay, D., 378
Boolean
    connectives, valuations, 10
    algebras, 139, 155
Boolos, G., 37, 72, 74n, 78, 82, 82n, 166n
    on Löb's Rule, 73
    producing finite countermodels, 73
Borowski, E. J., 237
"bouletic", *see* boulomaic
boulomaic (logic), 50, 203, 205n, 339, 370n, 446
    and Moore's Paradox, 358
    and normality, 358
Brady, R. T., 424
Braun, D., 300n
bringing about, 235, 237, 449
Brink, D., 249n
Broad, C. D., 252n
Brogaard, B., 398
Broome, J., 240n, 251
Brouwer, L. E. J.
    Bouwerian rule, 115, 446
    'Bouwersche' axiom (see also **B**), 34
Brown, C. D., 101
Brown, M. A., 37, 160n, 301
Broyles, J. E., 345
Bryant, J., 234, 250
Buckwalter, W., 204
Bull, R. A., 73, 340n
Bulygin, E., 248
Bunch, B. H., 30
Burgess, A., 380
Burgess, J. A., 302n, 376

Burgess, J. P., 166, 235n
    "it is true that" in propositional attitude ascriptions, 380
    modal probability logic, 384
    on the correct alethic modal logic, 166n
    on **S4.2**, 278n
    ontological commitments and modal locutions, 465
    tense logic, 195, 195n
Butchart, S., ii
Butterfield, J., 458n
Buvač, S., 50
Bybee, J., 300
Byrne, A., 422n

C (doxastic $\Diamond$), 203, 351
C2, C3, etc. (Lemmon's systems), 47, 49
Cameron, J. R., 240n
Cancellation (rule), 126
canonical
    — frame, 74
    — modal logic, 82, 105, 285, 286
    — model, 61
    canonically enforceable conditions, 195
    method of canonical accidents, 84
"can", *see* modal auxiliaries
Cantor, G., 126
Canty, J. T., 154
capacities, epistemic or recognitional, *see* recognitional capacities
Capone, A., 379
Cappelen, H., 379, 395
Cariani, F., 244, 301, 350, 351
Carlson, E., 242, 326
Carnap, R., 44n, 53, 299, 396
Carnielli, W., 5, 45, 245n
Carr, D., 351n
Carr, K., 242
Castañeda, H.-N., 234, 238, 240, 249n, 338n
    de se issues, 379
    agent-implicating deontic judgments, 251
    imperatives underlying *ought*-statements, 299
Casullo, A., 205
Cath, Y., 352n
Chagrov, A. V., 5, 39, 137, 361, 366, 369
Chalmers, D., 54n, 337, 432n
characteristic
    frame, 71, 82
    matrix, 112
    model, 61

Chellas, B. F., 20, 25, 34, 35, 41n, 45, 54, 78, 98, 161, 219, 243, 247, 287, 299, 362n
    errors remaining in *Modal Logic: An Introduction*, 133
    on conditional obligation, 241
    on locale semantics, 161n
    on neighbourhood semantics, 160, 161
    on U_O, 239n
Chisholm, R. M., 241, 242, 255, 379, 423
    Chisholm's Paradox, 234
Christenssn, D., 374
Church, A., 6n, 194
Cinque, G., 17
circularity (of definitions or analyses), 169, 225, 376
    non-vicious, 302
circumstances (in which an action is performed), 238, 243
Clark, A., 389
'classical' (= congruential), 20, 39n
closure-related conditions vs. closure conditions, 348
clusters, 404, 414, 419
Coates, D. J., 50
Cocchiarella, N. B., 28, 57
Coffa, J. A., 23
Cohen, B., 244
Cohen, L. J., 152n
Cohen, S., 394
Coliva, A., 355
collapsing (modalities), 42
Collingwood, R. G., 252n, 396
Collins, J., 345
Colonna, G., 155
coming about, 450–469
commutative, 11
"commutes with", 52n
comparatives, 458n, 461–466
compatibilism, 37
    and the conditional account free action, 50
complete
    non-modal version (of semantic completeness), 6
    w.r.t. a class of frames, 71
    w.r.t. a class of models, 60
composition
    compositional independence property, 299, 302, 304
    compositionality (and reverse compositional-

ity), 427
Comrie, B., 160
concept-possession issues, 205, 336, 343, 348, 351, 376, 384, 420–441
conditional obligation, 238, 241
conflicting obligations, 244–245, 248–250
congruential
    Congruentiality Rule, 29
    modal logics, 19, 25, 171–174
    operator (in a given logic), 31
Coniglio, M., 245n
connected frames, 75
    weak connectedness without transitivity, 77
    weakly connected, 75
consequence
    causal consequences of actions, 240n
    consequence relations, 8
    frame consequence, 23n, 80
    generalized consequence relations, 7, 14, 175n, 381
    global consequence, 148, 156
    local consequence, 49, 102, 146n, 156
    point consequence, 23
consequentialism (how to characterize), 240n
conservative extension, 135, 214, 316, 320
consistent
    — and inconsistent beliefs in doxastic logic, 392
    — and inconsistent beliefs in doxastic logic, 343, 375, 390
    logic, 38
    set of formulas, 61
constitutivism (about first-person authority), 359, 387
contexts (linguistic), 13
contextualism (epistemic), 300n, 394
contingency (and noncontingency), 17, 21, 22, 81, 154, 265, 326n
    as a truth-value in a three-valued setting, 152
    contingent truth and contingent falsity as truth-values in a four-valued setting, 153
    deontic analogue of, 236, 255, 287
        a complication, 238
    epistemic analogue of, see knowing whe-ther
    erroneous claim concerning Zolin's treatment of, 254n
    in connection with material and strict conditionals, 24
    Moisil's bizarre use of *la contingence*, 51
    necessary conclusions from contingent premisses, 23
    noncontingency and the rule of disjunction, 114
    Pizzi on, 254n
    the contingency of many moral claims, 243, 275, 279
    three notions of noncontingency in **S4**, 102, 214
contra-classical (modal logics), 38, 174
contracture, symmetric, 107
contrastivism (epistemic), 395
convergent frames, 75
Copeland, B. J., 54, 71
Cormack, A., 355
Correia, F., 213, 424
'correspondence' (ambiguity of), 219
da Costa, N., 245n
"could" (*see also* modal auxiliaries), 17
    epistemic interpretation of, 302
counterfactuals, 50, 242, 358
    in the analysis of knowledge, see knowledge, analysis of
    with monadic deontic operators instead of a special conditional obligation connective, 242
counterparts, 212n
cover (relation), 39n, 307
Crabbé, M., 74
Craig, E., 349
Cresswell, M. J., ii, 5, 16, 18, 26n, 31, 34, 37, 41, 44, 48, 49, 54, 54n, 69, 71n, 73, 82, 82n, 92, 96, 100, 123, 159, 209, 216n, 221, 237n, 246, 249n, 343, 360n, 362n, 363, 368, 369, 376n, 462
    on McKinsey, 165n
    on modal predicate logic, 57, 212n
    the "$\neg\Box p$" problem, 166
Cresswell, W. D'Arcy, 249n
Crisp, T. M., 458n
Cross, C., 331n
Curley, E. M., 364
Curry, H. B., 6n, 31n, 486n
    an early treatment of $\Diamond$, 50
Czermak, J., 288

D (*see also* **KD**), 33
    rule for natural deduction, 492
D2, D3, etc. (Lemmon's systems), 49
Dalla Chiara, M., 339

Danielsson, S., 238n, 244, 324–330
Davidson, D., 237, 451n
Davies, M. K., 258, 300, 352, 396, 427, 441, 447, 465
Davis, W. A., 394
Dawson, E. E., 266, 278, 279, 282, 290, 372, 434
de jure and *de facto* logics, *see* Leuenberger, S.
de re and *de dicto*, 237n, 442
de se attitude ascriptions, 203, 212, 356, 378, 379, 386n
Decew, J. W., 242
decidability
    of formulas (relative to a logic), 171
    of logics, 73
Dedekind, R., 194
Deduction Theorem, 49
defensibility (Hintikka's notion), 338
definability
    bimodal, 107
    explicit vs. implicit, 430, 437
    first-order or otherwise, 87, 105, 285
    modal, 83–87, 92, 96, 107, 138, 220, 267, 270
        local vs. global, 82, 185
    of connectives (*see also* definition), 312, 320
        K in terms of B, 371
    trimodal, 268
definition
    Bennett possibly confused about —, 214
    of $\Box$ in terms of noncontingency, 18
    of $\Diamond$ in terms of $\Box$, 16
    of ↔, 8
    Weingartner confused about —, 335
definitional
    $\Box$-definitional, 38, 266, 363
    equivalence, 373
    definitionally equivalent (logics), 416
    translation, *see* translation
Denecessitation (rule), 112
Denning, B., 235
dense (frames), 195
Denyer, N., 301
deontic logic, 1, 234–290
    "a dubious enterprise" (Pigden), 246
    and full modalization, 291
    Prior's claim of a basic difference with alethic modal logic, 304
    semi-paraconsistent, 370
DeRose, K., 339, 384, 394
Descartes, R.
    on dreaming, 104
    on God's choosing the necessary truths, 364
descending chain property, 111
desire, *see* boulomaic
determined
    by a class of frames, 71
    by a class of models, 60
    by a matrix, 154
Dever, J., 379
Df (in "$=_{Df}$"), 16
di-proposition, 275
diagrammatic representation of (certain) conditions on relations or frames, 65–68
difficulty (and ease), 301
dilemmas, moral, *see* conflicting obligations
Dipert, R., 194
direction of fit, 204, 205n, 356, 447
disbelief, 203, 433
discharge (of assumptions in a natural deduction proof), 472
disjoint union (of frames), 83
disjunction
    rule of —, *see* rule of disjunction
    the 'fatal disjunction', *see* Ross's Paradox, Danielsson's treatment
disjunctivism, 422
"distributes over" (ambiguity of), 52n
distributive triple, 51
van Ditmarsch, H., 104, 254n, 333, 333n, 397, 398
Divers, J., 55
domain-reflexive, 79, 90
    1-d.r. and 2-d.r., 190
    spurious 'alio' version, 94
Dorr, C., 356
Došen, K., 51, 155, 160n
doxastic logic (*see also* epistemic logic), 2, 125, 203
    and the modified rule of disjunction, 436
    quantified, 442
    semi-paraconsistent, 370
Drake, F. R., 165
Dreier, J., 240, 354
Dretske, F., 344–346, 389, 395
dual
    dual intuitionistic negation, 50
    form of a modal principle, 35–36
    of a connective, 35
Dubikajtis, L., 11n
Dubucs, J., 345

Dummett, M. A., 41n, 144, 269, 345n
    sequent notation, 7
Dunn, J. M., 6, 423
Dutant, J., 345, 378
Dutton, J. D., 332
duty (and duties) etc., 252n
Duží, M., 425
dynamic (logic); *see also under* epistemic logic, 237
dynatic (logic), 37, 50, 301

"**E**" (for **5**), 79
E, 25, 39
E1, **E5** (Lemmon's systems), 48, 360, 360n
E2, **E3**, etc. (further Lemmon systems), 47, 49, 227, 360, 363
van Eck, J. A., 243
Edelberg, W., 442n
Edelstein, R., 6
Edgington, D., 379
Egan, A., 300n
Égré, P., 378, 395
eliminativism (about belief etc.), 421n
EM, 160, 161, 246
    determined by the class of all locale frames, 161
embedding (*see also* translation), 142
EMN, 247
End, 33, 37, 38, 174
Enkratic Principle, 299
epistemic
    logic, 1, 2, 331–441
        dynamic, 333, 398
        possibility, 383, 428
equivalence relations, 15, 65
Esakia, L., 112, 137, 417n
euclidean, 65
Evans, G., 167
Ewing, A. C., 244
exclamatives, 331n
exclusive
    disjunction, 348
        as opposed to exclusionary disjunction, 455n
    interpretation of quantifiers, 68n, 95
explicit
    explicit definition, *see* definability, explicit vs. implicit
    explicit knowledge, belief, etc., *see* propositional attitudes, explicit or implicit

extensional (connectives, modal logics), 49n, 129, 174, 303, 322
    inappropriately extensional treatment of modality in Łukasiewicz and Törnebohm, 261, 363
    Zolin's terminology for, 304

F (future tense $\Diamond$), 177
f.m., *see* fully modalized
'factive', 353
Fagin, R., 341n
fair redescription (of beliefs), principle of, 341, 428
faithful (embedding), 142
Falcão, P. A., 155
Falsum
    sentential constant $\bot$, 10
    the 'Falsum' modal logic, 171n, 215, 303, 362, 364
        for deontic application, 253
family (of logics), 25
Fan, J., 104, 254n
Fara, M., 398
Feit, N., 379
Ferguson, T. M., 423
Fernández-Duque, D., 333
Feyerabend, P., 379
Feys, R., 48, 49
'field-reflexive' (relations), 90n
filter algebras, 154n
filtrations, 73, 85, 413
Fine, K., 112, 137, 195n, 209, 253, 283, 340
    first-order undefinability of the class of frames for a canonical logic, 285
    on analytic implication, 423
    on Aristotle's modal logic, 289
    on essence, 212
    on propositional quantifiers, 216, 340
    on **KM**, 287
finitary (consequence relation), 9
"finite-to-arbitrary" move, 117, 197
finite model property, 73, 111
    Bull's Theorem, 410n
    by-passing its use for a result on **S4.2**, 413
    failure of, 194
    possessed by **KM**, 287
Finlay, S., 242, 244, 251
Finte, K., 441n
Fiocco, M. O., 384

first-person examples: care needed, 300, 353, 356, 377, 377n, 432
Fishburn, P. C., 241, 461
Fisher, M., 152n
Fitch, F. B., 257, 449n, 475n
    Fitch derivation, 258, 340, 356, 397
Fitting, M., 6, 6n, 73n, 142, 143, 148n, 185, 246n, 288, 333
Flescher, A. M., 255
Floridi, L., 385
$Fm_S^\square$, $Fm_S^\lozenge$ (set of formulas S-equivalent to a $\square$-formula, to a $\lozenge$-formula, resp.), 36, 131, 213
$fm(S)$, the f.m. core of S, 297
f.m., *see* fully modalized
Folescu, M., 379
Føllesdal, D., 234, 241, 249n, 337
Font, J., 28, 29, 261
Forbes, G., 3, 352n, 458n, 467, 475n, 477n
Forrest, P., 427
Forrester, J. W., 238, 241, 249n, 251, 255n, 300
van Fraassen, B., 157, 240n, 248, 252n
    on conditional obligation, 241
Frajzyngier, Z., 301
frames, 55, 71–82
    frame for, 80
    frame-equivalent (formulas), 139
    general —, 83, 193
    locale —, 246
    neighbourhood —, 159
    Pargetter —, 221
    pointed —, 190
    with distinguished elements, 190
Frankfurt, H., 364
free choice permission, 253
Frege, G., 45, 54n, 305, 352n, 464
French, R., ii, 60n, 416, 417
Freund, M., 57
Fricker, E., 352
Fritz, P., 39n, 174n
Frost, K., 357
fully modalized
    formulas, 28, 56
        all modally invariant, 56
    deontic application, 314
    doxastic application, 441
    formulas, 476, 479
        modally invariant in semi-simplified models, 206
    logics, 290–304
functional (frames and relations), 68, 132, 157, 161, 307, 454
functional completeness (and strong functional completeness), 11

G (future tense $\square$), 177
Gabbay, D. M., 57, 89n, 159n, 235, 255, 277, 278, 308, 420n
Gailor, D., 300
Gallie, R. D., 81
Galton, A., 160
García-Carpintero, M., 394
Gärdenfors, P., 71, 160n, 161, 384
Garson, J. W., 6n, 57, 66, 68n, 79, 237n
Gasking, D., 386n
GD_n (Generalized Density schema), 196–202, 219, 297, 490
Geach, P. T., 78, 240n, 364, 379
    cancelling-out fallacy, 399
    merely Cambridge change, 457, 457n
    on comparatives, 462n
    on deontic logic, 251
    on intentional identity, 442n
    two kinds of intentionality, 349n, 352n
Gemes, K., 424
general frames, *see* frames, general
generality constraint, 427
generalized equivalence (condition), 443
    piecewise version, 445
generated
    Generation Theorem, 70, 83, 92, 184, 207, 298, 406
    subframe, 71
    submodel, 69, 184
Gensler, H. J., 244n, 354
Gentle Murder(er) Paradox (J. W. Forrester), 238, 300
Gentzen, G., 7, 471, 475n
Georgacarakos, G. N., 46, 66, 137, 414, 431
    Porte on, 414
George, B. R., 331n
Gerbrandy, J., 333
Gerla, G., 158
Gerson, M. S., 160n
Gettier, E. L., 344, 347–350, 392, 403n
    on Kaplan's problem, 258
Geurts, B., 353
GH rule, 180
Ghilardi, S., 308

Giannakidou, A., 204
Gibbard, A., 240
Gilabert, P., 251
Gillet, E., 219n, 335, 337, 366–368
Gindikin, S. G., 456
Giordani, A., 213
Girle, R., 3, 6n, 338n, 351
GL (for "Gödel–Löb" logic), *see* **KW**
Glick, E., 351n
Glivenko, V., 142
global consequence, *see* consequence, global
Goble, L., 5, 239n, 240–242, 248–250, 298n
    a picture of Goble's argument, 250
Gochet, P., 219n, 335, 337, 366–368
Gödel, K., 142
Goldblatt, R., 30n, 54, 57, 75n, 85n, 137, 177n, 185n, 245n
    on modal algebras and general frames, 83
    on modal definability, 85
    on non-normal logics between **S4** and **S5**, 194
    on **KM**, 286
Goldman, H. S., 242
Good Samaritan Paradox, 234, 238, 255n, 300
Goré, R., 6n, 486n
Goswick, D., 209, 252
Greco, J., 344
Green, C., 95n
Green, J. A., 45, 91
Green, M., 355
Greenough, P., 104, 378
Greenspan, P., 242
Gregory, D., 219
Grice, H. P., 243, 350
Grimes, T. R., 424
Groenendijk, J., 334
groupoids, 159
groups, 159
Grünbaum, A., 458n
Guerzoni, E., 331n
Guigon, G., 449n
Gunderson, D., 30
'gwalking' example, 352, 447

H (past tense $\Box$), 177
H, 87, 279, 411n, 418n
$\underline{\mathbf{H}}$, 133, 280
de Haan, F., 17
Hacking, I., 339, 384
Haddock, A., 422n

Hailperin, T., 91
Hájek, A., 355
Hájek, P., 261
Hakli, R., 50
Hale, B., ii, 277
Halldén, S., 48
Halldén completeness, 39, 119, 120, 134, 293
    and Halldén normality, 118, 415
    and strongly heterogeneous frames, 414
    Halldén unreasonable (disjunctions), 120n
    set of formulas valid at a point is Halldén complete, 185
Halpern, J. Y., 3, 125, 137, 178, 339, 341, 369, 374, 383, 385–403, 405, 430, 437–441
Hamblin, C. L.
    discreteness in tense logic, 196, 199, 200, 202
    modal probability logic, 384
Handfield, T., 339n
Hanfling, O., 394
Hansen, J., 241, 248, 253, 341n
Hanson, W. H., 60n, 262
"Hans(s)on" and similar names in deontic logic, 240
Hansson, B., 71, 160n, 161, 240, 251
Hansson, S. O., 240, 243, 251, 278
Harary, F., 93, 95, 95n
Hardy, G. H. (anecdote about), 379
Hare, R. M., 240n, 354
    moral judgments as imperatival, 299
    on 'inverted commas' uses of language, 204
    on universalizability, 239
Harrison, J., 246, 277, 345, 349, 350
Hart, H. L. A., 252n
Hart, S., 104, 395n
Hartland-Swann, J., 351n
Hartshorne, C., 101
Hasse diagram, 41n, 139, 307
Hawthorne, J., 345, 346, 356, 359, 395
Hayaki, R., 212n
Hazen, A. P., ii, 48, 95n, 305
Hazlett, A., 204n
Heathcote, A., 353, 378
Heifetz, A., 104, 395n
Heim, I., 334
Heller, D., 302
Helm, P., 457
Hempel, C. G., 226n
Hendricks, V. F., 333
Hendry, H. E., 11n

INDEX

Henle algebras, 154n
Henry, D. P., 251, 451n
    defining possibility using a sentential constant, 50
heterogeneous
    as opposed to homogeneous frames, 414n
    strongly heterogeneous frames, 414
Hetherington, S., 349, 351n
Heyd, D., 255n
Higginbotham, J., 379
Higgins, F. R., 302
Hill, D., 387
Hilpinen, R., 5, 234, 235, 241, 249n, 253, 333n, 337, 451n
    on being in a position to know, 339
    on the KK principle (4_K), 376
Hintikka, J., 3, 30, 59n, 68n, 203, 204, 205n, 235, 331n, 333, 333n, 334, 336, 338, 338n, 345, 352, 371, 375, 377, 379, 384, 424
    independence friendly logic, 333
    objecting to Kripke semantics, 166–167
    on *prima facie* duties, 244
    on iterated deontic operators, 239
    ought vs. *must*, 299
Hippler, H.-J., 252n
Hiż, H., 40
Hodes, H., 441n
Hodges, W., 39, 67
Hodkinson, I., 202, 287
van der Hoek, W., 79n, 122, 125, 333, 375, 380–385, 391, 396n, 398, 400, 404n, 421, 433
Hoffmann, A., ii
Hohfeld, W. N., 251
holds (sequent holds in a model), 481
Holliday, W. H., 212, 355, 394, 398, 403n
Holton, R., 204, 204n, 359, 379
homogeneous (frames), 414n, 415
hope – is a propositional attitude though one cannot 'hope' a proposition, 348
Horn formulas/conditions, 39, 67, 354n
Horn, A., 354n
Horn, L. R., 350, 354
Hornstein, N., 160
Horty, J. F., 20, 241, 248, 251, 452
    on *prima facie* duties, 244
Horwich, P., 458n
Hösli, B., 7
Huddleston, R., 331n, 332, 351n, 377
Hudson, H., 235

Huemer, M., 428
Hughes, G. E., 16, 26n, 31, 34, 37, 41, 44, 49, 54, 69, 71n, 73, 82, 82n, 85, 86n, 92, 96, 100, 122, 123, 209, 216n, 221, 223, 237n, 369
    (strong/weak) Hughes formulas, 64, 128
    claim about reflexive transitive relations, 89
    logic determined by the class of frames in which every point has a reflexive successor, 85
    logics with no minimal proper extensions, 112, 119
    on axiomatizations of **S5**, 92
    on modal predicate logic, 57
Hume, D.
    'Hume' (and 'Heimson') worlds, 59, 218–226
    Hume's Thesis (or Hume's Law), 268, 441
    Humean supervenience, 218
    on the conditional account of free action, 50
Humphreys, J. F., 89n
Hurley, S., 244
hybrids (of connectives), 261, 314

Iacona, A., 339n
Iatridou, S., 377
Icard, T. F., 355, 398
idempotent, 11, 320
identity
    element in an algebra, 159n
    relation of
        claimed not to exist, 68
        how to represent in diagrams, 67
Iliev, P., 398
imperatives, 248, 299
    to form questions, 331
implicature, conversational, 18, 243, 325, 350, 389, 392n
implicit
    implicit definition, *see* definability, explicit vs. implicit
    implicit knowledge, belief, etc., *see* propositional attitudes, explicit or implicit
"impossible"
    impossible worlds, *see* worlds, impossible
    syntactic differences between "possible" and — , 332
incorrigibility, 386–387
index (advice on using), 5
Indrzejczak, A., 486n, 493n
Innala, H.-P., 240n

Innermost Reduction Property, 42
Inoué, T., 142
instances (of a schema), *see also* substitution instances, 22
interdeducible (formulas), 34
internalism
    in epistemology, 403n
    in ethics, 354
    in philosophy of mind, 403n
interpolants *vs.* intermediaries, 308
interpolation, 308
intersection
    of sets in general, 4
intersections
    of accessibility relations, 107, 268, 277
    of modal logics, 39
    of neighbourhoods, 159
intransitive, 14
    K determined by the class of — frames, 270
    class of — frames not modally definable, 83, 138
    unfortunate uses of the term, 462n
introspection
    and incorrigibility, 387
    negative, *see* NI and 5_K
    positive, *see* PI and 4_K
    terminology not to be taken seriously, 370, 378
intuitionistic logic, 6, 50, 51, 473
irreflexive, 14
    class of — frames not modally definable, 83
Ivin, A., 242
Izvorski, R., 377

Jackson, F. C., 241–243, 337, 350, 386
    on desire as non-monotone, 358
de Jager, T., 395
Jäger, G., 7
Jago, M., 331n, 341n, 423n
Jansana, R., 28, 29
Jaspars, J., 380–385, 432, 433
Jennings, R. E., 21n, 30n, 160, 181, 240, 241, 245, 246, 248, 253, 340, 347, 393
    on pseudo-subjectivism, 244, 299
Jeřábek, E., 416
Jespersen, B., 425
Johansson, I. (*see also* Miminal Logic), 50
Johnson, K., 427
Jones, A. J. I., 229, 277, 300, 449
Jones, H. S., 160
Jónsson, B., 90n
Jørgensen, J., 240n
Jubien, M., ii
justification, 339n
    not entailing truth, 347

K (epistemic $\Box$), 203, 351
K
    axiom, 22, 26
    modal logic, 19
K.2, 75, 215
K.3, 74, 75
K.3$^+$, 75
K4, 44
    and transitive irreflexive frames, 271
    completeness result for, 72
K45
    and semi-simplified models, 206
    as a doxastic logic, 125
    faithfully embedding **S5** into —, 208
    how related to **KB4**, 90
K45$'$, 135
K4!, 143, 447
    sample (informal) proof in, 40
K4$_c$, 92
    determined by the class of dense frames, 195
    exercise concerning a generalized density formula in —, 136, 196
    on a revision exercise concerning —, 490
    sample proof in, 40
K5, 44
    completeness result for, 72
K5$'$, 135
K5$_c$, 136
K5! (= **K4!5!**), 43
Kahle, R., 22
Kajamies, T., 364
Kamp, J. A. W. ('Hans'), 202
Kane, R., 101
Kanger, S., 167, 257, 429
    on $\mathbf{U}_O$, 239n
Kant, I., 299, 424
Kapitan, T., 101
Kaplan, D., 28n, 340n
    'Kaplan's Problem', 258
Kaplan, M., 345
Karmo, T. E., 209–211
KB
    and symmetric irreflexive frames, 271
    completeness result for, 72

INDEX

KB (knowledge implying belief), 371, 373, 385
    sacrificed by Halpern, 389
KB4
    = **KB5**, 130, 140
    nothing between it and **S5**, 89, 417
KB4′ (logic of 'elsewhere"), 272
KB4′ (logic of 'elsewhere'), 96
KB5, see **KB4**
KB$^\Box$4, 98
KB$_c$ (= **KB!**), 42n, 92, 136, 403
KC$_n$, 92
KD, 52, 65
    KD ∩ **KVer**, 119
    K + D vs. **K ⊕ D**, 121
    a bimodal version for deontic logic, 229
    and various other modal logics are fully modalized, 295
    completeness result for, 72, 119
    decides all pure formulas, 172
    doxastic logics extending —, 432
    held not to be a plausible deontic logic, 245–250
    noncontingency in extensions of —, 253n
KD4
    completeness result for, 72
    every point in its canonical frame has a reflexive successor, 84
KD45
    and semi-simplified models, 206
    as a deontic or doxastic version of **S5**, 278, 297
    embedding — in **S4.2**, 282
KD$_c$, 52, 333
    completeness result for, 108
KDFi, 286–287
KDH, 280–282
KD!, 17, 52, 92, 123, 129, 130, 132, 157, 229, 307, 454
KD!45 (= **KD!4** = **KD!5**), 295
KDT!$^\Box$, 287
Keefe, R., 302n, 303
Kemmerling, A., 343
Kemp, G., 460–466
Kempson, R., 353
KEnd, 37, 38
    Halldén incompleteness of, 414
Kenny, A. J., 160, 203, 301, 370n
Keutsier, T., 244
KF, 66

KFi, 283–285
KH, 280
Khamara, E. J., 465n
Kibort, A., 160, 377
Kielkopf, C., 240, 278, 279n, 282, 287, 288, 290n
    correcting Porte, 49
Kim, J., 457
King, J., 54n, 348, 376n
Kiparsky, C., 204, 353
Kiparsky, P., 204, 353
KK principle, see **4$_K$**
KM, 86n, 286, 287
know
    "a capacity to know", 398
    "in a position to know", 339, 344, 378, 424n
    inverted commas use, 204, 333
    knowing *wh*-, 331n
    knowing how *vs.* knowing that, 351n
    knowing whether, 103, 104, 331, 332, 334–336, 353
knowability, principle of, *see* Fitch derivation
knowledge
    — claimed not to require belief
        David Lewis, 392
        Halpern, 374, 389–392
        Williams, Lemmon, Radford, McGinn, 392
    — claimed not to require justification, 372, 415
    — claimed not to require truth, 204
    analysis of —
        circularity issues, 376
        defeasibility, 403n
        justified true belief, 339n, 344, 347
        Lemmon's proposed, 376
        merely true belief, 372n, 415n
        not to be had, 349
        using counterfactuals, 344, 395
    inferable (as opposed to merely implicit), 425
Köhler, P., 39n
Kölbel, M., 394
Kolodny, N., 242
Kondo, M., 49
Kordig, C. R., 244
Körner, S., 349, 350
Kowalski, T., 98, 99, 102
Kracht, M., 48, 122, 415
Kratzer, A., 244, 300
Kripke, S. A., 53, 340, 346, 360n
    completeness issue for quantified **S4.2**, 57

his 'model structures', 414
his discussion of Lemmon's E5, 360
Kripke semantics, 2, 58–61
   model-theoretic vs. truth-theoretic, 167
Kripke-complete (modal logics), 82
on belief attribution, 343
on Halldén completeness, 39
on Nozick on knowledge, 344
Kroy, M., 239
KT
   KT (= **K** $\oplus$ **T**) versus **K** + **T**, 25, 191
   completeness result for, 63, 72
KTB
   completeness result for, 65, 72
KT$_c$
   — = **KT**$_c$**U** example, 261
   illustrating the relativity to a logic of the notion "nontrivial combination of variables and modalized formulas", 293
   point-generated frames for, 92
KTM, 105
KT!, 37, 42n, 174, 363
   — as Post complete, 38
   deontic fragment of, 261
   point-generated frames for, 92
KT!$^\square$, 288
KU (see also **U**), 79, 260, 274, 297, 298
   some extensions of, 420
Kudo, Y., 104
Kuhn, S. T., 18, 196, 199, 219n, 239, 240n, 277, 297
Kurosh, A. G., 111
von Kutschera, F., 258, 303, 398, 427
KVer, 37, 329
   — as Post complete, 38
   decides all pure formulas, 172
   point-generated frames for, 92
KW, 37, 38, 69, 82n, 83, 113, 120n, 173
   not canonical, 117
Kyburg, H., 374n

L and M (notation for $\square$ and $\Diamond$), 1
L-modal logic (of Łukasiewicz), 48, 49n, 50, 156, 261, 363
Lacey, H. M., 465
Laitinen, A., 447
Lakemeyer, G., 212n, 339, 341, 369, 412n, 423n
Lakoff, R., 240n
Lambert, K., 89
Langford, C. H., 216

lattices (of modal logics), 25, 39
Lazerowitz, M., 23
Le Loux-Schuringa, J. A., 160
Lebar, M., 334n
Leckey, M., 217
Leech, J., 277
Legg, C., 194
Lehrer, K., 205, 464n
Leibniz, G. W. von, 53
Leitgeb, H., 374n
Lemmon, E. J., 41n, 46, 81, 144, 269, 338, 360
   his natural deduction system, 471–475
   on 5_K, 371n
   on *knows/believes* contrasts, 351
   on 'must', 300
   on action sentences, 451n
   on Anderson's reduction (in deontic logic), 255
   on duties vs. obligations, 252n
   on talk of *prima facie* obligations, 244
   on the correct alethic modal logic, 166n
   on the KK principle (4_K), 376
Lenzen, W., ii, 3, 34n, 266, 278, 290, 336, 338n, 345, 351n, 371n, 391, 403, 408
   a picture of Lenzen's argument, 374
   Lenzen's Argument vs. Lenzen's Theorem, 374n
   on **S4.2**, 402
   on **S4.2** and **S4.4** as epistemic logics, 372, 407, 410, 431, 434, 438
   the case against 5_K, 371–375, 385, 389
Leonard, H. S., 152
Letz, R., 130
Leuenberger, S., 81, 240
Levesque, H., 212n, 333, 339, 341, 369, 412n, 423n
Lewis, C. I., 33, 46, 57, 260
   his Existence Postulate, 216
Lewis, D. K., ii, 57, 212, 212n, 226, 227n, 242, 245, 249n, 253, 334, 359, 379, 393, 403n, 423, 441n, 452, 466
   and the notion of a miracle, 217
   Humean supervenience, 218n
   on conditional obligation, 241
   on desire as belief, 447
Lewis, H. D., 244
'liberated' modal logics, 89
Lichtenberg, J., 235
Lindenbaum algebra (of a logic), 41n, 318
Lindström, S., 252n, 258, 451n

linear orders, 66
Linsky, B., 212
List, C., 21n
Litak, T., 197n
Lloyd, A. C., 336
Löb's Rule, 73, 127, 128
local consequence, *see* consequence, local
locale semantics, 161, 246, 248
locally finite/locally tabular, 44
Logue, H., 422n
Lokhorst, G.-J. C., 249n, 298n
López de Sa, D., 395
Lottery Paradox, 374
Lucas, J. R., 290, 304
Łukasiewicz, J. (*see also* L-modal logic), 31n
Łukasiewicz, J. (*see also* L-modal logic), 48, 152, 156, 261
luminous (conditions) – Williamson, 104, 378
Lycan, W. G., 237n, 331n

M (McKinsey's axiom), 34, 36, 81, 86n, 160, 167, 193, 285, 286, 288, 417, 418
    M_n (Lemmon–Scott generalized forms), 286
    not valid on the canonical frame for **KM**, 286
"M" (for Monotone) in "**EM**", 160
Ma, M., 376
Mac Lane, S., 90
MacFarlane, J., 242, 394
MacIntosh, J. J., 349, 352, 378
    MacIntosh logics, rule(s), 98, 127, 210, 388
Mackie, J. L., 343n
MacPherson, B., 21, 421
Macpherson, F., 422n
Maddux, R., 90n
Makinson, D., ii, 20, 71, 73, 155, 174, 176, 235, 240n, 241
    Makinson logics, 171–174, 215
    on deontic features of individual acts, 236
    on deontic properties for act types and tokens, 236
    on modal logic with and without ⊥ primitive ('Makinson's Warning'), 39, 162
    on special implication relations, 424
Maksimova, L., 112, 308
Malley, E., 249n
Mally, E., 249n, 337
Manley, D., 345
Mannison, D., 376, 392
Marcos, J., 229

Mares, E. D., 46, 57, 246, 255, 298n, 363
margins, (Williamson's) rule of, 127, 128, 378
Martin, M., 422n
Marx, Groucho, 377
Massey, G. J., 11n, 152, 155, 157, 239n
material (conditional or implication, biconditional or equivalence), 10
    "merely material" implication, 24
matrices, 155
Matsumoto, K., 142–152
    early modal sequent calculus, 486n
    Matsumoto normal (modal logic), 148–151
    Matsumoto's Theorem, 420, *see* **S4**, Matsumoto's Theorem concerning
maximal consistency, 61
Mayo, B., 235, 377n
McCawley, J., 377
McGinn, C., 351n, 392
McGinnis, C., 245n, 250, 370
McKenna, M., 50
McKinsey, J. C. C., 24, 34, 143, 193
    his substitutional semantics for modal logic, 165–169
McKinsey, M., 348
McNamara, P., 251, 255
McNaughton, D., 251
Meadows, T., 54
Medeiros, M., 475n
meet-reducible, meet-irreducible (in a lattice), 118
Melia, J., 241
Menzel, C., 209
Meredith, C. A., 46
Meskhi, V., 112, 137, 417n
van der Meyden, R., 235
Meyer, J.-J. C., 5, 235, 238, 421
Meyer, R. K., 373n
Meyer-Viol, W. P. M., 160
"might", *see* modal auxiliaries
Milne, P., 462n
Mind (the journal), 244, 255n
Minimal Logic (Johansson), 50, 473
    minimal negation alongside intuitionistic negation, 51
"minimal models" (Chellas's name for neighbourhood models), 160
Mints, G. E., 41n
miracles, 217
mirror image (of a tense-logical formula), 179
misprints, *see* typographical errors

Mittwoch, A., 251
Miyazaki, Y., 29
modal
    degree, 70
    logic, 1
        formal definition of a —, 18
        history of, 54
        modal auxiliaries (*must, might, should,...*), 1, 17, 205, 239n, 289, 299–302, 307, 350, 356, 395, 422
            agent implicating issues, 251, 301
            and negation, 355
        modal definability, *see* definability, modal
        modally define, 81
        modally invariant (formulas), 56, 206
        ontological argument, 101
        predicate logic, 5, 57, 101, 158, 212, 237n, 466
            Prior's Q, 209
        the modal fallacy, 345
modalities, 31
    affirmative, 31
modalized, *see* fully modalized
'model structures' terminology, 402n, 414
model-changing semantics, 390n
models
    $\mathcal{M}$ is a model for S, 60
    characteristic, *see* characteristic model
    simplified, 54
modular law, 309, 320
Moffett, M. A., 351n
Moisil, G. C., 31n, 50
Moltmann, F., 348, 421n, 462n, 467
    interpreting modals used with presentational pronouns, 301
monotone
    as opposed to *monotonic*, 20
    consequence relations, 22
    modal logics, 19
    operator (in a given logic), 31
monotonic, *see* non-monotonic (logic)
Montague, R., 218, 219n, 304n
Montgomery, H., 49
Moore, A. W., 258, 351
Moore, G. E., 384
    and disjunctivism, 422n
        Moore's Paradox, 355, 379, 398
            as wielded by Shoemaker against self-blindness, 360
        for desire, 358
    Moore's Rule, 357–360
    naturalistic fallacy, 254
    on the conditional account of free action, 50
    paradox of analysis, 376n
Moore, O. K., 229n, 254n
Moore, R. C., 125, 125n, 333
de Moraes, L., 11
moral
    — conflicts, *see* conflicting obligations
    — nihilism, 252
    — principles (as opposed to arbitrary moral judgments), 243, 279
    — vs. legal readings of deontic operators, 239n
    morally perfect worlds, 58
    nihilism, 252
Morato, V., 219n
Morgan, C. G., 89
Mortensen, C., 364, 365
Morton, A., 462n
Moses, Y., 125
Mott, P. L., 242
Mourelatos, A., 160
Mulac, A., 332
Mullen, J. D., 241
Müller, J., 425
Murai, T., 104
"must", *see* modal auxiliaries
Myers-Schulz, B., 392

N (operator for purely nomic/physical necessity), 226
$\mathcal{N}$
    neighbourhood assigning function, 160
    operator for 'necessitativity', 213
Nagel, E., 226n
Nair, S., 250n
Nakhnikian, G., 386n, 393
"nand", 11
vander Nat, A., 229
natural deduction (rules, systems), 7n, 56, 129, 471–493
    for a novel connective, 312
    for propositional quantifiers, 213
naturalistic fallacy, 254
NC ('Negative Certainty'), 371, 385
Necessitation (rule), 19
    "Necessitation" vs. "necessitation", 25
necessitative (vs. necessary), 23, 213

necessity
    absolute, 275, 277
        Smiley's sense, 257
        von Wright's sense, 250
    as now-unpreventability, 299, 326n
    physical or nomic, 58, 59, 203, 217–234, 257
    relative
        Smiley's sense, 257, 358
        von Wright's sense, 250
'neg-raising', 354
negative polarity items, 332
Negri, S., 50
neighbourhood semantics, 159, 171, 174, 181, 244, 248, 249, 340, 425n
    general neighbourhood frames, 160n
Nelkin, N., 341
Nelson, M., 60n
Nguyen, L. A., 486n
NI ('Negative Introspection' – but not 5_K), 371, 385
NI frames (non-identity frames), 92, 272
Nichols, S., 415n
Niiniluoto, I., 205n, 242
Ninan, D., 240n, 300
Nogina, E., 333
Nolan, D., 379
nomenclature (for modal logics), 33–36, 45
    Chellas's "!" notation, 37
    multimodal logics, 183
    non-normal logics, 46, 49
    Sobociński's, 287n
nomic (possibility, necessity), *see also* necessity, physical or nomic, 2, 59
non-monotonic (logic), 9n, 20, 125
    adaptive logic, 250n
non-normal worlds, *see* normal points or worlds
noncontingency, *see* contingency
Nordenfelt, L., 301
normal
    consequence relations, 22
    modal logics, 2, 18, 19
    operator (in a given logic), 31
    points or worlds, 360
notation
    $R(\cdot)$, 60
    R^*, 70
    $\Box$, 1
    $\Diamond$, 1
    $\Rightarrow$, 4n

$\|\cdot\|$ (for truth-sets), 54
$\succ$, 2, 7
$\bot$, 10
$\boxdot$, 416
$\cap$, 4
$\cup$, 4
$\dashv\vdash$, 9
η, γ, μ, ν (Moisil's operators), 50
$\in$, 4
$\wedge$, 6
$\langle$ and $\rangle$, 4
$\leftrightarrow$, 8
$\vee$, 6
$\neg$, 6
$\oplus$ (normal extension by adding), 76
$\rho(\cdot), \nu(\cdot)$ (Porte's mappings), 49
$\sim$ (operation from pairs of sets to binary relations), 15
$\sim_R$, 404
$\setminus$, 4
$\prec$ (strict implication), 46
$\subseteq$, 4
$\supset$, 7
$\times$, 4, 15
$\rightarrow$, 6
$\top$, 10
∇ (contingency), 18
$\varnothing$, 5
$\triangle$ (noncontingency), 21
$\vdash$, 8
$\vdash_{CL}$, 9
$|\cdot|$, 5
$\widetilde{O}_i$ (dual of modal operator O_i), 35
$\wp(\cdot)$, 5
$+$ (for modal logics), 81
$+$ (operation from pairs of sets to binary relations), 15, 406
$\&$, 4
$\Box\!\rightarrow$, 344n
$\nvdash$, 9
$\Vdash_{CL}$, 14
$\mathbb{N}$, 4, 85
Nat, 4
$\mathbb{Q}$, 4, 117, 202
$\mathbb{Z}$, 4, 84, 202
Nowak, M., 95n, 462n
Nowell-Smith, P. H., 255
Nozick, R., 238, 239, 344, 346
Nute, D., 250n

O (deontic $\Box$), 1, 205
$O_1, O_2, \ldots$: arbitrary occurrences of $\Box$, $\Diamond$, 31
O'Connor, D., 376n
objective
    — chance, 339
    — formulas in epistemic logic, 391, 397
    objective vs. subjective *ought*, 242, 244
    von Kutschera's objectivity operator O, 303, 398, 427
"obligation", "obliged", "obligatory", 252n
obviousness, 379
Odegaard, D., 102
Ohnishi, M., 143, 486n
Okasha, S., 376
Oksanan, M., 258
omniscience
    collective, 134, 333n
    logical or deductive, 162, 163, 205, 331–370, 422, 424, 432
    Weingartner on God's —, 387
ontological argument, modal, 101–102
Oppy, G., 101
Orayen, R., 236
Ore, O., 91n
    Ore-partitions, 395
Orłowska, E., 336
OS2, OS4, etc. (Smiley's systems), 260
Osborne, H., 244
"ought", *see* modal auxiliaries
Outermost Reduction Property, 42, 288

P (deontic $\Diamond$), 1, 234
P (epistemic $\Diamond$), 203, 351
P (past tense $\Diamond$), 177
p-morphisms, 84, 96, 107, 137, 267, 269, 272
Pacuit, E., 333
Palmer, F. R., 17
Palmigiano, A., 22
paraconsistent
    deontic logic, 245
    logic, 46
Paradox of the Knower (L. Åqvist), 300
Parent, X., 277
Pargetter, R., 217–225, 242, 257
Parisi, D., 251
Parry, W. T., 41n, 48, 423
    his "S4.5", 144
    on analytic implication, 423
Parsons, J., 395
Parsons, T., 348, 352n

Partee, B., 178
Partee, B. H., 343
partial
    partial orders (or partial orderings), 15
    strict partial orders (or strict partial orderings), 461
    partial functional (frames and relations), 68, 139
        alio- version of, 108
partitions, 90
    Ore-partitions, 395
PC ('Positive Certainty'), 371, 373, 385
    as Positive Confidence, 385
Peacocke, C., 359, 379, 458, 458n, 465
Peetz, V., 465n
Peirce, C. S., 93, 95, 194
Pelletier, J. F., 28, 29, 416n
perfect aspect, 160, 377, 452
Perloff, M., 451
permissibility, 1, 208, 236
    free choice permission, 244, 253
    legal vs. moral etc., 239n
    really dual to obligation?, 252
    relative, 241
Peron, N. M., 245
pertinent: $\Box(O, B,$ etc.$)$-pertinent points, 59n, 60, 205, 278, 369, 389
Perzanowski, J., 98–99, 262
Peterson, P. L., 352n
Pettit, P., 21n
PI ('Positive Introspection' – but not 4_K), 371, 385
    opposed by Hintikka, 378
piecewise
    piecewise convergent, connected etc., 74
    piecewise reflexive, universal, 79
    piecewise symmetric, 85, 138, 262
    piecewise version of a given condition, 76
Pigden, C. R., 246, 252, 268
Pigozzi, D., 23n
Pizzi, C., 5, 45, 254n
Plaza, J., 331n, 333, 397
Pledger, K., 48
Plumwood, V., 245n, 341, 351n, 364, 369, 370, 423n, 428
Poggiolesi, F., 493n
point-generated
    (sub)frame, 71
    (sub)model, 70

pointwise equivalent (models), 161n
Połacik, T., 347
Pollock, J., 90
Pörn, I., 229, 277, 300, 449
Porte, J., 48, 49, 191, 360n, 415
  on two kinds of non-normal worlds, 366
  on Zeman, Georgacarakos, and the semantics for **S4.4**, 413
Portner, P., 160, 300
possibilities (as opposed to possible worlds), 211, 421n
possibility
  physical (or nomic), 59
possible worlds
  philosophical issues surrounding, 57, 89, 227, 337
  semantics (*see also* Kripke semantics), 41, 53
    complaint by Hintikka discussed by Burgess, 167
  under a conceptualization, 157
Post completeness, 38, 151, 174, 242
Potocki, G., 249n
Potts, D. H., 93
Potts, T. C., 160, 251
Powers, L., 298, 345n
Prawitz, D., 471, 475n, 477n
pre-orders, 15
Predelli, S., 219n
preference (and preferability), 241
prenex normal form, 66n
Prest, M. Y., 89n
presupposition, 204, 352
pretabular (logics), 112
Price, A., 299
Priest, G., 6n, 46, 234, 402n, 425n
  on laws and physically impossible worlds, 59
prima facie obligation, 244, 299
'prime' (modal logics), 304
Prior, A. N., 37, 41n, 152n, 156, 177n, 243, 253n, 255n, 279, 287, 396
  branching time, 195n
  his system Q, 209
  history of tense logic, 195n
  many values as modal profiles, 154
  modal ontological argument, 101
  on Anderson's reduction of deontic logic, 253, 254n, 255
  on presentism, 205n
  on the absence of anything between obliga-
    tion and permissibility, 304–320
  on $\mathbf{U}_O$, 239n
  tense operators, 107, 178
Pritchard, D., 345, 422n
probability
  modally treated, 384
  subjective, 373, 374n
"problematic", 304
progressive (or continuous) aspect, 159, 301
prohairetic (logic), 241
propositional
  prop'l attitudes, 54n
    attitude toward the prop'n that p which is a belief but is not the belief that p, 447
    conventions governing their ascription, 343
    do not exist, 421n
    explicit or implicit, 205n, 339–342, 351, 424, 425, 432
    knowledge as not being one, 352
    logics of, 203, 358
    not knowing, *not* believing, etc., are not prop'l attitudes, 203, 421, 433
    veridical or otherwise, 205, 290
  prop'l constants, *see* sentential constants
  prop'l quantifiers, 213, 239, 257, 257n, 258, 277, 340, 370, 404n
    in non-modal classical propositional logic, 216n
    in spelling out an idea of Jonathan Bennett, 215
  prop'l variables, 6
    behaving in a special way, 397
    do not stand where the name of a proposition could stand, 349
propositions
  admissible — (in general frames), *see* admissible
  as sets of worlds, 54, 154, 317n
  informal reference to, 205
  no such thing as atomic —, 253
  propositional allomorphs (Dretske), 395
  referentialism concerning (Chalmers), 54n
  structured, 54, 341
  thought problematic, 421n
provability, modal logic of, 37, 166n
  Gödel–Löb or *GL* logic, 82n
pseudo-neighbourhood semantics, 162, 425
pseudo-subjectivism (Jennings), 244, 299
public announcements, logic of, 333

Pucella, R., 341n
Pullum, G. K., 331n, 332, 351n, 371, 377
pure (formulas), 37, 134, 171, 256
    none frame-equivalent to $\mathbf{D}_c$, 139
pure (rules), 485, 486
Purtill, R. L., 101

Q (Kanger–Smiley sentential constant), 257
Q (Prior's system), 209
quantified modal logic, *see* modal predicate logic
quantifiers
    individual, *see* modal predicate logic
    propositional, *see* propositional quantifiers
"quasi-" (prefix), 25
quasi-congruential (modal logics), 39
quasi-normal (modal logics), 25, 81, 191, 193
    McKinsey's $\mathbf{S4} + \mathbf{M}$, 34
"quasi-order", 15, 362
"quasi-reflexive", 79
quasi-regular (modal logics), 47
questions (*see also* knowing whether, etc.), 333
    and partitions, 396
    concealed, 334
    disjunctive, 350
    embedded, 204n, 331n, 334, 351
    presuppositions of, 353
Quine, W. V., 6n, 235n, 358

$\mathbf{R}$ here called $.4$, 97
R-chains, 69
Rabinovich, A., 202
Rabinowicz, W., 219n, 239, 397
Radford, C., 392
Ramachandran, M., 378
Ramsamujh, T. J., 30
range-reflexive, 79
    1-r.r. and 2-r.r., 190
    closure of a relation, 273
    spurious 'alio' version, 94
range-restriction (of a binary relation), 262
Rawling, P., 251
recognitional capacities (positive, negative, general), 103–104, 378
Reed, B., 384
reflexive, 14
    standard sense vs. "field-reflexive", 90n, 95
refraining, 433
regular
    modal logics, 18, 19
    canonical model completeness proofs for, 362
    operator (in a given logic), 31
Reichenbach, H., 160
Reicher, M., 402n
Reinhart, J., 360n
rejecting a modal principle (distinguished from accepting its negation), 217
relations (n-ary relations on a set), 4
relevant
    'relevant alternatives' epistemology, 395, 403n
    implication, 158, 298n, 423
    logic, 6, 46, 342, 474n
        first-degree entailments, 369
reportive vs. normative (interpretation of deontic vocabulary), 239
representable
    $\wedge$-representable (relations), 15, 457–460, 466
    $\leftrightarrow$-representable (relations), 15
    $\vee$-representable (relations), 15, 406, 407
    monadically representable (relations), 15, 366, 466
representative (instances of a schema), 22
Rescher, N., 195n, 337
Restall, G., 23, 153, 261, 305, 316n, 323, 493n
Reyolds, M., 202
Rickman, H. P., 255n
Rigel 7 example (M. Huemer), 428
de Rijke, M., 22
Rin, B. G., 305
Rini, A., ii
Rivière. C., 300n
Robbins, P., 427
Roberts, G. W., 423
Roelofsen, F., 331n
van Roojen, M., 240, 299
Rose, A., 11n
Rosefeldt, T., 351n
Rosenberg, S., 101
Rosenstein, J. G., 202n
Rosenthal, D. M., 355
Ross's Paradox, 243, 278
    Cariani's treatment, 350
    Danielsson's treatment, 324–330
    Körner's treatment, 350
Ross, A., 244
Ross, S., 244
Ross, W.D., 244
Routley, R., 49, 238, 239, 245n, 341, 351n, 364,

INDEX

369, 370, 402n, 423n, 428
    on animal belief, 392
    on S0.5, 368
Routley, V., *see* Plumwood, V.
Rubinstein, A., 300n
rule of disjunction, 82, 112–128
    conditional, 124
    conditional modified, 124
    in epistemic and doxastic logic, 436
    modified, 122
Rumfitt, I., 211, 345n
Rumsfeld, D., 371
Russell, B. (Russsell's Paradox), 96, 126
Russell, G., 23, 261
Rybakov, V., 143
Ryle, G., 351n

"S" with a numeral, for **KT**, 34, 49, 362
S0.5 (Lemmon's system), 47, 49, 366
    semantics for, 368
S1, 48, 362
S13 (Kaplan's system), 28n
S2, 33, 46, 47, 49, 260, 360, 362, 363
$S2^0$, $S3^0$, T^0 etc. (labels from Feys and Zeman), 49
S3, 33, 47, 360, 362
S3.5, 48
    semantics for, 369
S4, 23, 33, 41
    completeness result for, 63, 72
    logics between — and **S5**
        as epistemic logics, 402–441
        non-normal, 194
        Matsumoto's Theorem concerning, 142–149, 420, 435
    modalities in, 41
    non-equivalent one-variable formulas in, 44
    non-normal extensions of, 34
    rule of disjunction for, 113
S4.04, 137, 431, 434, 435
"S4.1" (an unfortunate name for **S4M**), 34, 133
S4.2
    Burgess argues against this as a logic of demonstrability, 278n
S4.2, 34, 148
    S4.2($\Diamond\Box$) = **KD45**, 434
    as an epistemic logic, 402, 416, 431
    cluster-based completeness result, 413
    completeness results for, 77, 282

    definability of belief in — as an epistemic logic (*see also* Lenzen), 410, 434
    embedding — in **S4.4**, 416
    embedding deontic logic in, 278
    natural deduction for, 486
    quantified, 57
    Stalnaker on — and internalism, 403n
S4.3, 34, 133
    all extensions normal, 34
    Bull's Theorem concerning, 73
    completeness results for, 77
    determined by $\langle \mathbb{Q}, \leq \rangle$, 117
    Stalnaker on — and defeasibility, 403n
"S4.3.2" (a name for **S4F**), 34n, 66
S4.4, 34, 97, 144
    an incorrect claim concerning, 133
    as an epistemic logic, 402
    completeness results for, 413
S4B$_2$, 435
S4F, 34n, 67n, 406–409, 413, 435
    Stalnaker on, 416n
S4Fi (= **S4M**), 287
S4M, 34, 86, 86n
    and 'McKinsey validity', 168
    completeness theorem for, 286
    versus **S4** + **M**, 34, 81, 193
S4Sch, and **Sch** (Schumm's axiom), 418
S5, 14
    alternative axiomatizations of, 44, 45, 91
        as **KD4** + **B** instead of **KD4** $\oplus$ **B**, 81
        as **KD4** with the Brouwerian rule, 116
    completeness result for, 65, 72
    non-equivalent one-variable formulas in, 44, 154
S6, 38, 48, 362
    Curley entertains — as a candidate Cartesian modal logic, 364
S7, S8, 48
"S9", 48
S_a, S_b, S_c (Porte's systems), 49
Saarinen, E., 421n, 441n, 442, 442n
Sadegh-Zadeh, K., 338n
Safe Return condition (on frames), 105
safety (as a requirement for knowledge), 344n, 344–345
Sahlqvist, H.
    Sahlqvist formulas, 78, 444
    unravelled frames, 269
Sainsbury, R. M., 345

Salerno, J., 398
Salmon, N., 219n, 458, 466
Salqvist, H.
    Sahlqvist formulas, 444
Samet, D., 104, 395n, 430, 437–441
Sartwell, C., 415n
Saunders, K. W., 252
Sayre-McCord, G., 241
Schaffer, J., 351n, 395
Scharle, T. W., 154
schematic letters, 2
    notation for, 3
Schiff, S., 252n
Schild, U., 235, 255
Schotch, P. K., 21n, 30n, 160, 245, 248, 340, 347n, 393
Schreider, J. A., 89n
Schumm, G. F., 6, 111n, 418n
    on S2, 48
    on Sobociński's question, 417
Schurz, G., 268, 396, 424
Schwartz, N., 252n
Schwarz, G., 333, 404, 406n, 408
Schwitzgebel, E., 392
Scott, D. S., 9, 159
    Scott's Rule, 22
Scroggs, S. J., 34, 72, 112, 120
Searle, J., 377n
Seg$_n$, 221
Segerberg, K., 20, 39n, 75, 219n, 252n, 404, 451n
    S4 determined by the class of posets, 72
    'quasi' terminology, 24
    all extensions of **S4.3** normal, 34
    axioms **Seg**$_n$, 221
    his cluster terminology, 419
    index frames, 409n
    labels for axioms and logics, 33, 34, 46, 66, 74, 189, 287, 418n
    logic of 'elsewhere', 93, 96
    logics with functional accessibility relations, 307
    nothing between **KB4** and **S5**, 89, 417n
    on **K.3**, 77
    on **S4.2**, 412
    on **S4F**, 408
    on 'Boolean atoms', 162, 292n
    on a restricted convergence axiom, 109n
    on completeness results w.r.t. classes of irreflexive frames, 195
    on consequence relations, 14
    on deontic applications of dynamic logic, 237
    on doxastic logic and belief revision, 333
    on extensions of **S4** non included in **S5**, 403
    on filtrations and the finite model property, 73
    on frames with distinguished elements and the quasi-normal logic **K** + **T**, 191
    on Macintosh logics, 98
    on Makinson's warning, 40
    on non-normal worlds, 361
    on p-morphisms, 96
    on S1 and relatives, 362n
    on Sobociński's logics, 87
    on the Falsum modal logic, 303, 362
    on the Fitch derivation, 397
    on the logic of action, 451
    on two-dimensional modal logic, 277
    on **S4Zem** (= **S4.04**), 431
    probability connective, 384
Segev, E., 430, 437–441
semi-euclidean (frames), 66, 405
'semi-paraconsistent'
    deontic logic, 245n, 250
    doxastic logic, 370
semi-simplified
    Kripke models, 58, 205–208, 266, 295, 385, 387, 404, 413
    neighbourhood/locale models, 248
semigroups, 43, 159
semilattices, 159
sense/reference distinction, 45
sensitivity (as a requirement for knowledge), 344n
sentence letters, *see* propositional variables
sentential constants
    a constant Ω with unusual logical powers, 315
    and contingency, 254n
    Anderson's sanction constant, 253
    for truth and falsity, 10
    in Kanger and Smiley, 257
    Porte's constant Ω for presenting the L-modal logic, 363
    signifying a reward, 255
    true at precisely the normal worlds, 363
    which cannot be construed as propositional constants, 217, 275, 429
separated (conditions), 444
sequents, 2, 7

holding on a valuation, 311
    natural deduction presented with sequent to sequent rules, 471
    plain vs. transitional, 307, 486
    sequent calculus, 312, 486, 493n
        vs. natural deduction, 7n
serial (frames and models), 65, 72
    a partly existential condition, 68
    class of — frames modally defined by **D**, 82
Setlur, R. V., 312
Seuren, P., 95, 354
Sharvit, Y., 331n
Sheffer stroke, 11
Shehtman, V. B., 160, 161n
Shoemaker, S., 358
Shoesmith, D. J., 14
Shope, R. K., 349
Shortt, J., 467n
"should", *see* modal auxiliaries
"shouldn't" in "I shouldn't be alive", 302
"shyster logicians" (Geach's complaint), 364
Silk, A., 242, 244
Simons, M., 253, 350
simplified (Kripke models), 53, 55, 205
Singer, M. G., 235, 239
Sinhababu, N., 203
Sinnott-Armstrong, W., 243
Sinnott-Armstrong. W., 234
Skinner, Q., 235
Slater, B. H., 96
Slavkovic, M., 277
Smiley, T. J., 14, 257, 260, 423, 429
    a redundant axiom, 261
Smith, C. S., 160
Smith, N., 355
Smithies, D., 359
Smullyan, R. M., 90n
Snedegar, J., 251
Snowdon, P., 351n
Sobel, J. H., 101
Sobociński, B., 11n, 87, 89
    his names for modal logics, 287
    his question about **S4.4** and **S5**, 89, 417
    Sobociński-regular (modal logics), 37, 38, 173
        multimodal case, 196n
"solid" (B. Williams), 386, 386n
Solovay, R., 235
Sorensen, R., 21n
    on blindspots, 258, 356, 398
    on collective omniscience, 135
    on Moore's Paradox, 355
Sosa, E., 241, 242, 255, 344
sound
    non-modal version, 6
    w.r.t. a class of frames, 71
    w.r.t. a class of models, 60
specificity
    act-specificity in deontic logic, 237
    failure of — not amounting to ambiguity, 251
    of desire, 446
    of propositional attitudes, 331
    without existence, 442–446
square of opposition, 17
Stalnaker, R,
    on impossible worlds, 364
Stalnaker, R., ii, 55, 93, 125, 280, 282, 333n, 341, 345, 371, 371n, 403n, 408, 410, 416, 435
    Stalnaker's assumption (concerning conditionals), 452
    Stalnaker's condition (on frames for doxastic logic), 406
Stanley, J., 351n
Steinacker, P., 246n
Steinsvold, C., 229, 429
Stenius, E., 248
Steup, M., 104, 378
Stich, S., 415n, 421n
Stine, G., 344n
Stirling, C., 458n
"stit", 452
Stojanovic, I., 219n
Stokhof, M., 334
Stoljar, D., 355, 359
strict
    equivalence, 153n
    implication, 10, 46, 257
    linear orders, 77, 178, 241
strict intermediaries (a logic provides), 310
structural constraint, 427
Sturch, R. L., 244
subjective, *see* objective
subjunctive mood, 300, 305
substitution instances (of a formula), 19
    vs. instances (of a schema), 55
substitution-invariant (consequence relations, rules), 311
subtraction, logical, 327
successor, R-successor, 59

supererogation, 229, 255
Suppe, F., 379
'surface implication' (Hintikka), 424
Sweetser, E., 300
Sylvan, R. (*see also* Routley, R.), 368
symmetric, 14
symmetric (relations, models, frames)
    1- and 2-symmetric points, 188, 191, 414
    remotely symmetric, 137, 414, 431
    symmetry in diagrammatic form, 66

T (*see also* **KT**), 33
    rule for natural deduction, 492
tableaux, 6
tabular (logics), 112
Talja, J., 451n
Tan, Y.-H., 241
Tarski, A., 34, 143, 193
tautologies, 6
    tautologous (sequent or formula), 7
Taylor, B. M., 160
$\mathbf{T}_n^\diamond$, 85
"tell", 204
Temkin, L., 462
tense logic, 1, 37, 177–202
    — vs. temporal logic, 195n, 415n
    nothing specifically to do with time (or tense), 178, 210, 437, 445
    unique characterization in, 437
TF (derived rule: truth-functional consequences), 26
"that" (complementizer), 331–332, 422
Thijsse, E., 161n, 339, 362n, 380–385, 421n, 433
Thomas, I., 416
Thomason, R. H., 195n, 243
Thomason, S. K., 85
Thompson, S. A., 332
Tichý, P., 425
Titelbaum, M., 299
Tkaczyk, M., 48
Tokarz, M., 154n, 295
    and fully modalized logics, 294
Tomberlin, J., 300
"too" (modal aspects of claims of sufficiency and excess), 301n
van der Torre, L., 241
tough-movement, 332
Tranøy, K. E.
    confused about iterated deontic operators, 239
    on ordinary language and obligation, 252n
    ought vs. *must*, 299
transitional (sequents), *see* sequents, plain vs. transitional
transitive, 14
    1-, 2- and 3-transitive points, 185–191
    class of — frames modally defined by **4**, 82
    negative transitivity, 461
    transitive closure (of a binary relation), 70
translation
    compositional, 265
    definitional, 265, 416
    particular translations
        $(\cdot)^{\mathbf{T!}}$, $(\cdot)^{\mathbf{Ver}}$, 38, 121, 134, 265
        $\tau_{\Box\diamond\Box}$, 418–420
        $\tau_{\Box\diamond}$, 287
        $\tau_{\diamond\Box}$, 266, 278, 281, 416
        $\tau_{\Box}$, 266, 416
        τ_{Th}, 416
        Fitting's $\Box\diamond$ translation, 288
    translational embedding, 142, 147, 259–261
    variable-fixed, 265
trivialism, 364
Truszczyński, M., 333, 404, 406n, 408
truth-functionality
    informal notion, 1
    precise explication, 10
Tsai, C.-C., 71
Tsohampidis, V., 54n
Tsohatzidis, S., 204n
two-dimensional (modal semantics), ii, 135, 217, 275, 334, 358, 441
    avoided under special conditions using a tense-logical expansion of the language, 443
Typanska, L., 242
typographical errors (remarked on), 7n, 15n, 122n, 166n, 241n, 287n, 301n, 305, 338n, 351n, 406n, 410, 427, 429n, 462n
Törnebohm, H., 261

U, 33, 47, 94, 107, 289, 419
    and range-reflexivity, 80
    for desire, 370n
    in deontic logic ($\mathbf{U}_O$), 204
        as synthetic *a priori*, 239n
        by contrast with $\Box(p \to \Box p)$, 287
        not well-formed on von Wright's approach, 239
    in doxastic logic ($\mathbf{U}_B$), 303, 370, 436
    in epistemic logic ($\mathbf{U}_K$) – objected to, 403

in tense logic, 190
label used rather than $\mathbf{T}^\square$, 37n
provable in **K5**, 80, 261, 360n
provable in $\mathbf{KT}_c$, 261
rule for natural deduction, 492
Uckelman, S., 451n
ultrafilter (extensions), 85
"unary", 6n
Unger, P., 352, 392n
uniform substitution, 18, 19, 44
internal, 86
must logics be closed under —?, 397
preserves validity, 57
unilateral (implication), 308
unique characterization (of $\square$, etc.), 96, 437n
uniquely satisfiable (condition on a relation), 92
universalizability, 239, 299
unravelling, 269–273, 296
Urmson, J. O., 255
Urquhart, A., 28, 29, 158, 195n, 416n, 423
US, *see* uniform substitution

valid
Gochet–Gillet valid, 367
McKinsey validity, *see* McKinsey, his substitutional semantics
real-world vs. general validity, 219n, 305
truth-functional validity, 6
validity according to the simplified Kripke semantics, 55
validity at a point, 185
validity of sequents, 7
validity on a frame, 71
validity versus truth, 82n, 185
valuation, 10
Boolean, 10
double use of the term, 54
Vendler, Z., 377
on objects of knowledge and belief, 352
on verbal aspect, 160
Venema, Y., 5, 22, 85n, 275, 277, 420n
Ver, 33, 37
veridical, 204
Verkuyl, H., 160
Verum
modal logic, *see* **KVer**
sentential constant $\top$, 10
Vikner, S., 160
Viney, D. W., 101
Vlach, F., 377

Vlach operators, 441
Volpe, G., 240
Voorbraak, F., 235, 374, 388, 405, 407, 410
Vorobej, M., 239n

W (*see also* **KW**), 33
von Wachter, D., 387
Wagner, C., 464n
Wajszczyk, J., 452n
Wall, D., 358
Walton, D., 301, 449n, 451n
Wang, H., 194
Wang, Y., 104, 254n
Wansing, H., 22, 239n, 339, 362n, 383n, 493n
want, *see* boulomaic
Watts, R. J., 332
w.d.e., *see* frames with distinguished elements
Weatherson, B., 300n, 394, 457
Wedgwood, R., 240n
Wehmeier, K. F., 60n, 68n, 300, 305, 458n
Weinberg, J. M., 415n
Weinberger, O., 240n
Weingartner, P.
confused about infallibility, 387
multiplying knowledge operators, 334–335, 368
on Parry-like relevance criteria, 423
Werner, T., 300
Wertheimer, R., 300
Wheatley, J. M., 377n
"whether", *see* knowing whether
White, A. R., 240n, 349, 384
White, R., 68
"why", 351
Williams, B., 354, 378, 386n
knowledge held to be 'at least as grand as belief', 392
on aggregation in deontic logic, 21n, 248
on anoxia and dreaming, 103
on beliefs as aiming at the truth, 356
Williams, J. N., 351n, 355, 358
Williamson, T., 46, 57, 92, 98, 122, 124, 126–128, 211, 212, 212n, 305n, 322n, 338n, 339, 340, 348, 351, 352n, 353, 371n, 376, 379, 398, 403n, 415n, 424n
and Smiley on coextensive vs. synonymous operators, 257
non-congruential vs. hyperintensional contexts, 376n
on contextualism, 394
on disjunctivism, 422

on logical neutrality, 241
on positive introspection (4_K) and luminosity, 378
on safety, 345
on the dim prospects for an analysis of the concept of knowledge, 349
symmetric accessibility relations in epistemic logic, 385
Williamson models for epistemic logic, 162–166

Wilson, D., 353
Wittgenstein, L., 68n
Wójcicki, R., 265
Wolter, F., 197n
Wolter, L., 302
worlds
    impossible, 364
    possible, *see* possible worlds
    world-mates, 461, 466n
Wright, C., 339n, 359
von Wright, G. H., 2, 25n, 34, 203, 249n, 304
    confused over epistemic analogues of possibility and contingency, 22
    his basic deontic logic as formulated by Danielsson, 324
    O and P as predicates rather than operators, 234, 324
    on *nullum crimen sine lege*, 252
    on bringing about, 451
    on deontic necessitation, 246
    on dyadic deontic and alethic operators, 250
    on the Ł-modal system, 261

Yablo, S., 253, 424
Yalcin, S., 384, 422

Zakharyaschev. M., 5, 39, 137, 361, 366, 369
Zalta, E. N., 60n, 212, 234
Zanardo, A., 195n
Zangwill, N., 204, 357
Zardini, E., 378
Zawadowski, M., 308
Zeman, J. J., 34, 49, 402, 408, 417
    S4Zem (= **S4.04**), 431
    Porte on, 414
    Zeman-modalized (formulas), 486
Ziegeler, D., 300, 302
Zolin, E., ii, 22, 38, 254n
    his embedding notation, 266, 288, 290, 418
    his use of the term *modality*, 31, 44n

on extensional modal logics, 304
on fully modalized logics, 290–295

www.ingramcontent.com/pod-product-compliance
Lightning Source LLC
Chambersburg PA
CBHW081412230426
43668CB00016B/2215